DEAN'S HANDBOOK OF ORGANIC CHEMISTRY

George W. Gokel, Ph.D.

Director, Program in Chemical Biology
Professor, Department of Molecular Biology and Pharmacology
Washington University School of Medicine
Professor, Department of Chemistry
Washington University
St. Louis Missouri

Second Edition

McGRAW-HILL

New York   Chicago   San Francisco   Lisbon   London   Madrid
Mexico City   Milan   New Delhi   San Juan   Seoul
Singapore   Sydney   Toronto

The *McGraw·Hill* Companies

Cataloging-in-Publication Data is on file with the Library of Congress

1 2 3 4 5 6 7 8 9 0   DOC/DOC   0 9 8 7 6 5 4 3

ISBN 0-07-137593-7

*The sponsoring editor for this book was Kenneth P. McCombs and the production supervisor
was Sherri Souffrance. It was set in Times Roman by Newgen Imaging Systems (P) Ltd. The
art director for the cover was Anthony Landi.*

Printed and bound by RR Donnelley.

McGraw-Hill books are available at special quantity discounts to use as premiums and sales
promotions, or for use in corporate training programs. For more information, please write
to the Director of Special Sales, McGraw-Hill Professional, Two Penn Plaza, New York,
NY 10121-2298. Or contact your local bookstore.

 This book is printed on recycled, acid-free paper containing
a minimum of 50% recycled, de-inked fiber.

CONTENTS

PREFACE

The first edition of the *Handbook of Organic Chemistry* was edited by Professor John A. Dean. It appeared in 1987 and has served as a widely used and convenient reference work for more than 15 years. When Professor Dean asked if I would work with him to develop a second edition, I was pleased to do so. I felt that as valuable as the first edition was, it would be more broadly useful if it contained discussions of the data, the means by which the data were acquired, and perhaps even how the data are applied in modern science. We thus began the revision with enhanced usability as the foremost goal. Sadly, just as we were beginning the effort, Professor Dean passed away. He will be sorely missed.

In following the original plan, many figures, structures, discussions of the methods, and illustrations of the data have been incorporated. Some tables have been reorganized. In some cases tables have been printed twice; although they contain the same data, they are arranged by different criteria. The intent is to make the data easier for the researcher to access and use. Some Internet addresses that can serve as a supplementary resource are included. Despite the numerous additions, the volume remains compact and accessible.

As Professor Dean was not involved in producing this edition, I take responsibility for errors of fact or omission. I hope the volume is error-free, but I would appreciate being informed of any mistakes that are found. Finally, I wish to express my thanks to Mrs. Jolanta Pajewska, who helped in improving the manuscript and the proofreading.

GEORGE W. GOKEL

SECTION 1

ORGANIC COMPOUNDS

NOMENCLATURE OF ORGANIC COMPOUNDS

The following synopsis of rules for naming organic compounds and the examples given in explanation are not intended to cover all the possible cases. For a more comprehensive and detailed description, see J. Rigaudy and S. P. Klesney, *Nomenclature of Organic Chemistry*, Sections A, B, C, D, E, F, and H, Pergamon Press, Oxford, 1979. This publication contains the recommendations of the Commission on Nomenclature of Organic Chemistry and was prepared under the auspices of the International Union of Pure and Applied Chemistry (IUPAC).

Hydrocarbons and Heterocycles

Alkanes. The saturated open-chain (acyclic) hydrocarbons (C_nH_{2n+2}) have names ending in -ane. The first four members have the trivial names *methane* (CH_4), *ethane* (CH_3CH_3 or C_2H_6), *propane* (C_3H_8), and *butane* (C_4H_{10}). For the remainder of the alkanes, the first portion of the name is derived from the Greek prefix (see Table 11.4) that cites the number of carbons in the alkane followed by -ane with elision of the terminal -a from the prefix, as shown in Table 1.1.

TABLE 1.1 Names of Straight-Chain Alkanes

n^*	Name	n^*	Name	n^*	Name	n^*	Name
1	Methane	11	Undecane‡	21	Henicosane	60	Hexacontane
2	Ethane	12	Dodecane	22	Docosane	70	Heptacontane
3	Propane	13	Tridecane	23	Tricosane	80	Octacontane
4	Butane	14	Tetradecane			90	Nonacontane
5	Pentane	15	Pentadecane	30	Triacontane	100	Hectane
6	Hexane	16	Hexadecane	31	Hentriacontane	110	Decahectane
7	Heptane	17	Heptadecane	32	Dotriacontane	120	Icosahectane
8	Octane	18	Octadecane			121	Henicosahectane
9	Nonane†	19	Nonadecane	40	Tetracontane		
10	Decane	20	Icosane§	50	Pentacontane		

$^*n$ = total number of carbon atoms.
† Formerly called enneane.
‡ Formerly called hendecane.
§ Formerly called eicosane.

For branching compounds, the parent structure is the longest continuous chain present in the compound. Consider the compound to have been derived from this structure by replacement of hydrogen by various alkyl groups. Arabic number prefixes indicate the carbon to which the alkyl group is attached. Start numbering at whichever end of the parent structure that results in the lowest-numbered locants. The arabic prefixes are listed in numerical sequence, separated from each other by commas and from the remainder of the name by a hyphen.

If the same alkyl group occurs more than once as a side chain, this is indicated by the prefixes di-, tri-, tetra-, etc. Side chains are cited in alphabetical order (before insertion of any multiplying prefix). The name of a complex radical (side chain) is considered to begin with the first letter of its complete name. Where names of complex radicals are composed of identical words, priority for citation is given to that radical which contains the lowest-numbered locant at the first cited point of difference in the radical. If two or more side chains are in equivalent positions, the one to be assigned the lowest-numbered locant is that cited first in the name. The complete expression for the side chain may be enclosed in parentheses for clarity or the carbon atoms in side chains may be indicated by primed locants.

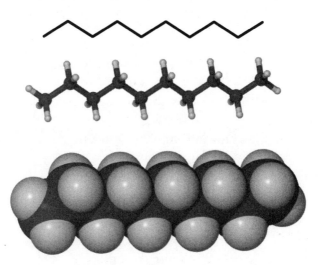

FIGURE 1.1   Projections for *n*-decane

If hydrocarbon chains of equal length are competing for selection as the parent, the choice goes in descending order to (1) the chain that has the greatest number of side chains, (2) the chain whose side chains have the lowest-numbered locants, (3) the chain having the greatest number of carbon atoms in the smaller side chains, or (4) the chain having the least-branched side chains.

These trivial names may be used for the unsubstituted hydrocarbons only:

Isobutane $(CH_3)_2CHCH_3$ Neopentane $(CH_3)_4C$

Isopentane $(CH_3)_2CHCH_2CH_3$ Isohexane $(CH_3)_2CHCH_2CH_2CH_3$

Univalent radicals derived from saturated unbranched alkanes by removal of hydrogen from a terminal carbon atom are named by adding -yl in place of -ane to the stem name. Thus the alkane *ethane* becomes the radical *ethyl*. These exceptions are permitted for unsubstituted radicals only:

Isopropyl $(CH_3)_2CH$—   Isopentyl $(CH_3)_2CHCH_2CH_2$—

Isobutyl $(CH_3)_2CHCH_2$—   Neopentyl $(CH_3)_3CCH_2$—

sec-Butyl $CH_3CH_2CH(CH_3)$—   *tert*-Pentyl $CH_3CH_2C(CH_3)_2$—

tert-Butyl $(CH_3)_3C$—   Isohexyl $(CH_3)_2CHCH_2CH_2CH_2$—

Note the usage of the prefixes iso-, neo-, *sec*-, and *tert*-, and note when italics are employed. Italicized prefixes are never involved in alphabetization, except among themselves; thus *sec*-butyl would precede isobutyl, isohexyl would precede isopropyl, and *sec*-butyl would precede *tert*-butyl.

Examples of alkane nomenclature are

$$\overset{4}{C}H_3-\overset{3}{C}H_2-\overset{2}{C}H-\overset{1}{C}H_3$$
$$\underset{|}{}CH_3$$

2-Methylbutane (or the trivial name, isopentane)

$$\overset{5}{C}H_3-\overset{4}{C}H_2-\overset{3}{C}H-CH_3$$
$$\overset{|}{C}H_2-CH_3$$
$$\underset{2}{}\quad\underset{1}{}$$

3-Methylpentane (not 2-ethylbutane)

$$\overset{8}{C}H_3-\overset{7}{C}H_2-\overset{6}{C}H_2-\overset{5}{C}H-\overset{4}{C}H_2-\overset{3}{C}H_2-\overset{2}{C}-\overset{1}{C}H_3$$
with CH$_3$ on carbon 2 and CH$_2$—CH$_3$ on carbon 5, CH$_3$ on carbon 2

5-Ethyl-2,2-dimethyloctane (note cited order)

$$\overset{8}{C}H_3-\overset{7}{C}H_2-\overset{6}{C}H-\overset{5}{C}H_2-\overset{4}{C}H_2-\overset{3}{C}H-\overset{2}{C}H_2-\overset{1}{C}H_3$$
with CH$_3$ on carbon 6 and CH$_2$—CH$_3$ on carbon 3

3-Ethyl-6-methyloctane (note locants reversed)

$$\overset{2}{C}H_3$$
$$CH_3-\overset{1}{C}-CH_3 \quad CH_3$$
$$\overset{8}{C}H_3-\overset{7}{C}H_2-\overset{6}{C}H_2-\overset{5}{C}H_2-\overset{4}{C}-\overset{3}{C}H_2-\overset{2}{C}H-\overset{1}{C}H_3$$
$$CH_3-C-CH_3$$
$$\underset{|}{CH_3}$$

4,4-Bis(1,1-dimethylethyl)-2-methyloctane
4,4-Bis-1',1'-dimethylethyl-2-methyloctane
4,4-Bis(*tert*-butyl)-2-methyloctane

Bivalent radicals derived from saturated unbranched alkanes by removal of two hydrogen atoms are named as follows: (1) If both free bonds are on the same carbon atom, the ending -ane of the hydrocarbon is replaced with -ylidene. However, for the first member of the alkanes it is methylene rather than methylidene. Isopropylidene, *sec*-butylidene, and neopentylidene may be used for the unsubstituted group only. (2) If the two free bonds are on different carbon atoms, the straight-chain group terminating in these two carbon atoms is named by citing the number of methylene groups comprising the chain. Other carbons groups are named as substituents. Ethylene is used rather than dimethylene for the first member of the series, and propylene is retained for $CH_3-CH-CH_2-$ (but trimethylene is $-CH_2-CH_2-CH_2-$).

Trivalent groups derived by the removal of three hydrogen atoms from the same carbon are named by replacing the ending -ane of the parent hydrocarbon with -ylidyne.

Alkenes and Alkynes. Each name of the corresponding saturated hydrocarbon is converted to the corresponding alkene by changing the ending -ane to -ene. For alkynes the ending is -yne. With more than one double (or triple) bond, the endings are -adiene, -atriene, etc. (or -adiyne, -atriyne, etc.). The position of the double (or triple) bond in the parent chain is indicated by a locant obtained by numbering from the end of the chain nearest the double (or triple) bond; thus $CH_3CH_2CH=CH_2$ is 1-butene and $CH_3C\equiv CCH_3$ is 2-butyne.

For multiple unsaturated bonds, the chain is so numbered as to give the lowest possible locants to the unsaturated bonds. When there is a choice in numbering, the double bonds are given the lowest locants, and the alkene is cited before the alkyne where both occur in the name. Examples:

CH$_3$CH$_2$CH$_2$CH$_2$CH=CH—CH=CH$_2$ 1,3-Octadiene

CH$_2$=CHC≡CCH=CH$_2$ 1,5-Hexadiene-3-yne

CH$_3$CH=CHCH$_2$C≡CH   4-Hexen-1-yne

CH≡CCH$_2$CH=CH$_2$ 1-Penten-4-yne

Unsaturated branched acyclic hydrocarbons are named as derivatives of the chain that contains the maximum number of double and/or triple bonds. When a choice exists, priority goes in sequence to (1) the chain with the greatest number of carbon atoms and (2) the chain containing the maximum number of double bonds.

These nonsystematic names are retained:

Ethylene   CH$_2$=CH$_2$

Allene   CH$_2$=C=CH$_2$

Acetylene   HC≡CH

An example of nomenclature for alkenes and alkynes is

$$HC\overset{6}{\equiv}C\overset{5}{-}\underset{4}{C}=\overset{2}{C}-\overset{3}{C}H=\overset{1}{C}H_2$$

with CH$_2$—CH$_2$—CH$_3$ and CH=CH$_2$ substituents   4-Propyl-3-vinyl-1,3-hexadien-5-yne

Univalent radicals have the endings -enyl, -ynyl, -dienyl, -diynyl, etc. When necessary, the positions of the double and triple bonds are indicated by locants, with the carbon atom with the free valence numbered as 1. Examples:

CH$_2$=CH—CH$_2$—   2-Propenyl

CH$_3$—C≡C—   1-Propynyl

CH$_3$—C≡C—CH$_2$CH=CH$_2$—   1-Hexen-4-ynyl

These names are retained:

Vinyl (for ethenyl)   CH$_2$=CH—

Allyl (for 2-propenyl)   CH$_2$=CH—CH$_2$—

Isopropenyl (for 1-methylvinyl but for unsubstituted radical only)   CH$_2$=C(CH$_3$)—

Should there be a choice for the fundamental straight chain of a radical, that chain is selected which contains (1) the maximum number of double and triple bonds, (2) the largest number of carbon atoms, and (3) the largest number of double bonds. These are in descending priority.

Bivalent radicals derived from unbranched alkenes, alkadienes, and alkynes by removing a hydrogen atom from each of the terminal carbon atoms are named by replacing the endings -ene, -diene, and -yne by -enylene, -dienylene, and -ynylene, respectively. Positions of double and triple bonds are indicated by numbers when necessary. The name *vinylene* instead of ethenylene is retained for —CH=CH—.

Monocyclic Aliphatic Hydrocarbons. Monocyclic aliphatic hydrocarbons (with no side chains) are named by prefixing cyclo- to the name of the corresponding open-chain hydrocarbon having the same number of carbon atoms as the ring. Radicals are formed as with the alkanes, alkenes, and alkynes. Examples:

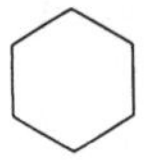

Cyclohexane — Cyclohexyl- (for the radical)

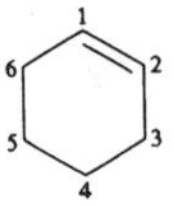

Cyclohexene — 1-Cyclohexenyl- (for the radical with the free valence at carbon 1)

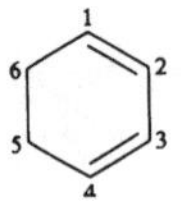

1,3-Cyclohexadiene — Cyclohexadienyl- (the unsaturated carbons are given numbers as low as possible, numbering from the carbon atom with the free valence given the number 1)

For convenience, aliphatic rings are often represented by simple geometric figures: a triangle for cyclopropane, a square for cyclobutane, a pentagon for cyclopentane, a hexagon (as illustrated) for cyclohexane, etc. It is understood that two hydrogen atoms are located at each corner of the figure unless some other group is indicated for one or both.

Monocyclic Aromatic Compounds.   Except for six retained names, all monocyclic substituted aromatic hydrocarbons are named systematically as derivatives of benzene. Moreover, if the substituent introduced into a compound with a retained trivial name is identical with one already present in that compound, the compound is named as a derivative of benzene. These names are retained:

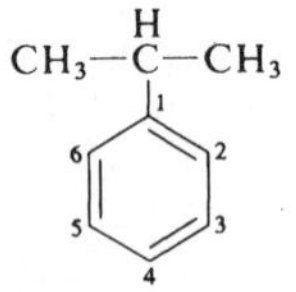

Cumene

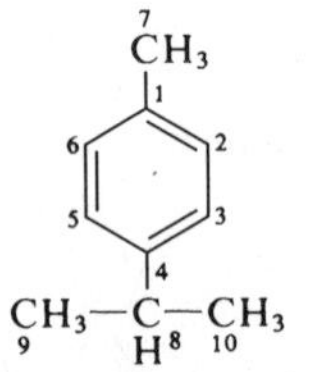

Cymene (all three forms; *para-* shown)

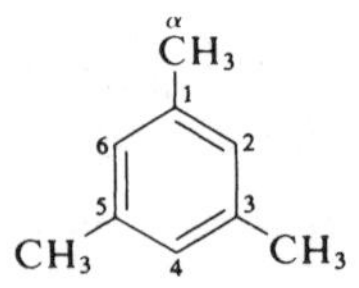

Mesitylene

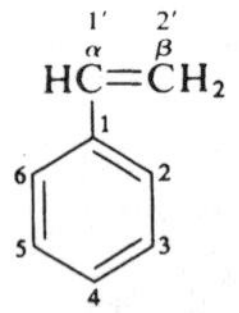

Styrene

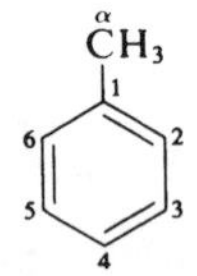

Toluene

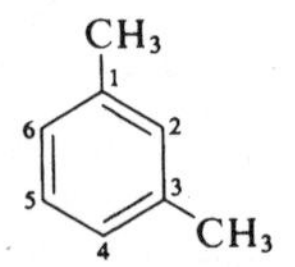

Xylene (all three forms; *meta-* shown)

The position of substituents is indicated by numbers, with the lowest locant possible given to substituents. When a name is based on a recognized trivial name, priority for lowest-numbered locants is given to substituents implied by the trivial name. When only two substituents are present on a benzene ring, their position may be indicated by o- (*ortho-*), m- (*meta-*), and p- (*para-*) (and alphabetized in the order given) used in place of 1,2-, 1,3-, and 1,4-, respectively.

Radicals derived from monocyclic substituted aromatic hydrocarbons and having the free valence at a ring atom (numbered 1) are named phenyl (for benzene as parent, since benzyl is used for the radical $C_6H_5CH_2-$), cumenyl, mesityl, tolyl, and xylyl. All other radicals are named as substituted phenyl radicals. For radicals having a single free valence in the side chain, these trivial names are retained:

Benzyl $C_6H_5CH_2-$

Benzhydryl (alternative to
  diphenylmethyl) $(C_6H_5)_2CH-$

Cinnamyl $C_6H_5CH=CH-CH_2-$

Phenethyl $C_6H_5CH_2CH_2-$

Styryl $C_6H_5CH=CH-$

Trityl $(C_6H_5)_3C-$

Otherwise, radicals having the free valence(s) in the side chain are named in accordance with the rules for alkanes, alkenes, or alkynes.

The name *phenylene* (*o*-, *m*-, or *p*-) is retained for the radical $-C_6H_4-$. Bivalent radicals formed from substituted benzene derivatives and having the free valences at ring atoms are named as substituted phenylene radicals, with the carbon atoms having the free valences being numbered 1,2-, 1,3-, or 1,4-, as appropriate.

Radicals having three or more free valences are named by adding the suffixes -triyl, -tetrayl, etc. to the systematic name of the corresponding hydrocarbon.

Fused Polycyclic Hydrocarbons.   The names of polycyclic hydrocarbons containing the maximum number of conjugated double bonds end in -ene. Here the ending does not denote one double bond. Names of hydrocarbons containing five or more fixed benzene rings in a linear arrangement are formed from a numerical prefix (see Table 11.4) followed by -acene. A partial list of the names of polycyclic hydrocarbons is given in Table 1.2. Many names are trivial.

Numbering of each ring system is fixed, as shown in Table 1.2, but it follows a systematic pattern. The individual rings of each system are oriented so that the greatest number of rings are (1) in a horizontal row and (2) the maximum number of rings are above and to the right (upper-right quadrant) of the horizontal row. When two orientations meet these requirements, the one is chosen that has the fewest rings in the lower-left quadrant. Numbering proceeds in a clockwise direction, commencing with the carbon atom not engaged in ring fusion that lies in the most counterclockwise position of the uppermost ring (upper-right quadrant); omit atoms common to two or more rings. Atoms common to two or more rings are designated by adding lowercase roman letters to the number of the position immediately preceding. Interior atoms follow the highest number, taking a clockwise sequence wherever there is a choice. Anthracene and phenanthrene are two exceptions to the rule on numbering. Two examples of numbering follow:

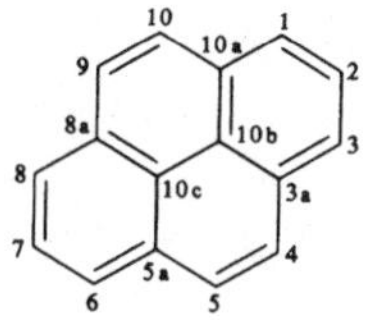

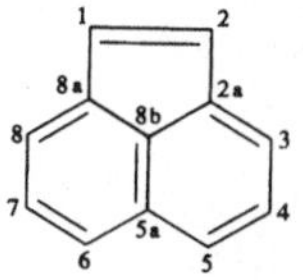

When a ring system with the maximum number of conjugated double bonds can exist in two or more forms differing only in the position of an "extra" hydrogen atom, the name can be made specific by indicating the position of the extra hydrogen(s). The compound name is modified with a locant followed by an italic capital *H* for each of these hydrogen atoms. Carbon atoms that carry an indicated hydrogen atom are numbered as low as possible. For example, 1*H*-indene is illustrated in Table 1.2; 2*H*-indene would be

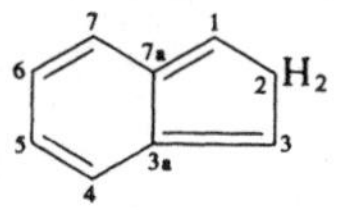

Names of polycyclic hydrocarbons with less than the maximum number of noncumulative double bonds are formed from a prefix dihydro-, tetrahydro-, etc., followed by the

name of the corresponding unreduced hydrocarbon. The prefix perhydro- signifies full hydrogenation. For example, 1,2-dihydronaphthalene is

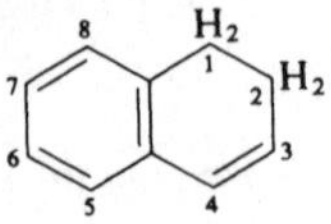

TABLE 1.2   Fused Polycyclic Hydrocarbons

Listed in order of increasing priority for selection as parent compound

Asterisk after a compound denotes exception to systematic numbering.

1. Pentalene	9. Acenaphthylene
2. Indene	10. Fluorene
3. Naphthalene	11. Phenalene
4. Azulene	12. Phenanthrene*
5. Heptalene	13. Anthracene*
6. Biphenylene	14. Fluoranthene
7. *asym*-Indacene	15. Acephenanthrylene
8. *sym*-Indacene	16. Aceanthrylene

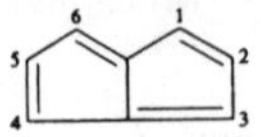

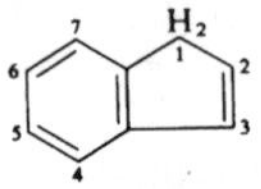

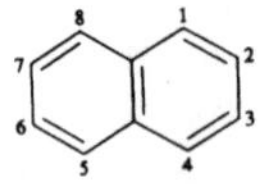

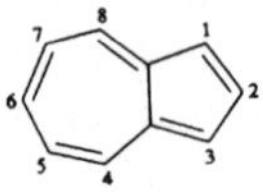

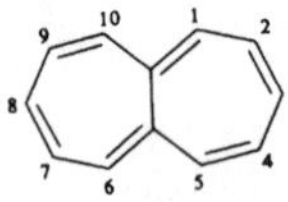

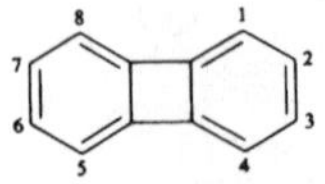

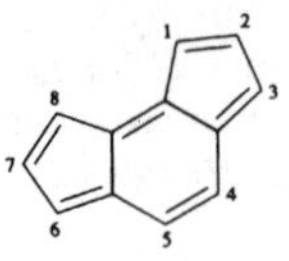

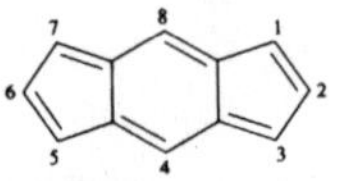

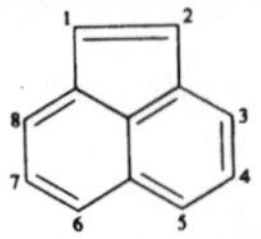

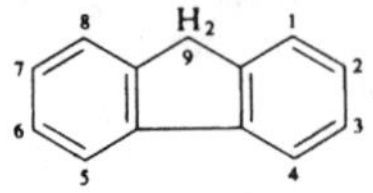

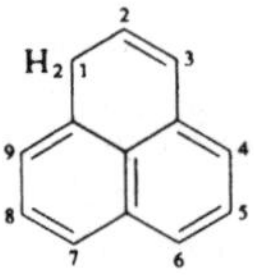

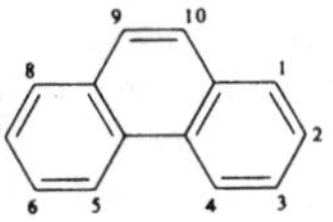

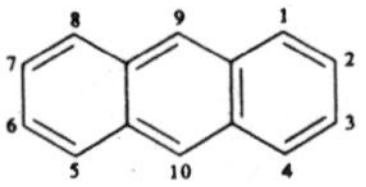

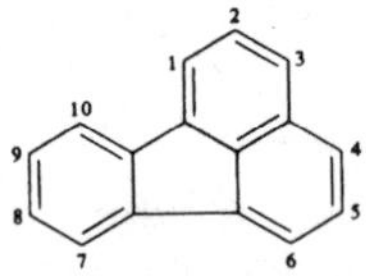

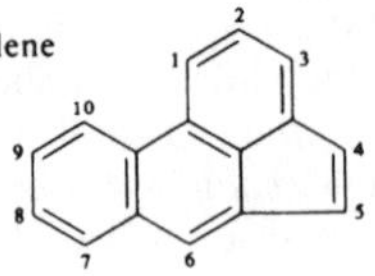

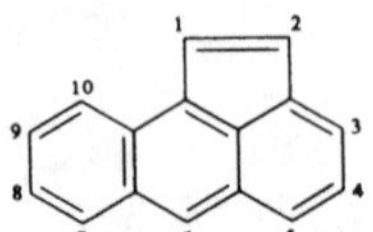

TABLE 1.2 Fused Polycyclic Hydrocarbons (*continued*)

Listed in order of increasing priority for selection as parent compound

Asterisk after a compound denotes exception to systematic numbering.

17. Triphenylene	19. Chrysene
18. Pyrene	20. Naphthacene

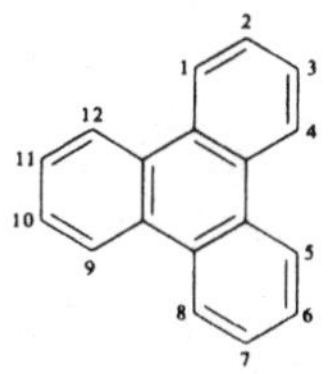

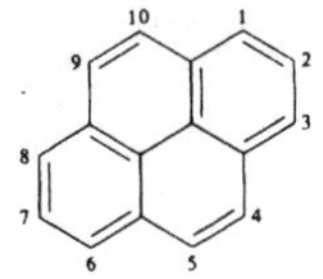

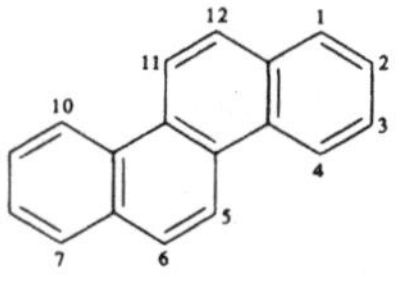

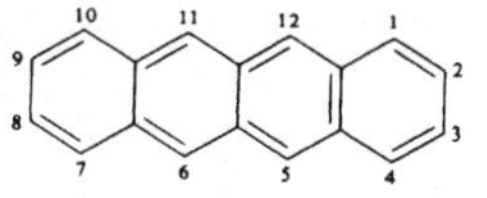

Examples of retained names and their structures are as follows:

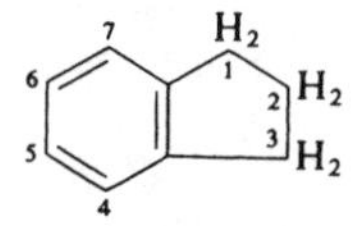

Indan 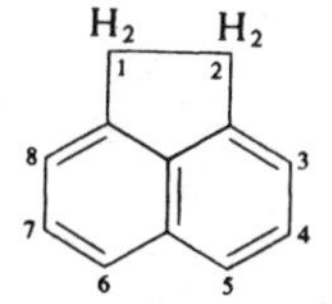   Acenaphthene 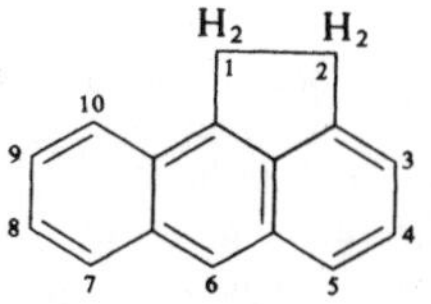   Aceanthrene

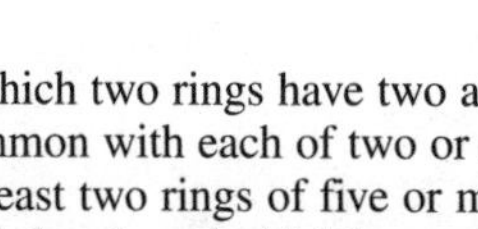

Polycyclic compounds in which two rings have two atoms in common or in which one ring contains two atoms in common with each of two or more rings of a contiguous series of rings and which contain at least two rings of five or more members with the maximum number of noncumulative double bonds and which have no accepted trivial name (Table 1.2) are named by prefixing to the name of the parent ring or ring system designations of the other components. The parent name should contain as many rings as possible (provided it has a trivial name) and should occur as far as possible from the beginning of the list in Table 1.2. Furthermore, the attached component(s) should be as simple as possible. For example, one writes dibenzo phenanthrene and not naphthophenanthrene because the attached component benzo- is simpler than naphtho-. Prefixes designating attached components are formed by changing the ending -ene into -eno-; for example, indeno- from indene. Multiple prefixes are arranged in alphabetical order. Several abbreviated prefixes are recognized; the parent is given in parentheses:

Acenaphtho-	(acenaphthylene)	Naphtho-	(naphthalene)
Anthra-	(anthracene)	Perylo-	(perylene)
Benzo-	(benzene)	Phenanthro-	(phenanthrene)

For monocyclic prefixes other than benzo-, the following names are recognized, each to represent the form with the maximum number of noncumulative double bonds: cyclopenta-, cyclohepta-, cycloocta-, etc.

Isomers are distinguished by lettering the peripheral sides of the parent beginning with *a* for the side 1,2-, and so on, lettering every side around the periphery. If necessary for clarity, the numbers of the attached position (1,2-, for example) of the substituent ring are also denoted. The prefixes are cited in alphabetical order. The numbers and letters are enclosed in square brackets and placed immediately after the designation of the attached component. Examples are

Benz[α]anthracene   Anthra[2,1-α]naphthacene

Bridged Hydrocarbons.   Saturated alicyclic hydrocarbon systems consisting of two rings that have two or more atoms in common take the name of the open-chain hydrocarbon containing the same total number of carbon atoms and are preceded by the prefix bicyclo-. The system is numbered commencing with one of the bridgeheads, numbering proceeding by the longest possible path to the second bridgehead. Numbering is then continued from this atom by the longer remaining unnumbered path back to the first bridgehead and is completed by the shortest path from the atom next to the first bridgehead. When a choice in numbering exists, unsaturation is given the lowest numbers. The number of carbon atoms in each of the bridges connecting the bridgeheads is indicated in brackets in descending order. Examples are

Bicyclo[3.2.1]octane   Bicyclo[5.2.0]nonane

Hydrocarbon Ring Assemblies.   Assemblies are two or more cyclic systems, either single rings or fused systems, that are joined directly to each other by double or single bonds. For identical systems naming may proceed (1) by placing the prefix bi- before the name of the corresponding radical or (2) for systems joined through a single bond, by placing the prefix bi- before the name of the corresponding hydrocarbon. In each case, the numbering of the assembly is that of the corresponding radical or hydrocarbon, one system being assigned unprimed numbers and the other primed numbers. The points of attachment are indicated by placing the appropriate locants before the name; an unprimed number is considered lower than the same number primed. The name *biphenyl* is used for the assembly consisting of two benzene rings. Examples are

1,1′-Bicyclopropyl or 1,1′-bicyclopropane   2-Ethyl-2′-propylbiphenyl

For nonidentical ring systems, one ring system is selected as the parent and the other systems are considered as substituents and are arranged in alphabetical order. The parent ring system is assigned unprimed numbers. The parent is chosen by considering the following characteristics in turn until a decision is reached: (1) the system containing the larger number of rings, (2) the system containing the larger ring, (3) the system in the lowest state of hydrogenation, and (4) the highest-order number of ring systems set forth in Table 1.2. Examples are given, with the deciding priority given in parentheses preceding the name:

(1) 2-Phenylnaphthalene
(2) and (4) 2-(2′-Naphthyl)azulene
(3) Cyclohexylbenzene

Radicals from Ring Systems. Univalent substituent groups derived from polycyclic hydrocarbons are named by changing the final *e* of the hydrocarbon name to -yl. The carbon atoms having free valences are given locants as low as possible consistent with the fixed numbering of the hydrocarbon. Exceptions are naphthyl (instead of naphthalenyl), anthryl (for anthracenyl), and phenanthryl (for phenanthrenyl). However, these abbreviated forms are used only for the simple ring systems. Substituting groups derived from fused derivatives of these ring systems are named systematically. Substituting groups having two or more free bonds are named as described in Monocyclic Aliphatic Hydrocarbons on p. 1.5.

Cyclic Hydrocarbons with Side Chains. Hydrocarbons composed of cyclic and aliphatic chains are named in a manner that is the simplest permissible or the most appropriate for the chemical intent. Hydrocarbons containing several chains attached to one cyclic nucleus are generally named as derivatives of the cyclic compound, and compounds containing several side chains and/or cyclic radicals attached to one chain are named as derivatives of the acyclic compound. Examples are

2-Ethyl-1-methylnaphthalene   Diphenylmethane
1,5-Diphenylpentane   2,3-Dimethyl-1-phenyl-1-hexene

Recognized trivial names for composite radicals are used if they lead to simplifications in naming. Examples are

1-Benzylnaphthalene   1,2,4-Tris(3-*p*-tolylpropyl)benzene

Fulvene, for methylenecyclopentadiene, and stilbene, for 1,2-diphenylethylene, are trival names that are retained.

Heterocyclic Systems. Heterocyclic compounds can be named by relating them to the corresponding carbocyclic ring systems by using replacement nomenclature. Heteroatoms are denoted by prefixes ending in *-a*, as shown in Table 1.3. If two or more replacement prefixes are required in a single name, they are cited in the order of their listing in the table. The lowest possible numbers consistent with the numbering of the corresponding carbocyclic system are assigned to the heteroatoms and then to carbon atoms bearing double

TABLE 1.3 Specialist Nomenclature for Heterocyclic Systems

Heterocyclic atoms are listed in decreasing order of priority

Element	Valence	Prefix	Element	Valence	Prefix
Oxygen	2	Oxa-	Antimony	3	Stiba-*
Sulfur	2	Thia-	Bismuth	3	Bisma-
Selenium	2	Selena-	Silicon	4	Sila-
Tellurium	2	Tellura-	Germanium	4	Germa-
Nitrogen	3	Aza-	Tin	4	Stanna-
Phosphorus	3	Phospha-*	Lead	4	Plumba-
Arsenic	3	Arsa-*	Boron	3	Bora-
			Mercury	2	Mercura-

* When immediately followed by -in or -ine, phospha- should be replaced by phosphor-, arsa- by arsen-, and stiba- by antimon-. The saturated six-membered rings corresponding to phosphorin and arsenin are named *phosphorinane* and *arsenane*. A further exception is the replacement of borin by borinane.

TABLE 1.4 Suffixes for Specialist Nomenclature of Heterocyclic Systems

Number of ring members	Rings containing nitrogen		Rings containing no nitrogen	
	Unsaturation*	Saturation	Unsaturation*	Saturation
3	-irine	-iridine	-irene	-irane
4	-ete	-etidine	-ete	-etane
5	-ole	-olidine	-ole	-olane
6	-ine†	‡	-in	-ane§
7	-epine	‡	-epin	-epane
8	-ocine	‡	-ocin	-ocane
9	-onine	‡	-onin	-onane
10	-ecine	‡	-ecin	-ecane

* Unsaturation corresponding to the maximum number of noncumulative double bonds. Heteroatoms have the normal valences given in Table 1.3.
† For phosphorus, arsenic, antimony, and boron, see the special provisions in Table 1.3.
‡ Expressed by prefixing perhydro- to the name of the corresponding unsaturated compound.
§ Not applicable to silicon, germanium, tin, and lead; perhydro- is prefixed to the name of the corresponding unsaturated compound.

TABLE 1.5 Trivial Names of Heterocyclic Systems Suitable for Use in Fusion Names

Listed in order of increasing priority as senior ring system

Asterisk after a compound denotes exception to systematic numbering.

Structure	Parent name	Radical name	Structure	Parent name	Radical name
	Thiophene	Thienyl		2H-Pyrrole	2H-Pyrrolyl
				Pyrrole	Pyrrolyl
	Thianthrene	Thianthrenyl		Imidazole	Imidazolyl
	Furan	Furyl		Pyrazole	Pyrazolyl
	Pyran (2H-shown)	Pyranyl		Isothiazole	Isothiazolyl
	Isobenzofuran	Isobenzofuranyl		Isoxazole	Isoxazolyl
	Chromene (2H-shown)	Chromenyl		Pyridine	Pyridyl
				Pyrazine	Pyrazinyl
	Xanthene*	Xanthenyl		Pyrimidine	Pyrimidinyl
	Phenoxathiin	Phenoxathiinyl		Pyridazine	Pyridazinyl

TABLE 1.5 Trivial Names of Heterocyclic Systems Suitable for Use in Fusion Names (*continued*)
Listed in order of increasing priority as senior ring system

Asterisk after a compound denotes exception to systematic numbering.

Structure	Parent name	Radical name	Structure	Parent name	Radical name
	Indolizine	Indolizinyl		Phthalazine	Phthalazinyl
	Isoindole	Isoindolyl		Naphthyridine (1,8-shown)	Naphthyridinyl
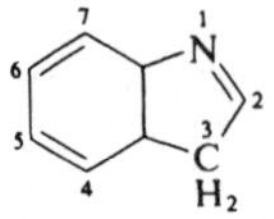	3*H*-Indole	3*H*-Indolyl		Quinoxaline	Quinoxalinyl
	Indole	Indolyl		Quinazoline	Quinazolinyl
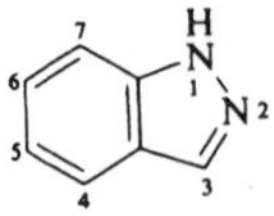	1*H*-Indazole	1*H*-Indazolyl		Cinnoline	Cinnolinyl
	Purine*	Purinyl		Pteridine	Pteridinyl
	4*H*-Quinolizine	4*H*-Quinolizinyl		4α*H*-Carbazole*	4α*H*-Carbazolyl
	Isoquinoline	Isoquinolyl		Carbazole*	Carbazolyl
	Quinoline	Quinolyl			

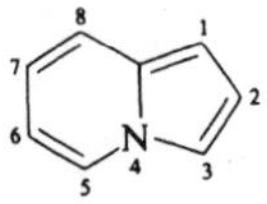

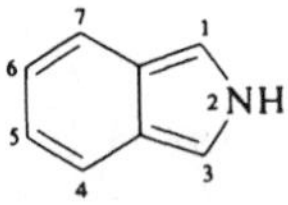

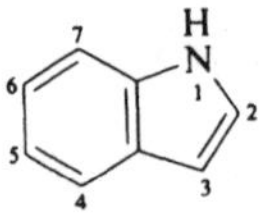

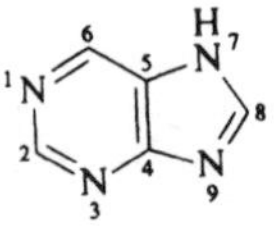

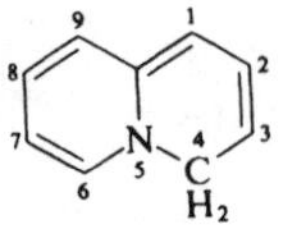

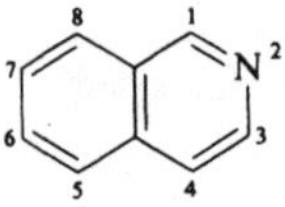

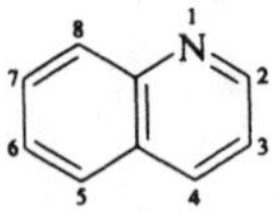

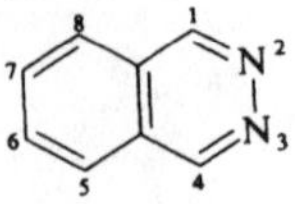

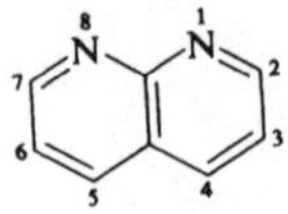

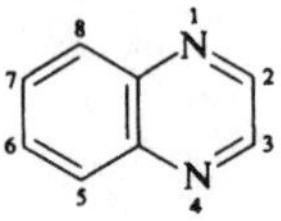

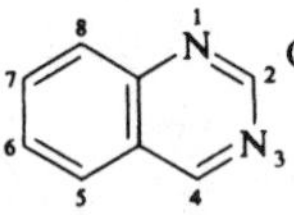

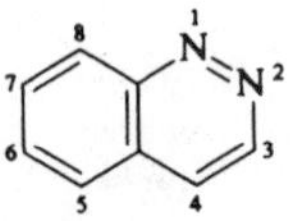

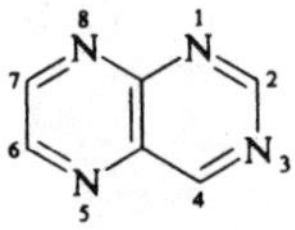

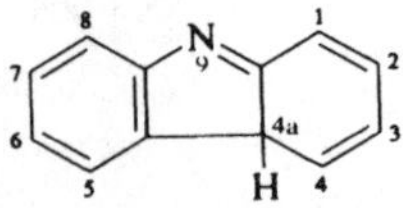

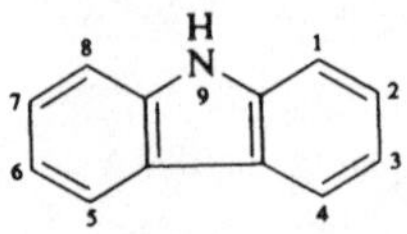

TABLE 1.5 Trivial Names of Heterocyclic Systems Suitable for Use in Fusion Names (*continued*)
Listed in order of increasing priority as senior ring system

Asterisk after a compound denotes exception to systematic numbering.

Structure	Parent name	Radical name	Structure	Parent name	Radical name
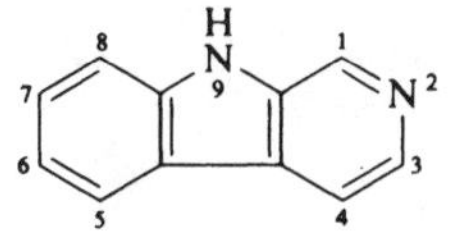	β-Carboline	β-Carbolinyl	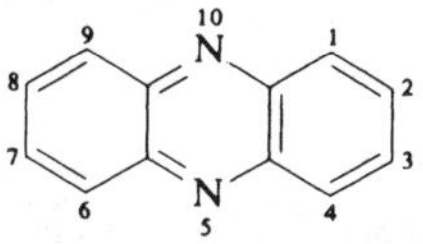	Phenazine	Phenazinyl
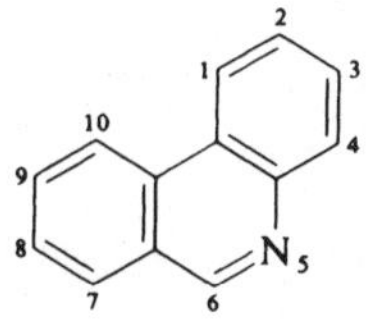	Phenanthri-dine	Phenanthri-dinyl	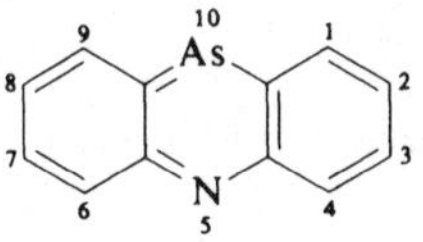	Phenarsazine	Phenarsazinyl
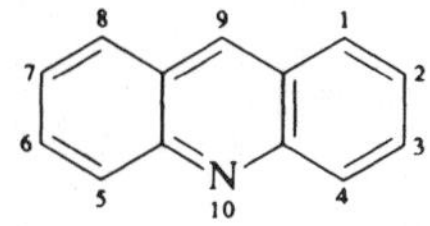	Acridine*	Acridinyl	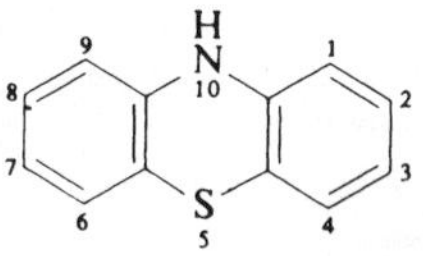	Phenothiazine	Phenothiazinyl
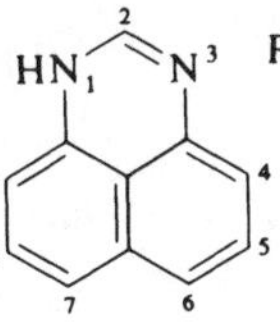	Perimidine	Perimidinyl	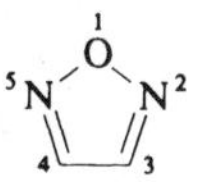	Furazan	Furazanyl
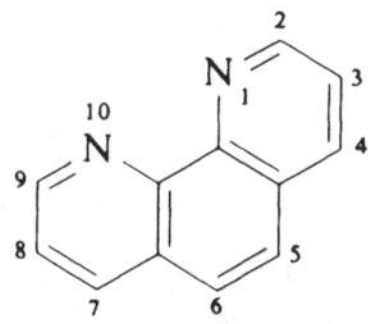	Phenanthroline (1,10-shown)	Phenanthrolinyl	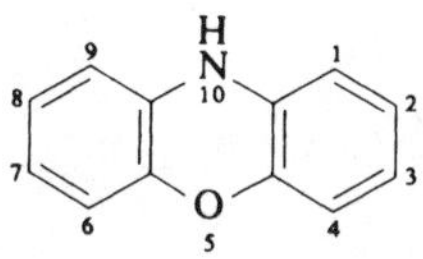	Phenoxazine	Penoxazinyl

or triple bonds. Locants are cited immediately preceding the prefixes or suffixes to which they refer. Multiplicity of the same heteroatom is indicated by the appropriate prefix in the series: di-, tri-, tetra-, penta-, hexa-, etc.

If the corresponding carbocyclic system is partially or completely hydrogenated, the additional hydrogen is cited using the appropriate *H*- or hydro- prefixes. A trivial name from Tables 1.5 and 1.6, if available, along with the state of hydrogenation may be used. In the specialist nomenclature for heterocyclic systems, the prefix or prefixes from

TABLE 1.6 Trivial Names for Heterocyclic Systems that are Not Recommended for Use in Fusion Names

Listed in order of increasing priority

Structure	Parent name	Radical name	Structure	Parent name	Radical name
	Isochroman	Isochromanyl		Pyrazoline (3-shown*)	Pyrazolinyl
	Chroman	Chromanyl		Piperidine	Piperidyl†
	Pyrrolidine	Pyrrolidinyl		Piperazine	Piperazinyl
	Pyrroline (2-shown*)	Pyrrolinyl		Indoline	Indolinyl
	Imidazolidine	Imidazolidinyl		Isoindoline	Isoindolinyl
	Imidazoline (2-shown*)	Imidazolinyl		Quinuclidine	Quinuclidinyl
	Pyrazolidine	Pyrazolidinyl		Morpholine	Morpholinyl‡

* Denotes position of double bond.
† For 1-piperidyl, use piperidino.
‡ For 4-morpholinyl, use morpholino.

Table 1.3 are combined with the appropriate stem from Table 1.4, eliding an *-a* where necessary. Examples of acceptable usage, including (1) replacement and (2) specialist nomenclature, are

(1) 1-Oxa-4-azacyclo-hexane	(1) 1,3-Diazacyclo-hex-5-ene	(1) Thiacyclopropane
(2) 1,4-Oxazoline Morpholine	(2) 1,2,3,4-Tetra-hydro-1,3-diazine	(2) Thiirane Ethylene sulfide

Radicals derived from heterocyclic compounds by removal of hydrogen from a ring are named by adding -yl to the names of the parent compounds (with elision of the final *e*, if present). These exceptions are retained:

Furyl (from furan)	Furfuryl (for 2-furylmethyl)
Pyridyl (from pyridine)	Furfurylidene (for 2-furylmethylene)
Piperidyl (from piperidine)	Thienyl (from thiophene)
Quinolyl (from quinoline)	Thenylidyne (for thienylmethylidyne)
Isoquinolyl	Furfurylidyne (for 2-furylmethylidyne)
Thenylidene (for thienylmethylene)	Thenyl (for thienylmethyl)

Also, piperidino- and morpholino- are preferred to 1-piperidyl- and 4-morpholinyl-, respectively.

If there is a choice among heterocyclic systems, the parent compound is decided in the following order of preference:

1. A nitrogen-containing component

2. A component containing a heteroatom, in the absence of nitrogen, as high as possible in Table 1.3

3. A component containing the greatest number of rings

4. A component containing the largest possible individual ring

5. A component containing the greatest number of heteroatoms of any kind

6. A component containing the greatest variety of heteroatoms

7. A component containing the greatest number of heteroatoms first listed in Table 1.3

If there is a choice between components of the same size containing the same number and kind of heteroatoms, choose as the base component that one with the lower numbers for the heteroatoms before fusion. When a fusion position is occupied by a heteroatom, the names of the component rings to be fused are selected to contain the heteroatom.

Common Names of Heterocycles Used Broadly in Biology. The naming of heterocycles by systematic methods is important but cumbersome for designating some of the most commonly occurring heterocycles. In particular, the bases that occur in ribonucleic acids (RNA) and deoxyribonucleic acids (DNA) have specific substitution patterns. Because they occur so commonly, they have been given trivial names that are invariably used when discussed or named in the biological literature.

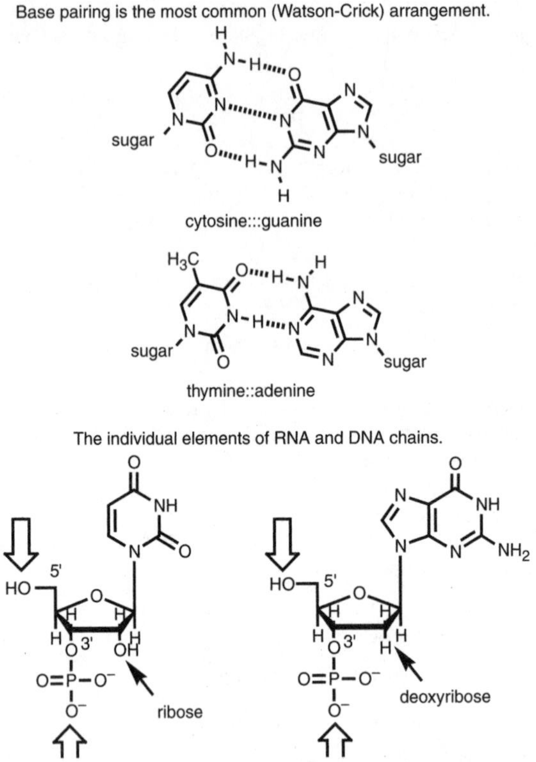

Base pairing is the most common (Watson-Crick) arrangement.

cytosine:::guanine

thymine::adenine

The individual elements of RNA and DNA chains.

ribose

deoxyribose

FIGURE 1.2 Base pairing in the most common (Watson–Crick) arrangement. The individual elements of RNA and DNA chains are shown in the lower panel of the figure. Hollow arrows indicate the points at which the 5′-hydroxyl group is esterified to the 3′-phosphate group to form the so-called "sugar–phosphate" backbone. Note the hydroxyl group (arrow) that is present on ribose but missing in deoxyribose.

The structural frameworks of DNA and RNA are organized by hydrogen bond formation between pairs of purine and pyrimidine bases. The pyrimidines are shown near the end of Table 1.5. Cytosine (C) and thymine (T) occur in DNA and form hydrogen-bonded pairs with the purines guanine (G) and adenine (A), respectively. The base pairs are abbreviated AT and GC, sometimes with dotted lines connecting them. The AT pair is held together by two hydrogen bonds and may be represented in shorthand as A::T. Three H-bonds hold together guanine and cytosine, giving G:::C. The so-called Watson–Crick base pairing is shown in Figure 1.2. In RNA, uracil replaces thymine but pairing still occurs with adenine to give A::U.

An alternative form of hydrogen bonding between base pairs is designated "Hoogsteen." This type of bonding cannot readily occur in nature because the purine and pyrimidine bases are constrained to long chains that must interact at numerous points.

Functionalized Compounds

There are several types of nomenclature systems that are recognized. Which type to use is sometimes obvious from the nature of the compound. Substitutive nomenclature, in general, is preferred because of its broad applicability, but radicofunctional, additive, and replacement nomenclature systems are convenient in certain situations.

Substitutive Nomenclature. The first step is to determine the kind of characteristic (functional) group for use as the principal group of the parent compound. A characteristic group is a recognized combination of atoms that confers characteristic chemical properties on the molecule in which it occurs. Carbon-to-carbon unsaturation and heteroatoms in rings are considered nonfunctional for nomenclature purposes.

Substitution means the replacement of one or more hydrogen atoms in a given compound by some other kind of atom or group of atoms, functional or nonfunctional. In substitutive nomenclature, each substituent is cited as either a prefix or a suffix to the name of the parent (or substituting radical) to which it is attached; the latter is denoted the parent compound (or parent group if a radical).

In Table 1.7 are listed the general classes of compounds in descending order of preference for citation as suffixes, that is, as the parent or characteristic compound. When oxygen is

TABLE 1.7 Characteristic Groups for Substitutive Nomenclature

Listed in order of decreasing priority for citation as principal group or parent name

Class	Formula*	Prefix	Suffix
1. Cations		-onio-	-onium
	H_4N^+	Ammonio-	-ammonium
	H_3O^+	Oxonio-	-oxonium
	H_3S^+	Sulfonio-	-sulfonium
	H_3Se^+	Selenonio-	-selenonium
	H_2Cl^+	Chloronio-	-chloronium
	H_2Br^+	Bromonio-	-bromonium
	H_2I^+	Iodonio-	-iodonium
2. Acids			
Carboxylic	—COOH	Carboxy-	-carboxylic acid
	—(C)OOH		-oic acid
	—C(=O)OOH		-peroxy $\cdots$ carboxylic acid
	—(C=O)OOH		-peroxy $\cdots$ oic acid
Sulfonic	—SO₃H	Sulfo-	-sulfonic acid
Sulfinic	—SO₂H	Sulfino-	-sulfinic acid
Sulfenic	—SOH	Sulfeno-	-sulfenic acid
Salts	—COOM		Metal $\cdots$ carboxylate
	—(C)OOM		Metal $\cdots$ oate
	—SO₃M		Metal $\cdots$ sulfonate
	—SO₂M		Metal $\cdots$ sulfinate
	—SOM		Metal $\cdots$ sulfenate
3. Derivatives of acids			
Anhydrides	—C(=O)OC(=O)—		-carboxylic anhydride
	—(C=O)O(C=O)—		-oic anhydride
Esters	—COOR	R-oxycarbonyl-	R $\cdots$ carboxylate
	—C(OOR)		R $\cdots$ oate
Acid halides	—CO—halogen	Haloformyl	-carbonyl halide
Amides	—CO—NH₂	Carbamoyl-	-carboxamide
	(C)O—NH₂		-amide

TABLE 1.7   Characteristic Groups for Substitutive Nomenclature (*continued*)

Listed in order of decreasing priority for citation as principal group or parent name

Class	Formula*	Prefix	Suffix
Hydrazides	$-CO-NHNH_2$ $-(CO)-NHNH_2$	Carbonyl-hydrazino-	-carbohydrazide -ohydrazide
Imides	$-CO-NH-CO-$	R-imido-	-carboximide
Amidines	$-C(=NH)-NH_2$ $-(C=NH)-NH_2$	Amidino-	-carboxamidine -amidine
4. Nitrile (cyanide)	$-CN$ $-(C)N$	Cyano-	-carbonitrile -nitrile
5. Aldehydes	$-CHO$ $-(C=O)H$ (then their analogs and derivatives)	Formyl- Oxo-	-carbaldehyde -al
6. Ketones	$>(C=O)$ (then their analogs and derivatives)	Oxo-	-one
7. Alcohols (and phenols)	$-OH$	Hydroxy-	-ol
Thiols	$-SH$	Mercapto-	-thiol
8. Hydroperoxides	$-O-OH$	Hydroperoxy-	
9. Amines	$-NH_2$	Amino-	-amine
Imines	$=NH$	Imino-	-imine
Hydrazines	$-NHNH_2$	Hydrazino-	-hydrazine
10. Ethers	$-OR$	R-oxy-	
Sulfides	$-SR$	R-thio-	
11. Peroxides	$-O-OR$	R-dioxy-	

* Carbon atoms enclosed in parentheses are included in the name of the parent compound and not in the suffix or prefix.

replaced by sulfur, selenium, or tellurium, the priority for these elements is in the descending order listed. The higher valence states of each element are listed before considering the successive lower valence states. Derivative groups have priority for citation as principal group after the respective parents of their general class.

In Table 1.8 are listed characteristic groups that are cited only as prefixes (never as suffixes) in substitutive nomenclature. The order of listing has no significance for nomenclature purposes.

Systematic names formed by applying the principles of substitutive nomenclature are single words except for compounds named as acids. First one selects the parent compound, and thus the suffix, from the characteristic group listed earliest in Table 1.7. All remaining functional groups are handled as prefixes that precede, in alphabetical order, the parent name. Two examples may be helpful:

Structure 1   Structure 2

TABLE 1.8 Characteristic Groups Cited Only as Prefixes in Substitutive Nomenclature

Characteristic group	Prefix	Characteristic group	Prefix
—Br	Bromo-	—IX$_2$	X may be halogen or a radical; dihalogenoiodo- or diacetoxyiodo-, e.g., —ICl$_2$ is dichloroido-
—Cl	Chloro-		
—ClO	Chlorosyl-		
—ClO$_2$	Chloryl-	=N$^\oplus$=N$^\ominus$	Diazo-
—ClO$_3$	Perchloryl-	—N$_3$, —N=N$^\oplus$=N$^\ominus$	Azido-
—F	Fluoro-	—N=O	Nitroso-
—I	Iodo-	—NO$_2$, —N$^\oplus\diagdown{O} \atop \diagup{O_\ominus}$	Nitro-
—IO	Iodosyl-	—N$^\oplus\diagdown{OH} \atop \diagup{O_\ominus}$	
		=N$^{\oplus\diagdown}$	aci-Nitro-
—IO$_2$	Iodyl*	—OR $O_\ominus$	R-oxy-; alkoxy- or aryloxy-
—I(OH)$_2$	Dihydroxyiodo-	—SR	R-thio-; alkylthio- or arylthio-
		—SeR (—TeR)	R-seleno- (R-telluro-)

* Formerly iodoxy-.

Structure 1 contains an ester group and an ether group. Since the ester group has higher priority, the name is ethyl 2-methoxy-6-methyl-3-cyclohexene-1-carboxylate. Structure 2 contains a carbonyl group, an hydroxy group, and a bromo group. The latter is never a suffix. Between the other two, the carbonyl group has higher priority, the parent has -one as suffix, and the name is 4-bromo-1-hydroxy-2-butanone.

Selection of the principal alicyclic chain or ring system is governed by the following selection rules:

1. For purely alicyclic compounds, the selection process proceeds successively until a decision is reached: (a) the maximum number of substituents corresponding to the characteristic group cited earliest in Table 1.7, (b) the maximum number of double and triple bonds considered together, (c) the maximum length of the chain, and (d) the maximum number of double bonds. Additional criteria, if needed for complicated compounds, are given in the IUPAC nomenclature rules.

2. If the characteristic group occurs only in a chain that carries a cyclic substituent, the compound is named as an aliphatic compound into which the cyclic component is substituted; a radical prefix is used to denote the cyclic component. This chain need not be the longest chain.

3. If the characteristic group occurs in more than one carbon chain and the chains are not directly attached to one another, then the chain chosen as parent should carry the largest number of the characteristic group. If necessary, the selection is continued as in rule 1.

4. If the characteristic group occurs only in one cyclic system, that system is chosen as the parent.

5. If the characteristic group occurs in more than one cyclic system, that system is chosen as parent which (a) carries the largest number of the principal group or, failing to reach a decision, (b) is the senior ring system.

6. If the characteristic group occurs both in a chain and in a cyclic system, the parent is that portion in which the principal group occurs in largest number. If the numbers are the same, that portion is chosen which is considered to be the most important or is the senior ring system.

7. When a substituent is itself substituted, all the subsidiary substituents are named as prefixes and the entire assembly   is regarded as a parent radical.

8. The seniority of ring systems is ascertained by applying the following rules successively until a decision is reached: (*a*) all heterocycles are senior to all carbocycles, (*b*) for heterocycles, the preference follows the decision process described under "Heterocyclic Systems," page 1–12, (*c*) the largest number of rings, (*d*) the largest individual ring at the first point of difference, (*e*) the largest number of atoms in common among rings, (*f*) the lowest letters in the expression for ring functions, (*g*) the lowest numbers at the first point of difference in the expression for ring junctions, (*h*) the lowest state of hydrogenation, (*i*) the lowest-numbered locant for indicated hydrogen, (*j*) the lowest-numbered locant for point of attachment (if a radical), (*k*) the lowest-numbered locant for an attached group expressed as a suffix, (*l*) the maximum number of substituents cited as prefixes, (*m*) the lowest-numbered locant for substituents named as prefixes, hydro prefixes, -ene, and -yne, all considered together in one series in ascending numerical order independent of their nature, and (*n*) the lowest-numbered locant for the substituent named as prefix which is cited first in the name.

Numbering of Compounds.   If the rules for aliphatic chains and ring systems leave a choice, the starting point and direction of numbering of a compound are chosen so as to give lowest-numbered locants to these structural factors, if present, considered successively in the order listed below until a decision is reached. Characteristic groups take precedence over multiple bonds.

1. Indicated hydrogen, whether cited in the name or omitted as being conventional.

2. Characteristic groups named as suffix following the ranking order of Table 1.7.

3. Multiple bonds in acyclic compounds; in bicycloalkanes, tricycloalkanes, and polycycloalkanes, double bonds having priority over triple bonds; and in heterocyclic systems whose names end in -etine, -oline, or -olene.

4. The lowest-numbered locant for substituents named as prefixes, hydro prefixes, -ene, and -yne, all considered together in one series in ascending numerical order.

5. The lowest locant for that substituent named as prefix which is cited first in the name.

For cyclic radicals, indicated hydrogen and thereafter the point of attachment (free valency) have priority for the lowest available number.

Prefixes and Affixes.   Prefixes are arranged alphabetically and placed before the parent name; multiplying affixes, if necessary, are inserted and *do not* alter the alphabetical order already attained. The parent name includes any syllables denoting a change of ring member or relating to the structure of a carbon chain. Nondetachable parts of parent names include

1. Forming rings: cyclo-, bicyclo-, spiro-;

2. Fusing two or more rings: benzo-, naphtho-, imidazo-;

3. Substituting one ring or chain member atom for another: oxa-, aza-, thia-;

4. Changing positions of ring or chain members: iso-, *sec*-, *tert*-, neo-;

5. Showing indicated hydrogen;

6. Forming bridges: ethano-, epoxy- and;

7. Hydro-.

Prefixes that represent complete terminal characteristic groups are preferred to those representing only a portion of a given group. For example, for the group $-C(=O)CH_3$, the prefix (formylmethyl-) is preferred to (oxoethyl-).

The multiplying affixes di-, tri-, tetra-, penta-, hexa-, hepta-, octa-, nona-, deca-, undeca-, and so on are used to indicate a set of *identical* unsubstituted radicals or parent compounds. The forms bis-, tris-, tetrakis-, pentakis-, and so on are used to indicate a set of identical radicals or parent compounds *each substituted in the same way*. The affixes bi-, ter-, quater-, quinque-, sexi-, septi-, octi-, novi-, deci-, and so on are used to indicate the number of identical rings joined together by a single or double bond.

Although multiplying affixes may be omitted for very common compounds when no ambiguity is caused thereby, such affixes are generally included throughout this handbook in alphabetical listings. An example would be ethyl ether for diethyl ether.

Conjunctive Nomenclature. Conjunctive nomenclature may be applied when a principal group is attached to an acyclic component that is directly attached by a carbon–carbon bond to a cyclic component. The name of the cyclic component is attached directly in front of the name of the acyclic component carrying the principal group. This nomenclature is not used when an unsaturated side chain is named systematically. When necessary, the position of the side chain is indicated by a locant placed before the name of the cyclic component. For substituents on the acyclic chain, carbon atoms of the side chain are indicated by Greek letters proceeding from the principal group to the cyclic component. The terminal carbon atom of acids, aldehydes, and nitriles is omitted when allocating Greek positional letters. Conjunctive nomenclature is not used when the side chain carries more than one of the principal group, except in the case of malonic and succinic acids.

The side chain is considered to extend only from the principal group to the cyclic component. Any other chain members are named as substituents, with appropriate prefixes placed before the name of the cyclic component.

When a cyclic component carries more than one identical side chain, the name of the cyclic component is followed by di-, tri-, etc., and then by the name of the acyclic component, and it is preceded by the locants for the side chains. Examples are

4-Methyl-1-cyclohexaneethanol

α-Ethyl-β,β-dimethylcyclohexaneethanol

When side chains of two or more different kinds are attached to a cyclic component, only the senior side chain is named by the conjunctive method. The remaining side chains are named as prefixes. Likewise, when there is a choice of cyclic component, the senior is chosen. Benzene derivatives may be named by the conjunctive method only when two or more identical side chains are present. Trivial names for oxo carboxylic acids may be used for the acyclic component. If the cyclic and acyclic components are joined by a double bond, the locants of this bond are placed as superscripts to a Greek capital delta that is inserted between the two names. The locant for the cyclic component precedes that for the acyclic component, e.g., indene-$\Delta^{1,\alpha}$-acetic acid.

Radicofunctional Nomenclature. The procedures of radicofunctional nomenclature are identical with those of substitutive nomenclature except that suffixes are never used. Instead, the functional class name (Table 1.9) of the compound is expressed as one word and the remainder of the molecule as another that precedes the class name. When the functional class name refers to a characteristic group that is bivalent, the two radicals attached to it are each named, and when different, they are written as separate words arranged in alphabetical order. When a compound contains more than one kind of group listed in Table 1.9, that kind is cited as the functional group or class name that occurs higher in the table, all others being expressed as prefixes.

Radicofunctional nomenclature finds some use in naming ethers, sulfides, sulfoxides, sulfones, selenium analogs of the preceding three sulfur compounds, and azides.

TABLE 1.9   Functional Class Names Used in Radicofunctional Nomenclature

Groups are listed in order of decreasing priority

Group	Functional class names
X in acid derivatives	Name of X (in priority order: fluoride, chloride, bromide, iodide; cyanide, azide; then the sulfur and selenium analogs)
—CN, —NC	Cyanide, isocyanide
$>$CO	Ketone; then S and Se analogs
—OH	Alcohol; then S and Se analogs
—O—OH	Hydroperoxide
$>$O	Ether or oxide
$>$S, $>$SO, $>$SO$_2$	Sulfide, sulfoxide, sulfone
$>$Se, $>$SeO, $>$SeO$_2$	Selenide, selenoxide, selenone
—F, —Cl, —Br, —I	Fluoride, chloride, bromide, iodide
—N$_3$	Azide

Replacement Nomenclature. Replacement nomenclature is intended for use only when other nomenclature systems are difficult to apply in the naming of chains containing heteroatoms. When no group is present that can be named as a principal group, the longest chain of carbon and heteroatoms terminating with carbon is chosen and named as though the entire chain were that of an acyclic hydrocarbon. The heteroatoms within this chain are identified by means of prefixes aza-, oxa-, thia-, etc., in the order of priority stated in Table 1.3. Locants indicate the positions of the heteroatoms in the chain. Lowest-numbered locants are assigned to the principal group when such is present. Otherwise, lowest-numbered locants are assigned to the heteroatoms considered together and, if there is a choice, to the heteroatoms cited earliest in Table 1.3. An example is

$$HO-\overset{13}{C}H_2-\overset{12}{O}-\overset{11}{C}H_2-\overset{10}{C}H_2-\overset{9}{O}-\overset{8}{C}H_2-\overset{7}{C}H_2-\overset{6}{N}-\overset{5}{C}H_2-\overset{4}{C}H_2-\overset{3}{N}-\overset{2}{C}H_2-\overset{1}{C}OOH$$

13-Hydroxy-9,12-dioxa-3,6-diazatridecanoic acid

Specific Functional Groups

Characteristic groups will now be treated briefly in order to expand the terse outline of substitutive nomenclature presented in Table 1.7. Alternative nomenclature will be indicated whenever desirable.

Acetals and Acylals.   Acetals, which contain the group $>C(OR)_2$, where R may be different, are named (1) as dialkoxy compounds or (2) by the name of the corresponding aldehyde or ketone followed by the name of the hydrocarbon radical(s) followed by the word *acetal*. For example, $CH_3-CH(OCH_3)_2$ is named either (1) 1,1-dimethoxyethane or (2) acetaldehyde dimethyl acetal.

A cyclic acetal in which the two acetal oxygen atoms form part of a ring may be named (1) as a heterocyclic compound or (2) by use of the prefix methylenedioxy for the group $-O-CH_2-O-$ as a substituent in the remainder of the molecule. For example,

(1) 1,3-Benzo[*d*]dioxole-5-carboxylic acid

(2) 3,4-Methylenedioxybenzoic acid

Acylals, $R^1R^2C(OCOR^3)_2$, are named as acid esters:

Butylidene acetate propionate

α-Hydroxy ketones, formerly called acyloins, had been named by changing the ending -ic acid or -oic acid of the corresponding acid to -oin. They are preferably named by substitutive nomenclature. For example,

$CH_3-CH(OH_3)-CO-CH_3$ 3-Hydroxy-2-butanone (formerly acetoin)

Acid Anhydrides.   Symmetrical anhydrides of monocarboxylic acids, when unsubstituted, are named by replacing the word *acid* by *anhydride*. Anhydrides of substituted monocarboxylic acids, if symmetrically substituted, are named by prefixing bis- to the name of the acid and replacing the word *acid* by *anhydride*. Mixed anhydrides are named by giving in alphabetical order the first part of the names of the two acids followed by the word *anhydride*, e.g., acetic propionic anhydride or acetic propanoic anhydride. Cyclic anhydrides of polycarboxylic acids, although possessing a heterocyclic structure, are preferably named as acid anhydrides. For example,

1,8;4,5-Napthalenetetracarboxylic dianhydride. (Note the use of a semicolon to distinguish the pairs of locants.)

Acyl Halides.   Acyl halides, in which the hydroxyl portion of a carboxyl group is replaced by a halogen, are named by placing the name of the corresponding halide after that of the acyl radical. When another group is present that has priority for citation as principal group or when the acyl halide is attached to a side chain, the prefix haloformyl- is used as, for example, in fluoroformyl-.

Alcohols and Phenols.   The hydroxyl group is indicated by a suffix -ol when it is the principal group attached to the parent compound and by the prefix hydroxy- when another group with higher priority for citation is present or when the hydroxy group is present in a side chain. When confusion may arise in employing the suffix -ol, the hydroxy group is indicated as a prefix; this terminology is also used when the hydroxyl group is attached to a heterocycle, as, for example, in the name 3-hydroxythiophene to avoid confusion with thiophenol (C_6H_5SH). Designations such as isopropanol, *sec*-butanol, and *tert*-butanol are incorrect because no hydrocarbon exists to which the suffix can be added. Many trivial names are retained. These structures are shown in Table 1.10. The radicals (RO—) are named by adding -oxy as a suffix to the name of the R radical, e.g., pentyloxy for $CH_3CH_2CH_2CH_2CH_2O$—. These contractions are exceptions: methoxy (CH_3O—), ethoxy (C_2H_5O—), propoxy (C_3H_7O—), butoxy (C_4H_9O—), and phenoxy (C_6H_5O—). For unsubstituted radicals only, one may use isopropoxy [$(CH_3)_2CH$—O—], isobutoxy [$(CH_3)_2CH_2CH$—O—], *sec*-butoxy [$CH_3CH_2CH(CH_3)$—O—], and *tert*-butoxy [$(CH_3)_3C$—O—].

TABLE 1.10   Retained Trivial Names of Alcohols and Phenols with Structures

Ally alcohol	$CH_2{=}CHCH_2OH$
tert-Butyl alcohol	$(CH_3)_3COH$
Benzyl alcohol	$C_6H_5CH_2OH$
Phenethyl alcohol	$C_6H_5CH_2CH_2OH$
Ethylene glycol	$HOCH_2CH_2OH$
1,2-Propylene glycol	$CH_3CHOHCH_2OH$
Glycerol	$HOCH_2CHOHCH_2OH$
Pentaerythritol	$C(CH_2OH)_4$
Pinacol	$(CH_3)_2COHCOH(CH_3)_2$
Phenol	C_6H_5OH

$$Xylitol \qquad HOCH_2CH{-}\overset{\displaystyle OH}{\underset{}{CH}}{-}\underset{\displaystyle OH}{CH}{-}\underset{\displaystyle OH}{CH}{-}CH_2OH$$

$$Geraniol \qquad (CH_3)_2C{=}CHCH_2CH_2\underset{\displaystyle CH_3}{C}{=}CHCH_2OH$$

Phytol

$$CH_2CH_2\overset{\displaystyle CH_3}{CH}CH_2CH_2CH_2CH(CH_3)_2$$
$$\underset{\displaystyle CH_3}{CH_2}CHCH_2CH_2CH_2\underset{\displaystyle CH_3}{C}{=}CHCH_2OH$$

TABLE 1.10 Retained Trivial Names of Alcohols and Phenols with Structures (*continued*)

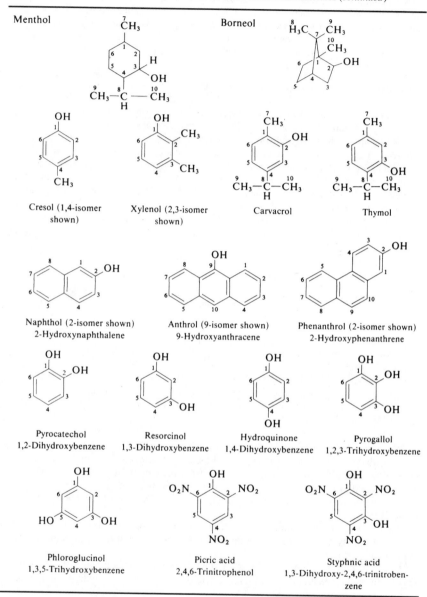

Menthol

Borneol

Cresol (1,4-isomer shown)

Xylenol (2,3-isomer shown)

Carvacrol

Thymol

Naphthol (2-isomer shown)
2-Hydroxynaphthalene

Anthrol (9-isomer shown)
9-Hydroxyanthracene

Phenanthrol (2-isomer shown)
2-Hydroxyphenanthrene

Pyrocatechol
1,2-Dihydroxybenzene

Resorcinol
1,3-Dihydroxybenzene

Hydroquinone
1,4-Dihydroxybenzene

Pyrogallol
1,2,3-Trihydroxybenzene

Phloroglucinol
1,3,5-Trihydroxybenzene

Picric acid
2,4,6-Trinitrophenol

Styphnic acid
1,3-Dihydroxy-2,4,6-trinitrobenzene

Bivalent radicals of the form O—Y—O are named by adding -dioxy to the name of the bivalent radicals except when forming part of a ring system. Examples are —O—CH$_2$—O—(methylenedioxy), —O—CO—O— (carbonyldioxy), and —O—SO$_2$—O— (sulfonyldioxy). Anions derived from alcohols or phenols are named by changing the final -ol to -olate.

Salts composed of an anion, RO—, and a cation, usually a metal, can be named by citing first the cation and then the RO anion (with its ending changed to -yl oxide), e.g., sodium benzyl oxide for $C_6H_5CH_2ONa$. However, when the radical has an abbreviated name, such as methoxy, the ending -oxy is changed to -oxide. For example, CH_3ONa is named sodium methoxide (not sodium methylate).

Aldehydes.   When the group —$C(=O)H$, usually written —CHO, is attached to carbon at one (or both) end(s) of a linear acyclic chain the name is formed by adding the suffix -al (or -dial) to the name of the hydrocarbon containing the same number of carbon atoms. Examples are butanal for $CH_3CH_2CH_2CHO$ and propanedial for $OHCCH_2CHO$.

Naming an acyclic polyaldehyde can be handled in two ways. (1) When more than two aldehyde groups are attached to an unbranched chain, the proper affix is added to -carbaldehyde, which becomes the suffix to the name of the longest chain carrying the maximum number of aldehyde groups. The name and numbering of the main chain do not include the carbon atoms of the aldehyde groups. (2) The name is formed by adding the prefix formyl- to the name of the -dial that incorporates the principal chain. Any other chains carrying aldehyde groups are named by the use of formylalkyl- prefixes. Examples are

$$
\begin{array}{c}
\text{CHO}\\
|\\
\text{OHC—CH}_2\text{—CH}_2\text{—CH}_2\text{—CH—CH}_2\text{—CHO}
\end{array}
$$

(1) 1,2,5-Pentanetricarbaldehyde
(2) 3-Formylheptanedial

$$
\begin{array}{cc}
\text{OHC—CH}_2\text{—CH}_2\text{—CH}_2 & \text{CHO}\\
& \searrow \qquad\qquad |\\
& \text{CH—CH—CH—CH}_2\text{—CHO}\\
& \nearrow \qquad\qquad |\\
\text{OHC—CH}_2\text{—CH}_2 & \text{CH}_2\text{—CHO}
\end{array}
$$

(1) 4-(2-Formylethyl)-3-(formylmethyl)-1,2,7-heptanetricarbaldehyde
(2) 3-Formyl-5-(2-formylethyl)-4-(formylmethyl)nonanedial

When the aldehyde group is directly attached to a carbon atom of a ring system, the suffix -carbaldehyde is added to the name of the ring system, e.g., 2-naphthalenecarbaldehyde. When the aldehyde group is separated from the ring by a chain of carbon atoms, the compound is named (1) as a derivative of the acyclic system or (2) by conjunctive nomenclature, for example, (1) (2-naphthyl)propionaldehyde or (2) 2-naphthalenepropionaldehyde.

An aldehyde group is denoted by the prefix formyl- when it is attached to a nitrogen atom in a ring system or when a group having priority for citation as principal group is present and part of a cyclic system.

When the corresponding monobasic acid has a trivial name, the name of the aldehyde may be formed by changing the ending -ic acid or -oic acid to -aldehyde. Examples are

Formaldehyde	Acrylaldehyde (not acrolein)
Acetaldehyde	Benzaldehyde
Propionaldehyde	Cinnamaldehyde
Butyraldehyde	2-Furaldehyde (not furfural)

The same is true for polybasic acids, with the proviso that all the carboxyl groups must be changed to aldehyde; then it is not necessary to introduce affixes. Examples are

Glyceraldehyde   Succinaldehyde

Glycolaldehyde   Phthalaldehyde (*o*-, *m*-, *p*-)

Malonaldehyde

These trivial names may be retained: citral (3,7-dimethyl-2,6-octadienal), vanillin (4-hydroxy-3-methoxybenzaldehyde), and piperonal (3,4-methylenedioxybenzaldehyde).

Amides. For primary amides the suffix -amide is added to the systematic name of the parent acid. For example, $CH_3—CO—NH_2$ is acetamide. Oxamide is retained for $H_2N—CO—CO—NH_2$. The name -carboxylic acid is replaced by -carboxamide.

For amino acids having trivial names ending in -ine, the suffix -amide is added after the name of the acid (with elision of -*e* for monoamides). For example, $H_2N—CH_2—CO—NH_2$ is glycinamide.

In naming the radical $R—CO—NH—$, either (1) the -yl ending of RCO— is changed to -amido or (2) the radicals are named as acylamino radicals. For example,

$CH_3—CO—NH$⟨ ⟩$—COOH$ (1) 4-Acetamidobenzoic acid
                             (2) 4-Acetylaminobenzoic acid

The latter nomenclature is always used for amino acids with trivial names.

N-substituted primary amides are named either (1) by citing the substituents as *N* prefixes or (2) by naming the acyl group as an *N* substituent of the parent compound. For example,

⟨ ⟩$—CO—NH—CH_3$ (1) *N*-Methylbenzamide
                       (2) Benzoylaminomethane

Amines. Amines are preferably named by adding the suffix -amine (and any multiplying affix) to the name of the parent radical. Examples are

$CH_3CH_2CH_2CH_2CH_2NH_2$ Pentylamine
$H_2NCH_2CH_2CH_2CH_2CH_2NH_2$ 1,5-Pentyldiamine
                                      or pentamethylenediamine

Locants of substituents of symmetrically substituted derivatives of symmetrical amines are distinguished by primes or else the names of the complete substituted radicals are enclosed in parentheses. Unsymmetrically substituted derivatives are named similarly or as *N*-substituted products of a primary amine (after choosing the most senior of the radicals to be the parent amine). For example,

$CH_2CH_2CH_2F$
    /
HN
    \
$CHF—CH_2CH_3$

(1) 1,3′-Diflurodipropylamine
(2) 1-Fluoro-*N*-(3-fluoropropyl)propylamine
(3) (1-Fluoropropyl)(3-fluoropropyl)amine

Complex cyclic compounds may be named by adding the suffix -amine or the prefix amino- (or aminoalkyl-) to the name of the parent compound. Thus three names are permissible for

(1) 4-Pyridylamine
(2) 4-Pyridinamine
(3) 4-Aminopyridine

Complex linear polyamines are best designated by replacement nomenclature. These trivial names are retained: aniline, benzidene, phenetidine, toluidine, and xylidine.

The bivalent radical —NH— linked to two identical radicals can be denoted by the prefix imino-, as well as when it forms a bridge between two carbon ring atoms. A trivalent nitrogen atom linked to three identical radicals is denoted by the prefix nitrilo-. Thus ethylenedi- aminetetraacetic acid (an allowed exception) should be named ethylenedinitrilotetraacetic acid.

Ammonium Compounds.   Salts and hydroxides containing quadricovalent nitrogen are named as a substituted ammonium salt or hydroxide. The names of the substituting radicals precede the word *ammonium*, and then the name of the anion is added as a separate word. For example, $(CH_3)_4N^+I^-$ is tetramethylammonium iodide.

When the compound can be considered as derived from a base whose name does not end in -amine, its quaternary nature is denoted by adding -ium to the name of that base (with elision of -*e*), substituent groups are cited as prefixes, and the name of the anion is added separately at the end. Examples are

$C_6H_5NH_3{}^+HSO_4^-$ Anilinium hydrogen sulfate
$[(C_6H_5NH_3)^+PtCl_6^{2-}$ Dianilinium hexachloroplatinate

The names *choline* and *betaine* are retained for unsubstituted compounds.

In complex cases, the prefixes amino- and imino- may be changed to ammonio- and iminio- and are followed by the name of the molecule representing the most complex group attached to this nitrogen atom and are preceded by the names of the other radicals attached to this nitrogen. Finally the name of the anion is added separately. For example, the name might be 1-trimethylammonioacridine chloride or 1-acridinyltrimethylammonium chloride.

When the preceding rules lead to inconvenient names, then (1) the unaltered name of the base may be used followed by the name of the anion or (2) for salts of hydrohalogen acids only the unaltered name of the base is used followed by the name of the hydrohalide. An example of the latter would be 2-ethyl-*p*-phenylenediamine monohydrochloride.

Azo Compounds.   When the azo group (—N=N—) connects radicals derived from identical unsubstituted molecules, the name is formed by adding the prefix azo- to the name of the parent unsubstituted molecules. Substituents are denoted by prefixes and suffixes. The azo group has priority for lowest-numbered locant. Examples are azobenzene for C_6H_5—N=N—C_6H_5, azobenzene-4- sulfonic acid for C_6H_5—N=N—$C_6H_5SO_3H$, and 2′,4-dichloroazobenzene-4′-sulfonic acid for ClC_6H_4—N=N—$C_6H_3ClSO_3H$.

When the parent molecules connected by the azo group are different, azo is placed between the complete names of the parent molecules, substituted or unsubstituted. Locants are placed between the affix azo and the names of the molecules to which each refers. Preference is given to the more complex parent molecule for citation as the first component, e.g., 2-aminonaphthalene-1-azo-(4'-chloro-2'-methylbenzene).

In an alternative method, the senior component is regarded as substituted by RN=N—, this group R being named as a radical. Thus 2-(7-phenylazo-2-naphthylazo)anthracene is the name by this alternative method for the compound named anthracene-2-azo-2'-naphthalene-7'-azobenzene.

Azoxy Compounds. Where the position of the azoxy oxygen atom is unknown or immaterial, the compound is named in accordance with azo rules, with the affix azo replaced by azoxy. When the position of the azoxy oxygen atom in an unsymmetrical compound is designated, a prefix *NNO-* or *ONN-* is used. When both the groups attached to the azoxy radical are cited in the name of the compound, the prefix *NNO-* specifies that the second of these two groups is attached directly to —N(O)—; the prefix *ONN-* specifies that the first of these two groups is attached directly to —N(O)—. When only one parent compound is cited in the name, the prefixed *ONN-* and *NNO-* specify that the group carrying the primed and unprimed substituents is connected, respectively, to the —N(O)— group. The prefix *NON-* signifies that the position of the oxygen atom is unknown; the azoxy group is then written as —N$_2$O—. For example,

2,2',4-Trichloro-*NNO*-azoxybenzene

Boron Compounds. Molecular hydrides of boron are called boranes. They are named by using a multiplying affix to designate the number of boron atoms and adding an Arabic numeral within parentheses as a suffix to denote the number of hydrogen atoms present. Examples are pentaborane(9) for B$_5$H$_9$ and pentaborane(11) for B$_5$H$_{11}$.

Organic ring systems are named by replacement nomenclature. Three- to ten-membered monocyclic ring systems containing uncharged boron atoms may be named by the specialist nomenclature for heterocyclic systems. Organic derivatives are named as outlined for substitutive nomenclature. The complexity of boron nomenclature precludes additional details; the text by Rigaudy and Klesney should be consulted.

Carboxylic Acids. Carboxylic acids may be named in several ways. (1) —COOH groups replacing CH$_3$— at the end of the main chain of an acyclic hydrocarbon are denoted by adding -oic acid to the name of the hydrocarbon. (2) When the —COOH group is the principal group, the suffix -carboxylic acid can be added to the name of the parent chain whose name and chain numbering *does not include* the carbon atom of the —COOH group. The former nomenclature is preferred unless use of the ending -carboxylic acid leads to citation of a larger number of carboxyl groups as suffix. (3) Carboxyl groups are designated by the prefix carboxy- when attached to a group named as a substituent or when another group is present that has higher priority for citation as principal group. In all cases, the principal chain should be linked to as many carboxyl groups as possible even though it might not be the longest chain present. Examples are

$$CH_3CH_2CH_2CH_2CH_2CH_2COOH$$

(1) Heptanoic acid
(2) 1-Hexanecarboxylic acid

$$C_6H_{11}COOH$$

(2) Cyclohexanecarboxylic acid

$$CH_3-CH_2-\overset{\overset{\displaystyle COOH}{|}}{CH}-CH_2-\overset{\overset{\displaystyle CH_2COOH}{|}}{CH}-CH_2-COOH$$

(3) 2-(Carboxymethyl)-1,4-hexanedicarboxylic acid

Removal of the OH from the —COOH group to form the acyl radical results in changing the ending -oic acid to -oyl or the ending -carboxylic acid to -carbonyl. Thus the radical $CH_3CH_2CH_2CH_2CO$— is named either pentanoyl or butanecarbonyl. When the hydroxyl has not been removed from all carboxyl groups present in an acid, the remaining carboxyl groups are denoted by the prefix carboxy-. For example, $HOOCCH_2CH_2CH_2CH_2CH_2CO$— is named 6-carboxyhexanoyl.

Many trivial names exist for acids: these are listed in Table 1.11. Generally, radicals are formed by replacing -ic acid by -oyl.* When a trivial name is given to an acyclic monoacid or diacid, the numeral 1 is always given as locant to the carbon atom of a carboxyl group in the acid or to the carbon atom with a free valence in the radical RCO—.

Ethers (R¹—O—R²). In substitutive nomenclature, one of the possible radicals, R—O—, is stated as the prefix to the parent compound that is senior from among R¹ or R². Examples are methoxyethane for $CH_3OCH_2CH_3$ and butoxyethanol for $C_4H_9OCH_2CH_2OH$.

When another principal group has precedence and oxygen is linking two identical parent compounds, the prefix oxy- may be used, as with 2,2′-oxydiethanol for $HOCH_2CH_2OCH_2CH_2OH$.

Compounds of the type RO—Y—OR, where the two parent compounds are identical and contain a group having priority over ethers for citation as suffix, are named as assemblies of identical units. For example, $HOOC-CH_2-O-CH_2CH_2-O-CH_2-COOH$ is named 2,2′-(ethylenedioxy)diacetic acid.

Linear polyethers derived from three or more molecules of aliphatic dihydroxy compounds, particularly when the chain length exceeds ten units, are most conveniently named by open-chain replacement nomenclature. For example, $CH_3CH_2-O-CH_2CH_2-O-CH_2CH_3$ could be 3,6-dioxaoctane or (2-ethoxy)ethoxyethane.

An oxygen atom directly attached to two carbon atoms already forming part of a ring system or to two carbon atoms of a chain may be indicated by the prefix epoxy-. For example, $CH_2-CH-CH_2Cl$ is named 1-chloro-2,3,-epoxypropane.

Symmetrical linear polyethers may be named (1) in terms of the central oxygen atom when there is an odd number of ether oxygen atoms or (2) in terms of the central hydrocarbon group when there is an even number of ether oxygen atoms. For example, $C_2H_5-O-C_4H_8-O-C_4H_8-O-C_2H_5$ is bis-(4-ethoxybutyl)ether, and 3,6-dioxaoctane (earlier example) could be named 1,2-bis(ethoxy)ethane.

Polyethers and Cyclic Polyethers. During the past several decades, linear and cyclic polyethers have gained considerable prominence. This is largely due to their remarkable ability to complex metallic and organic cations. The linear polyethers of the polyethylene

---

*Exceptions: formyl, acetyl, propionyl, butyryl, isobutyryl, valeryl, isovaleryl, oxalyl, malonyl, succinyl, glutaryl, furoyl, and thenoyl.

TABLE 1.11 Names of Some Carboxylic Acids

Systematic name	Trivial name	Systematic name	Trivial name
Methanoic	Formic	trans-Methylbutenedioic	Mesaconic*
Ethanoic	Acetic		
Propanoic	Propionic	1,2,2-Trimethyl-1,3-	Camphoric
Butanoic	Butyric	cyclopentanedicarboxylic	
2-Methylpropanoic	Isobutyric*	acid	
Pentanoic	Valeric	Benzenecarboxylic	Benzoic
3-Methylbutanoic	Isovaleric*	1,2-Benzenedicarboxylic	Phthalic
2,2-Dimethylpropanoic	Pivalic*	1,3-Benzenedicarboxylic	Isophthalic
Hexanoic	(Caproic)	1,4-Benzenedicarboxylic	Terephthalic
Heptanoic	(Enanthic)	Naphthalenecarboxylic	Naphthoic
Octanoic	(Caprylic)	Methylbenzenecarboxylic	Toluic
Decanoic	(Capric)	2-Phenylpropanoic	Hydratropic
Dodecanoic	Lauric*	2-Phenylpropenoic	Atropic
Tetradecanoic	Myristic*	trans-3-Phenylpropenoic	Cinnamic
Hexadecanoic	Palmitic*	Furancarboxylic	Furoic
Octadecanoic	Stearic*	Thiophenecarboxylic	Thenoic
		3-Pyridinecarboxylic	Nicotinic
Ethanedioic	Oxalic	4-Pyridinecarboxylic	Isonicotinic
Propanedioic	Malonic		
Butanedioic	Succinic	Hydroxyethanoic	Glycolic
Pentanedioic	Glutaric	2-Hydroxypropanoic	Lactic
Hexanedioic	Adipic	2,3-Dihydroxypropanoic	Glyceric
Heptanedioic	Pimelic*	Hydroxypropanedioic	Tartronic
Octanedioic	Suberic*	Hydroxybutanedioic	Malic
Nonanedioic	Azelaic*	2,3-Dihydroxybutanedioic	Tartaric
Decanedioic	Sebacic*	3-Hydroxy-2-phenylpropanoic	Tropic
Propenoic	Acrylic	2-Hydroxy-2,2-	Benzilic
Propynoic	Propiolic	diphenylethanoic	
2-Methylpropenoic	Methacrylic	2-Hydroxybenzoic	Salicylic
trans-2-Butenoic	Crotonic	Methoxybenzoic	Anisic
cis-2-Butenoic	Isocrotonic	4-Hydroxy-3-methoxybenzoic	Vanillic
cis-9-Octadecenoic	Oleic		
trans-9-Octadecenoic	Elaidic	3,4-Dimethoxybenzoic	Veratric
cis-Butenedioic	Maleic	3,4-Methylenedioxybenzoic	Piperonylic
trans-Butenedioic	Fumaric	3,4-Dihydroxybenzoic	Protocatechuic
cis-Methylbutenedioic	Citraconic*	3,4,5-Trihydroxybenzoic	Gallic

The names in parentheses are abandoned but are listed for reference to older literature.
* Systematic names should be used in derivatives formed by substitution on a carbon atom.

glycol type are discussed in Section 10. The cyclic polyethers are called crown ethers if they are monocyclic and cryptands if they are di- or multi-cyclic compounds.

Crown ethers are typically complicated structures and their names have evolved from the convenient, semi-systematic nomenclature developed by the pioneers in the field. The name "crown" was suggested because the cyclic polyethers "crown a cation" when they complex it. The most general naming system consists in identifying the largest cycle and then denoting the number and type of heteroatoms present. The most common repeating unit is ethyleneoxy or $-CH_2CH_2O-$, normally in the form 1,2-ethylenedioxy and this is presumed to be present unless otherwise noted. Ethylene oxide (oxirane) is the smallest cyclic compound containing this unit. Dioxane is formally its dimer. The trimer, called 9-crown-3, is known but the smallest compound normally considered to be a crown ether is 12-crown-4. 18-Crown-6 has six repeating ethyleneoxy units and is systematically named 1,4,7,10,13,16-hexaoxacyclooctadecane.

Examples of various crown ethers and cryptands are shown here. The top line of compounds may be named readily enough although the problem with this semi-systematic approach is obvious. If two methylenes were added to 18-crown-6, the compound could correctly be called 20-crown-6 but in the absence of unequivocal descriptors, the positions of the 3-carbon bridges would be unclear. The more cumbersome name 1,4,7,11,13,17-hexaoxacycloicosane tells clearly that the longer bridges are adjacent to each other. A similar problem is apparent in the last two entries of the second line. The designations dicyclohexano and dibenzo are clear as to the substituents but not their positions. The semi-systematic nomenclature is widely used, however, because it is so much less cumbersome for most purposes.

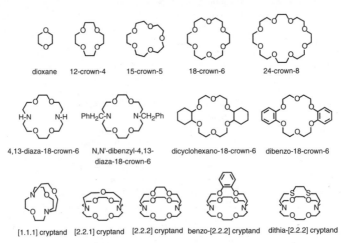

dioxane   12-crown-4   15-crown-5   18-crown-6   24-crown-8

4,13-diaza-18-crown-6   N,N'-dibenzyl-4,13-diaza-18-crown-6   dicyclohexano-18-crown-6   dibenzo-18-crown-6

[1.1.1] cryptand   [2.2.1] cryptand   [2.2.2] cryptand   benzo-[2.2.2] cryptand   dithia-[2.2.2] cryptand

An additional nuance in the nomenclature of these compounds concerns their complexes. The open-chained compounds are often referred to as podands and their complexes as podates. The cyclic ethers may also be called coronands and their complexes are therefore coronates. Complexed cryptands are cryptates. The even more complicated structures known as spherands, cavitands, or carcerands are called spherates, cavitates, or carcerates, respectively, when complexed. The combination of a macrocycle (crown ether or coronand) and a sidechain (podand) is typically called a lariat ether.

An alternate nomenclature system based upon IUPAC principles polymer systems has also been developed but it has not been adopted broadly. Using this method, 15-crown-5 would be called cyclo[pentakis(oxyethylene)] instead of 1,4,7,10,13-pentaoxacyclopentadecane. In this case, substituents and other heteroatoms make the names more complex.

Partial ethers of polyhydroxy compounds may be named (1) by substitutive nomenclature or (2) by stating the name of the polyhydroxy compound followed by the name of the etherifying radical(s) followed by the word *ether*. For example,

$$CH_2O - C_4H_9$$
$$HCOH$$
$$CH_2OH$$

(1) 3-Butoxy-1,2-propanediol

(2) Glycerol 1-buty ether; also, 1-O-butylglycerol

Cyclic ethers are named either as heterocyclic compounds or by specialist rules of hetero-cyclic nomenclature. Radicofunctional names are formed by citing the names of the radicals R^1 and R^2 followed by the word *ether*. Thus methoxyethane becomes ethyl methyl ether and ethoxyethane becomes diethyl ether.

Halogen Derivatives. Using substitutive nomenclature, names are formed by adding pre-fixes listed in Table 1.8 to the name of the parent compound. The prefix perhalo- implies the replacement of all hydrogen atoms by the particular halogen atoms.

Cations of the type $R^1R^2X^+$ are given names derived from the halonium ion, H_2X^+, by substitution, e.g., diethyliodonium chloride for $(C_2H_5)_2I^+Cl^-$.

These trivial names are retained: bromoform ($CHBr_3$), chloroform ($CHCl_3$), fluoroform (CHF_3), iodoform (CHI_3), phosgene ($COCl_2$), thiophosgene ($CSCl_2$), and dichlorocarbene radical ($>CCl_2$). Inorganic nomenclature leads to such names as carbonyl and thiocar-bonyl halides (COX_2 and CSX_2) and carbon tetrahalides (CX_4).

Hydroxylamines and Oximes. For RNH—OH compounds, prefix the name of the radi-cal R to hydroxylamine. If another substituent has priority as principal group, attach the prefix hydroxyamino- to the parent name. For example, C_6H_5NHOH would be named N-phenylhydroxylamine, but HOC_6H_4NHOH would be (hydroxyamino)phenol, with the point of attachment indicated by a locant preceding the parentheses.

Compounds of the type $R^1NH—OR_2$ are named (1) as alkoxyamino derivatives of com-pound R^1H, (2) as N,O-substituted hydroxylamines, (3) as alkoxyamines (even if R^1 is hydrogen), or (4) by the prefix aminooxy- when another substituent has priority for parent name. Examples of each type are as follows:

1. 2-(Methoxyamino)-8-naphthalenecarboxylic acid for $CH_3ONH—C_{10}H_6COOH$
2. O-phenylhydroxylamine for $H_2N—O—C_6H_5$ or N-phenylhydroxylamine for $C_6H_5NH—OH$
3. Phenoxyamine for $H_2N—O—C_6H_5$ (not preferred to O-phenylhydroxylamine)
4. Ethyl (aminooxy)acetate for $H_2N—O—CH_2CO—OC_2H_5$

Acyl derivatives, RCO—NH—OH and $H_2N—O—CO—R$, are named as N-hydroxy derivatives of amides and as O-acylhydroxylamines, respectively. The former may also be named as hydroxamic acids. Examples are N-hydroxyacetamide for $CH_3CO—NH—OH$ and O-acetylhydroxylamine for $H_2N—O—CO—CH_3$. Further substituents are denoted by prefixes with O- and/or N-locants. For example, $C_6H_5NH—O—C_2H_5$ would be O-ethyl-N-phenylhydroxylamine or N-ethoxyaniline.

For oximes, the word *oxime* is placed after the name of the aldehyde or ketone. If the carbonyl group is not the principal group, use the prefix hydroxyimino-. Compounds with the group $>N—OR$ are named by a prefix alkyloxyimino- as oxime O-ethers or as O-substituted oximes. Compounds with the group $>C = N(O)R$ are named by adding N-oxide after the name of the alkylideneamine compound. For amine oxides, add the word *oxide* after the name of the base, with locants. For example, $C_5H_5N—O$ is named pyridine N-oxide or pyridine 1-oxide.

Imines. The group $>C=NH$ is named either by the suffix -imine or by citing the name of the bivalent radical $R^1R^2C<$ as a prefix to amine. For example, $CH_3CH_2CH_2CH=NH$

could be named 1-butanimine or butylideneamine. When the nitrogen is substituted, as in $CH_2=N-CH_2CH_3$, the name is *N*-(methylidene) ethylamine.

Quinones are exceptions. When one or more atoms of quinonoid oxygen have been replaced by $>$NH or $>$NR, they are named by using the name of the quinone followed by the word *imine* (and preceded by proper affixes). Substituents on the nitrogen atom are named as prefixes. Examples are

O=⟨ ⟩=NH   *p*-Benzoquinone monoimine

HN=⟨ ⟩=NH   *p*-Benzoquinone diimine

Ketenes. Derivatives of the compound ketene, $CH_2=C=O$, are named by substitutive nomenclature. For example, $C_4H_9CH=C=O$ is butyl ketene. An acyl derivative, such as $CH_3CH_2-CO-CH_2CH=C=O$, may be named as a polyketone, 1-hexene-1,4-dione. Bis-ketene is used for two to avoid ambiguity with diketene (dimeric ketene).

Ketones. Acyclic ketones are named (1) by adding the suffix -one to the name of the hydrocarbon forming the principal chain or (2) by citing the names of the radicals R^1 and R^2 followed by the word *ketone*. In addition to the preceding nomenclature, acyclic mono-acyl derivatives of cyclic compounds may be named (3) by prefixing the name of the acyl group to the name of the cyclic compound. For example,

$$\overset{O}{\overset{\|}{C}}-CH_2CH_3$$

(1) 1-(2-Furyl)-1-propanone
(2) Ethyl 2-furyl ketone
(3) 2-Propionylfuran

When the cyclic component is benzene or naphthalene, the -ic acid or -oic acid of the acid corresponding to the acyl group is changed to -ophenone or -onaphthone, respectively. For example, $C_6H_5-CO-CH_2CH_2CH_3$ can be named either butyrophenone (or butanophenone) or phenyl propyl ketone.

Radicofunctional nomenclature can be used when a carbonyl group is attached directly to carbon atoms in two ring systems and no other substituent is present having priority for citation.

When the methylene group in polycarbocyclic and heterocyclic ketones is replaced by a keto group, the change may be denoted by attaching the suffix -one to the name of the ring system. However, when $\equiv$CH in an unsaturated or aromatic system is replaced by a keto group, two alternative names become possible. (1) The maximum number of noncumulative double bonds is added after introduction of the carbonyl group(s), and any hydrogen that remains to be added is denoted as indicated hydrogen with the carbonyl group having priority over the indicated hydrogen for lower-numbered locant. (2) The prefix oxo- is used, with the hydrogenation indicated by hydro prefixes; hydrogenation is considered to have occurred before the introduction of the carbonyl group. For example,

(1) 1(2*H*)-Naphthalenone
(2) 1-Oxo-1,2-dihydronaphthalene

When another group having higher priority for citation as principal group is also present, the ketonic oxygen may be expressed by the prefix oxo-, or one can use the name of the carbonyl-containing radical, as, for example, acyl radicals and oxo-substituted radicals. Examples are

$$HOOC-\!\!\!\overbrace{}-CH_2CH_2CH_2CCH_2CH_3$$

4-(4′-Oxohexyl)-1-benzoic acid

$$CH_3-C-\!\!\!\overbrace{}-C-CH_3$$

1,2,4-Triacetylbenzene

Diketones and tetraketones derived from aromatic compounds by conversion of two or four ⩾CH groups into keto groups, with any necessary rearrangement of double bonds to a quinonoid structure, are named by adding the suffix -quinone and any necessary affixes.

Polyketones in which two or more contiguous carbonyl groups have rings attached at each end may be named (1) by the radicofunctional method or (2) by substitutive nomenclature. For example,

(1) 2-Naphthyl 2-pyridyl diketone
(2) 1-(2-Naphthyl)-2-(2-pyridyl)ethanedione

Some trivial names are retained: acetone (2-propanone), biacetyl (2,3-butanedione), propiophenone (C_6H_5—CO—CH_2CH_3), chalcone (C_6H_5—CH=CH—CO—C_6H_5), and deoxybenzoin (C_6H_5—CH_2—CO—C_6H_5).

These contracted names of heterocyclic nitrogen compounds are retained as alternatives for systematic names, sometimes with indicated hydrogen. In addition, names of oxo derivatives of fully saturated nitrogen heterocycles that systematically end in -idinone are often contracted to end in -idone when no ambiguity might result. For example,

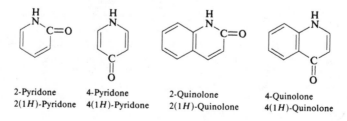

2-Pyridone
2(1*H*)-Pyridone

4-Pyridone
4(1*H*)-Pyridone

2-Quinolone
2(1*H*)-Quinolone

4-Quinolone
4(1*H*)-Quinolone

1-Isoquinolone
1(2H)-Isoquinolone

4-Oxazolone
4(5H)-Oxazolone

4-Pyrazolone
4(5H)-Pyrazolone

5-Pyrazolone
5(4H)-Pyrazolone

4-Isoxazoline
4(5H)-Isoxazolone

4-Thiazolone
4(5H)-Thiazolone

9-Acridone
9(10H)-Acridone

Lactones, Lactides, Lactams, and Lactims. When the hydroxy acid from which water may be considered to have been eliminated has a trivial name, the lactone is designated by substituting -olactone for -ic acid. Locants for a carbonyl group are numbered as low as possible, even before that of a hydroxyl group.

Lactones formed from aliphatic acids are named by adding -olide to the name of the nonhydroxylated hydrocarbon with the same number of carbon atoms. The suffix -olide signifies the change of $>CH \cdots CH_3$ into $>C \cdots C = O$ with $\lfloor O \rfloor$

Structures in which one or more (but not all) rings of an aggregate are lactone rings are named by placing -carbolactone (denoting the $-O-CO-$ bridge) after the names of the structures that remain when each bridge is replaced by two hydrogen atoms. The locant for $-CO-$ is cited before that for the ester oxygen atom. An additional carbon atom is incorporated into this structure as compared to the -olide.

These trivial names are permitted: γ-butyrolactone, γ-valerolactone, and δ-valerolactone. Names based on heterocycles may be used for all lactones. Thus, γ-butyrolactone is also tetrahydro-2-furanone or dihydro-2(3H)-furanone.

Lactides, intermolecular cyclic esters, are named as heterocycles. *Lactams* and *lactims,* containing a $-CO-NH-$ and $-C(OH)=N-$ group, respectively, are named as heterocycles, but they may also be named with -lactam or -lactim in place of -olide. For example,

(1) 2-Pyrrolidinone
(2) 4-Butanelactam

Nitriles and Related Compounds. For acids whose systematic names end in -carboxylic acid, nitriles are named by adding the suffix -carbonitrile when the $-CN$ group replaces the $-COOH$ group. The carbon atom of the $-CN$ group is excluded from the numbering of a chain to which it is attached. However, when the triple-bonded nitrogen atom is considered to replace three hydrogen atoms at the end of the main chain of an acyclic hydrocarbon, the suffix -nitrile is added to the name of the hydrocarbon. Numbering begins with the carbon attached to the nitrogen. For example, $CH_3CH_2CH_2CH_2CH_2CN$ is named (1) pentanecarbonitrile or (2) hexanenitrile.

Trivial acid names are formed by changing the endings -oic acid or -ic acid to -onitrile. For example, CH_3CN is acetonitrile. When the —CN group is not the highest priority group, the —CN group is denoted by the prefix cyano-.

In order of decreasing priority for citation of a functional class name, and the prefix for substitutive nomenclature, are the following related compounds:

Functional group	Prefix	Radicofunctional ending
—NC	Isocyano-	Isocyanide
—OCN	Cyanato-	Cyanate
—NCO	Isocyanato-	Isocyanate
—ONC	—	Fulminate
—SCN	Thiocyanato-	Thiocyanate
—NCS	Isothiocyanato-	Isothiocyanate
—SeCN	Selenocyanato-	Selenocyanate
—NCSe	Isoselenocyanato-	Isoselenocyanate

Peroxides.   Compounds of the type R—O—OH are named (1) by placing the name of the radical R before the word *hydroperoxide* or (2) by use of the prefix hydroperoxy- when another parent name has higher priority. For example, C_2H_5OOH is ethyl hydroperoxide.

Compounds of the type R^1O—OR^2 are named (1) by placing the names of the radicals in alphabetical order before the word *peroxide* when the group —O—O— links two chains, two rings, or a ring and a chain, (2) by use of the affix dioxy to denote the bivalent group —O—O— for naming assemblies of identical units or to form part of a prefix, or (3) by use of the prefix epidioxy- when the peroxide group forms a bridge between two carbon atoms, a ring, or a ring system. Examples are methyl propyl peroxide for CH_3—O—O—C_3H_7 and 2,2'- dioxydiacetic acid for HOOC—CH_2—O—O—CH_2—COOH.

Phosphorus Compounds.   Acyclic phosphorus compounds containing only one phosphorus atom, as well as compounds in which only a single phosphorus atom is in each of several functional groups, are named as derivatives of the parent structures listed in Table 1.12. Often these are purely hypothetical parent structures. When hydrogen attached to phosphorus is replaced by a hydrocarbon group, the derivative is named by substitution nomenclature. When hydrogen of an —OH group is replaced, the derivative is named by radicofunctional nomenclature. For example, $C_2H_5PH_2$ is ethylphosphine; $(C_2H_5)_2PH$, diethylphosphine; $CH_3P(OH)_2$, dihydroxy-methyl-phosphine or methylphosphonous acid; C_2H_5—$PO(Cl)(OH)$, ethylchlorophosphonic acid or ethylphosphonochoridic acid or hydrogen chlorodioxoethylphosphate(V); $CH_3CH(PH_2)COOH$, 2-phosphinopropionic acid; $HP(CH_2COOH)_2$, phosphinediyldiacetic acid; $(CH_3)HP(O)OH$, methylphosphinic acid or hydrogen hydridomethyldioxophosphate(V); $(CH_3O)_3PO$, trimethyl phosphate; and $(CH_3O)_3P$, trimethyl phosphite.

Salts and Esters of Acids.   Neutral salts of acids are named by citing the cation(s) and then the anion, whose ending is changed from -oic to -oate or from -ic to -ate. When different acidic residues are present in one structure, prefixes are formed by changing the anion ending -ate to -ato- or -ide to -ido-. The prefix carboxylato- denotes the ionic group —COO⁻. The phrase: (metal) salt of (the acid) is permissible when the carboxyl groups are not all named as affixes.

Acid salts include the word *hydrogen* (with affixes, if appropriate) inserted between the name of the cation and the name of the anion (or word *salt*).

TABLE 1.12 Parent Structures of Phosphorus-containing Compounds

Formula	Parent name	Substitutive prefix	Radicofunctional ending
H_3P	Phosphine	$H_2P—$ Phosphino-	Phosphide
H_5P	Phosphorane	$H_4P—$ Phosphoranyl-	—
		$H_3P<$ Phosphoroanediyl-	—
		$H_2P≶$ Phosphoranetriyl-	—
H_3PO	Phosphine oxide	—	—
H_3PS	Phosphine sulfide	—	—
H_3PNH	Phosphine imide	—	—
$P(OH)_3$	Phosphorous acid	—	Phosphite
$HP(OH)_2$	Phosphonous acid	—	Phosphonite
H_2POH	Phosphinous acid	—	Phosphinite
$P(O)(OH)_3$	Phosphoric acid	$P(O)≶$ Phosphoryl-	Phosphate(V)
$HP(O)(OH)_2$	Phosphonic acid	$HP(O)<$ Phosphonoyl-	Phosphonate
		$—P(O)OH_2$ Phosphinoyl-	—
$H_2P(O)OH$	Phosphinic acid	$H_2P(O)—$ Phosphinoyl-	Phosphinate
		$>P(O)OH$ Phosphinoco-	—
		Phosphinato-	—

Esters are named similarly, with the name of the alkyl or aryl radical replacing the name of the cation. Acid esters of acids and their salts are named as neutral esters, but the components are cited in the order: cation, alkyl or aryl radical, hydrogen, and anion. Locants are added if necessary. For example,

$$CH_2—CO—OC_2H_5$$
$$HOC—COO^-$$
$$CH_2—COO^-$$
$$K^+ \quad H^+$$

Potassium 1-ethyl hydrogen citrate

Ester groups in $R^1—CO—OR^2$ compounds are named (1) by the prefix alkoxycarbonyl- or aryloxycabonyl- for $—CO—OR^2$ when the radical R^1 contains a substituent with priority for citation as principal group or (2) by the prefix acyloxy- for $R^1—CO—O—$ when the radical R^2 contains a substituent with priority for citation as principal group. Examples are

$$CH_2CH_2CH_2CO—OCH_3$$

Methyl 3-methoxycarbonyl-2-naphthalenebutyrate

$$CO—OCH_3$$

$$[CH_3O—CO—CH_2CH_2\overset{+}{N}(CH_3)_3]\, Cl^-$$ [(2-Methoxycarbonyl)ethyl]trimethylammonium chloride

$$C_6H_5—CO—OCH_2CH_2COOH$$ 3-Benzoyloxypropionic acid

The trivial name *acetoxy* is retained for the $CH_3—CO—O—$ group. Compounds of the type $R^2C(OR^2)_3$ are named as R^2 esters of the hypothetical ortho acids. For example, $CH_3C(OCH_3)_3$ is trimethyl orthoacetate.

Silicon Compounds. SiH_4 is called silane; its acyclic homologs are called disilane, tri-silane, and so on, according to the number of silicon atoms present. The chain is numbered from one end to the other so as to give the lowest-numbered locant in radicals to the free valence or to substituents on a chain. The abbreviated form silyl is used for the radical SiH_3—. Numbering and citation of side chains proceed according to the principles set forth for hydrocarbon chains. Cyclic nonaromatic structures are designated by the prefix cyclo-.

When a chain or ring system is composed entirely of alternating silicon and oxygen atoms, the parent name *siloxane* is used with a multiplying affix to denote the number of silicon atoms present. The parent name *silazane* implies alternating silicon and nitrogen atoms; multiplying affixes denote the number of silicon atoms present.

The prefix sila- designates replacement of carbon by silicon in replacement nomenclature. Prefix names for radicals are formed analogously to those for the corresponding carbon-containing compounds. Thus silyl is used for SiH_3—, silyene for —SiH_2—, silylidyne for —$SiH<$.

Sulfur Compounds

Bivalent Sulfur. The prefix thio-, placed before an affix that denotes the oxygen-containing group or an oxygen atom, implies the replacement of that oxygen by sulfur. Thus the suffix -thiol denotes —SH, -thione denotes —(C)=S and implies the presence of an =S at a nonterminal carbon atom, -thioic acid denotes $[(C)=S]OH \rightleftharpoons [(C)=O]SH$ (that is, the *O*-substituted acid and the *S*-substituted acid, respectively), -dithioic acid denotes [—C(S)]SH, and -thial denotes —(C)HS (or -carbothialdehyde denotes —CHS). When -carboxylic acid has been used for acids, the sulfur analog is named -carbothioic acid or -carbodithioic acid.

Prefixes for the groups HS— and RS— are mercapto- and alkylthio-, respectively; this latter name may require parentheses for distinction from the use of thio- for replacement of oxygen in a trivially named acid. Examples of this problem are $4\text{-}C_2H_5—C_6H_4—CSOH$ named *p*-ethyl(thio)benzoic acid and $4\text{-}C_2H_5—S—C_6H_4—COOH$ named *p*-(ethylthio) benzoic acid. When —SH is not the principal group, the prefix mercapto- is placed before the name of the parent compound to denote an unsubstituted —SH group.

The prefix thioxo- is used for naming =S in a thioketone. Sulfur analogs of acetals are named as alkylthio- or arylthio-. For example, $CH_3CH(SCH_3)OCH_3$ is 1-methoxy-1-(methylthio)ethane. Prefix forms for -carbothioic acids are hydroxy(thiocarbonyl)- when referring to the *O*-substituted acid and mercapto(carbonyl)- for the *S*-substituted acid.

Salts are formed as with oxygen-containing compounds. For example, $C_2H_5—S—Na$ is named either sodium ethanethiolate or sodium ethyl sulfide. If mercapto- has been used as a prefix, the salt is named by use of the prefix sulfido- for —S^-.

Compounds of the type $R^1—S—R^2$ are named alkylthio- (or arylthio-) as a prefix to the name of R^1 or R^2, whichever is the senior.

Sulfonium Compounds. Sulfonium compounds of the type $R^1R^2R^3S^+X^-$ are named by citing in alphabetical order the radical names followed by -sulfonium and the name of the anion. For heterocyclic compounds, -ium is added to the name of the ring system. Replacement of $>$CH by sulfonium sulfur is denoted by the prefix thionia-, and the name of the anion is added at the end.

Organosulfur Halides. When sulfur is directly linked only to an organic radical and to a halogen atom, the radical name is attached to the word *sulfur* and the name(s) and number of the halide(s) are stated as a separate word. Alternatively, the name can be formed from R—SOH, a sulfenic acid whose radical prefix is sulfenyl-. For example, $CH_3CH_2—S—Br$ would be named either ethylsulfur monobromide or ethanesulfenyl bromide. When another principal group is present, a composite prefix is formed from the number and substitutive name(s) of the halogen atoms in front of the syllable thio. For example, BrS—COOH is (bromothio)formic acid.

Sulfoxides.   Sulfoxides, R^1—SO—R^2, are named by placing the names of the radicals in alphabetical order before the word *sulfoxide*. Alternatively, the less senior radical is named followed by sulfinyl- and concluded by the name of the senior group. For example, CH_3CH_2—SO—$CH_2CH_2CH_3$ is named either ethyl propyl sulfoxide or 1-(ethylsulfinyl)-propane.

When an $>$SO group is incorporated in a ring, the compound is named an oxide.

Sulfones.   Sulfones, R^1—SO_2—R^2, are named in an analogous manner to sulfoxides, using the word *sulfone* in place of *sulfoxide*. In prefixes, the less senior radical is followed by -sulfonyl-. When the $>SO_2$ group is incorporated in a ring, the compound is named as a dioxide.

Sulfur Acids.   Organic oxy acids of sulfur, that is, —SO_3H, —SO_2H, and —SOH, are named sulfonic acid, sulfinic acid, and sulfenic acid, respectively. In subordinate use, the respective prefixes are sulfo-, sulfino, and sulfeno-. The grouping —SO_2—O—SO_2— or —SO—O—SO is named sulfonic or sulfinic anhydride, respectively.

Inorganic nomenclature is employed in naming sulfur acids and their derivatives in which sulfur is linked only through oxygen to the organic radical. For example, $(C_2H_5O)_2SO_2$ is diethyl sulfate and C_2H_5O—SO_2—OH is ethyl hydrogen sulfate. Prefixes *O*- and *S*- are used where necessary to denote attachment to oxygen and to sulfur, respectively, in sulfur replacement compounds. For example, CH_3—S—SO_2—ONa is sodium *S*-methyl thiosulfate.

When sulfur is linked only through nitrogen, or through nitrogen and oxygen, to the organic radical, naming is as follows: (1) *N*-substituted amides are designated as *N*-substituted derivatives of the sulfur amides and (2) compounds of the type R—NH—SO_3H may be named as *N*-substituted sulfamic acids or by the prefix sulfoamino- to denote the group HO_3S—NH—. The groups —N$=$SO and —N$=SO_2$ are named sulfinylamines and sulfonylamines, respectively.

Sultones and Sultams.   Compounds containing the group —SO_2—O— as part of the ring are called -sultone. The —SO_2— group has priority over the —O— group for lowest-numbered locant.

Similarly, the —SO_2—N$=$ group as part of a ring is named by adding -sultam to the name of the hydrocarbon with the same number of carbon atoms. The —SO_2— has priority over —N$=$ for lowest-numbered locant.

Steroids.   Steroids are important natural products that have a special nomenclature. They typically consist of three fused 6-membered rings and a four fused 5-membered ring. The

Lanosterol   Cholesterol   Ergosterol

Androsterone   Estradiol   Testosterone

rings are designated A, B, C, and D as shown here. The most common, all *trans* ring fusion is illustrated in the right hand structure. Most steroids that occur naturally possess the two methyl groups ("bridgehead" or angular methyl groups) shown as carbons 18 and 19 at positions 10 and 13, respectively.

Cholesterol is the most common steroid of mammalian membranes. It is formed biologically from lanosterol, as shown. Ergosterol is the most common steroid of fungal membranes. It differs from cholesterol by the presence of two additional double bonds that affect its three dimensional structure. Also shown are three so-called steroid hormones, androsterone, estradiol, and testosterone. Note the presence of an aromatic A-ring in estradiol.

Vitamins. The vitamins are natural organic compounds of considerable diversity that occur widely. The name derives from the Latin *vita* (life) and "amin," a shortened form of amine. The name reflects the historical discovery of these substances, not all of which are amines. They are all of relatively low molecular weight, especially compared to peptides but in a range comparable to steroids. These substances are uniformly active and play various roles in biosynthesis and metabolism. The vitamins are too numerous to detail here but the most common examples are illustrated. They are classed using the common system, that is, water or fat soluble, depending on their approximate level of hydrophobicity or hydrophilicity. Their names are typically nonsystematic but the diversity of their structures requires that the trivial names be used.

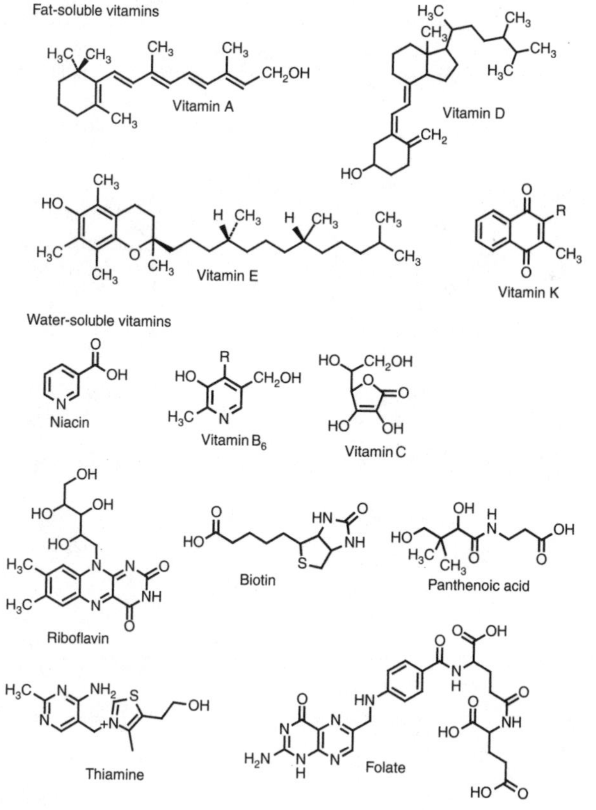

Fat-soluble vitamins

Vitamin A

Vitamin D

Vitamin E

Vitamin K

Water-soluble vitamins

Niacin

Vitamin B$_6$

Vitamin C

Riboflavin

Biotin

Panthenoic acid

Thiamine

Folate

Biological Nomenclature

The names assigned to compounds by organic chemists and biologists sometimes differ. Moreover, biological structures are made up of repeating components and rapidly become large and complex molecules. Thus, special terminology has been developed to assist in describing these compounds.

Amino Acids. An amino acid is any organic compound that possesses both amine (—NH$_2$) and carboxyl (—COOH) groups within the same organic framework. When both functional groups are attached to the same carbon atom, they are designated α-amino acids. These are of special significance in biology as they form the diverse monomer set from which peptides and proteins are built.

Aminoacetic acid is the simplest example of an α-amino acid. Among the 20 biologically most important amino acids, its structure is unique in two ways. First, it possesses no "sidechain" attached to the methylene carbon. Second, the lack of any substituent (other than hydrogen) means that aminoacetic acid, more commonly called glycine, is achiral. The other 19 "essential" or "common" amino acids possess sidechains attached in a stereochemically identical fashion.

The other 19 common α-amino acids have side chains attached at the position represented by R. Among these 19, proline is unique because its sidechain is attached at the other end to the amino nitrogen, which is therefore secondary rather than primary.

The twenty common α-amino acids may be named systematically. For example, when R is methyl, the compound may be called 2-methylaminoacetic acid. It may also correctly be called 2-aminopropanoic acid. By far, however, it is most commonly called alanine.

TABLE 1.13

Common	IUPAC	IUB	$-pK_{COOH}$	$-pK_{NH3}^+$	$pK_{Side\ chain}$	I_{pH}	
Alanine	Ala	A	2.34	9.69	—	6.01	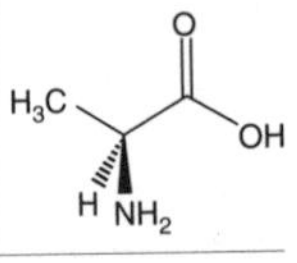
Arginine	Arg	R	2.17	9.04	12.84	10.76	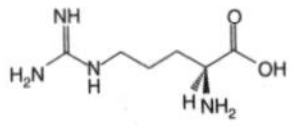
Asparagine	Asn	N	2.02	8.60	—	5.41	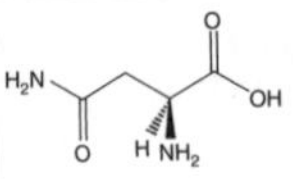

TABLE 1.13 (*continued*)

Common	IUPAC	IUB	$-pK_{COOH}$	$-pK_{NH_3^+}$	$pK_{Side\ chain}$	I_{pH}	
Aspartic acid	Asp	D	1.88	9.60	3.65	2.77	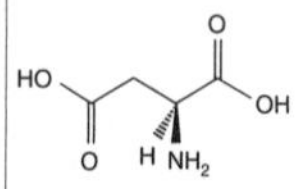
Cystine	Cys	C	1.71	8.18	10.28	5.02	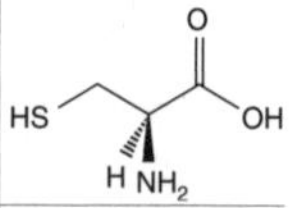
Glutamic acid	Glu	E	2.16	9.67	4.32	3.24	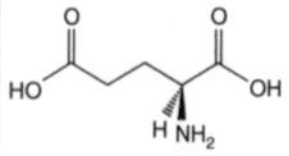
Glutamine	Gln	Q	2.17	9.13	—	5.65	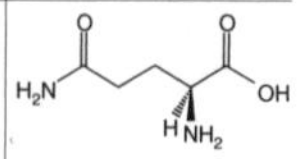
Glycine	Gly	G	2.34	9.60	—	5.97	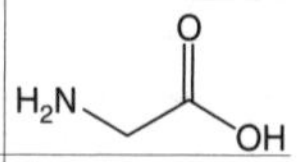
Histidine	His	H	1.82	9.17	6.00	7.59	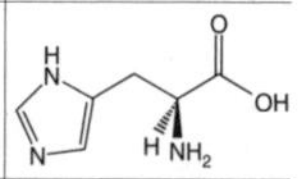
Isoleucine	Ile	I	2.36	9.68	—	6.02	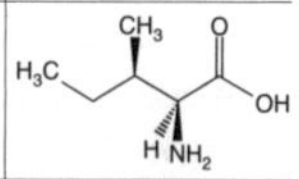
Leucine	Leu	L	2.36	9.60	—	5.98	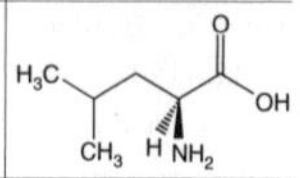
Lysine	Lys	K	2.18	9.12	10.53	9.82	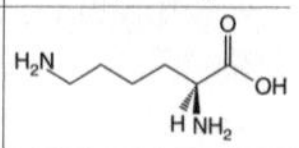
Methionine	Met	M	2.28	9.21	—	5.74	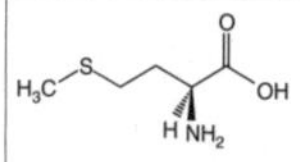
Phenylalanine	Phe	F	1.83	9.13	—	5.84	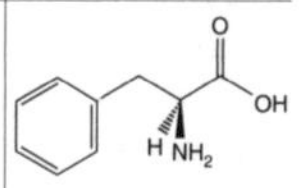
Proline	Pro	P	1.99	10.6	—	6.30	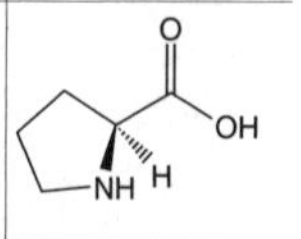
Serine	Ser	S	2.21	9.15	—	5.68	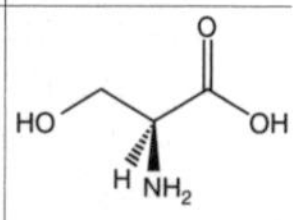

TABLE 1.13 (*continued*)

Common	IUPAC	IUB	$-pK_{COOH}$	$-pK_{NH_3^+}$	$pK_{\text{Side chain}}$	I_{pH}	
Threonine	Thr	T	2.71	9.62	—	6.16	
Tryptophan	Trp	W	2.38	9.39	—	5.89	
Tyrosine	Tyr	Y	2.20	9.11	10.07	5.66	
Valine	Val	V	2.32	9.62	—	5.96	

The names and structures of the twenty natural amino acids are given in Table 1.13. The International Union of Pure and Applied Chemistry (IUPAC) uses the three-letter abbreviations shown in the second column of Table 1.13 to describe amino acids. These are widely used in biological circles as well but are inappropriate when long peptide or protein sequences need to be described. For example, when proinsulin is cleaved, it forms the biologically important peptide insulin and another peptide usually called C-peptide. The human peptide consists of a linear chain of 31 amino acids that have the sequence, from amino to carboxyl, H_2N-Glu-Ala-Glu-Asp-Leu-Gln-Val-Glu-Gln-Glu-Leu-Gly-Gly-Gly-Pro-Gly-Ala-Gly-Ser-Leu-Gln-Pro-Leu-Ala-Leu-Glu-Gly-Ser-Leu-Gln-OH. This is readily comprehensible to most chemists because the abbreviations are typically the first three letters of the amino acid. Thus, alanine is Ala and arginine is Arg. Aspartic acid and asparagine cannot both be named Asp so the latter is distinguished as Asn.

Biologists often must compare peptides or proteins from different species. If the IUPAC nomenclature was used, the descriptor for this peptide would be about four-fold longer than if the International Union of Biology's (IUB) single-letter codes are used. Where possible, the first letter of the amino acid's name is used. As with the three-letter abbreviations, this is not always possible. Alanine is A and aspartic acid is arbitrarily assigned the letter D. Asparagine is called N. Glycine is G so glutamic acid is designated E. Although arbitrary, the name is logical for the homologue of aspartic acid. Glutamine is called Q. Arginine cannot use "A," which is taken by alanine, but the letter "R" is suggestive and serves as a mnemonic.

When the single-letter abbreviations are used, the human C-peptide sequence reduces to EAEDLQVGQVELGGGPGAGSLQPLALEGSLQ, often written in groups of five letters as EAEDL QVGQV ELGGG PGAGS LQPLA LEGSL Q so that the sequence can be more conveniently read and compared with other sequences.

Most of three-letter abbreviations are taken from the first three letters of the name of the corresponding amino acid and are pronounced as written (alanine–Ala, cysteine–Cys). The one-letter symbol for the amino acids is usually the first letter of the amino acid's name and is often used when comparing the amino acids sequences of several similar proteins.

Note that the single-letter abbreviations permit a fast comparison of the sequences so that their differences and similarities can quickly be discerned. The sequences for C-peptides from humans and from rats are compared below (differences are highlighted):

```
EAEDL QVGQV ELGGG PGAGS LQPLA LEGSL Q (Human)
EVEDP QVPQL ELGGG PEAGD LQTLA LEVAR Q (Rat)
```

Ten of the amino acids in each sequence differ. Thus, 21 of the 31 amino acids are identical within the sequence. The sequence homology is said to be $(21/31 \times 100 =)68\%$. It would have been much more difficult to make this comparison using the longer three-letter abbreviations. For a protein having 200 or 300 amino acids, the problem becomes correspondingly greater.

Despite the greater economy of using single letter amino acid abbreviations, it is the three-letter abbreviations that have become shortened names for the essential amino acids. For example, practicing scientists would identify the sequence GPAGW as "Gly-Pro-Ala-Gly-Try." A sequence containing both aspartic acid and asparagine such as GAGE would be referred to in conversation as "Gly-Asp-Gly-Asparagine."

Formally, however, the amino acids contained either in peptides or proteins are named in a fashion related to that used for alkanes. The "-ine" suffix is replaced by "-yl" to give, for GAG, glycyl-alanyl-glycine. Peptide and protein sequences are always written and named from the N-terminus to the C-terminus. The C-terminal amino acid retains its full name.

Stereochemistry

Concepts in stereochemistry, that is, chemistry in three-dimensional space, are in the process of rapid expansion. This section will deal with only the main principles. The compounds discussed will be those that have identical molecular formulas but differ in the arrangement of their atoms in space. *Stereoisomers* is the name applied to these compounds.

Stereoisomers can be grouped into three categories: (1) Conformational isomers differ from each other only in the way their atoms are oriented in space, but can be converted into one another by rotation about sigma bonds. (2) Geometric isomers are compounds in which rotation about a double bond is restricted. (3) Configurational isomers differ from one another only in configuration about a chiral center, axis, or plane. In subsequent structural representations, a broken line denotes a bond projecting behind the plane of the paper and a wedge denotes a bond projecting in front of the plane of the paper. A line of normal thickness denotes a bond lying essentially in the plane of the paper.

Conformational Isomers.   A molecule in a conformation into which its atoms return spontaneously after small displacements is termed a *conformer*. Different arrangements of atoms that can be converted into one another by rotation about single bonds are called *conformational isomers* (see Figure 1.3). A pair of conformational isomers can be but do not have to be mirror images of each other. When they are not mirror images, they are called *diastereomers*.

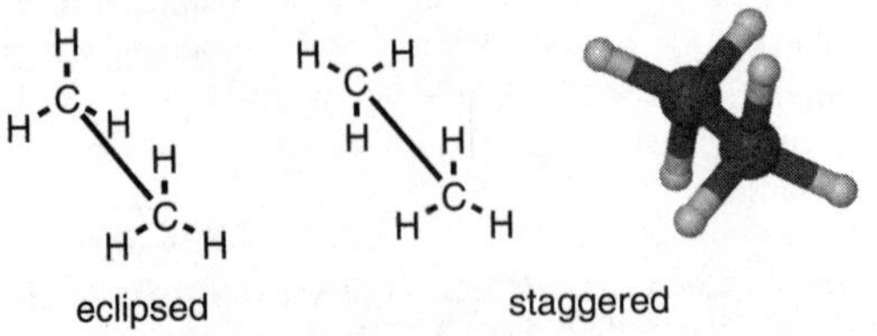

FIGURE 1.3   Eclipsed (left) and staggered (right) conformations of ethane. The ball and stick model is in the staggered conformation.

Acyclic Compounds.   Different conformations of acyclic compounds are best viewed by construction of ball-and-stick molecules or by use of Newman projections. Both types of representations are shown for ethane. Atoms or groups that are attached at opposite ends of a single bond should be viewed along the bond axis. If two atoms or groups attached at opposite ends of the bond appear directly one behind the other, these atoms or groups are described as eclipsed. That portion of the molecule is described as being in the eclipsed conformation. If not eclipsed, the atoms or groups and the conformation may be described as staggered. Newman projections show these conformations clearly.

Certain physical properties show that rotation about the single bond is not quite free. For ethane there is an energy barrier of about $3 \, kcal \cdot mol^{-1}$ ($12 \, kJ \cdot mol^{-1}$). The potential energy of the molecule is at a minimum for the staggered conformation, increases with rotation, and reaches a maximum at the eclipsed conformation. The energy required to rotate the atoms or groups about the carbon–carbon bond is called *torsional energy*. Torsional strain is the cause of the relative instability of the eclipsed conformation or any intermediate skew conformations.

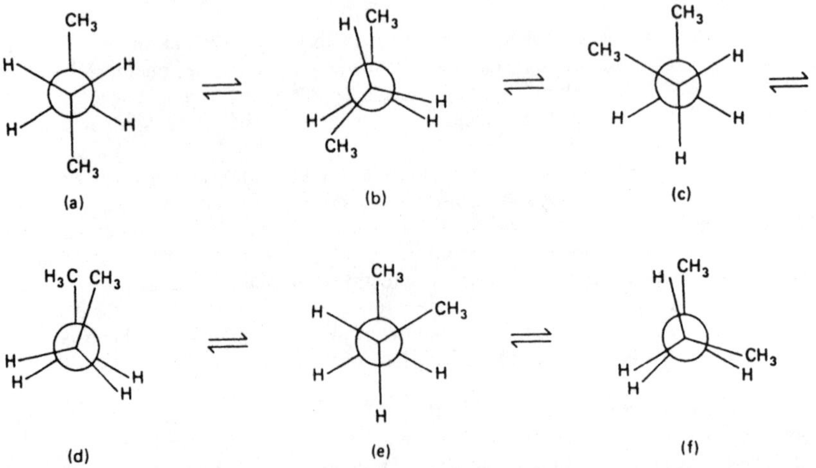

FIGURE 1.4   Conformations of butane. (*a*) *Anti*-staggered; (*b*) eclipsed; (*c*) *gauche*-staggered; (*d*) eclipsed; (*e*) *gauche*-staggered; (*f*) eclipsed. (Eclipsed conformations are slightly staggered for convenience in drawing; actually they are superimposed.)

In butane, with a methyl group replacing one hydrogen on each carbon of ethane, there are several different staggered conformations (see Figure 1.4). There is the *anti* conformation in which the methyl groups are as far apart as they can be (dihedral angle of 180°). There are two *gauche* conformations in which the methyl groups are only 60° apart; these are two nonsuperimposable mirror images of each other. The *anti* conformation is more stable than the *gauche* by about $0.9 \text{kcal} \cdot \text{mol}^{-1}$ ($4 \text{kJ} \cdot \text{mol}^{-1}$). Both are free of torsional strain. However, in a *gauche* conformation the methyl groups are closer together than the sum of their van der Waals' radii. Under these conditions van der Waals' forces are repulsive and raise the energy of conformation. This strain can affect not only the relative stabilities of various staggered conformations but also the heights of the energy barriers between them. The energy maximum (estimated at $4.8–6.1 \text{kcal} \cdot \text{mol}^{-1}$ or $20–25 \text{kJ} \cdot \text{mol}^{-1}$) is reached when two methyl groups swing past each other (the eclipsed conformation) rather than past hydrogen atoms.

Cyclic Compounds.   Although cyclic aliphatic compounds are often drawn as if they were planar geometric figures (a triangle for cyclopropane, a square for cyclobutane, and so on), their structures are not that simple. Cyclopropane does possess the maximum angle strain if one considers the difference between a tetrahedral angle (109.5°) and the 60° angle of the cyclopropane structure. Nevertheless the cyclopropane structure is thermally quite stable. The highest electron density of the carbon–carbon bonds does not lie along the lines connecting the carbon atoms. Bonding electrons lie principally outside the triangular internuclear lines and result in what are known as *bent bonds* (see Figure 1.5).

Cyclobutane has less angle strain than cyclopropane (only 19.5°). It is also believed to have some bent-bond character associated with the carbon–carbon bonds. The molecule exists in a nonplanar conformation in order to minimize hydrogen–hydrogen eclipsing strain.

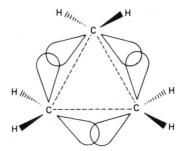

FIGURE 1.5   The bent bonds ("tear drops") of cyclopropane.

FIGURE 1.6   The conformations of cyclopentane.

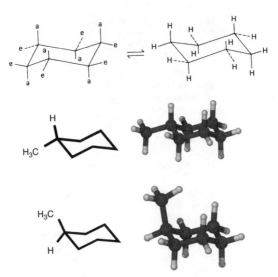

FIGURE 1.7 The two chair conformations of cyclohexane: a = axial hydrogen atom and e = equatorial hydrogen atom. The middle and bottom panels show methylcyclohexane in the chair form with the methyl group equatorial (middle) and axial (bottom).

Cyclopentane is nonplanar, with a structure that resembles an envelope (see Figure 1.6). Four of the carbon atoms are in one plane, and the fifth is out of that plane. The molecule is in continual motion so that the out-of-plane carbon moves rapidly around the ring.

The 12 hydrogen atoms of cyclohexane do not occupy equivalent positions. In the chair conformation six hydrogen atoms are perpendicular to the average plane of the molecule and six are directed outward from the ring, slightly above or below the molecular plane (see Figure 1.7). Bonds which are perpendicular to the molecular plane are known as *axial bonds*, and those which extend outward from the ring are known as *equatorial bonds*. The three axial bonds directed upward originate from alternate carbon atoms and are parallel with each other; a similar situation exists for the three axial bonds directed downward. Each equatorial bond is drawn so as to be parallel with the ring carbon–carbon bond once removed from the point of attachment to that equatorial bond. At room temperature, cyclohexane is interconverting rapidly between two chair conformations. As one chair form converts to the other, all the equatorial hydrogen atoms become axial and all the axial hydrogens become equatorial. The interconversion is so rapid that all hydrogen atoms on cyclohexane can be considered equivalent. Interconversion is believed to take place by movement of one side of the chair structure to produce the twist boat, and then movement of the other side of the twist boat to give the other chair form. The chair conformation is the most favored structure for cyclohexane. No angle strain is encountered since all bond angles remain tetrahedral. Torsional strain is minimal because all groups are staggered.

In the boat conformation of cyclohexane (Figure 1.8) eclipsing torsional strain is significant, although no angle strain is encountered. Nonbonded interaction between the two hydrogen atoms across the ring from each other (the "flagpole" hydrogens) is unfavorable. The boat conformation is about 6.5 kcal · mol^{-1} (27 kJ · mol^{-1}) higher in energy than the chair form at 25 °C.

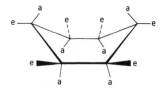

FIGURE 1.8   The boat conformation of cyclohexane. a = axial hydrogen atom and e = equatorial hydrogen atom.

FIGURE 1.9   Twist-boat conformation of cyclohexane.

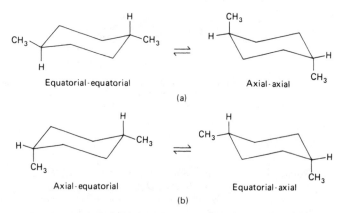

FIGURE 1.10   Two isomers of 1,4-dimethylcyclohexane. (*a*) *Trans* isomer; (*b*) *cis* isomer.

A modified boat conformation of cyclohexane, known as the twist boat (Figure 1.9), or skew boat, has been suggested to minimize torsional and nonbonded interactions. This particular conformation is estimated to be about $1.5\,\text{kcal}\cdot\text{mol}^{-1}$ ($6\,\text{kJ}\,\text{mol}^{-1}$) lower in energy than the boat form at room temperature.

The medium-size rings (7–12 ring atoms) are relatively free of angle strain and can easily take a variety of spatial arrangements. They are not large enough to avoid all nonbonded interactions between atoms.

Disubstituted cyclohexanes can exist as *cis–trans* isomers as well as axial–equatorial conformers. Two isomers are predicted for 1,4-dimethylcyclohexane (see Figure 1.10). For the *trans* isomer the diequatorial conformer is the energetically favorable form. Only one *cis* isomer is observed, since the two conformers of the *cis* compound are identical. Interconversion takes place between the conformational (equatorial-axial) isomers but not configurational (*cis–trans*) isomers.

The bicyclic compound decahydronaphthalene, or bicyclo[4.4.0]decane, has two fused six-membered rings. It exists in *cis* and *trans* forms (see Figure 1.11), as determined by the configurations at the bridgehead carbon atoms. Both *cis-* and *trans*-decahydronaphthalene can be constructed with two chair conformations.

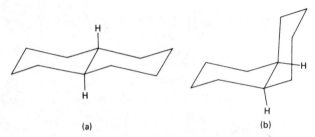

FIGURE 1.11 Two isomers of decahydronaphthalene, or bicyclo[4.4.0]decane. (*a*) *Trans* isomer; (*b*) *cis* isomer.

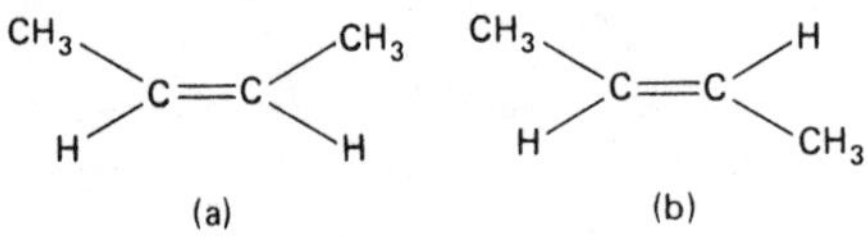

FIGURE 1.12 Two isomers of 2-butene. (*a*) *Cis* isomer, bp 3.8 °C, mp −138.9 °C, dipole moment 0.33 D; (*b*) *trans* isomer, bp 0.88 °C, mp −105.6 °C, dipole moment 0 D.

Geometrical Isomerism.   Rotation about a carbon–carbon double bond is restricted because of interaction between the *p* orbitals which make up the pi bond. Isomerism due to such restricted rotation about a bond is known as *geometric isomerism*. Parallel overlap of the *p* orbitals of each carbon atom of the double bond forms the molecular orbital of the pi bond. The relatively large barrier to rotation about the pi bond is estimated to be nearly 63 kcal · mol^{-1} (263 kJ · mol^{-1}).

When two different substituents are attached to each carbon atom of the double bond, *cis–trans* isomers can exist. In the case of *cis*-2-butene (Figure 1.12*a*), both methyl groups are on the same side of the double bond. The other isomer has the methyl groups on opposite sides and is designated as *trans*-2-butene (Figure 1.12*b*). Their physical properties are quite different. Geometric isomerism can also exist in ring systems; examples were cited in the previous discussion on conformational isomers.

For compounds containing only double-bonded atoms, the reference plane contains the double-bonded atoms and is perpendicular to the plane containing these atoms and those directly attached to them. It is customary to draw the formulas so that the reference plane is perpendicular to that of the paper. For cyclic compounds the reference plane is that in which the ring skeleton lies or to which it approximates. Cyclic structures are commonly drawn with the ring atoms in the plane of the paper.

Sequence Rules for Geometric Isomers and Chiral Compounds.   Although *cis* and *trans* designations have been used for many years, this approach becomes useless in complex systems. To eliminate confusion when each carbon of a double bond or a chiral center is

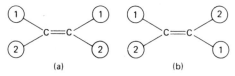

FIGURE 1.13 Configurations designated by priority groups. (*a*) Z (*cis*); (*b*) E (*trans*).

connected to different groups, the Cahn, Ingold, and Prelog system for designating configuration about a double bond or a chiral center has been adopted by IUPAC. Groups on each carbon atom of the double bond are assigned a first (1) or second (2) priority. Priority is then compared at one carbon relative to the other. When both first priority groups are on the *same side* of the double bond, the configuration is designated as Z (from the German *zusammen*, "together"), which was formerly *cis*. If the first priority groups are on *opposite sides* of the double bond, the designation is E (from the German *entgegen*, "in opposition to"), which was formerly *trans*. (See Figure 1.13.)

When a molecule contains more than one double bond, each E or Z prefix has associated with it the lower-numbered locant of the double bond concerned. Thus (see also the rules that follow)

aminoacetic acid or glycine

an α-amino acid

When the sequence rules permit alternatives, preference for lower-numbered locants and for inclusion in the principal chain is allotted as follows in the order stated: Z over E groups and *cis* over *trans* cyclic groups. If a choice is still not attained, then the lower-numbered locant for such a preferred group at the first point of difference is the determining factor. For example,

(2Z,5E)-2,5-Heptadienedioic acid

RULE 1. Priority is assigned to atoms on the basis of atomic number. Higher priority is assigned to atoms of higher atomic number. If two atoms are isotopes of the same element, the atom of higher mass number has the higher priority. For example, in 2-butene, the carbon atom of each methyl group receives first priority over the hydrogen atom connected to the same carbon atom. Around the asymmetric carbon atom in chloroiodomethanesulfonic acid, the priority sequence is I, Cl, S, H. In 1-bromo-1-deuteroethane, the priority sequence is Cl, C, D, H.

RULE 2. When atoms attached directly to a double-bonded carbon have the same priority, the second atoms are considered and so on, if necessary, working outward once again from the double bond or chiral center. For example, in 1-chloro-2-methylbutene, in CH_3 the second atoms are H, H, H and in CH_2CH_3 they are C, H, H. Since carbon has a higher atomic number than hydrogen, the ethyl group has the next highest priority after the chlorine atom.

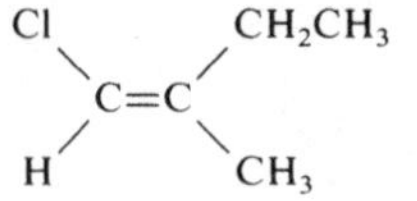

$$\begin{array}{c}
\text{Cl} \qquad\qquad \text{CH}_2\text{CH}_3 \\
\diagdown\qquad\diagup \\
\text{C}=\text{C} \\
\diagup\qquad\diagdown \\
\text{H} \qquad\qquad \text{CH}_3
\end{array}\qquad\qquad
\begin{array}{c}
\text{Cl} \qquad\qquad \text{CH}_3 \\
\diagdown\qquad\diagup \\
\text{C}=\text{C} \\
\diagup\qquad\diagdown \\
\text{H} \qquad\qquad \text{CH}_2\text{CH}_3
\end{array}$$

(Z)-1-Chloro-2-methylbutene 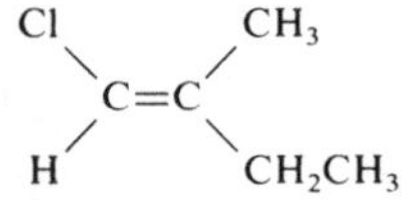 (E)-1-Chloro-2-methylbutene

RULE 3. When groups under consideration have double or triple bonds, the multiple-bonded atom is replaced conceptually by two or three single bonds to that same kind of atom. Thus, $=$A is considered to be equivalent to two A's, or $<^{\text{A}}_{\text{A}}$ and $\equiv$A equals $\Big\langle^{\text{A}}_{\substack{\text{A}\\\text{A}}}$.

However, a real $<^{\text{A}}_{\text{A}}$ has priority over $=$A; likewise a real $\Big\langle^{\substack{\text{A}\\\text{A}}}_{\text{A}}$ has priority over $\equiv$A.

Actually, both atoms of a multiple bond are duplicated, or triplicated, so that C$=$O is treated as $\begin{array}{c}\text{C}-\text{O}\\|\;\;|\\\text{O}\;\;\text{C}\end{array}$, that is $\begin{array}{c}\text{C}-\text{O}\\|\\(\text{O})\end{array}$ and $\begin{array}{c}\text{O}-\text{C}\\|\\(\text{C})\end{array}$, and C$\equiv$N is treated as $\begin{array}{c}\text{C}\diagdown\qquad\diagup\text{N}\\\diagup\quad\diagdown\quad\diagup\quad\diagdown\\(\text{N})\;\;(\text{N})\;\;(\text{C})\;\;(\text{C})\end{array}$. A phenyl carbon becomes $-\text{C}\big\langle^{\text{CH}}_{\text{CH}}$. Only the double-bonded atoms themselves are duplicated, not the atoms or groups attached to them. The duplicated atoms (or phantom atoms) may be considered as carrying atomic number zero. For example, among the groups OH, CHO, CH$_2$OH, and H, the OH group has the highest priority, and the C(O, O, H) of CHO takes priority over the C(O, H, H) of CH$_2$OH.

Chirality and Optical Activity. A compound is chiral (the term *dissymmetric* was formerly used) if it is not superimposable on its mirror image. A chiral compound does not have a plane of symmetry. Each chiral compound possesses one (or more) of three types of chiral element, namely, a chiral center, a chiral axis, or a chiral plane.

Chiral Center. The chiral center, which is the chiral element most commonly met, is exemplified by an asymmetric carbon with a tetrahedral arrangement of ligands about the carbon. The ligands comprise four different atoms or groups. One "ligand" may be a lone pair of electrons; another, a phantom atom of atomic number zero. This situation is encountered in sulfoxides or with a nitrogen atom. Lactic acid is an example of a molecule with an asymmetric (chiral) carbon. (See Figure 1.14.)

A simpler representation of molecules containing asymmetric carbon atoms is the Fischer projection, which is shown here for the same lactic acid configurations. A Fischer

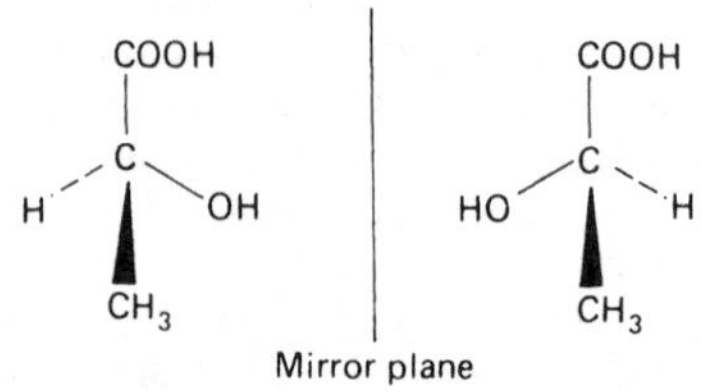

$$\begin{array}{cc}
\begin{array}{c}
\text{COOH}\\
|\\
\text{H}-\text{C}-\text{OH}\\
|\\
\text{CH}_3
\end{array}
&
\begin{array}{c}
\text{COOH}\\
|\\
\text{HO}-\text{C}-\text{H}\\
|\\
\text{CH}_3
\end{array}
\end{array}$$

Mirror plane

FIGURE 1.14A Stereo drawing of the lactic acid molecule.

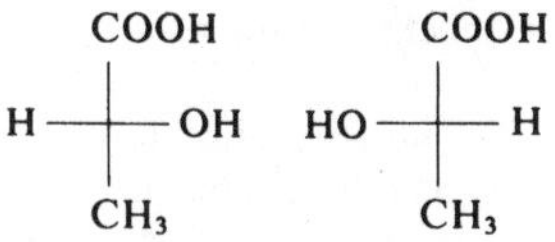

$$
\begin{array}{ccc}
\text{COOH} & & \text{COOH} \\
\text{H} \!-\!\!\!\!\!-\!\!\!\!\!- \text{OH} & \text{HO} \!-\!\!\!\!\!-\!\!\!\!\!- \text{H} \\
\text{CH}_3 & & \text{CH}_3
\end{array}
$$

FIGURE 1.14B   Fischer projection of the lactic acid molecule.

projection involves drawing a cross and attaching to the four ends the four groups that are attached to the asymmetric carbon atom. The asymmetric carbon atom is understood to be located where the lines cross. The horizontal lines are understood to represent bonds coming toward the viewer out of the plane of the paper. The vertical lines represent bonds going away from the viewer behind the plane of the paper as if the vertical line were the side of a circle. The principal chain is depicted in the vertical direction; the lowest-numbered (locant) chain member is placed at the top position. These formulas may be moved sideways or rotated through 180° in the plane of the paper, but they may not be removed from the plane of the paper (i.e., rotated through 90°). In the latter orientation it is essential to use thickened lines (for bonds coming toward the viewer) and dashed lines (for bonds receding from the viewer) to avoid confusion.

Enantiomers.   Two nonsuperimposable structures that are mirror images of each other are known as *enantiomers*. Enantiomers are related to each other in the same way that a right hand is related to a left hand. Except for the direction in which they rotate the plane of polarized light, enantiomers are identical in all physical properties. Enantiomers have identical chemical properties except in their reactivity toward optically active reagents.

Enantiomers rotate the plane of polarized light in opposite directions but with equal magnitude. If the light is rotated in a clockwise direction, the sample is said to be dextrorotatory and is designed as (+). When a sample rotates the plane of polarized light in a counterclockwise direction, it is said to be levorotatory and is designed as (−). Use of the designations *d* and *l* is discouraged.

Specific Rotation.   Optical rotation is caused by individual molecules of the optically active compound. The amount of rotation depends upon how many molecules the light beam encounters in passing through the tube. When allowances are made for the length of the tube that contains the sample and the sample concentration, it is found that the amount of rotation, as well as its direction, is a characteristic of each individual optically active compound.

Specific rotation is the number of degrees of rotation observed if a 1-dm tube is used and the compound being examined is present to the extent of 1 g per 100 mL. The density for a pure liquid replaces the solution concentration.

$$
\text{Specific rotation} = [\alpha] = \frac{\text{observed rotation (degrees)}}{\text{length (dm)} \times (\text{g}/100 \text{ mL})}
$$

The temperature of the measurement is indicated by a superscript and the wavelength of the light employed by a subscript written after the bracket; for example, $[\alpha]_{590}^{20}$ implies that the measurement was made at 20 °C using 590 nm radiation.

Optically Inactive Chiral Compounds.   Although chirality is a necessary prerequisite for optical activity, chiral compounds are not necessarily optically active. With an equal mixture of two enantiomers, no net optical rotation is observed. Such a mixture of enantiomers is said to be *racemic* and is designated as (±) and not as *dl*. Racemic mixtures usually have melting points higher than the melting point of either pure enantiomer.

A second type of optically inactive chiral compounds, *meso* compounds, will be discussed next.

Multiple Chiral Centers. The number of stereoisomers increases rapidly with an increase in the number of chiral centers in a molecule. A molecule possessing two chiral atoms should have four optical isomers, that is, four structures consisting of two pairs of enantiomers. However, if a compound has two chiral centers but both centers have the same four substituents attached, the total number of isomers is three rather than four. One isomer of such a compound is not chiral because it is identical with its mirror image; it has an internal mirror plane. This is an example of a diastereomer. The achiral structure is denoted as a *meso* compound. Diastereomers have different physical and chemical properties from the optically active enantiomers. Recognition of a plane of symmetry is usually the easiest way to detect a *meso* compound. The stereoisomers of tartaric acid are examples of compounds with multiple chiral centers (see Figure 1.15), and one of its isomers is a *meso* compound.

Stereochemistry is sometimes harder to discern in ring systems than in open-chained compounds. The smallest ring, cyclopropane, has six equivalent hydrogens, each of which may be substituted. When a substituent such as a methyl group replaces a hydrogen, the molecule remains achiral because a mirror plane is present that bisects the substituted carbon (and its substituent) and the opposite bond. One side of the cyclopropane therefore reflects the other. The presence of a substituent makes the other cyclopropane positions nonequivalent. Thus a second methyl group may be added on the same carbon (opposite side of the ring) or on one of the adjacent carbons on either the same or opposite sides of the ring. Figure 1.16 shows some of these possibilities. E-1,2-Dicarboxycyclopropane (a) exists in two distinct forms. They are nonsuperimposable mirror images; Z-1,2-Dicarboxycyclopropane (b) has an internal mirror plane and is therefore superimposable on its mirror image. Structure (d) in Figure 1.16 shows a molecular model of this compound. The carboxyl substituents do not appear to reflect each other but recall that they can rotate freely about the single bond.

A cyclic compound that has two differently substituted asymmetric carbons will have $2^2 = 4$ optical isomers. These will consist of pairs of *cis* and *trans* enantiomers. When the asymmetric centers have identical substituents, the *cis* isomer will have an internal reflection plane and is called a *meso* form. The *meso* forms of cis-1,2-dicarboxycyclopropane are shown in panels (b) and (d) of Figure 1.16 in a line angle drawing and as a molecular model.

Torsional Asymmetry. Rotation about single bonds of most acyclic compounds is relatively free at ordinary temperatures. There are, however, some examples of compounds in which nonbonded interactions between large substituent groups inhibit free rotation about a sigma bond. In some cases these compounds can be separated into pairs of enantiomers.

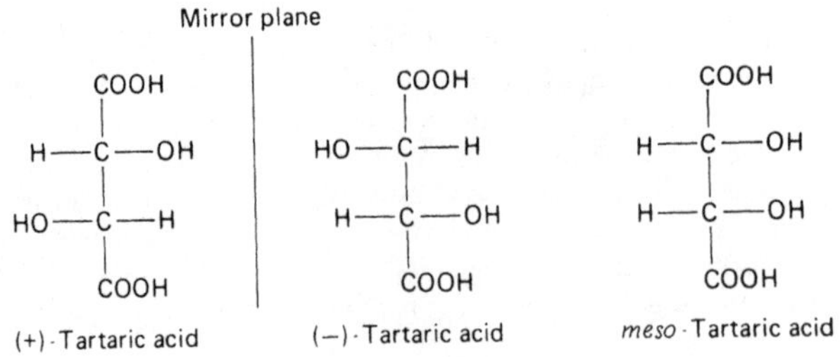

FIGURE 1.15 Isomers of tartaric acid.

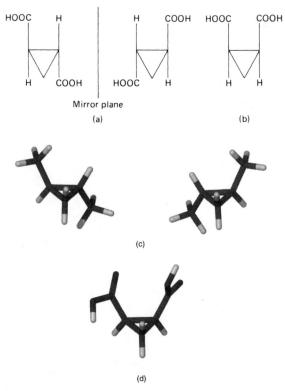

FIGURE 1.16 Isomers of cyclopropane-1,2-dicarboxylic acid. (*a*) *E*-1,2-Dicarboxycyclopropane (*trans* isomers); (*b*) *Z*-1,2-Dicarboxycyclopropane isomer. (*meso* isomer); (*c*) Molecular models of *E*- (*trans*-) 1,2-dimethylcyclopropane shown in the tube representation. Rotation of the right hand structure about a vertical axis through the center of the cyclopropane will superimpose the two methyl groups. The methylene of the rotated structure will be in the back, rather than the front, and not superimposed. (*d*) A mirror plane through the methylene and the back carbon–carbon bond is a plane of symmetry. The two carboxyl groups appear not to reflect each other in the model shown but they can rotate freely and will reflect each other on an instantaneous basis.

A *chiral axis* is present in chiral biaryl derivatives. When bulky groups are located at the *ortho* positions of each aromatic ring in biphenyl, free rotation about the single bond connecting the two rings is inhibited because of torsional strain associated with twisting rotation about the central single bond. Interconversion of enantiomers is prevented (see Figure 1.17).

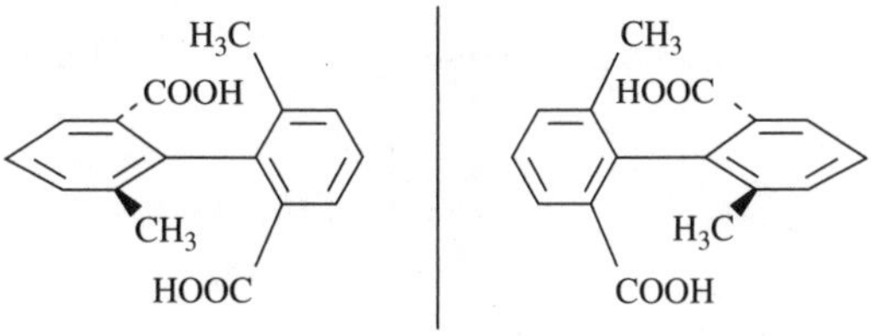

FIGURE 1.17 Isomers of biphenyl compounds with bulky groups attached at the *ortho* positions.

For compounds possessing a chiral axis, the structure can be regarded as an elongated tetrahedron to be viewed along the axis. In deciding upon the absolute configuration it does not matter from which end it is viewed; the nearer pair of ligands receives the first two positions in the order of precedence (see Figure 1.18). For the meaning of (*S*), see the discussion under "Absolute Configuration".

A *chiral plane* is exemplified by the plane containing the benzene ring and the bromine and oxygen atoms in the chiral compound shown in Figure 1.19. Rotation of the benzene ring around the oxygen-to-ring single bonds is inhibited when *x* is small (although no critical size can be reasonably established).

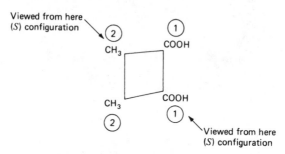

FIGURE 1.18   Example of a chiral axis.

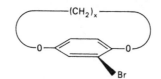

FIGURE 1.19   Example of a chiral plane.

Absolute Configuration.   The terms absolute stereochemistry and absolute configuration are used to describe the three-dimensional arrangement of substituents around a chiral element. A general system for designating absolute configuration is based upon the priority system and sequence rules. Each group attached to a chiral center is assigned a number, with number one the highest-priority group. For example, the groups attached to the chiral center of 2-butanol (see Figure 1.20) are assigned these priorities: 1 for OH, 2 for CH_2CH_3, 3 for CH_3, and 4 for H. The molecule is then viewed from the side opposite the group of lowest priority (the hydrogen atom), and the arrangement of the remaining groups are noted. If, in proceeding from the group of highest priority to the group of second priority and

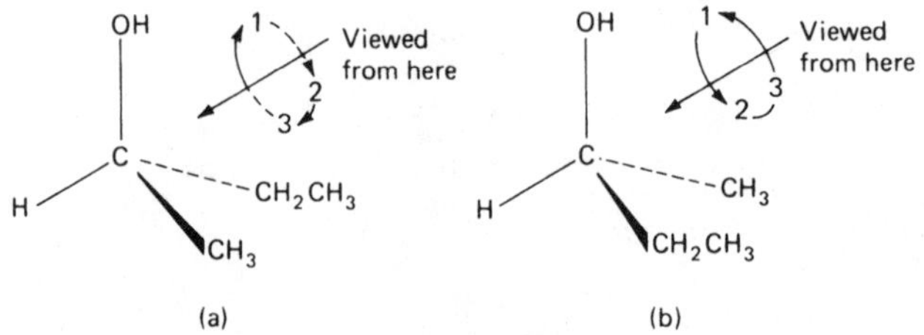

OH   OH
  Viewed   Viewed
  from here   from here

H — C ---- CH_2CH_3 H — C ---- CH_3
        CH_3 CH_2CH_3

(a)   (b)

FIGURE 1.20   Viewing angle as a means of designating the absolute configuration of compounds with a chiral axis. (*a*) (*R*)-2-butanol (sequence clockwise); (*b*) (*S*)-2-butanol (sequence counterclockwise).

thence to the third, the eye travels in a clockwise direction, the configuration is specified R (from the Latin *rectus*, "right"); if the eye travels in a counterclockwise direction, the configuration is specified S (from the Latin *sinister*, "left"). The complete name includes both configuration and direction of optical rotation, as for example, (S)-(+)-2-butanol.

The relative configurations around the chiral centers of many compounds have been established. One optically active compound is converted to another by a sequence of chemical reactions which are stereospecific; that is, each reaction is known to proceed spatially in a specific way. The configuration of one chiral compound can then be related to the configuration of the next in sequence. In order to establish absolute configuration, one must carry out sufficient stereospecific reactions to relate a new compound to another of known absolute configuration. Historically the configuration of D-(+)-2,3-dihydroxypropanal has served as the standard to which all configuration has been compared. The absolute configuration assigned to this compound has been confirmed by an X-ray crystallographic technique.

Stereochemistry in Biological Systems. Amino acids occur naturally in both D and L (R and S) enantiomeric configurations. Amino acids that occur in proteins almost always have the L configuration although amino acids that occur in bacterial peptides may have the enantiomeric D configuration. The two configurations are shown in Figure 1.21 for alanine.

The description of α-amino acids as D or L is a holdover from an older nomenclature system. In this system (S)-alanine is called L-alanine. The enantiomer would be D- or (R)-serine. The L (*laevo*, turned to the left; D = *dextro*, turned to the right) designation refers to the α-carbon in the essential amino acids. In alanine, there is a single α-carbon that is asymmetric. When two asymmetric centers are present as in L-threonine, the stereochemistry of both carbons must be considered. The common form of L-threonine is the 2S,3R stereoisomer.

Threonine (center) is shown in Figure 1.22 along with the simplest chiral amino acid, alanine. The only cyclic amino acid, proline, is pictured as well in the common L-configuration.

Extended Arrangements of Peptides and Proteins. Amino acids are linked from the carboxyl to the amine with formation of an amide bond, often referred to as the peptide link. The repeating ($-N-C-CO-$) unit is called the peptide or protein backbone. Peptides and proteins differ only in the number of amino acids present in the biopolymer chain. The cutoff is arbitrarily set. Often, but not always, a peptide is designated as having fewer than 100 amino acids and the protein possesses more. Backbone amide groups have been found to play a role in enzyme catalysis.

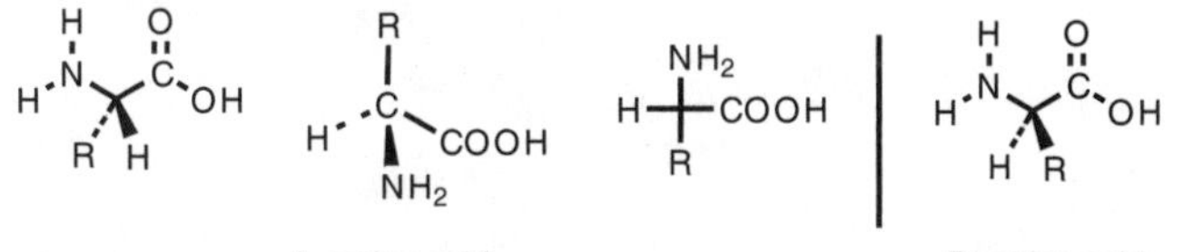

L-amino acid
(S configuration)

D-amino acid
(R configuration)

FIGURE 1.21 Stereochemistry of α-amino acids. The most common, L configuration is shown at the left.

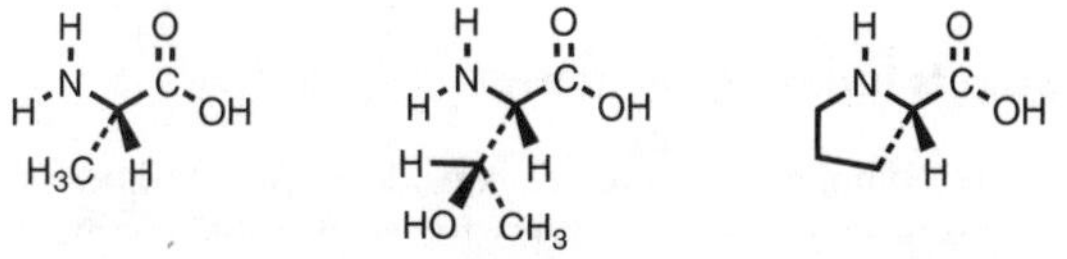

L-alanine
(S configuration)

L-threonine
(2S,3R configuration)

L-proline
(S configuration)

FIGURE 1.22 Structures of alanine, threonine, and proline.

The extended chains, that is, the backbones, may further organize into assemblies that have characteristic properties. The two most important of these are the α- helix and the β-sheet. The latter is illustrated in Figure 1.23. Panel (a) shows the extensive hydrogen bond organization of peptide chains that are oriented in opposite directions. The arrows indicate the nitrogen to carbonyl (arrowhead) direction. The lower panel (b) shows an alternate H-bond organization when the two peptide chains are parallel rather than antiparallel.

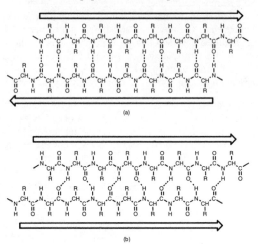

(a)

(b)

FIGURE 1.23   Hydrogen bonded interactions of peptide chains to form β-sheets. The chains are arranged antiparallel in panel (a) and parallel in panel (b).

An alternate organization for peptide chains is the α-helix. It is essentially a coil in which a carbonyl group H-bonds an amide nitrogen between every fourth residue. The resulting structure exhibits one full turn for each 3.6 amino acids, which spans 5.4 Å per turn. The resulting α-helix is a tight coil that lacks any significant interior space. A schematic representation of the H-bonded coil is shown in Figure 1.24.

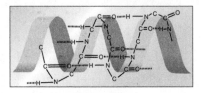

FIGURE 1.24

A number of other arrangements are possible for peptide or protein chains. A discussion of them is beyond the scope of this Handbook.

Chemical Abstracts Indexing System

When compounds of complex structure are considered, the number of name possibilities grows rapidly. To avoid having index entries for all possible names, Chemical Abstracts Service has developed what might be called the principle of inversion. The indexing system employs inverted entries to bring together related compounds in an alphabetically arranged index. The *index heading parent* from the Chemical Substance Index appears in the Formula Index in lightface before the "comma of inversion." The *substituents* follow the "comma of inversion" in alphabetical order. Any *name modification* appears on a separate line. If necessary, the chemical description is completed by citation of an associated ion, a functional derivative, a "salt with" or "compound with" terms and/or a stereochemical descriptor.

Quite naturally there is a certain amount of arbitrariness in this system, although the IUPAC nomenclature is followed. The preferred *Chemical Abstracts* index names for chemical substances have been, with very few exceptions, continued unchanged (since 1972) as set forth in the *Ninth Collective Index Guide* and in a journal article.* Any revisions appear in the updated *Index Guide*; new editions appear at 18-month intervals. Appendix VI is of particular interest to chemists. Reprints of the Appendix may be purchased from Chemical Abstracts Service, Marketing Division, P.O. Box 3012, Columbus, Ohio 43210.

PHYSICAL PROPERTIES OF PURE SUBSTANCES

TABLE 1.14   Empirical Formula Index for Organic Compounds

The alphanumeric designations are keyed to Table 1.15

Cl_2H_2Si: d226
Cl_3HSi: t247
Cl_6OSi_2: h28

C₁

$CBrClF_2$: b255
$CBrCl_3$: b358
$CBrF_3$: b360
CBr_2F_2: d75
$CClF_3$: c253
$CClNO_3S$: c240
CCl_2F_2: d170
CCl_3D: c127
CCl_3F: t232
CCl_3NO_2: t239
CCl_4O_2S: t236
CCl_4S: t235
CD_4O: m36
$CHBrCl_2$: b266
$CHBr_2Cl$: d71
$CHBr_3$: t206
$CHClF_2$: c85
$CHCl_2F$: d183
$CHCl_3$: c126
CHF_3: t295
CHF_3O_3S: t296
CHI_3: i36
CHN_3O_6: t385
CH_2BrCl: b256
CH_2Br_2: d88

CH_2Cl_2: d190
CH_2Cl_4Si: c165
CH_2I_2: d404
CH_2N_2: c285, d47
CH_2N_4: t136
CH_2O: f27
$(CH_2O)_x$: p1
CH_2O_2: f32
CH_2S_3: t434
CH_3Br: b300
CH_3Br_3Ge: m254
CH_3Cl: c137
CH_3ClHg: m295
CH_3ClO_2S: m32
CH_3Cl_3Ge: m437
CH_3Cl_3Si: t238
CH_3DO: m35
CH_3F: f18
CH_3I: i40
CH_3NO: f28
CH_3NO_2: m314, n56
CH_3NO_3: m313
CH_3N_5: a289
CH_4: m29
CH_4Cl_2Si: d199, m222
CH_4N_2O: f34, u12
$CH_4N_2O_2S$: f30
CH_4N_2S: t163
$CH_4N_4O_2$: n54
CH_4O: m34
CH_4O_2: m275
CH_4O_3S: m30

CH_4S: m33
CH_5AsO_3: m125
CH_5N: m115
CH_5NO_3S: a205
CH_5N_3: g29
CH_5N_3O: s3
CH_5N_3S: t162
CH_6N_2: m270
CH_6N_4: a180, a181
CH_6N_4O: c11
CN_4O_8: t126a

C₂

$C_2Br_2ClF_3$: d72
$C_2Br_2Cl_4$: d99
$C_2Br_2F_4$: d100
$C_2Br_2O_2$: o50
C_2ClF_3: c252
$C_2Cl_2F_3I$: d188
$C_2Cl_2F_4$: d227
$C_2Cl_2O_2$: o51
$C_2Cl_3F_3$: t251
C_2Cl_3N: t217
C_2Cl_4: t29
$C_2Cl_4F_2$: d347, d348, t26
C_2Cl_4O: t218
C_2Cl_6: h29
C_2D_3N: a30
$C_2D_4O_2$: a21
C_2D_6OS: d615

TABLE 1.14 Empirical Formula Index for Organic Compounds (*continued*)

The alphanumeric designations are keyed to table 1.15

C_2F_4: t63
C_2F_6: h44
$C_2F_6O_5S_2$: t297
$C_2HBrClF_3$: b258
$C_2HBr_2F_3$: d103
C_2HBr_2N: d63
C_2HBr_3: t205
C_2HBr_3O: t201
$C_2HBr_3O_2$: t202
$C_2HClF_2O_2$: c83
$C_2HCl_2F_3$: d232
C_2HCl_3: t230
C_2HCl_3O: d141
$C_2HCl_3O_2$: t216
C_2HCl_5: p9
$C_2HF_3O_2$: t287
C_2H_2: a41
C_2H_2BrClO: b224
$C_2H_2Br_2$: d80, d81
$C_2H_2Br_2F_2$: d74
$C_2H_2Br_2O$: b223
$C_2H_2Br_2O_2$: d62
$C_2H_2Br_4$: t9
$C_2H_2ClF_3$: c251
C_2H_2ClN: c27
$C_2H_2Cl_2$: d178, d179, d180
$C_2H_2Cl_2O$: c31
$C_2H_2Cl_2O_2$: d138
$C_2H_2Cl_4$: t27, t28
$C_2H_2F_3NO$: t286
C_2H_2O: k1
$C_2H_2O_2$: g27
$C_2H_2O_3$: g28
$C_2H_2O_4$: o48, o49
C_2H_3Br: b284
C_2H_3BrO: a35
$C_2H_3BrO_2$: b220
$C_2H_3Br_2Cl_3Si$: d82
$C_2H_3Br_3O$: t204
C_2H_3Cl: c109
$C_2H_3ClF_2$: c84
C_2H_3ClO: a37
$C_2H_3ClO_2$: c24, m187
$C_2H_3Cl_3$: t226, t227
$C_2H_3Cl_3O$: t228
$C_2H_3Cl_3Si$: t252

$C_2H_3Cl_5Si$: d182
$C_2H_3DO_2$: a20
C_2H_3FO: a43
$C_2H_3FO_2$: f6
$C_2H_3F_3$: t291
$C_2H_3F_3O$: t292
C_2H_3IO: a48
$C_2H_3IO_2$: i25
C_2H_3N: a29
C_2H_3NO: m287
C_2H_3NS: m289, m426
$C_2H_3N_3$: t199
$C_2H_3N_3S_2$: a290
C_2H_4BrCl: b254
C_2H_4BrNO: b218
$C_2H_4Br_2$: d77, d78
C_2H_4ClNO: c22
$C_2H_4Cl_2$: d176, d177
$C_2H_4Cl_2O$: d197
$C_2H_4Cl_6Si_2$: b204
C_2H_4FNO: f5
$C_2H_4F_2$: d346
C_2H_4INO: i24
$C_2H_4I_2$: d403
$C_2H_4N_2$: a106
$C_2H_4N_2O_2$: o54
$C_2H_4N_2O_4$: d632
$C_2H_4N_2O_6$: e126
$C_2H_4N_2S_2$: d710
$C_2H_4N_4$: a295, d235
$C_2H_4N_4O_2$: a324
C_2H_4O: a4, e129
C_2H_4OS: t142
$C_2H_4O_2$: a19, h86, m250
$C_2H_4O_2S$: m14
$C_2H_4O_3$: h87, p59
$C_2H_4O_5S$: s23
C_2H_4S: e130
$C_2H_5AlCl_2$: e57
C_2H_5Br: b277
$C_2H_5BrNaO_2S$: b278
C_2H_5BrO: b279, b308
C_2H_5Cl: c102
C_2H_5ClHg: e165
C_2H_5ClO: c103, c155
$C_2H_5ClO_2S$: e19

C_2H_5ClS: c156
$C_2H_5Cl_2OPS$: e117
$C_2H_5Cl_2O_2P$: e116
$C_2H_5Cl_3Si$: c153, t231
C_2H_5DO: e22
C_2H_5F: f17
$C_2H_5FO_3S$: e134
C_2H_5I: i34
C_2H_5IO: i35
C_2H_5N: e131
C_2H_5NO: a5, a6, m248
$C_2H_5NO_2$: e186, g25, m181, n53
$C_2H_5NO_3$: e185
C_2H_5NS: t141
$C_2H_5N_3O_2$: b215, o53
C_2H_6: e14
C_2H_6BrN: b281
C_2H_6Cd: d501
C_2H_6ClN: c106
$C_2H_6ClNO_2S$: d609
$C_2H_6ClO_2PS$: d504
$C_2H_6Cl_2Si$: d174
C_2H_6Hg: d546
$C_2H_6N_2$: a7
$C_2H_6N_2O$: a25, m444, n79
$C_2H_6N_2O_2$: m271
$C_2H_6N_2O_4S$: a107
$C_2H_6N_2S$: m431
$C_2H_6N_4O_2$: o52
C_2H_6O: d518, e21
C_2H_6OS: d614, m18
$C_2H_6O_2$: e16, e128
$C_2H_6O_2S$: d613
$C_2H_6O_3S$: d612, m297
$C_2H_6O_4S$: d610, h114
$C_2H_6O_5S_2$: m31
C_2H_6S: d611, e20
$C_2H_6S_2$: d516, e18
C_2H_6Te: d617
C_2H_6Zn: d624
$C_2H_7AsO_2$: d484
C_2H_7ClSi: c92
C_2H_7N: d461, e58
C_2H_7NO: a163, a164
$C_2H_7NO_3S$: a161

TABLE 1.14 Empirical Formula Index for Organic Compounds (*continued*)

The alphanumeric designations are keyed to Table 1.15

$C_2H_7NO_4S$: a169
C_2H_7NS: a162
$C_2H_7N_5$: b133
$C_2H_7O_3P$: d541
$C_2H_8N_2$: d539, d540, e15
$C_2H_8N_2O$: h120

$$C_3$$

$C_3Br_2F_6$: d85
$C_3Cl_3NO_2$: t219
$C_3Cl_3N_3$: t250
$C_3Cl_3N_3O_3$: t234
C_3Cl_6: h31
C_3Cl_6O: h23
C_3D_6O: a27
C_3HCl_5O: p7
C_3H_2ClN: c32
$C_3H_2Cl_2O_2$: m6
$C_3H_2Cl_4$: t33
$C_3H_2Cl_4O$: t21
$C_3H_2Cl_4O_2$: t229
$C_3H_2F_6O$: h45
$C_3H_2N_2$: m5
$C_3H_2N_2O_3$: i6
$C_3H_2O_2$: p241
C_3H_3Br: b344
C_3H_3Cl: c232
C_3H_3ClO: a65
$C_3H_3Cl_3O$: e13
C_3H_3N: a64
$C_3H_3NOS_2$: r3
$C_3H_3NO_2$: c287
C_3H_3NS: t140
$C_3H_3N_3O_2S$: a249
$C_3H_3N_3O_3$: c299
C_3H_4: a78, p240
C_3H_4BrClO: b340, b341
C_3H_4BrN: b339
$C_3H_4Br_2$: d95
$C_3H_4Br_2O_2$: d96
C_3H_4ClN: c220
$C_3H_4Cl_2$: d221, d222
$C_3H_4Cl_2O$: c221, c222, d139

$C_3H_4Cl_2O_2$: m219
$C_3H_4Cl_3NO$: m436
$C_3H_4F_4O$: t64
$C_3H_4N_2$: i4, p245
$C_3H_4N_2O$: c286
$C_3H_4N_2OS$: t152
$C_3H_4N_2O_2$: h84
$C_3H_4N_2S$: a291
C_3H_4O: p204, p242
$C_3H_4O_2$: a63, o59, p210
$C_3H_4O_3$: e124, o60
$C_3H_4O_4$: m3
C_3H_5Br: a85, b225, b335, b336
C_3H_5BrO: b276
$C_3H_5BrO_2$: b337, b338, m143
$C_3H_5Br_3$: t208
C_3H_5Cl: c216
C_3H_5ClO: c101, c215, p216
C_3H_5ClOS: e101
$C_3H_5ClO_2$: c218, c219, e98, m182
$C_3H_5Cl_3$: t244
$C_3H_5Cl_3O$: t245
$C_3H_5Cl_3Si$: a102
C_3H_5FO: f7
$C_3H_5F_3O_3S$: m438
C_3H_5I: a92, i50
C_3H_5N: p215
C_3H_5NO: a62, c290, h168, h169
$C_3H_5NO_2$: o55
C_3H_5NS: e161, m421
$C_3H_5N_3O$: c288
$C_3H_5N_3O_9$: g21
$C_3H_5N_3S$: c292
C_3H_6: c364, p205
C_3H_6BrCl: b257
$C_3H_6BrNO_4$: b316
$C_3H_6Br_2$: d92, d93
$C_3H_6Br_2O$: d94
C_3H_6ClNO: d502
$C_3H_6Cl_2$: d218, d219
$C_3H_6Cl_2O$: d220
$C_3H_6Cl_2Si$: d200

$C_3H_6Cl_4Si$: c229
$C_3H_6I_2$: d405
$C_3H_6N_2$: a274, d505
$C_3H_6N_2O$: i7
$C_3H_6N_2O_2$: m4, m269
$C_3H_6N_2S$: a292, i5
$C_3H_6N_2OS$: a58
$C_3H_6N_6$: t198
C_3H_6O: a26, a81, e10, m446, p211, p227, t345
C_3H_6OS: m420, t161
$C_3H_6O_2$: d647, e11, e135, h89, m111, p213
$C_3H_6O_2S$: m21, m293
$C_3H_6O_3$: d397, d398, d503, L1, L2, m38, m259, t388
$C_3H_6O_3S$: p198
C_3H_6S: p206, p228, t345a
$C_3H_6S_3$: t431
C_3H_7Br: b332, b333
C_3H_7BrO: b334
C_3H_7Cl: c210, c211
C_3H_7ClO: c111, c152, c213, c214
C_3H_7ClOS: c136
$C_3H_7ClO_2$: c212
$C_3H_7ClO_2S$: p197
$C_3H_7Cl_2OP$: d236
$C_3H_7Cl_3Si$: d194, p237
C_3H_7I: i48, i49
C_3H_7N: a82, p226
C_3H_7NO: a28, d522, m110, p212
$C_3H_7NO_2$: a73, a74, a75, a76, e91, m258, n73, n74
$C_3H_7NO_2S$: c370
$C_3H_7NO_3$: i105, n75, p233, s4
$C_3H_7NO_5S$: a288
C_3H_7NS: d620
$C_3H_7NS_2$: d517
$C_3H_7O_5P$: c17
C_3H_8: p191
C_3H_8ClN: c224
$C_3H_8Cl_2Si$: c75, c150

TABLE 1.14   Empirical Formula Index for Organic Compounds (*continued*)

The alphanumeric designations are keyed to Table 1.15

C_3H_8IN: d549
$C_3H_8N_2O$: d623, e230
$C_3H_8N_2O_2$: e92, f29
$C_3H_8N_2S$: d621
C_3H_8O: e171, p202, p203
$C_3H_8OS_2$: d425, m305
$C_3H_8O_2$: d441, m65, p194, p195
$C_3H_8O_2S$: m20
$C_3H_8O_3$: g16
C_3H_8S: e182, p199, p200
$C_3H_8S_2$: p196
C_3H_9Al: t327
$C_3H_8BO_3$: t319
$C_3H_9B_3O_6$: t320
C_3H_9BrGe: b363
C_3H_9BrSi: b364
C_3H_9ClGe: c254
C_3H_9ClSi: c255
C_3H_9IOS: t378
C_3H_9IS: t377
C_3H_9ISi: i55
C_3H_9N: i88, m246, p220, t328
C_3H_9NO: a269, a270, a271, a272, m69, m119, t329
$C_3H_9NO_2$: a268
$C_3H_9N_3Si$: a319
$C_3H_9O_3P$: d551, t364
$C_3H_9O_4P$: t363
$C_3H_{10}N_2$: m247, p192, p193
$C_3H_{10}N_2O$: d43
$C_3H_{11}Br_2N_3S$: a171

C_4

$C_4Cl_2F_6$: d185
$C_4Cl_2F_8$: d206
$C_4Cl_2O_3$: d189
$C_4Cl_3F_7$: h3
C_4Cl_6: h25
$C_4D_6O_3$: a23
$C_4F_6O_3$: t288

C_4HBrO_3: b299
$C_4HCl_3N_2$: t246
$C_4HF_7O_2$: h2
C_4H_2: b376
$C_4H_2Br_2S$: d101
$C_4H_2Cl_2N_2$: d223
$C_4H_2Cl_2O_2$: f38
$C_4H_2Cl_2O_3$: d208
$C_4H_2Cl_2S$: d228
$C_4H_2F_6O_2$: t294
$C_4H_2O_3$: m2
$C_4H_2O_4$: a42
C_4H_3BrS: b353
C_4H_3ClS: c242
$C_4H_3Cl_2N_3O$: d193
C_4H_3IS: i52
C_4H_4: b407
$C_4H_4BrNO_2$: b351
$C_4H_4Br_2O_2$: d69
$C_4H_4Br_2O_4$: d98
$C_4H_4ClNO_2$: c239
$C_4H_4Cl_2$: d168
$C_4H_4Cl_2O_2$: s19
$C_4H_4Cl_2O_3$: c25
$C_4H_4N_2$: b380, p244, p247, p267, s18
$C_4H_4N_2O_2$: d400, p268
$C_4H_4N_2O_2S$: d388
$C_4H_4N_2O_3$: b1
$C_4H_4N_2O_5$: a79
$C_4H_4N_4$: d40
C_4H_4O: f40
$C_4H_4O_2$: d422
$C_4H_4O_3$: s16
$C_4H_4O_4$: f37, ml
C_4H_4S: t154
$C_4H_5BrO_4$: b350
C_4H_5Cl: c63, c70
C_4H_5ClO: c283, c366, m28
$C_4H_5ClO_2$: a87
$C_4H_5ClO_3$: e191
$C_4H_5Cl_3O_2$: e226
$C_4H_5F_3O_2$: e227
C_4H_5N: b400, c365, m27, p269
C_4H_5NO: m290

$C_4H_5NO_2$: e106, m193, s17
$C_4H_5NO_2S$: e32
$C_4H_5NO_3$: h182
C_4H_5NS: a93
$C_4H_5N_3$: a284, i11
$C_4H_5N_3O$: a198
$C_4H_5N_3OS$: a191
$C_4H_5N_3O_2$: a154, a155, c289, m322
C_4H_6: b373, b374, b490, b491
$C_4H_6Br_2O_2$: d70
C_4H_6ClN: c73
$C_4H_6Cl_2$: d165, d166, d167
$C_4H_6Cl_2O$: c74
$C_4H_6Cl_2O_2$: m220
$C_4H_6Cl_3NSi$: c294
$C_4H_6N_2$: a151, m280, m281, m282
$C_4H_6N_2O_2$: e114
$C_4H_6N_2S$: a229
$C_4H_6N_4O$: d39
$C_4H_6N_4O_3$: a77
C_4H_6O: b406, c282, d356, d545a, m24, m396
$C_4H_6O_2$: b386, b401, b402, b403, b492, b497, b498, c367, m26, m114, v2
$C_4H_6O_2S$: d368
$C_4H_6O_3$: a22, a24, m334, o56, p225
$C_4H_6O_4$: d566, s14
$C_4H_6O_4S$: m23, t148
$C_4H_6O_5$: h180, h181, o61
$C_4H_6O_6$: tl, t2
C_4H_7Br: b240, b241, b242
$C_4H_7BrO_2$: b244, b280, b307, e75, m146
C_4H_7Cl: c68, c69, c163, c164
C_4H_7ClO: b501, c67, c115, i78
$C_4H_7ClO_2$: c71, c72, e94, m189
$C_4H_7Cl_2NSi$: c291
$C_4H_7Cl_3O$: t237

TABLE 1.14 Empirical Formula Index for Organic Compounds (*continued*)

The alphanumeric designations are keyed to Table 1.15

$C_4H_7Cl_3O_2Si$: c13
$C_4H_7FO_2$: e133
C_4H_7N: b499, i76
C_4H_7NO: h145, i98, m25, m334, p231, p275
$C_4H_7NO_2$: m331
$C_4H_7NO_3$: a46, e192, p12
$C_4H_7NO_4$: a314, i10
C_4H_7NS: m419
$C_4H_7N_3O$: c278
C_4H_8: b395, b396, b397, c300, m383
C_4H_8BrCl: b251
$C_4H_8Br_2$: d67, d68
$C_4H_8Br_2O$: b149
$C_4H_8Cl_2$: d162, d163, d164
$C_4H_8Cl_2O$: b158, d181
$C_4H_8Cl_2Si$: a89
$C_4H_8N_2O$: a105, a150
$C_4H_8N_2O_2$: d526, s13
$C_4H_8N_2O_3$: a313, g26
$C_4H_8N_2S$: a101, t79
C_4H_8O: b393, b404, b405, b493, e3, e232, i73, m96, m377, m385, t66
C_4H_8OS: e220, t107, t164
$C_4H_8O_2$: b398, b399, b495, d646, e51, h106, i75, m389, m390, p229
$C_4H_8O_2S$: e164, m294, t106
$C_4H_8O_3$: e23, e150, h116, h127, m64, m291, m298
$C_4H_8O_3S$: m384
C_4H_8S: a95, t81
$C_4H_8S_2$: d707
C_4H_9Br: b238, b239, b310, b311
C_4H_9BrO: b285
C_4H_9Cl: c64, c65, c161, c162
C_4H_9ClO: c66, c110
$C_4H_9ClO_2$: c89, c104, m67
C_4H_9ClSi: c93
$C_4H_9Cl_3Si$: b483, c225
$C_4H_9Cl_3Sn$: b481

C_4H_9F: f20
C_4H_9I: i30, i31, i43, i44
C_4H_9Li: b457, b458
C_4H_9N: p270
C_4H_9NO: a321, b394, b494, d458, e52, i74, m388, m448
$C_4H_9NO_2$: a138, a139, a222, b464, b465, h115, i71, n50
$C_4H_9NO_2S$: a204
$C_4H_9NO_3$: a187, a188, a189, a190, i70, n51
C_4H_9NSi: c298
$C_4H_9N_3O_2$: c277
C_4H_{10}: b378, m375
$C_4H_{10}ClN$: d467
$C_4H_{10}ClO_2PS$: d292
$C_4H_{10}ClO_3P$: d291
$C_4H_{10}Cl_2Si$: b160, m392
$C_4H_{10}N_2$: p179
$C_4H_{10}N_2O$: a231
$C_4H_{10}N_2O_4S$: a8
$C_4H_{10}O$: b391, b392, d300, m381, m382, m393
$C_4H_{10}OS$: e153
$C_4H_{10}OS_2$: b186
$C_4H_{10}O_2$: b381, b382, b383, b384, b385, b453, d438, d439, e34, m95
$C_4H_{10}O_2S$: m430, t149
$C_4H_{10}O_2S_2$: d424, h118
$C_4H_{10}O_3$: b181, b390, t351
$C_4H_{10}O_3S$: d338
$C_4H_{10}O_4S$: d336
$C_4H_{10}S$: b388, b389, d337, i104, m378, m379, m380, m395
$C_4H_{10}S_2$: b387, d294a
$C_4H_{10}S_3$: b187
$C_4H_{10}Zn$: d344
$C_4H_{11}ClSi$: c166
$C_4H_{11}N$: b377, b417, b418, d267, d268, d520, i63
$C_4H_{11}NO$: a136, a137, a221, d315, d465, e38, e62

$C_4H_{11}NO_2$: a165, a220, d245, d440
$C_4H_{11}NO_3$: t423
$C_4H_{11}O_2PS_2$: d296
$C_4H_{11}O_3P$: d314
$C_4H_{12}BrN$: t93
$C_4H_{12}ClN$: t94
$C_4H_{12}Ge$: t109
$C_4H_{12}IN$: t95
$C_4H_{12}N_2$: b379, b452, d521, m376, m377
$C_4H_{12}N_2O$: a166
$C_4H_{12}N_2S_2$: c369
$C_4H_{12}OSi$: m108
$C_4H_{12}O_3Si$: t326a
$C_4H_{12}O_4Si$: t92
$C_4H_{12}Pb$: t112
$C_4H_{12}Si$: t120
$C_4H_{12}Sn$: t123
$C_4H_{13}N_3$: d298
$C_4H_{14}OSi_2$: t105
$C_4H_{16}O_4Si_4$: t103

C_5

C_5Cl_5N: p12
C_5Cl_6: h27
C_5D_5N: p249
$C_5H_3Br_2N$: d97
$C_5H_3ClO_2$: f48
$C_5H_3Cl_2N$: d224
C_5H_4BrN: b345, b346
C_5H_4ClN: c233
C_5H_4FN: f23
$C_5H_4F_8O$: o18
$C_5H_4N_2O_3$: n76
$C_5H_4N_4O$: h186
$C_5H_4N_4O_3$: u13
C_5H_4OS: t156
$C_5H_4O_2$: f39
$C_5H_4O_2S$: t157
$C_5H_4O_3$: c271, f42
$C_5H_5ClN_2$: a149
$C_5H_5ClN_2O_2$: c167
$C_5H_5F_3O_2$: t293

TABLE 1.14 Empirical Formula Index for Organic Compounds (*continued*)

The alphanumeric designations are keyed to Table 1.15

C_5H_5N: p248

C_5H_5NO: h173, h174, h175, p262

$C_5H_5NO_2$: d401, h177

$C_5H_5NO_3S$: p263

$C_5H_5N_3O_2$: a248

$C_5H_5N_3O_4$: a160

$C_5H_5N_5$: a69

C_5H_6: m166

$C_5H_6Br_2N_2O_2$: d76

$C_5H_6Cl_2N_2O_2$: d173

$C_5H_6Cl_2O_2$: d195, g15

$C_5H_6Cl_4O_2$: t225

$C_5H_6N_2$: a281, a282, a283, g14, m397, v7

$C_5H_6N_2O$: a47, a197

$C_5H_6N_2OS$: h128

$C_5H_6N_2O_2$: d391

$C_5H_6N_2O_2$: e107

C_5H_6O: m59, m252

C_5H_6OS: f44

$C_5H_6O_2$: f46

$C_5H_6O_3$: g12

$C_5H_6O_4$: c270, m245

$C_5H_6O_4S_3$: b155

C_5H_6S: m427

$C_5H_7BrO_2$: m145

$C_5H_7BrO_3$: e79

$C_5H_7ClO_3$: m183, m188

C_5H_7N: m404

C_5H_7NO: f47

$C_5H_7NO_2$: e105

C_5H_7NS: t158

$C_5H_7N_3$: a228, d44

$C_5H_7N_3O$: a192

C_5H_8: c358, m147, m148, m171, p16, p17, p18, p19, p57

$C_5H_8Br_2O_2$: e115

$C_5H_8Br_4$: p21

$C_5H_8F_4O$: m414

$C_5H_8N_2$: d542, d603, e159, p272

$C_5H_8N_2O$: m449

$C_5H_8N_2O_2$: d538

$C_5H_8N_4O_{12}$: p22

C_5H_8O: c356, c368, d363, e8, m172, p51

$C_5H_8O_2$: a80, e56, g13, i84, m58, m161, m162, m163, m192, m217, m296, p31, p32, p40, p40a, p50, p208

$C_5H_8O_3$: e196, m112, o58

$C_5H_8O_4$: d545, g11, m274, m412

C_5H_9Br: b263

$C_5H_9BrO_2$: e81, e82, m144

C_5H_9Cl: c79

C_5H_9ClO: c192, d600, m179, p44

C_5H_9ClOS: b437, c229

$C_5H_9ClO_2$: b436, e99, e100, i65, m186

$C_5H_9F_3O_2Si$: t376

C_5H_9N: d602, m178, p33, t78

C_5H_9NO: b455, b456, c357, e193, m406

$C_5H_9NO_2$: m118, p273

$C_5H_9NO_4$: g9

$C_5H_9N_3$: i8

C_5H_{10} c352, m158, m159, m160, p47, p48, p49

$C_5H_{10}Br_2$: d91

$C_5H_{10}ClNO$: d288

$C_5H_{10}Cl_2$: d209

$C_5H_{10}Cl_2O_2Si$: c12

$C_5H_{10}Cl_2Si$: c351

$C_5H_{10}N_2$: d293, d474

$C_5H_{10}N_2O$: d543, p180

$C_5H_{10}N_2O_3$: g10

$C_5H_{10}O$: a91, c355, d596, i108, m157, m164, m165, m173, m174, m415, p27, p41, p42, t76

$C_5H_{10}OS$: m425

$C_5H_{10}O_2$: d454, d598, e208, h139, h156, h157, i66, i87, m102, m175, m176, m177, m286, p36, p219, t68

$C_5H_{10}O_2S$: e183, m306, m417

$C_5H_{10}O_3$: d289, d451, e162, m68, m278

$C_5H_{10}O_4$: b184

$C_5H_{10}O_5$: a310, r5, x8

$C_5H_{11}Br$: b305, b322, b323

$C_5H_{11}BrO_2$: b269

$C_5H_{11}BrO_2Si$: t370

$C_5H_{11}Cl$: c91, c148, c149, c191

$C_5H_{11}ClSi$: a86

$C_5H_{11}Cl_2N$: b159

$C_5H_{11}I$: i42, i47

$C_5H_{11}N$: a90, m405, p183

$C_5H_{11}NO$: d304, d597, h167, m307, t69

$C_5H_{11}NO_2$: a253, a254, b128, e231, i81, v1

$C_5H_{11}NO_2S$: m37

$C_5H_{11}NO_3$: n58

$C_5H_{11}NS_2$: d295

$C_5H_{11}O_5P$: t365

C_5H_{12}: d592, m149, p28

$C_5H_{12}ClN$: d475

$C_5H_{12}Cl_2O_2Si$: b157

$C_5H_{12}N_2$: a267, m368, m369

$C_5H_{12}N_2O$: b487, t124

$C_5H_{12}N_2O_2$: b430, o46

$C_5H_{12}N_2S$: t122

$C_5H_{12}N_2S_2$: p271

$C_5H_{12}O$: b460, d452, d595, e209, m153, m154, m155, m156, p37, p38, p39

$C_5H_{12}OSi$: t379

$C_5H_{12}O_2$: d594, m57, p30

$C_5H_{12}O_2S$: e221

$C_5H_{12}O_3$: h141, m66, t352, t425

$C_5H_{12}O_3S$: p34

$C_5H_{12}O_4$: p20, t114

$C_5H_{12}O_5$: x7

$C_5H_{12}S$: b463, e210, m150, m151, m152, p35

TABLE 1.14 Empirical Formula Index for Organic Compounds (*continued*)

The alphanumeric designations are keyed to Table 1.15

$C_5H_{12}Si$: t380
$C_5H_{13}N$: a251, a252, d601, m167, m168, m169, p53
$C_5H_{13}NO$: a213, a214, a255, d472, d473, e47, i89, p221
$C_5H_{12}NOSi$: t368
$C_5H_{13}NO_2$: a176, d442, d471, d523, m223
$C_5H_{13}N_3$: t110
$C_5H_{14}N_2$: d593, p29, t113
$C_5H_{14}OSi$: e50, t372
$C_5H_{14}O_2Si$: d255
$C_5H_{15}N_3$: a175

C_6

C_6BrD_5: b230
C_6BrF_5: b321
$C_6Cl_4O_2$: t24, t25
$C_6Cl_5NO_2$: p10
C_6Cl_6: h24
C_6D_6: b10
C_6D_{12}: c313
C_6F_6: h43
C_6HBr_5O: p6
$C_6HCl_4NO_2$: t30
C_5HCl_5: p8
C_6HCl_5O: p11
$C_6H_2BrFN_2O_4$: b272
$C_6H_2Cl_2O_4$: d172
$C_6H_2Cl_3NO_2$: t242a
$C_6H_2Cl_4$: t22, t23
$C_6H_3Br_2F$: d84
$C_6H_3Br_2NO_2$: d90
$C_6H_3Br_3O$, t207
$C_6H_3ClFNO_2$: c121
$C_6H_3ClN_2O_4$: c94, c95
$C_6H_3ClN_2O_4S$: d627
$C_6H_3Cl_2NO_2$: d203, d204, d205
$C_6H_3Cl_3$: t222, t223, t224
$C_6H_3Cl_3O$: t240, t241
$C_6H_3Cl_3O_2S$: d155
$C_6H_3FN_2O_4$: d633

$C_6H_3N_3O_6$: t382, t383
$C_6H_3N_3O_7$: p173
C_6H_4BrCl: b247, b248, b249
$C_6H_4BrClO_2S$: b231
C_6H_4BrF: b288, b289, b290
$C_6H_4BrNO_2$: b314
$C_6H_4BrN_3O_4$: b271
$C_6H_4Br_2$: d65
$C_6H_4Br_2N_2O_2$: d89
$C_6H_4Br_3N$: t203
C_6H_4ClF: c116, c117, c118
C_6H_4ClFO: c122
C_6H_4ClI: c135
$C_6H_4ClNO_2$: c175, c176, c177, c234, c235
$C_6H_4ClNO_3$: c186
$C_6H_4ClNO_4S$: n35
$C_6H_4ClO_2P$: c110
$C_6H_4Cl_2$: d152, d153, d154
$C_6H_4Cl_2N_2O_2$: d202
$C_6H_4Cl_2O$: d210, d211, d212, d213
$C_6H_4Cl_2O_2$: d171
$C_6H_4Cl_2O_2S$: c43
$C_6H_4Cl_3N$: t220, t221
$C_6H_4Cl_4Si$: c208
$C_6H_4FNO_2$: f21
$C_6H_4F_2$: d345
$C_6H_4INO_2$: i45
$C_6H_4I_2$: d402
$C_6H_4N_2$: c295, c296, c297
$C_6H_4N_2O_2$: b43
$C_6H_4N_2O_4$: d626
$C_6H_4N_2O_5$: d635
$C_6H_4N_4$: a273
$C_6H_4N_4O_6$: t381
$C_6H_4O_2$: b59
$C_6H_5BO_2$: c21
C_6H_5Br: b229
C_6H_5BrO: b325, b326
C_6H_5BrS: b354
C_6H_5Cl: c41
C_6H_5ClHg: p126
$C_6H_5ClN_2O_2$: c172, c173a, c173, c174

C_6H_5ClO: c194, c195, c196
$C_6H_5ClO_2$: c87, c88
$C_6H_5ClO_2S$: b23
C_6H_5ClS: c243
C_6H_5ClSe: p151
$C_6H_5Cl_2N$: d142, d143, d144, d145, d146, d147
$C_6H_5Cl_2OP$: p137
$C_6H_5Cl_2O_2P$: p105
$C_6H_5Cl_2P$: d216
$C_6H_5Cl_2PS$: p138
$C_6H_5Cl_3Si$: p155
C_6H_5D: b9
C_6H_5F: f11
C_6H_5FO: f22
$C_6H_5FO_2S$: b24
$C_6H_5F_7O_2$: e137
C_6H_5I: i27
C_6H_5NO: n78, p251, p252, p253
C_6H_5NOS: t153
$C_6H_5NO_2$: n30, n83, p255, p256, p257
$C_6H_5NO_3$: h176, n60, n61
$C_6H_5NO_4$: c272
$C_6H_5N_3$: b62
$C_6H_5N_3O$: h103
$C_6H_5N_3O_4$: d625
C_6H_6: b8a
$C_6H_6AsNO_6$: h153
C_6H_6BrN: b225, b226, b227
C_6H_6ClN: c33, c34, c35
C_6H_6ClNO: a148, c141
$C_6H_6ClNO_2S$: c42
$C_6H_6Cl_2N_2$: d215
$C_6H_6Cl_6$: h26
C_6H_6FN: f9
C_6H_6HgO: p127
C_6H_6IN: i26
$C_6H_6N_2O$: e43, p250, p254
$C_6H_6N_2O_2$: n24, n25, n26
$C_6H_6N_2O_3$: a244, a245, m84

TABLE 1.14   Empirical Formula Index for Organic Compounds (*continued*)

The alphanumeric designations are keyed to Table 1.15

$C_6H_6N_4O_4$: d637
C_6H_6O: p64
C_6H_6OS: a57, m428
$C_6H_6O_2$: a44, d377, d378, d379, m251
$C_6H_6O_2S$: b20, t155
$C_6H_6O_3$: h146, m253, t304, t305
$C_6H_6O_3S$: b22
$C_6H_6O_4$: d460
$C_6H_6O_5S$: d382
$C_6H_6O_6$: p207
$C_6H_6O_8S_2$: d381
C_6H_6S: t159
$C_6H_7AsO_3$: b11
$C_6H_7BO_2$: b12
$C_6H_7ClN_2$: c202, c203, c204, c205
C_6H_7N: a298, a299, m398, m399, m400
C_6H_7NO: a257, a258, a259, m101, m403, p264, p265
$C_6H_7NO_2S$: b21
$C_6H_7NO_3S$: a118, a119, a120, s23
$C_6H_7NO_6S_2$: a117
C_6H_7NS: a293
$C_6H_7N_3O$: p258
$C_6H_7N_3O_2$: n67, n68, n69
$C_6H_7O_2P$: p135
$C_6H_7O_3P$: p136
$C_6H_8AsNO_3$: a115, a116
$C_6H_8Cl_2O_2$: h62, m221
$C_6H_8N_2$: a223, a224, a225, a226, a227, d238, m121, m257, p107, p108, p109, p118
$C_6H_8N_2O$: a208, o63
$C_6H_8N_2O_2S$: b25, s22
$C_6H_8N_2O_3S$: d32
$C_6H_8N_4$: p181
C_6H_8O: c331, d525, h40, m216
$C_6H_8O_2$: b375, c322, d364, h42, m214, v4

$C_6H_8O_3$: a36, d365, f43, h183
$C_6H_8O_4$: d524, d544
$C_6H_8O_6$: a312, g8, i59
$C_6H_8O_7$: c273
C_6H_9Br: b262
C_6H_9ClO: c78
$C_6H_9ClO_3$: e95, e96
$C_6H_9F_3O_2$: b484
C_6H_9NO: v11
C_6H_9NOS: m418
$C_6H_9NO_2$: b438
$C_6H_9NO_6$: n21
$C_6H_9N_3$: a158
$C_6H_9N_3O_2$: a159, c284, h83
C_6H_{10}: c330, d488, h41, h82, m351
$C_6H_{10}N_2$: e172, p184
$C_6H_{10}N_2O_2$: c323
$C_6H_{10}N_2O_4$: d279
$C_6H_{10}N_2O_5$: a14
$C_6H_{10}N_4$: p26
$C_6H_{10}O$: c328, d26, d361, e5, e6, h78, m215, m350, m352
$C_6H_{10}O_2$: a96, c353, d359, e40, e104, e112, e166, h61, h71, h76, m349
$C_6H_{10}O_3$: d436, e53, e54, h121, p214
$C_6H_{10}O_4$: d325, d608, e17, h57, m272
$C_6H_{10}O_4S$: t151
$C_6H_{10}O_4S_2$: d709
$C_6H_{10}O_5$: d326
$C_6H_{10}O_6$: d616
$C_6H_{10}O_8$: t84
$C_6H_{10}S$: d27
$C_6H_{11}Br$: b261
$C_6H_{11}BrO_2$: b295, e76, e77, e78
$C_6H_{11}Cl$: c77
$C_6H_{11}ClO$: h73
$C_6H_{11}ClO_2$: b433, c151, e97

$C_6H_{11}Cl_3Si$: c344
$C_6H_{11}I$: i32
$C_6H_{11}N$: d25, h63, m339, m416
$C_6H_{11}NO$: c329, e217, f35, m376, o57, t352
$C_6H_{11}NO_2$: e61
C_6H_{12}: c312, d498, d499, e84, h75, m213, m347, m348
$C_6H_{12}Br_2$: d86
$C_6H_{12}ClN$: c160
$C_6H_{12}ClNO$: c112
$C_6H_{12}Cl_2$: d187
$C_6H_{12}Cl_2O$: b161
$C_6H_{12}Cl_2O_2$: b156, d169
$C_6H_{12}Cl_3O_3P$: t417
$C_6H_{12}Cl_3O_4P$: t416
$C_6H_{12}F_3NOSi$: m440
$C_6H_{12}NO_3P$: d293a
$C_6H_{12}N_2$: d45, t269
$C_6H_{12}N_2O_3$: s15
$C_6H_{12}N_2O_4S_2$: c371
$C_6H_{12}N_2S_4$: b174
$C_6H_{12}N_2Si$: t373
$C_6H_{12}N_4$: h52
$C_6H_{12}O$: a100, b488, c327, d497, d618, e87, h54, h72, h77, i72, m346, o47
$C_6H_{12}O_2$: b412, b413, b414, d500, e49, e88, e89, h66, h142, i62, m227, m302, m340, m341, m342, t77
$C_6H_{12}O_3$: d435, d457, d515, e37, e152, e154, i99, p2, p232, t67
$C_6H_{12}O_4Si$: d23
$C_6H_{12}O_6$: f36, g1, g6, i23, m11, s6
$C_6H_{12}O_7$: g4
$C_6H_{12}S$: c326
$C_6H_{13}Br$: b294
$C_6H_{13}BrO_2$: b267
$C_6H_{13}Cl$: c129
$C_6H_{13}ClO$: c130

TABLE 1.14   Empirical Formula Index for Organic Compounds (*continued*)

The alphanumeric designations are keyed to Table 1.15

$C_6H_{13}ClO_2$: c81
$C_6H_{13}ClO_3$: c105
$C_6H_{13}Cl_3O_3Si$: t415
$C_6H_{13}I$: i39
$C_6H_{13}N$: c334, h51, m371, m372, m373, m374
$C_6H_{13}NO$: d260, d553, e184, h144, p187
$C_6H_{13}NO_2$: a183, a184, h122, i79, L4, L5
$C_6H_{13}NO_4$: b182
$C_6H_{13}NO_4S$: m451
$C_6H_{13}NO_5$: g5, t428
C_6H_{14}: d489, d490, h55, m336, m337
$C_6H_{14}ClN$: d272
$C_6H_{14}Cl_4OSi_2$: b168
$C_6H_{14}N_2$: a182, a219, c318, c319
$C_6H_{14}N_2O$: a172, h123
$C_6H_{14}N_2O_2$: L12
$C_6H_{14}N_4O_2$: a311
$C_6H_{14}O$: b449, d417, d492, d492, d494, d495, d496, d701, e83, h68, h69, h70, m343, m344, m345
$C_6H_{14}OSi$: a97, e33, t374
$C_6H_{14}O_2$: b410, d251, d252, d491, e179, h58, h59, h60, i86, m338
$C_6H_{14}O_2S$: d704
$C_6H_{14}O_3$: b191, d253, e35, e156, h65, h172, t322
$C_6H_{14}O_4$: e127, t270
$C_6H_{14}O_4S$: d703
$C_6H_{14}O_6$: d738, m10, s5
$C_6H_{14}O_6S_2$: b188
$C_6H_{14}S$: b451, h64
$C_6H_{14}Si$: a104
$C_6H_{15}Al$: t263
$C_6H_{15}As$: t266
$C_6H_{15}B$: t268
$C_6H_{15}Bi$: t267
$C_6H_{15}ClO_2Si$: c154
$C_6H_{15}ClO_3Si$: c231
$C_6H_{15}ClSi$: b446

$C_6H_{15}Ga$: t274
$C_6H_{15}In$: t276
$C_6H_{15}N$: d411, d696, e85, e86, h80, m353a, t264
$C_6H_{15}NO$: a185, a216, a217, b419, b448, d270
$C_6H_{15}NOSi$: m439
$C_6H_{15}NO_2$: d254, e118
$C_6H_{15}NO_3$: t264
$C_6H_{15}NO_6S$: t424
$C_6H_{15}N_3$: a174
$C_6H_{15}O_3B$: t260
$C_6H_{15}O_3P$: d420, t282
$C_6H_{15}O_3PS$: t285
$C_6H_{15}O_4P$: t280
$C_6H_{15}P$: t281
$C_6H_{15}Sb$: t265
$C_6H_{16}Cl_2Si_2$: t104
$C_6H_{16}N_2$: d302, h56, t108
$C_6H_{16}OSi$: p218
$C_6H_{16}Br_2OSi_2$: b150
$C_6H_{16}O_2Si$: d249
$C_6H_{16}O_3SSi$: m22
$C_6H_{16}O_3Si$: t266b
$C_6H_{16}Si$: t284
$C_6H_{17}NO_3Si$: a280
$C_6H_{17}NO_5S$: b180
$C_6H_{17}N_3$: i9
$C_6H_{18}LiNSi_2$: L11
$C_6H_{18}N_2Si$: b172
$C_6H_{18}N_3ClSi$: c257a
$C_6H_{18}N_3OP$: h53
$C_6H_{18}N_4$: t272
$C_6H_{18}OSi_2$: h50
$C_6H_{18}O_3Si_3$: h48
$C_6H_{19}NOSi_2$: b210
$C_6H_{19}NSi_2$: h49
C_6N_4: t37

C₇

C_7F_5N: p23
$C_7H_3BrClF_3$: b250
$C_7H_3BrF_3NO_2$: b315

$C_7H_3ClF_3NO_2$: c182, c183, c184
$Cl_7H_3ClN_2O_5$: d630
$C_7H_3ClN_2O_6$: c96
$C_7H_3Cl_3O$: d160, d161
$C_7H_4BrF_3$: b233, b234
C_7H_4ClFO: f14
$C_7H_4ClF_3$: c51, c52, c53
C_7H_4ClN: c47, c48
C_7H_4ClNO: c206
$C_7H_4ClNO_3$: n41, n42
$C_7H_4ClNO_4$: c178, c179, c180, n66
$C_7H_4Cl_2O$: c55, c56, d150
$C_7H_4Cl_2O_2$: d156, d157, d158
$C_7H_4Cl_3F$: t233
$C_7H_4Cl_4S$: t34
$C_7H_4F_3NO_2$: n88, n89
$C_7H_4F_{12}O$: d718
$C_7H_4I_2O_3$: h111
$C_7H_4N_2O_2$: n40
$C_7H_4N_2O_6$: d628, d629
$C_7H_4N_2O_7$: d638
$C_7H_4O_3S$: h104
$C_7H_4O_4S$: s24
C_7H_5BrO: b66, b228
$C_7H_5BrO_2$: b232
$C_7H_5BrO_3$: b348
$C_7H_5ClF_3N$: a144, a145, a146
$C_7H_5ClN_2$: a141
C_7H_5ClO: b67, c38, c39
C_7H_5ClOS: p103
$C_7H_5ClO_2$: c45, c46, c46a, c237, c238, p102
$C_7H_5ClO_3$: c193
$C_7H_5Cl_2F$: c119
$C_7H_5Cl_2N$: d196
$C_7H_5Cl_2NO$: d151
$C_7H_5Cl_3$: t248, t249
C_7H_5FO: b69, f10
$C_7H_5FO_2$: f12, f13
$C_7H_5F_3$: t300
$C_7H_5F_3N_2O_2$: a241, a242
$C_7H_5F_3O$: t290

TABLE 1.14 Empirical Formula Index for Organic Compounds (*continued*)

The alphanumeric designations are keyed to Table 1.15

$C_7H_5F_4N$: a179
$C_7H_5IO_2$: i29
$C_7H_5IO_3$: i51
$C_7H_5I_2NO_2$: a156
C_7H_5N: b51
C_7H_5NO: b63, p121
$C_7H_5NO_3$: n27, n28
$C_7H_5NO_3S$: s1
$C_7H_5NO_4$: n37, n38, n39, p259, p260, p261
$C_7H_5NO_5$: h154
C_7H_5NS: b60, p122
$C_7H_5NS_2$: m17
$C_7H_5N_3O_2$: a238, n36, n55
$C_7H_5N_3O_2S$: a240
$C_7H_5N_3O_6$: t386
C_7H_6BrClO: b252
$C_7H_6BrNO_2$: n46
$C_7H_6BrNO_3$: h155
$C_7H_6Br_2$: b236, d102
C_7H_6ClF: c123, c124, c125, f16
C_7H_6ClNO: c40
$C_7H_6ClNO_2$: a140, c187, c188, c189, n47
$C_7H_6ClNO_3$: c140
$C_7H_6Cl_2$: c59, c60, d229, d230, d231
$C_7H_6Cl_2O$: d191, d192
$C_7H_6F_3N$: a129, a130, a131
$C_7H_6INO_2$: a203
$C_7H_6N_2$: a124, a125, a126, b38
$C_7H_6N_2O_3$: n29
$C_7H_6N_2O_4$: a237, d639, d640, d641
$C_7H_6N_2O_5$: d631, d33a
$C_7H_6N_2S$: a128, m15
C_7H_6O: b3
C_7H_6OS: t143
$C_7H_6O_2$: b44, h94, h95, h96, m240
$C_7H_6O_2S$: m16
$C_7H_6O_3$: d375, d376, f41, h99, h100, h101

$C_7H_6O_4$: d383, d384, d385
$C_7H_6O_5$: t306
$C_7H_6O_6S$: s28
C_7H_7Br: b85, b355, b356, b357
C_7H_7BrO: b235, b301, b302, b303
C_7H_7Cl: b89, c244, c245, c246
$C_7H_7ClN_4O_2$: c241
C_7H_7ClO: c57, c139, c158, c159
$C_7H_7ClO_2S$: t177
$C_7H_7ClO_3S$: m49
C_7H_7ClS: c248
$C_7H_7Cl_3Si$: b123, t192
C_7H_7F: f24, f25, f26
C_7H_7FO: f15, f19
$C_7H_7FO_2S$: t178
C_7H_7I: i53, i54
C_7H_7IO: i41
C_7H_7N: v9, v10
C_7H_7NO: a53, a54, a55, b4, f31
$C_7H_7NO_2$: a121, a122, a123, h97, h98, m401, m402, n85, n86, n87
$C_7H_7NO_3$: a286, a287, m81, m82, m323, m324, n44, n45
$C_7H_7NO_4S$: c16
$C_7H_7N_3$: a201, a202, m136
C_7H_8: b129, c310, t167
C_7H_8BrN: b304
C_7H_8ClN: c58, c142, c143, c144, c145, c146
C_7H_8ClNO: c138a, c138
$C_7H_8ClNO_2S$: c247
$C_7H_8Cl_2Si$: d198, m358
$C_7H_8N_2O$: a114, b72, p165
$C_7H_8N_2O_2$: d33, h165, m315, m316, m317
$C_7H_8N_2O_3$: m78, m79, m80
$C_7H_8N_2S$: p154
$C_7H_8N_4O_2$: t138

C_7H_8O: b78, c279, c280, c281, m48
C_7H_8OS: m429
$C_7H_8O_2$: d389, d390, h105, m87, m88, m89, m276
$C_7H_8O_2S$: t173
$C_7H_8O_3$: e136, f45, m304
$C_7H_8O_3S$: m127, t176
C_7H_8S: m367, p128, t147
C_7H_9ClSi: m357
C_7H_9N: b79, d604, d605, d606, d607, e211, e212, e213, m122, t180, t181, t182
C_7H_9NO: a218, b98, h126, m42, m43, m44
$C_7H_9NO_2$: d456
$C_7H_9NO_2S$: t174
$C_7H_9NO_3S$: a294
C_7H_9NS: m422, m423
$C_7H_9N_3O$: a133
C_7H_{10}: b130
$C_7H_{10}N_2$: a157, a177, a178, d476, m360, t168, t169, t170, t171
$C_7H_{10}N_2O$: m94
$C_7H_{10}N_2OS$: h129
$C_7H_{10}N_2O_2$: e173, m232
$C_7H_{10}N_2O_2S$: a210, t175
$C_7H_{10}O$: m61, m62, n108, t65
$C_7H_{10}O_2$: a40, c359
$C_7H_{10}O_3$: e12, h158, m333, t341
$C_7H_{10}O_4$: d550
$C_7H_{10}O_5$: d459
$C_7H_{10}Si$: m366
$C_7H_{11}Br$: b318
$C_7H_{11}BrO_4$: d286
$C_7H_{11}ClO$: c316
$C_7H_{11}ClO_4$: d290
$C_7H_{11}NO$: c340, h110
$C_3H_{11}NO_2$: a52
$C_7H_{11}NO_3$: m335
$C_7H_{11}NO_5$: a45
$C_7H_{11}NS$: c341

TABLE 1.14 Empirical Formula Index for Organic Compounds (*continued*)

The alphanumeric designations are keyed to Table 1.15

C_7H_{12}: c311, h22, m207, m208, n107
$C_7H_{12}O$: c309, c315, m204, m205, m206, m268
$C_7H_{12}O_2$: b417, c317, d355, e121
$C_7H_{12}O_3$: e168, e195
$C_7H_{12}O_4$: d317, d318, d552, d574, d575, d576, h8, m273, t126
$C_7H_{12}O_5$: g17, g18
$C_7H_{12}O_6Si$: m435
$C_7H_{12}O_7$: g3
$C_7H_{13}Br$: b260, b306
$C_7H_{13}BrO_2$: e80
$C_7H_{13}ClO$: h18
$C_7H_{13}N$: a250, d333, q5
$C_7H_{13}NO$: a317, c339
$C_7H_{13}NO_2$: a152
C_7H_{14}: c306, h19, m194
$C_7H_{14}ClN$: c113
$C_7H_{14}N_2$: d416
$C_7H_{14}N_2O$: a278
$C_7H_{14}N_2O_2$: e201
$C_7H_{14}O$: c308, c342, d569, d578, h5, h15, h16, h17, m197, m198, m199, m200, m201, m202, m203, m267
$C_7H_{14}O_2$: b479, c307, d258, e123, e170, e198, h10, i80, m75, m265, p52
$C_7H_{14}O_3$: i68
$C_7H_{14}O_6$: m256
$C_7H_{15}Br$: b291, b292
$C_7H_{15}Cl$: c128
$C_7H_{15}ClO_2$: c82
$C_7H_{15}Cl_3Si$: h21
$C_7H_{15}I$: i37
$C_7H_{15}N$: c324, d591, e202, e203, m209, m211, m211
$C_7H_{15}NO$: d469, e157, h125, m375, p185, p186
$C_7H_{15}NO_2$: p276
$C_7H_{15}NO_3$: c18, m452
$C_7H_{15}O_5P$: e119
C_7H_{16}: d570, d571, d572, d573, e197, h6, m264, t337
$C_7H_{16}BrNO_2$: a38
$C_7H_{16}ClNO_2$: a39
$C_7H_{16}N_2$: a215, m301, t366
$C_7H_{16}N_2O$: a277
$C_7H_{16}N_2O_2$: p182
$C_7H_{16}O$: d577, h12, h13, h14, m266, t338
$C_7H_{16}O_2$: d257, d331, m394
$C_7H_{16}O_2Si$: d256, e228
$C_7H_{16}O_3$: d700, t278, t321
$C_7H_{16}O_4$: t91
$C_7H_{16}S$: h9
$C_7H_{17}N$: h20, m269a
$C_7H_{17}NO$: d275
$C_7H_{17}NO_2$: b420, d274
$C_7H_{17}NO_5$: m255
$C_7H_{17}NO_6S$: t427
$C_7H_{17}NO_7S$: t426
$C_7H_{18}N_2$: d330, h7, i103, t118
$C_7H_{18}N_2O$: b173
$C_7H_{18}N_2O_2$: a276
$C_7H_{18}O_2Si$: b486
$C_7H_{18}O_3Si$: b485, t266a
$C_7H_{19}NOSi_2$: b209
$C_7H_{19}NSi$: d343, t371
$C_7H_{19}N_3$: d42, t419
$C_7H_{21}N_3Si$: t420

C_8

$C_8Br_4O_3$: t10
$C_8Cl_4O_3$: t31
C_8D_{10}: e67
$C_8HCl_4NO_2$: t32
$C_8H_3NO_5$: n72
$C_8H_4BrNO_2$: b296
$C_8H_4Cl_2O_2$: b14, b15, p171
$C_8H_4Cl_2O_4$: d217
$C_8H_4Cl_6$: b202
$C_8H_4F_3N$: t298
$C_8H_4F_6$: b206
$C_8H_4N_2$: d236, d237
$C_8H_4O_3$: p168
$C_8H_5Br_5$: p5
$C_8H_5ClO_4$: c209
$C_8H_5Cl_3O_3$: t242
$C_8H_5F_3O$: t289
$C_8H_5F_3O_2S$: t137
$C_8H_5F_6N$: b205
C_8H_5NO: b68
$C_8H_5NO_2$: i21, p170
$C_8H_5NO_3$: h166, i58
$C_8H_5NO_6$: n31, n32, n33, n34
C_8H_6: p82
C_8H_6BrClO: b246
C_8H_6BrN: b329
$C_8H_6Br_2O$: d64
$C_8H_6Br_4$: t11, t12
$C_8H_6ClF_3$: t299
C_8H_6ClN: c201
$C_8H_6ClNO_3$: c171
$C_8H_6Cl_2O$: d140
$C_8H_6Cl_2O_3$: d214
$C_8H_6Cl_4$: t35
$C_8H_6N_2$: q4
$C_8H_6N_2O_2$: a266, n65
$C_8H_6N_2O_6$: d636, m228
C_8H_6O: b42
$C_8H_6O_2$: b13, p169
$C_8H_6O_3$: b70, c14, f33, m239
$C_8H_6O_4$: b16, b17, m241, p167
C_8H_6S: b61
C_8H_7Br: b349
C_8H_7BrO: b221, b222
$C_8H_7BrO_2$: b327, b328
C_8H_7ClO: c28, c29, c30, p81, t187, t188, t189
C_8H_7ClOS: b91
$C_8H_7ClO_2$: b90, c200, m53, p69
$C_8H_7ClO_3$: c86, c197, m184, m185
$C_8H_7ClO_4$: c133
C_8H_7FO: f8
C_8H_7N: i18, p80, t184, t185, t186

TABLE 1.14   Empirical Formula Index for Organic Compounds (*continued*)

The alphanumeric designations are keyed to Table 1.15

C_8H_7NO: m9, m137, t190

$C_8H_7NO_2$: h133, n84

$C_8H_7NO_3$: n22, n23

$C_8H_7NO_3S$: t179

$C_8H_7NO_4$: a116a, m318, m319, m320, m321, n62, n63, n64

$C_8H_7NO_5$: m83

C_8H_7NS: b121, m135

$C_8H_7N_3O_2$: a153

C_8H_8: s11

C_8H_8BrNO: b219

$C_8H_8Br_2$: d79, d104, d105

C_8H_8ClNO: c23

$C_8H_8ClNO_3S$: a10

$C_8H_8Cl_2$: d233, d234

$C_8H_8Cl_2Si$: p166

$C_8H_8HgO_2$: p125

$C_8H_8N_2$: a260, m128

$C_8H_8N_2OS$: a207

C_8H_8O: a31, e9, m126, p76a

C_8H_8OS: m424, p153

$C_8H_8O_2$: b41, b97, h90, h91, h92, m45, m46, m129, m130, m131, m132, p78, p79

$C_8H_8O_2S$: t160

$C_8H_8O_3$: d370, d380, h130, h131, h137, h138, h160, m8, m50, m51, m52, m242, m277, m410, p68, t74

$C_8H_8O_4$: d21, h132

$C_8H_8O_4S$: a33

C_8H_9Br: b282, b283, b368, b369, b370, b371

C_8H_9BrO: b270, b286

$C_8H_9BrO_2$: b268

C_8H_9Cl: c107, c108, c258, c259, c260, c261

C_8H_9ClO: c90

C_8H_9N: b100, c360, i22, m447

C_8H_9NO: a18, a108, a109, a110, b96, m249

$C_8H_9NO_2$: a15, a16, a17, a211, a212, b88, d556, d557, d558, d559, e187, e214, e215, e216, m47, m116, m117, p115, t75

$C_8H_9NO_3$: a206, h163, h164, m85, n59

$C_8H_9NO_4$: d444

C_8H_{10}: e68, m244, x4, x5, x6

$C_8H_{10}N_2O$: d560

$C_8H_{10}N_4O_2$: c1, d240

$C_8H_{10}O$: b131, d579, d580, d581, d582, d583, d584, e28, e199, m105, m106, m107, m138, m139, m140, p112, p113

$C_8H_{10}O_2$: b18, d431, d432, d433, m54, p72, p111

$C_8H_{10}O_3$: c320, d446, h135, h159

$C_8H_{10}O_3S$: m434

$C_8H_{10}O_4$: d263

$C_8H_{10}S$: b105

$C_8H_{11}ClSi$: d585

$C_8H_{11}N$: b103, d477, d478, d479, d480, d481, d482, d483, e63, e64, e65, e180, e181, m141, m142, p114, t367

$C_8H_{11}NO$: a173, a256, a261, a262, a300, d470, e24, h117, m55, m71, m72, m73, p266

$C_8H_{11}NO_2$: d427, d428, d429

$C_8H_{11}NO_2S$: m433

$C_8H_{11}NO_3$: e132

$C_8H_{11}NO_3S$: d463

$C_8H_{11}N_5$: p93

C_8H_{12}: c345, v6

$C_8H_{12}N_2$: d239, d586, t119, x9

$C_8H_{12}N_2O_2$: d410

$C_8H_{12}N_2O_3$: d280

$C_8H_{12}N_4$: a323

$C_8H_{12}O$: e234

$C_8H_{12}O_2$: d508, e219, h185, n111

$C_8H_{12}O_3$: e194

$C_8H_{12}O_4$: d305, d316

$C_8H_{12}O_6Si$: t195

$C_8H_{12}Si$: d587

$C_8H_{13}N$: e235

C_8H_{14}: c349, d532, o17, o44, v5

$C_8H_{14}N_2$: p188

$C_8H_{14}O$: c348, d510, e7a, m262, o45

$C_8H_{14}O_2$: b459, c333, c363, d537, i69, m195

$C_8H_{14}O_3$: b415, b496, d712, e90

$C_8H_{14}O_4$: b447, d320, d335, d536a, e149, o24

$C_8H_{14}O_4S$: d619

$C_8H_{14}O_4S_2$: d708

$C_8H_{14}O_6$: d339, d340

$C_8H_{14}O_6Si$: t194

$C_8H_{15}ClO$: e145, o37

$C_8H_{15}N$: o27

$C_8H_{15}NO$: d367

$C_8H_{15}NO_2$: d468, e204, e205, e206

C_8H_{16}: c346, d506, d507, e108, o39, t357

$C_8H_{16}ClN$: c227

$C_8H_{16}O$: c347, d509, e109, e110, o34, o35, o36, o40

$C_8H_{16}O_2$: b431, c321, e142, e143, h79, i67, m261, o29, p234

$C_8H_{16}O_4$: e36, t125

$C_8H_{17}Br$: b320

$C_8H_{17}Cl$: c190

$C_8H_{17}Cl_3Si$: o43

$C_8H_{17}I$: i46

$C_8H_{17}N$: c350, d511

$C_8H_{17}NO_2$: p189

$C_8H_{17}NO_3S$: c335

$C_8H_{17}O_5P$: t283

TABLE 1.14 Empirical Formula Index for Organic Compounds (*continued*)

The alphanumeric designations are keyed to Table 1.15

C_8H_{18}: d533, e140, e174,
  e175, m260, o22, t100,
  t353, t354, t355
$C_8H_{18}ClNO_2$: a49
$C_8H_{18}Cl_2O_2Si_3$: d186
$C_8H_{18}Cl_2Si$: d184
$C_8H_{18}Cl_2Sn$: d136a
$C_8H_{18}F_3NOSi_2$: b212
$C_8H_{18}N_2$: c314
$C_8H_{18}N_2O$: m224, m367
$C_8H_{18}N_2O_4S$: h124
$C_8H_{18}O$: d115, d407, e144,
  o30, o31, o32, o33
$C_8H_{18}OSi_2$: d713
$C_8H_{18}OSn$: d137
$C_8H_{18}O_2$: d122, d535,
  e141, o25, o26, t356
$C_8H_{18}O_2S$: d135
$C_8H_{18}O_3$: b176, b411,
  d698, t277
$C_8H_{18}O_3S$: d134
$C_8H_{18}O_3Si$: t262
$C_8H_{18}O_4$: b189
$C_8H_{18}O_4S$: d131
$C_8H_{18}O_5$: t51
$C_8H_{18}S$: d132, d133, o28
$C_8H_{18}S_2$: b153, b154, d113,
  d114
$C_8H_{18}Si_2$: b208
$C_8H_{19}N$: d107, d406, d418,
  d536, e147, o41, t102
$C_8H_{19}NO$: d412
$C_8H_{19}NO_2$: b444, d247,
  d248
$C_8H_{19}NO_5$: b183
$C_8H_{19}O_3P$: d127
$C_8H_{20}BrN$: t48
$C_8H_{20}ClN$: t49
$C_8H_{20}Ge$: t56
$C_8H_{20}N_2$: d534, o23, t101,
  t273
$C_8H_{20}O_3SSi$: m19
$C_8H_{20}O_3Si$: t261
$C_8H_{20}O_4Si$: t47
$C_8H_{20}O_5P_2$: t59
$C_8H_{20}O_7P_2$: t58

$C_8H_{20}Pb$: t57
$C_8H_{20}Si$: t60
$C_8H_{20}Sn$: t62
$C_8H_{21}NO$: t50
$C_8H_{21}NOSi_2$: b207
$C_8H_{21}NO_2Si$: a275
$C_8H_{22}N_2O_3Si$: a167a, t324
$C_8H_{22}N_4$: b145
$C_8H_{22}O_2Si_2$: b211
$C_8H_{23}N_5$: t54
$C_8H_{24}Cl_2O_3Si_4$: d207
$C_8H_{24}O_2Si_3$: o21
$C_8H_{24}O_4Si_4$: o20
$C_8H_{28}N_4Si_4$: o19

C_9

$C_9F_{15}N_3$: t432
$C_9H_2Cl_6O_3$: h30
$C_9H_3Cl_3O_3$: b32
$C_9H_4O_5$: b31, c15
$C_9H_5BrClNO$: b255
$C_9H_5Br_2NO$: d87
C_9H_5ClINO: c132
$C_9H_5Cl_2N$: d225
C_9H_6BrN: b347
C_9H_6ClN: c236
C_9H_6ClNO: c134
$C_9H_6N_2O_2$: n77, t172
$C_9H_6O_2$: b56, c276
$C_9H_6O_3$: h108, h109
$C_9H_6O_4$: i16
$C_9H_6O_6$: b28, b29, b30
C_9H_7BrO: b259
C_9H_7ClO: c268
$C_9H_7ClO_2$: c76
$C_9H_7Cl_3O_3$: t243
C_9H_7N: i110, q3
C_9H_7NO: h178, i20
$C_9H_7NO_3$: h143, m285
$C_9H_7NO_4S$: h179
$C_9H_7N_3O_4S_2$: a247
C_9H_8: i17
$C_9H_8Cl_2O_2$: n109
$C_9H_8N_2$: m409

$C_9H_8N_2O_5$: n43
C_9H_8O: c266, i15
$C_9H_8O_2$: c267, d353
$C_9H_8O_3$: h107
$C_9H_8O_4$: a56, p124
C_9H_9BrO: b342
C_9H_9Cl: c217
C_9H_9ClO: c223
$C_9H_9ClO_3$: c199, c249
C_9H_9N: d486, m283, m284
C_9H_9NO: m93
$C_9H_9NO_2$: a9
$C_9H_9NO_2S$: t191
$C_9H_9NO_3$: a11, a12, b71
$C_9H_9N_3O$: a265
C_9H_{10}: a84, i13, m411, v3
$C_9H_{10}F_3NO_2$: m123
$C_9H_{10}N_2$: a301, p119a
$C_9H_{10}N_2O$: p147
$C_9H_{10}N_2O_2$: p83
$C_9H_{10}N_2O_3$: a132
$C_9H_{10}O$: a98, a99, c269,
  d360, i14, m113, p144,
  p145, p209, p217
$C_9H_{10}O_2$: b77, d485, e9a,
  e25, e26, e69, h170,
  h171, m39, m40, m41,
  m356, p74, p146
$C_9H_{10}O_2S$: b120
$C_9H_{10}O_3$: d430, e29, e30,
  e39, e48, e151, e218,
  m91, m279, m292, p75
$C_9H_{10}O_4$: d434, m288,
  m445
$C_9H_{10}O_8$: c354
$C_9H_{11}Br$: b297, b331,
  b361, b362
$C_9H_{11}BrO$: b343
$C_9H_{11}ClO_3S$: c114
$C_9H_{11}Cl_3Si$: c226, m354
$C_9H_{11}N$: a83, a199, a200,
  c332, t71, t80
$C_9H_{11}NO$: d462, m355,
  m432
$C_9H_{11}NO_2$: d464, e27, e59,
  e60, p84

TABLE 1.14 Empirical Formula Index for Organic Compounds (*continued*)

The alphanumeric designations are keyed to Table 1.15

$C_9H_{11}NO_3$: t437

C_9H_{12}: e158, i91, n91, p222, t333, t334, t335, v8

$C_9H_{12}Cl_2Si$: m353

$C_9H_{12}N_2O_4$: a246

$C_9H_{12}N_2O_6$: u14

$C_9H_{12}O$: b95, d547, d548, i106, i107, p142, p143, p235, t358, t359, t362

$C_9H_{12}O_2$: b110, e31, i85, n110, p73, p140, t348

$C_9H_{12}O_3$: m196, t315

$C_9H_{12}O_3S$: e222

$C_9H_{12}S$: p141

$C_9H_{13}N$: b480, d487, d622, e72, e169, e223, e224, i90, t330

$C_9H_{13}NO$: a264, b80, m86, n112

$C_9H_{13}NO_2$: a263

$C_9H_{13}N_3O_2$: t438

$C_9H_{14}BrN$: p159

$C_9H_{14}Br_3N$: p162

$C_9H_{14}ClN$: p160

$C_9H_{14}IN$: p161

$C_9H_{14}N_2$: n94

$C_9H_{14}O$: d527, d529, i82, t340

$C_9H_{14}OSi$: t375

$C_9H_{14}O_2Si$: d443

$C_9H_{14}O_3$: b192

$C_9H_{14}O_3Si$: p158

$C_9H_{14}O_5$: d262, d321

$C_9H_{14}O_6$: p201

$C_9H_{14}Si$: p163

$C_9H_{15}NO$: c361

$C_9H_{15}NO_2$: d568

$C_9H_{15}NO_5$: d261

$C_9H_{15}NSi$: t369

C_9H_{16}: h46

$C_9H_{16}Cl_2Si$: c333a

$C_9H_{16}N_2$: d46

$C_9H_{16}O$: d528

$C_9H_{16}O_2$: c325

$C_9H_{16}O_3$: b467

$C_9H_{16}O_4$: d303, d307, d322, d530, n95

$C_9H_{17}ClO$: n101

$C_9H_{17}N$: a88, n97

$C_9H_{17}NO$: m180

$C_9H_{17}NO_2$: e177, e178

C_9H_{18}: i94, n102, p224, t339

$C_9H_{18}NO$: t117

$C_9H_{18}N_2O_3Si$: t325

$C_9H_{18}O$: d531, n100, n103

$C_9H_{18}O_2$: e138, m329, n98

$C_9H_{18}O_3$: d111

$C_9H_{19}Br$: b317

$C_9H_{19}N$: i95, t332

$C_9H_{19}NO$: d116

$C_9H_{19}NO_2$: e120

$C_9H_{19}NO_3S$: c337

C_9H_{20}: n92, t346

$C_9H_{20}Cl_2Si$: m330

$C_9H_{20}N_2$: a296

$C_9H_{20}N_2S$: d136

$C_9H_{20}O$: n99, t347

$C_9H_{20}O_2$: b450, n96

$C_9H_{20}O_3$: d699, t279

$C_9H_{20}O_3Si$: a103

$C_9H_{20}O_4$: t408

$C_9H_{20}O_5$: t53

$C_9H_{21}BO_3$: t406

$C_9H_{21}ClO_3Si$: c230

$C_9H_{21}ClSi$: c257

$C_9H_{21}N$: n104, t407

$C_9H_{21}NO_3$: t309

$C_9H_{21}N_3$: t275

$C_9H_{21}O_3B$: t310

$C_9H_{21}O_3P$: t313

$C_9H_{22}N_2$: d327, n93

$C_9H_{22}O_3Si$: p238

$C_9H_{23}NO_3Si$: a279

$C_9H_{24}N_4$: b147

C_{10}

$C_{10}H_2O_6$: b27

$C_{10}H_4Cl_2O_2$: d201

$C_{10}H_6N_2$: b99

$C_{10}H_6N_2O_4$: d634

$C_{10}H_6N_2O_4S$: d48

$C_{10}H_6O_2$: n11

$C_{10}H_6O_3$: h152

$C_{10}H_6O_8$: b26

$C_{10}H_7Br$: b312

$C_{10}H_7BrO$: b313

$C_{10}H_7Cl$: c168, c169

$C_{10}H_7NO_2$: n57, n81, p123

$C_{10}H_7NO_8S_2$: n82

$C_{10}H_8$: a326, n2

$C_{10}H_8BrNO_2$: b287

$C_{10}H_8N_2$: d705

$C_{10}H_8O$: n9, n10

$C_{10}H_8O_2$: d392, d393, d394, d395, m191

$C_{10}H_8O_3$: h140

$C_{10}H_8O_3S$: n18

$C_{10}H_8O_7S_2$: h150, h151

$C_{10}H_8O_8S_2$: d396

$C_{10}H_9ClCrN_2O_3$: b143

$C_{10}H_9N$: m407, m408, n17

$C_{10}H_9NO$: a51, a236

$C_{10}H_9NO_2$: i19

$C_{10}H_9NO_3S$: a234

$C_{10}H_9NO_4S$: a193, a194, a195, a196

$C_{10}H_9NO_6$: d561

$C_{10}H_9NO_6S_2$: a232, a233

$C_{10}H_9N_3$: d706

$C_{10}H_{10}ClFO$: c120

$C_{10}H_{10}ClNO_2$: c26

$C_{10}H_{10}N_2$: a285, n4, n5

$C_{10}H_{10}N_2O$: m365

$C_{10}H_{10}O$: d362, m190, p96, p98

$C_{10}H_{10}O_2$: b64, s2

$C_{10}H_{10}O_3$: b73, m60

$C_{10}H_{10}O_4$: d588, d589, d590, h136, p152

$C_{10}H_{11}BrO$: b312a

$C_{10}H_{11}ClO_3$: c198

$C_{10}H_{11}ClO_4$: t317

$C_{10}H_{11}IO_4$: i28

$C_{10}H_{11}N$: p101

TABLE 1.14 Empirical Formula Index for Organic Compounds (*continued*)

The alphanumeric designations are keyed to Table 1.15

$C_{10}H_{11}NO_2$: a32, d448
$C_{10}H_{11}NO_4$: c10
$C_{10}H_{11}NO_6$: m226
$C_{10}H_{12}$: d244, t73
$C_{10}H_{12}NO$: b408
$C_{10}H_{12}N_2$: a170, b81, b102
$C_{10}H_{12}N_2O_2$: p77
$C_{10}H_{12}O$: a94, b500, e55, i77, m97, m386, m387, m391, p94, p95
$C_{10}H_{12}O_2$: e200, h112, h162, m70, m92, m98, m99, m100, p99, p100, p223
$C_{10}H_{12}O_3$: d426, e41, e163, m300, p71, p230
$C_{10}H_{12}O_4$: d447, m225, t314
$C_{10}H_{12}O_5$: d306, p239, t316
$C_{10}H_{12}O_6$: d514
$C_{10}H_{13}Br$: b298
$C_{10}H_{13}BrO$: b243
$C_{10}H_{13}Cl$: b434
$C_{10}H_{13}NO$: p129
$C_{10}H_{13}NO_2$: e45
$C_{10}H_{13}NO_2S$: b92
$C_{10}H_{13}N_5O_4$: a70
$C_{10}H_{14}$: b423, b424, b425, d282, d283, d284, i64, i100, i101, i102, t97, t98, t99
$C_{10}H_{14}NO_5PS$: p3
$C_{10}H_{14}N_2$: n20, p139
$C_{10}H_{14}N_2O$: d323, d334
$C_{10}H_{14}N_4O_4$: d399
$C_{10}H_{14}N_5O_7P$: a72
$C_{10}H_{14}O$: b469, b470, b471, b472, b473, b477, c20, c362, i92, i102a, p58, t116, t254
$C_{10}H_{14}O_2$: b432, b454, d450
$C_{10}H_{14}O_3$: c6, c9
$C_{10}H_{14}O_4$: m90, t318
$C_{10}H_{15}BrO$: b245

$C_{10}H_{15}N$: b421, d277, d278, d564, e233, i93, p97, t96
$C_{10}H_{15}NO$: d273, e1, e2
$C_{10}H_{15}NO_2$: d451, p106
$C_{10}H_{15}N_5O_{10}P_2$: a71
$C_{10}H_{16}$: a67, c2, d649, L6, L7, m453, p25, p175, p176, t5, t6, t253
$C_{10}H_{16}ClN$: b126
$C_{10}H_{16}Cl_2O_2$: d11
$C_{10}H_{16}N_2O_8$: e125
$C_{10}H_{16}O$: c3, c4, d352, d562, d563, L8, p177, p178, p243, t351
$C_{10}H_{16}OSi$: d519
$C_{10}H_{16}O_4$: c5, d266
$C_{10}H_{16}O_4S$: c7
$C_{10}H_{16}O_5$: d265, d301
$C_{10}H_{16}Si$: b127
$C_{10}H_{17}N$: a66, p274
$C_{10}H_{17}NO$: c343, m450
$C_{10}H_{18}$: d1, d2, p174
$C_{10}H_{18}N_2O_7$: h119
$C_{10}H_{18}O$: b216, b441, b442, c265, d3, g2, i60, i83, i109, L9, m13, p190, t7, t350
$C_{10}H_{18}O_2$: e112
$C_{10}H_{18}O_3$: d599, t70
$C_{10}H_{18}O_4$: b175, d9, d121, d332, d565
$C_{10}H_{18}O_4S$: d341
$C_{10}H_{18}O_6$: d421
$C_{10}H_{19}ClO$: d17
$C_{10}H_{19}N$: d12, t331
$C_{10}H_{19}NO_2$: e207
$C_{10}H_{20}$: c301, d18
$C_{10}H_{20}Br_2$: d73
$C_{10}H_{20}N_2S_4$: t61
$C_{10}H_{20}O$: b439, b440, c274, d6, d16, d357, d366, e148, m12, m303
$C_{10}H_{20}O_2$: d14, e146, e190, m63, m170
$C_{10}H_{20}O_5$: p45

$C_{10}H_{20}O_5Si$: t326
$C_{10}H_{21}Br$: b265
$C_{10}H_{21}Cl$: c80
$C_{10}H_{21}I$: i33
$C_{10}H_{21}N$: d294
$C_{10}H_{21}NO$: a230
$C_{10}H_{22}$: d7
$C_{10}H_{22}N_2$: d41
$C_{10}H_{22}O$: d15, d651, t72
$C_{10}H_{22}O_2$: d10, d106
$C_{10}H_{22}O_3$: d697, t413
$C_{10}H_{22}O_3S$: d13
$C_{10}H_{22}O_4$: t412
$C_{10}H_{22}O_5$: b190
$C_{10}H_{22}O_7$: d648
$C_{10}H_{23}N$: d19, d650
$C_{10}H_{23}NO$: d108
$C_{10}H_{23}NO_2$: d259
$C_{10}H_{24}N_2$: d8, t55, t111
$C_{10}H_{24}N_2O_2$: d645
$C_{10}H_{24}N_4$: b146
$C_{10}H_{24}OSi$: m109
$C_{10}H_{24}O_3Si$: m441
$C_{10}H_{24}O_6Si$: t429
$C_{10}H_{27}O_3N_3Si$: t323
$C_{10}H_{30}O_3Si_4$: d5
$C_{10}H_{30}O_5Si_5$: d4

C_{11}

$C_{11}H_4F_{20}O$: i1
$C_{11}H_7N$: c293
$C_{11}H_8O$: n1
$C_{11}H_8O_3$: h147, m310, n3
$C_{11}H_8O_3$: h148, h149
$C_{11}H_9Br$: b309
$C_{11}H_9Cl$: c157
$C_{11}H_9N$: p148
$C_{11}H_{10}$: m308, m309
$C_{11}H_{10}N_2S$: n19
$C_{11}H_{10}O$: m76, m77
$C_{11}H_{11}N$: n6
$C_{11}H_{12}N_2O$: a309
$C_{11}H_{12}N_2O_2$: t436
$C_{11}H_{12}O_2$: d358, e103, m104

TABLE 1.14 Empirical Formula Index for Organic Compounds (*continued*)

The alphanumeric designations are keyed to Table 1.15

$C_{11}H_{12}O_3$: e70
$C_{11}H_{13}ClO$: b428
$C_{11}H_{13}ClO_3$: c250
$C_{11}H_{13}NO$: b119
$C_{11}H_{13}NO_2$: t183
$C_{11}H_{13}NO_3$: a302, a303
$C_{11}H_{13}N_3O$: a113
$C_{11}H_{13}N_3O_3S$: d567
$C_{11}H_{14}O$: m103, p43
$C_{11}H_{14}O_2$: b426, b427, d455, e46
$C_{11}H_{14}O_3$: b409, b468, b476, e167
$C_{11}H_{14}O_4$: e155
$C_{11}H_{14}O_4Si$: d24
$C_{11}H_{15}NO$: d269
$C_{11}H_{15}NO_2$: d276, d466, e122
$C_{11}H_{16}$: b482, p24, p54
$C_{11}H_{16}N_2$: b114
$C_{11}H_{16}O$: b86, b461, b462, p56
$C_{11}H_{16}O_2$: a68
$C_{11}H_{16}O_3$: m299
$C_{11}H_{16}O_4$: d714
$C_{11}H_{17}N$: b429, e160
$C_{11}H_{17}NO$: e225
$C_{11}H_{17}NO_2$: b104
$C_{11}H_{17}O_3P$: b93
$C_{11}H_{18}O$: d308, n105, p4
$C_{11}H_{18}O_5$: d264
$C_{11}H_{19}ClO$: u11
$C_{11}H_{19}N$: a209
$C_{11}H_{20}O$: p55, u7
$C_{11}H_{20}O_2$: u9
$C_{11}H_{20}O_4$: d119, d287, d309
$C_{11}H_{21}BrO_2$: b367
$C_{11}H_{22}$: u8
$C_{11}H_{22}N_2$: d695
$C_{11}H_{22}O$: u1, u5, u6, u10
$C_{11}H_{22}O_2$: m218, u3
$C_{11}H_{22}O_4Si$: e7
$C_{11}H_{23}NO_2$: a297
$C_{11}H_{24}$: u2
$C_{11}H_{24}O$: d310, u4
$C_{11}H_{24}O_3Si$: t311

$C_{11}H_{24}O_4$: t410
$C_{11}H_{24}O_6$: p46
$C_{11}H_{24}O_6Si$: t430
$C_{11}H_{26}N_2$: d129
$C_{11}H_{26}N_2O_6$: b214

C_{12}

$C_{12}Br_{10}O$: b197
$C_{12}H_4Cl_6S_2$: b203
$C_{12}H_5ClO_3$: c170
$C_{12}H_6Br_4O_4S$: s25
$C_{12}H_6O_3$: n7
$C_{12}H_6O_{12}$: b19
$C_{12}H_7NO_2$: n8
$C_{12}H_8$: a3
$C_{12}H_8Br_2$: d66
$C_{12}H_8Cl_2OS$: b166
$C_{12}H_8Cl_2O_2S$: b165
$C_{12}H_8N_2$: p63
$C_{12}H_8N_2O_2$: a235
$C_{12}H_8N_2O_4S_2$: b194, b195
$C_{12}H_8O$: d50
$C_{12}H_8O_6$: b132
$C_{12}H_8S$: d52
$C_{12}H_9Br$: b237
$C_{12}H_9BrO$: b330
$C_{12}H_9ClO_2S$: c207
$C_{12}H_9N$: c8, d665, n16
$C_{12}H_9NO$: b74, b75, b76
$C_{12}H_9NO_2$: n48, n49
$C_{12}H_9NO_3$: n70, n71
$C_{12}H_9NS$: p66
$C_{12}H_{10}$: a2, b134
$C_{12}H_{10}ClN$: c61, c62
$C_{12}H_{10}ClO_3P$: d662
$C_{12}H_{10}ClP$: c99
$C_{12}H_{10}Cl_2Si$: d175
$C_{12}H_{10}Hg$: d675
$C_{12}H_{10}N_2$: a322
$C_{12}H_{10}N_2O$: n80, p89
$C_{12}H_{10}N_2O_2$: n52
$C_{12}H_{10}N_2O_2S$: a243
$C_{12}H_{10}N_3O_3P$: d682

$C_{12}H_{10}O$: d667, m311, m312, p131, p132
$C_{12}H_{10}OS$: d690
$C_{12}H_{10}O_2$: d387, h88, n14, n15
$C_{12}H_{10}O_2S$: d689, t150
$C_{12}H_{10}O_3$: n12
$C_{12}H_{10}O_3S$: b139
$C_{12}H_{10}O_4$: q1
$C_{12}H_{10}O_4S$: s27, t145
$C_{12}H_{10}S$: d688
$C_{12}H_{10}S_2$: d664
$C_{12}H_{10}Se_2$: d663
$C_{12}H_{11}ClNO_2P$: p134
$C_{12}H_{11}N$: a134, a135, b117, b118, d655
$C_{12}H_{11}NO$: n13, p70
$C_{12}H_{11}N_3$: p87
$C_{12}H_{11}O_3P$: d681
$C_{12}H_{12}$: d554, d555
$C_{12}H_{12}N_2$: b136, d673, p131
$C_{12}H_{12}N_2O$: o62
$C_{12}H_{12}N_2O_2$: b40
$C_{12}H_{12}N_2O_2S$: d36, d37
$C_{12}H_{12}N_4$: d31
$C_{12}H_{12}O$: e44
$C_{12}H_{12}O_2Si$: d687
$C_{12}H_{12}O_3$: t196
$C_{12}H_{12}O_6$: t193, t336
$C_{12}H_{13}N_3$: d34
$C_{12}H_{14}N_2O_3S$: a167
$C_{12}H_{14}N_4O_2S$: s21
$C_{12}H_{14}O_3$: e176
$C_{12}H_{14}O_4$: d329
$C_{12}H_{15}N$: d369
$C_{12}H_{15}NO$: b116
$C_{12}H_{15}N_3O_3$: t197
$C_{12}H_{16}$: c338, m212, p104
$C_{12}H_{16}O_2$: m364
$C_{12}H_{16}O_3$: d246
$C_{12}H_{17}N$: b115, c337
$C_{12}H_{17}NO$: d319, d342
$C_{12}H_{18}$: b489, c304, d414, d415, h47, p117, t435
$C_{12}H_{18}Cl_2N_4OS$: t139

TABLE 1.14 Empirical Formula Index for Organic Compounds (*continued*)

The alphanumeric designations are keyed to Table 1.15

$C_{12}H_{18}O$: d419, d513
$C_{12}H_{18}O_2$: b474, b475
$C_{12}H_{18}O_4$: b445
$C_{12}H_{19}N$: d413, h81
$C_{12}H_{20}O_2$: b185, b217, e102, L10
$C_{12}H_{20}O_3Si$: p157
$C_{12}H_{20}O_4$: d118
$C_{12}H_{20}O_4Si$: t8
$C_{12}H_{21}N$: t431
$C_{12}H_{21}N_3$: t422
$C_{12}H_{22}$: c305, d241
$C_{12}H_{22}O$: c303, e4
$C_{12}H_{22}O_3$: h67
$C_{12}H_{22}O_4$: d130, d324, d512, d702, d721
$C_{12}H_{22}O_{11}$: L3, m7, s20
$C_{12}H_{23}ClO$: d728
$C_{12}H_{23}N$: d242, d724
$C_{12}H_{23}NO$: a318
$C_{12}H_{24}$: d729
$C_{12}H_{24}N_2$: d694
$C_{12}H_{24}O$: c302, d731, m443, t349
$C_{12}H_{24}O_2$: d726, e113
$C_{12}H_{24}O_6$: h74
$C_{12}H_{25}Br$: b275
$C_{12}H_{25}Cl$: c100
$C_{12}H_{25}Cl_3Si$: d736
$C_{12}H_{26}$: d719
$C_{12}H_{26}O$: d350, d727, t354a
$C_{12}H_{26}O_2$: d722, d723
$C_{12}H_{26}O_3$: b151
$C_{12}H_{26}O_4$: t411
$C_{12}H_{26}O_4S$: d735
$C_{12}H_{26}S$: d725
$C_{12}H_{27}Al$: t307
$C_{12}H_{27}BO_3$: t209
$C_{12}H_{27}ClSn$: t215
$C_{12}H_{27}N$: d349, d732, t210
$C_{12}H_{27}O_3P$: t214
$C_{12}H_{27}O_4P$: t212
$C_{12}H_{27}P$: t213
$C_{12}H_{28}BrN$: t135
$C_{12}H_{28}N_2$: d720

$C_{12}H_{28}O_4Si$: t88, t134
$C_{12}H_{28}O_4Ti$: t166
$C_{12}H_{28}O_8Si$: t89
$C_{12}H_{36}O_4Si_4Ti$: t90

C_{13}

$C_{13}H_5N_3O_7$: t384
$C_{13}H_8ClNO_3$: c181
$C_{13}H_8ClNOS$: p67
$C_{13}H_8Cl_2O$: d159
$C_{13}H_8N_2O_7$: b193
$C_{13}H_8O$: f3
$C_{13}H_8OS$: t165
$C_{13}H_8O_2$: x3
$C_{13}H_9BrO$: b232
$C_{13}H_9ClO$: c49, c50
$C_{13}H_9ClO_2$: c131
$C_{13}H_9N$: a61
$C_{13}H_{10}$: f2
$C_{13}H_{10}ClNO$: a142, a143, d659
$C_{13}H_{10}Cl_2O_2$: m233
$C_{13}H_{10}N_2$: p90
$C_{13}H_{10}N_2O_3$: a239
$C_{13}H_{10}O$: b53, x1
$C_{13}H_{10}O_2$: b135, h102, p91
$C_{13}H_{10}O_3$: d386, d661, p150
$C_{13}H_{10}O_5$: t83
$C_{13}H_{11}Br$: b274
$C_{13}H_{11}Cl$: c97
$C_{13}H_{11}ClO$: c44
$C_{13}H_{11}NO$: a127, b5
$C_{13}H_{11}NO_2$: h161, p85
$C_{13}H_{11}NO_3$: p86
$C_{13}H_{12}$: d676
$C_{13}H_{12}N_2$: b54, d38, d670
$C_{13}H_{12}N_2O$: d693
$C_{13}H_{12}N_2S$: d692, t146
$C_{13}H_{12}N_4O$: p88
$C_{13}H_{12}N_4S$: d691
$C_{13}H_{12}O$: b138, d677, h113, m56, p76
$C_{13}H_{12}S$: b113

$C_{13}H_{13}ClSi$: c98
$C_{13}H_{13}N$: d678, m230, m93
$C_{13}H_{13}NO$: b106
$C_{13}H_{13}N_3$: d671
$C_{13}H_{14}N_2$: d35, m238, t344
$C_{13}H_{14}N_2O_3$: a59
$C_{13}H_{14}N_4O$: d660
$C_{13}H_{14}Si$: m231
$C_{13}H_{16}O_2$: m74
$C_{13}H_{16}O_3$: e71
$C_{13}H_{16}O_4$: d328
$C_{13}H_{17}NO_2$: e74
$C_{13}H_{20}$: p116
$C_{13}H_{20}N_2O_2$: d271
$C_{13}H_{20}O$: i56, i57
$C_{13}H_{22}ClN$: b125
$C_{13}H_{22}N_2$: d243
$C_{13}H_{22}O_2$: n106
$C_{13}H_{22}O_3Si$: b124
$C_{13}H_{26}$: t258
$C_{13}H_{26}N_2$: m236, t343
$C_{13}H_{26}O_2$: e229, t257
$C_{13}H_{27}Br$: b359
$C_{13}H_{28}$: t256
$C_{13}H_{28}O_4$: t409
$C_{13}H_{29}NO_4$: b169

C_{14}

$C_{14}H_6Cl_2O_2$: d148, d149
$C_{14}H_7ClO_2$: c36, c37
$C_{14}H_8ClNO_5$: c185
$C_{14}H_8O_2$: a305, p62
$C_{14}H_8O_3$: h93
$C_{14}H_8O_4$: d371, d372, d373, d374
$C_{14}H_8O_5S$: a308
$C_{14}H_8O_8$: a306, d312
$C_{14}H_9Br$: b324
$C_{14}H_9ClO_3$: c54
$C_{14}H_9Cl_5$: b167
$C_{14}H_9NO_2$: a111, a112
$C_{14}H_9NO_3$: a186

TABLE 1.14   Empirical Formula Index for Organic Compounds (*continued*)

The alphanumeric designations are keyed to Table 1.15

$C_{14}H_{10}$: a304, d654, p61
$C_{14}H_{10}Br_2O$: b273
$C_{14}H_{10}ClNO_3$: a147
$C_{14}H_{10}Cl_2O_4$: b162
$C_{14}H_{10}Cl_4$: b163
$C_{14}H_{10}N_2O_2$: d28, d29, d30
$C_{14}H_{10}O_2$: b34
$C_{14}H_{10}O_3$: b45, b65, x2
$C_{14}H_{10}O_4$: b137, d54, t82
$C_{14}H_{11}N$: d653, p120
$C_{14}H_{11}NOS$: a50
$C_{14}H_{12}$: d351, s9
$C_{14}H_{12}Cl_2O$: b164
$C_{14}H_{12}N_2O$: b37
$C_{14}H_{12}N_2O_2$: b35
$C_{14}H_{12}O$: a34, d22, m133, m134
$C_{14}H_{12}O_2$: b46, b83, b84, b107, b108, d652
$C_{14}H_{12}O_3$: b36, h35
$C_{14}H_{13}ClO$: c147
$C_{14}H_{13}N$: e93, i12
$C_{14}H_{13}NO$: b82
$C_{14}H_{13}NO_2$: b50
$C_{14}H_{14}$: d666
$C_{14}H_{14}N_2$: a168
$C_{14}H_{14}N_2O_3$: a325
$C_{14}H_{14}O$: d58
$C_{14}H_{14}OS$: b200
$C_{14}H_{14}O_2$: b109
$C_{14}H_{14}S_2$: b199, d57
$C_{14}H_{15}N$: d56, d668
$C_{14}H_{15}O_3P$: d61
$C_{14}H_{16}N_2$: d669
$C_{14}H_{16}O_2Si$: d437
$C_{14}H_{16}O_4$: d281
$C_{14}H_{18}O_4$: d285
$C_{14}H_{20}N_2O_6S$: m120
$C_{14}H_{20}O_5$: b39
$C_{14}H_{22}$: p130a
$C_{14}H_{22}O$: d123, d124, d125, d126
$C_{14}H_{22}O_2$: d112
$C_{14}H_{23}N$: d109, o42
$C_{14}H_{23}N_3O_{10}$: d299
$C_{14}H_{26}O_3$: h11
$C_{14}H_{26}O_4$: d408

$C_{14}H_{27}ClO$: t41
$C_{14}H_{28}$: t42, t43
$C_{14}H_{28}O_2$: t39
$C_{14}H_{29}Br$: b352
$C_{14}H_{29}Cl_3Si$: t46
$C_{14}H_{30}$: t38
$C_{14}H_{30}O$: t40
$C_{14}H_{31}N$: t44
$C_{14}H_{32}N_2O_4$: t87

C_{15}

$C_{15}H_{10}O_2$: b101, m124
$C_{15}H_{11}NO$: d679
$C_{15}H_{12}N_2O_2$: d672
$C_{15}H_{12}O$: d354, d685
$C_{15}H_{12}O_2$: d53
$C_{15}H_{13}NO$: a13
$C_{15}H_{14}O$: d684
$C_{15}H_{14}O_2$: b49, b141, d686
$C_{15}H_{14}O_3$: b111
$C_{15}H_{16}O$: m359
$C_{15}H_{16}O_2$: i97
$C_{15}H_{17}N_3$: d711
$C_{15}H_{18}OSi$: e42
$C_{15}H_{22}O_3$: d117
$C_{15}H_{24}$: t312
$C_{15}H_{24}O$: d120
$C_{15}H_{26}O$: h184
$C_{15}H_{26}O_6$: g19
$C_{15}H_{30}N_2$: t342
$C_{15}H_{30}N_3OP$: t405
$C_{15}H_{30}O$: p14
$C_{15}H_{30}O_2$: m413
$C_{15}H_{32}$: p13
$C_{15}H_{32}O_3Si_4$: p164
$C_{15}H_{32}O_{10}$: t389

C_{16}

$C_{16}H_{10}$: b52, fl, p246
$C_{16}H_{11}NO_2$: p149
$C_{16}H_{12}N_2O_5S$: a60
$C_{16}H_{12}N_4O_9S_2$: t3

$C_{16}H_{13}N$: p130
$C_{16}H_{14}$: d656, d657, e66
$C_{16}H_{14}O$: d658
$C_{16}H_{14}O_6S$: s26
$C_{16}H_{15}NO_4$: d445
$C_{16}H_{16}O_2$: b47, b112
$C_{16}H_{16}O_3$: d449
$C_{16}H_{18}ClN_3S$: m237
$C_{16}H_{19}ClSi$: b435
$C_{16}H_{20}N_2$: d59
$C_{16}H_{20}O_2Si$: d250
$C_{16}H_{22}O_4$: d128, d409
$C_{16}H_{22}O_{11}$: g7
$C_{16}H_{26}O_3$: d730
$C_{16}H_{26}O_7$: t52
$C_{16}H_{32}$: h37
$C_{16}H_{32}O_2$: h35
$C_{16}H_{33}Br$: b293
$C_{16}H_{33}I$: i38
$C_{16}H_{33}NO$: d297
$C_{16}H_{34}$: h4, h32
$C_{16}H_{34}O$: h36
$C_{16}H_{34}O_2$: h33
$C_{16}H_{34}S$: d644, h34
$C_{16}H_{35}N$: d643, h38
$C_{16}H_{35}O_4P$: b178
$C_{16}H_{36}BF_4N$: t19
$C_{16}H_{36}BrN$: t14
$C_{16}H_{36}ClN$: t15
$C_{16}H_{36}FN$: t16
$C_{16}H_{36}IN$: t18
$C_{16}H_{36}O_4Si$: t13
$C_{16}H_{36}Sn$: t20
$C_{16}H_{37}NO_4S$: t17

C_{17}

$C_{17}H_6O_7$: b55
$C_{17}H_{10}O$: b8
$C_{17}H_{12}O_3$: p119
$C_{17}H_{13}N_3O_5S_2$: p172
$C_{17}H_{16}O_4$: d60
$C_{17}H_{18}O_3$: b478
$C_{17}H_{20}N_2O$: b171
$C_{17}H_{20}N_4O_6$: r4

TABLE 1.14   Empirical Formula Index for Organic Compounds (*continued*)

The alphanumeric designations are keyed to Table 1.15

$C_{17}H_{21}NO_4$: c275
$C_{17}H_{22}N_2$: m234
$C_{17}H_{23}NO_3$: a315
$C_{17}H_{34}O_2$: m263
$C_{17}H_{36}$: h1
$C_{17}H_{37}N$: m229

C_{18}

$C_{18}H_9Cl_6O_4P$: t418
$C_{18}H_{10}O_6$: h85
$C_{18}H_{12}$: b6, b7, t396
$C_{18}H_{12}N_5O_6$: d683
$C_{18}H_{14}$: t4
$C_{18}H_{14}O$: d680
$C_{18}H_{14}O_8$: d55
$C_{18}H_{15}As$: t394
$C_{18}H_{15}N$: t392
$C_{18}H_{15}N_3Si$: a320
$C_{18}H_{15}O_3P$: t403
$C_{18}H_{15}O_4P$: t399
$C_{18}H_{15}P$: t400
$C_{18}H_{15}PS$: t402
$C_{18}H_{15}PSe$: t401
$C_{18}H_{15}Sb$: t393
$C_{18}H_{16}O_2$: b422
$C_{18}H_{16}Si$: t404
$C_{18}H_{18}O_3$: e73
$C_{18}H_{20}O_2$: b48
$C_{18}H_{25}NO_3$: i61
$C_{18}H_{30}O$: t211
$C_{18}H_{30}O_2$: o7
$C_{18}H_{31}N$: d733
$C_{18}H_{32}O_2$: o1
$C_{18}H_{32}O_{16}$: r1
$C_{18}H_{34}O_2$: o10, o11
$C_{18}H_{34}O_4$: d110
$C_{18}H_{36}$: d734, o8
$C_{18}H_{36}O$: o12
$C_{18}H_{36}O_2$: e139, o5
$C_{18}H_{37}Br$: b319
$C_{18}H_{37}Cl_3Si$: o15
$C_{18}H_{37}N$: o9
$C_{18}H_{37}NO$: o2
$C_{18}H_{38}$: o3
$C_{18}H_{38}O$: o6

$C_{18}H_{38}S$: o4
$C_{18}H_{39}ClSi$: t302
$C_{18}H_{39}N$: o13, t301
$C_{18}H_{39}O_7P$: t414
$C_{18}H_{40}Si$: t303

C_{19}

$C_{19}H_{15}Br$: b366
$C_{19}H_{15}Cl$: c256
$C_{19}H_{16}$: t397
$C_{19}H_{16}O$: t398
$C_{19}H_{18}BrP$: m442
$C_{19}H_{20}Br_4O_4$: i96
$C_{19}H_{20}O_4$: b87
$C_{19}H_{22}N_2O$: c264
$C_{19}H_{30}O_5$: m243
$C_{19}H_{32}$: p156
$C_{19}H_{34}ClN$: b122
$C_{19}H_{34}O_2$: m325
$C_{19}H_{36}O_2$: m327
$C_{19}H_{37}NO$: o14
$C_{19}H_{38}O_2$: m326
$C_{19}H_{40}$: n90, t115
$C_{19}H_{40}Cl_2Si$: m328

C_{20}

$C_{20}H_{10}Br_2O_5$: d83
$C_{20}H_{12}$: b57, b58, d49
$C_{20}H_{12}O_5$: f4
$C_{20}H_{14}O_4$: p65
$C_{20}H_{15}Br$: b365
$C_{20}H_{18}O_3Si$: t391
$C_{20}H_{19}N_3$: b2
$C_{20}H_{22}O_6$: t271
$C_{20}H_{24}N_2O_2$: q2
$C_{20}H_{24}O_6$: d51
$C_{20}H_{28}O_2P$: d674
$C_{20}H_{30}O_2$: a1
$C_{20}H_{31}N$: d20
$C_{20}H_{35}N$: t45
$C_{20}H_{36}O_2$: e188
$C_{20}H_{38}O_2$: e189
$C_{20}H_{40}$: i3

$C_{20}H_{40}O$: o16
$C_{20}H_{42}$: i2

C_{21}

$C_{21}H_{15}NO$: b142
$C_{21}H_{15}N_3O_3$: t390
$C_{21}H_{21}N$: t200
$C_{21}H_{22}N_2O_2$: s10
$C_{21}H_{24}O_2$: b144
$C_{21}H_{28}N_2O$: b170
$C_{21}H_{36}O$: p15
$C_{21}H_{39}N_3$: t255

C_{22}

$C_{22}H_{23}N_3O_9$: a316
$C_{22}H_{30}O_2S$: t144
$C_{22}H_{34}O_4$: b443
$C_{22}H_{39}N$: h39
$C_{22}H_{42}O_4$: d312
$C_{22}H_{44}O_2$: b466, d716
$C_{22}H_{46}$: d715
$C_{22}H_{46}O$: d717

C_{23}

$C_{23}H_{16}O_6$: m235
$C_{23}H_{26}N_2O_4$: b372

C_{24}

$C_{24}H_{16}N_2O_2$: b198
$C_{24}H_{18}$: t395
$C_{24}H_{20}BNa$: t128
$C_{24}H_{20}O_4Si$: t127
$C_{24}H_{20}Si$: t132
$C_{24}H_{20}Sn$: t133
$C_{24}H_{22}N_2O$: b140
$C_{24}H_{38}O_4$: b179, d313
$C_{24}H_{40}O_5$: c263
$C_{24}H_{46}O_4$: d643a
$C_{24}H_{50}$: t36
$C_{24}H_{51}N$: t387

TABLE 1.14 Empirical Formula Index for Organic Compounds (*continued*)

The alphanumeric designations are keyed to Table 1.15

$C_{24}H_{51}O_3P$: d423, t421	$C_{27}H_{46}O$: c262	$C_{32}H_{68}O_4Si$: t86
$C_{24}H_{52}O_4Si$: t85	$C_{27}H_{50}ClN$: b94	$C_{36}H_{75}O_3P$: d642
$C_{24}H_{54}OSn_2$: b201		$C_{38}H_{30}NiO_2P_2$: b213
	C_{28}	$C_{39}H_{74}O_6$: g20
C_{26}		$C_{40}H_{56}$: c19
		$C_{40}H_{82}O_6P_2$: b196
$C_{26}H_{20}$: t131	$C_{28}H_{22}$: t129	
$C_{26}H_{26}N_2O_2S$: b152	$C_{28}H_{31}ClN_2O_3$: r2	C_{45} to C_{57}
$C_{26}H_{26}OSi_2$: t130	$C_{28}H_{32}O_2Si_3$: t121	
$C_{26}H_{50}O_4$: b177, d311		$C_{45}H_{86}O_6$: g24
	C_{30} to C_{40}	$C_{48}H_{40}O_4Si_4$: o38
C_{27}		$C_{51}H_{98}O_6$: g23
	$C_{30}H_{50}$: s8	$C_{57}H_{104}O_6$: g22
	$C_{30}H_{62}$: s7	
$C_{27}H_{19}NO$: b148	$C_{30}H_{63}O_3P$: t308	
$C_{27}H_{42}ClNO_2$: b33	$C_{32}H_{66}$: d737	

TABLE 1.15 Physical Constants of Organic Compounds
See also the special tables of fats, oils, and waxes.

Names of the compounds in the table starting on p. 1.82 are arranged alphabetically. Usually substitutive nomenclature is employed; exceptions generally involve ethers, sulfides, sulfones, and sulfoxides. Each compound is given a number within its letter classification; thus compound c195 is 3-chlorophenol. The section "Nomenclature of Organic Compounds" should be consulted to familiarize oneself with present nomenclature systems.

Synonyms or *Alternate Names* are found at the bottom of each spread in their alphabetical listing; the number following the name refers to the numerical place of this compound in the table. For example, epichlorohydrin, c101, indicates that this compound is found listed under the name 1-chloro-2,3-epoxypropane.

Formulas are presented in a semistructural form when no ambiguity is possible. Complicated systems are drawn in complete structural form and located at the bottom of each page and keyed to the number of the entry.

Beilstein Reference. In the column so headed is found the reference to the volume and page numbers of the fourth edition of Beilstein: (Handbuch der Organischen Chemie) (Springer-Verlag, New York). Thus the entry 9, 202 refers to an entry in volume 9 appearing on page 202. When the volume number has a superscript attached, reference is made to the appropriate supplementary volume. For example, 12^2, 404 indicates that the compound will be found listed in the second supplement to volume 12 on page 404. The earliest Beilstein entry is listed. Supplementary information may be found in the supplements to the basic series; such coordinating references (series number, volume number, and page number of the main edition) along with the system number are found at the top of each *odd-numbered* page. Similarly, a back reference such as H 93; E II 64; E III 190 in a volume of Supplementary Series IV means that previous items on this compound are found in the same volume of the Basic Series on page 93, of Supplementary Series II on page 64, and of Supplementary Series III on page 190. The absence of a back reference implies that the compound involved is described *for the first time* in the series concerned.

Formula Weights are based on the International Atomic Weights of 1973 and are computed to the nearest hundredth.

Density values are given at room temperature unless otherwise indicated by the superscript figure; thus 0.9711^{112} indicates a density of 0.9711 for the substance at $112\,°C$. A densityof 0.899_4^{16} indicates a density of 0.899 for the substance at $16\,°C$ relative to water at $4\,°C$.

Refractive Index, unless otherwise specified, is given for the sodium line at 589.6 nm. The temperature at which the measurement was made is indicated by the superscript figure; otherwise it is assumed to be room temperature.

Melting Point is recorded in certain cases as 250 d and in some other cases as d 250, the distinction being made in this manner to indicate that the former is a melting point with decomposition at $250\,°C$, while the latter decomposition occurs only at $250\,°C$ and higher temperatures. Where a value such as $-2H_2O$, 120 is given, it indicates a loss of 2 mol of water per formula weight of the compound at a temperature of $120\,°C$.

Boiling Point is given at atmospheric pressure (760 mmHg) unless otherwise indicated; thus $82^{15\,mm}$ indicates that the boiling point is $82\,°C$ when the pressure is 15 mmHg. Also, subl 550 indicates that the compound sublimes at $550\,°C$.

Flash Point is given in degrees Celsius, usually closed up. Because values will vary with the specific procedure employed, and sometimes the method was not stated, the values listed for the flash point should be considered only as indicative. See also Table 4.13, Properties of Combustible Mixtures in Air.

Solubility is given in parts by weight (of the formula weight) per 100 parts by weight of the solvent and at room temperature. Other temperatures are indicated by the superscript. In the case of gases, the solubility is often expressed as $5^{10°}$ mL, which indicates that at $10\,°C$, 5 mL of the gas is soluble in 100 g of the solvent.

Abbreviations Used in the Table

abs, absolute	EtOH, ethanol, 95%	s, soluble
acet, acetone	expl, explodes	*sec*, secondary
alc, ethanol	glyc, glycerol	sl, slight or slightly
alk, alkali (i.e., aqueous NaOH	h, hot	soln, solution
or KOH)	HOAc, acetic acid	solv, solvent
anhyd, anhydrous	hyd, hydrolysis	subl, sublimes
aq, aqueous; water	hygr, hygroscopic	*s*, symmetrical
as, asymmetrical	i, insoluble	*sym*, symmetrical
atm, atmosphere	ign, ignites	*tert*, tertiary
BuOH, butanol	i-PrOH, isopropanol	v, very
bz, benzene	l (−), levorotatory	v s, very soluble
c, cold	L, designates configuration	v sl s, very slightly soluble
chl, chloroform, $CHCl_3$	*m*, meta position	vac, vacuo or vacuum
conc, concentrated	Me, methyl	vols, volumes
d, decomposes or decomposed	MeEtKe, methyl ethyl ketone	>, greater than
d (+), dextrorotatory	MeOH, methanol	<, less than
$-d_n$, deuterium substitution	misc, miscible; soluble in all	~, approximately
D, designates configuration	proportions	α, alpha position
deliq, deliquescent	NaOH, aqueous sodium	β, beta position
dil, dilute	hydroxide	γ, gamma position
diox, dioxane	*o*, ortho position	δ, delta position
DL (or *dl*), inactive (i.e., 50% D	org, organic	ε, epsilon position
and 50% L)	*p*, para position	ω, omega position (farthest
DMF, dimethylformamide	PE, petroleum ether	from parent functional group)
EtAc, ethyl acetate	pyr, pyridine	
eth, diethyl ether		

TABLE 1.15 Physical Constants of Organic Compounds (*continued*)

No.	Name	Formula	Formula weight	Beilstein reference	Density	Refractive index	Melting point	Boiling point	Flash point	Solubility in 100 parts solvent
a1	(−)-Abietic acid		302.44	9^2, 424	—		172–175			i aq; s alc, bz, chl, eth, acet, dil alk
a2	Acenaphthene		154.21	5, 586	1.069^{95}		93.45	279		i aq; 3.2 alc; 20 bz
a3	Acenaphthlyene		152.20	5, 625	0.899^{16}		80–83	280		i aq; v s alc, eth
a4	Acetaldehyde	CH_3CHO	44.05	1, 594	0.8053^0	1.3311^{20}	−123.5	20.2	−27	misc aq, alc
a5	Acetaldoxime	$CH_3CH{=}NOH$	59.07	1, 608	0.966	1.415^{20}	46.5	114.5	38	v s aq, alc; eth
a6	Acetamide	CH_3CONH_2	59.07	2^2, 177	0.9711^{112}	1.4158^{110}	80.1	221.15		70aq; 50alc; s chl, hot bz
a7	Acetamidine HCl	$CH_3({=}NH)NH_2 \cdot HCl$	94.54	2, 185			170–172			v s aq, alc; i acet, eth
a8	N-(2-Acetamido)-2-aminoethane-sulfonic acid	$H_2N(CO)CH_2NHCH_2\text{-}CH_2SO_3H$	182.20				>220 d			
a9	4-Acetamidobenzaldehyde	$CH_3CONHC_6H_4CHO$	163.18	14, 38			154–156			s aq, bz; sl s alc
a10	4-Acetamidobenzenesulfonyl chloride	$CH_3CONHC_6H_4SO_2Cl$	233.67	14, 439			149			d aq; v s alc, eth
a11	2-Acetamidobenzoic acid	$CH_3CONHC_6H_4COOH$	179.18	14, 337			185–187			sl s aq; v s alc, bz, eth, acet
a12	4-Acetamidobenzoic acid	$CH_3CONHC_6H_4COOH$	179.18	14, 432			260–262			i aq; s alc; sl s eth
a13	2-Acetamidofluorene		223.28	12, 1331			194			i aq; s alc, glycols
a14	N-(2-Acetamido)-iminodiacetic acid	$H_2NCOCH_2N(CH_2COOH)_2$	190.16				219 d			
a15	2-Acetamidophenol	$CH_3CONHC_6H_4OH$	151.17	13, 370			207–209			
a16	3-Acetamidophenol	$CH_3CONHC_6H_4OH$	151.17	13, 415	1.2933^{21}		146–149			
a17	4-Acetamidophenol	$CH_3CONHC_6H_4OH$	151.17	13, 460			170			s alc, acet

a18	Acetanilide	$CH_3CONHC_6H_5$	135.17	12, 237	1.219^{15}_{4}		114.2	304	173	$0.56aq^{25}$; 29 alc; 2bz; 27 chl; 25 acet; 5 eth
a19	Acetic acid	CH_3COOH	60.65	2, 96	1.0492^{20}_{4}	1.3716^{20}	16.63	117.90	40	misc aq, alc, eth, CCl_4
a20	Acetic acid-d	CH_3COOD	61.05		1.07	1.3715^{20}		115.5	40	misc aq, alc, eth, CCl_4
a21	Acetic-d_3, acid-d	CD_3COOD	64.08		1.11	1.3709^{20}		115.5	40	misc aq, alc, eth
a22	Acetic anhydride	$(CH_3CO)_2O$	102.09	2, 166	1.082^{15}_{4}	1.3904^{20}	−73.1	140.0	130	13 aq; s chl, eth
a23	Acetic anhydride-d_6	$(CD_3CO)_2O$	108.14			1.3875^{20}		65^{65mm}	54	d aq, alc
a24	Acetoacetic acid	CH_3COCH_2COOH	102.09	3, 630			36–37	d violently 100		misc aq, alc, eth
a25	Acetohydrazide	$CH_3CONHNH_2$	74.08	2, 191				129^{18mm}		
a26	Acetone	CH_3COCH_3	58.08	1, 635	0.7908^{20}_{4}	1.3588^{20}	−95.35	56.24	−20	misc aq, alc, chl
a27	Acetone-d^6	CD_3COCD_3	64.13		0.88	1.3554^{20}		55.5	−17	
a28	Acetone oxime	$(CH_3)_2C{=}NOH$	73.10	1, 649	0.901		60–63	135		v s aq, alc, eth

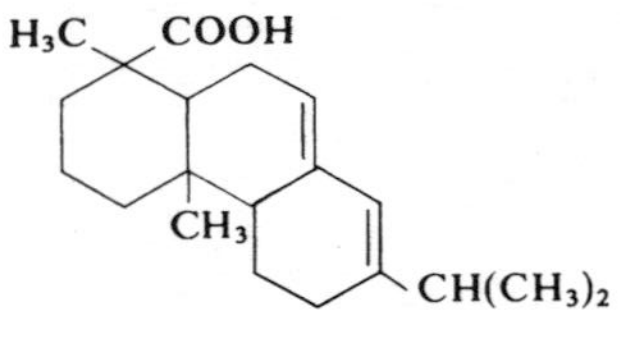

H_3C COOH CH_3 $CH(CH_3)_2$

a1

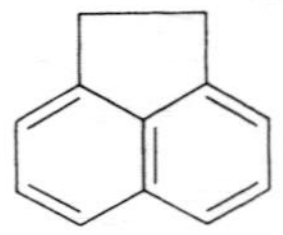

a2

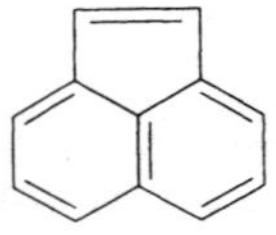

a3

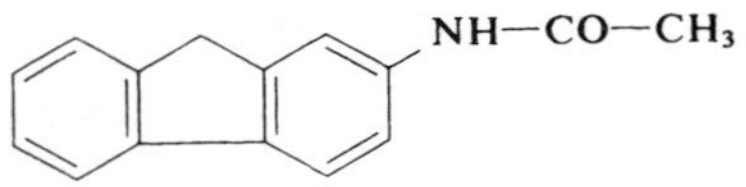

NH—CO—CH_3

a13

TABLE 1.15 Physical Constants of Organic Compounds (*continued*)

No.	Name	Formula	Formula weight	Beilstein reference	Density	Refractive index	Melting point	Boiling point	Flash point	Solubility in 100 parts solvent
a29	Acetonitrile	CH_3CN	41.05	2, 183	0.7857^{20}	1.3441^{20}	−43.8	81.60	5	misc aq, alc, chl
a30	Acetonitrile-d_3	CD_3CN	44.08		0.84	1.3420^{20}		80.7	5	misc aq, alc, chl
a31	Acetophenone	$C_6H_5COCH_3$	120.15	7, 271	1.0238^{25}	1.5322^{25}	19.62	202.08	82	0.55 aq; s alc, eth
a32	2-Acetylacetanilide	$C_6H_5NHCOCH_2COCH_3$	177.20	12, 518			85			sl s aq; s alc, hot bz, chl, eth, acids, alk
a33	4-Acetylbenzene-sulfonic acid, Na salt	$CH_3COC_6H_4SO_3{}^-Na^+$	222.02	11^2, 186			>300			i aq; v s alc, acet
a34	4-Acetylbiphenyl	$C_6H_5C_6H_4COCH_3$	196.25	7^2, 337			116–118	325–327		d aq, alc; misc bz, chl, eth
a35	Acetyl bromide	CH_3COBr	122.95	2, 174	1.663^{16}		−96	75–77	1	
a36	2-Acetylbutyrolactone		128.13	2, 173	1.1846_4^{20}	1.4585^{20}		107^{5mm}		21 aq
a37	Acetyl chloride	CH_3COCl	78.50		1.104_4^{20}	1.3886^{20}	−112.9	50.8	4	d aq, alc; misc bz, chl, eth
a38	Acetylcholine bromide	$(CH_3)_3NBrCH_2CH_2$—$OCOCH_3$	226.14	4^1, 428			114–116			v s aq (d hot aq); s alc; i eth
a39	Acetylcholine chloride	$(CH_3)_3NClCH_2CH_2$—$OCOCH_3$	181.66	4, 281			150–152			v s aq; alc; d hot aq; i eth
a40	2-Acetylcyclo-pentanone		126.16	7, 558	1.043	1.4905^{20}		$72–75^{80mm}$	72	v s aq, alc, eth
a41	Acetylene	$HC\equiv CH$	26.02	1, 228	0.90(g)		-81^{891mm}	−83.95 subl		90 aq; 14 alc; v s bz, eth; acet dissolves 25 acet^{15}
a42	Acetylenedicarboxylic acid	$HOOCC\equiv CCOOH$	114.06	2, 801			180 d			v s aq, alc, eth
a43	Acetyl fluoride	CH_3COF	62.04	2, 172	1.032		>−60	20		5 aq(d); misc alc, bz, eth

No.	Name	Formula	M.W.	Beilstein ref.	Density	n_D	mp, °C	bp, °C		Solubility
a44	2-Acetylfuran		110.11	17, 286	1.098	1.5065^{20}	29–30	$67^{10\,mm}$	71	
a45	N-Acetyl-L-glutamic acid	HOOCCH₂CH₂CH—(NHCOCH₃)COOH	189.17	4^{2}, 908			200–201			
a46	N-Acetylglycine	CH₃CONHCH₂COOH	117.10	4, 354			207–209			$2.7\ aq^{15}$; s alc; i eth
a47	N-Acetylimidazole		110.12	2, 174			93–96			
a48	Acetyl iodide	CH₃COI	169.96		2.0674^{20}_{4}	1.5491^{20}		108		d aq, alc; s bz, eth
a49	Acetyl-2-methyl-choline chloride	CH₃COOCH(CH₃)CH₂-NC1(CH₃)₃	195.69				171–173			
a50	2-Acetylphenothiazine		241.31				180–185			
a51	2-Acetylphenylaceto-nitrile	C₆H₅CH(CN)COCH₃	159.19	10, 699			89–92			v s aq, alc, chl; i eth
a52	N-Acetyl-4-piperidone		141.17		1.146	1.5026^{20}		218	>112	

Structures:

a36 — (lactone ring) COCH₃

a40 — (cyclopentanone ring) COCH₃

a44 — (furan ring) COCH₃

a47 — (imidazole) N—COCH₃

a50 — N(H)⋯S phenothiazine, COCH₃

a52 — N—COCH₃ (4-piperidone), O

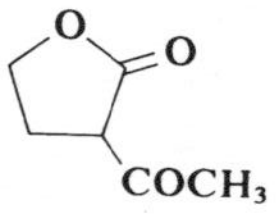

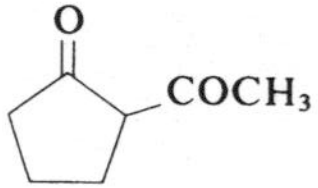

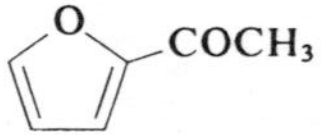

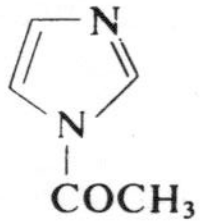

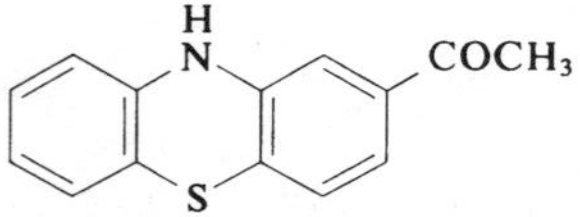

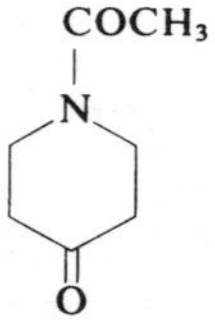

TABLE 1.15 Physical Constants of Organic Compounds (*continued*)

No.	Name	Formula	Formula weight	Beilstein reference	Density	Refractive index	Melting point	Boiling point	Flash point	Solubility in 100 parts solvent
a53	2-Acetylpyridine	$(C_5H_4N)COCH_3$	121.14	21, 279	1.080	1.5203^{20}		188–189	>112	v s alc, eth
a54	3-Acetylpyridine	$(C_5H_4N)COCH_3$	131.14	21, 279	1.102	1.5336^{20}		220	150	v s acids, alc, eth; s aq
a55	4-Acetylpyridine	$(C_5H_4N)COCH_3$	121.14	21, 279	1.095	1.5290^{20}		212	>112	0.33 aq^{25}; 20 alc; 5.9 chl; 5 eth; sl s bz
a56	Acetylsalicylic acid	$HOOCC_6H_4OOCCH_3$	180.16	10, 67	1.35		135			
a57	2-Acetylthiophene	$(C_4H_3S)COCH_3$	126.18	17, 287	1.1682^{22}	1.5564^{20}	10–11	214		sl s aq; misc alc; eth
a58	N-Acetylthiourea	$CH_3CONHC(S)NH_2$	118.16	3, 191			165–169			s hot aq, alc; sl s eth
a59	N-Acetyl-DL-tryptophan		246.27	22^2, 469			204–206			s aq, alc; v s eth
a60	Acid alizarin violet N		366.33	16^2, 127						
a61	Acridine		179.22	20, 459			107–110 subl 110	346		s alc, eth, CS_2, PE
a62	Acrylamide	$H_2C{=}CHCONH_2$	71.08	2, 400	1.122^{30}		84.5	125^{25mm}		215 aq^{30}, 86 alc^{30}; 63 acet; 2.7 chl; v s eth
a63	Acrylic acid	$H_2C{=}CHCOOH$	72.06	2, 397	1.0511^{20}	1.4224^{20}	13	140–141	54	misc aq, alc, bz, eth, chl, acet
a64	Acrylonitrile	$H_2C{=}CHCN$	53.06	2, 400	0.8060^{20}_4	1.3911^{20}	−83.7	77.4	0	7.3 aq; misc org solv
a65	Acryloyl chloride	$H_2C{=}CHCOCl$	90.51	2, 400	1.114	1.4350^{20}		72–76	16	d aq; v s chl
a66	1-Adamantanamine		151.25				206–208			sl s aq
a67	Adamantane		136.24		1.09	1.568	268 sealed tube	subl 205		

a68	1-Adamantane-carboxylic acid		180.25				174–175			
a69	Adenine		135.13	26, 420			>360 d	subl 220		0.05 aq; sl s alc; i chl, eth
a70	Adenosine		267.25	31, 27			234–236			s aq; i alc

N-Acetylsulfanilyl chloride, a10
Aconitic acid, p207
Acrolein, p204
Acrolein diethyl acetal, d258
Acrolein dimethyl acetal, d454
Acrylaldehyde, p207
1-Adamantanemethylamine, a209
Adenosine monophosphate, a72

a59

a60

a61

a66

a67

a68

a69

a70

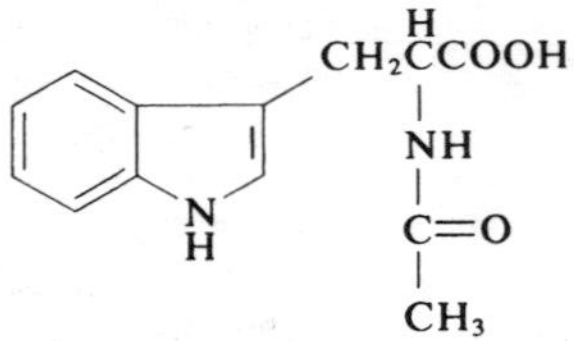

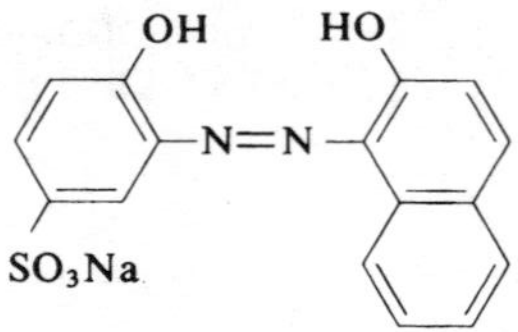

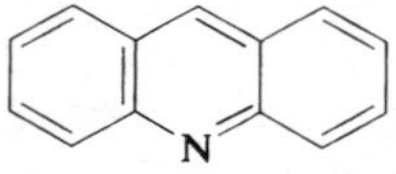

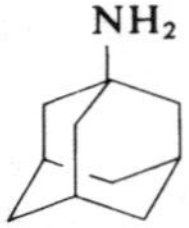

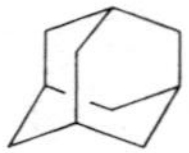

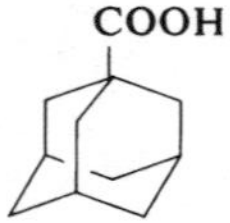

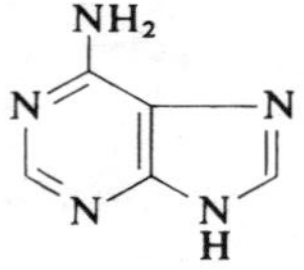

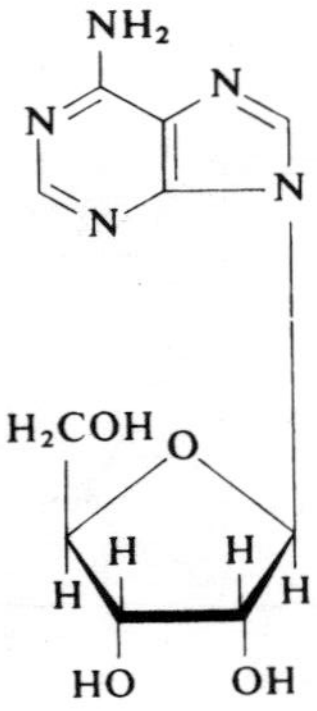

TABLE 1.15 Physical Constants of Organic Compounds (*continued*)

No.	Name	Formula	Formula weight	Beilstein reference	Density	Refractive index	Melting point	Boiling point	Flash point	Solubility in 100 parts solvent
a71	Adenosine-5'-diphosphoric acid		427.22							
a72	Adenosine-5'-phosphoric acid		347.22				200 d			v s hot aq, HCl
a73	D-α-Alanine	$CH_3CH(NH_2)COOH$	89.09	4, 385			291–293 d			16.7 aq^{25}; 8.7 alc^{25}; i eth
a74	DL-α-Alanine	$CH_3CH(NH_2)COOH$	89.09	4, 387	1.402		289 d	subl		16.7 aq^{25}; 8.7 alc^{25}; i eth
a75	L-α-Alanine	$CH_3CH(NH_2)COOH$	89.09	4, 381			315–316			v s aq; sl s alc; i eth
a76	β-Alanine	$H_2NCH_2CH_2COOH$	89.09	4, 401	1.437^{-5}		197–198 d			0.45 aq; 0.2 alc
a77	Allantoin		158.12	25, 474			238			
a78	Allene	$H_2C{=}C{=}CH_2$	40.06	1, 248		1.4168	−136.2	−34.5		
a79	Alloxan monohydrate		160.09	24, 500	1.787		253 d			s alc, acet, HOAc; sl s chl, PE, EtAc
a80	Allyl acetate	$H_2C{=}CHCH_2OCOCH_3$	100.12	2, 136	$0.928^{0.928}$	1.4040^{20}	−50	104	6	i aq; misc alc, eth
a81	Allyl alcohol	$H_2C{=}CHCH_2OH$	58.08	2, 436	0.8540^{20}_4	1.4127^{20}	glass	97.1	22	misc aq, alc, chl, eth
a82	Allylamine	$H_2C{=}CHCH_2NH_2$	57.10	4, 205	0.760^{20}_{20}	1.4205^{20}	−88.2	53.3	−28	misc aq, alc, chl, eth
a83	N-Allylaniline	$C_6H_5NHCH_2CH{=}CH_2$	133.19	12, 170	0.982^{25}	1.5630^{20}		218–220	89	i aq; s alc, eth
a84	Allylbenzene	$C_6H_5CH_2CH{=}CH_2$	118.18	5, 484	0.892^{20}_0	1.5122^{20}		156–157	33	i aq; s alc, eth
a85	Allyl bromide	$H_2C{=}CHCH_2Br$	120.98	1, 201	1.451^{25}_{25}	1.465^{25}	−50	70	7	i aq; misc org solv
a86	Allylchlorodimethylsilane	$H_2C{=}CHCH_2Si(CH_3)_2Cl$	134.7		0.8964^2	1.4195^{20}		110–112		
a87	Allyl chloroformate	$H_2C{=}CHCH_2OOCCl$	120.54		1.13	1.423		27	31	

			Formula wt	Beilstein ref.	Density	n_D	mp, °C	bp, °C	Solubility
a88	Allylcyclohexylamine	$C_6H_{11}NHCH_2CH{=}CH_2$	139.24		0.962	1.4664^{20}		66^{12mm}	
a89	Allyldichloromethyl-silane	$H_2C{=}CHCH_2Si(CH_3)Cl_2$	155.1		1.0758^{20}	1.4419^{20}		119–120	53
a90	*N*-Allyl-*N,N*-dimethylamine	$H_2C{=}CHCH_2N(CH_3)_2$	85.0			1.4010^{20}		63–64	
a91	Allyl ethyl ether	$H_2C{=}CHCH_2OCH_2CH_3$	86.13	1, 438	0.7651^{20}_4	1.3881^{20}		64–66	i aq; misc alc, eth
a92	Allyl iodide	$H_2C{=}CHCH_2I$	167.98	1, 202	1.846^{20}_4		−99.3	103.1	i aq; misc alc, eth

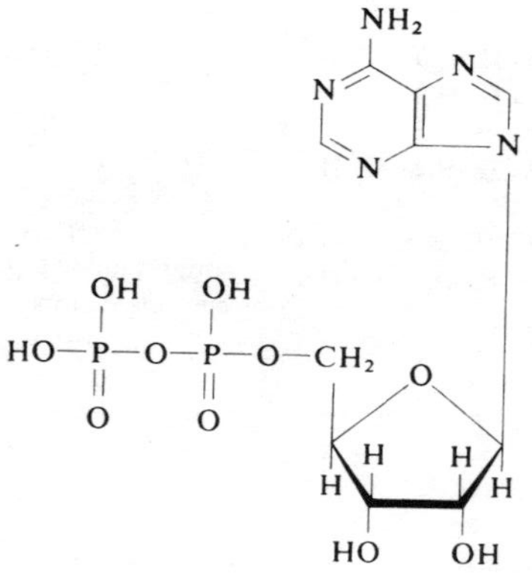

a71

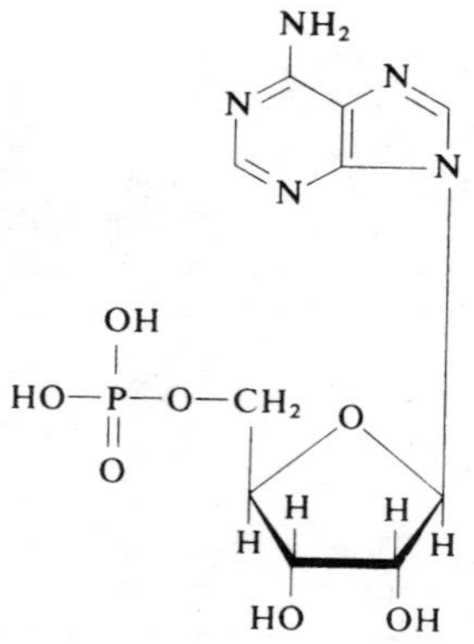

a72

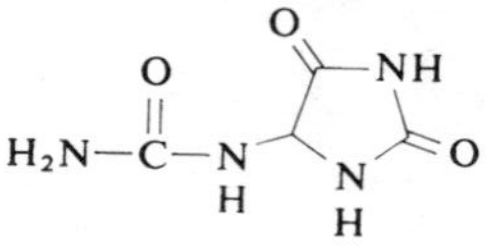

a77

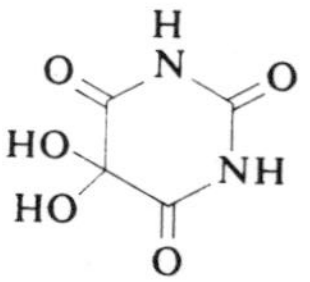

a79

TABLE 1.15 Physical Constants of Organic Compounds (*continued*)

No.	Name	Formula	Formula weight	Beilstein reference	Density	Refractive index	Melting point	Boiling point	Flash point	Solubility in 100 parts solvent
a93	Allyl isothiocyanate	$H_2C=CHCH_2NCS$	99.16	4, 214	1.013_4^{20}	1.5300^{20}	−80	150	46	0.2 aq; misc org solv
a94	1-Allyl-4-methoxybenzene	$H_2C=CHCH_2C_6H_4OCH_3$	148.21	6, 571	0.9645_4^{21}	1.5195^{20}		215–216		a slc, chl
a95	Allyl methyl sulfide	$H_2C=CHCH_2SCH_3$	88.17	1, 440	0.803	1.4714^{20}		91–93	18	
a96	1-Allyloxy-2,3-epoxypropane	$H_2C-CHCH_2OCH_2$ $\quad O \quad CH=CH_2$	114.14		0.962	1.4332^{20}		154	57	
a97	Allyloxytrimethylsilane	$H_2C=CHCH_2OSi(CH_3)_3$	130.3		0.7830	1.4075^{25}		100–101		
a98	2-Allylphenol	$H_2C=CHCH_2C_6H_4OH$	134.18	6, 572	1.0255_{15}^{15}	1.5455^{20}	−6	220	88	s alc, eth
a99	Allyl phenyl ether	$H_2C=CHCH_2OC_6H_5$	134.18	6, 144	0.9834^{15}	1.5200^{20}		192	62	i aq; s alc; misc eth
a100	Allyl propyl ether	$H_2C=CHCH_2OC_3H_7$	100.16	1^3, 1882	0.7670_4^{20}	1.3919^{20}		90–92	38	s alc; misc eth
a101	1-Allyl-2-thiourea	$H_2C=CHCH_2NHC(S)NH_2$	116.18	4, 211	1.219_{20}^{20}		78			3.3 aq; s alc; i bz; v sl s eth
a102	Allyltrichlorosilane	$H_2C=CHCH_2SiCl_3$	175.5		1.2011_4^{20}	1.4460^{20}		117.5		v sl s eth
a103	Allyltriethoxysilane	$H_2C=CHCH_2Si-(OC_2H_5)_3$	204.3		0.9030^{20}	1.4072^{20}		176^{740mm}		
a104	Allyltrimethylsilane	$H_2C=CHCH_2Si(CH_3)_3$	114.27	4, 209	0.7193_4^{20}	1.4074^{20}		85–86	7	v s aq, alc; v sl s eth
a105	Allylurea	$H_2C=CHCH_2NHCONH_2$	100.12	4, 344			78	58^{15mm} d		s acids, alc
a106	Aminoacetonitrile	H_2NCH_2CN	56.07	4, 344			101	d 165		v s aq; sl s alc; i eth
a107	Aminoacetonitrile hydrogen sulphate	$H_2NCH_2CN \cdot H_2SO_4$	154.14	4, 344				70^{3mm}		v sl s aq; s alc, eth
a108	2'-Aminoacetophenone	$H_2NC_6H_4COCH_3$	135.17	14, 41						
a109	3'-Aminoacetophenone	$H_2NC_6H_4COCH_3$	135.17	14, 45			98–99	289–290		s hot aq, alc, eth, HOAc; sl s bz
a110	4'-Aminoacetophenone	$H_2NC_6H_4COCH_3$	135.17	14, 46			106	293–295		

	Name	Formula	M.W.	Beil. ref.	m.p., °C		Solubility
a111	1-Aminoanthra-quinone		223.23	14, 177	253–255	subl	i aq; v s alc, bz, chl, eth, HOAc, HCl
a112	2-Aminoanthra-quinone		223.23	14, 191	295 d	subl	i aq, eth; s alc, bz
a113	4-Aminoantipyrine		203.25	24, 273	109		s aq, alc, bz; sl s eth
a114	2-Aminobenzamide	$H_2NC_6H_4CONH_2$	136.15	14, 320	110	300 sl d	v s hot aq, alc; i bz; sl s eth
a115	2-Aminobenzene-arsonic acid	$H_2NC_6H_4AsO(OH)_2$	217.06	16[1], 463	153		bz; sl s eth
a116	4-Aminobenzene-arsonic acid	$H_2NC_6H_4AsO(OH)_2$	217.06	16, 878	>300		s hot aq, alk CO_3, mineral acids
a116a	5-Aminobenzene-1,3-dicarboxylic acid	$H_2NC_6H_3(COOH)_2$	181.15	14[1], 636	>300		
a117	2-Aminobenzene-1,4-disulfonic acid	$H_2NC_6H_3(SO_3H)_2$	253.24				
a118	2-Aminobenzene-sulfonic acid	$H_2NC_6H_4SO_3H$	173.19	14, 681	d 325		1.5 aq[15]; v sl s alc, eth

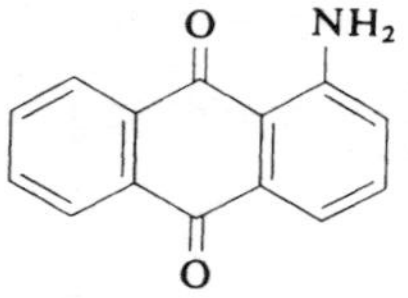

a111

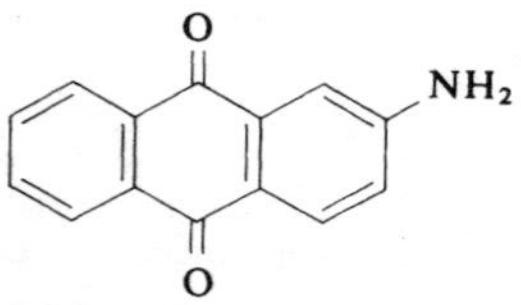

a112

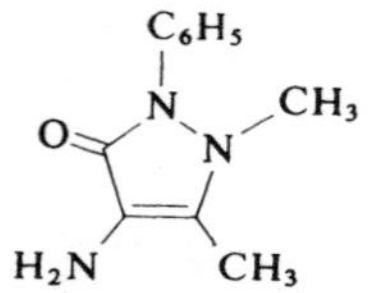

a113

TABLE 1.15 Physical Constants of Organic Compounds (*continued*)

No.	Name	Formula	Formula weight	Beilstein reference	Density	Refractive index	Melting point	Boiling point	Flash point	Solubility in 100 parts solvent
a119	3-Aminobenzene-sulfonic acid	$H_2NC_6H_4SO_3H$	173.19		1.69					2 aq[15]; sl s alc
a120	4-Aminobenzene-sulfonic acid	$H_2NC_6H_4SO_3H$	173.19	14, 695			d 288			1 aq[20]; sl s hot MeOH
a121	2-Aminobenzoic acid	$H_2NC_6H_4COOH$	137.14	14, 310	1.511^4		144–146	subl		v s hot aq, alc, eth
a122	3-Aminobenzoic acid	$H_2NC_6H_4COOH$	137.14	14, 383			172–174			sl s aq; v s alc; s eth
a123	4-Aminobenzoic acid	$H_2NC_6H_4COOH$	137.14	14, 418	1.374		187	268		0.59 aq; 5.6 alc; s alc, eth
a124	2-Aminobenzonitrile	$H_2NC_6H_4CN$	118.14	14, 322			49	288–290		s hot aq; v s alc, eth
a125	3-Aminobenzonitrile	$H_2NC_6H_4CN$	118.14	14, 391			53			v s hot aq, alc, eth
a126	4-Aminobenzonitrile	$H_2NC_6H_4CN$	118.14	14, 425			85	d		sl s aq; s alc, eth
a127	2-Aminobenzo-phenone	$H_2NC_6H_4COC_6H_5$	197.24	14, 76			108	223–226		
a128	2-Aminobenzothiazole		150.20	27, 182			132	d		v s alc, chl, eth
a129	2-Aminobenzotri-fluoride	$H_2NC_6H_4CF_3$	161.13	12^2, 453	1.290^{25}	1.4785^{25}	34	175	55	
a130	3-Aminobenzotri-fluoride	$H_2NC_6H_4CF_3$	161.13	12, 870	1.290	1.4800^{20}	6	187	85	
a131	4-Aminobenzotri-fluoride	$H_2NC_6H_4CF_3$	161.13	12^3, 2151	1.283^{27}	1.4815^{25}	38	107^{39mm}	85	
a132	N-(p-Aminobenzoyl)-glycine	$H_2NC_6H_4CONHCH_2COOH$	194.19	14^2, 258			198–199			i aq; s alc, bz, chl
a133	4-Aminobenzoyl hydrazide	$H_2NC_6H_4CONHNH_2$	151.17	14^1, 570			227			
a134	2-Aminobiphenyl	$H_2NC_6H_4C_6H_5$	169.23	12, 1317			53	299		sl s aq; s alc
a135	4-Aminobiphenyl	$H_2NC_6H_4C_6H_5$	169.23	12, 1318			54	191^{15mm}		s hot aq, alc, eth
a136	D-(+)-2-Amino-1-butanol	$CH_3CH_2CH(NH_2)CH_2OH$	89.14	4, 291	0.947^{20}	1.4521^{20}	−2	174	79	misc aq; s alc

a137	L-(−)-2-Amino-1-butanol	$CH_3CH_2CH(NH_2)CH_2OH$	89.14	4, 291	0.947^{20}	1.4525^{20}	−2	174	82	misc aq; s alc
a138	DL-2-Aminobutyric acid	$CH_3CH_2CH(NH_2)COOH$	103.12	4, 408			304	subl 300		21 aq; 0.2 hot alc
a139	4-Aminobutyric acid	$H_2NCH_2CH_2CH_2COOH$	103.12	4, 413			195.d			v s aq; i alc, eth
a140	2-Amino-4-chloro-benzoic acid	$H_2N(Cl)C_6H_3COOH$	171.58	14, 365			233			
a141	2-Amino-5-chloro-benzonitrile	$H_2N(Cl)C_6H_3CN$	152.58				99	$132^{0.5mm}$	>112	
a142	2-Amino-4′-chloro-benzophenone	$H_2NC_6H_4COC_6H_4Cl$	231.68	14^1, 389			104			
a143	2-Amino-5-chloro-benzophenone	$H_2N(Cl)C_6H_3COC_6H_5$	231.68	14, 79			100			
a144	2-Amino-5-chloro-benzotrifluoride	$H_2N(Cl)C_6H_3CF_3$	195.57	12^3, 1921	1.386	1.5069^{20}		$66–67^{3mm}$		
a145	3-Amino-4-chloro-benzotrifluoride	$H_2N(Cl)C_6H_3CF_3$	195.57		1.428	1.4975^{25}		$82–83^{9mm}$	none	
a146	5-Amino-2-chloro-benzotrifluoride	$H_2N(Cl)C_6H_3CF_3$	195.57				36			

Aminobenzenethiol, a293
Aminobenzoic acid hydrazide, a133
Aminobenzyl cyanide, a260
1-Aminobutane, b417
2-Aminobutane, b377
Aminobutanoic acids, a138, a139
3-Amino-2-butenamide, a150
4-Aminobutyraldehyde diethyl acetal, d247
α-Aminocaproic acid, a183
ε-Aminocaproic acid, a184

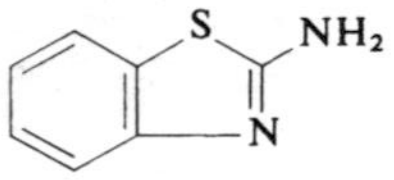

a128

TABLE 1.15 Physical Constants of Organic Compounds (*continued*)

No.	Name	Formula	Formula weight	Beilstein reference	Density	Refractive index	Melting point	Boiling point	Flash point	Solubility in 100 parts solvent
a147	2-(3-Amino-4-chloro-benzoyl)benzoic acid	$H_2N(Cl)C_6H_3CO-C_6H_4COOH$	275.69	14, 661			171–173			
a148	2-Amino-4-chloro-phenol	$H_2N(Cl)C_6H_3OH$	143.57	13, 383			138			
a149	2-Amino-5-chloro-pyridine	$H_2N(Cl)C_5H_3N$	128.56	22^2, 332			138	128^{11mm}		
a150	3-Aminocrotamide	$CH_3C(NH_2){=}CHCONH_2$	100.12	3, 660			102			
a151	3-Aminocrotononitrile	$CH_3C(NH_2){=}CHCN$	82.11							
a152	1-Amino-1-cyclo-hexanecarboxylic acid	$C_6H_{10}(NH_2)COOH$	143.19	14, 299			>300			
a153	5-Amino-2,3-dihydro-1,4-phthalazine-dione		177.16	25^1, 698			319–320			
a154	2-Amino-4,6-dihy-droxypyrimidine		127.10	24, 468			>300			
a155	4-Amino-2,6-dihy-droxypyrimidine		127.10	24, 469			>300			
a156	4-Amino-3,5-diiodo-benzoic acid	$I_2(NH_2)C_6H_2COOH$	388.93	14, 439			>300			i aq, alc
a157	2-Amino-4,6-di-methylpyridine	$(CH_3)_2(NH_2)(C_5H_2N)$	122.17	22, 435			64	235		
a158	4-Amino-2,6-di-methylpyridimide		123.16	24^2, 45			181			156 aq; 18.9 alc
a159	6-Amino-1,3-di-methyluracil		155.16	24, 471			295 d			
a160	5-Amino-2,6-dioxo-1,2,3,6-tetrahydro-4-pyrimidinecar-boxylic acid		171.11	25, 264			>300			

No.	Name	Formula	Formula wt	Beilstein ref.	Density	n_D	Melting point	Boiling point	Flash pt	Solubility
a161	2-Aminoethanesulfonic acid	$H_2NCH_2CH_2SO_3H$	125.15	4, 528			d > 300			6.45 aq[12]; i abs alc
a162	2-Aminoethanethiol	$HSCH_2CH_2NH_2$	77.14	4, 286			99–100			v s aq; s alc
a163	1-Aminoethanol	$CH_3CH(OH)NH_2$	61.08				97	110 d		s aq; sl s eth
a164	2-Aminoethanol	$H_2NCH_2CH_2OH$	61.08	4, 274	1.0158^{20}	1.4539^{20}	10.52	171		misc aq, org solv
a165	2-(2-Aminoethoxy)-ethanol	$H_2NCH_2CH_2OCH_2CH_2OH$	105.14	4^3, 642		1.460		218–224	93	
a166	2-(2-Aminoethyl-amino)ethanol	$H_2NCH_2CH_2NHCH_2CH_2OH$	104.15	4, 286	1.030	1.4861^{20}		241	129	v s aq, alc; sl s eth
a167	5-(2-Aminoethyl-amino)-1-naphtha-lenesulfonic acid	$H_2NCH_2CH_2NH\text{-}C_{10}H_6SO_3H$	266.32				>300			
a167a	3-(2-Aminoethyl-amino)propyl-trimethoxy-silane	$H_2NCH_2CH_2NHCH_2CH_2CH_2Si(OCH_3)_3$	222.1		1.01_4^{25}	1.4418^{25}		140^{15mm}	150	

4-Amino-*m*-cresol, a218
Aminocyclohexane, c334
Aminodecane, d19
2-Amino-2-deoxyglucose, g5
2-Amino-5-diethylaminopentane, d327

2-Amino-1,5-dihydro-1-methyl-4*H*-imidazol-4-one, c278
2-Aminodiphenylamine, p131
1-Amino-1,2-diphenylethane, d668
Aminodiphenylmethane, d678

Aminoethane, e58
1-(2-Aminoethyl)amino-2-[(2-aminoethyl)-aminoethyl]aminoethane, t54

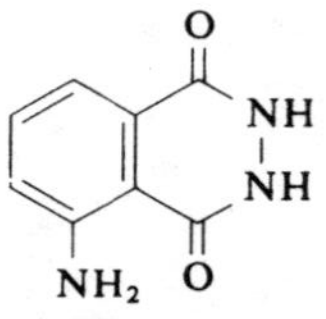

a153

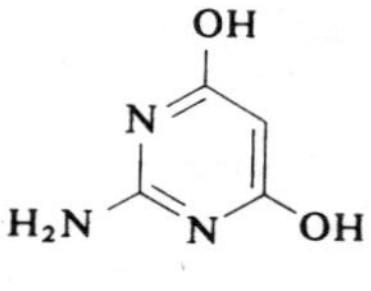

a154

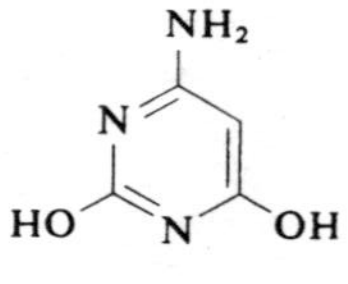

a155

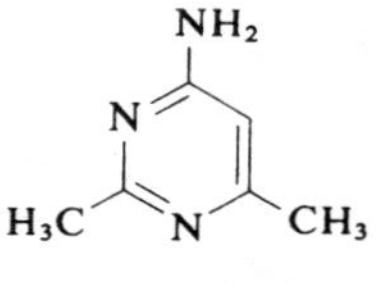

a158

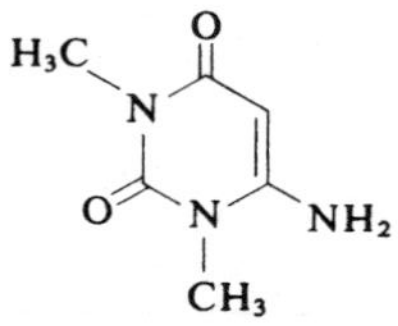

a159

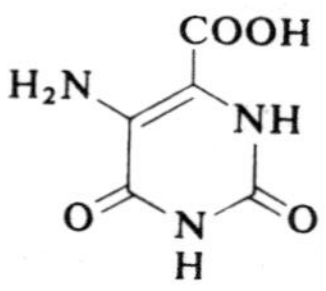

a160

1.95

TABLE 1.15 Physical Constants of Organic Compounds (*continued*)

No.	Name	Formula	Formula weight	Beilstein reference	Density	Refractive index	Melting point	Boiling point	Flash point	Solubility in 100 parts solvent
a168	3-Amino-9-ethyl-carbazole		210.28	22[1], 642			98–100			
a169	2-Aminoethyl hydrogen sulfate	$H_2NCH_2CH_2OSO_3H$	141.15	4, 276			280 d			
a170	3-(2-Aminoethyl)-indole		160.22	22[1], 636			118	$137^{0.15mm}$		i aq, bz, chl, eth; s alc, acet
a171	S-2-Aminoethyl-isothiouronium bromide HBr		281.02				194–195			
a172	N-(2-Aminoethyl)-morpholine		130.19		0.992	1.4755^{20}	25.6	205	175	
a173	p-(2-Aminoethyl)-phenol	$HOC_6H_4CH_2CH_2NH_2$	137.18	13, 625			161–163	175^{8mm}		
a174	N-(2-Aminoethyl)-piperazine		129.21		0.985	1.4983^{20}	−26	222	93	
a175	N-(2-Aminoethyl)-1,3-propanediamine	$H_2NCH_2CH_2CH_2NHCH_2CH_2NH_2$	117.20		0.928	1.4815^{20}			96	
a176	2-Amino-2-ethyl-1,3-propanediol	$HOCH_2C(NH_2)-(C_2H_5)CH_2OH$	119.16		1.099^{20}_{20}	1.490^{20}	38	152^{10mm}	74	misc aq; s alc
a177	2-(2-Aminoethyl)-pyridine	$H_2NCH_2CH_2(C_5H_4N)$	122.17	22, 434	1.021	1.5357^{20}		93^{12mm}		
a178	4-(2-Aminoethyl)-pyridine	$H_2NCH_2CH_2(C_5H_4N)$	122.17		1.012	1.5403^{20}		104^{9mm}		
a179	3-Amino-4-fluorobenzo-trifluoride	$H_2N(F)C_6H_3CF_3$	179.0			1.4608^{20}		81^{20mm}		
a180	Aminoguanidine H_2CO_3	$H_2NNHC(=NH)-NH_2 \cdot H_2CO_3$	136.11	3, 117			172 d			i aq; d hot aq

a181	Aminoguanidine nitrate	$H_2NNHC(=\!NH)\text{-}NH_2 \cdot HNO_3$	137.11	3, 117			137			
a182	N-Aminohexamethyl-eneimine	$C_6H_{12}N\text{—}NH_2$	114.19		0.984	1.4850^{20}		165	56	1.15 aq^{25}; 0.42 alc
a183	2-Aminohexanoic acid	$CH_3(CH_2)_3CH(NH_2)\text{-}COOH$	131.18	4, 433	1.172		d 327			v s aq; i alc
a184	6-Aminohexanoic acid	$H_2N(CH_2)_4CH_2COOH$	131.18	4, 434			204–206			
a185	6-Amino-1-hexanol	$H_2N(CH_2)_5CH_2OH$	117.19	4^2, 748			56–58	135^{30mm}		s eth
a186	1-Amino-4-hydroxy-anthraquinone		239.23	14, 268			207–209			v s aq; i alc, eth, chl
a187	L-2-Amino-3-hydroxy-butyric acid	$CH_3CH(OH)CH(NH_2)COOH$	119.12	4, 514			d 255–257			

NH₂ / CH₂CH₃ — a168

CH₂CH₂NH₂ / N–H — a170

$[H_3\overset{+}{N}CH_2CH_2SC(=\overset{+}{N}H_2)NH_2]\ 2Br^-$ — a171

O morpholine N–CH₂CH₂NH₂ — a172

HN piperazine N—CH₂CH₂NH₂ — a174

NH₂ ... OH anthraquinone — a186

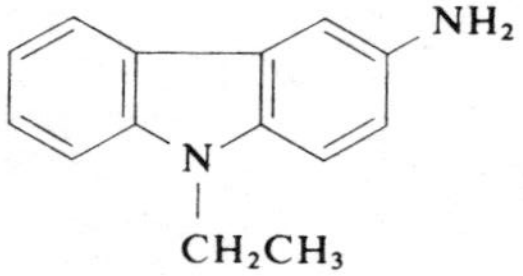

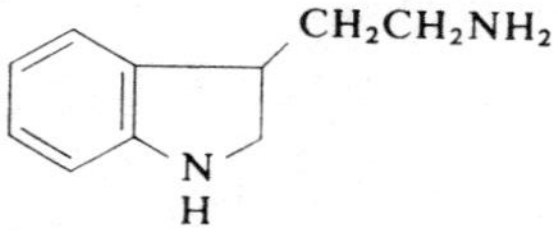

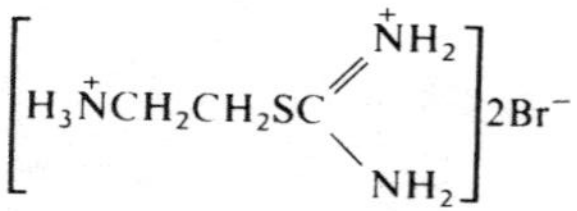

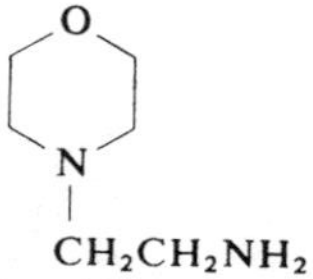

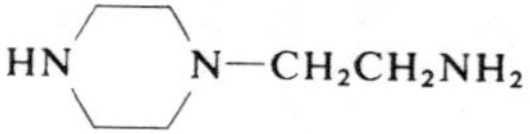

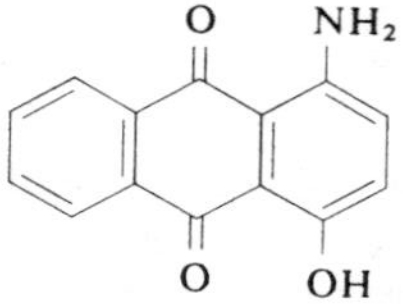

TABLE 1.15 Physical Constants of Organic Compounds (*continued*)

No.	Name	Formula	Formula weight	Beilstein reference	Density	Refractive index	Melting point	Boiling point	Flash point	Solubility in 100 parts solvent
a188	DL-2-Amino-4-hydroxy-butyric acid	$HOCH_2CH_2CH(NH_2)COOH$	119.12	4, 514			188–189			s alc
a189	L-2-Amino-4-hydroxy-butyric acid	$HOCH_2CH_2CH(NH_2)COOH$	119.12	4³, 1636			203 d			s aq; sl s alc, eth
a190	DL-4-Amino-3-hydroxy-butyric acid	$H_2NCH_2CH(OH)CH_2COOH$	119.12	4², 938			202 d			
a191	4-Amino-6-hydroxy-2-mercapto-pyrimidine hydrate		161.18	24, 476			>300			
a192	2-Amino-4-hydroxy-6-methylpyrimidine		125.13	24, 343			>300			i aq, alc, bz, eth
a193	4-Amino-3-hydroxy-1-naphthalenesulfonic acid		239.25	14, 846			295 d			sl s aq; i alc, eth
a194	4-Amino-5-hydroxyl-1-naphthalenesulfonic acid		239.25	14, 835						sl s hot aq; i eth
a195	5-Amino-6-hydroxy-2-naphthalenesulfonic acid		239.25							
a196	6-Amino-7-hydroxy-2-naphthalenesulfonic acid		239.25	14, 849			>300			
a197	2-Amino-3-hydroxy-pyridine	$H_2N(HO)(C_5H_3N)$	110.12	22², 408			172–174			0.77 aq; sl s alc
a198	4-Amino-2-hydroxy-pyrimidine		111.10	24, 314			>300			

			MW	Ref	d	d	mp	bp	Solubility
a199	1-Aminoindan		133.19	12, 1191		1.5613^{20}	1.5	97^{8mm}	sl s aq
a200	5-Aminoindan		133.19	12^1, 511			36	249^{745mm}	sl s aq
a201	5-Aminoindazole		133.15	25^2, 308			178		
a202	6-Aminoindazole		133.15	25, 317			206 d		
a203	2-Amino-5-iodobenzoic acid	$H_2N(I)C_6H_3COOH$	263.03	14, 373	1.038^{15}_4		221 d	94	sl s aq, PE; s alc
a204	DL-2-Amino-4-mercaptobutyric acid	$HSCH_2CH_2CH(NH_2)COOH$	135.19	4^3, 1647			232–233		
a205	Aminomethanesulfonic acid	$H_2NCH_2SO_3H$	111.12	1, 583			185 d		v s aq

2-Amino-2-(hydroxymethyl)-1,3-propanediol, t423
α-Amino-4-imidazolepropanoic acid, h83

Aminoiminomethanesulfinic acid, f30
N-(Aminoiminomethyl)-N-methylglycine, c277
2-Aminoisobutyric acid, a222

5-Aminoisophthalic acid, a116a
6-Amino-2,4-lutidine, a157
2-Amino-3-mercaptopropanoic acid, c370

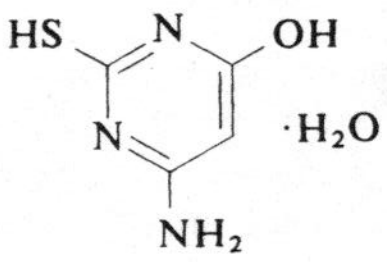

a191 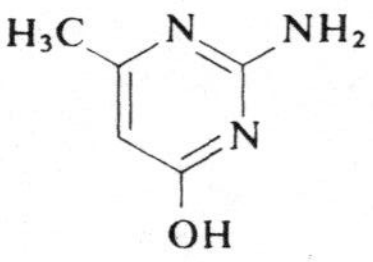 a192 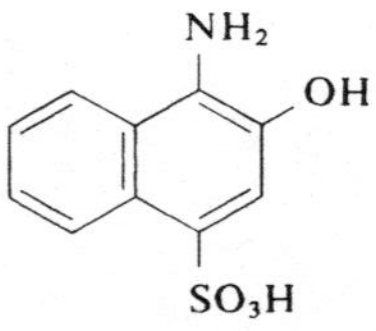 a193 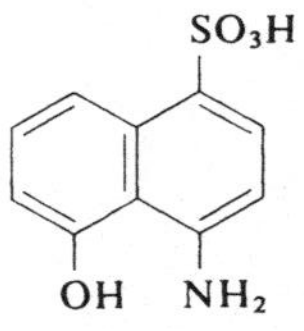 a194 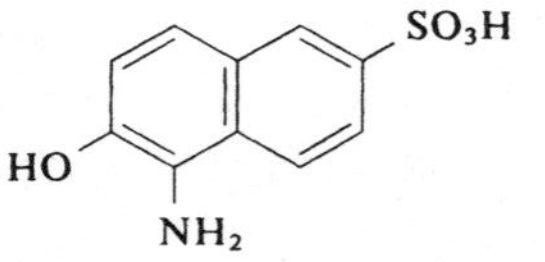 a195 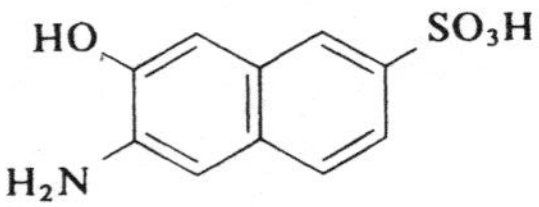 a196

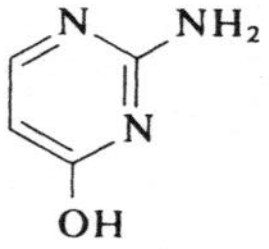

a198 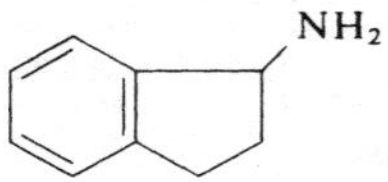 a199 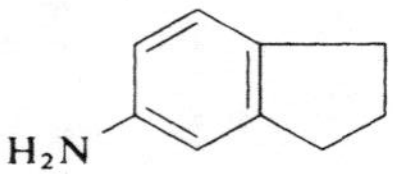 a200 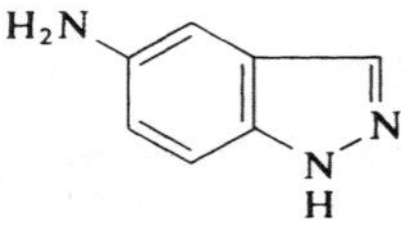 a201 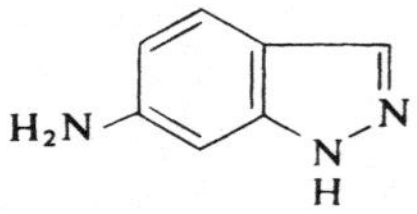 a202

TABLE 1.15 Physical Constants of Organic Compounds (*continued*)

No.	Name	Formula	Formula weight	Beilstein reference	Density	Refractive index	Melting point	Boiling point	Flash point	Solubility in 100 parts solvent
a206	3-Amino-4-methoxy-benzoic acid	$CH_3O(NH_2)C_6H_3COOH$	167.16	14^1, 657			241			
a207	2-Amino-6-methoxy-benzothiazole		180.23	27^2, 334			165–167			
a208	5-Amino-2-methoxy-pyridine	$CH_3O(NH_2)C_5H_3N$	124.14	22^2, 408		1.5745^{20}	31	90^{1mm}		
a209	1-(Aminomethyl)-adamantane		165.28		0.933	1.5137^{20}		$83–85^{0.3\,mm}$	92	
a210	4-(Aminomethyl)-benzenesulfonamide	$H_2NCH_2C_6H_4SO_2NH_2$	186.25				151–152			s dil alk, dil acid
a211	2-Amino-5-methyl-benzoic acid	$H_2N(CH_3)C_6H_3COOH$	151.17	14, 481			177 d			sl s aq; s alc, eth
a212	3-Amino-4-methyl-benzoic acid	$H_2N(CH_3)C_6H_3COOH$	151.17	14, 487			166			a aq
a213	DL-2-Amino-3-methyl-1-butanol	$(CH_3)_2CHCH(NH_2)$-CH_2OH	103.17			1.4543^{20}		77^{8mm}	83	
a214	L-2-Amino-3-methyl-1-butanol	$(CH_3)_2CHCH(NH_2)$-CH_2OH	103.17		0.926	1.4548^{20}		81^{8mm}	78	
a215	2-(Aminomethyl)-1-ethylpyrrolidine		128.22		0.887	1.4665^{20}		60^{16mm}		
a216	2-Amino-3-methyl-1-pentanol	$CH_3CH_2CH(CH_3)CH$-$(NH_2)CH_2OH$	117.19			1.4589^{20}	30	97^{14mm}		
a217	2-Amino-4-methyl-1-pentanol	$CH_3CH(CH_3)CH_2CH$-$(NH_2)CH_2OH$	117.19	4, 298	0.917	1.4511^{20}		200	90	
a218	4-Amino-3-methyl-phenol	$H_2N(CH_3)C_6H_3OH$	123.16	13, 593			179			
a219	4-(Aminomethyl)-piperidine		114.19			1.4900^{20}	25	200	78	
a220	2-Amino-2-methyl-1,3-propanediol	$HOCH_2C(CH_3)$-$(NH_2)CH_2OH$	105.14	4, 303			110	151^{10mm}		250 aq^{20}; s alc

			Mol. wt.	References	0.934^{20}_{20}	1.4480^{20}	30–31	165	67	misc aq; s alc, org solv
a221	2-Amino-2-methyl-1-propanol	$(CH_3)_2C(NH_2)CH_2OH$	89.14							
a222	2-Amino-2-methyl-propionic acid	$(CH_3)_2C(NH_2)COOH$	103.12	4, 414			335 sealed tube	280 subl		v s aq
a223	2-(Aminomethyl)-pyridine	$H_2NCH_2(C_5H_4N)$	108.14		1.049	1.5445^{20}		85^{12mm}		
a224	3-(Aminomethyl)-pyridine	$H_2NCH_2(C_5H_4N)$	108.14		1.062	1.5510^{20}	–21	74^{1mm}	100	
a225	2-Amino-3-methyl-pyridine	$H_2N(CH_3)(C_5H_3N)$	108.14	22[2], 342		1.5782^{20}	34	222		v s aq; s alc
a226	2-Amino-4-methyl-pyridine	$H_2N(CH_3)(C_5H_3N)$	108.14	22[2], 342			100	230		v s aq, alc, DMF
a227	2-Amino-6-methyl-pyridine	$H_2N(CH_3)(C_5H_3N)$	108.14	22[1], 633			45	209		v s aq
a228	2-Amino-4-methyl-pyrimidine		109.13	24, 84			160	subl		s hot aq; s alc
a229	2-Amino-4-methyl-thiazole		114.17	27, 159			45	232		v s aq, alc, eth
a230	2-Aminomethyl-3,5,5-trimethylcyclo-hexanol		171.29		0.969	1.4904^{20}	43–48	265	>112	

1-Amino-2-methoxyethane, m69
α-(Aminomethyl)benzyl alcohol, a262

3-Amino-α-methylbenzyl alcohol, a261
2-Amino-3-methylpentanoic acid, i79

2-Aminomethylthiophene, t158

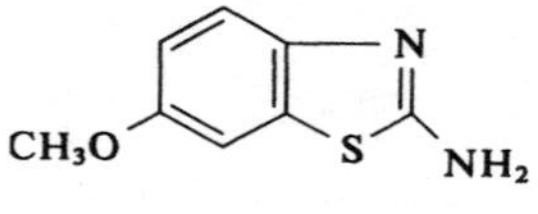

CH₃O ... N S NH₂   a207

CH₂NH₂ (adamantyl)   a209

CH₂NH₂ / CH₂CH₃ (pyrrolidine)   a215

CH₂NH₂ (piperidine, N–H)   a219

NH₂ ... N ... CH₃ (pyrimidine)   a228

H₃C ... N S NH₂ (thiazole)   a229

H₃C, H₃C, CH₃, CH₂NH₂, OH, H (cyclohexanol)   a230

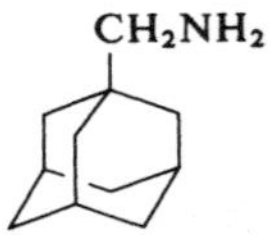

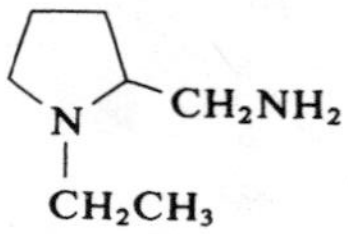

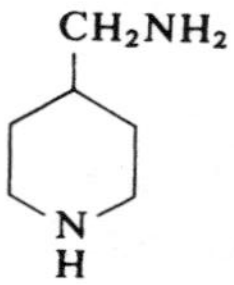

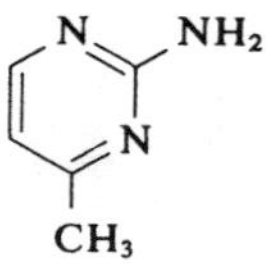

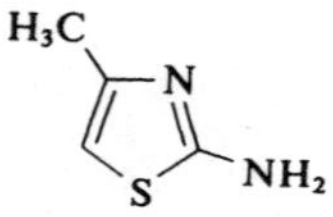

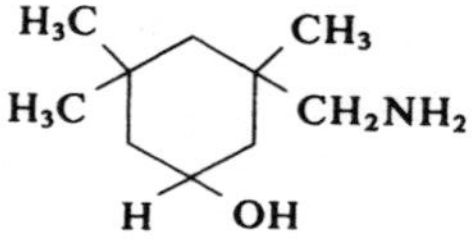

1.101

TABLE 1.15 Physical Constants of Organic Compounds (*continued*)

No.	Name	Formula	Formula weight	Beilstein reference	Density	Refractive index	Melting point	Boiling point	Flash point	Solubility in 100 parts solvent
a231	*N*-Aminomorpholine		102.14	27, 8	1.059	1.4772^{20}		168	58	
a232	2-Amino-1, 5-naphthalenedisulfonic acid		303.31	14, 786			>300			
a233	7-Amino-1,3-naphthalenedisulfonic acid		303.31	14, 784			>300			
a234	4-Amino-1-naphthalenesulfonic acid	$H_2NC_{10}H_6SO_3H$	223.26		1.670_4^{25}		d			0.031 aq; s dil alk
a235	4-Amino-1,8-naphthalimide		212.21	22^2, 452			360			
a236	3-Amino-2-naphthol	$H_2NC_{10}H_6OH$	159.19	13, 685			207			i aq; v s alc, eth
a237	2-Amino-4-nitrobenzoic acid	$H_2N(NO_2)C_6H_3COOH$	182.14	14, 374			270 d			
a238	2-Amino-5-nitrobenzonitrile	$H_2N(NO_2)C_6H_3CN$	163.14	14^2, 234			200–207			
a239	2-Amino-5-nitrobenzophenone	$C_6H_5COC_6H_3$-$(NH_2)NO_2$	242.23	14, 79			166–168			
a240	2-Amino-6-nitrobenzothiazole		195.20	27^2, 232			247–249			
a241	2-Amino-5-nitrobenzotrifluoride	$H_2N(NO_2)C_6H_3CF_3$	206.12				90–92			
a242	4-Amino-3-nitrobenzotrifluoride	$H_2N(NO_2)C_6H_3CF_3$	206.12				105–106			
a243	4-Amino-4′-nitrodiphenylsulfide	$O_2NC_6H_4SC_6H_4NH_2$	246.29	13, 534			142			
a244	2-Amino-4-nitrophenol	$O_2N(NH_2)C_6H_3OH$	154.13	13^2, 192			145			

No.	Name	Formula	Formula wt		mp	d / n_D / bp	Solubility
a245	4-Amino-2-nitrophenol	O2N(NH2)C6H3OH	154.13	13, 520	127		
a246	D-(−)-threo-2-Amino-1-(p-nitrophenyl)-1,3-propanediol	HOCH2CH(NH2)CH(OH)-C6H4NO2	212.21		163–165 163–165		
a247	2-Amino-5-(p-nitrophenylsulfonyl)-thiazole		285.30		222–226		
a248	2-Amino-5-nitropyridine	H2N(C5H3N)NO2	139.11	22^1, 631	188		sl s aq, bz, eth
a249	2-Amino-5-nitrothiazole		145.14		202 d		v sl s aq; 0.7 alc; 0.4 eth
a250	exo-2-Aminonorbornane		111.19	0.938	35	1.4807^{20} 49^{10mm}	

1-Aminonaphthalene, n17

a231

a232

a233

1-Amino-2-naphthol-4-sulfonic acid, a193

a235

1-Amino-2-naphthol-6-sulfonic acid, a195

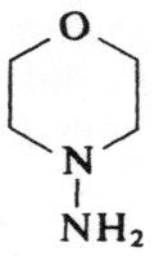

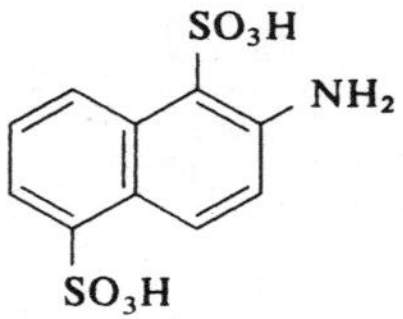

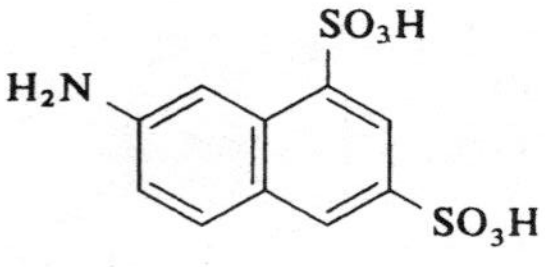

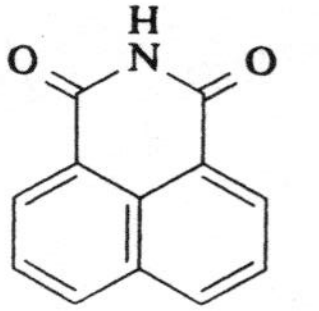

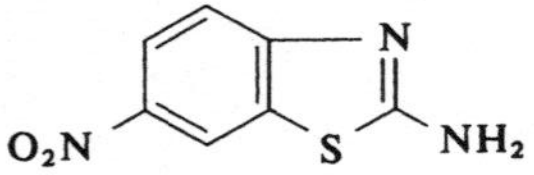

a240

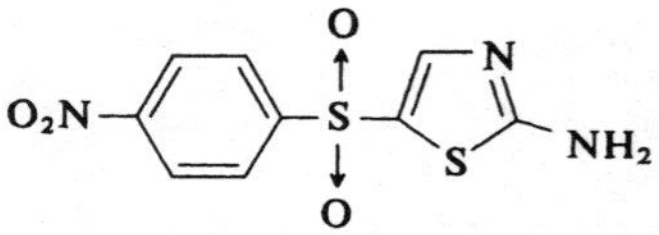

a247

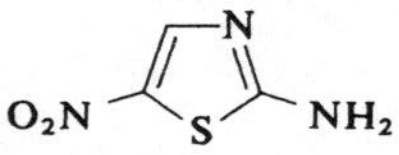

a249

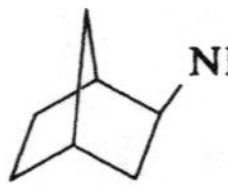

a250

TABLE 1.15 Physical Constants of Organic Compounds (*continued*)

No.	Name	Formula	Formula weight	Beilstein reference	Density	Refractive index	Melting point	Boiling point	Flash point	Solubility in 100 parts solvent
a251	2-Aminopentane	H(CH$_2$)$_3$CH(NH$_2$)CH$_3$	87.17	4, 177	0.739^{20}	1.4047^{20}		91–92	1	s aq, alc, eth, PE
a252	3-Aminopentane	C$_2$H$_5$CH(NH$_2$)C$_2$H$_5$	87.17	4, 178	0.749$^{20}_4$	1.4055^{20}		91		misc aq, alc, eth
a253	DL-2-Aminopentanoic acid	H(CH$_2$)$_3$CH(NH$_2$)COOH	117.15	4, 416			303	320 subl		5.5 aq^{18}, v sl s alc, chl, eth, PE
a254	5-Aminopentanoic acid	H$_2$N(CH$_2$)$_4$COOH	117.15	4, 418			158–161			v s aq; sl s alc; i eth
a255	5-Amino-1-pentanol	H$_2$N(CH$_2$)$_5$OH	103.17	4^1, 441		1.4615^{20}	37	122^{16mm}	65	
a256	2-Aminophenethyl alcohol	H$_2$NC$_6$H$_4$CH$_2$CH$_2$OH	137.18	13^3, 1679	1.045	1.5849^{20}		148^{4mm}	>112	
a257	2-Aminophenol	H$_2$NC$_6$H$_4$OH	109.13	13, 354			170–174			2 aq; 4.3 alc; v s eth; sl s bz
a258	3-Aminophenol	H$_2$NC$_6$H$_4$OH	109.13	13, 401			122–123	164^{11mm}		2.5 aq; v s alc, eth
a259	4-Aminophenol	H$_2$NC$_6$H$_4$OH	109.13	13, 427			190	284 d		0.65 aq; s alc, eth
a260	4′-Aminophenylacetonitrile	H$_2$NC$_6$H$_4$CH$_2$CN	132.17	14, 457			44	312		sl s hot aq; s alc
a261	1-(3-Aminophenyl)-ethanol	H$_2$NC$_6$H$_4$CH(CH$_3$)OH	137.18	13^3, 1654			68–71			
a262	2-Amino-1-phenyl-ethanol	C$_6$H$_5$CH(CH$_2$NH$_2$)OH	137.18	13^2, 361			56–57	160^{17mm}		v s aq; s alc
a263	1S,2S-(+)-2-Amino-1-phenyl-1,3-propanediol	C$_6$H$_5$CH(OH)CH(NH$_2$)-CH$_2$OH	167.21				109–113			
a264	L-2-Amino-3-phenyl-1-propanol	C$_6$H$_5$CH$_2$CH(NH$_2$)-CH$_2$OH	151.21	13^3, 1757			92–94			
a265	3-Amino-1-phenyl-2-pyrazolin-5-one		175.19				210–215			
a266	N-Aminophthalimide		162.15	20, 89	0.928	1.4750^{20}	200–202	146^{730mm}	36	
a267	N-Aminopiperidine		100.17	4, 301	1.175	1.4920^{20}		265^{739mm}	>112	
a268	3-Amino-1,2-propanediol	H$_2$NCH$_2$CH(OH)CH$_2$OH	91.11							

a269	DL-1-Amino-2-propanol	$CH_3CH(OH)CH_2NH_2$	75.11	4, 289	0.973	1.4483^{20}	−2	160	73	s aq, alc; i eth
a270	DL-2-Amino-1-propanol	$CH_3CH(NH_2)CH_2OH$	75.11	4[1], 432	0.943	1.4495^{20}		173–176		v s aq, alc, eth
a271	L-2-Amino-1-propanol	$CH_3CH(NH_2)CH_2OH$	75.11	4[1], 432	0.965	1.4495^{20}		176	62	v s aq, alc, eth
a272	3-Amino-1-propanol	$H_2NCH_2CH_2CH_2OH$	75.11	4, 288	0.982	1.4598^{20}	12	188	79	s aq, alc
a273	2-Amino-1-propene-1,1,3-tricarbonitrile	$NCC(CN){=}C(NH_2){-}CH_2CN$	132.13				171–173			s aq
a274	3-Aminopropionitrile	$H_2NCH_2CH_2CN$	70.09					185			
a275	3-Aminopropyl(diethoxy)methylsilane	$H_2N(CH_2)_3Si(CH_3)(OCH_2CH_3)_2$	191.4		0.916^{20}_{4}	1.427^{20}		$85\text{–}88^{8\,mm}$		
a276	N-(3-Aminopropyl)-iminodiethanol	$H_2N(CH_2)_3N{-}(CH_2CH_2OH)_2$	162.23		0.1071	1.4980^{20}		$170^{2\,mm}$	137	
a277	N-(3-Aminopropyl)-morpholine		144.22		0.9872^{20}_{20}	1.4761^{20}	−15	224	98	misc aq, alc, bz
a278	N-(3-Aminopropyl)-2-pyrrolidinone		142.20		1.014	1.5000^{20}		$120–123^{1\,mm}$	>112	

NH_2

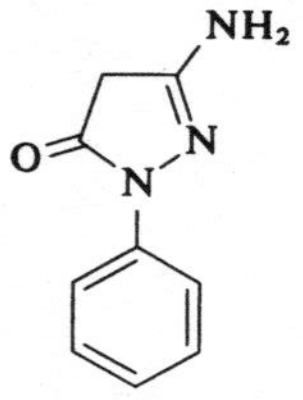

a265

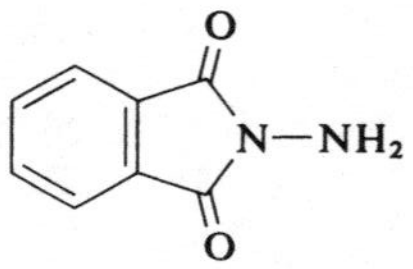

$N{-}NH_2$

a266

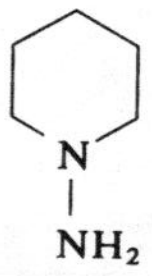

NH_2

a267

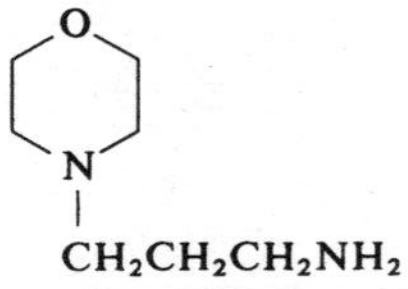

$CH_2CH_2CH_2NH_2$

a277

$CH_2CH_2CH_2NH_2$

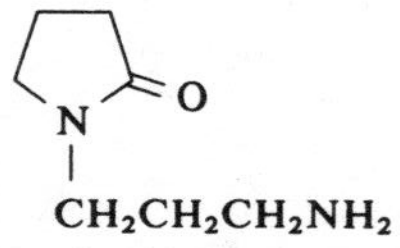

a278

TABLE 1.15 Physical Constants of Organic Compounds (*continued*)

No.	Name	Formula	Formula weight	Beilstein reference	Density	Refractive index	Melting point	Boiling point	Flash point	Solubility in 100 parts solvent
a279	3-Aminopropyl-triethoxysilane	$H_2N(CH_2)_3Si(OC_2H_5)_3$	221.37		0.9506_4^{20}	1.4225^{20}		217	96	s aq, alc, bz, eth
a280	3-Aminopropyl-trimethoxysilane	$H_2N(CH_2)_3Si(OCH_3)_3$	179.2		1.01_4^{25}	1.420^{25}		80^{8mm}	104	
a281	2-Aminopyridine	$(C_5H_4N)NH_2$	94.12	22, 428			58.1	210.6	92	s aq, alc, bz, eth
a282	3-Aminopyridine	$(C_5H_4N)NH_2$	94.12	22, 431			64	248		s aq, alc, bz, eth
a283	4-Aminopyridine	$(C_5H_4N)NH_2$	94.12	22, 433			155–158	273		s aq, alc; sl s bz, eth
a284	2-Aminopyrimidine		95.11	24, 80			123–126	subl		v s aq
a285	4-Aminoquinaldine		158.20	22, 453			169	333		sl s aq; v s alc, eth, acet; s hot bz
a286	4-Aminosalicyclic acid	$H_2NC_6H_3(OH)COOH$	153.14	14, 579			147 d			0.2 aq; 4.8 alc; s dil acid, alk
a287	5-Aminosalicyclic acid	$H_2NC_6H_3(OH)COOH$	153.14	14, 579			280 d			sl s aq, alc; s acid
a288	2-Amino-3-sulfopropionic acid	$HOOCCH(NH_2)-CH_2SO_3H$	187.17	4, 533			260 d			v s aq
a289	5-Amino-1,2,3,4-tetrazole hydrate		103.08	26, 403			204 d			
a290	5-Amino-1,3,4-thiadiazole-2-thiol		133.20	27, 674			235 d			
a291	2-Aminothiazole		100.14	27, 155			93			sl s aq, alc, eth
a292	2-Amino-2-thiazole		100.14	27, 136			91–93			
a293	2-Aminothiophenol	$H_2NC_6H_4SH$	125.19	13, 397		1.6405^{20}	26	234	79	1 aq^{12}; v s hot aq
a294	6-Amino-3-toluenesulfonic acid	$H_2NC_6H_3(CH_3)SO_3H$	187.22	14, 723			>300			
a295	3-Amino-1,2,4-triazole		84.08	26, 137			159			s aq, alc, chl

No.	Name	Formula	Formula wt.	Beilstein ref.	Density	n_D	m.p., °C	b.p., °C	Flash p., °C	Solubility
a296	5-Amino-2,2,4-trimethyl-1-cyclopentanemethylamine	H$_2$N(CH$_2$)$_{10}$COOH	156.27		0.901	1.4733^{20}		221	97	3.5 aq^{25}; s alc, CCl$_4$, eth, acids
a297	11-Aminoundecanoic acid		201.31				190–192			
a298	Aniline	C$_6$H$_5$NH$_2$	93.13	12, 59	1.0217^{20}	1.5855^{20}	−5.98	184.40	70	100 aq; v s alc
a299	Aniline hydrochloride	C$_6$H$_4$NH$_2$·HCl	129.59		1.222		198		193	
a300	2-Anilinoethanol	C$_6$H$_5$NHCH$_2$CH$_2$OH	137.18	12, 182	1.085	1.5793^{20}		150–152$^{10\text{mm}}$	>112	sl s aq; v s alc, chl, eth
a301	3-Anilinopropionitrile	C$_6$H$_5$NHCH$_2$CH$_2$CN	146.19				52–53			

6-Aminopurine, a69
2-Amino-3-pyridinol, a197
Aminopyrimidinediols, a154, a155
2-Aminosuccinamic acid, a313
Aminosuccinic acid, a314
6-Amino-2-thiouracil, a191
α-Amino-p-toluenesulfonamide, a210
2-Amino-1,1,3-tricyanopropene, a273
1-Aminotricyclo[3.3.1^{3,7}]decane, a66

Aminouracil, a155
2-Aminovaleric acid, a253
5-Aminovaleric acid, a254
Amyl compounds, see Pentyl
Amyl alcohol, p37
act-Amyl alcohol, m153
sec-Amyl alcohol, p38
tert-Amyl alcohol, m154
tert-Amylamine, d601

Amyl bromides, b322, b323
Amyl chloride, c191
Amyl iodide, i47
Amyl mercaptan, p35
Amyl methyl ketone, h15
Anethole, m97
Angelic acid, m162
Anilinesulfonic acids, a118, a119, a120
Aniline-2,5-disulfonic acid, a117

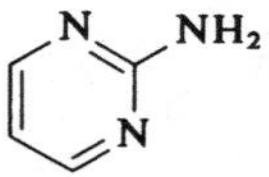

a284 — NH$_2$; a285 — NH$_2$, CH$_3$; a289 — H$_2$N, H·H$_2$O; a290 — H$_2$N, SH; a291 — NH$_2$; a292 — NH$_2$; a295 — NH$_2$; a296 — H$_3$C, CH$_3$, CH$_3$, CH$_2$NH$_2$, H$_2$N

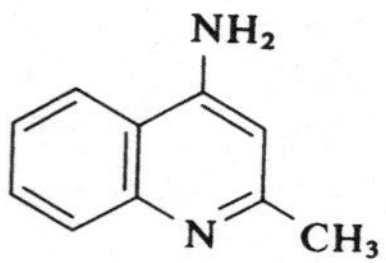

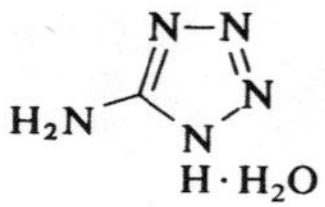

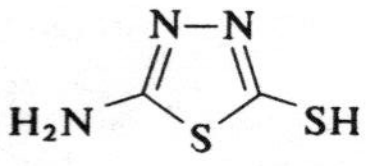

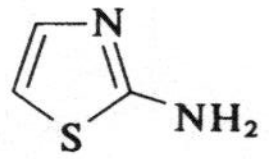

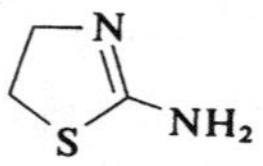

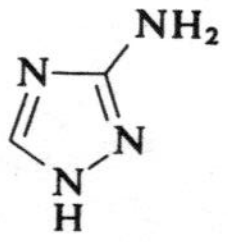

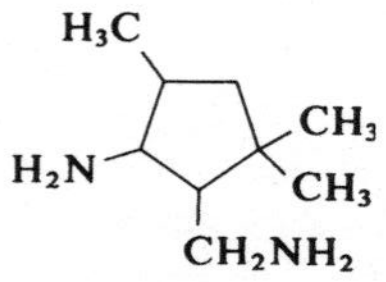

TABLE 1.15 Physical Constants of Organic Compounds (*continued*)

No.	Name	Formula	Formula weight	Beilstein reference	Density	Refractive index	Melting point	Boiling point	Flash point	Solubility in 100 parts solvent
a302	1-(o-Anisidino)-1,3-butanedione	$CH_3OC_6H_4NHCOCH_2COCH_3$	207.23	13^1, 117			84–85			
a303	1-(p-Anisidino)-1,3-butanedione	$CH_3OC_6H_4NHCOCH_2COCH_3$	207.23	13^1, 177			115–117			
a304	Anthracene		178.23	5, 657	1.25^{27}_4		216.3	340		i aq; 1.5 alc; 1.6 bz; 1.2 chl; 3.1 CS$_2$
a305	9,10-Anthracene-dione		208.22	7, 781	1.43^{20}_4		286	377	185	i aq; 0.44 alc; 0.26 bz; 0.61 chl; 0.11 eth
a306	9,10-Anthraquinone-1,5-disulfonic acid disodium salt		412.31	11, 340			>300			s aq
a307	9,10-Anthraquinone-2,6-disulfonic acid disodium salt		412.31	11, 342			>325			s aq
a308	9,10-Anthraquinone-2-sulfonic acid Na salt		310.26							
a309	Antipyrine		188.23	24, 27	1.088^{113}_4		114	319^{174mm}		100 aq; 77 alc; 100 chl; 2.3 eth
a310	L-(+)-Arabinose		150.13	31, 32			160–163 223 d			100 aq
a311	L-(+)-Arginine	$H_2NC(=NH)NH(CH_2)_3$-$CH(NH_2)COOH$	174.20	4, 420						17.6 aq; sl s alc
a312	L-(+)-Ascorbic acid		176.12				190–192 d			100 aq; 3.3 alc
a313	L-(+)-Asparagine hydrate	$H_2NCOCH_2CH(NH_2)$-$COOH \cdot H_2O$	150.14	4, 484			233–235			3.6 aq^{28}; s alk acids; i alc, bz, eth

| a314 | L-(+)-Aspartic acid | HOOCCH$_2$CH(NH$_2$)COOH | 133.10 | 4, 472 | | 0.45 aq; i alc, eth |
| a315 | Atropine | | 289.38 | 21, 27 | 270 sealed tube 114–116 | 0.22 aq; s bz, dil acid |

Anisaldehydes, m45, m46
Anisamide, m47
Anisic acids, m50, m51, m52
Anisidines, m42, m43, m44
Anisole, m48
p-Anisoyl chloride, m53

p-Anisyl alcohol, m54
Anthraflavic acid, d374
Anthranilamide, a114
Anthranilic acid, a121
Anthranionitrile, a124
9,10-Anthraquinone, a305

APDC, p271
Araboascorbic acid, i59
Aspirin, a56
Arsanilic acids, a115, a116

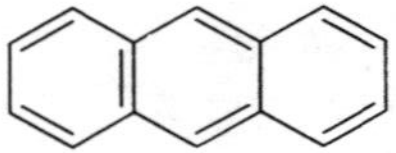

a304

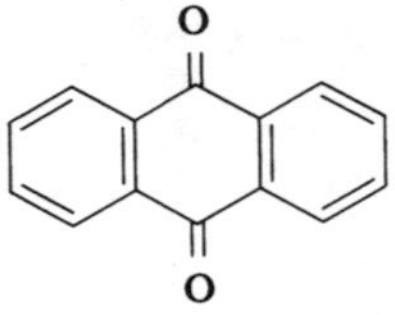

a305

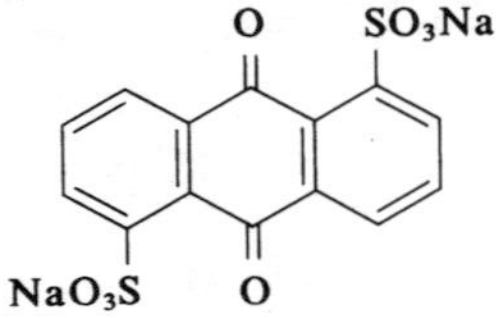

a306

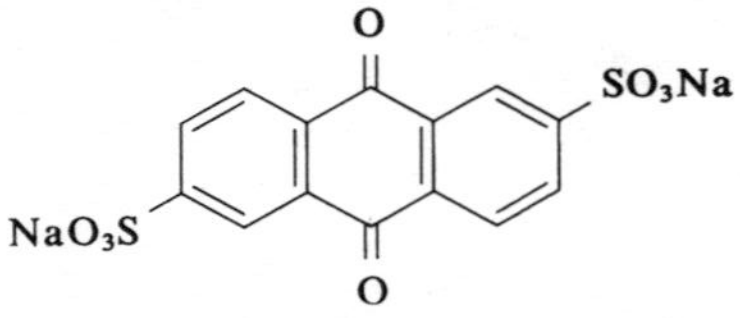

a307

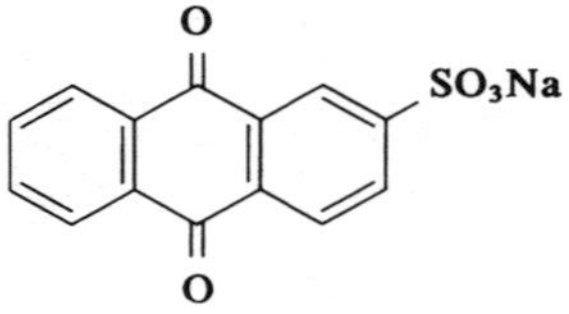

a308

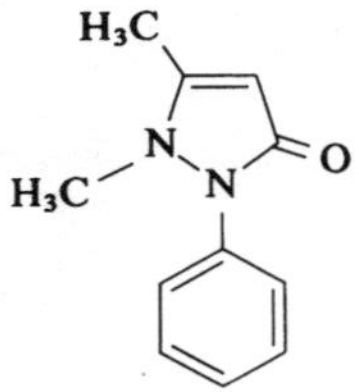

a309

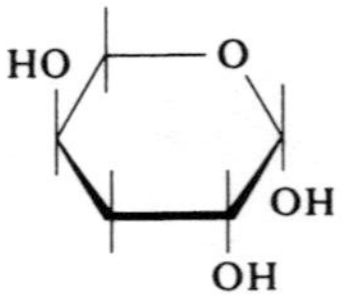

a310

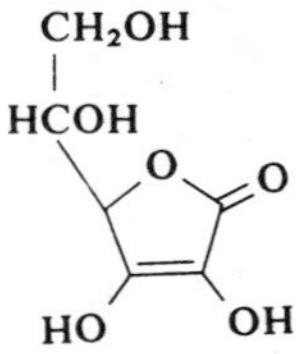

a312

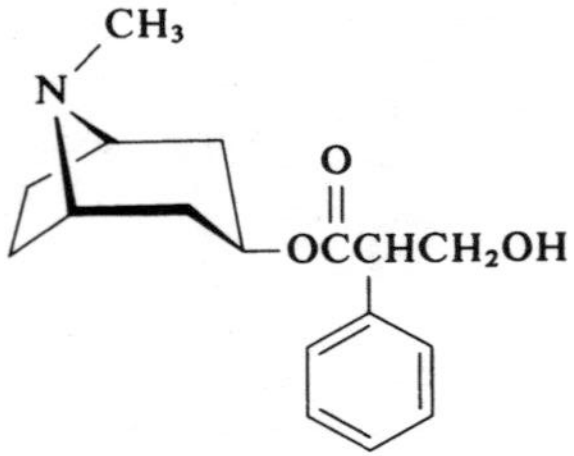

a315

TABLE 1.15 Physical Constants of Organic Compounds (*continued*)

No.	Name	Formula	Formula weight	Beilstein reference	Density	Refractive index	Melting point	Boiling point	Flash point	Solubility in 100 parts solvent
a316	Aurintricarboxylic acid, triammonium salt		473.44	10^2, 775			225 d			v s aq
a317	2-Azacyclooctanone		127.19	21, 242			35–38			
a318	2-Azacyclotridecanone		197.32				150–153	148^{10mm}		
a319	Azidotrimethylsilane	$(CH_3)_3SiN_3$	115.21		0.868	1.4142^{20}	−95	95–96	23	
a320	Azidotriphenylsilane	$(C_6H_5)_3SiN_3$	301.4				83–84	$100^{0.01mm}$	67	
a321	1-Aziridineethanol	$(CH_2)_2{=}NCH_2CH_2OH$	87.12	16, 8	1.088	1.4560^{20}		168		
a322	cis-Azobenzene	$C_6H_5N{=}NC_6H_5$	182.23		1.20		68.3	293		i aq; s alc, eth, HOAc
a323	2,2′-Azobis(2-methyl)-propionitrile	$(CH_3)_2C(CN)N{=}N{-}C(CN)(CH_3)_2$	164.21	4, 563				107 d		2 EtOH; 5 MeOH; can explode in acetone
a324	Azodicarbonamide	$H_2NCON{=}NCONH_2$	116.08	3, 123			225 d			s hot aq, dil acid
a325	4,4′-Azoxydianisole	$CH_3OC_6H_4N{=}N({\rightarrow}O){-}C_6H_4OCH_3$	258.28	16, 637			120			
a326	Azulene		128.17	5^2, 432			100.5	250		
b1	Barbituric acid		128.09	24, 467			248–252 d			s hot aq, dil acid
b2	Basic fuchsin		337.86	13, 765	1.22		d 186			0.3 aq; s alc, acids
b3	Benzaldehyde	C_6H_5CHO	106.12	7, 174	1.0447^{20}	1.5455^{20}	−26	178.9	62	0.3 aq; misc alc, eth
b4	Benzamide	$C_6H_5CONH_2$	121.14	9, 195	1.341^4		127.2	288		1.3 aq; 17 alc; 30 pyr
b5	Benzanilide	$C_6H_5CONHC_6H_5$	197.24	12, 262	1.315		163.1	117^{10mm}		i aq; 1.7 alc; sl s eth
b6	1,2-Benzanthracene		228.29	5, 718			155–157	437.6		sl s hot alc; s most other org solv

							subl	
b7	2,3-Benzanthracene		228.29	5², 628	1.35		341	
b8	7H-Benz[de]-anthracen-7-one		230.27	7, 518			170	
b8a	Benzene	C_6H_6	78.11	5, 179	0.8737^{25}	1.4979^{25}	5.53	80.10	−11	sl s most org solv 1.6 bz; 0.5 HOAc
b9	Benzene-d	C_6H_5D	79.12			1.4980^{20}		80	−11	0.17 aq; s most org solv
b10	Benzene-d_6	C_6D_6	84.16		0.95	1.4978^{20}		79.1	−11	

Azacyclopropane, e131
Azelaic acid, n95
Azelonitrile, n94
Aziridine, e131
Azobis(isobutyronitrile), a323

4,4'-Azoxyanisole, a325
Barbitol, d280
Behenic acid, d716
Behenyl alcohol, d717
Benzalacetone, p96

Benzal bromide, d102
Benzalphthalide, b101
Benzanthrone, b8
Benzeneacetaldehyde, p76a

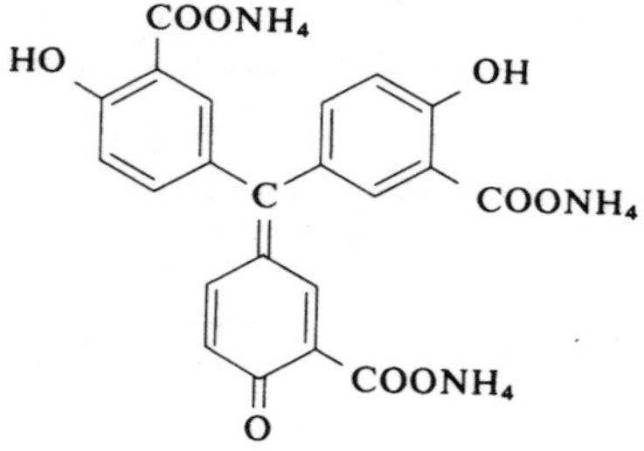

a316

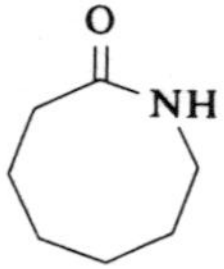

a317

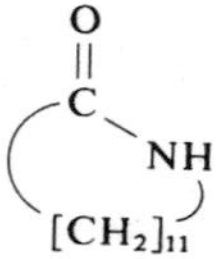

a318

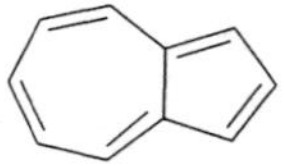

a326

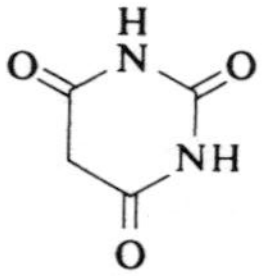

b1

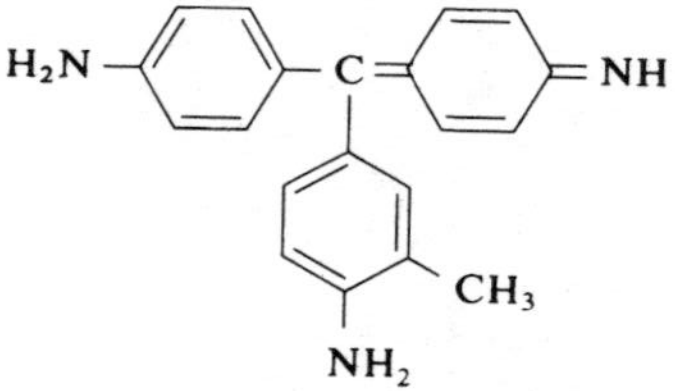

b2

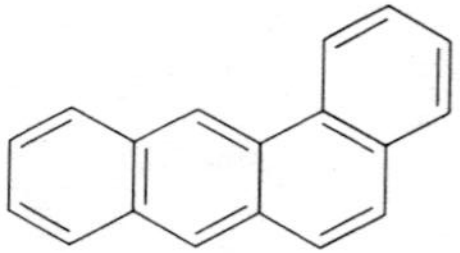

b6

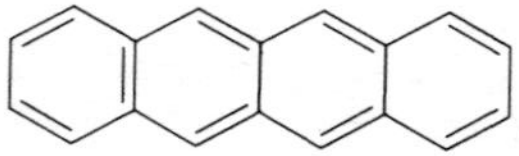

b7

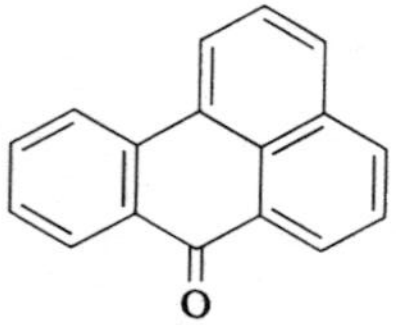

b8

1.111

TABLE 1.15 Physical Constants of Organic Compounds (*continued*)

No.	Name	Formula	Formula weight	Beilstein reference	Density	Refractive index	Melting point	Boiling point	Flash point	Solubility in 100 parts solvent
b11	Benzenearsonic acid	$C_6H_5AsO(OH)_2$	202.04	16, 868	1.760^{25}		163 d			2.5 aq; 2 alc
b12	Benzeneboronic acid	$C_6H_5B(OH)_2$	121.93	16, 920			217 to the anhydride	$-H_2O$ on standing in air		2.6 aq; 1.8 alc; 43 eth; s bz
b13	1,4-Benzenedicarb-aldehyde	$C_6H_4(CHO)_2$	134.13	7, 675			114	248		i aq; 6 bz; 17 acet; 2 eth; 14 diox; 46 MeOH
b14	1,3-Benzenedicarbonyl dichloride	$C_6H_4(COCl)_2$	203.02	9, 834			43–44	276	180	73 bz; 62 CCl_4
b15	1,4-Benzenedicarbonyl dichloride	$C_6H_4(COCl)_2$	203.02	9, 844			81	266	180	37 bz; 9 CCl_4
b16	1,3-Benzenedicarboxylic acid	$C_6H_4(COOH)_2$	166.13	9, 832			345–348	subl		0.012 aq; v s alc, HOAc; i bz, PE
b17	1,4-Benzenedicarboxylic acid	$C_6H_4(COOH)_2$	166.13	9, 841			subl without melting	subl		v sl s aq, chl, eth; sl s alc; s alk
b18	1,4-Benzenedimethanol	$C_6H_4(CH_2OH)_2$	138.17	6, 919	1.100^{17}		115	143^{1mm}	188	v s aq, alc, eth
b19	Benzenehexacarboxylic acid	$C_6(COOH)_6$	342.17	9, 1008			286 d			v s aq, alc
b20	Benzenesulfinic acid	$C_6H_5S(=O)OH$	142.16	11, 2			85	100 d		sl s aq; s alc, bz, eth
b21	Benzenesulfonamide	$C_6H_5SO_2NH_2$	157.19	11, 39			152			i aq; sl s alc; s eth
b22	Benzenesulfonic acid	$C_6H_5SO_2OH$	158.18	11, 26	1.3842^{15}_{15}		50–51			v s aq, alc; sl s bz
b23	Benzenesulfonyl chloride	$C_6H_5SO_2Cl$	176.62	11, 34	1.3286^{20}_{4}	1.5518	14.5	177^{100mm}	>112	i aq; s alc, eth
b24	Benzenesulfonyl fluoride	$C_6H_5SO_2F$	160.16	11^2, 23		1.4932^{18}		203–204		s alc, eth

No.	Name	Formula			m.p.	b.p.	Solubility
b25	Benzenesulfonyl hydrazide	$C_6H_5SO_2NHNH_2$	172.21	11, 52	101–103		flammable solid
b26	1,2,4,5-Benzenetetra-carboxylic acid	$C_6H_2(COOH)_4$	254.15	9, 997	276		1.5 aq; v s alc
b27	1,2,4,5-Benzenetetra-carboxylic anhydride		218.12	19, 196	283–286	397–400	
b28	1,2,3-Benzenetricar-boxylic acid dihydrate	$C_6H_3(COOH)_3 \cdot 2H_2O$	246.18	9, 976	192 d		sl s aq; v s eth
b29	1,2,4-Benzenetricar-boxylic acid	$C_6H_3(COOH)_3$	210.14	9, 977	321 d		2.1 aq; 25.3 alc; 7.9 acet; v s eth
b30	1,3,5-Benzenetricar-boxylic acid	$C_6H_3(COOH)_3$	210.14	9, 978	>330		sl s aq; v s alc; s eth
b31	1,2,4-Benzenetricar-boxylic anhydride		192.13	18, 468	161–164	245^{14mm}	50 acet; 22 EtAc

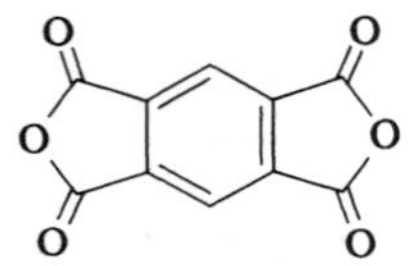

b27

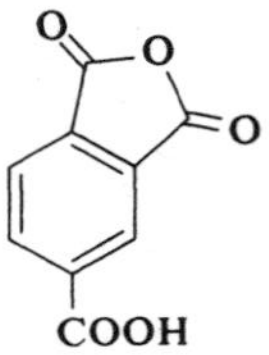

b31

TABLE 1.15 Physical Constants of Organic Compounds (*continued*)

No.	Name	Formula	Formula weight	Beilstein reference	Density	Refractive index	Melting point	Boiling point	Flash point	Solubility in 100 parts solvent
b32	1,3,5-Benzenetricarboxylic trichloride	$C_6H_3(COCl)_3$	265.48				35–36			
b33	Benzethonium chloride	$(CH_3)_3CCH_2C(CH_3)_2C_6H_4OCH_2CH_2OCH_2CH_2N^+(CH_3)_2CH_2C_6H_5Cl^-$	448.10				164–166			v s aq; s alc, acet
b34	Benzil	$C_6H_5COCOC_6H_5$	210.23	7, 747	1.23^{15}		94.9	346		i aq; s alc, eth
b35	Benzil-α-dioxime	$C_6H_5C(=NOH)C(=NOH)C_6H_5$		10, 342						s alk
b36	Benzilic acid	$(C_6H_5)_2C(OH)COOH$	228.25	7^1, 394			153			sl s aq; v s alc, eth
b37	Benzil monohydrazone	$C_6H_5C(=NNH_2)COC_6H_5$	224.26	23, 131			150–152			sl s aq, eth; v s alc
b38	Benzimidazole		118.14				170.5	>360		
b39	Benzo-15-crown-5		268.3				76–78			
b40	7,8-Benzo-1,3-diaza-spiro-[4.5]decane-2,4-dione		216.24				268–270			
b41	1,4-Benzodioxan		136.15	17, 54	1.142	1.5485^{20}		103^{6mm}	87	i aq; misc bz, eth, PE
b42	2,3-Benzofuran		118.14		1.072	1.5660^{20}	<–18	175		
b43	Benzofurazan-1-oxide		136.11	27^1, 740	1.080		69–71			
b44	Benzoic acid	C_6H_5COOH	122.13	9, 92			122.4	132.5^{10mm}	121	0.29 aq; 43 alc; 10 bz; 22 chl; 33 eth; 33 acet
b45	Benzoic anhydride	$(C_6H_5CO)_2O$	226.23	9, 164	1.199		39–40	360		i aq; s alc, acet, chl, bz, HOAc
b46	DL-Benzoin	$C_6H_5COCHOHC_6H_5$	212.25	8, 165	1.3100_4^{20}		134–136	344		s acet; 20 pyr
b47	Benzoin ethyl ether	$C_6H_5CH(OC_2H_5)COC_6H_5$	240.30	8, 174	1.1016_4^{17}	1.5727^{17}	61	195^{20mm}		s alc, bz, eth

	Name	Formula	Formula wt	Beilstein	Density	n_D	m.p., °C	b.p., °C	Solubility
b48	Benzoin isobutyl ether	$C_6H_5CH[OCH_2CH(CH_3)_2]COC_6H_5$	268.36		0.985	1.5485^{20}	85	$133^{0.5mm}$	v s alc, bz, eth
b49	Benzoin methyl ether	$C_6H_5CH(OCH_3)COC_6H_5$	226.28	8, 174	1.1278^{14}_4		48	189^{15mm}	sl s aq; s alc, NH$_4$OH
b50	α-Benzoinoxime	$C_6H_5CH(OH)C(=NOH)C_6H_5$	227.26	8, 175			151–152		0.2 aq; misc alc, bz, chl, eth
b51	Benzonitrile	C_6H_5CN	103.12	9, 275	1.0006^{25}	1.5257^{25}	−12.75	191.1; 71	i aq; s alc, eth
b52	Benzo[def]phenanthrene		202.26	5, 693	1.271^{23}		156	404	
b53	Benzophenone	$C_6H_5COC_6H_5$	182.22	7, 411	1.1108^{15}_4		48.1	305	i aq; 13.3 alc; 17 eth
b54	Benzophenone hydrazone	$C_6H_5C(=NNH_2)C_6H_5$	196.25	7, 417			98	230^{55mm}	

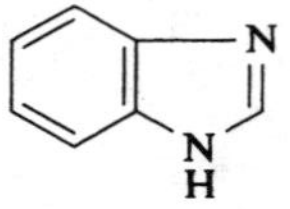

b38 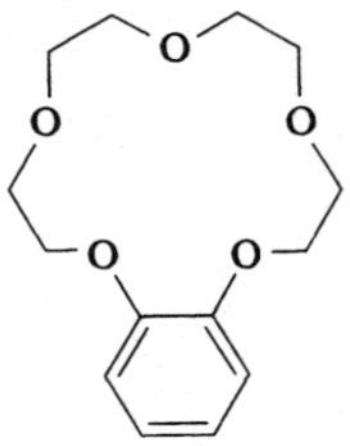   b39 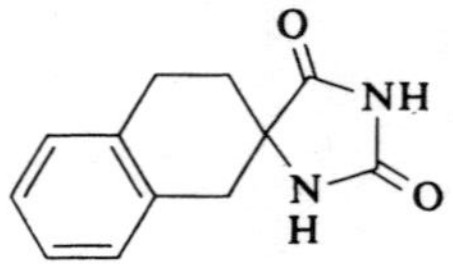   b40 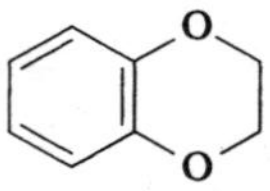   b41 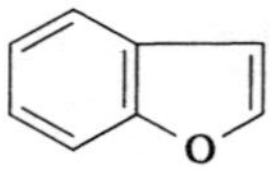   b42 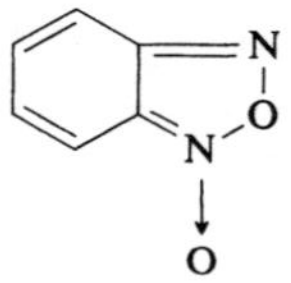   b43 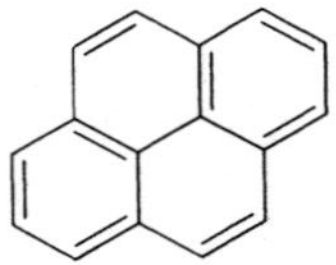   b52

TABLE 1.15 Physical Constants of Organic Compounds (*continued*)

No.	Name	Formula	Formula weight	Beilstein reference	Density	Refractive index	Melting point	Boiling point	Flash point	Solubility in 100 parts solvent
b55	3,3',4,4'-Benzophenonetetracarboxylic dianhydride		322.23				215–217			
b56	1-Benzopyran-4(4H)-one		146.15	17, 327			55–57	495		i aq; s bz; sl s alc
b57	1,2-Benzo[a]pyrene		252.32				179.3			i aq
b58	4,5-Benzo[e]pyrene		252.32				182			
b59	1,4-Benzoquinone	$O\!=\!C_6H_4\!=\!O$	108.10	7, 609	1.318_4^{20}		115.7			sl s aq; s alc, eth, hot bz, alk (with d)
b60	Benzothiazole		135.19		1.246_4^{20}	1.6379^{20}	2	231	>112	sl s aq; v s alc, CS$_2$
b61	Benzo[b]thiophene		134.20	17, 59	1.1937^{40}	1.6302^{40}	31.32	221		s alc, bz, chl, eth
b62	1,2,3-Benzotriazole		119.13	26, 38	1.238	1.6420^{20}	98.5	204^{15mm}		sl s aq; s alc, bz, chl
b63	Benzoxazole		119.12	27, 42		1.5594	30	182	58	sl s aq
b64	1-Benzoylacetone	$C_6H_5COCH_2COCH_3$	162.19	7, 680	1.090_{60}^{60}		60	260 sl d		sl s aq; v s alc, eth
b65	2-Benzoylbenzoic acid	$C_6H_5COC_6H_4COOH$	226.23	10, 747			129	265		sl s aq; v s alc, eth
b66	Benzoyl bromide	C_6H_5COBr	185.03	9, 195	1.5467^{20}			218–219	90	d aq, alc; misc eth
b67	Benzoyl chloride	C_6H_5COCl	140.57	9, 182	1.2114_4^{20}	1.5525^{20}	−1.0	197.2	68	d aq, alc; misc bz, CS$_2$, eth
b68	Benzoyl cyanide	C_6H_5COCN	131.13	10, 659			32	206		i aq
b69	Benzoyl fluoride	C_6H_5COF	124.11	9, 181	1.140	1.4960^{20}	−28	161	48	d hot aq; v s alc, eth
b70	Benzoylformic acid	$C_6H_5COCOOH$	150.13	10, 654			69			0.4 aq; 0.1 chl; 0.25 eth; sl s alc; i bz, PE
b71	N-Benzoylglycine	$C_6H_5CONHCH_2COOH$	179.18	9, 225			178–179			
b72	Benzoylhydrazine	$C_6H_5CONHNH_2$	136.15	9, 319			117			

No.	Name	Formula	Mol. wt.	Beilstein ref.	Density	n_D	mp (°C)	bp (°C)	Flash P (°C)	Solubility
b73	3-Benzoylpropionic acid	$C_6H_5COCH_2CH_2COOH$	178.19	10, 696			116			sl s aq; s alc
b74	2-Benzoylpyridine	$C_6H_5CO(C_5H_4N)$	183.21	21,330			44	317	150	s alc, bz, eth
b75	3-Benzoylpyridine	$C_6H_5CO(C_5H_4N)$	183.21	21,331			40	307	150	s alc, bz, eth
b76	4-Benzoylpyridine	$C_6H_5CO(C_5H_4N)$	183.21	21,331			71	315	150	sl s aq; misc alc, eth
b77	Benzyl acetate	$CH_3COOCH_2C_6H_5$	150.18	6, 435	1.0515^{25}	1.5232^{20}	−51.5	215.5	102	0.08 aq; misc alc, eth
b78	Benzyl alcohol	$C_6H_5CH_2OH$	108.13	6, 428	1.0413^{25}	1.5371^{25}	−15.3	205.45	100	misc aq, alc, eth
b79	Benzylamine	$C_6H_5CH_2NH_2$	107.16	12, 1013	0.9814^{19}	1.5424^{20}	10	185	60	
b80	2-Benzylaminoethanol	$C_6H_5CH_2NHCH_2CH_2OH$	151.21	12, 1040	1.065	1.5435^{20}		156^{12mm}	>112	
b81	(3-Benzylamino)-propionitrile	$C_6H_5CH_2NHCH_2CH_2CN$	160.22			1.5308^{20}				

Benzoresorcinol, d386
2-Benzothiazolethiol, m17
Benzotrichloride, t248

Benzotrifluoride, b4
Benzoylamide, b4
Benzoylbenzene, b53

Benzoyl peroxide, d54
1,2-Benzphenanthrene, b6
Benzylaniline, p92

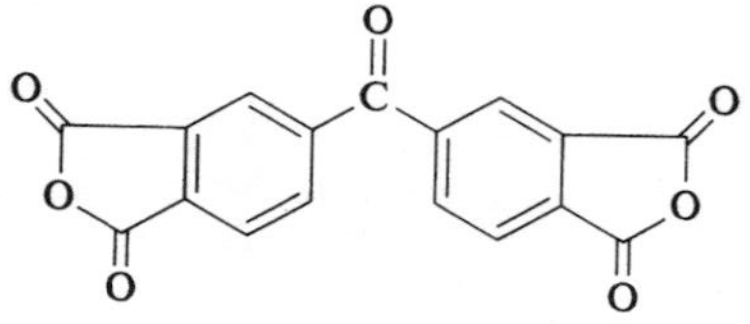

b55

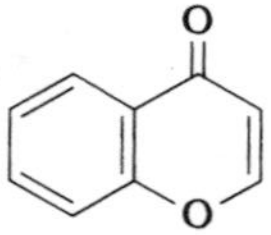

b56

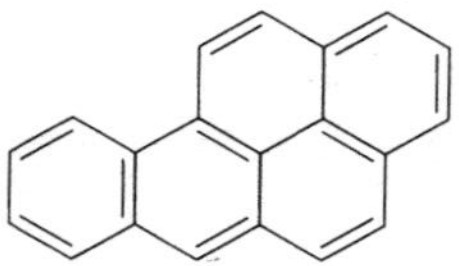

b57

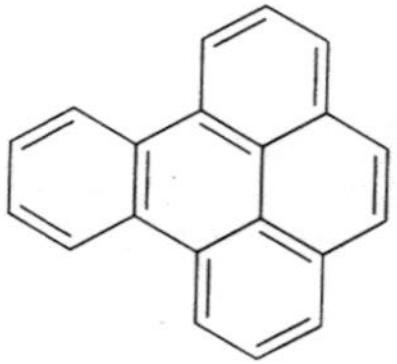

b58

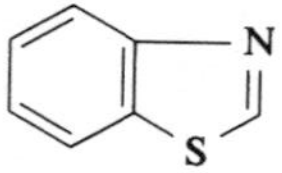

b60

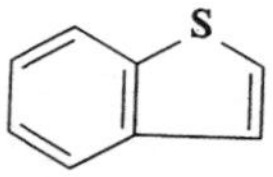

b61

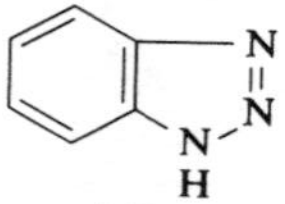

b62

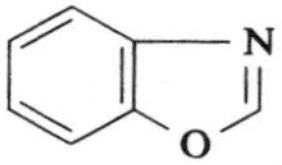

b63

TABLE 1.15 Physical Constants of Organic Compounds (*continued*)

No.	Name	Formula	Formula weight	Beilstein reference	Density	Refractive index	Melting point	Boiling point	Flash point	Solubility in 100 parts solvent
b82	N-Benzylbenzamide	$C_6H_5CONHCH_2C_6H_5$	211.26	9, 121			106			
b83	Benzyl benzoate	$C_6H_5COOCH_2C_6H_5$	212.25		1.118_4^{25}	1.5681^{21}	19.4	323.5	147	i aq; misc alc, chl, eth
b84	2-Benzylbenzoic acid	$C_6H_5CH_2C_6H_4COOH$	212.24	9^2, 471			110–113			sl s aq; s alc, bz, chl, eth
b85	Benzyl bromide	$C_6H_5CH_2Br$	171.04	5, 306	1.438_0^{22}	1.5752^{20}	–3.9	198–199	86	sl d aq
b86	Benzyl-*tert*-butanol	$C_6H_5CH_2CH_2C(CH_3)_2OH$	164.25	6, 548		1.5090^{20}	33	144^{85mm}	>112	
b87	Benzyl butyl-1,2-phthalate	$C_6H_5CH_2OOCC_6H_4\text{-}COOC_4H_9$	312.37		1.119_5^{25}				218	i aq; v s alc; sl s eth
b88	Benzyl carbamate	$C_6H_5CH_2OCONH_2$	151.17	6, 437			87–89	220 d		i aq; misc alc, chl, eth
b89	Benzyl chloride	$C_6H_5CH_2Cl$	126.59	5, 292	1.0993^{20}	1.5391^{20}	–43 to –48	179	73	d aq; s eth
b90	Benzyl chloroformate	$C_6H_5CH_2OC(O)Cl$	170.60	6, 437	1.195	1.5190^{20}		103^{20mm}	91	
b91	Benzyl chlorothiol formate	$C_6H_5CH_2S(COCl)$	186.5		1.237_4^{30}	1.5711^{30}		$80^{0.13mm}$	118	
b92	S-Benzyl-L-cysteine	$C_6H_5CH_2SCH_2CH(NH_2)COOH$	211.28	6, 465			214 d		>112	
b93	Benzyl diethyl phosphite	$C_6H_5CH_2P(O)(OC_2H_5)_2$	228.23	12^3, 2212	1.076	1.4930^{20}		110^{2mm}		
b94	Benzyldimethylstearylammonium chloride	$C_6H_5CH_2N[(CH_2)_{17}CH_3](CH_3)_2Cl \cdot H_2O$	442.18				67–69			
b95	Benzyl ethyl ether	$C_6H_5OC_2H_5$	136.20		0.9478^{20}	1.4958^{20}		185.0		i aq; misc alc, eth
b96	N-Benzylformamide	$C_6H_5CH_2NHCHO$	135.17	12, 1043			60–61			
b97	Benzyl formate	$C_6H_5CH_2OOCH$	136.15		1.081_4^{20}			203		i aq; s alc; misc eth
b98	O-Benzylhydroxylamine	$C_6H_5CH_2ONH_2$	123.16	6, 440				119^{30mm}		

b99	Benzylidenemalono-nitrile	$C_6H_5CH{=}C(CN)_2$	154.17	9, 895			83–85			
b100	N-Benzylidenemethyl-amine	$C_6H_5CH{=}NCH_3$	119.17	7, 213	0.967	1.5526^{20}		80^{18mm}	>112	
b101	3-Benzylidene-phthalide		222.24	17, 376			102			
b102	2-Benzyl-2-imid-azoline HCl		196.68				174			v s aq, alc; s chl; v sl s eth, EtAc
b103	Benzylmethylamine	$C_6H_5CH_2NHCH_3$	138.23	12, 1019	0.939	1.5224^{20}		184.189	77	
b104	3-(N-Benzyl-N-methylamino)-1,2-propanediol	$C_6H_5CH_2N(CH_3)\text{-}CH_2CH(OH)CH_2OH$	195.26		1.084	1.5341^{20}		206^{30mm}	>112	
b105	Benzyl methyl sulfide	$C_6H_5CH_2SCH_3$	138.23	6, 453	1.015	1.5620^{20}		195–198	73	
b106	3-Benzyloxyaniline	$C_6H_5CH_2OC_6H_4NH_2$	199.25	13, 404			63–67			
b107	3-Benzyloxybenz-aldehyde	$C_6H_5CH_2OC_6H_4CHO$	212.25	8, 73			56–58			
b108	4-Benzyloxybenz-aldehyde	$C_6H_5CH_2OC_6H_4CHO$	212.25	8, 73			73–74			

Benzyl cyanide, p80
Benzyl disulfide, d57
N-Benzylethanolamine, b80

Benzyl ether, d58
Benzylideneacetone, p96
Benzylideneacetophenone, d686

Benzyl mercaptan, p128
Benzyl methyl ketone, p144
Benzyloxyamine, b98

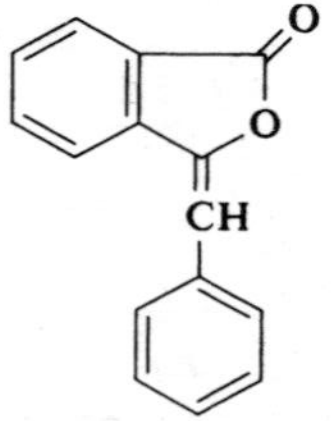

b101

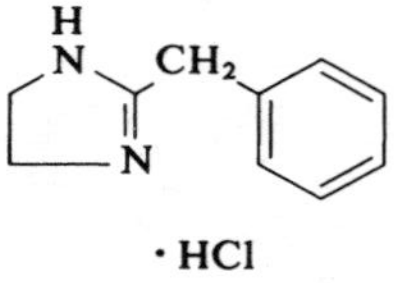

· HCl

b102

TABLE 1.15 Physical Constants of Organic Compounds (*continued*)

No.	Name	Formula	Formula weight	Beilstein reference	Density	Refractive index	Melting point	Boiling point	Flash point	Solubility in 100 parts solvent
b109	4-Benzyloxybenzyl alcohol	$C_6H_5CH_2OC_6H_4CH_2OH$	214.26				86–87			
b110	2-Benzyloxyethanol	$C_6H_5CH_2OCH_2CH_2OH$	152.19		1.07^{20}_{20}			255.9	129	0.4 aq
b111	4-Benzyloxy-3-methoxybenzaldehyde	$C_6H_5CH_2OC_6H_3\text{-}(OCH_3)CHO$	242.27							
b112	4'-Benzyloxypropiophenone	$C_6H_5CH_2OC_6H_4COC_2H_5$	240.30				100–102			
b113	Benzyl phenyl sulfide	$C_6H_5CH_2SC_6H_5$	200.30	6, 454	1.014	1.5467^{20}	43	197^{27mm}	>112	i aq; sl s alc; s eth
b114	1-Benzylpiperazine		176.26	20, 296	0.997	1.5379^{20}	7	279	>112	s aq, alc, eth
b115	4-Benzylpiperidine		175.28		1.021	1.5399^{20}		134^{7mm}	>112	
b116	1-Benzyl-4-piperidone		189.26							
b117	2-Benzylpyridine	$C_6H_5CH_2(C_5H_4N)$	169.23	20, 425	1.054	1.5785^{20}	10	276	125	i aq; v s alc, eth
b118	4-Benzylpyridine	$C_6H_5CH_2(C_5H_4N)$	169.23	20, 426	1.061^{20}_{0}	1.5818^{20}		287	115	s alc; v s eth
b119	1-Benzyl-2-pyrrolidinone		175.23		1.095	1.5525^{20}			>112	
b120	(Benzylthio)acetic acid	$C_6H_5CH_2SCH_2COOH$	182.24				59–63			
b121	Benzyl thiocyanate	$C_6H_5CH_2SCN$	149.22	6, 460			43	235		i aq; s alc; v s eth
b122	Benzyltributyl-ammonium chloride	$C_6H_5CH_2N(C_4H_9)_3^+Cl^-$	312.94				155 d			
b123	Benzyltrichlorosilane	$C_6H_5CH_2SiCl_3$	225.57		1.288^{20}_{4}	1.526^{20}		$140–142^{100mm}$		
b124	Benzyltriethoxysilane	$C_6H_5CH_2Si(OC_2H_5)_3$	254.40		0.986^{20}_{4}			$170–175^{70mm}$		
b125	Benzyltriethyl-ammonium chloride	$C_6H_5CH_2N(C_2H_5)_3^+Cl^-$	227.78				185 d			
b126	Benzyltrimethyl-ammonium chloride	$C_6H_5CH_2N(CH_3)_3^+Cl^-$	185.70	12, 1020						

No.	Name	Formula	M		Density	n_D	mp	bp	fp	Solubility
b127	Benzyltrimethyl-silane	$C_6H_5CH_2Si(CH_3)_3$	164.32		0.8933^{20}	1.4941^{20}				
b128	Betaine	$(CH_3)_3\overset{+}{N}CH_2COO^-$	117.15	4, 347			d > 310	190–191		160 aq; 55 MeOH; 6 alc
b129	Bicyclo[2.2.1]hepta-2,5-diene		92.14		0.909^{20}	1.4707^{20}	−20	89	−21	i aq; s PE
b130	Bicyclo[2.2.1]-2-heptene		94.16				46	96	−15	s eth
b131	Bicyclo[2.2.1]-5-heptene-2-carbaldehyde		122.16		1.018	1.4883^{20}		$67\text{–}70^{12mm}$	51	

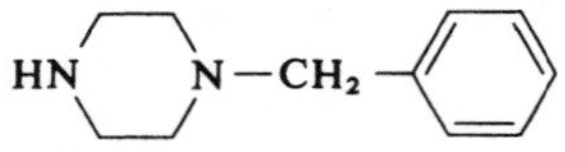

HN—N—CH₂—[phenyl]   b114

HN—[piperidine]—CH₂—[phenyl]   b115

O=[piperidine]—N—CH₂—[phenyl]   b116

[pyrrolidinone] O, N—CH₂—[phenyl]   b119

b129

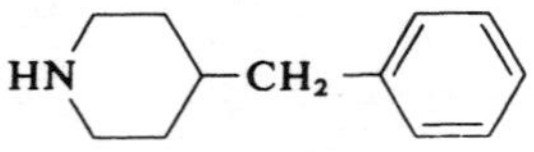

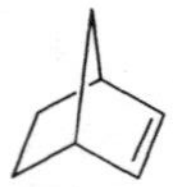

b130

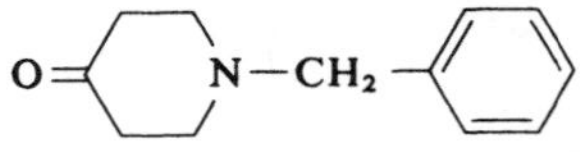

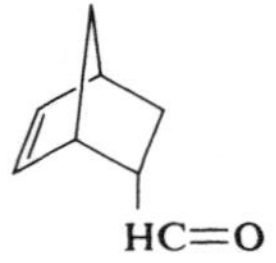

[bicycloheptene]—HC=O   b131

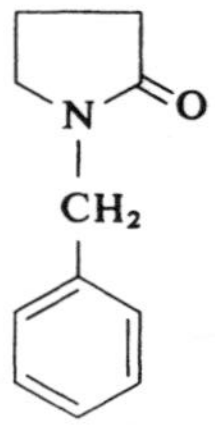

TABLE 1.15 Physical Constants of Organic Compounds (*continued*)

No.	Name	Formula	Formula weight	Beilstein reference	Density	Refractive index	Melting point	Boiling point	Flash point	Solubility in 100 parts solvent
b132	Bicyclo[2.2.2]oct-7-ene-2,3,5,6-tetracarboxylic-2,3,5,6-dianhydride		248.19				>300			
b133	Biguanide	$H_2NC(=NH)NHC(=NH)NH_2$	101.11	3, 93			130	d 142		s aq, alc; i bz, eth
b134	Biphenyl	$C_6H_5—C_6H_5$	154.20	5, 578	0.9939^{70}	1.5887^{77}	68.8	255.0		i aq; s alc, eth
b135	4-Biphenylcarboxylic acid	$C_6H_5C_6H_4COOH$	198.22	9, 671			226	subl		i aq; v s alc, eth; s bz
b136	(1,1'-Biphenyl)-4,4'-diamine	$H_2NC_6H_4C_6H_4NH_2$	184.23	13, 214			128	400^{740mm}		0.04 aq; s alc; 2 eth
b137	(1,1'-Biphenyl)-2,2'-dicarboxylic acid	$HOOCC_6H_4C_6H_4COOH$	242.23	9, 922			228–229			0.06 aq; s org solv
b138	4-Biphenylmethanol	$C_6H_5C_6H_4CH_2OH$	184.24	6^2, 636			101			
b139	4-Biphenylsulfonic acid	$C_6H_5C_6H_4SO_3H$	234.26				138			
b140	2-(4-Biphenylyl)-5-(4-*tert*-butylphenyl)-1,3,4-oxadiazole		354.46				138			
b141	o-Biphenylyl glycidyl ether		226.28				30–32	$120^{0.1mm}$		
b142	2-(4-Biphenylyl)-5-phenyloxazole		197.36				118			
b143	2,2'-Bipyridinium chlorochromate	$C_5H_4N—C_5H_4NH^+CrClO_3^-$	292.64							
b144	2,2-Bis[p-(allyloxy)phenyl]propane	$H_2C=CHCH_2OC_6H_4C(CH_3)_2C_6H_4OCH_2-CH=CH_2$	308.42		1.022	1.5636^{20}			>112	

b145	*N,N′*-Bis(3-amino-propyl)-ethylenediamine	$H_2N(CH_2)_3NHCH_2CH_2NH(CH_2)_3NH_2$	174.29					$118^{0.2mm}$		
b146	*N,N′*-Bis(3-amino-propyl)piperazine		200.33	23^2, 12	0.973	1.5015^{20}	15	152^{2mm}		
b147	*N,N′*-Bis(3-amino-propyl)-1,3-propanediamine	$H_2N(CH_2)_3NHCH_2CH_2CH_2NH(CH_2)_3NH_2$	188.32					$98–103^{1mm}$		
b148	2,5-Bis(4-biphenylyl)-oxazole		373.46				240			

Bicyclo[4.3.0]nonane, h46
Biphenol, d387
Biphenylamines, a134, a135
3-(*o*-Biphenylyloxy)-1,2-epoxypropane, b142
2,2′-Bipyridine, d705
Bis(4-aminophenyl)ether, o61
1,3-Bis(aminomethyl)cyclohexane, c314
1,2-Bis(benzylamino)ethane, d59

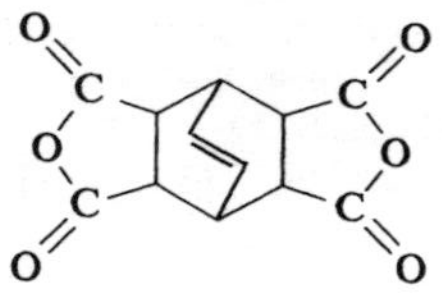

b132

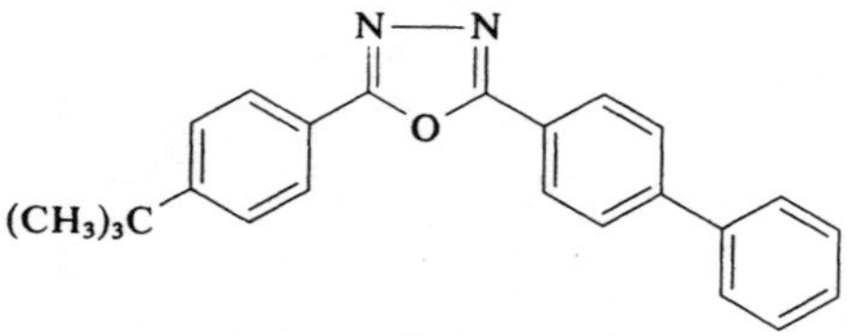

b140

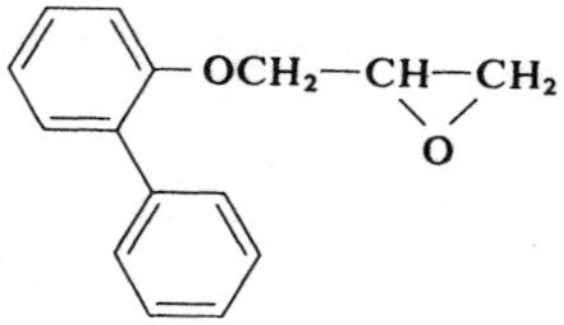

b141

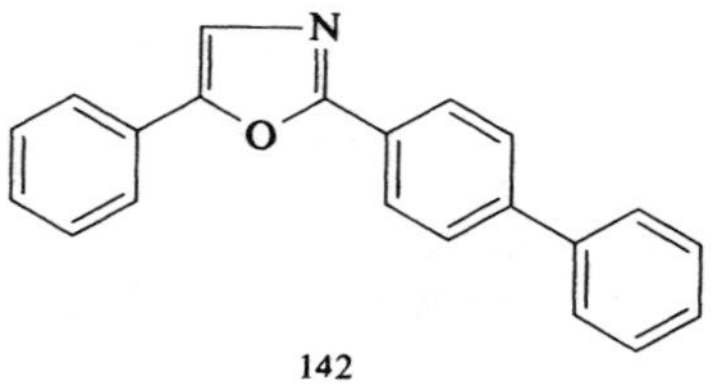

142

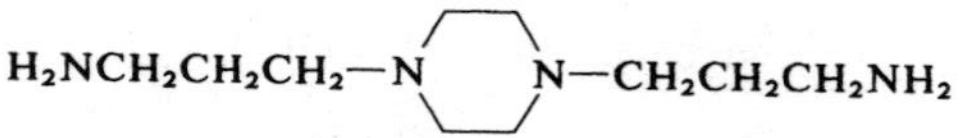

b146

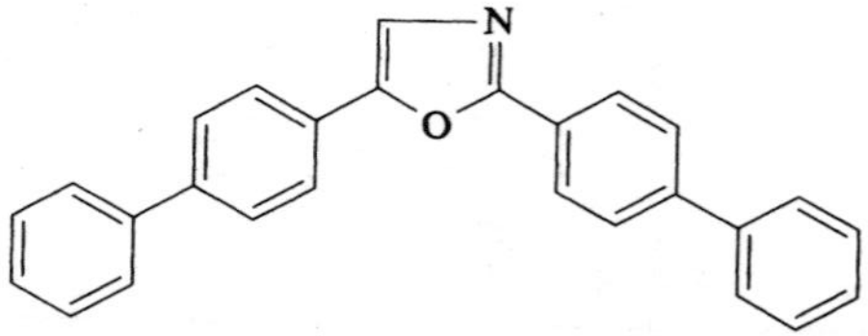

b148

TABLE 1.15 Physical Constants of Organic Compounds (*continued*)

No.	Name	Formula	Formula weight	Beilstein reference	Density	Refractive index	Melting point	Boiling point	Flash point	Solubility in 100 parts solvent
b149	Bis(2-bromoethyl) ether	$BrCH_2CH_2OCH_2CH_2Br$	231.92					$103–107^{20}$		
b150	1,3-Bis(bromoethyl)-tetramethyldi-siloxane	$[BrCH_2Si(CH_3)_2]_2O$	320.17		1.3918^{20}_4	1.4719^{20}		$103–104^{15mm}$		
b151	Bis(2-butoxyethyl) ether	$(C_4H_9OCH_2CH_2)_2O$	218.33		0.8853^{20}_{20}	1.4233^{20}	−60.2	254.6	47	0.3 aq; misc alc, eth, ketones, esters, CCl₄
b152	2,5-Bis(5-tert-butyl-2'-benzoxazolyl)-thiophene		430.57				201			
b153	Bis(sec-butyl) disulfide	$[CH_3CH_2CH(CH_3)]_2S_2$	178.36	1³, 1549	0.957	1.4920^{20}		164^{739mm}	>112	
b154	Bis(tert-butyl) disulfide	$(CH_3)_3CSSC(CH_3)_3$	178.36	1, 379	0.909	1.4930^{20}		204	79	
b155	Bis(carboxymethyl) trithiocarbonate	$HOOCCH_2SC(=S)-SCH_2COOH$	226.29	3, 252			172–175			
b156	1,2-Bis(2-chloro-ethoxy)ethane	$(ClCH_2CH_2OCH_2—)_2$	187.07		1.197^{20}_4	1.4617		108^{8mm}		
b157	Bis(2-chloroethoxy)-methylsilane	$H(CH_3)Si-(OCH_2CH_2Cl)_2$	203.1		1.1643^{20}_4	1.4431^{20}		$95–97^{18mm}$		
b158	Bis(2-chloroethyl) ether	$ClCH_2CH_2OCH_2CH_2Cl$	143.01	1², 335	1.2192^{20}	1.4575^{20}	−51.7	178.8	55	i aq; s most org solv
b159	Bis(2-chloroethyl)-N-methylamine	$CH_3N(CH_2CH_2Cl)_2$	156.07		1.118^{25}_4		−60	75^{10mm}		v sl s aq; misc most org solv
b160	Bis(chloromethyl)-dimethylsilane	$(CH_3)_2Si(CH_2Cl)_2$	157.12	4³, 1845	1.075^{20}_4	1.4600^{20}		160		

No.	Name	Formula	Formula wt	Beilstein ref	Density	n_D	mp, °C	bp, °C		Solubility
b161	Bis(2-chloro-1-methyl)ethyl ether	$ClCH_2CH(CH_3)OCH(CH_3)CH_2Cl$	171.07		1.1112^{20}_{20}			187.3	85	
b162	Bis(4-chlorophenoxy)acetic acid	$(ClC_6H_4O)_2CHCOOH$	313.14				142			
b163	2,2-Bis(p-chlorophenyl)-1,1-dichloroethane	$(ClC_6H_4)_2CHCHCl_2$	320.05				111			
b164	1,1-Bis(4'-chlorophenyl)ethanol	$(ClC_6H_4)_2C(OH)CH_3$	267.16	6^3, 3396			69			v sl s aq; s org solv
b165	Bis(4-chlorophenyl) sulfone	$ClC_6H_4SO_2C_6H_4Cl$	287.16	6, 327						
b166	Bis(4-chlorophenyl) sulfoxide	$ClC_6H_4S(O)C_6H_4Cl$	271.17	6^1, 149			144	250^{10mm}		
b167	1,1-Bis(p-chlorophenyl)-2,2,2-trichloroethane	$(ClC_6H_4)_2CHCCl_3$	354.49				109			
b168	1,3-Bis(dichloromethyl)tetramethyldisiloxane	$[Cl_2CH(CH_3)_2Si]_2O$	300.16		1.2213^{20}_{4}	1.4660^{20}		149^{40mm}		
b169	N,N-Bis(2,2-diethoxyethyl)methylamine	$[(C_2H_5O)_2CHCH_2]_2NCH_3$	263.38	4, 311	0.945	1.4259^{20}		222^{244mm}	60	i aq; 58 acet; 78 bz; 45 CCl₄; v s pyr, diox

Bis(3-*tert*-butyl-4-hydroxy-5-methylphenyl) sulfide, t144

Bis(2-cyanoethyl) ether, o63

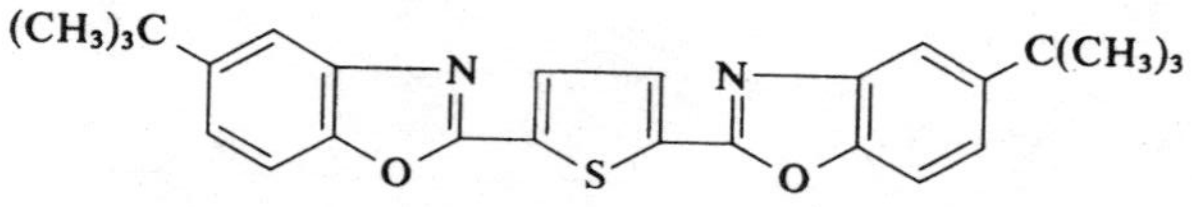

$(CH_3)_3C$ ⋯ $C(CH_3)_3$

b152

TABLE 1.15 Physical Constants of Organic Compounds (*continued*)

No.	Name	Formula	Formula weight	Beilstein reference	Density	Refractive index	Melting point	Boiling point	Flash point	Solubility in 100 parts solvent
b170	4,4′-Bis(diethyl-amino)benzo-phenone	$[(C_2H_5)_2NC_6H_4]_2C{=}O$	324.47	14, 98			95			
b171	4,4′-Bis(dimethyl-amino)benzo-phenone	$[(CH_3)_2NC_6H_4]_2C({=}O)$	268.36	14, 89				d 360		i aq; s alc, warm bz
b172	Bis(dimethylamino)-dimethylsilane	$[(CH_3)_2N]_2Si(CH_3)_2$	146.3		0.810^{22}	1.432^{22}	172–176	128–129		
b173	1,3-Bis(dimethyl-amino)-2-propanol	$[(CH_3)_2NCH_2]_2CHOH$	146.23	4, 290	0.897	1.4422^{20}	−98		>112	
b174	Bis(dimethylthio-carbamyl) disulfide	$[(CH_3)_2NC({=}S)S{-}]_2$	240.43	4, 76	1.29	1.4535^{20}	155–156	160^{11mm}		s alc, eth; sl s bz, acet; i aq
b175	1,4-Bis(2,3-epoxy-propoxy)butane	$H_2C\underset{O}{\diagdown}CHCH_2{-}OCH_2CH_2{-}]_2O$	202.25	1^2, 519	1.049	1.4110^{20}	−44.3		>112	
b176	Bis(2-ethoxyethyl) ether	$(C_2H_5OCH_2CH_2)_2O$	162.23		0.9072_4^{20}	1.4496^{25}		188.4	54	v s aq, alc, org solv
b177	Bis(2-ethylhexyl) decanedioate	$CH_3(CH_2)_3CH(C_2H_5)$-$CH_2OOC(CH_2)_8COOCH_2$-$CH(C_2H_5)(CH_2)_3CH_3$	426.66	1^4, 1786	1.9119_{25}^{25}					
b178	Bis(2-ethylhexyl) hydrogen phosphate	$[CH_3(CH_2)_3CH(C_2H_5)$-$CH_2O]_2P(O)OH$	322.43		0.965	1.4450^{20}	−60	209^{10mm}		
b179	Bis(2-ethylhexyl) o-phthalate	$[CH_3(CH_2)_3CH(C_2H_5)$-$CH_2OOC]_2C_6H_4$	390.57		0.9843^{20}	1.4859^{20}	−50	384	207	0.01 aq
b180	N,N-Bis(2-hydroxy-ethyl)-2-amino-ethanesulfonic acid	$(HOC_2H_4)_2$-$NCH_2CH_2SO_3H$	213.25	1, 468			152–154			
b181	Bis(2-hydroxyethyl) ether	$HOCH_2CH_2OCH_2CH_2OH$	106.12		1.118_{20}^{20}	1.4460^{20}	−10.45	245	143	misc aq, alc, acet, eth

No.	Name	Formula	Formula wt	Beilstein ref	Density	n_D	M.P., °C	B.P., °C	Flash	Solubility
b182	N,N-Bis(2-hydroxyethyl)glycine	$(HOCH_2CH_2)_2NCH_2COOH$	163.17				192 sl d			sl s aq
b183	Bis(2-hydroxyethyl)iminotris(hydroxymethyl)methane	$(HOCH_2CH_2)_2NC(CH_2OH)_3$	209.24				104			
b184	2,2-Bis(hydroxymethyl)propionic acid	$(HOCH_2)_2C(CH_3)COOH$	134.13	3, 401			189–191			
b185	4,8-Bis(hydroxymethyl)tricyclo[5.2.1^{2,6}]decane		196.29			1.5280^{20}			>112	
b186	Bis(2-mercaptoethyl)ether	$(HSCH_2CH_2)_2O$	138.25		1.114		−80	217		
b187	Bis(2-mercaptoethyl)sulfide	$(HSCH_2CH_2)_2S$	154.32		1.183	1.5982^{20}		136^{10mm}	90	
b188	1,4-Bis(methanesulfonoxy)butane	$(CH_3SO_2OCH_2CH_2{-})_2$	246.30				115–117			sl hyd aq; 0.1 alc; 1.4 acet
b189	1,2-Bis(methoxyethoxy)ethane	$(CH_3OCH_2CH_2OCH_2{-})_2$	178.23	1^3, 2107	0.990_4^{20}	1.4224^{20}	−45	216	110	misc aq
b190	Bis[2,(2-methoxyethoxy)ethyl] ether	$(CH_3OCH_2CH_2OCH_2{-}CH_2{-})_2O$	222.28		1.0087_4^{20}	1.4330^{20}	−27	275.3	140	s aq
b191	Bis(2-methoxyethyl)ether	$(CH_3OCH_2CH_2{-})_2O$	134.18		0.9440^{25}	1.4043^{25}	−68	162	70	misc aq

Bis(2-ethylhexyl) sebacate, b177
Bis(2-hydroxyethyl) sulfide, t149

2,2-Bis(hydroxymethyl)-2,2',2''-nitrilotriethanol, b183

Bis(4-hydroxyphenyl) sulfide, t150

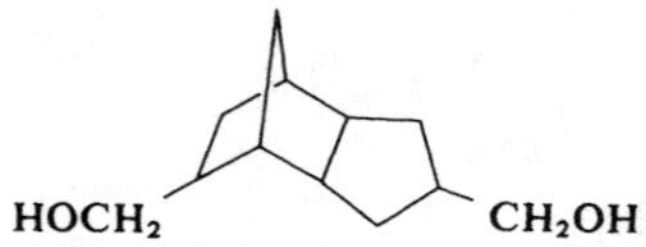

b185

TABLE 1.15 Physical Constants of Organic Compounds (*continued*)

No.	Name	Formula	Formula weight	Beilstein reference	Density	Refractive index	Melting point	Boiling point	Flash point	Solubility in 100 parts solvent
b192	Bis(2-methylallyl) carbonate	[H$_2$C=C(CH$_3$)CH$_2$O]$_2$C(=O)	170.21		0.943^{20}	1.4371		202	72	
b193	Bis(4-nitrophenyl) carbonate	(O$_2$NC$_6$H$_4$O)$_2$C(=O)	304.21	6^1, 120			141			
b194	Bis(3-nitrophenyl) disulfide	O$_2$NC$_6$H$_4$SSC$_6$H$_4$NO$_2$	308.22	6, 339			83			i aq; s alc, eth
b195	Bis(4-nitrophenyl) disulfide	O$_2$NC$_6$H$_4$SSC$_6$H$_4$NO$_2$	308.33	6, 340			181			sl s alc
b196	Bis(octadecyl)penta-erythritol diphosphite	[C$_{18}$H$_{37}$OP(OCH$_2$)$_2$]$_2$	721.01		0.925	1.457	40		261	
b197	Bis(pentabromophenyl) ether	C$_6$Br$_5$OC$_6$Br$_5$	969.22	6^1, 108			>300			
b198	1,4-Bis(5-phenyloxazol-2-yl)benzene		364.40				244			
b199	Bis(*p*-tolyl) disulfide	CH$_3$C$_6$H$_4$SSC$_6$H$_4$CH$_3$	246.39	6, 425			43–46			i aq; s alc; v s eth
b200	Bis(*p*-tolyl) sulfoxide	CH$_3$C$_6$H$_4$S(O)C$_6$H$_4$CH$_3$	230.33	6, 419			94–96			v s alc, bz, chl, eth
b201	Bis(tributyltin) oxide	(C$_4$H$_9$)$_3$SnOSn(C$_4$H$_9$)$_3$	596.08	5, 385	1.170	1.4864^{20}		180^{2mm}	>112	
b202	1,4-Bis(trichloro-methyl)benzene	Cl$_3$CC$_6$H$_4$CCl$_3$	312.84				108–110			i aq; 26 acet; 38 bz; 22 CCl$_4$; 33 eth; 3 MeOH
b203	Bis(2,4,5-trichloro-phenyl) disulfide	Cl$_3$C$_6$H$_4$SSC$_6$H$_4$Cl$_3$	425.01				140–144			
b204	1,2-Bis(trichloro-silyl)ethane	Cl$_3$SiCH$_2$CH$_2$SiCl$_3$	296.64		1.483$_4^{20}$	1.473^{20}	24.5	201–202		
b205	3,5-Bis(trifluoro-methyl)aniline	(F$_3$C)$_2$C$_6$H$_3$NH$_2$	229.13		1.467	1.4335^{20}		85^{15mm}	83	

b206	1,3-Bis(trifluoromethyl)benzene	$F_3CC_6H_4CF_3$	214.0	1.3790^{25}	1.3916^{25}		116	
b207	N,O-Bis(trimethylsilyl)acetamide	$CH_3C{=}N{-}Si(CH_3)_3$ | $O{-}Si(CH_3)_3$	203.43	0.832^{20}	1.4170^{20}		73^{35mm}	11
b208	Bis(trimethylsilyl)acetylene	$(CH_3)_3SiC{\equiv}CSi(CH_3)_3$	170.41	0.770^{20}_4	1.413^{20}		137	2
b209	Bis(trimethylsilyl)formamide	$HC{=}NSi(CH_3)_3$ | $O{-}Si(CH_3)_3$	189.41	0.885	1.4381^{20}		54–55^{13mm}	
b210	N,O-Bis(trimethylsilyl)hydroxylamine	$(CH_3)_3SiONHSi(CH_3)_3$	177.40	0.830	1.4112^{20}		78–80^{100mm}	28
b211	1,2-Bis(trimethylsilyloxy)ethane	$(CH_3)_3SiOCH_2CH_2{-}OSi(CH_3)_3$	206.43	0.842	1.4034^{20}		165–166	46
b212	N,O-Bis(trimethylsilyl)trifluoroacetamide	$CF_3C[{=}NSi(CH_3)_3]{-}OSi(CH_3)_3$	257.40	0.969	1.3939^{20}	−10	50^{14mm}	23
b213	Bis(triphenylphosphine)dicarbonylnickel	$[(C_6H_5)_3P]_2Ni(CO)_2$	639.32			209		

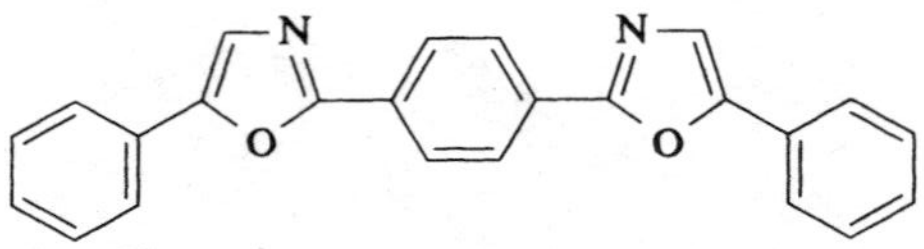

b198

1.129

TABLE 1.15 Physical Constants of Organic Compounds (*continued*)

No.	Name	Formula	Formula weight	Beilstein reference	Density	Refractive index	Melting point	Boiling point	Flash point	Solubility in 100 parts solvent
b214	1,3-Bis[tris(hydroxy-methyl)methyl-amino]propane	$CH_2[CH_2NHC(CH_2OH)_3]_2$	282.34	4^3, 859			170			
b215	Biuret	$H_2NCONHCONH_2$	103.08	3, 70	1.467_4^{-5}		110	d 190		v s alc; 2 aq
b216	1-Borneol		154.25	6, 72	1.011_4^{20}		204	212	65	i aq; 176 alc; s eth, bz, PE
b217	1-Bornyl acetate		196.29	6, 82	0.982	1.4626	27	224	84	sl s aq; s alc, eth
b218	N-Bromoacetamide	$CH_3CONBrH$	137.97	2, 181			102–105			sl s aq; v s eth
b219	p-Bromoacetanilide	$BrC_6H_4NHCOCH_3$	214.07	12, 642	1.717		168			i aq; s bz, chl, EtAc
b220	Bromoacetic acid	$BrCH_2COOH$	138.95	2, 213	1.934_4^{50}	1.4804^{50}	50	208		v s aq, alc, eth
b221	α-Bromoaceto-phenone	$C_6H_5COCH_2Br$	199.05	7, 283	1.647_4^{20}		50	135^{18mm}		i aq; v s alc, bz, chl, eth
b222	p-Bromoaceto-phenone	$BrC_6H_4COCH_3$	199.05	7, 283	1.647		54	255		s alc, bz, eth, HOAc
b223	Bromoacetyl bromide	$BrCH_2COBr$	201.86	2, 215	2.317_{22}^{22}	1.5480^{20}		150	none	d aq, alc
b224	Bromoacetyl chloride	$BrCH_2COCl$	157.40	2, 215	1.908	1.4960^{20}		128	none	d aq, alc
b225	2-Bromoaniline	$BrC_6H_4NH_2$	172.03	12, 631	1.578_4^{20}	1.6113^{20}	31	229		i aq; s alc, eth
b226	3-Bromoaniline	$BrC_6H_4NH_2$	172.03	12, 633	1.580_4^{20}	1.6250^{20}	16.8	251	>112	sl s aq; s alc, eth
b227	4-Bromoaniline	$BrC_6H_4NH_2$	172.03	12, 636	1.4970_4^{100}		66.3			i aq; v s alc, eth
b228	3-Bromobenzaldehyde	BrC_6H_4CHO	185.03	7, 238	1.587	1.5935^{20}		230	96	i aq; v s alc, eth
b229	Bromobenzene	C_6H_5Br	157.02	5, 206	1.4952_4^{20}	1.5580^{20}	−30.72	156.2	51	0.044 aq; 10.4 alc; misc bz, chl, PE; 71.6 eth
b230	Bromobenzene-d_5	C_6D_5Br	162.06					53^{23mm}		i aq; d alc; v s eth
b231	4-Bromobenzene-sulfonyl chloride	$BrC_6H_4SO_2Cl$	255.52	11, 57			74.5	153^{15mm}	65	i aq; s alc; eth
b232	4-Bromobenzo-phenone	$BrC_6H_4COC_6H_5$	261.12	7, 422			82	350		i alc; sl s bz, eth

b233	2-Bromobenzotri-fluoride	$BrC_6H_4CF_3$	225.01		1.652^{20}	1.4817^{20}		168	51	
b234	3-Bromobenzotri-fluoride	$BrC_6H_4CF_3$	225.01		1.613	1.4749^{20}		152	43	
b235	2-Bromobenzyl alcohol	$BrC_6H_4CH_2OH$	187.04	6, 445			82			s hot aq; v s alc, eth
b236	2-Bromobenzyl bromide	$BrC_6H_4CH_2Br$	249.94	5, 308		1.6193^{20}	31	129^{19mm}		d hot aq; s alc, eth
b237	4-Bromobiphenyl	$BrC_6H_4C_6H_5$	233.11	5, 580	9.9327_4^{25}		87	310		i aq; s alc, bz, eth
b238	1-Bromobutane	$CH_3CH_2CH_2CH_2Br$	137.02	1, 119	1.2686_4^{25}	1.4374^{25}	−112.4	101.6	23	i aq; s alc, bz, eth
b239	2-Bromobutane	$CH_3CH_2CHBrCH_3$	137.03	1, 119	1.2530_4^{25}	1.4360^{20}	−112.4	21		<0.1 aq; v s alc, eth
b240	1-Bromo-2-butene	$CH_3CH{=}CHCH_2Br$	135.01	1, 205	1.312	1.4765^{20}		99	11	
b241	2-Bromo-2-butene	$CH_3CH{=}C(Br)CH_3$	135.01	1, 205	1.328	1.4613^{20}		90^{740mm}	<1	
b242	4-Bromo-1-butene	$BrCH_2CH_2CH{=}CH_2$	135.01	1^1, 84	1.3230_4^{20}	1.4608^{30}		100	<1	i aq; s alc, eth

2-Bornanone, c3
Bromal, t201
Bromoacetaldehyde diethyl acetal, b267
Bromoanisoles, b301, b302, b303
p-Bromobenzenethiol, b354
4-Bromobenzyl cyanide, b329
exo-2-Bromobicyclo[2.2.1]heptane, b318
Bromobutanedioic acid, b350

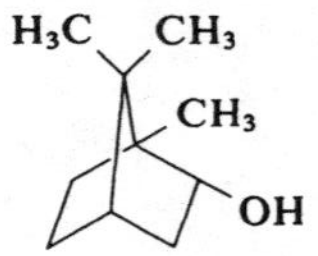
H3C
CH3
CH3
OH

b216

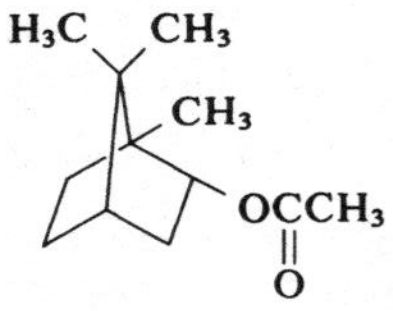
H3C
CH3
CH3
OCCH3
O

b217

TABLE 1.15 Physical Constants of Organic Compounds (*continued*)

No.	Name	Formula	Formula weight	Beilstein reference	Density	Refractive index	Melting point	Boiling point	Flash point	Solubility in 100 parts solvent
b243	4-Bromobutyl phenyl ether	C$_6$H$_5$OCH$_2$CH$_2$CH$_2$CH$_2$Br	229.12	6^2, 82		1.4720^{20}	41–42	153–156^{18mm}	>112	6.7 aq; s alc, eth
b244	2-Bromobutyric acid	CH$_3$CH$_2$CH(Br)COOH	167.01	2, 281	1.5669^{20}_{20}		−4	103^{10mm}		i aq; 15 alc; 200 chl; 62 eth
b245	*endo*-3-Bromo-D-camphor		231.14	7^2, 101	1.449		76–78	244		
b246	α-Bromo-*p*-chloroacetophenone	ClC$_6$H$_4$COCH$_2$Br	233.50	7, 285			96.5			
b247	2-Bromochlorobenzene	BrC$_6$H$_4$Cl	191.46	5, 209	1.6382^{25}_{4}	1.5789^{25}		204	79	i aq; v s bz
b248	3-Bromochlorobenzene	BrC$_6$H$_4$Cl	191.46	5, 209	1.6302^{20}_{4}	1.5771^{20}	−21	196	80	i aq; v s alc, eth
b249	4-Bromochlorobenzene	BrC$_6$H$_4$Cl	191.46	5, 209	1.576^{71}_{4}	1.5531^{70}	64.5	196		0.1 aq; misc MeOH, eth
b250	3-Bromo-4-chlorobenzotrifluoride	Br(Cl)C$_6$H$_3$CF$_3$	259.47	5^3, 294	1.743^{25}	1.4973^{25}	−22	191–192	60	i aq; s alc, chl, eth
b251	1-Bromo-4-chlorobutane	ClCH$_2$CH$_2$CH$_2$CH$_2$Br	171.47		1.488	1.4875^{20}		82^{30mm}		
b252	4-Bromo-6-chloro-*o*-cresol	Br(Cl)C$_6$H$_2$(OH)CH$_3$	221.49	6, 360			47			
b253	Bromochlorodifluoromethane	Br(Cl)CF$_2$	165.4	1, 89	1.83^{21}		−160.5	−4.01		
b254	1-Bromo-2-chloroethane	ClCH$_2$CH$_2$Br	143.43		1.7392^{20}_{4}	1.4917^{20}	−18.4	106.6	none	0.7 aq; misc org solv
b255	7-Bromo-5-chloro-8-hydroxyquinoline		258.51	21^1, 222			177–179			
b256	Bromochloromethane	ClCH$_2$Br	129.39	1, 67	1.923^{25}_{4}	1.480^{25}	−88	67.8	none	0.9 aq; misc MeOH, eth

No.	Name	Formula	Formula wt	Beilstein	Density	n_D	mp	bp	Flash pt	Solubility
b257	1-Bromo-3-chloropropane	$ClCH_2CH_2CH_2Br$	157.44	1, 109	1.472	1.486^{20}	<-50	143.5	none	0.1 aq; misc org solv
b258	2-Bromo-2-chloro-1,1,1-trifluoroethane	$HC(Br)ClCF_3$	197.4		1.8636^{25}	1.3738^{25}		50		
b259	α-Bromocinnamaldehyde	$C_6H_5CH=C(Br)CHO$	211.06	7, 358			66–68			
b260	Bromocycloheptane	BrC_7H_{13}	177.09	5, 29	1.2887^{22}_4	1.5052^{20}		72^{10mm}	68	i aq; v s chl, eth
b261	Bromocyclohexane	BrC_6H_{11}	163.06	5, 24	1.3264^{15}_4	1.4956^{15}		165.8	62	0.1 aq; 10 MeOH; 71 eth
b262	3-Bromocyclohexene		161.04	5^2, 40	1.3890^{20}_4	1.5292^{20}		$64–65^{15mm}$	35	
b263	Bromocyclopentane	BrC_5H_9	149.04	5, 19	1.3900^{20}_4	1.4881^{20}		137–139	2	
b264	Bromocyclopropane	BrC_3H_5	120.98			1.4605^{20}		69	94	
b265	1-Bromodecane	$CH_3(CH_2)_9Br$	221.19	1^2, 130	1.0658^{20}_4	1.4560^{20}	-30	238	none	i aq; v s chl, eth
b266	Bromodichloromethane	$HCBrCl_2$	163.83	1, 67	1.980^{20}	1.4964^{20}	-55	89.2		sl s aq; misc org solv
b267	2-Bromo-1,1-diethoxyethane	$BrCH_2CH(OC_2H_5)_2$	197.08	1, 625	1.310	1.4385^{20}		67^{18mm} / 180 d	51	s hot alc

2-Bromo-p-cumene, b298

β-Bromocumene, b297

4-Bromodiphenyl ether, b330

Bromoethene, b284

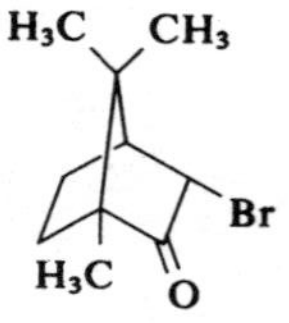

b245

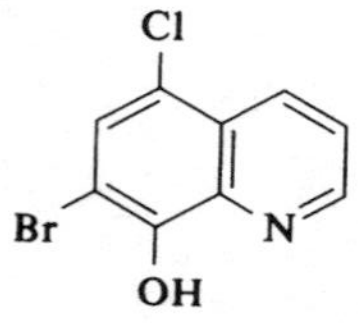

b255

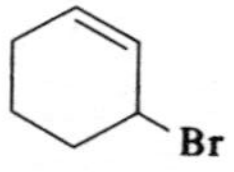

b262

TABLE 1.15 Physical Constants of Organic Compounds (*continued*)

No.	Name	Formula	Formula weight	Beilstein reference	Density	Refractive index	Melting point	Boiling point	Flash point	Solubility in 100 parts solvent
b268	4-Bromo-1,2-di-methoxybenzene	$BrC_6H_3(OCH_3)_2$	217.07	6, 784	1.702	1.5743^{20}		256	109	
b269	1-Bromo-2,2-di-methoxypropane	$CH_3C(OCH_3)_2CH_2Br$	183.05		1.355	1.4475^{20}		87^{80mm}	40	
b270	4-Bromo-2,6-di-methylphenol	$BrC_6H_2(CH_3)_2OH$	201.07	6, 485			78			v s hot alc, hot acet
b271	2-Bromo-4,6-di-nitroaniline	$BrC_6H_2(NO_2)_2NH_2$	262.02	12, 761			154	subl		
b272	3-Bromo-4,6-di-nitrofluorobenzene	$BrC_6H_2(NO_2)_2F$	264.9	9^1, 283			90–91			
b273	2-Bromo-2,2-di-phenylacetyl bromide	$BrC(C_6H_5)_2COBr$	354.05				63–65	184^{20mm}		
b274	α-Bromodiphenyl-methane	$C_6H_5CH(Br)C_6H_5$	247.14	5, 592	1.038	1.4580^{20}	40		110	0.1 aq; s alc, eth
b275	1-Bromododecane	$CH_3(CH_2)_{11}Br$	249.24	1^2, 133	1.601^{20}	1.4820^{20}	−9	135^{6mm}		i aq; sl s alc; s eth
b276	1-Bromo-2,3-epoxy-propane	$H_2C-CHCH_2Br$ (O)	136.98	17, 9			−40	134–136	56	
b277	Bromoethane	CH_3CH_2Br	108.97	1, 88	1.4708^{15}	1.4276^{15}	−118.6	38.4	none	0.91 aq
b278	2-Bromoethane-sulfonic acid, sodium salt	$BrCH_2CH_2SO_3^-Na^+$	211.02	4, 7			283–285 d			
b279	2-Bromoethanol	$BrCH_2CH_2OH$	124.97	1, 338	1.7629_4^{20}	1.4920^{20}		150	40	misc aq; s org solv
b280	2-Bromoethyl acetate	$CH_3COOCH_2CH_2Br$	167.01	2^1, 57	1.514_4^{20}	1.4547^{20}	−13.8	159	71	v s aq; misc alc, eth
b281	2-Bromoethylamine HBr	$BrCH_2CH_2NH_2 \cdot HBr$	204.90	4, 134			172–174			v s aq, alc

No.	Name	Formula	Formula wt		Density	n_D	mp	bp		Solubility
b282	o-Bromo(ethyl)-benzene	$CH_3CH_2C_6H_4Br$	185.07	5, 355	1.3566^{25}_{25}	1.5603^{20}		199	89	0.1 aq; misc org solv
b283	(2-Bromoethyl)-benzene	$C_6H_5CH_2CH_2Br$	185.07	5, 356	1.355	1.5563^{20}		221		i aq; s bz, eth
b284	Bromoethylene	$H_2C{=}CHBr$	106.96	1, 188	1.493^{20}	1.4350^{20}	−139.5	15.8		i aq; misc alc, eth
b285	2-Bromoethyl ethyl ether	$BrCH_2CH_2OCH_2CH_3$	153.02	1, 338	1.3572^{20}_4	1.4450^{20}		150	21	sl s aq; misc alc, eth
b286	2-Bromoethyl phenyl ether	$BrCH_2CH_2OC_6H_5$	201.07	6, 142			34	144^{40mm}	65	i aq; v s alc, eth
b287	N-(2-Bromoethyl)-phthalimide		254.09	21, 461			81–84			s hot aq; v s eth
b288	2-Bromofluoro-benzene	BrC_6H_4F	175.01		1.601	1.5337^{20}		156	43	
b289	3-Bromofluoro-benzene	BrC_6H_4F	175.01		1.567	1.5257^{20}		150	38	
b290	4-Bromofluoro-benzene	BrC_6H_4F	175.01	5, 209	1.593^{15}	1.5310^{15}	−17.4	151–152	60	
b291	1-Bromoheptane	$H(CH_2)_7Br$	179.11	1, 155	1.1384^{4}_4	1.4505^{20}	−58	180	60	i aq; v s alc, eth
b292	2-Bromoheptane	$H(CH_2)_5CH(Br)CH_3$	179.11	1, 155	1.142	1.4470^{20}		66^{21mm}	47	
b293	1-Bromohexadecane	$H(CH_2)_{16}Br$	305.35	$1^2, 138$	0.9991	1.4618	17.8	336	177	i aq; misc org solv
b294	1-Bromohexane	$H(CH_2)_6Br$	165.08	1, 144	1.1763^{20}_4	1.4475	−85	154–158	57	i aq; misc alc, eth

(Bromomethyl)benzene, b85

Bromoform, t206

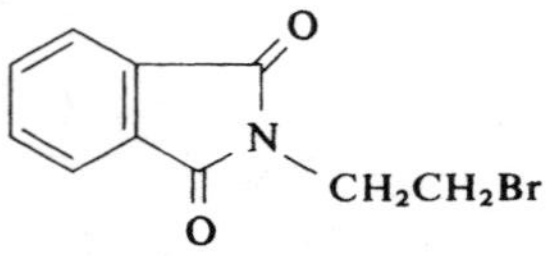

b287

TABLE 1.15 Physical Constants of Organic Compounds (*continued*)

No.	Name	Formula	Formula weight	Beilstein reference	Density	Refractive index	Melting point	Boiling point	Flash point	Solubility in 100 parts solvent
b295	DL-2-Bromohexanoic acid	$CH_3(CH_2)_3CH(Br)COOH$	195.06	2, 325	1.370	1.4720^{20}		$136\text{--}138^{18mm}$		
b296	5-Bromoisatin		226.03	21, 453				251–253		
b297	(2-Bromoisopropyl)-benzene	$C_6H_5CH(CH_3)CH_2Br$	199.10	5^1, 191	1.316	1.5480^{20}		108^{18mm}	91	
b298	2-Bromo-4-isopropyl-1-methylbenzene	$CH_3(Br)C_6H_3CH(CH_3)_2$	213.0		1.253^{25}_{25}	1.535^{25}	−20	120		i aq; 50 MeOH; misc org solv
b299	Bromomaleic anhydride		176.96	17, 435	1.905	1.5400^{20}		215	>112	
b300	Bromomethane	CH_3Br	94.94	1, 67	1.732^{0}_{0}	1.4234^{10}	−84	3.56	none	0.1 aq; s alc, chl, eth
b301	2-Bromo-1-methoxy-benzene	$BrC_6H_4OCH_3$	187.04	6, 197	1.5018^{25}_{4}	1.5737^{20}	2	223	96	i aq; v s alc, eth
b302	3-Bromo-1-methoxy-benzene	$BrC_6H_4OCH_3$	187.04	6, 198	1.477	1.5635^{20}	211	93		i aq; s alc, eth
b303	4-Bromo-1-methoxy-benzene	$BrC_6H_4OCH_3$	187.04	6, 199	1.4564^{20}_{4}	1.5630^{20}	10	223	94	sl s aq; v s alc, eth
b304	4-Bromo-2-methyl-aniline	$CH_3(Br)C_6H_3NH_2$	186.06	12, 838			56	240		sl s aq; v s alc
b305	1-Bromo-3-methyl-butane	$(CH_3)_2CHCH_2CH_2Br$	151.05	1, 136	1.210^{15}_{4}	1.4409^{20}	−112	119.7	32	0.02 aq; misc alc, eth
b306	(Bromomethyl)cyclo-hexane	$C_6H_{11}CH_2Br$	177.09	5^2, 18	1.269	1.4907^{20}		$76\text{--}77^{26mm}$	57	
b307	2-Bromomethyl-1,3-dioxalane		167.01	19^2, 8	1.613	1.4817^{20}		$80\text{--}82^{27mm}$	62	
b308	Bromomethyl methyl ether	$BrCH_2OCH_3$	124.97	1, 582	1.531	1.4550^{20}		87	26	

No.	Name	Formula	Formula wt	Beil. ref.	Density	n_D	mp, °C	bp, °C	Flash pt, °C	Solubility
b309	1-Bromo-2-methyl-naphthalene	BrC$_{10}$H$_6$CH$_3$	221.10	5, 568	1.418	1.6484^{20}		296	>112	0.06 aq; misc alc, eth
b310	1-Bromo-2-methyl-propane	(CH$_3$)$_2$CHCH$_2$Br	137.03	1, 126	1.2641^{20}	1.4362^{20}	−119	91.5	18	i aq; misc org solv
b311	2-Bromo-2-methyl-propane	(CH$_3$)$_3$CBr	137.03	1, 127	1.215^{25}_{25}	1.425^{25}	−16.2	73.1	18	i aq; misc org solv
b311a	α-Bromo-α-methyl-propiophenone	C$_6$H$_5$COC(CH$_3$)$_2$Br	227.11	7, 316	1.350	1.5561^{20}		148^{30mm}	>112	
b312	1-Bromonaphthalene	C$_{10}$H$_7$Br	207.08	5, 547	1.4834^{20}_4	1.6580^{20}	−1	281.1	>112	misc alc, bz, chl, eth
b313	1-Bromo-1-naphthol	BrC$_{10}$H$_6$OH	223.07	6, 650			78	130 d		i aq; s alc, bz, eth
b314	1-Bromo-2-nitro-benzene	BrC$_6$H$_4$NO$_2$	202.01	5[1], 247	09.9327^{25}_4		43	261		v alc; s bz, eth
b315	5-Bromo-2-nitro-benzotrifluoride	O$_2$N(Br)C$_6$H$_3$CF$_3$	270.02		1.7992^{25}	1.5180^{25}	40–44	99–100		
b316	2-Bromo-2-nitro-1,3-propanediol	(HOCH$_2$)$_2$C(Br)NO$_2$	199.99	1, 476			133			
b317	1-Bromononane	H(CH$_2$)$_9$Br	207.16	1[1], 63	1.084	1.4540^{20}		201	51	i aq; s chl, eth
b318	exo-2-Bromo-norbornane		175.07		1.363	1.5148^{20}		82^{29mm}	60	
b319	1-Bromooctadecane	H(CH$_2$)$_{18}$Br	333.41	1[1], 69			23	216^{12mm}		i aq; s alc, eth

α-Bromoisobutyrophenone, b311a
2-Bromomesitylene, b362

α-Bromo-4-nitro-o-cresol, h155

α-Bromo-p-nitrotoluene, n46

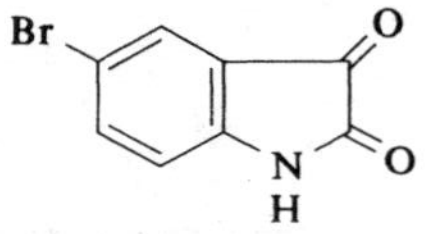

b296

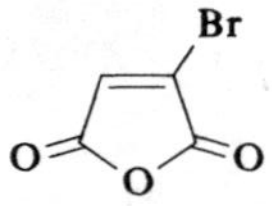

b299

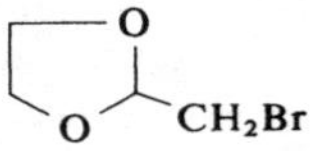

b307

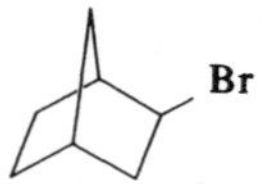

b318

TABLE 1.15 Physical Constants of Organic Compounds (*continued*)

No.	Name	Formula	Formula weight	Beilstein reference	Density	Refractive index	Melting point	Boiling point	Flash point	Solubility in 100 parts solvent
b320	1-Bromooctane	$H(CH_2)_8Br$	193.13	1, 160	1.1084^{25}	1.4503^{25}	−55	201	78	i aq; misc alc, eth
b321	Bromopentafluorobenzene	BrC_6F_5	246.97	1, 131	1.947^{20}	1.4490^{20}	−31	137	87	
b322	1-Bromopentane	$H(CH_2)_5Br$	151.05	1, 131	1.2337^{15}_4	1.4444^{20}	−88	129.6	31	i aq; s alc; misc eth
b323	2-Bromopentane	$CH_3CH_2CH_2CH(Br)CH_3$	151.05	1, 131	1.2039^{20}	1.4403^{20}		117	20	i aq; s alc, eth
b324	9-Bromophenanthrene		257.14	5, 671	1.409^{101}_4		54–58	190^{2mm}		s aq; misc chl, eth
b325	2-Bromophenol	BrC_6H_4OH	173.01	6, 197	1.492	1.5892^{20}	6	194	42	
b326	4-Bromophenol	BrC_6H_4OH	173.01	6, 198	1.5875^{80}		68	238		14 aq; v s alc, chl
b327	2-Bromo-2-phenylacetic acid	$C_6H_5CH(Br)COOH$	215.05	9, 451			83			
b328	p-Bromophenylacetic acid	$BrC_6H_4CH_2COOH$	215.05	9, 451			119			sl s aq; v s alc, eth
b329	p-Bromophenylacetonitrile	$BrC_6H_4CH_2CN$	196.05	9, 451			47–49			i aq; sl s alc; v s bz
b330	4-Bromophenyl phenyl ether	$BrC_6H_4OC_6H_5$	249.11	6^1, 105	1.423	1.6070^{20}	18	305	>112	
b331	1-Bromo-3-phenylpropane	$C_6H_5CH_2CH_2CH_2Br$	199.10	5, 391	1.310	1.5450^{20}		220	101	
b332	1-Bromopropane	$CH_3CH_2CH_2Br$	123.00	1, 108	1.3597^{15}	1.4370^{15}	−110.1	71.0	25	0.23 aq^{30}; misc alc
b333	2-Bromopropane	$CH_3CH(Br)CH_3$	123.00	1, 108	1.3222^{15}	1.4285^{15}	−89.0	59.5	19	0.3 aq^{18}; misc alc, bz, chl, eth
b334	3-Bromo-1-propanol	$BrCH_2CH_2CH_2OH$	139.00	1, 356	1.5374^{20}_4	1.4858^{20}		62^{5mm}	4	s aq; misc alc, eth
b335	1-Bromo-1-propene	$CH_3CH{=}CHBr$	120.98	1, 200	1.4133^4_4	1.4538^{20}	−116	63	4	i aq
b336	2-Bromo-1-propene	$CH_3C(Br){=}CH_2$	120.98	1, 200	1.362^{20}_0	1.4425^{20}	−125	49		
b337	2-Bromopropionic acid	$CH_3CH(Br)COOH$	152.98	2, 254	1.7000^{20}	1.4750^{20}	25.7	203	100	v s aq, alc, eth
b338	3-Bromopropionic acid	$BrCH_2CH_2COOH$	152.98	2, 256	1.480		62.5		65	s aq, alc, bz, chl, eth

No.	Name	Formula					mp	bp	fp	Solubility
b339	3-Bromopropionitrile	BrCH$_2$CH$_2$CN	133.98	2^2, 231	1.6152^{20}_4	1.4800^{20}		78^{10mm}	98	v s alc, eth
b340	2-Bromopropionyl chloride	CH$_3$CH(Br)COCl	171.43	2, 256	1.700^{11}	1.4800^{20}		133	51	d aq; s chl, eth
b341	3-Bromopropionyl chloride	BrCH$_2$CH$_2$COCl	171.43	2^2, 231	1.701	1.4968^{20}		57^{17mm}	79	
b342	α-Bromopropiophenone	C$_6$H$_5$COCHBrCH$_3$	213.08	7, 302	1.4304^{20}_4	1.5715^{20}		250	>112	s alc, bz, eth, acet
b343	3-Bromopropyl phenyl ether	C$_6$H$_5$OCH$_2$CH$_2$CH$_2$Br	215.10	6, 142	1.365	1.5464^{20}	10–11	130–134^{14mm}	96	
b344	3-Bromopropyne	BrCH$_2$C≡CH	118.97	1, 248	1.335	1.4905^{20}		88–90	18	i aq; s org solv
b345	2-Bromopyridine	BrC$_5$H$_4$N	158.00	20, 233	1.657^{18}	1.5720^{20}		194	54	s aq; v s alc, eth
b346	3-Bromopyridine	BrC$_5$H$_4$N	158.00	20, 233	1.645^0_4	1.5695^{20}	142–143	173	51	s HOAc
b347	3-Bromoquinoline	C$_9$H$_6$BrN	208.06	20, 363	1.533	1.6640^{20}	15	276	>112	
b348	5-Bromosalicylic acid	Br(HO)C$_6$H$_3$COOH	217.02	10, 107			166			0.3 aq^{80}, 85 alc^{25}; 70 eth^{25}
b349	β-Bromostyrene	C$_6$H$_5$CH=CHBr	183.05	5, 477	1.4224^{20}_4	1.6066^{20}	7	112^{20mm}	79	i aq; misc alc, eth
b350	Bromosuccinic acid	HOOCCH$_2$CH(Br)COOH	196.99	2, 621	2.073		172 d			18 aq; s alc

β-Bromophenetole, b286
3-Bromopropene, a85

3-(Bromopropyl)benzene, b330

5-Bromopseudocumene, b361

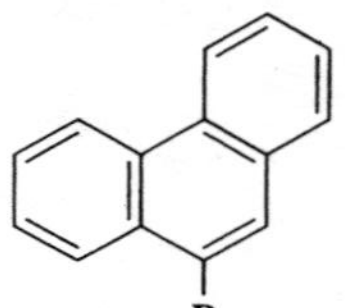

b324

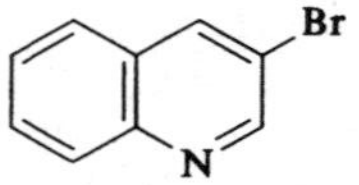

b347

TABLE 1.15 Physical Constants of Organic Compounds (*continued*)

No.	Name	Formula	Formula weight	Beilstein reference	Density	Refractive index	Melting point	Boiling point	Flash point	Solubility in 100 parts solvent
b351	*N*-Bromosuccinimide		177.99	21, 380	2.098		173 sl d			1.5 aq; 14.4 acet; 3.1 HOAc; 0.02 CCl_4
b352	1-Bromotetradecane	$H(CH_2)_{14}Br$	277.30	1^2, 136	1.0124_4^{25}	1.4600^{20}	6	178^{20mm}	>112	s alc; v s chl; misc bz, acet
b353	2-Bromothiophene	BrC_4H_3S	163.04	17, 33	1.684_4^{20}	1.5860^{20}		151	60	v s acet, eth
b354	4-Bromothiophenol	BrC_6H_4SH	189.08	6, 330			76	239	78	
b355	2-Bromotoluene	$BrC_6H_4CH_3$	171.04	5, 304	1.422_{25}^{25}	1.552^{25}	−26	181		0.1 aq; misc alc, bz, chl, eth
b356	3-Bromotoluene	$BrC_6H_4CH_3$	171.04	5, 305	1.4099^{20}	1.5517^{20}	−39.8	183.7	60	s alc, bz, eth
b357	4-Bromotoluene	$BrC_6H_4CH_3$	171.04	5, 305	1.3959_{25}^{35}	1.5490	28.5	184.5	85	s alc, bz, eth
b358	Bromotrichloromethane	$BrCCl_3$	198.28	1, 67	1.997_{25}^{25}	1.5063	−21	103.8	none	misc org solv
b359	1-Bromotridecane	$H(CH_2)_{13}Br$	263.27	1^2, 134	1.0262_4^{20}	1.4592^{20}	7	150^{10mm}	>112	v s chl
b360	Bromotrifluoromethane	$BrCF_3$	148.92	1^3, 83	1.5800_4^{20}			−57.8		v s chl
b361	5-Bromo-1,2,4-trimethylbenzene	$BrC_6H_2(CH_3)_3$	199.10	5, 403			73	235		i aq; s alc
b362	2-Bromo-1,3,5-trimethylbenzene	$BrC_6H_2(CH_3)_3$	199.10	5, 408	1.301	1.5511^{20}	2	225	96	i aq; s bz; v s eth
b363	Bromotrimethylgermane	$(CH_3)_3GeBr$	197.60		1.544^{18}	1.4705^{20}	−25	113.7		
b364	Bromotrimethylsilane	$(CH_3)_3SiBr$	153.10		1.160	1.4145^{20}		79	1	
b365	Bromotriphenylethylene	$(C_6H_5)_2C=C(Br)C_6H_5$	335.22				114–115			
b366	Bromotriphenylmethane	$(C_6H_5)_3CBr$	323.24	5, 704			152–154	230^{15mm}		
b367	11-Bromoundecanoic acid	$Br(CH_2)_{10}COOH$	265.20	2^2, 315			51	174^{2mm}		i aq; v s alc

No.	Name	Formula	Formula weight	Beilstein reference	Density	n_D	Melting point	Boiling point	Flash point	Solubility
b368	α-Bromo-o-xylene	BrCH$_2$C$_6$H$_4$CH$_3$	185.07	5, 365	1.381^{23}	1.5742^{20}	21	223–224	82	s alc, eth
b369	α-Bromo-m-xylene	BrCH$_2$C$_6$H$_4$CH$_3$	185.07	5, 374	1.370^{23}	1.5560^{20}		185^{340mm}	82	s alc, eth
b370	2-Bromo-p-xylene	BrC$_6$H$_3$(CH$_3$)$_2$	185.07	5, 385	1.340	1.5505^{20}	9–10	199–201	79	v s alc, eth
b371	4-Bromo-o-xylene	BrC$_6$H$_3$(CH$_3$)$_2$	185.07	5, 365	1.370^{15}	1.5560^{20}		215	80	77 alc; 1 bz; 20 chl
b372	Brucine		394.45	27^2, 797			178			
b373	1,2-Butadiene	CH$_3$CH=C=CH$_2$	54.09	1, 249	0.676_4^{10}	1.4205^{1}	−136.2	10.9		misc alc, eth
b374	1,3-Butadiene	CH$_2$=CHCH=CH$_2$	54.09	1, 249	0.650_4^{-6}	1.4293^{-25}	−108.9	−4.4		misc alc, eth
b375	1,3-Butadienyl acetate	CH$_3$C(=O)OCH=CH–CH=CH$_2$	112.13	2^3, 295	0.945	1.4690^{20}		60^{40mm}	33	
b376	1,3-Butadiyne	HC≡CC≡CH	50.06	1^3, 1056	0.7364_4^{0}	1.4189^{5}	−36	10.3		v s eth; s bz, acet
b377	2-Butanamine	CH$_3$CH$_2$CH(NH$_2$)CH$_3$	73.14	4, 160	0.7308_4^{15}	1.3963^{15}	−104.5	66	−19	misc aq, alc
b378	Butane	CH$_3$CH$_2$CH$_2$CH$_3$	58.12		0.6011^{0}	1.3562^{-13}	−138.3	−0.50		
b379	1,4-Butanediamine	H$_2$NCH$_2$CH$_2$CH$_2$CH$_2$NH$_2$	88.15	4, 264	0.877_4^{25}	1.4569^{20}	27–28	158–160	51	s aq
b380	Butanedinitrile	NCCH$_2$CH$_2$CN	80.09	2, 615	0.9867_4^{60}	1.4173^{60}	57.9	265–267		11.5 aq; s acet, chl, diox; sl s bz, eth
b381	1,2-Butanediol	CH$_3$CH$_2$CH(OH)CH$_2$OH	90.12	1, 477	1.006_0^{18}	1.4380^{20}	<−50	207.5	93	s aq, alc, acet
b382	1,3-Butanediol	CH$_3$CH(OH)CH$_2$CH$_2$OH	90.12	1, 477	1.00053^{20}	1.441^{20}		207.5	121	s aq, alc, acet; 9 eth

b351

CH$_3$O CH$_3$O

b372

TABLE 1.15 Physical Constants of Organic Compounds (*continued*)

No.	Name	Formula	Formula weight	Beilstein reference	Density	Refractive index	Melting point	Boiling point	Flash point	Solubility in 100 parts solvent
b383	1,4-Butanediol	HOCH$_2$CH$_2$CH$_2$CH$_2$OH	90.12	1,478	1.016_{25}^{25}	1.4452^{20}	120.9	230	>112	misc aq, alc, acet; 0.3 bz; 3.1 eth; 0.9 PE
b384	*meso*-2,3-Butanediol	CH$_3$CH(OH)CH(OH)CH$_3$	90.12	1,479	0.9939_4^{25}	1.4324^{35}	34.4	182	85	misc aq, alc
b385	D-(−)-2,3-Butanediol	CH$_3$CH(OH)CH(OH)CH$_3$	90.12	1^2,546	0.9869_4^{25}	1.4315^{25}	19.7	180^{715mm}	85	misc aq, alc; s eth
b386	2,3-Butanedione	CH$_3$C(O)C(O)CH$_3$	86.09	1,769	0.990_{15}^{15}	1.3951^{20}		88	26	25 aq; misc alc, eth
b387	1,4-Butanedithiol	HSCH$_2$CH$_2$CH$_2$CH$_2$SH	122.25	1,479	1.042	1.5290^{20}		106^{30mm}	70	i aq; v s alc
b388	1-Butanethiol	CH$_3$CH$_2$CH$_2$CH$_2$SH	90.19	1,370	0.8367_{25}^{25}	1.4403^{25}	−115.7	98.5	12	0.06 aq; v s alc, eth
b389	2-Butanethiol	CH$_3$CH$_2$CH(SH)CH$_3$	90.19	1,373	0.8246_{25}^{25}	1.4338^{25}	−165	85.0	21	sl s aq; v s alc, eth
b390	1,2,4-Butanetriol	HOCH$_2$CH$_2$CH(OH)-CH$_2$OH	106.12	1,519	1.018^{20}	1.4748^{20}		191^{18mm}	167	v s aq, alc
b391	1-Butanol	CH$_3$CH$_2$CH$_2$CH$_2$OH	74.12	1,367	0.8097^{20}	1.3993^{20}	−88.6	117.7	35	7.4 aq; misc alc, eth
b392	2-Butanol	CH$_3$CH$_2$CH(OH)CH$_3$	74.12	1,371	0.8069_4^{20}	1.3972^{20}	−114.7	99.5	26	12.5 aq; misc alc, eth
b393	2-Butanone	CH$_3$CH$_2$COCH$_3$	72.11	1^2,726	0.8049_4^{20}	1.3788^{20}	−86.7	79.6	−3	24 aq; misc alc, bz, eth
b394	2-Butanone oxime	CH$_3$CH$_2$C(=NOH)CH$_3$	87.12	1^2,730	0.9232_4^{20}	1.4428	−29.5	72^{25mm}		s aq; misc alc, eth
b395	1-Butene	CH$_3$CH$_2$CH=CH$_2$	56.10	1^3,715	0.6255_4^{-185}	1.3962^{20}	−185.3	−6.3		i aq; v s alc, eth
b396	cis-2-Butene	CH$_3$CH=CHCH$_3$	56.10	1^3,728	0.6213_4	1.3931^{-25}	−138.9	3.7		i aq; v s alc, eth
b397	trans-2-Butene	CH$_3$CH=CHCH$_3$	56.10	1^3,730	0.6041_4	1.3848^{-25}	−105.6	0.88		i aq; v s alc, eth
b398	cis-2-Butene-1,4-diol	HOCH$_2$CH=CHCH$_2$OH	88.11	1^2,567	0.0700_4^{20}	1.4793^{20}	12.5	234	128	s aq; v s alc
b399	trans-2-Butene-1,4-diol	HOCH$_2$CH=CHCH$_2$OH	88.11	1^3,2252	0.070_4^{20}	1.4779^{20}	27.3	132		v s aq, alc
b400	3-Butenenitrile	H$_2$C=CHCH$_2$CN	67.09	2,408	0.8341_4^{20}	1.4060^{20}	−87	119	21	sl s aq; misc alc, eth
b401	cis-2-Butenoic acid	CH$_3$CH=CHCOOH	86.09	2,412	1.0267_4^{20}	1.4482^{14}	14	168–169		v s aq; s alc

No.	Name	Formula	Formula wt	Beilstein ref.	Density	n_D	mp, °C	bp, °C	Flash pt, °C	Solubility
b402	*trans*-2-Butenoic acid	$CH_3CH=CHCOOH$	86.09	2, 408	0.964^{80}_{4}	1.4228^{77}	71.4	185.0	87	54.6 aq; v s EtOH, bz, acet
b403	3-Butenoic acid	$H_2C=CHCH_2COOH$	86.09	2, 407	1.0091^{20}_{20}	1.4249^{20}	−39	163	65	s aq; misc alc, eth
b404	*cis*-2-Buten-1-ol	$CH_3CH=CHCH_2OH$	72.11	1, 442	0.8662^{20}_{20}	1.4342^{20}	−89.4	123.6	56	16.6 aq; misc alc
b405	*trans*-2-Buten-1-ol	$CH_3CH=CHCH_2OH$	72.11	1, 442	0.8454^{20}_{4}	1.4289^{20}		121.2	56	16.6 aq; misc alc
b406	3-Buten-2-one	$H_2C=CHCOCH_3$	70.09	1, 728	0.8636^{20}_{4}	1.4086^{20}		81.4	−6	v s aq, alc, acet, eth
b407	1-Buten-3-yne	$HC{\equiv}CCH=CH_2$	52.07	1^3, 1032	0.7095^{5}_{4}	1.4161^{1}		5.1		
b408	4-Butoxyaniline	$CH_3(CH_2)_3OC_6H_4NH_2$	165.24	13^2, 226	0.992	1.5343^{20}		148–149^{13mm}		
b409	4-Butoxybenzoic acid	$CH_3(CH_2)_3OC_6H_4COOH$	194.23	10^2, 93			150			5 aq; s most org solv
b410	2-Butoxyethanol	$CH_3(CH_2)_3OCH_2CH_2OH$	118.18	1^2, 519	0.9012^{20}_{4}	1.4198^{20}	−40	170.2	60	misc aq, alc, bz, acet, PE, CCl_4
b411	2-(2-Butoxyethoxy)-ethanol	$HOCH_2CH_2OCH_2$-$CH_2OC_4H_9$	162.23	1^2, 521	0.9536^{20}_{20}	1.4306^{20}	−68.1	230.4	110	0.43 aq; misc alc, eth; s most org solv
b412	Butyl acetate	$C_4H_9OOCCH_3$	116.16	2, 130	0.8813_{4}	1.3941^{20}	−73.5	126.1	37	0.62 aq; s alc, eth
b413	DL-*sec*-Butyl acetate	$CH_3COOCH(CH_3)C_2H_5$	116.16	2^2, 141	0.865^{25}	1.3840^{25}		112.3	32	i aq; misc alc, eth
b414	*tert*-Butyl acetate	$(CH_3)_3COOCCH_3$	116.16	2, 131	0.8665^{20}_{4}	1.3853^{20}		97.8	15	
b415	*tert*-Butyl aceto-acetate	$(CH_3)_3COC(=O)CH_2$-$C(=O)CH_3$	158.20		0.954	1.4180^{20}			60	
b416	Butyl acrylate	$H_2C=CHCOOC_4H_9$	128.17	2^2, 388	0.894^{25}_{16}	1.4160		148	38	i aq; s alc, eth
b417	Butylamine	$CH_3CH_2CH_2CH_2NH_2$	73.14	4, 156	0.7327^{25}	1.3992^{25}	−50.5	77.9	−1	misc aq, alc, eth, PE
b418	*tert*-Butylamine	$(CH_3)_3CNH_2$	73.14				−67.5	44.4	−8	
b419	2-(*tert*-Butylamino)-ethanol	$(CH_3)_3CNHCH_2CH_2OH$	117.19	4, 173	0.6951^{20}_{4}	1.3788^{20}	42–45	90–92^{25mm}	68	misc aq, alc

TABLE 1.15 Physical Constants of Organic Compounds (*continued*)

No.	Name	Formula	Formula weight	Beilstein reference	Density	Refractive index	Melting point	Boiling point	Flash point	Solubility in 100 parts solvent
b420	3-(*tert*-Butylamino)-1,2-propanediol	$(CH_3)_3CNHCH_2CH(OH)\text{-}CH_2OH$	147.22				70	92^{1mm}		
b421	4-Butylaniline	$CH_3CH_2CH_2CH_2\text{-}C_6H_4NH_2$	149.24	12^1, 503	0.945	1.5350^{20}		120^{15mm}	101	
b422	2-*tert*-Butylanthraquinone		264.32				100			
b423	Butylbenzene	$CH_3CH_2CH_2CH_2C_6H_5$	134.22	5, 413	0.8604^{20}	1.4898^{20}	−88	183.3	59	misc alc, bz, eth
b424	*sec*-Butylbenzene	$CH_3CH_2CH(CH_3)C_6H_5$	134.22	5, 414	0.8608^{24}	1.4902^{20}	−82.7	173.3	45	misc alc, bz, eth
b425	*tert*-Butylbenzene	$(CH_3)_3CC_6H_5$	134.22	5, 415	0.8669^{20}	1.4927^{20}	−57.9	169.1	34	misc alc, bz, eth
b426	Butyl benzoate	$C_6H_5COOC_4H_9$	178.23	9, 112	1.000^{20}	1.496	−22	250		i aq; s alc, eth
b427	4-*tert*-Butylbenzoic acid	$(CH_3)_3CC_6H_4COOH$	178.23	9, 560			167			i aq; v s alc, bz
b428	4-*tert*-Butylbenzoyl chloride	$(CH_3)_3CC_6H_4COCl$	196.68	12, 1022	1.007	1.5364^{20}		135^{20mm}	87	
b429	N-(*tert*-Butyl)benzylamine	$C_6H_5CH_2NHC(CH_3)_3$	163.27		0.881	1.4968^{20}		80^{5mm}	80	i aq; misc alc, eth
b430	Butyl butyrate	$CH_3CH_2CH_2COOC_4H_9$	144.21	2, 271	0.8717^{20}_{20}	1.4035		156.9	51	
b431	*tert*-Butyl carbazate	$H_2NNHCOOC(CH_3)_3$	132.16				42	$65^{0.03mm}$	151	
b432	4-*tert*-Butylcatechol	$(CH_3)_3CC_6H_3(OH)_2$	166.22		1.049^{60}_{25}		55	285		0.2 aq^{80}; 240 eth^{25}; s alc; v s acet
b433	*tert*-Butyl chloroacetate	$ClCH_2COOC(CH_3)_3$	150.61	2^3, 444	1.053	1.4230^{20}		$48\text{-}49^{11mm}$	41	
b434	4-*tert*-Butyl-1-chlorobenzene	$(CH_3)_3CC_6H_4Cl$	168.67	5, 416	1.006	1.5108^{20}	23–25	217		
b435	*tert*-Butylchlorodiphenylsilane	$(CH_3)_3CSi(C_6H_5)_2Cl$	274.87	3^2, 11	1.057	1.5675^{20}		$90^{0.02mm}$	>112	
b436	Butyl chloroformate	$ClCOOC_4H_9$	136.58		1.074^{25}_{4}	1.4114^{20}		142	25	d aq; alc; misc eth

b437	S-*tert*-Butyl chlorothioformate	$ClC(=O)SC(CH_3)_3$	152.6		1.081_4^{30}	1.4691^{30}		42.0^{10mm}	46	
b438	*tert*-Butyl cyanoacetate	$NCCOOC(CH_3)_3$	141.17			1.4200^{20}		108		
b439	2-*tert*-Butylcyclohexanol	$(CH_3)_3CC_6H_{10}OH$	156.27		0.902		46			i aq
b440	4-*tert*-Butylcyclohexanol	$(CH_3)_3CC_6H_{10}OH$	156.27	6^1, 18	0.896		70	115^{15mm}	105	i aq
b441	2-*tert*-Butylcyclohexanone	$(CH_3)_3CC_6H_9(=O)$	154.25	7^3, 143		1.4565^{20}		62.5^{4mm}		
b442	4-*tert*-Butylcyclohexanone	$(CH_3)_3CC_6H_9(=O)$	154.25	7^1, 29			50	116^{20mm}	96	i aq
b443	Butyl decyl *o*-phthalate	$C_4H_9OOCC_6H_4COOC_{10}H_{21}$	362.51		0.994_{25}^{25}				202	
b444	*N*-Butyldiethanolamine	$C_4H_9N(CH_2CH_2OH)_2$	161.25	4, 285	0.986_{20}^{20}	1.4625^{20}	<-70	276	126	
b445	Butyl 3,4-dihydro-2,2-dimethyl-4-oxo-2*H*-pyran-6-carboxylate		226.27		1.054_{25}^{25}	1.4767^{20}		256–270	>112	misc alc, chl, eth

C(CH₃)₃

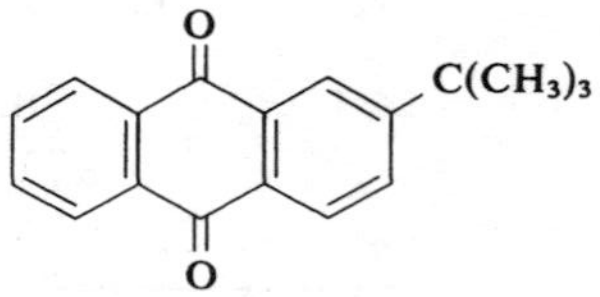

b422

CH₃ CH₃

C₄H₉O—C

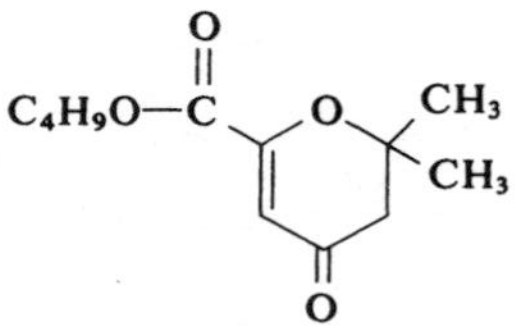

b445

TABLE 1.15 Physical Constants of Organic Compounds (*continued*)

No.	Name	Formula	Formula weight	Beilstein reference	Density	Refractive index	Melting point	Boiling point	Flash point	Solubility in 100 parts solvent
b446	*tert*-Butyldimethyl-chlorosilane	$(CH_3)_3CSi(CH_3)_2Cl$	150.7		1.028		91.5	124–126		v s aq; s alc
b447	1,3-Butylene diacetate	$CH_3CH(OOCCH_3)CH_2\text{-}CH_2OOCCH_3$	174.20	2, 143	0.89^{20}	1.4199^{20}		99^{8mm}	85	
b448	*N*-Butylethanolamine	$HOCH_2CH_2NHC_4H_9$	117.19	1^3, 1502	0.7495^{20}_4	1.444^{20}	-3.5	192	77	i aq; misc alc, eth
b449	Butyl ethyl ether	$C_4H_9OC_2H_5$	102.18		0.931^{50}_{20}	1.3818^{20}	-103	92.5		0.8 aq
b450	2-Butyl-2-ethyl-1,3-propanediol	$HOCH_2C(C_2H_5)(C_4H_9)\text{-}CH_2OH$	160.25			1.4587^{25}	41.4	195^{100mm}		
b451	Butyl ethyl sulfide	$C_4H_9SC_2H_5$	118.24	1^3, 1522	0.8376^{20}_4	1.4491^{20}	-95.1	144.2		s chl
b452	*tert*-Butylhydrazine HCl	$(CH_3)_3CNHNH_2 \cdot HCl$	124.61	4^3, 1734			191–174			
b453	*tert*-Butylhydroper-oxide	$(CH_3)_3C\text{—}O\text{—}OH$	90.12		0.896^{20}_4	1.4007^{20}	4–5	$33\text{-}4^{17mm}$	62	s aq, alc, chl, eth
b454	*tert*-Butylhydro-quinone	$(CH_3)_3CC_6H_3(OH)_2$	166.22				129			
b455	Butyl isocyanate	$CH_3CH_2CH_2CH_2NCO$	99.13	4, 175	0.880	1.4061^{20}		115	26	
b456	*tert*-Butyl isocyanate	$(CH_3)_3CNCO$	99.13		0.868	1.3865^{20}		86	26	
b457	Butyllithium	$CH_3CH_2CH_2CH_2Li$	64.06					$80^{0.0001mm}$	pyro-phoric	
b458	*tert*-Butyllithium	$(CH_3)_3CLi$	64.06					subl $70^{0.1mm}$	pyro-phoric	
b459	Butyl methacrylate	$H_2C\text{=}C(CH_3)COOC_4H_9$	142.19	1, 381	0.889^{25}_{15}	1.4220^{25}	-109	170	49	i aq; misc alc, eth
b460	*tert*-Butyl methyl ether	$(CH_3)_3COCH_3$	88.15		0.758	1.3685^{20}		56	-10	s aq; v s alc, eth
b461	2-*tert*-Butyl-4-methylphenol	$(CH_3)_3CC_6H_3(CH_3)OH$	164.25		0.9247^{75}_4	1.4969^{75}	51.7	237		i aq; s org solv

No.	Name	Formula	Formula weight	Beilstein reference	Density	n_D	Melting point, °C	Boiling point, °C		Solubility
b462	2-*tert*-Butyl-6-methylphenol	$(CH_3)_3CC_6H_3(CH_3)OH$	164.25			1.5195^{20}	32	230	107	v s alc
b463	Butyl methyl sulfide	$C_4H_9SCH_3$	104.21	1^3, 1521	0.8426^{20}_{4}	1.4477^{20}	−97.8	123.4	4	misc alc, eth
b464	Butyl nitrite	C_4H_9ONO	103.12	1, 369	0.9114^{0}_{4}	1.3768		78		sl s aq; v s alc, chl, eth, CS_2
b465	*tert*-Butyl nitrite	$(CH_3)_3CONO$	103.12	1^2, 415	0.8671^{20}_{4}	1.3687^{20}		63		
b466	Butyl octadecanoate	$CH_3(CH_2)_{16}COOC_4H_9$	340.60	2^2, 352	0.8551^{20}_{4}	1.4422^{25}	26.3	343	160	s alc, v s acet
b467	Butyl 4-oxopentanoate	$CH_3C({=}O)CH_2CH_2COOC_4H_9$	172.22		0.9735^{5}_{4}	1.4270^{20}		107^{6mm}	91	s alc, eth, acet
b468	*tert*-Butyl peroxybenzoate	$C_6H_5({=}O)O{-}OC(CH_3)_3$	194.23		1.021	1.4990^{20}		$76^{0.2mm}$	93	
b469	2-*sec*-Butylphenol	$CH_3CH_2CH(CH_3)C_6H_4OH$	150.22		0.982	1.5222^{20}	12	228	112	i aq; s alc; v s eth
b470	2-*tert*-Butylphenol	$(CH_3)_3CC_6H_4OH$	150.22	6^2, 489	0.9783^{3}_{4}	1.5228^{20}	−7	221–224	110	
b471	3-*tert*-Butylphenol	$(CH_3)_3CC_6H_4OH$	150.22				40–41	240		
b472	4-*sec*-Butylphenol	$CH_3CH_2CH(CH_3)C_6H_4OH$	150.22	6, 522	0.9694^{20}_{4}	1.5150	62	136^{25mm}	115	s hot aq, alc, eth
b473	4-*tert*-Butylphenol	$(CH_3)_3CC_6H_4OH$	150.22	6, 524	0.908^{114}_{4}	1.4787^{114}	100–101	237	149	i aq; s alc, eth
b474	2-(4-*sec*-Butylphenoxy)ethanol	$CH_3CH_2CH(CH_3)C_6H_4OCH_2CH_2OH$	194.2		1.008^{25}		<−20	158^{10mm}		0.1 aq
b475	2-(4-*tert*-Butylphenoxy)ethanol	$(CH_3)_3CC_6H_4OCH_2CH_2OH$	194.3		1.016^{25}		54	167^{10mm}	157	0.1 aq
b476	*tert*-Butyl phenyl carbonate	$C_6H_5OC({=}O)OC(CH_3)_3$	194.23		1.047	1.4805^{20}		$79^{0.8mm}$		
b477	Butyl phenyl ether	$CH_3CH_2CH_2CH_2OC_6H_5$	150.22	6, 143	0.9351^{20}_{4}	1.4970^{20}	−19	210.3	82	

TABLE 1.15 Physical Constants of Organic Compounds (*continued*)

No.	Name	Formula	Formula weight	Beilstein reference	Density	Refractive index	Melting point	Boiling point	Flash point	Solubility in 100 parts solvent
b478	4-*tert*-Butylphenyl salicylate	$HOC_6H_4COOC_6H_4-C(CH_3)_3$	270.31				62–64			0.1 aq; 79 alc; 153 EtAc; 158 toluene
b479	Butyl propionate	$CH_3CH_2COOC_4H_9$	130.19	2, 241	0.8818^{15}	1.3982^{25}		145.5		misc alc, eth
b480	4-*tert*-Butylpyridine	$(CH_3)_3C(C_5H_4N)$	135.21	20, 252	0.915	1.4952^{20}	−89.6	197	63	
b481	Butyltin chloride	$C_4H_9SnCl_3$	282.17		1.693	1.5229^{20}		93^{10mm}	81	
b482	4-*tert*-Butyltoluene	$(CH_3)_3C_6H_4CH_3$	148.25	5, 439	0.853	1.4897^{20}		192	54	d aq, hot alc; s eth
b483	Butyltrichlorosilane	$C_4H_9SiCl_3$	191.5	4^1, 582	1.1614^{20}	1.436^{20}		142–143		
b484	Butyl trifluoro-acetate	$CF_3COOC_4H_9$	170.1		1.0268^{22}	1.353^{22}		100.2		
b485	Butyltrimethoxysilane	$C_4H_9Si(OCH_3)_3$	178.3		0.9312_4^{20}	1.3979^{20}		164–165		s aq, alc, eth
b486	*tert*-Butyl tri-methylsilyl peroxide	$(CH_3)_3C-O-O-Si-(CH_3)_3$	162.3		0.8219_4^{20}	1.3935^{25}	d 135	41^{41mm}		0.3 aq
b487	Butyl urea	$C_4H_9NHCONH_2$	116.16	4^1, 371			93–95			s aq, alc, eth
b488	Butyl vinyl ether	$C_4H_9OCH=CH_2$	100.16		0.7792^{20}	1.4007^{20}	−112.7	94.2	−9	0.3 aq
b489	5-*tert*-Butyl-*m*-xylene	$(CH_3)_3CC_6H_3(CH_3)_2$	162.28	5, 447	0.867	1.4946^{20}		205–206	72	
b490	1-Butyne	$CH_3CH_2C\equiv CH$	54.09	1, 249	0.7110_4^{-31}	1.3962^{20}	−125.7	8.1		i aq; s alc, eth
b491	2-Butyne	$CH_3C\equiv CCH_3$	54.09	1, 249	0.6910_4^{20}	1.3920^{20}	−32.3	17.0		i aq; s alc, eth
b492	2-Butyne-1,4-diol	$HOCH_2C\equiv CCH_2OH$	86.09	1^1, 261		1.450^{25}	54–58	238	152	374 aq; 83 alc; 0.04 bz; 2.6 eth; 70 acet
b493	Butyraldehyde	$CH_3CH_2CH_2CHO$	72.11	1, 662	0.8016_4^{20}	1.3791^{20}	−96.4	74.8	−6.7	7.1 aq; misc alc, eth, acet, EtAc
b494	Butyramide	$CH_3CH_2CH_2CONH_2$	87.12	2, 275			116	216		16 aq; s alc
b495	Butyric acid	$CH_3CH_2CH_2COOH$	88.11	2, 264	0.9582_4^{20}	1.3980^{20}	−5.3	163.3	77	misc aq, alc, eth
b496	Butyric anhydride	$[CH_3CH_2CH_2C(O)]_2O$	158.20	2, 274	0.9668_4^{20}	1.4130^{20}	−65.7	199.5	87	s aq, alc(d), eth

b497	3-Butyrolactone		86.09	17^1, 130	1.056	1.4109	20	73^{29mm}	60	misc aq, alc, acet, bz, eth, CCl_4
b498	4-Butyrolactone		86.09	17, 234	1.124_4^{25}	1.4348^{25}	−43.5	204	98	3.3 aq; misc alc, eth
b499	Butyronitrile	$CH_3CH_2CH_2CN$	69.11	2^2, 252	0.7954_4^{15}	1.3860^{15}	−111.9	117.9	16	s aq, alc(d); misc eth
b500	Butyrophenone	$C_6H_5C(O)C_3H_7$	148.21	7, 313	1.021	1.5195^{20}	13	222	88	
b501	Butyryl chloride	$CH_3CH_2CH_2COCl$	106.55	2, 274	1.0263_4^{21}	1.4122^{20}	−89	102	21	
c1	Caffeine		194.19	26, 461	1.23_4^{18}		238	subl 178		2.1 aq; 1.5 alc; 18 chl; 0.19 eth; 1 bz
c2	DL-Camphene		136.24	5, 156	0.8422_4^{54}	1.4551^{54}	51–52	159		i aq; s alc, chl, eth
c3	D-(+)-Camphor		152.23	7, 101	0.9920_4^{25}		178.8	207.4	36	100 alc; 100 eth; 200 chl; 250 acet
c4	DL-Camphor		152.24	7, 135			177	204		4 aq; 100 alc; s chl, eth
c5	D-Camphoric acid		200.23	9, 745	1.1864_4^{20}		186–188		64	

Butyl o-phthalate, d128
Butyl propyl ketone, 036
Butyl stearate, b466
Butyl sulfate, d131

Butyl sulfides, d132, d133
Butyl sulfite, d134
Butyl sulfone, d135
Butyrolactam, p275

Cadaverine, p29
2-Camphanone, c3

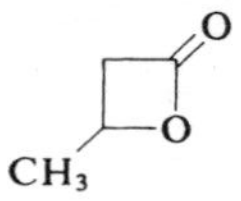

b497

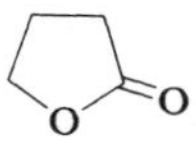

b498

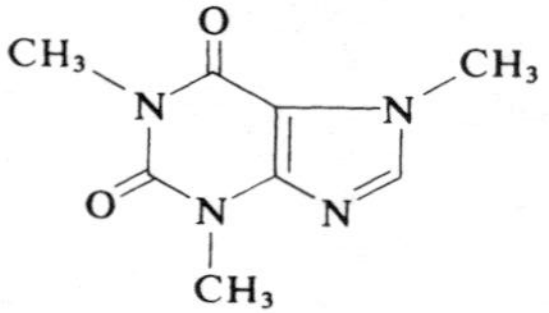

c1

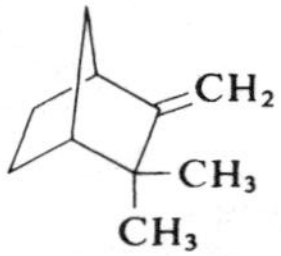

c2

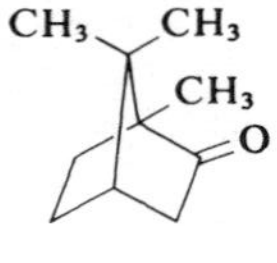

c3, c4

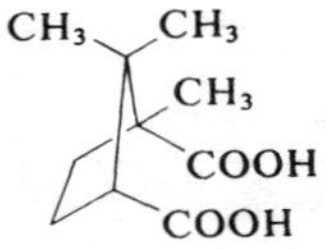

c5

TABLE 1.15 Physical Constants of Organic Compounds (*continued*)

No.	Name	Formula	Formula weight	Beilstein reference	Density	Refractive index	Melting point	Boiling point	Flash point	Solubility in 100 parts solvent
c6	DL-Camphoric anhydride		182.22	17, 455	1.194_4^{20}		225	270		s bz; sl s aq, alc, eth
c7	D-10-Camphor-sulfonic acid hydrate		250.32	11, 316			194 d			deliq moist air; sl s HOAc, EtAc; i eth
c8	Carbazole		167.21	20, 433			245–246	355		
c9	4-Carbethoxy-3-methyl-3-cyclohexen-1-one		182.22	10, 631	1.078	1.4880^{20}		268–272	>112	
c10	Carbobenzyloxy-glycine	$C_6H_5CH_2OC(=O)NH-CH_2COOH$	209.20				122			v s aq; i alc, bz, eth
c11	Carbohydrazide	$H_2NNHC(=O)NHNH_2$	90.09	3, 121			d 153			
c12	2-(Carbomethoxy)-ethylmethyl-dichlorosilane	$CH_3OC(=O)CH_2CH_2Si(CH_3)Cl_2$	201.1		1.187_4^{25}	1.4439^{25}		$98–99^{25mm}$		
c13	2-Carbomethoxyethyl-trichlorosilane	$CH_3OC(=O)CH_2CH_2SiCl_3$	221.6		1.3254_4^{20}	1.448^{20}		$88–89^{2mm}$		
c14	2-Carboxybenz-aldehyde	$HC(=O)C_6H_4COOH$	150.13	10, 666			96–98			
c15	4-Carboxy-1,2-ben-zenedicarboxylic anhydride		192.13	18, 468			161–164	$240–245^{14mm}$		15.5 DMF; 49.6 acet; 21.6 EtAc
c16	4-Carboxybenzene-sulfonamide	$HOOCC_6H_4SO_2NH_2$	201.20	11, 390			d 280			i aq, bz, eth; v s alc
c17	2-Carboxyethyl-phosphonic acid	$HOOCCH_2CH_2-P(O)(OH)_2$	154.06	4^2, 976						
c18	DL-Carnitine HCl	$(CH_3)_3NCH_2CHOH-CH_2COOH \cdot HCl$	197.66				197 d			v s aq; i acet, eth

No.	Name	Formula wt	Ref.	mp, °C	bp, °C	Density	n_D	Solubility
c19	trans-β-Carotene	536.89	30, 87	183				i aq; s bz, chl, CS$_2$
c20	D-(+)-Carvone	150.22	7, 153		230	0.965^{20}_{4}	1.4989^{20}	i aq; misc alc
c21	Catecholborane	119.92	88	12	$50^{50\text{mm}}$	1.000^{20}_{20}	1.5070^{20}	

Capric acid, d14
Caproaldehyde, h54
Caproic acid, h66
Caproic anhydride, h67
ε-Caprolactam, o57
ε-Caprolactone, h71
Capronitrile, h63
Caproyl chloride, h73
Caprylic acid, o29
Capryl alcohol, o30
Caprylaldehyde, o40

Caprylonitrile, o27
Capryloyl chloride, o37
CAPS, c337
N-(Carbamoylmethyl)iminodiacetic acid, a14
Carbamylurea, b215
Carbanilide, d693
Carbazole, d665
Carbitol, e35
Carbitol acetate, e36
Carbobenzoxy chloride, b90
4,4′-Carboxyldiphthalic anhydride, b55

N-Carbonylsulfamyl chloride, c240
Carboxybenzaldehyde, f33
(3-Carboxy-2-hydroxypropyl)trimethylammonium hydroxide, c18
3-Carbomethoxypropionyl chloride, m188
3-Carboxymethylimino)bis(ethylenenitrilo)-tetraacetic acid, d299
(Carboxylmethyl)trimethylammonium hydroxide, b128
3-Carboxypropyl disulfide, d708

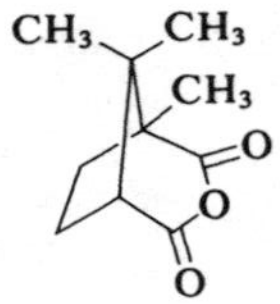

c6 (CH$_3$, CH$_3$, CH$_3$; O O O)

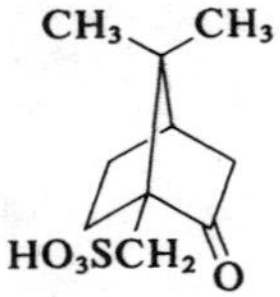

c7 (CH$_3$, CH$_3$, CH$_3$; HO$_3$SCH$_2$; O)

c8 (N–H)

c9 (CH$_3$; O=COCH$_2$CH$_3$; O)

c15 (HOOC; O O O)

c19 (CH$_3$ groups along polyene chain)
[Note: 9 successive
units]

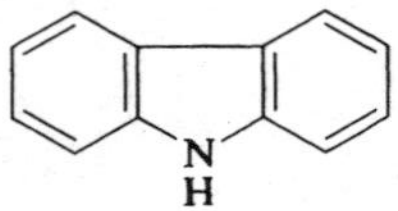

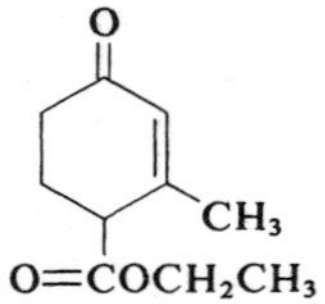

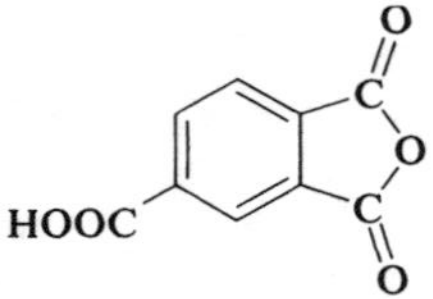

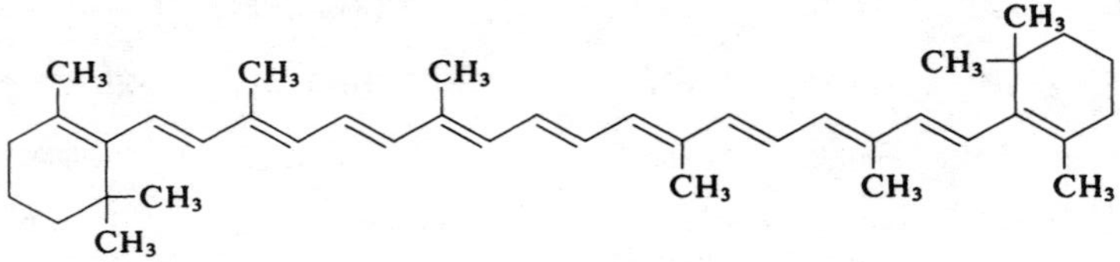

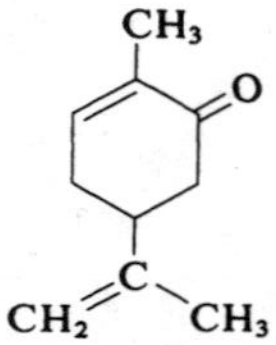

c20 (CH$_3$; O; CH$_2$=, CH–CH$_3$)

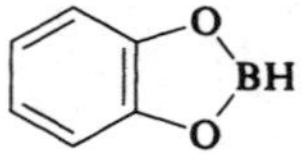

c21 (O–BH–O)

TABLE 1.15 Physical Constants of Organic Compounds (*continued*)

No.	Name	Formula	Formula weight	Beilstein reference	Density	Refractive index	Melting point	Boiling point	Flash point	Solubility in 100 parts solvent
c22	2-Chloroacetamide	$ClCH_2CONH_2$	93.51	2, 199			118	225 d		10 aq; 10 alc; sl s eth
c23	p-Chloroacetanilide	$ClC_6H_4NHCOCH_3$	169.61	12, 611	1.3854^{20}		179			i aq; v s alc, eth, CS_2
c24	Chloroacetic acid	$ClCH_2COOH$	94.50	2, 194	$1.580(c)$	1.4297^{65}	$63(\alpha)$	189		v s aq; s alc, bz, eth
c25	Chloroacetic anhydride	$[ClCH_2C(O)]_2O$	170.98	2, 199	1.5494^{20}_4		46	203		d aq; v s chl, eth
c26	p-Chloroacetoacetanilide	$CH_3COCH_2CONHC_6H_4Cl$	211.65				134			
c27	Chloroacetonitrile	$ClCH_2CN$	75.50	2, 201	1.193	1.4225^{20}		126	47	i aq; v s alc, bz, eth
c28	α-Chloroaceto-phenone	$C_6H_5COCH_2Cl$	154.60	7, 282	1.324^{15}		54	245		
c29	o-Chloroaceto-phenone	$ClC_6H_4COCH_3$	154.60	7^1, 151	1.188	1.5438^{20}		228^{738mm}	88	sl s aq; s eth
c30	p-Chloroaceto-phenone	$ClC_6H_4COCH_3$	154.60	7, 281	1.192^{20}_4	1.5549	20–21	237	90	i aq; misc alc, eth
c31	Chloroacetyl chloride	$ClCH_2COCl$	112.94	2, 199	1.4185^{25}_{25}	1.4530^{20}	−22.5	106	none	d aq, MeOH
c32	2-Chloroacrylo-nitrile	$H_2C{=}C(Cl)CN$	87.51		1.096	1.4290^{20}	−65	89	6	
c33	2-Chloroaniline	$ClC_6H_4NH_2$	127.57	12, 597	1.2125^{20}_4	1.5881^{20}	−1.94	208.8	97	0.88 aq; s alc, bz, eth
c34	3-Choroaniline	$ClC_6H_4NH_2$	127.57	12, 602	1.2150^{22}_4	1.5931^{20}	−10.4	230.5	123	i aq; s alc, bz, eth
c35	p-Chloroaniline	$ClC_6H_4NH_2$	127.57	12, 607	1.169^{77}_4	1.5546^{85}	72.5	232		s hot aq; v s alc, acet, eth, CS_2
c36	1-Chloroanthra-quinone		242.66	7, 787			160	subl		sl s alc; misc eth; s hot bz
c37	2-Chloroanthra-quinone		242.66	7, 787			211	subl		sl s alc, bz; i eth

No.	Name	Formula	Form. wt.	Ref.	Density	n_D	m.p.	b.p.		Solubility
c38	2-Chlorobenz-aldehyde	ClC_6H_4CHO	140.57	7, 233	1.2483^{20}_{4}	1.5658	11	215	87	sl s aq; s alc, bz, eth
c39	4-Chlorobenz-aldehyde	ClC_6H_4CHO	140.57	7, 235	1.1964^{61}	1.552^{61}	47	214	87	s aq; v s alc, bz, eth
c40	2-Chlorobenzamide	$ClC_6H_4CONH_2$	155.58	9, 336			142–144			s hot aq, hot alc, hot eth
c41	Chlorobenzene	C_6H_5Cl	112.56	5, 199	1.1063^{20}	1.5248^{20}	−45.3	131.7	23	0.049 aq^{30}; v s alc, bz, chl, eth
c42	4-Chlorobenzene-sulfonamide	$ClC_6H_4SO_2NH_2$	191.64	11, 55			146			
c43	4-Chlorobenzene-sulfonyl chloride	$ClC_6H_4SO_2Cl$	211.07	11, 55			55	141^{15mm}		d aq, alc; v s bz, eth
c44	4-Chlorobenzhydrol	$ClC_6H_4CH(OH)C_6H_5$	218.68	6, 680			58–60			
c45	2-Chlorobenzoic acid	ClC_6H_4COOH	156.57	9, 334	1.544^{25}_{4}		142			0.11 aq; v s alc, eth

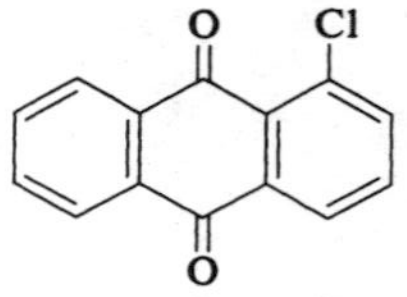

c36

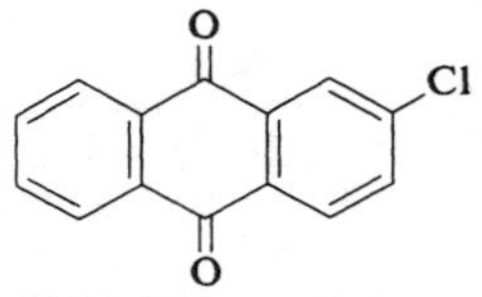

c37

TABLE 1.15 Physical Constants of Organic Compounds (*continued*)

No.	Name	Formula	Formula weight	Beilstein reference	Density	Refractive index	Melting point	Boiling point	Flash point	Solubility in 100 parts solvent
c46	3-Chlorobenzoic acid	ClC_6H_4COOH	156.57	9, 337	1.4964_4^{25}		157–158			0.04 aq; v s alc, eth
c46a	4-Chlorobenzoic acid	ClC_6H_4COOH	156.57	9, 340			241–243			0.02 aq; v s alc, eth
c47	2-Chlorobenzonitrile	ClC_6H_4CN	137.57	9, 336			46	232		s alc, eth
c48	4-Chlorobenzonitrile	ClC_6H_4CN	137.57	9, 341			93	22		s alc, bz, chl, eth
c49	2-Chlorobenzophenone	$ClC_6H_4COC_6H_5$	216.67	7, 419			44–47	300		s alc, acet, bz, eth
c50	4-Chlorobenzophenone	$ClC_6H_4COC_6H_5$	216.67	7, 419			77	196^{17mm}		
c51	2-Chlorobenzotrifluoride	$ClC_6H_4CF_3$	180.56		1.3540^{25}	1.4513^{25}	−6.4	152.3		
c52	3-Chlorobenzotrifluoride	$ClC_6H_4CF_3$	180.56		1.3311^{25}	1.4438^{25}	−56.7	137.7	36	
c53	4-Chlorobenzotrifluoride	$ClC_6H_4CF_3$	180.56		1.353^{20}	1.4463	−33.2	138.7	47	
c54	2-(4-Chlorobenzoyl)-benzoic acid	$ClC_6H_4COC_6H_4COCH$	260.68	10, 750			150			s alc, bz, eth
c55	2-Chlorobenzoyl chloride	ClC_6H_4COCl	175.01	9, 336	1.382	1.5718^{20}	−3	238	110	d aq, alc
c56	4-Chlorobenzoyl chloride	ClC_6H_4COCl	175.01	9, 341	1.377	1.5780^{20}	14	222	105	d aq, alc
c57	4-Chlorobenzyl alcohol	$ClC_6H_4CH_2OH$	142.59	6, 444			72	234		v s alc, eth
c58	4-Chlorobenzylamine	$ClC_6H_4CH_2NH_2$	141.60	12, 1074	1.164	1.5586^{20}		215	90	
c59	2-Chlorobenzyl chloride	$ClC_6H_4CH_2Cl$	161.03	5, 297	1.274	1.5591^{20}	−17	214	82	
c60	4-Chlorobenzyl chloride	$ClC_6H_4CH_2Cl$	161.03	5, 308			30	214	97	s alc; v s eth

c61	2(p-Chlorobenzyl)-pyridine	$ClC_6H_4CH_2—C_5H_4N$	203.67		1.390	1.5868^{20}		183^{20mm}	>112	
c62	4-(p-Chlorobenzyl)-pyridine	$ClC_6H_4CH_2—C_5H_4N$	203.67		1.167	1.5900^{20}			>112	v s chl
c63	1-Chloro-1,3-butadiene	$H_2C=CHCH=CHCl$	88.54	1[3], 949	0.9601^{20}_4	1.4712^{20}		68		0.11 aq; misc alc, eth
c64	1-Chlorobutane	$CH_3CH_2CH_2CH_2Cl$	92.57	1, 118	0.8864^{20}_4	1.4021^{20}	−123.1	78.44	−6	0.1 aq; misc alc, eth
c65	2-Chlorobutane	$CH_3CH_2CH(Cl)CH_3$	92.57	1, 119	0.8732^{20}_4	1.3971^{20}	−113.3	68.25	−15	0.1 aq; misc alc, eth
c66	4-Chloro-1-butanol	$ClCH_2CH_2CH_2CH_2OH$	108.56	1[2], 398	1.0883^{20}_4	1.4518^{20}		$86–89^{20mm}$	32	s alc, eth
c67	3-Chloro-2-butanone	$CH_3CH(Cl)COCH_3$	106.55	1, 669	1.055	1.4172^{20}		117	21	v s alc, eth
c68	cis-1-Chloro-2-butene	$CH_3CH=CHCH_2Cl$	90.55	1[2], 176	0.9426^{20}_4	1.4390^{20}		84.1	−15	s alc, acet
c69	3-Chloro-1-butene	$CH_3CH(Cl)CH=CH_2$	90.55	1[2], 174	0.90001^{20}_4	1.4155^{20}		62–65	−20	v s acet
c70	3-Chloro-1-butyne	$CH_3CH(Cl)C≡CH$	88.54	1[4], 970	0.961	1.4280^{20}		68–70	1	
c71	3-Chlorobutyric acid	$CH_3CH(Cl)CH_2COOH$	122.55	2, 277	1.186^{20}_4	1.4421^{20}	16.3	109^{17mm}		s alc, eth
c72	4-Chlorobutyric acid	$ClCH_2CH_2CH_2COOH$	122.55	2, 278	1.2336^{20}_4	1.4510^{20}	12–16	196^{22mm}	>112	sl s aq; v s eth
c73	4-Chlorobutyro-nitrile	$ClCH_2CH_2CH_2CN$	103.55	2, 278	1.158	1.4413^{20}		197	85	s alc, eth
c74	4-Chlorobutyryl chloride	$ClCH_2CH_2CH_2COCl$	141.00	2, 278	1.258	1.4609^{20}		174	72	d aq; alc; s eth
c75	Chloro(chloromethyl)-dimethylsilane	$ClCH_2Si(CH_3)_2Cl$	143.09	9, 594	1.086	1.4373^{20}		114^{752mm}	21	
c76	trans-p-Chloro-cinnamic acid	$ClC_6H_4CH=CHCOOH$	182.61				248–250			i aq; s alc, eth
c77	Chlorocyclohexane	ClC_6H_{11}	118.61	5, 21	1.000^{20}_4	1.4620^{20}	−44	142	28	
c78	2-Chlorocyclo-hexanone	$ClC_6H_9(=O)$	132.59	7, 10	1.161	1.4835^{20}	23	83^{10mm}	82	s bz, eth, diox

4-Chlorobenzyl mercaptan, c248

Chlorocresols, c158, c159

1.155

TABLE 1.15 Physical Constants of Organic Compounds (*continued*)

No.	Name	Formula	Formula weight	Beilstein reference	Density	Refractive index	Melting point	Boiling point	Flash point	Solubility in 100 parts solvent
c79	Chlorocyclopentane	ClC_5H_9	104.58	5, 19	1.0051_4^{20}	1.4512^{20}		114	15	i aq
c80	1-Chlorodecane	$CH_3(CH_2)_9Cl$	176.73	1, 168	0.868	1.4362^{20}	−34	223	83	i aq
c81	2-Chloro-1,1-diethoxyethane	$ClCH_2CH(OC_2H_5)_2$	152.62	1, 611	1.018	1.4157^{20}		157	29	
c82	3-Chloro-1,1-diethoxypropane	$ClCH_2CH_2CH(OC_2H_5)_2$	166.65	1, 632	0.995	1.4240^{20}		84^{25mm}	36	
c83	Chlorodifluoroacetic acid	$F_2C(Cl)COOH$	130.48	2, 201	1.118^{21}	1.3559^{20}	22.9	121.5		
c84	1-Chloro-1,1-difluoroethane	$CH_3C(Cl)F_2$	100.50				−131	−9		0.19 aq
c85	Chlorodifluoromethane	$HCClF_2$	86.47		1.209^{21}		−160	−40.8		0.30 aq
c86	α-Chloro-3',4'-dihydroxyacetophenone	$(HO)_2C_6H_3C(=O)CH_2Cl$	186.59	8, 273			176			
c87	1-Chloro-2,4-dihydroxybenzene	$ClC_6H_3(OH)_2$	144.56	6^2, 818			107	147^{18mm}		v s aq, alc, chl, eth
c88	2-Chloro-1,4-dihydroxybenzene	$ClC_6H_3(OH)_2$	144.56	6, 849			101–102	263		v s aq; i alc; s eth
c89	2-Chloro-1,1-dimethoxyethane	$ClCH_2CH(OCH_3)_2$	124.57		1.094_{20}^{20}	1.4148^{20}		130	28	
c90	4-Chloro-3,5-dimethylphenol	$Cl(CH_3)_2C_6H_2OH$	156.61	6^2, 463			115.5	246		0.1 aq; l alc; s bz, eth, alk
c91	1-Chloro-2,2-dimethylpropane	$(CH_3)_3CCH_2Cl$	106.59		0.866_4^{20}	1.4042^{20}	−20	84.4		
c92	Chlorodimethylsilane	$(CH_3)_2Si(Cl)H$	94.62		0.852_4^{20}	1.3827^{20}	−111	36	−28	
c93	Chlorodimethylvinylsilane	$(CH_3)_2Si(Cl)CH=CH_2$	120.7		0.884_4^{25}	1.414^{25}		82.5		

No.	Name	Formula	Mol. wt.	Beilstein ref.	Density	n_D	m.p., °C	b.p., °C	Flash pt., °C	Solubility
c94	1-Chloro-2,4-di-nitrobenzene	$ClC_6H_3(NO_2)_2$	202.55	5, 263	1.4982^{75}_{4}	1.5857^{60}	52–54	315	186	sl s alc; s hot alc, bz, eth
c95	1-Chloro-3,4-di-nitrobenzene	$ClC_6H_3(NO_2)_2$	202.55	5, 262	1.6867^{16}	1.5870^{20}			>112	v s eth; s alc
c96	2-Chloro-3,5-di-nitrobenzoic acid	$ClC_6H_2(NO_2)_2COOH$	246.56	9, 415			198	241 ex-plodes		0.3 aq
c97	α-Chlorodiphenyl-methane	$C_6H_5CH(Cl)C_6H_5$	202.68	5^2, 600	1.140^{20}_{4}	1.5951^{20}	17	140^{3mm}	>112	
c98	Chlorodiphenyl-methylsilane	$(C_6H_5)_2Si(Cl)CH_3$	232.8		1.1277^{20}_{4}	1.5742^{20}		295		
c99	Chlorodiphenyl-phosphine	$(C_6H_5)_2PCl$	220.64	16, 763	1.229	1.6338^{20}		320	>112	
c100	1-Chlorododecane	$CH_3(CH_2)_{11}Cl$	204.79		0.8673^{20}_{4}	1.4426	−9	116	93	v s alc; s bz
c101	1-Chloro-2,3-epoxy-propane	$H_2C{-}CHCH_2Cl$ (O)	92.53	17, 6	1.1812^{20}_{4}	1.4381^{20}	−57.2	116.1	33	5.9 aq; misc alc, chl,
c102	Chloroethane	CH_3CH_2Cl	64.52	1, 82	0.9214^{0}_{4}	1.3742^{10}	−136 to −138	12.3	−43	0.45 aq^0; 48 alc; misc eth
c103	2-Chloroethanol	$ClCH_2CH_2OH$	80.52	1, 337	1.197^{20}_{4}	1.4422^{20}	−67.5	128.6	60	misc aq. alc
c104	2-(2-Chloroethoxy)-ethanol	$ClCH_2CH_2OCH_2CH_2OH$	124.57	1, 467	1.180	1.4529^{20}		81^{5mm}	90	
c105	2-[2-(2-Chloroeth-oxy)ethoxy]ethanol	$ClCH_2CH_2OCH_2CH_2{-}OCH_2CH_2OH$	168.62	1, 468	1.160	1.4580^{20}		120^{5mm}	107	
c106	2-Chloroethylamine HCl	$ClCH_2CH_2NH_2{\cdot}HCl$	115.99	4, 133			146			
c107	1-Chloro-2-ethyl-benzene	$ClC_6H_4C_2H_5$	140.61		1.055^{25}_{25}		−81	179.2		i aq; misc alc, eth
c108	(2-Chloroethyl)-benzene	$C_6H_5CH_2CH_2Cl$	140.61	5, 354	1.069	1.5300^{20}		84^{16mm}	66	s alc, bz, eth

TABLE 1.15 Physical Constants of Organic Compounds (*continued*)

No.	Name	Formula	Formula weight	Beilstein reference	Density	Refractive index	Melting point	Boiling point	Flash point	Solubility in 100 parts solvent
c109	Chloroethylene	$H_2C{=}CHCl$	62.50	1, 186	0.97^{-14}	1.4125^{20}	−159.7	−13.9	15	sl s aq; s alc
c110	2-Chloroethyl ethyl ether	$ClCH_2CH_2OCH_2CH_3$	108.57	1, 337	0.989			107	15	
c111	2-Chloroethyl methyl ether	$ClCH_2CH_2OCH_3$	94.54	1, 337	1.035	1.4111^{20}		89–90		
c112	N-(2-Chloroethyl)-morpholine HCl		186.08				186			
c113	N-(2-Chloroethyl)-piperidine HCl		184.11	20, 17			236			
c114	2-Chloroethyl p-toluenesulfonate	$CH_3C_6H_4SO_3CH_2CH_2Cl$	234.70	11^2, 45	1.294	1.5290^{20}		$153^{0.3mm}$	>112	
c115	2-Chloroethyl vinyl ether	$H_2C{=}CHOCH_2CH_2Cl$	106.55	1^2, 473	1.048	1.4370^{20}	−69.7	110	16	0.6 aq
c116	1-Chloro-2-fluoro-benzene	ClC_6H_4F	130.55	5^1, 110	1.244	1.5010^{20}	−42.5	138.5	31	s alc, eth
c117	1-Chloro-3-fluoro-benzene	ClC_6H_4F	130.55		1.219	1.4944^{20}		126	20	s alc, eth
c118	1-Chloro-4-fluoro-benzene	ClC_6H_4F	130.55	5, 201	1.226^{20}_4	1.4967^{20}	−21.5	130–131		s alc, eth
c119	2-Chloro-6-fluoro-benzyl chloride	$Cl(F)C_6H_3CH_2Cl$	179.02		1.401	1.5372^{20}				
c120	4-Chloro-4′-fluoro-butyrophenone	$FC_6H_4C({=}O)CH_2CH_2CH_2Cl$	200.64		1.220	1.5255^{20}			110	
c121	3-Chloro-4-fluoro-nitrobenzene	$Cl(F)C_6H_3NO_2$	175.5		1.6028^{17}	1.5674^{17}	41.5	127^{17mm}		
c122	2-Chloro-4-fluoro-phenol	$Cl(F)C_6H_3OH$	146.5				23	88^{4mm}		
c123	2-Chloro-4-fluoro-toluene	$Cl(F)C_6H_3CH_3$	144.58		1.1972^{20}	1.4985^{25}		152–153		

No.	Name	Formula	Formula wt	Beilstein ref	Density	n_D	Melting point	Boiling point	Flash point	Solubility
c124	2-Chloro-6-fluoro-toluene	Cl(F)C6H3CH3	144.58		1.191	1.5026^{20}		156	46	0.82 aq
c125	4-Chloro-2-fluoro-toluene	Cl(F)C6H3CH3	144.58			1.4998^{20}		158		
c126	Chloroform	CHCl3	119.39	1, 61	1.4985^{15}	1.4486^{15}	−63.59	61.7	none	misc alc, eth
c127	Chloroform-d	CDCl3	120.39		1.50	1.4445^{20}		60.9	none	
c128	1-Chloroheptane	CH3(CH2)6Cl	134.65	1, 154	0.8810^{16}	1.4250^{20}		159–161	41	i aq
c129	1-Chlorohexane	CH3(CH2)5Cl	120.62		0.8780_4^{20}	1.4236^{20}	−69	134	38	i aq
c130	6-Chloro-1-hexanol	Cl(CH2)6OH	136.62		1.204			108^{14mm}	98	sl s aq; v s alc, eth
c131	4-Chloro-4'-hydroxy-benzophenone	ClC6H4C(=O)C6H4OH	232.67	8[2], 187		1.4557^{20}	175–178	257^{14mm}		
c132	5-Chloro-8-hydroxy-7-iodoquinoline		305.50				d 172			
c133	3-Chloro-4-hydroxy-mandelic acid	ClC6H3(OH)-CH(OH)COOH	202.60				145–147			i alc, eth; 0.8 chl; 0.6 HOAc
c134	5-Chloro-8-hydroxy-quinoline		179.61	21, 95			130			sl s aq HCl
c135	1-Chloro-4-iodo-benzene	ClC6H4I	238.46	5, 221	1.1864^{57}		53–54	226–227		s alc

2-Chloroethyl ether, b158
2-Chloro-6-fluorobenzal chloride, t233

c112

c113

α-Chloro-4-fluorotoluene, f16
2375-Chloro-2-hydroxyaniline, a148

Chlorohydroxybenzoic acids, c237, c238
1-Chloro-3-hydroxypropane, c214

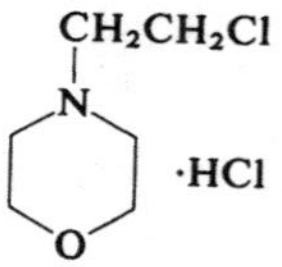

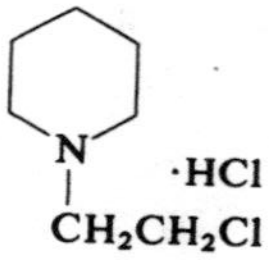

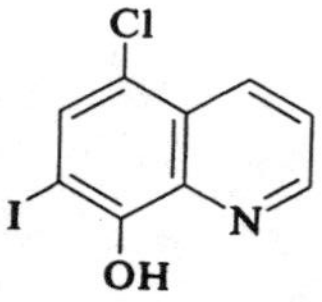

c132

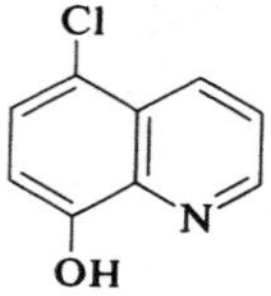

c134

TABLE 1.15 Physical Constants of Organic Compounds (*continued*)

No.	Name	Formula	Formula weight	Beilstein reference	Density	Refractive index	Melting point	Boiling point	Flash point	Solubility in 100 parts solvent
c136	1-Chloro-3-mercapto-2-propanol	HSCH$_2$CH(OH)CH$_2$Cl	126.61	1^3, 2156	1.277	1.5276^{20}		57$^{1.3mm}$	97	0.48 aq^{25}; s alc; misc chl, eth, HOAc
c137	Chloromethane	CH$_3$Cl	50.49	1, 59	0.92^{20}	1.3712^{-24}	−97.7	−24.22		
c137a	3-Chloro-4-methoxy-aniline	ClC$_6$H$_3$(OCH$_3$)NH$_2$	157.60	13, 511			50–55			
c138	5-Chloro-2-methoxy-aniline	ClC$_6$H$_3$(OCH$_3$)NH$_2$	157.60	13, 383			83–85			
c139	1-Chloro-2-methoxy-benzene	ClC$_6$H$_4$OCH$_3$	142.59	6, 184	1.123	1.5445^{20}		196	76	i aq; s alc, eth
c140	1-Chloro-4-methoxy-2-nitrobenzene	CH$_3$O(Cl)C$_6$H$_3$NO$_2$	187.58				45			s hot alc
c141	2-Chloro-6-methoxy-pyridine	CH$_3$O(Cl)(C$_5$H$_3$N)	143.57		1.207	1.5263^{20}		186		
c142	2-Chloro-6-methyl-pyridine	CH$_3$(Cl)C$_5$H$_3$NH$_2$	141.60	12^1, 388	1.152	1.5761^{20}	2	215	98	s alc
c143	3-Chloro-2-methyl-aniline	CH$_3$(Cl)C$_6$H$_3$NH$_2$	141.60	12, 836		1.5874^{20}	2	115–117^{10mm}	>112	
c144	3-Chloro-4-methyl-aniline	CH$_3$(Cl)C$_6$H$_3$NH$_2$	141.60	12, 988		1.5830^{20}	25	238	100	
c145	4-Chloro-2-methyl-aniline	CH$_3$(Cl)C$_6$H$_3$NH$_2$	141.60	12, 835		1.5848^{20}	27	241	99	s hot alc
c146	5-Chloro-2-methyl-aniline	CH$_3$(Cl)C$_6$H$_3$NH$_2$	141.60	12, 835		1.5840^{20}	22	237	160	
c147	DL-4-Chloro-2-(α-methylbenzyl)-phenol	C$_6$H$_5$CH(CH$_3$)-C$_6$H$_3$(Cl)OH	232.71	6^4, 4710				155^{2mm}		
c148	1-Chloro-3-methyl-butane	(CH$_3$)$_2$CHCH$_2$CH$_2$Cl	106.59	1, 135	0.8704$^{20}_4$	1.4084^{20}	−104	99	16	sl s aq; misc alc, eth

No.	Name	Formula	Formula weight	Beilstein ref.	Density	n_D	mp, °C	bp, °C	Flash pt, °C	Solubility
c149	2-Chloro-2-methyl-butane	$CH_3CH_2CCl(CH_3)_2$	106.59	1, 134	0.8650^{20}_4	1.4052^{20}	−73.7	85	16	i aq; s alc, eth
c150	Chloromethyldi-methylchlorosilane	$(CH_3)_2Si(Cl)CH_2Cl$	143.1		1.0865^{20}_4	1.4360^{20}		115–116		
c151	Chloromethyl 2,2-di-methylpropionate	$(CH_3)_3CCOOCH_2Cl$	150.61		1.045	1.4170^{20}			40	s alc; v s eth
c152	Chloromethyl ethyl ether	$ClCH_2OCH_2CH_3$	94.54	1^2, 645	1.04^{20}_4	1.4040^{20}		79–83		
c153	Chloromethylmethyl-dichlorosilane	$ClCH_2Si(CH_3)Cl_2$	163.5		1.2858^{20}_4	1.4500^{20}		121–122		
c154	Cloromethylmethyl-diethoxysilane	$ClCH_2Si(OC_2H_5)_2CH_3$	182.7		1.000^{20}_4	1.407^{25}		160–161		
c155	Chloromethyl methyl ether	$ClCH_2OCH_3$	80.51	1, 580	1.0703^{20}_4	1.3961^{20}	−103.5	57–59	15	d aq; s acet, CS$_2$
c156	Chloromethyl methyl sulfide	$ClCH_2SCH_3$	95.48		1.153	1.4963^{20}		105		
c157	1-(Chloromethyl)-naphthalene	$C_{10}H_7CH_2Cl$	176.65	5, 566		1.6380^{20}	32	169^{25mm}	>112	
c158	4-Chloro-2-methyl-phenol	$CH_3(Cl)C_6H_3OH$	142.59	6, 359			48	225		sl s aq
c159	4-Chloro-3-methyl-phenol	$CH_3(Cl)C_6H_3OH$	142.59	6, 381			68	235		i aq; s alc, bz, chl, eth, acet
c160	4-Chloro-N-methyl-piperidine HCl		170.08				164			

Chloromethylbenzenes, c244, c245, c246

(Chloromethyl)oxirane, c101

Chloromethyl pivalate, c151

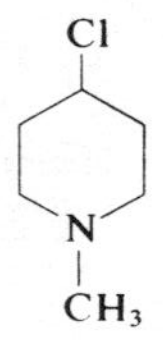

Cl —[piperidine ring]— N—CH₃

c160

TABLE 1.15 Physical Constants of Organic Compounds (*continued*)

No.	Name	Formula	Formula weight	Beilstein reference	Density	Refractive index	Melting point	Boiling point	Flash point	Solubility in 100 parts solvent
c161	1-Chloro-2-methyl-propane	$(CH_3)_2CHCH_2Cl$	92.57	1, 124	0.8829^{15}	1.4010^{15}	-130.3	68.9	21	0.09 aq; misc alc, eth
c162	2-Chloro-2-methyl-propane	$(CH_3)_3CCl$	92.57	1, 125	0.8474_4^{15}	1.3856^{20}	-25.4	50.8	18	sl s aq; misc alc, eth
c163	1-Chloro-2-methyl-propene	$(CH_3)_2C{=}CHCl$	90.55	1, 209	0.9186_4^{20}	1.4225^{20}		68.1	-1	misc alc, eth
c164	3-Chloro-2-methyl-propene	$ClCH_2C(CH_3){=}CH_2$	90.55	1, 209	0.9210_4^{15}	1.4272^{20}	-80	72	-10	misc alc, eth
c165	Chloromethyltri-chlorosilane	$ClCH_2SiCl_3$	183.9		1.465_4^{20}	1.4555^{20}		117–118		
c166	Chloromethyltri-methylsilane	$ClCH_2Si(CH_3)_3$	122.7	4^3, 1844	0.8861_4^{20}	1.4180^{20}		99	<1	
c167	6-(Chloromethyl)-uracil		160.56	23^1, 328			257 d			
c168	1-Chloronaphthalene	$C_{10}H_7Cl$	162.62	5, 541	1.1938_4^{20}	1.6332^{20}	-2.3	259.3	121	s alc, bz, PE
c169	2-Chloronaphthalene	$C_{10}H_7Cl$	162.62	17, 522	1.1377^{71}	1.6079^{71}	59.5	256		s alc, bz, chl, eth
c170	4-Chloro-1,8-naph-thalic anhydride		232.63				210			
c171	4'-Chloro-3'-nitro-acetophenone	$ClC_6H_3(NO_2)-C({=}O)CH_3$	199.60	7^3, 995			101			
c172	2-Chloro-4-nitro-aniline	$ClC_6H_3(NO_2)NH_2$	172.57	12, 733			109			sl s aq; v s alc, eth
c172a	2-Chloro-5-nitro-aniline	$ClC_6H_3(NO_2)NH_2$	172.57	12, 732			114			
c173	4-Chloro-2-nitro-aniline	$ClC_6H_3(NO_2)NH_2$	172.57	12, 729			119			v s alc, eth
c174	4-Chloro-3-nitro-aniline	$ClC_6H_3(NO_2)NH_2$	172.57	12, 731			101			v s alc; s eth

1.162

	Name	Formula								Solubility
c175	1-Chloro-2-nitrobenzene	$ClC_6H_4NO_2$	157.56	5, 241	1.348		32–33	246	123	s alc, bz, eth
c176	1-Chloro-3-nitrobenzene	$ClC_6H_4NO_2$	157.56	5, 243	1.534_4^{20}		46	236	103	sl s alc; v s eth, chl
c177	1-Chloro-4-nitrobenzene	$ClC_6H_4NO_2$	157.56	5, 243	1.520		82–84	242	110	sl s alc; v s eth, CS$_2$
c178	2-Chloro-4-nitrobenzoic acid	$ClC_6H_3(NO_2)COOH$	201.57	9, 404			141			s hot aq, hot bz
c179	2-Chloro-5-nitrobenzoic acid	$ClC_6H_3(NO_2)COOH$	201.57	9, 403	1.608^{18}		168			sl s aq; s alc, bz, eth
c180	4-Chloro-3-nitrobenzoic acid	$ClC_6H_3(NO_2)COOH$	201.57	9, 402	1.645^{18}		183			sl s alc; s hot aq
c181	4-Chloro-3-nitrobenzophenone	$ClC_6H_3(NO_2)C(=O)C_6H_5$	261.66	7^1, 230			104–105	235^{13mm}		
c182	2-Chloro-5-nitrobenzotrifluoride	$ClC_6H_3(NO_2)CF_3$	225.55		1.527	1.5083^{20}		231	98	
c183	4-Chloro-3-nitrobenzotrifluoride	$ClC_6H_3(NO_2)CF_3$	225.55		1.511	1.4893^{20}	–2.5	222	101	
c184	5-Chloro-2-nitrobenzotrifluoride	$ClC_6H_3(NO_2)CF_3$	225.55		1.526	1.4980^{20}	21–22	222–224	102	

Chloronicotinic acids, c234, c235

α-Chloronitrotoluene, n47

Chloronitro-α,α,α-trifluorotoluenes, c182, c183, c184

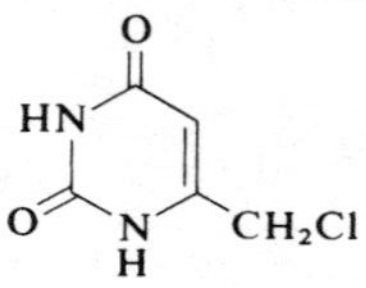

c167

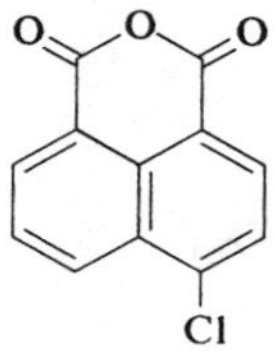

c170

TABLE 1.15 Physical Constants of Organic Compounds (*continued*)

No.	Name	Formula	Formula weight	Beilstein reference	Density	Refractive index	Melting point	Boiling point	Flash point	Solubility in 100 parts solvent
c185	o-(4-Chloro-3-nitro-benzoyl)benzoic acid	$ClC_6H_3(NO_2)COC_6H_4$-COOH	305.68	10, 752			201			
c186	2-Chloro-4-nitro-phenol	$ClC_6H_3(NO_2)OH$	173.56	6, 240			106			i aq; s alc, eth
c187	2-Chloro-4-nitro-toluene	$ClC_6H_3(NO_2)CH_3$	171.58	5, 329		1.5470^{70}	61	260		i aq
c188	2-Chloro-6-nitro-toluene	$ClC_6H_3(NO_2)CH_3$	171.58	5, 327		1.5377^{70}	36	238	125	i aq
c189	4-Chloro-3-nitro-toluene	$ClC_6H_3(NO_2)CH_3$	171.58	5, 329	1.297	1.5580^{20}	7	260^{745mm}	>112	i aq
c190	1-Chlorooctane	$CH_3(CH_2)_7Cl$	148.68	1, 159	0.875_4^{20}	1.4298^{20}	−61	183	54	i aq; v s alc, eth
c191	1-Chloropentane	$CH_3(CH_2)_4Cl$	106.60	1, 130	0.882_4^{20}	1.4118^{20}	−99.0	98.3	12	0.02 aq; misc alc, eth
c192	5-Chloro-2-pentanone	$ClCH_2CH_2CH_2COCH_3$	120.58	1^2, 738	1.0571_4^{18}	1.4375^{20}		72^{20mm}	62	s acet, eth
c193	3-Chloroperoxy-benzoic acid	$ClC_6H_4C(O)OOH$	172.57				94 d			
c194	2-Chlorophenol	ClC_6H_4OH	128.56	6, 183	1.2573_4^{25}	1.5579^{20}	9.3	175–176	63	sl s aq; v s alc, eth
c195	3-Chlorophenol	ClC_6H_4OH	128.56	6, 185	1.245_4^{25}	1.5565^{40}	33.5	214	>112	sl s aq; s alc, eth
c196	4-Chlorophenol	ClC_6H_4OH	128.56	6, 186	1.2238_4^{78}	1.5419^{45}	43.5	220	115	sl s aq; v s alc, chl, eth
c197	4-Chlorophenoxy-acetic acid	$ClC_6H_4OCH_2COOH$	186.59	6, 187			159			
c198	2-(4-Chlorophenoxy)-2-methylpropionic acid	$ClC_6H_4OC(CH_3)_2COOH$	214.65				122			
c199	DL-2-(4-Chlorophen-oxy)propionic acid	$ClC_6H_4OCH(CH_3)COOH$	200.62	6^3, 695			117			

No.	Name	Formula	Mol. wt.	Beilstein ref.	Density	n_D	m.p.	b.p.		Solubility
c200	4-Chlorophenylacetic acid	ClC$_6$H$_4$CH$_2$COOH	170.60	9, 448			105			v s aq, alc, eth; s bz
c201	p-Chlorophenylacetonitrile	ClC$_6$H$_4$CH$_2$CN	151.60	9, 448			30.5	267		
c202	2-Chloro-p-phenyl-enediamine sulfate	H$_2$NC$_6$H$_3$(Cl)NH$_2$· H$_2$SO$_4$	240.67	13, 117			253			
c202	4-Chloro-1,2-phenyl-enediamine	ClC$_6$H$_3$(NH$_2$)$_2$	142.59	13, 25			70			
c204	4-Chloro-1,3-phenyl-enediamine	H$_2$N(Cl)C$_6$H$_3$NH$_2$	142.59	13, 53			90			
c205	3-Chlorophenyl-hydrazine HCl	ClC$_6$H$_4$NHNH$_2$·HCl	179.05	15, 424			242 d			
c206	4-Chlorophenyl isocyanate	ClC$_6$H$_4$NCO	153.57	12, 616		1.5618^{20}	31	204	110	
c207	4-Chlorophenyl phenyl sulfone	ClC$_6$H$_4$SO$_2$C$_6$H$_5$	252.72	6^1, 149			94			74 acet; 44 bz; 5 CCl$_4$; 65 diox; 21 i-PrOH
c208	4-Chlorophenyltri-chlorosilane	ClC$_6$H$_4$SiCl$_3$	246.0		1.4316^{20}_4	1.5418^{20}		$115\text{–}117^{20mm}$		
c209	4-Chloro-o-phthalic acid	ClC$_6$H$_3$(COOH)$_2$	200.58	9, 816			148			
c210	1-Chloropropane	CH$_3$CH$_2$CH$_2$Cl	78.54	1, 104	0.8895^{15}	1.3880^{20}	−122.8	46.6	18	0.27 aq; misc alc, eth
c211	2-Chloropropane	CH$_3$CHClCH$_3$	78.54	1, 105	0.8563^{20}	1.3777^{20}	−117.2	35	−35	0.34 aq; misc alc, eth
c212	3-Chloro-1,2-propanediol	ClCH$_2$CH(OH)CH$_2$OH	110.54	1, 363	1.3218^{20}_4	1.4805^{20}		213	58	s aq, alc, eth
c213	1-Chloro-2-propanol	CH$_3$CH(OH)CH$_2$Cl	94.54	1, 363	1.115^{20}	1.4375^{20}		126–127	51	misc aq; s alc
c214	3-Chloro-1-propanol	ClCH$_2$CH$_2$CH$_2$OH	94.54	1, 356	1.1309^{20}_4	1.4460^{20}		160–162	73	

p-Chlorophenacyl bromide, b246
Chlorophenylamines, c33, c34, c35

4-Chlorophenyl sulfone, b165
4-Chlorophenyl sulfoxide, b166

Chloropicrin, t239
Chloroprene, c216

TABLE 1.15 Physical Constants of Organic Compounds (*continued*)

No.	Name	Formula	Beilstein reference	Formula weight	Density	Refractive index	Melting point	Boiling point	Flash point	Solubility in 100 parts solvent
c215	Chloro-2-propanone	$ClCH_2COCH_3$	1, 653	92.53	1.135^{15}	1.4350^{20}	-44.5	119.7	7	10 aq; misc alc, chl
c216	3-Chloro-1-propene	$ClCH_2CH=CH_2$	1, 198	76.53	0.939^{20}_4	1.4151^{20}	-134.5	45.2	-28	0.36 aq; misc alc, chl
c217	(3-Chloropropenyl)-benzene	$C_6H_5CH=CHCH_2Cl$	5^2, 372	152.62		1.5845^{20}	-19	108^{12mm}	79	
c218	2-Chloropropionic acid	$CH_3CH(Cl)COOH$	2, 248	108.52	1.182	1.4345^{20}		186	101	misc aq, alc, eth
c219	3-Chloropropionic acid	$ClCH_2CH_2COOH$	2, 249	108.52			41	205	>112	v s aq, alc, chl
c220	3-Chloropropio-nitrile	$ClCH_2CH_2CN$	2, 250	89.53	1.1443^{18}	1.4379^{20}	-50	176	75	d aq, alc
c221	2-Chloropropionyl chloride	$CH_3CH(Cl)COCl$	2, 248	126.97	1.308	1.4400^{20}		111	31	
c222	3-Chloropropionyl chloride	$ClCH_2CH_2COCl$	2, 250	126.97	1.3307^{13}	1.4570^{20}		145	61	i aq; d hot aq, hot alc; s alc; v s eth
c223	p-Chloropropio-phenone	$ClC_6H_4C(=O)CH_2CH_3$	7, 301	168.62			37	97^{1mm}		
c224	3-Chloropropylamine HCl	$ClCH_2CH_2CH_2NH_2 \cdot HCl$	4, 148	130.02			150			
c225	3-Chloropropyl-methyldichloro-silane	$Cl(CH_2)_3Si(CH_3)Cl_2$		191.6	1.2045^{20}	1.4580^{20}		70^{15mm}		
c226	2-Chloropropyl-(phenyl)dichloro-silane	$Cl(CH_2)_3SiCl_2(C_6H_5)$		253.6	1.241^{20}_4	1.5332^{20}		141^{10mm}		
c227	N-(3-Chloropropyl)-piperidine HCl		20, 18	198.14			220			

ID	Name	Formula	Formula wt	Beilstein ref.	Density	n_D	mp	bp		Solubility
c228	3-Chloropropyl thiolacetate	$CH_3C(=O)SCH_2CH_2CH_2Cl$	152.64	2^3, 493	1.159	1.4946^{20}		84^{10mm}	77	misc bz, alc, eth, EtAc
c229	3-Chloropropyltri-chlorosilane	$ClCH_2CH_2CH_2SiCl_3$	212.0		1.3590_4^{20}	1.4668^{20}		181.5	66	sl s aq; s alc, eth
c229	3-Chloropropyltri-ethoxysilane	$Cl(CH_2)_3Si(OC_2H_5)_3$	240.8		1.009_4^{20}	1.420^{20}		102^{10mm}		
c231	3-Chloropropyltri-methoxysilane	$Cl(CH_2)_3Si(OCH_3)_3$	198.72		1.077_4^{25}	1.4183^{25}		183	66	
c232	3-Chloropropyne	$ClCH_2C{\equiv}CH$	74.51	1, 248	1.0306_4^{25}	1.4349^{20}	−78	58	18	
c233	2-Chloropyridine	ClC_5H_4N	113.55	20, 230	1.205^{15}	1.5320^{20}		166^{714mm}	65	
c234	2-Chloro-3-pyridine-carboxylic acid	$C_5H_3N(Cl)COOH$	157.56	22^2, 35			d 175			
c235	6-Chloro-3-pyridine-carboxylic acid	$C_5H_3N(Cl)COOH$	157.56	22, 43			200 d			
c236	2-Chloroquinoline		163.61	20, 359	1.2464_4^{25}	1.6259^{25}	37	267		i aq; s alc, bz, eth
c237	4-Chlorosalicyclic acid	$HO(Cl)C_6H_3COOH$	172.57	10, 101			212			
c238	5-Chlorosalicylic acid	$HO(Cl)C_6H_3COOH$	172.57	10, 102	1.65		172			
c239	N-Chlorosuccinimide		133.53	21, 380			150–151			
c240	Chlorosulfonyl isocyanate	$ClSO_2NCO$	141.53		1.626	1.4467^{20}	−44	107		1.4 aq; 0.67 alc; 2 bz; sl s chl, eth

β-Chloropropionaldehyde diethyl acetal, c82

3-Chloropropylene-1,2-oxide, c102

1-Chloro-2,5-pyrrolidinedione, c239

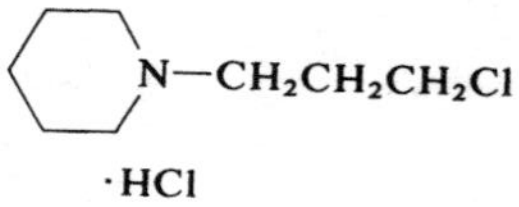

N–CH₂CH₂CH₂Cl · HCl, c227

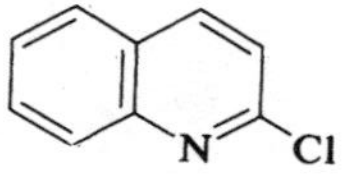

c236

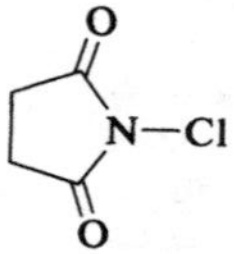

c231

TABLE 1.15 Physical Constants of Organic Compounds (*continued*)

No.	Name	Formula	Formula weight	Beilstein reference	Density	Refractive index	Melting point	Boiling point	Flash point	Solubility in 100 parts solvent
c241	8-Chlorotheophylline		214.61	26, 473			d 290			s alk
c242	2-Chlorothiophene	$Cl—C_4H_3S$	118.59	17, 32		1.5483^{20}	−72	129	22	i aq; misc alc, eth
c243	4-Chlorothiophenol	ClC_6H_4SH	144.62	6, 326			51	207		
c244	2-Chlorotoluene	$ClC_6H_4CH_3$	126.59	5, 290	1.0826^{20}_4	1.5250^2	−34	159.0	47	sl s aq; v s alc, bz, chl, eth
c245	3-Chlorotoluene	$ClC_6H_4CH_3$	126.59	5, 291	1.0760^{19}_4	1.5218^{20}	−48.9	161.8	50	s alc, bz, chl; misc eth
c246	4-Chlorotoluene	$ClC_6H_4CH_3$	126.59	5, 292	1.0697^{20}_4	1.5208^{20}	7.2	162.0	49	sl s aq; s alc, bz, eth
c247	N-Chloro-p-toluenesulfonamide, Na salt	$CH_3C_6H_4SO_2NCl^-Na^+$	227.67				167 d			s aq; i bz, chl, eth
c248	4'-Chloro-1-toluenethiol	$ClC_6H_4CH_2SH$	158.65	6, 466	1.202	1.5893^{20}	20		76	
c249	4-Chloro-o-tolyloxyacetic acid, Na salt	$ClC_6H_3(CH_3)O-CH_2COO^-Na^+$	222.61	6^3, 1265			220–225			
c250	4-(4-Chloro-o-tolyloxy)butyric acid	$ClC_6H_3(CH_3)O-(CH_2)_3COOH$	228.68				99–100			
c251	Chloro-2,2,2-trifluoroethane	CF_3CH_2Cl	118.5		1.389^0	1.3090^0	−105	6.9		
c252	Chlorotrifluoroethylene	$CF_2=CFCl$	116.48		1.315		−158.2	−27.9		
c253	Chlorotrifluoromethane	$ClCF_3$	104.46	1^3, 42			−181	−81.5		
c254	Chlorotrimethylgermane	$(CH_3)_3GeCl$	153.16		1.2382^{22}	1.4283^{20}	−13	102		
c255	Chlorotrimethylsilane	$(CH_3)_3SiCl$	108.64		0.8580^{20}_4	1.3885^{20}	−40	57	−40	

No.	Name	Formula	Formula weight	Beilstein reference	Density	Refractive index	Melting point	Boiling point		Solubility
c256	Chlorotriphenyl-methane	$(C_6H_5)_3CCl$	278.78	5, 700			110–112	230^{20mm}		v s bz, chl, eth
c257	Chlorotripropyl-silane	$(C_3H_7)_3SiCl$	192.8		0.882^{20}_4	1.440^{20}		199–201		
c257a	Chlorotris(di-methylamino)silane	$[(CH_3)_2N]_3SiCl$	195.8		0.975^{20}_4	1.442^{20}		$62–63^{12mm}$		
c258	α-Chloro-o-xylene	$CH_3C_6H_4CH_2Cl$	140.61	5, 364	1.063	1.5391^{20}		199	73	i aq; misc alc, eth
c259	α-Chloro-m-xylene	$CH_3C_6H_4CH_2Cl$	140.61	5, 373	1.064^{20}	1.5350^{20}		195–196	75	i aq; misc alc, eth
c260	α-Chloro-p-xylene	$CH_3C_6H_4CH_2Cl$	140.61	5, 384		1.5330^{20}		200	75	misc alc, bz, eth, acet
c261	4-Chloro-o-xylene	$ClC_6H_3(CH_3)_2$	140.61	5, 363	1.047	1.5283^{20}	4.5	223	66	misc alc, bz, eth, acet
c262	Cholesterol		386.66		1.067^{20}_4		148.5	360 sl d		1.29 alc; 35 eth; 22 chl; s bz, PE
c263	Cholic acid		408.58				198			0.028 aq; 0.06 alc; 2.8 acet; 0.036 bz; 0.5 chl

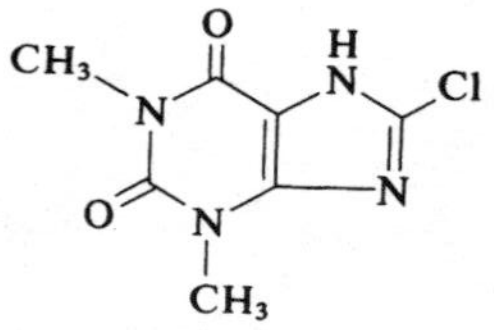

c241

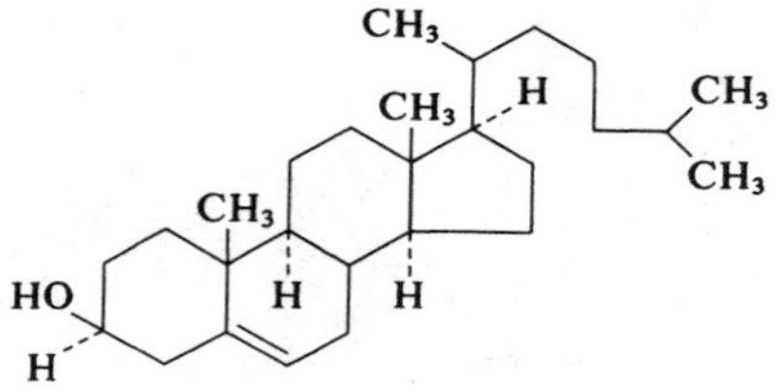

c262

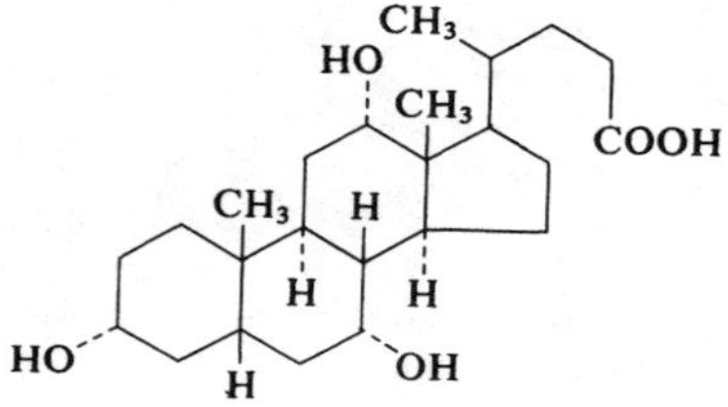

c263

TABLE 1.15 Physical Constants of Organic Compounds (*continued*)

No.	Name	Formula	Formula weight	Beilstein reference	Density	Refractive index	Melting point	Boiling point	Flash point	Solubility in 100 parts solvent
c264	Cinchonine		294.40	23², 369			~260			1.4 alc; 0.9 chl; 0.2 eth
c265	1,8-Cineole		154.25	17, 23	0.921^{25}_{25}	1.4572^{20}	1.5	174.4		misc alc, chl, eth
c266	*trans*-Cinnamaldehyde	$C_6H_5CH{=}CHCHO$	132.16	7, 348	1.050^{25}_{25}	1.6219^{20}	−7.5	246	71	0.014 aq; misc alc, chl, eth
c267	*trans*-Cinnamic acid	$C_6H_5CH{=}CHCOOH$	148.16	9, 573	1.2475^{4}		134	300		0.05 aq; 16 alc; 8 chl
c268	*trans*-Cinnamoyl chloride	$C_6H_5CH{=}CHCOCl$	166.61	9², 390	1.1617^{25}_{4}	1.614^{43}	35–36	258		s hot alc, CCl_4
c269	Cinnamyl alcohol	$C_6H_5CH{=}CHCH_2OH$	134.18	6, 570	1.0397^{35}_{35}	1.5758^{33}	33	250.0		s aq; v s alc, eth
c270	Citraconic acid	$CH_3C(COOH){=}CHCOOH$	130.10	2, 768	1.62		92 d			v s aq, alc, eth; sl s chl; i bz, PE
c271	Citraconic anhydride		112.08	17, 440	1.247	1.4712^{20}	8	214	101	i aq; s alk
c272	Citrazinic acid		155.11	22, 254			carbonizes without melting >300			
c273	Citric acid	$HOOCCH_2C(OH)(COOH)$-CH_2COOH	192.12	3, 556	1.665		154			59 aq
c274	Citronellol	$(CH_3)_2C{=}CHCH_2CH_2CH$-$(CH_3)CH_2CH_2OH$	156.27	1, 451	0.8570^{20}_{4}	1.4556^{20}		222	79	0.17 aq; 15 alc; 140 chl; 28 eth
c275	Cocaine		303.35	22², 150		1.5022^{98}	98	$187^{0.1mm}$		0.25 aq; v s alc, chl, eth
c276	Coumarin		146.15	17, 328	0.9354^{4}		69	298		0.25 aq; v s alc, chl, eth

c277	Creatine	$HOOCCH_2N(CH_3)$- $C({=}NH)NH_2$	131.14	4, 363			300			1.3 aq; 0.11 alc; ith
c278	Creatinine		113.12	24, 245			255 d			8 aq; sl s alc; i eth
c279	*o*-Cresol	$CH_3C_6H_4OH$	108.14	6, 349	1.0273^{41}	1.5361^{41}	30.9	190.8	81	$3.1\ aq^{40}$; misc alc, chl, eth; s alk
c280	*m*-Cresol	$CH_3C_6H_4OH$	108.14	6, 373	1.034^{20}_{4}	1.5438^{20}	12.2	202.7	86	$2.5\ aq^{40}$; misc alc, chl, eth; s alk
c281	*p*-Cresol	$CH_3C_6H_4OH$	108.14	6, 389	1.0179^{41}	1.5312^{41}	34.8	201.9	86	$2.3\ aq^{40}$; misc alc, chl, eth; s alk
c282	*trans*-Crotonaldehyde	$CH_3CH{=}CHCHO$	70.09	1, 728	0.8516^{20}	1.4373^{20}	−76.5	104.1	8	18.1 aq
c283	Crotonyl chloride	$CH_3CH{=}CHCOCl$	104.54	2, 411	1.091	1.4595^{20}		123	35	
c284	Cupferron	$C_6H_5N(NO)O^-NH_4^+$	155.16	16^1, 395			163–164			v s aq, alc

Chromone, b56
Chromotropic acid, d396
Chrysoldin, d31
Cinchophen, p149
Cinnamyl chloride, c217
Citral, d562, d563
Cleland's reagent, d424
2,4,6-Collidine, t367

p-Coumaric acid, h107
Cresotic acids, h137, h138
Cresylic acids, c279, c280, c281
Crotonic acid, b402
Crotononitrile, b400
Crotyl alcohols, b404, b405
Crotyl bromide, b240
Crotyl chloride, c68

12-Crown-4, t125
15-Crown-5, p45
18-Crown-6, h74
Cumene, i91
Cumic alcohol, i92
Cupron, b50

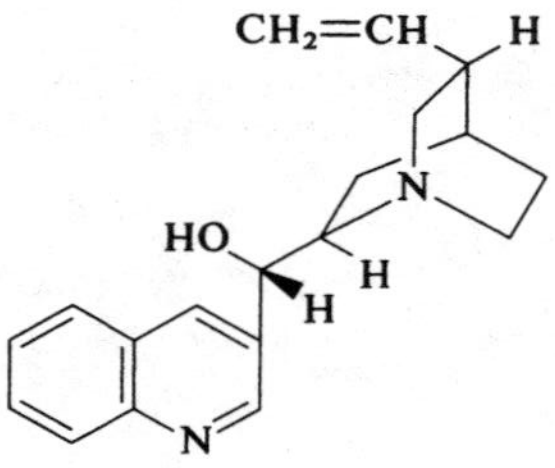

c264

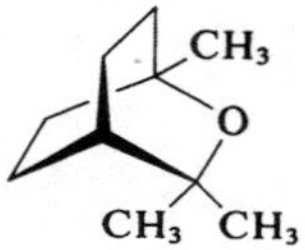

c265

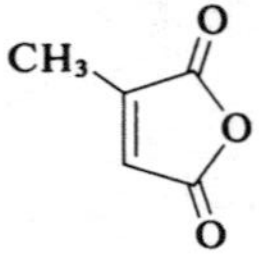

c271

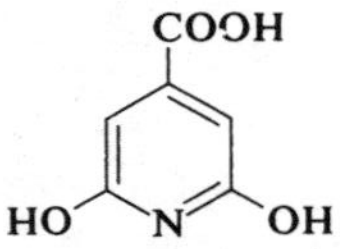

c272

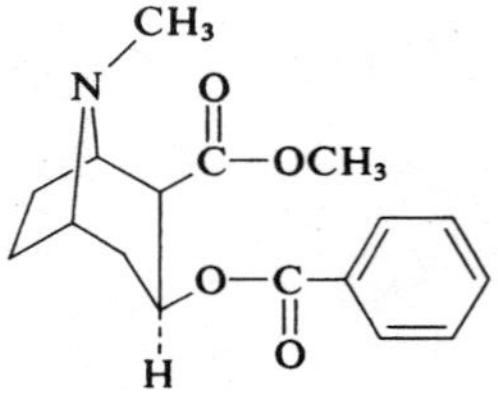

c275

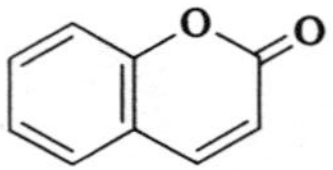

c276

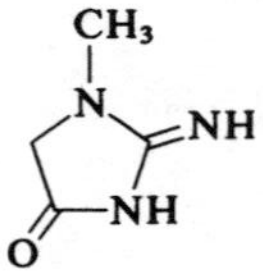

c278

TABLE 1.15 Physical Constants of Organic Compounds (*continued*)

No.	Name	Formula	Formula weight	Beilstein reference	Density	Refractive index	Melting point	Boiling point	Flash point	Solubility in 100 parts solvent
c285	Cyanamide	H_2NCN	42.04	3^2, 63	1.282^{20}_4		46	83^{380mm}		78 aq; 29 BuOH; 42 EtAc; s alc, eth
c286	2-Cyanoacetamide	$NCCH_2CONH_2$	84.08	2, 589			119.5		215	25 aq; 3.1 alc
c287	Cyanoacetic acid	$NCCH_2COOH$	85.06	2, 583			65–67	108^{15mm}	107	s aq, alc, eth; sl s bz
c288	Cyanoacetohydrazide	$NCCH_2C(=O)NHNH_2$	99.09	3, 66			110	d		v s aq; s alc; i eth
c289	Cyanoacetylurea	$NCCH_2C(=O)NHC(=O)NH_2$	127.10				214 d			
c290	2-Cyanoethanol	$NCCH_2CH_2OH$	71.08	3^2, 213	1.0588^{0}			$106-108^{11mm}$ 63^{4mm}		misc aq, alc; sl s eth
c291	2-Cyanoethyldi-chloromethylsilane	$NCCH_2CH_2Si(CH_3)Cl_2$	168.1	4, 71	1.202^{20}_4	1.455^{20}				
c292	1-Cyano-3-methyliso-thiourea, Na salt	$CH_3NHC(=NCN)S^-Na^+$	137.14				290 d			i aq; v s alc, eth
c293	1-Cyanonaphthalene	$C_{10}H_7CN$	153.18	9, 649	1.1113^{25}	1.6298^{18}	38	299		
c293	3-Cyanopropyltri-chlorosilane	$NCCH_2CH_2CH_2SiCl_3$	202.6		1.280^{25}	1.465^{25}		$93-$ 94^{8mm}		
c295	2-Cyanopyridine	$NC(C_5H_4N)$	104.11	22, 36		1.5288^{20}	28	215	89	s aq; v s alc, bz, eth
c296	3-Cyanopyridine	$NC(C_5H_4N)$	104.11	22, 41			52	240–245		v s aq, alc, bz, eth
c297	4-Cyanopyridine	$NC(C_5H_4N)$	104.11	22, 46			80			s aq; alc, bz, eth
c298	Cyanotrimethylsilane	$(CH_3)_3SiCN$	99.21	26, 239	0.7834^{20}	1.3924^{20}	11	114–117	1	0.5 aq; s hot alc, pyr; i acet, bz, chl, eth
c299	Cyanuric acid		129.08		1.768^{0}		d to HO-CN			i aq; v s alc, acet
c300	Cyclobutane	C_4H_8	56.10	5, 17	0.7038^{0}	1.3752^{0}	−90.7	12.5		i aq; v s alc, acet
c301	Cyclodecane	$C_{10}H_{20}$	140.27			1.4707^{20}		201	1	
c302	Cyclododecanol	$C_{12}H_{23}OH$	184.32				77			

No.	Name	Formula	Formula wt.	Beilstein ref.	Density	n_D^{20}	m.p., °C	b.p., °C	Flash point, °C	Solubility
c303	Cyclododecanone	$C_{12}H_{22}(=O)$	182.31	7^2, 48	0.906	1.5070^{20}	61	85^{1mm}		
c304	trans,trans,cis-1,5,9-cyclododeca-triene		162.28		0.8925_4^{20}		−18	231	87	
c305	trans-Cyclododecene		166.31	5, 29	0.863	1.4822^{20}		232–245	93	
c306	Cycloheptane	C_7H_{14}	98.18	6^3, 4086	0.811_4^{20}	1.4455^{20}	−8.0	118.8	6	v s alc, eth
c307	DL-trans-1,2-Cycloheptanediol	$C_7H_{12}(OH)_2$	130.19				61–63	$138–139^{15mm}$		
c308	Cycloheptanol	$C_7H_{13}OH$	114.19	6, 10	0.948^{20}	1.4760^{20}	2	185	71	sl s aq; v s alc, eth
c309	Cycloheptanone	$C_7H_{12}(=O)$	112.17	7, 13	0.9490_4^{20}	1.4611^{20}		179–181	55	i aq; v s alc; s eth
c310	1,3,5-Cyclohepta-triene		92.13	5, 280	0.888	1.5211^{20}	−75.3	115.5	26	s alc, eth; v s bz, chl
c311	Cycloheptene	C_7H_{12}	96.17	5, 65	0.824_4^{20}	1.4585^{20}		114.7	−6	s alc, eth
c312	Cyclohexane	C_6H_{12}	84.16	5, 20	0.7786_4^{20}	1.4262^{20}	6.5	80.7	−18	0.01 aq; misc alc, bz, acet, eth, CCl$_4$
c313	Cyclohexane-d$_{12}$	C_6D_{12}	92.26		0.89	1.4210^{20}		78	−18	

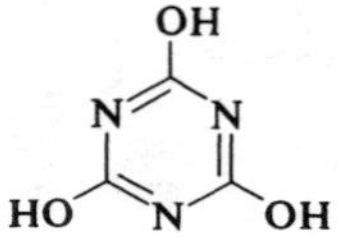

c299

c304

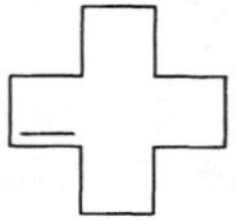

c305

c310

TABLE 1.15 Physical Constants of Organic Compounds (*continued*)

No.	Name	Formula	Formula weight	Beilstein reference	Density	Refractive index	Melting point	Boiling point	Flash point	Solubility in 100 parts solvent
c314	1,3-Cyclohexanebis-(methylamine)	$C_6H_{10}(NHCH_3)_2$	142.25						106	
c315	Cyclohexanecarb-aldehyde	$C_6H_{11}CHO$	112.17	7, 19	0.926	1.4500^{20}		163	40	
c316	Cyclohexanecarbonyl chloride	$C_6H_{11}COCl$	146.62	9, 9	1.096	1.4700^{20}		184	66	
c317	Cyclohexanecar-boxylic acid	$C_6H_{11}COOH$	128.17	7, 19	1.0480^{15}_4	1.4530^{20}	29	232.5		0.21 aq; s alc, bz, eth
c318	cis-1,2-Cyclohex-anediamine	$C_6H_{10}(NH_2)_2$	114.19	13, 1	0.931	1.4864^{20}		92^{18mm}		
c319	trans-1,2-Cyclohex-anediamine	$C_6H_{10}(NH_2)_2$	114.19	13, 1	0.931	1.4864^{20}		92^{18mm}		
c320	cis-1,2-Cyclohexane-dicarboxylic anhydride		154.17				34	158^{17mm}		
c320	cis-1,4-Cyclohexane-dimethanol	$C_6H_{10}(CH_2OH)_2$	144.21		0.978^{100}_4	1.4893^{20}	43	288	74	misc aq, alc; 2.5 eth
c322	1,3-Cyclohexanedione	$C_6H_8(=O)_2$	112.13	7, 554	1.0861^{91}	1.4576^{102}	103–105			s aq, alc, acet, chl
c323	1,2-Cyclohexanedione dioxime	$C_6H_8(=NOH)_2$	142.16	17^2, 526		super-cooled	185–188			s aq
c324	Cyclohexanemethyl-amine	$C_6H_{11}CH_2NH_2$	113.20	12, 12	0.870	1.4630^{20}		145–147	43	
c325	Cyclohexanepropionic acid	$C_6H_{11}CH_2CH_2COOH$	156.23	9, 82	0.912	1.4636^{20}	14–17	275.8		
c326	Cyclohexanethiol	$C_6H_{11}SH$	116.23	6, 8	0.950	1.4921^{20}		158–160	43	3.8 aq^{25}; misc alc, bz
c327	Cyclohexanol	$C_6H_{11}OH$	100.16	6, 5	0.9416^{30}	1.4629^{30}	25.2	161.1	67	

No.	Name	Formula	Mol. wt.		Density	n_D	mp, °C	bp, °C		Solubility
c328	Cyclohexanone	$C_6H_{10}(=O)$	98.15	7, 8	0.9478_4^{20}	1.4510^{20}	−45 to −47	155.7	46	15 aq[10.]; s alc, eth
c329	Cyclohexanone oxime	$C_6H_{10}(=NOH)$	113.16	7,10			89–91	206–210		s aq, eth; sl s alc
c330	Cyclohexene	C_6H_{10}	82.15	5, 63	0.8094_4^{20}	1.4464^{20}	−103.5	83.0	−12	0.02 aq; misc alc, bz, acet, eth
c331	2-Cyclohexen-1-one	$C_6H_8(=O)$	96.13	7[2], 55	0.993	1.4885^{20}	−53	168	61	v s alc
c332	2,3-Cyclohexeneo-pyridine		133.19	20[2], 176	1.025	1.5440		218	86	
c332a	[2-(3-Cyclohexenyl)-ethyl]methyldi-chlorosilane	$C_6H_9CH_2CH_2Si(CH_3)Cl_2$	223.2		1.0774_4^{20}	1.481^{25}		79–81[2mm]		
c333	Cyclohexylacetic acid	$C_6H_{11}CH_2COOH$	142.20	9[2], 9	1.007	1.4630^{20}	31–33	242–244	>112	sl s aq; s org solv
c334	Cyclohexylamine	$C_6H_{11}NH_2$	99.18	12, 5	0.8671^{20}	1.4593^{20}	−17.7	134.8	<32	misc aq, alc, eth, chl
c335	2-(Cyclohexylamino)-ethanesulfonic acid	$C_6H_{11}NHCH_2CH_2SO_3H$	207.29				>300			
c336	3-Cyclohexylamino-1-propanesulfonic acid	$C_6H_{11}NHCH_2CH_2-CH_2SO_3H$	221.32				>300			

Cyclohexanone cyanohydrin, h110
cis-4-Cyclohexene-1,2-dicarboximide, t75
cis-4-Cyclohexene-1,2-dicarboxylic anhydride, t74

Cyclohexene oxide, e5
N-(1-Cyclohexen-1-yl)morpholine, m450
N-(1-Cyclohexen-1-yl)pyrrolidine, p274

Cyclohexyl alcohol, c327

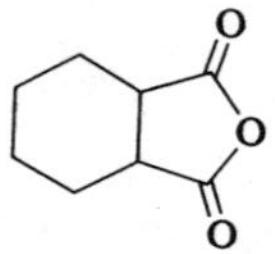

c320

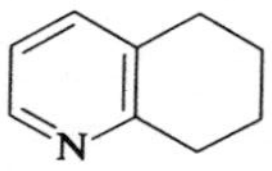

c332

TABLE 1.15 Physical Constants of Organic Compounds (*continued*)

No.	Name	Formula	Formula weight	Beilstein reference	Density	Refractive index	Melting point	Boiling point	Flash point	Solubility in 100 parts solvent
c337	4-Cyclohexylaniline	$C_6H_{11}C_6H_4NH_2$	175.28	12, 1209	0.9502^{20}	1.5258^{20}	53–56	166^{13mm}	98	i aq; v s alc, eth
c338	Cyclohexylbenzene	$C_6H_{11}C_6H_5$	160.26	5, 503			5–6	239–240		
c339	N-Cyclohexylformamide	$C_6H_{11}NHCHO$	127.18				38–40	137^{10mm}		
c340	Cyclohexyl isocyanate	$C_6H_{11}NCO$	125.17	12^2, 12	0.980	1.4551^{20}		168–170	48	
c341	Cyclohexyl isothiocyanate	$C_6H_{11}NCS$	141.24	12^2, 12	0.996	1.5350^{20}		219	71	s alc, eth
c342	Cyclohexylmethanol	$C_6H_{11}CH_2OH$	114.19	6, 14	0.9512^{25}	1.4640^{25}		181		
c343	N-Cyclohexyl-2-pyrrolidinone		167.25		1.026	1.495	12	284		
c344	Cyclohexyltrichlorosilane	$C_6H_{11}SiCl_3$	217.6		1.2222^4	1.477^{20}		90– 91^{10mm}		s CCl_4
c345	1,5-Cyclooctadiene		108.18	5, 116	0.8818^{25}	1.4905^{25}	−69	149–150	45	
c346	Cyclooctane	C_8H_{16}	112.22	5, 35	0.834	1.4574^{20}	14.8	151.1	30	
c347	Cyclooctanol	$C_8H_{15}OH$	128.22	6^2, 25	0.9740^{20}	1.4850^{20}	14–15	106– 108^{22mm}	86	
c348	Cyclooctanone	$C_8H_{14}(=O)$	126.20	7, 21	0.9584^{20}	1.6494^{20}	41–43	195–197	25	
c349	Cyclooctene	C_8H_{14}	110.20	5^1, 35	0.846	1.4698^{20}	−16	145–146	62	
c350	Cyclooctylamine	$C_8H_{15}NH_2$	127.23		0.928	1.4804^{20}	−48	190		
c351	Cyclopentamethylenedichlorosilane		169.1		1.5584	1.4679^{20}		169–170		
c352	Cyclopentane	C_5H_{10}	70.13	5, 19	0.7460^{20}	1.4065^{20}	−93.9	49.3	−37	i aq; misc alc, eth
c353	Cyclopentanecarboxylic acid	C_5H_9COOH	114.14	9, 6	1.0534	1.4540^{20}	4	216	93	sl s aq; s MeOH
c354	cis,cis,cis-1,2,3,4-Cyclopentanetetracarboxylic acid	$C_5H_6(COOH)_4$	246.17	9^2, 724			192– 195 d			

c355	Cyclopentanol	C_5H_9OH	86.13	6, 5	0.9488_4^{20}	1.4521^{20}	-19	140.9	51	sl s aq; s alc
c356	Cyclopentanone	$C_5H_8(=O)$	84.12	7, 5	0.9509_4^{18}	1.4366^{20}	-58	130.6	30	sl s aq; misc alc, eth
c357	Cyclopentanone oxime	$C_5H_8(=NOH)$	99.13	7, 7			53–55	196		s aq, alc, bz, chl, eth
c358	Cyclopentene	C_5H_8	68.11	5, 61	0.774	1.4228^{20}	-135.1	44.2	-28	
c359	2-Cyclopentene-1-acetic acid	$C_5H_7CH_2COOH$	126.16	9, 42	1.047	1.4675^{20}	19	$93-94^{2.5mm}$	>112	
c360	2,3-Cyclopenteneo-pyridine		119.17		1.018	1.5445^{20}		$87-88^{11mm}$	67	
c361	N-(1-Cyclopentene-1-yl)morpholine		153.23		0.957	1.5105^{20}		$105-106^{12mm}$	60	
c362	2-Cyclopentylidene-cyclopentanone		150.22		1.001	1.5231^{20}		140^{20mm}	103	
c363	3-Cyclopentylpro-pionic acid	$C_5H_9CH_2CH_2COOH$	142.20		0.996	1.4570^{20}		130^{12mm}	46	
c364	Cyclopropane	C_3H_6	42.08	5, 15	0.720_4^{-79}		-127.4	-32.8		37 mL per 100 mL aq[15]; v s alc, eth

Cyclohexylbenzene, p104
Cyclohexyl bromide, b261
Cyclohexyl chloride, c77
Cyclohexyl ketone, c328
Cyclohexyl mercaptan, c326

Cyclohexylmethane, m194
Cyclohexylmethyl bromide, b306
Cyclooctene oxide, e7a
Cyclopentanepropanoic acid, c363
Cyclopentene oxide, e37

Cyclopentyl bromide, b263
Cyclopentyl chloride, c79
Cyclopropyl bromide, b264
Cyclopropyl cyanide, c365

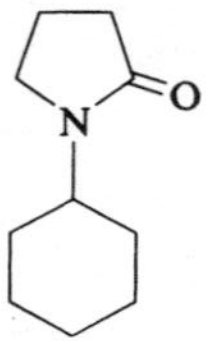

c343

c345

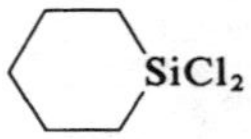

c351

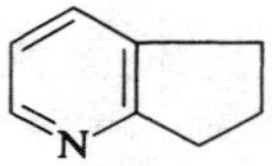

c360

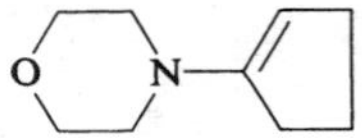

c361

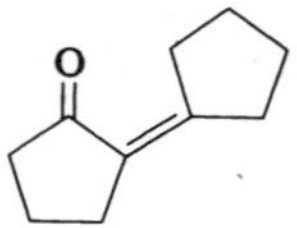

c362

TABLE 1.15 Physical Constants of Organic Compounds (*continued*)

No.	Name	Formula	Formula weight	Beilstein reference	Density	Refractive index	Melting point	Boiling point	Flash point	Solubility in 100 parts solvent
c365	Cyclopropanecarbonitrile	C_3H_5CN	67.09	9, 4	0.911^{16}	1.4207^{20}		135	32	s eth
c366	Cyclopropanecarbonyl chloride	C_3H_5COCl	104.54	9, 4	1.152	1.4522^{20}		119	23	sl s hot aq; s alc, eth
c367	Cyclopropanecarboxylic acid	C_3H_5COOH	86.09	9, 4	1.008	1.4380^{20}	17–19	182–184	71	s aq, alc, eth
c368	Cyclopropyl methyl ketone	$C_3H_5COCH_3$	84.12	7, 7	0.8993^{20}_4	1.4241^{20}		114	21	s aq, alc, eth
c369	Cystamine dihydrochloride	$H_2NCH_2CH_2SSCH_2$-$CH_2NH_2 \cdot 2HCl$	225.20	4, 287			217 d			v s aq, alc; i bz, eth
c370	L-(+)-Cysteine	$HSCH_2CH(NH_2)COOH$	121.16	4, 506			220 d			0.01 aq; s acid, alk; i alc
c371	L-Cystine	$HOOCCH(NH_2)CH_2$-$SSCH_2CH(NH_2)COOH$	240.30	4, 507			d 240			
d1	cis-Decahydronaphthalene	$C_{10}H_{18}$	138.26	5, 92	0.8963^{20}_4	1.4810^{20}	−43.0	195.8	58	v s alc, chl, eth; misc most ketones, esters
d2	trans-Decahydronaphthalene	$C_{10}H_{18}$	138.26	5^2, 56	0.8700^{20}_4	1.4697^{20}	−30.4	187.3	52	see under cis isomer
d3	Dehydro-2-naphthol	$C_{10}H_{17} \cdot OH$	154.25	6, 67	0.996	1.4992		109^{14mm}	>112	i aq
d4	Decamethylcyclopentasiloxane	$[-Si(CH_3)_2O-]_5$	370.8		0.9594^{20}	1.3982^{20}	−38	101^{20mm}		
d5	Decamethyltetrasiloxane	$(CH_3)_3SiO[Si(CH_3)_2O]_2$-$Si(CH_3)_3$	310.7		0.8536^{20}_4	1.3880^{20}	−70	194–195	86	sl s alc; s bz, PE
d6	Decanal	$H(CH_2)_9CHO$	156.27	1, 711	0.830^{15}	1.4280^{20}		207–209	85	i aq; s alc, eth
d7	Decane	$CH_3(CH_2)_8CH_3$	142.29	1, 168	0.7301^{20}_4	1.4119^{20}	−29.7	174.1	46	0.07 aq
d8	1,10-Decanediamine	$H_2N(CH_2)_{10}NH_2$	172.32	4, 273			62–63	140^{12mm}		0.1 aq; v s alc, esters, ketones
d9	Decanedioic acid	$HOOC(CH_2)_8COOH$	202.25	2, 718	1.207^{20}_4	1.422^{134}	134.5	295^{100mm}		sl s aq, eth; v s alc
d10	1,10-Decanediol	$HO(CH_2)_{10}OH$	174.28	1^2, 560			72–75	170^{8mm}		

No.	Name	Formula	Formula wt	Beilstein ref	Density	n_D	mp, °C	bp, °C	Flash pt, °C	Solubility
d11	Decanedioyl dichloride	$ClC(O)(CH_2)_8COCl$	239.14	2, 719	1.1212^{20}_{4}	1.4678^{20}		220^{75mm}	>112	d aq, alc
d12	Decanenitrile	$CH_3(CH_2)_8CN$	153.27	2, 356	0.8295^{15}_{4}	1.4295^{20}	−15	235–237		misc alc, chl, eth
d13	1-Decanesulfonic acid, Na salt	$CH_3(CH_2)_9SO_3^-Na^+$	244.33	4^3, 27			300			
d14	Decanoic acid	$CH_3(CH_2)_8COOH$	172.27	2^2, 309	0.8782^{50}	1.4288^{40}	31.4	270		0.015 aq; s alc, chl, bz, eth, CS_2
d15	1-Decanol	$CH_3(CH_2)_9OH$	158.29	1, 425	0.8297^{20}_{4}	1.4371^{20}	6.9	230.2	82	i aq; s alc, eth
d16	4-Decanone	$CH_3(CH_2)_5C(=O)(CH_2)_2CH_3$	156.27	1, 711	0.824^{20}_{20}	1.4237^{20}		207	71	i aq; misc alc, eth
d17	Decanoyl chloride	$CH_3(CH_2)_8C(=O)Cl$	190.71	2, 356	0.919	1.4410^{20}	−34.5	96^{5mm}	98	d aq; alc; s eth
d18	1-Decene	$CH_3(CH_2)_7CH=CH_2$	140.27	1^3, 858	0.7408^{20}_{4}	1.4215^{20}	−66.3	170.6	47	i aq; misc alc, eth
d19	Decylamine	$CH_3(CH_2)_9NH_2$	157.30	4, 199	0.787	1.4360^{20}	12–14	216–218	85	sl s aq; misc alc, bz, eth, acet
d20	Dehydroabietylamine		285.48							
d21	Dehydroacetic acid		168.15	17, 559		1.5460^{20}	111–113	269.9	>112	22 acet; 18 bz; 5 MeOH
d22	Deoxybenzoin	$C_6H_5CH_2C(=O)C_6H_5$	196.25	7, 431	1.201^{0}_{4}		55–56	320		i aq; v s alc, eth

Structure **d20** (labels: CH_3, CH_2NH_2, H_3C, $CH(CH_3)_2$)

Structure **d21** (labels: CH_3, CH_3, O, O, O)

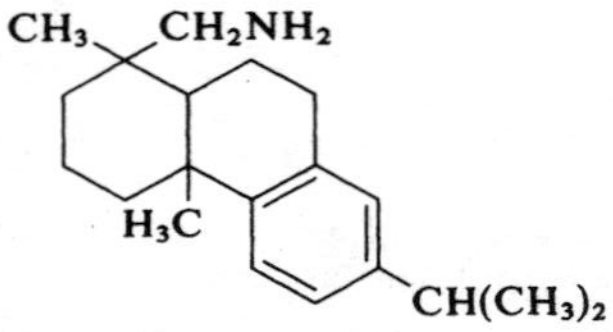

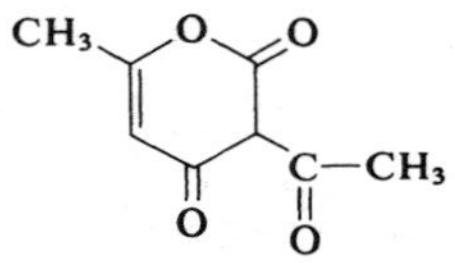

TABLE 1.15 Physical Constants of Organic Compounds (*continued*)

No.	Name	Formula	Formula weight	Beilstein reference	Density	Refractive index	Melting point	Boiling point	Flash point	Solubility in 100 parts solvent
d23	Diacetoxydimethylsilane	$(CH_3)_2Si(OOCCH_3)_2$	176.3		1.054^{20}_4	1.4030^{20}		164–166		
d24	Diacetoxymethylphenylsilane	$CH_3(C_6H_5)Si(OCOCH_3)_2$	238.3			1.487^{20}		127^{6mm}		
d25	Diallylamine	$(H_2C{=}CHCH_2)_2NH$	97.16	4, 208	0.787	1.4405^{20}	−88	111–112	15	i aq; misc alc, eth
d26	Diallyl ether	$(H_2C{=}CHCH_2)_2O$	98.15	1^2, 477	0.805^{18}	1.4240^{20}		94		sl s aq; misc alc, eth
d27	Diallyl sulfide	$(H_2C{=}CHCH_2)_2S$	114.21	1, 440	0.8877^{27}	1.4889^{20}	−83	138	46	sl s alc, eth
d28	1,2-Diaminoanthraquinone		238.25	14^1, 459			289–291			sl s alc, eth
d29	1,4-Diaminoanthraquinone		238.25	14, 197			265–268			sl s aq, alc; v s bz
d30	2,6-Diaminoanthraquinone		238.25	14, 215			>325			sl s hot aq. pyr
d31	2,5-Diaminoazobenzene HCl	$C_6H_5N{=}NC_6H_3-(NH_2)_2 \cdot HCl$	248.72	16, 383			235 d			sl s aq. alc
d32	2,5-Diaminobenzenesulfonic acid	$(H_2N)_2C_6H_3SO_3H$	188.21	14, 713			298 d			sl s aq; s alc, eth
d33	3,5-Diaminobenzoic acid	$(H_2N)_2C_6H_3COOH$	152.15	14, 453			228	−H$_2$O, 110		
d34	4,4′-Diaminodiphenylamine sulfate	$H_2NC_6H_4NHC_6H_4-NH_2 \cdot H_2SO_4$	297.33	13, 110			300			sl s aq; v s alc, bz, eth
d35	4,4′-Diaminodiphenylmethane	$H_2NC_6H_4CH_2C_6H_4NH_2$	198.27	13, 238			91–92	398	221	i aq; s alc, bz
d36	3,3′-Diaminodiphenyl sulfone	$H_2NC_6H_4SO_2C_6H_4NH_2$	248.30	13, 426			167–170			
d37	4,4′-Diaminodiphenyl sulfone	$H_2NC_6H_4SO_2C_6H_4NH_2$	248.30	13, 536			175–177			i aq; s alc, acet, HCl
d38	2,7-Diaminofluorene		196.25	13, 266			165–166			sl s aq; v s alc

								s aq
d39	2,4-Diamino-6-hydroxypyrimidine		126.12	24, 469			285 d	
d40	Diaminomaleonitrile	$NCC(NH_2){=}C(NH_2)CN$	108.10	4^2, 949		1.4805^{20}	178–179	107–125^{10mm}	93
d41	1,8-Diamino-p-menthane		170.30	13, 4	0.914		−45	110–112^{6mm}	102
d42	3,3'-Diamino-N-methyl-dipropyamine	$CH_3N[(CH_2)_3NH_2]_2$	145.25	4^4, 1279				

Diacetins, g17, g18
Diacetone acrylamide, d568
Diacetone alcohol, h142
Diacetonitrile, a151
(Diacetoxyiodo)benzene, i28
Diacetyl, b386

Diallyl, h41
2,5-Diaminoanisole, m94
1,4-Diaminobutane, b379
1,2-Diaminocyclohexanes, c318, c319
1,10-Diaminodecane, d8
p-Diaminodiphenyl, b136

3,3'-Diaminodipropylamine, i9
1,12-Diaminododecane, d720
1,2-Diaminoethane, e15
1,7-Diaminoheptane, h7
1,6-Diaminohexane, h56

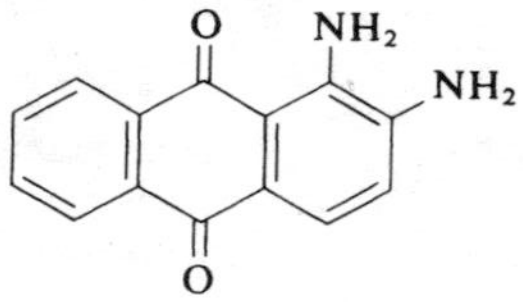

d28

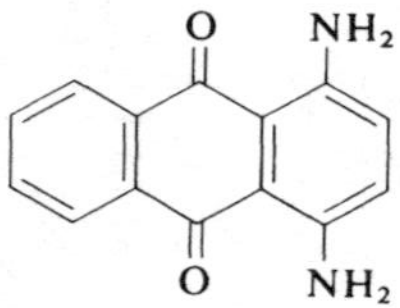

d29

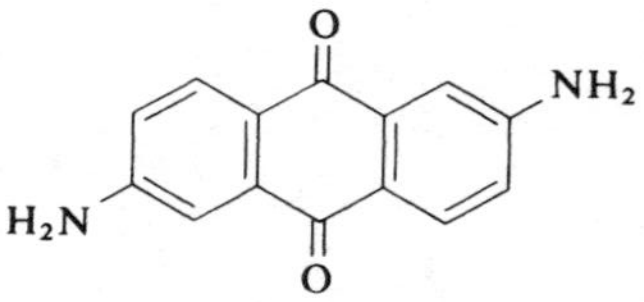

d30

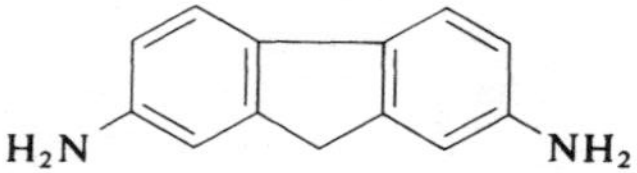

d38

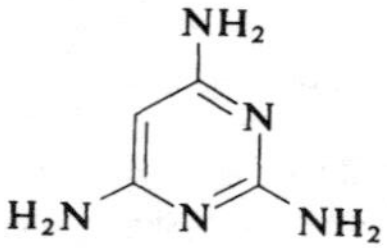

d39

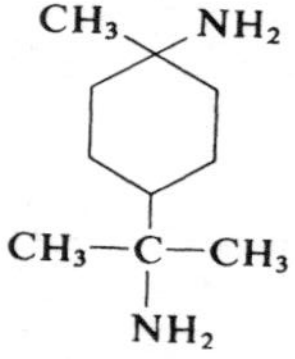

d41

TABLE 1.15 Physical Constants of Organic Compounds (*continued*)

No.	Name	Formula	Formula weight	Beilstein reference	Density	Refractive index	Melting point	Boiling point	Flash point	Solubility in 100 parts solvent
d43	1,3-Diamino-2-propanol	$H_2NCH_2CH(OH)CH_2NH_2$	90.13	4, 290			40–45	235		s aq, alc
d44	2,6-Diaminopyridine	$(H_2N)_2C_5H_3N$	109.13	22[1], 647			118–120	174		45 aq; 77 EtOH; 51 bz; 13 acet; 26 MeEtKe
d45	1,4-Diazabicyclo[2.2.2]octane		112.18				158			
d46	1,8-Diazabicyclo[5.4.0]undec-7-ene		152.24	23, 25	1.018	1.5219^{20}		$80^{0.6mm}$	>112	s eth, diox
d47	Diazomethane	$CH_2{=}N{=}N$	42.04				−145	−23	very explosive	
d48	1-Diazo-2-naphthol-4-sulfonic acid, Na salt		272.22	16, 595			166			
d49	Dibenz[de,kl]anthracene		252.32	5[1], 363	1.35		273–274	503		s bz; sl s alc, eth
d50	Dibenzofuran		168.20	17, 70	1.0886^{99}_4	1.6079^{99}	81–83	285		i aq; s alc, bz, eth
d51	2,3,11,12-Dibenzo-1,4,7,10,13-hexaoxacyclooctadeca-2,11-diene		360.41				162–164			
d52	Dibenzothiophene		184.26	17, 72			97,100	332–333		s aq; v s alc, bz
d53	Dibenzoylmethane	$C_6H_5C({=}O)CH_2{-}C({=}O)C_6H_5$	224.26	7, 769			78–79	220^{18mm}		s alc; v s eth
d54	Dibenzoyl peroxide	$C_6H_5C(O)O{-}OC(O)C_6H_5$	242.23				103–106	may explode when heated		sl s aq, alc; s bz, chl, eth

No.	Name	Formula	FW	Beil. ref.	Density	n_D	m.p., °C	b.p., °C	Solubility
d55	(—)-Dibenzoyl-L-tartaric acid hydrate	$[(C_6H_5COOCH(COOH){-}]_2 \cdot H_2O$	376.34	9,170			135	143	i aq; s alc, eth
d56	Dibenzylamine	$C_6H_5CH_2NHCH_2C_6H_5$	197.28	12, 1035	1.026	1.5731^{20}	−26	300	s hot alc, bz, eth
d57	Dibenzyl disulfide	$C_6H_5CH_2SSCH_2C_6H_5$	246.39	6, 465			69	d>270	misc alc, acet, chl, eth
d58	Dibenzyl ether	$C_6H_5CH_2OCH_2C_6H_5$	198.27	6, 434	1.0014_4^{20}	1.5610^{20}	3.5	298 d	

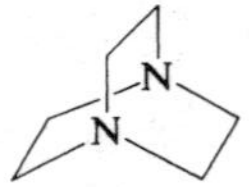

d45

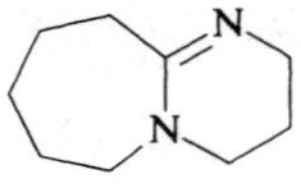

d46

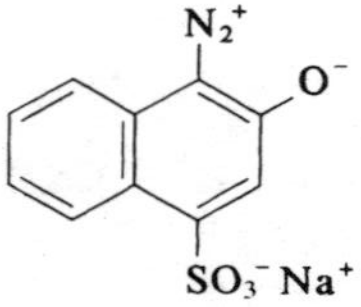

d48

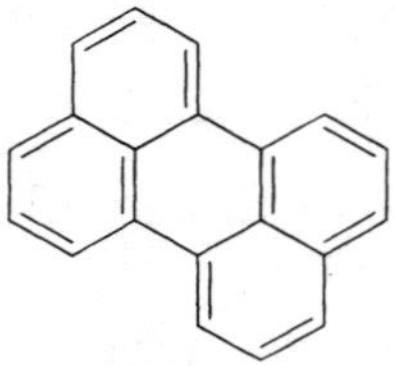

d49

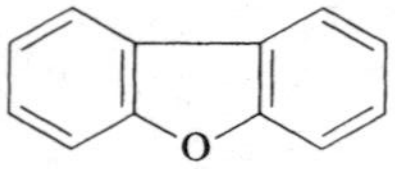

d50

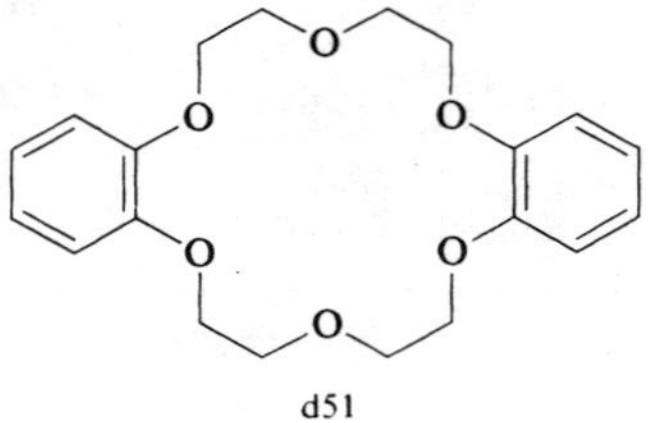

d51

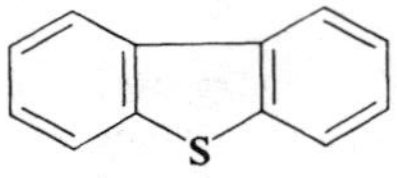

d52

TABLE 1.15 Physical Constants of Organic Compounds (*continued*)

No.	Name	Formula	Formula weight	Beilstein reference	Density	Refractive index	Melting point	Boiling point	Flash point	Solubility in 100 parts solvent
d59	*N,N'*-Dibenzylethylenediamine	$(C_6H_5CH_2NHCH_2-)_2$	240.35	12, 1067	1.024^{20}_{4}	1.5624^{20}	26	195^{4mm}	>112	v s alc, bz, chl, eth
d60	Dibenzyl malonate	$CH_2[COOCH_2C_6H_5]_2$	284.31	6, 436	1.137	1.5447^{20}	−5 to +5	$188^{0.2mm}$	>112	
d61	Dibenzyl phosphonate	$(C_6H_5CH_2O)_2P(O)H$	262.25		1.187	1.5540^{20}		$110^{0.01mm}$	>112	
d62	Dibromoacetic acid	$Br_2CHCOOH$	217.86	2, 218				$128-130^{16mm}$		
d63	Dibromoacetonitrile	Br_2CHCN	198.86	2, 219	2.296	1.5393^{20}		$67-69^{24mm}$	none	s warm alc, eth
d64	2,4'-Dibromoacetophenone	$BrC_6H_4C(=O)CH_2Br$	277.96	7, 285			108–110			
d65	1,4-Dibromobenzene	$C_6H_4Br_2$	235.92	5, 211	0.9641^{100}	1.5743^{100}	87.3	219	none	1.4 alc; s bz; 101 eth
d66	4,4'-Dibromobiphenyl	$BrC_6H_4C_6H_4Br$	312.00	5, 580			162–163	355–360	none	s bz; sl s hot alc
d67	1,3-Dibromobutane	$CH_3CH(Br)CH_2CH_2Br$	215.93	1, 120	1.800^{20}	1.5085^{20}		175	none	s chl, eth
d68	1,4-Dibromobutane	$BrCH_2CH_2CH_2CH_2Br$	215.93	1, 120	1.8080^{20}_{4}	1.5186^{20}	−20	198	>112	s chl
d69	1,4-Dibromo-2,3-butanedione	$BrCH_2C(=O)C(=O)CH_2Br$	243.89	1, 774			116–117			
d70	*trans*-2,3-Dibromo-2-butene-1,4-diol	$HOCH_2C(Br)=C(Br)CH_2OH$	245.91	1[1], 260			112–114			
d71	Dibromochloromethane	$HCClBr_2$	208.29	1, 67	2.451	1.5465^{20}	−22	120^{748mm}	none	misc alc, bz, eth
d72	1,2-Dibromo-2-chloro-1,1,2-trifluoroethane	$FCCl(Br)C(Br)F_2$	276.5		2.2478^{20}	1.4275^{20}		93–94		
d73	1,10-Dibromodecane	$Br(CH_2)_{10}Br$	300.09	1[1], 64	1.335^{30}	1.4912^{20}	27	160^{15mm}	>112	sl s alc; s eth
d74	1,2-Dibromo-1,1-difluoroethane	$CH_2BrC(Br)F_2$	223.87	1, 92	2.2238^{20}	1.4456^{20}	−61.3	93.4		i aq
d75	Dibromodifluoromethane	Br_2CF_2	209.81	1[1], 16	2.288^{15}_{4}	1.3999^{12}	−141.6	23–24	none	0.1 aq; misc alc, bz, chl, eth

No.	Name	Formula	Mol. wt.	Beilstein ref.	Density	n_D	mp, °C	bp, °C	Flash point	Solubility
d76	1,3-Dibromo-5,5-dimethylhydantoin		285.93				197 d			
d77	1,1-Dibromoethane	CH_3CHBr_2	187.87	1, 90	2.055^{20}	1.5379^{20}		113	none	i aq; v s alc, eth
d78	1,2-Dibromoethane	$BrCH_2CH_2Br$	187.87	1, 90	2.1802^{20}_4	1.5416^{15}	10.0	131.7		0.43 aq; misc alc, eth
d79	(1,2-Dibromoethyl)-benzene	$C_6H_5CH(Br)CH_2Br$	263.97	5, 356			70–74	140^{15mm}		
d80	cis-1,2-Dibromo-ethylene	$BrCH=CHBr$	185.86	1, 190	2.21^{17}_4	1.5431^{18}	−53	112.5		s alc, bz, chl, eth
d81	trans-1,2-Dibromo-ethylene	$BrCH=CHBr$	185.86	1, 190	2.246	1.5505^{18}	−6.5	108		
d82	1,2-Dibromoethyltri-chlorosilane	$BrCH_2CH(Br)SiCl_3$	321.3		2.046^{20}_4	1.537^{20}		90^{1mm}		
d83	4',5'-Dibromofluo-rescein		490.12	19, 228			270–273			s hot alc, HOAc
d84	2,4-Dibromo-1-fluoro-benzene	$Br_2C_6H_3F$	253.91		2.047^{20}	1.5840^{20}		105^{22mm}	92	
d85	1,2-Dibromohexa-fluoropropane	$CF_3CF(Br)C(Br)F_2$	309.83				72.8			

Dibenzyl ketone, d684

5,7-Dibromo-8-quinolinol, d87

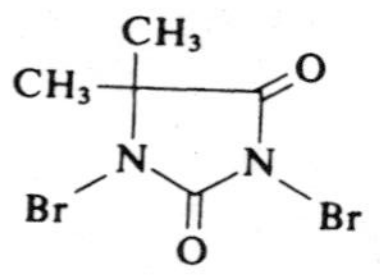

d76

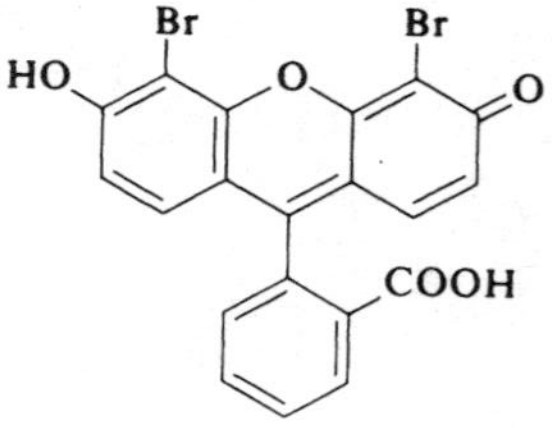

d83

TABLE 1.15 Physical Constants of Organic Compounds (*continued*)

No.	Name	Formula	Formula weight	Beilstein reference	Density	Refractive index	Melting point	Boiling point	Flash point	Solubility in 100 parts solvent
d86	1,6-Dibromohexane	$Br(CH_2)_6Br$	243.98	1, 145	1.586_4^{18}	1.5066^{20}	200–201	243	32	misc eth
d87	5,7-Dibromo-8-hydroxyquinoline		302.96	21, 97				subl		s alc, bz; v s eth
d88	Dibromomethane	CH_2Br_2	173.85	1, 67	2.4956_4^{20}	1.5419^{20}	−52.7	96.97	none	1.15 aq; misc alc, bz, acet, chl, eth; sl s aq; s HOAc
d89	2,6-Dibromo-4-nitroaniline	$Br_2C_6H_2(NO_2)NH_2$	295.93	12, 743			206–208			s bz, hot alc
d90	2,5-Dibromonitrobenzene	$Br_2C_6H_3NO_2$	280.91	5, 250	1.9581^{111}		82–84			
d91	1,5-Dibromopentane	$Br(CH_2)_5Br$	229.95	1, 131	1.6879_4^{15}	1.5092^{15}	−34	110^{15mm}	79	0.2 aq; misc alc, bz, chl, eth
d92	1,2-Dibromopropane	$CH_3CH(Br)CH_2Br$	201.90	1, 109	1.933^{20}	1.5203^{20}	−55.5	139.6	none	0.17 aq; s alc, eth
d93	1,3-Dibromopropane	$BrCH_2CH_2CH_2Br$	201.90	1, 110	1.9712_4^{25}	1.5233^{20}	−34	166.8	54	sl s aq; misc alc, bz, eth, acet
d94	2,3-Dibromopropanol	$BrCH_2CH(Br)CH_2OH$	217.90	1, 357	2.120_4^{20}	1.5599^{20}		$95–97^{10mm}$		
d95	2,3-Dibromopropene	$BrCH_2C(Br){=}CH_2$	199.88	1, 201	1.9336_4^{20}	1.5470^{20}	64–66	140–143	none	s aq, alc, bz
d96	2,3-Dibromopropionic acid	$BrCH_2CH(Br)COOH$	231.88	2, 258				160^{20mm}		
d97	2,6-Dibromopyridine	$Br_2(C_5H_3N)$	236.91	20^2, 153			118–119	255		v s aq, alc
d98	DL-2,3-Dibromosuccinic acid	$HOOCCH(Br)CH(Br)COOH$	275.89	2, 625			167		none	
d99	1,2-Dibromotetrachloroethane	$BrCCl_2CCl_2Br$	325.65	1, 93	2.713		220–222			
d100	1,2-Dibromotetrafluoroethane	$BrCF_2CF_2Br$	259.83		2.163^{25}	1.367^{25}	−110.5	47.3		s aq, alc
d101	2,5-Dibromothiophene	$Br_2C_4H_2S$	241.94	17, 33	2.147_{23}^{23}	1.6289^{20}	−6	221	110	i aq; v s alc, eth
d102	α,α-Dibromotoluene	$C_6H_5CHBr_2$	249.94	5, 308	1.510^{15}	1.6147^{20}		156^{23mm}		i aq; misc alc, eth
d103	1,2-Dibromo-1,1,2-trifluoroethane	$HC(Br)FC(Br)F_2$	241.8	1, 92	2.274^{27}	1.4191^{24}		76.5		

			MW	Beil. ref.	Density	n_D	m.p.	b.p.		Solubility
d104	α,α'-Dibromo-o-xylene	$C_6H_4(CH_2Br)_2$	263.97	5, 366	1.960		92–94			sl s alc, chl, eth
d105	α,α'-Dibromo-p-xylene	$C_6H_4(CH_2Br)_2$	263.97	5, 385	2.012^0		142–143	245		v s alc, chl; s eth
d106	1,2-Dibutoxyethane	$C_4H_9OCH_2CH_2OC_4H_9$	174.28		0.8374^{20}_{20}	1.4131^{20}	−69.1	203.6		0.2 aq; misc alc, acet
d107	Dibutylamine	$(C_4H_9)_2NH$	129.25	4, 157	0.7601^{20}_4	1.4177^{20}	−62	159.6	33	0.47 aq; s alc, acet, eth, EtAc, PE
d108	N,N-Dibutylamino-ethanol	$(C_4H_9)_2NCH_2CH_2OH$	173.29		0.860^{20}_{20}	1.444^{20}	<−70	227–230	93	i aq. MeOH; s acet, bz, EtOH, EtAc, PE
d109	N,N-Dibutylaniline	$C_6H_5N(C_4H_9)_2$	205.34	$12^2, 95$	0.904^{20}	1.5197^{20}		267–275	110	i aq; s acet, bz, EtOH, EtAc, eth
d110	Dibutyl decanedioate	$C_4H_9OOC(CH_2)_8COOC_4H_9$	314.45	2, 719	0.9366^{20}	1.4415^{20}	1.0	344–345	177	0.004 aq
d111	Di-tert-butyldi-carbonate	$(CH_3)_3COC(O)OC(CH_3)_3$	218.25		0.950	1.4103^{20}	23	$56^{0.5mm}$	37	
d112	2,5-Di-tert-butyl-1,4-dihydroxy-benzene	$[(CH_3)_3C]_2C_6H_2(OH)_2$	222.33				217–219			
d113	Dibutyl disulfide	$C_4H_9SSC_4H_9$	178.36	$1^2, 400$	0.9383^{20}_4	1.4920^{20}	−71	231.2	93	i aq: misc alc, eth

Dibutyl 1,2-benzenedicarboxylate, d128

Dibutyl butanedioate, d130

Dibutyl Cellosolve, d106

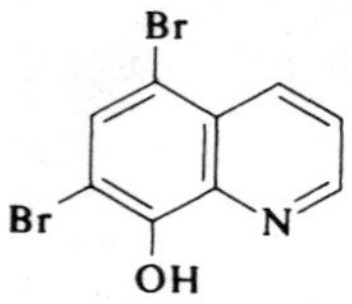

d87

TABLE 1.15 Physical Constants of Organic Compounds (*continued*)

No.	Name	Formula	Formula weight	Beilstein reference	Density	Refractive index	Melting point	Boiling point	Flash point	Solubility in 100 parts solvent
d114	Di-*tert*-butyl disulfide	$(CH_3)_3CSSC(CH_3)_3$	178.36		0.935	1.4920		229–33	93	
d115	Dibutyl ether	$C_4H_9OC_4H_9$	130.22	1, 369	0.7689^{20}	1.3992^{20}	−97.9	142.4	25	0.03 aq; misc alc, eth
d116	*N,N*-Dibutyl-formamide	$HC(=O)N(C_4H_9)_2$	157.26		0.864	1.4429^{20}		120^{15mm}	100	
d117	3,5-Di-*tert*-butyl-4-hydroxybenzoic acid	$[(CH_3)_3C]_2C_6H_2(OH)COOH$	250.34				206–209			
d118	Dibutyl maleate	$C_4H_9OOCCH=CHCOOC_4H_9$	228.28		0.9950^{20}	1.4454^{20}	<−80	d280	135	0.05 aq
d119	Di-*tert*-butyl malonate	$CH_2COOC(CH_3)_3$ | $COOC(CH_3)_3$	216.27			1.4184^{20}	−6.0	93^{10mm}		
d120	2,6-Di-*tert*-butyl-4-methylphenol	$[(CH_3)_3C]_2C_6H_2(CH_3)OH$	220.36	6^3, 2073	0.894_4^{75}	1.4859^{75}	70	265		i aq; s alc, bz, acet
d121	Dibutyl oxalate	$C_4H_9OOC{-}COOC_4H_9$	202.25	2, 540	0.986_4^{20}	1.4232^{20}	−30.0	239–240	108	misc alc, ketones, PE
d122	Di-*tert*-butyl peroxide	$(CH_3)_3CO{-}OC(CH_3)_3$	146.23		0.794^{20}	1.3890^{20}	−40	110		misc acet, octane
d123	2,4-Di-*tert*-butyl-phenol	$[(CH_3)_3C]_2C_6H_3OH$	206.33				56.5	263.5	115	s hot alc; i alk
d124	2,6-Di-*sec*-butyl-phenol	$[CH_3CH_2CH(CH_3)]_2{-}C_6H_3OH$	206.33	6^3, 2061	0.918	1.5100^{20}	−42	255–260	127	
d125	2,6-Di-*tert*-butyl-phenol	$[(CH_3)_3C]_2C_6H_3OH$	206.33				35–83	253	118	s hot alc; i alk
d126	3,5-Di-*tert*-butyl-phenol	$[(CH_3)_3C]_2C_6H_3OH$	206.33				87–89			
d127	Dibutyl phosphonate	$(C_4H_9O)_2P(O)H$	194.21	1, 187	0.9954_4^{20}	1.4231^{20}		118^{11mm}	121	sl s (hyd) aq; misc alc, acet, eth
d128	Dibutyl *o*-phthalate	$C_6H_4[COOC_4H_9]_2$	278.35	9^2, 586	1.0465_4^{20}	1.4926^{20}	−35	340	171	0.01 aq; v s alc, bz; acet, eth

No.	Name	Formula	Formula wt	Beilstein ref	Density	n_D	mp, °C	bp, °C	Flash pt	Solubility
d129	N,N-Dibutyl-1,3-propanediamine	$C_4H_9NHCH_2CH_2CH_2NHC_4H_9$	186.34		0.827	1.4463^{20}		205	103	i aq; s alc, eth
d130	Dibutyl succinate	$[C_4H_9OOCCH_2{-}]_2$	230.30	2^2, 551	0.9768^{20}_4	1.4299^{20}	−29.0	274.5		i aq; s alc, eth
d131	Dibutyl sulfate	$C_4H_9OSO_2OC_4H_9$	210.29		1.059^{25}_4	1.4213^{20}		$130–132^{11mm}$		i aq; v s alc, eth
d132	Dibutyl sulfide	$C_4H_9SC_4H_9$	146.30	1, 370	0.839^{16}_0	1.4530^{20}	−75.0	188.9	76	i aq; s alc, eth
d133	Di-tert-butyl sulfide	$(CH_3)_3CSC(CH_3)_3$	146.30		0.815	1.4506^{20}		151	48	
d134	Dibutyl sulfite	$(C_4H_9O)_2S(O)$	194.29	1^2, 397	0.9944^{22}_4	1.4310^{20}		108^{15mm}		i aq; s alc, eth
d135	Dibutyl sulfone	$(C_4H_9O)_2SO_2$	178.29	1, 371			46	295	143	
d136	N,N'-Dibutylthiourea	$C_4H_9NHC(=S)NHC_4H_9$	188.34				63.65			i aq; s alc; sl s eth
d136a	Dibutyltin dichloride	$(C_4H_9)_2SnCl_2$	303.83				39–41	135^{10mm}	>112	
d137	Dibutyltin oxide	$(C_4H_9)_2SnO$	248.92	4^1, 588			>300			
d138	Dichloroacetic acid	$Cl_2CHCOOH$	128.94	2, 202	1.563^{20}_4	1.4642^{20}	9–11	193–194	>112	misc aq, alc, eth
d139	1,1-Dichloroacetone	$CH_3C(O)CHCl_2$	126.97	1, 654	1.305^{18}_{15}			150		sl s aq; s alc; misc eth
d140	2',4'-Dichloroacetophenone	$Cl_2C_6H_3COCH_3$	189.04	7^2, 219		1.5635^{20}	33–34	145^{15mm}	>112	i aq
d141	Dichloroacetyl chloride	$Cl_2CHCOCl$	147.39	2, 204	1.5315^{16}_4	1.4603^{20}		107–108	66	d aq, alc; misc eth
d142	2,3-Dichloroaniline	$Cl_2C_6H_3NH_2$	162.02	12, 621		1.5969^{20}	23–24	252	>112	s alc; v s eth
d143	2,4-Dichloroaniline	$Cl_2C_6H_3NH_2$	162.02	12, 621	1.567^{20}		59.62	245		sl s aq; s alc, eth
d144	2,5-Dichloroaniline	$Cl_2C_6H_3NH_2$	162.02	12, 625			49–51	251		s alc, bz, eth

TABLE 1.15 Physical Constants of Organic Compounds (*continued*)

No.	Name	Formula	Formula weight	Beilstein reference	Density	Refractive index	Melting point	Boiling point	Flash point	Solubility in 100 parts solvent
d145	2,6-Dichloroaniline	$Cl_2C_6H_3NH_2$	162.02	12, 626			38.41	272		s alc, eth; sl s bz
d146	3,4-Dichloroaniline	$Cl_2C_6H_3NH_2$	162.02	12, 626			70–72			i aq; s alc, eth
d147	3,5-Dichloroaniline	$Cl_2C_6H_3NH_2$	162.02	12, 626			51–53	259^{741mm}		sl s alc, bz, acet
d148	1,5-Dichloroanthraquinone		277.11	7, 787			245–247			
d149	1,8-Dichloroanthraquinone		277.11	7, 788			202–203			sl s alc
d150	2,4-Dichlorobenzaldehyde	$Cl_2C_6H_3CHO$	175.01	7, 236			69–73	233		i aq; s alc
d151	2,4-Dichlorobenzamide	$Cl_2C_6H_3CONH_2$	190.03	9^3, 1376			191–194			
d152	1,2-Dichlorobenzene	$C_6H_4Cl_2$	147.01	5, 201	1.3059^{20}_{4}	1.5515	−17.0	180.4	65	misc alc, bz, eth
d153	1,3-Dichlorobenzene	$C_6H_4Cl_2$	147.01	5, 202	1.2884^{20}_{4}	1.5459	−24.8	173.1	63	0.01 aq; s alc, eth
d154	1,4-Dichlorobenzene	$C_6H_4Cl_2$	147.01	5, 203	1.2417^{60}	1.5285	53	174.1	65	s alc, bz, chl, eth
d155	2,5-Dichlorobenzenesulfonyl chloride	$Cl_2C_6H_3SO_2Cl$	245.51	11^1, 15			36–37			d hot aq, hot alc
d156	2,4-Dichlorobenzoic acid	$Cl_2C_6H_3COOH$	191.01	9, 342			157–160			s hot aq, alc, bz, chl
d157	2,5-Dichlorobenzoic acid	$Cl_2C_6H_3COOH$	191.01	9, 342			151–154	301		sl s aq; s alc, eth
d158	3,4-Dichlorobenzoic acid	$Cl_2C_6H_3COOH$	191.01	9, 343			207–209			s hot aq, eth; v s alc
d159	4,4′-Dichlorobenzophenone	$(ClC_6H_4)_2CO$	251.11	7, 420			144–146	353		s hot alc; v s chl, eth
d160	2,4-Dichlorobenzoyl chloride	$Cl_2C_6H_3COCl$	209.46	9, 342	1.494	1.5297^{20}	16–18	150^{34mm}	137	d aq, alc
d161	3,4-Dichlorobenzoyl chloride	$Cl_2C_6H_3COCl$	209.46	9, 344			30–33	242	142	d aq, alc
d162	1,2-Dichlorobutane	$CH_3CH_2CH(Cl)CH_2Cl$	127.01	1^1, 38	1.1182^{20}_{4}	1.4474^{15}		124		i aq; s chl, eth

d163	1,4-Dichlorobutane	$ClCH_2CH_2CH_2CH_2Cl$	127.01	1, 119	1.1598^{20}_{4}	1.4566^{20}	−38	155	40	i aq; s chl
d164	*meso*-2,3-Dichloro-butane	$CH_3CH(Cl)CH(Cl)CH_3$	127.01	1, 119	1.1025^{25}_{4}	1.4386^{25}	−80	115.9	18	i aq; s chl
d165	*cis*-1,4-Dichloro-2-butene	$ClCH_2CH{=}CHCH_2Cl$	125.00	1^3, 743	1.188^{25}_{4}	1.4887^{25}	−48	152	49	i aq; s org solv
d166	*trans*-1,4-Dichloro-2-butene	$ClCH_2CH{=}CHCH_2Cl$	125.00	1^3, 743	1.183^{25}_{4}	1.4861^{25}	1–3	$74–76^{40mm}$	56	i aq; s org solv
d167	3,4-Dichloro-1-butene	$ClCH_2CH(Cl)CH{=}CH_2$	125.00		1.150	1.4658^{20}	−61	123	28	
d168	1,4-Dichloro-2-butyne	$ClCH_2C{\equiv}CCH_2Cl$	122.98	1^3, 927	1.258^{20}_{4}	1.5048^{20}		165–168	160	
d169	1,1-Dichloro-2,2-di-ethoxyethane	$Cl_2CHCH(OC_2H_5)_2$	187.07	1, 614	1.138	1.4360^{20}		183–184	60	
d170	Dichlorodifluoro-methane	Cl_2CF_2	120.92	1, 61	1.486^{-30}		−158	−29.8		0.02 aq; 9 bz; 5.5 chl; 6 diox; s alc, eth
d171	4,6-Dichloro-1,3-di-hydroxybenzene	$Cl_2C_6H_2(OH)_2$	179.00	6^1, 403			104–106	254		sl s aq, bz; s eth
d172	2,5-Dichloro-3,6-di-hydroxy-*p*-benzo-quinone		208.98	8, 379			283–284			

2,6-Dichlorobenzyl chloride, t249
2,2′-Dichlorodiethyl ether, b158

5,5′-Dichloro-2,2′-dihydroxydiphenylmethane, m233

1,1-Dichlorodimethyl ether, d197
Dichlorohydrin, d220

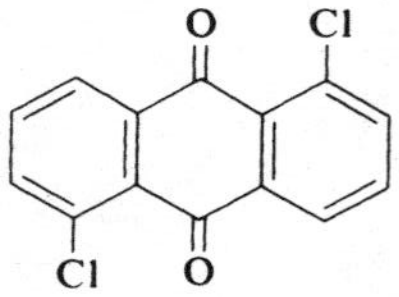

d148

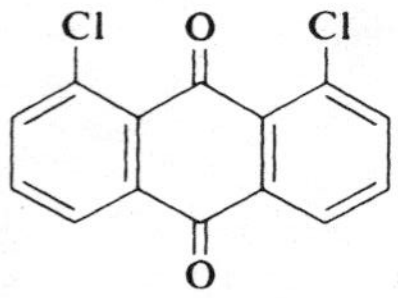

d149

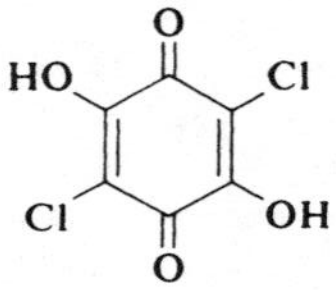

d172

TABLE 1.15 Physical Constants of Organic Compounds (*continued*)

No.	Name	Formula	Formula weight	Beilstein reference	Density	Refractive index	Melting point	Boiling point	Flash point	Solubility in 100 parts solvent
d173	1,3-Dichloro-3,5-dimethylhydantoin		197.02	24[2], 158			134–136			
d174	Dichlorodimethylsilane	$(CH_3)_2SiCl_2$	129.06		1.0644^{20}_4	1.4038^{20}	−16	70	−16	
d175	Dichlorodiphenylsilane	$(C_6H_5)_2SiCl_2$	253.20	16, 910	1.2224^{20}_4	1.582^{20}		308–309	157	d aq, alc
d176	1,1-Dichloroethane	CH_3CHCl_2	98.96	1, 83	1.17572^{20}_4	1.4164^{20}	−97.0	57.3	−5	0.51 aq; misc alc
d177	1,2-Dichloroethane	$ClCH_2CH_2Cl$	98.96	1, 84	1.25314^{20}_4	1.4448^{20}	−35.7	83.5	15	0.8 aq; misc alc, chl, eth
d178	1,1-Dichloroethylene	$H_2C=CCl_2$	96.94	1, 186	1.2129^{20}_4	1.4247^{20}	−122.6	31.6	−15	0.02 aq; s alc, bz, chl, eth
d179	cis-1,2-Dichloroethylene	$ClCH=CHCl$	96.94	1, 188	1.2818^{20}_4	1.4490^{20}	−80.1	60.7	6	0.7 aq; s alc, eth
d180	trans-1,2-Dichloroethylene	$ClCH=CHCl$	96.94	1, 188	1.2546^{20}_4	1.4462^{20}	−49.8	47.7	6	0.6 aq; s alc, eth
d181	2,2'-Dichloroethyl ether	$ClCH_2CH_2OCH_2CH_2Cl$	143.01	1[2], 335	1.2220^{20}_{20}	1.457^{20}		178.5	55	1.1 aq; s alc, bz, eth
d182	1,2-Dichloroethyltrichlorosilane	$ClCH_2CH(Cl)SiCl_3$	232.4		1.516^{25}_4	1.449^{25}		$82–84^{26mm}$		
d183	Dichlorofluoromethane	$FCHCl_2$	102.92	1, 61	1.345^{30}		−135	8.9		
d184	Dichloroheptylmethylsilane	$C_7H_{15}Si(CH_3)Cl_2$	225.2		0.9780^{20}_4	1.4396^{25}		207–208		
d185	1,2-Dichlorohexafluorocyclobutane	$F_6C_4Cl_2$	233.0			1.3342^{25}		59–60		
d186	1,5-Dichlorohexamethyltrisiloxane	$[ClC(CH_3)_2O]_2\text{-}Si_3(CH_3)_2$	277.4		1.0184^{20}_4	1.4071		184		
d187	1,6-Dichlorohexane	$Cl(CH_2)_6Cl$	155.07	1, 144	1.068	1.4568^{20}		87^{15mm}	73	s chl

No.	Name	Formula	M.W.	Beil. ref.	Density	n_D	m.p., °C	b.p., °C	Solubility
d188	1,2-Dichloro-2-iodo-1,1,2-trifluoroethane	F(I)C(Cl)C(Cl)F$_2$	278.9		2.200^{20}	1.4490^{20}		100–101	1.3 aq; misc alc, eth
d189	Dichloromaleic anhydride		166.95	17,434					none
d190	Dichloromethane	CH$_2$Cl$_2$	84.93	1, 60	1.3255_4^{20}	1.4246^{20}	−96.7	40.5	
d191	2,3-Dichloro-1-methoxybenzene	Cl$_2$C$_6$H$_3$OCH$_3$	177.03	6[1], 102			31–33		
d192	3,5-Dichloro-1-methoxybenzene	Cl$_2$C$_6$H$_3$OCH$_3$	177.03	6, 190			40–42		
d193	2,4-Dichloro-6-methoxy-1,3,5-triazine		179.99				86–88	132^{49mm}	
d194	(Dichloromethyl)dimethylchlorosilane	Cl$_2$CHSi(Cl)(CH$_3$)$_2$	177.5		1.237_4^{20}	1.461^{20}	−49	149	
d195	2,2-Dichloro-1-methylcyclopropanecarboxylic acid	Cl$_2$(C$_3$H$_2$)(CH$_3$)COOH	169.01				60–65	85^{8mm}	
d196	N-(Dichloromethylene)aniline	C$_6$H$_5$N=CCl$_2$	174.03	12, 447	1.265	1.5710^{20}		106^{30mm}	79
d197	Dichloromethyl ether	Cl$_2$CHOCH$_3$	114.96		1.271	1.4300^{20}		85	42

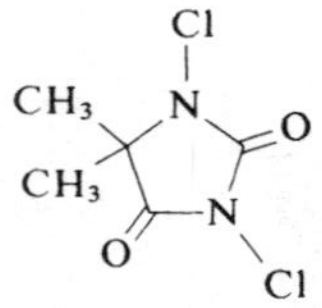

d173

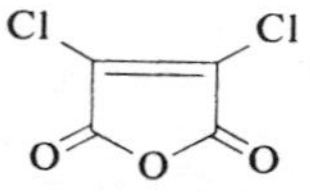

d189

d193

4,4'-Dichloro-α-methylbenzhydrol, b164

Dichloroisopropyl alcohol, d220

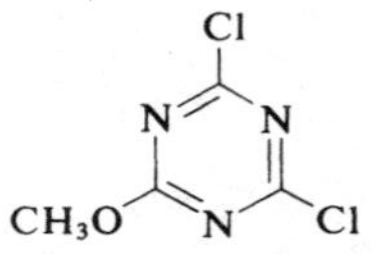

TABLE 1.15 Physical Constants of Organic Compounds (*continued*)

No.	Name	Formula	Formula weight	Beilstein reference	Density	Refractive index	Melting point	Boiling point	Flash point	Solubility in 100 parts solvent
d198	Dichloro(methyl)-phenylsilane	$C_6H_5Si(CH_3)Cl_2$	191.13		1.176	1.5190^{20}		205	82	sl s alc, bz, eth
d199	Dichloro(methyl)-silane	$HSi(CH_3)Cl_2$	115.04	4[1], 581	1.105		−93	41	−32	
d200	Dichloro(methyl)-vinylsilane	$H_2C{=}CHSi(CH_3)Cl_2$	141.07		1.0874^{20}_{4}	1.4300^{20}		92–93	4	
d201	2,3-Dichloro-1,4-naphthoquinone		227.05	7, 729			190–192			
d202	2,6-Dichloro-4-nitroaniline	$Cl_2C_6H_2(NO_2)NH_2$	207.02	12, 735			190–192			s PE
d203	2,3-Dichloronitro-benzene	$Cl_2C_6H_3NO_2$	192.00	5, 245	1.721^{14}		61–62	257–258		s hot alc; misc eth
d204	2,4-Dichloronitro-benzene	$Cl_2C_6H_3NO_2$	192.00	5, 245	1.439^{80}		29–32	258		
d205	3,4-Dichloronitro-benzene	$Cl_2C_6H_3NO_2$	192.00	5, 246	1.4567^{75}_{4}		41–42	255–256	123	
d206	2,3-Dichloroocta-fluorobutane	$CF_3CF(Cl)CF(Cl)CF_3$	271.0		1.6801^{20}	1.3100^{20}	−68	63		
d207	1,7-Dichloroocta-methyltetrasiloxane	$[Cl(CH_3)_2SiOSi-(CH_3)_2{-}]_2O$	351.6		1.0111^{20}_{4}	1.403^{20}	125–128	222	100	sl s aq; s alc, hot bz
d208	2,3-Dichloro-4-oxo-2-butenoic acid	$ClC(CHO){=}C(Cl)COOH$	168.96	3, 727	1.10584^{15}	1.4553^{20}			26	i aq; s alc, eth
d209	1,5-Dichloropentane	$Cl(CH_2)_5Cl$	141.04	1, 131			−72	63^{10mm}		s alc, eth
d210	2,3-Dichlorophenol	$Cl_2C_6H_3OH$	163.00	6[1], 102			58–60	206		
d211	2,4-Dichlorophenol	$Cl_2C_6H_3OH$	163.00	6, 189			42–43	210	113	v s alc, bz, chl, eth
d212	2,5-Dichlorophenol	$Cl_2C_6H_3OH$	163.00	6, 189			56–58	211		v s alc, bz, eth
d213	2,6-Dichlorophenol	$Cl_2C_6H_3OH$	163.00	6, 190			65–68	218–220		v s alc, eth
d214	2,4-Dichlorophenoxy-acetic acid	$Cl_2C_6H_3OCH_2COOH$	221.04				138	$160^{0.4mm}$		s alc, bz, chl, eth

No.	Name	Formula	Mol wt	Beilstein ref.	Density	n_D	mp, °C	bp, °C		Solubility
d215	2,5-Dichloro-p-phenylenediamine	$Cl_2C_6H_2(NH_2)_2$	177.03	13, 118			165 d			
d216	Dichlorophenyl-phosphine	$C_6H_5PCl_2$	178.99	16, 763	1.319	1.5980^{20}	−51	222	>112	s aq; v eth
d217	4,5-Dichloro-o-phthalic acid	$Cl_2C_6H_2(COOH)_2$	235.02	9^1, 366			193–195			
d218	1,2-Dichloropropane	$CH_3CH(Cl)CH_2Cl$	112.99	1, 105	1.1558^{20}	1.4390^{20}	−100.4	96.4	4	0.26 aq; misc alc, bz, chl, eth
d219	1,3-Dichloropropane	$ClCH_2CH_2CH_2Cl$	112.99	1, 105	1.878^{20}_{4}	1.4487^{20}	−99.5	120.5	32	v s alc, eth
d220	1,3-Dichloro-2-propanol	$ClCH_2CH(OH)CH_2Cl$	128.99	1, 364	1.3506^{17}_{4}	1.4835^{20}	−4	174.3	74	9.1 aq; misc alc, eth
d221	1,3-Dichloropropene	$ClCH_2CH{=}CHCl$	110.97	1, 199	1.217^{20}_{4}	1.470^{20}		112	10	i aq; s chl, eth
d222	2,3-Dichloro-1-propene	$ClCH_2C(Cl){=}CH_2$	110.97	1, 199	1.204^{25}_{25}	1.4611^{20}		94		misc alc; s eth
d223	3,6-Dichloropyri-dazine		148.98	20, 231			66–69			
d224	2,6-Dichloropyridine	$Cl_2(C_5H_3N)$	147.99				86–88			
d225	4,7-Dichloro-quinoline		198.05				84–86	148^{10mm}		
d226	Dichlorosilane	H_2SiCl_2	101.0				−122	8.3		
d227	1,2-Dichloro-1,1,2,2-tetrafluoro-ethane	$ClCF_2CF_2Cl$	170.93		1.470^{20}_{4}	1.290^{20}	−94	3.6		s alc, eth

1,1-Dichloro-2-propanone, d139

4,6-Dichlororesorcinol, d171

α,o-Dichlorotoluene, c59

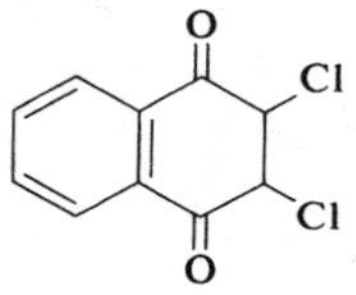

d201

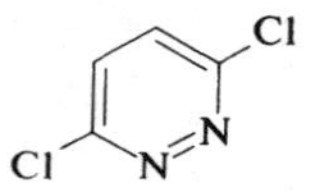

d223

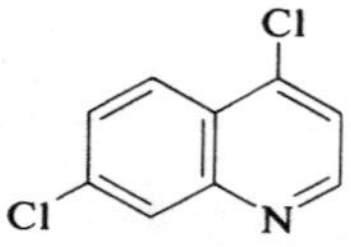

d225

TABLE 1.15 Physical Constants of Organic Compounds (*continued*)

No.	Name	Formula	Formula weight	Beilstein reference	Density	Refractive index	Melting point	Boiling point	Flash point	Solubility in 100 parts solvent
d228	2,5-Dichlorothiophene	$Cl_2(C_4H_2S)$	153.03	17,33	1.442	1.5621^{20}	-40.5	162		i aq; misc alc, eth
d229	2,4-Dichlorotoluene	$Cl_2C_6H_3CH_3$	161.03	5,295	1.2460^{20}_{20}	1.5454^{20}	-13	200.5	79	i aq
d230	2,6-Dichlorotoluene	$Cl_2C_6H_3CH_3$	161.03	5,296	1.254	1.5507^{20}		196–203	82	i aq; s chl
d231	3,4-Dichlorotoluene	$Cl_2C_6H_3CH_3$	161.03	5,296	1.251^{25}_{25}	1.5472^{20}	-14	201^{740mm}	85	i aq
d232	2,2-Dichloro-1,1,1-trifluoroethane	CF_3CHCl_2	152.9					28		
d233	α,α'-Dichloro-p-xylene	$C_6H_4(CH_2Cl)_2$	175.06	5,384			100	254		22.5 acet; 20 bz; 4.5 CCl_4; 11 eth; 18 EtAc
d234	2,5-Dichloro-p-xylene	$Cl_2C_6H_2(CH_3)_2$	175.06	5,384			71	222		27 acet; 44 bz; 39 eth 32 EtAc; 5 MeOH
d235	Dicyanodiamide	$H_2NC(=NH)NHCN$	84.08	3²,75	1.400^{25}_{4}		208–211			2.3 aq; 1.3 alc; i bz
d236	1,2-Dicyanobenzene	$C_6H_4(CN)_2$	128.13	9,815			139–141			v s bz, alc; s hot eth
d237	1,3-Dicyanobenzene	$C_6H_4(CN)_2$	128.13	9,836	0.951	1.4380^{20}	158–160			s alc, bz, chl, eth
d238	1,4-Dicyanobutane	$NC(CH_2)_4CN$	108.14	2,653			1–3	295	>112	
d239	1,6-Dicyanohexane	$NC(CH_2)_6CN$	136.20	2,694	0.954	1.4436^{20}	-3.5	185^{15mm}	>112	
d240	2,4-Dicyano-3-methylglutaramide	$CH_3CH[CH(CN)-CONH_2]_2$	194.19	2²,704			159–160			
d241	Dicyclohexyl	$C_6H_{11}\cdot C_6H_{11}$	166.31	5,108	0.864	1.4782^{20}	3–4	227	101	7 MeOH; misc bz, acet, eth
d242	Dicyclohexylamine	$(C_6H_{11})_2NH$	181.32	12,6	0.910	1.4842^{20}	-0.1	255.8	96	misc alc, bz, chl, eth
d243	N,N'-Dicyclohexylcarbodiimide	$C_6H_{11}N=C=NC_6H_{11}$	206.33				34–35	$122-124^{6mm}$		
d244	Dicyclopentadiene		132.21	5,495	0.930^{25}_{4}	1.5050^{25}	-1	170	26	s alc, eth

	Name	Formula								Solubility
d245	Diethanolamine	HOCH$_2$CH$_2$NHCH$_2$CH$_2$OH	105.14	4, 283	1.0883$^{30}_4$	1.4747^{30}	28.0	268.0	137	96 aq; 4 bz; 0.8 eth; misc MeOH, acet
d246	2,2-Diethoxyaceto-phenone	C$_6$H$_5$C(=O)CH(OC$_2$H$_5$)$_2$	208.26	7^1, 361	1.034	1.4995^{20}		131–134^{10mm}	110	
d247	4,4-Diethoxybutyl-amine	H$_2$N(CH$_2$)$_3$CH(OC$_2$H$_5$)$_2$	161.25	4, 319	0.933	1.4275^{20}		196	62	
d248	2,2-Diethoxy-N,N-dimethylethylamine	(C$_2$H$_5$O)$_2$CHCH$_2$N(CH$_3$)$_2$	161.25	4, 308	0.883	1.4129^{20}		170	45	
d249	Diethoxydimethyl-silane	(C$_2$H$_5$O)$_2$Si(CH$_3$)$_2$	148.28		0.840$^{20}_4$	1.3811^{20}	−87	114	11	
d250	Diethoxydiphenyl-silane	(C$_2$H$_5$O)$_2$Si(C$_6$H$_5$)$_2$	272.42		1.0329$^{20}_4$	1.5269^{20}		130^{2mm}		
d251	1,1-Diethoxyethane	CH$_3$CH(OC$_2$H$_5$)$_2$	118.18	1, 603	0.8254$^{20}_4$	1.3825^{20}	2.8	102.7	−21	5 aq; misc alc, eth
d252	1,2-Diethoxyethane	C$_2$H$_5$OCH$_2$CH$_2$OC$_2$H$_5$	118.18	1, 468	0.842	1.3922^{20}	−74	121.4	27	21 aq
d253	2,2-Diethoxyethanol	(C$_2$H$_5$O)$_2$CHCH$_2$OH	134.18	1, 818	0.888^{24}	1.4160^{20}		167	67	s alc, eth
d254	2,2-Diethoxyethyl-amine	(C$_2$H$_5$O)$_2$CHCH$_2$NH$_2$	133.19	4, 308	0.916	1.4170		162–163	45	
d255	Diethoxymethylsilane	(C$_2$H$_5$O)$_2$SiH(CH$_3$)	134.3		0.829$^{25}_4$	1.372^{25}		94–95		
d256	Diethoxymethylvinyl-silane	(C$_2$H$_5$O)$_2$Si-(CH$_3$)CH=CH$_2$	160.3		0.858$^{20}_4$	1.400^{20}		133–134		
d257	1,1-Diethoxypropane	CH$_3$CH$_2$CH(OC$_2$H$_5$)$_2$	132.20	1, 630	0.8232$^{20}_4$	1.3884^{20}		122.8	12	v s alc, eth
d258	3,3-Diethoxy-1-propene	(C$_2$H$_5$O)$_2$CHCH=CH$_2$	130.19	1, 727	0.854	1.4000^{20}		89–90	4	

α,p-Dichlorotoluene, c60

1,2-Dicyanoethane, b380

d244

TABLE 1.15 Physical Constants of Organic Compounds (*continued*)

No.	Name	Formula	Formula weight	Beilstein reference	Density	Refractive index	Melting point	Boiling point	Flash point	Solubility in 100 parts solvent
d259	2,2-Diethoxytri-ethylamine	$(C_2H_5O)_2CHCH_2N(C_2H_5)_2$	189.30	4, 309	0.850	1.4189^{20}		194–195	65	
d260	N,N-Diethyl-acetamide	$CH_3C(=O)N(C_2H_5)_2$	115.18	4, 110	0.925	1.4401^{20}		182–186	70	
d261	Diethyl acetamido-malonate	$C_2H_5OOCCH(NHCOCH_3)COOC_2H_5$	217.22	4^2, 891			97–98	185^{20mm}		
d262	Diethyl 1,3-acetone-dicarboxylate	$C_2H_5OOCCH_2CO-CH_2COOC_2H_5$	202.21	3, 791	1.113	1.4385^{20}		250	86	
d263	Diethyl acetylenedi-carboxylate	$C_2H_5OOCC{=}CCOOC_2H_5$	170.16	2, 803	1.063	1.4426^{20}		107^{11mm}	94	
d264	Diethyl 2-acetyl-glutarate	$C_2H_5OOCCH_2CH_2-(COCH_3)COOC_2H_5$	230.26		1.071	1.4386^{20}		154^{11mm}	>112	
d265	Diethyl acetyl-succinate	$C_2H_5OOCCH_2CH-(COCH_3)COOC_2H_5$	216.23	3, 801	1.081	1.4346^{20}		180–183^{50mm}	>112	
d266	Diethyl allyl-malonate	$C_2H_5OOCCH(CH_2-CH{=}CH_2)COOC_2H_5$	200.23	2, 776	1.015	1.4304^{20}		222–223	92	
d267	Diethylamine	$(C_2H_5)_2NH$	73.14	4, 95	0.7074^{20}_4	1.3864^{20}	−50.0	55.5	−28	misc aq, alc
d268	Diethylamine HCl	$(C_2H_5)_2NH \cdot HCl$	109.60	4, 95	1.048^{21}_4		39–41	320–330		s aq, alc, chl; i eth
d269	4-(Diethylamino)-benzaldehyde	$(C_2H_5)_2NC_6H_4CHO$	177.25	14^2, 25				174^{7mm}	48	s aq, alc, bz, eth
d270	2-Diethylamino-ethanol	$(C_2H_5)_2NCH_2CH_2OH$	117.19	4, 282	0.8800^{25}	1.4389^{25}	−70	163		
d271	2-(Diethylamino)-ethyl-4-amino-benzoate	$H_2NC_6H_4COOCH_2CH_2-N(C_2H_5)_2$	236.30	14, 424			61			0.5 aq; s alc, bz, eth
d272	2-Diethylaminoethyl chloride HCl	$ClCH_2CH_2N(C_2H_5)_2 \cdot HCl$	172.10	4^2, 618			208–210			
d273	3-(Diethylamino)-phenol	$(C_2H_5)_2NC_6H_4OH$	165.24	13, 408			65–69	170^{15mm}		s aq, alc, eth

No.	Name. Formula.	Formula	Formula wt.	Beilstein ref.	Density	n_D	mp, °C	bp, °C	Flash pt., °C	Solubility
d274	3-Diethylamino-1,2-propanediol	$(C_2H_5)_2NCH_2CH(OH)-CH_2OH$	147.22	4, 302	0.9732^{20}_{20}	1.4602^{20}		233–235	107	s aq, alc, chl, eth
d275	1-Diethylamino-2-propanol	$(C_2H_5)_2NCH_2CH(OH)-CH_3$	131.22	4^2, 737	0.889	1.4255^{20}	13.5	$55\text{–}59^{13mm}$	33	s alc
d276	4-(Diethylamino)-salicylaldehyde	$(C_2H_5)_2NC_6H_3(OH)CHO$	193.25	14, 234			62–64			
d277	N,N-Diethylaniline	$C_6H_5N(C_2H_5)_2$	149.24	12, 164	0.9302^{25}	1.5394^{25}	−34.4	216.3	97	1 aq; sl s alc, eth
d278	2,6-Diethylaniline	$(C_2H_5)_2C_6H_3NH_2$	149.24		0.906	1.5452^{20}	3	243	123	
d279	Diethyl azodicarboxylate	$C_2H_5OOCN{=}NCOOC_2H_5$	174.16	3, 123	1.106	1.4280^{20}		106^{13mm}	26	
d280	5,5-Diethylbarbituric acid		184.19	24^2, 279	1.220		188–192			0.7 aq; 7 alc; 1.3 chl; 3.2 eth; s acet, HOAc
d281	Diethyl benzalmalonate	$C_6H_5CH{=}C(COOC_2H_5)_2$	248.28	9, 892	1.107	1.5365^{20}		215^{30mm}	>112	
d282	1,2-Diethylbenzene	$C_6H_4(C_2H_5)_2$	134.22	5, 426	0.8800^{20}	1.5022^{20}	−31.3	183.4	49	s alc, eth
d283	1,3-Diethylbenzene	$C_6H_4(C_2H_5)_2$	134.22	5, 426	0.8640^{20}_{4}	1.4950^{20}	−83.9	181.1	50	s alc, eth
d284	1,4-Diethylbenzene	$C_6H_4(C_2H_5)_2$	134.22	5, 426	0.8620^{20}_{4}	1.4940^{20}	−42.85	183.8	56	s alc, eth
d285	Diethyl benzylmalonate	$C_6H_5CH_2CH(COOC_2H_5)_2$	250.29	9, 869	1.064	1.4868^{20}		162^{10mm}	>112	
d286	Diethyl bromomalonate	$BrCH(COOC_2H_5)_2$	239.07	2, 594	1.4022^{25}_{4}	1.4550^{20}	−54	233–235 d		i aq; misc alc, eth

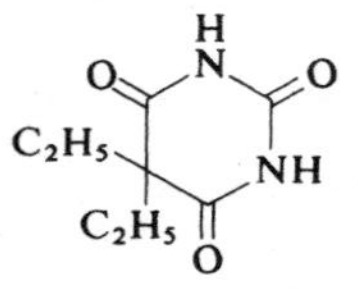

d280

TABLE 1.15 Physical Constants of Organic Compounds (*continued*)

No.	Name	Formula	Formula weight	Beilstein reference	Density	Refractive index	Melting point	Boiling point	Flash point	Solubility in 100 parts solvent
d287	Diethyl butyl-malonate	$C_4H_9CH(COOC_2H_5)_2$	216.28	2^1, 282	0.983	1.4220		235–240	93	v s alc, eth
d288	Diethylcarbamoyl chloride	$(C_2H_5)_2NCOCl$	135.59	4, 120		1.4515^{20}		187–190	75	d hot aq, hot alc
d289	Diethyl carbonate	$(C_2H_5O)_2C{=}O$	118.13	3, 5	0.9764_4^{20}	1.3843^{20}	−43.0	126.8	25	69 aq; misc alc, bz, eth, esters
d290	Diethyl chloro-malonate	$ClCH(COOC_2H_5)_2$	194.61	2^2, 537	1.2040_4^{20}	1.4310^{20}		222–223		misc alc, chl, eth
d291	Diethyl chloro-phosphate	$(C_2H_5O)_2P(O)Cl$	172.55	1, 332	1.194	1.4165^{20}		60^{2mm}		
d292	Diethyl chloro-thiophosphate	$(C_2H_5O)_2P(S)Cl$	188.61		1.200	1.4715^{20}		45^{3mm}		
d293	Diethylcyanamide	$(C_2H_5)_2NCN$	98.15	4, 121	0.846	1.4229^{20}		186–188	69	sl s aq; misc alc, eth
d293a	Diethyl cyanomethyl-phosphonate	$(C_2H_5O)_2P(O)CH_2CN$	177.14		1.095	1.4312^{20}		$101^{0.4mm}$	>112	
d294	N,N-Diethylcyclo-hexylamine	$C_6H_{11}N(C_2H_5)_2$	155.29	12, 6	0.850	1.4562^{20}		194–195	57	
d294a	Diethyl disulfide	$C_2H_5SSC_2H_5$	122.25	1, 347	0.9984_4^{20}	1.5063^{20}	−101.5	154.0		
d295	Diethyldithio-carbamic acid, Na salt	$(C_2H_5)_2NC({=}S)S^-\,Na^+ \cdot 3H_2O$	225.31	4^2, 613			95–99			
d296	Diethyl dithio-phosphate	$(C_2H_5O)_2P(S)SH$	186.23	1, 333	1.111	1.5120^{20}		60^{1mm}		
d297	N,N-Diethyldodecan-amide	$CH_3(CH_2)_{10}C(O)N(C_2H_5)_2$	255.45		0.847	1.4545^{20}		166^{2mm}	>112	
d298	Diethylenetriamine	$(H_2NCH_2CH_2)_2NH$	103.17	4, 255	0.9542_4^{20}	1.4826^{20}	−35	207.1	101	misc aq, alc, bz, eth

No.	Name	Formula	Formula weight		Density	n_D	Melting point	Boiling point	Flash point	Solubility
d299	Diethylenetriamine-pentaacetic acid	[(HOOCCH$_2$)$_2$NCH$_2$-CH$_2$]$_2$NCH$_2$COOH	393.35				220 d			
d300	Diethyl ether	C$_2$H$_5$OC$_2$H$_5$	74.12	1,314	0.7134$^{20}_4$	1.3527^{20}	−116.3	34.6	−40	6 aq; misc alc, bz, chl
d301	Diethyl ethoxymethyl-enemalonate	(C$_2$H$_5$OOC)$_2$C=CH-OC$_2$H$_5$	216.23	3,469	1.070	1.4620^{20}		279–281	155	
d302	N,N-Diethylethyl-enediamine	(C$_2$H$_5$)$_2$NCH$_2$CH$_2$NH$_2$	116.21	4,251	0.827	1.4360^{20}		145–147	30	
d303	Diethyl ethyl-malonate	C$_2$H$_5$CH(COOC$_2$H$_5$)$_2$	188.2	2,644	1.004$^{20}_{20}$	1.4158^{20}		75–77^{5mm}	88	sl s aq; v s alc, eth
d304	N,N-Diethyl-formamide	(C$_2$H$_5$)$_2$NCHO	101.15	4,109	0.908	1.4340^{20}		176–177	60	misc aq; v s alc, eth
d305	Diethyl fumarate	C$_2$H$_5$OOCCH=CH-COOC$_2$H$_5$	172.18	2,742	1.052$^{20}_4$	1.4406^{20}	1–2	218–219	91	
d306	Diethyl 3,4-furandi-carboxylate	(C$_2$H$_5$OOC)$_2$C$_4$H$_2$O	212.20		1.140	1.4717^{20}		155^{13mm}	82	
d307	Diethyl glutarate	C$_2$H$_5$OOCCH$_2$CH$_2$CH$_2$-COOC$_2$H$_5$	188.22	2,633	1.022	1.4240^{20}	−23.8	237	96	0.9 aq; v s alc, eth
d308	2,4-Diethyl-2,6-heptadienal	H$_2$C=CHCH(C$_2$H$_5$)-CH=C(C$_2$H$_5$)CHO	166.27					91^{12mm}		
d309	Diethyl heptane-dioate	C$_2$H$_5$OOC(CH$_2$)$_5$-COOC$_2$H$_5$	216.28	2,671	0.9945^{20}	1.4280^{20}	−24	192^{100mm}	>112	i aq; s alc, eth
d310	2,4-Diethyl-1-heptanol	CH$_3$CH$_2$CH$_2$CH(C$_2$H$_5$)-CH$_2$CH(C$_2$H$_5$)CH$_2$OH	172.31					109^{12mm}		

TABLE 1.15 Physical Constants of Organic Compounds (*continued*)

No.	Name	Formula	Formula weight	Beilstein reference	Density	Refractive index	Melting point	Boiling point	Flash point	Solubility in 100 parts solvent
d311	Di-(2-ethylhexyl) decanedioate	$C_4H_9CH(C_2H_5)CH_2OOC$-$(CH_2)_8COOCH_2CH$-$(C_2H_5)C_4H_9$	426.68		0.912^{25}	1.451^{25}		256^{5mm}	227	i aq; s alc, bz, acet
d312	Di-(2-ethylhexyl) hexanedioate	$C_4H_9CH(C_2H_5)CH_2OOC$-$(CH_2)_4COOCH_2CH$-$(C_2H_5)C_4H_9$	370.57		0.925^{25}_{25}	1.4474^{20}		214^{5mm}	193	s alc, eth, acet; iaq
d313	Di-(2-ethylhexyl) o-phthalate	$C_6H_4[COOCH_2CH$-$(C_2H_5)C_4H_9]$	390.56		0.981^{25}_{25}	1.4853^{20}	−50	384	207	
d314	Diethyl hydrogen phosphonate	$(C_2H_5O)_2P(O)H$	138.10	1, 330	1.079^{20}_{4}	1.4076^{20}		50–51^{2mm}	90	s aq (hyd), alc, eth
d315	N,N-Diethylhydroxyl-amine	$(C_2H_5)_2NOH$	89.14	4, 536	1.867	1.4195^{20}	−25	125–130	45	1.4 aq; s alc, eth
d316	Diethyl maleate	$C_2H_5OOCCH=CH$-$COOC_2H_5$	172.18	2, 751	1.0687^{20}	1.4400^{20}	−8.8	225.3	93	2.7 aq; misc alc, eth
d317	Diethyl malonate	$C_2H_5OOCCH_2COOC_2H_5$	160.17	2, 573	1.0550	1.4136^{20}	−48.9	199.3	100	v s aq, alc, eth
d318	Diethylmalonic acid	$HOOC(C_2H_5)_2COOH$	160.17	2, 686			127	d 170–180		s aq; v s alc, bz, eth
d319	N,N-Diethyl-3-methylbenzamide	$CH_3C_6H_4C(=O)N$-$(C_2H_5)_2$	191.27	9^2, 325	0.9962^{20}_{4}	1.5212^{20}		111^{1mm}		sl s aq; v s alc, eth
d320	Diethyl methyl-malonate	$C_2H_5OOCCH(CH_3)$-$COOC_2H_5$	174.20	2, 629	1.018^{20}_{4}	1.4130^{20}		198	76	
d321	Diethyl 2-methyl-2′-oxosuccinate	$C_2H_5OOCCH(CH_3)$-$C(=O)C(=O)OC_2H_5$	202.21	3, 794	1.073	1.4313^{20}		138^{23mm}	>112	
d322	Diethyl methyl-succinate	$C_2H_5OOCCH_2CH(CH_3)$-$COOC_2H_5$	188.22	2, 639	1.012	1.4199^{20}		217–218		
d323	N,N-Diethyl-4-nitrosoaniline	$C_6H_4(NO)N(C_2H_5)_2$	178.24	12, 684			82–84			
d324	Diethyl octanedioate	$C_2H_5OOC(CH_2)_6$-$COOC_2H_5$	230.30	2, 693	0.9822^{20}_{4}	1.4323^{20}	5.9	282	>112	i aq; s alc, eth

No.	Name	Formula	Formula weight	Beilstein ref.	Density	n_D	mp, °C	bp, °C	Flash pt	Solubility
d325	Diethyl oxalate	$C_2H_5OOCCOOC_2H_5$	146.14	2, 535	1.0785^{20}_4	1.4102	−40.6	185.4	75	3.6aq (gradual d); misc alc, eth
d326	Diethyl oxydiformate	$[C_2H_5OC(=O)]_2O$	162.14		1.12^{20}_4	1.3980^{20}		93^{18mm}	69	s alc, esters, ketones
d327	N^1,N^1-Diethyl-1,4-pentanediamine	$CH_3CH(NH_2)(CH_2)_3$-$N(C_2H_5)_2$	158.29		0.817	1.4429^{20}		200^{753mm}	68	s aq, alc, eth
d328	Diethyl phenylmalonate	$C_6H_5CH(COOC_2H_5)_2$	236.27	9, 854	1.0950^{20}_4	1.4913^{20}	16	170^{14mm}	>112	i aq; s alc
d329	Diethyl o-phthalate	$C_6H_4(COOC_2H_5)_2$	222.24	9, 798	1.232^{14}	1.5049^{14}	−3	295	140	i aq; misc alc, eth
d330	N,N-Diethyl-1,3-propanediamine	$(C_2H_5)_2NCH_2CH_2$-CH_2NH_2	130.24		0.826	1.4416^{20}		159	58	
d331	2,2-Diethyl-1,3-propanediol	$(C_2H_5)_2C(CH_2OH)_2$	132.20		1.052^{20}	1.4574^{25}	61.3	125^{10mm}		25 aq; v s alc, eth
d332	Diethyl propylmalonate	$C_2H_5OOCCH(C_3H_7)$-$COOC_2H_5$	202.25	2, 657	0.987	1.4185^{20}		221–222	91	
d333	1,1-Diethyl-2-propynylamine	$HC\equiv CC(C_2H_5)_2NH_2$	111.19		0.828	1.4409^{20}		71^{90mm}	21	
d334	N,N-Diethyl-3-pyridinecarboxamide	C_5H_4N—$C(=O)N$-$(C_2H_5)_2$	178.24	22^2, 34	1.060^{25}_4	1.5240^{20}	24–26	296–300	>112	
d335	Diethyl succinate	$C_2H_5OOCCH_2CH_2$-$COOC_2H_5$	174.20	2, 609	1.040^{20}_4	1.4200^{20}	−21	217.7	110	i aq; misc alc, eth
d336	Diethyl sulfate	$(C_2H_5O)_2SO_2$	154.18	1, 327	1.172^{25}	1.4004^{20}	−25	209 d	78	misc alc, eth
d337	Diethyl sulfide	$(C_2H_5)_2S$	90.19	1, 344	0.8367^{20}_4	1.4430^{20}	−103.9	92.1	−9	i aq; misc alc, eth
d338	Diethyl sulfite	$(C_2H_5O)_2S(O)$	138.19	1, 325	1.077^{25}_4			157.7		s aq(d), alc

TABLE 1.15 Physical Constants of Organic Compounds (*continued*)

No.	Name	Formula	Beilstein reference	Formula weight	Density	Refractive index	Melting point	Boiling point	Flash point	Solubility in 100 parts solvent
d339	(+)-Diethyl-L-tartrate	[—CH(OH)COOC$_2$H$_5$]$_2$	3, 512	206.19	1.204$_4^{20}$	1.4459^{20}	17	280	93	sl s aq; misc alc, eth
d340	(−)-Diethyl-D-tartrate	[—CH(OH)COOC$_2$H$_5$]$_2$	3^1, 181	206.19	1.205	1.4467^{20}		162^{19mm}	93	
d341	Diethyl 3,3'-thio-propionate	S(CH$_2$CH$_2$COOC$_2$H$_5$)$_2$	9^2, 325	234.32	1.095	1.4655^{20}		121^{2mm}		
d342	N,N-Diethyl-m-toluamide	CH$_3$C$_6$H$_4$C(=O)N-(C$_2$H$_5$)$_2$		191.27	0.996	1.5212^{20}		111^{1mm}		
d343	N,N-Diethyl-1,1,1-trimethylsilylamine	(C$_2$H$_5$)$_2$NSi(CH$_3$)$_3$		145.32	0.767	1.4081^{20}		125–126	10	
d344	Diethylzinc	(C$_2$H$_5$)$_2$Zn	5, 199	123.49	1.2065^{20}		−28	118		
d345	1,4-Difluorobenzene	C$_6$H$_4$F$_2$		114.09	1.1701^{20}	1.4415^{20}	−23.7	88.9	2	0.32 aq
d346	1,1-Difluoroethane	CH$_3$CHF$_2$		66.05	0.909^{21}	1.413	−117	−24.7	none	sl s alc; v s eth
d347	1,1-Difluorotetrachloroethane	Cl$_3$CCClF$_2$	1, 86	203.83	1.649		41	91		i aq; s alc, eth
d348	1,2-Difluorotetrachloroethane	FCl$_2$CCCl$_2$F	1^3, 365	203.83	1.6447$_4^{25}$	1.413^{25}	23.8	203.8		
d349	Dihexylamine	(C$_6$H$_{13}$)$_2$NH	4^1, 384	185.36	0.795	1.4320^{20}		192–195	95	s alc, eth
d350	Dihexyl ether	(C$_6$H$_{13}$)$_2$O	1^3, 1656	186.34	0.7936$_4^{20}$	1.4204^{20}		226.2	77	i aq; s eth
d351	9,10-Dihydroanthracene		5, 641	180.25	0.880		108–110	312		i aq; s alc, bz, eth
d352	(+)-Dihydrocarvone		7^3, 337	152.24	0.929^{19}	1.4718^{20}	25	221–222	81	sl s alc, eth; s chl
d353	Dihydrocoumarin		17, 315	148.16	1.169^{18}	1.5563^{20}	32–34	272	>112	
d354	10,11-Dihydro-5H-dibenzo-[a,d]cyclohepten-5-one			208.26	1.156	1.6332^{20}		148$^{0.3mm}$		
d355	3,4-Dihydro-2-ethoxy-2H-pyran			128.17	0.957	1.4394^{20}		42^{16mm}	24	

ID	Compound	Formula / MW		Density	n_D	bp/mp (°C)	
d356 d537	2,3-Dihydrofuran Dihydrolinalool	70.09 156.27	17^3, 141	0.927 0.925^{25}	1.4239^{20} 1.433^{20}	54–55 171^{11mm}	<1 178
d358	3,4-Dihydro-1(2H)-6-methoxynaphthalenone	176.22	9^2, 889			80	
d359	3,4-Dihydro-2-methoxy-2H-pyran	114.14			1.4425^{20}		16
d360	2,3-Dihydro-2-methylbenzofuran	134.18	17^1, 23	1.061	1.5308^{20}	197–198	62
d361	5,6-Dihydro-4-methyl-2H-pyran	98.15	17^3, 160	0.912	1.4495^{20}	117–118	21

Formula (d537, Dihydrolinalool): $(CH_3)_2C{=}CHCH_2CH_2{-}C(OH)(CH_3)CH_2CH_3$

Structures:

d351

d352 — CH_3 (ketone); $CH_3{-}C{=}CH_2$

d353 — (lactone, O)

d354 — (O)

d355 — OC_2H_5

d356 — (O)

d358 — (O), OCH_3

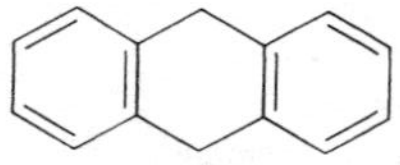

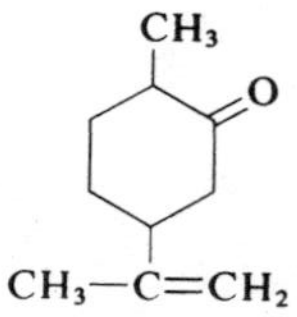

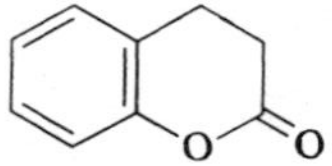

d359 — OCH_3, O

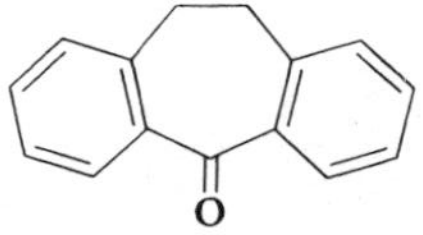

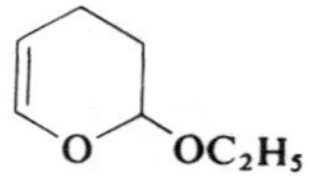

d360 — CH_3, O

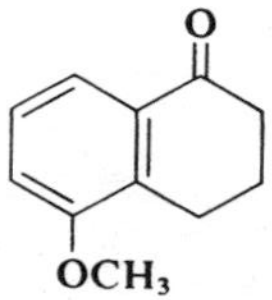

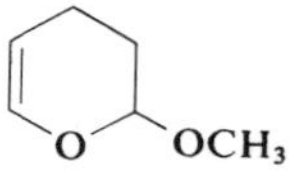

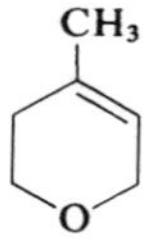

d361 — CH_3, O

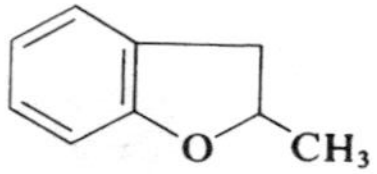

TABLE 1.15 Physical Constants of Organic Compounds (*continued*)

No.	Name	Formula	Formula weight	Beilstein reference	Density	Refractive index	Melting point	Boiling point	Flash point	Solubility in 100 parts solvent
d362	3,4-Dihydro-1(2H)-naphthalenone		146.19	7, 370	1.099	1.5685^{20}	5–6	116^{6mm}	>112	s aq, alc
d363	Dihydropyran		84.12		0.922^{19}_{15}	1.4410^{20}	–70	86	–15	
d364	5,6-Dihydro-2H-pyran-3-carbaldehyde		112.13		1.100	1.4980^{20}		78^{12mm}	77	
d365	3,4-Dihydro-2H-pyran-2-carboxylic acid, Na salt		150.11				242–244			
d366	Dihydroterpineol		256.27		0.907^{25}	1.4670^{20}		48^{17mm}	88	
d367	5,6-Dihydro-2,4,4,6-tetramethyl-4H-1,3-oxazine		141.21		0.886	1.4410^{20}				
d368	2,5-Dihydrothiophene-1,1-dioxide		118.15				64–66		>112	s aq, alc, bz, chl, eth
d369	1,2-Dihydro-2,2,4-trimethylquinoline		173.26	8, 266	0.934	1.5895^{20}	145–147	$90^{0.02mm}$	101	s warm alc, pyr, HOAc; i bz, chl, eth
d370	2',4',-Dihydroxyacetophenone	$(HO)_2C_6H_3C(=O)CH_3$	152.15		1.180			430		s alc, bz, chl, HOAc
d371	1,2-Dihydroxyanthraquinone		240.21	8, 439			287–289	430		s alc, bz, chl, HOAc
d372	1,4-Dihydroxyanthraquinone		240.21	8, 450			196			s alc, alk, eth
d373	1,8-Dihydroxyanthraquinone		240.21	8, 458			193–197	subl		0.005 alc; 0.2 eth; s chl
d374	2,6-Dihydroxyanthraquinone		240.21	8, 463			360 d			sl s aq, alc
d375	2,4-Dihydroxybenzaldehyde	$(HO)_2C_6H_3CHO$	138.12	8, 241			135–136	226^{22mm}		v s aq, alc, chl, eth

No.	Name	Formula							Solubility
d376	3,4-Dihydroxybenz-aldehyde	$(HO)_2C_6H_3CHO$	138.12	8, 246		153			5 aq; 79 hot alc; v s eth
d377	1,2-Dihydroxybenzene	$C_6H_4(OH)_2$	110.11	6, 759	1.344^4	104–106	245.5	137	43 aq; s alc, bz, chl, eth; v s pyr, alk
d378	1,3-Dihydroxybenzene	$C_6H_4(OH)_2$	110.11	6^2, 802	1.272^{15}	109–110	276	171	110 aq; 110 alc; v s eth, glyc; sl s chl
d379	1,4-Dihydroxybenzene	$C_6H_4(OH)_2$	110.11	6, 836	1.332^{15}	170–171	285–287		7 aq; v s alc, eth

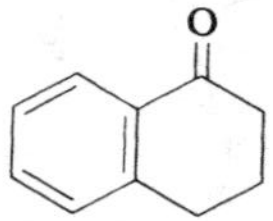

d362

d363

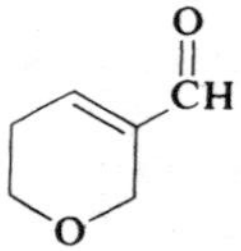

d364

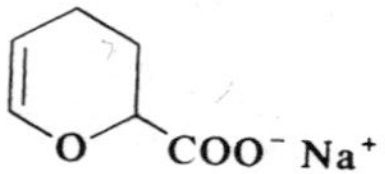

d365 — COO⁻ Na⁺

d366 — CH₃CHOH, CH₃, CH₃

d367 — CH₃, CH₃, N, CH₃, CH₃

d368 — S, O, O

d369 — CH₃, CH₃, CH₃, N, H

d371 — OH, OH, O, O

d372 — OH, OH, OH, OH, O, O

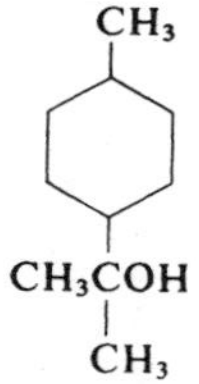

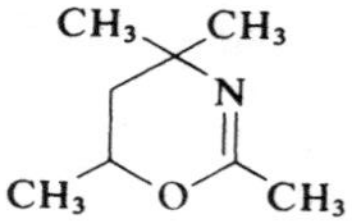

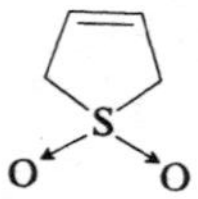

d373 — OH, OH, O, O

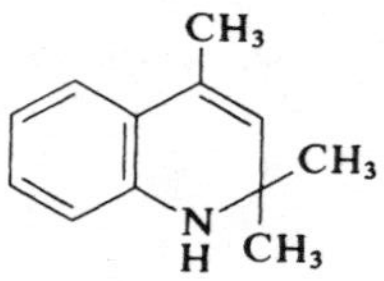

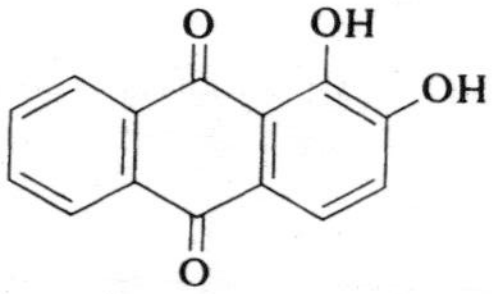

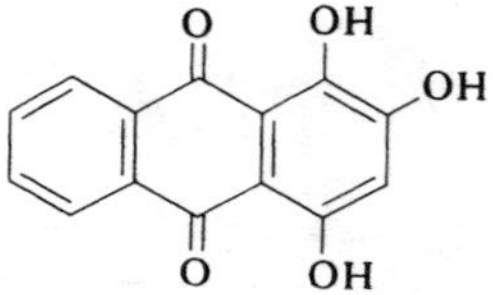

d374 — OH, O, O, HO

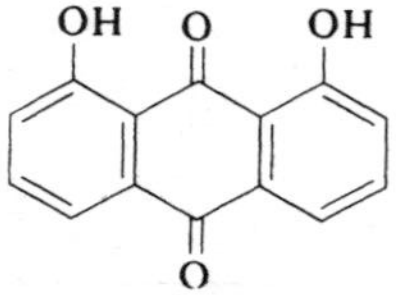

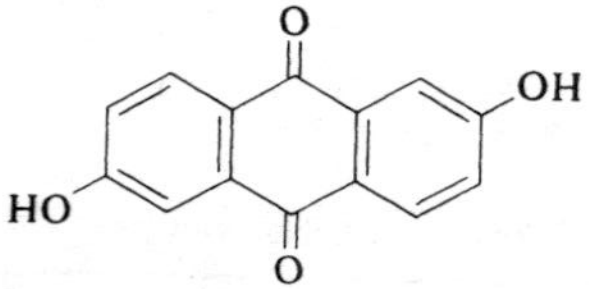

TABLE 1.15 Physical Constants of Organic Compounds (*continued*)

No.	Name	Formula	Formula weight	Beilstein reference	Density	Refractive index	Melting point	Boiling point	Flash point	Solubility in 100 parts solvent
d380	1,3-Dihydroxybenzene monoacetate	$HOC_6H_4OOCCH_3$	152.15	6, 816		1.5350^{20}		283	>112	
d381	2,5-Dihydroxy-p-benzenedisulfonic acid, K salt	$(HO)_2C_6H_2(SO_3^-K^+)_2$	346.43	11, 300			>300			v s aq
d382	2,5-Dihydroxy-benzenesulfonic acid, K salt	$(HO)_2C_6H_3SO_3^-K^+$	228.27	11, 300			251 d			v s aq
d383	2,4-Dihydroxybenzoic acid	$(HO)_2C_6H_3COOH$	154.12	10, 377			213			s hot aq, alc, eth
d384	2,5-Dihydroxybenzoic acid	$(HO)_2C_6H_3COOH$	154.12	10, 384			199–200			0.5 aq; s alc, eth
d385	3,5-Dihydroxybenzoic acid	$(HO)_2C_6H_3COOH$	154.12	10, 404			236 d			sl s aq; s alc, eth
d386	2,4-Dihydroxybenzo-phenone	$(HO)_2C_6H_3C(=O)C_6H_5$	214.22	8, 312			144–145			v s alc, eth, HOAc
d387	2,2'-Dihydroxybi-phenyl	$HOC_6H_4C_6H_4OH$	186.21	6, 989			110	315		s alc, bz, eth; sl s aq
d388	4,6-Dihydroxy-2-mercaptopyrimidine		144.15	24, 476			236			v s aq, alc, eth
d389	1,2-Dihydroxy-4-methylbenzene	$(HO)_2C_6H_3CH_3$	124.14	6, 878	1.129_4^{74}	1.5425^{74}	67–69	251		s aq, alc, bz, eth
d390	1,3-Dihydroxy-2-methylbenzene	$(HO)_2C_6H_3CH_3$	124.14	6, 878			115–118	264		s aq, alc, bz, eth
d391	2,4-Dihydroxy-6-methylpyrimidine		126.12	24, 342			318 d			
d392	1,5-Dihydroxynaph-thalene	$C_{10}H_6(OH)_2$	160.17	6, 980			259 d			sl s aq; s alc; v s eth

No.	Name	Formula	M. wt.	Beil. ref.	Density	B.p.	M.p.	Solubility
d393	1,7-Dihydroxynaphthalene	$C_{10}H_6(OH)_2$	160.17	6, 981			177–180	v s alc, eth
d394	2,3-Dihydroxynaphthalene	$C_{10}H_6(OH)_2$	160.17	6, 982			162–164	v s alc, eth
d395	2,7-Dihydroxynaphthalene	$C_{10}H_6(OH)_2$	160.17	6, 985			187 d	sl s aq; v s alc, eth
d396	4,5-Dihydroxynaphthalene-2,7-disulfonic acid	$(HO)_2C_{10}H_4(SO_3H)_2$	296.26	11, 307				v s aq; i alc, eth
d397	1,3-Dihydroxy-2-propanone	$HOCH_2C(=O)CH_2OH$	90.08	1, 846			65–71	v s aq, alc, acet, eth
d398	2,3-Dihydroxypropionaldehyde	$HOCH_2CHOHCHO$	90.08	1, 845	1.455^{18}_{18}	$140^{0.8mm}$ >112	145	3 aq; i bz, PE
d399	7-(2,3-Dihydroxypropyl)theophylline		254.25	24, 312			158	33 aq; 2 alc; 1 chl
d400	3,6-Dihydroxypyridazine		112.09				d 260	sl s hot alc; s hot aq
d401	2,3-Dihydroxypyridine	$(HO)_2C_5H_3N$	111.10	21^2, 107			245 d	

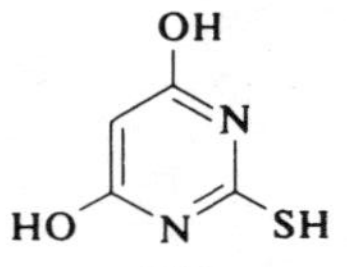

d388

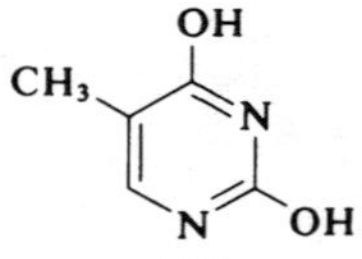

d391

d396

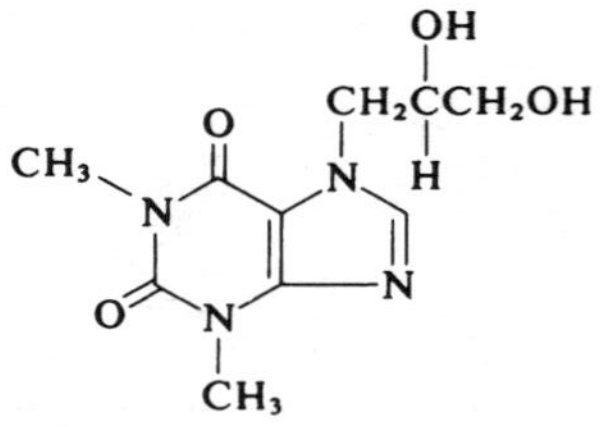

d399

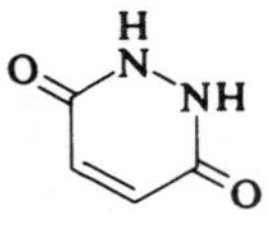

d400

TABLE 1.15 Physical Constants of Organic Compounds (*continued*)

No.	Name	Formula	Formula weight	Beilstein reference	Density	Refractive index	Melting point	Boiling point	Flash point	Solubility in 100 parts solvent
d402	1,4-Diiodobenzene	$C_6H_4I_2$	329.91	5, 227	2.132^{10}		131–133	285		sl s alc; v s eth
d403	1,2-Diiodoethane	ICH_2CH_2I	281.86	1, 99	3.325^{20}_{4}	1.7411^{20}	81	200		sl s aq; s alc, eth
d404	Diiodomethane	CH_2I_2	267.84	1, 71	2.5755^{20}_{4}		5.6	181		0.12 aq; misc alc, bz, eth, PE
d405	1,3-Diiodopropane	$ICH_2CH_2CH_2I$	295.88	1, 115	0.740	1.6423^{20}	-13	222	29	i aq; s chl, eth
d406	Diisobutylamine	$[(CH_3)_2CHCH_2]_2NH$	129.25	4, 166		1.4081^{20}	-77	137–139		s alc, acet, eth, EtAc
d407	Diisobutyl ether	$[(CH_3)_2CHCH_2]_2O$	130.22		0.761^{15}			122–124		i aq; misc alc, eth
d408	Diisobutyl hexanedioate	$[(CH_3)_2CHCH_2OOC-CH_2-]_2$	258.36		0.950^{25}_{25}				160	
d409	Diisobutyl-o-phthalate	$C_6H_4[COOCH_2CH(CH_3)_2]_2$	278.35		1.038^{25}_{25}				174	
d410	1,6-Diisocyanatohexane	$OCN(CH_2)_6NCO$	168.20	4^2, 711	1.040	1.4525^{20}		255	140	
d411	Diisopropylamine	$[(CH_3)_2CH]_2NH$	101.19	4, 154	0.7169^{20}	1.3924^{20}	-96.3	83.5	-6	11 aq
d412	2-(Diisopropylamino)-ethanol	$[(CH_3)_2CH]_2NCH_2CH_2OH$	145.25	4^1, 430	0.826	1.4417^{20}		187–192	57	
d413	2,6-Diisopropylaniline	$[(CH_3)_2CH]_2C_6H_3NH_2$	177.29	12, 168	0.940	1.5332^{20}	-45	257	123	
d414	1,3-Diisopropylbenzene	$C_6H_4[CH(CH_3)_2]_2$	162.28	5, 447	0.856^{20}_{4}	1.4980^{20}	-63	203	76	misc alc, bz, eth, acet
d415	1,4-Diisopropylbenzene	$C_6H_4[CH(CH_3)_2]_2$	162.28	5^2, 339	0.857^{20}_{4}	1.4889^{20}		203	76	misc alc, bz, eth, acet
d416	Diisopropylcyanamide	$[(CH_3)_2CH]_2NCN$	126.20	4^3, 279	0.839	1.4270^{20}		93^{25mm}	78	
d417	Diisopropyl ether	$[(CH_3)_2CH]_2O$	102.17	1, 362	0.7258^{20}_{4}	1.3689^{20}	-86.9	68.4	-12	1.2 aq; misc alc, bz, chl, eth
d418	N,N-Diisopropyl-ethylamine	$[(CH_3)_2CH]_2NC_2H_5$	129.25		0.742	1.4133^{20}		127	10	

d	Name	Formula								
d419	2,6-Diisopropylphenol	[(CH₃)₂CH]₂C₆H₃OH	178.28	6¹, 272	0.962	1.5140²⁰	18	256	>112	v s aq, alc, chl, eth
d420	Diisopropyl phosphite	[(CH₃)₂CHO]₂P(O)H	166.16	1, 363	0.997	1.4070²⁰		72–75¹⁰mm	>112	
d421	(+)-Diisopropyl L-tartrate	[—CH(OH)COOCH(CH₃)₂]₂	234.25	3, 517	1.114	1.4387²⁰		152¹²mm	109	
d422	Diketene		84.07		1.073	1.4330²⁰		127	33	
d423	Dilauryl phosphite	[CH₃(CH₂)₁₁O]₂P(O)H	418.64		0.946	1.4520²⁰	42–43		>112	
d424	threo-1,4-Dimercapto-2,3-butanediol	HSCH₂CH(OH)CH(OH)CH₂SH	154.25							
d425	2,3-Dimercapto-1-propanol	HSCH₂CH(SH)CH₂OH	124.22		1.2385²⁵	1.5720²⁵		120¹⁵mm	>112	8 aq(d); s alc, eth
d426	3′,4′-Dimethoxyacetophenone	(CH₃O)₂C₆H₃COCH₃	180.20	8², 298			49–51	286–288		sl s aq, alc, eth
d427	2,4-Dimethoxyaniline	(CH₃O)₂C₆H₃NH₂	153.18	13, 784			34–37	270 s l d		s alc, bz, eth
d428	2,5-Dimethoxyaniline	(CH₃O)₂C₆H₃NH₂	153.18	13, 788			80–82	176²²mm		s aq, alc
d429	3,4-Dimethoxyaniline	(CH₃O)₂C₆H₃NH₂	153.18	13, 780			88	281		s hot eth
d430	3,4-Dimethoxybenzaldehyde	(CH₃O)₂C₆H₃CHO	166.18	8, 255			42–43			v s alc, eth
d431	1,2-Dimethoxybenzene	C₆H₄(OCH₃)₂	138.17	6, 771	1.0819²⁵	1.5232²⁵	22.5	206.3	87	sl s aq; s alc, eth
d432	1,3-Dimethoxybenzene	C₆H₄(OCH₃)₂	138.17	6, 813	1.055	1.5240	−55	85–87⁷mm	87	s alc, bz, eth; sl s aq
d433	1,4-Dimethoxybenzene	C₆H₄(OCH₃)₂	138.17	6, 843	1.0364⁶⁸		55–60	213		s alc; v s bz, eth

Dihydroxytoluene, d390
3,5-Diiodosalicylic acid, h111
2Diisobutyl adipate, d408

Diisobutylene, t357
Diisobutyl ketone, d531
Diisopropyl ketone, d578

Diisopropylmethane, d572
Dimedone, d508
1,1-Dimethoxytrimethylamine, d523

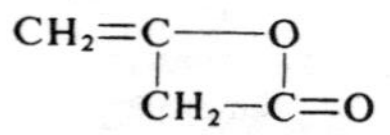

$$CH_2{=}C{-}O$$
$$\hphantom{CH_2=C-}|\hphantom{xx}|$$
$$CH_2{-}C{=}O$$

d422

TABLE 1.15 Physical Constants of Organic Compounds (*continued*)

No.	Name	Formula	Formula weight	Beilstein reference	Density	Refractive index	Melting point	Boiling point	Flash point	Solubility in 100 parts solvent
d434	3,4-Dimethoxybenzoic acid	$(CH_3O)_2C_6H_3COOH$	182.18	10^1, 188			180–181			0.047 aq; v s alc, eth
d435	1,1-Dimethoxy-3-butanone	$(CH_3O)_2CHCH_2COCH_3$	132.16		0.993	1.4150^{20}			49	
d436	2,5-Dimethoxy-2,5-dihydrofuran		130.14		1.073	1.4339^{20}		160–162	47	
d437	Dimethoxydiphenyl-silane	$(C_6H_5)_2Si(OCH_3)_2$	244.4		1.0771^{20}_4	1.5447^{20}		161^{15mm}		
d438	1,1-Dimethoxyethane	$CH_3CH(OCH_3)_2$	90.12	1, 603	0.8502^{20}	1.3796^{20}	−113	64.5	1	s aq, alc, chl, eth
d439	1,2-Dimethoxyethane	$CH_3OCH_2CH_2OCH_3$	90.12	1, 467	0.8629^{20}_4	1.4170^{20}	−68	85.2	1	misc aq, alc; s PE
d440	(2,2-Dimethoxy)-ethylamine	$H_2NCH_2CH(OCH_3)_2$	105.14	4^2, 758	0.965			135^{95mm}	53	
d441	Dimethoxymethane	$CH_2(OCH_3)_2$	76.10	1, 574	0.8601^{20}_{20}	1.3534^{20}	−104.8	42.3	−17	32 aq
d442	1,1-Dimethoxy-2-methylaminoethane	$CH_3NHCH_2CH(OCH_3)_2$	119.16	4^2, 759	0.928	1.4115^{20}		140	29	
d443	Dimethoxymethyl-phenylsilane	$(CH_3O)_2Si(CH_3)C_6H_5$	182.3		0.993^{20}_4	1.469^{20}		199–200		
d444	1,2-Dimethoxy-4-nitrobenzene	$(CH_3O)_2C_6H_3NO_2$	183.16	6, 789	1.1888^{133}		95–98	230^{17mm}		v s alc, eth; s chl
d445	2,5-Dimethoxy-4'-nitrostilbene	$(CH_3O)_2C_6H_3CH{=}CH{-}C_6H_4NO_2$	285.30	6^2, 987			117–119			
d446	2,6-Dimethoxyphenol	$(CH_3O)_2C_6H_3OH$	154.17	6, 1081			53–56	261		s alc; alk; v s eth
d447	(3,4-Dimethoxy)-phenylacetic acid	$(CH_3O)_2C_6H_3CH_2COOH$	196.20	10, 409			96–98			s aq; v s alc, eth
d448	(3,4-Dimethoxy)-phenylacetonitrile	$(CH_3O)_2C_6H_3CH_2CN$	177.20	10^1, 198			62–63	$171{-}178^{10mm}$		
d449	2,2-Dimethoxy-2-phenylacetophenone	$C_6H_5C(O)C(OCH_3)_2{-}C_6H_5$	256.30				67–70			

No.	Name	Formula	Formula wt	Beilstein ref.	Density	n_D	mp, °C	bp, °C	Flash pt, °C	Solubility
d450	1,1-Dimethoxy-2-phenylethane	$C_6H_5CH_2CH(OCH_3)_2$	166.22	7, 293	1.004	1.4950^{20}		221	83	misc aq, alc, bz, eth
d451	β-(3,4-Dimethoxy)-phenylethylamine	$(CH_3O)_2C_6H_3CH_2CH_2NH_2$	181.24	13, 800	1.074	1.5464^{20}		188^{15mm}		
d452	2,2-Dimethoxypropane	$(CH_3)_2C(OCH_3)_2$	104.15	1, 648	0.847	1.3780		83	4	
d453	1,1-Dimethoxy-2-propanone	$CH_3C(O)CH(OCH_3)_2$	118.13	1^1, 395	0.976	1.3978^{20}		143–147	37	
d454	3,3-Dimethoxy-1-propene	$(CH_3O)_2CHCH{=}CH_2$	102.13	1^1, 378	0.862	1.3954^{20}		89–90		
d455	1,2-Dimethoxy-4-propenylbenzene	$CH_3CH{=}CHC_6H_3(OCH_3)_2$	178.23	6, 956	1.055	1.5680^{20}		262–264	>112	
d456	2,6-Dimethoxypyridine	$(CH_3O)_2C_5H_3N$	139.15		1.053	1.5029^{20}		178–180	61	
d457	2,5-Dimethoxytetrahydrofuran	$(CH_3O)_2C_4H_6O$	132.16		1.020	1.4180^{20}		145–147	35	
d458	N,N-Dimethylacetamide	$CH_3C(O)N(CH_3)_2$	87.12	4, 59	0.9366^{25}	1.4356^{25}	−20	165.5	70	
d459	Dimethyl 1,3-acetonedicarboxylate	$[CH_3OOCCH_2]_2C{=}O$	174.15	3, 790	1.185	1.4434^{20}		150^{25mm}	>112	
d460	Dimethyl acetylenedicarboxylate	$CH_3OOCC{\equiv}CCOOCH_3$	142.11	2, 803	1.156	1.4470^{20}		$95–98^{19mm}$	86	
d461	Dimethylamine	$(CH_3)_2NH$	45.09	4, 39	0.680_4^0		−92.2	6.9		v s aq; s alc, eth

CH_3O—O—OCH_3

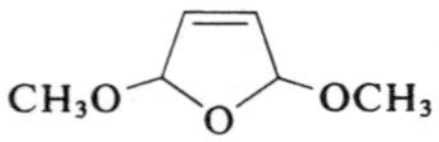

d436

TABLE 1.15 Physical Constants of Organic Compounds (*continued*)

No.	Name	Formula	Formula weight	Beilstein reference	Density	Refractive index	Melting point	Boiling point	Flash point	Solubility in 100 parts solvent
d462	4-Dimethylamino-benzaldehyde	$(CH_3)_2NC_6H_4CHO$	149.19	14, 31			74	176^{17mm}		s alc, chl, eth, HOAc
d463	p-(Dimethylamino)-benzenesulfonic acid, Na salt	$(CH_3)_2NC_6H_4SO_3^-Na^+$	223.23	14^3, 2023			>300			
d464	4-Dimethylamino-benzoic acid	$(CH_3)_2NC_6H_4COOH$	165.19	14, 426			241 d			s alc; sl s eth
d465	2-(Dimethylamino)-ethanol	$(CH_3)_2NCH_2CH_2OH$	89.14	4, 276	0.8876^{20}_4	1.4294^{20}		135	40	misc aq, alc, eth
d466	2-(Dimethylamino)-ethyl benzoate	$C_6H_5COOCH_2CH_2-N(CH_3)_2$	193.26		1.014	1.5077^{20}		155^{20mm}		
d467	2-Dimethylamino-ethyl chloride HCl	$(CH_3)_2NCH_2CH_2Cl \cdot HCl$	144.05	4, 133			205–208			
d468	2-(Dimethylamino)-ethyl methacrylate	$H_2C=C(CH_3)COOCH_2-CH_2N(CH_3)_2$	157.22	4^3, 649	0.933	1.4391^{20}		182–192	70	
d469	4-Dimethylamino-3-methyl-2-butanone	$(CH_3)_2NCH_2CH(CH_3)-COCH_3$	129.20	4^1, 452	0.841	1.4250^{20}		73^{35mm}	38	
d470	3-Dimethylamino-phenol	$(CH_3)_2NC_6H_4OH$	137.18	13, 405	1.5895^{26}		82–84	265–268		v s alc, bz, eth, acet
d471	3-(Dimethylamino)-1,2-propanediol	$(CH_3)_2NCH_2CH(OH)-CH_2OH$	119.16	4, 302	1.004	1.4609^{20}		216–217	105	s aq, alc, chl, eth
d472	1-Dimethylamino-2-propanol	$CH_3CH(OH)CH_2N(CH_3)_2$	103.17		0.837	1.4193^{20}		121–127	35	
d473	3-Dimethylamino-1-propanol	$(CH_3)_2NCH_2CH_2CH_2OH$	103.17	4^1, 433	0.872	1.4360^{20}		163–164	36	
d474	3-(Dimethylamino)-propionitrile	$(CH_3)_2NCH_2CH_2CN$	98.15	4^3, 1265	0.870	1.4258^{20}	−43	171^{750mm}	62	
d475	3-Dimethyl-aminopropyl chloride HCl	$(CH_3)_2NCH_2CH_2-CH_2Cl \cdot HCl$	158.07	4, 148			141–144		35	

d476	4-(Dimethylamino)-pyridine	$(CH_3)_2N(C_5H_4N)$	122.17	22², 341			108–110			v s aq, alc, bz, chl
d477	N,N-Dimethyl-aniline	$C_6H_5N(CH_3)_2$	121.18	12, 141	0.9559_4^{20}	1.5584^{20}	2.5	194.2	62	v s alc, chl, eth
d478	2,3-Dimethyl-aniline	$(CH_3)_2C_6H_3NH_2$	121.18	12, 1101	0.9931^{20}	1.5685^{20}	2.5	221–222	96	sl s aq; s alc, eth
d479	2,4-Dimethyl-aniline	$(CH_3)_2C_6H_3NH_2$	121.18	12, 1111	0.9804_4^{20}	1.5586^{20}		218	90	s alc, bz, eth
d480	2,5-Dimethyl-aniline	$(CH_3)_2C_6H_3NH_2$	121.18	12, 1135	0.9790_4^{21}	1.5592^{20}	11.5	218	93	sl s aq; s alc, eth
d481	2,6-Dimethyl-aniline	$(CH_3)_2C_6H_3NH_2$	121.18	12, 1107	0.984^{20}	1.5601^{20}	10–12	216	91	sl s aq; s alc, eth
d482	3,4-Dimethylaniline	$(CH_3)_2C_6H_3NH_2$	121.18	12, 1103	1.076^{18}		49–51	226	93	sl s aq; s alc
d483	3,5-Dimethylaniline	$(CH_3)_2C_6H_3NH_2$	121.18	12, 1131	0.972_4^{20}	1.5578^{20}	195–196	104^{14mm}		sl s aq; s alc
d484	Dimethylarsinic acid	$(CH_3)_2As(O)OH$	137.99							v s alc; 200 aq; i eth
d485	3,4-Dimethylbenzoic acid	$(CH_3)_2C_6H_3COOH$	150.18	9², 353			165–167	subl		s alc, bz
d486	2,5-Dimethylbenzo-nitrile	$(CH_3)_2C_6H_3CN$	131.18	9, 535	0.957	1.5284^{20}	13–14	223^{730mm}	92	
d487	N,N-Dimethylbenzyl-amine	$C_6H_5CH_2N(CH_3)_2$	135.21	12, 1019	0.900	1.5011^{20}	–75	183	54	
d488	2,3-Dimethyl-1,3-butadiene	$H_2C{=}C(CH_3)C(CH_3){=}CH_2$	82.15	1³, 991	0.7222_4^{25}	1.4362^{25}	–76.0	69.2	<1	
d489	2,2-Dimethylbutane	$CH_3CH_2C(CH_3)_3$	86.18	1, 150	0.6492^{20}	1.3688^{20}	–99.9	49.7	–28	
d490	2,3-Dimethylbutane	$(CH_3)_2CHCH(CH_3)_2$	86.18	1, 151	0.6616^{20}	1.3750^{20}	–128.5	58.0	–28	
d491	2,3-Dimethyl-2-3-butanediol	$(CH_3)_2C(OH)C(OH)(CH_3)_2$	118.18	1, 487			41.1	174.4		v s hot aq, alc, eth

3-Dimethylaminopropylamine, d593
Dimethylanisoles, d547, d548
2,4-Dimethyl-3-azapentane, d411

Dimethylbenzenes, x4, x5, x6
6,6-Dimethylbicyclo[3.1.1]hept-2-ene-2-ethanol, n105

Dimethyl (Z)-butenedioate, d544

TABLE 1.15 Physical Constants of Organic Compounds (*continued*)

No.	Name	Formula	Formula weight	Beilstein reference	Density	Refractive index	Melting point	Boiling point	Flash point	Solubility in 100 parts solvent
d492	2,2-Dimethyl-1-butanol	$CH_3CH_2C(CH_3)_2CH_2OH$	102.18	$1^3, 1675$	0.8286_4^{20}	1.4208^{20}	<-15	136.8		sl s aq; s alc, eth
d493	2,3-Dimethyl-1-butanol	$(CH_3)_2CHCH(CH_3)-CH_2OH$	102.18	$1^3, 1677$	0.8300_4^{20}	1.4205^{20}		149		s alc, eth
d494	2,3-Dimethyl-2-butanol	$(CH_3)_2CHC(CH_3)_2OH$	102.18	1, 413	0.8236_4^{20}	1.4176^{20}	-10.6	118.7	29	s aq; misc alc, eth
d495	3,3-Dimethyl-1-butanol	$(CH_3)_3CCH_2CH_2OH$	102.18	$1^3, 1677$	0.8147_4^{20}	1.4120^{20}	-60	143	47	s alc, eth
d496	3,3-Dimethyl-2-butanol	$(CH_3)_3CCH(OH)CH_3$	102.18	1, 412	0.8185_4^{20}	1.4151^{20}	5.3	120.4	28	s alc; misc eth
d497	3,3-Dimethyl-2-butanone	$(CH_3)_3CCOCH_3$	100.16	1, 694	0.7250_{25}^{25}	1.3939^{25}	-52.5	106.2	23	2.5 aq; s alc, eth
d498	2,3-Dimethyl-2-butene	$(CH_3)_2C{=}C(CH_3)_2$	84.16	1, 218	0.7081_4^{20}	1.4124^{20}	-74.3	73.2	-16	s alc, eth
d499	3,3-Dimethyl-1-butene	$(CH_3)_3CCH{=}CH_2$	84.16	1, 217	0.6531_4^{20}	1.3762^{20}	-115.2	41.3	-28	
d500	3,3-Dimethylbutyric acid	$(CH_3)_3CCH_2COOH$	116.16	2, 337	0.9124_4^{20}	1.4100^{20}	$6-7$	190	88	s alc, eth
d501	Dimethylcadmium	$(CH_3)_2Cd$	142.48		1.9846_4^{17}	1.5488	-4.5	105.5	≥ 150 ex-plo-des	d aq; s PE
d502	Dimethylcarbamyl chloride	$(CH_3)_2NCOCl$	107.54	4, 73	1.168	1.4540^{20}	-33	168	68	
d503	Dimethyl carbonate	$(CH_3O)_2C{=}O$	90.08	3, 4	1.065_4^{17}	1.3682^{20}	0.5	$90-91$	18	i aq; misc alc, eth
d504	Dimethyl chlorothio-phosphate	$(CH_3O)_2P(S)Cl$	160.56	$1^1, 143$	1.322	1.4819^{20}		67^{16mm}		
d505	Dimethylcyanamide	$(CH_3)_2NCN$	70.09	4, 74	0.867	1.4100^{20}		$161-163$	58	

			MW		density	n	mp	bp		solubility
d506	cis-1,2-Dimethyl-cyclohexane	$(CH_3)_2C_6H_{10}$	112.22	5, 36	0.7692_4^{20}	1.4335^{20}	−49.9	129.7	15	i aq; s alc, bz
d507	trans-1,2-Dimethyl-cyclohexane	$(CH_3)_2C_6H_{10}$	112.22	5, 36	0.7772_0^{20}	1.4273^{20}	−88.2	123.4	15	i aq; s alc, eth
d508	5,5-Dimethyl-1,3-cyclohexanedione		140.18	7, 559			d 149			0.4 aq; s alc, bz
d509	2,3-Dimethylcyclo-hexanol	$(CH_3)_2C_6H_9OH$	128.22		0.934	1.4653^{20}			65	
d510	2,6-Dimethylcyclo-hexanone		126.20	7, 23	0.925	1.4460^{20}		175	51	i aq; s alc, eth
d511	2,3-Dimethylcyclo-hexylamine	$(CH_3)_2C_6H_9NH_2$	127.23	7, 23	0.835	1.4595^{20}		160	51	
d512	Dimethyl decanedioate	$CH_3OOC(CH_2)_8COOCH_3$	230.30	2, 719	0.983_{20}^{30}	1.4335^{28}	23	144^{5mm}		i aq; s alc, eth
d513	5,7-Dimethyl-3,5,9-decatrien-2-one	$H_2C{=}CHCH_2CH(CH_3){-}CH{=}C(CH_3)CH{=}CH{-}COCH_3$	178.28					$79^{0.05mm}$		
d514	Dimethyl 2,5-dioxo-1,4-cyclohexanedi-carboxylate		228.20	10, 894			155–157			

Dimethyl 2-butynedioate, d460
Dimethyl Cellosolve, d439
Dimethylchlorosilane, c92

(Z)-2-Dimethylcrotonic acid, m162
Dimethyl 1,4-cyclohexanedione-2,5-dicarboxylic acid, d514

Dimethyl diphenyl sulfone 4,4'-dicarboxylate, s26

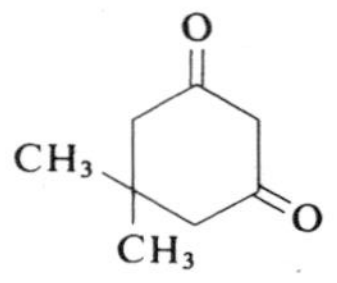

d508

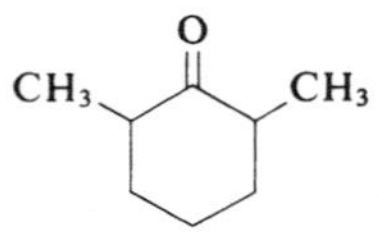

d510

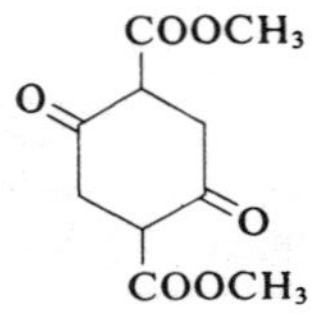

d514

TABLE 1.15 Physical Constants of Organic Compounds (*continued*)

No.	Name	Formula	Formula weight	Beilstein reference	Density	Refractive index	Melting point	Boiling point	Flash point	Solubility in 100 parts solvent
d515	2,3-Dimethyl-1,3-dioxolane-4-methanol		132.16		1.064_4^{20}	1.4383^{20}		188–189	80	misc aq, alc, eth
d516	Dimethyl disulfide	CH_3SSCH_3	94.20	1, 291	1.046	1.5253^{20}	−84.7	109.8	24	i aq; misc alc, eth
d517	Dimethyldithiocarbamic acid dihydrate, Na salt	$(CH_3)_2NCSS^-Na^+\cdot 2H_2O$	179.24	4, 75						
d518	Dimethyl ether	$(CH_3)_2O$	46.07	1, 281	0.661^{20}		−141.5	−24.9	−41	35% aq (5 atm); 15% bz; 11.8% acet
d519	Dimethylethoxyphenylsilane	$C_2H_5O(C_6H_5)Si(CH_3)_2$	180.3		0.9263_4^{20}	1.4799^{20}		93^{35mm}		
d520	N,N-Dimethylethylamine	$C_2H_5N(CH_3)_2$	73.14	4, 94	0.675	1.3720^{20}	−140	36–38	−36	
d521	N,N-Dimethylethylenediamine	$(CH_3)_2NCH_2CH_2NH_2$	88.15	$4^2, 690$	0.803	1.4260^{20}			23	misc aq, alc, bz, eth
d522	N,N-Dimethylformamide	$(CH_3)_2NCHO$	73.10	4, 58	0.9445_4^{25}	1.4282^{25}	−60.4	153.0	57	
d523	N,N-Dimethylformamide dimethyl acetal	$(CH_3)_2NCH(OCH_3)_2$	119.16		0.897	1.3972^{20}		103^{720mm}	7	sl s alc, eth
d524	Dimethyl fumarate	$CH_3OOCCH=CHCOOCH_3$	144.13	2, 741	1.045^{106}		105	193	<1	i aq; misc alc, eth
d525	2,5-Dimethylfuran	$(CH_3)_2(C_4H_2O)$	96.13	17, 41	0.9000_4^{20}	1.4414^{20}	−62	93		s alc, acet, eth, pyr
d526	Dimethylglyoxime	$CH_3C(=NOH)C(=NOH)CH_3$	116.12	1, 772			238–240			
d527	2,4-Dimethyl-2,6-heptadienal	$H_2C=CHCH_2CH(CH_3)CH=C(CH_3)CHO$	138.21					47^{2mm}		
d528	2,4-Dimethyl-2,6-heptadien-1-ol	$H_2C=CHCH_2CH(CH_3)CH=C(CH_3)CH_2OH$	140.23					88^{10mm}		

No.	Name	Formula	Formula wt.	Beilstein ref.	Density	n_D	mp, °C	bp, °C	Flash pt.	Solubility
d529	2,6-Dimethyl-2,5-heptadien-4-one	$(CH_3)_2C{=}CHC({=}O){-}CH{=}C(CH_3)_2$	138.21	1,751	0.8854^{20}	1.4968^{21}	28	198–199	79	sl s aq; s alc, eth
d530	Dimethyl heptanedioate	$CH_3OOC(CH_2)_5COOCH_3$	188.22	2^1,281	1.0625^{20}_4	1.4314^{20}	−21	122^{11mm}	>112	s alc
d531	2,6-Dimethyl-4-heptanone	$[(CH_3)_2CHCH_2]_2C{=}O$	142.24	1,710	0.806^{20}_{20}	1.4114^{20}	−41.5	168.1	48	0.06 aq; misc alc, bz, chl, eth
d532	2,5-Dimethyl-2,4-hexadiene	$(CH_3)_2C{=}CHCH{=}C(CH_3)_2$	110.20	1,259	0.7636^{20}_4	1.4741^{20}	12–14	132–134	29	i aq; s alc, eth
d533	2,5-Dimethylhexane	$(CH_3)_2CHCH_2CH_2CH(CH_3)_2$	114.24	1^3,283	0.6936^{20}_4	1.3925^{20}	−91.2	109.1	26	i aq; sl s alc; s eth
d534	2,5-Dimethyl-2,5-hexanediamine	$[(CH_3)_2C(NH_2)CH_2{-}]_2$	144.26		0.832	1.4459^{20}		64^{8mm}	62	
d534a	Dimethyl hexanedioate	$CH_3OOC(CH_2)_4COOCH_3$	174.20	1,652	1.0600^{20}_4	1.4285^{20}	8	112^{10mm}	107	i aq; s alc, eth
d535	2,5-Dimethyl-2,5-hexanediol	$[(CH_3)_2C(OH)CH_2{-}]_2$	146.23	1,492			86–90	214–215	126	
d536	1,5-Dimethylhexylamine	$(CH_3)_2CH(CH_2)_3CH(NH_2)CH_3$	129.25		0.767			154–156	48	
d537	2,5-Dimethyl-3-hexyne-2,5-diol	$(CH_3)_2CC{\equiv}CC(CH_3)_2$ OH OH	142.20	1,501		1.4209^{20}	94–95	205–206		

Dimethyleneimine, e131
Dimethylene oxide, e129

N,N-Dimethylethanolamine, d465
Dimethyl glutarate, d574

Dimethylglutaric acids, d575, d576

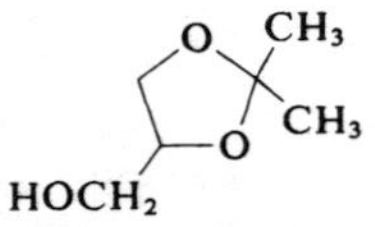

d515

TABLE 1.15 Physical Constants of Organic Compounds (*continued*)

No.	Name	Formula	Formula weight	Beilstein reference	Density	Refractive index	Melting point	Boiling point	Flash point	Solubility in 100 parts solvent
d538	5,5-Dimethyl-hydantoin		128.13	24, 289			176–178			v s aq, alc, bz, chl, eth, acet
d539	1,1-Dimethyl-hydrazine	$(CH_3)_2NNH_2$	60.10	4, 547	0.7911^{22}_4	1.4075^{20}	−58	63.9	1	misc aq, alc, eth, PE
d540	1,2-Dimethyl-hydrazine	$CH_3NHNHCH_3$	60.10	4, 547	0.8274^{20}_4	1.4209^{20}		81	flammable	misc aq, alc, eth, PE
d541	Dimethyl hydrogen phosphonate	$(CH_3O)_2P(=O)H$	110.05	1, 285	1.200^{20}_4	1.4009^{20}		170–171	96	s aq(hyd); misc alc, acet, eth
d542	1,2-Dimethyl-imidazole		96.13	23, 66	1.084		29–30	204	92	
d543	1,3-Dimethyl-2-imidazolidinone		114.15		1.044	1.4720^{20}		108^{17mm}	80	
d543a	Dimethylketene	$(CH_3)_2C=C=O$	70.09	1, 731			−97.5	34		d aq, alc; s eth
d544	Dimethyl maleate	$CH_3OOCCH=CHCOOCH_3$	144.13	2, 751	1.1513^{20}	1.4422^{20}	−17.5	200.4		8.7 aq
d545	Dimethyl malonate	$CH_3OOCCH_2COOCH_3$	132.12	2, 572	1.154^{20}_4	1.4135^{20}	−62	180–181	90	sl s aq; misc alc, eth
d546	Dimethylmercury	$(CH_3)_2Hg$	230.66	4, 678	3.1874^{20}	1.5452^{20}		92^{740mm}		i aq; s alc, eth
d547	3,4-Dimethyl-1-methoxybenzene	$(CH_3)_2C_6H_3OCH_3$	136.19	6, 481	0.9744^{14}_4	1.5198^{14}		200		i aq; s alc, bz, eth
d548	3,5-Dimethyl-1-methoxybenzene	$(CH_3)_2C_6H_3OCH_3$	136.19	6, 493	0.9627^{15}_4	1.5107^{15}		193	65	i aq; s alc, bz, eth
d549	N,N-Dimethylmethyleneammonium iodide	$H_2C=N(CH_3)_2^+I^-$	185.01	4^4, 153			219 d			
d550	Dimethyl methylene-succinate	$CH_3OOCCH_2C(=CH_2)COOCH_3$	158.15	2, 762	1.1241^{18}_4	1.4442^{20}	38	208		s alc, eth
d551	Dimethyl methyl-phosphonate	$(CH_3O)_2P(O)CH_3$	124.08	4^1, 572	1.145	1.4130^{20}		181	43	
d552	Dimethyl methyl-succinate	$CH_3OOCCH_2CH(CH_3)COOCH_3$	160.17	2^3, 1696	1.076	1.4200^{20}		196	83	

d553	2,6-Dimethyl-morpholine		115.18		0.9346^{20}	1.4470^{20}	−85	147	48	misc aq, alc, bz
d554	2,3-Dimethyl-naphthalene	$(CH_3)_2C_{10}H_6$	156.23	5, 571	1.0084^{20}		102–104	269		sl s alc; s bz, eth
d555	2,6-Dimethyl-naphthalene	$(CH_3)_2C_{10}H_6$	156.23	5, 570	1.142^0_4		110.2	262	107	i aq; sl s alc
d556	1,2-Dimethyl-3-nitrobenzene	$(CH_3)_2C_6H_3NO_2$	151.17	5, 367	1.129	1.5434^{20}	7–9	245		i aq; s alc
d557	1,2-Dimethyl-4-nitrobenzene	$(CH_3)_2C_6H_3NO_2$	151.17	5, 368	1.139		29–31	143^{20mm}		i aq; s alc
d558	1,3-Dimethyl-2-nitrobenzene	$(CH_3)_2C_6H_3NO_2$	151.17	5, 378	1.112	1.5220^{20}	14–16	225^{744mm}	87	i aq; s alc
d559	1,3-Dimethyl-4-nitrobenzene	$(CH_3)_2C_6H_3NO_2$	151.17	5, 378	1.117	1.5497^{20}	2	237–239	107	i aq; s alc
d560	N,N-Dimethyl-4-nitrosoaniline	$(CH_3)_2NC_6H_4NO$	150.18	12, 677			86	flammable		s alc, bz, chl, eth
d561	Dimethyl 2-nitro-1,4-phthalate	$O_2NC_6H_3(COOCH_3)_2$	239.18	9, 826			72–75	solid		i aq; s alc, eth

Dimethyl isophthalate, d589
1,4a-Dimethyl-7-isopropyl-1,2,3,4,4a,9,10,10a-octahydro-1-phenanthrenemethylamine, d20

Dimethyl itaconate, d550
2,2-Dimethyl-3-methylenenorbornane, c2
6,6-Dimethyl-2-methylenenorpinene, p176

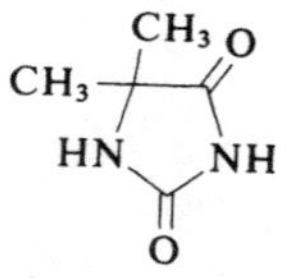

d538

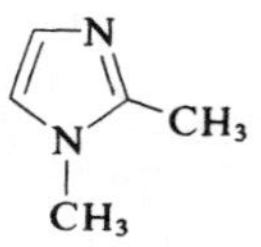

d542

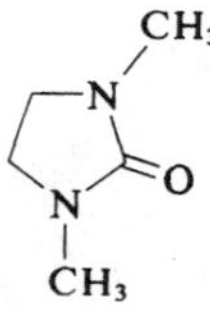

d543

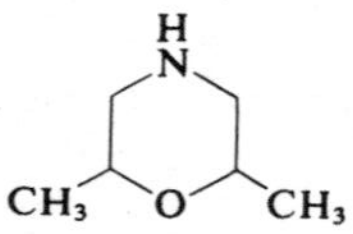

d553

TABLE 1.15 Physical Constants of Organic Compounds (*continued*)

No.	Name	Formula	Formula weight	Beilstein reference	Density	Refractive index	Melting point	Boiling point	Flash point	Solubility in 100 parts solvent
d562	cis-3,7-Dimethyl-2,6-octadienal		152.24		0.8888_4^{20}	1.4898^{20}		229	101	misc alc, eth, glyc
d563	trans-3,7-Dimethyl-2,6-octadienal		152.24		0.8869_4^{20}	1.4869^{20}		229	101	misc alc, eth, glyc
d564	3,7-Dimethyl-2,6-octadienenitrile		149.24		0.853	1.4753^{20}		268	>112	i aq; s alc
d565	Dimethyl octanedioate	CH₃OOC(CH₂)₆COOCH₃	202.25	2, 693	1.0210_4^{20}	1.4325^{20}	−4.8			i aq; s alc
d566	Dimethyl oxalate	CH₃OOCCOOCH₃	118.08	2, 534	1.148^{54}	1.379^{80}	50–54	163.5	75	6 aq; s alc, eth
d567	N¹-(4,5-Dimethyl-oxazol-2-yl)-sulfanilamide		267.31				193–194			s aq, acids, alk
d568	N-(1,1-Dimethyl-3-oxobutyl)acrylamide	CH₂=CHC(=O)NHC(CH₃)₂CH₂COCH₃	169.23				57–58	120^{8mm}		
d569	2,3-Dimethylpentanal	CH₃CH₂CH(CH₃)CH(CH₃)CHO	114.19		0.832	1.4132^{20}			58	
d570	2,2-Dimethylpentane	CH₃CH₂CH₂C(CH₃)₃	100.21	1, 157	0.6740^{20}	1.3824^{20}	−123.8	79.2	15	i aq; s alc, eth
d571	2,3-Dimethylpentane	CH₃CH₂CH(CH₃)-CH(CH₃)₂	100.21	1[2], 120	0.6951_4^{20}	1.3920^{20}	glass	89.8	−6	i aq; s alc, eth
d572	2,4-Dimethylpentane	(CH₃)₂CHCH₂CH(CH₃)₂	100.21	1, 158	0.6727_4^{20}	1.3815^{20}	−119.2	80.5	−6	s alc, eth
d573	3,3-Dimethylpentane	CH₃CH₂C(CH₃)₂CH₂CH₃	100.21	2, 633	0.6933^{20}	1.3905^{20}	−134.4	86.1		i aq; s alc, eth
d574	Dimethyl pentanedioate	CH₃OOC(CH₂)₃-COOCH₃	160.17		1.0934^{15}	1.4234^{20}		94–95^{13mm}	102	v s alc, eth
d575	2,2-Dimethyl-pentanedioic acid	HOOCC(CH₃)₂CH₂-CH₂COOH	160.17	2, 676			83–85			v s aq, alc, chl
d576	3,3-Dimethyl-pentanedioic acid	(CH₃)₂C(CH₂COOH)₂	160.17	2, 684			100–103			v s aq, alc, eth

No.	Name	Formula	M.W.	Beilstein	Density	n_D	m.p.	b.p.	Flash	Solubility
d577	2,4-Dimethyl-3-pentanol	$(CH_3)_2CHCH(OH)CH(CH_3)_2$	116.20	1, 417	0.829_4^{20}	1.4254^{20}	<70	140	37	sl s aq; s alc, eth
d578	2,4-Dimethyl-3-pentanone	$(CH_3)_2CHC(=O)CH(CH_3)_2$	114.19	1, 703	0.8062_4^{20}	1.3986^{20}	−80	124	15	misc alc, eth; s bz
d579	2,3-Dimethylphenol	$(CH_3)_2C_6H_3OH$	122.17	6, 480		1.5420^{20}	75	218	>112	v s alc, bz, chl, eth
d580	2,4-Dimethylphenol	$(CH_3)_2C_6H_3OH$	122.17	6, 486	1.0276^{14}	1.5390^{20}	27	210–212		v s alc, bz, chl, eth
d581	2,5-Dimethylphenol	$(CH_3)_2C_6H_3OH$	122.17	6, 494	0.965^{80}		74.5	211.5	73	v s alc, bz, chl, eth
d582	2,6-Dimethylphenol	$(CH_3)_2C_6H_3OH$	122.17	6, 485			49.0	203		v s alc, bz, chl, eth
d583	3,4-Dimethylphenol	$(CH_3)_2C_6H_3OH$	122.17	6, 480	1.064_4^{28}		62.5	225		v s alc, bz, chl, eth
d584	3,5-Dimethylphenol	$(CH_3)_2C_6H_3OH$	122.17	6, 480	1.008_4^{28}		64–68	219.5		v s alc, bz, chl, eth
d585	Dimethylphenyl-chlorosilane	$(CH_3)_2Si(Cl)C_6H_5$	170.7		1.032_4^{20}	1.508^{20}		192–193		
d586	4,5-Dimethyl-o-phenylenediamine	$(CH_3)_2C_6H_2(NH_2)_2$	136.20	13, 179			127–129			
d587	Dimethylphenyl-silane	$(CH_3)_2Si(H)C_6H_5$	136.3		0.8891_4^{20}	1.4995^{20}		156–157		

3,7-Dimethyl-6-octen-1-ol, c274
Dimethylolpropionic acid,

Dimethyl 3-oxoglutarate, d459
1,5-Dimethyl-2-phenyl-4-aminopyrazolone, a113

2,3-Dimethyl-1-phenyl-3-pyrazolin-5-one, a309

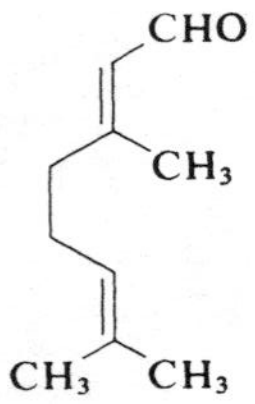

d562

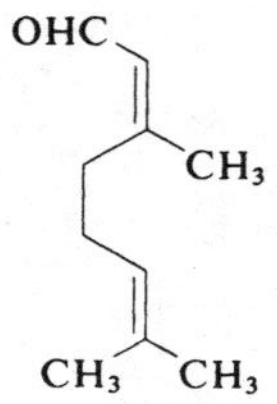

d563

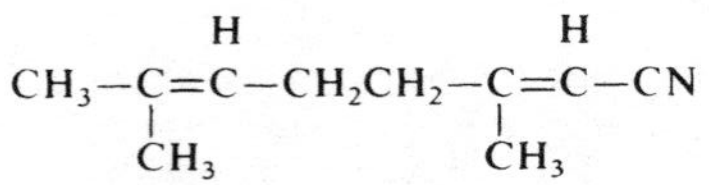

$CH_3-C=C-CH_2CH_2-C=C-C\equiv C-CN$ (with H and CH_3 substituents)

d564

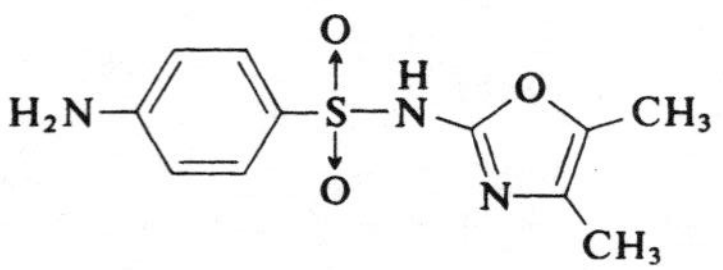

d567

TABLE 1.15 Physical Constants of Organic Compounds (*continued*)

No.	Name	Formula	Formula weight	Beilstein reference	Density	Refractive index	Melting point	Boiling point	Flash point	Solubility in 100 parts solvent
d588	Dimethyl-o-phthalate	$C_6H_4(COOCH_3)_2$	194.19	9, 797	1.1940^{20}_{20}	1.515^{21}	5.5	283.7	146	0.4 aq; misc alc, chl, eth; i PE
d589	Dimethyl-*m*-phthalate	$C_6H_4(COOCH_3)_2$	194.19	9, 834	1.1944^{20}	1.5168^{20}	67–68	282		i aq
d590	Dimethyl-*p*-phthalate	$C_6H_4(COOCH_3)_2$	194.19	4.3303	1.2		140–142	288		i aq, hot alc,eth
d591	2,6-Dimethyl-piperidine		113.20	20, 108	0.840	1.4394^{20}		127	11	
d592	2,2-Dimethylpropane	$(CH_3)_4C$	72.15		0.613^{0}	1.3476^{6}	−16.6	9.5		
d593	*N,N*-Dimethyl-1,3-propanediamine	$(CH_3)_2N(CH_2)_3NH_2$	102.18		0.812	1.4350^{20}		123	35	
d594	2,2-Dimethyl-1,3-propanediol	$(CH_3)_2C(CH_2OH)_2$	104.15	1, 483	1.11^{25}		127–128	208–210		180 aq; 12 bz; 60 acet; v s alc, eth
d595	2,2-Dimethyl-1-propanol	$(CH_3)_3CCH_2OH$	88.15	1, 406	0.812^{20}_{4}		52–54	113.1	36	3.6 aq; misc alc, eth
d596	2,2-Dimethylpropion-aldehyde	$(CH_3)_3CCHO$	186.25		0.793	1.3794^{20}	6	74^{730mm}	<1	
d597	2,2-Dimethyl propionamide	$(CH_3)_3CC(O)NH_2$	101.15	2, 320			154–157	212		
d598	2,2-Dimethyl-propionic acid	$(CH_3)_3CCOOH$	102.13	2, 319	0.905^{50}	1.3931^{37}	35.5	163.8	63	2.5 aq; v s alc, eth
d599	2,2-Dimethylpro-pionic anhydride	$[(CH_3)_3CC(O)]_2O$	186.25	2, 320	0.918	1.4092^{20}		193	57	
d600	2,2-Dimethyl-propionyl chloride	$(CH_3)_3CCOCl$	120.58	2, 320	0.979	1.4120^{20}		105–106	<1	d aq; alc; v s eth
d601	1,1-Dimethyl-propylamine	$CH_3CH_2C(CH_3)_2NH_2$	87.17	4, 179	0.731^{25}	1.3996^{20}	−105	77	65	misc aq, alc, eth
d602	1,1-Dimethyl-2-propynylamine	$HC{\equiv}CC(CH_3)_2NH_2$	83.13		0.790	1.4235^{20}		79–80	<1	

d603	3,5-Dimethyl-pyrazole		96.13	23, 74			108	218		s aq; v s bz, eth
d604	2,4-Dimethylpyridine	$(CH_3)_2(C_5H_3N)$	107.16	20, 244	0.9272^{25}_4	1.4991^{20}	<-60	158.3	37	17 aq[45]; v s alc, bz, eth
d605	2,6-Dimethylpyridine	$(CH_3)_2(C_5H_3N)$	107.16	20, 244	0.9200^{25}_4	1.4956^{25}	-6.0	143–144	33	43 aq[45]; s alc, eth
d606	3,4-Dimethylpyridine	$(CH_3)_2(C_5H_3N)$	107.16	20, 246	0.9395^{25}_4	1.5100^{25}	-12	164	53	sl s aq; s alc, eth
d607	3,5-Dimethylpyridine	$(CH_3)_2(C_5H_3N)$	107.16	20, 246	0.9396^{25}_4	1.5033^{25}	-9	170	53	s aq, alc, eth
d608	Dimethyl succinate	$CH_3OOCCH_2CH_2COOCH_3$	146.14	2, 609	1.202^{18}_4	1.4190^{20}	19.5	$195–200^{75mm}$	85	0.83 aq: 2.9 alc
d609	Dimethylsulfamoyl chloride	$(CH_3)_2NSO_2Cl$	143.59	4, 84	1.337	1.4518^{20}		114^{75mm}		
d610	Dimethyl sulfate	$(CH_3O)_2SO_2$	126.13	1, 283	1.3322^{20}_4	1.3874^{20}	-31.8	188 d	83	2.8 aq(hyd); s acet, bz, diox, eth
d611	Dimethyl sulfide	$(CH_3)_2S$	62.13	1, 288	0.8462^{21}_4	1.4354^{20}	-98.3	37.3	-36	2 aq; s alc, eth
d612	Dimethyl sulfite	$(CH_3O)_2SO$	110.13	1, 282	1.294	1.4083^{20}		126–127	30	
d613	Dimethyl sulfone	$(CH_3)_2SO_2$	94.13	1, 289			109	238	143	v s aq, alc, acet
d614	Dimethyl sulfoxide	$(CH_3)_2SO$	78.13	1, 289	1.100^{20}_4	1.4783^{20}	18.5	189.0	95	s alc, acet, bz, chl
d615	Dimethyl-d_6 sulfoxide	$(CD_3)_2SO$	84.18		1.18	1.4758^{20}		55^{5mm}	95	

Dimethyl phosphite, d541
Dimethyl pimelate, d530
Dimethyl propanedioate, d545

1,1-Dimethylpropargylamine, d602
N'-(4,6-Dimethyl-2-pyrimidinyl)sulfanilamide, s21

Dimethyl sebacate, d512
Dimethyl suberate, d565

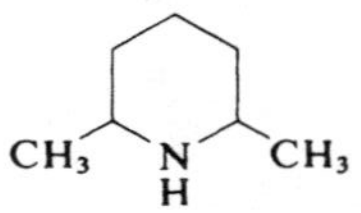

d591

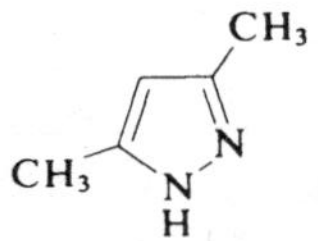

d603

TABLE 1.15 Physical Constants of Organic Compounds (*continued*)

No.	Name	Formula	Formula weight	Beilstein reference	Density	Refractive index	Melting point	Boiling point	Flash point	Solubility in 100 parts solvent
d616	(+)-Dimethyl-L-tartrate	$CH_3OOCCH(OH)CH(OH)COOCH_3$	178.14	3, 510	1.328^{20}_4		48–50	163^{23mm}		s aq; 200 alc^{-15}; v s bz
d617	Dimethyltelluride	$(CH_3)_2Te$	157.68	1, 291			–10	91–92		
d618	2,5-Dimethyltetrahydrofuran	$(CH_3)_2(C_4H_6O)$	100.16	17, 14	0.833	1.4041		90–92	26	d aq; v s alc; i eth
d619	Dimethyl-3,3′-thiodipropionate	$(CH_3OOCCH_2CH_2)_2S$	206.26		1.198	1.4740^{20}		148^{18mm}	>112	
d620	N,N-Dimethylthioformamide	$(CH_3)_2NC(S)H$	89.16	4, 70	1.047	1.5757^{20}		58^{1mm}	99	
d621	N,N′-Dimethylthiourea	$(CH_3NH)_2C{=}S$	104.18	4, 70			60–62			v s aq, alc, acet
d622	N,N-Dimethyl-p-toluidine	$CH_3C_6H_4N(CH_3)_2$	135.21	12, 902	0.937	1.5458^{20}		211	83	
d623	1,3-Dimethylurea	$(CH_3NH)_2C{=}O$	88.11	4, 65	1.386^{11}_4		101–104	268–270		v s aq, alc, i eth
d624	Dimethylzinc	$(CH_3)_2Zn$	95.45				–40	46	ignites in air	misc bz, PE; s eth
d625	2,4-Dinitroaniline	$(O_2N)_2C_6H_3NH_2$	183.12	12, 747	1.615^{14}		188			i aq; 0.75 alc
d626	1,3-Dinitrobenzene	$C_6H_4(NO_2)_2$	168.11	5, 258	1.575^{18}_4		89–90	300–303		0.05 aq; 2.7 alc; v s bz, chl, EtAc
d627	2,4-Dinitrobenzenesulfenyl chloride	$(O_2N)_2C_6H_3SCl$	234.62	6^2, 316			96			s bz, HOAc; d alc
d628	3,4-Dinitrobenzoic acid	$(O_2N)_2C_6H_3COOH$	212.12	9, 413			166	subl		0.7 aq; v s alc, eth
d629	3,5-Dinitrobenzoic acid	$(O_2N)_2C_6H_3COOH$	212.12	9, 413			207			1.9 hot aq; v s alc; sl s bz, eth
d630	3,5-Dinitrobenzoyl chloride	$(O_2N)_2C_6H_3COCl$	230.56	9, 414			69.5	196^{11mm}		d aq. alc; s eth
d631	2,6-Dinitro-p-cresol	$(O_2N)_2C_6H_2(OH)CH_3$	198.13	6, 414			77–79 (anhyd)			

No.	Name	Formula	Mol. wt.	Beil. ref.	Density	n_D	m.p. °C	b.p. °C	Flash pt.	Solubility
d631a	4,6-Dinitro-o-cresol	$(O_2N)_2C_6H_2(OH)CH_3$	198.13	6, 368			87.5			v s alc, acet, eth, alk
d632	1,1-Dinitroethane	$CH_3CH(NO_2)_2$	120.07	1, 102	1.3503^{24}			185–186		s alc, eth
d633	2,4-Dinitro-1-fluorobenzene	$FC_6H_3(NO_2)_2$	186.10	5, 262		1.5690^{20}	26	178^{25mm}	>112	s bz, eth, glyc
d634	1,5-Dinitronaphthalene	$C_{10}H_6(NO_2)_2$	218.17	5, 558			216–217	subl		s bz; v s eth; sl s alc
d635	2,4-Dinitrophenol	$(O_2N)_2C_6H_3OH$	184.11	6, 251	1.683		112–114			s alc, bz; 16 EtAc; 36 acet; 5 chl; 20 pyr
d636	2,4-Dinitrophenyl-acetic acid	$(O_2N)_2C_6H_3CH_2COOH$	226.15	9, 459			169–175			s alc, eth
d637	2,4-Dinitrophenyl-hydrazine	$(O_2N)_2C_6H_3NHNH_2$	198.14	15, 489			~200	flammable solid		sl s aq, alc; s acid
d638	3,5-Dinitrosalicyclic acid	$(O_2N)_2C_6H_2(OH)COOH$	228.12	10, 122			169–172			s aq; v s alc, eth
d639	2,4-Dinitrotoluene	$CH_3C_6H_3(NO_2)_2$	182.14	5, 339	1.321^{71}	1.442	64.66	300 sl d		1.2 alc; 9 eth
d640	2,6-Dinitrotoluene	$CH_3C_6H_3(NO_2)_2$	182.14	5, 341	1.2833^{111}	1.479	64–66			s alc
d641	3,4-Dinitrotoluene	$CH_3C_6H_3(NO_2)_2$	182.14	5, 341	1.2594^{111}		54–57		218	i aq; s alc
d641a	Dinonyl hexanedioate	$C_9H_{19}OOC(CH_2)_4$-$COOC_9H_{19}$	398.63		0.9172^{25}					
d642	Dioctadecyl phosphite	$(C_{18}H_{37}O)P(O)H$	586.97				57–59			
d643	Dioctylamine	$(C_8H_{17})_2NH$	241.46	4, 196	0.842	1.4610^{20}	14–16	298	>112	i aq; v s alc, eth
d644	Dioctyl sulfide	$(C_8H_{17})_2S$	258.51	1, 419	0.962	1.4609^{20}		180^{10mm}	>112	
d645	4,9-Dioxa-1,12-dodecanediamine	$H_2N(CH_2)_3O(CH_2)_2$-$O(CH_2)_3NH_2$	204.32					134-136^{4mm}	>112	

TABLE 1.15 Physical Constants of Organic Compounds (*continued*)

No.	Name	Formula	Formula weight	Beilstein reference	Density	Refractive index	Melting point	Boiling point	Flash point	Solubility in 100 parts solvent
d646	1,4-Dioxane		88.10	19, 3	1.0329_4^{20}	1.4224^{20}	11.7	101.2	12	misc aq, alc, bz, chl, eth, PE
d647	1,3-Dioxolane		74.08	19^2, 3	1.060_4^{20}	1.4000^{20}	−95	74–75	<1	misc aq; s alc, eth
d648	Dipentaerythritol	$(HOCH_2)_3CCH_2OCH_2-C(CH_2OH)_3$	254.28				215–218			
d649	Dipentene		136.24	5, 137	0.8402_4^{21}	1.4739^{20}		176	42	i aq; misc alc
d650	Dipentylamine	$(C_5H_{11})_2NH$	157.29	4^1, 378	0.777	1.4272		195–202	39	v s alc, eth
d651	Dipentyl ether	$(C_5H_{11})_2O$	158.29	1^1, 193	0.7833_4^{20}	1.4120^{20}	−69.4	186.8	63	misc alc, eth; s acet
d652	Diphenylacetic acid	$(C_6H_5)_2CHCOOH$	212.25	9, 673	1.258_{15}^{15}		148	195^{5mm}		s hot aq, alc, chl, eth
d653	Diphenylacetonitrile	$(C_6H_5)_2CHCN$	193.25	9, 674	0.990		76	181^{12mm}		s alc, eth
d654	Diphenylacetylene	$C_6H_5C{\equiv}CC_6H_5$	178.23	5, 656	1.160		60–61	300		v s hot alc, eth
d655	Diphenylamine	$(C_6H_5)_2NH$	169.23	12, 174			53–54	302	152	45 alc; v s bz, eth
d656	cis,cis-1,4-Diphenyl-1,3-butadiene	$C_6H_5CH{=}CH-CH{=}CHC_6H_5$	206.29	5, 676	0.9697_4^{101}	1.6347^{101} (He line)	70.5			s bz, chl, eth, PE
d657	cis,trans-1,4-Diphenyl-1,3-diene	$C_6H_5CH{=}CH-CH{=}CHC_6H_5$	206.29	5, 676	0.9974_4^{22}	1.6053^{22}	88	$133^{0.1mm}$		s alc, bz, eth, chl
d658	1,3-Diphenyl-2-buten-1-one	$C_6H_5C(O)CH{=}C(C_6H_5)CH_3$	222.27	7^2, 433	1.1080_4^{15}	1.6343^{20}	−30	246^{50mm}		i aq; s alc, eth
d659	Diphenylcarbamoyl chloride	$(C_6H_5)_2NCOCl$	231.68				glass 82–84			
d660	1,5-Diphenylcarbohydrazide	$(C_6H_5NHNH)_2C{=}O$	242.28	15, 292			168–171			s hot alc, acet, HOAc
d661	Diphenyl carbonate	$(C_6H_5O)_2C{=}O$	214.22	6, 158	1.296	1.5500^{20}	80–81	302–306	>112	s hot alc, bz, eth
d662	Diphenyl chlorophosphate	$(C_6H_5O)_2P(O)Cl$	268.64	6, 179				314^{272mm}		

No.	Name	Formula	Formula wt.	Beilstein ref.	Density	Refractive index	m.p., °C	b.p., °C		Solubility
d663	Diphenyl diselenide	$C_6H_5SeSeC_6H_5$	312.13	6, 346	1.557^{80}_{4}		61–64	310		s hot alc
d664	Diphenyl disulfide	$C_6H_5SSC_6H_5$	218.34	6, 323	1.353^{20}_{4}		58–60	355		s alc, bz, eth; i aq
d665	Diphenylenimine		167.21	20, 433	1.10^{18}		246			0.8 bz; 3 eth; 16 pyr; 11 acet; i aq
d666	1,2-Diphenylethane	$C_6H_5CH_2CH_2C_6H_5$	182.27	5, 598	0.995^{20}_{4}	1.5338	52.5	284		s alc; v s chl, eth
d667	Diphenyl ether	$C_6H_5OC_6H_5$	170.21	6, 146	1.0661^{20}_{4}	1.5763^{30}	26.9	258.3		s alc, bz, eth, HOAc
d668	1,2-Diphenylethyl-amine	$C_6H_5CH_2CH(C_6H_5)NH_2$	197.28	12, 1326	1.020	1.5802^{20}		311	115	
d669	N,N'-Diphenylethyl-enediamine	$C_6H_5NHCH_2CH_2NHC_6H_5$	212.30	12, 543			67.5	228–330	>112	v s alc, eth
d670	N,N'-Diphenyl-formamidine	$C_6H_5N=CHNHC_6H_5$	196.25	12, 236			138–141			s eth; v s chl
d671	1,3-Diphenyl-guanidine	$C_6H_5NHC(=NH)NHC_6H_5$	211.27	12, 369	1.13		150	d 170		s alc, hot bz, chl

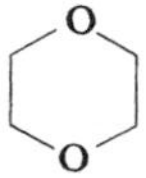

d646

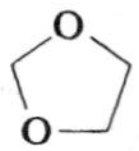

d647

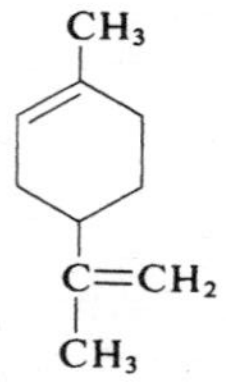

d649

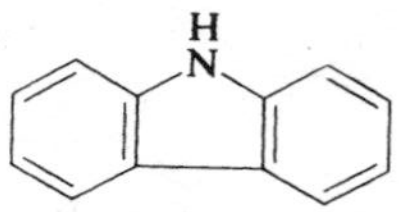

d665

TABLE 1.15 Physical Constants of Organic Compounds (*continued*)

No.	Name	Formula	Formula weight	Beilstein reference	Density	Refractive index	Melting point	Boiling point	Flash point	Solubility in 100 parts solvent
d672	5,5-Diphenyl-hydantoin		252.27	24, 410			295–298			i aq; 1.7 alc; 3.3 acet
d673	1,2-Diphenyl-hydrazine	$C_6H_5NHNHC_6H_5$	184.24	15, 123	1.158^{16}_{4}		123–126			v s alc; sl s bz
d674	Diphenyl isooctyl-phosphine	$(CH_3)_2CH(CH_2)_5PH\text{-}(OC_6H_3)_2$	346.40		1.044	1.522				
d675	Diphenylmercury	$(C_6H_5)_2Hg$	354.81	16, 946	2.318^{4}		124–125	d > 306	>112	s chl; sl s hot alc
d676	Diphenylmethane	$C_6H_5CH_2C_6H_5$	168.24	5^2, 498	1.3421^{10}_{4}	1.5768	25.9	264.5		v s alc, bz, chl, eth
d677	Diphenylmethanol	$(C_6H_5)_2CHOH$	184.24	6, 678			66.7	298		0.05 aq; v s alc, chl, eth
d678	1,1-Diphenylmethyl-amine	$C_6H_5CH(NH_2)C_6H_5$	183.25	12, 1323	1.0635^{22}_{22} super-cooled	1.5956^{99}	34	295	>112	sl s aq
d679	2,5-Diphenyloxazole		221.26	27, 78			72–73	360	176	
d680	2,6-Diphenylphenol	$(C_6H_5)_2C_6H_3OH$	246.31	6^3, 3631	1.223	1.5575^{20}	100–102	219^{26mm}		
d681	Diphenyl phosphite	$(C_6H_5O)_2P(=O)H$	234.19	6^1, 94	1.277	1.5518^{20}	12	$157^{0.17mm}$	>112	
d682	Diphenylphosphoryl azide	$(C_6H_5O)_2P(=O)N_3$	275.20							
d683	2,2-Diphenyl-1-picrylhydrazyl		394.32	16^2, 363			127 d			i aq; v s alc, eth
d684	1,3-Diphenyl-2-propanone	$C_6H_5CH_2C(=O)\text{-}CH_2C_6H_5$	210.28	7, 445	1.2		32–34	330		v s bz, chl, eth
d685	1,3-Diphenyl-2-propen-1-one	$C_6H_5CH=CHC(=O)\text{-}C_6H_5$	208.26	7, 478	1.0712^{62}_{4}	1.6458^{62}	57–58	208^{25mm}		s alc; v s bz, eth
d686	2,2-Diphenyl-propionic acid	$CH_3C(C_6H_5)_2COOH$	226.28	9^2, 474			175–177	300		s alc; v s bz, eth
d687	Diphenylsilanediol	$(C_6H_5)_2Si(OH)_2$	216.31	16, 909	1.118^{15}_{15}	1.6327^{20}	140 d	296	53	misc bz, eth, CS_2
d688	Diphenyl sulfide	$(C_6H_5)_2S$	186.28	6, 299			–40	379	>112	i aq; s hot alc, bz
d689	Diphenyl sulfone	$(C_6H_5)_2SO_2$	218.27	6, 300			128–129			

No.	Name	Formula	Formula wt.	Beilstein	Density	n_D	m.p., °C	b.p., °C		Solubility
d690	Diphenyl sulfoxide	$(C_6H_5)_2S{=}O$	202.28	6,300			69–71	207^{13mm}		i aq; v s chl, CCl_4
d691	Diphenylthiocarbazone	$C_6H_5N{=}NC(S)NH\text{-}NHC_6H_5$	256.33	16,26			168 d			i aq; v s alc, eth
d692	1,3-Diphenylthiourea	$C_6H_5NHC(S)NHC_6H_5$	228.32	12,394	1.32		154			
d693	1,3-Diphenylurea	$C_6H_5NHC(O)NHC_6H_5$	212.25	12,352	1.239		238	260 d		0.015 aq; s eth, HOAc
d694	1,2-Dipiperidinoethane		196.34	20[1],19	0.916	1.4876^{20}	−0.5	265		
d695	Dipiperidinomethane		182.31		0.915	1.4820^{20}		123^{15mm}	110	
d696	Dipropylamine	$(C_3H_7)_2NH$	101.19	4,138	0.7375^{20}_{4}	1.4043^{20}	−39.6	109.2	17	4 aq; v s alc, eth, PE
d697	Dipropylene glycol butyl ether	$CH_3CH(OH)CH_2OCH_2CH(OC_4H_9)CH_3$	190.3		0.917^{25}_{25}	1.425^{25}		229	113	

5,5-Diphenyl-2,4-imidazolidinedione, d672
Diphenyl ketone, b53
Diphenyl oxide, d667
Diphenylphosphorochloridate, d662

1,3-Diphenyl-1,3-propanedione, d53
sym-Diphenylthiourea, t146
Dipicolinic acid, p261
Di-2-propenylamine, d25

Dipropyl adipate, d702
Dipropylene glycol, h172

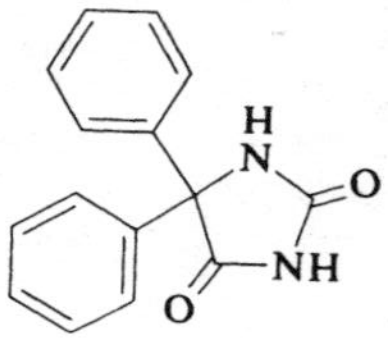

d672

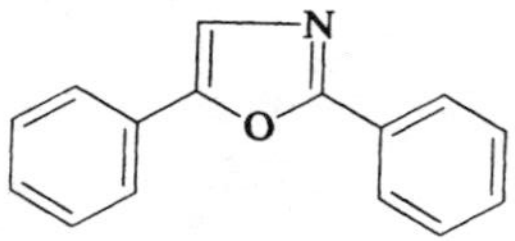

d679

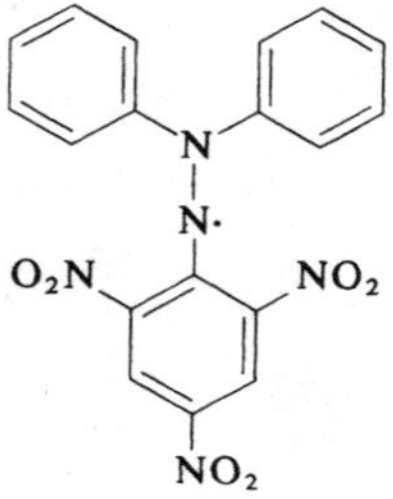

d683

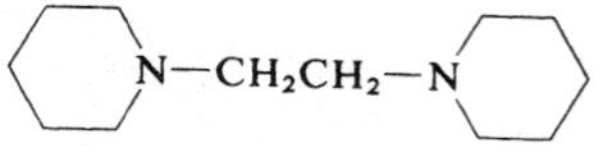

O_2N · · · NO_2 · · · NO_2

d694

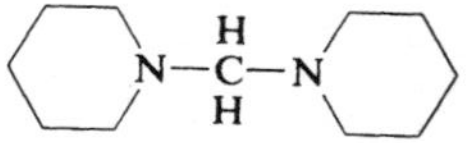

$N{-}CH_2CH_2{-}N$

d695

$N{-}\underset{H}{\overset{H}{C}}{-}N$

1.231

TABLE 1.15 Physical Constants of Organic Compounds (*continued*)

No.	Name	Formula	Beilstein reference	Formula weight	Density	Refractive index	Melting point	Boiling point	Flash point	Solubility in 100 parts solvent
d698	Dipropylene glycol ethyl ether	$CH_3CH(OH)CH_2OCH_2\text{-}CH(OC_2H_5)CH_3$		162.2	0.9302^{25}_{25}	1.419^{25}		388	90	
d699	Dipropylene glycol isopropyl ether	$CH_3CH(OH)CH_2OCH_2\text{-}CH[OCH(CH_3)_2]CH_3$		176.2	0.878^{25}_{25}	1.421^{25}		80.1	90	
d700	Dipropylene glycol methyl ether	$CH_3CH(OH)CH_2OCH_2\text{-}CH(OCH_3)CH_3$		148.2	0.951^{20}_{20}	1.419^{20}	−117	188.3	85	
d701	Dipropyl ether	$(C_3H_7)_2O$	1, 354	102.18	0.7466^{20}	1.3803^{20}	−123.2	89.6	4	0.4 aq i aq; s alc, eth
d702	Dipropyl hexane-dioate	$C_3H_7OOC(CH_2)_4\text{-}COOC_3H_7$	2^2, 574	230.30	0.9790^{20}_4	1.4314^{20}	−20	144^{10mm}		
d703	Dipropyl sulfate	$(C_3H_7O)_2SO_2$	1, 354	182.24	1.106^{20}_{20}		d 140	120^{20mm}	126	v s PE
d704	Dipropyl sulfone	$(C_3H_7)_2SO_2$	1, 359	150.24	1.028^{50}_4		28–30	270		0.5 aq; v s alc, bz, chl, eth, PE
d705	2,2′-Dipyridyl		23, 199	156.19			69.7	273		
d706	2,2′-Dipyridylamine		22^1, 630	171.20			89–90	222^{50mm}		
d707	1,3-Dithiane			120.24			53–55		90	
d708	4,4′-Dithiobutyric acid	$HOOC(CH_2)_3SS(CH_2)_3\text{-}COOH$	3, 312	238.32			110			
d709	3,3′-Dithiopropionic acid	$HOOCCH_2CH_2SSCH_2\text{-}CH_2COOH$		210.27			157–159			
d710	Dithiooxamide	$H_2NC(=S)C(=S)NH_2$	2, 565	120.20	1.10^{20}_4		170 d	subl		sl s aq; s alc; i eth
d711	1,3-Di-o-tolyl-guanidine	$(CH_3C_6H_4NH)_2C=NH$	12, 803	239.32			176–178			s hot alc, eth
d712	1,5-Di(vinyloxy)-3-oxapentane	$(CH_2=CHOCH_2CH_2)_2O$		158.20	0.975^{29}	1.445		81^{10mm}		
d713	1,3-Divinyltetra-methyldisiloxane	$[CH_2=CHSi(CH_3)_2]_2O$		186.39	0.811^{20}_4	1.412^{20}	−99.7	139		
d714	3,9-Divinyl-2,4,8,10-tetraoxaspiro[5.5]-undecane			212.25	1.251		40–54	120^{2mm}	110	

No.	Name	Formula	Mol. wt.	Beilstein	Density	n_D	mp	bp	Flash pt	Solubility
d715	Docosane	$CH_3(CH_2)_{20}CH_3$	310.61	1, 174	0.7782^{45}	1.4358^{45}	44.4	369		i aq; sl s alc; v s eth
d716	Docosanoic acid	$CH_3(CH_2)_{20}COOH$	340.60	2, 391	0.8221^{100}_{4}	1.4270^{100}	80–82	206^{60mm}	71	0.2 alc; 0.19 eth
d717	1-Docosanol	$CH_3(CH_2)_{21}OH$	326.61	1, 431			65–72	$180^{0.22mm}$	155	sl s eth; s alc, chl
d718	1H,1H,7H-Dodecafluoro-1-heptanol	$HCF_2(CF_2)_5CH_2OH$	332.0		1.7616^{20}	1.3180^{20}		169–170		
d719	Dodecane	$CH_3(CH_2)_{10}CH_3$	170.41	1, 171	0.7490^{20}_{4}	1.4216^{20}	−9.6	216.28		
d720	1,12-Dodecanediamine	$H_2N(CH_2)_{12}NH_2$	200.37	4, 273			62–65			
d721	Dodecanedioic acid	$HOOC(CH_2)_{10}COOH$	230.30	2, 729			128–130	245^{10mm}		misc alc, bz, chl, eth
d722	1,2-Dodecanediol	$CH_3(CH_2)_9CH(OH)\text{-}CH_2OH$	202.34	1^3, 2237			58–60			i aq; s alc, eth
d723	1,12-Dodecanediol	$HOCH_2(CH_2)_{10}CH_2OH$	202.34	1^2, 562			81–84	189^{12mm}	>112	i aq; s alc, eth
d724	Dodecanenitrile	$CH_3(CH_2)_{10}CN$	181.32	2, 363	0.827	1.4360^{20}		198^{100mm}		i aq; 100 alc; v s bz, eth
d725	1-Dodecanethiol	$CH_3(CH_2)_{11}SH$	202.40		0.845^{20}_{20}	1.4587^{20}		266–283	87	
d726	Dodecanoic acid	$CH_3(CH_2)_{10}COOH$	200.32	2, 359	0.869^{5}_{4}	1.4183^{82}	44	225^{100mm}		i aq; s alc, eth
d727	1-Dodecanol	$CH_3(CH_2)_{11}OH$	186.34	1, 428	0.8308^{25}_{4}	1.4413^{25}	23.8	259	>112	i aq; s alc, eth
d728	Dodecanoyl chloride	$CH_3(CH_2)_{10}COCl$	218.77	2, 363	0.946	1.4459^{20}		134^{11mm}	>112	i aq; s alc, eth
d729	1-Dodecene	$CH_3(CH_2)_9CH{=}CH_2$	168.32	1, 225	0.7584^{20}_{4}	1.4294^{20}	−35.2	213.4	77	s alc, eth, PE

d705

d706

d707

$CH_2{=}CH$ $CH{=}CH_2$

d714

TABLE 1.15 Physical Constants of Organic Compounds (*continued*)

No.	Name	Formula	Beilstein reference	Formula weight	Density	Refractive index	Melting point	Boiling point	Flash point	Solubility in 100 parts solvent
d730	2-Dodecen-1-yl-succinic anhydride			266.38				180^{5mm}	177	
d731	Dodecanal	$CH_3(CH_2)_{10}CHO$	1, 714	184.32	0.835	1.4344^{20}		185^{100mm}	101	misc alc, bz, chl, eth
d732	Dodecylamine	$CH_3(CH_2)_{11}NH_2$	4, 200	185.36			28–30	247–249	>112	
d733	4-Dodecylaniline	$CH_3(CH_2)_{11}C_6H_4NH_2$	12^3, 2776	261.46			40–41	220– 221^{15mm}		10 aq
d734	Dodecylcyclohexane	$CH_3(CH_2)_{11}C_6H_{11}$		252.50	0.8250	1.4580^{20}	12	$131^{0.8mm}$		
d735	Dodecyl sulfate, Na salt	$CH_3(CH_2)_{11}OSO_3^{-} Na^{+}$		288.38			204–207			
d736	Dodecyltrichloro-silane	$CH_3(CH_2)_{11}SiCl_3$		303.8		1.458^{20}		155^{10mm}		v s alc, eth
d737	Dotriacontane	$CH_3(CH_2)_{30}CH_3$	1, 177	450.88	0.8124^{20}	1.4364^{70}	68–70	467		sl s alc, bz, eth
d738	Dulcitol		1, 544	182.17	1.47^{20}		188–189	275^{1mm}		3.3 aq; sl s alc
e1	D-Ephedrine	$CH_3NHCH(CH_3)CH(OH)$-C_6H_5	13, 637	165.24			119	225		v s alc, eth
e2	L-Ephedrine	$CH_3NHCH(CH_3)CH(OH)$-C_6H_5	13, 636	165.24			34	255		5 aq; v s alc; s chl
e3	1,2-Epoxybutane	$CH_3CH_2CH-CH_2$ with O	17^2, 17	72.11	0.8297^{20}	1.3840^{20}	–150	63.2	–17	6 aq; misc alc, bz, chl, eth
e4	1,2-Epoxycyclo-dodecane			182.31	0.939	1.4773^{20}			27	
e5	1,2-Epoxycyclo-hexane		17, 21	98.15	0.970	1.4520^{2}		129–130		v s alc, bz, eth
e6	1,4-Epoxycyclohexane			98.15	0.969	1.4480^{20}		119^{713mm}	12	
e7	2-(3,4-Epoxycyclo-hexyl)ethyltri-methoxysilane			246.37	1.070^{25}_{4}	1.449^{25}		310	146	

No.	Name	Formula	Beilstein ref.	Formula wt	Density	n_D	Melting point, °C	Boiling point, °C	Flash point, °C	Solubility
e7a	1,2-Epoxycyclooctane			126.20			53–56	55^{5mm}	56	
e8	1,2-Epoxycyclopentane		17, 21	84.12	0.964	1.4336^{20}		102	10	i aq; s alc, eth
e9	1,2-Epoxyethyl-benzene	$C_6H_5CH{-}CH_2$ (O bridge)	17, 49	120.15	1.0523^{16}	1.5338^{20}	−37	194	79	
e9a	1,2-Epoxy-3-phenoxy-propane	$C_6H_5OCH_2CH{-}CH_2$ (O bridge)		150.18			2			
e10	1,2-Epoxypropane	$CH_3CH{-}CH_2$ (O bridge)	17, 6	58.08	0.859^{0}	1.3660^{20}	−112.1	34.2	−37	41 aq; misc alc, eth
e11	2,3-Epoxy-1-propanol	$H_2C{-}CHCH_2OH$ (O bridge)	17, 104	74.08	1.1143^{25}	1.4315^{20}		$66^{2.5mm}$	81	misc aq

Structures:

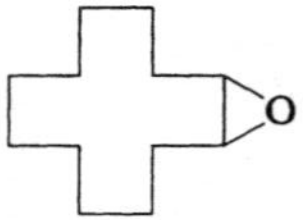

e4 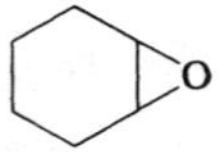 · e5 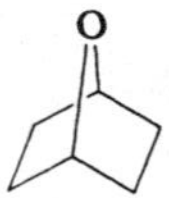 · e6 · e7a · e8

e7 — $CH_2CH_2Si(OCH_3)_2$

d730 — $CH_3(CH_2)_8CH{=}CHCH_2$—

d738
```
     H  OH OH H
     |  |  |  |
HOCH2—C—C—C—C—CH2OH
     |  |  |  |
     OH H  H  OH
```

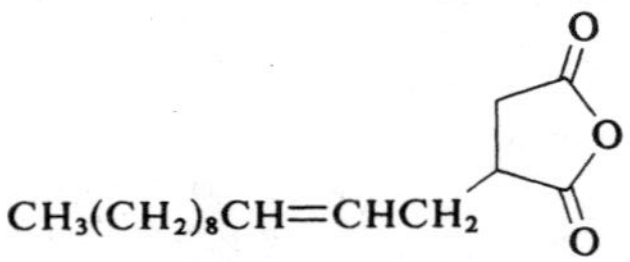

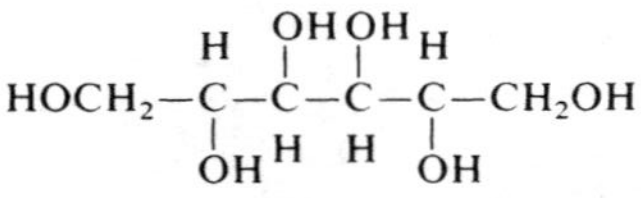

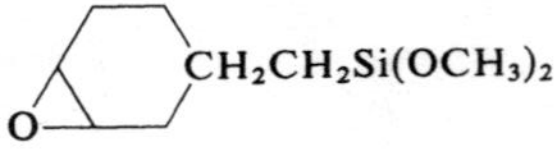

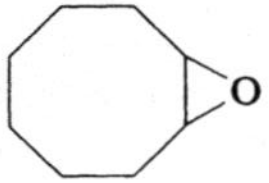

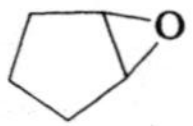

TABLE 1.15 Physical Constants of Organic Compounds (*continued*)

No.	Name	Formula	Formula weight	Beilstein reference	Density	Refractive index	Melting point	Boiling point	Flash point	Solubility in 100 parts solvent
e12	2,3-Epoxypropyl methacrylate	$H_2C{=}C(CH_3)COO\text{-}CH_2CH{-}CH_2$ (O)	142.15		1.042	1.4494^{20}		189	83	
e13	1,2-Epoxy-3,3,3-tri-chloropropane	$CH_2CH{-}CH_2$ (O)	161.42	$17^2, 14$	1.495	1.4778^{20}		151^{745mm}	66	
e14	Ethane	CH_3CH_3	30.07	1, 80	$0.5462,^{-88}$ $1.0493_4^0,$ g·L		-183.3	-88.6		4.7 mL aq; 46 mL alc^4
e15	1,2-Ethanediamine	$H_2NCH_2CH_2NH_2$	60.10	4, 230	0.8977^{20}	1.4568^{20}	8.5	117.3	33	misc aq, alc; i bz
e16	1,2-Ethanediol	$HOCH_2CH_2OH$	62.07	1, 465	1.11354^{20}	1.4318^{20}	-12.6	197.3	110	misc aq, alc, glyc, pyr
e17	1,2-Ethanediol diacetate	$CH_3COOCH_2CH_2OOCCH_3$	146.14	2, 142	1.1043^{20}	1.4150^{20}	-31	190.2	82	misc alc, eth
e18	1,2-Ethanedithiol	$HSCH_2CH_2SH$	94.20	1, 471	1.123^{24}	1.5580^{20}		146	50	v s alc, alk
e19	Ethanesulfonyl chloride	$CH_3CH_2SO_2Cl$	128.57	$4^2, 526$	1.357^{22}	1.4339^{20}		171		d aq, alc; v s eth
e20	Ethanethiol	CH_3CH_2SH	62.13	1, 340	0.8315^{25}	1.420^{25}	-147.9	35.0	-17	0.7 aq; s alc, eth
e21	Ethanol	CH_3CH_2OH	46.07	1, 292	0.7894_4^{20}	1.3614^{20}	-114.5	78.3	8	misc aq, alc, eth, chl
e22	Ethanol-*d*	CH_3CH_2OD	47.08	$1^3, 1287$	0.801	1.3595^{20}		78.8	12	misc aq, alc, eth
e23	Ethoxyacetic acid	$CH_3CH_2OCH_2COOH$	104.11	3, 233	1.1021^{20}	1.4190^{20}		97^{11mm}	97	s aq, alc, eth
e24	4-Ethoxyaniline	$CH_3CH_2OC_6H_4NH_2$	137.18	13, 436	1.06524^{20}	1.5609^{20}	4	250	115	i aq; s alc
e25	2-Ethoxybenzaldehyde	$CH_3CH_2OC_6H_4CHO$	150.18	8, 43	1.074	1.5422^{20}	20	136^{24mm}	107	misc alc, eth
e26	4-Ethoxybenzaldehyde	$CH_3CH_2OC_6H_4CHO$	150.18	8, 73	1.080_{25}^{25}	1.5584^{20}	13–14	255	>112	v s alc, bz, eth
e27	2-Ethoxybenzamide	$CH_3CH_2OC_6H_4CONH_2$	165.19	10, 93			132–133			sl s aq; s alc, eth
e28	Ethoxybenzene	$CH_3CH_2OC_6H_5$	122.17	6, 140	0.967_4^{20}	1.5074^{20}	-29.5	170.0		0.12 aq; misc alc, eth
e29	2-Ethoxybenzoic acid	$CH_3CH_2OC_6H_4COOH$	166.18	10, 64	1.105	1.5400^{20}	19.4	174^{15mm}	>112	sl s aq

	Name	Formula	MW	Ref	d	n_D	mp	bp		Solubility
e30	4-Ethoxybenzoic acid	$CH_3CH_2OC_6H_4COOH$	166.18	10, 156			197–199			sl s hot aq
e31	2-Ethoxybenzyl alcohol	$CH_3CH_2OC_6H_4CH_2OH$	152.19	6, 893		1.5321^{20}		265	50	
e32	Ethoxycarbonyl isothiocyanate	$CH_3CH_2OC(=O)NCS$	131.15	3^3, 279	1.112	1.5000^{20}		56^{18mm}		
e33	Ethoxydimethylvinyl-silane	$(CH_3)_2Si(OC_2H_5)CH=CH_2$	130.3		0.790_4^{20}	1.398^{20}		99^{710mm}		
e34	2-Ethoxyethanol	$CH_3CH_2OCH_2CH_2OH$	90.12	1, 467	0.9295^{20}	1.4075^{20}	−59	134.8	48	misc aq, alc, eth, acet
e35	2-(2-Ethoxyethoxy)-ethanol	$C_2H_5OCH_2CH_2OCH_2CH_2OH$	134.18	1^2, 520	0.9841_4^{25}	1.4254^{25}	−55	201.9	96	misc aq, alc, bz, chl, acet, pyr
e36	2-(2-Ethoxyethoxy)-ethyl acetate	$C_2H_5OCH_2CH_2OCH_2CH_2OOCCH_3$	176.21		1.0096^{20}	1.4213^{20}	−25	218.5	110	misc aq, alc, eth, oils
e37	2-Ethoxyethyl acetate	$CH_3COOCH_2CH_2OCH_2CH_3$	132.16	2^2, 155	0.9749_4^{20}	1.4023^{20}	−61.7	156.3	57	29 aq; misc alc, eth
e38	2-Ethoxyethylamine	$CH_3CH_2OCH_2CH_2CH_2NH_2$	89.14	4^2, 718	0.8512_4^{20}	1.4101^{20}		107	21	misc aq, alc, eth
e39	3-Ethoxy-4-hydroxy-benzaldehyde	$C_2H_5OC_6H_3(OH)CHO$	166.18	8, 256			76–78			s eth, glycols; 50 alc
e40	3-Ethoxymethacrolein	$C_2H_5OCH=C(CH_3)CHO$	114.15	1^4, 4082	0.960	1.4792^{20}		$78-81^{14mm}$	35	
e41	4-Ethoxy-3-methoxy-benzaldehyde	$C_2H_4OC_6H_3(OCH_3)CHO$	180.20	8, 256			59–60			s alc, bz, chl, eth
e42	Ethoxymethyldi-phenylsilane	$(C_6H_5)_2Si(CH_3)OC_2H_5$	242.4		1.018_4^{20}	1.544^{20}		$122^{0.3mm}$		
e43	Ethoxymethylene-malononitrile	$CH_3CH_2OCH=C(CN)_2$	122.13	3^1, 162			64–66	160^{12mm}		

TABLE 1.15 Physical Constants of Organic Compounds (*continued*)

No.	Name	Formula	Formula weight	Beilstein reference	Density	Refractive index	Melting point	Boiling point	Flash point	Solubility in 100 parts solvent
e44	1-Ethoxynaphthalene	$C_{10}H_7OCH_2CH_3$	172.23	6, 606	1.060_4^{20}	1.6040^{20}	5.5	280	>112	i aq; v s alc, eth
e45	N-(4-Ethoxyphenyl)-acetamide	$CH_3CH_2OC_6H_4NHCOCH_3$	179.21	13^2, 244			134–135			0.076 aq; 6.7 alc; 7.1 chl; 1.1 eth; s glyc
e46	*trans*-2-Ethoxy-5-(1-propenyl)phenol	$C_2H_5OC_6H_3(CH{=}CH{-}CH_3)OH$	178.23	6^2, 918						
e47	3-Ethoxypropylamine	$C_2H_5OCH_2CH_2CH_2NH_2$	103.17	4^3, 739	0.861	1.4178^{20}	86–88	136–138	32	
e48	3-Ethoxysalicyl-aldehyde	$C_2H_5OC_6H_3(OH)CHO$	166.18	8^2, 267				263–264		
e49	2-Ethoxytetrahydro-furan	$C_2H_5O(C_4H_7O)$	116.16	17^4, 1020	0.908	1.4140^{20}		170–172	16	
e50	Ethoxytrimethyl-silane	$(CH_3)_3SiOC_2H_5$	118.3		0.7573_4^3	1.3742^{20}				
e51	Ethyl acetate	$CH_3COOC_2H_5$	88.11	2, 125	0.9006_4^{20}	1.3724^{20}	−84	77.1	−3	9.7 aq; misc alc, acet, chl, eth
e52	Ethyl acetimidate HCl	$CH_3C({=}NH)OC_2H_5 \cdot$ HCl	123.58	2, 182			112–114			
e53	Ethyl acetoacetate (enol)	$CH_3COCH{=}C(OH)OC_2H_5$	130.15	3, 632	1.0119^{10}	1.4480^{10}	−44	180.8	84	1.9 aq; misc alc, chl
e54	Ethyl acetoacetate (keto)	$CH_3COCH_2COOC_2H_5$	130.15	3, 632	1.0368^{10}	1.4224^{10}	−39	180.8	84	12 aq; misc alc, chl
e55	p-Ethylacetophenone	$C_2H_5C_6H_4COCH_3$	148.21	7^4, 1101	0.993	1.5293^{20}	−20.6	114^{11mm}	90	1.5 aq; s alc, eth
e56	Ethyl acrylate	$CH_2{=}CHCOOCH_2CH_3$	100.12	2, 399	0.9405_4^{20}	1.4068^{20}	−71.2	99.5	15	
e57	Ethylaluminum dichloride	$CH_3CH_2AlCl_2$	126.95		1.207^{50}		32	113^{50mm}		
e58	Ethylamine	$CH_3CH_2NH_2$	45.09	4, 87	0.689_5^{15}		−81.0	16.6	−17	misc aq, alc, eth
e59	Ethyl 2-amino-benzoate	$H_2NC_6H_4COOCH_2CH_3$	165.19	14, 319	1.088^{15}		14	266–268		i aq; s alc, eth

No.	Name	Formula	M.W.	Ref.	Density	n	m.p.	b.p.	Flash	Solubility
e60	Ethyl 4-aminobenzoate	$H_2NC_6H_4COOCH_2CH_3$	165.19	14, 422	1.021_4^{20}		88–90	310		0.04 aq; 20 alc; 50 chl; 25 eth; s dil acid
e61	Ethyl 3-aminocrotonate	$CH_3C(NH_2)=CH-COOCH_2CH_3$	129.16	3, 654	0.914_4^{20}		33–35	210–215	71	i aq; s alc, bz, eth
e62	2-(Ethylamino)-ethanol	$CH_3CH_2NHCH_2CH_2OH$	89.14	4, 282		1.4402^{20}	–90	170		v s aq, alc, eth
e63	N-Ethylaniline	$C_6H_5NHCH_2CH_3$	121.18	12, 159	0.958_{25}^{25}	1.5559^{20}	–63	204.5	85	i aq; misc alc, eth
e64	2-Ethylaniline	$CH_3CH_2C_6H_4NH_2$	121.18	12[2], 584	0.983_4^{22}	1.5590^{20}	–44	210	97	sl s aq; v s alc, eth
e65	4-Ethylaniline	$CH_3CH_2C_6H_4NH_2$	121.18	12, 1090	0.975_4^{22}	1.5542^{20}	–5	216	85	sl s aq; v s alc, eth
e66	2-Ethylanthraquinone		236.27	7[1], 425			108–111			
e67	Ethylbenzene-d_{10}	$C_6D_5CD_2CD_3$	116.25		0.8670_4^{20}	1.4920^{20}		134.6	31	0.01 aq; misc alc, bz, chl, eth
e68	Ethylbenzene	$C_6H_5CH_2CH_3$	106.17	5[2], 274		1.4959^{20}	–95.0	136.2	20	
e69	Ethyl benzoate	$C_6H_5COOCH_2CH_3$	150.18	9, 110	1.0502_5^{25}	1.5052^{20}	–34.7	212.4	84	0.05 aq; misc alc, chl, bz, eth, PE
e70	Ethyl benzoylacetate	$C_6H_5C(=O)CH_2-COOCH_2CH_3$	192.21	10, 674	1.110	1.5338^{20}		265 d	140	i aq; misc alc, eth
e71	Ethyl 2-benzylacetoacetate	$CH_3COCH(CH_2C_6H_5)-COOC_2H_5$	220.27	10, 674	1.036	1.4996^{20}		276	>112	
e72	N-Ethylbenzylamine	$C_6H_5CH_2NHC_2H_5$	135.21	12, 1020	0.909	1.5117^{20}		194	66	
e73	Ethyl (2-benzyl)-benzoylacetate	$C_6H_5COCH(CH_2C_6H_5)-COOC_2H_5$	282.34	10, 764	1.110	1.5567^{20}		270^{80mm}	>112	

C_2H_5

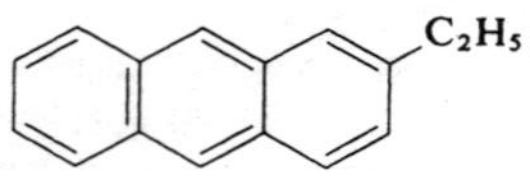

e66

TABLE 1.15 Physical Constants of Organic Compounds (*continued*)

No.	Name	Formula	Formula weight	Beilstein reference	Density	Refractive index	Melting point	Boiling point	Flash point	Solubility in 100 parts solvent
e74	Ethyl N-benzyl-N-cyclopropyl-carbamate	$C_6H_5CH_2N(C_3H_5)$-$COOCH_2CH_3$	219.28		0.997	1.5104^{20}			>112	
e75	Ethyl bromoacetate	$BrCH_2COOCH_2CH_3$	167.01	2, 214	1.506_{20}^{20}	1.4510^{20}		159	47	i aq; misc alc, eth
e76	Ethyl 2-bromo-butyrate	$CH_3CH_2CH(Br)$-$COOCH_2CH_3$	195.06	2^2, 255	1.329_{20}^{20}	1.4470^{20}		177 d	58	i aq; misc alc, eth
e77	Ethyl 4-bromo-butyrate	$BrCH_2CH_2CH_2$-$COOCH_2CH_3$	195.06	2, 283	1.363	1.4559^{20}		82^{10mm}	90	i aq; misc alc, eth
e78	Ethyl 2-bromoiso-butyrate	$(CH_3)_2C(Br)$-$COOCH_2CH_3$	195.06	2, 296	1.329_4^{20}	1.4446^{20}		67^{11mm}	60	i aq; misc alc, eth
e79	Ethyl 3-bromo-2-oxo-butyrate	$BrCH_2C(=O)$-$COOCH_2CH_3$	195.02	3^3, 409	1.554	1.4695^{20}		100^{10mm}	98	i aq; misc alc, eth
e80	Ethyl 2-bromo-pentanoate	$CH_3(CH_2)_2CH(Br)$-$COOCH_2CH_3$	209.09	2, 302	1.226	1.4486^{20}		190–192	77	i aq; misc alc, eth
e81	Ethyl 2-bromopro-pionate	$CH_3CH(Br)COOCH_2CH_3$	181.03	2, 255	1.447_{20}^{20}	1.4470^{20}	159–160	51		i aq; misc alc, eth
e82	Ethyl 3-bromopro-pionate	$BrCH_2CH_2COOCH_2CH_3$	181.03	2, 256	1.4123_4^{18}	1.4569^{18}		136^{50mm}	79	i aq; misc alc, eth
e83	2-Ethyl-1-butanol	$(C_2H_5)_2CHCH_2OH$	102.18	1, 412	0.8330^{20}	1.4224^{20}	−114.4	146.5	58	0.63 aq
e84	2-Ethyl-1-butene	$(C_2H_5)_2C=CH_2$	84.16	1^3, 814	0.6696_4^{20}	1.3967^{20}	−131.5	64.7		
e85	N-Ethylbutylamine	$CH_3(CH_2)_3NHCH_2CH_3$	101.19	4, 157	0.740_4^{20}	1.4050^{20}		108	18	s aq; alc, acet, eth
e86	2-Ethylbutylamine	$(C_2H_5)_2CHCH_2NH_2$	101.19	4, 192	0.776_{20}^{20}	1.4018^{20}	−89	121–125	21	0.31 aq
e87	2-Ethylbutyraldehyde	$(C_2H_5)_2CHCHO$	100.16	1, 693	0.8162_{20}^{20}	1.3928^{20}	−98.0	116.7	21	0.49 aq; misc alc, eth
e88	Ethyl butyrate	$CH_3CH_2CH_2COOCH_2CH_3$	116.16	2, 270	0.879_4^{20}			121.6	29	
e89	2-Ethylbutyric acid	$(C_2H_5)_2CHCOOH$	116.16	2, 333	0.9225_{20}^{20}	1.4133^{20}	−15	194.2	99	0.22 aq
e90	Ethyl butyrylacetate	$CH_3(CH_2)_2C(O)CH_2$-$COOC_2H_5$	158.20	3, 684	1.001	1.4295^{20}		104^{22mm}	78	

No.	Name	Formula	Mol. wt.	Beilstein	Density	n_D	mp	bp		Solubility
e91	Ethyl carbamate	$H_2NCOOCH_2CH_3$	89.09	3, 22	1.056		49–50	182–184		200 aq; 125 alc; 111 chl; 67 eth
e92	Ethyl carbazate	$H_2NNHCOOCH_2CH_3$	104.11	3, 98			44–47	110^{22mm}		
e93	N-Ethylcarbazole	(structure; N–C_2H_5)	195.27	20, 436			66–68			
e94	Ethyl chloroacetate	$ClCH_2COOCH_2CH_3$	122.55	2, 197	1.1498_4^{20}	1.4227^{20}		144–146	65	i aq; misc alc, eth
e95	Ethyl 2-chloroaceto-acetate	$CH_3C(=O)CH(Cl)-COOC_2H_5$	164.59	3, 662	1.190	1.4430^{20}	−26	107^{14mm}	50	i aq; s alc, eth
e96	Ethyl 4-chloroaceto-acetate	$ClCH_2C(=O)CH_2-COOC_2H_5$	164.59	3, 663	1.218_4^{17}	1.4520^{20}		115^{14mm}	96	i aq; misc alc, eth
e97	Ethyl 4-chloro-butyrate	$ClCH_2CH_2CH_2COOC_2H_5$	150.61	2, 278	1.0754_4^{20}	1.4306^{20}		186	51	s alc, acet, eth
e98	Ethyl chloroformate	$ClCOOC_2H_5$	108.52	3, 10	1.1403_4^{20}	1.3941^{20}	−81	95	2	misc alc, bz, chl, eth
e99	Ethyl 2-chloropropionate	$CH_3CH(Cl)COOC_2H_5$	136.58	2, 248	1.086_4^{20}	1.4185^{20}		147–148	38	i aq; misc alc, eth
e100	Ethyl 3-chloropropionate	$ClCH_2CH_2COOC_2H_5$	136.58	2, 250	1.1086_4^{20}	1.4249^{20}		163	54	misc alc, eth
e101	Ethyl chlorothio-formate	$ClC(=O)SCH_2CH_3$	124.59	3, 134	1.195	1.4820^{20}		132	30	
e102	Ethyl chrysanthemumate	(structure)	196.29	9^2, 45	0.906	1.4600^{20}		112^{10mm}		

e102 structure: CH_3, CH_3, $CH=C(CH_3)_2$, $C(=O)-OC_2H_5$

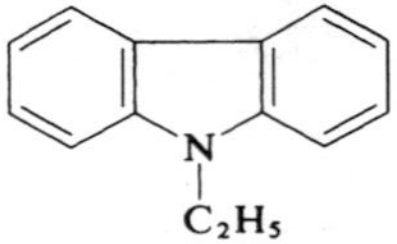

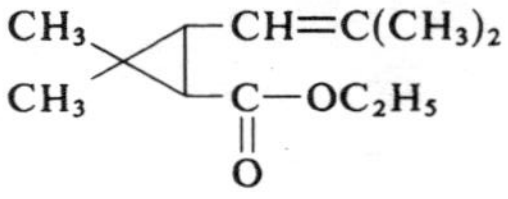

TABLE 1.15 Physical Constants of Organic Compounds (*continued*)

No.	Name	Formula	Formula weight	Beilstein reference	Density	Refractive index	Melting point	Boiling point	Flash point	Solubility in 100 parts solvent
e103	Ethyl *trans*-cinnamate	$C_6H_5CH=CHCOOCH_2CH_3$	176.22	9^2, 385	1.0495_4^{20}	1.5598^{20}	12	271.0		misc alc, eth; i aq
e104	Ethyl crotonate	$CH_3CH=CHCOOCH_2CH_3$	114.14	2, 411	0.9175_4^{20}	1.4248^{20}		138	2	i aq; s alc, eth
e105	Ethyl cyanoacetate	$NCCH_2COOCH_2CH_3$	113.12	2, 585	1.0564_4^{25}	1.4156^{20}	−22.5	206.0	110	i aq; misc alc, eth
e106	Ethyl cyanoformate	$NCCOOCH_2CH_3$	99.09	2, 547	1.003_4^{20}	1.3820^{20}		116	24	
e107	Ethyl cyano(hydroxy-imino)acetate	$NCC(=NOH)COOCH_2CH_3$	142.12	3, 775			130–132			
e108	Ethylcyclohexane	$C_6H_{11}CH_2CH_3$	112.22	5, 35	0.7879_4^{20}	1.4330^{20}	−111.3	131.8	18	
e109	*cis*-2-Ethylcyclo-hexanol	$CH_3CH_2C_6H_{10}OH$	128.22	6^2, 26	0.9274^{21}	1.4646^{20}		$74–79^{12mm}$	68	i aq
e110	4-Ethylcyclohexanol	$CH_3CH_2C_6H_{10}OH$	128.22	6^2, 26	0.889	1.4625^{20}		84^{10mm}	77	
e112	Ethyl cyclohexyl-acetate	$C_6H_{11}CH_2COOCH_2CH_3$	170.25	9, 14	0.948	1.4439^{20}		212	80	
e112	Ethyl cyclopropane-carboxylate	$C_3H_5COOCH_2CH_3$	114.14	9, 4	0.960	1.4197^{20}		129–133	18	
e113	Ethyl decanoate	$CH_3(CH_2)_8COOCH_2CH_3$	200.32	2, 356	0.862^{20}	1.4248^{20}		245	102	misc alc, chl, eth
e114	Ethyl diazoacetate	$N_2CH_2COOCH_2CH_3$	114.10	3^1, 211	1.0852_4^{18}	1.4588^{18}	−22	141^{720mm}	26	misc alc, bz, eth
								explodes when heated		
e115	Ethyl 2,3-dibromo-propionate	$BrCH_2CH(Br)COO-CH_2CH_3$	259.94	2, 259	1.7884_4^{16}	1.4986^{20}		214	91	s alc, eth
e116	Ethyl dichloro-phosphate	$CH_3CH_2OP(O)Cl_2$	162.94	1, 332	1.373	1.4338^{20}		65^{10mm}		
e117	Ethyl dichloro-thiophosphate	$CH_3CH_2OP(S)Cl_2$	179.01	1, 333	1.353	1.5040^{20}		$55–68^{10mm}$		
e118	*N*-Ethyldiethanol-amine	$CH_3CH_2N(CH_2CH_2OH)_2$	133.19	4, 284	1.014	1.4665^{20}	−50	246–252	123	
e119	Ethyl diethoxy-phosphinylformate	$(C_2H_5O)_2P(O)COOC_2H_5$	210.17	3^2, 103	1.110	1.4230^{20}		135^{13mm}		

No.	Name	Formula								Solubility
e120	Ethyl 3-(diethylamino)propionate	(C₂H₅)₂NCH₂CH₂-COOC₂H₅	173.26	4,404	0.881	1.4253^{20}		84^{12mm}	7	s alc, eth
e121	Ethyl 3,3-dimethylacrylate	(CH₃)₂C=CHCOOC₂H₅	128.17	2,433	0.9247^{20}_4	1.4350^{20}		155	33	
e122	Ethyl 2-dimethylaminobenzoate	(CH₃)₂NC₆H₄COOC₂H₅	193.25		1.061	1.5425^{20}			98	
e123	Ethyl 2,2-dimethylpropionate	(CH₃)₃CCOOCH₂CH₃	130.19	2², 280	0.8584^{18}_4	1.3922^{18}		118.2		s alc, eth
e124	Ethylene carbonate		88.06	19, 100	1.3208^{25}	1.4199^{40}	36.4	238	160	misc aq
e125	Ethylenediamine-N,N,N',N'-tetraacetic acid	(HOOCCH₂)₂NCH₂CH₂-N(CH₂COOH)₂	292.24				245 d			0.05 aq
e126	Ethylene dinitrate	O₂NOCH₂CH₂ONO₂	152.07		1.496^{15}_{15}	1.499^{15}	−22	106^{19mm}		misc aq, alc, bz
e127	2,2'-(Ethylenedioxy)-bisethanol	HOCH₂CH₂OCH₂CH₂-OCH₂CH₂OH	150.17		1.1274^{15}_4	1.4578^{15}	−72	285	166	
e128	Ethylene glycol	HOCH₂CH₂OH	62.07	1,465	1.1135^{20}_4	1.4319^{20}	−13	197.6	110	misc aq, alc, acet, glc, HOAc, pyr; sl s eth; i bz, chl
e129	Ethylene oxide	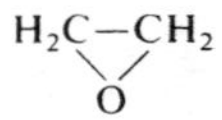 H₂C—CH₂ (O)	44.05		0.891^{0}_4	1.3597^{7}	−112.44	10.6	−18	misc aq; s alc, eth

H₂C—O
      C=O
H₂C—O

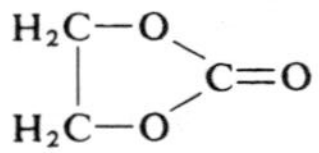

e124

TABLE 1.15 Physical Constants of Organic Compounds (*continued*)

No.	Name	Formula	Formula weight	Beilstein reference	Density	Refractive index	Melting point	Boiling point	Flash point	Solubility in 100 parts solvent
e130	Ethylene sulfide	$H_2C{-}CH_2$ with S	60.12	17^2, 12	1.010	1.4935^{20}		55–56	10	sl s alc, eth
e131	Ethylenimine	$H_2C{-}CH_2$ NH	43.07	3, 470	0.8321^{25}_4	1.4123^{25}	−78.0	56	−24	misc aq; sl s alc
e132	Ethyl (ethoxymethylene)cyanoacetate	$C_2H_5OCH{=}C(CN)\text{-}COOC_2H_5$	169.18				51–53	190^{30mm}		
e133	Ethyl fluoroacetate	$FCH_2COOC_2H_5$	106.10	2, 193	1.0926^{21}	1.3755^{20}		119^{753mm}	30	s aq
e134	Ethyl fluorosulfonate	$FSO_2OC_2H_5$	128.12					$23{-}25^{12mm}$	32	
e135	Ethyl formate	$HCOOC_2H_5$	74.08	2, 19	0.917^{20}_4	1.3599^{20}	−79.4	54.2	−28	12 aq; misc alc, eth
e136	Ethyl 2-furoate		140.14	18, 275	1.117^{20}_4	1.3030^{20}	33–36	196	70	i aq; s alc, eth
e137	Ethyl heptafluorobutyrate	$CF_3CF_2CF_2COOC_2H_5$	242.09		1.394^{20}			94–96		
e138	Ethyl heptanoate	$CH_3(CH_2)_5COOC_2H_5$	158.24	2^2, 295	0.8685^{20}_4	1.4144^{15}	−66	187		s alc, eth
e139	Ethyl hexadecanoate	$CH_3(CH_2)_{14}COOC_2H_5$	284.48	2^2, 336	0.8577^{25}_4	1.4347^{34}	22	191^{10mm}		s alc, eth
e140	3-Ethylhexane	$CH_3CH_2CH_2CH(C_2H_5)_2$	114.24	1^3, 478	0.7136^{20}_4	1.4016^{20}		118.5		sl s alc; s eth
e141	2-Ethyl-1,3-hexanediol	$C_3H_7CH(OH)CH\text{-}(C_2H_5)CH_2OH$	146.23			0.9325^{22}_2	1.4530^{22}	−40	244.2	129	0.6 aq; s alc
e142	Ethyl hexanoate	$CH_3(CH_2)_4COOC_2H_5$	144.21	2, 323	0.871^{20}_4	1.4075^{20}	−67	166–168	49	i aq; misc alc, eth
e143	2-Ethylhexanoic acid	$CH_3(CH_2)_3CH(C_2H_5)\text{-}COOH$	144.21	2, 349	0.9077^{20}_{20}	1.4241^{20}	−118.4	227.6	127	0.25 aq
e144	2-Ethyl-1-hexanol	$CH_3(CH_2)_3CH(C_2H_5)\text{-}CH_2OH$	130.23		0.9344^{20}_{20}	1.4231^{20}	−76	184.3	77	0.07 aq; s alc, bz, chl
e145	2-Ethylhexanoyl chloride	$CH_3(CH_2)_3CH(C_2H_5)\text{-}COCl$	162.66		0.939	1.4335^{20}		68^{11mm}	69	
e146	2-Ethylhexyl acetate	$CH_3(CH_2)_3CH(C_2H_5)\text{-}CH_2OOCCH_3$	172.27	2^2, 304	0.8718^{20}_{20}	1.4204^{20}	−93	198.6	82	0.03 aq; misc alc

			FW				mp	bp		Solubility
e147	2-Ethylhexylamine	$CH_3(CH_2)_3CH(C_2H_5)CH_2NH_2$	129.31		0.792^{20}_{20}	1.4273^{20}	<100 glass	165–169	57	i aq; s alc, acet, eth
e148	2-Ethylhexyl vinyl ether	$CH_3(CH_2)_3CH(C_2H_5)CH_2OCH{=}CH_2$	156.26		0.8102^{20}_{20}	1.4387^{20}		177.7		0.01 aq
e149	Ethyl hydrogen hexanedioate	$HOOC(CH_2)_4COOC_2H_5$	174.20	2^1, 277			28–29	180^{18mm}	>112	v s alc, eth
e150	Ethyl hydroxyacetate	$HOCH_2COOC_2H_5$	104.11	3, 236	1.0874^{15}_{4}			160		
e151	Ethyl 4-hydroxybenzoate	$HOC_6H_4COOC_2H_5$	166.18	10, 159			116	297 d		0.07 aq; v s alc, eth
e152	Ethyl 3-hydroxybutyrate	$CH_3CH(OH)CH_2COOC_2H_5$	132.16	3, 309	1.0172^{20}_{4}	1.4205^{20}		170	64	s aq, alc
e153	Ethyl 2-hydroxyethyl sulfide	$HOCH_2CH_2SCH_2CH_3$	106.19	1^2, 525	1.0200^{20}_{20}	1.4869^{20}		184.5		s eth
e154	Ethyl 2-hydroxyisobutyrate	$(CH_3)_2C(OH)COOC_2H_5$	132.16	3, 315	0.965	1.4078^{20}		150	44	d hot aq
e155	Ethyl 4-hydroxy-3-methoxyphenylacetate	$HOC_6H_3(OCH_3)CH_2COOC_2H_5$	210.23	10^1, 198			44–47	$180\text{–}185^{14mm}$		

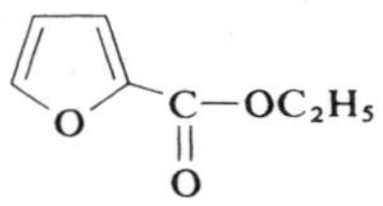

e136

TABLE 1.15 Physical Constants of Organic Compounds (*continued*)

No.	Name	Formula	Formula weight	Beilstein reference	Density	Refractive index	Melting point	Boiling point	Flash point	Solubility in 100 parts solvent
e156	2-Ethyl-2-(hydroxy-methyl)-1,3-propanediol	$CH_3CH_2C(CH_2OH)_3$	134.18	1^3, 2349			60–62	$159–161^{2mm}$		
e157	N-Ethyl-3-hydroxy-piperidine		129.20		0.970	1.4754^{20}		$93–95^{15mm}$	47	
e158	5-Ethylidene-2-norbornene		120.20		0.893	1.4895			38	
e159	2-Ethylimidazole		96.13				79–81			
e160	2-Ethyl-6-isopropyl-aniline	$(CH_3)_2CHC_6H_3-(C_2H_5)NH_2$	163.26		0.949			249		
e161	Ethyl isothiocyanate	CH_3CH_2NCS	87.14	4, 123	1.003_4^{18}	1.5142^{18}	−6	130–132	32	i aq; misc alc, eth
e162	Ethyl L-(+)-lactate	$CH_3CH(OH)COOC_2H_5$	118.13	3, 264	1.0328^{20}	1.4124^{20}	−26	154.5	70	misc aq, alc, eth, esters, PE
e163	Ethyl DL-mandelate	$C_6H_5CH(OH)COOC_2H_5$	180.21	10, 202			37	253–255		s alc, eth
e164	Ethyl 2-mercapto-acetate	$HSCH_2COOC_2H_5$	120.17	3, 255	1.0964^{15}	1.4571^{20}		54^{12mm}	47	
e165	Ethylmercury chloride	CH_3CH_2HgCl	165.13		3.5		192	subl		0.78 eth; 2.6 chl
e166	Ethyl methacrylate	$H_2C{=}C(CH_3)COOC_2H_5$	114.14	2, 423	0.909^{25}	1.4116^{25}		118	49	i aq; s alc, eth
e167	Ethyl 4-methoxy-phenylacetate	$CH_3OC_6H_4CH_2COOC_2H_5$	194.23	10^1, 83	1.097	1.5075^{20}		138^{7mm}	46	
e168	Ethyl 2-methylaceto-acetate	$CH_3C({=}O)CH(CH_3)-COOC_2H_5$	144.17	3, 679	1.019_4^{20}	1.4182^{20}		187	62	i aq; s alc, eth
e169	N-Ethyl-N-methyl-aniline	$C_6H_5N(CH_3)C_2H_5$	135.21	12, 162	0.9193_4^{55}	1.5474^{20}		203–205		i aq; misc alc, eth
e170	Ethyl 3-methyl-butyrate	$(CH_3)_2CHCH_2COOC_2H_5$	130.19	2^2, 275	0.868_{20}^{20}	1.3962^{20}	−99.3	134.7	26	0.2 aq; misc alc, bz
e171	Ethyl methyl ether	$CH_3CH_2OCH_3$	60.09	1, 314	0.725_0^0			10.8		s aq; misc alc, eth

No.	Name		Formula wt		Density	n_D	mp	bp		Solubility
e172	2-Ethyl-4-methyl-imidazole		110.16	23^2, 72	0.975	1.4995^{20}		292–295	137	i aq; sl s alc; s eth
e173	Ethyl 4-methyl-5-imidazolecarb-oxylate		154.17	25^1, 534			204–206			
e174	3-Ethyl-2-methyl-pentane	$(C_2H_5)_2CHCH(CH_3)_2$	114.24	1^3, 489	0.7193^{20}_4	1.4040^{20}	-115.0	115.7		
e175	3-Ethyl-3-methyl-pentane	$(C_2H_5)_3CCH_3$	114.24		0.7274^{20}_4	1.4078^{20}	-90.9	118.3		i aq; s eth
e176	Ethyl 3-methyl-3-phenylglycidate		206.24		1.09^{15}_4	1.508^{20}				
e177	Ethyl 1-methyl-2-piperidinecarb-oxylate		171.24	22^1, 485	0.975	1.4519^{20}		$92\text{–}96^{11mm}$	73	

Ethyl 2-hydroxypropionate, e162
Ethylidene bromide, d77
Ethylidene chloride, d176
Ethylidene dimethyl ether, d438
Ethylidene fluoride, d346
2,2'-Ethyliminodiethanol, e118
Ethyl iodide, i34

Ethyl isonicotinate, e216
Ethyl isonipecotate, e206
Ethyl isopropylacetate, e170
Ethyl isothiocyanatoformate, e32
Ethyl isovalerate, e170
Ethyl levulinate, e195
Ethyl linoleate, e188

Ethyl mercaptan, e20
Ethyl 3-methylcrotonate, e121
Ethyl methyl ketone, b393
Ethyl 1-methylnipecotate, e177
Ethyl 2-methyl-4-oxo-2-cyclohexene-1-carboxylate, c9
Ethyl 1-methylpipecolinate, e178

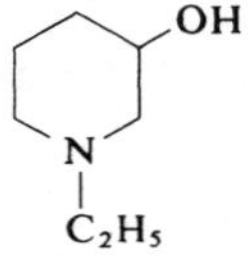

e157

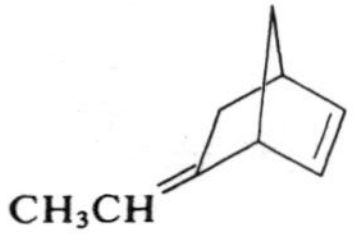

e158

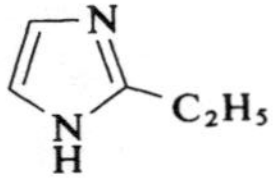

e159

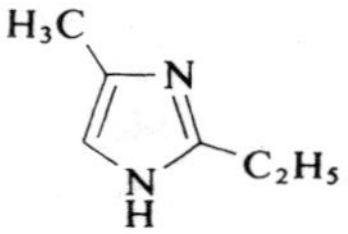

e172

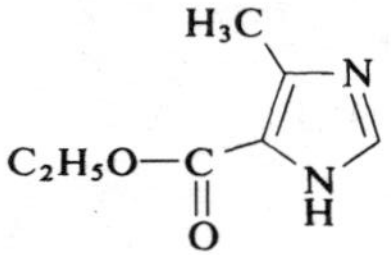

e173

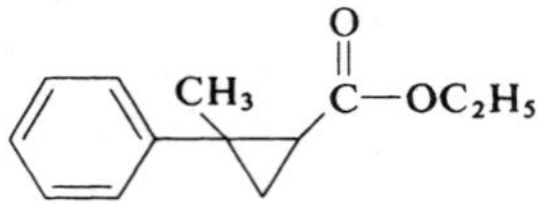

e176

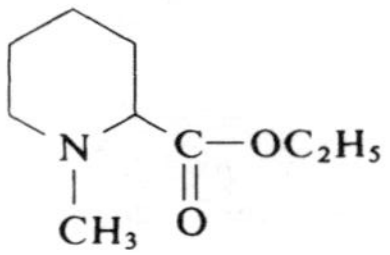

e177

TABLE 1.15 Physical Constants of Organic Compounds (*continued*)

No.	Name	Formula	Formula weight	Beilstein reference	Density	Refractive index	Melting point	Boiling point	Flash point	Solubility in 100 parts solvent
e178	Ethyl 1-methyl-3-piperidinecarboxylate		171.24		0.954	1.4510^{20}		89^{11mm}		
e179	2-Ethyl-2-methyl-1,3-propanediol	$HOCH_2C(C_2H_5)(CH_3)$-CH_2OH	118.18	1, 487			41–44	226		s alc, eth; sl s aq
e180	3-Ethyl-4-methylpyridine	$C_2H_5(CH_3)C_5H_3N$	121.18	20^2, 163	0.9286^{17}_{4}			198		s alc, bz, eth, acid
e181	5-Ethyl-2-methylpyridine	$C_2H_5(CH_3)C_5H_3N$	121.18	20, 248	0.9184^{23}_{4}	1.4974^{20}		178	66	i aq; misc alc, eth
e182	Ethyl methyl sulfide	$CH_3CH_2SCH_3$	76.15	1, 343	0.8422^{20}	1.4403^{20}	−105.9	66.7	49	
e183	Ethyl (methylthio)-acetate	$CH_3SCH_2COOC_2H_5$	134.20		1.043	1.4587^{20}			59	
e184	N-Ethylmorpholine		115.18	27^1, 203	0.9162^{20}_{20}	1.4410^{20}	−63	139	27	misc aq, alc, eth
e185	Ethyl nitrate	$CH_3CH_2ONO_2$	91.13	1, 329	1.100^{25}_{4}	1.3849^{22}	−94.6	87.7	flam-mable	1 aq; misc alc, eth
e186	Ethyl nitrite	CH_3CH_2ONO	75.07	1, 329	0.90^{15}_{1}			17		misc alc, eth
e187	4-Ethylnitrobenzene	$C_2H_5C_6H_4NO_2$	151.17	5, 358	1.118	1.5445^{20}	−32	245–246	>112	v s alc, eth
e188	Ethyl (Z,Z)-9,12-octadecadienoic acid	$H(CH_2)_5CH=CHCH_2$-$CH=CH(CH_2)_7COOC_2H_5$	308.51	2^2, 461	0.8846^{16}_{4}	1.4675^{20}		193^{6mm}	>112	misc DMF, oils
e189	Ethyl cis-9-octadecenoate	$CH_3(CH_2)_7CH=CH$-$(CH_2)_7COOC_2H_5$	310.52	2, 467	0.869^{20}_{4}	1.445^{25}	<−15	216^{15mm}		i aq; misc alc, eth
e190	Ethyl octanoate	$CH_3(CH_2)_6COOC_2H_5$	172.27	2, 348	0.878^{17}	1.4166^{20}	−47	206–208	75	i aq; misc alc, eth
e191	Ethyl oxalyl chloride	$CH_3CH_2OC(=O)COCl$	136.53	2, 541	1.2223^{20}_{4}	1.4164^{20}		135	41	d aq, alc; s bz, eth
e192	Ethyl oxamate	$CH_3CH_2OC(=O)CONH_2$	117.10	2, 544			114–116			s aq, eth; i bz
e193	2-Ethyl-2-oxazoline		99.13		0.982	1.4370^{20}	−62	128.4	29	

No.	Name	Formula	Formula wt	Beil. ref.	Density	n_D	mp, °C	bp, °C	Flash pt, °C	Solubility
e194	Ethyl 2-oxocyclopentanecarboxylate	$(O{=})C_5H_7COOC_2H_5$	156.18		1.054	1.4485^{20}		102^{11mm}	>112	v s aq; misc alc
e195	Ethyl 4-oxopentanoate	$CH_3C({=}O)CH_2CH_2COOC_2H_5$	144.17	3, 675	1.0112^{20}_{20}	1.4222^{20}		205–206	45	sl s aq; misc alc, eth
e196	Ethyl 2-oxopropionate	$CH_3C({=}O)COOC_2H_5$	116.12	3, 616	1.060^{16}_{4}	1.408^{16}		144		
e197	3-Ethylpentane	$(C_2H_5)_3CH$	100.20	1^3, 441	0.6982^{20}_{4}	1.3934^{20}	−118.6	93.5		i aq; s alc, eth
e198	Ethyl pentanoate	$CH_3(CH_2)_3COOC_2H_5$	130.19	2, 301	0.8774^{20}_{20}	1.3732^{20}	−91.3	145.5		0.2 aq; misc alc, eth
e199	4-Ethylphenol	$CH_3CH_2C_6H_4OH$	122.17	6, 472	1.0111^{25}_{4}	1.5239	47.0	218–219	77	i aq; misc alc, eth
e200	Ethyl phenylacetate	$C_6H_5CH_2COOC_2H_5$	164.20	9, 434	1.0333^{20}_{4}	1.4980^{20}		226		i aq; misc alc, eth
e201	Ethyl N-piperazinocarboxylate		158.20	23^2, 9	1.080	1.4765^{20}		273	>112	
e202	1-Ethylpiperidine		113.20	20, 17	0.8237^{20}_{4}	1.4440^{20}		131	18	
e203	2-Ethylpiperidine		113.20	20, 104	0.850	1.4510^{20}		143	31	
e204	Ethyl 2-piperidinecarboxylate		157.21	22, 7	1.006	1.4562^{20}		216–217	46	s aq
e205	Ethyl 3-piperidinecarboxylate		157.21		1.012	1.4601^{20}		104^{7mm}	90	

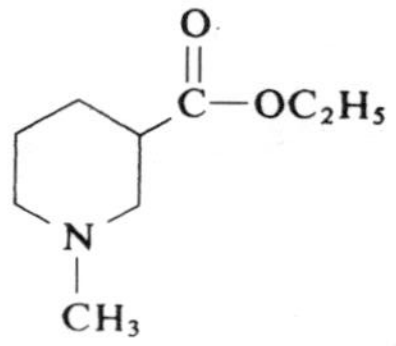

e178 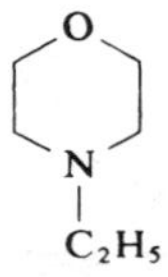 e184 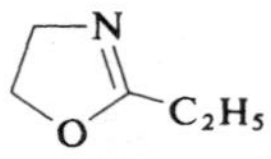 e193 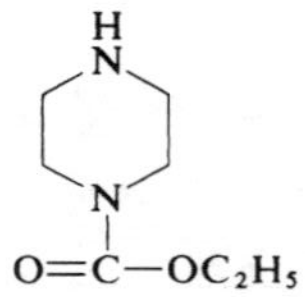 e201 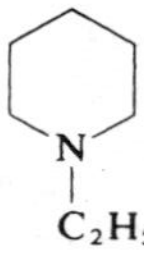 e202 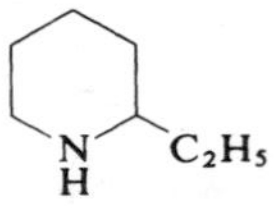 e203 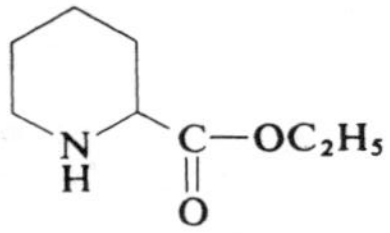 e204 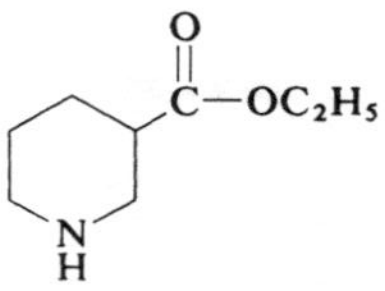 e205

TABLE 1.15 Physical Constants of Organic Compounds (*continued*)

No.	Name	Formula	Formula weight	Beilstein reference	Density	Refractive index	Melting point	Boiling point	Flash point	Solubility in 100 parts solvent
e206	Ethyl 4-piperidine-carboxylate		157.21		1.010	1.4591^{20}		204	80	s aq, alc, bz, eth
e207	Ethyl N-piperidine-propionate		185.27	20, 62	0.927	1.4545^{20}		217–219	87	sl s aq; misc alc, eth
e208	Ethyl propionate	$CH_3CH_2COOC_2H_5$	102.13	2, 240	0.891^{20}_4	1.3839^{20}	−73.9	99.1	12	2 aq; misc alc, eth
e209	Ethyl propyl ether	$CH_3CH_2OCH_2CH_2CH_3$	88.15	1, 354	0.739^{20}_4	1.3695^{20}	−79	62–63	32	sl s aq; misc alc, eth
e210	Ethyl propyl sulfide	$CH_3CH_2SCH_2CH_2CH_3$	104.21	1^3, 1432	0.8270^{20}_4	1.4462^{20}	−117.0	118.5	29	s alc
e211	2-Ethylpyridine	$CH_3CH_2C_5H_4N$	107.16	20, 241	0.937	1.4964^{20}		149	48	sl s aq; s alc, eth
e212	3-Ethylpyridine	$CH_3CH_2C_5H_4N$	107.16	20, 242	0.954	1.5015^{20}		162–165	47	v s alc, eth; sl s aq
e213	4-Ethylpyridine	$CH_3CH_2C_5H_4N$	107.16	20, 243	0.9404^{22}_4	1.5009^{20}		168		sl s aq; s alc, eth
e214	Ethyl 2-pyridine-carboxylate		151.17	22, 35	1.1194^{20}	1.5088^{20}	2	240–241	107	misc aq, alc, eth
e215	Ethyl 3-pyridine-carboxylate		151.17	22, 39	1.1070^{20}	1.5040^{20}	8–9	23–224	93	v s aq, alc, eth; s bz
e216	Ethyl 4-pyridine-carboxylate		151.17	22^2, 37	1.009^{15}_4	1.5009^{20}	23	220	87	i aq; s alc, bz, chl
e217	1-Ethyl-2-pyrrol-idinone		113.16		0.992	1.4652^{20}		97^{20mm}	76	misc alc, eth; sl s aq
e218	Ethyl salicylate	$C_6H_4(OH)COOC_2H_5$	166.18	10, 73	1.1314^{20}	1.5219^{20}	2–3	231–234	107	
e219	Ethyl sorbate	$CH_3CH=CHCH=CH-COOC_2H_5$	140.18	2, 484	0.959	1.4942^{20}		195.5	69	misc alc, eth; sl s aq
e220	S-Ethyl thioacetate	$CH_3C(=O)SCH_2CH_3$	104.16	2, 232	0.976^{28}_4	1.4503^{28}		116–117		i aq; v s alc, eth
e221	3-Ethylthio-1,2-propanediol	$C_2H_5SCH_2CH(OH)CH_2OH$	136.21		1.095	1.5065^{20}			>112	
e222	Ethyl 4-toluene-sulfonate	$CH_3C_6H_4SO_2OC_2H_5$	200.26	11, 99	1.166^{45}_4	1.5110^{20}	33	173^{15mm}	157	i aq; s alc, eth
e223	N-Ethyl-m-toluidine	$CH_3C_6H_4NHC_2H_5$	135.21	12, 857	0.957	1.5451^{20}		221	89	i aq; s alc, eth

			Formula wt	Beilstein	Density	n_D	mp	bp	fp	Solubility
e224	6-Ethyl-o-toluidine	$CH_3CH_2C_6H_3(CH_3)NH_2$	135.21		0.968	1.5525^{20}	-33	231	89	i aq; s alc, eth
e225	2-(N-Ethyl-m-toluidino)ethanol	$CH_3C_6H_4N(C_2H_5)CH_2CH_2OH$	179.26		1.019	1.5540^{20}		115^{1mm}	65	
e226	Ethyl trichloroacetate	$Cl_3CCOOC_2H_5$	191.44	2, 209	1.3834^{20}	1.4447^{20}		168	-1	
e227	Ethyl trifluoroacetate	$F_3CCOOC_2H_5$	142.08	2^2, 186	1.194	1.3068^{20}		60-62	35	
e228	Ethyl (trimethylsilyl)acetate	$(CH_3)_3SiCH_2COOC_2H_5$	160.29		0.876	1.4153^{20}		156-159		
e229	Ethyl undecanoate	$CH_3(CH_2)_9COOC_2H_5$	214.35	2, 358	0.859	1.4280^{20}		105^{4mm}	>112	i aq; s org solv
e230	Ethylurea	$CH_3CH_2NHC(=O)NH_2$	88.11	4, 115	1.213^{18}		93-96		75	v s aq; 80 alc; i eth
e231	N-Ethylurethane	$CH_3CH_2NHCOOC_2H_5$	117.15	4, 114	0.9814^{20}	1.4211^{20}		85^{20mm}		63 aq
e232	Ethyl vinyl ether	$CH_3CH_2OCH{=}CH_2$	72.11	1, 433	0.7531^{20}	1.3754^{20}	-115.8	35.7	-17	0.9 aq
e233	N-Ethyl-2,3-xylidine	$(CH_3)_2C_6H_3NHC_2H_5$	149.24	12, 1101	0.917	1.5468^{20}		227-228	71	
e234	1-Ethynyl-1-cyclohexanol	$C_6H_{10}(C{\equiv}CH)OH$	124.18	6^2, 100	0.967^{20}		30-31	180	62	2.4 aq; misc alc, bz, acet, ketones, PE

e206

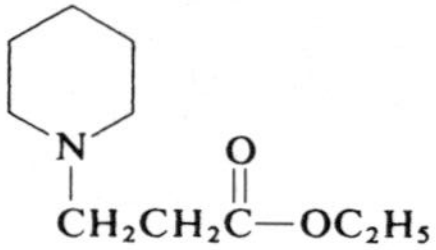

e207 $CH_2CH_2C{-}OC_2H_5$, O

e214 $C{-}OC_2H_5$, O, N

e215 $C{-}OC_2H_5$, O, N

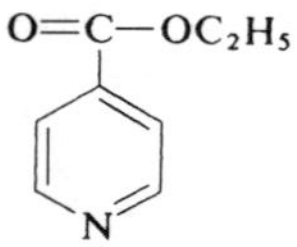

e216 $O{=}C{-}OC_2H_5$, N

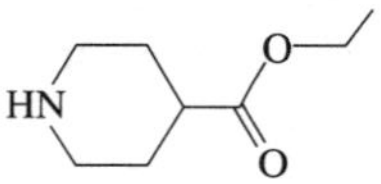

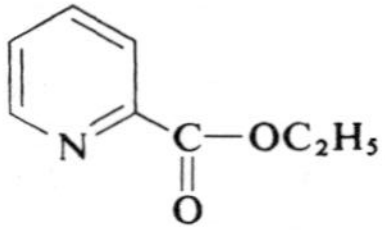

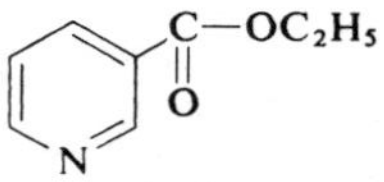

e217 O, $N{-}C_2H_5$

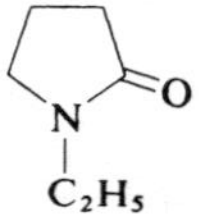

TABLE 1.15 Physical Constants of Organic Compounds (*continued*)

No.	Name	Formula	Formula weight	Beilstein reference	Density	Refractive index	Melting point	Boiling point	Flash point	Solubility in 100 parts solvent
e235	1-Ethynylcyclo-hexylamine	$C_6H_{10}(C{\equiv}CH)NH_2$	123.30		0.913	1.4817^{20}		66^{20mm}	42	sl s alc; s bz, eth
f1	Fluoranthene		202.26	5, 685	1.252^0		107–110	384		v s HOAc; s bz, eth
f2	Fluorene		166.22	5, 625	1.203^0		114.8	295		s alc, bz; v s eth
f3	9-Fluorenone		180.21	7, 465	1.1300^{99}_4	1.6369^{99}	82–85	342		s hot alc, hot HOAc, alk; i bz, chl, eth
f4	Fluorescein		332.31	19, 222			314 d			v s aq; s acet
f5	Fluoroacetamide	$FCH_2C(O)NH_2$	77.06	2, 193			107 subl			sl s aq, alc
f6	Fluoroacetic acid	FCH_2COOH	78.04	2, 193			33	165		
f7	Fluoroacetone	$CH_3C(O)CH_2F$	76.07		1.054	1.3700		75	7	
f8	p-Fluoroaceto-phenone	$FC_6H_4COCH_3$	138.14		1.138	1.5110^{20}		196	71	
f9	p-Fluoroaniline	$FC_6H_4NH_2$	111.12	12, 597	1.1725^{20}_4	1.5395^{20}	−1.9	187	73	sl s aq; s alc, eth
f10	o-Fluorobenzal-dehyde	FC_6H_4CHO	124.11	7[1], 132	1.178	1.5220^{20}	−44.5	91^{46mm}	55	
f11	Fluorobenzene	C_6H_5F	96.11	5, 198	1.0240^{20}_4	1.4657^{20}	−42.2	84.7	−12	0.15 aq; misc alc
f12	o-Fluorobenzoic acid	FC_6H_4COOH	140.11	9, 333	1.460^{25}_4		123–125			sl s aq; s alc, eth
f13	p-Fluorobenzoic acid	FC_6H_4COOH	140.11	9, 333	1.479^{25}_4		182.6			0.1 aq; s alc, eth
f14	p-Fluorobenzoyl chloride	FC_6H_4COCl	158.56	9[1], 137	1.342	1.5296^{20}	9	82^{20mm}	82	
f15	o-Fluorobenzyl alcohol	$FC_6H_4CH_2OH$	126.13	6[1], 222	1.173	1.5136^{20}			90	
f16	p-Fluorobenzyl chloride	$FC_6H_4CH_2Cl$	144.58		1.207	1.5130^{20}		82^{26mm}	60	
f17	Fluoroethane	CH_3CH_2F	48.06	1, 82	0.00220^0		−143.2	−37.7		198 mL aq; v s alc, eth

No.	Name	Formula	Formula weight	Beilstein reference	Density	n_D	Melting point, °C	Boiling point, °C	Flash point, °C	Solubility
f18	Fluoromethane	CH_3F	34.04	1, 59	1.1951 g·L		−141.8	−78.4		166 mL aq; v s alc, eth
f19	4-Fluoro-1-methoxybenzene	$FC_6H_4OCH_3$	126.13	6[1], 98	1.114	1.4877^{20}	−45	157	43	s eth
f20	2-Fluoro-2-methylpropane	$(CH_3)_3CF$	76.11	1[4], 286			−77	12.1	−12	
f21	1-Fluoro-4-nitrobenzene	$FC_6H_4NO_2$	141.10	5, 241	1.3300^{20}_4	1.5312^{20}	21	205	83	i aq; s alc, eth
f22	4-Fluorophenol	FC_6H_4OH	112.10	6, 183	1.128	1.4680^{20}	46–48	185	68	
f23	2-Fluoropyridine	FC_5H_4N	97.09	20[1], 80	1.0014^{17}	1.4716^{17}		126	28	
f24	o-Fluorotoluene	$FC_6H_4CH_3$	110.13	5, 290	0.9974^{20}	1.4691^{20}	−62.0	114.4	12	v s alc, eth
f25	m-Fluorotoluene	$FC_6H_4CH_3$	110.13	5, 290	0.9975^{20}	1.4688^{20}	−87.7	116.5	9	s alc, eth
f26	p-Fluorotoluene	$FC_6H_4CH_3$	110.13	5, 290			−56.7	116.6	40	s alc, eth
f27	Formaldehyde	$H_2C{=}O$	30.03	1, 558	0.8153^{-20}		−92	−19.5		
f28	Formamide	$HC(=O)NH_2$	45.04	2, 26	1.1334^{20}_4	1.4475^{20}	2.6	111^{20mm}	154	122 aq; s alc, eth
f29	Formamidine acetate	$HC(=NH)NH_2, HOOCCH_3$	104.11				158 d			misc aq, alc, acet
f30	Formamidinesulfinic acid	$H_2NC(=NH)S(O)OH$	108.12	3[1], 36			126 d			

Eugenol, m99
Fenchone, t351
Fenchyl alcohol, t350

Ferulic acid, h136
2,7-Fluorenediamine, d38
N-9H-2-(2-Fluorenyl)acetamide, a13

Fluorotrichloromethane, t232
Fluothane, b258
Formic acid hydrazide, f34

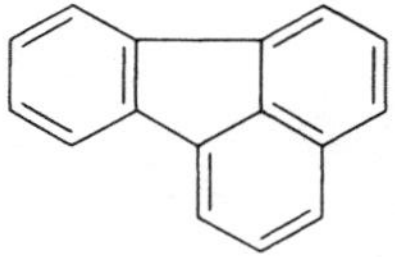

f1

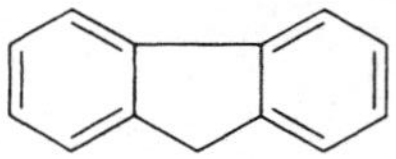

f2

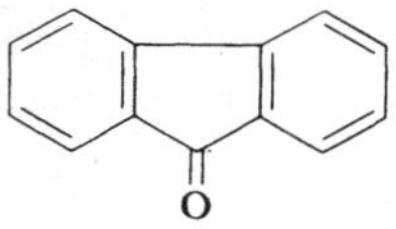

f3

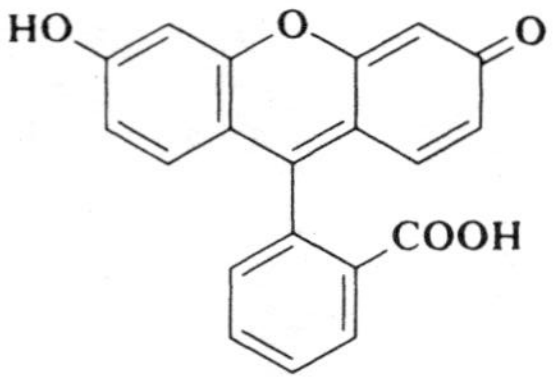

f4

TABLE 1.15 Physical Constants of Organic Compounds (*continued*)

No.	Name	Formula	Formula weight	Beilstein reference	Density	Refractive index	Melting point	Boiling point	Flash point	Solubility in 100 parts solvent
f31	Formanilide	C₆H₅NHCHO	121.14	12, 230	1.144		47	271		2.5 aq
f32	Formic acid	HCOOH	46.03	2, 8	1.220_4^{20}	1.3714^{20}	8.5	100.8	68	misc aq, alc, eth
f33	2-Formylbenzoic acid	C₆H₄(HCO)COOH	150.13	10, 666	1.404		98			s aq; v s alc, eth
f34	Formylhydrazine	HC(=O)NHNH₂	60.06	2, 93			54–56			v s alc, chl, eth; s bz
f35	N-Formylpiperidine		113.16	20, 45	1.019	1.4780^{20}		222	91	v s aq; 6.7 alc; s pyr
f36	D-(−)-Fructose		180.16	31, 321						
f37	Fumaric acid	HOOCCH=CHCOOH	116.07	2, 737	1.635_4^{20}		287	subl 200		0.6 aq; 9 alc; 0.7 eth
f38	Fumaroyl dichloride	ClC(=O)CH=CH-C(=O)Cl	152.96	2, 743	1.408^{20}	1.4988^{20}		161–164	73	d aq, alc
f39	2-Furaldehyde		96.09	17[2], 305	1.1598_4	1.5262^{20}	−36.5	161.8	68	8 aq; misc alc, eth
f40	Furan		68.07	17, 27	0.9371_4^{20}	1.4214^{20}	−85.6	31.4	−35	1 aq; misc alc, eth
f41	2-Furanacrylic acid		138.12	18, 300			141	286		0.2 aq; 1.1 bz; s alc, eth, HOAc
f42	2-Furancarboxylic acid		112.08	18, 272	1.132		133–134	230–232		4 aq; s alc; v s eth
f43	2,5-Furandimethanol		128.13	17[1], 90		1.5304^{20}	74–76	155	45	i aq; s alc, eth
f44	2-Furanmethanethiol		114.17	17[2], 116	1.1175_4	1.4618^{20}		175–177	65	misc aq(d); v s alc
f45	Furfuryl acetate		140.14	17[2], 115	1.1285_4	1.4868^{20}	−14.6	170.0	65	misc aq; s alc, eth
f46	Furfuryl alcohol		98.10	17, 112	1.0995_4	1.4900^{20}	−70	145–146	45	d aq; alc; s eth
f47	Furfurylamine		97.12	18, 584						
f48	2-Furoyl chloride		130.53	18, 276	1.324	1.5310^{20}	−2	170	85	200 aq; s pyr; sl s alc
g1	D-(+)-Galactose		180.16	31, 295			167			i aq; misc alc, eth
g2	Geraniol	(CH₃)₂C=CHCH₂CH₂-C(CH₃)=CHCH₂OH	154.25	1, 457	0.8894_4^{20}	1.4760^{20}		230	76	
g3	α-D-Glucoheptonic acid γ-lactone		208.17				145–148			s aq

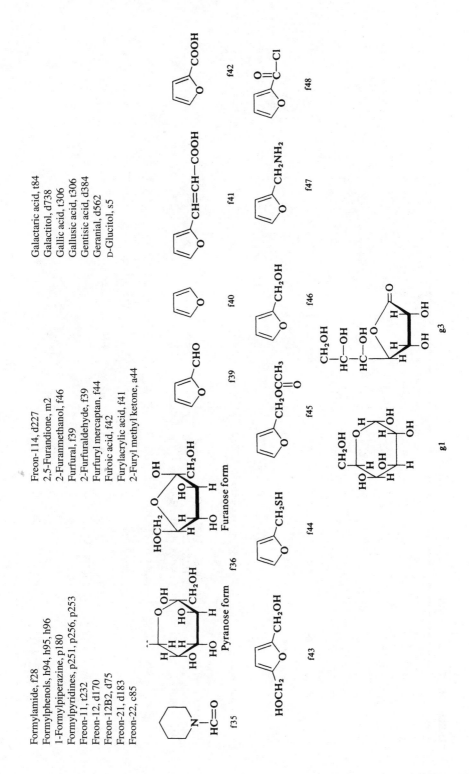

Formylamide, f28
Formylphenols, h94, h95, h96
1-Formylpiperazine, p180
Formylpyridines, p251, p256, p253
Freon-11, t232
Freon-12, d170
Freon-12B2, d75
Freon-21, d183
Freon-22, c85

Freon-114, d227
2,5-Furandione, m2
2-Furanmethanol, f46
Furfural, f39
2-Furfuraldehyde, f39
Furfuryl mercaptan, f44
Furoic acid, f42
Furylacrylic acid, f41
2-Furyl methyl ketone, a44

Galactaric acid, t84
Galactitol, d738
Gallic acid, t306
Gallusic acid, t306
Gentisic acid, d384
Geranial, d562
D-Glucitol, s5

TABLE 1.15 Physical Constants of Organic Compounds (*continued*)

No.	Name	Formula	Formula weight	Beilstein reference	Density	Refractive index	Melting point	Boiling point	Flash point	Solubility in 100 parts solvent
g4	D-Gluconic acid		196.16	3, 542			131			v s aq; sl s alc; i eth
g5	D-Glucosamine		179.17	1, 902			88(α)			v s aq; i chl, eth
g6	α-D-(+)-Glucose		180.16	31, 83	1.5620^{18}_4		146			91 aq; 0.83 MeOH; s pyr
g7	α-D-Glucose pentaacetate		390.34	31, 119			109–111			0.15 aq; 1.3 alc; 3 eth
g8	D-Glucurono-3,6-lactone		176.12				176–178			27 aq; 2.8 MeOH
g9	L-Glutamic acid		147.13	4, 488	1.538^{20}_4		d 247	subl 200		0.8 aq; i alc, eth
g10	L-Glutamine		146.15	4, 491			d 185			4 aq; 0.0035 MeOH; i bz, chl, eth, acet
g11	Glutaric acid	$HOOCCH_2CH_2CH_2COOH$	132.12	2, 631	1.429^{20}_4	1.4188^{106}	97.5	200^{20mm}		64 aq; v s alc, eth; s bz, chl; sl s PE
g12	Glutaric anhydride		114.10	17, 411			52–55			misc aq. alc
g13	Glutaric dialdehyde	$OCHCH_2CH_2CH_2CHO$	100.12	1, 776	1.3730^{20}		−6	150^{10mm}		s aq, alc, chl; i eth
g14	Glutaronitrile	$NCCH_2CH_2CH_2CN$	94.12	2, 635	0.9888^{23}	1.4345^{20}	−29	187–189 d	112	d aq, alc; s eth
g15	Glutaryl dichloride	$ClC(=O)CH_2CH_2CH_2C(=O)Cl$	169.01	2, 634	1.324	1.4720^{20}		216–218	106	
g16	Glycerol	$HOCH_2CH(OH)CH_2OH$	92.09	1, 502	1.2613^{20}	1.4746^{20}	18.18	182^{20mm}		misc aq. alc; 0.2 eth
g17	Glyceryl 1,2-diacetate	$HOCH_2CH(OOCCH_3)-CH_2OOCCH_3$	176.17	2, 147	1.184^{16}_4	1.1173^{15}	40	172^{40mm}		s aq, alc, bz, eth
g18	Glyceryl 1,3-diacetate	$CH_3COOCH_2CH(OH)-CH_2OOCCH_3$	176.17	2, 290	1.179^{15}	1.4395^{20}	42	172^{40mm}		s aq, alc, bz, chl
g19	Glyceryl tris-(butyrate)		302.37	2, 273	1.032^{20}_4	1.4359^{20}	−75	305–310	173	i aq; v s alc, eth
g20	Glyceryl tris-(dodecanoate)		639.02	2, 363	0.894^{60}_4	1.4404^{60}	46			v s bz, eth; sl s alc

g21	Glyceryl tris-(nitrate)	$O_2NOCH_2CH(ONO_2)$-CH_2ONO_2	227.09	1, 516	1.5944_4^{20}	1.4786^{12}	13.3	160^{5mm}	expl 270	0.18 aq; 54 alc; misc eth
g22	Glyceryl tris-(oleate)		885.46	4, 468	0.9154_4^{15}	1.4621^{40}	-4 to -5	235^{15mm}		s chl, eth, CCl_4
g23	Glyceryl tris-(palmitate)		807.35	2, 373	0.8663_4^{80}	1.4381^{80}	65-66	310-320		v s bz, chl, eth
g24	Glyceryl tris-(tetradecanoate)		723.18	2, 367	0.8854_4^{60}	1.4428^{60}	57			v s alc, bz, chl

Glycerol α-monochlorohydrin, c212
Glyceryl triacetate, p201
Glyceryl tris(laurate), g20

Glyceryl tris(myristate), g24

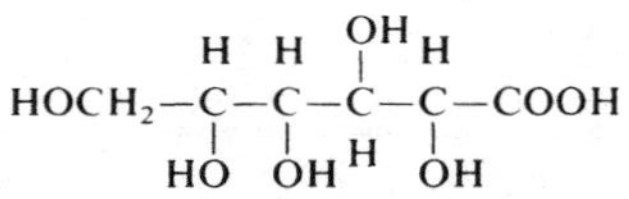

Glutaraldehyde, g13
Glyceraldehyde, d398
Glycerol dichlorohydrin, d220

g4, 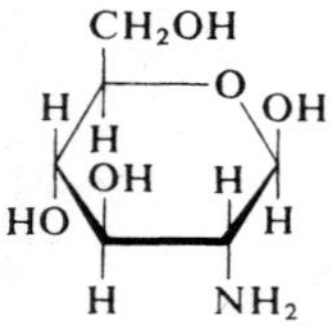 g5, 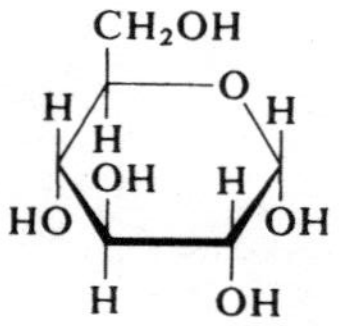 g6, 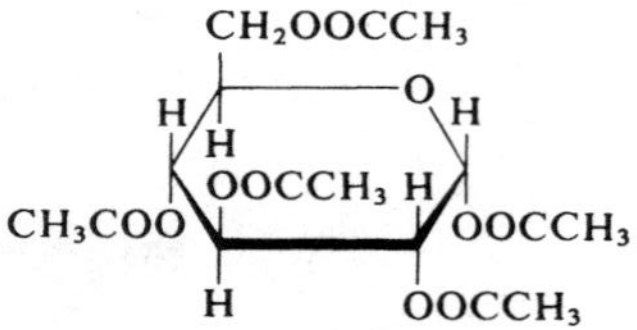 g7, 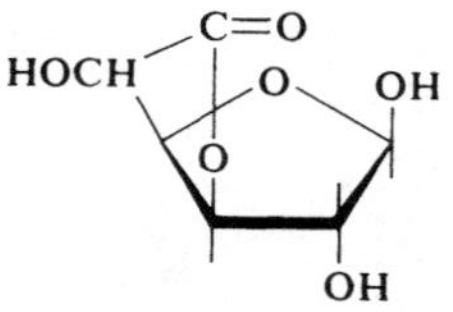 g8, 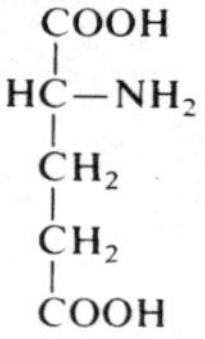 g9, 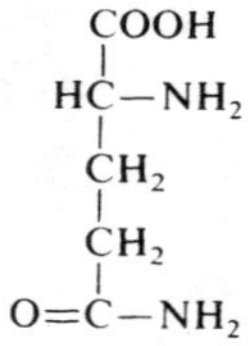 g10, 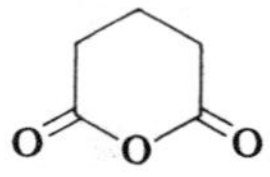 g12, g13, g19, 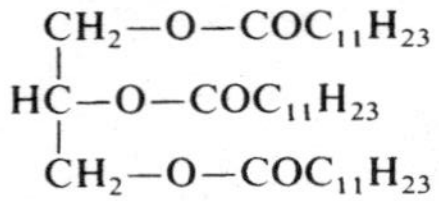 g20, 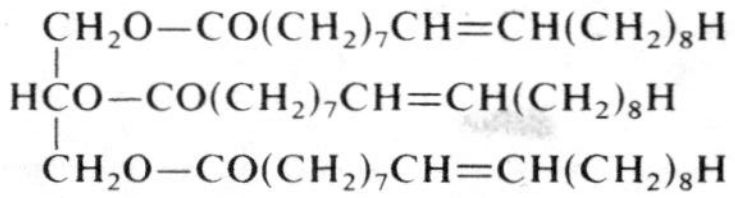 g22, 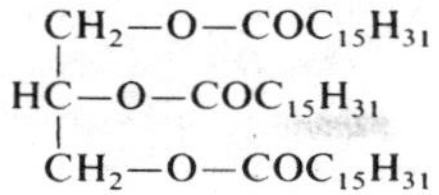 g23, 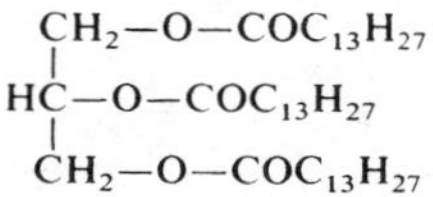 g24

g19
$CH_2-O-CO-C_3H_7$
$HC-O-CO-C_3H_7$
$CH_2-O-CO-C_3H_7$

g20
$CH_2-O-CO-C_{11}H_{23}$
$HC-O-CO-C_{11}H_{23}$
$CH_2-O-CO-C_{11}H_{23}$

g23
$CH_2-O-CO-C_{15}H_{31}$
$HC-O-CO-C_{15}H_{31}$
$CH_2-O-CO-C_{15}H_{31}$

g24
$CH_2-O-CO-C_{13}H_{27}$
$HC-O-CO-C_{13}H_{27}$
$CH_2-O-CO-C_{13}H_{27}$

g22
$CH_2O-CO(CH_2)_7CH=CH(CH_2)_8H$
$HCO-CO(CH_2)_7CH=CH(CH_2)_8H$
$CH_2O-CO(CH_2)_7CH=CH(CH_2)_8H$

g9
COOH
$HC-NH_2$
CH_2
CH_2
COOH

g10
COOH
$HC-NH_2$
CH_2
CH_2
$O=C-NH_2$

g4
$HOCH_2-C-C-C-C-COOH$ (with H, H, OH, H and HO, OH, H, OH substituents)

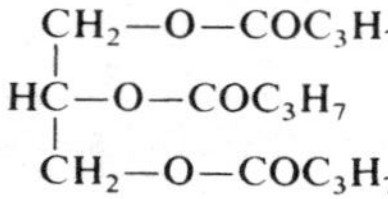

TABLE 1.15 Physical Constants of Organic Compounds (*continued*)

No.	Name	Formula	Formula weight	Beilstein reference	Density	Refractive index	Melting point	Boiling point	Flash point	Solubility in 100 parts solvent
g25	Glycine	H_2NCH_2COOH	75.07	4, 333	1.1607		d 233			25 aq; 0.6 pyr; i eth
g26	*N*-Glycylglycine	$H_2NCH_2C(=O)NHCH_2\text{-}COOH$	132.12	4, 371			d 262			s hot aq; sl s alc
g27	Glyoxal	$HC(=O)CHO$	58.04	1, 759	1.29^{20}_4	1.3826^{20}	15	51		violent reaction aq; s anhyd solv; mixtures with air may explode
g28	Glyoxylic acid	$HC(=O)COOH$	74.04	3, 594			98			v s aq; sl s alc, eth
g29	Guanidine	$H_2NC(=NH)NH_2$	59.07	3, 82			~60			v s aq, alc
h1	Heptadecane	$CH_3(CH_2)_{15}CH_3$	140.41	1, 173	0.7767^{22}	1.4360^{25}	22.0	302.2	148	s eth; sl s alc
h2	Heptafluorobutyric acid	$CF_3CF_2CF_2COOH$	214.04		1.645			120		
h3	Heptafluoro-2,3,3-trichlorobutane	$CF_3CCl_2CF(Cl)CF_3$	287.5		1.7484^{20}	1.3530^{20}	4	98		
h4	2,2,4,6,8,8-Heptamethylnonane	$(CH_3)_3CCH_2C(CH_3)_2CH_2\text{-}CH(CH_3)CH_2C(CH_3)_3$	226.45		0.793	1.4391^{20}		240		
h5	Heptanal	$CH_3(CH_2)_5CHO$	114.19	1^2, 750	0.8216^{15}_4	1.4285^{20}	-43	153	35	misc alc, eth; sl s aq
h6	Heptane	$CH_3(CH_2)_5CH_3$	100.21	1, 154	0.68384^{20}	1.3877^{20}	-90.6	98.4	-1	s alc, chl, eth
h7	1,7-Heptanediamine	$H_2N(CH_2)_7NH_2$	130.24	4, 271			27-29	147-149	87	
h8	Heptanedioic acid	$HOOC(CH_2)_5COOH$	160.17	2, 670	1.329^{15}		105.8	212^{10mm}	46	5 aq; v s alc, eth i aq
h9	1-Heptanethiol	$CH_3(CH_2)_6SH$	132.27	1, 415			-43.2	176.9	46	i aq
h10	Heptanoic acid	$CH_3(CH_2)_5COOH$	130.19	2, 338	0.9181^{20}_4	1.4221^{20}	-7.5	223.0	>112	0.25 aq; s alc, eth
h11	Heptanoic anhydride	$[CH_3(CH_2)_5CO]_2O$	242.36	2, 340	0.932^{20}_4	1.4332^{20}	-12.4	268	>112	i aq; s alc, eth
h12	1-Heptanol	$CH_3(CH_2)_6OH$	116.20	1, 414	0.8219^{20}_4	1.4242^{20}	-34.6	175.8	73	misc alc, eth
h13	2-Heptanol	$CH_3(CH_2)_4CH(OH)CH_3$	116.20	1, 415	0.8193^{20}_4	1.4210^{20}		160	41	0.35 aq; s alc, bz, eth

h14	3-Heptanol	$CH_3(CH_2)_3CH(OH)CH_2CH_3$	116.20	1^1, 205	0.818	1.4214^{20}	-35	66^{20mm}	54	sl s aq
h15	2-Heptanone	$CH_3(CH_2)_4COCH_3$	114.19	1,699	0.8197^{15}_4	1.4116^{15}	-35	151	47	s alc, eth
h16	3-Heptanone	$CH_3(CH_2)_3C(=O)CH_2CH_3$	114.19	1,699	0.8197^{20}_{20}	1.4085^{20}	-36.7	147.8	41	0.43 aq; s alc, eth
h17	4-Heptanone	$CH_3CH_2CH_2(O)CH_2CH_2CH_3$	114.19	1,699	0.821^{15}_4	1.4068^{20}	-32.1	143.7	48	0.53 aq; misc alc, eth
h18	Heptanoyl chloride	$CH_3(CH_2)_5COCl$	148.63	2,340	0.960^{20}	1.4300^{20}		173	57	d aq, alc; s eth
h19	1-Heptene	$CH_3(CH_2)_4CH=CH_2$	98.90	1,219	0.6970^{20}	1.3992^{20}	-118.9	93.6	-1	0.1 aq; s alc, eth
h20	1-Heptylamine	$CH_3(CH_2)_6NH_2$	115.22	4,193	0.777	1.4243^{20}	-23	154–156	35	s alc, acet, eth, PE
h21	Heptyltrichlorosilane	$CH_3(CH_2)_6SiCl_3$	233.7		1.087^{20}_4	1.4439^{25}		211–212		
h22	1-Heptyne	$CH_3(CH_2)_4C≡CH$	96.17	1,256	0.733	1.4075^{20}	-81	99–100	22	sl s aq; s acet
h23	Hexachloroacetone	$Cl_3CC(=O)CCl_3$	264.75	1,657	1.743	1.5112^{20}	-30	66^{6mm}	none	s bz, chl, eth
h24	Hexachlorobenzene	C_6Cl_6	284.78	5,205	2.044^{24}		231	323–326	none	s alc, eth
h25	Hexachloro-1,3-butadiene	$Cl_2C=CClCCl=CCl_2$	260.76	1,250	1.655	1.5550^{20}	-19	210–220	none	
h26	1,2,3,4,5,6-Hexachlorocyclohexane	$C_6H_6Cl_6$	290.83	5^2, 11	1.87^{20}		113			s bz, chl
h27	Hexachlorocyclo-1,3-pentadiene	C_6Cl_6	272.77		1.701^{25}_4	1.5644^{20}	-10	239	none	s bz, chl

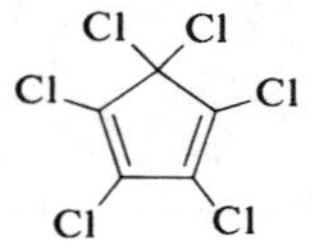

h27

TABLE 1.15 Physical Constants of Organic Compounds (*continued*)

No.	Name	Formula	Formula weight	Beilstein reference	Density	Refractive index	Melting point	Boiling point	Flash point	Solubility in 100 parts solvent
h28	Hexachlorodisiloxane	$Cl_3SiOSiCl_3$	284.9		2.091_4^{20}		−35	137		s alc, bz, chl, eth
h29	Hexachloroethane	Cl_3CCCl_3	236.74	1, 87			187–188		none	
h30	1,4,5,6,7,7-Hexa-chloro-5-norborn-ene-2,3-dicarbox-ylic anhydride		370.83				235–239	210	135	misc eth
h31	Hexachloropropene	$Cl_3CC(Cl)=CCl_2$	248.75	1, 200	1.765	1.5480^{20}		286.8		
h32	Hexadecane	$CH_3(CH_2)_{14}CH_3$	226.45	1, 172	0.7733_4^{20}	1.4345^{20}	18.2			sl s alc; s eth
h33	1,2-Hexadecanediol	$CH_3(CH_2)_{13}CH(OH)-CH_2OH$	258.45	1^3, 2244			72–74			s hot alc, chl, eth
h34	1-Hexadecanethiol	$CH_3(CH_2)_{15}SH$	258.51	1, 430	0.840	1.4720^{20}	18–20	184^{7mm}	101	s alc, chl, eth
h35	Hexadecanoic acid	$CH_3(CH_2)_{14}COOH$	256.43	2, 370	0.8524^{62}	1.4273^{80}	63–64	215^{15mm}	135	s alc, eth, PE
h36	1-Hexadecanol	$CH_3(CH_2)_{15}OH$	242.45	1, 429	0.8116^{60}	1.4355^{60}	49.3	344	132	v s alc, eth; s bz, chl
h37	1-Hexadecene	$CH_3(CH_2)_{13}CH=CH_2$	224.43	1, 226	0.7834^{20}	1.4401	4.1	274	140	
h38	1-Hexadecylamine	$CH_3(CH_2)_{15}NH_2$	241.46	4, 202			40–42	330		
h39	4-Hexadecylaniline	$CH_3(CH_2)_{15}C_6H_4NH_2$	317.56	12, 1186			51–52	254–255^{15mm}		s alc, eth
h40	2,4-Hexadienal	$CH_3CH=CHCH=CHCHO$	96.13	1^2, 809	0.898^{20}	1.5386^{20}		76^{30mm}	67	0.2 aq; 13 alc; 9 acet; 2.3 bz; 11 diox; 1 CCl_4
h41	1,5-Hexadiene	$H_2C=CHCH_2CH_2CH=CH_2$	82.15	1, 253	0.6923_4^{20}	1.4042^{20}	−140.7	59.5	<1	
h42	2,4-Hexadienoic acid	$CH_3CH=CHCH=CHCOOH$	112.13	2, 483			134.5	119^{10mm}	127	
h43	Hexafluorobenzene	C_6F_6	186.05	1^3, 132	1.6182^{20}	1.3781^{20}	5.1	80.3	10	sl s alc, eth
h44	Hexafluoroethane	F_3CCF_3	138.01		1.590^{-78}		−100.1	−78.3		s aq, bz, CCl_4
h45	1,1,1,3,3,3-Hexa-fluoro-2-propanol	$(CF_3)_2CHOH$	168.04		1.596^{25}	1.2750^{20}	−3	58.2	4	
h46	cis-Hexahydroindane		124.23	5, 82	0.876	1.4702	−53	167	23	s eth
h47	Hexamethylbenzene	$C_6(CH_3)_6$	162.28	5, 450			165.6	264		v s bz; s acet, eth

No.	Name	Formula	Formula weight	Beilstein reference	Density	n_D	Melting point	Boiling point	Flash point	Solubility
h48	Hexamethylcyclotrisiloxane	$[-Si(CH_3)_2O-]_3$	222.48	4^3, 1884			64	133–135	35	
h49	1,1,1,3,3,3-Hexamethyldisilazane	$(CH_3)_3SiNHSi(CH_3)_3$	161.40		0.774_4^{20}	1.4071^{20}		126	22	
h50	Hexamethyldisiloxane	$(CH_3)_3SiOSi(CH_3)_3$	162.38		0.764_4^{20}	1.3775^{20}	−67	101	−1	
h51	Hexamethyleneimine		99.18	20, 94	0.880	1.4631^{20}		$138^{749\text{mm}}$	18	67 aq; 8 alc; 10 chl
h52	Hexamethylenetetramine		140.19	1, 583	1.331^{-5}	subl 263			250	
h53	Hexamethylphosphoramide	$[(CH_3)_2N]_3P(O)$	179.20		1.027^{20}	1.4588^{20}	7.2	233	105	misc aq
h54	Hexanal	$CH_3(CH_2)_4CHO$	100.16	1^2, 745	0.8335_4	1.4035^{20}		131	32	v s alc, eth; sl s aq
h55	Hexane	$CH_3(CH_2)_4CH_3$	86.18	1, 142	0.6594_4^{20}	1.3749^{20}	−95.4	68.7	−23	misc alc, chl, eth
h56	1,6-Hexanediamine	$H_2N(CH_2)_6NH_2$	116.21	4, 269			42	205	81	v s aq; sl s alc, bz
h57	1,6-Hexanedioic acid	$HOOC(CH_2)_4COOH$	146.14	2, 649	1.360_4^{25}		152	337.5	196	1.4 aq; v s alc; s acet

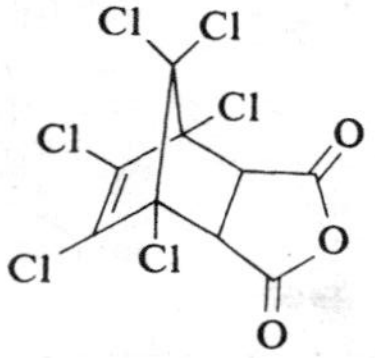

h30

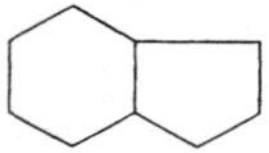

h46

$(CH_2)_6NH$ h51

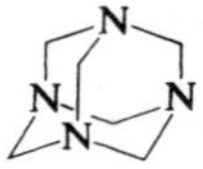

h52

TABLE 1.15 Physical Constants of Organic Compounds (*continued*)

No.	Name	Formula	Formula weight	Beilstein reference	Density	Refractive index	Melting point	Boiling point	Flash point	Solubility in 100 parts solvent
h58	DL-Hexanediol	$CH_3(CH_2)_3CH(OH)CH_2OH$	118.18	1[1], 251	0.951	1.4425^{20}		223–234	>112	v s aq, alc
h59	1,6-Hexanediol	$HO(CH_2)_6OH$	118.18	1, 484	0.958	1.4579^{25}	42.8	243–250	101	s aq, alc, eth
h60	2,5-Hexanediol	$CH_3CH(OH)CH_2CH_2CH(OH)CH_3$	118.18	1, 485	0.9617^{45}_{16}	1.4465^{20}	−50 glass	220.8	101	s aq, alc, eth
h61	2,5-Hexanedione	$CH_3COCH_2CH_2COCH_3$	114.14	1, 788	0.973	1.4260^{20}	−6	191.4	70	misc aq, alc, eth
h62	Hexanedioyl dichloride	$ClC(=O)(CH_2)_4COCl$	183.03	2, 653	1.259	1.4706^{20}		105^{2mm}	>112	
h63	Hexanenitrile	$CH_3(CH_2)_4CN$	97.16	2, 324	0.8052^{20}	1.4069^{20}	−80.3	163.6	43	i aq; s alc, eth
h64	1-Hexanethiol	$CH_3(CH_2)_5SH$	118.24	1[3], 1659	0.8424^{24}	1.4496^{20}	−80.5	152.7	20	i aq; v s alc, eth
h65	1,2,6-Hexanetriol	$HOCH_2CH(OH)(CH_2)_3\text{-}CH_2OH$	134.17		1.1063^{20}	1.4771	−32.8	178^{5mm}	79	misc alc, acet; i bz
h66	Hexanoic acid	$CH_3(CH_2)_4COOH$	116.16	2, 321	0.9265^{20}_4	1.4168^{20}	−4.0	205.7	104	1.1 aq; v s alc, eth
h67	Hexanoic anhydride	$[CH_3(CH_2)_4C(=O)]_2O$	214.31	2, 324	0.926	1.4280^{20}	−41	246–248	>112	s alc
h68	1-Hexanol	$CH_3(CH_2)_5OH$	102.18	1, 407	0.8186^{20}_4	1.4182^{20}	−51.6	157.5	60	8 aq; misc bz, eth; s alc
h69	2-Hexanol	$CH_3(CH_2)_3CH(OH)CH_3$	102.18	1, 408	0.8108^{25}_4	1.4128^{25}	−47	139.9	41	sl s aq; s alc, eth
h70	3-Hexanol	$CH_3CH_2CH_2CH(OH)\text{-}CH_2CH_3$	102.18	1, 408	0.8193^{24}_4	1.4160^{20}		135	41	
h71	6-Hexanolactone		114.14	17[2], 290	1.030	1.4630^{20}		97^{15mm}	109	
h72	2-Hexanone	$CH_3(CH_2)_3COCH_3$	100.16	1, 689	0.8209^{20}	1.4024^{20}	−56.9	127.2	35	v s alc, eth
h73	Hexanoyl chloride	$CH_3(CH_2)_4COCl$	134.61	2, 324	0.9754^{20}_4	1.4263^{20}	−87	153	79	d aq, alc; s eth
h74	1,4,7,10,13,16-Hexaoxacyclooctadecane		264.32				40			
h75	1-Hexene	$CH_3(CH_2)_3CH=CH_2$	84.16	1, 215	0.6732^{20}	1.3879^{20}	−139.8	63.5	−26	0.005 aq
h76	trans-3-Hexenoic acid	$CH_3CH_2CH=CHCH_2\text{-}COOH$	114.14	2, 435	0.963	1.4398^{20}	11–12	119^{22mm}	>112	
h77	trans-2-Hexen-1-ol	$CH_3CH_2CH_2CH=CH\text{-}CH_2OH$	100.16	1[2], 486	0.849	1.4343^{20}		158–160	54	

	Name	Formula	MW	Refs	Density	n	mp	bp	Solubility
h78	5-Hexen-2-one	$H_2C=CHCH_2CH_2COCH_3$	98.15	1, 734	0.847	1.4197^{20}		128–129 [23]	0.13 aq; v s alc, eth
h79	Hexyl acetate	$CH_3(CH_2)_5OOCCH_3$	144.21	2, 132	0.860^{20}_{20}	1.4090^{20}	−80	168–170 [37]	sl s aq; misc alc, eth
h80	Hexylamine	$CH_3(CH_2)_5NH_2$	101.19	4, 188	0.763^{25}_4	1.4180^{20}	−23	131–132 [8]	
h81	4-Hexylaniline	$CH_3(CH_2)_5C_6H_4NH_2$	177.29	12^3, 2759				146–148^{17mm}	
h82	1-Hexyne	$H(CH_2)_4C\equiv CH$	82.14	1^3, 977	0.7152^{20}_4	1.3989^{20}	−131.9	71.3	i aq; s alc, eth
h83	L-Histidine		155.16	25, 513			d 285		41 aq; v sl s alc
h84	Hydantoin		100.08	24, 242			220		s alc, alk; sl s eth
h85	Hydrindantin		322.27	8^1, 631			100		v sl s aq
h86	Hydroxyacetaldehyde	$HOCH_2CHO$	60.05	1, 817			93–94	d 252	v s aq, alc; sl s eth
h87	Hydroxyacetic acid	$HOCH_2COOH$	76.05	3, 228	1.366^{100}		80	110^{12mm}	

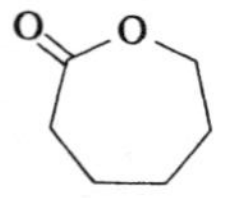

h71 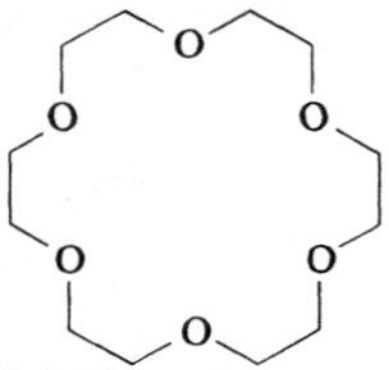   h74 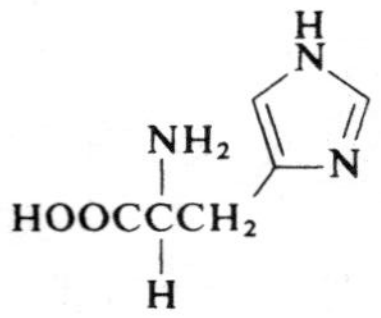   h83 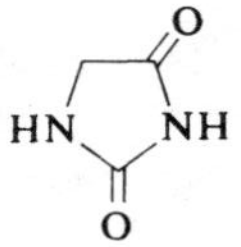   h84 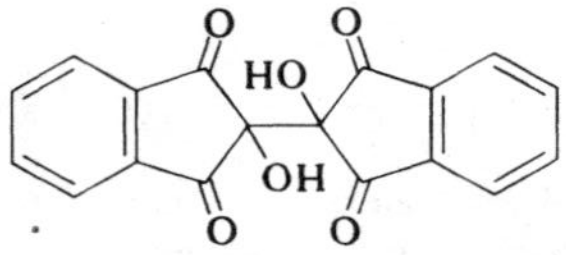   h85

TABLE 1.15 Physical Constants of Organic Compounds (*continued*)

No.	Name	Formula	Formula weight	Beilstein reference	Density	Refractive index	Melting point	Boiling point	Flash point	Solubility in 100 parts solvent
h88	1'-Hydroxy-2'-aceto-naphthone	$C_{10}H_6(OH)COCH_3$	186.21	8, 149			98–100	325 sl d		i aq; v s bz; s HOAc
h89	Hydroxyacetone	$HOCH_2COCH_3$	74.08	1[1], 84	1.082	1.4315^{20}	−17	145–146	56	misc aq, alc, eth
h90	o-Hydroxyaceto-phenone	$HOC_6H_4COCH_3$	136.15	8, 85	1.131^{21}	1.5584^{20}	4–6	213^{717mm}	>112	misc alc, eth; sl s aq
h91	m-Hydroxyaceto-phenone	$HOC_6H_4COCH_3$	136.15	8, 86	1.100^{100}	1.535^{100}	87–88	296		s aq; v s alc, bz, eth
h92	p-Hydroxyaceto-phenone	$HOC_6H_4COCH_3$	136.15	8, 87	1.109^{100}		106–107	147^{3mm}		v s alc, eth; sl s aq
h93	1-Hydroxyanthra-quinone		224.22	8, 338			196–198			
h94	2-Hydroxybenz-aldehyde	$C_6H_4(OH)CHO$	122.12	8, 31	1.167^{20}_4	1.5718^{20}	−7	196.7	76	1.7 aq^{86}; s alc, eth
h95	3-Hydroxybenz-aldehyde	$C_6H_4(OH)CHO$	122.12	8, 58			100–102	191^{50mm}		s alc, bz, eth; sl s aq
h96	4-Hydroxybenz-aldehyde	HOC_6H_4CHO	122.12	8, 64	1.129^{130}_4		117–119	subl		1 aq; 70 acet; 4 bz; v s alc, eth
h97	2-Hydroxybenz-aldehyde oxime	$C_6H_4(OH)CH{=}NOH$	137.14	8, 49			57	d		v s alc, bz, eth, acid
h98	2-Hydroxybenzamide	$C_6H_4(OH)CONH_2$	137.14	10, 87			140	d 270		0.2 aq; s alc, chl, eth
h99	2-Hydroxybenzoic acid	$C_6H_4(OH)COOH$	138.12	10, 43	1.443^{20}_4		157–159	211^{20mm}		0.2 aq; 37 alc; 33 eth; 33 acet; 2 chl; 0.7 bz
h100	3-Hydroxybenzoic acid	$C_6H_4(OH)COOH$	138.12	10, 134	1.473		201–203			0.8 aq; 10 eth
h101	4-Hydroxybenzoic acid	HOC_6H_4COOH	138.12	10, 149	1.468^4		214–215			0.2 aq; v s alc; 23 eth

No.	Name	Formula	Formula wt	Beilstein ref.	Density	n_D	mp, °C	bp, °C	Solubility
h102	p-Hydroxybenzophenone	$HOC_6H_4COC_6H_5$	198.22	8^2, 184			132–135		v s alc, eth; sl s aq
h103	1-Hydroxybenzotriazole		135.13	26, 41			155–158		
h104	6-Hydroxy-1,3-benzoxathiol-2-one		168.17	19^4, 2508			158–160		
h105	2-Hydroxybenzyl alcohol	$HOC_6H_4CH_2OH$	124.13	6, 891	1.161^{25}		86–87	subl 100	6.6 aq; v s alc, chl, eth; s bz
h106	3-Hydroxy-2-butanone	$CH_3COCH(OH)CH_3$	88.10	1, 827	0.997	1.4171^{20}	15	148	misc aq, alc; sl s eth
h107	p-Hydroxycinnamic acid	$HOC_6H_4CH{=}CHCOOH$	164.16	10, 297			210–213		s alc, eth; sl s aq
h108	4-Hydroxycoumarin		162.14	17, 488			213 d	50	s aq, alc, eth
h109	7-Hydroxycoumarin		162.14	18, 27			226–228	subl	v s alc, chl, alk, HOAc
h110	1-Hydroxy-1-cyclohexanecarbonitrile	$C_6H_{10}(OH)CN$	125.17	10, 5	1.031	1.4576^{20}	29	60	v s alc, eth; i bz, chl
h111	2-Hydroxy-3,5-diiodobenzoic acid	$I_2C_6H_2(OH)COOH$	389.91	10, 113			235 d		
h112	2′-Hydroxy-4′,6′-dimethylacetophenone	$(CH_3)_2C_6H_2(OH)COCH_3$	164.20				53–57		

2-Hydroxybenzenemethanol, h105
m-Hydroxybenzotrifluoride, t290

2-Hydroxybiphenyl, p131
4-Hydroxybiphenyl, p132

Hydroxybutanedioic acids, h180, h181

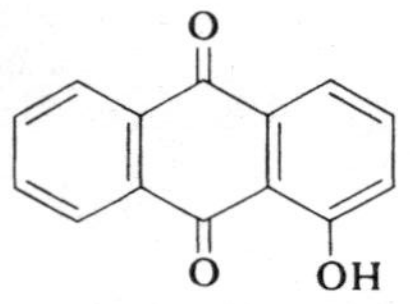

h93

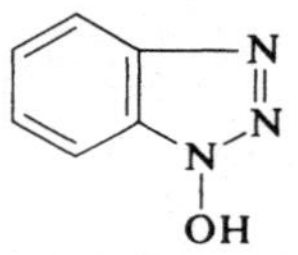

h103

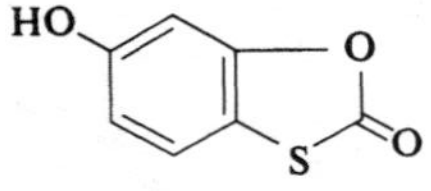

h104

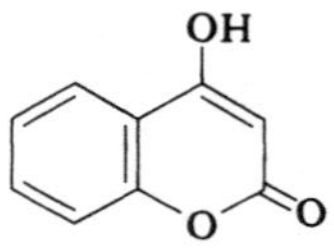

h108

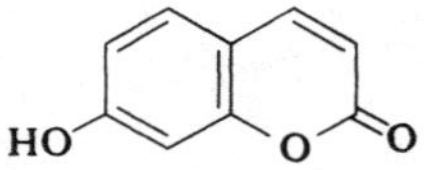

h109

TABLE 1.15 Physical Constants of Organic Compounds (*continued*)

No.	Name	Formula	Formula weight	Beilstein reference	Density	Refractive index	Melting point	Boiling point	Flash point	Solubility in 100 parts solvent
h113	2-Hydroxydiphenyl-methane	$C_6H_5CH_2C_6H_4OH$	184.24	6, 675			20.6	312		s alc, chl, eth, alk
h114	2-Hydroxyethane-sulfonic acid, Na salt	$HOCH_2CH_2SO_3^-\ Na^+$	148.11	4³, 42			191–194			v s aq
h115	N-(2-Hydroxyethyl)-acetamide	$HOCH_2CH_2NHCOCH_3$	103.12	4¹, 430	1.1233^{20}_{20}	1.4575^{20}	63–65	d	176	misc aq: sl s bz
h116	2-Hydroxyethyl acetate	$CH_3COOCH_2CH_2OH$	104.11	2, 141	1.108^{15}			181–186	102	misc aq, alc, chl, eth
h117	3-(α-Hydroxyethyl)-aniline	$CH_3CH(OH)C_6H_4NH_2$	137.18	13³, 1654			68–71			
h118	2-Hydroxyethyl disulfide	$HOCH_2CH_2SSCH_2CH_2OH$	154.25	1, 471	1.261	1.5655^{20}	25–27	$158^{3.5mm}$		
h119	N-(2-Hydroxyethyl)-ethylenediamine-N,N,N'-triacetic acid	$HOOCH_2N(CH_2CH_2OH)\text{-}CH_2CH_2N(CH_2COOH)_2$	278.26				212 d			
h120	2-Hydroxyethyl-hydrazine	$HOCH_2CH_2NHNH_2$	76.10	4¹, 562	1.119		−70	220	73	misc aq; s alc
h121	2-Hydroxyethyl methacrylate	$HOCH_2CH_2OOC\text{-}C(CH_3){=}CH_2$	130.14	27, 7	1.034	1.4515^{20}		$67^{3.5mm}$	97	
h122	N-(β-Hydroxyethyl)-morpholine		131.18	27, 7	1.083	1.4760^{20}		227	99	misc aq
h123	N-(β-Hydroxyethyl)-piperazine		130.19	23², 6	1.061	1.5065^{20}		246	>112	
h124	N-(2-Hydroxyethyl)-piperazine-N'-ethanesulfonic acid		238.31				234 d			

No.	Name	Formula							Solubility
h125	4'-(2-Hydroxyethyl)-piperidine		129.20	21[2], 10	1.0059[15/4]	1.5368[20]	199–202	92	v s aq, alc, chl
h126	2'-(2-Hydroxyethyl)-pyridine	HOCH$_2$CH$_2$C$_5$H$_4$N	123.16	21, 50	1.093		116[9mm]		v s aq, alc, eth
h127	2-Hydroxyisobutyric acid	(CH$_3$)$_2$C(OH)COOH	104.11	3, 313			84[1.5mm]	77–80	
h128	4-Hydroxy-2-mer-capto-6-methyl-pyrimidine		142.18	24[3], 1289				330 d	
h129	4-Hydroxy-2-mer-capto-6-propyl-pyrimidine		170.23					219–221	0.1 aq; 1.7 alc; 1.7 acet; v s alk; i bz
h130	2-Hydroxy-3-meth-oxybenzaldehyde	CH$_3$OC$_6$H$_3$(OH)CHO	152.15	8, 240				40–42	v s alc, eth; sl s aq
h131	4-Hydroxy-3-meth-oxybenzaldehyde	CH$_3$OC$_6$H$_3$(OH)CHO	152.15	8, 247	1.056			80–81	1 aq; s alc, chl, pyr
h132	4-Hydroxy-3-meth-oxybenzoic acid	CH$_3$OC$_6$H$_3$(OH)COOH	168.15	10, 392				210	0.12 aq; v s alc

Hydroxyethanal, h86
3-(α-Hydroxyethyl)aniline, a261
N-(2-Hydroxyethyl)-3-aza-1,5-pentanediol, t264

N-(2-Hydroxyethyl)ethyleneimine, a321
N-(2-Hydroxyethyl)piperidine, p185
2-(2-Hydroxyethyl)piperidine, p186

O-Hydroxyethylresorcinol, h159
2-Hydroxyisobutyronitrile, h145
2-Hydroxy-3-methyl-2-cyclopenten-1-one, m214

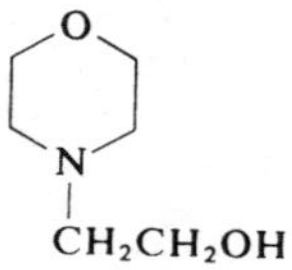

h122 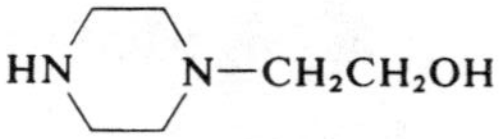 h123 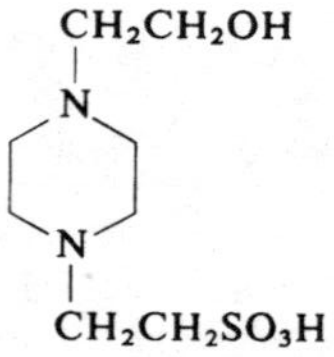 h124 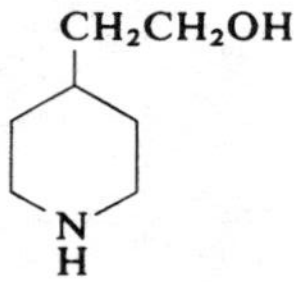 h125 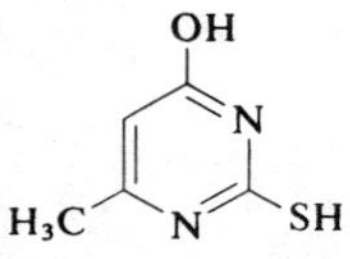 h128 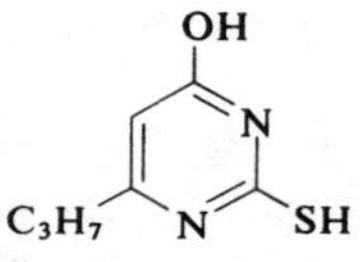 h129

TABLE 1.15 Physical Constants of Organic Compounds (*continued*)

No.	Name	Formula	Formula weight	Beilstein reference	Density	Refractive index	Melting point	Boiling point	Flash point	Solubility in 100 parts solvent
h133	4-Hydroxy-3-meth-oxybenzonitrile	$CH_3OC_6H_3(OH)CN$	149.15	10, 398			85–87			v s alc, chl, eth
h134	2-Hydroxy-4-meth-oxybenzophenone	$CH_3OC_6H_3(OH)COC_6H_5$	228.25	8, 312			66	155^{5mm}		
h135	4-Hydroxy-3-meth-oxybenzyl alcohol	$CH_3OC_6H_3(OH)CH_2OH$	154.17	6, 1113			113–115			s hot aq, alc, eth, EtAc; sl s bz, PE
h136	4-Hydroxy-3-meth-oxycinnamic acid	$CH_3OC_6H_3(OH)CH{=}CH{-}COOH$	194.19	10, 436			174			s alc, chl, eth, alk
h137	2-Hydroxy-3-methyl-benzoic acid	$CH_3C_6H_3(OH)COOH$	152.15	10, 220			165–166			s alc, chl, eth, alk
h138	2-Hydroxy-4-methyl-benzoic acid	$CH_3C_6H_3(OH)COOH$	152.15	10, 233			177			s alc, chl, eth, alk
h139	4-Hydroxy-3-methyl-2-butanone	$HOCH_2CH(CH_3)COCH_3$	102.13	1^1, 422	0.993	1.4340^{20}		92^{15mm}	78	s alc, HOAc; sl s eth
h140	7-Hydroxy-4-methyl-coumarin		176.17	18, 31			194–195			
h141	2-Hydroxymethyl-2-methyl-1,3-propanediol	$HOCH_2C(CH_3)(CH_2OH)_2$	120.09	1, 520			199–203			
h142	4-Hydroxy-4-methyl-2-pentanone	$(CH_3)_2C(OH)CH_2COCH_3$	116.16		0.9385^{20}	1.4235^{20}	−42.8	169	12	misc aq
h143	N-(Hydroxymethyl)-phthalimide		177.16	21, 475			142–145			sl s aq, alc, bz
h144	4-Hydroxy-N-methyl-piperidine		115.18	21^1, 188		1.4775^{20}	29–31	200	>112	
h145	2-Hydroxy-2-methyl-propanenitrile	$(CH_3)_2C(OH)CN$	85.10	3, 316	0.9267^{25}_4	1.3992^{20}	−19	95	63	s aq, alc, chl, eth

	Name	Formula	Mol. wt.	Beil. ref.	M.p., °C		Solubility
h146	3-Hydroxy-2-methyl-4-pyrone		126.11		161–162		1.2 aq; v s hot aq; s alc, alk; sl s bz, eth
h147	2-Hydroxy-1-naphthaldehyde	$C_{10}H_6(OH)CHO$	172.18	8, 143	82–85	192^{27mm}	v s alc, bz, eth, alk
h148	1-Hydroxy-2-naphthalenecarboxylic acid	$C_{10}H_6(OH)COOH$	188.18	10, 331	191–192		
h149	3-Hydroxy-2-naphthalenecarboxylic acid	$C_{10}H_6(OH)COOH$	188.18	10, 333	222–223		v s alc, eth; s bz, chl
h150	2-Hydroxy-3,6-naphthalenedisulfonic acid, disodium salt	$C_{10}H_5(OH)(SO_3^-Na^+)_2$	348.25	11, 288			v s aq, alc; i eth
h151	4-Hydroxy-2,7-naphthalenedisulfonic acid, disodium salt	$C_{10}H_5(OH)(SO_3^-Na^+)_2$	348.25	11, 227	>300		
h152	2-Hydroxy-1,4-naphthoquinone		174.16	8, 300	d 185		s HOAc

3-Hydroxymethylpiperidine, p187

h140

h143

1-Hydroxy-2-napthoic acid, h148

h144

3-Hydroxy-2-napthoic acid, h149

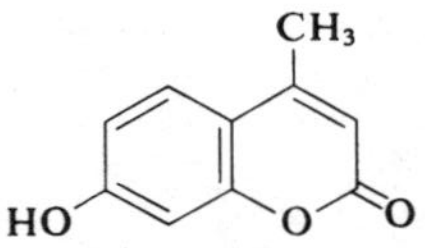

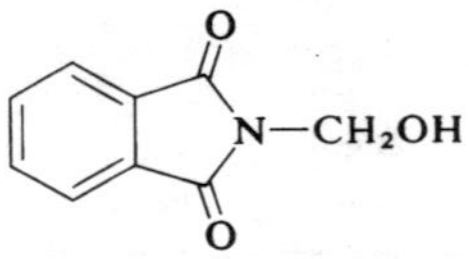

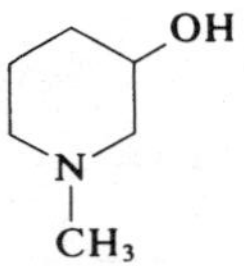

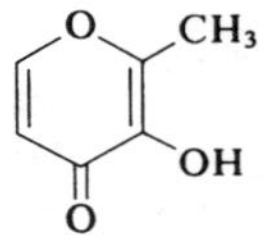

h146

h149

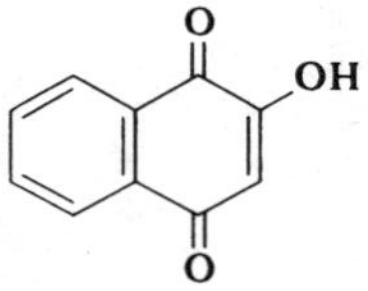

h152

TABLE 1.15 Physical Constants of Organic Compounds (*continued*)

No.	Name	Formula	Formula weight	Beilstein reference	Density	Refractive index	Melting point	Boiling point	Flash point	Solubility in 100 parts solvent
h153	4-Hydroxy-3-nitrobenzenearsonic acid	$HOC_6H_3(NO_2)-AsO(OH)_2$	263.04	16[1], 456			>300			v s alc, acet, HOAc, alk; sl s aq; i eth
h154	3-Hydroxy-4-nitrobenzoic acid	$HOC_6H_3(NO_2)COOH$	183.12	10, 146			229–231			
h155	2-Hydroxy-5-nitrobenzyl bromide	$HOC_6H_3(NO_2)CH_2Br$	232.04	6, 367			147–149			
h156	5-Hydroxy-1-pentanal	$HO(CH_2)_4CHO$	102.13		1.055	1.4530^{20}		115^{15mm}	>112	s aq
h157	5-Hydroxy-2-pentanone	$CH_3COCH_2CH_2CH_2OH$	102.13	1, 831	1.007^{20}_{4}	1.4372^{20}		144^{100mm}	93	misc aq; s alc, eth
h158	4-Hydroxy-3-penten-2-one acetate	$CH_3COOC(CH_3){=}CH{-}COCH_3$	142.15			1.4525^{20}		195	75	
h159	2-(m-Hydroxyphenoxy)ethanol	$HOC_6H_4OCH_2CH_2OH$	154.17				83–86			
h160	4-Hydroxyphenylacetic acid	$HOC_6H_4CH_2COOH$	152.15	10, 190			149–151	subl		v s alc, eth; sl s aq
h161	2-Hydroxy-N-phenylbenzamide	$HOC_6H_4CONHC_6H_5$	213.14	12, 500			136			v s alc, bz, chl, eth
h162	4-(p-Hydroxphenyl)-2-butanone	$HOC_6H_4CH_2CH_2COCH_3$	164.20				82–83			
h163	D-(−)-p-Hydroxyphenylglycine	$HOC_6H_4CH(NH_2)COOH$	167.16	14[1], 659			240 d			sl s aq, alc, bz, acet
h164	N-(p-Hydroxyphenyl)glycine	$HOC_6H_4NHCH_2COOH$	167.16	13, 488			220–248 d			s alk, acid; v sl s aq, alc, acet, bz, chl, eth
h165	1-(3-Hydroxyphenyl)urea	$HOC_6H_4NHCONH_2$	152.15	13, 417			182–184			

No.	Name	Formula	M.W.	Ref.	Density	n_D	M.P.	B.P.	F.P.	Solubility
h166	N-Hydroxyphthalimide		163.13	21, 500			233 d			misc aq, alc; s eth
h167	N-Hydroxypiperidine		101.15	20, 80			37–40	111^{55mm}	77	misc aq, alc; s eth
h168	2-Hydroxypropionitrile	$CH_3CH(OH)CN$	71.08	3^2, 209	0.9834^{25}	1.4027^{25}	−34	103^{50mm}	>112	misc aq, alc, acet; 2.3 eth; i bz, PE
h169	3-Hydroxypropionitrile	$HOCH_2CH_2CN$	71.08	3, 298	1.0404^{25}_4	1.4256^{20}	−46	228		v s alc, eth; sl s aq
h170	o-Hydroxypropiophenone	$HOC_6H_4COCH_2CH_3$	150.18	8, 102	1.094	1.5480^{20}		115^{15mm}	>112	v s alc, eth; sl s aq
h171	p-Hydroxypropiophenone	$HOC_6H_4COCH_2CH_3$	150.18	8, 102			148			misc aq, alc
h172	1-(2-Hydroxy-1-propoxy)-2-propanol	$CH_3CH(OH)CH_2OCH_2CH(OH)CH_3$	134.18		1.0252^{20}_{20}	1.4440^{20}		231.8	138	s aq, alc, bz; sl s eth
h173	2-Hydroxypyridine	HOC_5H_4N	95.10	21, 43			105–107	280–281		s aq, alc; sl s eth
h174	3-Hydroxypyridine	HOC_5H_4N	95.10	21, 46			126–129	151^{3mm}		v s aq, alc; sl s eth
h175	4-Hydroxypyridine	HOC_5H_4N	95.10	21, 48				230^{12mm}		v s aq; i alc, bz, eth
h176	2-Hydroxypyridine-5-carboxylic acid	$HO(C_5H_3N)COOH$	139.11	22, 215			>300			sl s aq, alc, eth

6-Hydroxynicotinic acid, h176
α-Hydroxy-α-phenylacetophenone, b46
3-(p-Hydroxyphenyl)alanine, t437

2-Hydroxy-2-phenylbenzeneacetic acid, b36
3-Hydroxy-1-propanesulfonic acid γ-sultone, p198

2-Hydroxypropanoic acids, L1, L2
1-Hydroxy-2-propanone, h89
3-Hydroxypropionitrile, c290

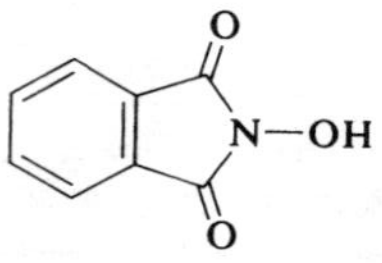

h166

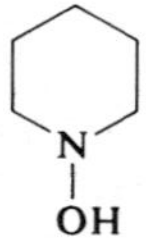

h167

TABLE 1.15 Physical Constants of Organic Compounds (*continued*)

No.	Name	Formula	Formula weight	Beilstein reference	Density	Refractive index	Melting point	Boiling point	Flash point	Solubility in 100 parts solvent
h177	3-Hydroxypyridine-*N*-oxide	$(HO)C_5H_4N{=}O$	111.10				190–192			
h178	8-Hydroxyquinoline		145.16	21, 91			76	267		v s alc, acet, bz, chl
h179	8-Hydroxyquinoline-5-sulfonic acid		225.22	22, 407			213 d			v s aq; sl s alc, eth
h180	DL-Hydroxysuccinic acid	$HOOCCH(OH)CH_2COOH$	134.09	3, 435			131–133			56 aq; 45 EtOH; 18 acet; 0.8 eth; 23 diox; i bz
h181	L-Hydroxysuccinic acid	$HOOCCH(OH)CH_2COOH$	134.09	3, 419			100			36 aq; 87 EtOH; 2.7 eth; 61 acet; 75 diox
h182	*N*-Hydroxy-succinimide		115.09	21, 380			93–95			v s aq
h183	6-Hydroxytetrahydro-pyran-2-carboxylic acid lactone		128.13		1.226	1.4593^{20}				
h184	3-Hydroxy-3,7,11-trimethyl-1,6,10-dodecatriene	$H_2C{=}CHC(OH)(CH_3)\text{-}CH_2CH_2CH{=}C(CH_3)\text{-}CH_2CH_2CH{=}C(CH_3)_2$	222.37		0.8760_4^{25}	1.4769^{25}		114^{1mm}	96	s abs alc
h185	3-Hydroxy-2,2,4-trimethyl-3-pentenoic acid β-lactone		140.18		0.947	1.4380^{20}	−18	170	62	
h186	Hypoxanthine		136.11	26, 416			d 150			0.25 aq; s alk, acid
i1	1*H*,1*H*,11*H*-Icosa-fluoro-1-undecanol	$HCF_2(CF_2)_9CH_2OH$	531.1		0.7777^{37}	1.4346^{40}	95–97	181^{200mm}	>112	
i2	Icosane	$CH_3(CH_2)_{18}CH_3$	282.56	1, 174			36.4	343.8		
i3	1-Icosene	$CH_3(CH_2)_{17}CH{=}CH_2$	280.54	1^3, 881			28.7	342.4		

No.	Name	Formula	Mol. wt.	Beil. ref.	Density	n_D	mp, °C	bp, °C		Solubility
i4	Imidazole		68.08	23, 45			90–91	257	145	v s aq, alc, chl, eth
i5	2-Imidazolidinethi- one		102.16	24, 4			203–204	subl 100		2 aq; s alc, pyr; i bz, acet, chl, eth
i6	Imidazolidinetrione		114.06	24, 16			230			5 aq; s alc
i7	2-Imidazolidone		86.09	25, 315			131	209^{18mm}		v s aq, hot alc
i8	2-(4-Imidazolyl)- ethylamine		111.15				83–84		118	v s aq, alc, hot chl
i9	3,3'-Iminobispropyl- amine	H₂NCH₂CH₂CH₂NHCH₂- CH₂CH₂NH₂	131.22		0.938	1.4810^{20}	−14	151^{50mm}		
i10	Iminodiacetic acid	HOOCCH₂NHCH₂COOH	133.10	4, 365			243 d			2 aq; v sl s bz, eth
i11	Iminodiacetonitrile	NCCH₂NHCH₂CN	95.11	4, 367			77			s aq, alc: sl s eth
i12	Iminodibenzyl		195.27				105–108			
i13	Indan		118.18	6, 575	0.9639^{20}_{4}	1.5360^{20}	−51.4	176.5	50	s alc, chl, eth; i aq
i14	5-Indanol		134.18	7, 360	1.1090^{45}_{4}	1.561^{45}	51–53	255		v s alc, eth; sl s aq
i15	1-Indanone		132.16				40–42	243–245		s alc, eth; sl s aq

5-Hydroxyvaleraldehyde, h156
Imidodicarbonic diamide, b215

Indalone, b445

Indanamines, a199, a200

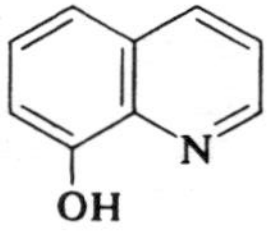

h178 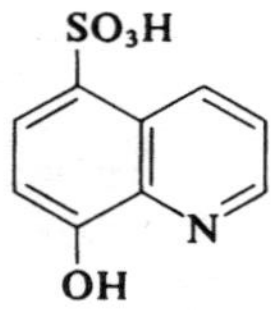   h179 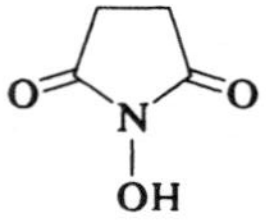   h182 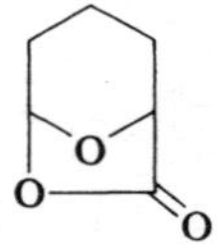   h183 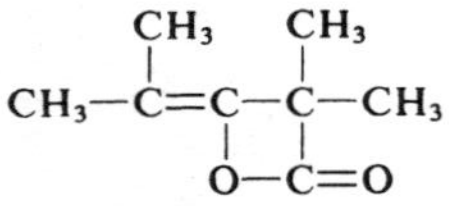   h185   i12

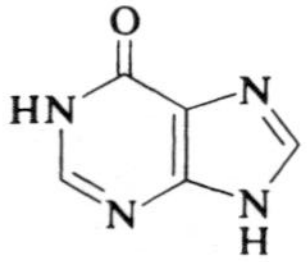

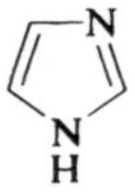

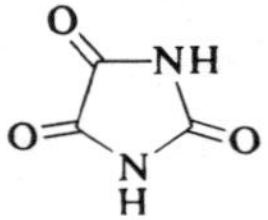

i6 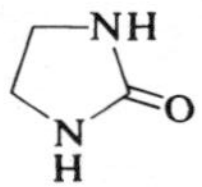   i7 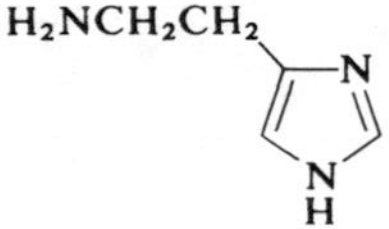   i8

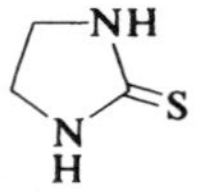

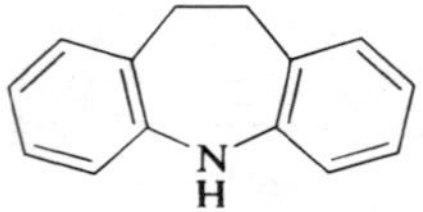

h186

i4   i5   i13 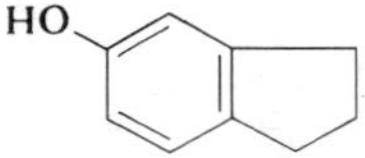   i14 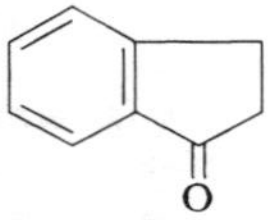   i15

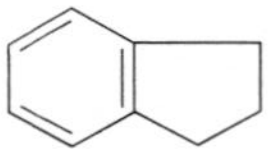

TABLE 1.15 Physical Constants of Organic Compounds (*continued*)

No.	Name	Formula	Formula weight	Beilstein reference	Density	Refractive index	Melting point	Boiling point	Flash point	Solubility in 100 parts solvent
i16	1,2,3-Indantrione hydrate		178.14				d 241			
i17	Indene		116.16	5, 515	0.9968^{20}_4	1.5762^{20}	−1.8	181.6	78	misc alc, bz, chl, eth
i18	Indole		117.15	20, 304	1.0643	1.609^{60}	52	253		s hot aq, bz, eth
i19	Indole-3-acetic acid		175.19	22, 66			168–170			v s alc; s acet, eth
i20	Indole-3-carbaldehyde		145.16	21, 313			195–198			s hot aq, hot alc, alk
i21	Indole-2,3-dione		147.13	21, 432			203.5 d			
i22	Indoline		119.17	20, 257	1.063	1.5906^{20}		221	92	sl s aq
i23	Inositol		180.16	$6^2, 1157$	1.752		225–227			14 aq; sl s aq; i eth
i24	Iodoacetamide	ICH_2CONH_2	184.96	2, 223			91–93			s hot aq
i25	Iodoacetic acid	ICH_2COOH	185.95	2, 222			82–83			s aq, alc; v sl s eth
i26	3-Iodoaniline	$IC_6H_4NH_2$	219.03	12, 670	1.821	1.6820^{20}	25	146^{15mm}	>112	i aq; s alc, eth
i27	Iodobenzene	C_6H_5I	204.01	5, 215	1.83834^{25}	1.621^{18}	−30	188.3	74	misc alc, chl, eth
i28	Iodobenzene diacetate	$C_6H_5I(OOCCH_3)_2$	322.10				163–165			
i29	2-Iodobenzoic acid	IC_6H_4COOH	248.02	9, 363	2.249^{25}_4		162			s alc, eth; sl s aq
i30	1-Iodobutane	$CH_3CH_2CH_2CH_2I$	184.02	1, 123	1.616^{20}_4	1.4999^{20}	−103.5	129–130	33	i aq; s alc, eth
i31	2-Iodobutane	$CH_3CH_2CH(I)CH_3$	184.02		1.592^{20}_4	1.4991^{20}	−104.0	118–120	28	i aq; s alc, eth
i32	Iodocyclohexane	$C_6H_{11}I$	210.06	$5^2, 13$	1.626^{15}_5	1.5472^{20}		180		i aq; s eth
i33	1-Iododecane	$CH_3(CH_2)_9I$	268.18	1, 168	1.257^{20}_4	1.4827^{20}		132^{15mm}		i aq; s alc, eth
i34	Iodoethane	CH_3CH_2I	155.97	1, 96	1.9358^{20}	1.5137	−110.9	72.4	none	0.4 aq; misc alc, bz, chl, eth
i35	2-Iodoethanol	ICH_2CH_2OH	171.97	1, 339	2.2197^{20}_4	1.5694^{20}		75^{5mm}	65	s aq; v s alc, eth
i36	Iodoform	CHI_3	393.73	1, 73	4.008		120–123		none	1.4 alc; 10 chl; 13 eth; v s bz, acet
i37	1-Iodoheptane	$CH_3(CH_2)_6I$	226.10	1, 155	1.3732^{20}_4	1.4900^{20}	−48.2	204	78	i aq; s alc, eth
i38	1-Iodohexadecane	$CH_3(CH_2)_{15}I$	352.35	1, 172	1.121	1.4806^{20}		206–207^{10mm}		i aq; s alc, eth

No.	Name	Formula	Formula wt		Density	n_D	mp	bp	Flash pt	Solubility
i39	1-Iodohexane	$CH_3(CH_2)_5I$	212.08	1, 146	1.4377^{20}_4	1.4926^{20}		179.5	61	i aq; misc alc, eth
i40	Iodomethane	CH_3I	141.94	1, 69	2.2789^{20}_4	1.5308^{20}	−66.5	42.4	none	1.4 aq; misc alc, eth
i41	4-Iodomethoxyben-zene	$IC_6H_4OCH_3$	234.04	6, 208			48–50	237^{726mm}		s hot alc, eth
i42	1-Iodo-3-methyl-butane	$(CH_3)_2CHCH_2CH_2I$	198.06	1^3, 367	1.509^{20}_4	1.4939^{20}		147.5		misc alc, eth; sl s aq
i43	1-Iodo-2-methyl-propane	$(CH_3)_2CHCH_2I$	184.02	1, 128	1.603^{20}_4			119		i aq; misc alc, eth
i44	2-Iodo-2-methyl-propane	$(CH_3)_3CI$	184.02	1^3, 326	1.571^0_0	1.4918^{20}	−38.2			d aq; misc alc, eth
i45	1-Iodo-3-nitrobenzene	$IC_6H_4NO_2$	249.01	5, 253	1.9477^{50}_4		36–38	280		i aq; s alc, eth
i46	1-Iodooctane	$CH_3(CH_2)_7I$	240.13	1, 160	1.330^{20}_4	1.4889^{20}	−45.9	221		s alc, eth
i47	1-Iodopentane	$CH_3(CH_2)_4I$	198.06	1, 133	1.512^{20}_0	1.4954^{20}	−85.6	154.5	79	sl s aq; s alc, eth
i48	1-Iodopropane	$CH_3CH_2CH_2I$	169.99	1, 113	1.7489^{20}	1.5058^{20}	−101	102.5	none	0.1 aq; misc alc, eth

i16

Indonaphthene, i17

i18

4-Iodoanisole, i41

i19

i20

5-Iodoanthranilic acid, a203

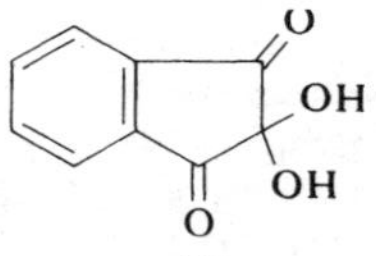

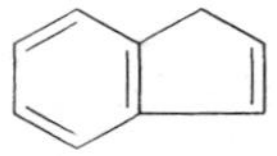

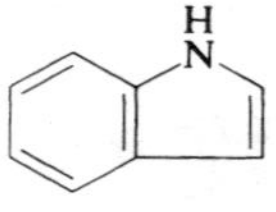

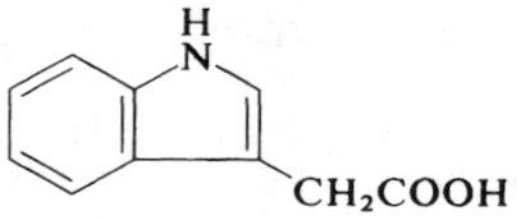

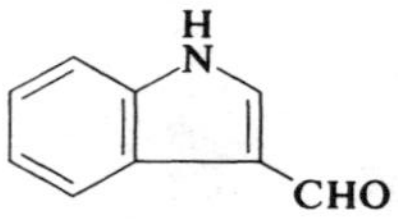

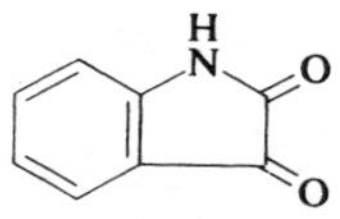

i21

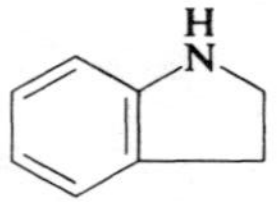

i22

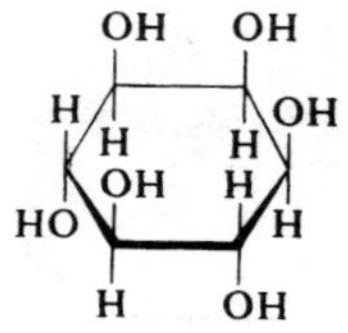

i23

TABLE 1.15 Physical Constants of Organic Compounds (*continued*)

No.	Name	Formula	Formula weight	Beilstein reference	Density	Refractive index	Melting point	Boiling point	Flash point	Solubility in 100 parts solvent
i49	2-Iodopropane	(CH₃)₂CHI	169.99	1, 114	1.7025_4^{20}	1.4992^{20}	−90.0	89.5	none	0.14 aq; misc alc, bz, chl, eth
i50	3-Iodo-1-propene	ICH₂CH=CH₂	167.97	1^3, 114	1.845_4^{22}	1.5540^{21}	−99	1–3		misc alc, chl, eth
i51	5-Iodosalicylic acid	IC₆H₃(OH)COOH	264.02	10, 112			189–191		71	v s alc; i bz, chl
i52	2-Iodothiophene		210.04	17, 34	1.902	1.6520^{20}	−40	73^{15mm}	90	v s eth
i53	2-Iodotoluene	IC₆H₄CH₃	218.04	5, 310	1.713	1.6079^{20}		211	82	i aq; s alc, eth
i54	3-Iodotoluene	IC₆H₄CH₃	218.04	5, 311	1.698	1.6040^{20}		80^{10mm}	<1	i aq; misc alc, eth
i55	Iodotrimethylsilane	(CH₃)₃SiI	200.10	7, 168	1.406_4^{20}	1.4710^{20}		106	104	s alc, bz, chl, eth
i56	α-Ionone		192.30	7, 168	0.932^{20}	1.4980^{20}		124^{11mm}	>112	s alc, bz, chl, eth
i57	β-Ionone		192.30	7, 167	0.946^{17}	1.521^{17}		140^{18mm}	>112	sl s aq; hot alc, acet
i58	Isatoic anhydride		163.13	27, 264			233 d			s aq, alc, acet, pyr
i59	D-(−)-Isoascrobic acid		176.12	6^2, 80			169 d			
i60	DL-Isoborneol		154.25				212	subl		v s alc, chl, eth
i61	2-Isobutoxy-1-isobutoxycarbonyl-1,2-dihydroquinoline		303.40		1.022	1.5230^{20}		$140^{0.2mm}$	>112	
i62	Isobutyl acetate	(CH₃)₂CHCH₂OOCCH₃	116.16	2, 131	0.8745^{20}	1.3902^{20}	−98.9	118.0	25	0.7 aq; v s alc
i63	Isobutylamine	(CH₃)₂CHCH₂NH₂	73.14	4, 163	0.724_4^{20}	1.3972^{20}	−84.6	67.7	−26	misc aq, alc, acet, eth
i64	Isobutylbenzene	C₆H₅CH₂CH(CH₃)₂	134.22	5, 414	0.8673_4^{20}	1.4855^{20}	−51.5	172.8	55	misc alc, eth
i65	Isobutyl chloroformate	ClCOOCH₂CH(CH₃)₂	136.58	3, 12	1.053	1.4070^{20}		128.8	26	misc bz, chl, eth
i66	Isobutyl formate	HCOOCH₂CH(CH₃)₂	102.13	2, 21	0.8854^{20}	1.3855^{20}	−94.5	98.4	10	1 aq; misc alc, eth
i67	Isobutyl isobutyrate	(CH₃)₂CHCH₂OOCCH(CH₃)₂	144.22	2, 291	0.8542^{20}	1.3999^{20}	−80.7	147.5		0.5 aq; misc alc
i68	Isobutyl lactate	CH₃CH(OH)COOCH₂CH(CH₃)₂	146.19	3^2, 188	0.971_{20}^{20}	1.4181^{25}		96^{40mm}		

No.	Name	Formula	Formula wt	Beil. ref	Density	n_D	mp, °C	bp, °C	Flash pt	Solubility
i69	Isobutyl methacrylate	$H_2C=C(CH_3)COOCH_2CH(CH_3)_2$	142.19		0.882^{25}_{15}	1.4170^{25}		155	45	misc alc, eth
i70	Isobutyl nitrate	$(CH_3)_2CHCH_2ONO_2$	119.12		1.015^{20}_{4}	1.4028^{20}		123–125		i aq; misc alc, eth
i71	Isobutyl nitrite	$(CH_3)_2CHCH_2ONO$	103.12	1,377	0.870^{22}_{4}	1.3715^{22}		67	4	misc alc; sl s aq(d)
i72	Isobutyl vinyl ether	$(CH_3)_2CHCH_2OCH=CH_2$	100.16		0.7702^{20}_{20}	1.3961^{20}	−132.3	83.4		0.2 aq
i73	Isobutyraldehyde	$(CH_3)_2CHCHO$	72.11	1,671	0.7988^{20}_{4}	1.3723^{20}	−65.9	63–64	−40	11 aq; misc alc, bz, acet, chl, eth
i74	Isobutyramide	$(CH_3)_2CHCONH_2$	87.12	2,293	1.013		127–129	216–220		
i75	Isobutyric acid	$(CH_3)_2CHCOOH$	88.11	2,288	0.950^{20}_{4}	1.3925^{20}	−46	154	55	17 aq; misc alc, chl, eth

Isatin, i21
Isethionic acid, h114
Isoamyl acetate, i80
Isoamyl alcohol, m155
sec-Isoamyl alcohol, m156
Isoamyl bromide, b305
Isoamyl iodide, i42
Isoamyl nitrite, i81

Isobutane, m375
Isobutene, m383
α-Isobutoxy-α-phenylacetophenone, b48
Isobutoxyacetylene, m352
Isobutyl alcohol, m381
Isobutyl bromide, b310
Isobutyl chloride, c161
Isobutyl chlorocarbonate, i65

Isobutyl 1,2-dihydro-2-isobutoxy-1-quinolinecarboxylate, i61
Isobutyl ether, d407
Isobutyl heptyl ketone, t349
Isobutyl mercaptan, m379
Isobutyraldehyde, m374
Isobutyramide, m388
Isobutyric acid, m390

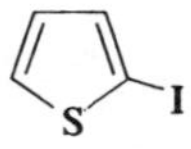

i52

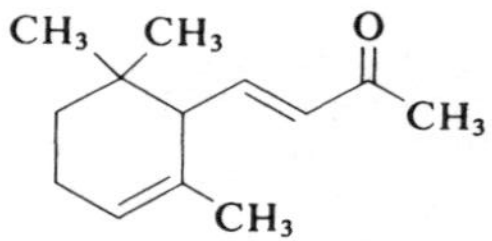

i56

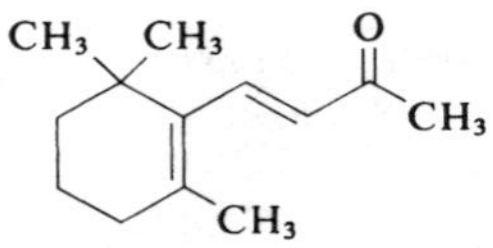

i57

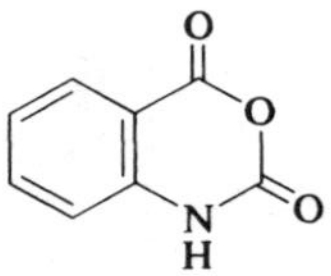

i58

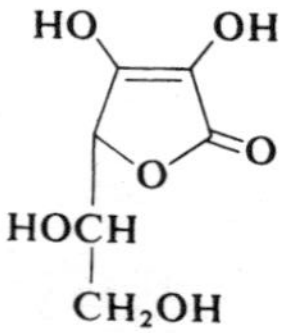

i59   HO   OH   HOCH   CH₂OH

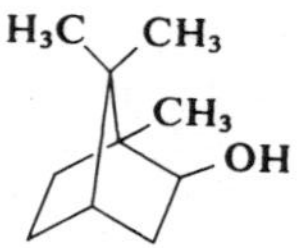

i60   H₃C   CH₃   CH₃   OH

i61   N   OCH₂CH(CH₃)₂   O=C—OCH₂CH(CH₃)₂

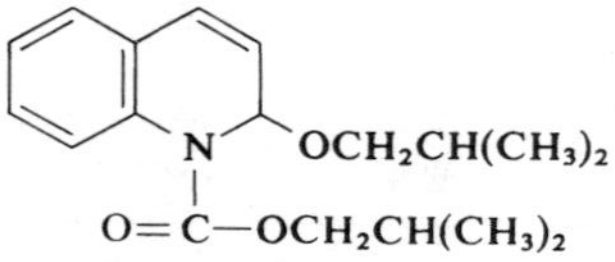

TABLE 1.15 Physical Constants of Organic Compounds (*continued*)

No.	Name	Formula	Formula weight	Beilstein reference	Density	Refractive index	Melting point	Boiling point	Flash point	Solubility in 100 parts solvent
i76	Isobutyronitrile	$(CH_3)_2CHCN$	69.11	2, 294	0.7704^{20}	1.3734^{20}	−71.5	103.8	3	v s alc, eth; sl s aq
i77	Isobutyrophenone	$C_6H_5COCH(CH_3)_2$	148.21	7, 316	0.9880^{20}	1.5172		217	84	d aq; d alc; s eth
i78	Isobutyryl chloride	$(CH_3)_2CHCOCl$	106.55	2, 293	1.017	1.4073^{20}	−90	91–93	1	4 aq; sl s hot alc
i79	L-Isoleucine	$CH_3CH_2CH(CH_3)CH(NH_2)COOH$	131.18	4, 454			d 284	subl 168		
i80	Isopentyl acetate	$CH_3COOCH_2CH_2CH(CH_3)_2$	130.19	2, 132	0.876^{15}_4	1.4007^{20}	−78.5	142.0	25	0.25 aq; misc alc, eth
i81	Isopentyl nitrite	$(CH_3)_2CHCH_2CH_2ONO$	117.15	1, 402	0.872	1.3860^{20}		99	10	misc alc, eth; sl s aq
i82	Isophorone		138.21	7, 65	0.923	1.4759^{20}	−8.1	215.2	84	1.2 aq
i83	DL-Isopinocampheol		154.25	6, 67			35–36	217		
i84	Isopropenyl acetate	$CH_3COOC(CH_3){=}CH_2$	100.12	2^2, 278	0.909	1.4005^{20}		94	18	
i85	2-Isopropoxyphenol	$(CH_3)_2CHOC_6H_4OH$	152.19	6^3, 4209	1.030	1.5157^{20}		100–102^{11mm}		
i86	1-Isopropoxy-2-propanol	$CH_3CH(OH)CH_2OCH(CH_3)_2$	118.1		0.879^{25}_{25}	1.407^{25}		47.9	49	
i87	Isopropyl acetate	$(CH_3)_2CHOOCCH_3$	102.13	2, 130	0.870^{20}_4	1.3773^{20}	−73.4	88.2	16	3 aq; misc alc, eth
i88	Isopropylamine	$(CH_3)_2CHNH_2$	59.11	4, 152	0.686^{25}_{25}	1.3711^{25}	−101	32.4	−17	misc aq, alc, eth
i89	2-Isopropylaminoethanol	$(CH_3)_2CHNHCH_2CH_2OH$	103.17	4, 282	0.8970^{20}_4	1.4395^{20}		75^{11mm}		misc aq, alc, eth
i90	2-Isopropylaniline	$(CH_3)_2CHC_6H_4NH_2$	135.2		0.966	1.4915^{20}		222	46	s alc, bz, eth
i91	Isopropylbenzene	$C_6H_5CH(CH_3)_2$	120.20	5, 393	0.864^{20}_4	1.5206^{20}	−96.0	152.4		misc alc, eth; i aq
i92	4-Isopropylbenzyl alcohol	$(CH_3)_2CHC_6H_4CH_2OH$	150.22	6^3, 1911	0.982^{15}		28	248.4	>112	
i93	N-Isopropylbenzylamine	$C_6H_5CH_2NHCH(CH_3)_2$	149.24		0.892	1.5025^{20}		200	87	
i94	Isopropylcyclohexane	$C_6H_{11}CH(CH_3)_2$	126.24	5, 41	0.8023^{20}_4	1.4399^{20}	−90	155	35	v s alc, eth
i95	N-Isopropylcyclohexylamine	$C_6H_{11}NHCH(CH_3)_2$	141.26		0.859	1.4480^{20}		60^{12mm}	33	

No.	Name	Formula	Mol. wt.	Beilstein ref.	m.p., °C	b.p., °C	Density	n_D	Solubility
i96	4,4'-Isopropylidene-bis[2-(2,6-dibromophenoxy)-ethanol]	$(CH_3)_2C[C_6H_2(Br)_2-OCH_2CH_2OH]_2$	632.01		107				
i97	4,4'-Isopropylidene-diphenol	$(CH_3)_2C[C_6H_4OH]_2$	228.29	6, 1011	153–156	220^{4mm}			
i98	Isopropyl isocyanate	$(CH_3)_2CHCNO$	85.11	4, 155	−2	74–75	0.866	1.3825^{20}	s aq, alc, eth
i99	Isopropyl S-(−)-lactate	$(CH_3)_2CHOOC-CH(OH)CH_3$	132.16	3, 282		166–168	0.998^{20}	1.4082^{25}	

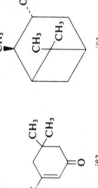

i82

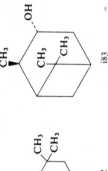

i83

TABLE 1.15 Physical Constants of Organic Compounds (*continued*)

No.	Name	Formula	Formula weight	Beilstein reference	Density	Refractive index	Melting point	Boiling point	Flash point	Solubility in 100 parts solvent
i100	2-Isopropyl-1-methyl-benzene	$CH_3C_6H_4CH(CH_3)_2$	134.21	5, 419	0.8766_4^{20}	1.5006^{20}	−71.5	178.2		misc alc, eth
i101	3-Isopropyl-1-methyl-benzene	$CH_3C_6H_4CH(CH_3)_2$	134.21	5, 419	0.8610_4^{20}	1.4930^{20}	−63.75	175.1		misc alc, eth
i102	4-Isopropyl-1-methyl-benzene	$CH_3C_6H_4CH(CH_3)_2$	134.21	5, 420	0.8573_4^{20}	1.4909^{20}	−67.9	177.1	47	misc alc, eth
i102a	2-Isopropyl-5-methyl-phenol	$CH_3C_6H_3(OH)$- $CH(CH_3)_2$	150.22	6, 532	0.9258_4^{80}		49–51	232		i aq; v s alc, chl, eth
i103	N^1-Isopropyl-2-methyl-1,2-propanedi-amine	$(CH_3)_2C(NH_2)CH_2NH$- $CH(CH_3)_2$	130.24		0.822	1.4269^{20}		147–149	90	
i104	Isopropyl methyl sulfide	$(CH_3)_2CHSCH_3$	90.18	1, 367			−101.5	84.7		
i105	Isopropyl nitrate	$(CH_3)_2CHONO_2$	105.09	$1^3, 1465$	1.036_{19}^{19}	1.3912^{16}		102.1		misc alc, eth
i106	2-Isopropylphenol	$(CH_3)_2CHC_6H_4OH$	136.19	6, 504	1.012^{20}	1.5259^{20}	15–16	212–213	107	316 alc; 350 eth
i107	4-Isopropylphenol	$(CH_3)_2CHC_6H_4OH$	136.19	6, 505	0.990^{20}		59–61	212		
i108	Isopropyl vinyl ether	$(CH_3)_2CHOCH{=}CH_2$	86.13		0.7534^{20}	1.3849^{20}	−140	5–6		
i109	Isopulegol		154.25	6, 65	0.911	1.4725^{20}		91^{12mm}	78	v sl s aq
i110	Isoquinoline		129.16	20, 380	1.0910_4^{30}	1.6208^{30}	26.5	243.2	107	sl s aq; s acid
k1	Ketene	$H_2C{=}C{=}O$	42.04	1, 724			−151	−41		s acet, eth; d aq
L1	DL-Lactic acid	$CH_3CH(OH)COOH$	90.08	3, 268	1.249_4^{15}		16.8	122^{14mm}		s aq, alc; i chl
L2	L-(+)-Lactic acid	$CH_3CH(OH)COOH$	90.08	3, 261	1.2060_4^{25}	1.4392^{20}	53	119^{12mm}	>112	v s aq, alc, eth
L3	α-Lactose		342.30	31, 408	1.525^{20}		219 d	subl 293		17 aq; i alc, eth
L4	DL-Leucine	$(CH_3)_2CHCH_2$- $CH(NH_2)COOH$	131.18	4, 447			d 332			1 aq; 0.13 alc; i eth
L5	L-Leucine	$(CH_3)_2CHCH_2$- $CH(NH_2)COOH$	131.18	4, 437	1.293^{18}		d 293	subl 145		2.4 aq; 0.07 alc; 1 HOAc; i eth

L6	(+)-Limonene		136.24	5, 133	0.8411^{20}_{4}	1.4715	−96.5	175–176	53	misc alc, eth
L7	(−)-Limonene		136.24	5, 136	0.844	1.4706^{20}	−96.5	175–176	48	misc alc, eth
L8	(+)-Limonene oxide		152.24	17, 44	0.929	1.4661^{20}		114^{50mm}	65	
L9	Linalool		154.25	1, 462	0.865^{15}	1.4615^{20}		199	76	misc alc, eth
L10	Linalyl acetate		196.29	2, 141	0.895^{20}	1.451		220 d	84	misc alc, eth
L11	*N*-Lithiohexamethyl-disilazane	$(CH_3)_3SiN(Li)Si(CH_3)_3$	167.3				70–72	115		

Isopropyl mercaptan, p200
1-Isopropyl-4-methyl-1,3-cyclohexadiene, t5
1-Isopropyl-4-methyl-1,4-cyclohexadiene, t6
Isopropyl methyl ketone, m157
Isopropyltolueness, i100, i101, i102
Isopseudocumenol, t358
Isovaleraldehyde, m174
Isovaleric acid, m177
Isovaleronitrile, m178
Isovaleryl chloride, m179
Itaconic acid, m245

Keto compounds, *see* Oxo
2-Ketobutyric acid, o56
5-Keto-1,7,7-trimethylnorcamphane, c3
4-Ketovaleric acid, o58
Koshland's reagent I, h155
Lactonitrile, h168
Lauraldehyde, d731
Lauric acid, d726
Lauronitrile, d724
Lauroyl chloride, d728
Lauryl alcohol, d727

Laurylamine, d732
Lauryl bromide, b275
Lauryl mercaptan, d725
Lauryl sulfate, d735
Lepidine, m408
Leucinol, a217
Levulinic acid, o58
Linoleic acid, o1
Linolenic acid, o7

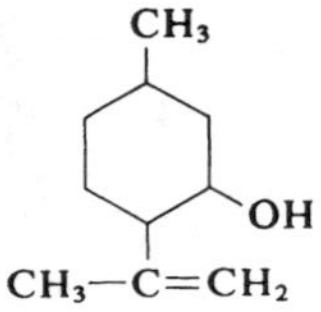

i109

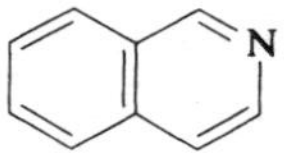

i110

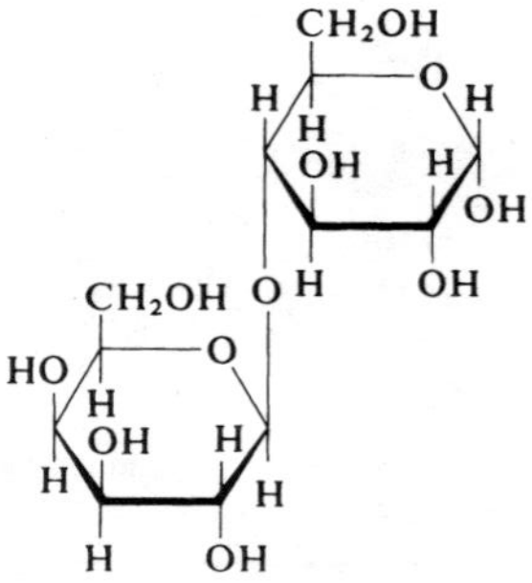

L3

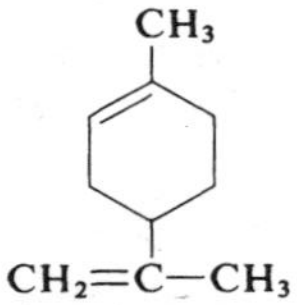

L6, L7

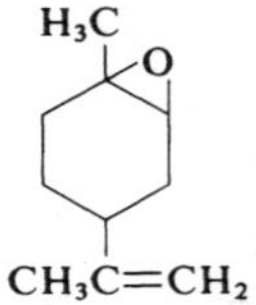

L8

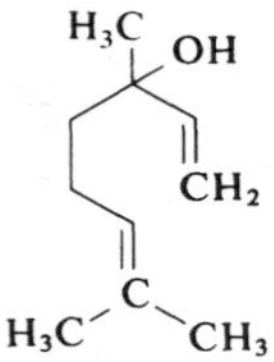

L9

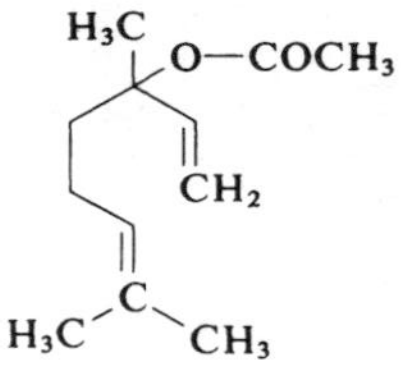

L10

TABLE 1.15 Physical Constants of Organic Compounds (*continued*)

No.	Name	Formula	Formula weight	Beilstein reference	Density	Refractive index	Melting point	Boiling point	Flash point	Solubility in 100 parts solvent
L12	L-(+)-Lysine	$H_2N(CH_2)_4$-$CH(NH_2)COOH$	146.19	4, 435			d 224			v s aq; sl s alc; i eth
m1	Maleic acid	HOOCCH=CHCOOH	116.07	2, 748	1.590		138–139			79 aq; 70 alc; 8 eth
m2	Maleic anhydride		98.06	17, 432	1.48		52.8	202.0	103	s aq (to acid), alc (to ester); 227 acet; 53 chl; 50 bz; 112 EtAc
m3	Malonic acid	$HOOCCH_2COOH$	104.06	2, 566	1.63		135 d			154 aq; 42 alc; 8 eth
m4	Malonodiamide	$H_2NCOCH_2CONH_2$	102.09	2, 582			168–170			9 aq; i alc, eth
m5	Malononitrile	$NCCH_2CN$	66.06	2, 589	1.049		32–34	220	112	13 aq; 40 alc; 20 eth
m6	Malonyl dichloride	$ClCOCH_2COCl$	140.95	21, 252	1.4486^{19}	1.4620^{20}		53^{19mm}	47	d hot aq; s eth
m7	D-(+)-Maltose hydrate		342.30	31, 386	1.540^{17}		102–103	d 130		v s aq; sl s alc; i eth
m8	D-Mandelic acid	$C_6H_5CH(OH)COOH$	152.15	10, 197	1.300^{20}_4		119	d		16 aq; 100 alc; s eth
m9	Mandelonitrile	$C_6H_5CH(OH)CN$	133.15	10, 193	1.117	1.5315^{20}	−10	d 170		v s alc, chl, eth; i aq
m10	Mannitol		182.17	1, 534	1.52^{20}		166–168			18 aq; 1.2 alc; i eth
m11	D-(+)-Mannose		180.16	31, 284	1.54^{20}		128–130	$290^{3.5mm}$		250 aq; 28 pyr; 0.8 alc
m12	L-Menthol		156.27	6, 28	0.890^{15}_{15}	1.458^{25}	43–45	212	93	v s alc, chl, eth, PE
m13	L-Menthone		154.25	7, 38	0.895^{20}_4	1.4510^{20}	−6	207	69	misc alc, eth; sl s aq
m14	Mercaptoacetic acid	$HSCH_2COOH$	92.12	3, 245	1.325	1.5030^{20}	−16.5	96^{5mm}	>112	misc aq, alc, bz, eth
m15	2-Mercaptobenzimidazole		150.20	24, 119			303–304			sl s aq; s alc

m16	o-Mercaptobenzoic acid	HSC_6H_4COOH	154.19	10, 125			164–165	v s alc, HOAc
m17	2-Mercaptobenzothiazole		167.25	27, 185	1.42_4^{20}		180–181 d	2 alc; 1 eth; 10 acet; 1 bz; s alk; i aq
m18	2-Mercaptoethanol	$HSCH_2CH_2OH$	78.13	1, 470	1.1143_4^{20}	1.5006^{20}	156.9	73	misc aq, alc, bz, eth
m19	2-Mercaptoethyltriethoxysilane	$HSCH_2CH_2Si(OC_2H_5)_3$	224.38		0.988_4^{20}	1.432^{20}	210	104	eth

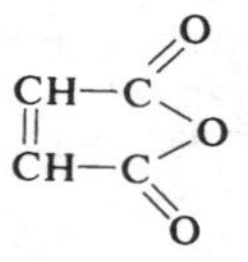

m2

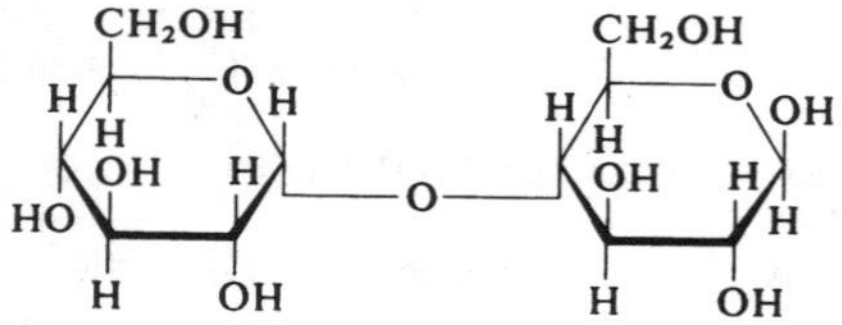

m7

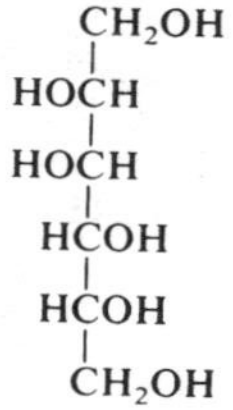

m10

CH₂OH
HOCH
HOCH
HCOH
HCOH
CH₂OH

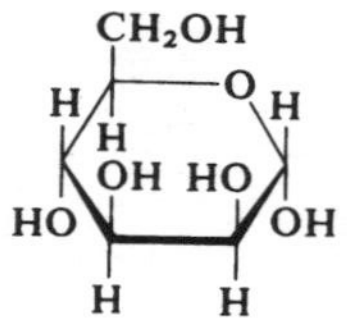

m11

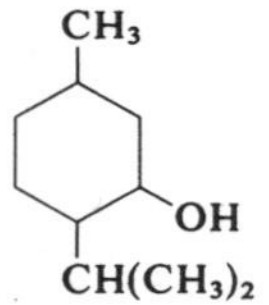

m12

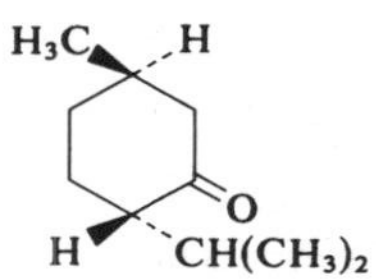

m13

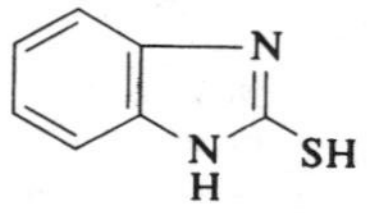

m15

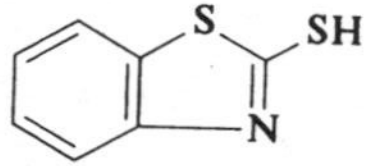

m17

TABLE 1.15 Physical Constants of Organic Compounds (*continued*)

No.	Name	Formula	Formula weight	Beilstein reference	Density	Refractive index	Melting point	Boiling point	Flash point	Solubility in 100 parts solvent
m20	3-Mercapto-1,2-propanediol	$HSCH_2CH(OH)CH_2OH$	108.16	1, 519	1.2951^{14}_{14}	1.5243^{20}		118^{5mm}	>112	misc alc; v s acet
m21	2-Mercaptopropionic acid	$CH_3CH(SH)COOH$	106.14	3, 289	1.2201^{15}	1.4809^{20}	10	117^{16mm}	87	misc aq, alc, eth, acet
m22	(3-Mercaptopropyl)-trimethoxysilane	$HS(CH_2)_3Si(OCH_3)_3$	196.34		1.0394^{20}	1.4416^{20}		93^{40mm}	48	
m23	Mercaptosuccinic acid	$HOOCCH_2CH(SH)COOH$	150.15	3, 439			152–154			50 aq; 50 alc; s eth
m24	Methacrylaldehyde	$H_2C{=}C(CH_3)CHO$	70.09	1, 731	0.8304^{20}_4	1.4160^{20}	−81	69	−15	6 aq; misc alc, eth
m25	Methacrylamide	$H_2C{=}C(CH_3)CONH_2$	85.11	2^2, 399			109–111			s alc; sl s eth
m26	Methacrylic acid	$H_2C{=}C(CH_3)COOH$	86.09	2, 421	1.0153^{20}_{20}	1.4314^{20}	16	163	76	9 aq; misc alc, eth
m27	Methacrylonitrile	$H_2C{=}C(CH_3)CN$	67.91	2, 423	0.8001^{20}_4	1.4007^{20}	−35.8	90.3	12	2.6 aq; misc acet, bz
m28	Methacryloyl chloride	$H_2C{=}C(CH_3)COCl$	104.54	2^2, 394	1.070	1.4447^{20}		95–96	2	3.3 mL aq; 47 mL alc
m29	Methane	CH_4	16.04	1, 56	0.4240^{bp} 0.7168 g/L		−182.5	−161.5		v s aq(d)
m30	Methanesulfonic acid	CH_3SO_3H	96.10	4, 4	1.4812^{18}_4	1.4303^{20}	20	167^{10mm}	>112	1.5 bz; misc aq
m31	Methanesulfonic anhydride	$(CH_3SO_2)_2O$	174.19				71	138^{10mm}		v s aq(d)
m32	Methanesulfonyl chloride	CH_3SO_2Cl	114.55	4, 5	1.4805^{18}_4	1.4518^{20}	−32	161	110	s alc, eth
m33	Methanethiol	CH_3SH	48.11	1, 288	0.8665^{20}_{20}		−123.0	6.0		2.3 aq; v s alc, eth
m34	Methanol	CH_3OH	32.04	1, 273	0.7913^{20}_4	1.3284^{20}	−97.7	64.7	11	misc aq, alc, bz, chl, eth

No.	Name	Formula								
m35	Methanol-*d*	CH_3OD	33.05	1^3, 1186	0.8127^{20}_4	1.3270^{20}	−110	65.5	11	misc aq, alc, eth
m36	Methanol-*d₄*	CD_3OD	36.07	1^3, 1187	0.888	1.3256^{20}		65.4	11	misc aq, alc, eth
m37	DL-Methionine	$CH_3SCH_2CH_2CH(NH_2)COOH$	149.21	4^2, 938	1.340		281 d			3 aq; i eth; v sl s alc
m38	Methoxyacetic acid	CH_3OCH_2COOH	90.08	3, 232	1.174	1.4158^{20}		202–204	>112	misc aq, alc, eth
m39	*o*-Methoxyaceto-phenone	$CH_3OC_6H_4COCH_3$	150.18	8, 85	1.090^{20}_4	1.5393^{20}		131^{18mm}	108	
m40	*m*-Methoxyaceto-phenone	$CH_3OC_6H_4COCH_3$	150.18	8, 86	1.094	1.5410^{20}		239–241	110	s aq
m41	*p*-Methoxyaceto-phenone	$CH_3OC_6H_4COCH_3$	150.18	8, 87	1.082^{41}_4	1.5335^{20}	36–38	154^{26mm}		v s alc, eth
m42	2-Methoxyaniline	$CH_3OC_6H_4NH_2$	123.16	13, 358	1.098^{15}_{15}	1.5730^{20}	5	225	98	i aq; misc alc, eth
m43	3-Methoxyaniline	$CH_3OC_6H_4NH_2$	123.16	13, 404	1.096	1.5794^{20}	1	251	>112	s alc, acid; sl s aq
m44	4-Methoxyaniline	$CH_3OC_6H_4NH_2$	123.16	13, 435	1.087		60	243		v s alc; sl s aq
m45	2-Methoxybenz-aldehyde	$CH_3OC_6H_4CHO$	136.15	8, 43	1.127	1.560^{20}	35–36	236	117	sl s alc, bz; i eth
m46	4-Methoxybenz-aldehyde	$CH_3OC_6H_4CHO$	136.15	8, 67	1.119	1.5713^{20}	−1	248	108	misc alc
m47	4-Methoxybenzamide	$CH_3OC_6H_4CONH_2$	151.17	10^2, 100			164–167	295		s aq; v s alc; sl s eth
m48	Methoxybenzene	$C_6H_5OCH_3$	108.14	6, 138	0.9942^{20}	1.5170^{20}	−37.5	153.8	51	1 aq; misc alc, eth
m49	4-Methoxybenzene-sulfonyl chloride	$CH_3OC_6H_4SO_2Cl$	206.65	11, 243			40–43			d aq; s alc, eth
m50	2-Methoxybenzoic acid	$CH_3OC_6H_4COOH$	152.15	10, 64	1.180		100	200		0.5 aq; v s alc, eth

TABLE 1.15 Physical Constants of Organic Compounds (*continued*)

No.	Name	Formula	Beilstein reference	Formula weight	Density	Refractive index	Melting point	Boiling point	Flash point	Solubility in 100 parts solvent
m51	3-Methoxybenzoic acid	$CH_3OC_6H_4COOH$	10, 137	152.15			104	172^{10mm}		s hot aq, alc, eth
m52	4-Methoxybenzoic acid	$CH_3OC_6H_4COOH$	10, 154	152.15	1.385^4		185	275–280		0.04 aq; v s alc, chl
m53	4-Methoxybenzoyl chloride	$CH_3OC_6H_4COCl$	10, 163	170.60		1.5810^{20}	22	145^{14mm}	87	i aq(d); s alc(d); s bz, acet
m54	4-Methoxybenzyl alcohol	$CH_3OC_6H_4CH_2OH$	6, 897	138.17	1.109^{25}_4	1.5442^{20}	23–25	259	>112	i aq; s alc, eth
m55	4-Methoxybenzyl-amine	$CH_3OC_6H_4CH_2NH_2$	13, 606	137.18	1.050^{15}	1.5462^{20}		236–237	>112	v s aq, alc, eth
m56	2-Methoxybiphenyl	$CH_3OC_6H_4C_6H_5$	6, 672	184.24	1.023	1.6105^{20}		274	>112	misc aq
m57	3-Methoxy-1-butanol	$CH_3OCH(CH_3)CH_2\text{-}CH_2OH$		104.15	0.9229^{20}_{20}	1.4145^{20}	−85	161.1	46	
m58	4-Methoxy-3-buten-2-one	$CH_3OCH=CHCOCH_3$		100.12	0.982	1.4660^{20}		200	63	v s org solv
m59	1-Methoxy-1-buten-3-yne	$CH_3OCH=CHC\equiv CH$		82.10	0.906^{20}_4	1.4818^{20}		122–125	8	s CCl_4
m60	4-Methoxycinnamic acid	$CH_3OC_6H_4CH=CHCOOH$	10, 298	178.19			172–187			
m61	1-Methoxy-1,3-cyclo-hexadiene		6^3, 367	110.16	0.929	1.4885^{20}		40^{15mm}	26	
m62	1-Methoxy-1,4-cyclo-hexadiene		6^3, 367	110.16	0.940	1.4819^{20}		148–150	36	
m63	7-Methoxy-3,7-di-methyloctanal	$(CH_3)_2C(OCH_3)CH_2\text{-}CH_2CH(CH_3)CH_2CHO$	19^4, 617	186.30	0.877	1.4374^{20}		$60^{0.45mm}$	98	
m64	2-Methoxy-1,3-dioxolane			104.11	1.092	1.4091^{20}		129–130	31	
m65	2-Methoxyethanol	$CH_3OCH_2CH_2OH$	1, 467	76.10	0.9646^{20}	1.4021^{20}	−85.1	124.6	46	misc aq

m66	2-(2-Methoxyethoxy)-ethanol	$CH_3OCH_2CH_2OCH_2CH_2OH$	120.15		1.0354_4^{20}	1.4264^{20}	−50	194.1	83	misc aq, alc, bz, eth, ketones
m67	2-Methoxyethoxy-methyl chloride	$CH_3OCH_2CH_2OCH_2Cl$	124.57		1.091	1.4270^{20}		$50^{13\text{mm}}$	>112	misc aq
m68	2-Methoxyethyl acetate	$CH_3COOCH_2CH_2OCH_3$	118.13	2, 141	1.0049^{20}	1.4022^{20}	−65.1	144.5	43	misc aq
m69	2-Methoxyethylamine	$CH_3OCH_2CH_2NH_2$	75.11	4^2, 718	0.864	1.4054^{20}		95	9	v s aq, alc
m70	1-Methoxy-2-indanol		164.20	6, 970		1.5482^{20}		$146^{11\text{mm}}$	>112	
m71	2-Methoxy-5-methyl-aniline	$CH_3OC_6H_3(CH_3)NH_2$	137.18	13^2, 388			52–54	235		s aq; v s alc, bz, eth
m72	3-Methoxy-4-methyl-aniline	$CH_3OC_6H_3(CH_3)NH_2$	137.18	13, 574	1.065	1.5647^{20}	51–54	250–252		s alc
m73	4-Methoxy-2-methyl-aniline	$CH_3OC_6H_3(CH_3)NH_2$	137.18	13^2, 330		1.5647^{20}	13–14	248–249	>112	s alc
m74	(4S,5S)-(−)-4-Methoxymethyl-2-methyl-5-phenyl-2-oxazoline		205.26			1.5155^{20}		$79^{0.05\text{mm}}$		
m75	4-Methoxy-4-methyl-2-pentanone	$(CH_3)_2C(OCH_3)CH_2COCH_3$	130.18		0.906	1.4181^{25}			61	misc aq
m76	1-Methoxynaphthalene	$C_{10}H_7OCH_3$	158.20	6, 606	1.090	1.6220^{20}		$135^{12\text{mm}}$	>112	

Methoxyethane, e171

2-Methoxyethoxychloromethane, m67

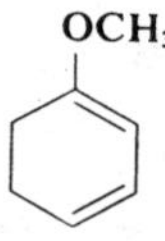

m61

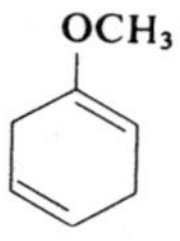

m62

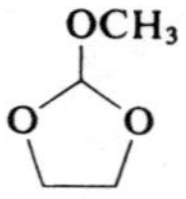

m64

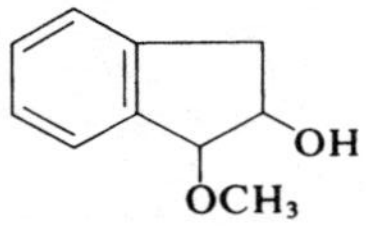

m70

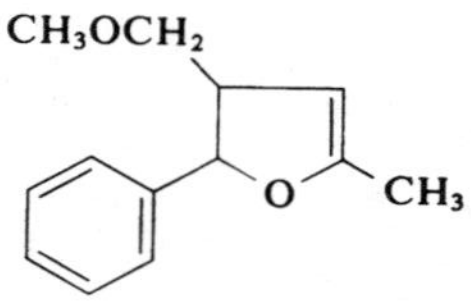

m74

TABLE 1.15 Physical Constants of Organic Compounds (*continued*)

No.	Name	Formula	Formula weight	Beilstein reference	Density	Refractive index	Melting point	Boiling point	Flash point	Solubility in 100 parts solvent
m77	2-Methoxynaphthalene	$C_{10}H_7OCH_3$	158.20	6, 640			72	272		s bz, eth, CS_2
m78	2-methoxy-4-nitroaniline	$CH_3OC_6H_3(NO_2)NH_2$	168.15	13, 390			138–140			
m79	2-Methoxy-5-nitroaniline	$CH_3OC_6H_3(NO_2)NH_2$	168.15	13, 389	1.207^{156}		117–119			s alc, hot bz, HOAc
m80	4-Methoxy-2-nitroaniline	$CH_3OC_6H_3(NO_2)NH_2$	168.15	13, 521			123–126			sl s aq; s alc, eth
m81	2-Methoxynitrobenzene	$CH_3OC_6H_4NO_2$	153.14	6, 217	1.2527^{20}_4	1.5619^{20}	9.4	277	>112	0.17 aq; s alc, eth
m82	4-Methoxynitrobenzene	$CH_3OC_6H_4NO_2$	153.14	6, 230	1.233		54	260		i aq; v s alc, eth
m83	4-Methoxy-3-nitrobenzoic acid	$CH_3OC_6H_3(NO_2)COOH$	197.15	10, 181			186–189			
m84	2-Methoxy-5-nitropyridine	$CH_3OC_5H_3N(NO_2)$	154.13	21^2, 33			108–109			
m85	4-Methoxy-2-nitrotoluene	$CH_3OC_6H_3(NO_2)CH_3$	167.16	6, 411	1.207	1.5525^{20}	17	267	>112	
m86	p-Methoxyphenethylamine	$CH_3OC_6H_4CH_2CH_2NH_2$	151.21	13, 626		1.5379^{20}		138^{20mm}		
m87	2-Methoxyphenol	$CH_3OC_6H_4OH$	124.14	6, 768	1.112 (liquid)	1.5429	28	205	82	1.5 aq; misc alc, eth
m88	3-Methoxyphenol	$CH_3OC_6H_4OH$	124.14	6, 813	1.131	1.5510^{20}	<−17.5	115^{5mm}	>112	misc alc, eth; sl s aq
m89	4-Methoxyphenol	$CH_3OC_6H_4OH$	124.14	6, 843			55–57	243		v s bz; s alk
m90	3-(4-Methoxyphenoxy)-1,2-propanediol	$CH_3OC_6H_4OCH_2CH(OH)CH_2OH$	198.22	6^3, 4411			76–80			

No.	Name	Formula	Formula wt	Beilstein ref	Density	n_D	mp	bp	Flash point	Solubility
m91	4-Methoxyphenyl-acetic acid	CH₃OC₆H₄CH₂COOH	166.18	10, 190			86–88	140^{3mm}		i aq; v s alc; s eth
m92	o-Methoxyphenyl-acetone	CH₃OC₆H₄CH₂COCH₃	164.20	8^3, 397	1.054	1.5250^{20}		130^{10mm}	>112	s alc, eth
m93	(o-Methoxyphenyl)-acetonitrile	CH₃OC₆H₄CH₂CN	147.18	10, 188			65–68	143^{15mm}		s hot bz
m94	2-Methoxy-p-phenylenediamine sulfate	CH₃OC₆H₃(NH₂)₂·H₂SO₄	236.26	13^3, 1349			283 d			
m95	1-Methoxy-2-propanol	CH₃OCH₂CH(OH)CH₃	90.1		0.919^{20}_{20}	1.4021^{20}	−97	120.1	38	misc aq, acet, bz, eth
m96	2-Methoxypropene	CH₃C(OCH₃)=CH₂	72.11	1, 435	0.753	1.3820^{20}		34–36	−18	
m97	trans-1-Methoxy-4-(1-propenyl)benzene	CH₃OC₆H₄CH=CHCH₃	148.21	6, 566	0.9883^{20}_4	1.5615^{20}	21.4	237	90	misc chl, eth; 50 alc; s bz, EtAc
m98	2-Methoxy-4-propenylphenol	CH₃OC₆H₃(OH)-CH=CHCH₃	164.20	6, 955	1.087^{20}_4	1.5748^{20}	−10	266	>112	misc alc, eth; sl s aq
m99	2-Methoxy-4-(2-propenyl)phenol	CH₃OC₆H₃(OH)-CH₂CH=CH₂	164.20	6, 961	1.0664^{20}_4	1.5408^{20}	−9.2	255	>112	misc alc, chl, eth; s HOAc, alk; i aq
m100	p-Methoxypropiophenone	CH₃OC₆H₄COCH₂CH₃	164.20	8, 103	1.071	1.5465^{20}	27–29	273–275	>112	
m101	2-Methoxypyridine	CH₃OC₅H₄N	109.13	21, 44	1.038	1.5029^{29}		142	32	misc aq
m102	2-Methoxytetrahydrofuran		102.13	17^4, 1019	0.972	1.4119^{20}		105–107	7	

α-Methoxy-α-phenylacetophenone, b49

6-Methoxytetralin, m103

Methoxy-1-tetralone, d358

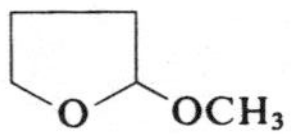

m102

TABLE 1.15 Physical Constants of Organic Compounds (*continued*)

No.	Name	Formula	Formula weight	Beilstein reference	Density	Refractive index	Melting point	Boiling point	Flash point	Solubility in 100 parts solvent
m103	6-Methoxy-1,2,3,4-tetrahydro-naphthalene		162.23	6^2, 537		1.5402^{20}		90^{1mm}	>112	
m104	6-Methoxy-1-tetralone		176.22	9^2, 889			77–79	171^{11mm}		
m105	2-Methoxytoluene	$CH_3OC_6H_4CH_3$	122.17	6, 352	0.9851^{15}_{15}	1.5161^{20}		170–172	51	i aq; v s alc, eth
m106	3-Methoxytoluene	$CH_3OC_6H_4CH_3$	122.17	6, 376	0.9697^{25}_{25}	1.5131^{20}		175–176	54	s alc, bz, eth; i aq
m107	4-Methoxytoluene	$CH_3OC_6H_4CH_3$	122.17	6, 392	0.969^{25}_{25}	1.5112^{20}		174	53	s alc, eth; i aq
m108	Methoxytrimethyl-silane	$CH_3OSi(CH_3)_3$	104.2		0.7560^{20}_4	1.3678^{20}		57–58		
m109	Methoxytripropyl-silane	$CH_3OSi(C_3H_7)_3$	188.4		0.822^{20}_4	1.428^{20}		83^{12mm}		s aq
m110	N-Methylacetamide	$CH_3CONHCH_3$	73.10	4, 58	0.9460^{35}	1.4253^{35}	30.6	206		24 aq; misc alc, eth
m111	Methyl acetate	CH_3COOCH_3	74.08	2, 224	0.9342^{20}_4	1.3619^{20}	−98.1	56.3	−16	50 aq; misc alc
m112	Methyl acetoacetate	$CH_3COCH_2COOCH_3$	116.12	3, 632	1.0747^{20}	1.4186^{20}	−80	171.7	70	i aq; v s alc, eth
m113	p-Methylaceto-phenone	$CH_3C_6H_4COCH_3$	134.18	7, 307	1.0051	1.5328^{20}	22–24	226	92	
m114	Methyl acrylate	$H_2C\!=\!CHCOOCH_3$	86.09	2, 399	0.9561^{20}_4	1.4117^{18}	−76.5	80.2	6	6 aq; s alc, eth
m115	Methylamine	CH_3NH_2	31.06	4, 32	0.699^{-11}_4		−93.5	−6.3	0	959 mL aq; 10.5 bz; s alc; misc eth
m116	Methyl 2-amino-benzoate	$H_2NC_6H_4COOCH_3$	151.17	14, 317	1.68^{19}_4	1.5820^{20}	24	256	104	sl s aq; v s alc, eth
m117	2-(N-Methylamino)-benzoic acid	$CH_3NHC_6H_4COOH$	151.17	14, 323			170–172 d			0.2 aq; s alc, eth
m118	Methyl 3-amino-crotonate	$CH_3C(NH_2)\!=\!CHCOOCH_3$	115.13	3, 632			81–83			

No.	Name	Formula	Formula weight	Beilstein reference	Density	Refractive index	Melting point	Boiling point	Flash point	Solubility
m119	2-(Methylamino)-ethanol	$CH_3NHCH_2CH_2OH$	75.11	4, 276	0.9375^{20}	1.4387^{20}		155–156	72	misc aq, alc, eth
m120	4-Methylamino-phenol sulfate	$(CH_3NHC_6H_4OH)_2 \cdot H_2SO_4$	344.39	13, 442			260 d			4 aq; sl s alc; i eth
m121	2-(Methylamino)-pyridine	$CH_3NHC_5H_4N$	108.14	22^1, 629	1.052^{29}_{29}	1.5785^{20}	15	201	87	s aq; v s alc, eth
m122	N-Methylaniline	$C_6H_5NHCH_3$	107.16	12, 135	0.989^{20}_{4}	1.5704^{20}	−57	196	73	sl s aq; s alc, eth
m123	N-Methylanilinium trifluoroacetate	$C_6H_5NHCH_3 \cdot HOOCCF_3$	221.18				65–66			
m124	2-Methylanthra-quinone		222.24	7, 809			177	subl		v s bz; s alc, eth
m125	Methylarsonic acid	$CH_3AsO(OH)_2$	139.96	4, 613			161			v s aq; s alc
m126	4-Methylbenz-aldehyde	$CH_3C_6H_4CHO$	120.15	7, 297	1.0194^{17}_{4}	1.5447^{20}	−4	205	80	misc alc, eth; sl s aq
m127	Methyl benzene-sulfonate	$C_6H_5SO_2OCH_3$	172.20	11^2, 20	1.2889^{0}_{4}	1.5151^{20}		154^{20mm}		v s alc, chl, eth
m128	2-Methylbenz-imidazole		132.17	23, 145			176–177			s alk, hot aq; sl s alc
m129	Methyl benzoate	$C_6H_5COOCH_3$	136.15	9, 109	1.0933^{15}_{4}	1.5205^{15}	−12.1	199.5	82	0.2 aq; misc alc, eth

Methylal, d441
Methyl alcohol, m34
Methylaminoacetaldehyde dimethyl acetal, d442
α-(1-Methylaminoethyl)benzyl alcohols, e1, e2

2-Methyl-p-anisidine, m71
4-Methyl-m-anisidine, m72
5-Methyl-o-anisidine, m73
Methylanisoles, m105, m106, m107
Methyl anthranilate, m116

Methylanthranilic acids, a211, a212
N-Methylanthranilic acid, m117
Methylbenzene, t167
4-Methylbenzenesulfonic acid, t176

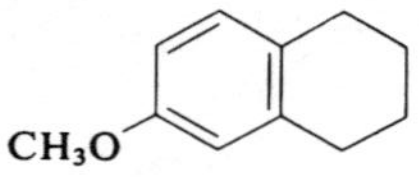

CH_3O

m103

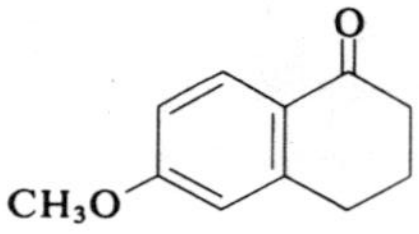

CH_3O　O

m104

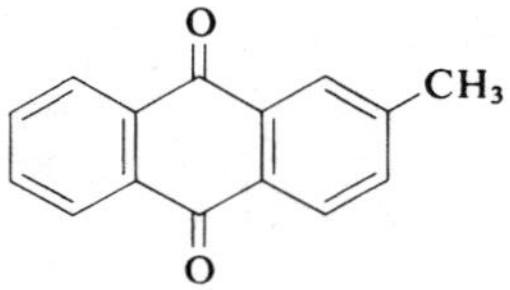

CH_3　O　O

m124

N　CH_3　N–H

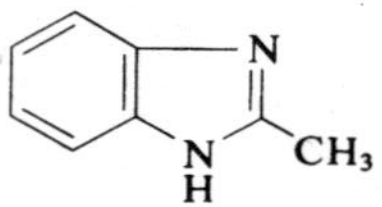

m128

TABLE 1.15 Physical Constants of Organic Compounds (*continued*)

No.	Name	Formula	Formula weight	Beilstein reference	Density	Refractive index	Melting point	Boiling point	Flash point	Solubility in 100 parts solvent
m130	2-Methylbenzoic acid	$CH_3C_6H_4COOH$	136.15	9, 462	1.062		107–108	258–259		sl s aq; v s alc
m131	3-Methylbenzoic acid	$CH_3C_6H_4COOH$	136.15	9, 475	1.054		111–113	263		0.09 aq; v s alc
m132	4-Methylbenzoic acid	$CH_3C_6H_4COOH$	136.15	9, 483			180–182	274–275		v s alc, eth
m133	2-Methylbenzophenone	$CH_3C_6H_4COC_6H_5$	196.25	7, 439	1.083	1.5958^{20}	<–18	309–311	>112	v s alc, org solv
m134	4-Methylbenzophenone	$CH_3C_6H_4COC_6H_5$	196.25	7, 440			59–60	326		v s bz, eth
m135	2-Methylbenzothiazole		149.22	27, 46	1.173	1.6170^{20}	12–14	238	102	s alc, HOAc; i aq
m136	5-Methyl-1 *H*-benzotriazole		133.15	26, 58			80–82	$210–212^{12mm}$	75	
m137	2-Methylbenzoxazole		133.15	27, 46	1.121	1.5497^{20}	8.5–10	178	75	
m138	α-Methylbenzyl alcohol	$C_6H_5CH(CH_3)OH$	122.17	6, 475	1.0191^{13}	1.5211^{20}	21	204	85	v s alc; s bz, chl
m139	3-Methylbenzyl alcohol	$CH_3C_6H_4CH_2OH$	122.17	6, 494	0.916^{17}	1.5334^{20}	<–20	217		5 aq; s alc, eth
m140	4-Methylbenzyl alcohol	$CH_3C_6H_4CH_2OH$	122.17	6, 498			59–61	217		s alc, eth; sl s aq
m141	DL-α-Methylbenzylamine	$C_6H_5CH(CH_3)NH_2$	121.18	12, 1094	0.940	1.5254^{20}	12–13	185	79	
m142	4-Methylbenzylamine	$CH_3C_6H_4CH_2NH_2$	121.18	12, 1141	0.952	1.5340^{20}		195	75	s alc
m143	Methyl bromoacetate	$BrCH_2COOCH_3$	152.98	2, 213	1.616	1.4586^{20}		52^{15mm}	62	
m144	DL-Methyl-2-bromobutyrate	$CH_3CH_2CH(Br)COOCH_3$	181.04	2, 282	1.573			$137–138^{50mm}$		

m145	Methyl 4-bromo-crotonate	$BrCH_2CH{=}CHCOOCH_3$	179.02		1.522	1.4980^{20}		85^{13mm}	91	s alc
m146	Methyl 2-bromo-propionate	$CH_3CH(Br)COOCH_3$	167.01	2, 253	1.497	1.5420^{20}		51^{19mm}	51	misc alc, eth
m147	2-Methyl-1,3-butadiene	$H_2C{=}C(CH_3)CH{=}CH_2$	68.12	1, 252	0.6814^{20}_4	1.4216^{20}	-145.9	34.1	-53	misc alc, eth
m148	3-Methyl-1,2-butadiene	$CH_3C(CH_3){=}C{=}CH_2$	68.12	1, 252	0.6944^{20}_4	1.4179^{20}	-113.6	40.9	-12	
m149	2-Methylbutane	$CH_3CH_2CH(CH_3)_2$	72.15	1, 134	0.6197^{20}_4	1.3537^{20}	-159.9	27.9	-56	0.005 aq; misc alc
m150	2-Methyl-1-butanethiol	$CH_3CH_2CH(CH_3)CH_2SH$	104.22	1^2, 421	0.848	1.4465^{20}		119.0	19	s alc, eth; i aq
m151	2-Methyl-2-butanethiol	$CH_3CH_2C(CH_3)_2SH$	104.22	1^1, 196	0.842	1.4385^{20}	-103.9	99.1	-1	s alc, eth; i aq
m152	3-Methyl-1-butanethiol	$(CH_3)_2CHCH_2CH_2SH$	104.22	1, 405	0.8354^{20}_4	1.4432^{20}	-133.5	118.4	18	misc alc, chl, eth
m153	2-Methyl-1-butanol	$CH_3CH_2CH(CH_3)CH_2OH$	88.15	1, 388	0.8164^{20}_4	1.4100^{20}	<-70	128	50	3 aq; misc alc, eth
m154	2-Methyl-2-butanol	$CH_3CH_2C(CH_3)_2OH$	88.15	1, 388	0.8090^{20}	1.4050^{20}	-9.0	102.0	21	11 aq; misc alc, bz, chl, eth
m155	3-Methyl-1-butanol	$(CH_3)_2CHCH_2CH_2OH$	88.15	1, 392	0.8129^{15}_4	1.4085^{15}	-117.2	132.0	45	2 aq; misc alc, bz, chl, eth, PE, HOAc

α-Methylbenzyl alcohol, p112
N-Methylbenzylamine, b103
Methylbenzyl bromides, b368, b369

Methylbenzyl chlorides, c258, c259, c260
Methylbis(2-chloroethoxy)silane, b157
N-Methylbis(2-chloroethyl)amine, b159

Methyl bromide, b300
3-Methyl-1-buten-1-carboxylic acid, m349

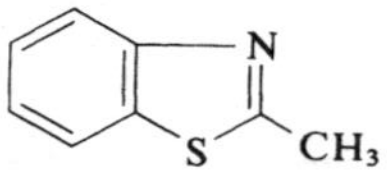

m135

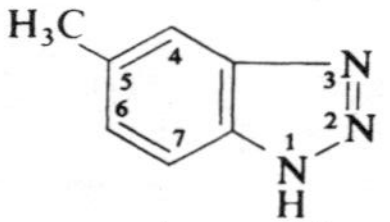

m136

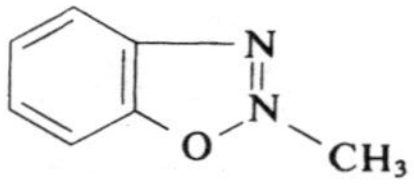

m137

TABLE 1.15 Physical Constants of Organic Compounds (*continued*)

No.	Name	Formula	Formula weight	Beilstein reference	Density	Refractive index	Melting point	Boiling point	Flash point	Solubility in 100 parts solvent
m156	3-Methyl-2-butanol	$(CH_3)_2CHCH(OH)CH_3$	88.15	1, 391	0.8179^{20}	1.4096^{20}		111.5	26	2.8 aq; misc alc, eth
m157	3-Methyl-2-butanone	$(CH_3)_2CHCOCH_3$	86.13	1, 682	0.802^{20}_4	1.3890	92	94–95		misc alc, eth
m158	2-Methyl-1-butene	$CH_3CH_2C(CH_3)=CH_2$	70.14	1, 211	0.6504^{20}_4	1.3777^{20}	−137.6	31.2		misc alc, eth
m159	2-Methyl-2-butene	$CH_3CH=C(CH_3)_2$	70.14	1, 211	0.6620^{20}_4	1.3878^{20}	−133.8	38.6	−45	misc alc, eth; i aq
m160	3-Methyl-1-butene	$(CH_3)_2CHCH=CH_2$	70.14	1^3, 797	0.6272^{20}_4	1.3638^{20}	−168.5	20.1		misc alc, eth
m161	(E)-2-Methyl-2-butenoic acid	$CH_3CH=C(CH_3)COOH$	100.12	2, 430	0.969	1.4342^{81}	64	198.5		s alc, eth; v s hot aq
m162	(Z)-2-Methyl-2-butenoic acid	$CH_3CH=C(CH_3)COOH$	100.12	2, 428	0.9834^7	1.4437^{47}	45	185		s alc, eth; v s hot aq
m163	3-Methyl-2-butenoic acid	$(CH_3)_2C=CHCOOH$	100.12	2, 432	1.006^{24}		69	194–195		s aq, alc, eth
m164	2-Methyl-3-buten-2-ol	$(CH_3)_2C(OH)CH=CH_2$	86.13	1, 444	0.8672^{20}_{20}	1.4160^{20}	2.6	98–99	13	
m165	3-Methyl-3-buten-1-ol	$H_2C=C(CH_3)CH_2CH_2OH$	86.13		0.853	1.4337^{20}			36	
m166	2-Methyl-1-buten-3-yne	$H_2C=C(CH_3)C≡CH$	66.10	1^1, 126		1.4140^{20}	−113	32	−6	
m167	N-Methylbutylamine	$CH_3CH_2CH_2CH_2NHCH_3$	87.17	4, 157	0.736	1.3995^{20}		91	<1	misc aq, alc, eth
m168	1-Methylbutylamine	$CH_3CH_2CH_2CH(CH_3)NH_2$	87.17	4, 177	0.7384^{20}_4	1.4029^{20}	−75	91	35	
m169	2-Methylbutylamine	$CH_3CH_2CH(CH_3)CH_2NH_2$	87.17	4^3, 342	0.738	1.4116^{20}		94–97	3	misc alc, eth
m170	3-Methylbutyl-3-methylbutyrate	$(CH_3)_2CHCH_2CH_2OOCCH_2CH(CH_3)_2$	172.27	2, 312	0.8541^{25}	1.4100^{25}		194.0		misc alc, eth
m171	3-Methyl-1-butyne	$(CH_3)_2CHC≡CH$	68.12	1, 251	0.666^{20}_4	1.3740^{20}	−89.8	26.4		misc alc, eth
m172	2-Methyl-3-butyn-2-ol	$(CH_3)_2C(OH)C≡CH$	84.12	1^1, 235	0.8672^{20}_{20}	1.4209^{20}	2.6	104–105	25	misc aq, acet, bz

m173	2-Methylbutyraldehyde	$CH_3CH_2CH(CH_3)CHO$	86.13	$1^1, 352$	0.804	1.3919^{20}		90–92	4	misc alc, eth; sl s aq
m174	3-Methylbutyraldehyde	$(CH_3)_2CHCH_2CHO$	86.13	1, 684	0.7852^{20}	1.3882^{20}	−51	92–93	19	1.4 aq; misc alc, eth
m175	Methyl butyrate	$CH_3CH_2CH_2COOCH_3$	102.13	$2^4, 786$	0.898_4	1.3879^{20}	−85	102	14	
m176	2-Methylbutyric acid	$CH_3CH_2CH(CH_3)COOH$	102.13	$2^4, 888$	0.936	1.4055^{20}		176.5	>112	4 aq; s alc, chl, eth
m177	3-Methylbutyric acid	$(CH_3)_2CHCH_2COOH$	102.13	2, 309	0.9308^{20}_4	1.4033^{20}	−30.0	176.5	70	misc alc, eth
m178	3-Methylbutyronitrile	$(CH_3)_2CHCH_2CN$	83.13	$2^2, 278$	0.7925^{19}_4	1.3927^{20}	−101	129		d aq; alc; s eth
m179	3-Methylbutyryl chloride	$(CH_3)_2CHCH_2COCl$	120.58	2, 315	0.985^{20}	1.4161^{20}		115–117	18	
m180	1-(3-Methylbutyryl)-pyrrolidine		155.24		0.938	1.4710^{20}			104	
m181	Methyl carbamate	$H_2NCOOCH_3$	75.07	3, 21	1.136^{56}_4		52–54	177		220 aq; 73 alc; s eth
m182	Methyl chloroacetate	$ClCH_2COOCH_3$	108.52	2, 197	1.238^{20}_{20}	1.4220^{20}	−33	130–132	57	i aq; misc alc, eth

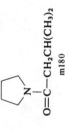

m180

TABLE 1.15 Physical Constants of Organic Compounds (*continued*)

No.	Name	Formula	Formula weight	Beilstein reference	Density	Refractive index	Melting point	Boiling point	Flash point	Solubility in 100 parts solvent
m183	Methyl 2-chloroacetoacetate	$CH_3COCH(Cl)COOCH_3$	150.56		1.236	1.4465^{20}	−32.7	137	71	
m184	Methyl *m*-chlorobenzoate	$ClC_6H_4COOCH_3$	170.60	9, 338		1.4923^{20}	21	101^{12mm}		
m185	Methyl *p*-chlorobenzoate	$ClC_6H_4COOCH_3$	170.60	9, 340	1.382^{20}		44			s alc
m186	Methyl 4-chlorobutyrate	$ClCH_2CH_2CH_2COOCH_3$	136.58	2, 278	1.1268^{14}	1.4321^{20}		176	59	v s eth; s alc, acet
m187	Methyl chloroformate	$ClCOOCH_3$	94.50	3, 9	1.223^{20}_4	1.3865^{20}		71	<1	misc alc, bz, chl, eth
m188	Methyl 3-(chloroformyl)propionate	$CH_3OOCCH_2CH_2COCl$	150.56	2^2, 553	1.223	1.4402^{20}		65^{3mm}	73	
m189	Methyl 2-chloropropionate	$CH_3CH(Cl)COOCH_3$	122.55	2, 248	1.075	1.4193^{20}		132–133	36	s alc
m190	2-Methylcinnamaldehyde	$C_6H_5CH=C(CH_3)CHO$	146.19	7, 369	1.047^{17}_4	1.6045^{20}		149^{27mm}	79	
m191	6-Methylcoumarin		160.17	17, 337			75–76	303^{725mm}	4	v s alc, eth; i aq
m192	Methyl crotonate	$CH_3CH=CHCOOCH_3$	100.12	2, 410	0.9444^{20}_4	1.4242^{20}		121		misc alc, eth
m193	Methyl cyanoacetate	$NCCH_2COOCH_3$	99.09	2, 584	1.1225^{25}	1.4166^{25}	−13.1	205.1	110	
m194	Methylcyclohexane	$C_6H_{11}CH_3$	98.19	5, 29	0.7694^{20}	1.4231^{20}	−126.6	100.9	−3	i aq; s alc, eth
m195	Methyl cyclohexanecarboxylate	$C_6H_{11}COOCH_3$	142.20	9^1, 5	0.9954^{16}_4	1.4445^{20}		183	60	
m196	4-Methyl-1,2-cyclohexanedicarboxylic anhydride		168.19		1.162	1.4774^{20}				
m197	1-Methylcyclohexanol	$C_6H_{10}(CH_3)OH$	114.19	6, 11	0.9251^{25}	1.4587^{25}	26	168	67	i aq; s bz, chl
m198	(Z)-2-Methylcyclohexanol	$C_6H_{10}(CH_3)OH$	114.19	6^2, 17	0.9340^{20}_4	1.4654^{20}	7	165	58	misc alc, eth

No.	Name	Formula	M.W.		Density	n_D	m.p.	b.p.		Solubility
m199	(E)-2-Methylcyclohexanol	$C_6H_{10}(CH_3)OH$	114.19	6, 11	0.9247_4^{20}	1.4616^{20}	−4	165.5	58	misc alc; s eth
m200	(Z)-3-Methylcyclohexanol	$C_6H_{10}(CH_3)OH$	114.19	6, 12	0.9155^{20}	1.4572^{20}	−6	94	62	misc alc, eth
m201	(E)-3-Methylcyclohexanol	$C_6H_{10}(CH_3)OH$	114.19	6, 12	0.9214^{20}	1.4580^{20}	−1	84	62	misc alc, eth
m202	(Z)-4-Methylcyclohexanol	$CH_3C_6H_{10}OH$	114.19	6, 14	0.9122_4^{20}	1.4614^{20}		171	70	misc alc, eth
m203	(E)-4-Methylcyclohexanol	$CH_3C_6H_{10}OH$	114.19	6, 14	0.9118_4^{21}	1.4559^{20}		173–175	70	misc alc; s eth
m204	2-Methylcyclohexanone	$CH_3C_6H_9(=O)$	112.17	7, 14	0.925_4^{20}	1.4478^{20}		162–163	46	i aq; s alc, eth
m205	3-Methylcyclohexanone	$CH_3C_6H_9(=O)$	112.17	7, 15	0.9155_4^{20}	1.4460^{20}		168–169	51	i aq; s alc, eth
m206	4-Methylcyclohexanone	$CH_3C_6H_9(=O)$	112.17	7, 18	0.916_4^{20}	1.4455^{20}		169–171	40	i aq; s alc, eth
m207	1-Methyl-1-cyclohexene		96.17	5, 66	0.809_4^{20}	1.4502^{20}	−121	111	−3	i aq; s alc, eth
m208	4-Methyl-1-cyclohexene		96.17	5, 67	0.799	1.4412^{20}	−115.5	102	−1	i aq; s alc, eth
m209	N-Methylcyclohexylamine	$C_6H_{11}NHCH_3$	113.20	12, 6	0.868	1.4560^{20}		149	29	i aq; s alc, eth

Methyl chloroform, t226

(E)-2-Methylcrotonic acid, m161

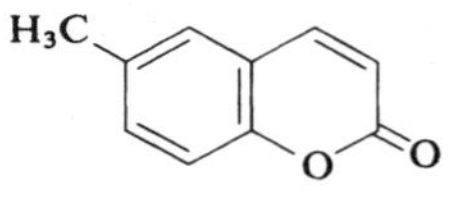

m191

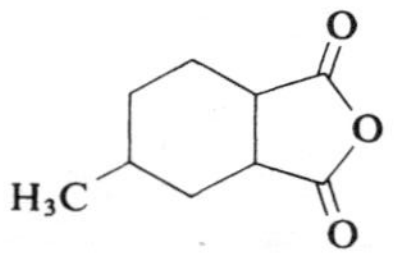

m196

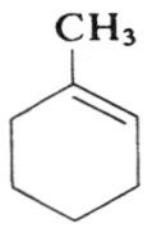

m207

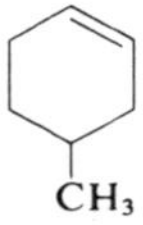

m208

TABLE 1.15 Physical Constants of Organic Compounds (*continued*)

No.	Name	Formula	Formula weight	Beilstein reference	Density	Refractive index	Melting point	Boiling point	Flash point	Solubility in 100 parts solvent
m210	3-Methylcyclo-hexylamine	$C_6H_{10}(CH_3)NH_2$	113.20	12, 10	0.855	1.4525^{20}		$150^{730\text{mm}}$	22	
m211	4-Methylcyclo-hexylamine	$C_6H_{10}(CH_3)NH_2$	113.20	12, 12	0.855	1.4531^{20}		151–154	26	
m212	Methylcyclopenta-diene dimer		160.26		0.941	1.4976^{20}	−51	200	26	
m213	Methylcyclopentane	$C_5H_9CH_3$	84.16	5, 27	0.7487^{20}	1.4097^{20}	−142.4	71.8	−27	0.013 aq
m214	3-Methyl-1,2-cyclo-pentanedione		112.13	$7^1, 310$			105–107			
m215	2-Methylcyclo-pentanone		98.15	$7^2, 13$	0.9200^{20}_4	1.4347^{20}	−76	139–140		s aq; v s alc, eth
m216	3-Methyl-2-cyclo-penten-1-one		96.13	$7^1, 46$	0.971	1.4780^{20}		$74^{15\text{mm}}$	65	
m217	2-Methyl-propanecarboxylic acid		100.12	9, 6	1.027	1.4395^{20}		$191^{745\text{mm}}$	87	
m218	Methyl decanoate	$CH_3(CH_2)_8COOCH_3$	186.30	2, 356	1.3808^{19}	1.4421^{20}	−18	223–224	80	i aq; misc alc, eth
m219	Methyl dichloro-acetate	$Cl_2CHCOOCH_3$	142.97	2, 203			−52	143		i aq; s alc
m220	Methyl 2,2-dichloro-1-methylcyclopro-panecarboxylate		183.03		1.245	1.4639^{20}		$74^{8\text{mm}}$	74	
m221	Methyl 2,3-dichloro-propionate	$ClCH_2CH(Cl)COOCH_3$	157.00	$2^1, 111$	1.3282^{20}_4	1.4447^{20}		$92^{50\text{mm}}$	42	s alc
m222	Methyldichloro-silane	CH_3SiHCl_2	115.0		1.1047^{20}_4	1.4222^{20}	−93	41	−25	
m223	N-Methyldiethanol-amine	$CH_3N(CH_2CH_2OH)_2$	119.16	4, 284	1.0377^{20}	1.4685^{20}		246–248	126	misc aq, alc

No.	Name	Formula	Formula wt.	Beilstein ref.	Density	n_D	mp, °C	bp, °C	Solubility
m224	O-Methyl-N,N′-diisopropylurea	$(CH_3)_2CHNHC(OCH_3){=}NCH(CH_3)_2$	158.25		0.871	1.4358^{20}	35	$50\text{–}52^{0.1mm}$	misc alc, eth; sl s aq
m225	Methyl 3,4-dimethoxybenzoate	$(CH_3O)_2C_6H_3COOCH_3$	196.20	10, 396			57–60	283	
m226	Methyl 4,5-dimethoxy-2-nitrobenzoate	$(CH_3O)_2C_6H_2(NO_2)COOCH_3$	241.20	10, 403			141–144		
m227	Methyl 2,2-dimethylpropionate	$(CH_3)_3CCOOCH_3$	116.16	2^1, 139	0.873	1.3880^{20}	−1	101–103	
m228	2-Methyl-3,5-dinitrobenzoic acid	$CH_3C_6H_2(NO_2)_2COOH$	226.15	9, 474			205–207		
m229	N-Methyldioctylamine	$(C_8H_{17})_2NCH_3$	255.49	4^3, 381		1.4424^{20}	−30.1	165^{15mm}	
m230	N-Methyldiphenylamine	$(C_6H_5)_2NCH_3$	183.26	12, 180	1.048_4^{20}	1.6193^{20}	−7.6	135^{6mm}	i aq; s alc, eth
m231	Methyldiphenylsilane	$(C_6H_5)_2Si(H)CH_3$	198.3		0.9974^{20}	1.569^{20}			
m232	N,N′-Methylenebisacrylamide	$H_2C{=}CHCONHCH_2NHCOCH{=}CH_2$	154.17				>300		
m233	2,2′-Methylenebis(4-chlorophenol)	$CH_2[C_6H_3(Cl)OH]_2$	269.13				177–178		100 EtOH; 100 eth; s PE

Methyl 4,6-dimethyl-2-oxo-2H-pyran-5-carboxylate, m288

Methyldinitrophenols, d631, d633a
Methyl enanthate, m261

Methylene bromide, d88

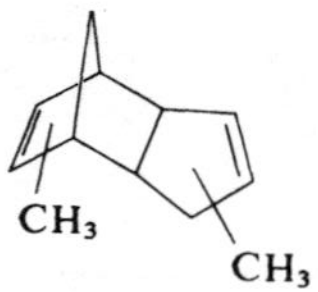

Structures:

m212

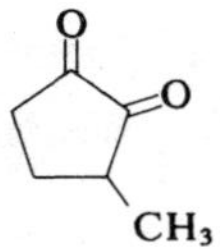

m214 — O, CH₃

m215 — O, CH₃

m216 — O, CH₃

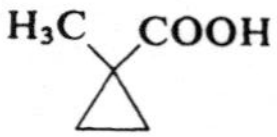

m217 — H₃C—⟨cyclopropane⟩—COOH

m220 — H₃C, COOCH₃, Cl, Cl (cyclopropane)

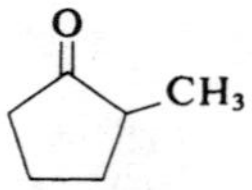

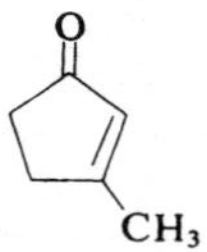

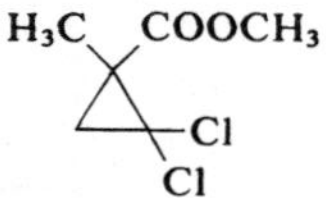

TABLE 1.15 Physical Constants of Organic Compounds (*continued*)

No.	Name	Formula	Formula weight	Beilstein reference	Density	Refractive index	Melting point	Boiling point	Flash point	Solubility in 100 parts solvent
m234	4,4′-Methylenebis-(N,N-dimethyl-aniline)	$CH_2[C_6H_4N(CH_3)_2]_2$	254.38	13, 239			90			i aq, alc, eth, bz; sl s chl; s pyr
m235	4,4′-Methylenebis-(3-hydroxy-2-naphthoic acid	$CH_2[C_{10}H_5(OH)COOH]_2$	388.38	10, 575			d > 280			
m236	1,1′-Methylenebis(3-methylpiperidine)	$CH_2(CH_3 \cdot C_5H_9N)_2$	210.37		0.887	1.4734^{20}		160^{50mm}	110	4 aq; 1.3 alc; s chl
m237	Methylene blue		373.90	27, 393			190 d			
m238	4,4′-Methylenedi-aniline	$CH_2(C_6H_4NH_2)_2$	198.26	13, 238			92	221		0.2 aq; v s alc, eth
m239	3,4-Methylenedi-oxybenzaldehyde		150.13	19, 115			37	264		
m240	1,2-Methylenedi-oxybenzene		122.12	19, 20	1.064	1.5398		173	55	sl s aq, chl, alc, eth
m241	3,4-Methylenedi-oxybenzoic acid		166.13	19, 269			229	subl 210		
m242	3,4-Methylenedi-oxybenzyl alcohol		152.14	19, 67			53–55			misc alc, bz, freons
m243	3,4-Methylenedioxy-6-propylbenzyldi-ethyleneglycol butyl ether		338.45		1.05	1.50^{20}		180^{1mm}	171	
m244	5-Methylene-2-nor-bornene		106.17		0.981	1.4819^{20}			4	
m245	Methylenesuccinic acid	$H_2C=C(COOH)CH_2-COOH$	130.10	2, 760	1.573		162 d			8.2 aq; 20 alc; v s bz, chl, eth, PE
m246	N-Methylethylamine	$CH_3CH_2NHCH_3$	59.11	4^2, 589	0.690	1.3760		35	−12	v s aq, alc

m247	N-Methylethylenedi-amine	$CH_3NHCH_2CH_2NH_2$	74.13	4^{1}, 415		0.841	1.4395^{20}	114–116	41	misc aq
m248	N-Methylformamide	$HCONHCH_3$	59.07	4, 58	–3.8	0.9988^{25}	1.4300^{25}	180–185	98	misc aq
m249	N-Methylformanilide	$C_6H_5N(CH_3)CHO$	135.17	12, 234	8–13	1.095	1.5593^{20}	244	126	23 aq; misc alc
m250	Methyl formate	$HCOOCH_3$	60.05	2, 18	–99.0	0.9815^{15}	1.3465^{15}	31.5	–32	s aq; v s alc; misc eth
m251	5-Methylfurfuraldehyde		110.11	17, 289		1.1072^{18}_{4}	1.5263^{20}	187	72	
m252	2-Methylfuran		82.10	17, 36	–88	0.915^{20}_{4}	1.4332^{20}	63–66	–26	0.3 aq
m253	Methyl furoate		126.11	18, 274		1.179^{20}	1.4862^{20}	181	181	s alc, eth; sl s aq
m254	Methylgermanium tribromide	CH_3GeBr_3	327.35			2.6337^{4}	1.5770^{20}	168	73	

$(CH_3)_2N$ — [phenothiazine ring with N^+, S] — $N(CH_3)_2$
$Cl^- \cdot 3H_2O$
m237

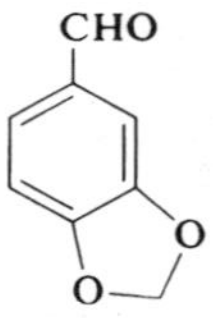

CHO — m239

m240

COOH — m241

CH_2OH — m242

C_3H_7 / $CH_2OCH_2CH_2OCH_2CH_2OC_4H_9$ — m243

CH_2 — m244

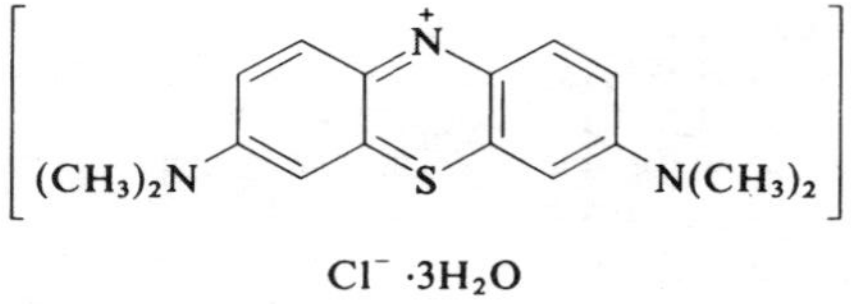

CH_3 — CHO — m251

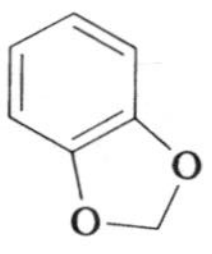

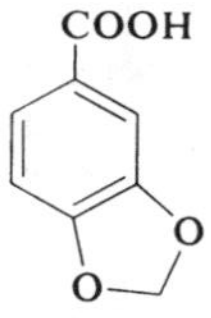

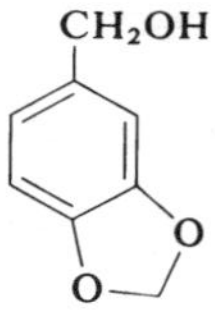

CH_3 — m252

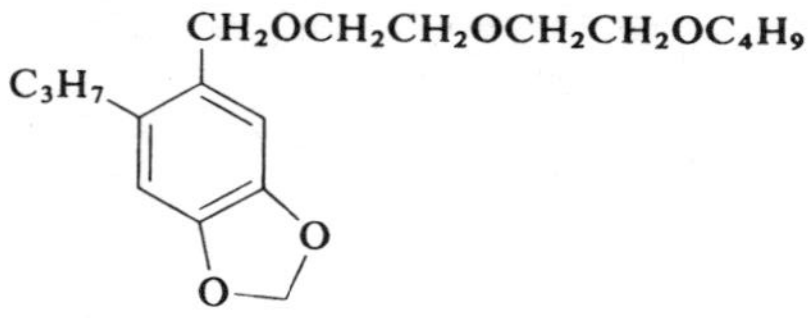

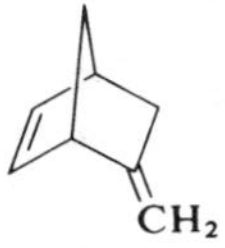

$COOCH_3$ — m253

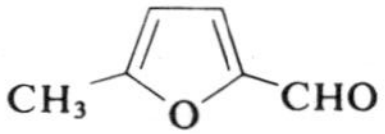

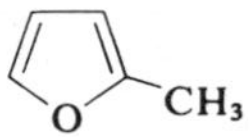

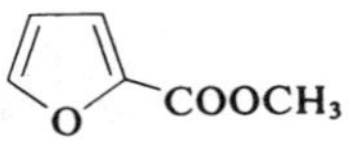

TABLE 1.15 Physical Constants of Organic Compounds (continued)

No.	Name	Formula	Formula weight	Beilstein reference	Density	Refractive index	Melting point	Boiling point	Flash point	Solubility in 100 parts solvent
m255	N-Methyl-D-glucamine		195.22				128–129			100 aq; 1.2 alc
m256	α-Methylglucoside		194.19	31, 179	1.46^{30}_{4}		168	$200^{0.2\text{mm}}$ $125^{10\text{mm}}$	126	63 aq; i alc, eth
m257	DL-2-Methyl-glutaronitrile	$NCCH_2CH_2CH(CH_3)CN$	108.14	2, 656	0.950					
m258	N-Methylglycine	CH_3NHCH_2COOH	89.09	4, 345	1.168^{18}		d 212			42 aq; sl s alc
m259	Methyl glycolate	$HOCH_2COOCH_3$	90.08	3, 236			74	151	4	s aq; misc alc, eth
m260	2-Methylheptane	$CH_3(CH_2)_4CH(CH_3)_2$	114.23	1, 161	0.6978^{20}_{4}	1.3974^{20}	−109.0	117.7		s eth; sl s alc
m261	Methyl heptanoate	$CH_3(CH_2)_5COOCH_3$	144.22	2, 339	0.8815^{20}_{4}	1.4115^{20}	−55.8	173.8	52	s alc, eth; sl s aq
m262	6-Methyl-5-hepten-2-one	$(CH_3)_2C{=}CHCH_2CH_2COCH_3$	126.20	1^2, 797	0.855^{16}_{4}	1.4392^{20}	−67	$73^{18\text{mm}}$	50	misc alc, eth
m263	Methyl hexadecanoate	$CH_3(CH_2)_{14}COOCH_3$	270.46	2, 372			28	$196^{15\text{mm}}$		s alc, chl, eth
m264	2-Methylhexane	$CH_3(CH_2)_3CH(CH_3)_2$	100.21	1, 156	0.6786^{20}	1.3849^{20}	−118.3	90.1	−3	s alc; misc eth
m265	Methyl hexanoate	$CH_3(CH_2)_4COOCH_3$	130.19	2, 323	0.9038^{0}_{4}	1.4038^{23}	−71	151	54	v s alc, eth
m266	5-Methyl-2-hexanol	$(CH_3)_2CHCH_2CH_2CH(OH)CH_3$	116.20	1, 416	0.814^{20}_{4}	1.4176^{20}		150	46	s alc, eth; i aq
m267	5-Methyl-2-hexanone	$(CH_3)_2CHCH_2CH_2COCH_3$	114.19	1^2, 756	0.888^{20}_{4}	1.4062^{20}		141	41	0.5 aq; misc alc, eth
m268	5-Methyl-3-hexen-2-one	$(CH_3)_2CHCH{=}CHCOCH_3$	112.17			1.4400^{20}				
m268a	1-Methylhexylamine	$H(CH_2)_5CH(NH_2)CH_3$	115.22	4, 194	0.7665^{18}	1.4175^{20}		144	54	sl s aq; s alc, eth
m269	1-Methylhydantoin		114.10	24, 244			157	subl		s aq, alc; 3 eth
m270	Methylhydrazine	CH_3NHNH_2	46.07	4^2, 957	0.866	1.4235^{20}	−52.4	87.5	21	misc aq, alc; s PE
m271	Methyl hydrazino-carboxylate	$H_2NNHCOOCH_3$	90.08	3^1, 46			70–73	$108^{12\text{mm}}$		
m272	Methyl hydrogen glutarate	$HOOCCH_2CH_2COOCH_3$	146.14	2^2, 565	1.169	1.4381^{20}		$151^{10\text{mm}}$	>112	

		Mol. wt	Beilstein	Density	n_D	mp	bp	fp	Solubility
m273	Methyl hydrogen hexanedioate — $HOOC(CH_2)_4COOCH_3$	160.17	2, 652	1.081	1.4401^{20}	8–9	162^{10mm}	>112	s alc
m274	Methyl hydrogen succinate — $HOOCCH_2CH_2COOCH_3$	132.12	2, 608			56–59	151^{20mm}		v s aq, alc, eth
m275	Methyl hydroperoxide — CH_3OOH	48.04	1^2, 270	1.997_4^{15}	1.3642^{15}		38^{65mm}		misc aq, alc, eth; s bz
m276	Methylhydroquinone	124.14	6, 874			125–128			
m277	Methyl 4-hydroxybenzoate — $HOC_6H_4COOCH_3$	152.15	10, 158			126–128	270 d		v s alc, eth, acet
m278	Methyl 2-hydroxyisobutyrate — $(CH_3)_2C(OH)COOCH_3$	118.13	3^2, 223	1.023	1.4112^{20}		137	42	v s aq, alc
m279	Methyl 4-hydroxyphenylacetate — $HOC_6H_4CH_2COOCH_3$	166.18	10, 191	1.030	1.4970^{20}	57–60	162–163^{5mm}		
m280	1-Methylimidazole	82.11	23, 46			–60	198	92	misc aq
m281	2-Methylimidazole	82.11	23, 65			143	268		
m282	4-Methylimidazole	82.11	23, 69			46–48	263	>112	

N-Methylguanidine acetic acid, c277
4-Methylhexahydrophthalic anhydride, m196
Methyl hydroxyacetate, m259

Methyl 4-hydroxy-3-methoxybenzoate, m445
Methyl 2-hydroxypropionate, m291
2,2′-Methyliminodiethanol, m223

2,2′-Methyliminobis(acetaldehyde diethyl acetal), b169

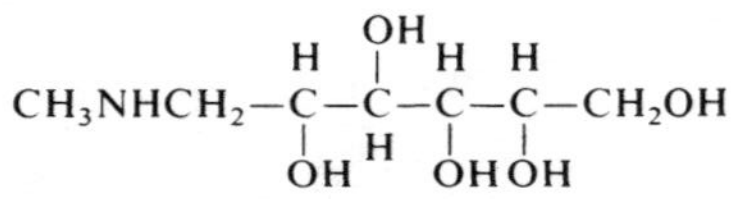

m255 — CH_3NHCH_2–C–C–C–C–CH_2OH (with H, OH, H, H and OH, H, OH, OH substituents)

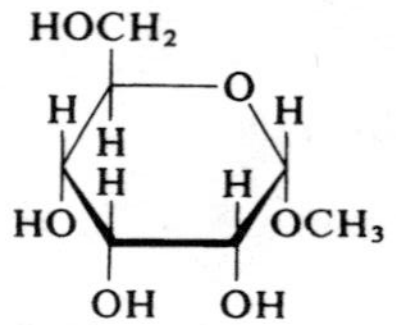

m256

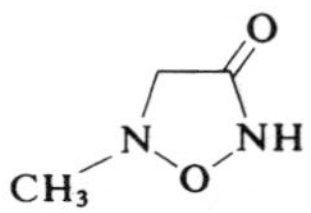

m269

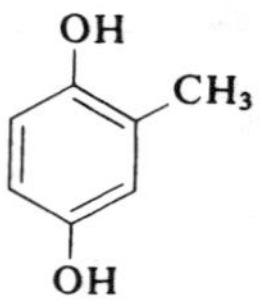

m276

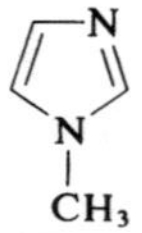

m280

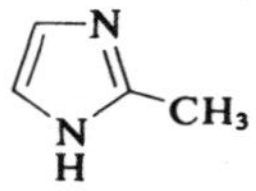

m281

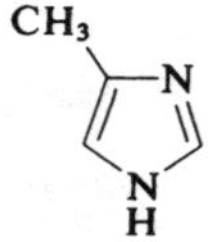

m282

TABLE 1.15 Physical Constants of Organic Compounds (*continued*)

No.	Name	Formula	Formula weight	Beilstein reference	Density	Refractive index	Melting point	Boiling point	Flash point	Solubility in 100 parts solvent
m283	2-Methyl-1*H*-indole		131.18	20, 311	1.07^{20}_4		58–60	273		v s alc, eth; s hot aq
m284	3-Methyl-1*H*-indole		131.18	20, 315			95	266		s hot aq, alc, bz
m285	*N*-Methylisatoic anhydride		177.16	27, 265			165 d			
m286	Methyl isobutyrate	(CH$_3$)$_2$CHCOOCH$_3$	102.13	2, 290	0.891^{20}	1.3840^{20}	–84	93	<1	misc alc, eth; sl s aq
m287	Methyl isocyanate	CH$_3$NCO	57.05	4, 77	0.967	1.3695^{20}	–17	37–39	–18	s aq
m288	Methyl isodehydacetate		182.18	18, 410			60–63	167^{14mm}		
m289	Methyl isothiocyanate	CH$_3$NCS	73.12	4, 77	1.069	1.5258^{37}	35–36	119	32	v s alc, eth; sl s aq
m290	5-Methylisoxazole		83.09	27, 16	1.018	1.4386^{20}	~66	122	30	misc aq(d), alc, eth
m291	Methyl lactate	CH$_3$CH(OH)COOCH$_3$	104.10	3, 280	1.0882^{20}_4	1.4131^{20}		144.8	52	
m292	Methyl mandelate	C$_6$H$_5$CH(OH)COOCH$_3$	166.18	10, 202	1.1756^{20}		51–54	135^{12mm}		s aq, alc, bz, chl
m293	Methyl mercaptoacetate	HSCH$_2$COOCH$_3$	106.14		1.187	1.4657^{20}		43^{10mm}	30	s alc, eth
m294	Methyl 3-mercaptopropionate	HSCH$_2$CH$_2$COOCH$_3$	120.17	3², 214	1.085	1.4640^{20}		55^{14mm}	60	
m295	Methylmercury chloride	CH$_3$HgCl	251.10		4.06^{25}		170			
m296	Methyl methacrylate	H$_2$C=C(CH$_3$)COOCH$_3$	100.12	2², 398	0.9433^{20}	1.4146^{20}	–48.2	100.3	10	1.6 aq; s ketones, esters, CCl$_4$
m297	Methyl methanesulfonate	CH$_3$SO$_2$OCH$_3$	110.13	4, 4	1.2943^{20}_4	1.4138^{20}		202–203	104	20 aq; 100 DMF
m298	Methyl methoxyacetate	CH$_3$OCH$_2$COOCH$_3$	104.11	3, 236	1.0511^{20}_4	1.3964^{20}		130	35	v s alc, eth; sl s aq

m299	Methyl 1-methoxybicyclo[2.2.2]oct-5-ene-2-carboxylate		196.25		1.086	1.4886[20]	105[17mm]	103
m300	Methyl 4-methoxyphenylacetate	CH$_3$OC$_6$H$_4$CH$_2$COOCH$_3$	180.20	10, 191	1.135	1.5165[20]	158[19mm]	36
m301	1-Methyl-4-(methylamino)piperidine		128.22		0.882	1.4672[20]		55
m302	Methyl 3-methylbutyrate	(CH$_3$)$_2$CHCH$_2$COOCH$_3$	116.16	2[2], 274	0.881$_4$[20]	1.3800[25]	116–117	
m303	2-Methyl-6-methylene-2-octanol	C$_2$H$_5$C(=CH$_2$)(CH$_2$)$_3$-C(CH$_3$)$_2$OH	156.27		0.784	1.4431[20]	84[10mm]	76	sl s aq; misc alc, eth

Methyl iodide, i40
Methyl isoamyl ketone, m267
Methyl isobutenyl ketone, m350
Methyl isobutyl ketone, m346
Methyl isonicotinate, m402
Methyl isopentyl ketone, m267

Methyl isovalerate, m302
2-Methyllactic acid, h127
Methyl linoleate, m325
Methyl mercaptan, m33
Methylmercaptoanilines, m422, m423
4-Methylmercaptobenzaldehyde, m424

Methylmercaptophenols, m431a, m429
7-Methyl-3-methylene-1,6-octadiene, m453
1-Methyl-4-(1-methylethenyl)cyclohexane, d649
5-Methyl-2-(1-methylethyl)cyclohexanol, m12
5-Methyl-2-(1-methylethyl)cyclohexanone, m13

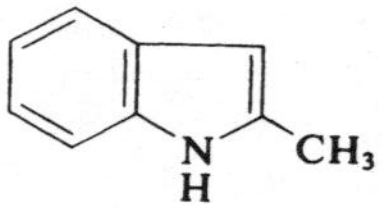

m283

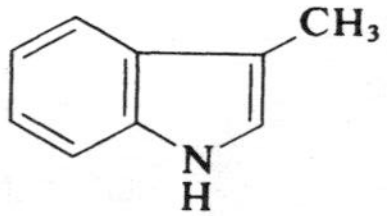

m284

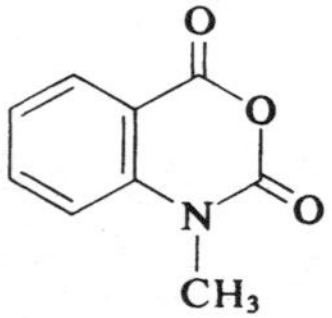

m285

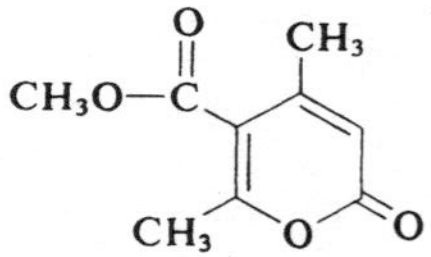

m288

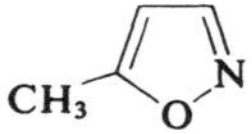

m290

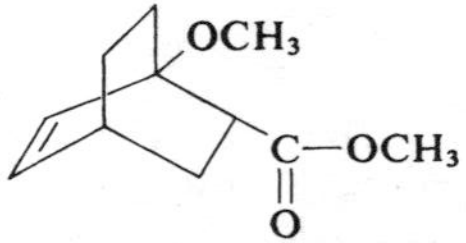

m299

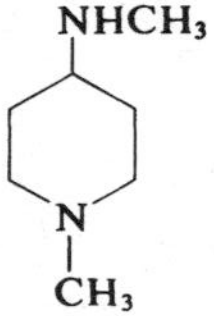

m301

TABLE 1.15 Physical Constants of Organic Compounds (*continued*)

No.	Name	Formula	Formula weight	Beilstein reference	Density	Refractive index	Melting point	Boiling point	Flash point	Solubility in 100 parts solvent
m304	Methyl 2-methyl-3-furancarboxylate		140.14		1.116	1.4730^{20}		75^{20mm}	63	
m305	Methyl S-methyl-thiomethyl sulfoxide	$CH_3S(=O)CH_2SCH_3$	124.22		1.191	1.5487^{20}		$95^{2.5mm}$	>112	
m306	Methyl 3-(methyl-thio)propionate	$CH_3SCH_2CH_2COOCH_3$	134.20		1.077	1.4650^{20}		75^{13mm}	72	s aq, alc, eth
m307	N-Methylmorpholine		101.15	27, 6	0.920	1.4349^{20}	−66	116	23	v s alc, eth
m308	1-Methylnaphthalene	$C_{10}H_7CH_3$	142.20	5, 566	1.025^{14}_4	1.6159^{20}	−30.5	244.7	82	v s alc, eth
m309	2-Methylnaphthalene	$C_{10}H_7CH_3$	142.20	5, 567	1.029^{20}_4	1.6026^{40}	34.6	241.4		1.4 alc; 10 bz; s chl
m310	2-Methyl-1,4-naphthoquinone		172.18	7^2, 656			105–107			s alc, eth; i aq
m311	Methyl 1-naphthyl ketone	$C_{10}H_7COCH_3$	170.21	7, 401	1.1336^0_4	1.6284^{20}	12	296–298		sl s alc; s CS_2
m312	Methyl 2-naphthyl ketone	$C_{10}H_7COCH_3$	170.21	7, 402			53–55	300–301		sl s aq; s alc, eth
m313	Methyl nitrate	CH_3ONO_2	77.04	1, 284	1.2075^{20}_4	1.3748^{20}	−83.0	64 explodes		s alc, eth
m314	Methyl nitrate	CH_3ONO	61.04	1, 284	0.991 (liquid)			−17.35		v s alc; s bz
m315	2-Methyl-4-nitro-aniline	$CH_3C_6H_3(NO_2)NH_2$	152.15	12, 846	1.586^{140}_4		131–133			s alc, acet, eth
m316	2-Methyl-5-nitro-aniline	$CH_3C_6H_3(NO_2)NH_2$	152.15	12, 844			104–107			v s alc; s eth
m317	4-Methyl-2-nitro-aniline	$CH_3C_6H_3(NO_2)NH_2$	152.15	12, 100			115–116			s alc, eth
m318	Methyl 2-nitro-benzoate	$O_2NC_6H_4COOCH_3$	181.15	9, 372	1.280	1.5350^{20}	−13	$106^{0.1mm}$	>112	

No.	Name	Formula	Mol wt	Beil. ref	Density	n_D	m.p., °C	fl. p.	b.p., °C	Solubility
m319	2-Methyl-3-nitro-benzoic acid	$CH_3C_6H_3(NO_2)COOH$	181.15	9, 471			182–184			
m320	4-Methyl-3-nitro-benzoic acid	$CH_3C_6H_3(NO_2)COOH$	181.15	9, 502			187–190			
m321	5-Methyl-2-nitro-benzoic acid	$CH_3C_6H_3(NO_2)COOH$	181.15	9, 482			134–136			
m322	2-Methyl-5-nitro-imidazole		127.10	$23^1, 23$			252–254			
m323	3-Methyl-2-nitro-phenol	$CH_3C_6H_3(NO_2)OH$	153.14	6, 385			35–39			
m324	4-Methyl-2-nitro-phenol	$CH_3C_6H_3(NO_2)OH$	153.14	6, 412	1.240^{20}_4	1.574^{40}	32–35		125^{22mm}	v s alc, eth
m325	Methyl 9,12-octa-decadienoate	$CH_3(CH_2)_4CH=CHCH_2-CH=CH(CH_2)_7COOCH_3$	294.46		0.8886^{18}_4	1.4593^{25}	–35		212^{16mm}	misc DMF
m326	Methyl octa-decanoate	$CH_3(CH_2)_{16}COOCH_3$	298.51	2, 379			38–39	>112	215^{15mm}	s alc, eth
m327	Methyl cis-9-octa-decenoate	$CH_3(CH_2)_7CH=CH-(CH_2)_7COOCH_3$	296.50	2, 467	0.879^{18}_4	1.4521^{20}	19.9		168^{2mm}	misc abs alc, eth
m328	Methyloctadecyldi-chlorosilane	$C_{18}H_{37}Si(CH_3)Cl_2$	367.5		0.930^{20}_4				$185^{2.5mm}$	

Methyl 2-methyllactate, m278
Methyl methyl-2-propenoate, m296
Methyl methylsulfinylmethyl sulfide, m305

Methyl myristate, m413
Methyl nicotinate, m404
4-Methyl-3-nitroanisole, m85

Methyl 6-nitrovertrate, m226
Methyl nonyl ketone, u5
Methyl oleate, m327

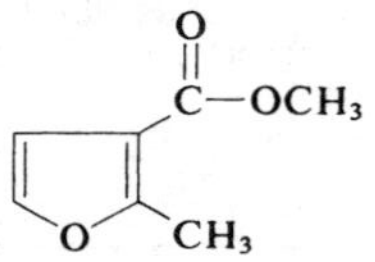

m304

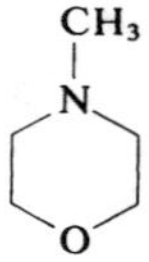

m307

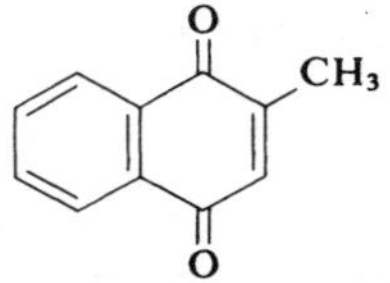

m310

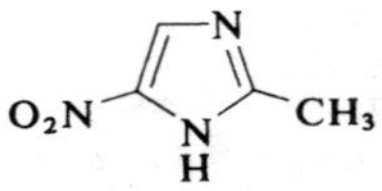

m322

TABLE 1.15 Physical Constants of Organic Compounds (*continued*)

No.	Name	Formula	Formula weight	Beilstein reference	Density	Refractive index	Melting point	Boiling point	Flash point	Solubility in 100 parts solvent
m329	Methyl octanoate	$CH_3(CH_2)_6COOCH_3$	158.24	2, 348	0.8775^{20}_4	1.4160^{25}	-40	192.9		v s alc, eth; i aq
m330	Methyloctyldichlorosilane	$C_8H_{17}Si(CH_3)Cl_2$	227.3		0.9764^{20}_4	1.444^{20}		94^{6mm}	>112	
m331	3-Methyl-2-oxazolidinone		101.11	27, 11	1.170	1.4541^{20}	15	$87-90^{1mm}$	20	
m332	2-Methyl-2-oxazoline		85.11	10, 597	1.005	1.4340^{20}		110	>112	
m333	Methyl 2-oxocyclopentanecarboxylate		142.15	3, 616	1.145	1.4560^{20}		105^{19mm}	39	misc alc, eth; sl s aq
m334	Methyl 2-oxopropionate	$CH_3C(=O)COOCH_3$	102.09		1.130	1.4065^{20}		134-137		
m335	Methyl 2-oxo-1-pyrrolidineacetate		157.17		1.131	1.4719^{20}			110	
m336	2-Methylpentane	$CH_3CH_2CH_2CH(CH_3)_2$	86.18	1, 148	0.6532^{20}	1.3725^{20}	-153.7	60.3	-23	
m337	3-Methylpentane	$(CH_3CH_2)_2CHCH_3$	86.18	1, 149	0.6643^{20}	1.3765^{20}	<-50 glass	63.3	-6	
m338	2-Methyl-2,4-pentanediol	$(CH_3)_2C(OH)CH_2CH(OH)CH_3$	118.18	1, 486	0.9216^{20}_4	1.4270^{20}	<-50 glass	198.3	101	misc aq
m339	4-Methylpentanenitrile	$(CH_3)_2CHCH_2CH_2CN$	97.16	2^2, 290	0.8035^{20}_4	1.4061^{20}	-51.1	153.5		s alc; misc eth
m340	Methyl pentanoate	$CH_3(CH_2)_3COOCH_3$	116.16	2, 301	0.875	1.3962^{20}		128	22	sl s aq; misc alc, eth
m341	2-Methylpentanoic acid	$CH_3CH_2CH_2CH(CH_3)COOH$	116.16	2^2, 288	0.9242^{20}_{20}	1.4135^{20}	-85 glass	196.4	107	1.3 aq
m342	3-Methylpentanoic acid	$CH_3CH_2CH(CH_3)CH_2COOH$	116.16	2, 331	0.9262^{20}	1.4159^{20}	-42	196-198	85	s alc, eth
m343	2-Methyl-1-pentanol	$CH_3CH_2CH_2CH(CH_3)CH_2OH$	102.18	1, 409	0.8242^{20}	1.4190^{20}		148.0	50	s alc, eth

m344	3-Methyl-3-pentanol	$(CH_3CH_2)_2C(CH_3)OH$	102.18	1, 411	0.8281^{20}	1.4186^{20}	<-38	122.4	46	misc alc, eth; sl s aq
m345	4-Methyl-2-pentanol	$(CH_3)_2CHCH_2$-$CH(OH)CH_3$	102.18	1,410	0.8080^{20}	1.4112^{20}	-90	131.7	41	1.6 aq
m346	4-Methyl-2-pentanone	$(CH_3)_2CHCH_2COCH_3$	100.16	1,691	0.8006^{20}_{4}	1.3958^{20}	-83.5	115.7	13	1.7 aq; misc alc, bz, eth
m347	2-Methyl-1-pentene	$CH_3CH_2CH_2$-$C(CH_3)=CH_2$	84.16	1^1, 90	0.6799^{20}_{4}	1.3920^{20}	-135.7	62.1	-26	s alc
m348	2-Methyl-2-pentene	$CH_3CH_2CH=C(CH_3)_2$	84.16	1, 217	0.6865^{20}_{4}	1.4003^{20}	-135.1	67.3	-23	s alc
m349	4-Methyl-2-pentenoic acid	$(CH_3)_2CHCH=$CHCHCOOH	114.14	2^2, 406	0.9529	1.4489	35	115^{20mm}	46	i aq; v s alc
m350	4-Methyl-3-penten-2-one	$(CH_3)_2C=CHCOCH_3$	98.15	1,736	0.854^{20}_{4}	1.4458^{20}	-42	129.5	30	3.1 aq
m350a	1-Methylpentylamine	$CH_3(CH_2)_3$-$CH(NH_2)CH_3$	101.19	4, 190	0.767^{20}_{4}	1.3930^{20}	-19	116–118	13	s aq, alc, PE
m351	4-Methyl-1-pentyne	$(CH_3)_2CHCH_2C≡CH$	82.15	1^2, 506	0.7041^{20}_{4}	1.3930^{20}	-104.8	61.2		13 aq; misc bz, acet, PE, EtAc; s eth
m352	3-Methyl-1-pentyn-3-ol	$CH_3CH_2C(CH_3)(OH)$-$C≡CH$	98.15		0.8688^{20}_{4}	1.4318^{20}	-30.6	121–122	38	
m353	Methyl-(2-phenylethyl)dichlorosilane	$C_6H_5CH_2CH_2Si(CH_3)Cl_2$	219.2		1.111^{20}_{4}	1.510^{20}		99^{6mm}		

o-Methylolphenol, h105
2-Methyloxacyclopropane, p227

Methyl oxirane, p227
Methyl palmitate, m263

Methyl pentyl ketone, h15

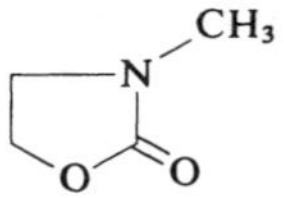

m331

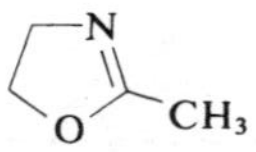

m332

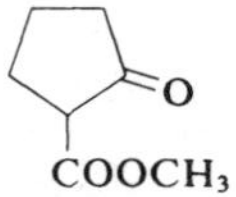

m333

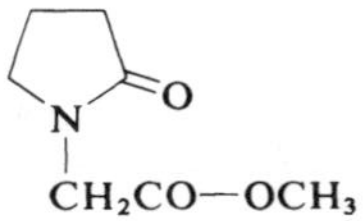

m335

TABLE 1.15 Physical Constants of Organic Compounds (*continued*)

No.	Name	Formula	Formula weight	Beilstein reference	Density	Refractive index	Melting point	Boiling point	Flash point	Solubility in 100 parts solvent
m354	(1-Methylphenethyl)-trichlorosilane	$C_6H_5CH(CH_3)CH_2SiCl_3$	253.6		1.226_4^{20}	1.515^{20}		116^{10mm}		s alc, EtAc, HOAc
m355	N-(4-Methylphenyl)-acetamide	$CH_3C_6H_4NHCOCH_3$	149.19	12[2], 501	1.212^{15}		153	307	90	
m356	Methyl phenylacetate	$C_6H_5CH_2COOCH_3$	150.18	9, 434	1.044	1.5075^{20}		215		i aq; misc alc, eth
m357	Methylphenylchloro-silane	$C_6H_5(CH_3)Si(H)Cl$	156.7		1.1054_4^{20}	1.571^{20}		113^{100mm}		
m358	Methylphenyldi-chlorosilane	$C_6H_5Si(CH_3)Cl_2$	191.1		1.187_4^{20}			205–206		
m359	p-(1-Methyl-2-phenylethyl)phenol	$C_6H_5CH_2CH(CH_3)-$ C_6H_4OH	212.29	15, 117			73	335		
m360	1-Methyl-1-phenyl-hydrazine	$C_6H_5N(CH_3)NH_2$	122.17		1.038^{22}	1.5834^{20}		118^{21mm}	96	misc alc, bz, chl, eth
m361	1-Methyl-3-phenyl-propyl acetate	$C_6H_5CH_2CH_2CH(CH_3)-$ $OOCCH_3$	192.26	6[1], 258	0.991			$74^{0.05mm}$	>112	
m362	3-Methyl-1-phenyl-2-pyrazolin-5-one		174.20	24, 20			130	287^{265mm}		
m363	Methylphenylsilane	$C_6H_5Si(CH_3)H_2$	122.1		0.889_4^{20}	1.506^{20}		139–240		
m364	Methyl phenyl sulfide	$C_6H_5SCH_3$	124.21	6, 297	1.058	1.5852^{20}	−15	188		i aq; s alc
m365	N-Methylpiperazine		100.17		0.903	1.4655^{20}		138	42	v s aq; alc, eth
m366	2-Methylpiperazine		100.17	23, 17			65–67	155.6	22	78 aq; 37 acet; 32 bz
m367	4-Methyl-1-piper-azinepropanol		158.25	23[3], 123		1.4835^{20}	28–30	120–121[9mm]		
m368	N-Methylpiperidine	$C_5H_{10}N{-}CH_3$	99.19	20, 19	0.816	1.4378^{20}		106–107	<1	v s aq; misc alc, eth
m369	2-Methylpiperidine	$CH_3C_5H_9NH$	99.19	20, 95	0.844	1.4459^{20}	−5	119	8	v s aq; misc alc, eth

m370	3-Methylpiperidine	$CH_3C_5H_9NH$	99.19	20, 100	0.845	1.4470^{20}		126	<1	v s aq
m371	4-Methylpiperidine	$CH_3C_5H_9NH$	99.19	20, 101	0.838	1.4458^{20}		124	7	v s aq
m372	1-Methyl-3-piperidinemethanol		129.20	21^2, 8	1.013	1.4772^{20}		140–145	94	
m373	1-Methyl-4-piperidone		113.16	21^2, 215	0.920	1.4614^{20}			60	9 aq; misc alc, bz, chl, eth
m374	2-Methylpropanal	$(CH_3)_2CHCHO$	72.11	1, 671	0.7891^{20}	1.3727^{20}	−65	64.1		13 mL aq; 1320 mL alc; 2890 mL eth
m375	2-Methylpropane	$(CH_3)_3CH$	58.12	1, 124	0.557^{20}		−159.6	−11.7		
m376	N-Methyl-1,3-propanediamine	$H_2NCH_2CH_2CH_2NHCH_3$	88.15	4^1, 419	0.844	1.4468^{20}		139–141	35	
m377	2-Methyl-1,2-propanediamine	$(CH_3)_2C(NH_2)CH_2NH_2$	88.15	4, 266	0.841	1.4410^{20}			23	
m378	1-Methyl-1-propanethiol	$CH_3CH_2CH(SH)CH_3$	90.19	1, 373	0.8246^{25}_{4}	1.4338^{25}	−165	84–85	21	sl s aq; v s alc, eth

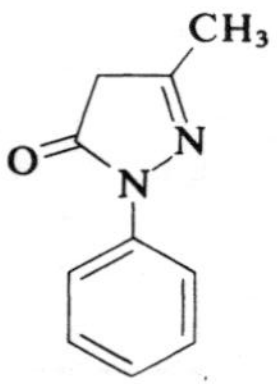

m362

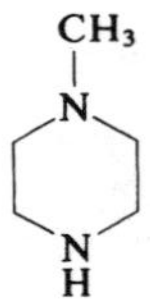

m365

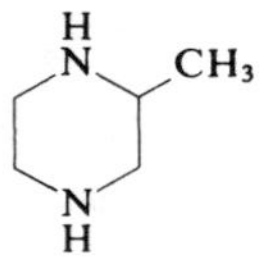

m366

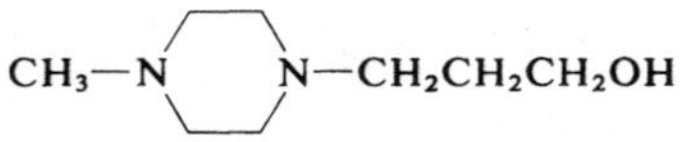

m367 CH_3-N (piperazine) $N-CH_2CH_2CH_2OH$

m372 CH_2OH ring, $N-CH_3$

m373 $O=$ ring, $N-CH_3$

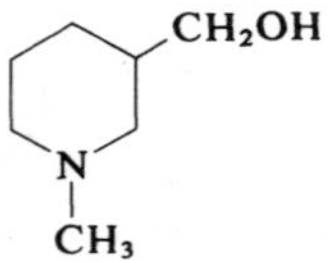

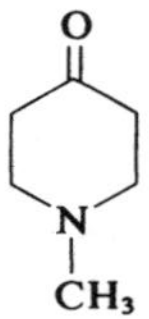

TABLE 1.15 Physical Constants of Organic Compounds (*continued*)

No.	Name	Formula	Formula weight	Beilstein reference	Density	Refractive index	Melting point	Boiling point	Flash point	Solubility in 100 parts solvent
m379	2-Methyl-1-propane-thiol	$(CH_3)_2CHCH_2SH$	90.19	1, 378	0.8357_4^{20}	1.4396^{20}	−79	88.5	−9	v s alc, eth
m380	2-Methyl-2-propane-thiol	$(CH_3)_3CSH$	90.19	1, 383	0.7943_4^{25}	1.4198^{25}	1.1	64.2	−26	i aq
m381	2-Methyl-1-propanol	$(CH_3)_2CHCH_2OH$	74.12	1, 373	0.8016^{20}	1.3958^{20}	−108	107.9	27	10 aq; misc alc, eth
m382	2-Methyl-2-propanol	$(CH_3)_3COH$	74.12	1, 379	0.7858_4^{20}	1.3877^{20}	25.8	82.4	15	misc aq, alc, eth
m383	2-Methylpropene	$(CH_3)_2C{=}CH_2$	56.10	1, 207	0.6266_4^{-140}		−140.4	−6.9		v s alc, eth
m384	2-Methyl-2-propene-1-sulfonic acid, Na salt	$H_2C{=}C(CH_3)CH_2SO_3^-{\cdot}Na^+$	158.15				>300			
m385	2-Methyl-2-propen-1-ol	$H_2C{=}C(CH_3)CH_2OH$	72.11	1, 443	0.857	1.4250^{20}		113–115	33	
m386	4-Methyl-2-(2-pro-penyl)phenol	$CH_3C_6H_3{-}$ $(CH_2CH{=}CH_2)OH$	148.21	6¹, 287	0.980	1.5385^{20}		238	101	
m387	6-Methyl-2-(2-pro-penyl)phenol	$CH_3C_6H_3{-}$ $(CH_2CH{=}CH_2)OH$	148.21	6¹, 287	0.992	1.5381^{20}		231–233	94	
m388	N-Methylpropion-amide	$CH_3CH_2CONHCH_3$	87.12		0.9305^{25}	1.4345^{25}	−30.9	148		
m389	Methyl propionate	$CH_3CH_2COOCH_3$	88.11	2, 239	0.915_4^{20}	1.3770^{20}	−88	79.7	−2	6 aq; misc alc, eth
m390	2-Methylpropionic acid	$(CH_3)_2CHCOOH$	88.11	2, 288	0.950_4^{20}	1.3930^{20}	−46.1	154.7	55	23 aq; misc alc, chl, eth
m391	4'-Methylpropio-phenone	$CH_3C_6H_4COCH_2CH_3$	148.21	7, 317	0.993	1.5280^{20}	7.2	239	96	
m392	Methylpropyldi-chlorosilane	$CH_3CH_2CH_2Si(CH_3)Cl_2$	157.1		1.04_4^{25}	1.425^{25}		125		
m393	Methyl propyl ether	$CH_3CH_2CH_2OCH_3$	74.12	1, 354	0.738_4^{20}			39.1		sl s aq; misc alc, eth

No.	Name	Formula	Formula wt	Beilstein	Density	n_D	mp, °C	bp, °C	Solubility	Solubility
m394	2-Methyl-2-propyl-1,3-propanediol	$C_3H_7C(CH_3)(CH_2OH)_2$	132.20	1[1], 254	0.8424[20]	1.4442[20]	53–55	230		s aq
m395	Methyl propyl sulfide	$CH_3SCH_2CH_2CH_3$	90.18	1[3], 1432	0.830	1.3961[20]	–113.0	95.5		v s aq, alc, eth
m396	Methyl 2-propynyl ether	$CH_3OCH_2C{\equiv}CH$	70.09	1, 454				61–62	<1	
m397	2-Methylpyrazine	$CH_3C_5H_4N$	94.12	23, 94	1.030	1.5042[20]	–29	135	50	v s aq, alc, eth
m398	2-Methylpyridine	$CH_3C_5H_4N$	93.13	20, 234	0.950[15]	1.5010[20]	–67	128–129	26	v s aq; s alc, eth
m399	3-Methylpyridine	$CH_3C_5H_4N$	93.13	20, 239	0.961[4]	1.5068[20]	–18.3	143.5	36	s aq, alc, eth
m400	4-Methylpyridine	$CH_3C_5H_4N$	93.13	20, 240	0.957[15]	1.5058	3.8	143–145	56	s aq, alc, eth
m401	Methyl 3-pyridine-carboxylate	$(C_5H_4N)COOCH_3$	137.14	22, 39			39	209		s aq, alc, bz
m402	Methyl 4-pyridine-carboxylate	$(C_5H_4N)COOCH_3$	137.14	22, 46	1.001	1.5122[20]	8.5	207–209	82	
m403	1-Methyl-2-pyridone		109.13	21, 268	1.112	1.5690[20]	7	250[740mm]		
m404	N-Methylpyrrole		81.2	20, 163	0.914	1.4875[20]	–57	113	15	i aq; misc alc, eth

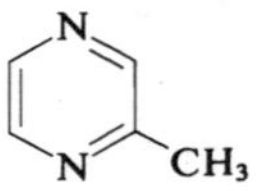

m397 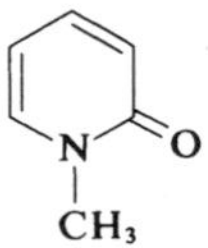   m403    m404

TABLE 1.15 Physical Constants of Organic Compounds (*continued*)

No.	Name	Formula	Formula weight	Beilstein reference	Density	Refractive index	Melting point	Boiling point	Flash point	Solubility in 100 parts solvent
m405	N-Methylpyrrolidine		85.15	20, 4	0.819_4^{20}	1.4247^{20}		80–81	−21	misc aq; eth
m406	N-Methyl-2-pyrrolidinone		99.13	21^2, 213	1.0279^{25}	1.4680^{25}	−24.4	202	95	misc aq, alc, bz, eth
m407	2-Methylquinoline		143.19	20, 387	1.058	1.6108^{20}	−2	248	79	i aq; s chl, eth
m408	4-Methylquinoline		143.19	20, 395	1.0826_4^{20}	1.6200^{20}	9–10	261–263	>112	misc alc, bz, eth
m409	2-Methylquinoxaline		144.18	23^1, 44	1.118	1.6156^{20}	180–181	245–247	107	misc aq
m410	Methyl salicylate	$HOC_6H_4COOCH_3$	152.15	10, 70	1.1831^{20}	1.5240^{20}	−8.6	223.0	110	0.7 aq; s chl, eth; misc alc, HOAc
m411	α-Methylstyrene	$C_6H_5C(CH_3){=}CH_2$	118.18	5, 484	0.909	1.5375^{20}	−23.2	165.5	45	
m412	Methylsuccinic acid	$HOOCCH_2CH(CH_3)COOH$	132.12	2, 636	1.411	1.4303	110–112	d		66 aq; v s alc, eth
m413	Methyl tetradecanoate	$CH_3(CH_2)_{12}COOCH_3$	242.40	2^2, 326	0.855	1.4362^{20}	18.4	323	>112	misc alc, bz, eth
m414	2-Methyl-3,3,4,4-tetrafluoro-2-butanol	$HCF_2CF_2C(CH_3)_2OH$	160.11		1.282	1.3524^{20}		117	73	
m415	2-Methyltetrahydrofuran		86.13	17, 12	0.860	1.4056^{20}		78–80	−11	
m416	1-Methyl-1,2,3,6-tetrahydropyridine		97.16		0.837	1.4570^{20}		113–114	8	
m417	3-Methyltetrahydrothiophene-1,1-dioxide		134.20		1.191	1.4772^{20}		276	>112	
m418	4-Methyl-5-thiazoleethanol		143.21		1.196	1.5508^{20}		135^{7mm}		
m419	2-Methyl-2-thiazoline		101.17	27, 13	1.067	1.5200^{20}	−101	145	37	
m420	Methyl thioacetate	CH_3COSCH_3	90.14			1.4628		98	10	s alc, eth
m421	(Methylthio)acetonitrile	CH_3SCH_2CN	87.14		1.039	1.4826^{20}		63^{15mm}	67	

No.	Name	Formula								
m422	2-(Methylthio)-aniline	$CH_3SC_6H_4NH_2$	139.22	13, 399	1.111	1.6239^{20}		234	>112	i aq; misc alc, eth
m423	3-(Methylthio)-aniline	$CH_3SC_6H_4NH_2$	139.22	13^1, 141	1.130	1.6423^{20}		165^{16mm}	>112	i aq; misc alc, eth
m424	4-(Methylthio)-benzaldehyde	$CH_3SC_6H_4CHO$	152.22	8^1, 533	1.144	1.6452^{20}		90^{1mm}	44	
m425	3-(Methylthio)-2-butanone	$CH_3CH(SCH_3)COCH_3$	118.20	1^4, 3993	0.975	1.4710^{20}		50–54^{20mm}		
m426	Methyl thiocyanate	CH_3SCN	73.12	3, 175	1.068^{20}	1.4697^{20}	−51	130–133	38	
m427	3-Methylthiophene		98.17	17, 38	1.016	1.5180^{20}	−69.0	115.4	11	
m428	5-Methyl-2-thio-phenecarbaldehyde		126.18	17^1, 151	1.170	1.5825^{20}		114^{25mm}	87	
m429	4-(S-Methylthio)-phenol	$CH_3SC_6H_4OH$	140.20	6^1, 419			83–85	153–156^{20mm}		

1-Methyl-2-(3-pyridyl)pyrrolidine, n20
Methylresorcinol, d390
Methylsalicyclic acids, h137, h138
Methyl stearate, m326

Methylsuccinyl chloride, m188
methylsulfonic acid, m30
Methyl theobromine, c1
3-Methyl-2-thiabutane, i104

Methyl thienyl ketone, a57
Methyl thioglycolate, m293

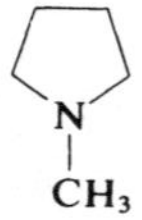

m405 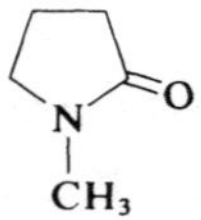 m406 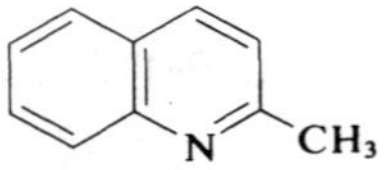 m407 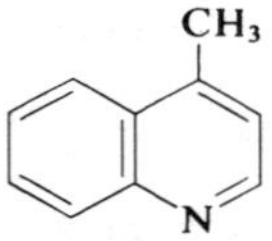 m408 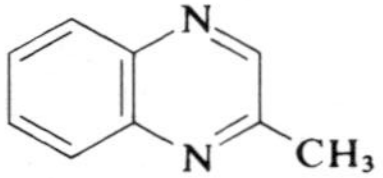 m409

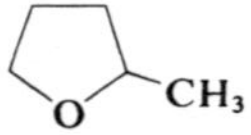

m415 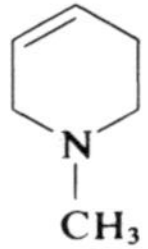 m416 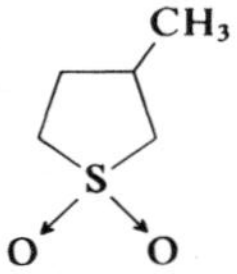 m417

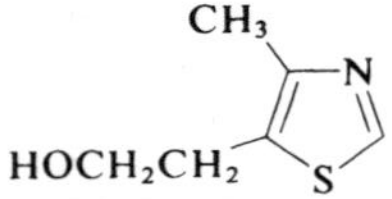

m418 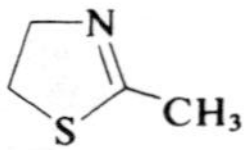 m419 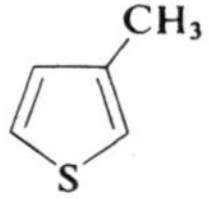 m427 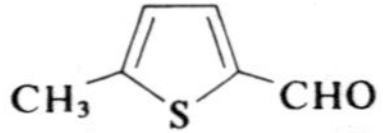 m428

TABLE 1.15 Physical Constants of Organic Compounds (*continued*)

No.	Name	Formula	Formula weight	Beilstein reference	Density	Refractive index	Melting point	Boiling point	Flash point	Solubility in 100 parts solvent
m430	3-Methylthio-1,2-propanediol	$CH_3SCH_2CH(OH)CH_2OH$	122.19		1.164	1.5160^{20}			>112	
m431	N-Methylthiourea	$CH_3NHC(=S)NH_2$	90.15	4, 70			119–121			v s aq, alc
m432	N-Methyl-o-toluamide	$CH_3C_6H_4CONHCH_3$	149.19	9, 465	1.168^{15}		69–71			
m433	N-Methyl-p-toluenesulfonamide	$CH_3C_6H_4SO_2NHCH_3$	185.25	11, 105			76–79			
m434	Methyl p-toluenesulfonate	$CH_3C_6H_4SO_2OCH_3$	186.23	11, 99			27.5			
m435	Methyltriacetoxysilane	$CH_3Si(OOCCH_3)_3$	220.3	4^3, 1896	1.175^{20}_4	1.408^{20}		88^{3mm}		
m436	Methyl 2,2,2-trichloroacetimidate	$Cl_3CC(=NH)OCH_3$	176.43	2, 212	1.425	1.4780^{20}		149	none	
m437	Methyltrichlorogermane	CH_3GeCl_3	193.98		1.730			111		
m438	Methyl trifluoromethanesulfonate	$CF_3SO_2OCH_3$	164.10		1.450	1.3244^{20}		94–99	38	
m439	N-Methyl-N-trimethylsilylacetamide	$CH_3CON(CH_3)-Si(CH_3)_3$	145.3	4^4, 4011	1.439^{20}_4	0.901^{20}		154		
m440	N-Methyl-N-(trimethylsilyl)-trifluoroacetamide	$CF_3CON(CH_3)-Si(CH_3)_3$	199.25		1.075	1.3802^{20}		132	25	
m441	Methyltripropoxysilane	$CH_3Si(OC_3H_7)_3$	220.4		0.88^{20}_4	1.4085^{20}		83^{13mm}		
m442	(Methyl)triphenylphosphonium bromide	$[CH_3P(C_6H_5)_3]^+Br^-$	357.24				230–233			

No.	Name	Formula	Formula wt	Beilstein ref	Density	n_D	m.p., °C	b.p., °C	Flash pt, °C	Solubility
m443	2-Methylundecanal	$CH_3(CH_2)_8CH(CH_3)-$ CHO	184.32		0.830_4^{15}	1.4321^{20}		271	93	s alc, eth
m444	Methyl urea	$CH_3NHCONH_2$	74.08	4, 64	1.204		101–102	d		v s aq, alc; i eth
m445	Methyl vanillate	$CH_3OC_6H_3(OH)COOCH_3$	182.18	10, 396			64–65	285–287		s hot alc, hot PE
m446	Methyl vinyl ether	$CH_3OCH{=}CH_2$	58.08	1^3, 1857	0.7511_4^{20}	1.3947	−112	5.5	−56	0.8 aq; v s alc
m447	2-Methyl-5-vinyl-pyridine		119.17		0.898	1.5437^{20}		100^{50mm}	65	
m448	Morpholine		87.12	27, 5	1.007_4^{20}	1.4542^{20}	−4.9	128.9	35	misc aq, alc, bz, eth
m449	4-Morpholine-carbonitrile		112.12		1.109	1.4730^{20}		$73^{0.6mm}$	104	
m450	N-Morpholino-1-cyclohexene		167.25		0.995	1.5128^{20}		117–122	68	
m451	2-(N-Morpholino)-ethanesulfonic acid		195.24				>300			

$CH_2{=}CH$— (2-methylpyridine ring, N, CH_3) m447

morpholine, N–H m448

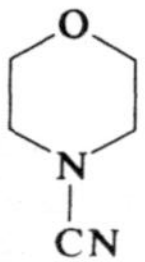

morpholine, N–CN m449

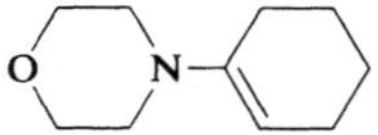

morpholine, N–(cyclohexenyl) m450

morpholine, $N-CH_2CH_2SO_3H$ m451

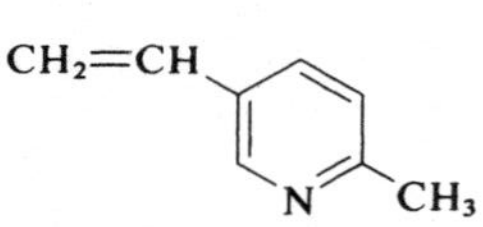

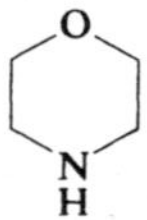

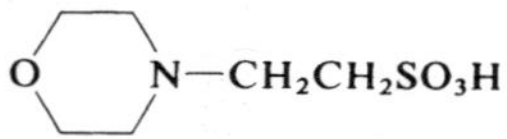

TABLE 1.15 Physical Constants of Organic Compounds (*continued*)

No.	Name	Formula	Formula weight	Beilstein reference	Density	Refractive index	Melting point	Boiling point	Flash point	Solubility in 100 parts solvent
m452	3-(*N*-Morpholino)-1,2-propanediol		161.20		1.157		37–38	191^{30mm}	>112	s alc, chl, eth, HOAc
m453	β-Myrcene	$(CH_3)_2C=CHCH_2CH_2-C(=CH_2)CH=CH_2$	136.24	1, 264	0.794^{20}_{4}	1.4709^{20}		166–168	39	s alc, eth
n1	1-Naphthaldehyde	$C_{10}H_7CHO$	156.18	7, 400	1.150^{20}_{4}	1.6520^{20}	1–2	161^{15mm}	>112	s alc, eth
n2	Naphthalene	$C_{10}H_8$	128.17	5, 531	1.162^{20}_{4}	1.5821^{100}	80.2 subl above mp	217.7	78	0.3 aq; 7 alc; 33 bz; 50 chl
n3	1-Naphthalenecarboxylic acid	$C_{10}H_7COOH$	172.18	9, 647			160–162	300		v s hot alc, eth
n4	1,5-Naphthalenediamine	$C_{10}H_6(NH_2)_2$	158.20	13, 203			185–187			s hot aq, hot alc
n5	1,8-Naphthalenediamine	$C_{10}H_6(NH_2)_2$	158.20	13, 204	1.1265^{99}_{4}	1.6828^{99}	66.5	205^{12mm}		sl s aq; s alc, eth
n6	1-Naphthalenemethylamine	$C_{10}H_7CH_2NH_2$	157.22	12, 1316	1.073	1.6429^{20}		290–293	>112	
n7	1,8-Naphthalic anhydride		198.18	17, 521			267–269			sl s HOAc
n8	1,8-Naphthalimide		197.19	21, 527			300			sl s alc; i bz, eth, aq
n9	1-Naphthol	$C_{10}H_7OH$	144.17	6, 596	1.0954^{99}_{4}	1.6206^{99}	96	288		v s alc, bz, chl, eth
n10	2-Naphthol	$C_{10}H_7OH$	144.17	6, 627	1.217^{4}		121–123	285–286	161	0.1 aq; 125 alc; 6 chl; 77 eth; s alk
n11	1,4-Naphthoquinone		158.16	7, 724	1.422		128	subl <100		s bz, chl, eth, alk
n12	(2-Naphthoxy)acetic acid	$C_{10}H_7OCH_2COOH$	202.21	6, 645			155–157			
n13	2-(1-Naphthyl)-acetamide	$C_{10}H_7CH_2CONH_2$	185.23	9, 666			181–183			i aq; s bz, CS_2

	Name	Formula	M	Density	n	mp	bp	Solubility
n14	1-Naphthyl acetate	$C_{10}H_7OOCCH_3$	186.21			43–46		s alc, eth
n15	1-Naphthylacetic acid	$C_{10}H_7CH_2COOH$	186.21			135	d	3.3 alc; v s chl, eth
n16	1-Naphthylaceto-nitrile	$C_{10}H_7CH_2CN$	167.21		1.6192^{20}	33–35	194^{18mm}	s alc
n17	1-Naphthylamine	$C_{10}H_7NH_2$	143.18	1.1232^{25}	1.6703	50	301	0.2 aq; v s alc, eth
n18	2-Naphthylsulfonic acid	$C_{10}H_7SO_3H$	208.23	1.441^{25}		91	d	77 aq; s alc, eth
n19	1-(1-Naphthyl)-2-thiourea	$C_{10}H_7NHC(=S)NH_2$	202.28			198		0.6 aq; 2.4 acet; s alc
n20	Nicotine		162.24	1.0097^{20}_{4}	1.5282^{20}	−79	123^{17mm}	misc aq; v s alc, eth, PE

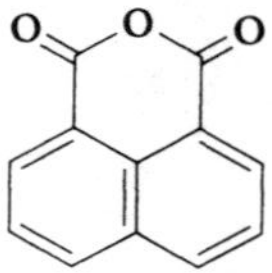

n7

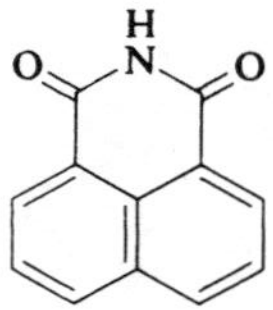

n8

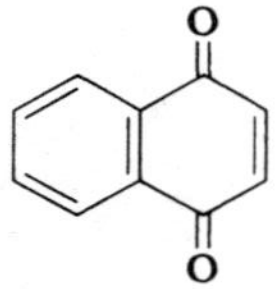

n11

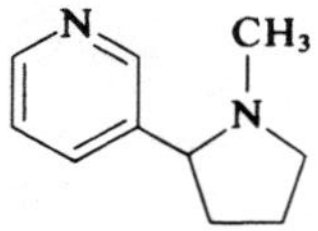

n20

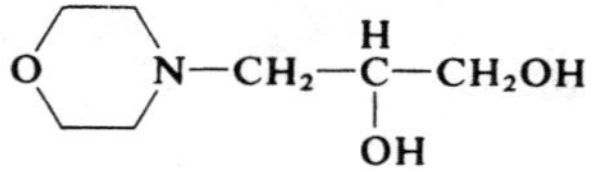

m452

TABLE 1.15 Physical Constants of Organic Compounds (*continued*)

No.	Name	Formula	Formula weight	Beilstein reference	Density	Refractive index	Melting point	Boiling point	Flash point	Solubility in 100 parts solvent
n21	Nitrilotriacetic acid	$N(CH_2COOH)_3$	191.14	4, 369			246 d			0.1 aq; s hot alc
n22	*m*-Nitroacetophenone	$O_2NC_6H_4COCH_3$	165.15	7, 288			76–78	202		s alc, eth
n23	*p*-Nitroacetophenone	$O_2NC_6H_4COCH_3$	165.15	7, 288			78–80	202		s alc
n24	*o*-Nitroaniline	$O_2NC_6H_4NH_2$	138.13	12, 687	1.442^{15}		69–70	284		s hot aq, alc, chl
n25	*m*-Nitroaniline	$O_2NC_6H_4NH_2$	138.13	12, 698	1.43		114	306		0.1 aq; 5 alc; 6 eth
n26	*p*-Nitroaniline	$O_2NC_6H_4NH_2$	138.13	12, 711	1.437^{14}		146	260^{100mm}	165	4 alc; 3.3 eth; s bz
n27	3-Nitrobenzaldehyde	$O_2NC_6H_4CHO$	151.12	7, 250	1.2792_4^{20}		58	164^{23mm}		s alc, chl, eth
n28	4-Nitrobenzaldehyde	$O_2NC_6H_4CHO$	151.12	7, 256	1.496		106–107			s alc, bz, HOAc
n29	2-Nitrobenzamide	$O_2NC_6H_4CONH_2$	166.12	9, 373	1.462_4^{32}		174–178	317		s hot aq, hot alc, eth
n30	Nitrobenzene	$C_6H_5NO_2$	123.11	5, 233	1.205_4^{15}	1.5546^{15}	5.8	210.8	87	v s alc, bz, eth
n31	2-Nitrobenzene-1,4-dicarboxylic acid	$O_2NC_6H_3(COOH)_2$	211.13	9, 851			270–272			
n32	3-Nitrobenzene-1,2-dicarboxylic acid	$O_2NC_6H_3(COOH)_2$	211.13	9, 823			216 d			2 aq; v s hot alc
n33	4-Nitrobenzene-1,2-dicarboxylic acid	$O_2NC_6H_3(COOH)_2$	211.13	9, 828			163–166			v s aq, alc; s eth
n34	5-Nitrobenzene-1,3-dicarboxylic acid	$O_2NC_6H_3(COOH)_2$	211.13	9, 840			260–261			0.15 aq; v s alc, eth
n35	2-Nitrobenzene-sulfonyl chloride	$O_2NC_6H_4SO_2Cl$	221.62	11, 67			65–67			s eth; d hot aq, alc
n36	6-Nitrobenzimidazole		163.14	23, 135	1.58		207–209			s alc, acid
n37	2-Nitrobenzoic acid	$O_2NC_6H_4COOH$	167.12	9, 370			146–148			0.7 aq; 33 alc; 22 eth
n38	3-Nitrobenzoic acid	$O_2NC_6H_4COOH$	167.12	9, 376	1.494		142			0.3 aq; 33 alc; 40 acet
n39	4-Nitrobenzoic acid	$O_2NC_6H_4COOH$	167.12	9, 389	1.58		242.8			9 alc; 2 eth; 5 acet

No.	Name	Formula	M.W.	Beilstein ref.	Density	n_D	m.p., °C	b.p., °C		Solubility
n40	4-Nitrobenzonitrile	$O_2NC_6H_4CN$	148.12	9, 397			146–149			s HOAc; sl s aq, alc
n41	3-Nitrobenzoyl chloride	$O_2NC_6H_4COCl$	185.57	9, 381			32–35	275–278		d aq, alc; v s eth
n42	4-Nitrobenzoyl chloride	$O_2NC_6H_4COCl$	185.57	9, 394			75	$205^{105\text{mm}}$		d aq, alc; s eth
n43	N-(p-Nitrobenzoyl)-glycine	$O_2NC_6H_4CONHCH_2COOH$	224.17	9, 395			131–133			
n44	3-Nitrobenzyl alcohol	$O_2NC_6H_4CH_2OH$	153.14	6, 449			30–32	$180^{3\text{mm}}$		s aq, alc, eth
n45	4-Nitrobenzyl alcohol	$O_2NC_6H_4CH_2OH$	153.14	6, 450			92–94	$185^{12\text{mm}}$		v s alc, eth; sl s aq
n46	4-Nitrobenzyl bromide	$O_2NC_6H_4CH_2Br$	216.04	5, 334			98–100			2 alc; v s eth
n47	4-Nitrobenzyl chloride	$O_2NC_6H_4CH_2Cl$	171.58	5, 329			70–73			8 alc; s eth
n48	2-Nitrobiphenyl	$O_2NC_6H_4C_6H_5$	199.21	5, 582	1.44^{25}_{4}	1.613^{25}	36.7	325	179	s alc, acet, CCl_4
n49	4-Nitrobiphenyl	$O_2NC_6H_4C_6H_5$	199.21	5, 583			112–114	340		sl s alc; v s eth
n50	1-Nitrobutane	$CH_3CH_2CH_2CH_2NO_2$	103.18	1, 123	0.9752^{20}_{20}	1.4112	−81.3	152.8		sl s aq; misc alc, eth
n51	3-Nitro-2-butanol	$CH_3CH(NO_2)CH(OH)CH_3$	119.12	1, 373	1.1296^{25}_{4}	1.4414^{20}	76–78	$92^{10\text{mm}}$		
n52	2-Nitrodiphenylamine	$O_2NC_6H_4NHC_6H_5$	214.22	12, 690			91			i aq; s alc

Ninhydrin, i16
Nioxime, c323
2,2',2''-Nitrilotriethanol, t264

1,1',1''-Nitrilotris(2-propanol), t309
2-Nitro-p-anisidine, m79
5-Nitro-o-anisidine, m78

Nitroanisoles, m81, m82
4-Nitrobenzyl cyanide, n65
Nitrocresols, m323, m324

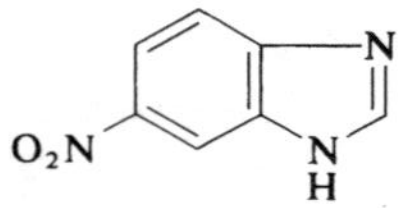

n36

TABLE 1.15 Physical Constants of Organic Compounds (*continued*)

No.	Name	Formula	Formula weight	Beilstein reference	Density	Refractive index	Melting point	Boiling point	Flash point	Solubility in 100 parts solvent
n53	Nitroethane	$CH_3CH_2NO_2$	75.07	1, 99	1.0528^{20}_{20}	1.3920^{20}	−90	114.1	30	4.5 aq; misc alc, eth; s alk, chl
n54	1-Nitroguanidine	$O_2NNHC(=NH)NH_2$	104.07	3, 126			d 225			0.4 aq; sl s MeOH
n55	5-Nitro-1*H*-indazole		163.14	23, 129			207–209			s alc, bz, eth, acet
n56	Nitromethane	CH_3NO_2	61.04	1, 74	1.1322^{25}_{4}	1.3795^{25}	−28.4	101.2	35	11 aq; s alc, eth
n57	1-Nitronaphthalene	$C_{10}H_7NO_2$	173.17	5, 553	1.223		59–60	304	90	s alc; v s chl, eth
n58	3-Nitro-2-pentanol	$CH_3CH_2CH(NO_2)$-$CH(OH)CH_3$	133.15	1, 385	1.08184^{25}_{4}	1.4430^{20}		100^{10mm}		
n59	2-Nitrophenethyl alcohol	$O_2NC_6H_4CH_2CH_2OH$	167.16	6, 218	1.190	1.5637^{20}	2	267	>112	
n60	2-Nitrophenol	$O_2NC_6H_4OH$	139.11	6, 213	1.495		44–45	214–216		s alc, bz, eth, alk
n61	4-Nitrophenol	$O_2NC_6H_4OH$	139.11	6, 226	1.495		112–114	279		s aq; v s alc, chl, eth
n62	4-Nitrophenyl acetate	$O_2NC_6H_4OOCCH_3$	181.15	6, 233			77–79			s aq; v s alc, bz, eth
n63	2-Nitrophenylacetic acid	$O_2NC_6H_4CH_2COOH$	181.15	9, 454			139–142			s hot aq, alc
n64	4-Nitrophenylacetic acid	$O_2NC_6H_4CH_2COOH$	181.15	9, 455			153			s alc, bz, eth
n65	4-Nitrophenylaceto-nitrile	$O_2NC_6H_4CH_2CN$	162.15	9, 456			117			s alc, eth
n66	4-Nitrophenyl chloroformate	$O_2NC_6H_4OOCCl$	201.57	6^1, 120			77–79	162^{19mm}		
n67	2-Nitro-*p*-phenylene-diamine	$O_2NC_6H_3(NH_2)_2$	153.14	13, 120			137–140			
n68	4-Nitro-*o*-phenylene-diamine	$O_2NC_6H_3(NH_2)_2$	153.14	13, 29			199–201			s acid
n69	4-Nitrophenyl-hydrazine	$O_2NC_6H_4NHNH_2$	153.14	15, 468			156 d			s alc, chl, eth, hot bz

#	Name	Formula	M.W.		Density	n_D	M.P.	B.P.		Solubility
n70	2-Nitrophenyl phenyl ether	$O_2NC_6H_4OC_6H_5$	215.21	6[2], 222	1.2539[22]	1.575[20]	<−20	184[8mm]		s alc, eth
n71	4-Nitrophenyl phenyl ether	$O_2NC_6H_4OC_6H_5$	215.21	6, 232			53–56	320		s bz, eth
n72	3-Nitrophthalic anhydride		193.11	17,486			163–165			sl s aq, bz
n73	1-Nitropropane	$CH_3CH_2CH_2NO_2$	89.09	1,115	1.0009[20]	1.4016[20]	−104.0	131.2	33	1.4 aq; misc alc, eth
n74	2-Nitropropane	$(CH_3)_2CHNO_2$	89.09	1,116	0.9876[20]	1.3949[20]	−91.3	120.3	37	1.7 aq; misc alc, eth
n75	2-Nitro-1-propanol	$CH_3CH(NO_2)CH_2OH$	105.09	1,358	1.1841[25][4]	1.4379[20]	159–162	99[10mm]	100	s aq, alc, eth
n76	4-Nitropyridine-N-oxide	$O_2NC_5H_4N(O)$	140.10							
n77	8-Nitroquinoline		174.16	20,373			89–91			s alc, bz, eth; i aq
n78	Nitrosobenzene	C_6H_5NO	107.11	5,230			67–69	59[18mm]		i aq; s alc
n79	N-Nitrosodimethylamine	$(CH_3)_2NNO$	74.08	8,84	1.0048[20][4]	1.4368[20]		151–153	61	v s aq, alc, eth
n80	p-Nitrosodiphenylamine	$C_6H_5NHC_6H_4NO$	198.22				144–145			v s alc, bz, chl, eth
n81	1-Nitroso-2-naphthol	$C_{10}H_6(NO)OH$	173.16	7,712			109–110			3 alc; s bz, eth, alk

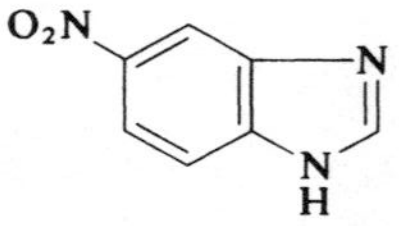

n55

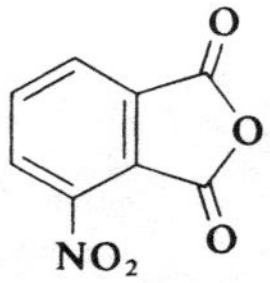

n72

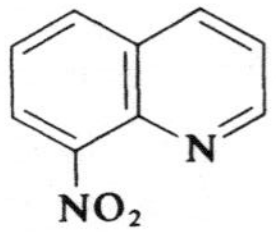

n77

TABLE 1.15 Physical Constants of Organic Compounds (*continued*)

No.	Name	Formula	Formula weight	Beilstein reference	Density	Refractive index	Melting point	Boiling point	Flash point	Solubility in 100 parts solvent
n82	1-Nitroso-2-naphthol-3,6-di-sulfonic acid, di-Na salt hydrate		377.26	11^2, 190			>300			2.5 aq; sl s alc
n83	4-Nitrosophenol	ONC_6H_4OH	123.11	7, 622			d 126			s aq; v s alc, eth; explodes on contact with conc acid, alk, or fire
n84	β-Nitrostyrene	$C_6H_5CH=CHNO_2$	149.15	5, 478	1.1622^{19}	1.5472^{20}	58	250	106	s alc; v s eth
n85	2-Nitrotoluene	$CH_3C_6H_4NO_2$	137.14	5, 318	1.1581^{20}_4	1.5459^{20}	−10	222	101	s alc, bz
n86	3-Nitrotoluene	$CH_3C_6H_4NO_2$	137.14	5, 321	1.392		15.5	231.9	106	misc alc, eth; s bz
n87	4-Nitrotoluene	$CH_3C_6H_4NO_2$	137.14	5, 323			53–54	238		s alc, bz, chl, eth
n88	2-Nitro-α,α,α-trifluorotoluene	$CF_3C_6H_4NO_2$	191.11	5^2, 251			31–32	105^{20mm}		v s alc, bz
n89	3-Nitro-α,α,α-trifluorotoluene	$CF_3C_6H_4NO_2$	191.11	5, 327	1.436^{16}_4	1.4715^{20}	−2.4	200–205	87	s alc, eth
n90	Nonadecane	$CH_3(CH_2)_{17}CH_3$	268.51	1, 174	0.7776^{32}_4	1.4335^{38}	31.9	330.6	168	s eth; sl s alc
n91	1,8-Nonadiyne	$HC\equiv C(CH_2)_5C\equiv CH$	120.20	1^2, 248	0.8159^{4}_{1}	1.4492^{20}	−21	55^{13mm}	41	s abs alc, eth
n92	Nonane	$CH_3(CH_2)_7CH_3$	128.26	1, 165	0.7176^{20}_4	1.4054^{20}	−53.5	150.8	31	v s alc, bz, eth
n93	1,9-Nonanediamine	$H_2N(CH_2)_9NH_2$	158.29	4, 272	0.929	1.4460^{20}	37–38	258	>112	
n94	Nonanedinitrile	$NC(CH_2)_7CN$	150.23	2, 709	1.029^{20}_4			176^{11mm}		
n95	1,9-Nonanedioic acid	$HOOC(CH_2)_7COOH$	188.22	2, 707			106.5	286^{100mm}		0.24 aq; v s alc; 3 eth
n96	1,9-Nonanediol	$HO(CH_2)_9OH$	160.26	1, 493			45–47	177^{15mm}	81	s alc, eth
n97	Nonanenitrile	$CH_3(CH_2)_7CN$	139.24	2, 354	0.821^{15}	1.4260^{20}	−34.2	224.0	100	
n98	Nonanoic acid	$CH_3(CH_2)_7COOH$	158.24	2, 352	0.906^{20}	1.4330^{20}	12.5	254	75	s alc, chl, eth
n99	1-Nonanol	$CH_3(CH_2)_8OH$	144.26	1, 423	0.8274^{20}_4	1.4338^{20}	−5.5	213.1		0.6 aq; misc alc, eth
n100	5-Nonanone	$(C_4H_9)_2CO$	142.24	1, 710	0.8064^{20}	1.4190^{20}	−50	187	60	misc alc, eth

n101	Nonanoyl chloride	$CH_3(CH_2)_7COCl$	176.69	2, 353	0.9464^{15}	1.4377^{20}	−60.5	215.4	81	d aq, alc: s eth
n102	1-Nonene	$H(CH_2)_7CH{=}CH_2$	126.24	1^2, 202	0.7292^{20}	1.4157^{20}	−81.4	146.9	46	
n103	Nonyl aldehyde	$CH_3(CH_2)_7CHO$	142.24	1, 708	0.827^{19}	1.4240^{20}		185	63	sl s aq; s alc, eth
n104	Nonylamine	$CH_3(CH_2)_8NH_2$	143.27	4, 198	0.782	1.4330^{20}		201	62	
n105	Nopol		166.26		0.973	1.4930^{20}		230–240	98	
n106	Nopyl acetate		210.3		0.9805^{25}	1.4721^{20}				
n107	Norbornane		96.17	5^2, 45			82–84	168–172	33	s alc
n108	2-Norbornanone		110.16	7, 57			88–91	118^{11mm}	110	
n109	trans-5-Norbornene-2,3-dicarbonyl dichloride		219.07		1.349	1.5165^{20}				

Nitroso-R-salt, n82
Nitroterephthalic acid, n31
2-Nitro-p-toluidine, m317
4-Nitro-o-toluidine, m315

5-Nitro-o-toluidine, m316
4-Nitroveratrole, d444
Nitroxylenes, d556, d557, d558, d559
Nonyl alcohol, n99

2,5-Norbornadiene, b129
exo-2-Norbornanamine, a250
5-Norbornen-2-carbaldehyde, b131
Norbornene, b130

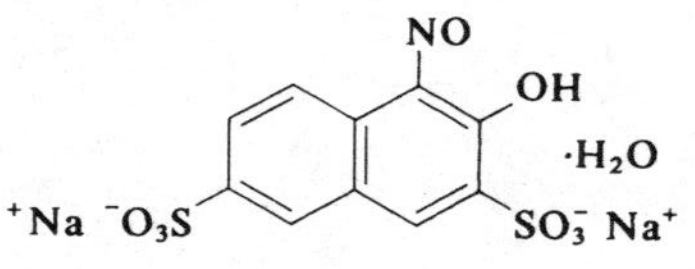

n82

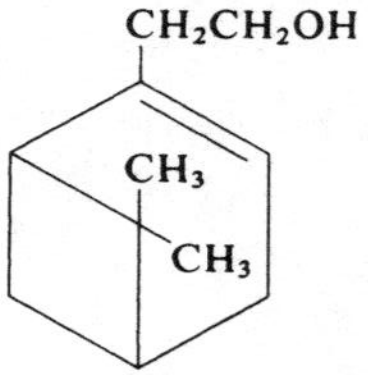

CH_2CH_2OH

n105

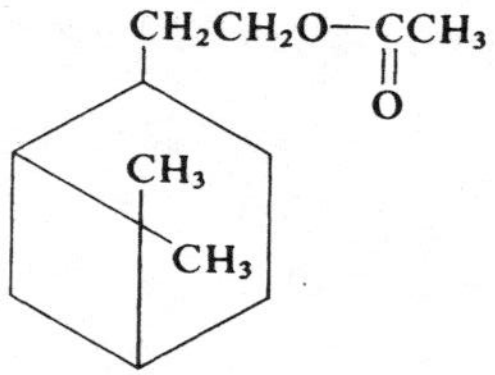

$CH_2CH_2O{-}CCH_3$

n106

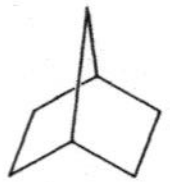

n109

n107

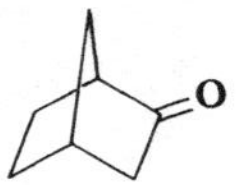

n108

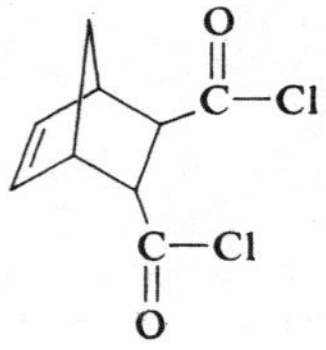

TABLE 1.15 Physical Constants of Organic Compounds (*continued*)

No.	Name	Formula	Formula weight	Beilstein reference	Density	Refractive index	Melting point	Boiling point	Flash point	Solubility in 100 parts solvent
n110	5-Norbornen-2-yl acetate		152.19		1.044	1.4700^{20}		76^{14mm}	62	
n111	*exo*-2-Norbornyl formate		140.18		1.048	1.4622^{20}		67^{16mm}	53	
n112	(+)-Norephedrine HCl		187.67	13[2], 371			174–176			
o1	(Z,Z)-9,12-Octadecadienoic acid	$CH_3(CH_2)_4CH{=}CHCH_2{-}CH{=}CH(CH_2)_7COOH$	280.44	2, 496	0.9025^{20}_{4}	1.4699^{20}	−5	230^{16mm}		v s eth; misc PE; s abs alc
o2	Octadecanamide	$CH_3(CH_2)_{16}CONH_2$	283.50	2, 384			108–109	251^{12mm}		s hot alc, hot eth
o3	Octadecane	$CH_3(CH_2)_{16}CH_3$	254.50	1, 173	0.7767^{28}_{4}	1.4367^{28}	28.2	316.7	165	s acet, eth; sl s alc
o4	1-Octadecanethiol	$CH_3(CH_2)_{17}SH$	286.57			1.4648	29–31	360	185	s eth; sl s alc
o5	Octadecanoic acid	$CH_3(CH_2)_{16}COOH$	284.48	2, 377	0.847^{70}	1.4299^{80}	70	383		4.9 alc; 20 bz; 50 chl; 3.9 acet
o6	1-Octadecanol	$CH_3(CH_2)_{17}OH$	270.50	1, 431	0.8123^{58}	1.4388^{20}	57.9	203^{10mm}		s alc, eth
o7	9,12,15-Octadecatrienoic acid	$CH_3CH_2CH{=}CH{-}CH_2{-}CH{=}CHCH_2{-}(CH_2)_6COOH$	278.44	2, 499	0.791^{18}_{4}	1.4800^{20}		230^{17mm}	>112	s alc, bz, eth
o8	1-Octadecene	$CH_3(CH_2)_{15}CH{=}CH_2$	252.49	1, 226	0.791^{18}_{4}	1.4439^{20}	17.7	314.9	148	s hot acet
o9	9-Octadecen-1-amine	$CH_3(CH_2)_7CH{=}CH{-}(CH_2)_8NH_2$	267.50		0.813	1.4578^{20}			154	
o10	(Z)-9-Octadecenoic acid	$CH_3(CH_2)_7CH{=}CH{-}(CH_2)_7COOH$	282.47	2, 463	0.8906^{20}_{4}	1.4571^{20}	4	286^{100mm}		misc alc, eth; s bz, chl
o11	(E)-9-Octadecenoic acid	$CH_3(CH_2)_7CH{=}CH{-}(CH_2)_7COOH$	282.47	2[2], 441	0.851^{79}	1.4308^{99}	44–45	288^{100mm}		s bz, chl, eth
o12	(Z)-9-Octadecen-1-ol	$CH_3(CH_2)_7CH{=}CH{-}(CH_2)_8OH$	268.49	1, 453	0.849^{20}_{4}	1.4610^{20}	13–19	195^{8mm}	>112	s alc, eth
o13	Octadecylamine	$CH_3(CH_2)_{17}NH_2$	269.52	4, 196	0.777^{27}		50–52	232^{32mm}	148	s alc, bz, eth
o14	Octadecyl isocyanate	$CH_3(CH_2)_{17}NCO$	295.51		0.847	1.4501^{20}		170^{2mm}	185	

No.	Name	Formula	Formula wt.	Beilstein ref.	Density	Refractive index	Melting point, °C	Boiling point, °C	Flash point, °C	Solubility
o15	Octadecyltrichlorosilane	$CH_3(CH_2)_{17}SiCl_3$	387.94		0.984	1.4602^{20}		223^{10mm}	89	
o16	Octadecyl vinyl ether	$CH_3(CH_2)_{17}OCH{=}CH_2$	296.54		0.821^{30}_4	1.4440^{30}	28	187^{5mm}	177	
o17	1,7-Octadiene	$H_2C{=}CH(CH_2)_4CH{=}CH_2$	110.20		0.746	1.4221^{20}		114–121	9	
o18	1H,1H,5H-Octafluoro-1-pentanol	$HCF_2CF_2CF_2CF_2CH_2OH$	232.08		1.6647^{20}	1.3190^{20}		140–141	74	
o19	Octamethylcyclotetrasilazane	$[-(CH_3)_2SiNH-]_4$	292.7		0.95^{22}	1.458^{25}		224–225		
o20	Octamethylcyclotetrasiloxane	$[-(CH_3)_2SiO-]_4$	296.62		0.9558^{20}	1.3968^{20}	17.5	175	90	s bz, PE; sl s alc
o21	Octamethyltrisiloxane	$[(CH_3)_3SiO]_2Si(CH_3)_2$	236.0		0.8200^{20}	1.3848^{20}	~−80	152–153	38	
o22	Octane	$CH_3(CH_2)_6CH_3$	114.23	1, 159	0.7025^{20}	1.3974^{20}	−56.8	125.7	15	s eth; sl s alc
o23	1,8-Octanediamine	$H_2N(CH_2)_8NH_2$	144.26	4, 271			50–52	225	165	
o24	1,8-Octanedioic acid	$HOOC(CH_2)_6COOH$	174.20	2, 691			140–144	230^{15mm}		0.16 aq; 0.6 eth; s alc
o25	1,2-Octanediol	$CH_3(CH_2)_5CH(OH)CH_2OH$	146.23	1^3, 2217			36–38	132^{10mm}	>112	

n110 — bicyclic (norbornene) ester, $O{-}C({=}O)CH_3$

n111 — bicyclic (norbornane) ester, $O{-}CH({=}O)$

n112 — $C_6H_5{-}CH(OH){-}CH(NH_2){-}CH_3$ (H, OH, NH₂, CH₃)

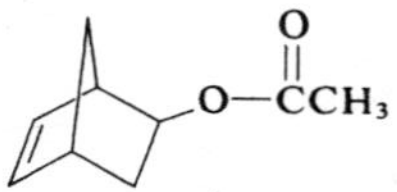

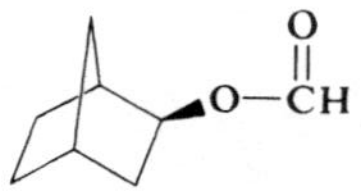

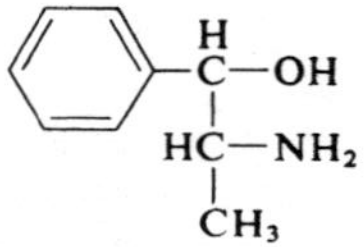

TABLE 1.15 Physical Constants of Organic Compounds (continued)

No.	Name	Formula	Formula weight	Beilstein reference	Density	Refractive index	Melting point	Boiling point	Flash point	Solubility in 100 parts solvent
o26	1,8-Octanediol	HO(CH$_2$)$_8$OH	146.23	1, 490			59–61	172^{20mm}		v s alc; sl s aq, eth
o27	Octanenitrile	CH$_3$(CH$_2$)$_6$CN	125.22	2, 349	0.8135^{20}	1.4202^{20}	−45.6	205.2	73	s eth; sl s alc
o28	1-Octanethiol	CH$_3$(CH$_2$)$_7$SH	146.30	1^3, 1710	0.843	1.4525^{20}	−49.2	199.0	68	s alc
o29	Octanoic acid	CH$_3$(CH$_2$)$_6$COOH	144.21	2, 347	0.9088_4^{20}	1.4279^{20}	16.6	239.3	110	0.07 aq; v s alc, chl, eth, PE
o30	1-Octanol	CH$_3$(CH$_2$)$_7$OH	130.23	1, 418	0.8258_4^{20}	1.4296^{20}	−15.0	195.2	81	0.06 aq; misc alc, chl, eth
o31	DL-2-Octanol	CH$_3$(CH$_2$)$_5$CH(OH)CH$_3$	130.23	1, 419	0.8207_4^{20}	1.4202^{20}	−38.6	179–180	71	0.08 aq; misc alc, eth
o32	DL-3-Octanol	CH$_3$(CH$_2$)$_4$CH(OH)-CH$_2$CH$_3$	130.23	1^1, 208	0.8216^{20}	1.4262^{20}		174–176	65	
o33	4-Octanol	CH$_3$(CH$_2$)$_3$CH(OH)-CH$_2$CH$_2$CH$_3$	130.23		0.8192^{20}	1.425^{20}		176.6	71	
o34	2-Octanone	CH$_3$(CH$_2$)$_5$COCH$_3$	128.22	1, 704	0.819_4^{20}	1.4150^{20}	−16	173	62	i aq; misc alc, eth
o35	3-Octanone	CH$_3$(CH$_2$)$_4$COCH$_2$CH$_3$	128.22	1, 706	0.8220_4^{20}	1.4150^{20}		167–168	46	i aq; misc alc, eth
o36	4-Octanone	CH$_3$(CH$_2$)$_3$COCH$_2$-CH$_2$CH$_3$	128.22	1, 706	0.809	1.4139^{20}		164	45	
o37	Octanoyl chloride	CH$_3$(CH$_2$)$_6$COCl	162.66	2, 348	0.955_{15}^{15}	1.4350^{20}	<−70	195	75	d aq; alc; s eth
o38	Octaphenylcyclo-tetrasiloxane	[—(C$_6$H$_5$)$_2$SiO—]$_4$	793.2		1.185			340^{1mm}		s alc, bz, HOAc
o39	1-Octene	CH$_3$(CH$_2$)$_5$CH=CH$_2$	112.22	1, 221	0.7149_4^{20}	1.4087^{20}	−101.7	121.3	21	i aq; misc alc, eth
o40	Octyl aldehyde	CH$_3$(CH$_2$)$_6$CHO	128.22	1, 704	0.821_4^{20}	1.4183^{20}	12–15	163.4	51	sl s aq; misc alc
o41	Octylamine	CH$_3$(CH$_2$)$_7$NH$_2$	129.25	4, 196	0.782	1.4290^{20}	−5 to −1	175–177	62	i aq; s alc, eth
o42	4-Octylaniline	CH$_3$(CH$_2$)$_7$C$_6$H$_4$NH$_2$	205.35	12, 1185	1.07^{20}	1.447^{20}		175^{13mm}		
o43	Octyltrichlorosilane	CH$_3$(CH$_2$)$_7$SiCl$_3$	247.7					226^{30mm}		
o44	1-Octyne	CH$_3$(CH$_2$)$_5$C≡CH	110.19	1, 258	0.7457_4^{20}	1.4159^{20}	−79.3	126.2	63	i aq; s alc, eth
o45	1-Octyn-3-ol	CH$_3$(CH$_2$)$_4$CH(OH)-C≡CH	126.20		0.864	1.4410^{20}				

No.	Name	Formula	Formula weight	Beilstein reference	Density	n_D	Melting point, °C	Boiling point, °C	Solubility
o46	L-(+)-Ornithine	$H_2N(CH_2)_3CH(NH_2)-COOH$	132.16	4, 420			142		v s aq, alc; sl s eth
o47	Oxacycloheptane		100.16		0.890	1.440^{20}	10	122	
o48	Oxalic acid	HOOCCOOH	90.04	2, 502	1.90^{17}		189 d	none	9.5 aq; 24 alc; 1.3 eth
o49	Oxalic acid dihydrate	$HOOCCOOH \cdot 2H_2O$	126.07	2, 502	1.653^{19}_{4}		$-2H_2O$, 102	none	14 aq; 40 alc; 1 eth
o50	Oxalyl bromide	BrCO—COBr	215.84	2, 542	1.667^{20}_{4}	1.5220		103^{720mm}	s eth; violent d aq, alc
o51	Oxalyl chloride	ClCO—COCl	126.93		1.488^{13}_{4}	1.4340^{13}	−12	64	
o52	Oxalyl dihydrazide	$H_2NNHCO—CONHNH_2$	118.10	2, 559			240 d		s hot aq; sl s alc, eth
o53	Oxamic hydrazide	$H_2NCO—CONHNH_2$	103.08	2, 559			218 d		s alk; sl s aq; i eth
o54	Oxamide	$H_2NCO—CONH_2$	88.07	2, 545			d 350		sl s hot aq, alc
o55	2-Oxazolidone		87.08	27, 135			86–89	220^{48mm}	
o56	2-Oxobutyric acid	$CH_3CH_2C(=O)COOH$	102.09	3, 629	1.200^{17}_{4}	1.3972^{20}	32–34	82^{16mm}	v s aq, alc; v sl s eth

o47

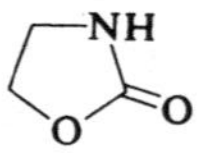

o55

TABLE 1.15 Physical Constants of Organic Compounds (continued)

No.	Name	Formula	Formula weight	Beilstein reference	Density	Refractive index	Melting point	Boiling point	Flash point	Solubility in 100 parts solvent
o57	2-Oxohexamethyl-eneimine		113.16	21^2, 216	1.024^{25}_4	1.4935	69.2	180^{50mm}		84 aq
o58	4-Oxopentanoic acid	$CH_3COCH_2CH_2COOH$	116.12	3, 671	1.1447^{25}_4	1.4396^{20}	33–35	245.8	137	v s aq, alc, bz, eth
o59	2-Oxopropional-dehyde	CH_3COCHO	72.06	1, 762	1.0455^{24}	1.4209^{20}		72	none	s aq, alc
o60	2-Oxopropionic acid	$CH_3COCOOH$	88.06	3, 608	1.267^{15}_4	1.4315^{20}	11.8	165 d	82	misc aq, alc, eth
o61	2,2'-Oxydiacetic acid	$HOOCCH_2OCH_2COOH$	134.09	3, 234			142–145	d		v s aq, alc; sl s eth
o62	4,4'-Oxydianiline	$H_2NC_6H_4OC_6H_4NH_2$	200.24	13, 441			190 d			
o63	3,3'-Oxydipropio-nitrile	$NCCH_2CH_2OCH_2CH_2CN$	124.14		1.043	1.4405^{20}		$112^{0.5mm}$	>112	
p1	Paraformaldehyde	$(CH_2O)_x$	132.16	1, 566			156 d		71	slowly s aq; s alk; i alc, eth
p2	Paraldehyde	$[—CH(CH_3)O—]_3$	132.16	19, 385	0.9984^{15}_4	1.4049^{20}	12.5	124		11 aq; misc alc, chl
p3	Parathion	$(C_2H_5O)_2P(=S)(O)-C_6H_4NO_2$	291.27		1.26^{25}_4	1.5370^{25}	6	375		v s alc, bz, eth
p4	DL-Patchenol		166.26	6^2, 64	0.987	1.5045^{20}	137–139	234–238	107	
p5	Pentabromoethyl-benzene	$CH_3CH_2C_6Br_5$	500.67	5, 357						
p6	Pentabromophenol	C_6Br_5OH	488.62	6, 206	1.690		223–226	subl		sl s alc, eth
p7	Pentachloroacetone	$Cl_2CHCOCCl_3$	230.31	1, 656		1.4967^{20}	21 anhyd	192	none	i aq; v s acet
p8	Pentachlorobenzene	C_6HCl_5	250.34	5, 205	1.8342^{16}		82–85	275–277		v s bz, chl, eth
p9	Pentachloroethane	Cl_2CHCCl_3	202.30	1, 87	1.6712^{25}	1.5030^{20}	−29.0	160.5	none	0.05 aq; misc alc, eth
p10	Pentachloronitro-benzene	$C_6Cl_5NO_2$	295.34	5, 247	1.718^{25}_4		140–143			s bz, chl
p11	Pentachlorophenol	C_6Cl_5OH	266.34	6, 194	1.978^{22}_4		190–191	310 d		v s alc; s bz; 148 eth
p12	Pentachloropyridine	C_5Cl_5N	251.33	20, 232			124–126			

p13	Pentadecane	$CH_3(CH_2)_{13}CH_3$	212.42	1, 172	0.7684^{20}_{4}	1.4319^{20}	9.9	270.6	132	v s alc, eth
p14	8-Pentadecanone	$[CH_3(CH_2)_6]_2CO$	226.40	1, 717			41–43	178		s alc
p15	3-Pentadecylphenol	$C_{15}H_{31}C_6H_4OH$	304.52				45–48	195^{1mm}		
p16	1,2-Pentadiene	$CH_3CH_2CH{=}C{=}CH_2$	68.12	1, 251	0.6926^{20}_{4}	1.4209^{20}	−137.3	44.9		
p17	(*E*)-1,3-Pentadiene	$CH_3CH{=}CHCH{=}CH_2$	68.12	1, 251	0.6760^{20}	1.4301^{20}	−87.5	42.0	−28	
p18	(*Z*)-1,3-Pentadiene	$CH_3CH{=}CHCH{=}CH_2$	68.12	1, 251	0.6910^{20}	1.4363^{20}	−140.8	44.1	−28	
p19	1,4-Pentadiene	$H_2C{=}CHCH_2CH{=}CH_2$	68.12	1, 251	0.6608^{22}_{4}	1.3888^{20}	−148.3	26.0	4	
p20	Pentaerythritol	$C(CH_2OH)_4$	136.15	1, 528	1.38^{25}_{4}	1.548	260	subl		6 aq; v sl s alc; i eth
p21	Pentaerythrityl tetrabromide	$C(CH_2Br)_4$	387.76	1, 142			158–160	305–306		
p22	Pentaerythrityl tetranitrate	$C(CH_2ONO_2)_4$	316.15	1^2, 602	1.773^{20}_{4}		140		sensitive to shock; explodes on percussion	acet; sl s eth, alc

3-Oxo-*N*-phenylbutanamide, a32
2,2′-Oxybis(chloroethane), b158
1,1′-Oxybis(2-methylpropane), d407
1,1′-Oxybis(pentane), d651
3,3′-Oxybis(1-propene), d26
2,2′-Oxydiethanethiol, b186
2,2′-Oxydiethanol, b181
Palmitic acid, h35
Pamoic acid, m235
Parabanic acid, i6
Pelargonaldehyde, n103
Pelargonic acid, n98
Pelargonoyl chloride, n101
Pelargononitrile, n97
Pentabromophenyl ether, b197
Pentaerythritol diformal, t126

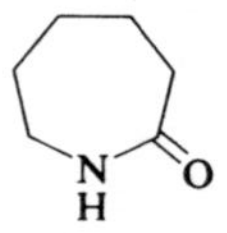

o57

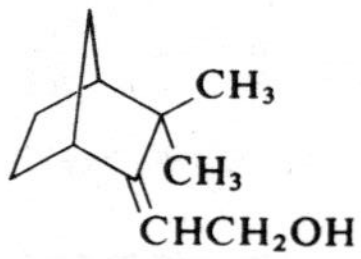

p4

TABLE 1.15 Physical Constants of Organic Compounds (*continued*)

No.	Name	Formula	Formula weight	Beilstein reference	Density	Refractive index	Melting point	Boiling point	Flash point	Solubility in 100 parts solvent
p23	Pentafluorobenzonitrile	C_6F_5CN	193.07		1.532	1.4425^{20}		185–190	29	
p24	Pentamethylbenzene	$C_6H(CH_3)_5$	148.25	5, 443	0.9174^{20}	1.527^{20}	54.4	231		v s alc, bz
p25	1,2,3,4,5-Pentamethylcyclopentadiene		136.24		0.870	1.4733^{20}		58^{13mm}	44	
p26	1,5-Pentamethylenetetrazole		138.17	26^2, 213			59–61	194^{12mm}		
p27	Pentanal	$CH_3CH_2CH_2CH_2CHO$	86.13	1, 676	0.8095^{20}	1.3942^{20}	–92	102–103	12	1.4 aq; misc alc, eth
p28	Pentane	$CH_3CH_2CH_2CH_2CH_3$	72.15	1, 130	0.6262^{20}	1.3575^{20}	–129.7	36.1	–49	misc alc, eth
p29	1,5-Pentanediamine	$H_2N(CH_2)_5NH_2$	102.18	4, 266	0.873^{25}	1.4591^{20}	–129.7	178–180	62	s aq, alc; sl s eth
p30	1,5-Pentanediol	$HO(CH_2)_5OH$	104.15	1, 481	0.9941^{20}	1.4494^{20}	–15.6	242.5	125	s aq; alc; sl s eth
p31	2,3-Pentanedione	$CH_3CH_2COCOCH_3$	100.11	1, 776	0.957	1.4068^{20}	–52	110–112	19	
p32	2,4-Pentanedione	$CH_3COCH_2COCH_3$	100.11	1, 777	0.9721^{25}	1.4510^{20}	–23.1	140.6	40	17 aq; misc alc, eth
p33	Pentanenitrile	$CH_3CH_2CH_2CH_2CN$	83.13	2, 301	0.8035^{15}	1.3991^{15}	–96.8	141.3	40	i aq; s alc, eth
p34	1-Pentanesulfonic acid, Na salt	$CH_3(CH_2)_4SO_3^-Na^+$	174.19	4^3, 23			>300			4 aq
p35	1-Pentanethiol	$CH_3(CH_2)_4SH$	104.22	1, 384	0.840	1.4460^{20}	–75.7	126.6	18	i aq; misc alc, eth
p36	Pentanoic acid	$CH_3(CH_2)_3COOH$	102.13	2, 299	0.9390^{20}	1.4080^{20}	–33.7	185.5	88	2.4 aq; v s alc, eth
p37	1-Pentanol	$CH_3(CH_2)_4OH$	88.15	1, 383	0.8148^4	1.4100^{20}	–78.9	137.8	32	2.7 aq; misc alc, eth
p38	2-Pentanol	$CH_3CH_2CH_2CH(OH)CH_3$	88.15	1, 384	0.8393^{20}	1.4064^{20}	glass	119.0	40	16.6 aq; misc alc, eth
p39	3-Pentanol	$CH_3CH_2CH(OH)CH_2CH_3$	88.15	1, 385	0.8150^{25}	1.4079^{25}	–69	115.6	40	5.2 aq; s alc, eth
p40	γ-Pentanolactone		100.12	17, 235	1.057	1.4330	–31	207–208	81	
p40a	δ-Pentanolactone		100.12	17, 235	1.079	1.4575^{20}		$60^{0.5mm}$	100	

ID	Name	Formula	MW	Ref	Density	n	mp	bp	flash	Solubility
p41	2-Pentanone	$CH_3CH_2CH_2COCH_3$	86.13	1,676	0.8095^{20}	1.3903	−77.8	101.7	7	misc acet, bz, eth, PE
p42	3-Pentanone	$CH_3CH_2COCH_2CH_3$	86.13	1,679	0.8143^{20}	1.3923^{20}	−39.0	102.0	12	3.4 aq
p43	Pentanophenone	$C_6H_5CO(CH_2)_3CH_3$	162.23	7,327	0.988	1.5143^{20}		107^{5mm}	102	s alc, eth
p44	Pentanoyl chloride	$CH_3CH_2CH_2CH_2COCl$	120.58	2,301	1.016	1.4216^{20}		125–127	23	
p45	1,4,7,10,13-Penta-oxacyclo-pentadecane	$[-CH_2CH_2O-]_5$	220.27			1.4615^{20}		$135^{0.2mm}$		
p46	3,6,9,12,15-Penta-oxahexadecanol	$CH_3O(CH_2CH_2O)_4CH_2CH_2OH$	252.31		0.933	1.4500^{20}		$133^{0.005mm}$	>112	
p47	1-Pentene	$CH_3CH_2CH_2CH=CH_2$	70.14	1,210	0.6410^{20}	1.3714^{20}	−165.2	30.0		misc alc, bz, eth
p48	(E)-2-Pentene	$CH_3CH_2CH=CHCH_3$	70.14	1,210	0.6482^{20}	1.3793^{20}	−140.2	36.3	−45	misc alc, eth
p49	(Z)-2-Pentene	$CH_3CH_2CH=CHCH_3$	70.14	1,210	0.6503^{20}	1.3830^{20}	−151.4	36.9		misc alc, eth
p50	4-Pentenoic acid	$H_2C=CHCH_2CH_2COOH$	100.11	2,425	0.9843^{18}	1.4341^{18}	<−18	187–189		sl s aq; s alc, eth
p51	3-Penten-2-one	$CH_3CH=CHCOCH_3$	84.12	1,732	0.8624^{20}	1.4405^{20}		121–124	21	s aq
p52	Pentyl acetate	$CH_3(CH_2)_4OOCCH_3$	130.19	2,131	0.8753^{20}	1.4028^{20}	<−100	149.2	23	0.17 aq
p53	Pentylamine	$CH_3(CH_2)_4NH_2$	87.17	4,175	0.752	1.4110^{20}	−55	104	4	v s aq; misc alc, eth
p54	Pentylbenzene	$CH_3(CH_2)_4C_6H_5$	148.25	5,434	0.8594^{20}_{4}	1.4885^{20}	−78.3	202.2	65	s alc; misc bz, eth

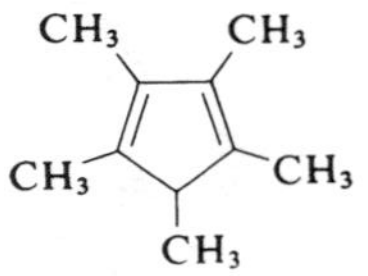

p25

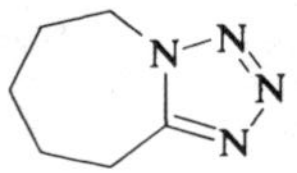

p26

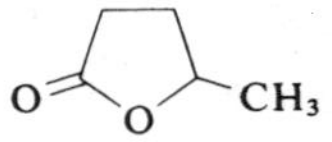

p40

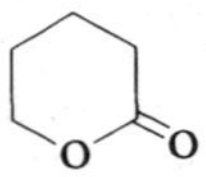

p40a

TABLE 1.15 Physical Constants of Organic Compounds (*continued*)

No.	Name	Formula	Formula weight	Beilstein reference	Density	Refractive index	Melting point	Boiling point	Flash point	Solubility in 100 parts solvent
p55	4-*tert*-Pentylcyclohexanone		168.28	7^3, 173	0.920	1.4677^{20}		125^{16mm}	104	
p56	4-*tert*-Pentylphenol	$CH_3CH_2C(CH_3)_2$-C_6H_4OH	164.25	6, 548	0.9622^{20}		93	262.2		s alc, eth
p57	1-Pentyne	$CH_3CH_2CH_2C{\equiv}CH$	68.11	1, 250	0.6901^{20}_4	1.3852^{20}	-105.7	40.2		v s alc; misc eth
p58	L-Perillaldehyde		150.22	7, 158	0.9645^{20}_4	1.5072^{20}		105^{10mm}	95	
p59	Peroxyacetic acid	$CH_3C({=}O)OOH$	76.05		1.226^{15}_4		0.1	105 explodes 110		v s aq, alc, eth
p60	Petroleum ether	principally pentanes and hexanes			0.640			35–80	-40	misc bz, chl, eth, CCl_4
p61	Phenanthrene		178.23	5, 667	1.179^{25}		100	340		1.6 alc; 50 bz; 30 eth
p62	9,10-Phenanthrenedione		208.22	7, 796	1.405^4		209–211			s bz, eth, hot alc
p63	1,10-Phenanthroline		180.21	23, 227			117			1.4 bz; s alc, acet
p64	Phenol	C_6H_5OH	94.11	6, 110	1.0576^{41}_4	1.5418^{41}	40.9	181.8	79	6.7 aq; 8.2 bz; v s alc, chl, eth, alk
p65	Phenolphthalein		318.33	18, 143	1.299^{25}_4		258–262			8.2 alc; 1 eth
p66	Phenothiazine		199.28	27, 63			185.1	371		v s bz; s eth; sl s alc
p67	Phenothiazine-10-carbonyl chloride		261.73	27, 66			168–171			
p68	Phenoxyacetic acid	$C_6H_5OCH_2COOH$	152.15	6, 161			98	285 sl d		1.3 aq; v s alc, bz, HOAc, CS_2, eth
p69	Phenoxyacetyl chloride	$C_6H_5OCH_2COCl$	170.60	6, 162	1.235	1.5340^{20}		225–256		d aq, alc; s eth
p70	*p*-Phenoxyaniline	$C_6H_5OC_6H_4NH_2$	185.23	13, 438			82–84	189^{14mm}		s hot aq; v s alc, eth

p71	2-Phenoxybutyric acid	CH$_3$CH$_2$CH(OC$_6$H$_5$)COOH	180.20	6, 163			79–83	258	sl s aq
p72	2-Phenoxyethanol	C$_6$H$_5$OCH$_2$CH$_2$OH	138.17	6, 146	1.102^{22}_{4}	1.5370^{20}	14	245.2	110
p73	1-Phenoxy-2-propanol	C$_6$H$_5$OCH$_2$CH(OH)CH$_3$	152.19	6', 85	1.063^{25}_{4}	1.523^{20}	13–18	240	135
p74	Phenoxy-2-propanone	C$_6$H$_5$OCH$_2$COCH$_3$	150.18	6, 151	1.097	1.5210^{20}		230	85
p75	DL-2-Phenoxy-propionic	CH$_3$CH(OC$_6$H$_5$)COOH	166.18	6, 163			116–119	265	

s aq; v s alc, eth

s alc; sl s aq

Peracetic acid, p59
Perdeuterocyclohexane, c313
Perylene, d49
Phenacetin, e45
Phenacyl bromide, b221
Phenacyl chloride, c28

9,10-Phenanthraquinone, p62
Phenazone, a309
1,2,4-Phenenyl triacetate, t193
Phenethyl alcohol, p113
sec-Phenethyl alcohol, m138
Phenethylamine, p114

Phenethyl bromide, b283
Phenethyl chloride, c108
p-Phenetidine, e24
Phenetole, e28
Phenoxyacetone, p74
4-Phenoxybutyl bromide, b243

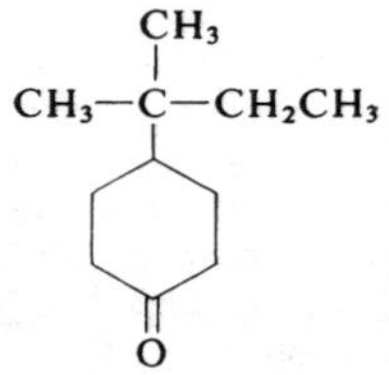

CH$_3$—C(CH$_3$)—CH$_2$CH$_3$ p55

CH$_3$—C=CH$_2$ CHO   p58

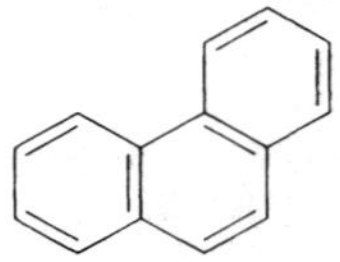

p61

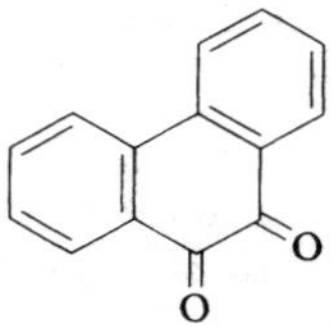

p62

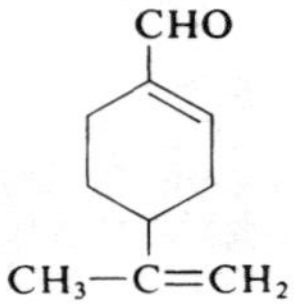

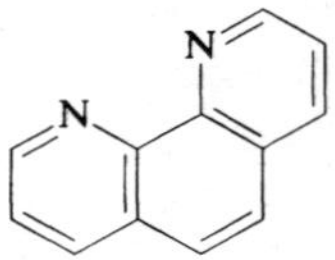

p63

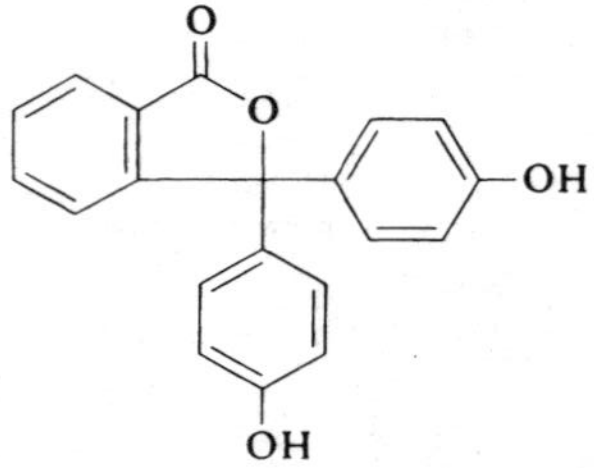

OH   OH   p65

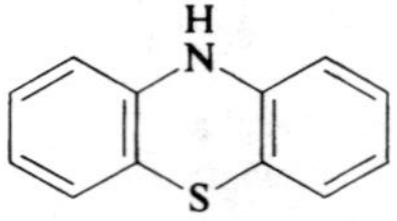

p66

COCl   p67

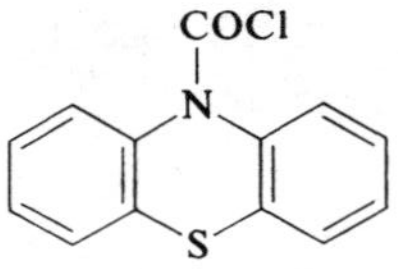

TABLE 1.15 Physical Constants of Organic Compounds (*continued*)

No.	Name	Formula	Formula weight	Beilstein reference	Density	Refractive index	Melting point	Boiling point	Flash point	Solubility in 100 parts solvent
p76	3-Phenoxytoluene	$C_6H_5OC_6H_4CH_3$	184.24	6, 377	1.051	1.5727^{20}	33–34	271–273	>112	sl s aq; s alc, eth
p76a	Phenylacetaldehyde	$C_6H_5CH_2CHO$	120.15	7, 292	1.027^{25}	1.5273^{20}		195	86	
p77	2-(2-Phenyl-acetamido)-acetaldoxime	$C_6H_5CH_2CONHCH_2$-$CH{=}NOH$	192.22				147–151			
p78	Phenyl acetate	$C_6H_5OOCCH_3$	136.15	6, 152	1.073^{20}_{20}	1.5030^{20}		196	76	misc alc, eth, chl
p79	Phenylacetic acid	$C_6H_5CH_2COOH$	136.15	9, 431	1.091^{77}_{4}		76.5	265.5		s hot aq, alc, eth
p80	Phenylacetonitrile	$C_6H_5CH_2CN$	117.15	9, 441	1.0214^{15}	1.5233^{20}	−23.8	233.5	101	i aq; misc alc, eth
p81	Phenylacetyl chloride	$C_6H_5CH_2COCl$	154.60	9, 436	1.169	1.5325^{20}		95^{12mm}		d aq, alc
p82	Phenylacetylene	$C_6H_5C{\equiv}CH$	102.14	5, 511	0.9300^{20}_{4}	1.5470^{20}	−44.9	142.4	31	misc alc, eth
p83	Phenylacetylurea	$C_6H_5CH_2CONHCONH_2$	178.19	14, 495			212–216 d 283			sl s alc, bz, chl, eth
p84	L-3-Phenyl-α-alanine	$C_6H_5CH_2CH(NH_2)COOH$	165.19							3 aq; s hot alc; i eth
p85	2-(Phenylamino)-benzoic acid	$C_6H_5NHC_6H_4COOH$	213.24	14, 327			185 d			s hot alc
p86	Phenyl 4-amino-salicylate	$H_2NC_6H_3(OH)COOC_6H_5$	229.24				153			0.7 mg aq
p87	p-Phenylazoaniline	$C_6H_5N{=}NC_6H_4NH_2$	197.24	16^1, 310			128	>360		v s alc, bz, chl, eth
p88	Phenylazoformic acid 2-phenylhydrazide	$C_6H_5N{=}NCONHNHC_6H_5$	240.27	16, 24			156–159 d			
p89	p-Phenylazophenol	$C_6H_5N{=}NC_6H_4OH$	198.23	16, 96			155–157	230^{20mm}		v s alc, eth
p90	2-Phenylbenzimidazole		194.24	23, 230			291			s abs alc; sl s bz, chl
p91	Phenyl benzoate	$C_6H_5COOC_6H_5$	198.22	9, 116	1.235		70	314		v s hot alc; sl s eth
p92	N-Phenylbenzylamine	$C_6H_5CH_2NHC_6H_5$	183.25	12, 1023	1.061		27–38	306–307		s alc, chl, eth
p93	1-Phenylbiguanide	$C_6H_5NHC({=}NH)NH$-$C({=}NH)NH_2$	177.21				144–146			v s aq, alc

No.	Name	Formula	M.W.	Beil. Ref.	Density	n_D	M.P.	B.P.	Flash P.	Solubility
p94	1-Phenyl-2-butanone	$CH_3CH_2COCH_2C_6H_5$	148.21	7, 314	0.998	1.5122^{20}		112^{15mm}	90	s alc; misc eth; i aq
p95	4-Phenyl-2-butanone	$C_6H_5CH_2CH_2COCH_3$	148.21	7, 314	0.989	1.5122^{20}		235	98	s alc, eth
p96	(E)-4-Phenyl-3-buten-2-one	$C_6H_5CH=CHCOCH_3$	146.19	7, 364	1.0097^{45}	1.5836^{45}	41.5	261	65	v s alc, bz, chl, eth
p97	4-Phenylbutylamine	$C_6H_5CH_2CH_2CH_2-CH_2NH_2$	149.24	12, 1165	0.944	1.5196^{20}		124^{17mm}	101	
p98	2-Phenyl-3-butyn-2-ol	$CH_3C(OH)(C_6H_5)C{\equiv}CH$	146.19	6^2, 559			51–52	217–218		0.8 aq: s alc, bz, acet
p99	2-Phenylbutyric acid	$CH_3CH_2CH(C_6H_5)COOH$	164.20	9^2, 356			42–44	270–272		s bz, eth
p100	4-Phenylbutyric acid	$C_6H_5CH_2CH_2CH_2COOH$	164.20	9, 539			50–52	165^{10mm}		s alc, eth
p101	DL-2-Phenylbutyronitrile	$CH_3CH_2CH(C_6H_5)CN$	145.21	9, 541	0.974	1.5086^{20}		$114\text{—}115^{15mm}$ 71^{9mm}	>112	
p102	Phenyl chloroformate	C_6H_5OOCCl	156.57							
p103	S-Phenyl chlorothioformate	C_6H_5SCOCl	172.6		1.269_4^{30}	1.5786^{30}	–14	101^{10mm}	116	

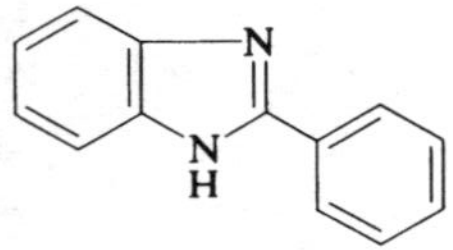

p90

TABLE 1.15 Physical Constants of Organic Compounds (*continued*)

No.	Name	Formula	Formula weight	Beilstein reference	Density	Refractive index	Melting point	Boiling point	Flash point	Solubility in 100 parts solvent
p104	Phenylcyclohexane	$C_6H_5C_6H_{11}$	160.26	5, 503	0.9427^{20}	1.5263^{20}	7.0	240.1	98	v s alc, eth
p105	Phenyl dichlorophosphate	$C_6H_5OP(O)Cl_2$	210.98	6, 179	1.412	1.5230^{20}		241–243	>112	
p106	N-Phenyldiethanol-amine	$C_6H_5N(CH_2CH_2OH)_2$	181.24	12, 183	1.120^{60}_{20}		56–58	350 sl d		5 aq; v s alc; 29 eth; 25 bz
p107	o-Phenylenediamine	$C_6H_4(NH_2)_2$	108.14	13, 6			103–104	256–258		v s alc, chl, eth
p108	m-Phenylenediamine	$C_6H_4(NH_2)_2$	108.14	13^3, 10	1.139^{15}_{15}		62–63	234–237		s aq, alc, acet, chl
p109	p-Phenylenediamine	$H_2NC_6H_4NH_2$	108.14	13, 61			145–147	267	68	1 aq; s alc; chl, eth
p110	o-Phenylene phosphorochloridite		174.52	27, 809	1.466	1.5712^{20}		80^{20mm}	>112	
p111	1-Phenyl-1,2-ethanediol	$C_6H_5CH(OH)CH_2OH$	138.17	6, 907			66–68	272–274		v s aq, alc, bz, eth, chl, HOAc
p112	1-Phenylethanol	$CH_3CH(C_6H_5)OH$	122.17	6, 475	1.0150^{20}_{20}	1.5211^{20}	21.4	203.9		2.3 aq
p113	2-Phenylethanol	$C_6H_5CH_2CH_2OH$	122.17	6, 478	1.0185^{25}	1.5317^{20}	−27	221	102	2 aq; misc alc, eth
p114	2-Phenylethylamine	$C_6H_5CH_2CH_2NH_2$	212.28	12, 1096	0.9640^{25}_{4}	1.5332^{20}		195	90	s aq; v s alc, eth
p115	D-(−)-α-Phenylglycine	$C_6H_5CH(NH_2)COOH$	151.17	14, 460			305–310			
p116	1-Phenylheptane	$C_6H_5(CH_2)_6CH_3$	176.30	5, 451	0.860	1.4842^{20}		233	95	misc eth
p117	1-Phenylhexane	$C_6H_5(CH_2)_5CH_3$	162.28	5^2, 337	0.861	1.4865^{20}	−61	226	83	misc alc, bz, chl, eth
p118	Phenylhydrazine	$C_6H_5NHNH_2$	108.14	15^2, 44	1.0978^{20}_{4}	1.6070^{20}	19.5	243.5 d	88	
p119	Phenyl 3-hydroxy-2-naphthoate	$C_{10}H_6(OH)COOC_6H_5$	264.28	10, 335			129–132	261^{160mm}		
p119a	2-Phenyl-2-imidazoline		146.19	23, 154			94–99			
p120	2-Phenylindole		193.25	20, 467			17	250^{10mm}		d aq, alc; s eth
p121	Phenyl isocyanate	C_6H_5NCO	119.12	12, 437	1.0956^{20}	1.5350^{20}	−30	162–163	55	
p122	Phenyl isothiocyanate	C_6H_5NCS	135.19	12, 453	1.1288^{25}_{4}	1.6497^{20}	−21	221	87	i aq; s alc, eth

No.	Name	Formula	Formula wt	Beilstein reference	Density	n	mp/°C	bp/°C	Solubility
p123	*N*-Phenylmaleimide		173.17	21, 400			89–90	$163^{12\text{mm}}$	s alc, chl, eth
p124	Phenylmalonic acid	$C_6H_5CH(COOH)_2$	180.16				155 d		0.17 aq; s alc, bz, acet
p125	Phenylmercury(II) acetate	$C_6H_5HgOOCCH_3$	336.74				149		s bz, eth, pyr
p126	Phenylmercury(II) chloride	C_6H_5HgCl	313.15	16, 952			250–252		
p127	Phenylmercury(II) hydroxide	C_6H_5HgOH	294.70				190 d		
p128	Phenylmethanethiol	$C_6H_5CH_2SH$	124.21	6, 453	1.058^{20}			70; 194–195	1.0 aq; v s hot alc
p129	*N*-Phenylmorpholine		163.22	27, 6			57	268	s alc, bz, chl, eth
p130	*N*-Phenyl-1-naphthylamine	$C_{10}H_7NHC_6H_5$	219.29	12, 1224			60–62	$226^{15\text{mm}}$	
p130a	1-Phenyloctane	$C_6H_5(CH_2)_7CH_3$	190.33	5, 453	0.8572^{20}_{4}	1.4840^{20}	−36	261–263	misc eth
p131	2-Phenylphenol	$C_6H_5C_6H_4OH$	170.21	6^2, 623	1.213		57	282	s alc, chl, eth, alk
p132	4-Phenylphenol	$C_6H_5C_6H_4OH$	170.21	6, 674			164–165	305	s alc, chl, eth, alk

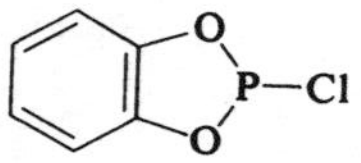

p110 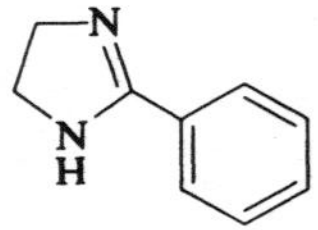   p119a 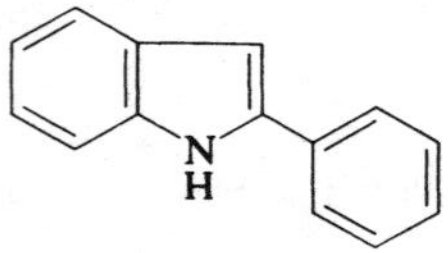   p120 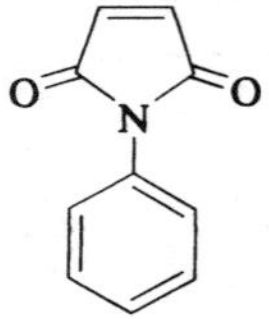   p123 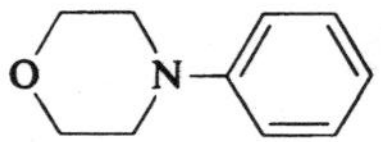   p129

TABLE 1.15 Physical Constants of Organic Compounds (*continued*)

No.	Name	Formula	Formula weight	Beilstein reference	Density	Refractive index	Melting point	Boiling point	Flash point	Solubility in 100 parts solvent
p133	N-phenyl-p-phenyl enediamine	$C_6H_5NHC_6H_4NH_2$	184.24	13, 76			73–75			
p134	Phenyl N-phenyl-phosphoramido-chloridate	$C_6H_5NHP(=O)(Cl)OC_6H_5$	267.66	12, 588			132–134			
p135	Phenylphosphinic acid	$C_6H_5PH(O)OH$	142.09	16, 791			83–85			
p136	Phenylphosphonic acid	$C_6H_5P(O)(OH)_2$	158.09	16, 803			163–166			
p137	Phenylphosphonic dichloride	$C_6H_5P(O)Cl_2$	194.99	16, 804	1.375	1.5600^{20}	3	258	>112	
p138	Phenylphosphono-thioic dichloride	$C_6H_5P(S)Cl_2$	211.05	16, 807	1.360	1.6244^{20}		205^{130mm}		
p139	N-Phenylpiperazine		162.24	6, 930	1.0621^{20}_4	1.5875^{20}	44–45	286	>112	i aq; misc alc
p140	2-Phenyl-1,2-propanediol	$CH_3C(C_6H_5)(OH)CH_2OH$	152.19	6^1, 253	1.010	1.5494^{20}		160–162^{26mm}	>112	
p141	3-Phenyl-1-propanethiol	$C_6H_5CH_2CH_2CH_2SH$	152.26	6, 502				109^{10mm}	90	
p142	1-Phenyl-1-propanol	$C_6H_5CH(OH)CH_2CH_3$	136.19	6, 503	0.9915^{25}_4	1.5169^{23}	−18	219		misc alc, bz
p143	3-Phenyl-1-propanol	$C_6H_5CH_2CH_2CH_2OH$	136.19	7^2, 233	1.008	1.5257^{20}		235	109	s aq; misc alc, eth
p144	1-Phenyl-2-propanone	$C_6H_5CH_2COCH_3$	134.18	7^2, 237	1.01574^{20}_4	1.5160^{20}	27	100^{13mm}	84	v s alc, eth; misc bz
p145	2-Phenylpropion-aldehyde	$CH_3CH(C_6H_5)CHO$	134.18	9, 508	1.009^{20}_4	1.5175^{20}		202–205	69	i aq; s alc
p146	3-Phenylpropionic acid	$C_6H_5CH_2CH_2COOH$	150.18	24, 2	1.047^{100}_4		47–48	280		0.6 aq; s bz, alc, chl, eth, HOAc, PE
p147	1-Phenyl-3-pyrazol-idinone		162.19				121			10 hot aq; hot alc; s alk, acid

No.	Name	Formula	Mol. wt.		Density	n_D	M.P., °C	B.P., °C		Solubility
p148	2-Phenylpyridine	$C_6H_5C_5H_4N$	155.20	20, 424		1.6242^{20}		268–270		s alc, eth
p149	2-Phenyl-4-quinolinecarboxylic acid		249.27	22, 103			214–215		>112	0.8 alc; 1 eth; 0.3 chl
p150	Phenyl salicylate	$C_6H_4(OH)COOC_6H_5$	214.22	1.25 · 10, 76	1.25		41–43	173^{12mm}		17 alc; 66 bz; s acet, chl, eth; 0.015 aq
p151	Phenylselenenyl chloride	C_6H_5SeCl	191.52	6^3, 1110			63–65	120^{20mm}		
p152	Phenylsuccinic acid	$HOOCCH_2-$ $CH(C_6H_5)COOH$	194.19	9, 865			167–169	$-H_2O$, >168 100^{6mm}		s hot aq, alc, eth
p153	S-Phenyl thioacetate	$C_6H_5SCOCH_3$	152.22			1.5720^{20}			79	
p154	1-Phenyl-2-thiourea	$C_6H_5NHC(S)NH_2$	152.22	12, 388	1.3		154			0.25 aq; s alc, alk
p155	Phenyltrichloro-silane	$C_6H_5SiCl_3$	211.56	16, 911	1.329^{20}	1.5230^{20}		201	91	

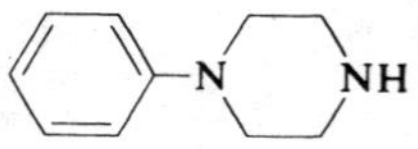

p139

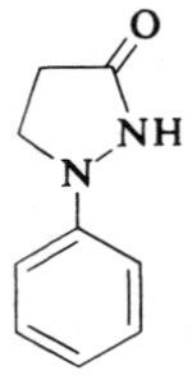

p147

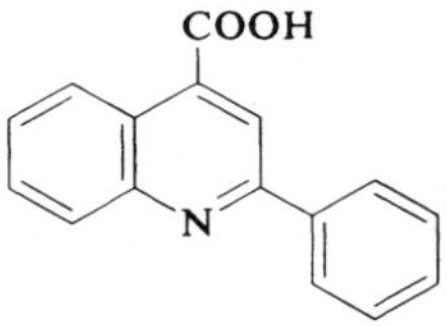

p149

TABLE 1.15 Physical Constants of Organic Compounds (*continued*)

No.	Name	Formula	Formula weight	Beilstein reference	Density	Refractive index	Melting point	Boiling point	Flash point	Solubility in 100 parts solvent
p156	1-Phenyltridecane	$C_6H_5(CH_2)_{12}CH_3$	260.47	16, 911	0.8555_4^5	1.4814^{20}	10	346	>112	
p157	Phenyltriethoxysilane	$C_6H_5Si(OC_2H_5)_3$	240.38		0.996	1.4604^{20}		$113^{1.0mm}$	42	
p158	Phenyltrimethoxysilane	$C_6H_5Si(OCH_3)_3$	198.3		1.064_4^{20}	1.4734^{20}		211		
p159	Phenyltrimethylammonium bromide	$[C_6H_5N(CH_3)_3]^+Br^-$	216.13	12^2, 88			210 d			v s aq; s hot alc
p160	Phenyltrimethylammonium chloride	$[C_6H_5N(CH_3)_3]^+Cl^-$	171.67	12, 158			237 subl			s aq; v s alc; sl s chl
p161	Phenyltrimethylammonium iodide	$[C_6H_5N(CH_3)_3]^+I^-$	263.12	12^2, 88			175			s aq, alc; sl s acet
p162	Phenyltrimethylammonium tribromide	$[C_6H_5N(CH_3)_3]^+Br_3^-$	375.95				114–116			
p163	Phenyltrimethylsilane	$C_6H_5Si(CH_3)_3$	150.30	16^1, 525	0.873	1.4907^{20}		168–170	44	s hot aq, hot alc, eth
p164	Phenyltris(trimethylsiloxy)silane	$[(CH_3)_3SiO]_3SiC_6H_5$	372.8		0.970_4^{25}	1.459^{25}		264–266	121	
p165	Phenylurea	$C_6H_5NHCONH_2$	136.15	12, 346	1.302		145–147	238		
p166	Phenylvinyldichlorosilane	$H_2C{=}CH(C_6H_5)SiCl_2$	203.2		1.196_4^{25}	1.534^{25}		$87^{1.5mm}$		
p167	o-Phthalic acid	$C_6H_4(COOH)_2$	166.13	9, 791	1.593_4^{20}		206–208			0.6 aq; 10 alc; 0.5 eth; v sl s chl
p168	Phthalic anhydride		148.12	17, 469	1.53		130.8	285 subl		0.6 aq(d); s alc
p169	Phthalide		134.13	17, 310			72–74	290		s alc
p170	Phthalimide		147.13	21, 458	1.164_4^{99}		238	subl		v s alk; v sl s bz, PE
p171	o-Phthaloyl dichloride	$C_6H_4(COCl)_2$	203.02	9, 805	1.409^{20}	1.5684^{20}	15–16	280–282	>112	d aq, alc; s eth

No.	Name	Formula	MW	Density	Ref.	mp	bp	n_D		Solubility
p172	Phthalylsulfa-thioazole		403.44							
p173	Picric acid	$(O_2N)_3C_6H_2OH$	229.11	1.763^{20}	6, 265	272 d	explodes >300			s alk; sl s alc; i chl
p174	Pinane		138.3	0.839^{20}_4	5, 93	−50	167–168	1.4616^{20}		
p175	(+)-α-Pinene		136.24	0.8591^{20}_{14}	5, 146	−55	155–156	1.4660^{20}	32	1.3 aq; 8.2 alc; 10 bz; 2.9 chl; 1.6 eth
p176	(−)-β-Pinene		136.24	0.8590^{20}	5, 154	−61.5	166	1.4666^{20}	32	
p177	α-Pinene oxide		152.24	0.964	5, 152		103^{50mm}	1.4690^{20}	65	misc alc, eth
p178	β-Pinene oxide		152.24	0.976	17^2, 44		100^{27mm}	1.4765^{20}	66	

Phloroglucinol, t305
Phorone, d529
Phthalaldehydic acid, f33
m-Phthalic acid, b16
p-Phthalic acid, b17
Phthalonitrile, d236

Picolinaldehyde, p251
Picolines, m398, m399, m400
Picolinic acids, p255, p257
Picolinonitrile, c295
Picolylamines, a223, a224
Picramide, t381

Pimelic acid, h8
Pinacol, d491
Pinacolone, d497
Pinacolyl alcohol, d496
3-Pinanol, i83

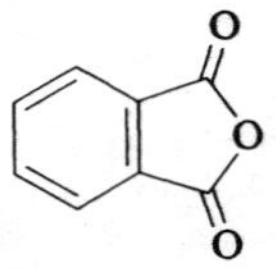

p168

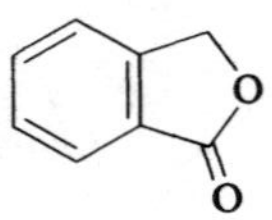

p169

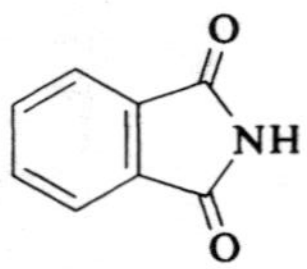

p170   NH

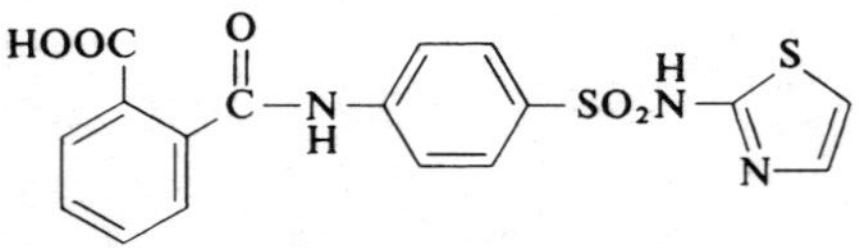

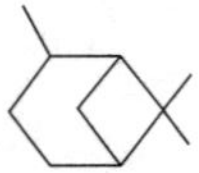

p174 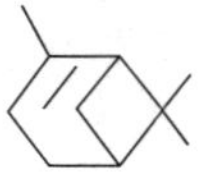   p175 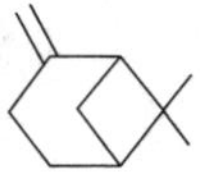   p176

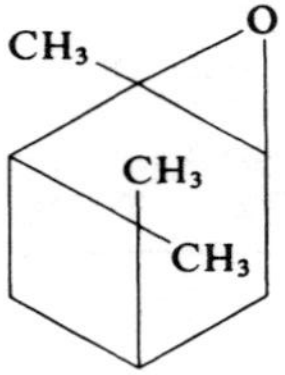

HOOC   N H   SO$_2$N H   S   N
p172

p177   CH$_3$ O   CH$_3$ CH$_3$

p178   CH$_2$—O   CH$_3$ CH$_3$ CH$_3$

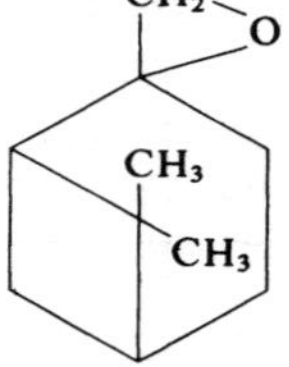

TABLE 1.15 Physical Constants of Organic Compounds (*continued*)

No.	Name	Formula	Formula weight	Beilstein reference	Density	Refractive index	Melting point	Boiling point	Flash point	Solubility in 100 parts solvent
p179	Piperazine		86.14	23, 4		1.446^{113}	108–110	145–146 $97^{0.5mm}$	109	v s aq; 50 alc; i eth
p180	1-Piperazinecarb-aldehyde		114.15		1.107	1.5094^{20}			101	
p181	1,4-Piperazinedi-carbonitrile		136.16	23^1, 5			167–170			
p182	3-(1-Piperazinyl)-1,2-propanediol		160.22				73–77	$133^{0.1mm}$	4	misc aq; s alc, bz, chl
p183	Piperidine		85.15	20, 6	0.8659^{15}	1.4525^{20}	−10.5	106.4		
p184	1-Piperidine-carbonitrile		110.16	20, 56	0.951	1.4705^{20}		102^{10mm}	97	misc aq; s alc
p185	N-Piperidineethanol		129.20	20, 25	0.9732^{25}_{25}	1.4804^{20}	38–40	200–202	68	v s aq, alc, eth
p186	2-Piperidineethanol		129.20	21, 2	1.010^{17}			234	102	
p187	3-Piperidinemethanol		115.18	21^2, 8	1.026			$107^{3.5mm}$	>112	
p188	1-Piperidinepropio-nitrile		138.21		0.933	1.4695^{20}		111^{16mm}		
p189	3-Piperidino-1,2-propanediol		159.23	20, 34			77–80			
p190	trans-Piperitol		154.3		0.9178^{25}	1.4729^{-42}				
p191	Propane	$CH_3CH_2CH_3$	44.10	1, 104	0.5842^{-42}	1.3397^{-42}	−187.7	−42.1		6.5 mL aq; 790 mL alc; 926 mL eth; 1300 mL chl; 1450 mL bz
p192	1,2-Propanediamine	$CH_3CH(NH_2)CH_2NH_2$	74.13	4, 257	0.878^{15}	1.4460^{20}		119.7	33	misc aq, bz; s alc, eth
p193	1,3-Propanediamine	$H_2NCH_2CH_2CH_2NH_2$	74.13	4, 261	0.8844^{25}	1.4575^{20}	−12	140	48	misc alc, eth; s aq
p194	1,2-Propanediol	$CH_3CH(OH)CH_2OH$	76.10	1, 472	1.0364^{20}_{4}	1.4331^{20}	−60	188	107	misc aq, acet, chl; s alc, eth
p195	1,3-Propanediol	$HOCH_2CH_2CH_2OH$	76.10	1, 475	1.0597^{20}_{4}	1.4396^{20}	−26.7	214.4	79	misc aq, alc

p196	1,3-Propanedithiol	$HSCH_2CH_2CH_2SH$	108.23	1, 476	1.0772_4^{20}	1.5405^{20}	−79	169	40	misc alc, bz, eth, chl
p197	1-Propanesulfonyl chloride	$CH_3CH_2CH_2SO_2Cl$	142.60	4, 8	1.2864_4^{15}			66^{8mm}		d hot aq, hot alc
p198	1,3-Propane sultone		122.14		1.392		30–33	180^{30mm}		

Pipecolines, m369, m370, m371
1-Piperazineethanol, h123
1-Piperidinecarboxyaldehyde, f35
Piperonal, m239
Piperonyl alcohol, m242
Piperonyl butoxide, m243
Piperonylic acid, m241
Pivalaldehyde, d596

Pivalamide, d597
Pivalic acid, d598
Pivalic anhydride, d599
Pivaloyl chloride, d600
Pivaloyloxymethyl chloride, c151
POPOP, b198
PPO, d679
Prehnitene, t97

Procaine, d271
Proline, p273
Propadiene, a78
1-Propanal, p211
1,3-Propanedicarboxylic acid, g11
Propanedioic acid, m3
1,2-Propanediol cyclic carbonate, p225

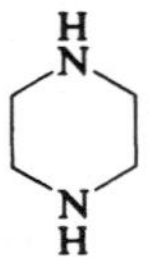

p179

p180

p181

p182

p183

p184

p185

p186

p187

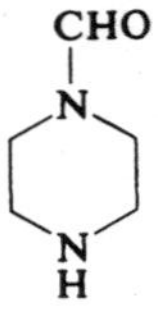

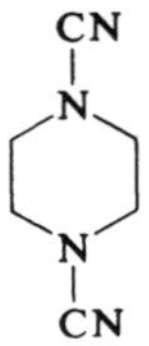

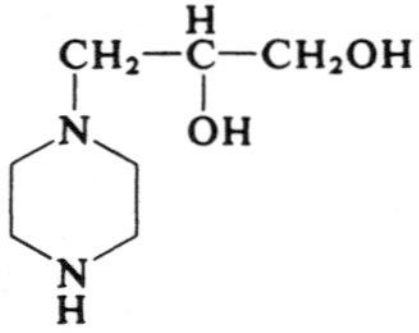

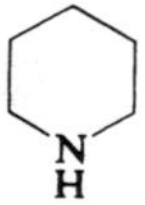

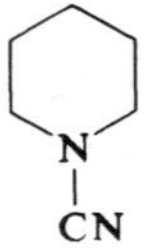

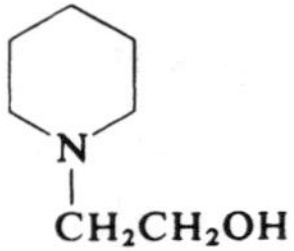

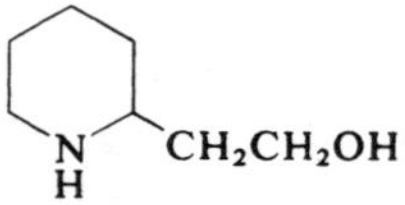

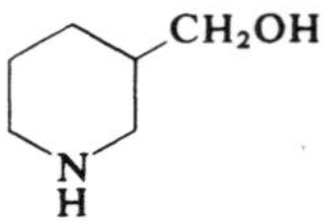

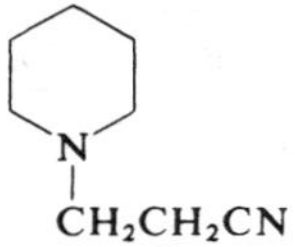

p188

p189

p190

p198

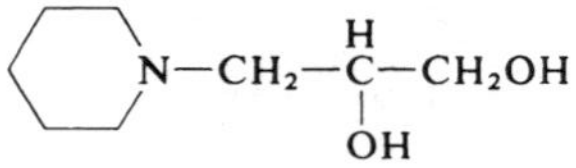

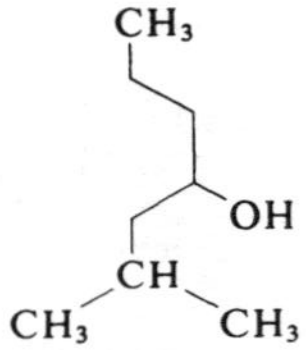

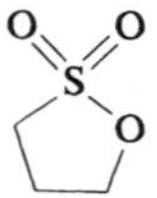

TABLE 1.15 Physical Constants of Organic Compounds (*continued*)

No.	Name	Formula	Formula weight	Beilstein reference	Density	Refractive index	Melting point	Boiling point	Flash point	Solubility in 100 parts solvent
p199	1-Propanethiol	$CH_3CH_2CH_2SH$	76.16	1, 359	0.8364^{25}	1.4380^{20}	−113.1	67.7	−20	s alc, eth
p200	2-Propanethiol	$CH_3CH(SH)CH_3$	76.16	1, 367	0.809_4^{25}	1.4255^{20}	−130.5	52.6	−34	misc alc, eth; sl s aq
p201	1,2,3-Propanetriol triacetate	$H_3CCOO-CH(CH_2OOCCH_3)_2$	218.21	2, 147	1.596_4^{20}	1.4302^{20}	−78	258–260	148	7.2 aq; misc alc, bz, chl, eth
p202	1-Propanol	$CH_3CH_2CH_2OH$	60.10	1, 350	0.8037_4^{20}	1.3856^{20}	−126.2	97.2	15	misc aq, alc, eth
p203	2-Propanol	$(CH_3)_2CHOH$	60.10	1, 360	0.7855_4^{20}	1.3772^{20}	−89.5	82.4	22	misc aq, alc, chl, eth
p204	2-Propenal	$H_2C=CHCHO$	56.07	1, 725	0.8389^{20}	1.4017^{20}	−87.0	52.7	−18	21 aq; s alc, eth
p205	Propene	$H_2C=CHCH_3$	42.08	1, 196	0.6104_4^{-48}	1.3567^{-40}	−185.2	−47.7		45 mL aq; 1200 mL alc; 500mL acet
p206	2-Propene-1-thiol	$H_2C=CHCH_2SH$	74.15	1, 440	0.925_4^{23}		d 200	67–68	21	misc alc, eth
p207	(Z)-1,2,3-Propene-tricarboxylic acid		174.11	2, 849						50 aq; s alc; sl s eth
p208	1-Propen-2-yl acetate	$H_2C=C(OOCCH_3)CH_3$	100.12		0.909	1.4000^{20}		97	18	
p209	o-Propenylphenol	$CH_3CH=CHC_6H_4OH$	134.18	6¹, 279	1.044	1.5754^{20}	−33.4	230–231	90	37 aq(hyd); misc alc (reacts), bz, eth, acet
p210	β-Propiolactone		72.06		1.1460_4^{20}	1.4131^{20}		162.3	70	
p211	Propionaldehyde	CH_3CH_2CHO	58.08	1, 629	0.8071_4^{20}	1.3646^{19}	−81	48–49	−9	30 aq; misc alc, eth
p212	Propionamide	$CH_3CH_2CONH_2$	73.10	2, 243	0.9597_4^{20}	1.4160^{110}	79	222.2		v s aq, alc, chl, eth
p213	Propionic acid	CH_3CH_2COOH	74.09	2, 234	0.9934_4^{20}	1.3865^{20}	−21	140.8	51	misc aq; s alc, chl, eth
p214	Propionic anhydride	$[CH_3CH_2C(=O)]_2O$	130.14	2, 242	1.0125_4^{20}	1.4047^{20}	−45	167	73	d aq; s alc, chl, eth
p215	Propionitrile	CH_3CH_2CN	55.08	2, 245	0.7818_4^{20}	1.3658^{20}	−92.8	97.2	6	10 aq; misc alc, eth

No.	Name	Formula	M.W.	Ref.	Density	n_D	m.p.	b.p.	f.p.	Solubility
p216	Propionyl chloride	CH_3CH_2COCl	92.53	2, 243	1.065_4^{20}	1.4051^{20}	−94	80	11	d aq, alc
p217	Propiophenone	$C_6H_5COCH_2CH_3$	134.18	7^2, 231	1.0105_4^{20}	1.5258^{20}	18.6	218.0	87	misc bz, eth, abs alc
p218	Propoxytrimethyl-silane	$CH_3CH_2CH_2OSi(CH_3)_3$	132.3		0.768_4^{20}	1.384^{20}		100^{735mm}		2.3 aq; misc alc, eth
p219	Propyl acetate	$CH_3CH_2CH_2OOCCH_3$	102.13	2, 129	0.836_4^{20}	1.3844^{20}	−92	101.6	12	misc aq, alc, eth
p220	Propylamine	$CH_3CH_2CH_2NH_2$	59.11	4, 136	0.7173^{20}	1.3882^{20}	−83.0	47.9	−37	s alc, eth
p221	2-(Propylamino)-ethanol	$C_3H_7NHCH_2CH_2OH$	103.17	4, 282	0.900	1.4415^{20}		182^{746mm}	78	i aq; s alc, eth
p222	Propylbenzene	$CH_3CH_2CH_2C_6H_5$	120.20	5, 390	0.8621^{20}	1.4912^{20}	−99.6	159.2		s bz, eth
p223	Propyl benzoate	$C_6H_5COOCH_2CH_2CH_3$	164.20	9, 112	1.0232^{20}	1.5003^{20}	−51.6	231.2	47	v s aq, alc, bz, eth
p224	Propylcyclohexane	$CH_3CH_2CH_2C_6H_{11}$	126.24	5^2, 23	0.7929_4^{20}	1.4370^{20}	−94.9	156.7		misc aq, alc, PE
p225	Propylene carbonate		102.09		1.2041_4^{20}	1.4210^{20}	−55	240	132	
p226	Propyleneimine		57.09		0.8017^{25}	1.4084^{25}		66.0		

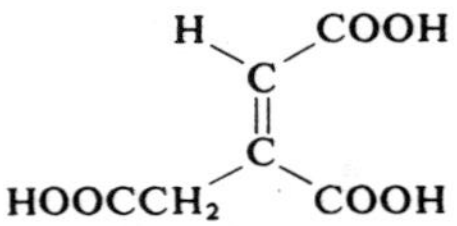

p207

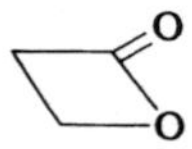

p210

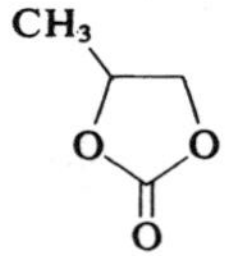

p225

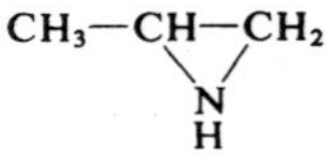

p226

$CH_3-CH-CH_2$
$\diagdown\;N\;\diagup$
    H
p226

TABLE 1.15 Physical Constants of Organic Compounds (*continued*)

No.	Name	Formula	Formula weight	Beilstein reference	Density	Refractive index	Melting point	Boiling point	Flash point	Solubility in 100 parts solvent
p227	Propylene oxide	CH$_3$CH—CH$_2$ (O)	58.08	17, 6	0.8287^{20}	1.3660^{20}	−112.1	37–38	−37	41 aq; misc alc, eth
p228	Propylene sulfide	CH$_3$CH—CH$_2$ (S)	102.18	1, 354	0.736	1.3800^{20}	−123	88–90	4	
p229	Propyl formate	CH$_3$CH$_2$CH$_2$OOCH	88.10	2, 21	0.9006^{20}_4	1.3769^{20}	−92.9	80.9	−3	2 aq; misc alc, eth
p230	Propyl 4-hydroxybenzoate	HOC$_6$H$_4$COOCH$_2$CH$_2$CH$_3$	180.20	10, 160			86–87			0.05 aq; v s alc, eth
p231	Propyl isocyanate	CH$_3$CH$_2$CH$_2$NCO	85.11	4^1, 366	0.908	1.3970^{20}		83–84	26	s aq, alc, eth
p232	Propyl lactate	CH$_3$CH(OH)COOC$_3$H$_7$	132.16	3, 265	0.9962^{20}_{20}	1.4167^{25}		86^{40mm}		s alc, eth
p233	Propyl nitrate	CH$_3$CH$_2$CH$_2$ONO$_2$	105.09	1, 355	1.0538^{20}_4	1.3976^{20}	−100	110.1	23	s alc, eth
							may explode on heating			
p234	2-Propylpentanoic acid	(CH$_3$CH$_2$CH$_2$)$_2$CHCOOH	144.21	2, 350	0.921	1.4250^{20}		220		
p235	o-Propylphenol	CH$_3$CH$_2$CH$_2$C$_6$H$_4$OH	136.19	6, 499	1.015^{20}	1.5279^{20}		224–226	93	s alc, eth
p236	Propylphosphonic dichloride	CH$_3$CH$_2$CH$_2$P(O)Cl$_2$	160.97	4, 596	1.290	1.4643^{20}		$88–90^{50mm}$	>112	
p237	Propyltrichlorosilane	CH$_3$CH$_2$CH$_2$SiCl$_3$	177.53	4, 630	1.1851^{20}_4	1.429^{20}		123–124	2	
p238	Propyltriethoxysilane	C$_3$H$_7$Si(OC$_2$H$_5$)$_3$	206.4		0.892^{20}_4	1.396^{20}		179–180		
p239	Propyl 3,4,5-trihydroxybenzoate	(HO)$_3$C$_6$H$_2$COOC$_3$H$_7$	212.20	1, 246			150			0.35 aq; 1 alc; 83 eth
p240	Propyne	CH$_3$C≡CH	40.06		0.691^{-20}_4	1.3725^{-20}	−102.8	−23.2		v s alc; 3000 mL eth
p241	2-Propynoic acid	HC≡CCOOH	70.05	2, 477	1.138^{20}_4	1.4320^{20}	9	102^{200mm}	58	s aq, alc, eth
p242	2-Propyn-1-ol	HC≡CCH$_2$OH	56.06	1, 454	0.9715^{20}_4	1.4320^{20}	−51.8	113.6	33	misc aq, alc, bz, chl
p243	(+)-Pulegone		152.24	7, 81	0.9346^{15}_4	1.4850^{20}		224	82	misc alc, chl, eth

No.	Name	Formula	M.W.	Beilstein	Density	n	m.p., °C	b.p., °C		Solubility
p244	Pyrazine		80.09	23, 91	1.031_4^{61}	1.4953^{61}	53	115–116		v s aq, alc, eth
p245	Pyrazole		68.08	23, 39		1.4203	70	186–188	85	s aq, alc, bz, eth
p246	Pyrene		202.26	5, 693			150–151		20	misc aq, bz; v s alc, eth
p247	Pyridazine		80.09	23, 89	1.1035_4^{25}	1.5230^{23}	−8	208	20	misc aq, alc, eth
p248	Pyridine	C_5H_5N	79.10	20, 181	0.9782_4^{25}	1.5067^{25}	−41.6	115.2		
p249	Pyridine-d_5	C_5D_5N	84.14		1.05	1.5079^{20}		114.4		
p250	2-Pyridinealdoxime	$(C_5H_4N)CH{=}NOH$	122.13	21[1], 288			110–112			
p251	2-Pyridinecarb-aldehyde	$(C_5H_4N)CHO$	107.11	21[1], 287	1.126	1.5370^{20}		181	54	s aq, eth
p252	3-Pyridinecarb-aldehyde	$(C_5H_4N)CHO$	107.11	21[1], 288	1.135	1.5493^{20}		97^{15mm}	60	
p253	4-Pyridinecarb-aldehyde	$(C_5H_4N)CHO$	107.11	21, 287	1.172	1.5440^{20}		78^{12mm}	54	s aq, eth
p254	3-Pyridinecarbamide	$(C_5H_4N)CONH_2$	122.13	22, 40			130–133			100 aq; 66 alc
p255	Pyridine-2-carboxylic acid	$(C_5H_4N)COOH$	123.11	22, 33	1.400	1.466	134–136	subl		s aq, alc, bz
p256	Pyridine-3-carboxylic acid	$(C_5H_4N)COOH$	123.11	22, 38	1.473		236.6	subl		1.4 aq: s alk

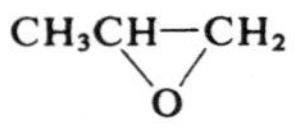

$CH_3CH{-}CH_2$ / O

p227

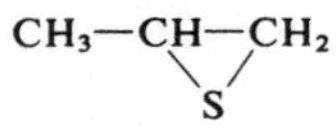

$CH_3{-}CH{-}CH_2$ / S

p228

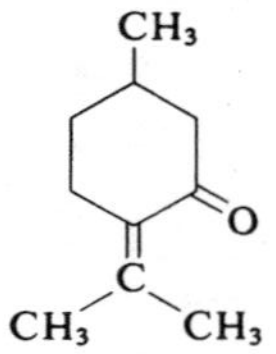

(CH_3, =O, CH_3, CH_3)

p243

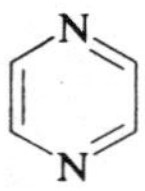

p244

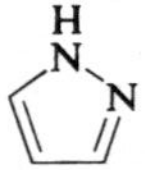

p245

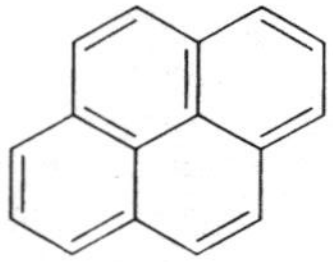

p246

p247

TABLE 1.15 Physical Constants of Organic Compounds (*continued*)

No.	Name	Formula	Formula weight	Beilstein reference	Density	Refractive index	Melting point	Boiling point	Flash point	Solubility in 100 parts solvent
p257	Pyridine-4-carboxylic acid	(C$_5$H$_4$N)COOH	123.11	22, 45			319	260^{15mm}		0.52 aq; i alc, bz, eth
p258	4-Pyridinecarboxylic hydrazide	(C$_5$H$_4$N)CONHNH$_2$	137.14	22^1, 504			171.4			14 aq; 2 alc; 0.1 chl
p259	2,3-Pyridinedicarboxylic acid	(C$_5$H$_3$N)(COOH)$_2$	167.12	22, 150			190 d			0.56 aq; s alk
p260	2,5-Pyridinedicarboxylic acid	(C$_5$H$_3$N)(COOH)$_2$	167.12	22, 153			236–237	subl d		s hot acid
p261	2,6-Pyridinedicarboxylic acid	(C$_5$H$_3$N)(COOH)$_2$	167.12	22, 154			250 d			sl s aq; v sl s alc
p262	Pyridine-*N*-oxide	C$_5$H$_5$(NO)	95.10	20^2, 131			66	270		v s aq
p263	3-Pyridinesulfonic acid	(C$_5$H$_4$N)SO$_3$H	159.16	22, 387			>300			
p264	2-Pyridylmethanol	(C$_5$H$_4$N)CH$_2$OH	109.13	21^1, 203	1.131	1.5420^{20}		113^{16mm}		v s aq, alc, eth
p265	3-Pyridylmethanol	(C$_5$H$_4$N)CH$_2$OH	109.13	21, 50	1.124	1.5445^{20}		154^{28mm}		v s aq, eth
p266	3-(3-Pyridyl)-1-propanol	(C$_5$H$_4$N)CH$_2$CH$_2$CH$_2$OH	137.18		1.045	1.5295^{20}				
p267	Pyrimidine		80.09	23, 89	1.016	1.5035^{20}	20–22	123–124	31	misc aq; s alc, eth
p268	2,4(1H,3H)-Pyrimidinedione		112.09	24, 312			335			0.3 aq; s alk
p269	Pyrrole		67.09	20, 159	0.9691_4^{20}	1.5102^{20}	−23.4	129.8	38	4.5 aq; v s alc, eth
p270	Pyrrolidine		71.12	20, 4	0.8520_4^{20}	1.4431^{20}	−57.8	88–89	2	misc aq; s alc, chl, eth
p271	1-Pyrrolidinecarbodithioic acid, ammonium salt		164.29				153–155			
p272	1-Pyrrolidinecarbonitrile		96.13		0.954	1.4690^{20}		77$^{1.8mm}$	107	

p273	L-(−)-2-Pyrrolidinecarboxylic acid	115.13	22, 2			d 220			162 aq; 66 abs alc
p274	1-Pyrrolidino-1-cyclohexene	151.25	21, 236	0.940	1.5225^{20}		115^{15mm}	39	
p275	2-Pyrrolidinone	85.11		1.116_4^{25}	1.486^{25}	25	245	145	misc aq, alc, bz, chl, eth, EtAc
p276	3-(N-Pyrrolidino)-1,2-propanediol	145.20	20^1, 4			46–48	158^{30mm}		
q1	Quinhydrone	218.20	7, 617	1.401_4^{20}		171			s hot aq, alc, eth

Pyridinols, h173, h174, h175
3-Pyridinol N-oxide, h177
2(1H)-Pyridone, h173
2-(2-Pyridyl)pyridine, d705
Pyrocatechol, d377

Pyrrolidinedithiocarbamate, p271
Pyruvic acid, o60
Pyruvic aldehyde, o59
Pyruvic aldehyde dimethyl acetal, d451
Quinaldine,

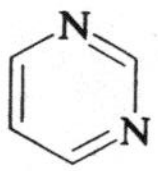

Pyrogallol, t304
Pyromellitic acid, b26
Pyromellitic dianhydride, b27
Pyromucic acid, f42
Pyromucic aldehyde, f39

p267 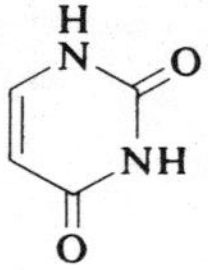   p268 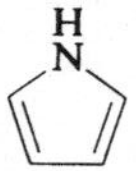   p269 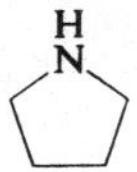   p270 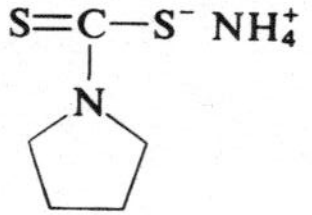   p271 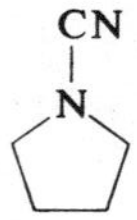   p272 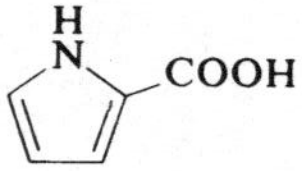   p273 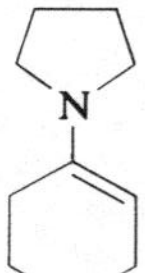   p274 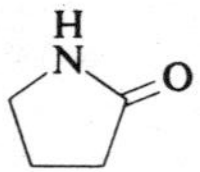   p275 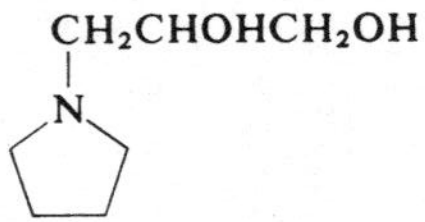   p276

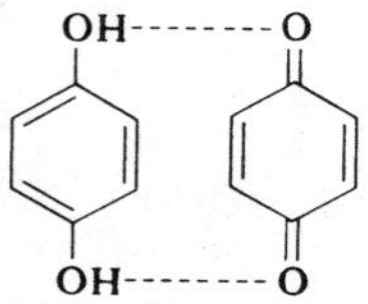

$CH_2CHOHCH_2OH$

q^1

TABLE 1.15 Physical Constants of Organic Compounds (*continued*)

No.	Name	Formula	Formula weight	Beilstein reference	Density	Refractive index	Melting point	Boiling point	Flash point	Solubility in 100 parts solvent
q2	Quinine		324.44			1.625	177 d			125 alc; 1.2 bz; 83 chl
q3	Quinoline		129.16	20, 339	1.095_4^{20}	1.6273^{20}	−14.9	237	101	0.6 aq; misc alc, eth
q4	Quinoxaline		130.15	23, 176	1.1334_4^{48}	1.6231^{48}	29–30	229.5		v s aq, alc, bz, eth
q5	Quinuclidine		111.19	20, 144			156 sealed tube			v s aq, alc, eth
r1	D-Raffinose pentahydrate		594.52	31, 462			80	d 118		14 aq; 10 MeOH
r2	Rhodamine B		479.02	19, 346			165			v s aq, alc
r3	Rhodanine		133.19	27, 242	0.868		170			v s hot aq, alc, eth
							may explode on rapid heating			
r4	Riboflavin		376.37	1^1, 434			d 278			v s alk(d); i eth
r5	D-(−)-Ribose		150.13				87			s aq; sl s alc
s1	Saccharin		183.19	27, 168			229–230			0.34 aq; 3 alc; 8 acet
s2	Safrole		162.19	19, 39	1.095^{20}	1.5370^{20}	11.2	232–234	97	v s alc; misc chl eth
s3	Semicarbazide	$H_2NNHCONH_2$	75.07	3, 98			96			v s aq, alc; i eth

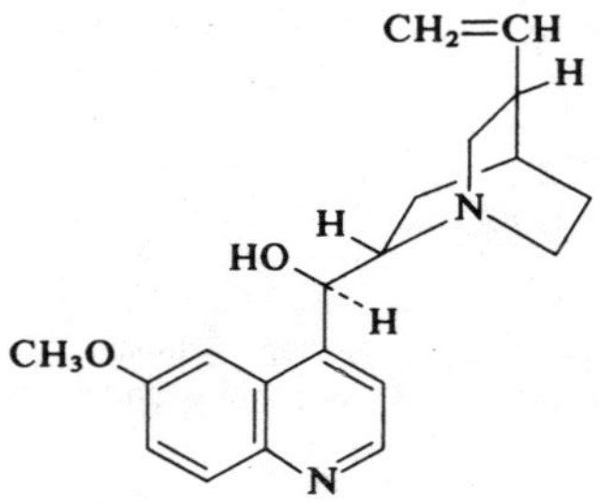

r1

s1

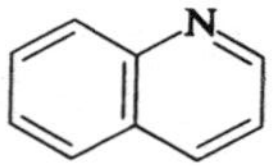

r5

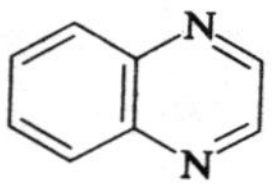

q5

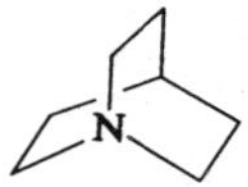

q4

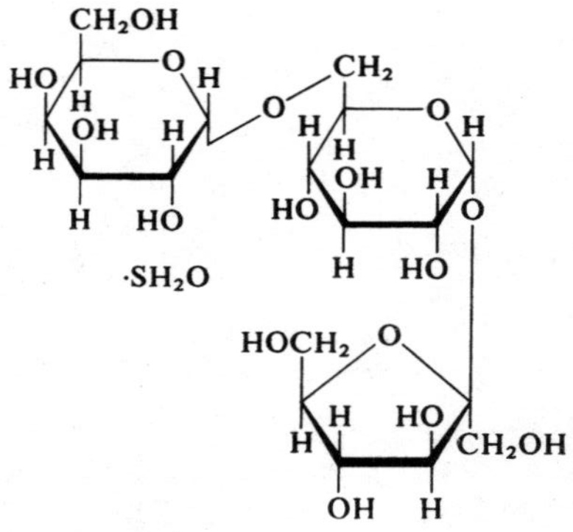

q3

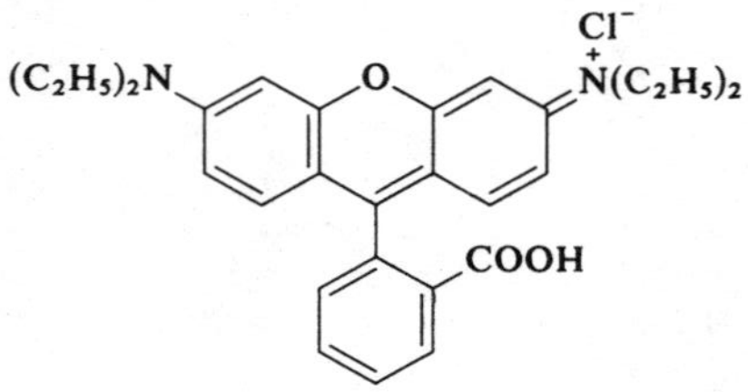

r4

s2

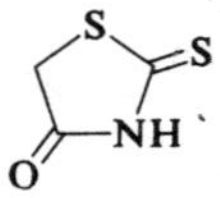

r3

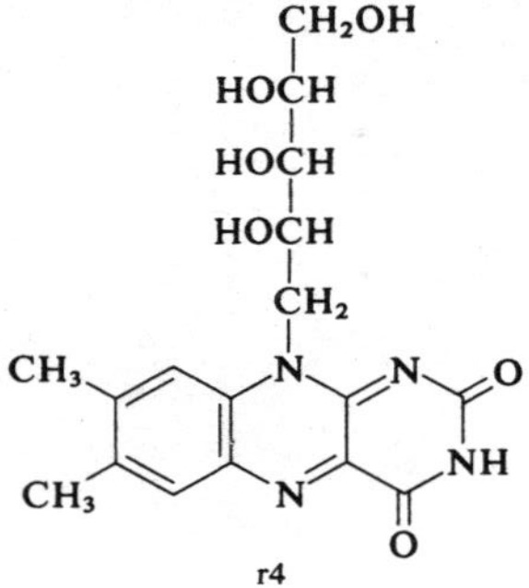

q2

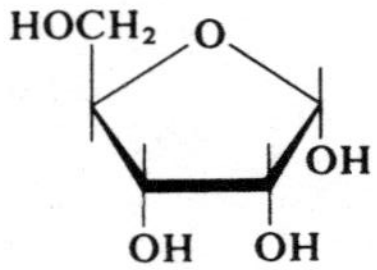

r2

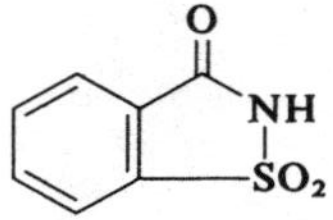

1.353

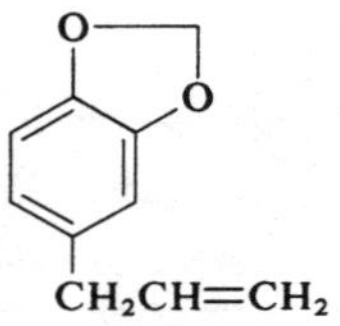

TABLE 1.15 Physical Constants of Organic Compounds (*continued*)

No.	Name	Formula	Formula weight	Beilstein reference	Density	Refractive index	Melting point	Boiling point	Flash point	Solubility in 100 parts solvent
s4	L-Serine	$HOCH_2CH(NH_2)COOH$	105.09	4, 505			222 d			s aq; v sl s alc, eth
s5	D-Sorbitol		182.17	1, 533	1.472^{-5}		110–112			83 aq; s hot alc, acet
s6	L-(−)-Sorbose		180.16	1, 927	1.65^{15}		165			55 aq; v sl s alc
s7	Squalane	$[(CH_3)_2CH(CH_2)_3$-$CH(CH_3)(CH_2)_3$-$CH(CH_3)CH_2CH_2—]_2$	422.80	1^1, 72	0.810	1.4530^{15}	−38	350	218	s bz, chl, eth, PE
s8	Squalene	$\{CH_3[C(CH_3)=CHCH_2$-$CH_2]_2C(CH_3)=CHCH_2—]_2$	410.73	1^1, 130	0.8584^{20}_4	1.4965^{20}	−75	285^{25mm}	200	v s eth, acet, PE
s9	trans-Stilbene	$C_6H_5CH=CHC_6H_5$	180.25	5, 630	0.970		124	206–207		v s bz, eth
s10	L-Strychnine		334.42	27^2, 723	1.36^{20}_4		284–286	270^{mm}		6.2 alc; 20 chl; 0.55 bz; 15 mg aq
s11	Styrene	$C_6H_5CH=CH_2$	104.15	5, 474	0.9060^{20}	1.5468^{20}	−30.6	145.1	31	s alc, acet, eth
s12	Succinamic acid	$H_2NCOCH_2CH_2COOH$	117.10	2, 614			153–156			s aq; sl s alc; i eth
s13	Succinamide	$H_2NCOCH_2CH_2CONH_2$	116.12	2, 614			260 d	125 subl		0.45 aq; i alc, eth
s14	Succinic acid	$HOOCCH_2CH_2COOH$	118.09	2, 601	1.552		187–190	235 d		7.7 aq; 5.4 alc; 2.8 acet; 0.88 eth; i bz
s15	Succinic acid 2,2-dimethylhydrazide	$HOOCCH_2CH_2CONH$-$N(CH_3)_2$	160.17				154–155			11 aq; 2.5 acet; 5 MeOH
s16	Succinic anhydride		100.07	17, 407	1.41		119.6	261		s alc, chl; v sl s eth
s17	Succinimide		99.09	21, 369	0.985		125–127	287		33 aq; 4 alc; i eth
s18	Succinonitrile	$NCCH_2CH_2CN$	80.09	2, 615	1.395^{15}	1.473^{15}	46–48	265–267	>112	d aq, alc; s bz
s19	Succinyl chloride	$ClCOCH_2CH_2COCl$	154.98	2, 613	1.5874^{25}		17	192–193	76	200 aq; 0.59 alc
s20	Sucrose		342.30	31, 424			192 d			0.15 aq; s alk
s21	Sulfamethazine		278.34				198–201			
s22	Sulfanilamide	$H_2NC_6H_4SO_2NH_2$	172.21	14, 698			164–166			0.76 aq; 2.7 alc; 20 acet; s acid, alk

No.	Name	Formula	Formula wt	Ref	mp/bp	Solubility
s23	Sulfoacetic acid	HO$_3$SCH$_2$COOH	140.11	4, 21	84–86	s aq, alc; i eth, chl
s24	o-Sulfobenzoic acid cyclic anhydride		184.17	19, 110	245 d, 186^{18mm}	s bz, chl, eth; i aq
s25	4,4′-Sulfonylbis-(2,6-dibromophenol)	[HO(Br)$_2$C$_6$H$_2$]$_2$SO$_2$	565.88	6, 865	289–292	

Senecioic acid, m163
Skatole, m284
Sodium tetraphenylborate, t128
Solketal, d515
Sorbic acid, h42
Sorbic aldehyde, h40
Stearamide, o2
Stearic acid, o5

Stearyl bromide, b319
Styrene dibromide, d79
Styrene glycol, p111
Styrene oxide, e9
Suberic acid, o24
Suberonitrile, d239
Succinic acid monoamide, s12
Succinonitrile, b380

Succinyl dihydrazide, s15
Sulfanilic acid, a120
N-Sulfinylaniline, t152
3-Sulfoalanine, a288
Sulfolane, t106
3-Sulfolene, d368
Sulfonyldianilines, d36, d37

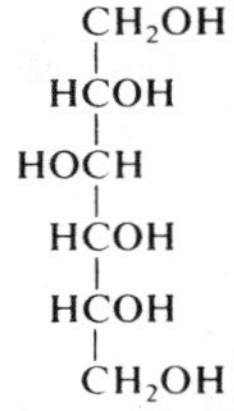

s5

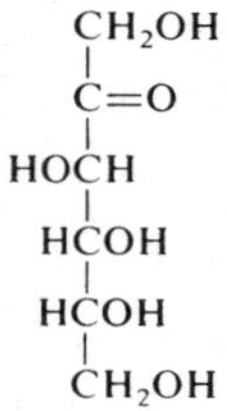

s6

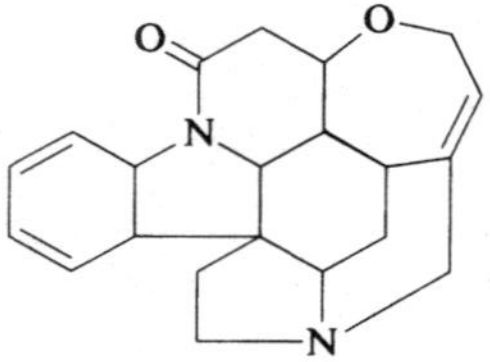

s10

s16

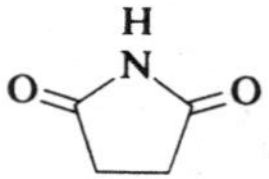

s17

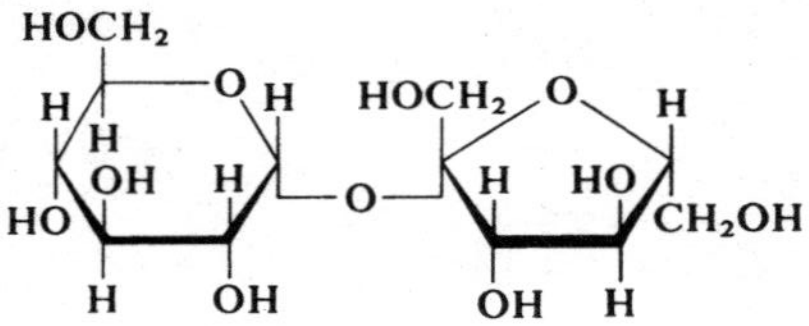

s20

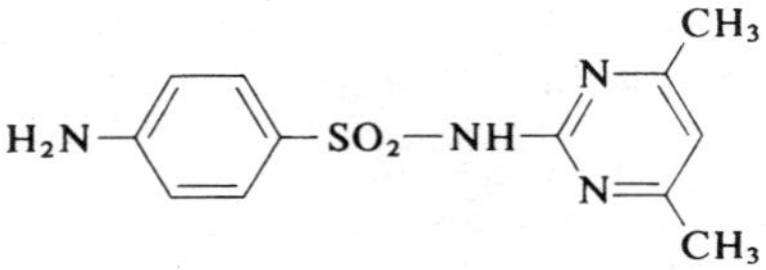

s21

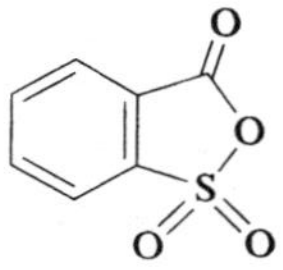

s24

TABLE 1.15 Physical Constants of Organic Compounds (*continued*)

No.	Name	Formula	Formula weight	Beilstein reference	Density	Refractive index	Melting point	Boiling point	Flash point	Solubility in 100 parts solvent
s26	4,4'-Sulfonylbis-(methyl benzoate)	$(CH_3OOCC_6H_4)_2SO_2$	334.35	10^2, 109			195–196			s alc, eth, acet; i aq
s27	4,4'-Sulfonyl-diphenol	$(HOC_6H_4)_2SO_2$	250.27	6, 861	1.3663^{15}		245–247			v s aq. alc; s eth
s28	5-Sulfosalicyclic acid	$HO_3SC_6H_3(OH)COOH$	254.21	11, 411			120			
t1	D-(−)-Tartaric acid	$HOOCCH(OH)-CH(OH)COOH$	150.09	3, 520	1.7598^{20}_4		168–170			139 aq; 33 alc; 0.4 eth
t2	meso-Tartaric acid hydrate	$HOOCCH(OH)-CH(OH)COOH \cdot xH_2O$	150.09	3, 528	1.666^{20}_4		140			125 aq
t3	Tartrazine		534.37	25, 252						v s aq
t4	p-Terphenyl	$C_6H_5C_6H_4C_6H_5$	230.31	5, 695			212–213	383		misc alc, eth
t5	α-Terpinene		136.24	5, 126	0.8375^{20}_4	1.4775^{20}		174	46	
t6	γ-Terpinene		136.24	5, 128	0.853^{15}_4	1.4754^{16}		183	51	
t7	Terpinen-4-ol		154.25	6, 55	0.9338^{20}_4	1.4820^{20}	36.4	219	79	v s alc, eth
t8	Tetraallyloxysilane	$(H_2C=CHCH_2O)_4Si$	256.4		0.9824^{20}_4	1.4336^{20}		114^{12mm}		
t9	1,1,2,2,-Tetrabromo-ethane	$Br_2CHCHBr_2$	345.67	1, 94	2.9529^{25}	1.6323^{25}	0.0	243.5	none	misc alc, eth; 0.07 aq
t10	Tetrabromophthalic anhydride		463.72	17, 485			274–276			sl s bz; i aq, alc
t11	α,α,α',α'-Tetrabromo-o-xylene	$C_6H_4(CHBr_2)_2$	421.77	5, 367			114–116			v s chl
t12	α,α,α',α'-Tetrabromo-m-xylene	$C_6H_4(CHBr_2)_2$	421.77	5, 375			105–108			v s bz, chl
t13	Tetrabutoxysilane	$(C_4H_9O)_4Si$	320.5		0.899^{20}_4	1.413^{20}		115^{3mm}		
t14	Tetrabutylammonium bromide	$(C_4H_9)_4N^+Br^-$	322.38				103–104			
t15	Tetrabutylammonium chloride	$(C_4H_9)_4N^+Cl^-$	277.92	4^3, 292			83–86			

No.	Name	Formula	Mol. wt.	Beilstein ref.	Density	n_D	mp, °C	bp, °C	Flash pt, °C	Solubility
t16	Tetrabutylammonium fluoride trihydrate	$(C_4H_9)_4N^+F^-\cdot 3H_2O$	315.52	4^3, 292			62–63			sl s aq; s alc, eth
t17	Tetrabutylammonium hydrogen sulfate	$(C_4H_9)_4N^+HSO_4^-$	339.54				169–171			
t18	Tetrabutylammonium iodide	$(C_4H_9)_4N^+I^-$	369.38	4, 157			145–148			
t19	Tetrabutylammonium tetrafluoroborate	$(C_4H_9)_4N^+BF_4^-$	329.28	4^3, 293			160–162			
t20	Tetrabutyltin	$(C_4H_9)_4Sn$	347.15		1.057	1.4742^{20}	−97	145^{10mm}	107	v s acet, chl
t21	1,1,3,3-Tetrachloroacetone	$Cl_2CHCOCHCl_2$	195.86	1, 656	1.624^{15}_4	1.497^{18}		182^{745mm}	none	
t22	1,2,3,4-Tetrachlorobenzene	$C_6H_2Cl_4$	215.89	5, 204			46–47	254	>112	v s eth; sl s alc
t23	1,2,4,5-Tetrachlorobenzene	$C_6H_2Cl_4$	215.89	5, 205	1.858^{22}		138–140	240–246	>112	s bz, chl, eth

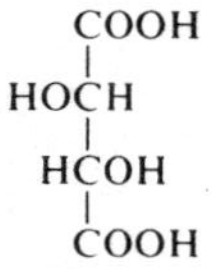

t1

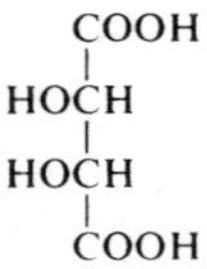

t2

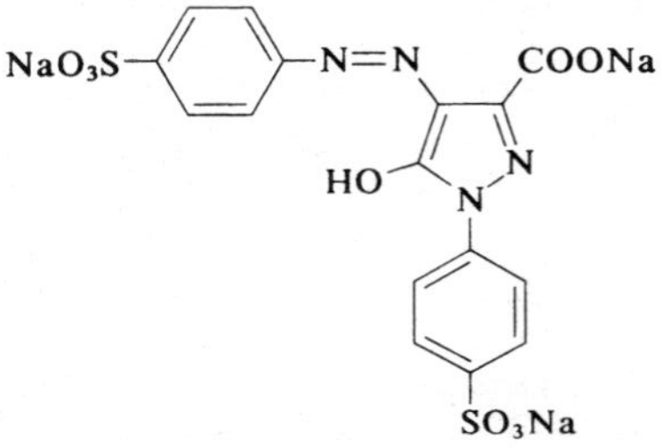

t3

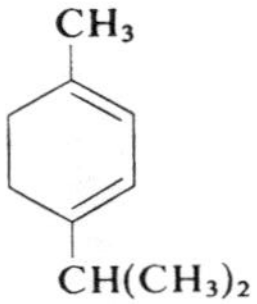

t5

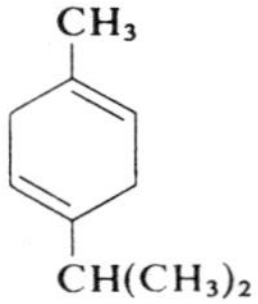

t6

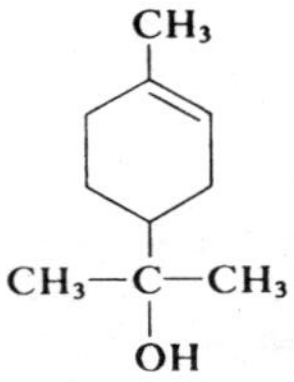

t7

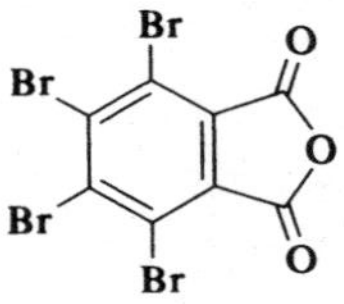

t10

TABLE 1.15 Physical Constants of Organic Compounds (*continued*)

No.	Name	Formula	Beilstein reference	Formula weight	Density	Refractive index	Melting point	Boiling point	Flash point	Solubility in 100 parts solvent
t24	Tetrachloro-*o*-benzoquinone	$C_6Cl_4(=O)_2$	7, 602	245.88			127–129			s eth; sl s chl; i aq
t25	Tetrachloro-*p*-benzoquinone	$C_6Cl_4(=O)_2$	7, 602	245.88			290	subl		
t26	Tetrachloro-1,2-difluoroethane	$Cl_2CFCFCl_2$		203.83	1.6447^{25}	1.4130^{25}	26.0	92.8		0.012 aq
t27	1,1,1,2-Tetrachloroethane	$ClCH_2CCl_3$	1, 86	167.85	1.598	1.4819^{20}		130	none	0.02 aq; misc alc
t28	1,1,2,2-Tetrachloroethane	$Cl_2CHCHCl_2$	1, 86	167.85	1.5866^{25}_4	1.4910^{25}	−43.8	146.3	none	0.3 aq; misc alc, chl, eth, PE
t29	Tetrachloroethylene	$Cl_2C=CCl_2$	1, 187	165.83	1.6230^{20}_4	1.5057^{20}	−22.4	121.1	none	misc alc, chl, eth
t30	2,3,5-Tetrachloronitrobenzene	$HC_6Cl_4NO_2$	5, 247	260.89	1.744^{25}_4		98–101	304		s alc, bz, chl
t31	Tetrachlorophthalic anhydride		17, 484	285.90			254–258	371		d hot aq; sl s eth
t32	3,4,5,6-Tetrachlorophthalimide		21, 505	284.91			>300			
t33	1,1,2,3-Tetrachloro-2-propene	$ClCH=C(Cl)CHCl_2$	1^1, 83	179.86	1.530	1.5163^{20}	165		none	
t34	2,3,5,6-Tetrachlorothioanisole	$HC_6Cl_4SCH_3$		262.0			59–61			
t35	2,4,5,6-Tetrachloro-*m*-xylene	$C_6Cl_4(CH_3)_2$	5, 373	243.95			220–222			
t36	Tetracosane	$CH_3(CH_2)_{22}CH_3$	1, 175	338.66	0.7786^{51}	1.4283^{70}	51.1	391		9.4 chl; s eth
t37	Tetracyanoethylene	$(NC)_2C=C(CN)_2$		128.09			200	subl 120		
t38	Tetradecane	$CH_3(CH_2)_{12}CH_3$	1, 171	198.40	0.7627^{20}_4	1.4290^{20}	5.9	253.5		v s alc, eth
t39	Tetradecanoic acid	$CH_3(CH_2)_{12}COOH$	2, 365	228.38	0.8528^4	1.4273^{70}	58.5	250^{100mm}		v s bz, chl, eth; s alc

No.	Name	Formula	Mol. wt.	Beilstein	Density	n_D	mp (°C)	bp (°C)	Flash (°C)	Solubility
t40	1-Tetradecanol	$CH_3(CH_2)_{13}OH$	214.39	1, 428	0.8151^{50}	1.4358^{50}	37.8	264		s eth; sl s alc
t41	Tetradecanoyl chloride	$CH_3(CH_2)_{12}COCl$	246.82	2, 368			−1	168^{15mm}		d aq, alc; s eth
t42	1-Tetradecene	$CH_3(CH_2)_{11}CH{=}CH_2$	196.38	1, 226	0.7775^{15}	1.4351^{20}	−12.9	251.2	115	v s alc; s eth
t43	7-Tetradecene	$CH_3(CH_2)_5CH{=}CH{-}(CH_2)_5CH_3$	196.38		0.764	1.4351^{20}		250	99	
t44	1-Tetradecylamine	$CH_3(CH_2)_{13}NH_2$	213.41	4, 201			40–42	162^{15mm}		
t45	4-Tetradecylaniline	$CH_3(CH_2)_{13}C_6H_4NH_2$	213.41	12^3, 2780			46–49	156^{3mm}		
t46	Tetradecyltrichloro-silane	$CH_3(CH_2)_{13}SiCl_3$	331.8			1.382				
t47	Tetraethoxysilane	$(CH_3CH_2O)_4Si$	208.33		0.9344^{20}	1.383^{20}	−77	165.8	46	d aq; s alc
t48	Tetraethylammonium bromide	$(CH_3CH_2)_4N^+Br^-$	210.16	4, 104	1.3974^{20}		287 d			v s aq, alc, acet, chl
t49	Tetraethylammonium chloride	$(CH_3CH_2)_4N^+Cl^-$	165.71	4, 104	1.0801^{21}_{4}		37.5			141 aq; s alc; 8.2 chl
t50	Tetraethylammonium hydroxide	$(CH_3CH_2)_4N^+OH^-$	147.26	4, 103						misc aq
t51	Tetraethylene glycol	$(HOCH_2CH_2OCH_2CH_2)_2O$	194.23	1, 468	1.1252^{20}_{20}	1.4590^{20}	−6	307.8	176	misc aq, alc, bz, eth
t52	Tetraethylene glycol dimethacrylate	$[H_2C{=}C(CH_3)COOCH_2CH_2OCH_2CH_2]_2O$	330.37		1.08			220^{1mm}	62	
t53	Tetraethylene glycol monomethyl ether	$CH_3O(CH_2CH_2O)_3CH_2CH_2OH$	208.26	1.08...		0.987	1.4453^{20}		166^{11mm}	>112

Tetraethyl orthosilicate, t47

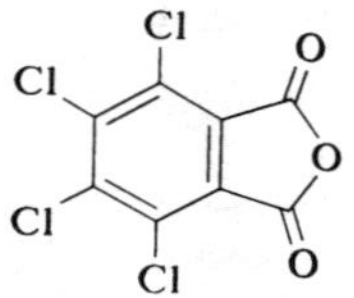

t31

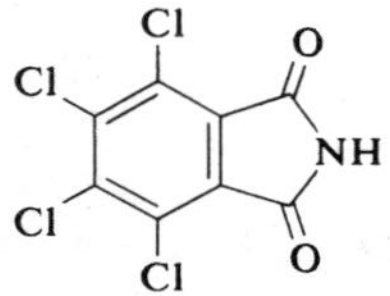

t32

1.359

TABLE 1.15 Physical Constants of Organic Compounds (*continued*)

No.	Name	Formula	Formula weight	Beilstein reference	Density	Refractive index	Melting point	Boiling point	Flash point	Solubility in 100 parts solvent
t54	Tetraethylenepentamine	(H₂NCH₂CH₂NHCH₂CH₂)₂NH	189.31		0.999^{20}_{20}	1.5055^{20}	−40	340	185	misc aq, alc, eth
t55	N,N,N',N'-Tetraethylethylenediamine	(C₂H₅)₂NCH₂CH₂N(C₂H₅)₂	172.32	4, 251	0.808	1.4343^{20}		189–192	58	
t56	Tetraethylgermanium	(C₂H₅)₄Ge	188.84	4, 631	1.1989	1.5198^{20}	−90	165.5		s alc, eth; i aq
t57	Tetraethyllead	(C₂H₅)₄Pb	323.45	4, 639	1.6534^{20}	1.4196^{20}	−136	152^{291mm}		s bz; misc eth
t58	Tetraethyl pyrophosphate	[(C₂H₅O)₂P(O)]₂O	290.20		1.1854		d 170			d aq; misc alc, bz, chl
t59	Tetraethyl pyrophosphite	[(C₂H₅O)₂P]₂O	258.19		1.057	1.4341^{20}		81^{1mm}	>112	i aq
t60	Tetraethylsilane	(C₂H₅)₄Si	144.34	4[2], 1007	0.7624^{20}	1.4246^{20}	70	153–155		3.8 alc; 7.1 eth; s bz, acet, chl; 0.02 aq
t61	Tetraethylthiuram disulfide	[(C₂H₅)₂NC(=S)S—]₂	296.54	4, 122	1.30					i aq; s eth
t62	Tetraethyltin	(C₂H₅)₄Sn	234.94	4, 632	1.1994^{20}		−112	181		i aq
t63	Tetrafluoroethylene	F₂C=CF₂	100.02	1[3], 638	1.1507^{-40}		−131.2	−75.6		
t64	2,2,3,3-Tetrafluoro-1-propanol	HCF₂CF₂CH₂OH	132.06		1.4853^4	1.3197^{20}	−15	109–110	49	
t65	1,2,3,6-Tetrahydrobenzaldehyde	C₆H₉CHO	110.16	7[1], 48	0.940	1.4745^{20}		163–164	57	
t66	Tetrahydrofuran		72.11	17, 10	0.8892^4	1.4072^{20}	−108.5	66	−17	misc aq, alc eth, PE
t67	2,5-Tetrahydrofurandimethanol		132.16		1.1542^{25}	1.4766^{25}	<−50	265		misc aq, alc, bz, chl; s eth
t68	Tetrahydro-2-furanmethanol		102.13	17[2], 106	1.0524^{20}	1.4520^{20}	<−80	178	83	misc aq, alc, bz, chl, eth, acet
t69	Tetrahydro-2-furanmethylamine		101.15	18[2], 415	0.980	1.4560^{20}		154^{744mm}	45	

No.	Name	Formula	Formula wt		Density	n_D	mp	bp	fp	Solubility
t70	2-(Tetrahydrofuryl-oxy)tetrahydro-pyran		186.25		1.030	1.4606^{20}			97	
t71	1,2,3,4-Tetrahydro-isoquinoline		133.19	20, 275	1.064	1.5668^{20}	-30	232–233	98	misc alc, bz, chl, eth, acet, PE
t72	Tetrahydrolinalool	$(CH_3)_2CHCH_2CH_2CH_2$-$C(CH_3)(OH)CH_2CH_3$	158.28		0.925^{25}	1.433^{20}				
t73	1,2,3,4-Tetrahydro-naphthalene	$C_{10}H_{12}$	132.21	5, 491	0.9702^{20}_{4}	1.5414^{20}	-35.8	207.6	77	
t74	cis-1,2,3,6-Tetra-hydrophthalic anhydride		152.15	17, 462			101–102			
t75	cis-1,2,3,6-Tetra-hydrophthalimide		151.17				134–138			
t76	Tetrahydropyran		86.14		0.8814^{20}_{4}	1.4211^{20}	-45	88	-20	misc aq, alc, eth
t77	Tetrahydropyran-2-methanol		116.16		1.0254^{20}_{4}	1.4580^{20}	-70 glass	187	93	misc aq, alc, bz, eth

Tetraglyme, b190
1,2,3,4-Tetrahydrobenzene, c330

Tetrahydrodicyclopentadiene, t253
Tetrahydro-2,5-dimethoxyfuran, d457

Tetrahydrofurfurylamine, t69
Tetrahydrofurfuryl alcohol, t68

t66

$HOCH_2$ CH_2OH t67

CH_2OH t68

CH_2NH_2 t69

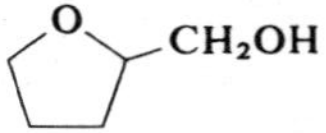

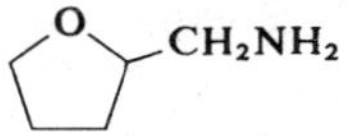

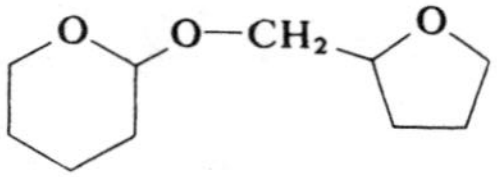

t70

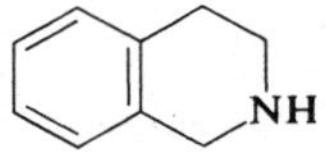

t71

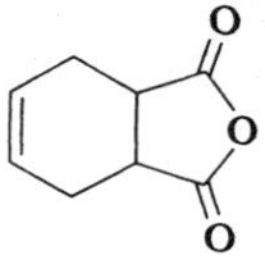

t74

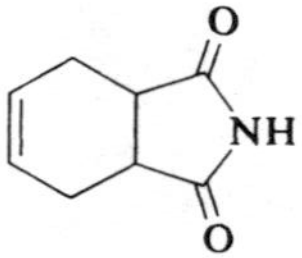

t75

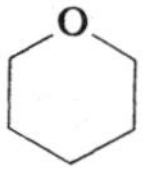

t76

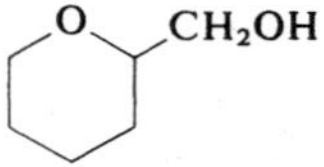

CH_2OH t77

TABLE 1.15 Physical Constants of Organic Compounds (*continued*)

No.	Name	Formula	Formula weight	Beilstein reference	Density	Refractive index	Melting point	Boiling point	Flash point	Solubility in 100 parts solvent
t78	1,2,3,6-Tetrahydropyridine		83.13	20^3, 1912	0.911	1.4800^{20}	−48	108	16	
t79	3,4,5,6-Tetrahydropyrimidinethiol		116.19	24, 5			210–212			
t80	1,2,3,4-Tetrahydroquinoline		133.19	20, 262	1.061	1.5924	15–16	249	100	s aq; misc alc, eth
t81	Tetrahydrothiophene		88.17	17^1, 5	0.9987^{20}	1.5048^{20}	−96.2	120.9	12	misc alc, eth; i aq
t82	1,4,9,10-Tetrahydroxyanthracene		242.23	8, 431			147–149			
t83	2,2′,4,4′-Tetrahydroxybenzophenone	$[(HO)_2C_6H_3]_2C{=}O$	246.22	8, 496			200–203			
t84	Tetrahydroxyhexanedioic acid	$HOOC[CH(OH)]_4COOH$	210.14	3, 581			230 d			0.003 aq; s alk
t85	Tetrakis(2-ethylbutoxy)silane	$[CH_3CH_2CH(C_2H_5)CH_2O]_4Si$	432.8		0.8924^{20}_4	1.430^{20}		171^{2mm}		
t86	Tetrakis(2-ethylhexoxy)silane	$[CH_3(CH_2)_3CH(C_2H_5)CH_2O]_4Si$	549.95		0.8804^{20}_4	1.4388^{20}		194^{1mm}	190	
t87	N,N,N′,N′-Tetrakis(p-hydroxypropyl)ethylenediamine	$\{[CH_3CH(OH)CH_2]_2NCH_2{-}\}_2$	292.42	4^4, 1685	1.013	1.4812^{20}		$175{-}181^{0.8mm}$		
t88	Tetrakis(isopropoxy)silane	$[(CH_3)_2CHO]_4Si$	264.4		0.877^{20}_4	1.385^{20}		64^{5mm}		
t89	Tetrakis(2-methoxyethoxy)silane	$(CH_3OCH_2CH_2O)_4Si$	328.4		1.079^{20}_4	1.422^{20}		182^{10mm}		
t90	Tetrakis(trimethylsiloxy)titanium	$[(CH_3)_3SiO]_4Ti$	404.7		0.900^{20}_4	1.427^{20}		110^{10mm}		

No.	Name	Formula	Formula weight	Beilstein reference	Density	n_D	mp, °C	bp, °C		Solubility
t91	1,1,3,3-Tetramethoxy-propane	$[(CH_3O)_2CH]_2CH_2$	164.20	1, 287	0.997	1.4081^{20}		183	54	55 aq
t92	Tetramethoxysilane	$(CH_3O)_4Si$	152.2		1.052_4^{20}	1.368^{20}		121–122	20	
t93	Tetramethyl-ammonium bromide	$(CH_3)_4N^+Br^-$	154.06	4, 51	1.56		d > 230	subl > 360		s aq, hot alc
t94	Tetramethyl-ammonium chloride	$(CH_3)_4N^+Cl^-$	109.60	4, 51	1.169_4^{20}		d > 230	subl > 300		s aq, hot alc
t95	Tetramethyl-ammonium iodide	$(CH_3)_4N^+I^-$	201.06		1.829		d 230			sl s aq; v s abs alc
t96	N,N,3,5-Tetra-methylaniline	$(CH_3)_2C_6H_3N(CH_3)_2$	149.24	12, 1131	0.913	1.5443^{20}		226–228	90	
t97	1,2,3,4-Tetramethyl-benzene	$C_6H_2(CH_3)_4$	134.22	5, 430	0.905_4^{20}	1.5187^{20}	−6.2	205.0	68	misc alc, eth
t98	1,2,3,5-Tetramethyl-benzene	$C_6H_2(CH_3)_4$	134.22	5, 430	0.8906_4^{20}	1.5134^{20}	−23.7	198.0	63	s alc; v s eth
t99	1,2,4,5-Tetramethyl-benzene	$C_6H_2(CH_3)_4$	134.22	5, 431	0.8384^{81}		79.2	196.8	73	v s alc, bz, eth

Structures:

t78 — N H

t79 — NH, SH, N

t80 — N H

t81 — S

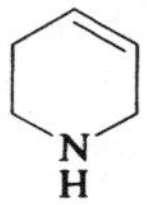

t82 — OH, OH / OH, OH

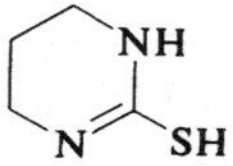

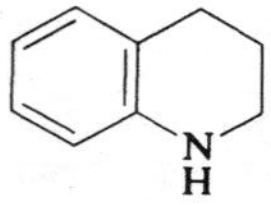

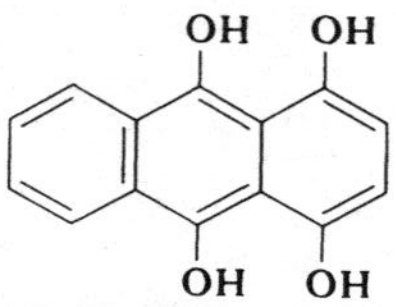

TABLE 1.15 Physical Constants of Organic Compounds (*continued*)

No.	Name	Formula	Formula weight	Beilstein reference	Density	Refractive index	Melting point	Boiling point	Flash point	Solubility in 100 parts solvent
t100	2,2,3,3-Tetramethyl-butane	$(CH_3)_3CC(CH_3)_3$	114.23	1, 165	0.656^{-120}		−120.7	106.5	<1	s aq, alc, eth
t101	N,N,N',N'-Tetra-methyl-1,4-butane-diamine	$(CH_3)_2N(CH_2)_4N(CH_3)_2$	144.26	4, 265	0.786^{20}	1.4280^{20}		169	46	
t102	1,1,3,3-Tetramethyl-butylamine	$(CH_3)_3CCH_2C(CH_3)_2NH_2$	129.25	4, 198	0.805	1.4240^{20}		137–143	32	s alc, eth, PE; i aq
t103	1,3,5,7-Tetramethyl-cyclotetrasiloxane	$[-SiH(CH_3)O-]_4$	240.5		0.9912^{20}_4	1.3870^{20}	−69	134–135		
t104	1,1,4,4-Tetramethyl-1,4-dichlorodisilyl-ethylene	$[(CH_3)_2Si(Cl)CH_2-]_2$	215.3				37	198^{734mm}	68	
t105	Tetramethyldi-siloxane	$[(CH_3)_2SiH]_2O$	134.3	$17^1, 5$	0.757^{20}_4	1.370^{20}		71^{731mm}		
t106	Tetramethylene sulfone		120.71		1.2614^{30}_4	1.4820^{30}	27.6	285	165	misc aq, acet, bz
t107	Tetramethylene sulfoxide		104.17		1.158	1.5200^{20}			>112	
t108	N,N,N',N'-Tetra-methylethylene diamine	$(CH_3)_2NCH_2CH_2N(CH_3)_2$	116.21	4, 250	0.770	1.4179^{20}	−55	120–122	10	
t109	Tetramethyl-germanium	$(CH_3)_4Ge$	132.73		1.006^0	1.3871^{20}	−88	43.4		
t110	1,1,3,3-Tetramethyl-guanidine	$[(CH_3)_2N]_2C=NH$	115.18					163		
t111	N,N,N',N'-Tetra-methyl-1,6-hexane-diamine	$[(CH_3)_2N(CH_2)_3-]_2$	172.32	$4^1, 423$	0.806	1.4359^{20}		209–210	73	
t112	Tetramethyllead	$(CH_3)_4Pb$	267.33	4, 639	1.995^{20}_4		−27.5	110		misc alc, eth

t113	*N,N,N′,N′*-Tetramethylmethanediamine	$(CH_3)_2NCH_2N(CH_3)_2$	102.18	4, 54	0.749^{20}	1.4005		85	< 1	
t114	Tetramethyl orthocarbonate	$C(OCH_3)_4$	136.15	3^2, 4	1.023	1.3845^{20}	−5	114	6	
t115	2,6,10,14-Tetramethylpentadecane	$[(CH_3)_2CH(CH_2)_3$-$CH(CH_3)CH_2]_2CH_2$	268.53		0.7827^{20}_{4}	1.4379^{20}	−100	167^{11mm}		s bz, chl, eth, PE
t116	2,3,5,6-Tetramethylphenol	$(CH_3)_4C_6HOH$	150.22	6, 547			108–110	250		
t117	2.2,6,6-Tetramethylpiperidino-N-oxy-(free radical)		156.25				36–38		67	
t118	*N,N,N′,N′*-Tetramethyl-1,3-propanediamine	$(CH_3)_2N(CH_2)_3N(CH_3)_2$	130.24	4, 262		1.4234^{20}		145–146	31	
t119	Tetramethylpyrazine		136.20	23, 99			84–86	190		
t120	Tetramethylsilane	$(CH_3)_4Si$	88.23	4, 625	0.6411^{20}_{4}	1.3585^{20}	−99.5	26.5	−27	v s alc, eth
t121	1,2,2,3-Tetramethyl-1,1,3,3-tetraphenyltrisiloxane	$[(C_6H_5)_2Si(CH_3)O]_2$-$Si(CH_3)_2$	484.8		1.07^{20}_{4}	1.551^{25}		$235^{0.5mm}$	221	

2,2,5,5-Tetramethyl-3,4-dithiahexane, d114
Tetramethylene chlorobromide, b251
Tetramethylethylene glycol, d491
Tetramethylene oxide, t66
Tetramethylene sulfide, t81
Tetramethylolmethane, p20
2,2,4,4-Tetramethyl-3-thiapentane, d133
Tetramethylthiuram disulfide, b174

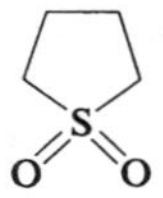
t106

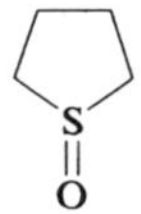
t107

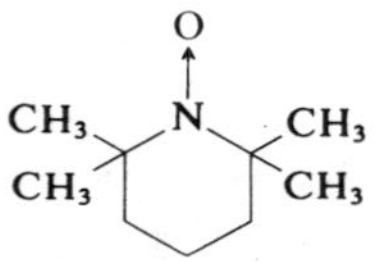
t117

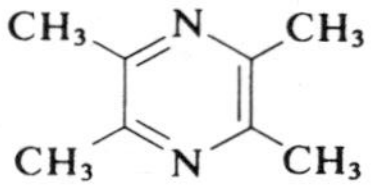
t119

TABLE 1.15 Physical Constants of Organic Compounds (*continued*)

No.	Name	Formula	Formula weight	Beilstein reference	Density	Refractive index	Melting point	Boiling point	Flash point	Solubility in 100 parts solvent
t122	1,1,3,3-Tetramethyl-2-thiourea	$(CH_3)_2NC(=S)N(CH_3)_2$	132.23	4^1, 336	1.3149^{25}	1.5201	75–77	245		
t123	Tetramethyltin	$(CH_3)_4Sn$	178.83	4, 631	0.9687^{20}_4	1.4493^{25}	−54.8	78	65	misc aq, alc, chl, eth
t124	1,1,3,3-Tetramethyl-urea	$(CH_3)_2NC(=O)N(CH_3)_2$	116.16	4, 74			−1.2	176		v s alc, eth, alk
t124a	Tetranitromethane	$C(NO_2)_4$	196.03	1, 80	1.6229^{25}_4	1.4358^{25}	13.5	126	> 112	
t125	1,4,7,10-Tetraoxa-cyclododecane		176.21		1.089	1.4621^{20}	16		> 112	
t126	2,4,8,10-Tetraoxa-spiro[5.5]undecane		160.17	19, 436			52–55	$83^{1.5mm}$		
t127	Tetraphenoxysilane	$(C_6H_5O)_4Si$	400.5		1.1141^{60}	1.554^{60}	48–49	237^{1mm}		v s aq, acet; s chl
t128	Tetraphenylboron sodium	$(C_6H_5)_4B^-Na^+$	342.23				> 300			
t129	1,1,4,4-Tetraphenyl-1,3-butadiene	$(C_6H_5)_2C=CHCH=C-(C_6H_5)_2$	358.49	5, 750			207–209			
t130	1,1,3,3-Tetraphenyl-1,3-dimethyldi-siloxane	$[(C_6H_5)_2Si(CH_3)]_2O$	410.7		1.076^{25}_4	1.5866^{26}	50	$215^{0.5mm}$	193	
t131	Tetraphenylethylene	$(C_6H_5)_2C=C(C_6H_5)_2$	332.45	5, 743			222–224	420		
t132	Tetraphenylsilane	$(C_6H_5)_4Si$	336.5		1.078^{20}_4		236–237	228^{3mm}		
t133	Tetraphenyltin	$(C_6H_5)_4Sn$	427.11		1.490^0		226	> 420	110	
t134	Tetrapropoxysilane	$(C_3H_7O)_4Si$	264.4		0.9162^{20}_4	1.401^{20}		94^{5mm}		
t135	Tetrapropylam-monium bromide	$(CH_3CH_2CH_2)_4N^+Br^-$	266.27	4^1, 364			270 d			s aq
t136	1H-Tetrazole		70.06	26, 346			156–158	subl		s aq, alc, acet
t137	2-Thenoyltrifluoro-acetone		222.18				40–44	98^{8mm}		

No.	Name	Formula	Mol. wt.	Beilstein ref.	Density	n_D	mp/°C	bp/°C	Flash pt/°C	Solubility
t138	Theobromine		180.17	26, 457			357	subl 290		0.05 aq; 0.045 alc; s alk; i bz, chl, eth
t139	Thiamine HCl		337.27				d 248			100 aq; 1 alc
t140	Thiazole		85.13	27, 15	1.200^{17}	1.5375^{20}		117–118	22	s alc, eth; sl s aq
t141	Thioacetamide	CH$_3$C(=S)NH$_2$	75.13	2, 232			112–114			16 aq; sl s alc, eth
t142	Thioacetic acid	CH$_3$CO—SH	76.12	2, 230	1.065	1.4630^{20}	<−17	88–91	<1	s aq; misc alc, eth
t143	Thiobenzoic acid	C$_6$H$_5$CO—SH	138.19	9, 419	1.174	1.6020^{20}	15–18	d	>112	misc eth; v s alc; i aq

Tetrantoin, b40
2,5,8,13-Tetraoxadodecane, b189
3,6,9,12-Tetraoxatridecanol, t53
Tetraphene, b6
2-Thenoic acid, t157
2-Thiabutane, e182

Thiacyclobutane, t345a
1-Thia-3-cyclopentene, 1,1-dioxide, d368
3-Thiaheptane, b451
2-Thiahexane, b463
3-Thiahexane, e210
Thianaphthene, b61

5-Thianonane, d132
2-Thiapentane, m395
3-Thiapentane, d337
Thioanisole, m364
2-Thiobarbituric acid, d388

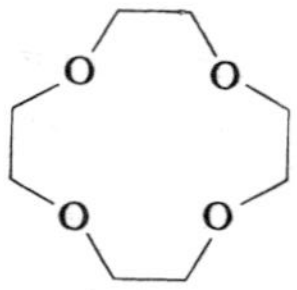

t125

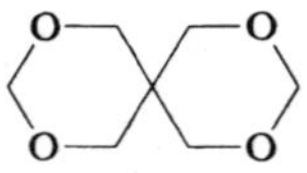

t126

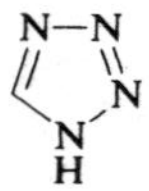

t136

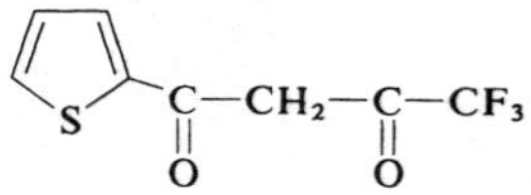

t137

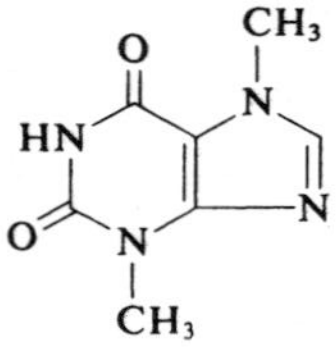

t138

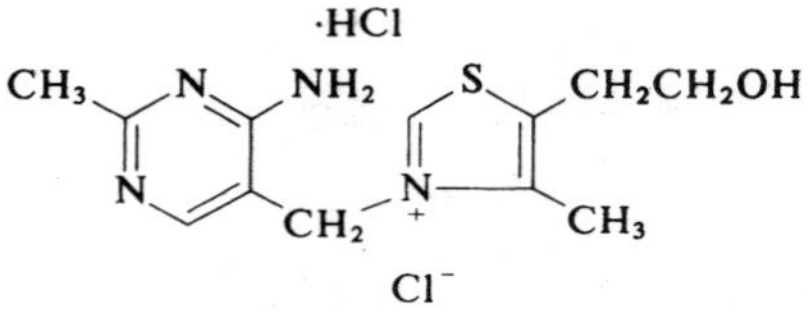

t139 ·HCl

t140

TABLE 1.15 Physical Constants of Organic Compounds (*continued*)

No.	Name	Formula	Formula weight	Beilstein reference	Density	Refractive index	Melting point	Boiling point	Flash point	Solubility in 100 parts solvent
t144	4,4'-Thiobis(2-*tert*-butyl-6-methyl-phenol)	$[(HO)_2C_6H_3]_2S$	358.54				127	316^{40mm}	240	
t145	4,4'-Thiobis(1,3-dihydroxybenzene)		250.27	6^3, 6291			175–177			
t146	Thiocarbanilide	$C_6H_5NHCSNHC_6H_5$	228.32	12, 394	1.32^{24}		154			v s alc, eth
t147	*p*-Thiocresol	$HSC_6H_4CH_3$	124.21	6, 416			43–44	195	68	s alc, eth; i aq
t148	2,2'-Thiodiacetic acid	$(HOOCCH_2)_2S$	150.15	3, 253			129			s aq, alc
t149	2,2'-Thiodiethanol	$(HOCH_2CH_2)_2S$	122.19	1, 470	1.1824^{20}_4	1.5203^{20}	–16	282	110	misc aq, alc; sl s eth
t150	4,4'-Thiodiphenol	$(HOC_6H_4)_2S$	218.27	6, 860			150–155			
t151	3,3'-Thiodipropionic acid	$(HOOCCH_2CH_2)_2S$	178.21				134			3.4 aq; v s alc
t152	2-Thiohydantoin		116.14	24, 260			231 d			sl s aq; i alc, eth
t153	N-Thionylaniline	$C_6H_5N=SO$	139.18	12, 578	1.236	1.6270^{20}		200		
t154	Thiophene	C_4H_4S	84.14	17, 29	1.05734^{25}	1.5257^{25}	–38.2	84.2	–1	misc alc, eth; i aq
t155	2-Thiopheneacetic acid	$(C_4H_3S)CH_2COOH$	142.18	18, 293			63–67	160^{22mm}		
t156	2-Thiophenecarbaldehyde	$(C_4H_3S)CHO$	112.15	17, 285	1.200	1.5900^{20}		198	77	s eth
t157	2-Thiophenecarboxylic acid	$(C_4H_3S)COOH$	128.15	18, 289			128.5	260		s aq, chl; v s alc, eth
t158	2-Thiophenemethylamine	$(C_4H_3S)CH_2NH_2$	113.19	18^4, 7096	1.103	1.5569^{20}		99^{28mm}	73	
t159	Thiophenol	C_6H_5SH	110.18	6, 294	1.0766^{20}	1.5897^{20}	–14.9	169.1	50	v s alc; misc bz, eth
t160	Thiophenoxyacetic acid	$C_6H_5SCH_2COOH$	168.21	6, 313			64–66			

No.	Name	Formula	M.W.	Ref.	Density	n_D	M.P.	B.P.		Solubility
t161	Thiopropionic acid	CH$_3$CH$_2$CO—SH	90.14	2, 264	1.014	1.4640^{20}	182–184	108–110	11	s aq, alc
t162	3-Thiosemicarbazide	H$_2$NC(=S)NHNH$_2$	91.14	3, 195			176–178			9 aq; s alc; sl s eth
t163	Thiourea	H$_2$NC(=S)NH$_2$	76.12	3, 180	1.045				42	
t164	1,4-Thioxane		104.17	19, 3	1.114	1.5095^{20}		147		v s bz, chl, hot HOAc
t165	Thioxanthen-9-one		212.27	17, 357			211	273^{715mm}		
t166	Titanium(IV) iso-propoxide	Ti[OCH(CH$_3$)$_2$]$_4$	284.26	1^2, 382	0.955	1.4654^{20}	18–20	218^{10mm}	22	misc alc, chl, eth, acet, HOAc
t167	Toluene	C$_6$H$_5$CH$_3$	92.14	5, 280	0.8660^{20}_4	1.4969^{20}	−95.0	110.6	7	s hot aq, alc, eth
t168	2,4-Toluenediamine	CH$_3$C$_6$H$_3$(NH$_2$)$_2$	122.17	13, 124			97–99	283.5		v s aq, alc, eth
t169	2,5-Toluenediamine	CH$_3$C$_6$H$_3$(NH$_2$)$_2$	122.17	13, 144			64	273–274		v s aq, alc, eth
t170	2,6-Toluenediamine	CH$_3$C$_6$H$_3$(NH$_2$)$_2$	122.17	13, 148			104–106			s aq, alc
t171	3,4-Toluenediamine	CH$_3$C$_6$H$_3$(NH$_2$)$_2$	122.17	13, 148			88–90	156^{18mm}		v s aq
t172	Toluene-2,4-diiso-cyanate	CH$_3$C$_6$H$_3$(NCO)$_2$	174.16	13, 138	1.2244^{20}_4	1.5689^{20}	20–21	251	121	d aq, alc; misc bz, acet, eth

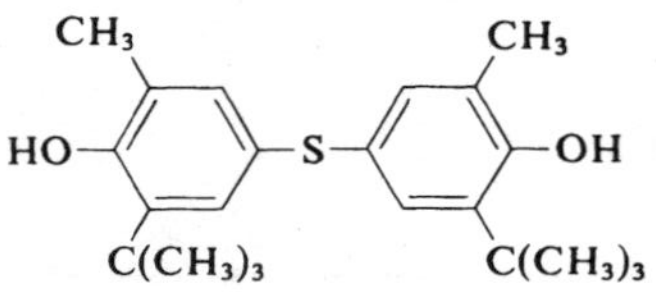

t144 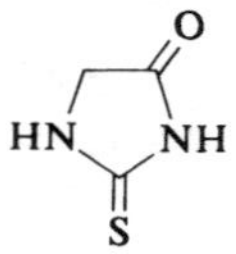 t152 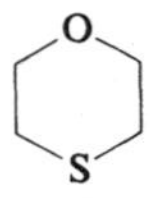 t164 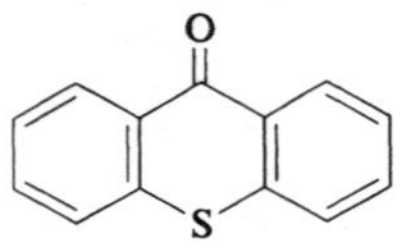 t165

TABLE 1.15 Physical Constants of Organic Compounds (*continued*)

No.	Name	Formula	Formula weight	Beilstein reference	Density	Refractive index	Melting point	Boiling point	Flash point	Solubility in 100 parts solvent
t173	p-Toluenesulfinic acid	$CH_3C_6H_4SO_2H$	172.20	11, 9			85			v s alc, eth
t174	p-Toluenesulfonamide	$CH_3C_6H_4SO_2NH_2$	171.22	11, 104			137–140			0.2 aq; 3.6 alc
t175	p-Toluenesulfonyl-hydrazide	$CH_3C_6H_4SO_2NHNH_2$	186.23	11^2, 66			110 d			
t176	p-Toluenesulfonic acid	$CH_3C_6H_4SO_3H$	172.20	11, 97				140^{20mm}		67 aq; s alc, eth
t177	p-Toluenesulfonyl chloride	$CH_3C_6H_4SO_2Cl$	190.65	11, 103			69–71	134^{10mm}		v s alc, bz, eth; i aq
t178	p-Toluenesulfonyl fluoride	$CH_3C_6H_4SO_2F$	174.19	11^2, 54			41–42	112^{16mm}		
t179	p-Toluenesulfonyl isocyanate	$CH_3C_6H_4SO_2NCO$	197.21			1.4355^{20}		144^{10mm}		
t180	o-Toluidine	$CH_3C_6H_4NH_2$	107.16	12, 772	0.9984^{20}	1.5725^{20}	−16.1	200.4	85	1.7 aq; s alc, eth
t181	m-Toluidine	$CH_3C_6H_4NH_2$	107.16	12, 853	0.989^{20}_4	1.5681^{20}	−30.4	203.4	85	misc alc, eth
t182	p-Toluidine	$CH_3C_6H_4NH_2$	107.16	12, 880	1.046^{20}_4	1.5532^{59}	43.8	200.6	88	7.4 aq; v s alc, eth
t183	1-(o-Toluidino)-1,3-butanedione	$CH_3C_6H_4NHCOCH_2$-$COCH_3$	191.23	12, 823			104–106	143		
t184	o-Tolunitrile	$CH_3C_6H_4CN$	117.15	9, 466	0.9955^{20}	1.5279^{20}	−13	205.2	84	i aq; misc alc, eth
t185	m-Tolunitrile	$CH_3C_6H_4CN$	117.15	9, 477	0.976^{15}	1.5256^{20}	−23	210	86	0.09 aq; v s alc, eth
t186	p-Tolunitrile	$CH_3C_6H_4CN$	117.15	9, 489	0.9785^{30}_4		29.5	217.6		i aq; v s alc, eth
t187	o-Toluoyl chloride	$CH_3C_6H_4COCl$	154.60	9, 464	1.185	1.5549^{20}		90^{12mm}	76	
t188	m-Toluoyl chloride	$CH_3C_6H_4COCl$	154.60	9, 477	1.173	1.5485^{20}		86^{5mm}	76	
t189	p-Toluoyl chloride	$CH_3C_6H_4COCl$	154.60	9, 484	1.169	1.5535^{20}	−2	225–257	82	s alc, eth; i aq
t190	m-Tolyl isocyanate	$CH_3C_6H_4NCO$	133.15	12, 864	1.033	1.5305^{20}		76^{12mm}	65	
t191	(p-Tolylsulfonyl)-methyl isocyanide	$CH_3C_6H_4SO_2CH_2NC$	195.24				114–115			
t192	p-Tolyltrichloro-silane	$CH_3C_6H_4SiCl_3$	225.6		1.3^{20}_4			218–220		

No.	Name	Formula	Formula weight	Beilstein ref.	Density	n_D	mp, °C	bp, °C	Flash pt, °C	Solubility
t193	1,2,4-Triacetoxybenzene	$C_6H_3(OOCCH_3)_3$	252.22	6, 1089			98–100			sl s aq; i alc, eth
t194	Triacetoxyethylsilane	$C_2H_5Si(OOCCH_3)_3$	234.3		1.1428^{20}_4	1.4123^{20}		$107\text{–}108^{8mm}$		
t195	Triacetoxyvinylsilane	$(CH_3COO)_3SiCH{=}CH_2$	232.3		1.167^{20}_4	1.423^{20}		113^{1mm}	104	
t196	1,3,5-Triacetylbenzene	$C_6H_3(COCH_3)_3$	204.23	7, 866			160–162		17	
t197	Triallyl-s-triazine-2,4,6(1H,3H,5H)-trione		249.27			1.5129^{20}		152^{4mm}	>112	
t198	2,4,6-Triamino-1,3,5-triazine		126.12	26, 245	1.573^{250}		>250	subl		
t199	1H-1,2,4-Triazole		69.07	26, 13			119–121	260 d		
t200	Tribenzylamine	$(C_6H_5CH_2)_3N$	287.41	12, 1038	0.991^{95}		91–94		65	s aq, alc
t201	Tribromoacetaldehyde	Br_3CCHO	280.76	1, 626	2.665	1.5850^{20}		174	65	s hot alc, eth
t202	Tribromoacetic acid	Br_3CCOOH	296.76	2, 220			130–133	245 d		s aq, alc, chl, eth
t203	2,4,6-Tribromoaniline	$Br_3C_6H_2NH_2$	329.83	12, 663	2.35		120–122	300		s aq, alc, eth
t204	2,2,2-Tribromoethanol	Br_3CCH_2OH	282.77	1^2, 338			80–81	93^{10mm}		s hot alc, chl, eth
t205	1,1,2-Tribromoethylene	$BrCH{=}CBr_2$	264.74	1, 191	1.708^{21}	1.6247^{25}		162.5		2 aq; s alc, bz, eth

$CH_2{=}CHCH_2$—N, N—$CH_2CH{=}CH_2$, $CH_2CH{=}CH_2$ t197

H_2N, NH_2 t198

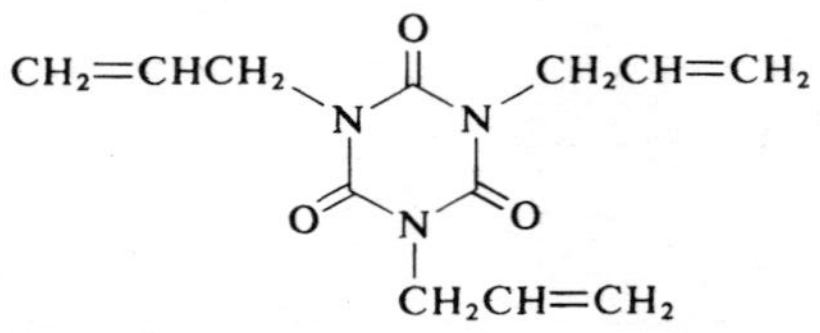

H, N, N, N   t199

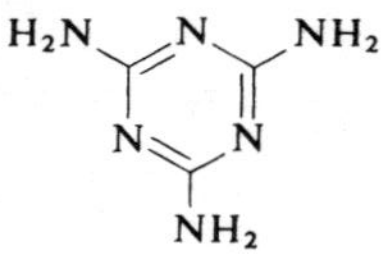

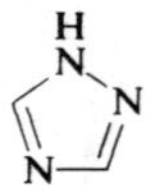

TABLE 1.15 Physical Constants of Organic Compounds (*continued*)

No.	Name	Formula	Formula weight	Beilstein reference	Density	Refractive index	Melting point	Boiling point	Flash point	Solubility in 100 parts solvent
t206	Tribromomethane	$CHBr_3$	252.77	1, 68	2.9031^{15}	1.6005^{15}	8.1	149.6	none	0.3 aq; misc eth, MeOH
t207	2,4,6-Tribromophenol	$Br_3C_6H_2OH$	330.82	6, 203	2.55		94–96	244		s alc, chl, eth; i aq
t208	1,2,3-Tribromo-propane	$BrCH_2CH(Br)CH_2Br$	280.78	1, 112	2.4114^{15}		16–17	219–221		s alc, eth
t209	Tributoxyborane	$(C_4H_9O)_3B$	230.16	1^2, 398	0.8580^{20}	1.4092^{20}	−70	233.5	93	hyd aq
t210	Tributylamine	$(C_4H_9)_3N$	185.36	4, 157	0.7784^{20}_4	1.4283^{20}	−70	216–217	63	v s alc, eth; s acet
t211	2,4,6-Tri-*tert*-butylphenol	$[(CH_3)_3C]_3C_6H_2OH$	262.44		0.864^{27}		131	278		
t212	Tributyl phosphate	$(C_4H_9O)_3P(O)$	266.32	1^2, 397	0.972^{25}	1.4226^{25}	< -80	289 d	146	0.04 aq; misc org solv
t213	Tributylphosphine	$(C_4H_9)_3P$	202.32	4^2, 971	0.812	1.4619^{20}		150^{50mm}	40	misc alc, bz, eth, PE
t214	Tributyl phosphite	$(C_4H_9O)_3P$	250.32	1^1, 187	0.9254^{20}	1.4326^{20}		125^{7mm}	121	
t215	Tributyltin chloride	$(C_4H_9)_3SnCl$	325.49	2, 206	1.200	1.4905^{20}		173^{25mm}	>112	120 aq; v s alc, eth
t216	Trichloroacetic acid	Cl_3CCOOH	163.39	2, 206	1.629^{61}		57–58	196–197	none	
t217	Trichloroaceto-nitrile	Cl_3CCN	144.39	2, 212	1.4403^{25}_4	1.4409^{20}		85.7	none	
t218	Trichloroacetyl chloride	Cl_3CCOCl	181.83	2, 210	1.629	1.4689^{20}		114–116	none	
t219	Trichloroacetyl isocyanate	$Cl_3CC(=O)NCO$	188.40			1.4809^{20}		85^{20mm}	65	
t220	2,4,5-Trichloro-aniline	$Cl_3C_6H_2NH_2$	196.46	12, 627			93–95	270		s alc
t221	2,4,6-Trichloro-aniline	$Cl_3C_6H_2NH_2$	196.46	12, 627			73–75	262		s alc, eth
t222	1,2,3-Trichloro-benzene	$C_6H_3Cl_3$	181.45	5, 203	1.69^{25}_{25}		52.6	221	113	v s bz, CS_2
t223	1,2,4-Trichloro-benzene	$C_6H_3Cl_3$	181.45	5, 204	1.446^{25}	1.5707^{20}	17	214	110	misc bz, eth, PE

No.	Name	Formula	Formula wt	Beilstein ref.	Density	n_D	mp, °C	bp, °C	Flash pt, °C	Solubility in 100 parts solvent
t224	1,3,5-Trichlorobenzene	$C_6H_3Cl_3$	181.45	5, 204		1.5662^{19}	63.4	208.5	107	v s bz, eth, PE
t225	2,2,2-Trichloro-1,1-dimethylethyl chloroformate	$ClCOOC(CH_3)_2CCl_3$	239.92				30–32	83–84^{14mm}	none	
t226	1,1,1-Trichloroethane	CH_3CCl_3	133.41	1, 85	1.3376^{20}_4	1.4379^{20}	−30.4	74.0	none	0.13 aq; s bz, eth
t227	1,1,2-Trichloroethane	$ClCH_2CHCl_2$	133.41	1, 85	1.4416^{20}_4	1.4711^{20}	−36.6	113.5	none	0.4 aq; misc alc, eth
t228	2,2,2-Trichloroethanol	Cl_3CCH_2OH	149.40	1, 338	1.557^{20}_{20}	1.4885^{20}	17.8	151		8 aq; misc alc, eth
t229	2,2,2-Trichloroethyl chloroformate	$ClCOOCH_2CCl_3$	211.86		1.539	1.4703^{20}		171–172	none	
t230	1,1,2-Trichloroethylene	$ClCH{=}CCl_2$	131.39	1, 187	1.4649^{20}_4	1.4775^{20}	−84.8	86.7	none	0.1 aq; misc alc, chl, eth
t231	Trichloroethylsilane	$C_2H_5SiCl_3$	163.5		1.2373^{20}_4	1.4256^{20}	−106	100.5	27	
t232	Trichlorofluoromethane	Cl_3CF	137.4		1.485^{21}	1.384^{20}	−111	23.8		0.14 aq; s alc, eth
t233	α,α,2-Trichloro-6-fluorotoluene	$ClC_6H_3(F)CHCl_2$	213.47	5^3, 701	1.446	1.5506^{20}		228–230	>112	
t234	Trichloroisocyanuric acid		232.41	25, 256			249–251			

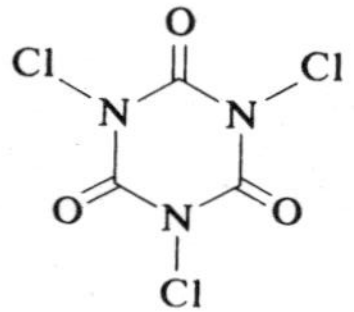

t234

TABLE 1.15 Physical Constants of Organic Compounds (*continued*)

No.	Name	Formula	Formula weight	Beilstein reference	Density	Refractive index	Melting point	Boiling point	Flash point	Solubility in 100 parts solvent
t235	Trichloromethane-sulfenyl chloride	Cl_3CSCl	185.89	3, 135	1.700_4^{20}	1.5436^{20}		146–148	none	
t236	Trichloromethane sulfonyl chloride	Cl_3CSO_2Cl	217.88	3^2, 16			139			s alc, eth
t237	1,1,1-Trichloro-2-methyl-2-propanol	$(CH_3)_2C(OH)CCl_3$	177.46	1, 382			99	167		s alc, bz, chl, eth
t238	Trichloromethyl-silane	CH_3SiCl_3	149.48		1.2754^{20}	1.4108^{20}	−90	66	5	
t238a	1,2,4-Trichloro-5-nitrobenzene	$Cl_3C_6H_2NO_2$	226.45	5, 246	1.790^{20}		49–55	288		v s bz, eth
t239	Trichloronitro-methane	Cl_3CNO_2	164.38	1, 76	1.6558_4^{20}	1.4611^{20}	−64	112		misc alc, bz; s eth
t240	2,4,5-Trichloro-phenol	$Cl_3C_6H_2OH$	197.45	6^2, 180			67	253		615 acet; 163 bz; 525 eth; s alc; i aq
t241	2,4,6-Trichloro-phenol	$Cl_3C_6H_2OH$	197.45	6, 190	1.4901_4^{75}		69	246	none	525 acet; 113 bz; 354 eth; v s alc; i aq
t242	(2,4,5-Trichloro-phenoxy)acetic acid	$Cl_3C_6H_2OCH_2COOH$	255.49	6^3, 702			153			s alc; v sl s aq
t243	2-(2,4,5-Trichloro-phenoxypropionic acid	$Cl_3C_6H_2O-CH(CH_3)COOH$	269.51				181.6			0.14 aq; 16 acet; 0.16 bz; 7.1 eth
t244	1,2,3-Trichloro-propane	$ClCH_2CH(Cl)CH_2Cl$	147.43	1, 106	1.3880^{20}	1.4834^{20}	−14.7	156.9	82	misc alc, eth; i aq
t245	1,1,1-Trichloro-2-propanol	$CH_3CH(OH)CCl_3$	163.43	1, 365			50	162	82	2.9 aq; v s alc, eth

No.	Name, Formula	Formula	Formula weight	Beilstein reference	Density	Refractive index	mp, °C	bp, °C	Flash point, °C	Solubility
t246	2,4,6-Trichloro-pyrimidine		183.43	23, 90		1.5700^{20}	23–25	210–215	>112	d aq; s bz, chl
t247	Trichlorosilane	$HSiCl_3$	135.45		1.3417^{20}_4	1.400^{20}	−1.28	31–32	−20	s alc, bz, eth
t248	α,α,α-Trichlorotoluene	$C_6H_5CCl_3$	195.48	5, 300	1.3756^{20}_4	1.5570^{20}	−5.0	220.8	97	
t249	α,2,6-Trichloro-toluene	$Cl_2C_6H_3CH_2Cl$	195.48	26, 35			36–39	119^{14mm}		v s alc, eth
t250	2,4,6-Trichloro-1,3,5-triazine		184.41				148	190^{720mm}		i aq; s alc
t251	1,1,2-Trichloro-trifluoroethane	$Cl_2CFCClF_2$	187.38	1^3, 157	1.5635^{25}	1.3557^{25}	−36.4	47.6	none	0.017 aq
t252	Trichlorovinyl-silane	$H_2C\!=\!CHSiCl_3$	161.49		1.243^{20}_4	1.4300^{20}	−95	90–93	−9	
t253	Tricyclo[5.2.1.0^{2,6}]-decane		136.24	5, 164			77–79	193	40	
t254	Tricyclo[5.2.1.0^{2,6}]-decan-8-one		150.22	7^2, 133	1.063	1.5025^{20}	74–75	132^{30mm}		
t255	1,3,5-Tricyclohexyl-hexahydro-s-triazine		333.57					97^{6mm}		

Tricine, t428

Tricyclo[3.3.1.1^{3,7}]decane, a67

Tricyclo[5.2.1.0^{2,6}]decane-4,8-dimethanol, b185

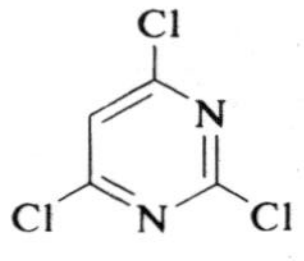

t246

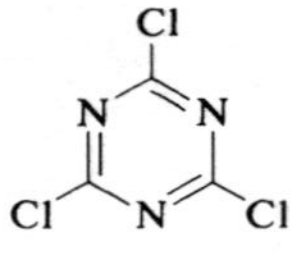

t250

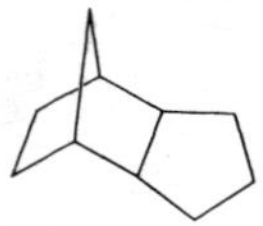

t253

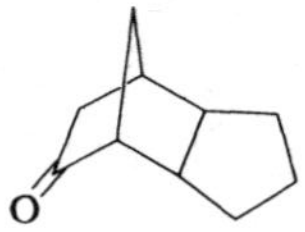

t254

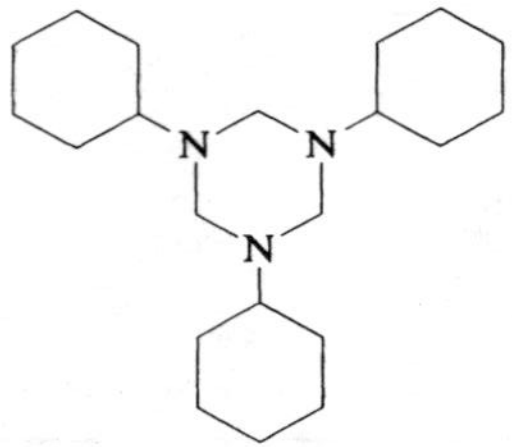

t255

TABLE 1.15 Physical Constants of Organic Compounds (*continued*)

No.	Name	Formula	Formula weight	Beilstein reference	Density	Refractive index	Melting point	Boiling point	Flash point	Solubility in 100 parts solvent
t256	Tridecane	$CH_3(CH_2)_{11}CH_3$	184.37	1, 171	0.7563_4^{20}	1.4256^{20}	−5.4	235.4	79	v s alc, eth
t257	Tridecanoic acid	$CH_3(CH_2)_{11}COOH$	214.35	2, 364			41–42	236^{100mm}		v s alc, eth; i aq
t258	1-Tridecene	$CH_3(CH_2)_{10}CH{=}CH_2$	182.35	1, 225	0.7653_4^{20}	1.4334^{20}	−23.1	232.8	79	s alc; v s eth
t259	Triethanolamine	$(HOCH_2CH_2)_3N$	149.19	4, 285	1.1242_4^{20}	1.4835^{25}	21.6	335.4	185	misc aq, alc, acet; 4.5 bz; 1.6 eth; s chl
t260	Triethoxyborane	$(CH_3CH_2O)_3B$	145.99	1, 335	0.864_{20}^{20}	1.3740^{20}		117–118	11	d aq
t261	Triethoxyethyl-silane	$(C_2H_5O)_3SiC_2H_5$	192.3		0.8963_4^{20}	1.3955^{20}		158–159		
t261a	Triethoxymethyl-silane	$CH_3Si(OC_2H_5)_3$	178.30	4, 629	0.895_4^{20}	1.3845^{20}		141–143	23	s alc
t261b	Triethoxysilane	$(C_2H_5O)_3SiH$	164.28	1, 334	0.875_4^{20}	1.3762		131.5	26	
t262	Triethoxyvinyl-silane	$(C_2H_5O)_3SiCH{=}CH_2$	190.32		0.903_4^{20}	1.3978^{20}		160–161	34	
t263	Triethylaluminum	$(C_2H_5)_3Al$	114.17	4, 643	0.832^{25}	1.3980^{25}	−58	194		d aq, air
t264	Triethylamine	$(C_2H_5)_3N$	101.19	4, 99	0.7326_4^{25}		−114.7	89.6	−6	5.5 aq; misc alc, eth; s acet, EtAc
t265	Triethylantimony	$(C_2H_5)_3Sb$	208.94	4, 618	1.324^{16}	1.42	−29	159.5		i aq; misc alc, eth
t266	Triethylarsine	$(C_2H_5)_3As$	162.11	4, 602	1.150_4^{20}			140^{736mm}		i aq; v s alc, eth
t267	Triethylbismuthine	$(C_2H_5)_3Bi$	296.17	4, 622	1.82			107^{79mm}		explodes when heated in air
t268	Triethylborane	$(C_2H_5)_3B$	98.00	4, 641	0.6961^{23}		−92.9	95		i aq; d air
t269	Triethylenediamine		112.18				158	174		45 aq; 13 acet; 77 alc; 51 bz
t270	Triethylene glycol	$(HOCH_2CH_2OCH_2{-})_2$	150.17	1, 468	1.1274_4^{15}	1.4578^{15}	−4.3	285	165	misc aq, alc, bz
t271	Triethylene glycol dibenzoate	$(C_6H_5COOCH_2\,CH_2OCH_2{-})_2$	358.39		1.2715^{30}	1.5252^{50}	47			
t272	Triethylenetetramine	$(H_2NCH_2CH_2NHCH_2{-})_2$	146.24	4, 255	0.982	1.4971^{20}	12	266–267	143	

No.	Name	Formula	M.W.		Density	n_D	m.p.	b.p.	Flash	Solubility
t273	*N,N,N'*-Triethylethylenediamine	$(C_2H_5)_2NCH_2CH_2NHC_2H_5$	144.26	4^2, 691	0.804	1.4311^{20}		55^{13mm}	32	
t274	Triethylgallium	$(C_2H_5)_3Ga$	156.91	26, 2	1.0576^{30}	1.4595^{20}	−82.3	142.6		
t275	1,3,5-Triethylhexahydro-s-triazine		171.29		0.894			207–208		
t276	Triethylindium	$(C_2H_5)_3In$	202.01		1.260^{20}	1.538^{20}	−32	144	55	misc alc, chl, eth
t277	Triethyl orthoacetate	$CH_3C(OC_2H_5)_3$	162.23	2, 129	0.8847^{25}_{4}	1.3950^{25}		142		d aq; s alc, eth
t278	Triethyl orthoformate	$HC(OC_2H_5)_3$	148.20	2, 20	0.891^{20}_{4}	1.3919^{20}	−76	146	30	
t279	Triethyl orthopropionate	$CH_3CH_2C(OC_2H_5)_3$	176.26	2, 240	0.876	1.3995^{20}		155–160	60	v s alc, eth
t280	Triethyl phosphate	$(C_2H_5O)_3P(O)$	182.16	1, 332	1.0725^{19}	1.4045^{20}		215–216		s aq(d), alc, eth
t281	Triethylphosphine	$(C_2H_5)_3P$	118.16	4, 582	0.800^{15}_{4}		−88	129	pyrophoric	i aq; misc alc, eth
t282	Triethyl phosphite	$(C_2H_5O)_3P$	166.2	1, 330	0.969^{20}_{4}	1.4131^{20}		65^{24mm}	55	
t283	Triethyl phosphonoacetate	$(CH_3CH_2O)_2P(O)CH_2COOC_2H_5$	224.19	4^1, 573	1.130	1.4310^{20}		145^{9mm}	>112	i aq(hyd); misc alc, acet, bz, eth, PE

Triethylenediamine, d45
Triethylene glycol, e127
Triethylene glycol dimethyl ether, b189

O,O,O-Triethyl phosphorothioate, t285

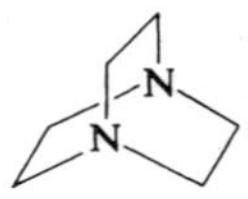

t269

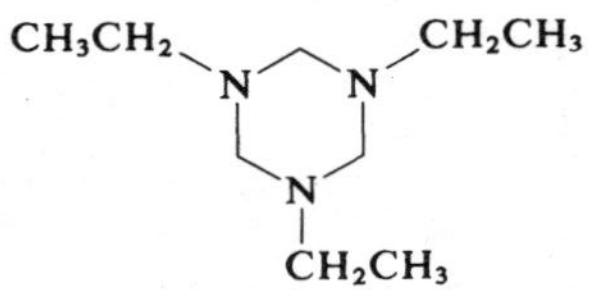

t275

Tridecylbenzene, p156
3-Triethoxysilylpropylamine, a279
Triethyl borate, t260

TABLE 1.15 Physical Constants of Organic Compounds (*continued*)

No.	Name	Formula	Formula weight	Beilstein reference	Density	Refractive index	Melting point	Boiling point	Flash point	Solubility in 100 parts solvent
t284	Triethylsilane	$(C_2H_5)_3SiH$	116.28	4, 625	0.731_4^{20}	1.412^{20}		107–108	107	i aq; misc alc, eth
t285	Triethyl thiophosphate	$(C_2H_5O)_3P(S)$	198.22	1, 333	1.082	1.4480^{20}		100^{16mm}		
t286	2,2,2-Trifluoroacetamide	CF_3CONH_2	113.04	2², 186			75	162.5		
t287	Trifluoroacetic acid	CF_3COOH	114.02	2², 186	1.4890_4^{20}	1.2850^{20}	−15.3	71.8	41	misc aq
t288	Trifluoroacetic anhydride	$[CF_3C(O)]_2O$	210.03	2², 186	1.487	>1.30	−65	39	73	
t289	α,α,α-Trifluoroacetophenone	$C_6H_5COCF_3$	174.12		1.240	1.4595^{20}		165–166		
t290	α,α,α-Trifluoro-*m*-cresol	$CF_3C_6H_4OH$	162.11	6¹, 187	1.333	1.4588^{20}	−1.8	178–179		
t291	1,1,1-Trifluoroethane	CH_3CF_3	84.04				−111.3	−47.3		
t292	2,2,2-Trifluoroethanol	CF_3CH_2OH	100.04		1.3842_4^{20}	1.2907^{22}	−43.5	74.1	29	
t293	2,2,2-Trifluoroethyl acrylate	$CF_3CH_2OOCCH{=}CH_2$	154.0		2.142_4^{25}	1.3981^{25}		46^{125mm}		
t294	2,2,2-Trifluoroethyl trifluoroacetate	$CF_3CH_2OOCCF_3$	196.0		1.4725_4^{18}	1.2812^{18}	−65.5	55		
t295	Trifluoromethane	HCF_3	70.01	1, 59	1.52^{-100}		−155.2	−82.2	none	75 mL aq; 500 mL alc
t296	Trifluoromethanesulfonic acid	CF_3SO_3H	150.07	3³, 34	1.695^{25}	1.3250^{25}	34	162		v s aq; misc eth alc
t297	Trifluoromethanesulfonic anhydride	$(CF_3SO_2)_2O$	282.13	3⁴, 35	1.677	1.3212^{20}		84		d aq, alc
t298	3-(Trifluoromethyl)-benzonitrile	$CF_3C_6H_4CN$	171.12	9, 478	1.2813^{20}	1.4505^{20}	14.5	189	72	

No.	Name	Formula	Formula wt	Beilstein ref	Density	n_D	mp, °C	bp, °C	Flash pt, °C	Solubility
t299	3-(Trifluoromethyl)-benzyl chloride	CF₃C₆H₄CH₂Cl	194.59		1.254	1.4605		70^{12mm}		
t300	α,α,α-Trifluoro-toluene	C₆H₅CF₃	146.11	5, 290	1.1886^{20}	1.4145^{20}	−29	102	12	v s alc, eth; i aq
t301	Trihexylamine	[CH₃(CH₂)₅]₃N	269.52	4, 188	0.871^{20}_4	1.456^{20}		263–265	>112	
t302	Trihexylchloro-silane	[CH₃(CH₂)₅]₃SiCl	319.1					155^{5mm}		
t303	Trihexylsilane	[CH₃(CH₂)₅]₃SiH	284.60	6, 1071	1.45	1.448^{20}		160^{5mm}		
t304	1,2,3-Trihydroxy-benzene	C₆H₃(OH)₃	126.11				131–133			59 aq; 77 alc; 62 eth
t305	1,3,5-Trihydroxy-benzene	C₆H₃(OH)₃	126.11	6, 1092			218–220	subl d		1 aq; 10 alc; s eth
t306	3,4,5-Trihydroxy-benzoic acid	(HO)₃C₆H₂COOH	170.12	10, 470			d 235			1.1 aq; 17 alc; 1 eth; 20 acet; i bz, chl, PE
t307	Triisobutylaluminum	[(CH₃)₂CHCH₂]₃Al	198.33		0.781^{25}		6	86^{10mm}	pyro-phoric 235	
t308	Triisodecyl phosphite	[(CH₃)₂CH(CH₂)₇O]₃P	502.80		0.886^{25}_{15}	1.454^{25}	<0	$180^{0.1mm}$		
t309	Triisopropanolamine	[CH₃CH(OH)CH₂]₃N	191.27	1, 363	0.9996^{50}_{20}	1.3764^{20}	46	305.4	152	v s aq
t310	Triisopropoxyborane	[(CH₃)₂CHO]₃B	188.08		0.815			139–141	17	
t311	Triisopropoxyvinyl-silane	[(CH₃)₂CHO]₃Si-CH=CH₂	232.4		0.863^{25}_4	1.396^{25}		179–181		
t312	1,3,5-Triisopropyl-benzene	C₆H₃[CH(CH₃)₂]₃	204.36	5, 458	0.845	1.4884^{20}		232–236	86	
t313	Triisopropyl phosphite	[(CH₃)₂CHO]₃P	208.24	1, 363	0.914^{20}_4	1.4101^{20}		64^{11mm}	73	i aq (sl hyd)

TABLE 1.15 Physical Constants of Organic Compounds (*continued*)

No.	Name	Formula	Formula weight	Beilstein reference	Density	Refractive index	Melting point	Boiling point	Flash point	Solubility in 100 parts solvent
t314	3,4,5-Trimethoxy-benzaldehyde	$(CH_3O)_3C_6H_2CHO$	196.20	8, 391			73–75	165^{10mm}		
t315	1,2,3-Trimethoxy-benzene	$C_6H_3(OCH_3)_3$	168.19	6, 1081	1.112		43–45	241		v s alc, eth; s chl
t316	3,4,5-Trimethoxy-benzoic acid	$(CH_3O)_3C_6H_2COOH$	212.20	10, 481			168–171	227^{10mm}		
t317	3,4,5-Trimethoxy-benzoyl chloride	$(CH_3O)_3C_6H_2COCl$	230.65	10, 487			79–81	185^{18mm}		
t318	3,4,5-Trimethoxy-benzyl alcohol	$(CH_3O)_3C_6H_2CH_2OH$	198.22	6, 1159	1.233	1.5459^{20}		228^{25mm}	>112	
t319	Trimethoxyborane	$(CH_3O)_3B$	103.91	1, 287	0.920_4^{23}	1.3568^{20}	–34	67–68	–1	hyd aq; misc alc, eth
t320	Trimethoxyboroxine	$[-OB(OCH_3)-]_3$	173.53	1^3, 3214	1.195	1.3996^{20}	10	130	10	
t321	1,3,3-Trimethoxy-butane	$(CH_3O)_2C(CH_3)CH_2-CH_2OCH_3$	148.20		0.940	1.4096^{20}		63^{20mm}	45	
t321a	Trimethoxy(methyl)-silane	$CH_3Si(OCH_3)_3$	136.23		0.9548_4^{20}	1.3696^{20}		102–103	21	
t322	1,3,3-Trimethoxy-propane	$CH_3OCH_2CH_2-CH(OCH_3)_2$	134.18	1, 820	0.942	1.4004^{20}		$45-$ 46^{17mm}		
t323	(Trimethoxysilyl)-propyldiethylene-triamine	$(CH_3O)_3Si(CH_2)_3NH-CH_2CH_2NHCH_2CH_2NH_2$	265.4		1.03_4^{20}	1.463^{20}				
t324	N-[3-Trimethoxy-silyl)propyl]-ethylenediamine	$(CH_3O)_3Si(CH_2)_3NH-CH_2CH_2NH_2$	222.4		1.010	1.4450^{20}		146^{15mm}	>112	
t325	N-(Trimethoxysilyl-propyl)imidazole		230.3		1.00_4^{20}	1.45^{25}				
t326	3-(Trimethoxysilyl)-propyl methacrylate	$(CH_3O)_3SiCH_2CH_2-CH_2OOCC(CH_3)=CH_2$	249.3		1.045_4^{20}	1.429^{25}		190	92	

No.	Name	Formula	Formula wt	Beil. ref.	Density	n_D	mp, °C	bp, °C	Flash pt	Solubility
t327	Trimethylaluminum	$(CH_3)_3Al$	72.09	4, 643	0.752^{20}	1.432^{12}	15.4	20^{8mm}	pyrophoric	s alk; v sl s alc
t328	Trimethylamine	$(CH_3)_3N$	59.11	4, 43	0.636		−117.1	2.9	−6	41 aq; misc alc; s bz, chl, eth
t329	Trimethylamine-N-oxide	$(CH_3)_3N(O)$	75.11				257			s aq. MeOH
t330	2,4,6-Trimethyl-aniline	$(CH_3)_3C_6H_2NH_2$	135.21	12, 1160	0.963	1.5510^{20}		233	96	
t331	1,3,3-Trimethyl-6-azabicyclo[3.2.1]-octane		153.27		0.902	1.4716^{20}		194	75	
t332	3,3,5-Trimethyl-1-azacycloheptane		141.26		0.852	1.4563^{20}		180	67	
t333	1,2,3-Trimethyl-benzene	$C_6H_3(CH_3)_3$	120.20	5, 399	0.894_4^{20}	1.5139^{20}	−25.4	176.1	48	i aq; s alc, eth
t334	1,2,4-Trimethyl-benzene	$C_6H_3(CH_3)_3$	120.20	5, 400	0.8756_4^{20}	1.5048^{20}	−43.9	169.4	48	s alc, bz, eth
t335	1,3,5-Trimethyl-benzene	$C_6H_3(CH_3)_3$	120.20	5, 406	0.8637_4^{20}	1.4994^{20}	−44.7	164.7	44	misc alc, bz, eth

Trimellitic acid, b29
Trimesic acid, b30
Trimesoyl chloride, b32
Trimethylacetaldehyde, d596

Trimethylacetamide, d597
Trimethylacetic acid, d598
Trimethylacetic anhydride, d599
Trimethylacetyl chloride, d600

endo-1,7,7-Trimethylbicyclo[2.2.1]heptan-2-ol, b216

t325: imidazol-1-yl—$CH_2CH_2CH_2Si(OCH_3)_2$

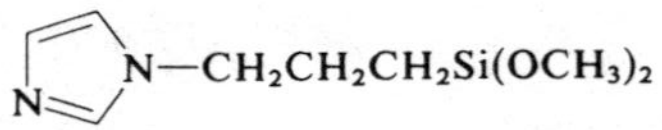

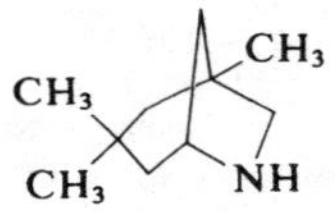

t331

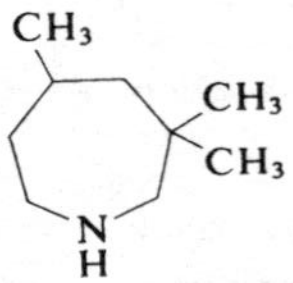

t332

TABLE 1.15 Physical Constants of Organic Compounds (*continued*)

No.	Name	Formula	Formula weight	Beilstein reference	Density	Refractive index	Melting point	Boiling point	Flash point	Solubility in 100 parts solvent
t336	Trimethyl 1,2,4-benzenetricarboxylate	$C_6H_3(COOCH_3)_3$	252.22	9^1, 429	1.261	1.5214^{20}	38–40	194^{12mm}	>112	
t337	2,2,3-Trimethylbutane	$(CH_3)_2CHC(CH_3)_3$	100.20	1^2, 121	0.6901^{20}_4	1.3894^{20}	−24.9	80.9		s alc, eth
t338	2,3,3-Trimethyl-2-butanol	$(CH_3)_3CC(CH_3)_2OH$	116.20	1^2, 447	0.8380^{25}_4	1.4233^{22}	15–17	130.5		misc alc, eth
t339	1,1,3-Trimethylcyclohexane	$C_6H_9(CH_3)_3$	126.24	7, 65		1.4296^{20}		136.6		
t340	3,5,5-Trimethylcyclohex-2-ene-1-one		138.2	19^3, 1604	0.9252^{20}_{20}	1.478^{20}	−8.1	215.2	96	1.2 aq
t341	2,2,6-Trimethyl-1,3-dioxen-4-one		142.15		1.088	1.4622^{20}	12–13	$65–67^{2mm}$		
t342	4,4'-Trimethylenebis-(1-methylpiperidine)		238.42		0.896	1.4820^{20}	13	215^{50mm}	110	
t343	4,4'-Trimethylenedipiperidine		210.37				65–68			
t344	4,4'-Trimethylenedipyridine		198.27				57–60			
t345	Trimethylene oxide		58.08	17, 6	0.8930^{25}_4	1.3895^{25}		50	<1	misc aq
t345a	Trimethylene sulfide		74.15	17^1, 3	1.025^{20}	1.5102^{20}	−73.3	95.0	<1	v s org solv
t346	2,2,5-Trimethylhexane	$(CH_3)_2CHCH_2CH_2\text{-}C(CH_3)_3$	128.26	1^3, 516	0.7072^{20}	1.3997^{20}	−105.8	124.1		
t347	3,5,5-Trimethyl-1-hexanol	$(CH_3)_3CCH_2CH(CH_3)\text{-}CH_2CH_2OH$	144.25		0.8236^{20}_4	1.4300^{25}	<−70	194		s alc, eth
t348	Trimethylhydroquinone	$(CH_3)_3C_6H(OH)_2$	152.19	6,931			172–174			s aq; v s alc, bz, eth

No.	Name	Formula							
t348a	2,6,8-Trimethyl-4-nonanol	$(CH_3)_2CHCH_2CH(CH_3)$-$CH_2CH(OH)CH_2$-$CH(CH_3)_2$	186.33	0.8193			225	93	
t349	2,6,8-Trimethyl-4-nonanone	$(CH_3)_2CHCH_2CH(CH_3)$-$CH_2C(O)CH_2CH(CH_3)_2$	184.31	0.818^{20}_{20}	−75		218.4		
t350	α-(−)-1,3,3-Tri-methyl-2-norbornanol		154.25	0.9641^{20}_{4}	6, 70	48	201	73	s alc, eth

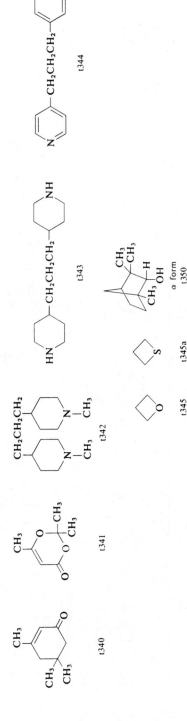

TABLE 1.15 Physical Constants of Organic Compounds (*continued*)

No.	Name	Formula	Formula weight	Beilstein reference	Density	Refractive index	Melting point	Boiling point	Flash point	Solubility in 100 parts solvent
t351	(+)-1,3,3-Tri-methyl-2-norbor-nanone		152.24	7, 96	0.948^{18}	1.4635^{18}	5–6	192–193	52	v s alc, eth
t352	Trimethyl orthoacetate	$CH_3C(OCH_3)_3$	120.15	1^2, 128	0.9428^{25}_4	1.3859^{25}		105		v s alc, eth
t353	Trimethyl orthoformate	$HC(OCH_3)_3$	106.12	2, 19	0.9676^{20}_4	1.3790^{20}		100.6	15	
t354	2,4,4-Trimethyl-1-oxazoline		113.16		0.887	1.4213^{20}		112–113	12	
t355	2,2,3-Trimethyl-pentane	$(CH_3)_3CCH(CH_3)$-CH_2CH_3	114.23	1^1, 62	0.7160^{20}_4	1.4030^{20}	−112.3	109.8		s eth; sl s alc
t356	2,2,4-Trimethyl-pentane	$(CH_3)_2CHCH_2C(CH_3)_3$	114.23	1^2, 127	0.6919^{20}_4	1.3915^{20}	−107.4	99.2	−7	s bz, chl, eth
t357	2,3,4-Trimethyl-pentane	$(CH_3)_2CH[CH(CH_3)]_2CH_3$	114.24	1^3, 500	0.7190^{20}_4	1.4042^{20}	−109.2	113.5		s alc, org solv
t358	2,2,4-Trimethyl-1,3-pentanediol	$(CH_3)_2CHCHCH(OH)$-$C(CH_3)_2CH_2OH$	146.22	1^3, 2225	0.9285^{55}_{15}	1.4513^{15}	46	229	113	1.8 aq; 75 alc; 22 bz; 25 acet
t359	2,4,4-Trimethyl-1-pentene	$(CH_3)_3CCH_2$-$C(CH_3)\!=\!CH_2$	112.22	1^3, 849	0.7150^{20}_4	1.4112^{20}	−93	101.4	<1	
t360	2,3,5-Trimethyl-phenol	$(CH_3)_3C_6H_2OH$	136.19	6, 518			92–95	230–231		
t361	2,3,6-Trimethyl-phenol	$(CH_3)_3C_6H_2OH$	136.19				62–64			
t362	2,4,6-Trimethyl-phenol	$(CH_3)_3C_6H_2OH$	136.19	6, 158			68–71	220		
t363	Trimethyl phosphate	$(CH_3O)_3P(O)$	140.08	1, 286	1.197	1.3958^{20}	−46	197	none	100 aq; s alc
t364	Trimethyl phosphite	$(CH_3O)_3P$	124.08	1, 285	1.046^{20}_4	1.4080^{20}	<−78	111–112	40	d aq; misc alc, acet, bz, PE

	Name	Formula	Formula wt		Density	n_D	mp, °C	bp, °C		Solubility
t365	Trimethyl phosphonoacetate	$(CH_3O)_2P(O)CH_2COCH_3$	182.11		1.125	1.4370^{20}		$118^{0.85mm}$	>112	s aq, alc, acet, bz
t366	1,2,4-Trimethyl-piperazine		128.22		0.8512^{25}_{25}	1.4480^{25}	<−50	151^{746mm}		
t367	2,4,6-Trimethyl-pyridine	$(C_5H_2N)(CH_3)_3$	121.18	20, 250	0.9166^{22}_{4}	1.4979^{20}	−43	170.5	57	3.5 aq; misc eth; s alc, bz, chl
t368	N-(Trimethylsilyl)-acetamide	$CH_3CONHSi(CH_3)_3$	131.25				52–54	185–186	57	
t369	N-(Trimethylsilyl)-aniline	$(CH_3)_3SiNHC_6H_5$	165.3		0.940^{20}_{4}	1.522^{20}		207–208		
t370	Trimethylsilyl bromoacetate	$BrCH_2COOSi(CH_3)_3$	211.14		1.284	1.4421^{20}		$57–58^{9mm}$	28	
t371	N-(Trimethylsilyl)-diethylamine	$(CH_3)_3SiN(C_2H_5)_2$	145.33					127^{738mm}		
t372	2-(Trimethylsilyl)-ethanol	$(CH_3)_3SiCH_2CH_2OH$	118.25		0.825	1.4246^{20}		$71–73^{35mm}$	50	
t373	N-(Trimethylsilyl)-imidazole		140.26		0.956	1.4751^{20}		99^{14mm}	80	
t374	3-(Trimethyl-silyloxy)allene	$(CH_3)_3SiOCH_2CH{=}CH_2$	130.3		0.7830^{30}_{4}	1.4075^{25}		100–102		

Trimethylolpropane, e156
Trimethylsilyl cyanide, c298

Trimethylsilyldiethylamine, d343
Trimethylsilyl iodide, i55

Trimethylsilylnitrile, c298

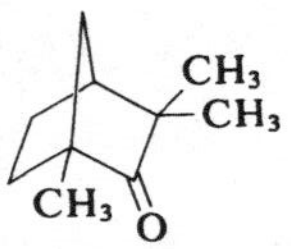

t351

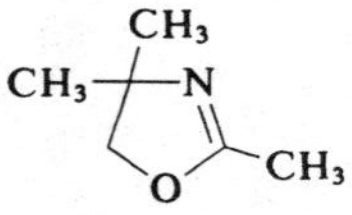

t354

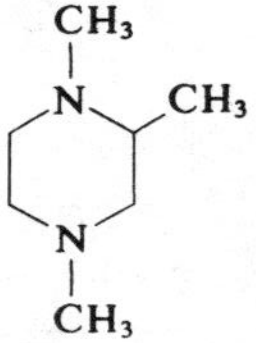

t366

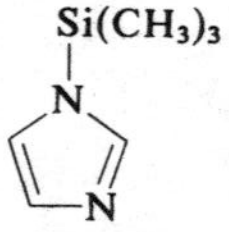

t373

TABLE 1.15 Physical Constants of Organic Compounds (*continued*)

No.	Name	Formula	Formula weight	Beilstein reference	Density	Refractive index	Melting point	Boiling point	Flash point	Solubility in 100 parts solvent
t375	Trimethylsilyl-phenoxide	$(CH_3)_3SiOC_6H_5$	166.3		0.9256^{20}_4	1.4782^{20}	−55	81^{23mm}		
t376	Trimethylsilyl trifluoroacetate	$(CH_3)_3SiOOCCF_3$	186.2		1.077^{20}_4	1.3880^{20}		89–90		
t377	Trimethylsulfonium iodide	$[(CH_3)_3S]I$	204.07				215–220	subl		
t378	Trimethylsulf-oxonium iodide	$[(CH_3)_3S(O)]I$	220.07				175 d			
t379	Trimethylvinyl-oxysilane	$(CH_3)_3SiOCH=CH_2$	116.2		0.772^{20}_4	1.389^{20}		74–75		
t380	Trimethylvinyl-silane	$(CH_3)_3SiCH=CH_2$	100.2		0.690^{20}_4	1.3920^{20}		55	<1	
t381	2,4,6-Trinitro-aniline	$(O_2N)_3C_6H_2NH_2$	228.12	12, 763	1.762^{14}		188–190	explodes		s hot acet; sl s alc
t382	1,2,4-Trinitro-benzene	$C_6H_3(NO_2)_3$	213.11	5, 271	1.73^{16}		61–62	explodes		5.5 alc; 7.1 eth; i aq
t383	1,3,5-Trinitro-benzene	$C_6H_3(NO_2)_3$	213.11	5, 271	1.688^{20}_4		122.5	explodes		0.035 aq; 1.9 alc; 1.5 eth; 6.2 bz
t384	2,4,7-Trinitro-9-fluorenone		315.20	7^2, 410			175–176	explodes		v s bz, acet; sl s aq
t385	Trinitromethane	$HC(NO_2)_3$	151.04	1, 79	1.597^{24}_4		15	47^{22mm}		s aq, alk
t386	2,4,6-Trinitro-toluene	$(O_2N)_3C_6H_2CH_3$	227.13	5, 347	1.654^{20}_4		80.1	explodes		1.5 alc; 4 eth; s bz, acet; 0.01 aq
t387	Trioctylamine	$[CH_3(CH_2)_7]_3N$	353.68	4, 196	0.809	1.4485^{20}		365–367	>112	17.2 aq; v s alc, bz, eth, EtAc
t388	s-Trioxane		90.08	19, 381	1.170^{65}		64	115	45	
t389	Tripentaerythritol		372.41				245 d			
t390	2,4,6-Triphenoxy-s-triazine		357.37				232–234			

No.	Name	Formula	Formula wt	Beilstein	Density	n_D	mp, °C	bp, °C	Solubility
t391	Triphenoxyvinyl-silane	$(C_6H_5O)_3SiCH=CH_2$	334.5		1.1304_4^{25}	1.562^{25}		210^{7mm}	s acet, eth; sl s alc
t392	Triphenylamine	$(C_6H_5)_3N$	245.33	12,181	0.7740_0^0		125–127	347–348	v s bz, eth; sl s alc
t393	Triphenylantimony	$(C_6H_5)_3Sb$	353.07	16,891	1.4343^{25}		52–54	377	v s bz, eth; s alc
t394	Triphenylarsine	$(C_6H_5)_3As$	306.24	16,828	1.2225^{48}	1.6139^{48}	60–62	233^{14mm}	v s bz, eth; s alc
t395	1,3,5-Triphenyl-benzene	$(C_6H_5)_3C_6H_3$	306.41	5,737	1.205		172–174	460	v s bz; s abs alc, eth
t396	Triphenylene		228.29	5,720	1.302		199	425	s alc; v s bz, eth
t397	Triphenylmethane	$(C_6H_5)_3CH$	244.34	5,698	1.01434_4^{99}		93.4	360	v s hot alc, eth; 49 chl; 7 bz; s PE; i aq
t398	Triphenylmethanol	$(C_6H_5)_3COH$	260.34	6,713	1.1994_4^0		164.2	360	v s alc, bz, eth; i aq
t399	Triphenyl phosphate	$(C_6H_5O)_3P(O)$	326.29	6,179			49–51	244^{10mm}; 223	misc alc; s bz, acet, chl, eth; i aq

2,4,6-Trinitrophenol, p173
Triolein, g22

Trioxymethylene, t388
Tripalmitin, g23

Triphenylmethyl bromide, b366

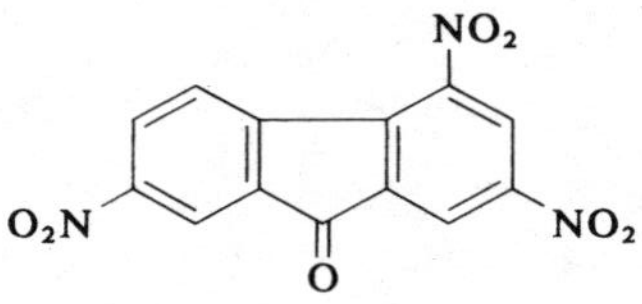

t384

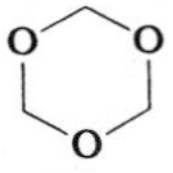

t388

CH₂OH
(HOCH₂)₃CCH₂OCH₂CCH₂OCH₂CCH₂OCH₂C(CH₂OH)₃
CH₂OH

t389

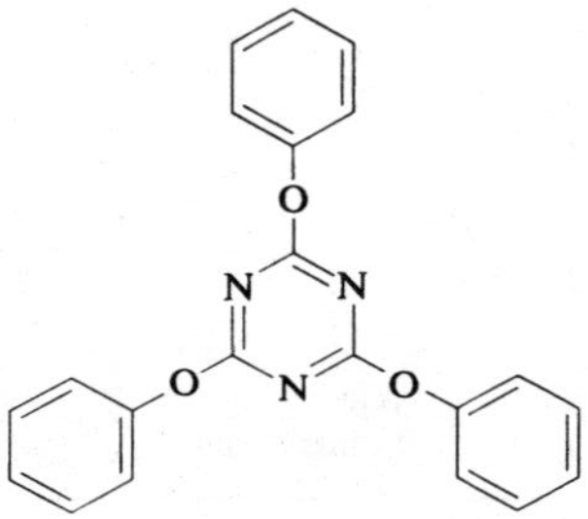

t390

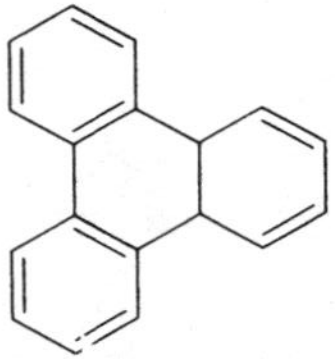

t396

TABLE 1.15 Physical Constants of Organic Compounds (*continued*)

No.	Name	Formula	Beilstein reference	Formula weight	Density	Refractive index	Melting point	Boiling point	Flash point	Solubility in 100 parts solvent
t400	Triphenylphosphine	$(C_6H_5)_3P$	16, 759	262.29	1.0758_4^{81}		80.5	377	181	v s eth; s bz, chl, HOAc; sl s alc; i aq
t401	Triphenylphosphine selenide	$(C_6H_5)_3P(Se)$		341.25			187–189			
t402	Triphenylphosphine sulfide	$(C_6H_5)_3P(S)$	16, 784	294.36			162–164			
t403	Triphenyl phosphite	$(C_6H_5O)_3P$	6, 177	310.29	1.184_{15}^{25}	1.5903^{20}	22–24	360	218	s alc, bz, chl, eth
t404	Triphenylsilane	$(C_6H_5)_3SiH$	16², 605	260.41			42–44	152^{2mm}		
t405	Tripiperidino-phosphine oxide		20, 88	299.40			40–42	273^{50mm}		
t406	Tripropoxyborane	$(CH_3CH_2CH_2O)_3B$	1², 369	188.08	0.8576_4^{20}	1.3948^{20}		175		v s alc; misc eth
t407	Tripropylamine	$(CH_3CH_2CH_2)_3N$	4, 139	143.27	0.753	1.4160^{20}	−93	155–158	36	s aq, alc, eth
t408	Tripropylene glycol	$H(OCH_2CH_2CH_2)_3OH$		192.3	1.018	1.442^{25}		267.2	141	s aq
t409	Tripropylene glycol butyl ether	$HO(CH_2CH_2CH_2O)_3$-$(CH_2)_3CH_3$		248.4	0.934_{25}^{25}	1.430^{25}		276	135	
t410	Tripropylene glycol ethyl ether	$HO(CH_2CH_2CH_2O)_3$-CH_2CH_3		220.3	0.0948_{25}^{25}	1.427^{25}		486	132	
t411	Tripropylene glycol isopropyl ether	$HO(CH_2CH_2CH_2O)_3$-$CH(CH_3)_2$		234.8	0.0942_{25}^{25}	1.428^{25}		112.7	124	
t412	Tripropylene glycol-methyl ether	$HO(CH_2CH_2CH_2O)_3CH_3$		206.3	0.967_{25}^{25}	1.428^{25}	−42	242.4	127	misc aq, alc, eth
t413	Tripropyl orthoformate	$HC(OCH_2CH_2CH_3)_3$		190.28	0.8805_4^{20}	1.4072^{20}		108^{5mm}		
t414	Tris(butoxyethyl) phosphate	$(C_4H_9OCH_2CH_2O)_3P(O)$		398.48	1.006	1.4359^{20}		228^{4mm}	> 112	
t415	Tris(2-chloro-ethoxy)silane	$(ClCH_2CH_2O)_3SiH$		267.6	1.2886_4^{20}	1.4577^{20}		118^{2mm}		

No.	Name	Formula	M.W.		Density	n	mp, °C	bp, °C			Solubility
t416	Tris(2-chloro-ethyl) phosphate	$(ClCH_2CH_2O)_3P(O)$	285.49	1^2, 337	1.390	1.4721^{20}		330		232	misc alc, bz, eth
t417	Tris(2-chloro-ethyl) phosphite	$(ClCH_2CH_2O)_3P$	269.49		1.3534^{20}	1.4863^{20}		115^{2mm}		190	
t418	Tris(2,6-dichloro-phenyl) phosphate	$(Cl_2C_6H_3O)_3P(O)$	533.09				208–210				
t419	Tris(dimethylamino)-methane	$CH[N(CH_3)_2]_3$	145.25		1.4360^{20}		42–	43^{12mm}			
t420	Tris(dimethylamino)-methylsilane	$[(CH_3)_2N]_3SiCH_3$	175.4		0.850_4^{22}	1.432^{22}	–11	56^{17mm}			
t421	Tris(2-ethylhexyl) phosphite	$[CH_3(CH_2)_3CH(CH_2CH_3)-CH_2O]_3P$	418.6		0.902_4^{20}	1.4494^{20}		$164^{0.3mm}$		185	i aq
t422	Tris(heptafluoro-propyl)-s-triazine		585.1		1.7158^{25}	1.7158^{25}		165			
t423	Tris(hydroxy-methyl)amino-methane	$(HOCH_2)_3CNH_2$	121.14	4, 303			172	220^{10mm}			

Triphenylsilyl azide, a320
Tripropyl borate, t406

Tripropylsilyl chloride, c258
TRIS, t423

Tris(dimethylamino)silyl chloride, c257a
1,1,1-Tris(hydroxymethyl)ethane, h141

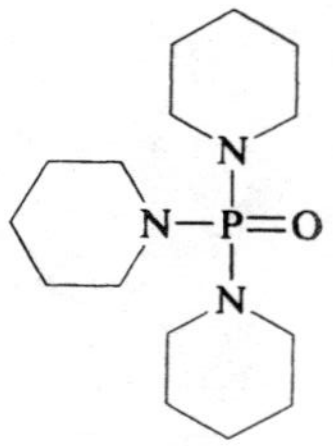

t405

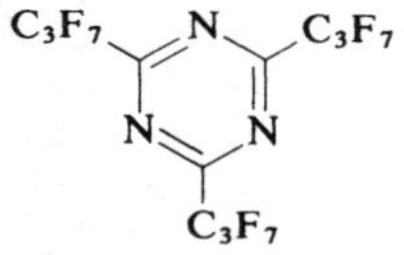

t422

1.389

TABLE 1.15 Physical Constants of Organic Compounds (*continued*)

No.	Name	Formula	Formula weight	Beilstein reference	Density	Refractive index	Melting point	Boiling point	Flash point	Solubility in 100 parts solvent
t424	2-[Tris(hydroxy-methyl)methyl-amino]-1-ethane-sulfonic acid	$(HOCH_2)_3CNHCH_2CH_2SO_3H$	229.25				223–225			
t425	1,1,1-Tris(hydroxy-methyl)ethane	$CH_3C(CH_2OH)_3$	120.15	1, 520						
t426	3-[N-Tris(hydroxy-methyl)methyl-amino]-2-hydroxypropane-sulfonic acid	$(HOCH_2)_3CNHCH_2CH(OH)CH_2SO_3H$	259.3				226			
t427	3-[Tris(hydroxy-methyl)-methyl-amino]-1-propanesulfonic acid	$(HOCH_2)_3CNHCH_2CH_2CH_2SO_3H$	243.28				240 d			
t428	N-[Tris(hydroxy-methyl)methyl]-glycine	$(HOCH_2)_3CNHCH_2COOH$	179.17				184 d			
t429	Tris(2-methoxy-ethoxy)methyl-silane	$CH_3Si(OCH_2CH_2OCH_3)_3$	268.4		1.045_4^{20}	1.420^{20}		145^{16mm}		
t430	Tris(2-methoxy-ethoxy)vinylsilane	$H_2C\!=\!CHSi(OCH_2CH_2OCH_3)_3$	280.38		1.034_4^{25}	1.427^{25}		284–286	65	
t431	Tris(2-methylallyl)-amine	$[H_2C\!=\!C(CH_3)CH_2]_3N$	173.91	$4^3, 462$	0.794	1.4575^{20}		$83–85^{15mm}$	53	
t432	Tris(pentafluoro-ethyl)-s-triazine		435.1		1.6506^{25}	1.3131^{25}		121–122		

t433	1,3,5-Trithiane		138.27	19, 382			216–218			s bz; sl s alc, eth
t434	Trithiocarbonic acid	$(HS)_2C(S)$	110.21	3, 221	1.483_4^{20}	1.8225^{20}	−26.9	57.8		d aq, alc; sl s eth
t435	1,2,4-Trivinylcyclo-hexane	$(H_2C{=}CH)_3C_6H_9$	162.28		0.836	1.4780^{20}		88^{20mm}	68	
t436	L-(−)-Tryptophan		204.23	22, 546			280–285 d			1 aq; s hot alc, alk; i eth, chl
t437	L-Tyrosine	$(HO)C_6H_4CH_2$-$CH(NH_2)COOH$	181.19	14, 605			> 300 d			0.03 aq; 0.01 alc; s alk; i eth
t438	L-Tyrosine hydrazide	$HOC_6H_4CH_2CH(NH_2)$-$CONHNH_2$	195.22	14^1, 665			196–198			
u1	Undecanal	$CH_3(CH_2)_9CHO$	170.30	1, 712	0.825	1.4322^{20}	−4	115^{5mm}	96	i aq; s alc, eth
u2	Undecane	$CH_3(CH_2)_9CH_3$	156.31	1, 170	0.7402_4^{20}	1.4173^{20}	−25.6	195.9	60	i aq; misc alc, eth
u3	Undecanoic acid	$CH_3(CH_2)_9COOH$	186.30	2, 358	0.8907	1.4294^{45}	28.5	228^{160mm}		s alc, chl, eth; i aq
u4	1-Undecanol	$CH_3(CH_2)_{10}OH$	172.31	1, 427	0.8324^{20}	1.4402^{20}	15.9	242.8	> 112	0.02 aq; s alc
u5	2-Undecanone	$CH_3(CH_2)_8COCH_3$	170.30	1, 173	0.829	1.4280^{20}	11–12	231–232	88	s alc, bz, chl, eth, acet; i aq
u6	6-Undecanone	$CH_3(CH_2)_4$-$CO(CH_2)_4CH_3$	170.30	1, 174	0.831	1.4280^{20}	14.6	228	88	i aq; v s alc, eth
u7	10-Undecenal	$H_2C{=}CH(CH_2)_8CHO$	168.28		0.810	1.4427^{20}			92	
u8	1-Undecene	$CH_3(CH_2)_8CH{=}CH_2$	154.29	1, 225	0.763_4^{20}	1.4261^{20}	−49.2	192.7		i aq; misc alc, eth

Tris(7-methylnonyl) phosphite, t308
Trityl alcohol, t398
Tryptamine, a170
Tyramine, a173
Umbelliferone, h109
Undecyl alcohol, u4
Undecyl-10-en-1-oic acid, u9
Undecylenic aldehyde, u7
Undecylic aldehyde, u1

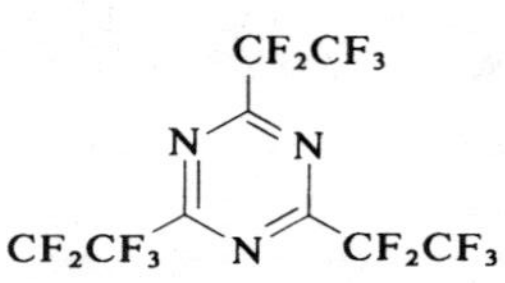

t432

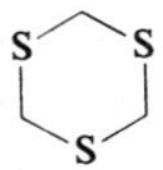

t433

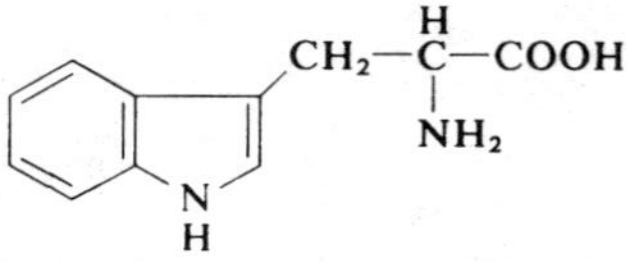

t436

TABLE 1.15 Physical Constants of Organic Compounds (*continued*)

No.	Name	Formula	Formula weight	Beilstein reference	Density	Refractive index	Melting point	Boiling point	Flash point	Solubility in 100 parts solvent
u9	10-Undecenoic acid	$H_2C=CH(CH_2)_8COOH$	184.28	2, 458	0.907_4^{24}	1.4493^{20}	24.5	137^{2mm}	148	s alc, chl, eth; i aq
u10	10-Undecen-1-ol	$H_2C=CH(CH_2)_9OH$	170.30	1, 452	0.850^{15}	1.4500^{20}	−2	245	93	
u11	10-Undecenoyl chloride	$H_2C=CH(CH_2)_8COCl$	202.73	2, 459	0.944	1.4532^{20}		122^{10mm}	93	
u12	Urea	$(H_2N)_2CO$	60.06	3, 42	1.32_1^{18}		132.7	d > mp		100 aq; 20 alc
u13	Uric acid		168.11	26, 513	1.893^{20}		>300	d		s alk; i aq, alc, eth
u14	Uridine		244.20	31, 23			165			s aq; hot alc, pyr
v1	L-Valine	$(CH_3)_2CH-$ $CH(NH_2)COOH$	117.15	4, 427	1.230		315	subl		8.8 aq; v sl s alc, eth
v2	Vinyl acetate	$H_2C=CHOOCCH_3$	86.09	2^1, 63	0.9318_4^{20}	1.3959^{20}	−92.8	72.5	−6	2 aq; misc alc, eth
v3	5-Vinylbicyclo-[2.2.1]-2-heptene		120.19		0.84	1.4802	−80	141		
v4	Vinyl crotonate	$CH_3CH=CHCOOCH=CH_2$	112.13	2^3, 1263	0.940	1.4488^{20}		50^{10mm}	27	
v5	Vinylcyclohexane	$C_6H_{11}CH=CH_2$	110.20	5^1, 35		1.4463^{20}		128	21	
v6	4-Vinyl-1-cyclohexene		108.18	5^1, 63	0.830_4^{20}	1.4640^{20}	−101	126–127	20	
v7	1-Vinylimidazole		94.12	23^4, 569	1.039	1.5308^{20}		78– 79^{13mm}	81	
v8	5-Vinyl-2-norbornene		120.20		0.8411	1.4802^{20}	−80	141	27	
v9	2-Vinylpyridine	$(C_5H_4N)CH=CH_2$	105.14	20, 256	0.975	1.5490^{20}		158–159	43	v s alc, chl, eth
v10	4-Vinylpyridine	$(C_5H_4N)CH=CH_2$	105.14	20^2, 170	0.975	1.5500^{20}		65^{15mm}	51	sl s hot aq, hot alc
v11	N-Vinyl-2-pyrrolidinone		111.14		0.980	1.5120^{20}		93^{13mm}	93	
x1	Xanthene		182.22	17, 73			101	310–312		s bz, eth; sl s alc, aq
x2	Xanthen-9-carboxylic acid		226.23	12^2, 279			217 d			s hot alc, eth
x3	9-Xanthenone		196.21	17, 354			174	350^{730mm}		0.5 alc; v s chl

Uracil, p268
5-Ureidohydantoin, a77
Urethane, e91
Valeraldehyde, p27
Valeric acid, p36
γ-Valerolactone, p40
Valerone, d531
Valeronitrile, p33
Valeryl chloride, p44
Valinols, a213, a214

Vanillic acid, h132
Vanillin, h131
o-Vanillin, h130
Vanillyl alcohol, h135
Veratraldehyde, d430
Veratric acid, d434
Veratrole, d431
Veronal, d280
Vinylacetic acid, b403
Vinyl bromide, b284

Vinyl 2-butenoate, v4
Vinyl chloride, c109
Vinylidene chloride, d178
Vinyltrimethylsilane, t380
Vinyltris (2-methoxyethoxy)silane, t430
Vitamin B₁, t139
Vitamin B₂, r4
Vitamin C, a312
Xanthone, x3

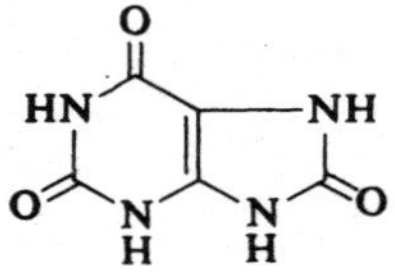

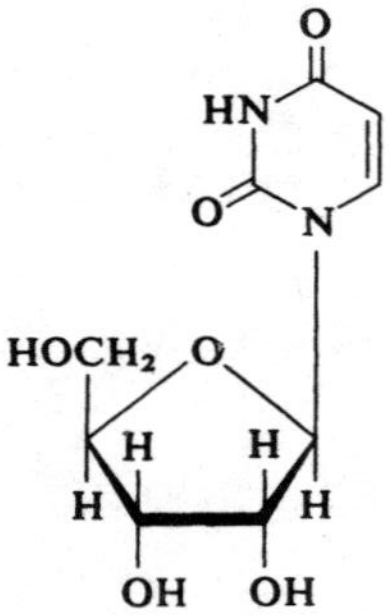

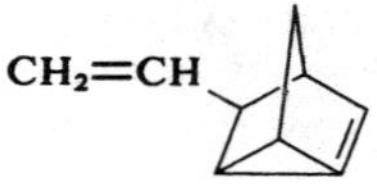

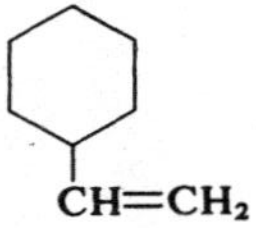

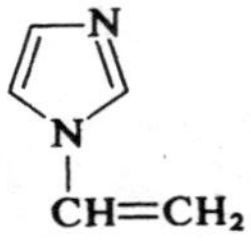

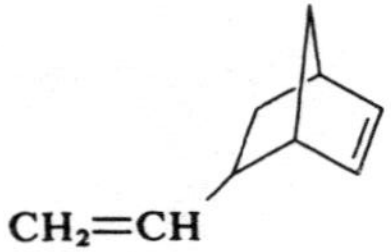

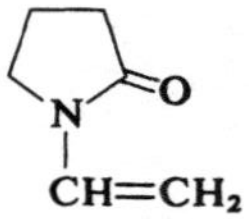

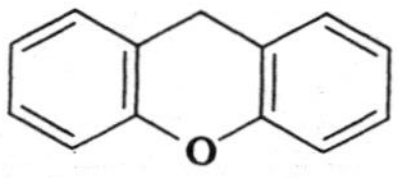

1.393

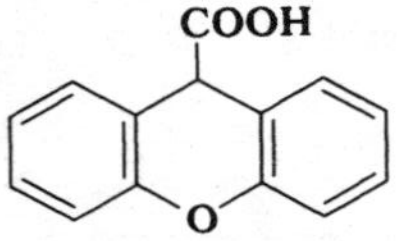

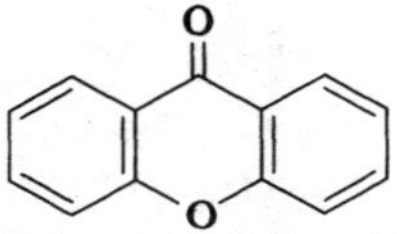

TABLE 1.15 Physical Constants of Organic Compounds (*continued*)

No.	Name	Formula	Formula weight	Beilstein reference	Density	Refractive index	Melting point	Boiling point	Flash point	Solubility in 100 parts solvent
x4	o-Xylene	$C_6H_4(CH_3)_2$	106.17	5, 362	0.8802_4^{20}	1.5054^{20}	−25.2	144.4	32	misc alc, eth; 0.017 aq
x5	m-Xylene	$C_6H_4(CH_3)_2$	106.17	5, 370	0.8684_4^{15}	1.4972^{20}	−47.9	139.1	25	misc alc, eth; 0.02 aq
x6	p-Xylene	$C_6H_4(CH_3)_2$	106.17	5, 382	0.8611_4^{20}	1.4958^{20}	13.3	138.4	30	v s eth; s alc; 0.02 aq
x7	Xylitol	$HOCH_2(CHOH)_3CH_2OH$	152.15	1, 531			95–97			s aq
x8	D-(+)-Xylose		150.13	31, 47	1.535^0		144–145			117 aq; s hot alc, pyr
x9	m-Xylylenediamine	$C_6H_4(CH_2NH_2)_2$	136.20	13, 186	1.032	1.5709^{20}		265^{745mm}	>112	

Xylene-α,α′-diol, b18
Xylenols, d579, d580, d581, d582, d583, d584

o-Xylyl bromide, b368
Xylyl chlorides, c258, c259, c260

p-Xylylene glycol, b18

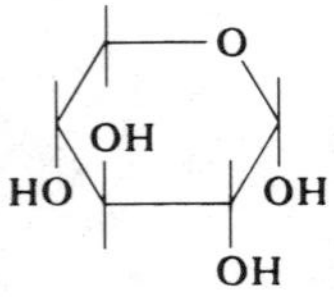

SECTION 2

INORGANIC AND ORGANOMETALLIC COMPOUNDS

This section summarizes the properties of various chemical compounds commonly encountered in organic chemistry that are often referred to either as inorganic or organometallic. In some cases, the distinction is artificial and the reader should refer to Section 1 if the expected compound is not presented here. Most compounds containing metals are collected here, rather than in Section 1.

TABLE 2.1   Physical Constants of Inorganic Compounds

Explanation of column headings

Names, while following the IUPAC nomenclature, are generally alphabetized by the central atom to facilitate their location. An example of the table organization is given below for Al_3C_4, aluminum tetracarbide. It is entered in Table 2.1 as follows:

Main heading	**Aluminum**
Subgrouping	carbide
Actual table listing	Aluminum
	(tetra-) carbide (tri-)

Solvates are listed under the entry for the anhydrous salt. Hydrazine hydrate, $H_2N\!-\!NH_2 \cdot H_2O$ is listed under hydrazine. Magnesium sulfate heptahydrate (epsom salt) is listed under **Magnesium**, using the subgroup sulfate 7-water. Inorganic acids are entered under hydrogen, For example, HF is listed under **Hydrogen**, using the subgroup fluoride. Where an elemental designation would be confusing or inappropriate, the compound is listed alphabetically as in the case of **hydroxylamine**, $HONH_2$.

Abbreviations used in the Table

>, greater than	ca, approximately	g, gas	pyr, pyridine
α, alpha position	chl, chloroform	glyc, glycerol	s, soluble
a, acid	conc, concentrated	h, hot	satd, saturated
abs, absolute	cub, cubic	hex, hexagonal	sl, slightly
acetone, acet	d, decompose(s)	hyd, hydrolysis	soln, solution
alc, alcohol	dil, dilute	i, insoluble	solv, solvent(s)
alk, alkali, (aq NaOH)	DMF,	ign, ignites	subl, sublimes
anhyd, anhydrous	dimethylformamide	lq, liquid	tetr, tetragonal
aq, aqueous	eth, diethyl ether	MeOH, methanol	THF, tetrahydrofuran
aq reg, aqua regia	EtOH, ethanol	min, mineral	tr, transition
atm, atmosphere	expl, explodes,	misc, miscible	v, very
bz, benzene	explosive	org, organic	vac, *vacuo* or vacuum
c, solid state	fcc, face-centered cubic	PE, petroleum ether	viol, violently

Formula Weights are based on the International Atomic Weights of 1973 and are computed to the nearest hundredth of an a.m.u.

Density values are given at room temperature unless otherwise indicated by a superscript figure indicating a temperature in °C. Thus, 2.487^{15} indicates a density of 2.487 for the named substance at 15 °C. For gases density values are given in grams per liter $(g \cdot L^{-1})$.

Melting Point values are recorded in °C. In certain cases decomposition is indicated with the letter "d" that either precedes or follows the number. The value 250d indicates that the substance melts at 250 °C with decomposition. The value d 250 indicates that decomposition only occurs at 250 °C and higher temperatures. Where a value such as "$-6H_2O$, 150" is given, it indicates a loss of 6 moles of water per formula weight of the compound at a temperature of 150 °C.

Boiling Point values are given at atmospheric pressure (760 mm of mercury) unless otherwise indicated; a value of 82 means that the boiling point is 82 °C at 760 mm Hg. A value of 82^{15} means that the boiling point is 82 °C when the pressure is 15 mmHg. The specification "subl 550" indicates that the compound sublimes at 550 °C.

Solubility is given in parts by weight (of the formula weight) per 100 parts by weight of the solvent. If the solvent is unspecified, it is water. If no temperature is specified, the solubility is for the substance at room temperature. Other temperatures (in °C) are indicated by superscript. The symbols of the common mineral acids represent aqueous solutions of those acids.

TABLE 2.1 Physical Constants of Inorganic Compounds

Name	Formula	Formula weight	Density	Melting point, °C	Boiling point, °C	Solubility in 100 parts solvent
Aluminum	Al	26.98	2.70	660.1	2450	s HCl, H_2SO_4, alk
acetylacetonate	$Al(C_5H_7O_2)_3$	324.31	1.27	subl 193 (vac)	314	i aq; v s alc; s bz, eth
ammonium bis(sulfate) 12-water	$AlNH_4(SO_4)_2 \cdot 12H_2O$	453.33	1.64	$-12\ H_2O$, 250	d >280	15 aq; i alc
bis(acetylsalicylate)	$Al(OOCC_6H_4O\text{-}COCH_3)_2OH$	402.30				v sl s aq, alc, eth
bromide	$AlBr_3$	266.71	2.64^{10}	97.5	253.3	d viol aq; s alc, acet, bz CS_2
butoxide, *sec-*	$Al(C_4H_9O)_3$	246.33	0.967		$200\text{--}206^{30\ mm}$	v s org solv (flash point 27°C)
butoxide, *tert-*	$Al(C_4H_9O)_3$	246.33	1.025^{20}_0	subl 180		v s org solv
(tetra-) carbide, tri-	Al_4C_3	143.96	2.36	2100	d >2200	d to CH_4 in aq (fire hazard)
chlorate	$Al(ClO_3)_3$	277.35				v s aq; s alc
chloride	$AlCl_3$	133.34	2.44	$194^{2.5\ atm}$	subl 181	70 aq (viol); 100^{12} abs alc; s CCl_4, eth; sl s bz
chloride 6-water	$AlCl_3 \cdot 6H_2O$	241.43	2.40	d 100		83^{20} aq; 25 abs alc; s eth
ethoxide	$Al(C_2H_5O)_3$	162.14	1.42^{20}_0	134	$205^{14\ mm}$	s hot aq (d); v sl a alc, eth
fluoride	$AlF3$	83.98	2.882^{25}_4	1040	subl 1276	0.562^5 aq; i a, alk, alc, acet
hydroxide	$Al(OH)_3$	78.00	2.42	$-H_2O$, 300		i aq; s a,alk
iodide	AlI_3	407.71	3.98^{25}	191	360	s aq(d); s alc, CS^2, eth
isopropoxide	$Al(C_3H_7O)_3$	204.25	1.0346^{20}_0	118.5	$135^{10\ mm}$	d aq; s alc, bz, chl, PE
nitrate 9-water	$Al(NO_3)_3 \cdot 9H_2O$	375.13		73	d 135	64^{25} aq; 100 alc; s acet
oxide	Al_2O_3	101.96	3.965	2054	2980	i aq; v sl s a, alk
phenoxide	$Al(C_6H_5O)_3$	306.27	1.23	d 265		d aq; s alc, chl, eth
potassium bis (sulfate) 12-water	$AlK(SO_4)_2 \cdot 12H_2O$	474.39	1.757^{20}	$-9H_2O$, 92	$-12H_2O$, 200	11.4^{20} aq
propoxide	$Al(C_3H_7O)_3$	204.25	1.0578^{20}_0	106	$248^{14\ mm}$	d aq; s alc
sodium bis(sulfate) 12-water	$AlNa(SO_4)_2 \cdot 12H_2O$	458.28	1.675^{20}	61		110^{15} aq

TABLE 2.1 Physical Constants of Inorganic Compounds (*continued*)

Name	Formula	Formula weight	Density	Melting point, °C	Boiling point, °C	Solubility in 100 parts solvent
Aluminum						
stearate	$Al(C_{18}H_{35}O_2)_3$	877.42	1.010	103		i aq; s alc, bz, alk
sulfate	$Al_2(SO_4)_3$	342.15	2.710	d 770		36.4[20] aq; sl s alc
sulfate 18-water	$Al_2(SO_4)_3 \cdot 18H_2O$	666.45	1.69[17]	d 86.5		87[0] aq; i alc
tetrahydroborate	$Al(BH_4)_3$	71.53		−64.5	44.5	d aq
Amidosulfuric acid	H_2NSO_3H	97.09	2.126	205	d	14.7 aq
Ammonia	NH_3	17.03	0.7188[20] g·L^{-1}	−77.75	−33.42	89.9 aq; 13.2[20] alc; s eth, org solv
$-d_3$ or [^{2}H]	ND_3 or N^2H_3	20.05	0.8437[20] g·L^{-1}	−74.33	−31.05	
Ammonium						
acetate	$NH_4C_2H_3O_2$	77.08	1.17[20]	114	d	148[4] aq; 7.9[15] MeOH; s alc
benzoate	$NH_4C_7H_5O_2$	139.16	1.260	d 198	subl 160	20[15] aq; 2.8 alc; s glyc; i eth
boranate, tetrafluoro-	NH_4BF_4	104.84	1.87[15]	subl		25[16] aq
bromide	NH_4Br	97.95	2.429	452 (under pressure)	d 397 (vac)	76[20] aq; s acet, alc, eth
carbamate	NH_2COONH_2	78.07		subl 60		v s aq; sl s alc; i eth
carbonate 1-water	$(NH_4)_2CO_3 \cdot H_2O$	114.10		d 20		100[15] aq; i alc
cerate(IV), hexanitrato-	$(NH_4)_2[Ce(NO_3)_6]$	548.23				135[20] aq; s alc, HNO_3
chloride	NH_4Cl	53.49	1.527	subl 340		26[15] aq; 0.6[19] abs alc; i acet, eth
chromate	$(NH_4)_2CrO_4$	152.08	1.91[12]	d 180		34[20] aq; sl s MeOH, acet; i alc
chromium(III) bis(sulfate) 12-water	$NH_4Cr(SO_4)_2 \cdot 12H_2O$	478.34	1.72	94		7.2[20] aq
citrate	$(NH_4)_3C_6H_5O_7$	243.22	1.48	d		100 aq; sl s alc
copper(II) tetrachloride 2-hydrate	$Cu(NH_4)_2Cl_4 \cdot 2H_2O$	277.46	1.993	−2H$_2$O, 110	d>120	40[20] aq; s alc
dichromate(VI)	$(NH_4)_2Cr_2O_7$	252.06	2.1554$_4^{25}$	d 170		36[20] aq; s alc (flammable)

Name	Formula	Formula wt	Density	mp (°C)	bp (°C)	Solubility
dithiocarbamate	$NH_4S—CS—NH_2$	110.19	1.4512^{20}	d 99		v s aq; s alc; sl s eth
diuranate(VI)	$(NH_4)_2U_2O_7$	624.22				v sl s aq, alk; s acids
fluoride	NH_4F	37.04	1.009^{25}	subl		100^0 aq; s alc
formate	NH_4OOCH	63.06	1.280	116	d 180	143^{20} aq; s alc, eth
hexadecanoate	$NH_4OOC(CH_2)_{14}CH_3$	273.45		21–22		s aq; sl s bz; i alc, acet
hexafluoroaluminate	$(NH_4)_3AlF_6$	195.10		d>100		v s aq
hydrogen carbonate	NH_4HCO_3	79.06	1.78	d 35	subl	22^{20} aq; i alc, acet
hydrogen citrate	$(NH_4)_2HC_6H_5O_7$	226.19	1.58			100 aq; sl s alc
hydrogen difluoride	NH_4HF_2	57.04	1.50	125.6		v s aq; sl s alc
hydrogen oxalate 1-water	$NH_4HC_2O_4·H_2O$	125.08	1.556	$-H_2O$, 170		s aq; i bz, eth
hydrogen phosphate	$(NH_4)_2HPO_4$	132.05	1.619	d 155		69^{20} aq; i alc, acet
hydrogen phosphate, di-	$NH_4H_2PO_4$	115.03	1.803^{19}	d 190		37^{20} aq; sl s alc; i acet
hydrogen sulfate	NH_4HSO_4	115.11	1.78	146.9		100 aq; i alc, acet
hydrogen sulfide	NH_4HS	51.11	1.17	d 25	d 350	128^0 aq; s alc; sl s acet; i bz
hydrogen sulfite	NH_4HSO_3	99.10	2.03	subl 150 (in N_2)		72^0 aq
hydroxide	NH_4OH	35.05		−77		misc aq
iodide	NH_4I	144.95	2.514^{25}	subl 551	220 (vac)	172^{20} aq; v s alc, acet
iron(II) bis(sulfate) 6-water	$Fe(NH_4)_2(SO_4)_2·6H_2O$	392.14	1.864^{20}	d 100		36^{20} aq; i alc
molybdate(VI)(6-) 4-water, hepta-	$(NH_4)_6Mo_7O_{24}·4H_2O$	1235.86	2.498	$-H_2O$, 90	d 190	43 aq; s a; i alc
nitrate	NH_4NO_3	80.04	1.725^{25}	169.6	$210^{11\,mm}$	192^{20} aq; 3.8^{20} alc; 17^{20} MeOH
octadecanoate	$NH_4OOC(CH_2)_{16}CH_3$	301.50		21–22		sl s aq; s alc; i acet
octanoate	$NH_4OOCC_7H_{15}$	161.24		d on standing		v s aq, alc, acet; sl s eth
oxalate 1-water	$(NH_4)_2C_2O_4·H_2O$	142.11	1.50	d 70		5.1^{20} aq
palladate(II) tetrachloro-	$(NH_4)_2PdCl_4$	284.29	2.170	d		v s aq; i abs alc
perchlorate	NH_4ClO_4	117.50	1.95	d 240		22^{20} aq; s MeOH; sl s alc, acet
peroxodisulfate	$(NH_4)_2S_2O_8$	228.18	1.982	d 120	expl 180	58^0 aq
phosphate, hexafluoro-	NH_4PF_6	163.00	2.180^{12}	d		75^{20} aq; s alc, acet
phosphinate	$NH_4PH_2O_2$	83.03	1.634	200	d 240	100 aq; 5 alc; i acet
picrate	$NH_4C_6H_2N_3O_7$	246.14	1.719	d	expl 423	1.1^{20} aq; sl s alc
platinate(IV), hexachloro-	$(NH_4)_2PtCl_6$	443.89	3.065	d		0.5^{20} aq
silicate, hexafluoro-	$(NH_4)_2SiF_6$	178.14	2.011	d		18.6^{20} aq; i alc, acet

TABLE 2.1 Physical Constants of Inorganic Compounds (*continued*)

Name	Formula	Formula weight	Density	Melting point, °C	Boiling point, °C	Solubility in 100 parts solvent
Ammonium						
sulfamate	$NH_4SO_3NH_2$	114.13		131	d 160	v s aq; sl s alc
sulfate	$(NH_4)_2SO_4$	132.14	1.769^{20}	d > 280		43.5^{25} aq; i alc, acet
sulfide	$(NH_4)_2S$	68.14		d		v s aq; s alc
DL-tartrate	$(NH_4)_2C_4H_4O_6$	184.15	1.601	d		58^{15} aq; sl s alc
tetraborate 4-water	$(NH_4)_2B_4O_7 \cdot 4H_2O$	263.44				s aq; i alc
thiocyanate	NH_4SCN	76.12	1.305	149.6	d 170	128^0 aq; v s alc; s acet
thiosulfate	$(NH_4)_2S_2O_3$	148.20	1.679	d 150		v s aq
vanadate(V)(1-)	NH_4VO_3	116.98	2.326	d 200		0.48^{20} aq
Antimony						
(III) chloride	$SbCl_3$	228.11	3.14^{20}	73.4	223.5	10^{20} aq; s alc, bz, chl
(V) chloride	$SbCl_5$	299.02	2.336_4^{20}	3.5	140	d aq; s HCl, chl, CCl_4
(III) fluoride	SbF_3	178.75	4.379^{20}	292	376	444^{20} aq
(V) fluoride	SbF_5	216.74	2.99^{23}	8.3	141	d viol aq; s HOAc; forms solids with alc, bz, CS_2, eth
hydride	SbH_3	124.77	4.36^{15}	-91.5	-18.4	20^0 mL aq; s CS_2
(III) oxide	Sb_2O_3	291.50	5.2	655	1425	v sl s aq; s HCl, KOH
(V) oxide	Sb_2O_5	323.50	2.78	$-O_2$, >300		v sl s aq; sl s warm KOH, eth
potassium oxide tartrate 0.5-water	$K(SbO)C_4H_4O_6 \cdot 0.5H_2O$	333.93	2.607	d 100		8.3^{20} aq; 6.7 glyc; i alc
(III) sulfide	Sb_2S_3	339.69	4.64	546		0.002^{20} aq d; s H_2SO_4
(V) sulfide	Sb_2S_5	403.82				i aq; s HCl, d NaOH
Argon	Ar	39.95	1.7824 $g \cdot L^{-1}$	-189.38	-185.87	3.36^{20} mL aq
Arsenic	As	74.92	5.72	$817^{28\,atm}$	subl 612	i aq; s HNO_3
(III) chloride	$AsCl_3$	181.28	2.1497_4^{25}	-16	130.2	d aq; misc chl, CCl_4, eth; s alc
(III) oxide dimer	As_4O_6	395.68	4.15	313	465	1.8^{20} aq; s alc

<table type="page" style="rotated 90 degrees">

Name	Formula	Form. wt.	Density	m.p., °C	b.p., °C	Solubility
(V) oxide	As_2O_5	229.84	4.32	d 800		66^{20} aq; s alc
(III) sulfide	As_2S_3	246.04	3.46	300–325	707	i aq; s alk; slowly s hot HCl
Barium						
acetate 1-water	$Ba(C_2H_3O_2)_2 \cdot H_2O$	273.46	2.19	d 150		76^{20} aq; 0.14 alc
benzenesulfonate	$Ba(O_3SC_6H_5)_2$	451.70				s aq; sl s alc
carbonate	$BaCO_3$	197.35	4.43	d 1360		0.002 aq; s a
chlorate 1-water	$Ba(ClO_3)_2 \cdot H_2O$	322.26	3.18	$-H_2O$, 120		34^{20} aq
chloride	$BaCl_2$	208.25	3.856	962	2029	36^{20} aq
fluoride	BaF_2	175.34	4.89	1368	2272	0.16^{20} aq
hydrogen phosphate	$BaHPO_4$	233.32	4.165^{15}	d 410		0.01 aq; s a
hydroxide 8-water	$Ba(OH)_2 \cdot 8H_2O$	315.48	2.18^{16}	78		3.9^{20} aq
manganate(VI)(2-)	$BaMnO_4$	256.28	4.85			v sl s aq
nitrate	$Ba(NO_3)_2$	261.35	3.24	575	d	9^{20} aq
nitrite 1-water	$Ba(NO_2)_2 \cdot H_2O$	247.37	3.173^{20}	d 115		73^{20} aq; i alc
oxide	BaO	153.34	5.72	2013	3088	3.5^{20} aq
perchlorate 3-water	$Ba(ClO_4)_2 \cdot 3H_2O$	390.29	2.74	d 400		198^{25} aq; s MeOH; sl s alc, acet
permanganate	$Ba(MnO_4)_2$	375.21	3.77	d 200		62^{11} aq
peroxide	BaO_2	169.34	4.96	450	$-O_2$, 800	1.5^0 aq
sulfate	$BaSO_4$	233.40	4.50^{15}	1580		0.0002 aq
sulfide	BaS	169.40	4.25^{15}	2227		7.9^{22} aq d
sulfite	$BaSO_3$	217.40		d		0.02^{20} aq
thiocyanate 2-water	$Ba(SCN)_2 \cdot 2H_2O$	289.53	2.286^{18}	d 160		170^{25} aq
thiosulfate 1-water	$BaS_2O_3 \cdot H_2O$	267.48	3.5^{18}	d 220		0.21^{20} aq
Beryllium	Be	9.01	1.86	1277	2484	i q; s a, alk
bromide	$BeBr_2$	168.83	3.465^{25}	506–509	521	v s aq; s alc; 19 pyr
chloride	$BeCl_2$	79.92	1.899^{25}	399	482	42 aq; s alc, eth, CS_2, pyr; i bz
fluoride	BeF_2	47.01	1.986_4^{25}	552	1175	v s aq but slow
hydride	BeH_2	11.03		$-H_2$, 220		d slowly aq; d rapidly a
hydroxide	$Be(OH)_2$	43.03	1.92	134 d		s hot conc a, alk
iodide	BeI_2	262.82	4.2	480	482	hyd aq; s alc, eth, CS_2
oxide	BeO	25.01	3.01	$2408(\alpha)$	3787	s conc H_2SO_4
sulfate 4-water	$BeSO_4 \cdot 4H_2O$	177.14	1.713^{11}	$-4H_2O$, 270	d 580	39^{20} aq; i alc

TABLE 2.1 Physical Constants of Inorganic Compounds (*continued*)

Name	Formula	Formula weight	Density	Melting point, °C	Boiling point, °C	Solubility in 100 parts solvent
Bismuth						
chloride, tri-	$BiCl_3$	315.34	4.75	ca 232	447	d aq; s HCl, alc, eth, acet
fluoride, penta-	BiF_5	303.98	5.4^{25}	151	230	d viol aq giving O_3
hydroxide	$Bi(OH)_3$	260.00	4.36	$-H_2O$, 100		d aq; s a
(III) nitrate 5-water	$Bi(NO_3)_3 \cdot 5H_2O$	485.07	2.83	d 30		d aq; s, acet
(III) oxide	Bi_2O_3	495.96	8.76	817	1890	i aq; s a
Boron						
bromide, tri-	BBr_3	250.57	2.695^{0}	-46.0	91.3	d aq
chloride, tri-	BCl_3	117.19	1.354^{12}	-107	18	d aq, alc
fluoride, tri-	BF_3	67.81	$2.99 \text{ g} \cdot \text{L}^{-1}$	-127.1	-100.4	105^{0} mL aq; s bz, chl, CCl_4
fluoride-1-diethyl ether	$BF_3 \cdot O(C_2H_5)_2$	141.94	1.125^{25}_4	-60.4	125.7	d aq
fluoride-1-methanol	$BF_3 \cdot CH_3OH$	131.89	1.203		$59^{4\,mm}$	
oxide	B_2O_3	69.62	2.46	450	2065	sl s aq
Bromine						
	Br_2	159.81	3.1028	-7.3	58.75	3.6^{20} aq; v s alc, chl, eth, CS_2
fluoride, tri-	BrF_3	136.90	2.803^{25}	8.77	125.74	d viol aq; d alk
Cadmium						
acetate	$Cd(C_2H_3O_2)_2$	230.50	2.341	256	d	v s aq
chloride	$CdCl_2$	183.32	4.047	568	961	120^{25} aq
iodide	CdI_2	366.21	5.670^{30}	387	796	85^{20} aq; s alc, acet, eth
oxide	CdO	128.40	8.15	subl 1497		i aq; s a
sulfate-water (3/8)	$3CdSO_4 \cdot 8H_2O$	769.56	3.09	$-H_2O$, 40	forms mono-hydrate 80	94.4^{25}; i alc
sulfide	CdS	144.46	4.82 hex		sub 1380 (in N_2)	3.13^{18} aq; s a
Calcium						
acetate	$Ca(C_2H_3O_2)_2$	158.17	d > 160			37^{0} aq; i alc, acet, bz
arsenate(V)	$Ca_3(AsO_4)_2$	398.08	3.620			0.013^{25} aq

bromide	$CaBr_2$	199.90	3.353	765	806–812	143_{20} aq; v s alc, acet
carbide, di-	CaC_2	64.10	2.22		2300	d aq giving C_2H_2
carbonate	$CaCO_3$	100.09	2.930	d 900		0.0013^{20}; s a
chlorate	$Ca(ClO_3)_2$	206.99		340		178 aq; s alc, acet
chloride	$CaCl_2$	110.99	2.15	772	1940	75^{20}; s alc, acet
chloride 6-water	$CaCl_2 \cdot 6H_2O$	219.08	1.71	$-6H_2O$, 200		536^{20} aq; s alc
citrate 4-water	$Ca(C_6H_6O_7) \cdot 4H_2O$	570.51		$-4H_2O$, 120		0.85^{18} aq; 0.0065^{18} alc
cyanamide	$CaCN_2$	80.11	2.29_4^{20}	1340		i aq; no known solv
cyanide	$Ca(CN)_2$	92.12		d 350		d aq
diphosphate	$Ca_2P_2O_7$	254.10	3.09	1230		i aq; s a
fluoride	CaF_2	78.08	3.180	1418	2510	0.002^{20} aq; sl s a
formate	$Ca(OOCH)_2$	130.12	2.015	d		16.6^{20} aq; i alc
glycerophosphate	$Ca[C_3H_5(OH)_2]PO_4$	210.16		d > 170		1.7^{20} aq, i alc
hydrogen phosphate, di-1-water	$Ca(H_2PO_4)_2 \cdot H_2O$	252.07	2.220_4^{18}	$-H_2O$, 109	d 203	1.8^{30} aq
hydroxide	$Ca(OH)_2$	74.09	2.24	$-H_2O$, 522		0.17^{10} aq; s a
hypochlorite	$Ca(OCl)_2$	142.99	2.35	100 d		d aq evolving Cl_2; i alc
iodate 6-water	$Ca(IO_3)_2 \cdot 6H_2O$	497.98		d 35		0.24^{20} aq; i alc
lactate 5-water	$Ca(C_3H_5O_3)_2 \cdot 5H_2O$	308.30		$-3H_2O$, 100	$-5H_2O$, 120	5.4^{15} aq; v sl s alc
nitrate	$Ca(NO_3)_2$	164.09	2.504^{18}	561		152^{30} aq
nitrite 4-water	$Ca(NO_2)_2 \cdot 4H_2O$	204.5	1.674_0^0	$-2H_2O$, 44		84.5^{18} aq; sl s alc
oleate	$Ca(C_{18}H_{33}O_2)_2$	603.01		83–84	d 140	0.04 aq; s bz, chl; v sl s alc
oxide	CaO	56.08	3.25	2927	3500	0.13^{25} aq; s a
palmitate	$Ca(C_{16}H_{31}O_2)_2$	550.93		d 155		0.003 aq; sl s bz, chl; i alc, eth
pantothenate (vitamin B_3)	$Ca[O_2CH_2CH_2CH_2HO\text{-}CH(OH)C(CH_3)_2\text{-}CH_2OH]_2$	476.55		d 195–196		35 aq; sl s alc, acet
peroxide	CaO_2	72.08	2.92_4^{25}	d 275		sl s aq; s a
phenoxide	$Ca(OC_6H_5)_2$	226.28				sl s aq, alc
phosphate	$Ca_3(PO_4)_2$	310.18	3.14	1730		0.03^{25}, s a; i alc
salicylate 2-water	$Ca(C_7H_5O_3)_2 \cdot 2H_2O$	350.34		$-2H_2O$, 120	d 240	2.8^{15} aq; 0.015^{16} EtOH
selenate 2-water	$CaSeO_4 \cdot 2H_2O$	219.07	2.68_4^{20}	$-2H_2O$, 200	d 698	9.2^{25} aq
stearate	$Ca(C_{18}H_{35}O_2)_2$	607.04		179–180		0.004^{15} aq; s hot pyr; i chl, eth

TABLE 2.1 Physical Constants of Inorganic Compounds (*continued*)

Name	Formula	Formula weight	Density	Melting point, °C	Boiling point, °C	Solubility in 100 parts solvent
Calcium						
succinate 3-water	$CaC_4H_6O_4 \cdot 3H_2O$	212.22				1.28^{20} aq; s a; i alc
sulfate	$CaSO_4$	136.14	2.960	1400		0.20 aq; s a
sulfate hemihydrate	$CaSO_4 \cdot 0.5H_2O$	145.15		$-H_2O$, 163		0.3^{20} aq; s a, glyc
sulfate 2-water	$CaSO_4 \cdot 2H_2O$	172.17	2.32	$-2H_2O$, 163		0.26^{20} aq; s a, glyc
sulfite 2-water	$CaSO_3 \cdot 2H_2O$	156.17		$-2H_2O$, 100		0.004 aq; s a; sl s alc
DL-tartrate 4-water	$CaC_4H_4O_6 \cdot 4H_2O$	260.21		$-4H_2O$, 200		0.0045^{25} aq; sl s alc
tetrahydridoaluminate	$Ca(AlH_4)_2$	102.10		d 160		ign moist air; d viol aq, alc
thiocyanate 3-water	$Ca(SCN)_2 \cdot 3H_2O$	210.29				150 aq; v s alc
Carbon						
(graphite)	C	12.01	2.25^{20}	$4000^{63.5\ atm}$	3930	i aq; alc
bromide, tetra-	CBr_4	331.65	3.42	90.1	190	i aq; s alc, chl, eth
chloride, tetra-	CCl_4	153.82	1.5867^{20}_{20}	-22.9	76.7	i aq; s alc, chl, eth
hydride, tetra-	CH_4	16.04	0.415^{-164}	-182.48	-161.49	i aq; s bz
iodide, tetra-	CI_4	519.63	4.34	d 171		sl hyd aq; s alc, bz, eth
oxide, mono-	CO	28.01	0793 (lq) 1.250 $g \cdot L^{-1}$ (gas)	-205.05	-191.49	2.1 mL aq; s alc, bz
oxide di-	CO_2	44.01	1.56^{-79} (c) 1.975 $g \cdot L^{-1}$	-56.2 solid subl	-78.44	31^{15} mL aq
(tri-) oxide, di-	C_3O_2	68.03	1.114^0_4	-112.19	6.4	d aq to malonic acid
selenide, di-	CSe_2	169.93	2.663^{25}_4	-43	125.1	i aq; d alc, pyr; misc CCl_4; s acet, eth
sulfide, di-	CS_2	76.14	1.261^{22}	-111.6	46.26	0.29^{20} aq; s alc, eth

			$g \cdot L^{-1}$			known in soln only
Carbonic acid	$H_2CO_3(CO_2 + H_2O)$	62.03				
Carbonyl						
chloride	$COCl_2$	98.92	1.392	-127.8	7.6	hyd aq; s bz
fluoride	COF_2	66.01	1.139^{-114}	-114.0	-83.3	hyd aq
sulfide	COS	60.07	1.073^0	-138.81	-50.23	54[20] mL aq; s alc, CS_2
Cerium						
(III) chloride	$CeCl_3$	246.48	3.92	8.7	1730	100[20] aq; 30 alc; s acet
(IV) fluoride	CeF_4	216.12	4.80	>650	d>550	i aq; s a
(IV) oxide	CeO_2	172.13				i aq; s a
(IV) sulfate	$Ce(SO_4)_2$	332.24	3.91	d 195		hyd aq; s H_2SO_4
Cesium						
bromide	$CsBr$	212.81	4.44	635	1300	107[18] aq
carbonate	Cs_2CO_3	325.82		d 610		260[15] aq; 11[20] alc; s eth
chloride	$CsCl$	168.36	3.988	645	1324	187[20] aq; 34[25] MeOH; v s alc
fluoride	CsF	151.90	4.115	703	1231	322[18] aq
hydroxide	$CsOH$	149.91	3.675	272	990	386[15] aq; s alc
iodide	CsI	259.81	4.510	621	ca 1280	77[20] aq; s EtOH; i acet
nitrate	$CsNO_3$	194.91	3.685_4^{20}	414	d 849	23[20] aq; s acet; v sl s alc
oxalate	$Cs_2C_2O_4$	353.82	3.230^{15}			313 aq
selenate	Cs_2SeO_4	408.77	4.4528_4^{20}			244[12] aq
sulfate	Cs_2SO_4	361.87	4.243	1019		179[20] aq; i alc, acet, pyr
Chlorine						
fluoride, tri-	ClF_3	92.45	1.825^{11}	-76.28	11.74	hyd viol aq; glass wool and org matter ign
(di-) oxide	Cl_2O	86.91	3.02^2	-120.6	2.1	3.5[20] (hyd to HClO); s CCl_4
oxide, di-	ClO_2	67.46	1.642^0	-59.6	10.9	11.2[10] aq
(di-) oxide, hepta-	Cl_2O_7	182.90	1.805^{25}	-91.5	83.6	d aq; expl on concussion or contact with flame or I_2
Chlorosulfonic acid	HSO_3Cl	116.52	1.753_4^{20}	-80	158	d viol aq to $HCl + H_2SO_4$
Chromium						
(II) acetate	$Cr(C_2H_3O_2)_2$	170.10	1.77^{18}			sl s aq, alc; i eth
carbonyl, hexa-	$Cr(CO)_6$	220.06		d 130	expl 210	i aq, alc, eth

TABLE 2.1 Physical Constants of Inorganic Compounds (*continued*)

Name	Formula	Formula weight	Density	Melting point, °C	Boiling point, °C	Solubility in 100 parts solvent
Chromium						
(II) chloride	$CrCl_2$	122.90	2.878	815	1300	v s aq
(III) chloride	$CrCl_3$	158.35	2.76^{15}	877	subl 947	i aq, alc, acet, eth
(III) fluoride	CrF_3	108.99	3.8	1100	subl	i aq, alc; s HF
(III) nitrate 9-water	$Cr(NO_3)_3 \cdot 9H_2O$	400.15		60	d 100	208^{15} aq; s alc
(III) oxide	Cr_2O_3	152.02	5.21	2330	3000	i aq, alc
(IV) oxide	CrO_2	83.99	4.89	$-O_2$, 300		i aq; s HNO_3
(VI) oxide	CrO_3	99.99	2.70	198	d 250	167^{20} aq; may ign org materials
(III) phosphate 6-water	$CrPO_4 \cdot 6H_2O$	255.06	2.121^{14}	100		i aq; v s a, alk; sl s HOAc
(III) sulfate 7-water	$CrSO_4 \cdot 7H_2O$	274.17				23^0 aq
(III) sulfate 18-water	$Cr_2(SO_4)_3 \cdot 18H_2O$	716.45	1.7	d 100		220^{20} aq
Chromyl						
chloride	CrO_2Cl_2	154.90	1.92	−96.5	117	d aq; s eth
Cobalt						
(II) acetate 4-water	$Co(C_2H_3O_2)_2 \cdot 4H_2O$	249.08	1.705^{19}	$-4H_2O$, 140		s aq; 2.1^{15} MeOH
(III) acetate	$Co(C_2H_3O_2)_3$	236.07		d 100		s aq, alc, HOAc
(II) bromide	$CoBr_2$	218.75	4.909^{25}_4	678 (in N_2)		112^{20} aq
(II) carbonate	$CoCO_3$	118.94	4.13	d		0.18^{15} aq; s a
(II) chloride	$CoCl_2$	129.84	3.356	740	1087	53^{20} aq
(II) fluoride	CoF_2	96.93	4.46	1127	1739	1.36^{20} aq; s a
(III) fluoride	CoF_3	115.93	3.88	d		d aq; i alc, bz, eth
(II) hydroxide	$Co(OH)_2$	92.95	3.597^{15}_4	d		0.0018 aq; s a
(II) iodide	CoI_2	312.74	5.68	505 d	570 (vac)	203 aq
(II) nitrate 6-water	$Co(NO_3)_2 \cdot 6H_2O$	291.04	1.87	55	d 74	155^{30} aq; v s alc
(II) oxalate	CoC_2O_4	146.95	3.021^{25}_4	d 250		0.002^{18} aq; s a
(II) oxide	CoO	74.93	6.45	1805		i aq; s a

(II,III) oxide	Co_3O_4	240.80	6.07	d 900		i aq; v sl s a
(II) sulfate 7-water	$CoSO_4 \cdot 7H_2O$	281.10	2.03^{25}_4	96.8	$-7H_2O$, 420	65^{20} aq; sl s alc
Copper						
(II) acetate hydrate	$Cu(C_2H_3O_2)_2 \cdot H_2O$	199.65	1.882	115	d 240	8 aq; 0.48 MeOH; sl s eth, glyc
(II) acetate–metaarsenite	$Cu(C_2H_3O_2)_2 \cdot 3Cu(AsO_2)_2$	1013.77				i aq; s a, NH_4OH
(I) bromide	$CuBr$	143.45	4.98	488	1318	v sl s aq; s a
(II) bromide	$CuBr_2$	223.31	4.710^{20}_4	498	d 100	126 aq; s alc, acet, pyr; i bz, eth
(II) chlorate 6-water	$Cu(ClO_3)_2 \cdot 6H_2O$	338.53		65		242^{18} aq; v s alc; s acet
(I) chloride	$CuCl$	98.99	4.14	430	1212	0.024 aq; s HCl
(II) chloride	$CuCl_2$	134.44	3.386^{25}_4	d 300		73^{20} aq; s alc, acet
(II) chloride 2-water	$CuCl_2 \cdot 2H_2O$	170.47	2.54	$-2H_2O$, 100	d	76^{25} aq; v s alc; s acet
(I) cyanide	$CuCN$	89.56	2.92	473 (in N_2)	d	0.00026 aq; s HCl, KCN
(II) fluoride	CuF_2	101.54	4.23	770	1449	0.075 aq; s a
formate	$Cu(OOCH)_2$	153.55	1.831			12.5 aq
hydroxide	$Cu(OH)_2$	97.55	3.368	160		i aq; s a
(I) iodide	CuI	190.44	5.62	588	1207	i aq; s HCl, KI
(II) nitrate 3-water	$Cu(NO_3)_2 \cdot 3H_2O$	241.60	2.05	114.5	d 170	138^0 aq; v s alc
(II) oleate	$Cu(OOCC_{17}H_{33})_2$	626.43				i aq; sl s alc; s eth
(I) oxide	Cu_2O	143.08	6.0	1236	$-O_2$, 1800	i aq; s HCl
(II) oxide	CuO	79.54	6.315^{14}	d 1122		i aq, alc; s a
(II) perchlorate	$Cu(ClO_4)_2$	262.43	2.225^{23}	82.3	d 130	146^{30} aq; s eth; i bz, CCl_4
(II) stearate	$Cu(OOCC_{17}H_{35})_2$	630.46		ca 250		i aq, alc, eth; s pyr, hot bz
(II) sulfate	$CuSO_4$	159.61	3.603	805 d		14.3^0 aq
(II) sulfate 5-water	$CuSO_4 \cdot 5H_2O$	249.68	2.284^{16}_4	$-5H_2O$, 150	-21.15	32^{20} aq; s MeOH, glyc; sl s EtOH
Cyanogen	$NC—CN$	52.04	$2.335\ \text{g·L}^{-1}$	-27.84		420^{20} mL aq; 230 mL alc
azide	$NC—N_3$	68.04		detonates		s acetonitrile; can be handled safely only in solv
bromide	$CNBr$	105.93	2.015^{20}_4	51.4	61.35	v s aq, alc, eth
chloride	$CNCl$	61.48	1.186	-6.90	13.0	s aq, alc, eth
iodide	CNI	152.92		146–147	subl 140	s aq, alc, eth
Deuterium	D_2 or $^2H_2$	4.03	0.169^{mp} (lq)	-252.89	-248.24	sl s aq
oxide	D_2O or ^2H_2O	20.03	1.1056^{20}	3.82	101.43	misc aq

TABLE 2.1 Physical Constants of Inorganic Compounds (*continued*)

Name	Formula	Formula weight	Density	Melting point, °C	Boiling point, °C	Solubility in 100 parts solvent
Disulfuryl dichloride	$S_2O_5Cl_2$	215.03	1.818^{11}	−37.5	152.5	d aq, a
Diphosphoric(V) acid	$H_4P_2O_7$	117.98		61		s aq
Fluorine	F_2	38.00	1.554^{25} g·L⁻¹	−219.70	−188.20	d aq viol
Fluoroboric acid	HBF_4	87.81		d 130		v s aq
Fluorosulfonic acid	HSO_3F	100.07	1.743^{15}	−87.3	165.5	s aq
-d or [²H]	DSO_3F or $^2HSO_3F$	101.08	1.523^{-142}	−89	163	s aq
Germane	GeH_4	76.62		−165.9	−88.5	sl s hot HCl
Gold						
(III) chloride	$AuCl_3$	303.33	3.9	254 d	subl 265	68^{20} aq
Helium	He	4.00	0.1784^0 g · L⁻¹ 0.1249 (lq)	$-272.2^{25\,\text{atm}}$	−268.935	0.861^{20} mL aq
Hydrazine	H_2NNH_2	32.05	1.0083^{20}	1.54	113.8	misc aq, alc
hydrate	$H_2NNH_2 \cdot H_2O$	50.16	1.038^{21}	−51.7	119.4	misc aq, alc
Hydrazinium						
(1+) chloride	H_2NNH_3Cl	68.51	1.4226^{20}	92.6	d 240	v s aq
(2+) chloride	ClH_3NH_3Cl	104.97	1.378	198	d 200	v s aq; sl s alc
(2+) sulfate	$(H_3NNH_3)SO_4$	130.13		254	d	3.4^{20} aq; i alc
Hydrogen	H_2	2.02	0.0899 g·L⁻¹	−259.76	−252.76	1.9 mL aq
azide	HN_3	43.03	1.126	−80	37	v s aq (v expl)
borate (1−)	HBO_2	43.83	2.486	236		v sl s aq
borate(3−), ortho-	H_3BO_3	61.83	1.435^{15}	171.0	d 300	6.4^{30} aq
bromide	HBr	80.92	3.388^{20} g·L⁻¹ 2.160^{-66} (lq)	−86.81	−66.71	193^{25} aq; s alc
bromide	48% HBr + H_2O		1.49 g·L⁻¹	−11	126	v s aq (constant boiling)
bromide-*d*	DBr or ^2HBr	81.92	3.39^{20} g·L⁻¹	−87.46	−66.5	v s aq

Name	Formula	Formula wt	Density	Melting point, °C	Boiling point, °C	Solubility
chloride	HCl	36.46	1.526^{20} g·L^{-1}; 1.187^{-85} (lq)	−114.18	−85.00	72^{20} aq
chloride	20.24% HCl + H$_2$O		1.097		110	v s aq (constant boiling)
cyanide	HCN	27.06		−13.24	25.70	v s aq
fluoride	HF	20.01	0.901 g·L^{-1}; 1.2675^{10} (lq)	−83.57	19.52	v s aq
fluoride	35.35% HF + H$_2$O		0.922^{0} g·L^{-1}; 0.957^{19} (lq)		120	v s aq (constant boiling)
iodide	HI	127.92	5.37^{20} g·L^{-1}; 2.799^{-35} (lq)	−50.79	−35.35	70^{0} aq
iodide	57% HI + H$_2$O		1.70^{15}		127	v s aq (constant boiling)
nitrate	HNO$_3$	63.02	1.5027	−41.59	83	v s aq
nitrate	69% HNO$_3$ + H$_2$O		1.41^{20}		120.5	misc aq (constant boiling)
oxide	H$_2$O	18.02	1.000^{4}	0.00	100.00	misc aq
oxide-d_2	D$_2$O or ^{2}H$_2$O	20.03	1.1045	3.82	101.43	v s aq (commercial 72% a)
perchlorate 2-water	HClO$_4$·2H$_2$O	136.49	1.67^{20}	−17.8	203	440^{25} aq
periodate(1−)	HIO$_4$	191.91		subl 110	d 138	113 aq
periodate(5−)	H$_5$IO$_6$	227.94		130	d 140	misc aq; s alc, eth
peroxide	H$_2$O$_2$	34.02	1.4649^{0}	−0.40	151.2	s aq
phosphate(V) (1−)	HPO$_3$	79.98	2.2–2.5	subl	d 213	v s aq (commercial 85% a)
phosphate(V) (3−)	H$_3$PO$_4$	98.00	1.88	42.3		26^{17} mL aq; s alc, eth
phosphide	PH$_3$	34.00	1.529	−133.81	−87.78	
selenide	H$_2$Se	80.98	2.12^{-42} g·L^{-1}	−65.73	−42	9.5^{20} mL aq
sulfide	H$_2$S	34.08	1.1906 g·L^{-1}	−85.52	−60.33	0.334^{25} mL aq
telluride	H$_2$Te	129.63	6.234 g·L^{-1}	−49	−2	d aq
tungstate(VI) (2−)	H$_2$WO$_4$	249.86	5.5	−H$_2$O, 100	$58^{22\,mm}$	i aq; s alk, HF
Hydroxylamine	HONH$_2$	33.03	1.332	33.1		s aq, alc
Hydroxylammonium						
chloride	HONH$_3$Cl	69.49	1.680^{20}	150.5	d	83^{17} aq; 4.4^{20} alc
sulfate	(HONH$_3$)$_2$SO$_4$	164.14		d 170		69^{20} aq

TABLE 2.1 Physical Constants of Inorganic Compounds (*continued*)

Name	Formula	Formula weight	Density	Melting point, °C	Boiling point, °C	Solubility in 100 parts solvent
Iodic acid	HIO_3	175.91	4.629^0	d 110 to H_5IO_6	d 195 to I_2O_5	310^{16} aq
Iodine	I_2	53.82	4.660^{20}	113.60	184.24	0.029^{20} aq; s alc, bz, chl, CS_2, CCl_4, eth
bromide	IBr	206.81	4.4157^0	42	116.d	s aq, alc, eth
chloride	ICl	162.36	3.20	27.38	97.8	d aq; s alc, eth
chloride, tri-	ICl_3	233.26	3.202	101 d	102	d aq; s alc, bz, eth
fluoride, penta-	IF_5	221.90	3.252	8.5	102	d aq
fluoride, hepta-	IF_7	259.89	2.8^6 g $\cdot$ L^{-1}	4.5	5.5	d aq
(di-) oxide, penta-	I_2O_5	333.81	4.799^{25}	d 275		187^{13} aq
Iron	Fe	55.85	7.86	1537	2872	i aq; s a
(II) bromide	$FeBr_2$	215.67	4.636	691	934	117^{20} aq
(III) bromide	$FeBr_3$	295.57		subl		s aq
carbonyl, penta-	$Fe(CO)_5$	195.00	1.49	−21	103	i aq; s alc, bz, eth
(II) chloride	$FeCl_2$	126.75	3.16^{25}	677	1024	63^{20} aq; v s alc, acet;i eth
(III) chloride	$FeCl_3$	162.21	2.898	304	332	74^0 aq
(III) ferrate(II), hexacyano-	$Fe_4[Fe(CN)_6]_3$	859.25	1.80	d		i aq; s HCl
(II) fluoride	FeF_2	93.84	4.09	1100	1837	sl s aq; s a
(III) fluoride	FeF_3	112.84	3.87	subl 927		0.091^{25} aq; s a; i alc, bz
(II) iodide	FeI_2	309.66	5.315	587	1093	s aq
(III) nitrate 9-water	$Fe(NO_3)_3 \cdot 9H_2O$	404.02	1.684^{21}	47	d 100	138^{20}
(II) oxalate 2-water	$FeC_2O_4 \cdot 2H_2O$	179.90	2.28	d 150–160		0.044^{18} aq; s a
(II) oxide	FeO	71.85	5.7	1377		i aq; s a
(III) oxide	Fe_2O_3	159.69	5.24	1462 d		i aq; s HCl
(II, III) oxide	Fe_3O_4	231.54	5.1	1597		i aq; s a
(II) sulfate 7-water	$FeSO_4 \cdot 7H_2O$	278.04	1.89		d 3414	48^{20} aq

			g · L^{-1}			
(III) sulfate	Fe$_2$(SO$_4$)$_3$	399.88	3.097^{18}	d 1178		sl s aq (hyd); sl s alc
(III) sulfate 9-water	Fe$_2$(SO$_4$)$_3$·9H$_2$O	562.01	2.1	d 175		440 aq
Krypton	Kr	83.80	3.736	−157.2	−153.4	5.94^{20} mL aq
Lead						
Pb	Pb	207.21	11.34 (fcc)	327.50	1753	i aq; s HNO$_3$
(II) acetate 3-water	Pb(C$_2$H$_3$O$_2$)$_2$·3H$_2$O	379.33	2.55	d 200		46^{15} aq
(IV) acetate	Pb(C$_2$H$_3$O$_2$)$_4$	443.37	2.228^{17}	175		d aq; s chl
(II) azide	Pb(N$_3$)$_2$	291.23		expl 350		0.023^{18} aq; s HOAc
(II) carbonate	PbCO$_3$	267.20	6.6	d 340		i aq; s a, alk
(II) chromate(VI) (2−)	PbCrO$_4$	323.18	6.12^{15}	844	d	i aq; s a
(IV) fluoride	PbF$_4$	283.21	6.7	600		hyd aq
(II) nitrate	Pb(NO$_3$)$_2$	331.23	4.53^{20}	d 200		56^{20} aq; 1.3 MeOH
(II) oleate	Pb(C$_{18}$H$_{33}$O$_2$)$_2$	770.12				i aq; s alc, bz, eth
(II) oxide	PbO	223.21	9.53	886	1516	0.0017^{20}, s HNO$_3$
(IV) oxide	PbO$_2$	239.21	9.375	d 752		i aq; s HCl
(II) phosphate	Pb$_3$(PO$_4$)$_2$	811.59	6.9	1014		i aq; s HNO$_3$, alk
(II) stearate	Pb(C$_{18}$H$_{35}$O$_2$)$_2$	774.15		ca 125		0.05^{35} aq; s hot alc
(II) sulfate	PbSO$_4$	303.28	6.2	1090		0.004 aq
Lithium	Li	6.94	0.535^{20}	180.6	1340	d aq to LiOH
aluminate, tetrahydrido-	LiAlH$_4$	37.95	0.917	d 125	d 430	d aq; alc; 30 eth (flammable)
amide	LiNH$_2$	22.96	1.178^{18}	374		d aq; i bz, eth
benzoate	LiC$_7$H$_5$O$_2$	128.05		>300		33 aq; 7.7 alc
boronate	LiBH$_4$	21.79	0.666	268	d 380	d aq; s eth, THF
bromate	LiBrO$_3$	134.85	3.62			179^{20} aq
bromide	LiBr	86.84	3.464	550	1289	164 aq; s alc, eth
carbonate	Li$_2$CO$_3$	73.89	2.11^{0}	720	d	1.3^{20} aq; i alc; s a
chloride	LiCl	42.40	2.068	610	1383	77^{20} aq; s alc, acet
fluoride	LiF	25.94	2.640^{20}	846	1717	0.13^{25} aq; s a
hydride	LiH	7.95	0.780	688.7	d 950	d aq; no known solv (flammable)
hydroxide	LiOH	23.95	2.54	471.2	1626	12.4^{20} aq
iodide	LiI	133.84	4.061	467	1178	165^{20} aq; v s alc
iodide 3-water	LiI·3H$_2$O	187.89	3.5	73	−3H$_2$O, 300	200 aq; 200 alc
nitrate	LiNO$_3$	68.94	2.38	261		70^{20} aq; s alc

TABLE 2.1 Physical Constants of Inorganic Compounds (*continued*)

Name	Formula	Formula weight	Density	Melting point, °C	Boiling point, °C	Solubility in 100 parts solvent
Lithium						
perchlorate	$LiClO_4$	106.40	2.43^{25}	236	d 400	56^{20} aq
sulfate	Li_2SO_4	109.88	2.22	860	1105	34.5^{20} aq
Magnesium						
	Mg	24.31	1.74^{20}	650		i aq; s a
amide	$Mg(NH_2)_2$	56.37	1.39^{25}_4	ign in air		d viol aq giving NH_3
bromide	$MgBr_2$	184.13	3.72	711	1158	101^{20} aq
bromide 6-water	$MgBr_2 \cdot 6H_2O$	292.22	2.00	165 d		160^{20} aq; s alc
carbonate	$MgCO_3$	84.32	2.958	d 402		0.01 aq; s a
chloride	$MgCl_2$	95.23	2.41	714	1437	54.6^{20} aq
hydride	MgH_2	26.34	1.45	d 287 (vac)	ign air	d viol aq, alc
hydroxide	$Mg(OH)_2$	58.33	2.36	268 d		i aq; s a
oleate	$Mg(C_{18}H_{33}O_2)_2$	293.61				i aq; s alc, eth, PE
oxide	MgO	40.52	3.58	2825	3260	i aq; s a
perchlorate	$Mg(ClO_4)_2$	223.23	2.21^{20}	d 251		49.6 aq
sulfate 7-water	$MgSO_4 \cdot 7H_2O$	246.49	1.67	$-6H_2O$, 120	$-7H_2O$, 250	27.2 aq; s alc
sulfite 6-water	$MgSO_3 \cdot 6H_2O$	212.47	1.725	$-6H_2O$, 200	d	66^{25} aq
Manganese						
acetate 4-water	$Mn(C_2H_3O_2)_2 \cdot 4H_2O$	245.08	1.589	54d		38^{50} aq; s alc
bromide 4-water	$MnBr_2 \cdot 4H_2O$	286.82		d		200 aq; s alc
carbonate	$MnCO_3$	114.94	3.125			0.0065^{25} aq; s a
(di-) carbonyl, deca-	$Mn_2(CO)_{10}$	389.99	1.75^{25}	155 (CO atm)	d 110	i aq; s org solv
chloride 4-water	$MnCl_2 \cdot 4H_2O$	197.91	2.01	$-4H_2O$, 198		143 aq; s alc; i eth
(III) fluoride	MnF_3	111.93	3.54	d 600		hyd aq; s a
nitrate 6-water	$Mn(NO_3)_2 \cdot 6H_2O$	287.05	1.8	25.8		v s aq, alc
(IV) oxide	MnO_2	86.94	5.026	d 530		i aq; s HCl
sulfate hydrate	$MnSO_4 \cdot H_2O$	169.01	2.95	$-H_2O$, 400		70^{20} aq

Name	Formula		Density	mp	bp	Solubility
Mercury	Hg	200.59	13.594^{20}	-38.86	356.60	i aq; s HNO_3
(II) acetate	$Hg(C_2H_3O_2)_2$	318.70	3.28	178	subl > 241	25^{10} aq; 7.5^{15} MeOH
(II) bromide	$HgBr_2$	360.44	6.05	241	d	0.56^{20} aq; 20^{25} alc
(I) chloride	Hg_2Cl_2	472.09	7.150	subl 382		0.00027 aq; s aqua regia
(II) chloride	$HgCl_2$	271.52	5.44	277	304	6.6^{20} aq; 33 alc; 4 eth
(II) cyanide	$Hg(CN)_2$	252.65	3.996	d 320		9.3^{20} aq; 8 alc; 25 MeOH
(II) fluoride	HgF_2	238.61	8.95^{15}	645	647	hyd aq; s HF
(II) iodide	HgI_2	454.45	6.28	259	350	0.006^{25} aq; 1 alc; 1.7 acet
(II) nitrate	$Hg(NO_3)_2$	324.63	4.3	79	d	v s aq; s acet
(II) oxide	HgO	216.61	11.14	d 476		0.005^{25} aq; s a
(I) sulfate	Hg_2SO_4	497.29	7.56	d		0.06^{25} aq; s HNO_3
(II) sulfate	$HgSO_4$	296.68	6.47	d		d aq; s a
(II) sulfide, red	HgS	232.68	8.10	subl 583		i aq; s aqua regia
Molybdenum						
carbonyl, hexa-	$Mo(CO)_6$	264.02	1.96	subl 102	156.4	s bz
(V) chloride	$MoCl_5$	273.21	2.928	194	264	hyd aq; s conc a
(VI) oxide	MoO_3	143.95	4.696^{26}	801	1155	0.22^{28} aq; s alk, NH_3
sulfide, di-	MoS_2	160.08	5.06^{15}	2375	subl 450	i aq; s aqua regia
Molybdic acid hydrate	$H_2MoO_4 \cdot H_2O$	179.97	3.124^{15}	$-H_2O$, 70		0.133^{18}; s alk
Molybdic phosphoric acid	$H_3[P(Mo_2O_7)_6] \cdot 28H_2O$	2365.71	2.53	78		hyd aq
Neon	Ne	20.18	0.8899^0 g·L^{-1}	-248.6	-246.1	1.05^{20} mL aq
Nickel	Ni	58.71	8.90	1455	2920	i aq; s HNO_3
acetylacetonate	$Ni(C_5H_7O_2)_2$	256.93	1.455^{17}	229	235	s aq, alc, bz, chl
bromide	$NiBr_2$	218.53	5.098	963	subl	131^{20} aq
chloride 6-water	$NiCl_2 \cdot 6H_2O$	237.70				111^{20} aq
dimethylglyoxime	$Ni(HC_4H_6N_2O_2)_2$	288.91		subl 250		i aq; s abs alc, a
formate 2-water	$Ni(OOCH)_2 \cdot 2H_2O$	184.78	2.154^{20}	$-2H_2O$, 130	d 180	s aq; i alc
nitrate 6-water	$Ni(NO_3)_2 \cdot 6H_2O$	290.81	2.05	56.7		150^{20} aq
sulfate 6-water	$NiSO_4 \cdot 6H_2O$	262.86	2.07	53.3	136.7	40^{20} aq

TABLE 2.1 Physical Constants of Inorganic Compounds (*continued*)

Name	Formula	Formula weight	Density	Melting point, °C	Boiling point, °C	Solubility in 100 parts solvent
Niobium						
(V) chloride	$NbCl_5$	270.20	2.75	204	250	s HCl, CCl_4
(V) fluoride	NbF_5	187.91	2.70_4^{80}	80	235	hyd aq, alc
(V) oxide	Nb_2O_5	265.82	4.6	1512		i aq; s HF, hot H_2SO_4
Nitrogen	N_2	28.01	1.165^{20} $g \cdot L^{-1}$	-210.00	-195.81	1.52^{20} mL aq
[^{15}N]	$^{15}N_2$	30.01	1.25^{20} $g \cdot L^{-1}$	-209.95	-195.73	
chloride, tri-	NCl_3	120.37	1.653^{20}	-27	71	i aq; s bz, CS_2, CCl_4
(di-) oxide	N_2O	44.02	1.8433^{20} $g \cdot L^{-1}$	-90.85	-88.47	130^0 mL aq; s alc
oxide	NO	30.01	1.2488^{20} $g \cdot L^{-1}$	-163.64	-151.76	7^0 mL aq
(di-) oxide, tetra-	N_2O_4	92.02	1.447_4^{20}	-9.3	21.10 d	d aq; s HNO_3, H_2SO_4, chl
(di-) oxide, penta-	N_2O_5	108.01	2.05^{15}	30	47.0	s aq, chl
Nitrosyl						
chloride	$NOCl$	65.47	1.592^{-5}	-61.5	-5.5	hyd aq
fluoride	NOF	49.01	2.788^{20} $g \cdot L^{-1}$	-132.5	-59.9	hyd aq
Nitryl						
chloride	NO_2Cl	81.46	2.81^{100} $g \cdot L^{-1}$	-145	-13.5	d aq
fluoride	NO_2F	65.00	2.7^{20} $g \cdot L^{-1}$	-166.0	-72.4	d aq
Osmium oxide, tetra-	OsO_4	254.20	4.91	40.6	130.0	7.24^{25} aq; 375^{25} CCl_4
Oxygen	O_2	32.00	1.331^{20} $g \cdot L^{-1}$	-218.75	-182.96	36^{25} mL aq

			Density			
Ozone	O_3	48.00	1.998^{20} $g \cdot L^{-1}$	-192.5	-110.50	49.4^0 mL aq
Palladium	Pd	106.4	12.023	1550	2940	s hot HNO_3, H_2SO_4
acetate	$Pd(C_2H_3O_2)_2$	224.49		205d		i aq, alc; s acet, chl
chloride	$PdCl_2$	177.30	4.0^{18}	680	d 680	s aq
nitrate	$Pd(NO_3)_2$	230.42		d		hyd aq; s HNO_3
oxide	PdO	122.40	8.70^{20}	870 d		i aq, a
Perchloryl fluoride	ClO_3F	102.46	0.637	-147.74	-46.67	26^{17} mL aq; s alc, eth
Phosphine	PH_3	34.00	1.529	-133.81	-87.78	
Phosphinic acid	HPH_2O_2	66.00	1.493^{19} $g \cdot L^{-1}$	26.5	d 50	s aq
Phosphonic acid	H_2PHO_3	82.00	1.651^{21}	ca 73	d 180	v s aq, alc
Phosphoric acid						
meta-	HPO_3	79.98	2.2–2.5			slowly hyd aq; s alc
ortho-	H_3PO_4	98.00	1.88	42.35	to $H_4P_2O_7$ ca 200; to $HPO_3 > 300$	
commercial 85% acid			1.685	anhyd 150		v s aq
fluoro-	H_2PO_3F	99.99	1.818	-80		
Phosphorus (white)	P (P₄ molecules)	30.97	1.828	44.2	280.3	i aq; 0.025 alc; 1 eth; 2.5 chl, bz; 1.25 CS_2
(red)	P	30.97	2.34	597	subl 416	i aq (ign in air 260°C)
bromide, tri-	PBr_3	270.73	2.85^{15}	-40.5	173.2	d aq, alc; s acet, CS_2
bromide, penta-	PBr_5	430.56	3.46^{20}	d 100		d aq; s CCl_4, CS_2
chloride, tri-	PCl_3	137.35	1.575^{20}	-91	75	d aq, alc; s bz, chl
chloride, penta-	PCl_5	208.27	2.119^{20}	subl 100	166 d	hyd aq; s CCl_4, CS_2
fluoride, penta-	PF_5	125.98	5.805	-93.8	-84.6	hyd aq
(tetra-) oxide, hexa-	P_4O_4	219.90	2.136_4^{20} $g \cdot L^{-1}$	24	175 (in N_2)	hyd aq; s bz, CS_2
(tetra-) oxide, deca-	P_4O_{10}	283.88	2.30	340	subl 360	d aq; s H_2SO_4
(tetra-) selenide, tri-	P_4Se_3	360.80	1.31	245	360–400	hyd aq; s bz, chl, acet
(tetra-) sulfide, deca-	P_4S_{10}	444.54	2.09	288	514	hyd aq; s alk, CS_2

TABLE 2.1 Physical Constants of Inorganic Compounds (*continued*)

Name	Formula	Formula weight	Density	Melting point, °C	Boiling point, °C	Solubility in 100 parts solvent
Phosphoryl chloride, tri-	$POCl_3$	153.35	1.645^{25}	2	105	d aq, alc
Platinic(IV) acid 6-water, hexachloro-	$H_2PtCl_6 \cdot 6H_2O$	517.92	2.431	60		v s aq, alc
Platinum	Pt	195.09	21.45^{20}	1770	3824	i aq; s aqua regia, fused alk
Platinum (II) chloride	$PtCl_2$	266.00	6.05	d 581		i aq; s HCl, NH$_4$OH
(IV) oxide	PtO_2	227.09	10.2	450		i aq, aqua regia
Potassium	K	39.10	0.856^{20}	63.7	765.5	d to KOH aq; s a
acetate	$KC_2H_3O_2$	98.14	1.57^{25}	292		256^{20} aq; 34 alc
bismuthate(4—), heptaiodo-	K_4BiI_7	1253.82				d aq; s alk iodide soln
borate, tetrahydrido-	KBH_4	53.95	1.11	d 497		21^{25} aq; 3.5^{20} MeOH
bromate	$KBrO_3$	167.01	3.27^{17}	350	d 370	6.9^{20} aq
bromide	KBr	119.01	2.75	734	1398	65^{20} aq; 0.4 alc
carbonate	K_2CO_3	138.20	2.29	901	d	111^{20} aq; i alc
chlorate	$KClO_3$	122.55	2.238^{20}	368	d 368	7.3^{20} aq; 2 glyc
chloride	KCl	74.56	1.988	771	1437	34^{20} aq; 7 glyc
chromate(VI)	K_2CrO_4	194.20	2.732^{18}	975		64^{20} aq; i alc
citrate hydrate	$K_3C_6H_5O_7 \cdot H_2O$	324.42	1.98	d 230		167^{15} aq
cobaltate(III) 1.5-water, hexanitrito-	$K_3[Co(NO_2)_6] \cdot 1.5H_2O$	479.30		d 200		0.089^{17} aq; v sl s alc
cyanate	KOCN	81.11	2.048	d 700–900		s aq; sl s alc
cyanide	KCN	65.12	1.52^{16}	622	1625	50 aq
dichromate(VI)	$K_2Cr_2O_7$	294.19	2.676^{25}	398	d 500	12.3^{20} aq
disulfate(IV)	$K_2S_2O_5$	222.32				s aq (flammable if ground)

ethyldithiocarbonate	KC_2H_5OCSS	160.30	1.558[22]	d 200		v s aq
ferrate(III), hexacyano-	$K_3[Fe(CN)_6]$	329.26	1.89	d		84[20] aq (slow)
fluoride	KF	58.10	2.481	858	1517	95[20] aq
formate	KOOCH	84.10	1.91	167.5	d 168	337[20] aq
gluconate	$KC_6H_{11}O_7$	234.24		d 180		v s aq; i alc, bz, chl
hydride	KH	40.11	1.43	417 d		d aq
hydrogen arsenate, di-	KH_2AsO_4	180.02	2.867	288		19[6] aq; 63 gly; i alc
hydrogen carbonate	$KHCO_3$	100.11	2.17	d 100–200		34[20] aq
hydrogen difluoride	KHF_2	78.11	2.37	238.7	d 478	30[20] aq; s alc
hydrogen bisiodate	$KH(IO_3)_2$	389.92		d		1.3[15] aq
hydrogen oxalate	KHC_2O_4	128.11	2.044	d		2.5 aq
hydrogen bisoxalate dihydrate, tri-	$KH_3(C_2O_4)_2 \cdot 2H_2O$	254.20	1.836	d		1.8[13] aq
hydrogen phosphate	K_2HPO_4	174.18		d		150 aq
hydrogen phosphate, di-	KH_2PO_4	136.09	2.338	400	d	22.6[20] aq
hydrogen phthalate	$KHC_8H_4O_4$	204.22	1.636[25]	d		10.2 aq; sl s alc
hydrogen sulfate	$KHSO_4$	136.17	2.24	197	d	48[20] aq
hydrogen tartrate	$KHC_4H_4O_6$	188.18	1.956			0.5[20] aq
hydroxide	KOH	56.11	2.044	406	1320	112[20] aq; 33 alc
iodate	KIO_3	214.02	3.89[25]	560 d		8.1[20] aq; i alc
iodide	KI	166.02	3.12	681	1345	144[20] aq; 4,5 alc; 1.2 acet
manganate(VI)	K_2MnO_4	197.12		d 190		s aq (stable in KOH)
nitrate	KNO_3	101.10	2.109[16]	334.3	d 400	32[20] aq; 0.16 alc; s glyc
nitrite	KNO_2	85.10	1.915	441	d 250	306[20] aq
oxalate hydrate	$K_2C_2O_4 \cdot H_2O$	184.24	2.127[4]	$-H_2O$, 160		36[20] aq
periodate	KIO_4	230.01	3.618[15]	582 d	d	0.42[20] aq
permanganate	$KMnO_4$	158.03	2.703	d 240		6.34[20] aq
peroxide	K_2O_2	110.20		490		d
peroxodisulfate	$K_2S_2O_8$	270.32	2.477	d 100		5.3[20] aq
phenolsulfonate hydrate	$KC_6H_4(OH)SO_3 \cdot H_2O$	240.28	1.87			s aq, alc
phosphate	K_3PO_4	212.28	2.564[17]	1340		92[20] aq
selenocyanate	KSeCN	144.08		d 100		s aq

TABLE 2.1 Physical Constants of Inorganic Compounds (*continued*)

Name	Formula	Formula weight	Density	Melting point, °C	Boiling point, °C	Solubility in 100 parts solvent
Potassium						
silicate(2−)	K_2SiO_3	154.29	2.27	976		s aq
silicate, hexafluoro-	K_2SiF_6	220.25		d		sl s aq
sodium tartrate 4-water	$KNaC_4H_4O_6 \cdot 4H_2O$	282.23	1.790	70–80	d 220	54[15] aq
sorbate	$KC_6H_7O_2$	150.22	1.363	d 270		110[20] aq
stannate(IV) 3-water	$K_2SnO_3 \cdot 3H_2O$	298.94	3.197	−3H$_2$O, 140		100[20] aq
sulfate	K_2SO_4	174.27	2.662	1067	1670	11[20] aq; 1.3 glyc; i alc
sulfite dihydrate	$K_2SO_3 \cdot 2H_2O$	194.30		d		106[20] aq
thiocarbonate	K_2CS_2	186.41		d		v s aq
thiocyanate	KSCN	97.18	1.886[14]	173	d 500	217[20] aq; 200 acet; 8 alc acet; 8 alc
thiosulfate	$K_2S_2O_3$	190.33		d 400		155[20] aq
titanate(IV), oxobis-(oxalato)diaqua-	$K_2[TiO(C_2O_4)_2](H_2O)_2]$	354.18				v s aq
Rhenium(VII) sulfide	Re_2S_7	596.85	4.866	d 460	subl 850	i aq; s HNO$_3$
Rhodium(III) chloride	$RhCl_3$	209.28		d 450		i aq; s KOH, KCN
Rubidium						
chloride	RbCl	120.94	2.76	715	1381	91[20] aq; 1.1 MeOH
iodide	RbI	212.37	3.55	640	1304	144[18] aq
nitrate	$RbNO_3$	147.47	3.11	310		53[20] aq
sulfate	Rb_2SO_4	267.03	3.613[20]	1060		48[20] aq
Ruthenium						
(III) chloride	$RuCl_3$	207.47	3.11	d 500		i aq; s HCl, alc
(IV) oxide	RuO_2	133.07	6.97	d		i aq; s fused alk
Selenic acid	H_2SeO_4	144.98	2.9508[15]	58	260	567[20] aq (viol)
Selenium	Se	78.96	4.81[20]	221	685	s CS$_2$, KOH, KCN
(IV) oxide	SeO_2	110.96	3.954[15]	340	subl 315	38[14] aq; 10[12] MeOH

		Formula wt	Density	mp	bp	Solubility
(di-) sulfide, hexa-	Se_2S_6	350.28	2.44	121.5		i aq; 1.2 bz; s CS_2
(tetra-) sulfide, tetra-	Se_4S_4	444.08	3.20	113		i aq; 0.04 bz; s CS_2
Silane	SiH_4	32.09	0.68^{-185}	−184.7	−111.9	d aq slowly
Silicon	Si	28.09	2.33^{25}	1415	2680	s HF + HNO_3, fused alk oxides
carbide	SiC	40.07	3.217	subl 2700	d 2972	s fused alk
chloride	$SiCl_4$	169.89	1.48^{20}	−70	57.6	hyd aq; s bz, CCl_4, eth
isothiocyanate, tetra-	$Si(NCS)_4$	260.40		143.8	314.2	d aq
oxide, di- (quartz)	SiO_2	60.08	2.64–2.66	1423	2230	i aq; s HF
oxide-tungsten trioxide-water (1/12/26) (silico-tungstic acid)	$SiO_2 \cdot 12WO_3 \cdot 26H_2O$	3310.66				v s aq, alc
telluride, tri-	Si_2Te_3	438.97				
Silver	Ag	107.87	10.49^{15}	892	2164	i aq; s HNO_3
acetate	$AgC_2H_3O_2$	166.92	3.259^{15}	960.15		1.04^{20} aq
azide	AgN_3	149.89		d	297	i aq; s KCN, HNO_3 (expl)
carbonate	Ag_2CO_3	275.77	6.077	252		0.003^{30} aq
chlorate	$AgClO_3$	191.34	4.430^{20}	d 220	d 270	15.3^{20} aq
chloride	AgCl	143.34	5.56	231	1564	0.00019 aq; s NH_4OH
chromate(VI)	Ag_2CrO_4	331.77	5.625^{25}	455		0.002^{20} aq; s HNO_3, NH_4OH
cyanide	AgCN	133.90	3.95	d 320		i aq; s KCN
fluoride	AgF	126.88	5.852^{16}	435	1150	172^{20} aq
(II) fluoride	AgF_2	145.87	4.57	690	d 700	hyd viol aq
iodate	$AgIO_3$	282.80	5.525^{20}	>200	d	0.004^{20} aq
iodide	AgI	234.80	5.683^{30}	558	1505	i aq; s KCN
nitrate	$AgNO_3$	169.89	4.352^{19}	210	d 440	216^{20} aq
nitrite	$AgNO_2$	153.89	4.453	d 140		0.41^{25} aq
oxide	Ag_2O	231.76	7.22^{25}	d 200		0.002^{25} aq
(II) oxide	AgO	123.88	7.483^{25}	d 100		i aq; s alk
permanganate	$AgMnO_4$	226.81	4.49	d		0.9 aq; d alc
phosphate, ortho-	Ag_3PO_4	418.62	6.370	849		0.006 aq
sulfate	Ag_2SO_4	311.83	5.45^{30}	660	d 1085	0.80^{20} aq

TABLE 2.1 Physical Constants of Inorganic Compounds (*continued*)

Name	Formula	Formula weight	Density	Melting point, °C	Boiling point, °C	Solubility in 100 parts solvent
Sodium	Na	22.99	0.968^{20}	97.82	881.4	d aq to NaOH
acetate	$NaC_2H_3O_2$	82.04	1.528	324		46.5^{20} aq
aluminate, tetrachloro-	$NaAlCl_4$	191.80		151		s aq
amide	$NaNH_2$	39.02		210	subl 400	d viol aq
aurate(III) dihydrate, tetrachloro-	$NaAuCl_4 \cdot 2H_2O$	397.80	1.6	d 100		166^{20} aq
azide	NaN_3	65.01	1.846^{20}	d		41^{20} aq; 0.3 alc
benzoate	$NaC_6H_5O_2$	144.11				63^{25} aq; 1.3 alc
bismuthate(V)(1−)	$NaBiO_3$	280.00		d		i aq; d a
boranate	$NaBH_4$	37.84	1.074	497 d	d	55^{25} aq; 4 alc; 1.4 pyr; 5 DMF
borate, tetra-	$Na_2B_4O_7$	201.27	2.367	742.5		2.6^{20} aq
borate, tetrafluoro-	$NaBF_4$	109.82	2.47^{20}	384	d	108^{27} aq
bromate	$NaBrO_3$	150.91	3.339^{17}	380 d		36^{20} aq
bromide	$NaBr$	102.91	3.205^{18}	747	1447	90^{20} aq; 6 alc; 16 MeOH
carbonate	Na_2CO_3	106.00	2.533	850.0	d	21.5^{20}; s glyc
carbonate 10-water	$Na_2CO_3 \cdot 10H_2O$	286.14	1.46	34		50 aq; s glyc
chlorate	$NaClO_3$	106.45	2.489	248	d 350	96^{20} aq; 0.77 alc; 25 glyc
chloride	$NaCl$	58.45	2.164^{20}	801	1465	36^{20} aq; 10 glyc
chlorite	$NaClO_2$	90.45		d 180–200		34^{17} aq
chromate(VI)	Na_2CrO_4	161.97	2.723	792		84^{20}
citrate 2-water	$Na_3C_6H_5O_7 \cdot 2H_2O$	294.10		−2H₂O, 150		77^{25} aq
cobaltate(III), hexanitrito-	$Na_3[Co(NO_2)_6]$	403.98				v s aq
cyanate	$NaOCN$	65.01	1.893^{20}	550		s aq d; 0.22^0 alc
cyanide	$NaCN$	49.02		562	1530	58.7^{20} aq
cyanoborohydride	$NaBH_3CN$	62.84		d 242		(flammable solid)
dichromate(VI) 2-water	$Na_2Cr_2O_7 \cdot 2H_2O$	298.00	2.348^{25}_4	356 anhyd	d 400	208^{20} aq
diethyldithiocarbamate	$NaS_2CN(C_2H_5)_2$	225.31		94 anhyd		s aq, alc

2.26

Name	Formula	M	Density	mp	bp	Solubility
dimethylarsonate 3-water	NaO$_2$As(CH$_3$)$_2$·3H$_2$O	214.03		60	−3H$_2$O, 120	200 aq; 40 alc
diphosphate(V)	Na$_4$P$_2$O$_7$	265.90	2.45	988		2.26^{0} aq
dithionate 2-water	Na$_2$S$_2$O$_6$·2H$_2$O	242.13	2.189	−2H$_2$O, 110	d 267	6.05^{20} aq
dithionate(III) (hydrosulfite)	Na$_2$S$_2$O$_4$	174.13		d		22^{20} aq
dodecylsulfate (laurate)	NaO$_3$SOC$_{12}$H$_{25}$	288.38				10 aq
ethoxide	NaOC$_2$H$_5$	68.06		>300		d aq; s abs alc
ethylenebis (aminodiacetate) (EDTA)	Na$_4$C$_2$H$_4$N$_2$(C$_2$H$_3$O$_2$)$_4$	380.20				103 aq
ethylsulfate	NaO$_3$SOC$_2$H$_5$	148.11				18.8^{20} aq
ferrate(II) 10-water, hexacyano-	Na$_4$[Fe(CN)$_6$]·10H$_2$O	484.07	1.458	−10H$_2$O, 82	d 435	
ferrate(III) 2-water, pentacyanonitrosyl-(nitroprusside)	Na$_2$[Fe(CN)$_5$NO]·2H$_2$O	297.65	1.72			40^{16} aq
fluoride	NaF	41.99	2.78	996	1787	4^{20} aq; i alc
formate	NaOOCH	68.02	1.919	253 d		81^{20} aq; s glyc; sl s alc
gluconate	NaC$_6$H$_{11}$O$_7$	218.13				59^{25} aq; sl s alc; i eth
glycerophosphate	NaC$_3$H$_5$(OH)$_2$PO$_4$	216.03	1.396	d 130		60 aq; i alc
hydride	NaH	24.00		d 425		d viol aq, alc
hydrogen carbonate	NaHCO$_3$	84.01	2.20	−CO$_2$, 270		9.6^{20} aq; i alc
hydrogen phosphate hydrate, di-	NaH$_2$PO$_4$·H$_2$O	137.99	2.040	−H$_2$O, 100		71^{0} aq
hydrogen phosphate 7-water	Na$_2$HPO$_4$·H$_2$O	268.07	1.679	d	d 200	185^{40} aq
hydrogen sulfate	NaHSO$_4$	120.07	2.435	315		28.5^{25} aq; d alc
hydrogen sulfite	NaHSO$_3$	104.06	1.48	d		29 aq; 1.4 alc
hydrogen sulfide 2-water	NaHS·2H$_2$O	92.09		55		s aq, alc, eth
hydroxide	NaOH	40.01	2.130^{25}	322	1557	108^{20} aq; 14 abs alc; 24 MeOH; s glyc
hydroxymethanesulfinate dihydrate	NaO$_2$SCH$_2$OH·2H$_2$O	154.12		63–64		v s aq; i abs alc, bz, eth
hypochlorite	NaClO	74.44				53^{20} aq (anhyd v expl)

TABLE 2.1 Physical Constants of Inorganic Compounds (*continued*)

Name	Formula	Formula weight	Density	Melting point, °C	Boiling point, °C	Solubility in 100 parts solvent
Sodium						
iodate	$NaIO_3$	197.90	4.277^{20}	d		8.1^{20} aq
iodite	NaI	149.92	3.667^{0}	660	1304	178^{20}
lactate	$NaOOCCHOHCH_3$	112.07		d		misc aq, alc
methoxide	$NaOCH_3$	54.03		>300		d aq; s alc
molybdate dihydrate	$Na_2MoO_4 \cdot 2H_2O$	241.95	3.28	687	$-2H_2O$, 100	65^{20} aq
nitrate	$NaNO_3$	85.01	2.257	308	d 380	88^{20} aq
nitrite	$NaNO_2$	69.00	2.168^{0}	271	d 320	81^{20} aq
oxalate	$Na_2C_2O_4$	134.01	2.27			3.4^{20} aq
oxide	Na_2O	61.98	2.27	1132	d 1950	d aq to NaOH
perchlorate	$NaClO_4$	122.44	2.499	468		201^{20}
periodate	$NaIO_4$	213.91	3.865^{16}	d 300	d	10.3^{20} aq
peroxide	Na_2O_2	77.99	2.805	675		v s aq (d)
peroxoborate 4-water	$NaBO_3 \cdot 4H_2O$	153.88		d 60		2.5 aq
peroxodisulfate(VI)	$Na_2S_2O_8$	238.13		d		55 aq
phosphate 12-water	$Na_3PO_4 \cdot 12H_2O$	380.12	1.62	73.4	$-11H_2O$, 100	28.3^{15} aq
platinate(IV) 6-water, hexachloro-	$Na_2PtCl_6 \cdot 6H_2O$	561.88	2.50	$-6H_2O$, 110		v s aq; s alc
propionate	$NaOOCCH_2CH_3$	96.07				100^{25} aq; 4.1^{25} alc
salicylate	$NaC_7H_5O_3$	160.11				95^{20} aq; 11 alc; 25 glyc
selenate(VI)	Na_2SeO_4	188.94	3.098			27^{20} aq
silicate, hexafluoro-	Na_2SiF_6	188.05	2.679	red heat		0.44^{0} aq; i alc
stannate(IV) 3-water	$Na_2SnO_3 \cdot 3H_2O$	266.71		d 140		50^{0} aq
stearate	$NaOOCC_{17}H_{35}$	306.47		d		sl s aq
sulfate	Na_2SO_4	142.06	2.664	884		19.5^{20}
sulfate 10-water	$Na_2SO_4 \cdot 10H_2O$	322.19	1.464	32.4	$-10H_2O$, 100	36^{15} aq

sulfide	Na_2S	78.05	1.856^{14}	950		15.7^{20} aq
sulfite	Na_2SO_3	126.06	2.633^{15}	d		26^{20} aq
tartrate dihydrate	$Na_2C_4H_4O_6 \cdot 2H_2O$	230.08	1.818	$-2H_2O, 120$		29^6 aq
tetraphenylborate	$NaB(C_6H_5)_4$	342.24				s aq, acet
thiocyanate	$NaSCN$	81.07		287		134^{20} aq
thiosulfate	$Na_2S_2O_3$	158.11	2.345			s aq; i alc
thiosulfate 5-water	$Na_2S_2O_3 \cdot 5H_2O$	248.18	1.685	$-5H_2O, 100$		70^{20} aq (d slowly)
tungstate(VI) dihydrate	$Na_2WO_4 \cdot 2H_2O$	329.86	3.245	$-2H_2O, 100$		88^0
Strontium						
carbonate	$SrCO_3$	147.64	3.70	$-CO_2, 1172$		0.001^{25} aq; s a
chloride	$SrCl_2$	158.52	3.052	874	2058	52.9^{20} aq
chromate(VI)	$SrCrO_4$	203.64	3.895^{15}			0.09^{20} aq; s HCl
hydroxide	$Sr(OH)_2$	121.64	3.625	375 (in H_2)	$-H_2O, 710$	1.77^{20} aq
Sulfamic acid	H_2NSO_3H	97.09	2.126	d 200		14.7 aq
Sulfinyl						
bromide	$SOBr_2$	207.88	2.67	-49.5	139.7	d aq
chloride	$SOCl_2$	118.98	1.656^{15}	-104.5	75.8	hyd aq
fluoride	SOF_2	86.06	3.0^{-44}	-110	-43.8	d aq; s bz, chl, eth
Sulfonyl						
chloride	SO_2Cl_2	134.98	1.6674^{20}	-46	69.3	d aq; s bz
fluoride	SO_2F_2	102.07	$3.72 \, g \cdot L^{-1}$	-135.8	-55.38	4 mL aq; 24 mL alc; 136 mL CCl_4; 210 mL toluene
Sulfur	S	32.07	1.92	106.8	444.60	i aq; 23^0 CS_2; s alc, bz
	S_8	256.51	1.96^{20}	115.21	444.60	i aq; 23^0 CS_2; s alc, bz
(di-) chloride, di-	S_2Cl_2	135.03	1.688^{15}	-80	138.1	hyd aq
fluoride, tetra-	SF_4	108.07	1.919^{-73}	-121	-38	d viol aq; s bz
fluoride, hexa-	SF_6	146.07	1.88^{-50} $g \cdot L^{-1}$	-50.8	subl 63.8	sl s aq; s alc, KOH
oxide, di	SO_2	64.07	2.716^{20} $g \cdot L^{-1}$	-75.47	-10.01	3937^{20} mL aq; 25 mL alc
oxide, tri(III)	SO_3	80.07	1.46^{-10} (lq) 1.9225^{20}	16.86	43.4	slowly v s aq

TABLE 2.1 Physical Constants of Inorganic Compounds (*continued*)

Name	Formula	Formula weight	Density	Melting point, °C	Boiling point, °C	Solubility in 100 parts solvent
Sulfuric acid	H_2SO_4	98.08	1.8318^{20}	10.38	335.5	v s aq
chloro-	$HOSO_2Cl$	116.52	1.753^{20}	−80	152	d viol aq
fluoro-	FSO_2OH	100.07	1.726^{25}	−88.98	162.6	d viol aq
Tantalum	Ta	180.95	16.69	2985	5513	i aq; s HF, fused alk
(V) fluoride	TaF_5	275.95	4.74^{20}	95–97	229	s aq
Tellurium	Te	127.60	6.24^{20}	450	1009	i aq; s HNO$_3$, KOH
Thallium	Tl	204.37	11.85	303.5	1487	i aq; s HNO$_3$
(III) acetate sesquihydrate	$Tl(C_2H_3O_2)_3 \cdot 1.5H_2O$	408.53		182 d		
(I) bromide	$TlBr$	284.31	7.54	460	825	0.05^{20} aq; s alc
(I) chloride	$TlCl$	239.85	7.004^{30}	429	816	0.33^{20} aq
(I) ethoxide	$TlOC_2H_5$	249.43	3.493^{20}	−3	d 130	sl s alc; s eth
(I) fluoride	TlF	223.39	8.23^{4}	322	700	78^{15} aq
(I) nitrate	$TlNO_3$	266.40	5.556	206	430	9.6^{20} aq
(III) nitrate 3-water	$Tl(NO_3)_3 \cdot 3H_2O$	444.43		102–103		s aq
(I) oxide	Tl_2O	424.78	9.52^{16}	300	1080	v s aq (d); s a
(III) oxide	Tl_2O_3	456.78	10.19^{22} (hex)	717	−O$_2$, 875	i aq; s a
(I) sulfate	Tl_2SO_4	504.85	6.77	632	d	4.9^{20} aq
Thiocarbonyl chloride	$CSCl_2$	114.98	1.509^{15}	ca −2	73.5	d aq; s eth
Thiocyanogen	$(SCN)_2$	116.16				d aq; s alc, CS$_2$, eth
Thionyl. *see* **Sulfinyl**						
Tin (silver-white, tetr) (gray, cub)	Sn	118.69	7.28 5.75	231.89 stable −161 to 13.2	2623	i aq; s HCl, H$_2$SO$_4$
(IV) bromide	$SnBr_4$	438.36	3.35^{33}	30	207	hyd aq; s acet
(II) chloride	$SnCl_2$	189.61	3.95	247	652	84^{0} aq; s alc, eth

Name	Formula	Formula wt	Density	Melting point	Boiling point	Solubility
(IV) chloride	$SnCl_4$	260.53	2.226	−34	115	s aq, eth
(II) diphosphate(V)	$Sn_2P_2O_7$	411.32	4.009^{16}	213		i aq; s conc a
(II) fluoride	SnF_2	156.70	4.57^{25}	subl 705		30 aq
(IV) fluoride	SnF_4	194.70	4.780^{19}	1630	subl 1900	hyd aq
(IV) oxide	SnO_2	150.70	6.95	881	1210	i aq
(II) sulfide	SnS	150.77	5.08	765		i aq; s conc HCl
(IV) sulfide	SnS_2	182.83	4.5			i aq; d by aqua regia
(II) zirconate (IV), hexafluoro-	$SnZrF_6$	323.92	4.21			s aq
Titanium	Ti	47.90	4.507	1660	3318	s hot a, HF
(III) chloride	$TiCl_3$	154.27	2.71	subl 831 (vac)	d 5000	s aq, alc
(IV) chloride	$TiCl_4$	189.73	1.726	−24.10	136.4	s cold aq, alc
hydride, di	TiH_2	49.92	3.752	d 400		
(IV) isopropoxide	$Ti[OCH(CH_3)_2]_4$	284.26	0.9711^{20}	ca 20	$220^{10\,mm}$	
(IV) oxide (rutile)	TiO_2	79.90	4.23	1857		s HF
(III) sulfate	$Ti_2(SO_4)_3$	384.00				s HCl
Trisulfuryl dichloride	$ClSO_2OSO_2OSO_2Cl$	295.09	1.90^{20}	18.7	$61^{3\,mm}$	
Tungsten						
(VI) chloride	WCl_6	396.57	2.721^{282}	281.5	340.5	hyd aq; s CS_2, CCl_4
(VI) oxide	WO_3	231.86	7.16	1472	1837	i aq; s hot alk
sulfide, di-	WS_2	247.98	7.5^{10}	d 1250		s HNO_3 + HF
Uranyl						
(VI) acetate 2-water	$UO_2(C_2H_3O_2)_2 \cdot 2H_2O$	422.13	2.893^{15}	−2H$_2$O, 110	d 275	7.7^{15} aq
nitrate 6-water	$UO_2(NO_3)_2 \cdot 6H_2O$	502.13	2.807^{13}	60.2	d 100	155^{20} aq
Vanadium						
(III) oxide	V_2O_3	149.00	4.87	2067		i aq; s HNO_3 + HF
(V) oxide	V_2O_5	181.90	3.35	670	1690	0.80 aq; s a, alk
(IV) oxide sulfate	$VOSO_4$	163.00				v s aq
Xenon	Xe	131.30	5.8971^{0} g·L^{-1}	−111.8	−108.10	10.8^{20} mL aq
fluoride, di-	XeF_2	169.30	3.13^{25}	129.0	subl 114	2.5^{0} aq

TABLE 2.1 Physical Constants of Inorganic Compounds (*continued*)

Name	Formula	Formula weight	Density	Melting point, °C	Boiling point, °C	Solubility in 100 parts solvent
Xenon						
fluoride, tetra-	XeF_4	207.30	3.03^{25}	117.1	subl 116	hyd aq; s F_3CCOOH
fluoride, hexa-	XeF_6	245.30	3.411^{25}	49.5	75.6	hyd aq
Zinc	Zn	65.37	7.14^{25}	419.6	911	i aq; s a, alk
acetate dihydrate	$Zn(C_2H_3O_2)_2 \cdot 2H_2O$	219.49	1.735	237		41.6^{20} aq; 3.3 alc
bromide	$ZnBr_2$	225.21	4.22	402	650	446^{20} aq; 200 alc; s eth
carbonate	$ZnCO_3$	125.38	4.398	$-CO_2$, 300		0.02^{25} aq; s a, alk
chloride	$ZnCl_2$	136.29	2.907^{25}	318	732	395^{20} aq; 77 alc; 50 glyc
chromate(VI)	$ZnCrO_4$	181.36	3.40			i aq; s a
cyanide	$Zn(CN)_2$	117.42	1.852	d 800		0.058^{18} aq; s KCN, alk
fluoride	ZnF_2	103.38	5.00^{25}	872	1500	1.6^{20} aq
iodide	ZnI_2	319.22	4.736^{25}	446	730	432^{20} aq; 50 glyc
nitrate 6-water	$Zn(NO_3)_2 \cdot 6H_2O$	297.47	2.065^{14}	36.4	$-6H_2O$, 131	146^0 aq
oxide	ZnO	81.37	5.67	1970		i aq; s a, alk
peroxide	ZnO_2	97.38	3.00	d 150		i aq; d slowly
p-phenolsulfonate 8-water	$Zn[C_6H_4(OH)SO_3]_2 \cdot 8H_2O$	555.83		$-8H_2O$, 120		63 aq; 56 alc
phosphate(V)	$Zn_3(PO_4)_2$	386.05	3.998^{15}	900		i aq; s a, NH_4OH
phosphide	Zn_3P_2	258.09	4.55	>420	subl 1100 (in H_2)	d aq; s bz, CS_2; d viol HCl
propionate	$Zn(OOCCH_2CH_3)_2$	211.52				32 aq; 2.8 alc
silicate 6-water, hexafluoro-	$ZnSiF_6 \cdot 6H_2O$	315.54	2.104	d 100		v s aq
stearate	$Zn(OOCC_{17}H_{35})_2$	632.33		ca 120		i aq, alc, eth; s bz
sulfate	$ZnSO_4$	161.44	3.54	1200		53.8^{20} aq
sulfate 7-water	$ZnSO_4 \cdot 7H_2O$	287.54	1.957	$-7H_2O$, 280		96^{20} aq; 40 glyc; i alc
sulfide	ZnS	97.43	4.087	1722		i aq; s a
thiocyanate	$Zn(SCN)_2$	181.53			d 500	0.14^{18} aq; s alc

Zirconium						
(IV) chloride	Zr	91.22	6.52^{30}	1852	4504	s aqua regia
chloride oxide 8-water	$ZrCl_4$	233.05	2.803^{15}	437	subl 334	hyd aq; s alc, eth
hydroxide	$ZrCl_2O \cdot 8H_2O$	322.25	1.91	$-8H_2O, 210$		s aq
(IV) oxide	$Zr(OH)_4$	159.25	3.25	$-2H_2O, 500$		s a
silicate(4$-$)	ZrO_2	123.22	5.85	2677	4275	s hot H_2SO_4, HF slowly
sulfate 4-water	$ZrSiO_4$	183.31	4.56	d 1538		very inert
	$Zr(SO_4)_2 \cdot 4H_2O$	355.41	3.22^{16}	anhyd 380		52.5^{18} aq

SECTION 3
PROPERTIES OF ATOMS, RADICALS, AND BONDS

NUCLIDES

TABLE 3.1 Table of Nuclides

Explanation of column headings

Nuclide. Each nuclide is identified by its atomic number Z, equal to the number of protons in the nucleus; the corresponding symbol for that element; and the mass number A, equal to the sum of the numbers of protons Z and neutrons N in the nucleus. Thus, $A = Z + N$, or $N = A - Z$. The m following the mass number (e.g., $^{69\,m}Zn$) indicates an isomer of that nuclide.

Half-Life. For the radioactive nuclides this time period corresponds to that during which loss by disintegration of 50% of the nuclide occurs. The units of time are designated by year (yr), day (d), hour (h), minute (min), and second (s).

Natural Abundance. The isotopic abundances listed are on an "atom percent" basis for the stable nuclides present in naturally occurring elements in the earth's crust.

Thermal Neutron Absorption Cross Section. The ease with which a given nuclide can absorb a thermal neutron (energy $\leq \frac{1}{40}$ eV) and become of a different nuclide is indicated by the cross section, given here in units of barns (1 barn = 10^{-24} cm²). If the mode of reaction is other than (n, γ), it is so indicated, for example, (n, p) or (n, α), where n = neutron, p = proton, γ = gamma ray, and α = alpha particle ($^4_2$He).

Major Radiations. In this column are listed the principal mode(s) of decay and the energies of the emanating radiations in million electronvolts (MeV). The gamma-ray (γ) intensities, where given, are given to the nearest whole percentage in parentheses following the numerical energy value for that particular γ. In most cases the radiations listed should be sufficient for identification of the particular nuclide. The following designations are used: negatron (β^-), positron (β^+), conversion electron (e^-), gamma ray (γ), and alpha particle (α).

Nuclide			Natural abundance, %	Thermal neutron absorption cross section, barns	Major radiations
Symbol	Mass	Half-life			
^{1}H	1.007 825		99.985	0.332	
^{2}H	2.014 102		0.015	0.000 5	
^{3}H	3.016 050	12.26 yr			β^-, 0.018 6; no γ
^{6}Li	6.015 125		7.42	953(n, α)	
^{7}Li	7.016 004		92.58	0.037	
^{7}Be	7.016 929	53.6 d		54,000(n, p)	γ, 0.477(10)
^{9}Be	9.012 186		100	0.009	
^{10}Be	10.013 534	2.5×10^6 yr			β^-, 0.555; no γ
^{10}B	10.012 939		19.7	3837(n, α)	
^{11}B	11.009 305		80.3	0.005	
^{11}C	11.011 432	20.34 min			β^+, 0.97; γ, 0.511
^{13}C	13.003 354		1.108	0.000 9	
^{14}C	14.003 242	5730 yr			β^-, 0.156; no γ
^{13}N	13.005 738	9.96 min			β^+, 1.20; γ, 0.511
^{14}N	14.003 074		99.635	1.81(n, p)	
^{19}O	19.003 578	29.1 s			β^-, 4.60; γ, 0.197(97), 1.37(59)
^{18}F	18.000 937	109.7 min			β^+, 1.74; γ, 0.511
^{22}Na	21.994 437	2.62 yr			β^+, 1.820, 0.545; γ, 0.511, 1.275(100)

TABLE 3.1 Table of Nuclides (*continued*)

Nuclide		Half-life	Natural abundance, %	Thermal neutron absorption cross section, barns	Major radiations
Symbol	Mass				
^{23}Na	22.934 473		100	0.53	
^{24}Na	23.990 962	14.96 h			β^-, 4.17, 1.389; γ, 0.511, 1.275(100)
^{25}Mg	24.985 839		10.11	0.3	
^{28}Mg	27.983 875	21.2 h			β^-, 0.46; e^-, 0.03; γ, 0.031(96), 0.40(30), 0.95(30), 1.35(70)
^{26}Al	25.986 891	7.4×10^5 yr			β^-, 8.5; γ, 0.511, 1.12(4), 1.81(100)
^{27}Al	26.981 539		100	0.235	
^{28}Al	27.981 905	2.31 min			β^-, 2.85; γ, 1.780 (100)
^{30}Si	29.973 763		3.12	0.11	
^{31}Si	30.975 349	2.62 h			β^-, 1.48; γ, 1.26
^{31}P	30.973 765		100	0.19	
^{32}P	31.973 909	14.28 d			β^-, 1.710
^{33}P	32.971 728	24.4 d			β^-, 0.248; no γ
^{34}S	33.967 865		4.22	0.27	
^{35}S	34.969 031	87.9 d			β^-, 0.167; no γ
^{38}S	37.971 230	2.87 h			β^-, 3.0, 1.1; γ, 1.88(95)
^{35}Cl	34.968 851		75.53	44	
^{36}Cl	35.968 309	3.08×10^5 yr		100	β^-, 0.714; γ, 0.511
^{37}Cl	36.965 898		24.47	0.4	
^{38}Cl	37.968 005	37.29 min			β^-, 4.91; γ, 1.60(38)
^{39}Cl	38.968 008	55.5 min			β^-, 3.45, 2.18, 1.91; γ, 0.246(44)
^{37}Ar	32.966 772	35.1 d			Cl X rays
^{40}K	39.964 000	1.26×10^9 yr	0.118	70	β^-, 1.314; β^+, 0.483; γ, 1.460(11)
^{41}K	40.961 832		6.77	1.2	
^{42}K	41.962 406	12.36 h			β^-, 3.52; γ, 0.31, 1.524(18)
^{44}Ca	43.955 490		2.06	0.7	
^{45}Ca	44.956 189	165 d			β^-, 0.252
^{47}Ca	46.954 538	4.535 d			β^-, 1.98, 0.67; γ, 0.49(5), 0.815(5), 1.308(74)
^{46}Sc	45.955 919	83.9 d			β^-, 1.48, 0.357, γ, 0.889(100), 1.120(100)

TABLE 3.1   Table of Nuclides (*continued*)

Nuclide		Half-life	Natural abundance, %	Thermal neutron absorption cross section, barns	Major radiations
Symbol	Mass				
^{44}Ti	43.959 572	48 yr			γ, 0.068(90), 0.078(98); e^-, 0.065, 0.073
^{48}V	47.952 259	16.0 d			β^+, 0.696; γ, 0.511, 0.945(10), 0.983(100), 1.312(97), 2.241(3)
^{49}V	48.949 522	330 d			Ti X rays
^{50}Cr	49.946 054		4.31	17	
^{51}Cr	50.944 768	27.8 d			γ, 0.320(9); e^-, 0.315
^{54}Mn	53.940 362	303 d			γ, 0.835(100); e^-, 0.829
^{55}Mn	54.938 050		100	13.3	
^{56}Mn	55.938 910	2.576 h			β^-, 2.85; γ, 0.847(99), 1.811(29), 2.110(15)
^{54}Fe	53.939 617		5.84	2.9	
^{55}Fe	54.938 299	2.60 yr			Mn X rays
^{58}Fe	57.933 282		0.31	1.1	
^{59}Fe	58.934 878	45.6 d			β^-, 1.57, 0.475; γ, 0.143(1), 0.192(3), 1.095(56), 1.292(44)
^{57}Co	56.936 296	270 d			γ, 0.014(9), 0.122(87), 0.136(11), 0.692; e^-, 0.115, 0.129
^{58}Co	57.935 761	71.3 d			β^+, 0.474; γ, 0.511, 0.810(99), 0.865(1), 1.67(1)
^{59}Co	58.933 189		100	19	
^{60}Co	59.933 813	5.263 yr		6	β^-, 1.48, 0.314; γ, 1.173(100), 1.332(100)
^{62}Ni	61.928 342		3.66	15	
^{63}Ni	62.929 664	92 yr			β^-, 0.067; no γ
^{64}Ni	63.927 958		1.16	1.5	
^{65}Ni	64.930 072	2.564 h			β^-, 2.13; γ, 0.368(5), 1.115(16), 1.481(25)
^{63}Cu	62.929 592		69.1	4.5	
^{64}Cu	63.929 759	12.80 h			β^-, 0.573; β^+, 0.656; e^-, 1.33; γ, 0.511, 1.34(1)

TABLE 3.1 Table of Nuclides (*continued*)

Nuclide		Half-life	Natural abundance, %	Thermal neutron absorption cross section, barns	Major radiations
Symbol	Mass				
^{64}Zn	63.929 145	$> 8 \times 10^{15}$ yr	48.89	0.46	
^{65}Zn	64.929 234	245 d			β^+, 0.327; e^-, 1.106; γ, 0.511, 1.115(49)
^{68}Zn	67.924 857		18.56	1.0	
$^{69\,m}$Zn		13.8 h			γ, 0.439(95); e^-, 0.429
^{71}Ge	70.924 956	11.4 d			Ga X rays
^{75}As	74.921 595		100	4.5	
^{76}As	75.922 397	26.4 h			β^-, 2.97; γ, 0.559(43), 0.657(6), 1.22(5) 1.44(1), 1.789, 2.10(1)
^{77}As	76.920 645	38.7 h			β^-, 0.68; γ, 0.086, 0.239(3), 0.522(1)
^{75}Se	74.922 525	120.4 d			γ, 0.066(1), 0.097(1), 0.121(17), 0.136(57), 0.265(60), 0.280(25), 0.401(12); e^- 0.085, 0.095, 0.109, 0.124, 0.253
^{79}Br	78.918 329		50.52	8.5	
^{80}Br	79.918 536	17.6 min			β^-, 2.00; β^+, 0.87; γ, 0.511, 0.618(7), 0.666(1)
^{81}Br	80.916 292		49.48	3	
^{82}Br	81.916 802	35.34 h			β^-, 0.444; γ, 0.554(66), 0.619(41), 0.698(27), 0.777(83), 0.828(25), 1.044(29), 1.317(26), 1.475(17)
^{85}Kr	84.912 523	10.76 yr		< 15	β^-, 0.67; γ, 0.514
^{86}Rb	85.911 193	18.66 d			β^-, 1.78; γ, 1.078(9)
^{85}Sr	84.912 989	64.0 d			γ, 0.514(100); e^-, 0.499
^{90}Y	89.907 163	64.0 h			β^-, 2.27; no γ
^{95}Nb	94.906 832	35.0 d		~ 7	β^-, 0.160; γ, 0.765(100)

TABLE 3.1   Table of Nuclides (*continued*)

Nuclide		Half-life	Natural abundance, %	Thermal neutron absorption cross section, barns	Major radiations
Symbol	Mass				
^{99}Mo	98.907 720	66.7 h			β^-, 1.23; γ, 0.041(12), 0.181(7), 0.372(1), 0.740(12), 0.780(4)
$^{99\,m}$Tc		6.049 h			γ, 0.140(90); e^-, 0.110
^{103}Ru	102.906 306	39.5 d			β^-, 0.70, 0.21; γ, 0.497(88), 0.610(6)
^{108}Pd	107.903 891		26.7	12	
^{109}Pd	108.905 954	13.47 h			β^-, 1.028; γ, 0.088(5), 0.129, 0.31, 0.41, 0.60, 0.64
^{109}Ag	108.904 756		48.65	89	
$^{110\,m}$Ag		39.2 s			γ, 0.088(5)
^{111}Ag	110.905 316	7.5 d			β^-, 1.05; γ, 0.247(1), 0.342(6)
^{109}Cd	108.904 928	453 d			γ, 0.088; e^-, 0.062
^{115}Cd	114.905 431	53.5 h			β^-, 1.11; γ, 0.230(1), 0.262(2), 0.49(10), 0.53(26)
$^{113\,m}$In		99.8 min			γ, 0.393(64); e^-, 0.365, 0.389
^{114}In	113.904 905	72 s			β^-, 1.988; β^+, 0.42; γ, 1.299
^{113}Sn	112.905 187	115 d			γ, 0.255(2)
^{121}Sb	120.903 816		57.25	6	
^{122}Sb	121.905 183	2.80 d			β^-, 1.97; β^+, 0.56; γ, 0.584(66), 0.686(3), 1.14(1) 1.26(1)
^{123}Sb	122.904 213	1.3×10^{16} yr	42.75	3.3	
^{124}Sb	123.905 973	60.4 d		2000	β^-, 2.31; γ, 0.603(97), 0.644(7), 0.72(14), 0.967(2), 1.048(2), 1.31(3), 1.37(5), 1.45(2), 1.692(50), 2.088(7)
^{125}Sb	124.905 232	2.71 yr		<20	β^-, 0.61; e^-, 0.114, 0.395; γ, 0.176(6), 0.427(31), 0.463(10), 0.599(24), 0.634(11), 0.66(3)

TABLE 3.1 Table of Nuclides (*continued*)

Nuclide			Natural abundance, %	Thermal neutron absorption cross section, barns	Major radiations
Symbol	Mass	Half-life			
^{132}Te	131.908 523	77.7 h			β^-, 0.22; e^-, 0.197; γ, 0.053(17), 0.230(90)
^{125}I	124.904 578	60.2 d		9	γ, 0.035(7); e^-, 0.030
^{127}I	126.904 470		100	6.4	
^{128}I	127.905 838	24.99 min			β^-, 2.12; γ, 0.441(14), 0.528(1), 0.743, 0.969
^{131}I	130.906 127	8.05 d		~0.7	β^-, 0.806, 0.606; e^-, 0.330; γ, 0.080(3), 0.284(5), 0.364(82), 0.637(7), 0.723(2)
^{132}I	131.907 981	2.26 h			β^-, 2.12; γ, 0.24(1), 0.52(20), 0.67(44), 0.773(89), 0.955(22) 1.14(6), 1.28(7), 1.40(14), 1.45(1), 1.91(1), 1.99(1)
^{133}Xe	132.905 815	5.270 d		190	β^-, 0.346; e^-, 0.045, 0.075; γ, 0.081(37)
^{131}Cs	130.905 466	9.70 d			Xe X rays
^{134}Cs	133.906 823	2.046 yr		136	β^-, 0.662; γ, 0.57(23), 0.605(98), 0.796(99), 1.038(1), 1.168(2), 1.365(3)
^{137}Cs	136.906 770	30.0 yr		0.11	β^-, 1.176, 0.514; e^-, 0.624, 0.656; γ, 0.662(85)
^{131}Ba	130.906 716	12.0 d			γ, 0.124(28), 0.216(19), 0.25(5), 0.373(13), 0.496(48), 0.60(3); e^-, 0.118, 0.180, 0.460
^{133}Ba	132.905 879	7.2 yr			γ, 0.080(36), 0.276(7), 0.302(14), 0.356(69), 0.382(8); e^-, 0.266, 0.319

TABLE 3.1   Table of Nuclides (*continued*)

Nuclide		Half-life	Natural abundance, %	Thermal neutron absorption cross section, barns	Major radiations
Symbol	Mass				
^{137m}Ba		2.554 min			γ, 0.662(89); e^-, 0.624, 0.656
^{140}Ba	139.910 565	12.80 d		< 20	β^-, 1.02; γ, 0.030(11), 0.163(6), 0.305(6), 0.438(5), 0.537(34)
^{141}Ce	140.908 219	32.5 d		30	β^-, 0.581; e^-, 0.104, 0.139; γ, 0.145(48)
^{144}Ce	143.913 591	284 d		1.0	β^-, 0.31; γ, 0.080(2), 0.134(11)
^{197}Au	196.966 541		100	98.8	
^{198}Au	197.968 231	2.697 d		26,000	β^-, 0.962; e^-, 0.329, 0.398; γ, 0.412(95), 0.676(1), 1.088
^{199}Au	198.968 773	3.15 d		~30	β^-, 0.46, 0.30; γ, 0.158(37), 0.208(8); e^-, 0.125, 0.145
^{197}Hg	196.967 360	65 h			γ, 0.77(18), 0.191(2), 0.268
^{203}Hg	202.972 880	46.9 d			β^-, 0.214; e^-, 0.194, 0.264, 0.275; γ, 0.279(77)
^{203}Tl	202.972 353		29.50	11	
^{205}Tl	203.973 865	3.81 yr			β^-, 0.766
^{210}Pb	209.984 187	20.4 yr			β^-, 0.061; γ, 0.047(4); α, 3.72
^{207}Bi	206.978 438	30.2 yr			γ, 0.570(98), 1.063(77), 1.771(9); e^- 0.482, 0.975, 1.048
^{210}Po	209.982 876	138.40 d		< 0.03	α, 5.305; γ, 0.803
^{226}Ra	226.025 360	1602 yr		20	α, 4.78, 4.60; γ, 0.186(4), 0.26, 0.42, 0.61; e^-, 0.170
^{241}Am	241.056 714	433 yr		700	α, 5.49, 5.44; γ, 0.060(36), 0.101, 0.208, 0.335, 0.37, 0.663, 0.722

ELECTRONEGATIVITY

According to Pauling, electronegativity χ is the relative attraction of an atom for the valence electrons in a covalent bond. It is proportional to the effective nuclear charge and inversely proportional to the covalent radius.

$$\chi = \frac{0.31(n + 1 \pm c)}{r} + 0.50$$

where n is the number of valence electrons, c is any formal valence charge on the atom and the sign before it corresponds to the sign of this charge, and r is the covalent radius. Because electronegativity is concerned with atoms in molecules rather than atoms in isolation, it is not possible to define precise electronegativity values. Pauling determined his set of values from bond energy data based on experimentally measured heats of dissociation and formation. Originally the element fluorine, whose atoms have the greatest attraction for electrons, was given an arbitrary electronegativity of 4.0. A revision of Pauling's values based on newer heat data assigns 3.9 to fluorine. A unit positive charge changes the χ value for an atom by about two-thirds of the electronegativity difference between it and the atom next on its right in the Periodic Table, and a unit negative charge similarly decreases the χ value.

The greater the difference in electronegativity, the greater is the ionic character of the bond. The amount of ionic character I is given by the expression

$$I = 1 - e^{-0.25(\chi_A - \chi_B)^2}$$

The bond is fully covalent when $(\chi_A - \chi_B) < 0.5$ (and $I < 6\%$). A different expression was proposed by Hannay-Smyth.*

$$I = 0.46|\chi_A - \chi_B| + 0.035(\chi_A - \chi_B)^2$$

Other sets of electronegativities of the elements have been proposed. The rather direct, but somewhat limited, method of Mulliken makes use of the ionization potential IP and electron-affinity data (Table 3.3). Numerical values are obtained that coincide with values from other methods if electronegativities are calculated from

$$\chi = \frac{IP + A}{5.6}$$

Electronegativities on the Allred-Rochow scale† are given by

$$\chi = 0.359\frac{Z_{eff}}{r^2} + 0.744$$

where Z_{eff} is the effective nuclear charge and r is the atomic radius.

Using Pauling's values, electronegativities of the elements are arranged in periodic order in Table 3.2A.

---

* Hannay-Smyth, *J. Am. Chem. Soc.*, **68**: 171 (1946).
† J. Inorg. Nucl, *Chem.*, **5**: 264, 269 (1958).

TABLE 3.2A Electronegativities of the Elements

H 2.2												
Li 1.0	Be 1.5							B 2.0	C 2.5	N 3.0	O 3.5	F 4.0
Na 0.9	Mg 1.2							Al 1.5	Si 1.8	P 2.1	S 2.4	Cl 2.8
K 0.9	Ca 1.0	Sc 1.3	Ti–V 1.6	Cr–Mn 1.6	Fe–Ni 1.8	Cu 1.9	Zn 1.7	Ga 1.6	Ge 1.8	As 2.0	Se 2.4	Br 2.7
Rb 0.8	Sr 1.0	Y 1.2	Zr–Nb 1.6	Mo–Tc 1.8	Ru–Pd 2.2	Ag 1.9	Cd 1.5	In 1.7	Sn 1.8	Sb 1.9	Te 2.1	I 2.2
Cs 0.7	Ba 0.9	La–Lu 1.1	Hf–Ta 1.3	W–Re 1.8	Os–Pt 2.2	Au 2.4	Hg 1.4	Tl 1.8	Pb 1.8	Bi 1.9		

TABLE 3.2B Electronegativities of the Groups

Group	χ	Group	χ	Group	χ
F	4.0	OH	3.7	C$\equiv$N	3.3
Cl	2.8	OCH$_3$	3.7	C$\equiv$CH	3.3
Br	2.7	NO$_2$	3.4	CH$=$CH$_2$	3.0
I	2.2	NH$_2$	3.4	C$_6$H$_5$	3.0
CF$_3$	3.4	N(CH$_3$)$_2$	3.0	COOH	2.8
CCl$_3$	3.0			SiH$_3$	2.2
CHCl$_2$	2.8			PH$_2$	2.3
				SH	2.8

Electronegativities have important uses in chemistry in addition to predicting the amount of ionic character in a bond. The bond stretching force constant k (in units of 10^5 dynes $\cdot$ cm^{-1}) can be estimated for stable molecules exhibiting their normal covalences by the expression:

$$k = 1.67 \, N \left(\frac{\chi_A \chi_B}{d^2} \right)^{3/4} + 0.30$$

where N is the bond order (i.e., the effective number of covalent or ionic bonds acting between the two atoms A and B) and d is the internuclear distance in angstroms.

An estimate of the percent ionic character may be made for organometallic compounds of the type alkyl-metal for metals in common use in organic synthesis. Among the alkali metals (row 1 of the Periodic Table, these are Li (43%), Na (47%), and K (51%)). The percent ionic character for an organomagnesium compound (typically a Grignard reagent), the bond is estimated to be 34% ionic. The more covalent organozinc and organocadmium compounds have correspondingly less ionic character: Zn (18%) and Cd (15%).

Electronegativities have also been estimated for various common substituent groups. They are arranged in Table 3.2B in clusters of related residues. The values for the individual halogens are from Table 3.2A.

Electronegativity is proportional to the work function ϕ, which is the energy necessary to just remove an electron from the metal surface in thermoelectric or photoelectric emission.

$$\chi = 0.44\,\phi - 0.15$$

ELECTRON AFFINITY

The *electron affinity* of an atom A is defined as the energy released when an atom and an electron react to form a negative ion in the gas phase at $0\,\mathrm{K}$.

$$A(g) + e^- = A^-(g)$$

An example is the capture of an electron by chlorine to give chloride anion.

$$Cl(g) + e^- = Cl^-(g).$$

Conceptually related to this is the *ionization potential*, which is the energy for the process

$$A \rightarrow A^+ + e^-.$$

An example is the loss of an electron by an alkali metal to give the alkali metal cation.

$$Na \rightarrow Na^+ + e^-$$

The second ionization potential carried this process further, that is, $A^+ \rightarrow A^{2+} + e^-$ and so on for the third or more ionization potentials. Both electron affinities and ionization potentials are typically expressed in electron volts (eV). Data for electron affinities are given in Table 3.3. Uncertainty in the final data figures is given in parentheses.

Source: H. Hotop and W. C. Lineberger, *J. Phys. Chem. Ref. Data*, **4**: 539 (1975). Data for ionization potentials are available in C. E. Moore, *National Bureau of Standards U. S. Publication NSRDS-NBS*, **34** (1970).

TABLE 3.3   Electron Affinities of Elements, Molecules and Radicals

A. Elements

Element	Electron affinity, eV*	Element	Electron affinity, eV*
Aluminum	0.46(3)	Beryllium	<0
Antimony	1.05(5)	Bismuth	1.1(2)
Argon	<0	Boron	0.28(1)
Arsenic	0.80(5)	Bromine	3.364(4)
Astatine	2.8(2)	Cadmium	<0
Barium	<0	Calcium	<0

TABLE 3.3   Electron Affinities of Elements, Molecules and Radicals (*continued*)

A. Elements

Element	Electron affinity, eV*	Element	Electron affinity, eV*
Carbon	1.268(5)	Oxygen	1.462(3)
Cesium	0.4715(5)	Palladium	0.6(3)
Chlorine	3.615(4)	Phosphorus	0.743(10)
Chromium	0.66(5)	Platinum	2.128
Cobalt	0.7(2)	Polonium	1.9(3)
Copper	1.226(10)	Potassium	0.5012(5)
Fluorine	3.399(3)	Radon	<0
Francium	(0.456)	Rare earths	≤0.5 (estimate)
Gallium	0.30(15)	Rhenium	0.15(10)
Germanium	1.2(1)	Rhodium	1.2(3)
Gold	2.3086(7)	Rubidium	0.4860(5)
Hafnium	>0	Ruthenium	1.1(3)
Helium	<0	Scandium	<0
Hydrogen	0.754 209(3)	Selenium	2.0206(3)
Indium	0.30(15)	Silicon	1.385(5)
Iodine	3.061(4)	Silver	1.303(7)
Iridium	1.6(2)	Sodium	0.546(5)
Iron	0.25(20)	Strontium	<0
Krypton	<0	Sulfur	2.0772(5)
Lanthanum	0.5(3)	Tantalum	0.6(4)
Lead	1.1(2)	Technetium	0.7(3)
Lithium	0.620(7)	Tellurium	1.9708(3)
Magnesium	<0	Thallium	0.3(2)
Manganese	<0	Tin	1.25(10)
Mercury	<0	Titanium	0.2(2)
Molybdenum	1.0(2)	Tungsten	0.6(4)
Neon	<0	Vanadium	0.5(2)
Nickel	1.15(10)	Xenon	<0
Niobium	1.0(3)	Yttrium	0.0(3)
Nitrogen	−0.07(8)	Zinc	≈0
Osmium	1.1(3)	Zirconium	0.5(3)

B. Molecules

Molecule	Electron affinity, eV*	Molecule	Electron affinity, eV*
BF_3	2.65	SF_6	1.43
p-Benzoquinone	1.34	2,3,5,6-Tetrachloro-	2.40
NO_2	3.91	benzoquinone	
O_2	0.45	Tetracyanoethylene	2.88

* To convert into $kJ \cdot mol^{-1}$ multiply by 96.48. To convert into $kcal \cdot mol^{-1}$ multiply by 23.06.

TABLE 3.3 Electron Affinities of Elements, Radicals and Molecules (*continued*)

C. Radicals

Radical	Electron affinity, eV*	Radical	Electron affinity, eV*
CH_3	1.08	OH	1.83
C_2H_5	0.89	CF_3O	1.35
C_6H_5	2.20	CH_3O	0.38
CCl_3	1.22	PH_2	1.60
CF_3	1.85	SH	2.19
CN	3.17	CH_3S	1.32
NH_2	1.12	SCN	2.17
C_6H_5NH	1.55	SeCN	2.64
$(C_6H_5)_2N$	1.19	SiF_3	3.35

* To convert into $kJ \cdot mol^{-1}$ multiply by 96.48. To convert into $kcal \cdot mol^{-1}$ multiply by 23.06.

BOND LENGTHS AND STRENGTHS

The bonds most commonly encountered in organic chemistry are those between carbons and between carbon and heteroatoms of the first row of the Periodic Table. Generally and very approximately, single bonds between two carbon atoms are about 1.5 Å in length and have strengths near to, but usually less than, 100 kcal/mole (418 kJ/mol). Likewise, carbon–carbon double bonds are stronger (~150 kcal/mol) and shorter (~1.35 Å). Triple bonds are shorter still ($C\equiv C \approx 1.2$ Å, ~200 kcal/mol). Bonds between carbon and either nitrogen or oxygen are somewhat shorter owing to the heteroatom's electronegativity. A few general examples are shown below and detailed data may be found in the Table 3.4.

General Trends in the Length of Common Organic Chemical Bonds

Bond	Length (Å)	Bond	Length (Å)	Bond	Length (Å)
C—C	1.54	C—O	1.43	$H—C(sp^3)$	1.12
C=C	1.33	C=O	1.20	N—H	1.03
C≡C	1.21	C—N	1.47	O—H	0.97
		C=N	1.28	C—Cl	1.76
		C≡N	1.15		

TABLE 3.4A   Bond Lengths between Carbon and other Elements

The numbers in parentheses following a numerical value represent the standard deviation of that value in terms of the final listed digit.

To convert the bond length from angstroms into nanometers, multiply by 0.1; to convert angstroms into picometers, multiply by 100.

Bond type	Bond length, Å*
Carbon–carbon	
Single bond	
Paraffinic: —C—C—	1.541(3)
In presence of —C=C— or of aromatic ring	1.53(1)
In presence of —C=O bond	1.516(5)
In presence of two carbon–oxygen double bonds	1.49(1)
In presence of two carbon–carbon double bonds	1.426(5)
Aryl—C=O	1.47(2)
In presence of one carbon–carbon triple bond: —C—C≡C—	1.460(3)
In presence of one carbon-nitrogen triple bond: —C—C≡N	1.464(5)
In compounds with tendency to dipole formation, e.g., C=C—C=O	1.44(1)
In aromatic compounds	1.395(3)
In presence of carbon–carbon double and triple bonds: —C=C—C≡C—	1.426(5)
In presence of two carbon-carbon triple bonds: —C≡C—C≡C—	1.373(4)
Double bond	
Single: —C=C—	1.337(6)
Conjugated with a carbon–carbon double bond: —C=C—C=C—	1.336(5)
Conjugated with a carbon-oxygen double bond: —C=C—C=O	1.36(1)
Cumulative: —C=C=C— or —C=C=O	1.309(5)
Triple bond	
Simple: —C≡C—	1.204(2)
Conjugated: —C≡C—C=C—, —C≡C—C=O, or —C≡C—aryl	1.206(4)

Bond type	Bond length, Å			
Carbon–halogen				
	Fluorine	Chlorine	Bromine	Iodine
Paraffinic: R—X	1.379(5)	1.767(2)	1.938(5)	2.139(1)
Olefinic: —C=C—X	1.333(5)	1.719(5)	1.89(1)	2.092(5)
Aromatic: Ar—X	1.328(5)	1.70(1)	1.85(1)	2.05(1)
Acetylenic: —C≡C—X	(1.27)	1.635(5)	1.795(10)	1.99(2)

Bond type	Bond length, Å
Carbon–hydrogen	
Paraffinic	
In methane	1.092
In CD_4	1.094
In monosubstituted carbon: H—C—Y	1.096(5)
In disubstituted carbon: H—C— (with X above and Y below)	1.073(5)

TABLE 3.4A   Bond Lengths between Carbon and Other Elements (*continued*)

Bond type	Bond length, Å
Carbon–hydrogen (*continued*)	

Paraffinic (*continued*)

In trisubstituted carbon: $H-\overset{\overset{\displaystyle X}{|}}{\underset{\underset{\displaystyle Y}{|}}{C}}-Y$ 1.070(7)

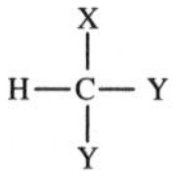

Olefinic

Simple: $H-C{=}C-$	1.083(5)
Cumultative carbon–carbon double bonds: $H-C{=}C{=}C-$	1.07(1)
Cumulative carbon–carbon–oxygen double bonds: $H-C-C{=}C{=}O$	1.08(1)

Aromatic	1.084(5)
Acetylenic (in C_2H_2, 1.059)	1.055(5)
In small rings	1.081(5)
In presence of a carbon triple bond: $H-C{\equiv}C-$	1.115(4)

Carbon–nitrogen	

Single bond

Paraffinic:

3 covalent nitrogen: RNH_2, R_2NH, R_3N	1.472(5)
4 covalent nitrogen: RNH_3^+, R_3N-BX_3	1.479(5)
In $-C-N{=}$	1.475(10)
In aromatic compounds	1.43(1)
In conjugated heterocyclic systems (partial double bond)	1.353(5)
In $-N-C{=}O$ (partial double bond)	1.322(5)

Double bond: $-C{=}N-$	1.32
Triple bond (in CN radical, 1.1774): $-C{\equiv}N$	1.157(5)

Carbon–oxygen	

Single bond

Paraffinic and saturated heterocyclic: $-C-O-$	1.426(5)

Strained, as in epoxides: $-\overset{|}{\underset{|}{C}}\overset{\diagdown\qquad\diagup}{\underset{O}{\qquad}}C-$ 1.435(5)

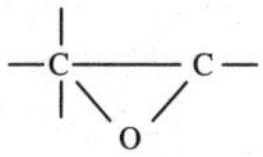

In aromatic compounds, as $Ar-OH$	1.36(1)
Longer bond in carboxylic acids and esters (HCOOH, 1.312)	1.358(5)
In conjugated heterocyclics, as furan	1.371(16)

TABLE 3.4A Bond Lengths between Carbon and Other Elements (*continued*)

Bond type	Bond length, Å
Carbon–oxygen (*continued*)	
Double bond	
In CO^+	1.115
In CO	1.128
In CO_2^+	1.177
In HCO	1.198(8)
In carbonyls	1.145(10)
In aldehydes and ketones	1.215(5)
In acyl halides: R—CO—X	1.171(4)
Shorter bond in carboxylic acids and esters	1.233(5)
In zwitterion forms	1.26(1)
In O=C=	1.160(1)
In isocyanates: RN=C=O	1.17(1)
In conjugated systems, as in partial triple bond: O=C—C=C	1.215(5)
In *p*-quinones	1.15(2)
In metal acetylacetonates	1.28(2)
In calcite: $CaCO_3$	1.29(1)
Carbon–selenium	
Single bond	
Paraffinic: —C—Se—	1.98(2)
In presence of fluorine, as in perfluoro compounds: —CF—Se—	1.95(2)
Double bond	
In Se=C=, as SeCS and SeCO	1.709(3)
In CSe radical	1.67
Carbon–silicon	
Alkyl substituent: H_3C—Si or H_2C—Si	1.870(5)
Aryl substituent: aryl—Si	1.843(5)
Electronegative substituent: R—Si—X	1.854(5)
Carbon–sulfur	
Single bond	
Paraffinic: —C—S—	1.817(5)
In presence of fluorine, as in perfluoro-compounds: —CF—S—	1.835(1)
In heterocyclic systems: partial double bonds	1.718(5)
Double bond	
In S=C: thiophene, S=CR_2	1.71(1)
In sulfoxides and sulfones	1.80(1)
In presence of second carbon–carbon double bond: S=C—C=C—	1.555(1)
In SC radical [in CS_2^+, 1.554(5)]	1.5349(2)

TABLE 3.4A Bond Lengths between Carbon and Other Elements (*continued*)

Bond type	Bond length, Å	Bond type	Bond length, Å
		Other elements and carbon	
C—Al	2.24(4)	C—In	2.16(4)
C—As (paraffinic)	1.98(1)	C—Mo	2.08(4)
C—B	1.56(1)	C—Ni	2.107(5)
C—Be	1.93	C—Pb (alkyl)	2.30(1)
C—Bi	2.30	C—Pd	2.27(4)
C—Co	1.83(2)	C—Sb (paraffinic)	2.202(16)
C—Cr	1.92(4)	C—Sn	
C—Fe	1.84(2)	alkyl	2.143(5)
C—Ge		electronegative	
Alkyl	1.98(3)	substituent	2.18(2)
Aryl	1.945(5)	C—Te	1.904
C—Hg	2.07(1)	C—Tl	2.705(5)
in Hg(CN)$_2$	1.99(2)	C—W	2.06

TABLE 3.4B Bond Lengths between Elements Other than Carbon

Elements	Bond type	Bond length, Å	Elements	Bond type	Bond length, Å
	Boron			Hydrogen (*continued*)	
B—B	B$_2$H$_6$	1.77(1)	H—Mg	MgH	1.731
B—Br	BBr$_3$	1.87(2)	H—Na	NaH	1.887
B—Cl	BCl$_3$	1.72(1)	H—Sb	H$_3$Sb	1.707
B—F	BF$_3$, R$_2$BF	1.29(1)	H—Se	H$_2$Se	1.460
B—H	Boranes	1.21(2)	H—Sn	SnH$_4$	1.701
	Bridge	1.39(2)	D—Br	DBr	1.4144
B—N	Borazoles	1.42(1)	D—Cl	DCl	1.2746
B—O	B(OH)$_3$,	1.362(5)	D—I		1.6165
	(RO)$_3$B		T—Br		1.4144
			T—Cl		1.2740
	Hydrogen			Nitrogen	
H—Al	AlH	1.646			
H—As	AsH$_3$	1.519	N—Cl	NO$_2$Cl	1.79(2)
H—Be	BeH	1.343	N—F	NF$_3$	1.36(2)
H—Br	HBr	1.408	N—H	NH$_4$$^+$	1.034(3)
H—Ca	CaH	2.002		NH$_3$, RNH$_2$	1.012
H—Cl	HCl	1.274		H$_2$NNH$_2$	1.038
H—F	HF	0.917		R—CO—NH$_2$	0.99(3)
H—Ge	GeH$_4$	1.53		HN=C=S	1.013(5)
H—I	HI	1.609	N—D	ND	1.041
H—K	KH	2.244	N—N	HN$_3$	1.02(1)
H—Li	LiH	2.595		R$_2$NNH$_2$	1.451(5)

TABLE 3.4B   Bond Lengths between Elements Other than Carbon (*continued*)

Elements	Bond type	Bond length, Å	Elements	Bond type	Bond length, Å
	Nitrogen (*continued*)			Phosphorus (*continued*)	
	N_2O	1.126(2)	P—H	PH_3, PH_4^+	1.424(5)
	N_2^+	1.116	P—I	PI_3	2.52(1)
N—O	NO_2Cl	1.24(1)	P—N	Single bond	1.491
	RO—NO_2	1.36(2)	P—O	Single bond	1.447
	NO_2	1.188(5)		p^3 bonding	1.67
N=O	N_2O	1.186(2)		sp^3 bonding	1.54(4)
	RNO_2	1.22(1)	P—S	p^3 bonding	2.12(5)
	NO^+	1.0619		sp^3 bonding	2.08(2)
N—Si	SiN	1.572		In rings	2.20(2)
			P—C	Single bond	1.562
	Oxygen			p^3 bonding	1.87(2)
O—H	H_2O	0.958		Silicon	
	ROH	0.97(1)			
	OH^+	1.0289	Si—Br	$SiBr_4$, R_3SiBr	2.16(1)
	HOOH	0.960(5)	Si—Cl	$SiCl_4$, R_3SiCl	2.019(5)
	D_2O	0.9575	Si—F	SiF_4, R_3SiF	1.561(3)
	OD	0.9699		SiF_6	1.58
O—O	HO—OH	1.48(1)	Si—H	SiH_4	1.480(5)
	O_2^+	1.227		R_3SiH	1.476(5)
	O_2^-	1.26(2)	Si—I	SiI_4	2.34
	O_2^{2-}	1.49(2)		R_3SiI	2.46(2)
	O_3	1.278(5)	Si—O	R_3SiOR	1.633(5)
O—Al	AlO	1.618	Si—Si	H_3SiSiH_3	2.30(2)
O—As	As_4O_6 (bridges)	1.79			
O—Ba	BaO	1.940		Sulfur	
O—Cl	ClO_2	1.484			
	OCl_2	1.68	S—Br	$SOBr_2$	2.27(2)
O—Mg	MgO	1.749	S—Cl	S_2Cl_2	1.585(5)
O—Os	OsO_4	1.66	S—F	SOF_2	1.585(5)
O—Pb	PbO	1.934	S—H	H_2S	1.333
				RSH	1.329(5)
	Phosphorus			D_2S	1.345
			S—O	SO_2	1.4321
P—Br	PBr_3	2.23(1)		$SOCl_2$	1.45(2)
P—Cl	PCl_3	2.00(2)	S—S	RSSR	2.05(1)
P—F	$PFCl_2$	1.55(3)			

TABLE 3.5 Bond Strengths

The quantity $D_0(A\!-\!B)$ corresponds to the bond dissociation energy at 0 K, all species considered to be ideal gases, for a bond $A\!-\!B$ which is broken through the reaction: Eq. $AB \rightarrow A + B$

where

$$D_0 = \Delta H f_0^\circ(A) + \Delta H f_0^\circ(B) - \Delta H f_0^\circ(AB)$$

D_0 at 298 K, or $\Delta H f_{298}$ is greater than D_0 at 0 K by an amount which lies between RT and $3/2\,RT$, or between 0.6 and 0.9 kcal·mol^{-1}. In polyatomic molecules this difference may be somewhat greater. It is important to note that the bond dissociation energy refers to the enthalpy change $\Delta H f$ in the dissociation process.

The strengths of carbon–carbon bonds are greatest for triple bonds and typically decrease with bond order. Other factors play a role as well. A carbon–carbon single bond between two multiple bonds will share some additional electron density and be correspondingly strengthened. Bonds are weaker than "normal" when cleavage of a bond gives a particularly stable species. For example, cleavage of a single bond in ethylbenzene to give benzyl radical and methyl radical, $C_6H_5CH_2\!-\!CH_3 \rightarrow C_6H_5CH_2\bullet + \bullet CH_3$, is favorable because benzyl radical is stable. Thus, the $C\!-\!C$ bond strength in this case is only 72 kcal/mol compared to the single bond of propane, which is 85 kcal/mol. Hexamethylethane may dissociate into two t-butyl radicals $[(CH_3)_3C\bullet + \bullet CH_3]$. The combination of steric crowding and the formation of stabilized radicals makes the $C\!-\!C$ bond strength in $(CH_3)_3C\!-\!C(CH_3)_3$ only about 68 kcal/mol. A few selected examples follow:

The numbers in parentheses following a numerical value represent the standard deviation of that value in terms of the final listed digit(s).

To convert the tabulated values (in kcal·mol^{-1}) to kJ·mol^{-1}, multiply by 4.184.

Source: T. L. Cottrell, *The Strengths of Chemical Bonds*, 2nd ed., Butterworth, London, 1958; B. deB. Darwent, National Standard Reference Data Series, National Bureau of Standards, no. 31, Washington, 1970; S. W. Benson, *J. Chem. Educ.*, **42**: 502 (1965); and J. A. Kerr, *Chem. Rev.*, **66**: 465 (1966).

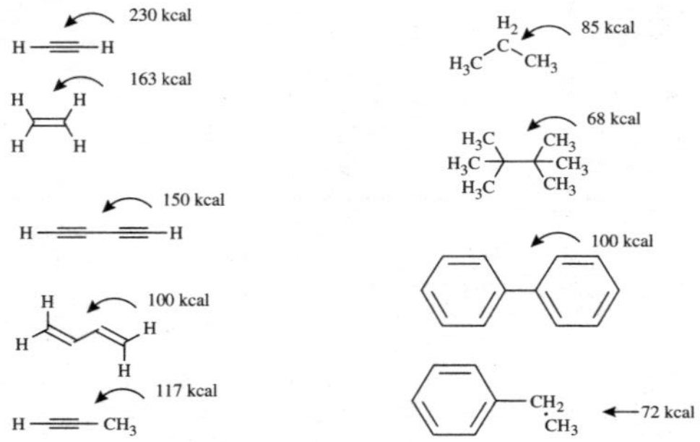

TABLE 3.5 Bond Strengths (*continued*)

Boron / Bromine

Bond	D_0°, kcal · mol^{-1}	ΔHf_{298}, kcal · mol^{-1}
Boron		
H$_3$B—BH$_3$		35
F$_2$B—F	133(20)	
Bromine		
Br—Br	45.45(1)	46.10(1)
Br—CH$_3$	67(2)	68(2)
Br—CH$_2$Br		61(3)
Br—CHBr$_2$		62(4)
Br—CBr$_3$	49(3)	50(3)
Br—CCl$_3$	51(3)	52(3)
Br—CF$_3$		68(3)
Br—CF$_2$CF$_3$		68.7(15)
Br—CF$_2$CF$_2$CF$_3$		66.5(15)
Br—CHF$_2$		69
Br—Cl	51.6(1)	52.3(1)
Br—F	67.2	68.1
Br—CN		91
Br—CO—C$_6$H$_5$		64
Br—N	68(5)	
Br—NF$_2$		53

Carbon (*continued*)

Bond	D_0°, kcal · mol^{-1}	ΔHf_{298}, kcal · mol^{-1}
CH$_3$—CH$_2$CN		73(2)
CH$_3$—CH(CH$_3$)CN		79(2)
CH$_3$—C(C$_6$H$_5$)CN(CH$_3$)		60
C$_2$H$_5$—CH$_2$CN		76.9(17)
CH$_3$—CF$_3$		101.2(11)
CH$_2$F—CH$_2$F		88(2)
CF$_3$—CF$_3$		97(2)
CF$_2$=CF$_2$		76(3)
CF$_3$—CN		120
CH$_3$—CO	80.6	81.9
CH$_3$—CHO		75
CH$_3$CO—CF$_3$		73.8
CH$_3$CO—COCH$_3$		67(2)
C$_2$H$_5$CO—COC$_6$H$_5$		66.4
C$_6$H$_5$CH$_2$CO—CH$_2$C$_6$H$_5$		65.4
C$_6$H$_5$CH$_2$—COOH		68.1
(C$_6$H$_5$CH$_2$)$_2$CH—COOH		59.4
NC—CN	143(5)	144(5)
CF$_3$—NF$_2$		65(3)
CH$_3$—NH$_2$		79(3)
C$_6$H$_5$CH$_2$—NH$_2$		72(1)

Bond		
Br—NO	27.8(15)	28.7(15)
Br—O	55.3(1)	56.2(1)

Carbon

Compound		
$HC{\equiv}CH$		230(2)
$H_2C{=}CH_2$		163
$CH_3{-}CH_3$		88
$CH_3{-}C(CH_3)_2CH_3$		69(2)
$CH_3{-}C(CH_3)_3$		80
$CH_3{-}C_6H_5$		93
$CH_3{-}CH_2C_6H_5$		72
$CH_3{-}CH(CH_3)C_6H_5$		71
$C_2H_5{-}CH_2C_6H_5$		71
$C_3H_7{-}CH_2C_6H_5$		67(2)
$CH_3{-}(CH{=}CH_2)$		29
$CH_3{-}(CH_2CH{=}CH_2)$		72
$CH_3{-}(C{\equiv}CH)$		117
$(CH_3)_3C{-}C(CH_3)_3$		67.5
$(CH_3)_3C{-}C(C_6H_5)_3$		15
$(CH_2{=}CH){-}(CH{=}CH_2)$		100
$C_6H_5{-}C_6H_5$		100
$(HC{\equiv}C){-}(C{\equiv}CH)$		150
$CH_3{-}CN$	119(5)	121(5)

Compound	Value
$CH_3{-}NHC_6H_5$	68
$C_6H_5{-}N(CH_3)C_6H_5$	65
$C_6H_5CH_2{-}NHCH_3$	69(1)
$C_6H_5CH_2{-}N(CH_3)_2$	61(1)
$CH_3{-}(N{=}NCH_3)$	52.5
$C_2H_5{-}(N{=}NC_2H_5)$	50.0
$(CH_3)_3C{-}[N{=}NC(CH_3)_3]$	43.5
$C_6H_5CH_2{-}(N{=}NCH_2C_6H_5)$	37.6
$CF_3{-}(N{=}NCF_3)$	55.2
$H_2C{=}NH$	154(5)
$HC{\equiv}N$	224
$CH_3{-}NO$	41.8(9)
$C_2H_5{-}NO$	42.0(13)
$C_3H_7{-}NO$	40.1(18)
$(CH_3)_2CH{-}NO$	41.0(13)
$C_4H_9{-}NO$	51.5(10)
$C_6H_5{-}NO$	51.5(10)
$Cl_3C{-}NO$	32
$F_3C{-}NO$	31
$C_6F_5{-}NO$	50.5(10)
$NC{-}NO$	29(3)
$CH_3{-}NO_2$	59(3)
$C_2H_5{-}NO_2$	62
$CH_3{-}OCH_3$	80

TABLE 3.5 Bond Strengths (*continued*)

Carbon (*continued*)

Bond	D_0°, kcal · mol⁻¹	ΔH_{f298}, kcal · mol⁻¹
$CH_3-OC_6H_5$		91
$CH_3-OCH_2C_6H_5$		67
$C_2H_5-OC_6H_5$		51
$C_6H_5CH_2-OCOCH_3$	67	
$C_6H_5CH_2-OCOC_6H_5$		69
$CH_3CO-OCH_3$		97
$CH_3-O-SOCH_3$		67
$CH_2{=}CHCH_2-OSOCH_3$		50
$C_6H_5CH_2-OSOCH_3$		53
$C{=}O$	256.2(1)	257.3(1)
$H_2C{=}O$		175
$OC{=}O$	125.7(1)	127.2(1)
$SC{=}O$	148	150
$C{\equiv}O$		257
CH_3-SH	71(3)	73(3)
$CH_3-SC_6H_5$		68(2)
$CH_3-SCH_2C_6H_5$		59(2)
$OC-S$	72.9	74.2
CH_2-CH_3		96
$CH_2CH_2-CH_3$		25.5
$(CH_3)_2C-CH_3$		51
$CHCH-CH_3$		32
(methylcyclohexane, CH_3)		27.5

Chlorine (*continued*)

Bond	D_0°, kcal · mol⁻¹	ΔH_{f298}, kcal · mol⁻¹
$Cl-COC_6H_5$		74(3)
$Cl-Cl^+$		94
$Cl-Cl$	57.3(1)	
$Cl-ClO$	33.3(10)	
$O_3Cl-ClO_4$		
$Cl-F$	59.5(5)	58
O_3Cl-F		61
$Cl-N$		62
$Cl-NCl$		67
$Cl-NCl_2$		91
$Cl-NF_2$		ca 32
$Cl-NH_2$		60(6)
$Cl-NO$	37.0(15)	38.0(15)
$Cl-NO_2$	33(1)	34(1)
$Cl-O$	64(1)	
$OCl-O$	58(3)	
O_2Cl-O		48(1)
$Cl-SiCl_3$		111
$Cl-CH_3^+$		51
$Cl-Cl^+$		94

Fluorine

Bond	D_0°, kcal · mol⁻¹	ΔH_{f298}, kcal · mol⁻¹
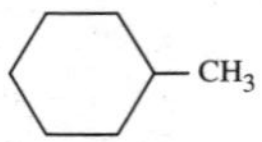 $F-CH_3$		108(5)
$F-C(CH_3)_3$		105

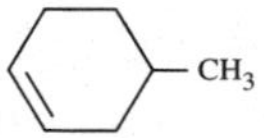

Fluorine (continued)

Bond	D°	D°
$F\!-\!C_6H_5$		116
$F\!-\!CCl_3$		106(5)
$F\!-\!CCl_2F$		110(6)
$F\!-\!CClF_2$		117(6)
$F\!-\!CF_3$		125(4)
$F\!-\!COCH_3$		119
$F\!-\!F$	37(1)	
$OF\!-\!F$	64(3)	
$O_2F\!-\!F$	18.4	
$F\!-\!N$	71(10)	72(10)
$F\!-\!NF$	75(5)	76(5)
$F\!-\!NF_2$	57(2)	58(2)
$F\!-\!NO$	55.2(10)	56.3(10)
$F\!-\!NO_2$	46(5)	
F^+	>60	

Gallium

Bond	D°
$CH_3\!-\!Ga(CH_3)_3$	59.5

Hydrogen

Bond	D°	D°
$H\!-\!Br$	86.6(1)	87.5(1)
$H\!-\!C$	80	81.0(5)
$H\!-\!CH$		108(6)
$H\!-\!CH_2$		113(1)
$H\!-\!CH_3$	112.3(1)	103(2)
$D\!-\!CD_3$	102(2)	
$H\!-\!(C\!\equiv\!CH)$	104.92(5)	125(1)
$H\!-\!(C\!=\!CH_2)$		102
$H\!-\!CH_2CH_3$		98(1)
$H\!-\!CH_2C\!\equiv\!CH$		93.9(12)
$H\!-\!CH_2CH\!=\!CH_2$		85

Radicals and ions

Bond	D°
[methyl-substituted cyclohexenyl structure]	35
[methyl-substituted cyclohexadienyl structure]	11.5
$(CH_3)_2C(\dot CH_2)\!-\!CH_3$	20
$\dot OCH_2\!-\!CH_3$	12
$(CH_3)_2C(\dot O)\!-\!CH_3$	7
$CH_2CO\!-\!CH_3$	30
$\dot OC\!-\!CH_3$	11
$O_2C\!-\!CH_3$	−20
$CH_3\!-\!CH_3^+$	46
$CH_2\!-\!CH_3^+$	119
$CH_2\!=\!CH_2^+$	162
$HC\!\equiv\!CH^+$	223

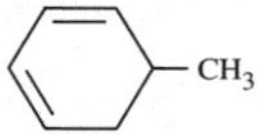

Chlorine

Bond	D°	D°
$Cl\!-\!C$	80.8	80(10)
$Cl\!-\!CH_3$	81(5)	
$Cl\!-\!C(CH_3)_3$	78.5	
$Cl\!-\!CH_2Cl$	74(3)	
$Cl\!-\!CCl_3$	70(5)	
$Cl\!-\!CF_3$	86.1(8)	
$Cl\!-\!CCl_2F$	73(2)	
$Cl\!-\!CF_2Cl$	76(2)	
$Cl\!-\!CF_2CF_3$	82.7(17)	
$Cl\!-\!(CH\!=\!CH_2)$	84	
$Cl\!-\!CN$	105	
$Cl\!-\!COCl$	78.5	
$Cl\!-\!COCH_3$	83.5	

TABLE 3.5 Bond Strengths (*continued*)

Hydrogen (*continued*)

Bond	D°_0, kcal · mol⁻¹	ΔHf_{298}, kcal · mol⁻¹
H—cyclopropyl		101(3)
H—CH₂CH₂CH₃		98(2)
H—CH(CH₃)₂		94.5
H—cyclobutyl		95(3)
H—CH₂CH(CH₃)₂		86
H—CH(CH₃)CH₂CH₃		95(1)
H—C(CH₃)₃		91
(cyclopentenyl, H—)		81(1)
(CH=CH₂ / H—CH / CH=CH₂)		80(1)
(cyclopentenyl, H—)		82(1)
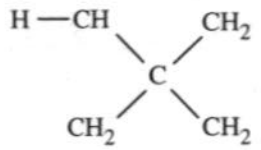 (H—CH / C / CH₂ / CH₂ / CH₂)		99(1)
H—C(CH₃)₂CH=CH₂		79
H—cyclopentyl		94.5(10)
H—CH₂C(CH₃)₃		100(1)
H—C₆H₅		103
H—CH₂C₆H₅		85(1)

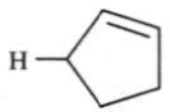

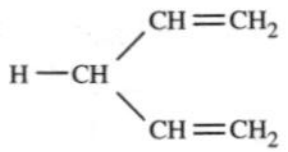

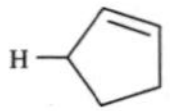

Hydrogen (*continued*)

Bond	D°_0, kcal · mol⁻¹	ΔHf_{298}, kcal · mol⁻¹
H—Cl	102.3(1)	30(2)
H—CO		87(1)
H—CHO		90
H—COOH		87(1)
H—COCH₃		87(1)
H—COCH₂CH₃		
(tetrahydrofuranyl, H—)		92(1)
H—COC₆H₅		87(1)
H—COCF₃		91(2)
H—F	135(1)	135.8
H—H	103.25	104.19
H—D	104.07(1)	105.00
D—D	105.05(1)	105.96
H—I	70.4(1)	71.3(1)
H—N	85(2)	85(2)
H—NH	89(2)	90(2)
H—NH₂	103(2)	104(2)
H—NHCH₃		103(2)
H—N(CH₃)₂		95(2)
H—NHC₆H₅		80(3)
H—N(CH₃)C₆H₅		74(3)
H—NF₂		76(3)
H—N₃		85

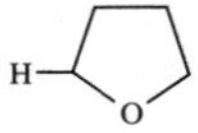

3.24

H—C(C_6H_5)$_3$		75	H—NO		< 49
			H—O	101.3(5)	102.3(5)
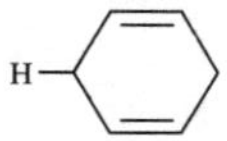		74	H—OH	118.0(2)	119.2(2)
			H—OCH_3		104.4(10)
			H—OCH_2CH_3		104.2
H—cyclohexyl		95.5(10)	H—$OC(CH_3)_3$		105(1)
H—cycloheptyl		92.5(10)	H—$OCH_2C(CH_3)_3$		102.3(15)
H—norbornyl		97(3)	H—OC_6H_5		88(5)
H—CH_2Br	97(5)		H—ONO		78.3(5)
H—$CHBr_2$		104	H—ONO_2		101.2(5)
H—CBr_3	88(2)	90(2)	H—OOH	88.5(20)	89.5(20)
H—CH_2Cl		101	H—$OOCCH_3$		112(4)
H—$CHCl_2$		99.0	H—$OOCCH_2CH_3$		110(4)
H—CCl_3	89(3)	90(3)	H—$OOCC_3H_7$		103(4)
H—CCl_2CHCl_2		94(2)	H—SH	90(1)	91(1)
H—CCl_2CCl_3		95(2)	H—SCH_3		ca 88
H—CH_2F		101(2)	H—SiH_3		94(3)
H—CHF_2		101(2)	H—$Si(CH_3)_3$		90(3)
H—CF_3	105(3)	106(3)	H—$SiCl_3$		91.3(14)
H—CF_2Cl		104(1)	$\dot{C}H_2$—H		106
H—CH_2CF_3		106.7(11)	$\dot{C}H$—H		106
H—CF_2CH_3		99.5(1)	$\dot{C}$—H		81
H—CF_2CF_3		103.1(15)	$\dot{C}H_2CH_2$—H		39
H—$CF_2CF_2CF_3$		104(2)	$\dot{O}CH_2$—H		22
H—CH_2I		103(2)	$\dot{C}O$—H		19
H—CHI_2		103(2)	$\dot{C}HCH$—H		43
H—CN	127(5)	129(5)	H—$\dot{O}$		102
H—CH_2CN		ca 93	H—$O\dot{O}$		47
H—$CH(CH_3)CN$		90(2)	H—$O\dot{C}H_2$		31
H—$C(CH_3)_2CN$		87(2)	H—$OO\dot{C}$		31
H—CH_2NH_2		95(2)	$\dot{C}H$—H		ca 125
H—$CH_2Si(CH_3)_3$		99(1)	$\dot{C}OCH_2$—H		43.5
H—CH_2COCH_3		98.3(18)			

TABLE 3.5 Bond Strengths (*continued*)

Hydrogen (*continued*)

Bond	D_0°, kcal·mol^{-1}	$\Delta H f_{298}$, kcal·mol^{-1}
CH_2CO—H		36
(cyclohexadienyl ring)—H		40
(cyclohexene ring)—H		47.5
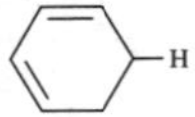 (cyclohexadiene ring)—H		24
C^+—H		85
CH_3—H		30
$CH_3CH_2^+$—H		29
$CH_2CH_3^+$—H		79
H—H$^+$		62

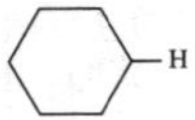

Iodine

Bond	D_0°, kcal·mol^{-1}	$\Delta H f_{298}$, kcal·mol^{-1}
I—Br	41.9(1)	42.5(1)
I—CH_3	54(3)	55.5(30)
I—CH_2CH_3		53.5
I—$CH(CH_3)_2$		53
I—$C(CH_3)_3$		49.5
I—CH_2CF_3		56(1)
I—CF_2CH_3		52(1)

Nitrogen

Bond	D_0°, kcal·mol^{-1}	$\Delta H f_{298}$, kcal·mol^{-1}
N—N	225.07(1)	225.96(1)
F_2N—NF_2	20(1)	21(1)
H_2N—NH_2		71(2)
H_2N—$NHCH_3$		65
H_2N—$N(CH_3)_2$		63
H_2N—NHC_6H_5		51
HN—N_2		9(1)
ON—N	113.5(10)	114.9(10)
ON—NO_2	8.4(2)	9.5(2)
O_2N—NO_2	12.7(5)	13.7(5)
NN—O		40
ON—O		73
HN=NH		109(10)
HN=O		115
N≡N		226
N—N$^+$		200
N—NO$^+$		155
ON—O$^+$		56
ON$^+$—O		56

Osmium

Bond	D_0°, kcal·mol^{-1}	$\Delta H f_{298}$, kcal·mol^{-1}
O_3Os—O		72(5)

Oxygen

Bond	D_0°, kcal·mol^{-1}	$\Delta H f_{298}$, kcal·mol^{-1}
HO—CH_3	88.5(30)	90(3)
HO—$(CH{=}CH_2)$		87

Bond		
I—CF$_2$CF$_3$		51(1)
I—C$_3$F$_7$		50(1)
I—(CH=CHCH$_3$)		41
I—CH$_3^+$		62
I—C$_6$H$_5$		64(1)
I—C$_6$F$_5$		66
I—Cl	49.7(1)	50.5(1)
I—COCH$_3$		52.5
I—CN		73(1)
I—F	66.4(10)	67(1)
I$^+$—H		70
I—I	35.60(1)	36.15
I—I$^+$		61
I—NO		17(1)
I—NO$_2$		18(1)

Lead

Bond		
CH$_3$—Pb(CH$_3$)$_3$		49.4(10)

Lithium

Bond	
Li—H	58

Mercury

Bond		
Hg—Br	16.4(10)	17.4(10)
CH$_3$—HgCH$_3$		57.5
C$_2$H$_5$—HgC$_2$H$_5$		43.7(10)
C$_3$H$_7$—HgC$_3$H$_7$		47.1
(CH$_3$)$_2$CH—HgCH(CH$_3$)$_2$		40.7
C$_6$H$_5$—HgC$_6$H$_5$		68

Bond		
HO—CH$_2$CH=CH$_2$		109
HO—C$_6$H$_5$		103
HO—CH$_2$C$_6$H$_5$		77
HO—CHO		96(3)
HO—COCH$_3$		108(5)
HO—COCH$_2$CH$_3$		43
HO—Cl		60(3)
HO—I		56(3)
HO—NCH$_3$		50
HO—OC(CH$_3$)$_3$		46(2)
O—O	117.97(10)	119.11
CF$_3$O—OCF$_3$	49.5(5)	51.1(5)
CH$_3$O—OCH$_3$		46
C$_2$H$_5$O—OC$_2$H$_5$		37.6(2)
C$_3$H$_7$O—OC$_3$H$_7$		38
O—OF		37
O—O$_2$ClF		58
FO—OF	110.7	62(20)
O=PBr$_3$		119(5)
O=PCl$_3$		122(5)
O=PF$_3$		130(5)
O—O$^+$		168
HO—CH$_3^+$		67

Phosphorus

Bond		
P—Br		63.7
P—Cl		78.5
P—F		117
P—H		79(1)
P—O	141.5(10)	142.3(10)
P—P	115(2)	116(2)
P=S	82	

TABLE 3.5 Bond Strengths (continued)

Bond	D°_0, kcal·mol⁻¹	$\Delta H_{f\,298}$, kcal·mol⁻¹
Ruthenium		
O—RuO₃	104	
Selenium		
Se—Cl		
Se—F		
Se—O	81(23)	58
Se—Se	65	68
Silicon		
Si—Br	69(14)	
Si—Cl	76(12)	
Si—F	135	
Si—H	74(6)	
Si—I	56	
Si—N	ca 104	
Si—O	185(7)	
Si—S	147(3)	
Si—Se	134(6)	
Si—Si	42	
H₃Si—SiH₃	81(4)	
(CH₃)₃Si—Si(CH₃)₃	81	
(C₆H₅)₃Si—Si(C₆H₅)₃	88(7)	
Si—Te	122(9)	

Bond	D°_0, kcal·mol⁻¹	$\Delta H_{f\,298}$, kcal·mol⁻¹
Sulfur		
S—Cl	16	61
O₂S—F	115	
S—N	123.6(20)	
S—O	130.8(20)	
OS—O	81.9(10)	83.2(10)
O₂S—O	101.5(15)	102.5(15)
S—S		65(5)
HS—SH		
S—Te	60	
HS⁺—H		104
HS—H⁺		161
OS—O⁺		155
Tin		
BrSn—Br		78
Br₃Sn—Br		65
C₂H₅Sn—(C₂H₅)₃		ca 57
Sn—Cl		76
Sn—H		61.0(7)
Sn—I		65
Sn—O	130(5)	131(5)
Sn—S	111(5)	112(5)
Xenon		
Xe—F		31(1)

Sodium		Zinc	
Na—H	47	Zn—H	19.6(5)
Na—K	14.3	C₂H₅Zn—C₂H₅	
Na—Na	17.3		
Na—OH	91(3)		ca 48

$$Na-H \quad 47$$
$$Na-K \quad 14.3$$
$$Na-Na \quad 17.3$$
$$Na-OH \quad 91(3)$$

$$Zn-H \quad 19.6(5)$$
$$C_2H_5Zn-C_2H_5 \quad ca\ 48$$

BOND AND GROUP DIPOLE MOMENTS

All bonds between equal atoms are given zero values. Because of their symmetry, methane and ethane molecules are nonpolar. The principle of bond moments thus requires that the CH_3 group moment equal one H—C moment. Hence the substitution of any aliphatic H by CH_3 does not alter the dipole moment, and all saturated hydrocarbons have zero moments as long as the tetrahedral angles are maintained.

The group moment always includes the C—X bond. When the group is attached to an aromatic system, the moment contains the contributions through resonance of those polar structures postulated as arising through charge shifts around the ring.

All values for bond and group dipole moments in Tables 3.6 and 3.7 were obtained in benzene solution.

TABLE 3.6 Bond Dipole Moments

Bond	Moment, D*	Bond	Moment, D*
H—C		Se—C	0.7
Aliphatic	0.3	Si—C	1.2
Aromatic	0.0	Si—H	1.0
C—C	0.0	Si—N	1.55
C≡C	0.0	H—Sb	−0.08
C—O		G—As	−0.10
Ether, aliphatic	0.74	H—P	0.36
Alcohol, aliphatic	0.7	H—I	0.38
C=O		H—Br	0.78
Aliphatic	2.4	H—Cl	1.08
Aromatic	2.65	H—F	1.94
O—H	1.51	C—Te	0.6
C—S	0.9	N—F	0.17
C=S	2.0	P—I	0.3
S—H	0.65	P—Br	0.36
S—O	(0.2)	P—Cl	0.81
S=O		As—I	0.78
Aliphatic	2.8	As—Br	1.27
Aromatic	3.3	As—Cl	1.64
C—N, aliphatic	0.45	As—F	2.03
C=N	1.4	Sb—I	0.8
C≡N (nitrile)	3.6	Sb—Br	1.9
NC (isonitrile)	3.0	Sb—Cl	2.6
N—H	1.31	S—Cl	0.7
N—O	0.3	Cl—O	0.7
N=O	2.0	I—Br	1.2
N: lone pair on sp^3N	1.0	I—Cl	1
C—P, aliphatic	0.8	Br—Cl	0.57
P—O	(0.3)	Br—F	1.3
P=O	2.7	Cl—F	0.88
P—S	0.5	Li—C	1.4
P=S	2.9	K—Cl	10.6
B—C, aliphatic	0.7	K—F	7.3
B—O	0.25		

TABLE 3.6 Bond Dipole Moments (*continued*)

Bond	Moment, D*	Bond	Moment, D*
Cs—Cl	10.5	Dative (coordination) bonds (*continued*)	
Cs—F	7.9		
		$P \rightarrow O$	2.9
Dative (coordination) bonds		$S \rightarrow O$	3.0
		$As \rightarrow O$	4.2
$N \rightarrow B$	2.6	$Se \rightarrow O$	3.1
$O \rightarrow B$	3.6	$Te \rightarrow O$	2.3
$S \rightarrow B$	3.8	$P \rightarrow S$	3.1
$P \rightarrow B$	4.4	$P \rightarrow Se$	3.2
$N \rightarrow O$	4.3	$Sb \rightarrow S$	4.5

*To convert debye units D into coulomb-meters, multiply by 3.33564×10^{-30}.

TABLE 3.7 Group Dipole Moments

	Moment, D*	
Group	Aromatic C—X	Aliphatic C—X
C—CH$_3$	0.37	0.0
C—C$_2$H$_5$	0.37	0.0
C—C(CH$_3$)$_3$	0.5	0.0
C—CH=CH$_2$	<0.4	0.6
C—C≡CH	0.7	0.9
C—F	1.47	1.79
C—Cl	1.59	1.87
C—Br	1.57	1.82
C—I	1.40	1.65
C—CH$_2$F	1.77(g)	
C—CF$_3$	2.54	2.32
C—CH$_2$Cl	1.85	1.95
C—CHCl$_2$	2.04	1.94
C—CCl$_3$	2.11	1.57
C—CH$_2$Br	1.86	1.96
C—C≡N	4.05	3.4
C—NC	3.5	3.5
C—CH$_2$CN	1.86	2.0
C—C=O	2.65	2.4
C—CHO	2.96	2.49
C—COOH	1.64	1.63
C—CO—CH$_3$	2.96	2.75
C—CO—OCH$_3$	1.83	1.75
C—CO—OC$_2$H$_5$	1.9	1.8
C—OH	1.6	1.7
C—OCH$_3$	1.28	1.28
C—OCF$_3$	2.36	

TABLE 3.7 Group Dipole Moments (*continued*)

Group	Moment, D*	
	Aromatic C—X	Aliphatic C—X
C—OCOCH$_3$	1.69	
C—OC$_6$H$_5$	1.16	1.16
C—CH$_2$OH	1.68	1.68
C—NH$_2$	1.53	1.46
C—NHCH$_3$	1.71	
C—N(CH$_3$)$_2$	1.58	0.86
C—NHCOCH$_3$	3.69	
C—N(C$_6$H$_5$)$_2$	(0.3)	−0.3
C—NCO	2.32	2.8
C—N$_3$	1.44	
C—NO	3.09	
C—NO$_2$	4.01	2.70
C—CH$_2$NO$_2$	3.3	3.4
C—SH	1.22	1.55
C—SCH$_3$	1.34	1.40
C—SCF$_3$	2.50	
C—SCN	3.59	3.6
C—NCS	2.9	3.3
C—SC$_6$H$_5$	1.51	1.5
C—SF$_5$	3.4	
C—SOCF$_3$	3.88	
(C—)$_2$SO$_2$	5.05	4.53
(C—)$_2$SO$_2$CH$_3$	4.73	
(C—)$_2$SO$_2$CF$_3$	4.32	
C—SeH	1.08	
C—SeCH$_3$	1.31	1.32
C—Si(CH$_3$)$_3$	0.44	0.4

*To convert debye units D into coulomb-meters, multiply by 3.33564×10^{-30}.

SECTION 4
PHYSICAL PROPERTIES

SOLUBILITIES

TABLE 4.1 Solubility of Gases in Water

Explanation of the column headings

α, *Volume of gas in milliliters (mL).* The column or line entry headed "α" gives the volume of gas (in milliliters) at standard conditions (0°C and 760 mm or 101.325 kN·m^{-2}) dissolved in 1 mL of water at the temperature stated (in degrees Celsius) and when the pressure of the gas without the water vapor is 760 mm.

A, *Volume of gas in milliliters (mL).* The line entry "A" indicates the same quantity as "α" except that the gas itself is at the uniform pressure of 760 mm when in equilibrium with water.

l, *Volume of gas in milliliters (mL).* The column headed "l" gives the volume of the gas (in milliliters) dissolved in 1 mL of water when the pressure of the gas plus that of the water vapor is 760 mm.

q, *Weight of the gas in grams (g).* The column headed "q" gives the weight of gas (in grams) dissolved in 100 g of water when the pressure of the gas plus that of the water vapor is 760 mm

Temp. °C	Acetylene		Air*		Ammonia		Bromine	
	α	q	$\alpha(\times 10^3)$	% oxygen in air	α	q	α	q
0	1.73	0.200	29.18	34.91	1130	89.5	60.5	42.9
1	1.68	0.194	28.42	34.87	—	—	—	—
2	1.63	0.188	27.69	34.82	—	—	54.1	38.3
3	1.58	0.182	26.99	34.78	—	—	—	—
4	1.53	0.176	26.32	34.74	1047	79.6	48.3	34.2
5	1.49	0.171	25.68	34.69	—	—	—	—
6	1.45	0.167	25.06	34.65	—	—	43.3	30.6
7	1.41	0.162	24.47	34.60	—	—	—	—
8	1.37	0.157	23.90	34.56	947	72.0	38.9	27.5
9	1.34	0.154	23.36	34.52	—	—	—	—
10	1.31	0.150	22.84	34.47	870	68.4	35.1	24.8
11	1.27	0.146	22.34	34.43	—	—	—	—
12	1.24	0.142	21.87	34.38	857	65.1	31.5	22.2
13	1.21	0.138	21.41	34.34	837	63.6	—	—
14	1.18	0.135	20.97	34.30	—	—	28.4	20.0
15	1.15	0.131	20.55	34.25	770	—	—	—
16	1.13	0.129	20.14	34.21	775	58.7	25.7	18.0
17	1.10	0.125	19.75	34.17	—	—	—	—
18	1.08	0.123	19.38	34.12	—	—	23.4	16.4
19	1.05	0.119	19.02	34.08	—	—	—	—
20	1.03	0.117	18.68	34.03	680	52.9	21.3	14.9
21	1.01	0.115	18.34	33.99	—	—	—	—
22	0.99	0.112	18.01	33.95	—	—	19.4	13.5
23	0.97	0.110	17.69	33.90	—	—	—	—
24	0.95	0.107	17.38	33.86	639	48.2	17.7	12.3
25	0.93	0.105	17.08	33.82	—	—	—	—
26	0.91	0.102	16.79	33.77	—	—	16.3	11.3
27	0.89	0.100	16.50	33.73	—	—	—	—
28	0.87	0.098	16.21	33.68	586	44.0	15.0	10.3
29	0.85	0.095	15.92	33.64	—	—	—	—
30	0.84	0.094	15.64	33.60	530	41.0	13.8	9.5
35	—	—	—	—	—	—	—	—
40	—	—	14.18	—	400	31.6	9.4	6.3
45	—	—	—	—	—	—	—	—
50	—	—	12.97	—	290	23.5	6.5	4.1
60	—	—	12.16	—	200	16.8	4.9	2.9
70	—	—	—	—	—	11.1	3.8	1.9
80	—	—	11.26	—	—	6.5	3.0	1.2
90	—	—	—	—	—	3.0	—	—
100	—	—	11.05	—	—	0.0	—	—

*Free from NH$_3$ and CO$_2$; total pressure of air + water vapor is 760 mm.

TABLE 4.1 Solubility of Gases in Water (*continued*)

Temp. °C	Carbon dioxide		Carbon monoxide		Chlorine		Ethane		Ethylene		Hydrogen	
	α	q	α	q	l	q	α	q	α	q	α	q
0	1.713	0.3346	0.03537	0.004397	—	—	0.09874	0.01317	0.226	0.0281	0.02148	0.0001922
1	1.646	0.3213	0.03455	0.004293	—	—	0.09476	0.01263	0.219	0.0272	0.02126	0.0001901
2	1.584	0.3091	0.03375	0.004191	—	—	0.09093	0.01212	0.211	0.0262	0.02105	0.0001881
3	1.527	0.2978	0.03297	0.004092	—	—	0.08725	0.01162	0.204	0.0253	0.02084	0.0001862
4	1.473	0.2871	0.03222	0.003996	—	—	0.08372	0.01114	0.197	0.0244	0.02064	0.0001843
5	1.424	0.2774	0.03149	0.003903	—	—	0.08033	0.01069	0.191	0.0237	0.02044	0.0001824
6	1.377	0.2681	0.03078	0.003813	—	—	0.07709	0.01025	0.184	0.0228	0.02025	0.0001806
7	1.331	0.2589	0.03009	0.003725	—	—	0.07400	0.00983	0.178	0.0220	0.02007	0.0001789
8	1.282	0.2492	0.02942	0.003640	—	—	0.07106	0.00943	0.173	0.0214	0.01989	0.0001772
9	1.237	0.2403	0.02878	0.003559	—	—	0.06826	0.00906	0.167	0.0207	0.01972	0.0001756
10	1.194	0.2318	0.02816	0.003479	3.148	0.9972	0.06561	0.00870	0.162	0.0200	0.01955	0.0001740
11	1.154	0.2239	0.02757	0.003405	3.047	0.9654	0.06328	0.00838	0.157	0.0194	0.01940	0.0001725
12	1.117	0.2165	0.02701	0.003332	2.950	0.9346	0.06106	0.00808	0.152	0.0188	0.01925	0.0001710
13	1.083	0.2098	0.02646	0.003261	2.856	0.9050	0.05894	0.00780	0.148	0.0183	0.01911	0.0001696
14	1.050	0.2032	0.02593	0.003194	2.767	0.8768	0.05694	0.00753	0.143	0.0176	0.01897	0.0001682
15	1.019	0.1970	0.02543	0.003130	2.680	0.8495	0.05504	0.00727	0.139	0.0171	0.01883	0.0001668
16	0.985	0.1903	0.02494	0.003066	2.597	0.8232	0.05326	0.00703	0.136	0.0167	0.01869	0.0001654
17	0.956	0.1845	0.02448	0.003007	2.517	0.7979	0.05159	0.00680	0.132	0.0162	0.01856	0.0001641
18	0.928	0.1789	0.02402	0.002947	2.440	0.7738	0.05003	0.00659	0.129	0.0158	0.01844	0.0001628
19	0.902	0.1737	0.02360	0.002891	2.368	0.7510	0.04858	0.00639	0.125	0.0153	0.01831	0.0001616
20	0.878	0.1688	0.02319	0.002838	2.299	0.7293	0.04724	0.00620	0.122	0.0149	0.01819	0.0001603
21	0.854	0.1640	0.02281	0.002789	2.238	0.7100	0.04589	0.00602	0.119	0.0146	0.01805	0.0001588
22	0.829	0.1590	0.02244	0.002739	2.180	0.6918	0.04459	0.00584	0.116	0.0142	0.01792	0.0001575
23	0.804	0.1540	0.02208	0.002691	2.123	0.6739	0.04335	0.00567	0.114	0.0139	0.01779	0.0001561
24	0.781	0.1493	0.02174	0.002646	2.070	0.6572	0.04217	0.00551	0.111	0.0135	0.01766	0.0001548

TABLE 4.1 Solubility of Gases in Water (*continued*)

Temp. °C	Carbon dioxide		Carbon monoxide		Chlorine		Ethane		Ethylene		Hydrogen	
	α	q	α	q	l	q	α	q	α	q	α	q
25	0.759	0.144 9	0.021 42	0.002 603	2.019	0.641 3	0.041 04	0.005 35	0.108	0.013 1	0.017 54	0.000 153 5
26	0.738	0.140 6	0.021 10	0.002 560	1.970	0.625 9	0.039 97	0.005 20	0.106	0.012 9	0.017 42	0.000 152 2
27	0.718	0.136 6	0.020 80	0.002 519	1.923	0.611 2	0.038 95	0.005 06	0.104	0.012 6	0.017 31	0.000 150 9
28	0.699	0.132 7	0.020 51	0.002 479	1.880	0.597 5	0.037 99	0.004 93	0.102	0.012 3	0.017 20	0.000 149 6
29	0.682	0.129 2	0.020 24	0.002 442	1.839	0.584 7	0.037 09	0.004 80	0.100	0.012 1	0.017 09	0.000 148 4
30	0.665	0.125 7	0.019 98	0.002 405	1.799	0.572 3	0.036 24	0.004 68	0.098	0.011 8	0.016 99	0.000 147 4
35	0.592	0.110 5	0.018 77	0.002 231	1.602	0.510 4	0.032 30	0.004 12	—	—	0.016 66	0.000 142 5
40	0.530	0.097 3	0.017 75	0.002 075	1.438	0.459 0	0.029 15	0.003 66	—	—	0.016 44	0.000 138 4
45	0.479	0.086 0	0.016 90	0.001 933	1.322	0.422 8	0.026 60	0.003 27	—	—	0.016 24	0.000 134 1
50	0.436	0.076 1	0.016 15	0.001 797	1.225	0.392 5	0.024 59	0.002 94	—	—	0.016 08	0.000 128 7
60	0.359	0.057 6	0.014 88	0.001 522	1.023	0.329 5	0.021 77	0.002 39	—	—	0.016 00	0.000 117 8
70	—	—	0.014 40	0.001 276	0.862	0.279 3	0.019 48	0.001 85	—	—	0.016 0	0.000 102
80	—	—	0.014 30	0.000 980	0.683	0.222 7	0.018 26	0.001 34	—	—	0.016 0	0.000 079
90	—	—	0.014 2	0.000 57	0.39	0.127	0.017 6	0.000 8	—	—	0.016 0	0.000 046
100	—	—	0.014 1	0.000 00	0.00	0.000	0.017 2	0.000 0	—	—	0.016 0	0.000 000

TABLE 4.1 Solubility of Gases in Water (*continued*)

Temp. °C	Hydrogen sulfide α	Hydrogen sulfide q	Methane α	Methane q	Nitric oxide α	Nitric oxide q	Nitrogen* α	Nitrogen* q	Oxygen α	Oxygen q	Sulfur dioxide l	Sulfur dioxide q
0	4.670	0.7066	0.055 63	0.003 959	0.073 81	0.009 833	0.023 54	0.002 942	0.048 89	0.006 945	79.789	22.83
1	4.522	0.6839	0.054 01	0.003 842	0.071 84	0.009 564	0.022 97	0.002 869	0.047 58	0.006 756	77.210	22.09
2	4.379	0.6619	0.052 44	0.003 728	0.069 93	0.009 305	0.022 41	0.002 798	0.046 33	0.006 574	74.691	21.37
3	4.241	0.6407	0.050 93	0.003 619	0.068 09	0.009 057	0.021 87	0.002 730	0.045 12	0.006 400	72.230	20.66
4	4.107	0.6201	0.049 46	0.003 513	0.066 32	0.008 816	0.021 35	0.002 663	0.043 97	0.006 232	69.828	19.98
5	3.977	0.6001	0.048 05	0.003 410	0.064 61	0.008 584	0.020 86	0.002 600	0.042 87	0.006 072	67.485	19.31
6	3.852	0.5809	0.046 69	0.003 312	0.062 98	0.008 361	0.020 37	0.002 537	0.041 80	0.005 918	65.200	18.65
7	3.732	0.5624	0.045 39	0.003 217	0.061 40	0.008 147	0.019 90	0.002 477	0.040 80	0.005 773	62.973	18.02
8	3.616	0.5446	0.044 13	0.003 127	0.059 90	0.007 943	0.019 45	0.002 419	0.039 83	0.005 632	60.805	17.40
9	3.505	0.5276	0.042 92	0.003 039	0.058 46	0.007 747	0.019 02	0.002 365	0.038 91	0.005 498	58.697	16.80
10	3.399	0.5112	0.041 77	0.002 955	0.057 09	0.007 560	0.018 61	0.002 312	0.038 02	0.005 368	56.647	16.21
11	3.300	0.4960	0.040 72	0.002 879	0.055 87	0.007 393	0.018 23	0.002 263	0.037 18	0.005 246	54.655	15.64
12	3.206	0.4814	0.039 70	0.002 805	0.054 70	0.007 233	0.017 86	0.002 216	0.036 37	0.005 128	52.723	15.09
13	3.115	0.4674	0.038 72	0.002 733	0.053 57	0.007 078	0.017 50	0.002 170	0.035 59	0.005 014	50.849	14.56
14	3.028	0.4540	0.037 79	0.002 665	0.052 50	0.006 930	0.017 17	0.002 126	0.034 86	0.004 906	49.033	14.04
15	2.945	0.4411	0.036 90	0.002 599	0.051 47	0.006 788	0.016 85	0.002 085	0.034 15	0.004 802	47.276	13.54
16	2.865	0.4287	0.036 06	0.002 538	0.050 49	0.006 652	0.016 54	0.002 045	0.033 48	0.004 703	45.578	13.05
17	2.789	0.4169	0.035 25	0.002 478	0.049 56	0.006 524	0.016 25	0.002 006	0.032 83	0.004 606	43.939	12.59
18	2.717	0.4056	0.034 48	0.002 422	0.048 68	0.006 400	0.014 97	0.001 970	0.032 20	0.004 514	42.360	12.14
19	2.647	0.3948	0.033 76	0.002 369	0.047 85	0.006 283	0.015 70	0.001 935	0.031 61	0.004 426	40.838	11.70

TABLE 4.1 Solubility of Gases in Water (*continued*)

Temp. °C	Hydrogen sulfide		Methane		Nitric oxide		Nitrogen*		Oxygen		Sulfur dioxide	
	α	q	α	q	α	q	α	q	α	q	α	q
20	2.582	0.3846	0.033 08	0.002 319	0.047 06	0.006 173	0.015 45	0.001 901	0.031 02	0.004 339	39.374	11.28
21	2.517	0.3745	0.032 43	0.002 270	0.046 25	0.006 059	0.015 22	0.001 869	0.030 44	0.004 252	37.970	10.88
22	2.456	0.3648	0.031 80	0.002 222	0.045 45	0.005 947	0.014 98	0.001 838	0.029 88	0.004 169	36.617	10.50
23	2.396	0.3554	0.031 19	0.002 177	0.044 69	0.005 838	0.014 75	0.001 809	0.029 34	0.004 087	35.302	10.12
24	2.338	0.3463	0.030 61	0.002 133	0.043 95	0.005 733	0.014 54	0.001 780	0.028 81	0.004 007	34.026	9.76
25	2.282	0.3375	0.030 06	0.002 091	0.043 23	0.005 630	0.014 34	0.001 751	0.028 31	0.003 931	32.786	9.41
26	2.229	0.3290	0.029 52	0.002 050	0.042 54	0.005 530	0.014 13	0.001 724	0.027 83	0.003 857	31.584	9.06
27	2.177	0.3208	0.029 01	0.002 011	0.041 88	0.005 435	0.013 94	0.001 698	0.027 36	0.003 787	30.422	8.73
28	2.128	0.3130	0.028 52	0.001 974	0.041 24	0.005 342	0.013 76	0.001 672	0.026 91	0.003 718	29.314	8.42
29	2.081	0.3055	0.028 06	0.001 938	0.040 63	0.005 252	0.013 58	0.001 647	0.026 49	0.003 651	28.210	8.10
30	2.037	0.2983	0.027 62	0.001 904	0.040 04	0.005 165	0.013 42	0.001 624	0.026 08	0.003 588	27.161	7.80
35	1.831	0.2648	0.025 46	0.001 733	0.037 34	0.004 757	0.012 56	0.001 501	0.024 40	0.003 315	22.489	6.47
40	1.660	0.2361	0.023 69	0.001 586	0.035 07	0.004 394	0.011 84	0.001 391	0.023 06	0.003 082	18.766	5.41
45	1.516	0.2110	0.022 38	0.001 466	0.033 11	0.004 059	0.011 30	0.001 300	0.021 87	0.002 858	—	—
50	1.392	0.1883	0.021 34	0.001 359	0.031 52	0.003 758	0.010 88	0.001 216	0.020 90	0.002 657	—	—
60	1.190	0.1480	0.019 54	0.001 144	0.029 54	0.003 237	0.010 23	0.001 052	0.019 46	0.002 274	—	—
70	1.022	0.1101	0.018 25	0.000 926	0.028 10	0.002 668	0.009 77	0.000 851	0.018 33	0.001 856	—	—
80	0.917	0.0765	0.017 70	0.000 695	0.027 00	0.001 984	0.009 58	0.000 660	0.017 61	0.001 381	—	—
90	0.84	0.041	0.017 35	0.000 40	0.0265	0.001 13	0.009 5	0.000 38	0.0172	0.000 79	—	—
100	0.81	0.000	0.017 0	0.000 00	0.0263	0.000 00	0.009 5	0.000 00	0.0170	0.000 00	—	—

*Atmospheric nitrogen containing 98.815% N_2 by volume + 1.185% inert gases.

TABLE 4.1 Solubility of Gases in Water (*continued*)

Substance		0°	10°	20°	30°	40°	60°	80°
Argon	α	0.052 8	0.041 3	0.033 7	0.028 8	0.025 1	0.0209 9	0.018 4
Helium	A	0.009 8	0.009 11	0.008 6	0.008 39	0.008 41	0.009 02	0.0009 42$^{70°}$
Hydrogen bromide	l	612	582		533$^{25°}$	385	469$^{50°}$	406$^{75°}$
Hydrogen chloride	α	512	475	442	412		339	
Krypton	α	0.110 5	0.081 0	0.062 6	0.051 1	0.043 3	0.035 7	
Neon	A		0.0117 9$^{9°}$	0.010 6	0.010 0	0.009 48$^{42°}$		0.009 84$^{73°}$
Nitrous oxide	A		0.88	0.63				
Ozone	$g \cdot L^{-1}$	0.039 4	0.029 9$^{12°}$	0.021 0$^{19°}$	0.0139$^{27°}$	0.004 2	0	
Radon	α	0.510	0.326	0.222	0.162	0.126	0.085	
Xenon	α	0.242	0.174	0.123	0.098	0.082		

VAPOR PRESSURES

TABLE 4.2   Vapor Pressure of Mercury

Temp. °C	mm of Hg	Temp. °C	mm of Hg	Temp. °C	mm of Hg
0	0.000 185	78	0.078 89	158	3.873
		80	0.088 80	160	4.189
2	0.000 228				
4	0.000 276	82	0.100 0		
6	0.000 335	84	0.112 4	162	4.528
8	0.000 406	86	0.126 1	164	4.890
10	0.000 490	88	0.141 3	166	5.277
		90	0.1582	168	5.689
12	0.000 588			170	6.128
14	0.000 706	92	0.1769		
16	0.000 846	94	0.1976	172	6.596
18	0.001 009	96	0.2202	174	7.095
20	0.001 201	98	0.2453	176	7.626
		100	0.2729	178	8.193
22	0.001 426			180	8.796
24	0.001 691	102	0.3032		
26	0.002 000	104	0.3366	182	9.436
28	0.002 359	106	0.3731	184	10.116
30	0.002 777	108	0.4132	186	10.839
		110	0.4572	188	11.607
32	0.003 261	112	0.5052	190	12.423
34	0.003 823	114	0.5576		
36	0.004 471	116	0.6150	192	13.287
38	0.005 219	118	0.6776	194	14.203
40	0.006 079	120	0.7457	196	15.173
		122	0.8198	198	16.200
42	0.007 067	122	0.8198	200	17.287
44	0.008 200	124	0.9004		
46	0.009 497	126	0.9882		
48	0.010 98	128	1.084	202	18.437
50	0.012 67	130	1.186	204	19.652
				206	20.936
52	0.014 59	132	1.298	208	22.292
54	0.016 77	134	1.419	210	23.723
56	0.019 25	136	1.551		
58	0.022 06	138	1.692	212	25.233
60	0.025 24	140	1.845	214	26.826
				216	28.504
62	0.028 83	142	2.010	218	30.271
64	0.032 87	144	2.188	220	32.133
66	0.037 40	146	2.379		
68	0.042 51	148	2.585		
70	0.048 25	150	2.807	222	34.092
				224	36.153
72	0.054 69	152	3.046	226	38.318
74	0.061 89	154	3.303	228	40.595
76	0.069 93	156	3.578	230	42.989

TABLE 4.2 Vapor Pressure of Mercury (*continued*)

Temp. °C	mm of Hg	Temp. °C	mm of Hg	Temp. °C	mm of Hg
232	45.503	302	257.78	372	994.34
234	48.141	304	269.17	374	1028.9
236	50.909	306	280.98	376	1064.4
238	53.812	308	293.21	378	1100.9
240	56.855	310	305.89	380	1138.4
242	60.044	312	319.02	382	1177.0
244	63.384	314	332.62	384	1216.6
246	66.882	316	346.70	386	1257.3
248	70.543	318	361.26	388	1299.1
250	74.375	320	376.33	390	1341.9
252	78.381	322	391.92	392	1386.1
254	82.568	324	408.04	394	1431.3
256	86.944	326	424.71	396	1477.7
258	91.518	328	441.94	398	1525.2
260	96.296	330	459.74	400	1574.1
262	101.28	332	478.13	430	2464
264	106.48	334	497.12	460	3715
266	111.91	336	516.74	490	5420
268	117.57	338	537.00		
270	123.47	340	557.90	520	7691
				550	10650
272	129.62	342	579.45	600	22.87 atm
274	136.02	344	601.69	650	35.49 atm
276	142.69	346	624.64	700	52.51 atm
278	149.64	348	648.30		
280	156.87	350	672.69	750	74.86 atm
				800	103.31 atm
282	164.39	352	697.83	850	138.42 atm
284	172.21	354	723.73	900*	180.92 atm
286	180.34	356	750.43	950	226.58 atm
288	188.79	358	777.92		
290	197.57	360	806.23	1000	290.5 atm
				1050	358.1 atm
292	206.70	362	835.38	1100	437.3 atm
294	216.17	364	865.36	1150	521.3 atm
296	226.00	366	896.23	1200	616.8 atm
298	236.21	368	928.02	1250	721.4 atm
300	246.80	370	960.66	1300	835.9 atm

*Critical point.

TABLE 4.3 Vapor Pressure of Water for Temperatures from -10 to $120\,°C$

The values in the table are for water in contact with its own vapor. Where the water is in contact with air at a temperature t in degrees Celsius, the following correction must be added: Correction (when $t \leqslant 40\,°C$) = $p(0.775 - 0.000\,313\,t)/100$; correction (when $t > 50\,°C$) = $p(0.0652 - 0.000\,087\,5\,t)/100$.

t, °C	p, mmHg	t, °C	p, mmHg	t, °C	p, mmHg	t, °C	p, mmHg
−10.0	2.149	11.5	10.176	22.2	20.070	30.8	33.312
−9.5	2.236	12.0	10.518	22.4	20.316	31.0	33.695
−9.0	2.326	12.5	10.870	22.6	20.565	31.2	34.082
−8.5	2.418	13.0	11.231	22.8	20.815	31.4	34.471
−8.0	2.514	13.5	11.604	23.0	21.068	31.6	34.864
−7.5	2.613	14.0	11.987	23.2	21.324	31.8	35.261
−7.0	2.715	14.5	12.382	23.4	21.583	32.0	35.663
−6.5	2.822	15.0	12.788	23.6	21.845	32.2	36.068
−6.0	2.931	15.2	12.953	23.8	22.110	32.4	36.477
−5.5	3.046	15.4	13.121	24.0	22.387	32.6	36.891
−5.0	3.163	15.6	13.290	24.2	22.648	32.8	37.308
−4.5	3.284	15.8	13.461	24.4	22.922	33.0	37.729
−4.0	3.410	16.0	13.634	24.6	23.198	33.2	38.155
−3.5	3.540	16.2	13.809	24.8	23.476	33.4	38.584
−3.0	3.673	16.4	13.987	25.0	23.756	33.6	39.018
−2.5	3.813	16.6	14.166	25.2	24.039	33.8	39.457
−2.0	3.956	16.8	13.347	25.4	24.326	34.0	39.898
−1.5	4.105	17.0	14.530	25.6	24.617	34.2	40.344
−1.0	4.258	17.2	14.715	25.8	24.912	34.4	40.796
−0.5	4.416	17.4	14.903	26.0	25.209	34.6	41.251
0.0	4.579	17.6	15.092	26.2	25.509	34.8	41.710
0.5	4.750	17.8	15.284	26.4	25.812	35.0	42.175
1.0	4.926	18.0	15.477	26.6	26.117	35.2	42.644
1.5	5.107	18.2	15.673	26.8	26.426	35.4	43.117
2.0	5.294	18.4	15.871	27.0	26.739	35.6	43.595
2.5	5.486	18.6	16.071	27.2	27.055	35.8	44.078
3.0	5.685	18.8	16.272	27.4	27.374	36.0	44.563
3.5	5.889	19.0	16.477	27.6	27.696	36.2	45.054
4.0	6.101	19.2	16.685	27.8	28.021	36.4	45.549
4.5	6.318	19.4	16.894	28.0	28.349	36.6	46.050
5.0	6.543	19.6	17.105	28.2	28.680	36.8	46.556
5.5	6.775	19.8	17.319	28.4	29.015	37.0	47.067
6.0	7.013	20.0	17.535	28.6	29.354	37.2	47.582
6.5	7.259	20.2	17.753	28.8	29.697	37.4	48.102
7.0	7.513	20.4	17.974	29.0	30.043	37.6	48.627
7.5	7.775	20.6	18.197	29.2	30.392	37.8	49.157
8.0	8.045	20.8	18.422	29.4	30.745	38.0	49.692
8.5	8.323	21.0	18.650	29.6	31.102	38.2	50.231
9.0	8.609	21.2	18.880	29.8	31.461	38.4	50.774
9.5	8.905	21.4	19.113	30.0	31.824	38.6	51.323
10.0	9.209	21.6	19.349	30.2	32.191	38.8	51.879
10.5	9.521	21.8	19.587	30.4	32.561	39.0	52.442
11.0	9.844	22.0	19.827	30.6	32.934	39.2	53.009

TABLE 4.3 Vapor Pressure of Water for Temperatures from −10 to 120°C (*continued*)

t, °C	p, mmHg	t, °C	p, mmHg	t, °C	p, mmHg	t, °C	p, mmHg
39.4	54.580	58.5	139.34	78.5	334.2	96.4	667.31
39.6	54.156	59.0	142.60	79.0	341.0	96.6	672.20
39.8	54.737	59.5	145.99	79.5	348.1	96.8	677.12
40.0	55.324	60.0	149.38	80.0	355.1	97 0	682.07
40.5	56.81	60 5	152.91	80.5	362.4	97.2	687.04
41.0	58 34	61.0	156.43	81.0	369.7	97.4	692.05
41.5	59.90	61.5	160.10	81.5	377.3	97.6	697.10
42.0	61.50	62.0	163.27	82.0	384.9	97.8	702.17
42.5	63.13	62.5	167.58	82.5	392.8	98.0	707.27
43.0	64.80	63.0	171.38	83.0	400.6	98.2	712.40
43.5	66.51	63.5	175.35	83.5	408.7	98.4	717.56
44.0	68.26	64.0	179.31	84.0	416.8	98.6	722.75
44.5	70.05	64.5	183.43	84.5	425.2	98.8	727.98
45.0	71.88	65.0	187.54	85.0	433.6	99.0	733.24
45.5	73.74	65.5	191.82	85.5	442.3	99.2	738.53
46.0	75.65	66.0	196.09	86.0	450.9	99.4	743.85
46.5	77.61	66.5	200.53	86.5	459.8	99.6	749.20
47.0	79.60	67.0	204.96	87.0	468.7	99.8	754.58
47.5	81.64	67.5	209.57	87.5	477.9	100.0	760.00
48.0	83.71	68.0	214.17	88.0	487.1	101.0	787.57
48.5	85.85	68.5	218.95	88.5	496.6	102.0	815.86
49.0	88.02	69.0	223.73	89.0	506.1	103.0	845.12
49.5	90.24	69.5	228.72	89.5	515.9	104.0	875.06
50.0	92.51	70.0	233.7	90.0	525.76	105.0	906.07
50.5	94.86	70.5	238.8	90.5	535.83	106.0	937.92
51.0	97.20	71.0	243.9	91.0	546.05	107.0	970.60
51.5	99.65	71.5	249.3	91.5	556.44	108.0	1004.42
52.0	102.09	72.0	254.6	92.0	566.99	109.0	1038.92
52.5	104.65	72.5	260.2	92.5	577.71	110.0	1074.56
53.0	107.20	73.0	265.7	93.0	588.60	111.0	1111.20
53.5	109.86	73.5	271.5	93.5	599.66	112.0	1148.74
54.0	112.51	74.0	277.2	94.0	610.90	1130	1187.42
54.5	115.28	74.5	283.2	94.5	622.31	114.0	1227.25
55.0	118.04	75.0	289.1	95.0	633.90	115.0	1267.98
55.5	120.92	75.5	295.3	95.2	638.59	116.0	1309.94
56.0	123.80	76.0	301.4	95.4	643.30	117.0	1352.95
56.5	126.81	76.5	307.7	95.6	648.05	118.0	1397.18
57.0	129.82	77.0	314.1	95.8	652.82	119.0	1442.63
57.5	132.95	77.5	320.7	96.0	657.62	120.0	1489.14
58.0	136.08	78.0	327.3	96.2	662.45		

TABLE 4.4 Vapor Pressure of Deuterium Oxide

t, °C	p, mmHg	t, °C	p, mmHg	t, °C	p, mmHg
0	3.65	20	15.2	80	331.6
1	3.93	30	28.0	90	495.5
2	4.29	40	49.3	100	722.2
3	4.65	50	83.6	101.43	760.0
3.8	5.05	60	136.6		
10	7.79	70	216.1		

BOILING POINTS

TABLE 4.5A Boiling Points for Common Organic Solvents

Arranged in order of increasing boiling point

Compound name	bp (°C)	Other name	Compound name	bp (°C)	Other name
Ethylene oxide	10.6	Oxirane	Benzene	80.1	
Chloroethane	12.3	Ethyl chloride	Cyclohexane	80.7	
Furan	31.4		Propyl formate	80.9	
Methyl formate	31.5		Acetonitrile	81.6	
Diethyl ether	34.6	Ethyl ether	2-Propanol	82.4	Isopropyl alcohol
Propylene oxide	34.5		1,1-Dimethylethanol	82.4	*t*-Butanol
Pentane	36.1		Cyclohexene	83.0	
Bromoethane	38.4	Ethyl bromide	Diisopropyl amine	83.5	
Dichloromethane	39.8	Methylene chloride	1,2-Dichloroethane	83.7	
Dimethoxyethane	42.3	DME, glyme	Thiophene	84.2	
Carbon disulfide	46.3		Trichloroethylene	87.2	
1-Isopropoxy- 2-propanol	47.9		Isopropyl acetate	88.2	
Ethyl formate	54.2		1-Bromo- 2-methylpropane	91.5	Isobutyl bromide
Acetone	56.2	Dimethyl ketone	2,5-Dimethylfuran	93–94	
Methyl acetate	56.3		Ethyl chloroformate	94.0	
1,1-Dichloroethane	57.3		Allyl alcohol	96.6	
Dichloroethylene	60.6		1,2-Dichloropropane	96.8	
Chloroform	61.2		1-Propanol	97.2	*n*-Propyl alcohol
Methanol	64.7		Heptane	98.4	
Tetrahydrofuran	66.0	THF	1-Chloro- 3-methylbutane	99.0	
Diisopropyl ether	68.0	Isopropyl ether	Ethyl propanoate	99.1	Ethyl propionate
Hexane	68.7		2-Butanol	99.6	*sec*-Butanol
1-Chloro- 2-methylpropane	68.9	Isobutyl chloride	Formic acid	100.8	
1,1,1-Trichloroethane	74.0		Methylcyclohexane	100.9	
1,3-Dioxolane	74–75		1,4-Dioxane	101.2	
Carbon tetrachloride	76.7		Nitromethane	101.2	
Ethyl acetate	77.1		Propyl acetate	101.5	
1-Chlorobutane	77.9	Butyl chloride	2-Pentanone	101.7	Methyl propyl ketone
Ethanol	78.3		3-Pentanone	102.0	Diethyl ketone
2-Butanone	79.6	Methyl ethyl ketone	2-Methyl-2-butanol	102.0	*t*-Pentanol
2-Methyltetrahydrofuran	80.0		1,1-Diethoxyethane	102.7	

TABLE 4.5A   Boiling Points for Common Organic Solvents (*continued*)

Arranged in order of increasing boiling point

Compound name	bp (°C)	Other name	Compound name	bp (°C)	Other name
Butyl formate	106.6		4-Heptanone	143.7	Dipropyl ketone
2-Methyl-1-propanol	107.9	Isobutanol	*o*-Xylene	144.4	1,2-Dimethylbenzene
Toluene	110.6	Methylbenzene	2-Methoxyethyl acetate	144.5	
sec-Butyl acetate	112.3				
1,1,2-Trichloroethane	113.5		1,1,2,2-Tetrachloroethane	146.3	
Nitroethane	114.1				
Pyridine	115.2		3-Heptanone	147.8	Ethyl butyl ketone
3-Pentanol	115.6		Tribromomethane	149.6	Bromoform
4-Methyl-2-pentanone	115.7	Methyl isobutyl ketone	Nonane	150.8	
			2-Heptanone	151	Methyl pentyl ketone
1-Chloro-2,3-epoxypropane	116.1	Epichlorohydrin			
			Isopropylbenzene	152.4	Cumene
1-Butanol	117.7		*N,N*-Dimethylformamide	153.0	DMF
Acetic acid	117.9				
Isobutyl acetate	118.0		Methoxybenzene	153.8	Anisole
2-Pentanol	119.3	*sec*-Pentanol	Ethyl lactate	154.5	
1-Bromo-3-methylbutane	119.7	Isopentyl bromide	Cyclohexanone	155.7	
			Bromobenzene	156.2	
1-Methoxy-2-propanol	120.1		1,2,3-Trichloropropane	156.9	
2-Nitropropane	120.3	Nitroisopropane			
Tetrachloroethylene	121.1		1-Hexanol	157.5	
Ethyl butanoate	121.6	Ethyl butyrate	Propylbenzene	159.2	
3-Hexanone	123	Ethyl propyl ketone	Cyclohexanol	161.1	
2,4-Dimethyl-3-pentanone	124	Diisopropyl ketone	Bis(2-methoxyethyl) ether	160	Diglyme
2-Methoxyethanol	124.6		Isopentyl propanoate	160.2	Isopentyl propionate
Octane	125.7				
Butyl acetate	126.1		2-Heptanol	160.4	
Diethyl carbonate	126.8		Pentachloroethane	160.5	
2-Hexanone	127.2	Methyl butyl ketone	2-Furaldehyde	161.8	Furfural
1-Chloro-2-propanol	127.4		2,6-Dimethyl-4-heptanone	168.1	Diisobutyl ketone
2-Chloroethanol	128.6				
1-Nitropropane	131.2		4-Hydroxy-4-methyl-2-pentanone	169.2	
Chlorobenzene	131.7				
1,2-Dibromoethane	131.7		2-Furanmethanol	170.0	2-Hydroxymethylfuran
4-Methyl-2-pentanol	131.7				
3-Methyl-1-butanol	132.0		Ethoxybenzene	170	Phenetole
Cyclohexylamine	134.8	Aminocyclohexane	2-Butoxyethanol	170.2	Butyl cellosolve
2-Ethoxyethanol	134.8		Diisopentyl ether	173.4	
Ethylbenzene	136.2		Decane	174.2	
1-Pentanol	138		1,3-Dichloro-2-propanol	174.3	
p-Xylene	138.4	1,4-Dimethylbenzene	Cyclohexyl acetate	174–175	
m-Xylene	139.1	1,3-Dimethylbenzene	1-Heptanol	175.8	
Acetic anhydride	140.0		Furfuryl acetate	175–177	
2,4-Pentanedione	140.6	Acetylacetone	4-Isopropyl-1-methylbenzene	177.1	*p*-Cymene
Isopentyl acetate	142				
Dibutyl ether	142.4				

TABLE 4.5A   Boiling Points for Common Organic Solvents (*continued*)

Arranged in order of increasing boiling point

Compound name	bp (°C)	Other name	Compound name	bp (°C)	Other name
Isopentyl butanoate	178.6	Isopentyl butyrate	*o*-Chloroaniline	208.8	1-Amino-
Bis(2-chloroethyl) ether	178.8	Dichloro diethyl ether			2-chlorobenzene
2-Octanol	179		Nitrobenzene	210.8	
1,2-Dichlorobenzene	180.4	*o*-Dichlorobenzene	Ethyl benzoate	212.4	
Ethyl acetoacetate	180.8		Isophorone	215.2	3,5, 5-
Phenol	181.8	Hydroxybenzene			Trimethylcyclo-
2-Ethyl-1-hexanol	184.3				hex-2-en-1-one
Aniline	184.4	Aminobenzene	Naphthalene	217.7	
Benzyl ethyl ether	185.0		2-(2-Ethoxyethoxy) ethyl acetate	218.5	
Diethyl oxalate	185.4				
1,2-Propanediol	188	Propylene glycol	Acetamide	221.2	
Bis(2-ethoxyethyl) ether	188.4		Methyl salicylate	223.0	
			Diethyl maleate	225.3	
Dimethylsulfoxide	189.0	DMSO	1,4-Butanediol	230	
1,2-Ethanediol diacetate	190.2	Diacetoxyethane	Propyl benzoate	231.2	
			1-Decanol	230.2	
Benzonitrile	191.0	Cyanobenzene	Phenylacetonitrile	233.5	Cyano-
2,5-Hexanedione	191.4				methylbenzene
2-(2-Methoxyethoxy) ethanol	194.1		Quinoline	237	
			Tributyl borate	238.5	
N,*N*-Dimethylaniline	194.2	Dimethyl- aminobenzene	Propylene carbonate	240	
			2-Phenoxyethanol	240	
1-Octanol	195.2		Bis(2-hydroxyethyl) ether	245	Diethylene glycol
1,2-Ethanediol	197.3	Ethylene glycol			
Diethyl malonate	199.3		Dibutyl oxalate	245.5	
Methyl benzoate	199.5		Butyl benzoate	250	
o-Toluidine	200.4	1-Amino- 2-methylbenzene	1,2,3-Propanetriol triacetate	258– 259	Glycerol triacetate
p-Toluidine	200.6	1-Amino- 4-methylbenzene	1-Chloronaphthalene	259.3	
			Isopentyl benzoate	262	
2-(2-Ethoxyethoxy) ethanol	202		*trans*-Ethyl cinnamate	271.0	
			Bis (2-(2-methoxy- ethoxy)- ethyl) ether	275.3	Triglyme
Acetophenone	202.1	Methyl phenyl- ketone			
			1-Methoxy- 2-nitrobenzene	277	
1,2-Dibutoxyethane	203.6				
1,2-Phenylethanol	203.9	Phenethyl alcohol	Isopentyl salicylate	277– 278	
m-Toluidine	203.4	1-Amino- 3-methylbenzene	1-Bromonaphthalene	281.1	
			Dimethyl *o*-phthalate	283.7	1,2-Bis(carbo- methoxy) benzene
Benzyl alcohol	205.5	Hydroxy- methylbenzene			
Camphor	207	1,7,7-Trimethyl- bicyclo[2.2.1] heptan-2-one	2,2′-(Ethylenedioxy) bisethanol	285	
			Glycerol	290	
1,3-Butanediol	207.5		Diethyl *o*-phthalate	295	
1,2,3,4-Tetrahy- dronaphthalene	207.6	Tetralin	Benzyl benzoate	323.5	
			Dibutyl *o*-phthalate	340.0	
γ-Valerolactone	207– 208		Dibutyl decanedioate	344–345	

TABLE 4.5B Boiling Points for Common Organic Solvents

Arranged alphabetically by solvent name

Compound name	bp (°C)	Other name	Compound name	bp (°C)	Other name
Acetamide	221.2		Chlorobenzene	131.7	
Acetic acid	117.9		1-Chlorobutane	77.9	Butyl chloride
Acetic anhydride	140.0		1-Chloro-2,	116.1	Epichlorohydrin
Acetone	56.2	Dimethyl ketone	3-epoxypropane		
Acetonitrile	81.6		1-Chloro-2-propanol	127.4	
Acetophenone	202.1	Methyl phenyl	Chloroethane	12.3	Ethyl chloride
		ketone	2-Chloroethanol	128.6	
Allyl alcohol	96.6		Chloroform	61.2	
Aniline	184.4	Aminobenzene	1-Chloro-		
Benzene	80.1		3-methylbutane	99.0	
Benzonitrile	191.0	Cyanobenzene	1-Chloro-	68.9	Isobutyl chloride
Benzyl alcohol	205.5	Hydroxy-	2-methylpropane		
		methylbenzene	1-Chloronaphthalene	259.3	
Benzyl benzoate	323.5		Cyclohexane	80.7	
Benzyl ethyl ether	185.0		Cyclohexanol	161.1	
Bis(2-(2-methoxy-	275.3	Triglyme	Cyclohexanone	155.7	
ethoxy)-			Cyclohexene	83.0	
ethyl) ether			Cyclohexyl acetate	174–	
Bis(2-chloroethyl)	178.8	Dichloro		175	
ether		diethyl ether	Cyclohexylamine	134.8	Aminocyclohexane
Bis(2-ethoxyethyl)	188.4		Decane	174.2	
ether			1-Decanol	230.2	
Bis(2-hydroxyethyl)	245	Diethylene glycol	1,2-Dibromoethane	131.7	
ether			1,2-Dibutoxyethane	203.6	
Bis(2-methoxyethyl)	160	Diglyme	Dibutyl decanedioate	344–	
ether				345	
Bromobenzene	156.2		Dibutyl ether	142.4	
Bromoethane	38.4	Ethyl bromide	Dibutyl o-phthalate	340.0	
1-Bromo-	91.5	Isobutyl bromide	Dibutyl oxalate	245.5	
2-methylpropane			1,2-Dichlorobenzene	180.4	o-Dichlorobenzene
1-Bromo-	119.7	Isopentyl bromide	Dichloroethylene	60.6	
3-methylbutane			Dichloromethane	39.8	Methylene chloride
1-Bromonaphthalene	281.1		1,1-Dichloroethane	57.3	
1,3-Butanediol	207.5		1,2-Dichloroethane	83.7	
1,4-Butanediol	230		1,2-Dichloropropane	96.8	
1-Butanol	117.7		1,3-Dichloro-	174.3	
2-Butanol	99.6	*sec*-Butanol	2-propanol		
2-Butanone	79.6	Methyl ethyl ketone	1,1-Diethoxyethane	102.7	
2-Butoxyethanol	170.2	Butyl cellosolve	Diethyl carbonate	126.8	
Butyl acetate	126.1		Diethyl ether	34.6	Ethyl ether
sec-Butyl acetate	112.3		Diethyl maleate	225.3	
Butyl benzoate	250		Diethyl malonate	199.3	
Butyl formate	106.6		Diethyl o-phthalate	295	
Camphor	207	1,7,7-Trimethylbi-	Diethyl oxalate	185.4	
		cyclo [2.2.1]	Diisopentyl ether	173.4	
		heptan-2-one	Diisopropyl amine	83.5	
Carbon disulfide	46.3		Diisopropyl ether	68.0	Isopropyl ether
Carbon tetrachloride	76.7		Dimethoxyethane	42.3	DME, glyme
o-Chloroaniline	208.8	1-Amino-			
		2-chlorobenzene			

TABLE 4.5B Boiling Points for Common Organic Solvents (*continued*)

Arranged alphabetically by solvent name

Compound name	bp (°C)	Other name	Compound name	bp (°C)	Other name
N,N-Dimethylaniline	194.2	Dimethyl-aminobenzene	2-Heptanone	151	Methyl pentyl ketone
1,1-Dimethylethanol	82.4	t-Butanol	3-Heptanone	147.8	Ethyl butyl ketone
N,N-Dimethyl-formamide	153.0	DMF	4-Heptanone	143.7	Dipropyl ketone
			Hexane	68.7	
2,5-Dimethylfuran	93–94		2,5-Hexanedione	191.4	
2,6-Dimethyl-4-heptanone	168.1	Diisobutyl ketone	1-Hexanol	157.5	
			2-Hexanone	127.2	Methyl butyl ketone
2,4-Dimethyl-3-pentanone	124	Diisopropyl ketone	3-Hexanone	123	Ethyl propyl ketone
Dimethyl o-phthalate	283.7	1,2-Bis(carbo-methoxy)benzene	4-Hydroxy-4-methyl-2-pentanone	169.2	
			Isobutyl acetate	118.0	
Dimethylsulfoxide	189.0	DMSO	Isopentyl acetate	142	
1,4-Dioxane	101.2		Isopentyl benzoate	262	
1,3-Dioxolane	74–75		Isopentyl butanoate	178.6	Isopentyl butyrate
1,2-Ethanediol	197.3	Ethylene glycol	Isopentyl propanoate	160.2	Isopentyl propionate
1,2-Ethanediol diacetate	190.2	Diacetoxyethane	Isopentyl salicylate	277–278	
Ethoxybenzene	170	Phenetole	Isophorone	215.2	3,5,5-Trimethylcycl-ohex-2-en-1-one
2-Ethoxyethanol	134.8				
2-(2-Ethoxyethoxy)ethanol	202		Isopropyl acetate	88.2	
			Isopropylbenzene	152.4	Cumene
2-(2-Ethoxyethoxy)ethyl acetate	218.5		1-Isopropoxy-2-propanol	47.9	
Ethyl acetate	77.1		4-Isopropyl-1-methylbenzene	177.1	p-Cymene
Ethyl acetoacetate	180.8				
Ethyl benzoate	212.4		Methanol	64.7	
Ethyl butanoate	121.6	Ethyl butyrate	Methoxybenzene	153.8	Anisole
Ethyl chloroformate	94.0		Methyl acetate	56.3	
trans-Ethyl cinnamate	271.0		Methyl benzoate	199.5	
Ethyl formate	54.2		Methyl formate	31.5	
Ethyl lactate	154.5		Methyl salicylate	223.0	
Ethyl propanoate	99.1	Ethyl propionate	Methylcyclohexane	100.9	
Ethylbenzene	136.2		1-Methoxy-2-nitrobenzene	277	
Ethylene oxide	10.6	Oxirane			
2,2'-(Ethylenedioxy)bisethanol	285		1-Methoxy-2-propanol	120.1	
			2-(2-Methoxyethoxy)ethanol	194.1	
2-Ethyl-1-hexanol	184.3				
Formic acid	100.8		2-Methoxyethanol	124.6	
2-Furaldehyde	161.8	Furfural	2-Methoxyethyl acetate	144.5	
Furan	31.4		2-Methyl-1-propanol	107.9	Isobutanol
2-Furanmethanol	170.0	2-Hydroxy-methylfuran	2-Methyl-2-butanol	102.0	t-Pentanol
			2-Methyltetrahy-drofuran	80.0	
Furfuryl acetate	175–177		3-Methyl-1-butanol	132.0	
Glycerol	290		4-Methyl-2-pentanol	131.7	
Heptane	98.4		4-Methyl-2-pentanone	115.7	Methyl isobutyl ketone
1-Heptanol	175.8				
2-Heptanol	160.4				

TABLE 4.5B   Boiling Points for Common Organic Solvents (*continued*)

Arranged alphabetically by solvent name

Compound name	bp (°C)	Other name	Compound name	bp (°C)	Other name
Naphthalene	217.7		Propyl formate	80.9	
Nitrobenzene	210.8		Propylbenzene	159.2	
Nitroethane	114.1		Pyridine	115.2	
Nitromethane	101.2		Quinoline	237	
1-Nitropropane	131.2		1,1,2,2-Tetra-	146.3	
2-Nitropropane	120.3	Nitroisopropane	chloroethane		
Nonane	150.8		Tetrachloroethylene	121.1	
Octane	125.7		Tetrahydrofuran	66.0	THF
1-Octanol	195.2		Thiophene	84.2	
2-Octanol	179		Toluene	110.6	Methylbenzene
Pentachloroethane	160.5		*o*-Toluidine	200.4	1-Amino-
Pentane	36.1				2-methylbenzene
2,4-Pentanedione	140.6	Acetylacetone	*m*-Toluidine	203.4	1-Amino-
1-Pentanol	138				3-methylbenzene
2-Pentanol	119.3	*sec*-Pentanol	*p*-Toluidine	200.6	1-Amino-
3-Pentanol	115.6				4-methylbenzene
2-Pentanone	101.7	Methyl propyl	Tribromomethane	149.6	Bromoform
		ketone	Tributyl borate	238.5	
3-Pentanone	102.0	Diethyl ketone	1,1,1-Trichloroethane	74.0	
Phenol	181.8	Hydroxybenzene	1,1,2-Trichloroethane	113.5	
2-Phenoxyethanol	240		Trichloroethylene	87.2	
Phenylacetonitrile	233.5	Cyano-	1,2,3,4-Tetrahy-	207.6	Tetralin
		methylbenzene	dronaphthalene		
1-Phenylethanol	203.9	Phenethyl alcohol	1,2,3-Trichloro	156.9	
1,2-Propanediol	188	Propylene glycol	propane		
1,2,3-Propanetriol	258–	Glycerol	γ-Valerolactone	207–	
triacetate	259	triacetate		208	
1-Propanol	97.2	*n*-Propyl alcohol	*o*-Xylene	144.4	1,2-Dimethyl-
2-Propanol	82.4	Isopropyl alcohol			benzene
Propylene carbonate	240		*m*-Xylene	139.1	1,3-Dimethyl-
Propylene oxide	34.5				benzene
Propyl acetate	101.5		*p*-Xylene	138.4	1,4-Dimethyl-
Propyl benzoate	231.2				benzene

TABLE 4.5C   Boiling Points for Common Organic Solvents

Arranged by compound type in order of increasing boiling point

	Compound name	(°C)	Other name
acid	Formic acid	100.8	
acid	Acetic acid	117.9	
alcohol	1-Isopropoxy-2-propanol	47.9	
alcohol	Methanol	64.7	
alcohol	Ethanol	78.3	
alcohol	2-Propanol	82.4	Isopropyl alcohol
alcohol	1,1-Dimethylethanol	82.4	*t*-Butanol
alcohol	Allyl alcohol	96.6	

TABLE 4.5C   Boiling Points for Common Organic Solvents (*continued*)

Arranged by compound type in order of increasing boiling point

	Compound name	(°C)	Other name
alcohol	1-Propanol	97.2	*n*-Propyl alcohol
alcohol	2-Butanol	99.6	*sec*-Butanol
alcohol	2-Methyl-2-butanol	102.0	*t*-Pentanol
alcohol	2-Methyl-1-propanol	107.9	Isobutanol
alcohol	3-Pentanol	115.6	
alcohol	1-Butanol	117.7	
alcohol	2-Pentanol	119.3	*sec*-Pentanol
alcohol	1-Methoxy-2-propanol	120.1	
alcohol	2-Methoxyethanol	124.6	
alcohol	1-Chloro-2-propanol	127.4	
alcohol	2-Chloroethanol	128.6	
alcohol	4-Methyl-2-pentanol	131.7	
alcohol	3-Methyl-1-butanol	132.0	
alcohol	2-Ethoxyethanol	134.8	
alcohol	1-Pentanol	138	
alcohol	1-Hexanol	157.5	
alcohol	Cyclohexanol	161.1	
alcohol	2-Heptanol	160.4	
alcohol	4-Hydroxy-4-methyl-2-pentanone	169.2	
alcohol	2-Furanmethanol	170.0	2-Hydroxymethylfuran
alcohol	2-Butoxyethanol	170.2	Butyl cellosolve
alcohol	1,3-Dichloro-2-propanol	174.3	
alcohol	1-Heptanol	175.8	
alcohol	2-Octanol	179	
alcohol	Phenol	181.8	Hydroxybenzene
alcohol	2-Ethyl-1-hexanol	184.3	
alcohol	1,2-Propanediol	188	Propylene glycol
alcohol	2-(2-Methoxyethoxy) ethanol	194.1	
alcohol	1-Octanol	195.2	
alcohol	1,2-Ethanediol	197.3	Ethylene glycol
alcohol	2-(2-Ethoxyethoxy) ethanol	202	
alcohol	1-Phenylethanol	203.9	Phenethyl alcohol
alcohol	Benzyl alcohol	205.5	Hydroxymethylbenzene
alcohol	1,3-Butanediol	207.5	
alcohol	Methyl salicylate	223.0	
alcohol	1,4-Butanediol	230	
alcohol	1-Decanol	230.2	
alcohol	2-Phenoxyethanol	240	
alcohol	Bis(2-hydroxyethyl) ether	245	Diethylene glycol
alcohol	Isopentyl salicylate	277–278	
alcohol	2,2'-(Ethylenedioxy)bisethanol	285	
alcohol	Glycerol	290	
aldehyde	2-Furaldehyde	161.8	Furfural
amide	*N,N*-Dimethylformamide	153.0	DMF
amide	Acetamide	221.2	
amine	Diisopropylamine	83.5	
amine	Pyridine	115.2	

TABLE 4.5C Boiling Points for Common Organic Solvents (*continued*)

Arranged by compound type in order of increasing boiling point

	Compound name	(°C)	Other name
amine	Cyclohexylamine	134.8	Aminocyclohexane
amine	Aniline	184.4	Aminobenzene
amine	*N,N*-Dimethylaniline	194.2	Dimethylaminobenzene
amine	*o*-Toluidine	200.4	1-Amino-2-methylbenzene
amine	*p*-Toluidine	200.6	1-Amino-4-methylbenzene
amine	*m*-Toluidine	203.4	1-Amino-3-methylbenzene
amine	*o*-Chloroaniline	208.8	1-Amino-2-chlorobenzene
amine	Quinoline	237	
anhydride	Acetic anhydride	140.0	
bromide	Bromoethane	38.4	Ethyl bromide
bromide	1-Bromo-2-methylpropane	91.5	Isobutyl bromide
bromide	1-Bromo-3-methylbutane	119.7	Isopentyl bromide
bromide	1,2-Dibromoethane	131.7	
bromide	Tribromomethane	149.6	Bromoform
bromide	Bromobenzene	156.2	
bromide	1-Bromonaphthalene	281.1	
chloride	Chloroethane	12.3	Ethyl chloride
chloride	Dichloromethane	39.8	Methylene chloride
chloride	1,1-Dichloroethane	57.3	
chloride	Dichloroethylene	60.6	
chloride	Chloroform	61.2	
chloride	1-Chloro-2-methylpropane	68.9	Isobutyl chloride
chloride	1,1,1-Trichloroethane	74.0	
chloride	Carbon tetrachloride	76.7	
chloride	1-Chlorobutane	77.9	Butyl chloride
chloride	1,2-Dichloroethane	83.7	
chloride	Trichloroethylene	87.2	
chloride	1,2-Dichloropropane	96.8	
chloride	1-Chloro-3-methylbutane	99.0	
chloride	1,1,2-Trichloroethane	113.5	
chloride	1-Chloro-2,3-epoxypropane	116.1	Epichlorohydrin
chloride	Tetrachloroethylene	121.1	
chloride	1-Chloro-2-propanol	127.4	
chloride	2-Chloroethanol	128.6	
chloride	Chlorobenzene	131.7	
chloride	1,1,2,2-Tetrachloroethane	146.3	
chloride	1,2,3-Trichloropropane	156.9	
chloride	Pentachloroethane	160.5	
chloride	1,3-Dichloro-2-propanol	174.3	
chloride	Bis(2-chloroethyl) ether	178.8	Dichloro diethyl ether
chloride	1,2-Dichlorobenzene	180.4	*o*-Dichlorobenzene
chloride	*o*-Chloroaniline	208.8	1-Amino-2-chlorobenzene
chloride	1-Chloronaphthalene	259.3	
ester	Methyl formate	31.5	
ester	Ethyl formate	54.2	

TABLE 4.5C Boiling Points for Common Organic Solvents (*continued*)

Arranged by compound type in order of increasing boiling point

	Compound name	(°C)	Other name
ester	Methyl acetate	56.3	
ester	Ethyl acetate	77.1	
ester	Propyl formate	80.9	
ester	Isopropyl acetate	88.2	
ester	Methyl chloroformate	94.0	
ester	Ethyl propanoate	99.1	Ethyl propionate
ester	Propyl acetate	101.5	
ester	Butyl formate	106.6	
ester	*sec*-Butyl acetate	112.3	
ester	Isobutyl acetate	118.0	
ester	Ethyl butanoate	121.6	Ethyl butyrate
ester	Butyl acetate	126.1	
ester	Diethyl carbonate	126.8	
ester	Isopentyl acetate	142	
ester	2-Methoxyethyl acetate	144.5	
ester	Ethyl lactate	154.5	
ester	Isopentyl propanoate	160.2	Isopentyl propionate
ester	Cyclohexyl acetate	174–175	
ester	Furfuryl acetate	175–177	
ester	Isopentyl butanoate	178.6	Isopentyl butyrate
ester	Ethyl acetoacetate	180.8	
ester	Diethyl oxalate	185.4	
ester	1,2-Ethanediol diacetate	190.2	Diacetoxyethane
ester	Diethyl malonate	199.3	
ester	Methyl benzoate	199.5	
ester	γ-Valerolactone	207–208	
ester	Ethyl benzoate	212.4	
ester	2-(2-Ethoxyethoxy)ethyl acetate	218.5	
ester	Methyl salicylate	223.0	
ester	Diethyl maleate	225.3	
ester	Propyl benzoate	231.2	
ester	Propylene carbonate	240	
ester	Dibutyl oxalate	245.5	
ester	Butyl benzoate	250	
ester	1,2,3-Propanetriol triacetate	258–259	Glycerol triacetate
ester	Isopentyl benzoate	262	
ester	*trans*-Ethyl cinnamate	271.0	
ester	Isopentyl salicylate	277–278	
ester	Dimethyl *o*-phthalate	283.7	1,2-Bis(carbomethoxy) benzene
ester	Diethyl *o*-phthalate	295	
ester	Benzyl benzoate	323.5	
ester	Dibutyl *o*-phthalate	340.0	
ester	Dibutyl decanedioate	344–345	
ether	Ethylene oxide	10.6	Oxirane
ether	Furan	31.4	

TABLE 4.5C   Boiling Points for Common Organic Solvents (*continued*)
Arranged by compound type in order of increasing boiling point

	Compound name	(°C)	Other name
ether	Diethyl ether	34.6	Ethyl ether
ether	Propylene oxide	34.5	
ether	Dimethoxyethane	42.3	DME, glyme
ether	1-Isopropoxy-2-propanol	47.9	
ether	Tetrahydrofuran	66.0	THF
ether	Diisopropyl ether	68.0	Isopropyl ether
ether	1,3-Dioxolane	74–75	
ether	2-Methyltetrahydrofuran	80.0	
ether	2,5-Dimethylfuran	93–94	
ether	1,4-Dioxane	101.2	
ether	1,1-Diethoxyethane	102.7	
ether	1-Chloro-2,3-epoxypropane	116.1	Epichlorohydrin
ether	1-Methoxy-2-propanol	120.1	
ether	2-Methoxyethanol	124.6	
ether	2-Ethoxyethanol	134.8	
ether	Dibutyl ether	142.4	
ether	2-Methoxyethyl acetate	144.5	
ether	Methoxybenzene	153.8	Anisole
ether	Bis(2-methoxyethyl) ether	160	Diglyme
ether	2-Furanmethanol	170.0	2-Hydroxymethylfuran
ether	Ethoxybenzene	170	Phenetole
ether	2-Buroxyethanol	170.2	Butyl cellosolve
ether	Diisopentyl ether	173.4	
ether	Bis(2-chloroethyl) ether	178.8	Dichloro diethyl ether
ether	Benzyl ethyl ether	185.0	
ether	Bis(2-ethoxyethyl)ether	188.4	
ether	2-(2Methoxyethoxy)ethanol	194.1	
ether	2-(2-Ethoxyethoxy) ethanol	202	
ether	1,2 Dibutoxyethane	203.6	
ether	2-(2-Ethoxyethoxy)ethyl acetate	218.5	
ether	2-Phenoxyethanol	240	
ether	Bis(2-hydroxyethyl) ether	245	Diethylene glycol
ether	Bis(2-(2-methoxyethoxy)-ethyl) ether	275.3	Triglyme
ether	1-Methoxy-2-nitrobenzene	277	
ether	2,2'-(Ethylenedioxy) bisethanol	285	
hydrocarbon	Pentane	36.1	
hydrocarbon	Hexane	68.7	
hydrocarbon	Benzene	80.1	
hydrocarbon	Cyclohexane	80.7	
hydrocarbon	Cyclohexene	83.0	
hydrocarbon	Heptane	98.4	
hydrocarbon	Methylcyclohexane	100.9	
hydrocarbon	Toluene	110.6	Methylbenzene
hydrocarbon	Octane	125.7	
hydrocarbon	Ethylbenzene	136.2	
hydrocarbon	*p*-Xylene	138.4	1,4-Dimethylbenzene
hydrocarbon	*m*-Xylene	139.1	1,3-Dimethylbenzene

TABLE 4.5C   Boiling Points for Common Organic Solvents (*continued*)

Arranged by compound type in order of increasing boiling point

	Compound name	(°C)	Other name
hydrocarbon	*o*-Xylene	144.4	1,2-Dimethylbenzene
hydrocarbon	Nonane	150.8	
hydrocarbon	Isopropylbenzene	152.4	Cumene
hydrocarbon	Propylbenzene	159.2	
hydrocarbon	Decane	174.2	
hydrocarbon	4-Isopropyl-l-methylbenzene	177.1	*p*-Cymene
hydrocarbon	1,2,3,4-Tetrahydronaphthalene	207.6	Tetralin
hydrocarbon	Naphthalene	217.7	
ketone	Acetone	56.2	Dimethyl ketone
ketone	2-Butanone	79.6	Methyl ethyl ketone
ketone	2-Pentanone	101.7	Methyl propyl ketone
ketone	3-Pentanone	102.0	Diethyl ketone
ketone	4-Methyl-2-pentanone	115.7	Methyl isobutyl ketone
ketone	3-Hexanone	123	Ethyl propyl ketone
ketone	2,4-Dimethyl-3-pentanone	124	Diisopropyl ketone
ketone	2-Hexanone	127.2	Methyl butyl ketone
ketone	2,4-Pentanedione	140.6	Acetylacetone
ketone	4-Heptanone	143.7	Dipropyl ketone
ketone	3-Heptanone	147.8	Ethyl butyl ketone
ketone	2-Heptanone	151	Methyl pentyl ketone
ketone	Cyclohexanone	155.7	
ketone	2,6-Dimethyl-4-heptanone	168.1	Diisobutyl ketone
ketone	4-Hydroxy-4-methyl-2-pentanone	169.2	
ketone	2,5-Hexanedione	191.4	
ketone	Acetophenone	202.1	Methyl phenyl ketone
ketone	Camphor	207	1,7,7-Trimethylbicyclo-[2.2.1] heptan-2-one
ketone	Isophorone	215.2	3,5,5-Trimethylcyclo-hex-2-en-1-one
miscellaneous	Carbon disulfide	46.3	
miscellaneous	Tributyl borate	238.5	
nitrile	Acetonitrile	81.6	
nitrile	Benzonitrile	191.0	Cyanobenzene
nitrile	Phenylacetonitrile	233.5	Cyanomethylbenzene
nitro compound	Nitromethane	101.2	
nitro compound	Nitroethane	114.1	
nitro compound	2-Nitropropane	120.3	Nitroisopropane
nitro compound	1-Nitropropane	131.2	
nitro compound	Nitrobenzene	210.8	
nitro compound	1-Methoxy-2-nitrobenzene	277	
sulfide	Thiophene	84.2	
sulfoxide	Dimethylsulfoxide	189.0	DMSO

TABLE 4.6   Molecular Elevation of the Boiling Point

Ebullioscopic constants

Molecular weights can be determined with the relation

$$M = K_b \frac{1000 \, w_2}{w_1 \Delta T_b}$$

where ΔT_b is the elevation of the boiling point brought about by the addition of w_2 grams of solute to w_1 grams of solvent and K_b is the ebullioscopic constant. In the column headed "Barometric correction" is given the number of degrees for each millimeter of difference between the barometric reading and 760 mmHg to be subtracted from K_b if the pressure is lower, or added if higher, than 760 mm. In general, the effect is within experimental error if the pressure is within 10 mm of 760 mm.

Compound	Barometric correction	K_b
Acetic acid	0.000 8	3.07
Acetic anhydride		3.53
Acetone	0.000 4	1.71
Acetonitrile		1.30
Acetophenone		5.65
Aniline	0.000 9	3.52
Benzene	0.000 7	2.53
Benzonitrile		3.87
Bromobenzene	0.001 6	6.26
Bromoethane		2.53
2-Butanone		2.28
cis-2-Butene-1,4-diol		2.86
D-(+)-Camphor	0.001 5	5.611
Carbon disulfide	0.000 6	2.34
Carbon tetrachloride	0.001 3	5.03
Chlorobenzene	0.001 1	4.15
Chloroethane		1.95
Chloroform	0.000 9	3.63
Cyclohexane	0.000 7	2.79
1,2-Dibromoethane	0.001 6	6.608
1,1-Dichloroethane		3.13
1,2-Dichloroethane		3.44
Dichloromethane		2.60
Diethyl ether	0.000 5	2.02
Diethyl sulfide		3.23
Dimethoxymethane		2.125
N,N-Dimethylacetamide		3.22
Dimethyl sulfide		1.85
1,4-Dioxane		3.270
Ethanol	0.000 3	1.22
Ethoxybenzene		5.0
Ethyl acetate	0.000 7	2.77
Formic acid		2.4
Glycerol		6.52
Heptane	0.000 8	3.43
Hexane		2.75
2-Hydroxybenzaldehyde		4.96

TABLE 4.6   Molecular Elevation of the Boiling Point (*continued*)

Compound	Barometric correction	K_b
Iodoethane		5.16
Iodomethane		4.19
4-Isopropyl-1-methylbenzene		5.52
Methanol	0.000 2	0.83
Methoxybenzene		4.502
Methyl acetate	0.000 5	2.15
2-Methyl-2-butanol		2.255
3-Methyl-1-butanol		2.65
3-Methylbutyl acetate		4.83
Methyl formate		1.649
2-Methyl-1-propanol		2.166
2-Methyl-2-propanol		1.745
Naphthalene	0.001 4	5.80
Nitrobenzene		5.24
Nitroethane		2.60
Nitromethane		1.86
Octane		4.02
Pentyl acetate		4.83
Phenol	0.000 9	3.60
Piperidine		2.84
1-Propanol		1.59
Propionic acid		3.51
Propionitrile		1.87
Pyridine		2.710
Quinoline		5.84
1,1,2,2-Tetrachloroethylene		5.50
1,2,3,4-Tetrahydronaphthalene		5.582
Toluene	0.000 8	3.33
p-Toluidine		4.14
Trichloroethylene		4.43
1,1,2-Trichloro-1,2,2-trifluoroethane		5.75
Triethylamine		3.45
Water	0.000 1	0.512

Distillation is an important historical method for the separation of liquids. A mixture or solution of two liquids is heated until the vapor pressure of the lower boiling compound reaches the pressure of the surroundings. This may be ambient pressure or a lowered pressure caused by application of a vacuum. In either event, vaporization occurs, the vapors are condensed on a cold surface, and the condensed liquid is collected. If the boiling points of the two liquids are sufficiently different given the pressure and efficiency of the apparatus, separation may be achieved. A zeotrope is a mixture that can be separated by distillation.

In contrast, certain mixtures of two (binary) or three (ternary) components form constant boiling mixtures that cannot be separated by distillation. In such cases, each component contributes a fixed amount and the boiling point of the mixture is characteristic of the components. Such a system is called an azeotrope. The boiling point of an azeotrope may be higher or lower than that of the individual components. Common binary azeotropes are listed in Table 4.7 and ternary azeotropes are listed in Table 4.8.

TABLE 4.7 Binary Azeotropic (Constant Boiling) Mixtures

A zeotrope is a mixture that can be separated by distillation.

A. Binary azeotropes containing water

System	BP of azeotrope, °C	Composition, wt %	
		Water	Other component
Inorganic acids			
Hydrogen bromide	126	52.5	47.5
Hydrogen chloride	108.58	79.78	20.22
Hydrogen fluoride	111.35	64.4	35.6
Hydrogen iodide	127	43	57
Hydrogen peroxide	zeotrope		
Nitric acid	120.7	32.6	67.4
Perchloric acid	203	28.4	71.6
Organic acids			
Formic acid	107.2	22.6	77.4
Acetic acid	zeotrope		
Propionic acid	99.9	82.3	17.7
Isobutyric acid	99.3	79	21
Butyric acid	99.4	81.6	18.4
Pentanoic acid	99.8	89	11
Isopentanoic acid	99.5	81.6	18.4
Perfluorobutyric acid	97	71	29
Crotonic acid	99.9	97.8	2.2
Alcohols			
Ethanol	78.17	4	96
Allyl alcohol	88.9	27.7	72.3
1-Propanol	71.7	71.7	28.3
2-Propanol	80.3	12.6	87.4
1-Butanol	92.7	42.5	57.5
2-Butanol	87.0	26.8	73.2
2-Methyl-2-propanol	79.9	11.7	88.3
1-Pentanol	95.8	54.4	45.6
2-Pentanol	91.7	36.5	63.5
3-Pentanol	91.7	36.0	64.0
2,2-Dimethyl-2-propanol	87.35	27.5	72.5
1-Hexanol	97.8	67.2	32.8
1-Octanol	99.4	90	10
Cyclopentanol	96.25	58	42
1-Heptanol	98.7	83	17
Phenol	99.52	90.8	9.2
2-Methoxyphenol	99.5	87.5	12.5

TABLE 4.7 Binary Azeotropic (Constant Boiling) Mixtures (*continued*)

System	BP of azeotrope, °C	Composition, wt %	
		Water	Other component
Alcohols (continued)			
1-Phenylphenol	99.95	98.75	1.25
Benzyl alcohol	99.9	91	9
2,3-Dimethyl-2, 3-butanediol	zeotrope		
Furfuryl alcohol	98.5	80	20
Aldehydes			
Propionaldehyde	47.5	2	98
Butyraldehyde	68	6	94
Pentanal	83	19	81
Paraldehyde	90	28.5	71.5
Furaldehyde	97.5	65	35
Amines			
N-Methylbutylamine	82.7	15	85
Furfurylamine	99	74	26
Piperidine	92.8	35	65
Pyridine	93.6	41.3	58.7
2-Methylpyridine	93.5	48	52
3-Methylpyridine	97	60	40
4-Methylpyridine	97.35	62.8	37.2
2,6-Dimethylpyridine	96.02	51.8	48.2
Dibutylamine	97	50.5	49.5
Dihexylamine	99.8	92.8	7.2
Triallylamine	95	38	62
Tributylamine	99.65	79.7	20.3
Aniline	98.6	80.8	19.2
N-Ethylaniline	99.2	83.9	16.1
1-Methyl-2-(2-pyridyl) pyrrolidine	99.85	97.5	2.5
Halogenated hydrocarbons			
Chloroform	56.1	2.8	97.2
Carbon tetrachloride	42.6	2.8	97.2
Trichloroethylene	73.4	17	83
Tetrachloroethylene	88.5	17.2	82.8
1,2-Dichloroethane	72	8.3	91.7
1-Chloropropane	44	2.2	97.8
1,2-Dichloropropane	78	12	88
Chlorobenzene	90.2	28.4	71.6

TABLE 4.7 Binary Azeotropic (Constant Boiling) Mixtures (*continued*)

System	BP of azeotrope, °C	Composition, wt %	
		Water	Other component
Esters			
Ethyl formate	52.6	5	95
Isopropyl formate	65.0	3	97
Propyl formate	71.6	2.3	97.7
Isobutyl formate	80.4	7.8	92.2
Butyl formate	83.8	14.5	85.5
Isopentyl formate	90.2	21	79
Pentyl formate	91.6	28.4	71.6
Benzyl formate	99.2	80	20
Ethyl acetate	70.38	8.47	91.53
Allyl acetate	83	14.7	85.3
Isopropyl acetate	76.6	10.6	89.4
Propyl acetate	87.4	16.5	83.5
Isobutyl acetate	82.4	14	86
Butyl acetate	90.2	28.7	71.3
Isopentyl acetate	93.8	36.3	63.7
Pentyl acetate	95.2	41	59
Hexyl acetate	97.4	61	39
Phenyl acetate	98.9	75.1	24.9
Benzyl acetate	99.6	87.5	12.5
Methyl propionate	71.4	3.9	96.1
Ethyl propionate	81.2	10	90
Isopropyl propionate	85.2	19.9	80.1
Propyl propionate	88.9	23	77
Isobutyl propionate	92.75	52.2	47.8
Isopentyl propionate	96.55	48.5	51.5
Methyl butyrate	82.7	11.5	88.5
Ethyl butyrate	87.9	21.5	78.5
Propyl butyrate	94.1	36.4	63.6
Isobutyl butyrate	96.3	46	54
Butyl butyrate	97.2	53	47
Isopentyl butyrate	98.05	63.5	36.5
Methyl isobutyrate	77.7	6.8	93.2
Ethyl isobutyrate	85.2	15.2	84.8
Propyl isobutyrate	92.2	30.8	69.2
Isobutyl isobutyrate	95.5	39.4	60.6
Isopentyl isobutyrate	97.4	56.0	44.0
Methyl isopentanoate	87.2	19.2	80.8
Ethyl isopentanoate	92.2	30.2	69.8
Propyl isopentanoate	96.2	45.2	54.8
Isobutyl isopentanoate	97.4	55.8	44.2
Isopentyl isopentanoate	98.8	74.1	25.9

TABLE 4.7   Binary Azeotropic (Constant-Boiling) Mixtures (*continued*)

System	BP of azeotrope, °C	Composition, wt %	
		Water	Other component
Esters (*continued*)			
Ethyl pentanoate	94.5	40	60
Ethyl hexanoate	97.2	54	46
Methyl benzoate	99.08	79.2	20.8
Ethyl benzoate	99.4	84.0	16.0
Propyl benzoate	99.7	90.9	9.1
Butyl benzoate	99.9	94	6
Isopentyl benzoate	99.9	95.6	4.4
Ethyl phenylacetate	99.7	91.3	8.7
Methyl cinnamate	99.9	95.5	4.5
Methyl phthalate	99.95	97.5	2.5
Diethyl *o*-phthalate	99.98	98.0	2.0
Ethyl chloroacetate	95.2	45.1	54.9
Butyl chloroacetate	98.12	75.5	24.5
Methyl acrylate	71	7.2	92.8
Isobutyl carbonate	98.6	74	26
Ethyl crotonate	93.5	38	62
Methyl lactate	99	80	20
1,2-Ethanediol diacetate	99.7	84.6	15.4
Ethyl nitrate	74.35	22	78
Propyl nitrate	84.8	20	80
Isobutyl nitrate	89.0	25	75
Methyl sulfate	98.6	73	27
Ethers			
Ethyl vinyl ether	34.6	1.5	98.5
Diethyl ether	34.2	1.3	98.7
Ethyl propyl ether	59.5	4	96
Diisopropyl ether	62.2	4.5	95.5
Butyl ethyl ether	76.6	11.9	88.1
Diisobutyl ether	88.6	23	77
Dibutyl ether	92.9	33	67
Diisopentyl ether	97.4	54	46
1,1-Diethoxyethane	82.6	14.5	85.5
Diphenyl ether	99.33	96.75	3.25
Methoxybenzene	95.5	40.5	59.5
Hydrocarbons			
Pentane	34.6	1.4	98.6
Hexane	61.6	5.6	94.4
Heptane	79.2	12.9	87.1
2,2,4-Trimethylpentane	78.8	11.1	88.9

TABLE 4.7 Binary Azeotropic (Constant Boiling) Mixtures (*continued*)

System	BP of azeotrope, °C	Composition, wt %	
		Water	Other component
Hydrocarbons (continued)			
Nonane	94.8	82	18
Undecane	98.85	96.0	4.0
Dodecane	99.45	98	2
Acrolein	52.4	2.6	97.4
Cyclohexene	70.8	8.93	91.07
Cyclohexane	69.5	8.4	91.6
1-Octene	88.0	28.7	71.3
Benzene	69.25	8.83	91.17
Toluene	84.1	13.5	86.5
Ethylbenzene	92.0	33.0	67.0
m-Xylene	92	35.8	64.2
Isopropylbenzene	95	43.8	56.2
Naphthalene	98.8	84	16
Ketones			
Acetone	zeotrope		
2-Butanone	73.5	11	89
2-Pentanone	83.3	19.5	80.5
Cyclopentanone	94.6	42.4	57.6
4-Methyl-2-pentanone	87.9	24.3	75.7
2-Heptanone	95	48	52
3-Heptanone	94.6	42.2	57.8
4-Heptanone	94.3	40.5	59.5
4-Hydroxy-4-methyl-2-pentanone	98.8	87.3	12.7
4-Methyl-3-penten-2-one	91.8	34.8	65.2
Nitriles			
Acetonitrile	76.5	16.3	83.7
Isobutyronitrile	82.5	23	77
Butyronitrile	88.7	32.5	67.5
Acrylonitrile	70.6	14.3	85.7
Miscellaneous			
Hydrazine	120	32.3	67.7
Acetamide	zeotrope		
Nitromethane	83.59	23.6	76.4
Nitroethane	87.22	28.5	71.5
2,5-Dimethylfuran	77.0	11.7	88.3
Trioxane	91.4	30	70
Carbon disulfide	42.6	2.8	97.2

TABLE 4.7 Binary Azeotropic (Constant-Boiling) Mixtures (*continued*)

B. Binary azeotropes containing organic acids

System	BP of azeotrope, °C	Composition, wt %	
		Acid	Other component
Formic acid			
2-Methylbutane	27.2	4	96
Pentane	34.2	20	80
Hexane	60.6	28	72
Methylcyclopentane	63.3	29	71
Cyclohexane	70.7	70	30
Methylcyclohexane	80.2	46.5	53.5
Heptane	78.2	56.5	43.5
Octane	90.5	63	37
Benzene	71.05	31	69
Toluene	85.8	50	50
o-Xylene	95.5	74	26
m-Xylene	92.8	71.8	28.2
Styrene	97.8	73	27
Iodomethane	42.1	6	94
Chloroform	59.15	15	85
Carbon tetrachloride	66.65	18.5	81.5
Trichloroethylene	74.1	25	75
Tetrachloroethylene	88.2	50	50
Bromoethane	38.2	3	97
1,2-Dibromoethane	94.7	51.5	48.5
1,2-Dichloroethane	77.4	14	86
1-Bromopropane	64.7	27	73
2-Bromopropane	77.4	14	86
1-Chloropropane	45.6	8	92
2-Chloropropane	34.7	1.5	98.5
1-Chloro-2-methylpropane	63.0	19	81
Bromobenzene	98.1	68	32
Chlorobenzene	93.7	59	41
Fluorobenzene	73.0	27	73
o-Chlorotoluene	100.2	83	17
Pyridine	127.43	61.4	38.6
2-Methylpyridine	158.0	25	75
2-Pentanone	105.3	32	68
3-Pentanone	105.4	33	67
Nitromethane	97.07	45.5	54.5
Diethyl sulfide	82.2	35	65
Diisopropyl sulfide	93.5	62	38
Dipropyl sulfide	98.0	83	17
Carbon disulfide	42.55	17	83

TABLE 4.7 Binary Azeotropic (Constant Boiling) Mixtures (*continued*)

System	BP of azeotrope, °C	Composition, wt %	
		Acid	Other component
Acetic acid			
Hexane	68.3	6.0	94.0
Heptane	91.7	23	67
Octane	105.7	53.7	46.3
Nonane	112.9	69	31
Decane	116.75	79.5	20.5
Undecane	117.9	95	5
Cyclohexane	78.8	9.6	90.4
Methylcyclohexane	96.3	31	69
Benzene	80.05	2.0	98.0
Toluene	100.6	28.1	71.9
o-Xylene	116.6	78	22
m-Xylene	115.35	72.5	27.5
p-Xylene	115.25	72	28
Ethylbenzene	114.65	66	34
Styrene	116.8	85.7	14.3
Isopropylbenzene	116.0	84	16
Triethylamine	163	67	33
Nitromethane	101.2	96	4
Nitroethane	112.4	30	70
Pyridine	138.1	51.1	48.9
2-Methylpyridine	144.1	40.4	59.6
3-Methylpyridine	152.5	30.4	69.6
4-Methylpyridine	154.3	30.3	69.7
2,6-Dimethylpyridine	148.1	22.9	77.1
Carbon tetrachloride	76	98.46	1.54
Trichloroethylene	86.5	96.2	3.8
Tetrachloroethylene	107.4	61.5	38.5
1,2-Dibromoethane	114.4	55	45
2-Iodopropane	88.3	9	91
1-Bromobutane	97.6	18	82
1-Bromo-2-methylpropane	90.2	12	88
Chlorobenzene	114.7	58.5	41.5
Trichloronitromethane	107.65	80.5	19.5
1,4-Dioxane	119.5	77	23
Diisopropyl sulfide	111.5	48	52
Propanoic (Propionic) acid			
Heptane	97.8	2	98
Octane	120.9	21.5	78.5

TABLE 4.7 Binary Azeotropic (Constant Boiling) Mixtures (*continued*)

System	BP of azeotrope, °C	Composition, wt %	
		Acid	Other component
Acetic acid (continued)			
Nonane	134.3	54.0	46.0
Decane	139.8	80.5	19.5
o-Xylene	135.4	43	57
p-Xylene	132.5	34	66
1,3,5-Trimethylbenzene	139.3	77	23
Isopropylbenzene	139.0	65	35
Propylbenzene	139.5	75	25
Camphene	138.0	65	35
α-Pinene	136.4	58.5	41.5
Methoxybenzene	140.8	96	4
Pyridine	148.6	67.2	32.8
2-Methylpyridine	154.5	55.0	45.0
1,2-Dibromoethane	127.8	17.5	82.5
1-Iodo-2-methylpropane	119.5	9	91
Chlorobenzene	128.9	18	82
Dipropyl sulfide	136.5	45	55
Butyric (butanoic) acid			
Undecane	162.4	84.4	15.5
o-Xylene	143.0	10	90
m-Xylene	138.5	6	94
p-Xylene	137.8	5.5	94.5
Ethylbenzene	135.8	4	96
Styrene	143.5	15	85
1,2,4-Trimethylbenzene	159.5	45	55
1,3,5-Trimethylbenzene	158.0	38	62
Isopropylbenzene	149.5	20	80
Propylbenzene	154.5	28	72
Butylbenzene	162.5	75	25
Naphthalene	zeotrope		
Indene	163.7	84	16
Camphene	152.3	2.8	97.2
Methoxybenzene	152.9	12	88
Pyridine	163.2	92.0	8.0
2-Furaldehyde	159.4	42.5	57.5
1,2-Dibromoethane	131.1	3.5	96.5
1-Iodobutane	129.8	2.5	97.5
Chlorobenzene	131.75	2.8	97.2
1,4-Dichlorobenzene	162.0	57	43
o-Bromotoluene	163.0	72	28
m-Bromotoluene	163.6	79.5	20.5
p-Bromotoluene	161.5	75	25
α-Chlorotoluene	160.8	65	35
Ethyl bromoacetate	157.4	84	16
Propyl chloroacetate	160.5	40	60

TABLE 4.7 Binary Azeotropic (Constant Boiling) Mixtures (*continued*)

System	BP of azeotrope, °C	Composition, wt%	
		Acid	Other component
Isobutyric (2-methylpropanoic) acid			
2,7-Dimethyloctane	148.6	48	52
o-Xylene	141.0	22	78
m-Xylene	139.9	15	85
p-Xylene	136.4	13	87
Styrene	142.0	27	73
1,2,4-Trimethylbenzene	152.3	63	37
Isopropylbenzene	146.8	35	65
Propylbenzene	149.3	49	51
Camphene	148.1	45	55
D-Limonene	152.5	78	22
Methoxybenzene	149.0	42	58
Ethyl bromoacetate	153.0	40	60
Ethyl 2-oxopropionate	153.0	60	40
1,2-Dibromoethane	130.5	6.5	93.5
1-Iodobutane	128.8	7	93
1-Bromohexane	148.0	35	65
Bromobenzene	148.6	35	65
Chlorobenzene	131.5	8	92
o-Bromotoluene	153.9	85	15
α-Chlorotoluene	153.5	80	20
Diisopentyl ether	154.2	93	7
Ethyl bromoacetate	153.0	40	60

C. Binary azeotropes containing alcohol

System	BP of azeotrope, °C	Composition, wt %	
		Alcohol	Other component
Methanol			
Pentane	30.9	7	93
Cyclopentane	38.8	14	86
Cyclohexane	53.9	36.4	63.6
Methylcyclohexane	59.2	54	46
Heptane	59.1	51.5	48.5
Octane	62.8	67.5	32.5
Nonane	64.1	83.4	16.6
Benzene	57.5	39.1	60.9
Fluorobenzene	59.7	32	68
Toluene	63.5	72.5	27.5
Bromomethane	3.55	99.55	0.45
Iodomethane	37.8	95.5	4.5

TABLE 4.7 Binary Azeotropic (Constant Boiling) Mixtures (*continued*)

System	BP of azeotrope, °C	Composition, wt %	
		Alcohol	Other component
Methanol (*continued*)			
Bromodichloromethane	63.8	60	40
Chloroform	53.4	87.4	12.6
Carbon tetrachloride	55.7	79.44	20.56
Bromoethane	34.9	5.3	94.7
1,2-Dichloroethane	61.0	32	68
Trichloroethylene	59.3	38	62
1-Bromopropane	54.5	21	79
2-Bromopropane	48.6	15.0	85.0
1-Chloropropane	40.5	9.5	90.5
2-Chloropropane	33.4	6	94
2-Iodopropane	61.0	38	62
1-Chlorobutane	57.0	27	73
Isobutyl formate	64.6	95	5
Methyl acetate	53.5	19	81
Methyl acrylate	62.5	54	46
Methyl nitrate	52.5	73	27
Acetone	55.5	12.1	87.9
1,4-Dioxane	zeotrope		
Dipropyl ether	63.8	72	28
Methyl *tert*-butyl ether	51.3	14.3	85.7
Diethyl sulfide	61.2	62	38
Carbon disulfide	39.8	71	29
Thiophene	59.7	16.4	83.6
Nitromethane	64.4	9.1	90.9
Ethanol			
Pentane	34.3	5	95
Cyclopentane	44.7	7.5	92.5
Hexane	58.7	21	79
Cyclohexane	64.8	29.2	70.8
Heptane	70.9	49	51
Octane	77.0	78	22
Benzene	67.9	31.7	68.3
Fluorobenzene	70.0	75	25
Toluene	76.7	68	32
Bromodichloromethane	75.5	72	28
Iodomethane	41.2	96.8	3.2
Chloroform	59.3	93	7
Trichloronitromethane	77.5	34	66
Carbon tetrachloride	65.0	84.2	15.8
1,2-Dichloroethane	70.5	37	63
3-Chloro-1-propene	44	5	95

TABLE 4.7 Binary Azeotropic (Constant Boiling) Mixtures (*continued*)

System	BP of azeotrope, °C	Composition, wt %	
		Alcohol	Other component
Ethanol (*continued*)			
1-Bromopropane	62.8	20.5	79.5
2-Bromopropane	55.6	10.5	89.5
1-Chloropropane	45.0	6	94
2-Chloropropane	35.6	2.8	97.2
1-Iodopropane	75.4	44	56
2-Iodopropane	71.5	27	73
1-Bromobutane	75.0	43	57
1-Chlorobutane	65.7	20.3	79.7
2-Butanone	74.8	40	60
1,1-Diethoxyethane	78.0	76	24
Dipropyl ether	74.5	44	56
Acetonitrile	72.5	44	56
Acrylonitrile	70.8	41	59
Nitromethane	76.1	29	71
Carbon disulfide	42.6	91	9
Diethyl sulfide	72.6	56	44
1-Propanol			
Hexane	65.7	4	96
Cyclohexane	74.7	18.5	81.5
Methylcyclohexane	87.0	34.7	65.3
Heptane	84.6	34.7	65.3
Octane	93.9	70	30
Benzene	77.1	16.9	83.1
Toluene	92.5	51.2	48.8
o-Xylene	zeotrope		
m-Xylene	97.1	94	6
p-Xylene	96.9	92.2	7.8
Styrene	97.0	8	92
Propyl formate	80.7	3	97
Butyl formate	95.5	64	36
Propyl acetate	94.7	51	49
Ethyl propionate	93.4	48	52
Methyl butyrate	94.4	49	51
Dipropyl ether	85.7	30	70
1,1-Diethoxyethane	92.4	37	63
1,4-Dioxane	95.3	55	45
Chloroform	zeotrope		
Carbon tetrachloride	73.4	92.1	7.9
Trichloronitromethane	94.1	58.5	41.5
Iodoethane	70	93	7
1,2-Dichloroethane	80.7	19	81

TABLE 4.7 Binary Azeotropic (Constant Boiling) Mixtures (*continued*)

System	BP of azeotrope, °C	Composition, wt %	
		Alcohol	Other component
1-Propanol (*continued*)			
Tetrachloroethylene	94.0	52	48
1-Bromopropane	69.7	9	91
1-Chlorobutane	74.8	18	82
Chlorobenzene	96.5	80	20
Fluorobenzene	80.2	18	82
Nitromethane	89.1	48.4	51.6
1-Nitropropane	97.0	8.8	91.2
Carbon disulfide	45.7	94.5	5.5
2-Propanol			
Pentane	35.5	6	94
Hexane	62.7	23	77
Cyclohexane	69.4	32	68
Heptane	76.4	50.5	49.5
Octane	81.6	84	16
Benzene	71.7	33.7	66.3
Fluorobenzene	74.5	30	70
Toluene	80.6	69	31
Chloroform	60.8	4.2	95.8
Trichloronitromethane	81.9	35	65
Carbon tetrachloride	69.0	18	82
1,2-Dichloroethane	74.7	43.5	56.5
Iodoethane	67.1	15	85
3-Bromo-1-propene	66.5	20	80
1-Chloropropane	46.4	2.8	97.2
1-Bromopropane	66.8	20.5	79.5
2-Bromopropane	57.8	12	88
1-Iodopropane	79.8	42	58
2-Iodopropane	76.0	32	68
1-Chlorobutane	70.8	23	77
Ethyl acetate	75.3	25	75
Isopropyl acetate	81.3	60	40
Methyl propionate	76.4	37	63
Acrylonitrile	71.7	56	44
Butylamine	74.7	60	40
2-Butanone	77.5	32	68
1,1-Diethoxyethane	81.3	63	37
Ethyl propyl ether	62.0	10	90
Diisopropyl ether	66.2	14.1	85.9

TABLE 4.7 Binary Azeotropic (Constant Boiling) Mixtures (*continued*)

System	BP of azeotrope, °C	Composition, wt %	
		Alcohol	Other component
1-Butanol			
Cyclohexane	79.8	9.5	90.5
Cyclohexene	82.0	5	95
Hexane	68.2	3.2	96.8
Methylcyclohexane	95.3	20	80
Heptane	93.9	18	82
Octane	108.5	45.2	54.8
Nonane	115.9	71.5	28.5
Toluene	105.5	27.8	72.2
o-Xylene	116.8	75	25
m-Xylene	116.5	71.5	28.5
p-Xylene	115.7	68	32
Ethylbenzene	115.9	65.1	34.9
Butyl formate	105.8	23.6	76.4
Isopentyl formate	115.9	69	31
Butyl acetate	117.2	47	53
Isobutyl acetate	114.5	50	50
Ethyl butyrate	115.7	64	36
Ethyl isobutyrate	109.2	17	83
Methyl isopentanoate	113.5	40	60
Ethyl borate	113.0	52	48
Ethyl carbonate	116.5	63	37
Isobutyl nitrate	112.8	45	55
Dibutyl ether	117.8	82.5	17.5
Diisobutyl ether	113.5	48	52
1,1-Diethoxyethane	101.0	13	87
Carbon tetrachloride	76.6	97.6	2.4
Tetrachloroethylene	110.0	68	32
2-Bromo-2-methylpropane	90.2	7	93
2-Iodo-2-methylpropane	110.5	30	70
Chlorobenzene	115.3	56	44
Paraldehyde	115.8	52	48
Hexaldehyde	116.8	77.1	22.9
Ethylenediamine	124.7	35.7	64.3
Pyridine	118.6	69	31
1-Nitropropane	115.3	32.2	67.8
Butyronitrile	113.0	50	50
Diisopropyl sulfide	112.0	45	55
2-Methyl-2-propanol			
Cyclohexene	80.5	14.2	85.8
Cyclohexane	78.3	14	86

TABLE 4.7   Binary Azeotropic (Constant Boiling) Mixtures (*continued*)

System	BP of azeotrope, °C	Composition, wt %	
		Water	Other component
2-Methyl-2-propanol (*continued*)			
Methylcyclopentane	71.0	5	95
Hexane	68.3	2.5	97.5
Methylcyclohexane	92.6	32	68
Heptane	90.8	27	73
2,5-Dimethylhexane	98.7	42	58
1,3-Dimethylcyclohexane	102.2	56	44
2,2,4-Trimethylpentane	92.0	27	73
Benzene	79.3	7.4	92.6
Chlorobenzene	107.1	63	37
Fluorobenzene	84.0	9	91
Toluene	101.2	45	55
Ethylbenzene	107.2	80	20
p-Xylene	107.1	88.6	11.4
Butyl formate	103.0	40	60
Isobutyl formate	97.4	12	88
Propyl acetate	101.0	17	83
Isobutyl acetate	107.6	92	8
Methyl butyrate	101.3	25	75
Ethyl isobutyrate	105.5	52	48
Methyl chloroacetate	107.6	12	88
Dipropyl ether	89.5	10	90
Isobutyl vinyl ether	82.7	6.2	93.8
1,1-Diethoxyethane	98.2	20	80
2-Pentanone	101.8	19	81
3-Pentanone	107.7	20	80
1,2-Dichloroethane	83.5	6.5	93.5
1-Bromobutane	95.0	21	79
1-Chlorobutane	77.7	4	96
2-Bromo-2-methylpropane	88.8	12	88
2-Iodo-2-methylpropane	104.0	36	64
1-Nitropropane	105.3	15.2	84.8
Isobutyl nitrate	105.6	36	64
Diisopropyl sulfide	105.8	73	27
3-Methyl-1-butanol			
Heptane	97.7	7	93
Octane	117.0	30	70
Toluene	109.7	10	90
Ethylbenzene	125.7	49	51
Isopropylbenzene	131.6	94	6
Camphene	130.9	24	76

TABLE 4.7   Binary Azeotropic (Constant Boiling) Mixtures (*continued*)

System	BP of azeotrope, °C	Composition, wt %	
		Alcohol	Other component
3-Methyl-1-butanol (*continued*)			
Bromobenzene	131.7	85	15
o-Fluorotoluene	112.1	14.0	86.0
Butyl acetate	125.9	16.5	83.5
Paraldehyde	123.5	22.0	78.0
Dibutyl ether	129.8	65	35
Cyclohexanol			
o-Xylene	143.0	14	86
m-Xylene	138.9	5	95
Propylbenzene	153.8	40	60
Indene	160.0	75	25
Camphene	151.9	41	59
Cineole	160.6	92	8
Allyl alcohol			
Methylcyclohexane	85.0	42	58
Hexane	65.5	4.5	95.5
Cyclohexane	74.0	58	42
2,5-Dimethylhexane	89.3	50	50
Octane	93.4	68	32
Benzene	76.75	17.36	82.64
Toluene	92.4	50	50
Propyl acetate	94.2	53	47
Methyl butyrate	93.8	55	45
1,2-Dichloroethane	79.9	18	82
3-Iodo-1-propene	89.4	28	72
Chlorobenzene	96.2	85	15
Diethyl sulfide	85.1	45	55
Phenol			
2,7-Dimethyloctane	159.5	6	94
Decane	168.0	35	65
Tridecane	180.6	83.1	16.9
Butylbenzene	175.0	46	54
1,2,4-Trimethylbenzene	166.0	25	75
1,3,5-Trimethylbenzene	163.5	21	79
Indene	177.8	47	53
Camphene	156.1	22	78
Benzaldehyde	175.6	51.0	49.0

TABLE 4.7 Binary Azeotropic (Constant Boiling) Mixtures (*continued*)

System	BP of azeotrope, °C	Composition, wt %	
		Alcohol	Other component
Phenol (*continued*)			
1-Octanol	195.4	13	87
2-Octanol	184.5	50	50
Dipentyl ether	180.2	78	22
Diisopentyl ether	172.2	15	85
2-Methylpyridine	185.5	75.4	24.6
3-Methylpyridine	188.9	71.2	29.8
4-Methylpyridine	190.0	67.5	32.5
2,4-Dimethylpyridine	193.4	57.0	43.0
2,6-Dimethylpyridine	185.5	72.5	27.5
2,4,6-Trimethylpyridine	195.2	52.3	47.7
Aniline	185.8	41.9	58.1
Ethylene diacetate	195.5	39.2	60.8
Iodobenzene	177.7	53	47
Benzyl alcohol			
Naphthalene	204.1	60	40
D-Limonene	176.4	11	89
1,3,5-Triethylbenzene	203.2	57	43
o-Cresol	zeotrope		
m-Cresol	207.1	61	39
p-Cresol	206.8	62	38
N-Methylaniline	195.8	30	70
N,N-Dimethylaniline	193.9	6.5	93.5
N-Ethylaniline	202.8	50	50
N, N-Diethylaniline	204.2	72	28
Iodobenzene	187.8	12	88
Nitrobenzene	204.0	58	42
o-Bromotoluene	181.3	7	93
Borneol	205.1	85.8	14.2
2-Ethoxyethanol			
Methylcyclohexane	98.6	15	85
Heptane	96.5	14	86
Octane	116.0	38	62
Toluene	110.2	10.8	89.2
Ethylbenzene	127.8	48	52
p-Xylene	128.6	50	50
Styrene	130.0	55	45
Propylbenzene	134.6	80	20
Isopropylbenzene	133.2	67	33
Camphene	131.0	65	35
Propyl butyrate	133.5	72	28

TABLE 4.7 Binary Azeotropic (Constant Boiling) Mixtures (*continued*)

System	BP of azeotrope, °C	Composition, wt %	
		Alcohol	Other component
2-Butoxyethanol			
Dipentene	164.0	53	47
1,3,5-Trimethylbenzene	162.0	32	68
Butylbenzene	169.6	73.4	26.6
Camphene	154.5	30	70
o-Cresol	191.6	15	85
Phenetole	167.1	52	48
Cineole	168.9	58.5	41.5
Benzaldehyde	171.0	91	9
Diisobutyl sulfide	163.8	42	58
1,2-Ethanediol			
Heptane	97.9	3	97
Decane	161.0	23	77
Tridecane	188.0	55	45
Toluene	110.1	2.3	97.7
Styrene	139.5	16.5	83.5
Stilbene	196.8	87	13
m-Xylene	135.1	6.55	93.45
p-Xylene	134.5	6.4	93.6
1,3,5-Trimethylbenzene	156	13	87
Propylbenzene	152	19	81
Isopropylbenzene	147.0	18	82
Naphthalene	183.9	51	49
1-Methylnaphthalene	190.3	60.0	40.0
2-Methylnaphthalene	189.1	57.2	42.8
Anthracene	197	98.3	1.7
Indene	168.4	26	74
Acenaphthene	194.65	74.2	25.8
Fluorene	196.0	82	18
Camphene	152.5	20	80
Camphor	186.2	40	60
Biphenyl	192.3	66.5	33.5
Diphenylmethane	193.3	68.5	31.5
Benzyl alcohol	193.1	56	44
2-Phenylethanol	194.4	69	31
o-Cresol	189.6	27	73
m-Cresol	195.2	60	40
3,4-Dimethylphenol	197.2	89	11
Menthol	188.6	51.5	48.5
Ethyl benzoate	186.1	46.5	53.5
o-Bromotoluene	166.8	25	75
Dibutyl ether	139.5	6.4	93.6

TABLE 4.7   Binary Azeotropic (Constant Boiling) Mixtures (*continued*)

System	BP of azeotrope, °C	Composition, wt %	
		Alcohol	Other component
1,2-Ethanediol (continued)			
Methoxybenzene	150.5	10.5	89.5
Diphenyl ether	193.1	60	40
Benzyl phenyl ether	195.5	87	13
Acetophenone	185.7	52	48
2,4-Dimethylaniline	188.6	47	53
N,N-Dimethylaniline	175.9	33.5	66.5
m-Toluidine	188.6	42	58
2,4,6-Trimethylpyridine	170.5	9.7	90.3
Quinoline	196.4	79.5	20.5
Tetrachloroethylene	119.1	94	6
1,2-Dibromoethane	129.8	4	96
Chlorobenzene	130.1	94.4	5.6
α-Chlorotoluene	167.0	30	70
Nitrobenzene	185.9	59	41
o-Nitrotoluene	188.5	48.5	51.5
1,2-Ethanediol monoacetate			
Indene	180.0	20	80
1-Octanol	189.5	71	29
Phenol	197.5	65	35
o-Cresol	199.5	51	49
m-Cresol	206.5	31	69
p-Cresol	206.0	33	67
Dipentyl ether	180.8	42	58
Diisopentyl ether	170.2	28	72
m-Bromotoluene	182.0	32	68

D. Binary azeotropes containing ketones

System	BP of azeotrope, °C	Composition, wt %	
		Ketone	Other component
Acetone			
Cyclopentane	41.0	36	64
Pentane	32.5	20	80
Cyclohexane	53.0	67.5	32.5
Hexane	49.8	59	41

TABLE 4.7 Binary Azeotropic (Constant Boiling) Mixtures (*continued*)

System	BP of azeotrope, °C	Composition, wt %	
		Ketone	Other component
Acetone (continued)			
Heptane	55.9	89.5	10.5
Diethylamine	51.4	38.2	61.8
Methyl acetate	55.8	48.3	51.7
Diisopropyl ether	54.2	61	39
Chloroform	64.4	78.1	21.9
Carbon tetrachloride	56.1	11.5	88.5
Carbon disulfide	39.3	67	33
Ethylene sulfide	51.5	57	43
2-Butanone			
Cyclohexane	71.8	40	60
Hexane	64.2	28.6	71.4
Heptane	77.0	70	30
2,5-Dimethylhexane	79.0	95	5
Benzene	78.33	44	56
2-Methyl-2-propanol	78.7	69	31
Butylamine	74.0	35	65
Ethyl acetate	77.1	11.8	88.2
Methyl propionate	79.0	60	40
Butyl nitrite	76.7	30	70
1-Chlorobutane	77.0	38	62
Fluorobenzene	79.3	75	25

E. Miscellaneous binary azeotropes

System	BP of azeotrope, °C	Composition, wt %	
		Solvent	Other component
Solvent: acetamide			
Dipentene	169.2	18	82
Biphenyl	213.0	50.5	49.5
Diphenylmethane	215.2	56.5	43.5
1,2-Diphenylethane	218.2	68	32
o-Xylene	142.6	11	89
m-Xylene	138.4	10	90
p-Xylene	137.8	8	92
Styrene	144	12	88
4-Isopropyl-1-methylbenzene	170.5	19	81

TABLE 4.7 Binary Azeotropic (Constant Boiling) Mixtures (*continued*)

System	BP of azeotrope, °C	Composition, wt %	
		Solvent	Other component
Solvent: acetamide (*continued*)			
Naphthalene	199.6	27	73
1-Methylnaphthalene	209.8	43.8	56.2
2-Methylnaphthalene	208.3	40	60
Indene	177.2	17.5	82.5
Acenaphthene	217.1	64.2	35.8
Camphene	155.5	12	88
Camphor	199.8	23	77
Benzaldehyde	178.6	6.5	93.5
3,4-Dimethylphenol	221.1	96	4
2-Methoxy-4-(2-propenyl)-phenol	220.8	88	12
N-Methylaniline	193.8	14	86
N-Ethylaniline	199.0	18	82
N,N-Diethylaniline	198.1	24	76
Diphenyl ether	214.6	52	48
Safrole	208.8	32	68
Tetrachloroethylene	120.5	97.4	2.6
Solvent: aniline			
Nonane	149.2	13.5	86.5
Decane	167.3	36	64
Undecane	175.3	57.5	42.5
Dodecane	180.4	71.5	28.5
Tridecane	182.9	86.2	13.8
Tetradecane	183.9	95.2	4.8
Butylbenzene	177.8	46	54
1,2,4-Trimethylbenzene	168.6	13.5	86.5
1,3,5-Trimethylbenzene	164.3	12.0	88.0
Indene	179.8	41.5	58.5
1-Octanol	183.9	83	17
o-Cresol	191.3	8	92
Dipentyl ether	177.5	55	45
Diisopentyl ether	169.3	28	72
Hexachloroethane	176.8	66	34
Solvent: pyridine			
Heptane	95.6	25.3	74.7
Octane	109.5	56.1	43.9
Nonane	115.1	89.9	10.1
Toluene	110.1	22.2	77.8
Phenol	183.1	13.1	86.9
Piperidine	106.1	8	92

TABLE 4.7 Binary Azeotropic (Constant Boiling) Mixtures (*continued*)

System	BP of azeotrope, °C	Composition, wt %	
		Solvent	Other component
Solvent: thiophene			
Methylcyclopentane	71.5	14	86
Cyclohexane	77.9	41.2	58.8
Hexane	68.5	11.2	88.8
Heptane	83.1	83.2	16.8
2,3-Dimethylpentane	80.9	64	36
2,4-Dimethylpentane	76.6	42.7	57.3
Solvent: benzene			
Methylcyclopentane	71.7	16	84
Cyclohexene	78.9	64.7	35.3
Cyclohexane	77.6	51.9	48.1
Hexane	68.5	4.7	95.3
Heptane	80.1	99.3	0.7
2,2-Dimethylpentane	75.9	46.3	53.7
2,3-Dimethylpentane	79.4	78.8	21.2
2,4-Dimethylpentane	75.2	48.3	51.7
2,2,4-Trimethylpentane	80.1	97.7	2.3
Solvent: bis(2-hydroxyethyl) ether			
Biphenyl	232.7	48	52
Diphenylmethane	236.0	52	48
1,3,5-Trimethylbenzene	210.0	22	78
Naphthalene	212.6	22	78
1-Methylnaphthalene	277.0	45	55
2-Methylnaphthalene	225.5	39	61
Acenaphthene	239.6	62	38
Fluorene	243.0	80	20
Benzyl acetate	214.9	7	93
Bornyl acetate	223.0	18	82
Ethyl fumarate	217.1	10	90
Dimethyl *o*-phthalate	245.4	96.3	3.7
Methyl salicylate	220.6	15	85
2-Hydroxy-1-isopropyl-4-methylbenzene	232.3	13	87
1,2-Dihydroxybenzene	259.5	46	54
Safrole	225.5	33	67
Isosafrole	233.5	46	54
Benzyl phenyl ether	241.5	80	20
Nitrobenzene	210.0	10	90
m-Nitrotoluene	224.2	25	75
o-Nitrophenol	216.0	10.5	89.5
Quinoline	233.6	29	71
p-Dibromobenzene	212.9	13	87

TABLE 4.8   Ternary Azeotropic Mixtures

A. Ternary azeotropes containing water and alcohols

System	BP of azeotrope, °C	Composition, wt %		
		Water	Alcohol	Other component
Methanol				
Chloroform	52.3	1.3	8.2	90.5
2-Methyl-1,3-butadiene	30.2	0.6	5.4	94.0
Methyl chloroacetate	67.9	6.3	81.2	13.5
Ethanol				
Acetonitrile	72.9	1	55	44
Acrylonitrile	69.5	8.7	20.3	71.0
Benzene	64.9	7.4	18.5	74.1
Butylamine	81.8	7.5	42.5	50.0
Butyl methyl ether	62	6.3	8.6	85.1
Carbon disulfide	41.3	1.6	5.0	93.4
Carbon tetrachloride	62	4.5	10.0	85.5
Chloroform	55.3	2.3	3.5	94.2
Crotonaldehyde	78.0	4.8	87.9	7.3
Cyclohexane	62.6	4.8	19.7	75.5
1,2-Dichloroethane	66.7	5	17	78
1,1-Diethoxyethane	77.8	11.4	27.6	61.0
Diethoxymethane	73.2	12.1	18.4	69.5
Ethyl acetate	70.2	9.0	8.4	82.6
Heptane	68.8	6.1	33.0	60.9
Hexane	56.0	3	12	85
Toluene	74.4	12	37	51
Trichloroethylene	67.0	5.5	16.1	78.4
Triethylamine	74.7	9	13	78
1-Propanol				
Benzene	67	7.6	10.1	82.3
Carbon tetrachloride	65.4	5	11	84
Cyclohexane	66.6	8.5	10.0	81.5
1,1-Dipropoxyethane	87.6	27.4	51.6	21.0
Dipropoxymethane	86.4	8.0	44.8	47.2
Dipropyl ether	74.8	11.7	20.2	68.1
3-Pentanone	81.2	20	20	60
Propyl acetate	82.5	17.0	10.0	73.0
Propyl formate	70.8	13	5	82
Tetrachloroethylene	81.2	12.5	20.7	66.8
2-Propanol				
Benzene	66.5	7.5	18.7	73.8
Butylamine	83	12.5	40.5	47.0

TABLE 4.8 Ternary Azeotropic Mixtures (*continued*)

System	BP of azeotrope, °C	Composition, wt %		
		Water	Alcohol	Other component
2-Propanol (*continued*)				
Cyclohexane	64.3	7.5	18.5	74.0
Toluene	76.3	13.1	38.2	48.7
Trichloroethylene	69.4	7	20	73
1-Butanol				
Butyl acetate	89.4	37.3	27.4	35.3
Butyl formate	83.6	21.3	10.0	68.7
Dibutyl ether	90.6	29.9	34.6	35.5
Heptane	78.1	41.4	7.6	51.0
Hexane	61.5	19.2	2.9	77.9
Nonane	90.0	69.9	18.3	11.8
Octane	86.1	60.0	14.6	25.4
2-Butanol				
Carbon tetrachloride	65	4.05	4.95	91.00
Cyclohexane	69.7	8.9	10.8	80.3
Isooctane	76.3	9	19	72
2-Methyl-1-propanol				
Isobutyl acetate	86.8	30.4	23.1	46.5
Isobutyl formate	80.2	17.3	6.7	76.0
Toluene	81.3	17.9	16.4	65.7
2-Methyl-2-propanol				
Benzene	67.3	8.1	21.4	70.5
Carbon tetrachloride	64.7	3.1	11.9	85.0
Cyclohexane	65.0	8	21	71
3-Methyl-1-butanol				
Isopentyl acetate	93.6	44.8	31.2	24.0
Isopentyl formate	89.8	32.4	19.6	48.0
Allyl alcohol				
Benzene	68.2	8.6	9.2	82.2
Carbon tetrachloride	65.2	5	11	84
Cyclohexane	66.2	8	11	81
Hexane	59.7	8.5	5.1	86.4

TABLE 4.8 Ternary Azeotropic Mixtures (*continued*)

B. Other ternary azeotropes

System	BP of azeotrope, °C	Composition, wt %	System	BP of azeotrope, °C	Composition, wt %
Water Acetone 2-Methyl-1,3-butadiene	32.5	0.4 7.6 92.0	Water Nitromethane Heptane	71.4	7.9 29.7 62.4
Water Acetonitrile Benzene	66	8.2 23.3 68.5	Water Nitromethane Nonane	80.7	17.4 58.3 24.3
Water Acetonitrile Trichloroethylene	67	6.4 20.5 73.1	Water Nitromethane Octane	77.4	12.4 44.3 43.3
Water Acetonitrile Triethylamine	68.6	3.5 9.6 86.9	Water Nitromethane Pentane	33.1	2.1 6.5 91.4
Water 2-Butanone Cyclohexane	63.6	5 35 60	Water Nitromethane Undecane	82.8	20.6 73.3 6.1
Water Butyraldehyde Hexane	55.0	4 21 75	Water Pyridine Dodecane	93.5	40.5 54.5 5.0
Water Formic acid Isopentanoic acid	107.6	21.3 76.3 2.4	Water Pyridine Undecane	93.1	38.5 51.0 10.5
Water Formic acid Isobutyric acid	107.0	15.5 66.8 17.7	Water Pyridine Decane	92.3	35.5 45.5 19.0

Component	BP, °C	Wt %
Water	107.6	19.5
Formic acid		75.9
Butyric acid		4.6
Water	107.2	18.6
Formic acid		71.9
Propionic acid		9.5
Water	105	11.0
Hydrogen bromide		10.4
Chlorobenzene		78.6
Water	96.9	20.2
Hydrogen chloride		5.3
Chlorobenzene		74.5
Water	107.3	64.8
Hydrogen chloride		15.8
Phenol		19.4
Water	116.1	54
Hydrogen fluoride		10
Fluorosilic acid		36
Water	75.1	11.5
Nitroethane		75.1
Heptane		64.0
Water	59.5	8.4
Nitroethane		9.3
Hexane		82.3
Water	82.4	19.1
Nitromethane		68.1
Decane		12.8
Water	83.1	21.5
Nitromethane		75.3
Dodecane		3.2

Component	BP, °C	Wt %
Water	90.5	30.5
Pyridine		37.0
Nonane		32.5
Water	86.7	22.4
Pyridine		25.5
Octane		52.0
Water	78.6	14.0
Pyridine		15.5
Heptane		70.5
Acetic acid	134.4	23
Pyridine		55
Acetic anhydride		22
Acetic acid	134.1	31.4
Pyridine		38.2
Decane		30.4
Acetic acid	129.1	13.5
Pyridine		25.2
Ethylbenzene		61.3
Acetic acid	98.5	3.4
Pyridine		10.6
Heptane		86.0
Acetic acid	128.0	20.7
Pyridine		29.4
Nonane		49.9
Acetic acid	115.7	10.4
Pyridine		20.1
Octane		69.5
Acetic acid	132.2	17.7
Pyridine		30.5
o-Xylene		51.8

TABLE 4.8 Ternary Azeotropic Mixtures (*continued*)

System	BP of azeotrope, °C	Composition, wt%	System	BP of azeotrope, °C	Composition, wt%
Acetic acid Pyridine p-Xylene	129.2	10.2 22.5 67.3	Methanol Methyl acetate Hexane	47.4	14.6 36.8 48.6
Acetic acid 2,6-Dimethylpyridine Undecane	163.0	75.0 13.8 11.2	Ethanol Acetone Chloroform	63.2	10.4 24.3 65.3
Acetic acid 2,6-Dimethylpyridine Decane	147.0	12.6 74.3 13.1	Ethanol Acetonitrile Triethylamine	70.1	8 34 58
Acetic acid 2-Methylpyridine Decane	141.3	19.9 46.8 33.3	Ethanol Benzene Cyclohexane	64.7	29.6 12.8 57.6
Acetic acid 2-Methylpyridine Nonane	135.0	12.8 38.4 48.8	Ethanol Chloroform Hexane	57.3	9.5 56.1 34.4
Acetic acid 2-Methylpyridine Octane	121.3	3.6 24.8 71.6	1-Propanol Benzene Cyclohexane	73.8	15.5 30.4 54.2
Acetic acid Benzene Cyclohexane	77.2	7.6 34.4 58.0	2-Propanol Benzene Cyclohexane	69.1	31.1 15.0 53.9
Acetic acid 2-Methyl-1-butanol Isopentyl acetate	132	15 54 31	1-Butanol Benzene Cyclohexane	77.4	4 48 48
Propionic acid 2-Methylpyridine Decane	149.3	29.5 32.0 38.5	1-Butanol Pyridine Toluene	108.7	11.9 20.7 76.4

Component		
Propionic acid	140.1	16.5
2-Methylpyridine		21.5
Nonane		42.0
Propionic acid	123.7	4.5
2-Methylpyridine		10.5
Octane		85.0
Propionic acid	153.4	43.0
2-Methylpyridine		40.0
Undecane		17.0
Propionic acid	147.1	55.5
Pyridine		26.4
Undecane		18.1
Methanol	57.5	23
Acetone		30
Chloroform		47
Methanol	47	14.6
Acetone		30.8
Hexane		59.6
Methanol	53.7	17.4
Acetone		5.8
Methyl acetate		76.8
Methanol	50.8	17.8
Methyl acetate		48.6
Cyclohexane		33.6
1,2-Ethanediol	185.0	8.7
Phenol		74.6
2,6-Dimethylpyridine		16.7
1,2-Ethanediol	185.1	5.9
Phenol		79.1
2-Methylpyridine		15.0
1,2-Ethanediol	186.4	15.9
Phenol		67.7
3-Methylpyridine		16.4
1,2-Ethanediol	188.6	29.5
Phenol		54.8
2,4,6-Trimethylpyridine		15.7
Acetone	60.8	3.6
Chloroform		68.8
Hexane		27.6
Acetone	49.7	51.1
Methyl acetate		5.6
Hexane		43.3
Chloroform	62.0	79.7
Ethyl formate		5.3
2-Bromopropane		15.7
1,4-Dioxane	101.8	44.3
2-Methyl-1-propanol		26.7
Toluene		29.0

FREEZING POINTS

Crystalline organic compounds typically have a characteristic melting point. In some cases, a compound may have more than one arrangement in the crystal and such polymorphs will exhibit differences in melting behavior. For the most part, however, the melting point of a compound is characteristic and invariant for a pure sample. When contaminated by a second substance, however, the melting (or freezing) point is typically lowered.

A classical test used to determine if two samples that have the same melting point are identical is to intimately mix them and record the melting point of the mixture. If the melting point of the blend is identical to that of each individual compound prior to mixing, the two substances are judged to be identical. When the two compounds are not the same, the melting point of the mixture is typically lowered and broadened.

The melting or freezing point depression can be used to determine an approximate molecular weight for a given substance. The cryoscopic constant K_f gives the depression of the melting point ΔT (in degrees Celsius) produced when 1 mol of solute is dissolved in 1000 g of a solvent. It is applicable only to dilute solutions for which the number of moles of solute is negligible in comparison with the number of moles of solvent. Because camphor is conveniently available and has a large K_f value, it is often used in this application. A known amount of the substance whose molecular weight is to be determined (the solute) is mixed with a larger amount of camphor (the solvent). Melting the mixture permits intimate mixing. The lowering or depression of the freezing point is then used to calculate the molecular weight from the following equation

$$M_2 = \frac{1000 w_2 K_f}{w_1 \Delta T}$$

where M_2 is the molecular weight to be determined of the solute, w_1 is the exact weight of the solvent, w_2 is the exact weight of the added compound of unknown molecular mass, ΔT is the change in temperature, and K_f is the cryoscopic constant given in the Table 4.9A.

In Table 4.9A, a range of compounds and their cryoscopic constants are recorded. All of the same data are included in Table 4.9B but they are rearranged in order of increasing melting point at the left and decreasing K_f at the right.

TABLE 4.9A Molecular Lowering of the Melting of Freezing Point

Compound	K_f	MP	Compound	K_f	MP
Acetamide	4.04	80.1	Camphene	31.08	51–52
Acetic acid	3.90	16.63	Camphorquinone	45.7	199
Acetone	2.40	−95.35	D-(+)-Camphor	39.7	178.8
Ammonia	0.957	−77.75	Carbon tetrachloride	29.8	−22.9
Aniline	5.87	−5.98	o-Cresol	5.60	30.9
Antimony(III) chloride	17.95	73.4	p-Cresol	6.96	34.8
Benzene	5.12	5.53	Cyclohexane	20.0	6.5
Benzonitrile	5.34	−12.75	Cyclohexanol	39.3	25.2
Benzophenone	9.8	48.1	Cyclohexylcyclohexane	14.52	3–4
Bicyclohexane	14.52	3–4	Cyclopentadecanone	21.3	64–66
Biphenyl	8.0	68.8	Z-Decahydronaphthalene	19.47	−43.0
Borneol	35.8	204	E-Decahydronaphthalene	20.81	−30.4
Bornylamine	40.6	163	Dibenz[de,kl]anthracene	25.7	273–274
Butanedinitrile	18.26	57.9	Dibenzyl ether	6.27	3.5

TABLE 4.9A Molecular Lowering of the Melting of Freezing Point (*continued*)

Compound	K_f	MP	Compound	K_f	MP
1,2-Dibromoethane	12.5	10.0	Nitrobenzene	6.852	5.8
Diethyl ether	1.79	−116.3	Octadecanoic acid	4.50	70
1,2-Dimethoxybenzene	6.38	22.5	2-Oxohexa-		
N, N-Dimethylacetamide	4.46	−20.0	methyleneimine	7.30	69.2
2,2-Dimethyl-1-propanol	11.0	52–54	Phenol	7.40	40.9
Dimethyl sulfoxide	4.07	18.5	Pyridine	4.75	−41.6
1,4-Dioxane	4.63	11.7			
Diphenylamine	8.60	53–54	Quinoline	1.95	−14.9
Diphenyl ether	7.88	26.9	Succinonitrile	18.26	46–48
1,2-Ethanediamine	2.43	8.5	Sulfuric acid	1.86	10.38
Ethoxybenzene	7.15	−29.5	1,1,2,2,-Tetrabro-		
Formamide	3.85	2.6	moethane	21.7	0.0
Formic acid	2.77	8.5	1,1,2,2,-Tetrachloro1,2-	37.7	26.0
Glycerol	3.3–3.7	18.18	difluoroethane		
Hexamethylphos-			Tetramethylene sulfone	64.1	27.6
phoramide	6.93	7.2	p-Toluidine	5.372	43.8
N-Methylacetamide	6.65	30.6	Tribromomethane	14.4	8.1
2-Methyl-2-butanol	10.4	−9.0	1,3,3-Trimethyl-2-	6.7	1–2
Methylcyclohexane	14.13	−126.6	oxabicyclo[2.2.2]octane		
Methyl Z-9-octadecenoate	3.4	19.9	Triphenylmethane	12.45	93.4
2-Methyl -2-propanol	8.37	25.8	Water	1.86	0.000
Naphthalene	6.94	80.2	p-Xylene	4.3	13.3

The same data are presented in Table 4.9B but the information is rearranged at the left in order of increasing melting point and at the right in order of decreasing K_f.

TABLE 4.9B Molecular Lowering of the Melting of Freezing Point

Compound	K_f	MP	Compound	K_f	MP
Methylcyclohexane	14.13	−126.6	Tetramethylene sulfone	64.1	27.6
Diethyl ether	1.79	−116.3	Camphorquinone	45.7	199
Acetone	2.4	−95.35	Bornylamine	40.6	163
Ammonia	0.957	−77.75	D-(+)-Camphor	39.7	178.8
Z-Decahydronaphthalene	19.47	−43	Cyclohexanol	39.3	25.2
Pyridine	4.75	−41.6	1,1,2,2-Tetrachloro-1,2-	37.7	26
E-Decahydronaphthalene	20.81	−30.4	difluoroethane		
Ethoxybenzene	7.15	−29.5	Borneol	35.8	204
Carbon tetrachloride	29.8	−22.9	Camphene	31.08	51–52
N, N- Dimethylacetamide	4.46	−20	Carbon tetrachloride	29.8	−22.9
Quinoline	1.95	−14.9	Dibenz[de,kl]anthracene	25.7	273–274
Benzonitrile	5.34	−12.75	1,1,2,2-Tetrabromoethane	21.7	0
2-Methyl-2-butanol	10.4	−9	Cyclopentadecanone	21.3	64–66
Aniline	5.87	−5.98	E- Decahydronaphthalene	20.81	−30.4
1,1,2,2-Tetrabromoethane	21.7	0	Cyclohexane	20	6.5
Water	1.86	0	Z-Decahydronaphthalene	19.47	−43
1,3,3-Trimethyl-2-	6.7	1–2	Succinonitrile	18.26	46–48
oxabicyclo[2.2.2]octane			Butanedinitrile	18.26	57.9

TABLE 4.9B Molecular Lowering of the Melting of Freezing Point (*continued*)

Compound	K_f	MP	Compound	K_f	MP
Formamide	3.85	2.6	Antimony(III) chloride	17.95	73.4
Bicyclohexane	14.52	3–4	Bicyclohexane	14.52	3–4
Dibenzyl ether	6.27	3.5	Cyclohexylcyclohexane	14.52	6.5
Benzene	5.12	5.53	Tribromomethane	14.4	8.1
Nitrobenzene	6.852	5.8	Methylcyclohexane	14.13	−126.6
Cyclohexane	20	6.5	1,2-Dibromoethane	12.5	10
Cyclohexylcyclohexane	14.52	6.5	Triphenylmethane	12.45	93.4
Hexamethyl-phosphoramide	6.93	7.2	2,2-Dimethyl-1-propanol	11	52–54
Tribromomethane	14.4	8.1	2-Methyl-2-butanol	10.4	−9
1,2-Ethanediamine	2.43	8.5	Benzophenone	9.8	48.1
Formic acid	2.77	8.5	Diphenylamine	8.6	53–54
1,2-Dibromoethane	12.5	10	2-Methyl-2-propanol	8.37	25.8
Sulfuric acid	1.86	10.38	Biphenyl	8	68.8
1,4-Dioxane	4.63	11.7	Diphenyl ether	7.88	26.9
p-Xylene	4.3	13.3	Phenol	7.4	40.9
Acetic acid	3.9	16.63	2-Oxohexamethylene imine	7.3	69.2
Glycerol	3.3–3.7	18.18	Ethoxybenzene	7.15	−29.5
Dimethyl sulfoxide	4.07	18.5	p-Cresol	6.96	34.8
Methyl Z-9-octadecenoate	3.4	19.9	Naphthalene	6.94	80.2
1,2-Dimethoxybenzene	6.38	22.5	Hexamethylphosphoramide	6.93	7.2
Cyclohexanol	39.3	25.2	Nitrobenzene	6.852	5.8
2-Methyl-2-propanol	8.37	25.8	1,3,3-Trimethyl-2-oxabicyclo[2.2.2]octane	6.7	1–2
1,1,2,2-Tetrachloro-1,2-difluoroethane	37.7	26	N-Methylacetamide	6.65	30.6
Diphenyl ether	7.88	26.9	1,2-Dimethoxybenzene	6.38	22.5
Tetramethylene sulfone	64.1	27.6	Dibenzyl ether	6.27	3.5
N-Methylacetamide	6.65	30.6	Aniline	5.87	−5.98
o-Cresol	5.6	30.9	o-Cresol	5.6	30.9
p-Cresol	6.96	34.8	p-Toluidine	5.372	43.8
Phenol	7.4	40.9	Benzonitrile	5.34	−12.75
p-Toluidine	5.372	43.8	Benzene	5.12	5.53
Succinonitrile	18.26	46–48	Pyridine	4.75	−41.6
Benzophenone	9.8	48.1	1,4-Dioxane	4.63	11.7
Camphene	31.08	51–52	Octadecanoic acid	4.5	70
2,2-Dimethyl-1-propanol	11	52–54	N, N- Dimethylacetamide	4.46	−20
Diphenylamine	8.6	53–54	p-Xylene	4.3	13.3
Butanedinitrile	18.26	57.9	Dimethyl sulfoxide	4.07	18.5
Cyclopentadecanone	21.3	64–66	Acetamide	4.04	80.1
Biphenyl	8	68.8	Acetic acid	3.9	16.63
2-Oxohexamethylene imine	7.3	69.2	Formamide	3.85	2.6
Octadecanoic acid	4.5	70	Glycerol	3.3–3.7	18.18
Antimony(III) chloride	17.95	73.4	Methyl Z-9-octadecenoate	3.4	19.9
Acetamide	4.04	80.1	Formic acid	2.77	8.5
Naphthalene	6.94	80.2	1,2-Ethanediamine	2.43	8.5
Triphenylmethane	12.45	93.4	Acetone	2.4	−95.35
Bornylamine	40.6	163	Quinoline	1.95	−14.9
D-(+)-Camphor	39.7	178.8	Water	1.86	0
Camphorquinone	45.7	199	Sulfuric acid	1.86	10.38
Borneol	35.8	204	Diethyl ether	1.79	−116.3
Dibenz[de,kl] anthracene	25.7	273–274	Ammonia	0.957	−77.75

VISCOSITY, DIELECTRIC CONSTANT, DIPOLE MOMENT, SURFACE TENSION, AND REFRACTIVE INDEX

Several additional physical properties are summarized in Table 4.10. These are viscosity, dielectric constant, dipole moment, and surface tension for selected common organic compounds.

Viscosity

Viscosity may be thought of as a fluid's resistance to flow or internal friction. It is characterized by the property called viscosity, which is designated η (eta). A fluid passing through a tube will flow more freely if its viscosity is low and more slowly if it is more viscous. Viscosity is given in Table 4.10 in units of milliNewton·second·meter^{-2} or mN·s·m^{-2}. In fluid mechanics, the unit "poise" is equal to a force of 1 dyne·cm^{-2} for two fluids passing each other at a rate of 1 cm·s^{-1}. Viscosity is typically greater at lower temperatures; temperatures (in °C) are given in parentheses in the table.

Dielectric Constant

The dielectric constant (or relative permittivity) is usually expressed using the symbol ε. The dielectric ε is defined as the ratio of electric fields E_0/E for a vacuum and a substance placed between the plates of a capacitor. The dielectric constant of a vacuum is 1 and substances that can orient to greater or lesser extents in the applied field will have higher dielectric constants. The dielectric constant of heptane at 20°C is ~1.9. Acetonitrile, $CH_3C\equiv N:$, has a dielectric constant at 20°C of 37.5. The dielectric constant for water is near 80.

The choice of a solvent for a particular reaction will usually depend on more than one variable. These include the liquid temperature range, the dielectric constant, and whether or not the solvent is reactive in the chemical reaction. The most important consideration in the latter context is often the presence of an easily transferred proton. Certain solvents, such as acetone, are considered aprotic but may transfer a proton under basic conditions. Thus the designations given below are general and approximate. They are intended to guide the reader to the more detailed information contained in the full tables.

Common Solvents Listed in Order of Increasing Dielectric Constant (Ascending Polarity)

Aprotic Solvents		Protic Solvents
Dielectric constants < 15	Dielectric constants > 15	Dielectric range 6–80
carbon disulfide	acetone	acetic acid
carbon tetrachloride	benzonitrile	trifluoroacetic acid
tetrachloroethylene	hexamethylphosphoramide	phenols
1,2-dichloroethylene	N-methylpyrrolidone	butanols
chlorobenzene	nitrobenzene	propanols
dichloromethane	nitromethane	ethanol
chloroform	N,N-dimethylformamide (DMF)	2,2,2-trifluoroethanol
1,4-dioxane	acetonitrile	methanol
diphenyl ether	dimethyl sulfoxide (DMSO)	various glycols
diethyl ether		water

	Aprotic Solvents	Protic Solvents
Dielectric constants < 15	Dielectric constants > 15	Dielectric range 6–80

tetrahydrofuran (THF)
1,2-dimethoxyethane (DME)
ethyl acetate
butyl acetate
N,N-dimethylaniline
pyridine

Dipole Moment

The dipole moment of a molecule is the vectorial sum of the individual dipoles within it. Bond dipoles are usually represented using the symbol μ and are expressed in units of Debye. The dipoles result from charge separation. The carbon–carbon bond in ethane, H_3C—CH_3 is symmetrical and not expected to have a dipole moment. The carbon–oxygen bond of methanol (CH_3OH), on the other hand, links two elements of differing electronegativity and is expected to have a significant molecular dipole. Methanol's molecular dipole is ~1.7 whereas, for ethane, it is 0.

Experimentally, the molecular dipole can be measured. Individual bond dipoles cannot be measured but they have been inferred from experimental data for a variety of compounds. Estimates of the molecular dipole can be made by vector addition of individual moments. Such estimates (calculations) are indicative but may differ significantly from the measured values. The latter are recorded in Table 4.10.

Four examples of dipole moments are instructive. First, the dipoles for chloromethane and dichloromethane are 1.87 D and 1.60 D, respectively. Although two chlorine–carbon bonds are present in the latter, the dipole is not along either but rather bisects the angle between them. This is illustrated schematically using the stylized arrow with its positive end in the form of a cross. The orientation question is shown clearly in the rigid dichlorobenzene framework. The dipole is 2.13 D for the *ortho*-isomer and 0 D when the dipoles exactly oppose each other.

The orientation of dipoles can be assessed by comparing otherwise identical molecules. For example, amino is electron releasing and nitro is electron withdrawing. The molecular dipoles (1.53 D for aniline and 4.22 D for nitrobenzene) add to give an overall molecular dipole of 6.3 D for 4-nitroaniline. The importance of two polar C—O bonds should be large

1.87 D   1.60 D   2.13 D   1.50 D   0 D

1.53 D   4.22 D   6.3 D   2.20 D

if they are added and small if they oppose. The dynamics of the system must be taken into account for structures such as 1,2-dihydroxyethane (ethylene glycol), which is illustrated in the eclipsed and staggered conformers.

Surface Tension

The surface tension is the force that acts on the surface of a liquid that tends to minimize the surface area of the liquid. Surface tension is also sometimes referred to as interfacial force or interfacial tension. The property of surface tension is temperature dependent. For the majority of compounds the dependence of the surface tension γ on the temperature can be given as

$$\gamma = a - bt$$

where a and b are constants and t is the temperature in °C.

TABLE 4.10 Viscosity, Dielectric Constant, Dipole Moment, and Surface Tension of Selected Organic Substances

States:	[g], gas	Solvents:	[B], benzene,	C_6H_6
	[lq], liquid		[C], carbon tetrachloride,	CCl_4
			[D], 1,4-dioxane,	$C_4H_8O_2$
			[H], hexane,	C_6H_{14}
			[cH], cyclohexane,	C_6H_{12}

The temperature in degrees Celsius at which the viscosity, dielectric constant, dipole moment, and surface tension of a substance were measured is shown in this table in parentheses after the value. The solvent used or the physical state of the substance are also shown in parentheses after the temperature in square brackets, for example, [g] or [b]

Alternate names for entries are listed in Table 1-14 at the bottom of each double page.

Substance	Viscosity η, mN·s·m^{-2}	Dielectric constant ε	Dipole moment, D	Surface tension, dyn·cm^{-1}	
				a	b
Acetaldehyde	0.280 (0)	21.8(10)	2.71 [g]	23.90	0.1360
	0.256 (10)	21.1 (21)			
	0.22 (20)				
Acetaldoxime	1.415 (20)	3 (23)	0.830 (20) [lq]	30.1 (35)	
			0.90 (25) [B]		
Acetamide	1.32 (105)	59.2 (83)	3 90 (25) [B]	47.66	0.102 1
	1.06 (120)	60.6 (94)	2.44 (30) [B]		
Acetanilide	2.22 (120)	3.65 (25) [B]	46.21	0.091 2	
	1.90 (130)				
Acetic acid	1.232 (20)	6.15 (20)	1 76[g]	29.58	0.009 4
	0.796 (50)	6.29 (40)	1.92 (20) [B]		

TABLE 4.10 Viscosity, Dielectric Constant, Dipole Moment, and Surface Tension of Selected Organic Substances (*continued*)

Substance	Viscosity η, mN·s·m^{-2}	Dielectric constant, ε	Dipole moment, D	Surface tension, dyn·cm^{-1}	
				a	*b*
Acetic anhydride	0.907 (20) 0.699 (40)	23.3 (0) 21.2 (20)	2.8 [g] 3.15 (20) [B]	35.52	0.143 6
Acetone					
[lq]	0.391 (0) 0.318 (20)	20.7 (25) 17.6 (56)	2.77 (22) [B]	26.26	0.112
[g]	0.009 33 (100) 0.012 8 (225)	1.015 9 (100)	2.87		
Acetonitrile	0.397 (10) 0.329 (30)	37.5 (20) 26.6 (82)	3.97 [g] 3.47 (20) [B]	29.58	0.117 8
Acetophenone	2.015 (15) 1.511 (30)	17.39 (25) 8.64 (202)	2.96 (30) [B]	41.92	0.115 4
Acetyl bromide		16.2 (20)	2.45 (20) [B]		
Acetyl chloride					
[lq]		16.9 (2) 15.8 (22)	2.47 (20) [B]	26.7 (15)	
[g]		1.0217 (20)	2.71		
Acetylene [g]	0.010 2 (30) 0.012 6 (101)	1.001 34 (0)	0	3.42	0.193 5 [lq]
Acrylic acid	1.3 (20) 1.16 (25)			28.1 (30)	
Acrylonitrile	0.35 (20) 0.34 (25)	33.0 (20)	3.91 [g] 3.54 (25) [B]	29.58	0.117 8
Allyl acetate	0.207 (30)			28.73	0.118 6
Allylamine	0.375 (25)		1.3 (25) [B]	27.49	0.128 7
Allyl isothiocyanate		17.2 (18)	3.2 (20) [B]	36.76	0.107 4
2-Aminoethanol	30.85 (15) 19.35 (25)	37.72 (25)	2.59 (25) [D]	51.11	0.111 7
Aniline	5.30 (15) 4.40 (20) 3.18 (30)	6.89 (20) 5.93 (70)	1.53 (20) [B]	44.83	0.108 5
Benzaldehyde	1.321 (25)	19.7 (0) 17.8 (20)	2.77 (20) [lq]	40.72	0.109 0
Benzaldehyde oxime (mp 30) (mp 128)		3.8 (20)	1.2 (25) [B] 1.5 (25) [B]		
Benzene	0.649 (20) 0.566 (30) 0.395 (60)	2.292 (15) 2.274 (25) 1.002 8 [g]	0	28.88 (20)	27.56 (30)
Benzamide			3.42 (25) [B]	47.26	0.070 5
Benzenesulfonyl chloride			4.50 (20) [B]	45.48	0.111 7
Benzenethiol	1.239 (20) 1.144 (25)	4.38 (25)	1.13 (25) [lq] 1.19 (20) [B]	41.41	0.120 2

TABLE 4.10 Viscosity, Dielectric Constant, Dipole Moment, and Surface Tension of Selected Organic Substances (*continued*)

Substance	Viscosity η, mN·s·m^{-2}	Dielectric constant, ε	Dipole moment, D	Surface tension, dyn·cm^{-1}	
				a	b
Benzonitrile	1.447 (15)	26.5 (20)	4.40 [g]	41.69	0.115 9
	1.111 (30)	24.0 (40)	3.9 (20) [B]		
Benzophenone	4.79 (55)	14.60 (18)	3.09 (50) [lq]	46.31	0.112 8
	1.38 (120)	11.4 (50)	2.98 (25) [B]		
Benzoyl bromide	1.956 (20)	21.33 (20)	3.40 (20) [B]	45.85	0.139 7
	1.798 (25)	20.74 (25)			
Benzoyl chloride		29 (0)	3.16 (25) [B]	41.34	0.108 4
		23 (20)			
Benzyl acetate	1.399 (45)	5.1 (21)	1.80 (25) [B]		
Benzyl alcohol	5.58 (20)	13.0 (20)	1.67 (25) [B]	38.25	0.138 1
	4.65 (30)	9.5 (70)			
	3.01 (45)				
Benzylamine	1.59 (25)	5.5 (1)	1.15 (20) [lq]	42.33	0.121 3
		4.6 (21)	1.38 (25) [B]		
Benzyl benzoate	8.51 (25)	4.9 (20)	2.06 (30) [B]	48.07	0.106 5
Benzyl butyl o-phthalate	65 (20)				
Benzyl chloride	1.400 (20)	7.0 (13)	1.83 (20) [B]	39.92	0.122 7
	1.290 (25)				
Benzylethylamine		4.3 (20)			
Benzyl ethyl ether		3.9 (20)		32.83 (20)	29.97 (40)
Biphenyl		2.53 (75)	0	41.52	0.093 1
Bis(2-ethoxyethyl)ether			1.92 (25) [B]	29.74	0.117 6
Bis(2-hydroxyethyl)ether	38.0 (20)	31.69 (20)	2.31 (20) [B]	46.97	0.088 0
	30.0 (25)				
1,2-Bis(methoxyethoxy)-ethane	3.76 (20)				
Bis(2-methoxyethyl) ether	1.99 (20)		1.97 (25) [B]	32.47	0.116 4
DL-Bornyl acetate	0.981 (25)	4.6 (21)	1.89 (22)		
3-Bromoaniline	6.81(20)	13.0 (19)	2.67 (20) [B]		
	3.70 (40)				
4-Bromoaniline	1.81 (80)	7.06 (30)	2.88 (25) [B]		
Bromobenzene	1.196 (15)	5.40 (25)	1.70 [g]	38.14	0.116 0
	0.985 (30)		1.50 (20) [lq]		
1-Bromobutane	0.633 (20)	7.88 (−10)	2.17 [g]	28.71	0.112 6
	0.597 (25)	7.07 (20)	2.04 (20) [lq]		
DL-2-Bromobutane	1.434 (20)	8.64 (25)	2.22 [g]	27.48	0.110 7
			2.14 (25) [lq]		
1-Bromo-2-chlorobenzene		6.80 (20)	2.15 (20) [B]		
1-Bromo-3-chlorobenzene		4.58 (20)	1.52 (22) [B]		
1-Bromo-4-chlorobenzene			0.1 (25) [B]	40.03	0.100 2

TABLE 4.10 Viscosity, Dielectric Constant, Dipole Moment, and Surface Tension of Selected Organic Substances (*continued*)

Substance	Viscosity η, mN·s·m^{-2}	Dielectric constant, ε	Dipole moment, D	Surface tension, dyn·cm^{-1}	
				a	b
Bromochloro-methane	0.670 (20)	7.79	1.66 (25) [B]	33.32 (20)	
Bromocyclo-hexane	2.0 (25)	11 (−65) 7.9 (25)	1.08 (25) [lq] 2.3 (25) [B]	36.13	0.111 7
1-Bromodecane		4.75 (1) 4.44 (25)	2.08 (20) [lq] 1.90 (25) [lq]	31.26	0.085 6
Bromodichloro-methane			1.31 (25) [B]	35.11	0.129 4
1-Bromododecane		4.07 (25)	2.01 (25) [lq] 1.89 (25) [B]	32.58	0.088 2
Bromoethane	0.397 (20) 0.348 (30)	13.6 (−60) 9.39 (20)	2.03 [g] 2.04 (20) [lq]	26.52	0.115 9
1-Bromo-2-ethoxyethane				31.98	0.112 9
1-Bromo-2-ethoxypentane		6.45 (25)	2.32 (25) [B]		
2-Bromo-2-ethoxypentane		6.40 (25)	2.07 (25) [B]		
3-Bromo-3-ethoxypentane		8.24 (25)	2.15 (25) [B]		
Bromoethylene		4.78 (25)	1.42 [g]		
Bromoform	2.152 (15) 1.741 (30)	4.39 (20)	1.00 [g] 0.92 (25) [lq]	48.14	0.130 8
1-Bromoheptane		5.33 (25) 4.48 (90)	2.17 [g] 2.02 (20) [lq]	30.74	0.098 2
2-Bromoheptane		6.46 (22)	2.08 (20) [B]		
3-Bromoheptane		6.93 (22)	2.06 (20) [B]		
4-Bromoheptane		6.81 (22)	2.06 (20) [B]		
1-Bromohexa-decane		3.71 (25)	1.98 (20) [lq] 1.96 (25) [C]	33.37	0.086 1
1-Bromohexane		6.30 (1) 5.82 (25)	2.06 (20) [lq]	29.81	0.096 7
Bromomethane		9.82 (0) 1.006 8 (100) [g]	1.79 [g]	26.52	0.115 9
1-Bromo-3-methylbutane		8.04 (−56) 6.05 (20)	1.95 (20) [B]	28.10	0.099 6
2-Bromo-3-methylbutyric acid		6.5 (20)			
1-Bromo-2-methylpropane	0.643 (20) 0.518 (40) 3.26 (90)	7.70 (0) 7.2 (25)	1.92 (25) [lq] 1.99 (20) [B]	26.96	0.105 9
1-Bromonaph-thalene	5.99 (15) 3.20 (40)	5.83 (25) 5.12 (20)	1.29 (25) [lq]	46.44	0.101 8
1-Bromononane		5.42 (−20) 4.74 (25)	1.95 (25) [lq]	31.36	0.089 4

TABLE 4.10 Viscosity, Dielectric Constant, Dipole Moment, and Surface Tension of Selected Organic Substances (*continued*)

Substance	Viscosity η, mN·s·m^{-2}	Dielectric constant, ε	Dipole moment, D	Surface tension, dyn·cm^{-1}	
				a	b
1-Bromooctane		6.35 (−50)	1.99 (20) [lq]	31.00	0.092 8
			1.88 (25) [lq]		
1-Bromopenta-decane		3.9 (20)			
1-Bromopentane		9.9 (−90)	2.21 [g]	29.51	0.104 9
		6.32 (25)	2.09 (20) [lq]		
p-Bromophenol				48.88	0.107 0
1-Bromopropane	0.539 (15)	8.09 (25)	2.17 [g]	28.30	0.121 8
	0.459 (30)		3.16 (20) [lq]		
2-Bromopropane	0.536 (15)	9.46 (25)	2.21 [g]	26.21	0.118 3
	0.437 (30)		2.10 (25) [lq]		
1-Bromotetra-decane		3.84 (25)	1.92 (20) [lq]	32.93	0.087 8
			1.83 (25) [lq]		
o-Bromotoluene	1.3 (25)	4.28 (58)	1.45 (20) [B]	36.62	0.099 8
m-Bromotoluene		5.36 (58)	1.77 (20) [B]		
p-Bromotoluene		5.49 (58)	1.95 (20) [B]	36.40	0.099 7
Bromotrifluoro-methane	0.15 (25)		0.65 [g]	4 (25)	
1-Bromoundecane		4.73 (−9)		31.94	0.086 1
Butane	0.007 39 (20) [g]		0	14.87	0.120 6
	0.008 39 (60) [g]				
1,3-Butanediol	130.3 (20)	28.8 (25)		37.8 (25)	
	89 (25)				
1,4-Butanediol	65–70 (25)	33 (15)	3.93 (20) [lq]		
		30 (30)	2.4 (15) [D]		
2,3-Butanediol	121 (25)			36 (25)	
Butanesulfonyl chloride			3.94 (25) [D]	37.33	0.097 7
1-Butanethiol	0.501 (20)	5.07 (25)	1.54 (25) [lq]	28.07	0.114 2
	0.450 (30)	4.59 (50)	or [B]		
1,2,4-Butanetriol	1227 (25)				
1-Butanol	2.948 (20)	17.8 (20)	1.66 [g]	27.18	0.089 8
	1.782 (40)	8.2 (118)	20 [B]		
DL-2-Butanol	3.907 (20)	16.6 (25)	1.66 (30) [B]	23.47 (20)	22.62 (30)
	0.527 (100)				
2-Butanone	0.428 (20)	18.5 (20)	3.2 (30) [lq]	26.77	0.112 2
	0.349 (40)	15.3 (60)	2.76 (25) [B]		
2-Butanone oxime		3.4 (20)		31.89	0.102 2
1-Butene [g]	0.007 6 (20)	1.003 2 (20)	0.30	15.19	0.132 3[lq]
	0.010 0 (120)				
2-Butene			0.33 [g, *cis*]	16.11	0.128 9
			0 [g, *trans*]		
3-Butenenitrile		28.1 (20)	4.53 [g]	31.40	0.108 5
2-Butoxyethanol	3.15 (25)	9.30 (25)	2.08 (25) [B]	28.18	0.081 6
	1.51 (60)				
Butoxyethyne		6.62 (25)	2.05 (25) [lq]		

TABLE 4.10 Viscosity, Dielectric Constant, Dipole Moment, and Surface Tension of Selected Organic Substances (*continued*)

Substance	Viscosity η, mN·s·m^{-2}	Dielectric constant, ε	Dipole moment, D	Surface tension, dyn·cm^{-1}	
				a	*b*
2-(2-Butoxyeth-oxy)ethanol	4.76 (25)			30.0 (25)	
1-Butoxy-2-propanol	2.55 (25)			26.5 (25)	
Butyl acetate	0.734 (20) 0.688 (25)	6.85 (−73) 5.01 (20)	1.86 (22) [B]	27.55	0.106 8
DL-*sec*-Butyl acetate				23.33 (22)	21.24 (42)
tert-Butyl acetate			1.91 (25) [B]	24.69	0.110 2
Butylamine	0.681 (20)	4.88 (20)	1.00 [g] 1.22 (20) [lq]	26.24	0.112 2
sec-Butylamine		4.4 (21)	1.28 (25) [B]	23.75	0.105 7
tert-Butylamine			1.29 (25) [B]	19.44	0.102 8
Butylbenzene	1.035 (20) 0.960 (25)	2.36 (20)	0.36 (20) [lq]	31.28	0.102 5
sec-Butylbenzene		2.36 (20)	0.37 (20) [lq]	30.48	0.097 9
tert-Butylbenzene		2.37 (20)	0.36 (20) [lq]	30.10	0.098 5
Butyl butyrate	0.84 (25)			27.65	0.096 5
Butyl decyl *o*-phthalate	55 (20)				
N-Butyldietha-nolamine	55 (25)				
4-*tert*-Butyl-2,5-dimethylphenol	8.30 (80)				
4-*tert*-Butyl-2,6-dimethylphenol	2.72 (80)				
6-*tert*-Butyl-2,4-dimethylphenol	2.10 (80)				
6-*tert*-butyl-3,4-dimethylphenol	3.50 (80)				
N-Butylethano-lamine	17.4 (25)				
Butyl ethyl ether	0.421 (20) 0.397 (25)		1.24	22.75	0.104 9
Butyl formate	0.691 (20) 0.940 (0)	2.43 (80)	2.08 (26) [lq] 2.03 (25) [B]	27.08	0.102 6
Butyl methyl ether			1.25 (25) [B]	22.17	0.105 7
2-*tert*-Butyl-4-methylphenol	2.55 (80)		1.31 (20) [B]		
Butyl nitrate		13 (20)	2.99 (20) [B]	30.35	0.112 6
2-(2-*sec*-Butylph-enoxy)ethanol	65.1 (25)				
2-(4-*tert*-Butylph-enoxy)ethanol	122.5 (25)				
Butyl propionate			1.79 (22) [B]	27.37	0.099 3

TABLE 4.10 Viscosity, Dielectric Constant, Dipole Moment, and Surface Tension of Selected Organic Substances (*continued*)

Substance	Viscosity η, mN·s·m^{-2}	Dielectric constant, ε	Dipole moment, D	Surface tension, dyn·cm^{-1}	
				a	b
4-*tert*-Butyl-pyridine	1.495 (20)		2.87 (25) [C]	35.48	0.095 1
Butyl stearate	8.26 (25) 4.9 (50)	3.11 (30)	1.88 (24) [B]	33.0 (25)	32.7 (30)
Butyl vinyl ether	0.5 (20)		1.25 (25) [H]	21.99 (20)	
Butyraldehyde	0.455 (20) 0.367 (39)	13.4 (26)	2.45 (40) [lq]	26.67	0.092 5
Butyric acid	1.540 (20) 0.980 (40)	2.97 (20)	1.65 (30) [B]	28.35	0.092 0
Butyric anhydride	1.615 (20) 1.486 (25)	13 (20)		28.93 (20)	28.44 (25)
4-Butyrolactone	1.75 (25)	39.1 (20)	4.12 (25) [B]		
Butyronitrile	0.624 (15) 0.515 (30)	20.3 (21)	4.07 [g] 3.6 (20) [B]	29.51	0.103 7
Camphor		11.35 (20)	2.91 (20) [B] 3.10 (25) [B]		
Carbon disulfide	0.363 (20)	3.0 (−112) 2.64 (20)	0 [g] 0.12 (20) [lq]	35.29	0.148 4
Carbon tetra-chloride	0.965 (20) 0.793 (25)	2.24 (20) 2.23 (25)	0	29.49	0.122 4
Carbon tetra-fluoride	0.020 (25)	1.000 6 (25) [g]	0	14 (−73)	
Carvone		11 (22)	2.8 (15) [B]	36.54	0.092 0
Chloroacetic acid	3.15 (50) 1.92 (75)	20 (20) 12.3 (60)	2.31 (30) [B]	43.27	0.111 7
o-Chloroaniline	0.925 (25)	13.4 (25)	1.78 (20) [B]	43.41	0.090 4
m-Chloroaniline		13.4 (19)	2.68 (20) [B]		
p-Chloroaniline			2.99 (25) [B]	48.69	0.109 9
Chlorobenzene	0.799 (20) 0.631 (40)	5.71 (20) 4.2 (120)	1.72 [g] 1.56 (20) [lq]	35.97	0.119 1
1-Chlorobutane	0.469 (15)	9.07 (−30) 7.39 (20)	2.13 [g] 2.0 (20) [B]	25.97	0.111 7
2-Chlorobutane	0.439 (15)	7.09 (30)	2.14 [g] 2.1 (20) [B]	24.40	0.111 8
Chlorocy-clohexane		10.9 (−47) 7.6 (25)	2.2 (25) [B]	33.90	0.110 1
Chlorodifluoro-methane	0.23 (25) 0.013 (25) [g]	6.11 (24)	1.4 [g]	8 (25)	
1-Chlorododecane		4.2 (20)	2.11 (25) [lq] 1.94 (20) [B]	31.56	0.090 4
1-Chloro-2-2,3-epoxypropane	1.03 (25)	25.6 (1) 22.6 (22)	1.8 (25) [C]	39.76	0.136 0
Chloroethane	0.279 (10)	1.013 (19) [g]	2.0 [g] 1.96 (20) [lq]	21.18 (5)	20.58 (10)
2-Chloroethanol	3.913 (15)	25.8 (25) 13 (132)	1.77 [g] 1.90 (25) [B]	38.9 (20)	

TABLE 4.10 Viscosity, Dielectric Constant, Dipole Moment, and Surface Tension of Selected Organic Substances (*continued*)

Substance	Viscosity η, mN·s·m^{-2}	Dielectric constant, ε	Dipole moment, D	Surface tension, dyn·cm^{-1} a	b
Chloroform	0.596 (15)	4.81 (20)	1.1 [g]	29.91	0.129 5
	0.514 (30)	4.31 (50)	1.1 (25) [lq]		
1-Chloroheptane		4.48 (20)	1.86 (22) [B]	28.94	0.096 1
2-Chloroheptane		6.52 (22)	2.05 (22) [B]		
3-Chloroheptane		6.70 (22)	2.06 (22) [B]		
4-Chloroheptane		6.54 (22)	2.06 (22) [B]		
1-Chlorohexane			1.94 (20) [B]	28.32	0.103 8
Chloromethane					
[g]	0.0106 (20)	1.006 9 (100)	1.87		
	0.012 9 (80)				
[lq]		12.6 (−20)	1.86 (20)	19.5	0.165 0
1-Chloro-3-methylbutane		7.63 (−70) 6.05 (20)	1.94 (20) [B]	25.51	0.107 6
Chloromethyl methyl ether			1.88 [C]		
1-Chloro-2-methylpropane	0.462 (20) 0.373 (40)	7.87 (−38) 6.49 (14)	2.06 [g] 2.0 (25) [B]	24.40	0.109 9
2-Chloro-2-methylpropane	0.543 (15)	10.95 (0) 9.96 (20)	2.11 [g] 2.13 (25) [B]	20.06 (15)	18.35 (30)
1-Chloronaph-thalene	2.940 (25)	5.04 (25)	1.33 (25) [lq] 1.52 (25) [B]	44.12	0.103 5
o-Chloronitro-benzene		38 (50) 32 (80)	4.62 [g] 6.22 (50) [lq]	48.10	0.117 1
m-Chloronitro-benzene		21 (50) 18 (80)	3.72 [g] 3.30 (50) [lq]	49.71	0.141 7
p-Chloronitro-benzene		8 (120)	2.81 [g] 2.83 (90) [lq]	45.84	0.104 6
1-Chlorooctane		5.05 (25)	2.14 (25) [lq]	29 64	0.096 1
Chloropenta-fluoroethane	0.26 (25) 0.013 (25) [g]		0.5 [g]	5 (25)	
1-Chloropentane	0.580 (20)	6.6 (11)	2.14 [g] 1.94 (20) [B]	27.09	0.107 6
o-Chlorophenol	2.250 (45) 4.11 (25)	6.31 (25)	2.19 [g] 1.46 (20) [lq]	42.5	0.112 2
m-Chlorophenol	4.722 (45) 11.55 (25)		2.19 (25) [B]	43.7	0.100 9
p-Chlorophenol	4.99 (50)		2.09 (20) [B]	19.51	0.087 5
1-Chloropropane	0.372 (15) 0.318 (30)	7.7 (20)	2.05 [g] 1.96 (20) [B]	24.41	0.124 6
2-Chloropropane	0.335 (15) 0.299 (30)	9.82 (20)	2.17 [g] 2.1 (20) [B]	21.37	0.088 3
1-Chloro-2-propanone		30 (19)	2.22 [g] 2.37 (20), [H]		
3-Chloro-1-propene	0.347 (15)	8.2 (20)	2.0 [g] 1.8 (20) [B]	25.50	0.094 6

TABLE 4.10 Viscosity, Dielectric Constant, Dipole Moment, and Surface Tension of Selected Organic Substances (*continued*)

Substance	Viscosity η, mN·s·m^{-2}	Dielectric constant, ε	Dipole moment, D	Surface tension, dyn·cm^{-1}	
				a	b
o-Chlorotoluene		4.45 (20)	1.57 [g]		
		4.2 (55)	1.41 (20) [lq]		
m-Chlorotoluene		5.5 (20)	1.77 (20) [lq]		
		5.0 (60)	1.8 (22) [B]		
p-Chlorotoluene		6.08 (20)	2.21 [g]	34.93	0.108 2
		5.6 (55)	1.90 (20) [lq]		
Chlorotrifluoro-methane	0.016 (25)	1.001 3 (29) [g]	0.50 [g]	14 (−73)	
Chlorotrimethy-lsilane			2.09 (20) [B]	19.51	0.087 5
Cinnamaldehyde		17 (20)	3.74 [g]		
		16.9 (24)	3.30 (30) [lq]		
o-Cresol	3.506 (46)	11.5 (25)	2.32 (25) [lq]	39.43	0.101 1
			1.45 (25) [B]		
m-Cresol	18.42 (20)	11.8 (25)	2.39 (20) [lq]	38.00	0.092 4
	5.057 (45)		1.61 (25) [B]		
p-Cresol	5.607 (45)	9.91 (58)	2.35 (20) [lq]	38.58	0.096 2
			1.54 (20) [B]		
Crotonic acid			2.13 (30) [B]		
Cyanoacetic acid		33.4 (19)			
Cycloheptanol				35.02	0.092 3
1,3-Cyclohex-adiene		2.6 (−89)	0.38 (20) [B]		
Cyclohexane	0.980 (20)	2.05 (15)	0	27.62	0.118 8
	0.534 (60)	2.02 (25)			
Cyclohexane-carboxylic acid		2.6 (31)			
1,4-Cyclohex-anedione		15.0 (25)	1.41 [g]		
			1.3 (30) [B]		
Cyclohexanol	41.07 (30)	15.0 (25)	1.86 (25) [C]	35.33	0.096 6
	17.19 (45)	7.24 (100)			
Cyclohexanone	2.453 (15)	20 (−40)	3.11 (20) [B]	37.67	0.124 2
	1.803 (30)	18.2 (20)	3.01 (25) [B]		
Cyclohexanone oxime		3.0 (89)	0.83 (25) [B]		
Cyclohexene	0.650 (20)	2.6 (−105)	0.61 [g]	29.23	0.122 3
		2.22 (25)	0.28 (20) [lq]		
Cyclohexylamine	1.662 (20)	4.73 (20)	1.22 (20) [lq]	34.19	0.118 8
	1.16 (49)		1.26 (20) [B]		
Cyclohexyl-benzene	3.681 (0)		0.62 (20) [B]		
Cyclohexyl-methanol		9.7 (60)	1.68 (20) [B]		
		8.1 (80)			
o-Cyclohexyl-phenol		3.97 (55)			

TABLE 4.10 Viscosity, Dielectric Constant, Dipole Moment, and Surface Tension of Selected Organic Substances (*continued*)

Substance	Viscosity η, mN·s·m^{-2}	Dielectric constant, ε	Dipole moment, D	Surface tension, dyn·cm^{-1}	
				a	b
p-Cyclohexyl-phenol		4.42 (131)			
Cyclooctane			0	32.02	0.109 0
Cyclopentane	0.439 (20)	1.965 (20)	0	25.53	0.146 2
Cyclopentanol		25 (−20) 18 (20)	1 72 (25) [C]	35.04	0.101 1
Cyclopentanone		16 (−51)	3.30 [g] 2.93 (25) [B]	35.55	0.110 0
Cyclopentene			0.98 (25) [H]	25.94	0.149 5
p-Cymene	3.402 (20)	2.243 (20)	0 [lq]	28.83	0.087 7
cis-Decahydro-naphthalene	3.381 (20)	2.18 (20)	0	32.18 (20)	31.01 (30)
trans-Decahydro-naphthalene	2.128 (20)	2.17 (20)	0	29.89 (20)	28.87 (30)
Decamethylcy-clopentasiloxane		2.5 (20)		19.56	0.056 5
Decamethylte-trasiloxane	1.28 (20)	2.4 (20)	0.79 (25) [lq]	86.20 (25)	
Decane	0.928 (20) 0.775 (22)	1.991 (20) 1.844 (130)	0	25.67	0.092 0
1-Decanol		8.1 (20)	1.71 (20) [B] 1.62 (25) [B]	30.34	0.073 2
1-Decene	0.805 (20)		0.42 (20) [B]	25.84	0.091 9
Diallyl sulfide		4.9 (20)	1.33 (25) [B]		
Dibenzofuran		3.0 (100)	0.88 (25) [B]		
Dibenzylamine		3.6 (20)	0.97 (20) [lq] 1.02 (20) [B]	43.27	0.108 6
Dibenzyl decanedioate		4.6 (25)			
Dibenzyl ether	3.711 (25)		1.39 (21) [B]	38.2 (35)	
o-Dibromobenzene		7.35 (20)	2.13 (20) [B]		
m-Dibromobenzene		3.80 (20)	1.5 (20) [B]		
p-Dibromobenzene		2.57 (95)	0	41.84	0.100 7
1,4-Dibromobutane			2.16 (20) [lq] 2.06 (20) [B]	48.24	0.119 0
2,3-Dibromobutane		5.75 (25)	2.20 [g] 1.7 (25) [lq]		
1,2-Dibromoethane	1.721 (20) 1.286 (40)	4.78 (25) 4.09 (131)	1.11 [g] 1.14 (20) [lq]	35.43	0.142 8
cis-1,2-Dibro-moethylene		7.7 (0) 7.08 (25)	1.35 (B)		
trans-1,2-Dibro-moethylene		2.9 (0) 2.88 (25)	0		
1,2-Dibro-moheptane		3.8 (25)	1.78 (25) [D]		

TABLE 4.10 Viscosity, Dielectric Constant, Dipole Moment, and Surface Tension of Selected Organic Substances (*continued*)

Substance	Viscosity η, mN·s·m^{-2}	Dielectric constant, ε	Dipole moment, D	Surface tension, dyn·cm^{-1}	
				a	b
2,3-Dibro-moheptane		5.1 (25)	2.15 (25) [B]		
3,4-Dibro-moheptane		4.7 (25)	2.15 (25) [B]		
Dibromomethane		7.77 (10)	1.43 [g]	42.77	0.148 8
		6.7 (40)	1.85 (20) [lq]		
1,2-Dibromo-propane	1.5 (25)	4.3 (20)	1.43 (25) [B]	36.81	0.115 5
Dibromotetra-fluoroethane	0.72 (25)	2.34 (25)		18.9 (20)	18.1 (25)
Dibutylamine	0.95 (20)	2.978 (20)	1.06 (20) [lq]	26.50	0.095 2
			1.05 (20) [B]		
Dibutyl decanedioate	9.03 (25)	4.54 (30)	2.64 (25) [B]		
Dibutyl ether	0.602 (30)	3.06 (25)	1.18 [g]	24.78	0.093 4
			1.19 (20) [lq]		
Dibutyl maleate	5.62 (20)		2.70 (25) [B]	32.46	0.086 5
	4.76 (25)				
2,6-Di-*tert*-butyl-4-methylphenol	3.47 (80)		1.68 (20) [B]		
Dibutyl o-phthalate	19.91 (20)	6.436 (30)	2.97 (20) [lq]	33.40 (20)	
	7.85 (45)	5.99 (45)	2.85 (30) [B]		
Dichloroacetic acid	3.23(50)	8.2 (22)		37.8	0.092 7
	1.92 (75)	7.8 (61)			
o-Dichlorobenzene	1.324 (25)	9.93 (25)	2.51 [g]	26.84 (20)	35.55 (30)
		7.10 (90)	2.26 (24) [B]		
m-Dichlorobenzene	1.045 (23)	5.04 (25)	1.68 [g]	38.30	0.114 7
	0.955 (33)	4.22 (90)	1.38 (24) [B]		
p-Dichlorobenzene	0.839 (55)	2.41 (50)	0	34.66	0.087 9
	0.668 (79)				
1,4-Dichlorobutane		8.9 (25)	2.22 [g]	37.79	0.117 4
			2.13 (25) [lq]		
Dichlorodifluoro-methane	0.26 (25)	2.13 (29)	0.51 [g]	9 (25)	
	0.013 (25) [g]				
1,1-Dichloroethane	0.505 (25)	10.1 (18)	2.06 [g]	27.03	0.118 6
	0.430 (30)	10.86 (16)	2.00 (25) [B]		
1,2-Dichloroethane	0.887 (15)	12.7 (−10)	1.48 [g]	35.43	0.142 8
	0.730 (30)	10.65 (20)	1.7 (20) [B]		
1,1-Dichloroe-thylene	0.442 (0)	4.67 (16)	1.30 (25) [B]		
	0.358 (20)				
cis-1,2-Dichlo-roethylene	0.467 (20)	9.20 (25)	2.95 [g]	28 (20)	
	0.444 (25)		1.90 (25) [B]		
trans-1,2-Dichlo-roethylene	0.423 (15)	2.14 (25)	0.70 (25) [B]	25 (20)	
	0.404 (20)				

TABLE 4.10 Viscosity, Dielectric Constant, Dipole Moment, and Surface Tension of Selected Organic Substances (*continued*)

Substance	Viscosity η, mN·s·m^{-2}	Dielectric constant, ε	Dipole moment, D	Surface tension, dyn·cm^{-1}	
				a	*b*
2,2′-Dichloro-ethyl ether	2.41 (20) 2.065 (25)	21.2 (20)	2.61 (20) [B]	40.57	0.130 6
Dichlorofluoro-methane	0.34 (25) 0.011 (25) [g]	5.34 (28)	1.3 [g]	18 (25)	
Dichloromethane	0.449 (15) 0.393 (30)	9.14 (20) 1.006 5 (100)	1.60 [g] 1.90 (20) [B]	30.41	0.128 4
2,4-Dichloro-phenol		[g]	1.60 (25) [B]	46.59	0.122 1
1,2-Dichloro-propane	0.865 (20) 0.700 (25)	8.925 (26) 7.90 (35)	1.87 (25) [B]	31.42	0.124 0
1,3-Dichloro-propane			2.08 [g] 2.2 (25) [B]	36.40	0.123 3
2,2-Dicloro-propane	0.769 (15) 0.619 (30)	11.37 (20)	2.62 [g] 2.20 (25) [B]	23.60 (20)	22.53 (30)
1,1-Dichloro-2-propanone		14 (20)			
1,2-Dichlorotetra-fluoroethane	0.38 (25) 0.011 (25) [g]	2.26 (25)	0.53 [g]	12 (25)	
α,α-Dichloro-toluene		6.9 (20)	2.07 (20) [B] 2.05 (25) [B]	41.26	0.103 5
Diethanolamine	368 (30) 196 (40)	2.81 (25)	2.84 (25) [B]		
1,1-Diethoxyethane		3.80 (25)	1.08 [g]	23.46	0.103 0⁻
1,2-Diethoxyethane	0.65 (20)		1.99 (20) [B] 1.65 (25) [B]		
Diethoxymethane				23.87	0.129 1
Diethylamine	0.388 (10) 0.273 (38)	3.6 (22)	0.92 [g] 1.11 (25) [lq]	22.71	0.114 3
N, N-Diethy-laniline	1.15 (30) 0.750 (75)	5.5 (19)	1.40 (20) [lq] 1.80 (20) [B]	36.59	0.104 0
Diethyl carbonate	0.868 (15) 0.748 (25)	2.82 (20)	1.07 [g] 0.91 (25) [B]	28.62	0.110 0
Diethyl decanedioate		5.0 (30)	2.38 (20) [lq] 2.52 (20) [B]	34.68	0.095 9
Diethyl ether	0.247 (15) 0.245 (20)	4.335 (20) 3.97 (40)	1.15 [g] 1.22 (16) [lq]	18.92	0.090 8
Diethyl ethyl phosphonate	1.627 (15) 0.969 (45)	11.00 (15) 9.86 (45)	2.95 (32) [lq] 2.91 (20) [C]	30.63	0.097 5
Diethyl fumarate		6.5 (23)	2.40 (20) [B]		
Diethyl glutarate		6.7 (30)	2.46 (30) [lq]	34.34	0.101 0
Di(2-ethylhexyl)-2-ethylhexyl-phosphonate	6.00 (45) 3.61 (65)	4.09 (45) 3.94 (65)			
Di(2-ethylhexyl) *o*-phthalate	33.67 (35) 21.40 (45)	4.91 (35) 4.77 (45)			

TABLE 4.10 Viscosity, Dielectric Constant, Dipole Moment, and Surface Tension of Selected Organic Substances (*continued*)

Substance	Viscosity η, mN·s·m^{-2}	Dielectric constant, ε	Dipole moment, D	Surface tension, dyn·cm^{-1}	
				a	b
Diethyl maleate	3.57 (20) 3.14 (25)	8.58 (23)	2.56 (25) [B]	34.67	0.103 9
Diethyl malonate	2.15 (20) 1.94 (25)	8.03 (25)	2.49 (20) [lq] 2.54 (25) [B]	33.91	0.104 2
Diethyl nonanedioate		5.13 (30)			
Diethyl oxalate	2.311 (15) 1.618 (30)	8.1 (21)	2.49 (20) [D]	34.32	0.111 9
Diethyl o-phthalate	9.18 (35) 6.41 (45)	7.34 (35) 7.13 (45)	2.8 (25) [B]	38.47	0.096 3
Diethyl succinate		6.64 (30)	2.3 [g] 2.37 (30) [lq]	33.97	0.104 1
Diethyl sulfate		29 (20)	4.46 (25) [D]	35.47	0.097 6
Diethyl sulfide	0.446 (20) 0.422 (25)	5.72 (25) 5.24 (50)	1.52 [g] 1.58 (20) [B]	27.33	0.110 6
Diethyl sulfite		16 (20) 14 (50)			
Diethylzinc		2.5 (20)	0.62 (25) [B]		
1,1-Difluoroethane	0.243 (21)	2.30 [g]			
1,2-Dihydroxy-benzene		2.6 (−89)	2.60 (25) [B]	47.6	0.084 9
1,3-Dihydroxy-benzene		3.2 (18)	2.09 (44) [B]	54.8	0.071 7
1,4-Dihydroxy-benzene			1.4 (44) [B]		
1,2-Diiodobenzene		5.7 (20)	1.70 (20) [B]		
1,3-Diiodobenzene		4.3 (25)	1.22 (20) [B]		
1,4-Diiodobenzene		2.9 (120)	0.19 (20) [B]		
cis-1,2-Diiodo-ethylene		4.46 (83)	0.71 [B]		
trans-1,2-Diiodo-ethylene		2.19 (83)	0		
Diiodomethane	3.043 (15) 2.392 (30)	5.316 (25)	1.08 (25) [B]	70.21	0.161 3
Diisobutylamine		2.7 (22)	1.10 (25) [B]	24.00	0.091 2
Diisobutyl o-phthalate	30 (20)				
Diisopentylamine		2.5 (18)	1.48 (30) [B]	26.04	0.085 8
Diisopentyl ether	1.40 (11) 1.012 (20)	2.82 (20)	0.98 (20) [lq] 1.23 (25) [B]	24.76	0.087 1
Diisopropylamine	0.40 (25)		1.26 (25) [B]	21.83	0.107 7
Diisopropyl ether	0.379 (25)	3.88 (25)	1.13 [g] 1.26 (25) [B]	19.89	0.104 8
1,2-Dimethoxy-benzene	3.281 (25) 2.184 (40)	4.09 (25)	1.32 (25) [B]	34.4	0.064 2

TABLE 4.10 Viscosity, Dielectric Constant, Dipole Moment, and Surface Tension of Selected Organic Substances (*continued*)

Substance	Viscosity η, mN·s·m^{-2}	Dielectric constant, ε	Dipole moment, D	Surface tension, dyn·cm^{-1}	
				a	b
1,1-Dimetho-xyethane				23.90	0.115 9
1,2-Dimetho-xyethane	0.530 (10) 0.455 (25)	7.60 (10) 7.20 (25)	1.71 (25) [B]	48.0 (25)	
Dimethoxy-methane	0.340 (15) 0.325 (20)	2.65 (20)	0.74 [g]	23.59	0.119 9
N,N-Dimethy lacetamide	2.141 (20) 0.838 (30)	37.78 (25)	3.80 [g] 4.60 (20) [lq]	32.43 (30)	29.50 (50)
Dimethylamine	0.207 (15) 0.186 (25)	6.32 (0) 5.26 (25)	1.03 [g] 1.14 (25) [lq]	29.50	0.126 5
N,N-Dimethy-laniline	1.285 (25) 0.91 (50)	4.9 (20) 4.4 (70)	1.61 [g] 1.55 (25) [B]	38.14	0.104 9
2,4-Dimethy-laniline			1.40 (25) [B]	39.34	0.099 6
2,2-Dimethyl-butane	0.351 (25) 0.330 (30)	1.873 (25)	0	18.29	0.099 0
2,3-Dimethyl-butane	0.361 (25) 0.342 (30)	1.890 (25)	0	19.38	0.100 0
2,3-Dimethyl-1-butanol				26.22	0.099 2
N,N-Dimethyl-butyramide	1.271	2.00			
Dimethyl carbonate			0.90 [g] 0.96 (25) [B]	31.94	0.134 3
1,1-Dimethyl-cyclopentane			0	23.78	0.101 6
2,2-Dimethyl-1, 3-dioxolane-4-methanol	11 (20)				
Dimethyl ether	0.010 4 (60)	5.02 (25) 2.97 (110)	1.30 [g] 1.25 (25) [B]	14.97	0.147 8
N,N-Dimethyl-formamide	0.845 (20) 0.598 (50)	38.3 (20) 36.71 (25)	3.86 (25) [B]	36.76 (20)	34.40 (40)
2,4-Dimethyl-heptane		1.9 (20)	0	23.21	0.092 9
2,5-Dimethyl-heptane		1.9 (20)	0	23.21	0.092 9
2,6-Dimethyl-heptane		2 (20)	0	22.77	0.088 7
2,6-Dimethyl-4-heptanone	1.03 (20)		2.66 (25) [C]		
Dimethyl hexanedioate	14 (20)		2.28 (20) [B]	38.26	0.113 8
Dimethyl hydrogen phosphonate	1.08 (25)				

TABLE 4.10 Viscosity, Dielectric Constant, Dipole Moment, and Surface Tension of Selected Organic Substances (*continued*)

Substance	Viscosity η, mN·s·m^{-2}	Dielectric constant, ε	Dipole moment, D	Surface tension, dyn·cm^{-1}	
				a	b
Dimethyl maleate	3.54 (20) 3.21 (25)		2.48 (25) [C]	40.73	0.122 0
Dimethyl malonate		10 (20)	2.41 (20) [B]	39.72	0.120 8
2,2-Dimethyl-pentane		1.91 (20)	0	19.94	0.095 7
2,3-Dimethyl-pentane	0.406 (20)	1.939 (20)	0	21.96	0.099 5
2,4-Dimethyl-pentane	0.361 (20)	1.914 (20)	0	20.09	0.097 2
3,3-Dimethyl-pentane		1.94 (20)	0	21.59	0.099 6
2,4-Dimethyl-phenol			1.48 (20) [B] 1.98 (60) [B]	34.57	0.086 9
2,5-Dimethyl-phenol	1.55 (80)		1.43 (20) [B] 1.52 (60) [B]	36.72	0.085 0
3,4-Dimethyl-phenol	3.00 (80)	4.8 (17)	1.77 (20) [B]	35.75	0.091 0
3,5-Dimethyl-phenol	2.42 (80)		1.76 (20) [B]	34.09	0.080 7
Dimethyl o-phthalate	17.2 (25) 6.41 (45)	8.25 (25) 8.11 (45)	2.8 (25) [B]		
2,2-Dimethyl-propane	0.328 (0) 0.303 (5)	1.80 (20) 1.678 (98)	0	12.05 (20)	10.98 (30)
N,N-Dimethyl-propionamide	0.935	33.1			
2,5-Dimethyl-pyrazine		2.43 (20)	0		
2,3-Dimethyl-quinoxaline		2.3 (25)	0		
Dimethyl succinate		5.1 (20)	2.09 (20) [B]	39.00	0.119 1
Dimethyl sulfate		48.3 (20) 46.4 (20)	4.31 (25) [D]	41.26	0.116 3
Dimethyl sulfide	0.289 (20) 0.265 (36)	6.2 (20)	1.45 (25) [B]	26.07	0.080 5
Dimethyl sulfite	0.715 (30) 0.436 (80)	22.5 (23)	2.93 (20) [B]	36.48	0.125 3
Dimethyl sulfoxide	2.47 (20) 1.192 (55)	48.9 (20) 41.9 (55)	3.9 (25) [B]	43.54 (20)	42.41 (30)
2,4-Dimethyl-tetrahydrothiophene-1, 1-dioxide	9.04	29.5			

TABLE 4.10 Viscosity, Dielectric Constant, Dipole Moment, and Surface Tension of Selected Organic Substances (*continued*)

Substance	Viscosity η, mN·s·m^{-2}	Dielectric constant, ε	Dipole moment, D	Surface tension, dyn·cm^{-1}	
				a	*b*
N,N-Dimethyl-*o*-toluidine		3.4 (20)	0.88 (25) [B]		
N,N-Dimethyl-*p*-toluidine		3.9 (20)	1.29 (25) [B]		
Dinonyl hexanedioate	37 (20)		2.53 (25) [B]		
Dinonyl *o*-phthalate		4.65 (35) 4.52 (45)			
Dioctyl decanedioate		4.0 (27)			
Dioctyl *o*-phthalate		5.1 (25)	3.06 (25) [C]		
1,4-Dioxane	1.439 (15) 1.087 (30)	2.24 (20) 2.21 (25)	0	36.23	0.139 1
Dipentyl ether	1.188 (15) 0.922 (30)	2.77 (25)	0.98 (20) [lq] 1.24 (25) [B]	26.66	0.092 5
Dipentyl *o*-phthalate	17.03 (35) 11.51 (45)	5.79 (35) 5.62 (45)	2.71 (20) [lq]	32.56	0.073 9
Dipentyl sulfide		3.83 (25)	1.59 (25) [B]	29.55	0.087 6
Diphenylamine	4.66 (55) 1.04 (130)	3.3 (52)	1.31 (20) [C] 1.01 (25) [B]	45.36	0.101 7
1,2-Diphenylethane		2.4 (110)	0 (110) [lq] 0.45 (25) [B]		
Diphenyl ether	2.61 (40) 2.09 (50)	3.65 (30)	1.16	28.70	0.078 0
Diphenylmethane		2.7 (18) 2.57 (26)	0.26 (30) [lq] 0.3 (25) [B]		
1,1-Dipropoxyethane				25.03	0.097 2
Dipropoxymethane				25.17	0.095 3
Dipropylamine	0.534 (20) 0.427 (37)	3.07 (20)	1.01 (20) [lq] 1.03 (20) [B]	24.86	0.102 2
Dipropyl carbonate				28.94	0.101 5
Dipropylene glycol butyl ether	4.23 (25)			28.2 (25)	
Dipropylene glycol ethyl ether	3.11 (25)			27.7 (25)	
Dipropylene glycol isopropyl ether	386 (25)			25.9 (25)	

TABLE 4.10 Viscosity, Dielectric Constant, Dipole Moment, and Surface Tension of Selected Organic Substances (*continued*)

Substance	Viscosity η, mN·s·m^{-2}	Dielectric constant, ε	Dipole moment, D	Surface tension, dyn·cm^{-1}	
				a	b
Dipropylene glycol methyl ether	3.1 (25)			28.8 (25)	
Dipropyl ether	0.448 (15) 0.376 (30)	3.39 (26)	1.21 [g] 1.17 (30) [H]	22.60	0.104 7
Divinyl ether		3.9 (20)	1.07 (20) [lq]		
Dodecamethyl-cyclohexasil-oxane		2.6 (20)			
Dodecamethyl-pentasiloxane		2.5 (20)		17.08 (25)	
Dodecane	1.508 (20) 1.378 (25)	2.05 (−10) 2.01 (20)	0	27.12	0.088 4
1-Dodecanol		5.15 (20) 6.5 (25)	1.52 (20) [B]	31.25	0.0748
6-Dodecyne		2.17 (25)			
1,2-Epoxybutane	0.41 (20) 0.40 (25)		2.01 (20) [B]	23.9 (20)	
Erythritol		28 (128)			
Ethane [g]	0.009 0 (20) 0.011 4 (100)	1.001 5 (0)	0	1.24	0.166 0 [lq]
1,2-Ethanedia-mine	1.54 (20) 1.226 (30)	16.8 (18) 14.2 (20)	1.96 [g] 1.92 (25) [B]	44.77	0.139 8
1,2-Ethanediol	26.09 (15) 13.55 (30)	38.66 (20) 37.7 (25)	2.28 [g] 2.3 (25) [D]	50.21	0.089 0
1,2-Ethanediol diacetate	3.13 (20)	13 (30)	2.34 (30) [B]		
Ethanesulfonic acid				45.74	0.082 4
Ethanesulfonyl chloride			3.89 (25) [B]	43.43	0.117 7
Ethanethiol	0.003 16 [g]	6.9 (15)	1.57 [g] 1.40 (20) [B]	25.06	0.079 3
Ethanol	1.209 (19) 0.991 (30)	25.00 (20) 20.21 (55)	1.69 [g] 1.71 (25) [B]	24.05	0.083 2
Ethoxybenzene	1.364 (15) 1.040 (30)	4.22 (20)	1.41 [g] 1.36 (25) [CS$_2$]	35.17	0.110 4
2-Ethoxyethanol	2.04 (20) 1.85 (25)	29.6 (24)	2.24 (30) [B]	30.59	0.089 7
2-(2-Ethoxyeth-oxy)ethanol	3.71 (25)			31.8 (25)	27.2 (75)
2-Ethoxyethyl acetate	1.025 (25)	7.567 (30)	2.25 (30) [B]	31.8 (25)	

TABLE 4.10 Viscosity, Dielectric Constant, Dipole Moment, and Surface Tension of Selected Organic Substances (*continued*)

Substance	Viscosity η, mN·s·m^{-2}	Dielectric constant, ε	Dipole moment, D	Surface tension, dyn·cm^{-1}	
				a	b
1-Ethoxy-2-methylbutane		3.96 (20)			
1-Ethoxynaph-thalene		3.3 (19)			
1-Ethoxypentane		3.6 (23)			
1-Ethoxy-2-propanol	1.68 (25)			25.9 (25)	
α-Ethoxytoluene		3.9 (20)			
Ethyl acetate	0.473 (15)	6.11 (20)	1.78 [g]	26.29	0.116 1
	0.426 (25)	5.30 (77)	1.84 (25) [lq]		
Ethyl acetoacetate	1.419 (20)	15.7 (22)	3.22 (18) [B keto form]	34.42	0.101 5
	1.508 (25)		2.04 (−80) [CS₂, enol form]		
Ethylamine		6.94 (10)	1.40 (25) [B]	22.63	0.137 2
2-(Ethylamino)ethanol	12.40 (25)				
N-Ethylaniline	2.04 (25)	5.76 (20)		39.00	0.107 0
	1.08 (55)				
Ethylbenzene	0.669 (20)	2.41 (20)	0.37 (25) [lq]	31.48	0.109 4
	0.531 (40)				
Ethyl benzoate	2.407 (15)	6.02 (20)	1.95 [g]	37.16	0.105 9
	1.751 (30)		1.93 (25) [B]		
Ethyl α-bromobutyrate		8 (20)	2.40 (25) [B]		
2-Ethyl-1-butanol	8.021 (15)	6.19 (90)		25.06 (15)	24.32 (25)
	5.892 (25)				
Ethyl butyrate	0.771 (15)	5.10 (18)	1.74 (22) [B]	26.55	0.104 5
	0.613 (25)				
2-Ethylbutyric acid	3.3 (20)			26.3 (20)	
Ethyl carbamate	0.916 (105)	14.2 (50)	2.59 (30) [D]		
	0.715 (120)				
Ethyl chloroacetate		11.4 (21)	2.65 (25) [B]	34.18	0.117 7
Ethyl chloroformate		11 (20)	2.56 (35) [B]	28.90	0.108 4
Ethyl cinnamate	8.7 (20)	6.1 (18)	1.86 (20) [B]	39.99	0.104 5
Ethyl crotonate		5.4 (20)	1.95 (24) [B]	29.31	0.106 6
Ethyl cyanoacetate	3.256 (15)	26.9 (20)	4.04 (30) [B]	38.80	0.109 2
	2.148 (30)				
Ethylcyclohexane	0.843 (20)	2.054 (20)	0 [g]	27.78	0.105 4
	0.787 (25)				
Ethyl dichloroacetate		12 (2)	2.63 (25) [B]	34.89	0.115 8
		10 (22)			

TABLE 4.10   Viscosity, Dielectric Constant, Dipole Moment, and Surface Tension of Selected Organic Substances (*continued*)

Substance	Viscosity η, mN·s·m^{-2}	Dielectric constant, ε	Dipole moment, D	Surface tension, dyn·cm^{-1}	
				a	b
N-Ethyldiethanolamine	53 (25)				
Ethyl dodecanoate		3.4 (20)	1.3 (20) [lq]	30.05	0.086 3
		2.7 (143)			
Ethylene		1.001 44 (0)	0 [g]	−2.7	0.185 4
Ethylene carbonate	1.85 (40)	89.6 (40)	4.87 (25) [B]		
		69.4 (91)			
Ethylenediamine	1.540 (18)	16.0 (18)	1.98 [g]	44.77	0.139 8
		14.2 (20)	1.92 (25) [B]		
Ethylene dinitrate		28.3 (20)	3.58 (25) [B]	49.1 (0)	46.7 (45)
2,2'-(Ethylenedioxy)diethanol	38 (20)	23.69 (20)	5.58 [lq]	47.33	0.088 0
Ethylene glycol	26.09 (15)	41.2 (20)	2.27 [g]	50.21	0.089 0
	13.35 (30)	37.7 (25)	2.20 (15) [lq]		
Ethylene oxide	0.3 (0)	14 (−1)	1.88 [g]	27.66	0.166 4
			1.92 (20) [lq]		
Ethyleneimine	0.418 (25)	18.3 (25)	1.89 [g]	7.9 (20)	
			1.77 (25) [B]		
Ethyl formate	0.419 (15)	7.16 (25)	1.94 [g]	26.47	0.131 5
	0.358 (30)		1.96 (25) [lq]		
Ethyl fumarate		6.5 (23)		33.90	0.105 6
Ethyl hexadecanoate		3.2 (20)	1.2 [lq]	32.86	0.085 9
		2.71 (104)			
2-Ethyl-1, 3-hexanediol	323 (20)				
Ethyl hexanoate			1.80 (20) [B]	27.73	0.096 0
2-Ethylhexanoic acid	7.7 (20)				
2-Ethyl-1-hexanol	9.8 (20)	4.41 (90)	1.74 (25) [B]	30.0 (22)	
2-Ethylhexyl acetate	1.5 (20)				
Ethyl isobutyrate				25.33	0.104 6
Ethyl isopentyl ether		3.96 (20)			
Ethyl isothiocyanate		19.5 (21)	3.67 (20) [B]	38.69	0.132 6
Ethyl lactate	2.44 (25)	13.1 (25)	2.4 (20) [B]	30.72	0.098 3
Ethyl maleate		8.6 (23)			
Ethyl 3-methylbutyrate		4.71 (18)		25.79	0.100 6
Ethyl methyl ether			1.22 [g]	18.56	0.131 7
Ethyl methyl sulfide	0.373 (20)			27.63	0.128 6
	0.354 (25)				
Ethyl nitrate		19.4 (20)	2.93 (20) [B]	30.81	0.134 5

TABLE 4.10   Viscosity, Dielectric Constant, Dipole Moment, and Surface Tension of Selected Organic Substances (*continued*)

Substance	Viscosity η, mN·s·m^{-2}	Dielectric constant, ε	Dipole moment, D	Surface tension, dyn·cm^{-1}	
				a	*b*
Ethyl 9-octadecenoate		3.2 (25)	1.83 (20) [lq]		
Ethyl 4-oxopentanoate		12 (21)			
3-Ethylpentane		1.94 (20)	0	22.52	0.103 2
Ethyl pentanoate	0.847 (20)	4.7 (18)	1.76 (28) [B]	27.15	0.099 9
Ethyl pentyl ether		3.6 (23)	1.2 (20) [B]	24.19	0.099 2
Ethyl phenylacetate	5.3 (21)	1.82 (30)			
Ethyl phenyl sulfide			4.08 (25) [B]	39.30	0.113 1
Ethyl propionate	0.564 (15) 0.473 (30)	5.65 (19)	1.75 (22) [B]	26.72	0.116 8
Ethyl propyl ether	0.323 (20) 0.225 (60)		1.16 (25) [B]	21.92	0.105 4
Ethyl salicylate	1.772 (45)	7.99 (30)	2.85 (25) [B]	31.00	0.109 1
Ethyl stearate		2.98 (40) 2.69 (100)	1.65 (40) [lq]		
Ethyl thiocyanate		29.3 (21)	3.33 (20) [B]	37.28	0.122 6
o-Ethyltoluene				32.33	0.106 0
p-Ethyltoluene		2.24 (25)	0	30.98	0.107 5
Ethyl trichloroacetate		7.8 (20)	2.56 (25) [B]	32.97	0.107 3
Ethyl vinyl ether	0.2		1.26 (20) [B]	19.00 (20)	
Ethynyl acetate				32.81 (20)	30.20 (40)
Fluorobenzene	0.620 (15) 0517 (30)	5.42 (25) 4.7 (60)	1.61 [g]	29.67	0.120 4
1-Fluorohexane				23.41	0.100 1
2-Fluoro-2-methylbutane		5.89 (20)	1.92 (25) [B]		
1-Fluoropentane		4.24 (20)	1.85 (25) [B]	22.81	0.131 5
o-Fluorotoluene	0.680 (20) 0.601 (30)	4.22 (30) 3.9 (60)	1.35 [g] 1.26 (30) [lq]		
m-Fluorotoluene	0.608 (20) 0.534 (30)	5.42 (30) 4.9 (60)	1.86 [g] 1.66 (30) [lq]	32.31	0.125 7
p-Fluorotoluene	0.622 (20) 0.522 (30)	5.86 (30) 5.3 (60)	2.00 [g] 1.76 (30) [lq]	30.44	0.110 9
Formamide	4.320 (15) 2.296 (30)	111.0 (20) 103.5 (40)	3.73 [g]	59.13	0.084 2
Formanilide	1.65 (120)		3.37 (25) [C]	44.30	0.087 5
Formic acid	1.966 (15) 1.219 (40)	58.5 (15) 57.0 (21)	1.35 [g] 1.20 (25) [B]	39.87	0.109 8
2-Furaldehyde	2.475 (0) 1.494 (25)	41.9 (20) 34.9 (50)	2.13 (25) [lq] 3.63 (25) [B]	46.41	0.132 7
Furan	0.380 (20) 0.361 (25)	2.95 (25)	0.66 [g] 0.67 (20) [B]	24.10 (20)	23.38 (25)

TABLE 4.10 Viscosity, Dielectric Constant, Dipole Moment, and Surface Tension of Selected Organic Substances (*continued*)

Substance	Viscosity η, mN·s·m^{-2}	Dielectric constant, ε	Dipole moment, D	Surface tension, dyn·cm^{-1}	
				a	b
Furfuryl alcohol	4.62 (25)		1.92 (25) [lq]	ca 38 (20)	
Glycerol	945 (25)	42.5 (25)	2.68 (25) [D]	63.14 (17)	62.5 (25)
	134 (50)				
Glycerol triacetate		7.2 (20)	2.73 (25) [B]	37.88	0.081
Glycerol trinitrate	36.0 (20)	19 (20)	3.38 (25) [B]	55.74	0.250 4
	13.6 (40)				
Glycerol trioleate		3.2 (26)	3.11 (23) [B]	36.03	0.069 9
Glycerol tripalmitate		2.9 (65)	2.80 (23) [B]	32.26	0.067 2
Glycerol tristearate		2.8 (70)	2.86 (23) [B]	32.73	0.068 5
Heptanaldehyde	0.977 (15)	9.1 (20)	2.26 (40) [lq]	28.64	0.092 0
			2.58 (22) [B]		
Heptane	0.416 (20)	1.924 (20)	0	22.10	0.098 0
	0.341 (40)	1.85 (70)			
Heptanoic acid	3.40 (30)	2.6 (71)		29.88	0.084 8
1-Heptanol	7.014 (20)	12.1 (22)	1.73 (20) [B]		
	8.53 (15)				
DL-2-Heptanol	5.06 (25)	9.21 (22)	1.73 (20) [B]		
DL-3-Heptanol		6.9 (22)	1.73 (20) [B]		
4-Heptanol		6.2 (22)	1.72 (20) [B]		
2-Heptanone	0.854 (15)	11.95 (20)	2.61 (22) [B]	28.76	0.105 6
	0.686 (30)	8.27 (100)			
3-Heptanone		12.9 (22)	2.81 (22) [B]	28.24	0.101 5
4-Heptanone	0.736 (20)	12.60 (20)	2.74 (20) [B]	28.11	0.106 0
		9.46 (80)			
1-Heptene	0.35 (20)	2.07 (20)	0.34 (20) [lq]	22.28	0.099 1
	0.34 (25)				
Hexadeca-methylcyclo-octasiloxane		2.7 (20)			
Hexadecane	3.591 (22)		0	29.18	0.085 4
1-Hexadecanol		3.8 (50)	1.67 (25) [B]		
1,5-Hexadiene	0.275 (20)				
	0.244 (36)				
2,4-Hexadiene		2.2 (25)	0.31 (25) [B]		
Hexafluoro-benzene			0	22.6 (20)	
Hexamethyl-disiloxane		2.2 (20)	0.37 (25) [lq]	17.01	0.076 3
Hexamethyl-phosphoramide	3.47 (20)	30 (20)	4.31 (25) [lq]	33.8 (20)	
Hexane	0.313 (20)	1.904 (15)	0	20.44	0.102 2
	0.271 (40)	1.890 (20)			
Hexanedinitrile	5.99	32.45	3.8 (25) [B]	47.88	0.097 3

TABLE 4.10   Viscosity, Dielectric Constant, Dipole Moment, and Surface Tension of Selected Organic Substances (*continued*)

Substance	Viscosity η, mN·s·m^{-2}	Dielectric constant, ε	Dipole moment, D	Surface tension, dyn·cm^{-1}	
				a	*b*
2,4-Hexanedione				32.22	0.100 2
Hexanenitrile	1.041 (15)	17.26 (25)		29.64	0.090 7
	0.830 (30)				
Hexanoic acid	3.525 (15)	2.63 (71)	1.13 (25) [lq]	28.05 (20)	27.55 (25)
	2.511 (3)				
1-Hexanol	6.203 (15)	13.3 (25)	1.55 (20) [B]	27.81	0.080 1
	3.872 (30)	8.5 (75)			
2-Hexanone	0.584 (25)	14.6 (15)	2.68 (22) [B]	28.18	0.109 2
1-Hexene	0.26 (20)	2.051 (20)	0.34 (20) [lq]	20.47	0.102 7
	0.25 (25)				
Hexyl acetate				28.44	0.097 0
4-Hydroxy-4-methyl-2-pentanone	2.9 (25)	18.2 (25)	3.24 (20) [B]	31.0 (20)	
Iodobenzene	1.774 (17)	4.62 (20)	1.71 [g]	41.52	0.112 3
	0.488 (149)		1.3 (20) [B]		
1-Iodobutane		6.22 (20)	2.10 [g]	30.82	0.013 1
		4.52 (130)	1.90 (20) [B]		
2-Iodobutane			2.06 (20) [B]	30.32	0.105 6
1-Iodododecane		3.9 (20)	1.87 (20) [C]		
Iodoethane	0.617 (15)	10.2 (−50)	1.91 [g]	31.67	0.128 6
	0.540 (30)	7.82 (20)	1.69 (20) [lq]		
1-Iodoheptane		4.9 (22)	1.86 (22) [B]	32.18	0.088 7
3-Iodoheptane		6.4 (22)	1.95 (22) [B]		
1-Iodohexadecane		3.5 (20)		34.49	0.088 0
1-Iodohexane		5.37 (20)	1.94 (20) [C]	31.63	0.084 5
Iodomethane	0.500 (20)	7.00 (20)	1.64 [g]	33.42	0.123 4
	0.424 (40)		1.42 (20) [B]		
1-Iodo-3-methylbutane		5.6 (19)	1.85 (20) [B]	30.37	0.091 5
2-Iodo-2-methylbutane		8.19 (20)	2.20 (20) [B]		
1-Iodo-2-methylpropane	0.875 (20)	6.5 (20)	1.89 (20) [B]	30.26	0.017 2
	0.697 (40)				
1-Iodooctane		4.6 (25)	1.80 (25) [lq]	32.51	0.091 5
			1.90 (20) [C]		
2-Iodooctane		5.8 (20)	2.07 (20) [C]		
1-Iodopentane		5.81 (20)	1.90 (20) [B]	31.41	0.101 4
1-Iodopropane	0.837 (15)	7.00 (20)	2.03 [g]	31.64	0.113 6
	0.670 (30)		1.86 (20) [B]		
2-Iodopropane	0.732 (15)	7.87 (20)	2.01 (20) [B]	29.35	0.110 7
	0.620 (30)				
p-Iodotoluene		4.4 (35)	1.72 (22) [B]	39.23	0.096 5
α-Ionone		11 (18)		34.10	0.094 9
β-Ionone		12 (20)		35.36	0.095 0

TABLE 4.10 Viscosity, Dielectric Constant, Dipole Moment, and Surface Tension of Selected Organic Substances (*continued*)

Substance	Viscosity η, mN·s·m^{-2}	Dielectric constant, ε	Dipole moment, D	Surface tension, dyn·cm^{-1}	
				a	b
Iron pentacarbonyl		2.6 (20)			
Isobutyl acetate		5.29 (20)	1.87 (22) [B]	25.59	0.101 3
Isobutylamine	0.553 (25)	4.43 (21)	1.27 (25) [B]	24.48	0.109 2
Isobutylbenzene		2.319 (20)	0.31 (20) [lq]	29.39	0.096 1
		2.298 (30)			
Isobutyl butyrate		4.1 (20)		24.47	0.084 3
Isobutyl formate	0.680 (20)	6.41 (19)	1.89 (20) [B]	26.14	0.112 2
Isobutyl isobutyrate				30.92	0.127 0
Isobutyl nitrate		2.7 (20)			
Isobutyl pentanoate		3.8 (19)			
Isobutyl propionate				28.97	0.116 6
Isobutyric acid	1.44 (15)	2.7 (20)	1.09 (25) [lq]	26.88	0.092 0
Isobutyric anhydride	14 (20)				
Isobutyronitrile	0.551 (15)	20.4 (24)	3.61 (25) [B]	24.93 (20)	23.84 (30)
	0.456 (30)				
Isopentyl acetate	0.872 (20)	4.81 (20)	1.84 (22) [B]	26.75	0.098 9
	0.790 (25)	4.63 (30)	1.76 (30) [lq]		
Isopentyl butyrate		4.0 (20)		27.32	0.091 8
Isopentyl pentanoate		3.6 (19)	1.8 (28) [B]		
Isopentyl propionate	4.2 (20)				
Isopropyl acetate	0.559 (20)		1.86 (22) [B]	24.44	0.107 2
Isopropylamine	0.36 (25)	5.45 (20)	1.45 (25) [B]	19.91	0.097 2
Isopropylbenzene	0.791 (20)	2.39 (20)	0.65 [g]	30.32	0.105 4
	0.739 (25)		0.39 (20) [lq]		
Isopropyl formate	0.512 (20)			24.56	0.114 7
1-Isopropyl-4-methylbenzene	3.402 (20)	2.24 (20)	0	29.44 (20)	
	1.600 (30)				
Isoquinoline	3.253 (30)	10.7 (20)	2.75 [g]		
			2.55 (25) [B]		
Lactamide					
Lactic acid	40.33 (25)	22 (17)			
Lactonitrile	2.01 (30)	38 (20)		38.31	0.096 0
D-Limonene		2.4 (20)	1.57 (25) [B]	29.50	0.092 9
DL-Limonene		2.3 (20)	0.63 (25) [B]	29.11	0.091 3
DL-Mandelonitrile		17.8 (23)		45.90	0.098 8
Menthol	6.89 (35)		1.55 (20) [B]		
2-Mercaptoethanol	3.4 (20)				
Methacrylic acid	1.32 (20)		1.65	26.5 (25)	
Methacrylonitrile	0.392 (20)		3.69 [g]	24.4 (20)	
Methane [g]	0.010 9 (20)	1.000 94 (0)	0	*	
	0.013 3 (100)				
Methanesulfonic acid				52.28	0.089 3

* $38.618 - 0.1873T - 0.000356T^2$

TABLE 4.10 Viscosity, Dielectric Constant, Dipole Moment, and Surface Tension of Selected Organic Substances (*continued*)

Substance	Viscosity η, mN·s·m^{-2}	Dielectric constant, ε	Dipole moment, D	Surface tension, dyn·cm^{-1} a	b
Methanethiol			1.26 [g]	28.09	0.169 6
Methanol	0.676 (10)	41.8 (−20)	1.69 [g]	24.00	0.077 3
	0.544 (25)	33.62 (20)	1.68 (22) [B]		
o-Methoxy-benzaldehyde			4.34 (20) [B]	45.34	0.110 5
p-Methoxy-benzaldehyde		22.3 (22)	3.26 (35) [B]	44.69	0.104 7
		10.4 (248)			
Methoxybenzene	1.152 (15)	4.33 (25)	1.36 [g]	38.11	0.120 4
	0.789 (30)	3.9 (70)	1.24 (20) [B]		
2-Methoxyethanol	1.72 (20)	16.93 (25)	2.04 (25) [B]	33.30	0.098 4
	1.60 (25)	16.0 (30)			
2-(2-Methoxy-ethoxy)ethanol	3.48 (25)			34.8 (25)	29.9 (75)
	1.61 (60)				
2-Methoxyethyl acetate		8.25 (20)	2.13 (30) [B]		
1-Methoxy-2-nitrobenzene			4.83 [g]	48.62	0.118 5
o-Methoxyphenol		12 (25)		41.2	0.094 3
2-Methoxy-4-(2-propenyl) phenol	6.931 (25)		2.46 (25) [B]		
o-Methoxytoluene		3.5 (20)			
m-Methoxytoluene		3.5 (20)			
p-Methoxytoluene		4.0 (20)		36.20	0.107 1
N-Methyl-acetamide	3.88 (30)	178.9 (30)	4.39 (20) [D]	33.67 (30)	30.62 (50)
	2.54 (45)	138.6 (60)			
Methyl acetate	0.388 (20)	7.03 (20)	1.70 [g]	27.95	0.128 9
	0.320 (40)	6.68 (25)	1.75 (25) [B]		
Methyl acetoacetate	1.704 (20)			34.98	0.094 4
Methyl acrylate	1.398 (20)		1.77 (25) [B]		
Methylamine	0.285 (15)	11.4 (−10)	1.29 [g]	22.87	0.148 8
	0.236 (0)	10.0 (18)			
N-Methylaniline	2.02 (25)		1.67 (25) [B]	39.32	0.097 0
	1.084 (55)				
Methyl benzoate	2.298 (15)	6.59 (20)	1.86 (25) [B]	40.10	0.117 1
	1.673 (30)				
2-Methyl-1, 2-butadiene	0.266 (0.3)	2.1 (25)	0.15 [g]		
	0.223 (20)				
2-Methylbutane	0.237 (15)	1.871 (0)	0.13 [g]	17.20	0.110 3
	0.215 (25)	1.845 (20)			
2-Methyl-1-butanol	5.50 (20)	14.7 (25)		21.5 (25)	
	1.44 (60)				
2-Methyl-2-butanol	5.48 (15)	5.82 (25)	1.72 (20) [B]	24.18	0.074 8
	2.81 (30)				

TABLE 4.10   Viscosity, Dielectric Constant, Dipole Moment, and Surface Tension of Selected Organic Substances (*continued*)

Substance	Viscosity η, mN·s·m^{-2}	Dielectric constant, ε	Dipole moment, D	Surface tension, dyn·cm^{-1}	
				a	b
3-Methyl-1-butanol	4.81 (15) 2.96 (30)	14.7 (25) 5.82 (130)	1.82 (25) [B]	25.76	0.082 0
3-Methyl-2-butanol	3.51 (25)			23.0 (25)	
2-Methyl-1-butene		2.20 (20)	0.52 (20) [lq]	18.81	0.1148
2-Methyl-2-butene			0.11 (25) [lq] 0.34 (25) [B]	19.70	0.127 1
3-Methyl-1-butene		1.002 8 (100) [g]	0.25 [g]	16.42	0.103 1
2-Methylbutyl acetate	0.872 (20)	4.63 (30)	1.82 (22)	26.75	0.098 9
Methyl butyrate	0.580 (20) 0.459 (40)	5.6 (20)	1.72 (22) [B]	27.48	0.114 5
3-Methylbutyric acid	2.731 (15) 2.411 (20)	2.64 (20)	0.63 (25)	27.28	0.088 6
3-Methylbutyronitrile		18 (220)	3.62 (25) [C]	27.58	0.082 7
Methyl chloroacetate		12.9 (21)		37.90	0.130 4
Methyl cyanoacetate	3.82 (50) 2.69 (65)	19.23 (50) 17.57 (65)		41.32	0.107 4
Methylcyclohexane	0.734 (20) 0.685 (25)	2.02 (20) 2.07 (25)	0	26.11	0.113 0
cis-2-Methylcyclohexanol	18.08 (25) 13.60 (30)	13.3*	2.58 (30) [lq]* 1.95 (25) [B]*	32.45	0.077 0*
trans-2-Methylcyclohexanol	37.13 (25) 25.14 (30)				
cis-3-Methylcyclohexanol	19.7 (25) 17.23 (30)	16.47 (20)	1.91	29.08	0.062 9*
trans-3-Methylcyclohexanol	25.52 (16) 15.60 (30)	8.05	1.75	28.80 (30)	
cis-4-Methylcyclohexanol	0.247 (25)	13.3*	2.70 (30) [lq]* 1.9 (25) [B]*	29.07	0.069 0*
trans-4-Methylcyclohexanol	0.385 (25)				
2-Methylcyclohexanone		16 (−15) 14 (20)	2.98 (25) [B]	34.06	0.102 7
3-Methylcyclohexanone		18 (−80) 12 (20)	3.06 (25) [B]	33.06	0.092 5
4-Methylcyclohexanone		15 (−41) 12 (20)	3.07 (25) [B]	32.83	0.093 5
Methylcyclopentane	0.507 (20) 0.478 (25)	1.985 (20)	0	24.63	0.116 3
Methyl decanoate			1.65 (20) [H]	30.33	0.091 2
Methyl dichloroacetate				37.00	0.121 9

* Mixed isomers.

TABLE 4.10 Viscosity, Dielectric Constant, Dipole Moment, and Surface Tension of Selected Organic Substances (*continued*)

Substance	Viscosity η, mN·s·m^{-2}	Dielectric constant, ε	Dipole moment, D	Surface tension, dyn·cm^{-1}	
				a	b
Methyl dodecanoate			1.70 (20) [H]	31.37	0.089 3
N-Methyl-formamide	1.99 (15) 1.65 (25)	200.1 (15) 182.4 (25)	3.86 (25) [B]	37.96 (30)	35.02 (50)
Methyl formate	0.360 (15) 0.319 (29)	8.5 (20)	1.77 [g]	28.29	0.157 2
Methyl heptanoate				28.29	0.157 2
2-Methyl-2-heptanol		3.38 (−7) 2.46 (25)			
2-Methyl-3-heptanol		3.37 (20) 3.75 (60)	1.63 (20) [B]		
2-Methyl-4-heptanol		3.30 (20) 3.65 (60)			
3-Methyl-3-heptanol		3.74 (20) 2.89 (60)			
3-Methyl-4-heptanol		9.1 (−20) 7.4 (20)			
4-Methyl-3-heptanol		5.25 (20) 4.62 (55)			
4-Methyl-4-heptanol		2.87 (20) 3.27 (60)			
Methyl hexadecanoate				31.50	0.077 5
2-Methylhexane	0.378 (20)	1.92 (20)	0	21.22	0.096 64
3-Methylhexane	0.372 (20)	1.93 (20)	0	21.73	0.097 0
Methyl hexanoate			1.70 (20) [H]	28.47	0.104 5
Methyl isobutyrate	0.523 (20) 0.419 (40)		1.98 (20) [B]	25.99	0.113 1
Methyl methacrylate	0.632 (20)	2.9 (20)	1.68 (25) [B]	28–29 (30)	
Methyl o-methoxybenzoate		7.7 (21)			
Methyl p-methoxybenzoate		4.3 (33)			
1-Methylnaphthalene		2.7 (20)	0.23 (20) [B]	39.96	0.0934
Methyl o-nitrobenzoate		28 (25)	3.67 (30) [B]		
Methyl octadecanoate				32.20	0.77 5
2-Methyloctane		1.97 (20)	0	23.76	0.094 0
4-Methyloctane		1.97 (20)	0	24.22	0.094 0
Methyl octanoate				29.93	0.100 2
Methyl oleate	4.88 (20)	3.211 (20)		31.3 (25)	25.4 (100)
2-Methylpentane	0.310 (20) 0.295 (25)	1.88 (20)	0	19.37	0.099 7

TABLE 4.10 Viscosity, Dielectric Constant, Dipole Moment, and Surface Tension of Selected Organic Substances (*continued*)

Substance	Viscosity η, mN·s·m^{-2}	Dielectric constant, ε	Dipole moment, D	Surface tension, dyn·cm^{-1}	
				a	b
3-Methylpentane	0.307 (25) 0.292 (30)	1.895 (20)	0	20.26	0.106 0
2-Methyl-2, 4-pentanediol	34.4 (20)		2.9 (0)	33.1 (20)	
4-Methylpen- tanenitrile	0.980 (20) 9.843 (30)	15.5 (22)	3.53 (25) [B]	28.89	0.091 7
Methyl pentanoate	0.713 (20)	4.3 (19)	1.62 (22) [B]	27.85	0.104 4
2-Methyl-1- pentanol				26.98	0.081 9
3-Methyl-1- pentanol				26.92	0.078 9
4-Methyl-1- pentanol				25.93	0.074 3
2-Methyl-2- pentanol				25.07	0.086 1
3-Methyl-2- pentanol				27.14	0.091 9
4-Methyl-2- pentanol	4.074 (25)			24.67	0.082 1
2-Methyl-3- pentanol				26.43	0.091 4
3-Methyl-3- pentanol				25.48	0.088 8
4-Methyl-2- pentanone	0.585 (20) 0.522 (30)	13.11 (20) 11.78 (40)		23.64 (20)	19.62 (60)
4-Methyl-3- penten-2-one	0.879 (25)	15.6 (0) 15.1 (20)	3.20 (25) [B]		
1-Methyl-1- phenylhy- drazine		7.3 (19)	1.84 (15) [B]		
Methyl phenyl sulfide			1.38 (20) [B]	42.81	0.123 8
2-Methylpropane	0.007 44 (20) [g]		0	12.83	0.123 6
2-Methylpro- panenitrile	0.551 (15) 0.456 (30)	20.2	4.07 [g] 3.60 (20) [B]		
2-Methyl-1- propanol	4.70 (15) 2.876 (30)	26 (−34) 17.93 (25)	2.96 (30) [lq] 1.78 (20) [B]	24.53	0.079 5
2-Methyl-2- propanol	3.316 (20) 2.039 (40)	10.9 (30) 8.49 (50)	1.67 (22) [B]	20.02 (15)	19.10 (30)
2-Methylpropene			0.50 [g]	14.84	0.131 9
N-Methylpro- pionamide	6.06 (20) 3.56 (40)	185 (20) 151 (40)	3.59 [g]	31.20 (20)	29.12 (50)
Methyl propionate	0.477 (15)	6.21	1.70 (22) [B]	27.58	0.125 8
2-Methylpro- pionic acid	1.213 (25) 1.126 (30)	2.73 (40)	1.08 (25) [lq]	25.55 (20)	25.13 (25)

TABLE 4.10   Viscosity, Dielectric Constant, Dipole Moment, and Surface Tension of Selected Organic Substances (*continued*)

Substance	Viscosity η, mN·s·m^{-2}	Dielectric constant, ε	Dipole moment, D	Surface tension, dyn·cm^{-1}	
				a	*b*
1-Methylpropyl acetate				25.72	0.105 4
2-Methylpropyl acetate	0.702 (20) 0.366 (78)	5.29 (20)	1.87 (22) [B]	25.59	0.101 3
2-Methylpropy-lamine	21.7 (25)	4.43 (21)	1.27 (27)	24.48	0.109 2
2-Methylpropyl formate	0.680 (20)	6.41 (19)	1.88 (22)	26.14	0.112 2
Methyl propyl ketoxime		3.3 (20)			
2-Methylpyridine	0.805 (20) 0.710 (30)	9.8 (20)	1.96 (25) [B]	36.11	0.124 3
3-Methylpyridine			2.41 (25) [B]	37.35	0.115 3
4-Methylpyridine			2.60 (25) [B]	37.71	0.114 1
N-Methyl-2-pyrrolidinone	1.666 (25)	32.0 (25)	4.09 (30) [B]		
Methyl salicylate		9.41 (30)	2.47 (25) [B]	42.15	0.117 4
Methyl tetrade-canoate			1.62 (25) [B]	31.00	0.080 0
2-Methyltetra-hydrofuran	0.601 (0) 0.536 (10)	6.92 (0) 6.63 (10)			
Methyl thiocyanate	64.3 (0)	4.3 (19)	3.34 (20) [B]	40.66	0.130 5
Morpholine	2.53 (15) 1.79 (30)	7.33 (25)	1.75 (25) [lq] 1.52 (25) [B]	37.63 (20)	36.24 (30)
Naphthalene	0.780 (100) 0.967 (80)	2.54 (85)	0	42.84	0.110 7
1-Naphthonitrile		16 (70)			
2-Naphthonitrile		17 (70)			
o-Nitroaniline		34.5 (90)	4.28 (20) [B]		
p-Nitroaniline		56.3 (160)	6.3 (25) [B]	60.62	0.092 3
o-Nitroanisole			4.83 [g]	48.62	0.118 5
Nitrobenzene	2.165 (15) 1.55 (35)	34.82 (25) 24.9 (90)	4.22 [g] 3.96 (25) [B]	46.34	0.115 7
m-Nitrobenzyl alcohol		22 (20)			
2-Nitrobiphenyl	12 (45)		3.82 (20) [B]		
Nitroethane	0.677 (20) 0.63 (35)	28.06 (30) 27.4 (35)	3.61 [g]	35.27	0.125 5
Nitromethane	0.692 (15) 0.596 (30)	35.87 (30) 35.1 (35)	3.46 [g]	40.72	0.167 8
1-Nitro-2-methoxy-benzene			4.83 [g]	48.62	0.118 5
o-Nitrophenol	2.343 (45)	17 (50)	3.14 (25) [B]	47.35	0.117 4
1-Nitropropane	0.798 (25) 0.70 (35)	23.24 (30) 22.7 (35)	3.60 [g]	32.62	0.100 9

TABLE 4.10 Viscosity, Dielectric Constant, Dipole Moment, and Surface Tension of Selected Organic Substances (*continued*)

Substance	Viscosity η, mN·s·m^{-2}	Dielectric constant, ε	Dipole moment, D	Surface tension, dyn·cm^{-1}	
				a	b
2-Nitropropane	0.750 (25)	25.52 (30)	3.76 [g]	32.18	0.115 8
N-Nitrosodime-thylamine		53 (20)	4.01 (20) [B]		
o-Nitrotoluene	2.37 (20)	27.4 (20)	3.72 (20) [B]	44.10	0.117 4
	1.63 (40)	22.0 (58)			
m-Nitrotoluene	2.33 (20)	24 (20)	4.20 (20) [B]	43.54	0.111 8
	1.60 (40)	22 (58)			
p-Nitrotoluene	1.20 (60)	22 (52)	4.47 (25) [B]	42.26	0.097 4
Nonane	0.713 (20)	1.972 (20)	0	24.72	0.093 5
	0.666 (25)	1.85 (110)			
1-Nonanol	14.3 (20)		1.72 (20) [B]	29.79	0.078 9
1-Nonene	0.620 (20)		0.59 (20) [B]	24.90	0.093 8
	0.586 (25)				
(Z,Z)-9,12-Octadeca-dienoic acid		2.70 (70)	1.40 (18) [Hx]		
		2.60 (120)			
Octamethylcy-clotetrasiloxane	2.20 (20)	2.4 (20)	0.42 (25) [lq]	20.19	0.081 1
			0.67 (25) [B]		
Octamethyltri-siloxane	0.82 (20)	2.3 (20)	0.64 (25) [lq]	67.56 (25)	
Octane	0.546 (20)	1.95 (20)	0	23.52	0.095 1
	0.433 (40)	1.83 (110)			
Octanenitrile	1.811 (15)	13.90 (25)		29.61	0.080 2
	1.356 (30)				
Octanoic acid	5.828 (20)	2.45 (20)	1.15 (25) [lq]	29.2 (20)	28.7 (25)
	4.690 (25)				
1-Octanol	10.64 (15)	11.3 (10)	1.72 (20) [B]	29.09	0.079 5
	6.125 (30)	10.34 (20)			
2-Octanol		8.20 (20)	1.65 (20) [B]	27.96	0.082 0
		6.52 (40)			
2-Octanone		10.39 (20)	2.72 (15) [B]		
		7.42 (100)			
1-Octene	0.470 (20)	2.084 (20)	0.34 (20) [lq]	23.68	0.095 8
	0.447 (25)				
Oleic acid	38.80 (20)	2.46 (20)	1.44 (25) [lq]	32.80 (20)	27.94 (90)
	27.64 (25)	2.45 (60)			
Oxalyl chloride		3.5 (21)	0.93 (20) [B]		
2-Oxohexame-thyleneimine	9 (78)		3.88 (25) [B]		
4-Oxopentanoic acid				41.69	0.076 3
Palmitic acid		2.3 (70)			
Paraldehyde		13.9 (25)	1.91 (25) [lq]	28.28	0.106 2
Parathion	15.30 (25)		4.98 (25) [B]	39.2 (25)	

TABLE 4.10 Viscosity, Dielectric Constant, Dipole Moment, and Surface Tension of Selected Organic Substances (*continued*)

Substance	Viscosity η, mN·s·m^{-2}	Dielectric constant, ε	Dipole moment, D	Surface tension, dyn·cm^{-1}	
				a	b
Pentachloroethane	2.741 (15) 2.070 (30)	3.73 (20)	0.92 [g] 0.98 (25) [lq]	37.09	0.117 8
Pentadecane	2.814 (22)		0	28.78	0.085 7
cis-1,3-Pentadiene		2.32 (25)	0.50 (25) [B]		
Pentanaldehyde		10.1 (17)	2.59 (20) [B]	27.96	0.101 0
Pentane	0.237 (15) 0.215 (25)	2.011 (−90) 1.84 (20)	0	18.25	0.112 1
1,5-Pentanediol	128 (20)		2.45 (20) [D]	43.2 (20)	
2,4-Pentanedione	0.6 (20)	25.7 (20) 17.39 (25)	3.03 [g] 2.5 (20) [B]	33.28	0.1144
Pentanenitrile	0.779 (15) 0.637 (30)	17.4 (21)	3.57 (25) [B]	27.44 (20)	26.33 (30)
1-Pentanethiol		4.55 (25) 4.23 (50)	1.54 (25) [lq]		
Pentanoic acid	2.359 (15) 1.774 (30)	2.66 (20)	1.61 (20) [D]	28.90	0.088 7
1-Pentanol	4.650 (15) 2.987 (20)	16.9 (20) 13.9 (25)	1.71 (20) [B]	27.54	0.087 4
2-Pentanol	5.130 (15) 2.780 (30)	13.82 (22)	1.66 (22) [B]	25.96	0.100 4
3-Pentanol	7.337 (15) 3.306 (30)	13.02 (22)	1.64 (22) [B]	24.60 (20)	23.76 (30)
2-Pentanone	0.473 (25)	15.45 (20) 11.73 (80)	2.72 (22) [B]	24.89	0.065 5
3-Pentanone	0.493 (15) 0.423 (30)	19.4 (−20) 17.00 (20)	2.72 (20) [B]	27.36	0.104 7
1-Pentene	0.24 (0)	2.10 (20)	0.34 (20) [lq]	18.20	0.109 9
cis-2-Pentene				19.73	0.117 2
trans-2-Pentene				18.90	0.099 7
Pentyl acetate	0.924 (20) 0.862 (25)	4.75 (20)	1.72 [g] 1.91 (25) [B]	27.66	0.099 4
Pentylamine	1.018 (20)	4.5 (22)	1.55 (30) [B]	24.4 (13)	
Pentyl formate		6.5 (20)		28.09	0.102 3
Pentyl nitrate		9 (18)			
Phenanthrene		2.8 (20)	0		
Phenol	6.024 (35) 3.421 (50)	9.78 (60)	1.53 (20) [B]	43.54	0.106 8
Phenoxyacetal-dehyde		4.8 (20)			
Phenoxyacetylene		4.8 (20)	1.42 (25) [lq]		
2-Phenyl-acetamide				46.26	0.078 8
Phenyl acetate	1.799 (45)	5.23 (20)	1.54 (22) [B]		
Phenylacetonitrile	1.93 (25)	19.0 (25) 8.5 (234)	3.47 (27) [B]	44.57	0.115 5

TABLE 4.10 Viscosity, Dielectric Constant, Dipole Moment, and Surface Tension of Selected Organic Substances (*continued*)

Substance	Viscosity η, mN·s·m^{-2}	Dielectric constant, ε	Dipole moment, D	Surface tension, dyn·cm^{-1}	
				a	b
Phenylacetylene		3.0 (20)	0.72 (20) [B]		
1-Phenylethanol		13 (20)	1.51 (20) [B]	42.88	0.103 8
		7.6 (90)			
Phenylhydrazine		7.2 (21)	1.67 (25) [B]	48.14	0.129 2
Phenyl isocyanate		8.8 (20)			
Phenyl isothio-cyanate		10 (20)		42.73	0.108 6
1-Phenylpropene		2.7 (20)			
2-Phenylpropene		2.3 (20)			
3-Phenylpropene		2.6 (20)			
Phenyl propyl ether				34.27	0.105 6
Phenyl salicylate		6.3 (50)		45.20	0.097 6
Phosgene		4.7 (0)			
		4.3 (22)			
Phthalide		36 (75)			
DL-α-Pinene	1.61 (25)	2.64 (25)	0.60 (25) [B]	28.35	0.094 4
L-β-Pinene	1.70 (20)	2.76 (20)		28.26	0.093 4
	1.41 (25)				
Piperidine	1.679 (15)	5.8 (20)	1.19 (25) [B]	31.79	0.115 3
	1.224 (30)				
Propane [g]	0.008 1 (20)	1.6 (0)	0	9.22	0.087 4 [lq]
	0.010 7 (125)				
1,2-Propane-diamine	1.46	10.2			
1,3-Propane-diamine	17.85	9.55	1.96 (25) [B]		
1,2-Propanediol	56.0 (20)	32.0 (20)	2.27 (25) [D]	72.0 (25)	
	18.0 (40)				
1,3-Propanediol	56.0 (20)	35.0 (20)	2.52 (25) [D]	47.43	0.090 3
	18.0 (40)				
1-Propanethiol			1.55 (25) [lq]	27.38	0.127 2
2-Propanethiol			1.64 (25) [lq]	24.26	0.117 4
1-Propanol	2.522 (15)	22.2 (20)	1.67 [g]	25.26	0.077 7
	1.722 (30)	20.33 (25)	1.75 (25) [B]		
2-Propanol	2.859 (15)	18.3 (25)	1.69 [g]	22.90	0.078 9
	1.765 (30)	16.24 (40)	1.66 (30) [B]		
2-Propenaldehyde			3.04 [g]		
			2.90 (25) [B]		
Propene [g]	0.008 43 (20)	1.88 (20)	0.35 [g]	9.99	0.142 7 [lq]
	0.009 33 (50)	1.44 (90)			
2-Propen-1-ol	1.363 (20)	21.6 (15)	1.63 [g]	27.53	0.090 2
	0.914 (40)				
Propionaldehyde	0.357 (15)	18.5 (17)	2.75 [g]		
	0.317 (27)		2.57 (20) [B]		

TABLE 4.10   Viscosity, Dielectric Constant, Dipole Moment, and Surface Tension of Selected Organic Substances (*continued*)

Substance	Viscosity η, mN·s·m^{-2}	Dielectric constant, ε	Dipole moment, D	Surface tension, dyn·cm^{-1}	
				a	b
Propionamide			3.4 (30) [B]	39.05	0.090 9
Propionic acid	1.175 (15)	3.30 (10)	1.76 [g]	28.68	0.099 3
	0.956 (30)	3.44 (40)	1.77 (25) [D]		
Propionic anhydride	1.144 (20)	18.3 (16)		30.30 (20)	29.70 (25)
	1.061 (25)				
Propionitrile	0.454 (15)	22.2 (20)	4.06 [g]	29.63	0.115 3
	0.389 (30)	24.2 (50)	3.60 (20) [B]		
Propyl acetate	0.585 (20)	5.69 (19)	1.86 (25) [B]	26.60	0.112 0
	0.460 (40)				
Propylamine	0.343 (25)	5.31 (20)	1.17 [g]	24.86	0.124 3
			1.36 (20) [B]		
Propylbenzene		2.37 (20)	0.35 (25) [lq]	31.13	0.107 5
		2.351 (30)			
Propyl benzoate				36.55	0.106 9
Propyl butyrate	0.831 (20)	4.3 (20)		27.06	0.100 0
Propyl chloroacetate				32.91	0.108 3
Propylene carbonate	2.53	64.4			
Propylene oxide	0.327 (20)		2.00 [g]		
	0.28 (25)				
Propyleneimine	0.491 (25)		1.77 [g, *cis*]		
			1.60 [g, *trans*]		
Propyl formate	0.574 (20)	7.72 (19)	1.91 (22) [B]	26.77	0.111 9
	0.417 (40)				
Propyl isobutyrate	0.831 (20)			25.83	0.101 5
Propyl nitrate		14 (18)	3.01 (20) [B]	29.67	0.123 7
Propyl pentoate	1.053 (20)	4 (19)		27.72	0.098 4
Propyl propionate	0.673 (20)	4.7 (20)	1.79 (22) [B]	26.85	0.105 9
Propyne			0.75 [g]	14.51	0.148 2
2-Propyn-1-ol	1.68 (20)	24.5 (20)	1.78 (25) [B]	38.59	0.127 0
Pulegone		9.5 (20)	2.00 (25) [B]		
Pyradazine			3.97 (35) [D]	50.55	0.103 6
Pyrazine		2.8 (54)	0		
Pyridine	1.130 (10)	12.3 (25)	2.20 (20) [B]	39.82	0.130 6
	0.829 (30)	9.4 (116)	2.25 [g]		
Pyrimidine			2.44 (35) [D]	32.85	0.101 0
Pyrrole	1.352 (20)	7.48 (18)	1.80 (25) [B]	39.81	0.110 0
	1.233 (25)	8.13 (25)			
Pyrrolidine			1.58 (20) [B]	31.48	0.090 0
2-Pyrrolidone	13.3 (25)		3.55 (25) [B]		
Quinoline	4.354 (15)	9.00 (25)	2.18 (25) [B]	45.25	0.106 3
	3.37 (25)				
Safrole	2.294 (25)	3.1 (21)			
Salicylaldehyde	2.90 (20)	13.9 (20)	3.1 (30) [lq]	45.38	0.124 2
	1.67 (45)		2.86 (20) [B]		

TABLE 4.10 Viscosity, Dielectric Constant, Dipole Moment, and Surface Tension of Selected Organic Substances (*continued*)

Substance	Viscosity η, mN·s·m^{-2}	Dielectric constant, ε	Dipole moment, D	Surface tension, dyn·cm^{-1}	
				a	b
Squalane	6.08 (20)		0		
Squalene	12 (25)		0.68 (25) [B]		
D-Sorbitol		33 (80)			
Stearic acid	11.6 (70)	2.29 (70)	1.76 (25) [D]		
		2.26 (100)			
Styrene	0.751 (20)	2.43 (25)	0.13 (25) [lq]	32.0 (20)	30.98 (30)
	0.696 (25)	2.32 (75)			
Succinonitrile	2.591 (60)	56.5 (57)	3.68 (30)	53.26	0.107 9
	2.008 (75)	54 (68)	[toluene]		
1, 1,2,2-Tetrabromo-ethane	13.950 (11)	8.6 (3)	1.29 (20) [H]	52.37	0.146 3
	9.797 (20)	7.0 (22)			
1, 1,2,2-Tetrachloro-difluoroethane	1.21 (25)	2.52 (25)		26.13	0.113 3
	1.208 (30)				
1, 1,2,2-Tetrachloroethane	1.844 (15)	8.20 (20)	1.29 [g]	38.75	0.126 8
	1.456 (30)		1.45 (25) [H]		
Tetrachloroethylene	1.932 (15)	2.30 (25)	0	32.86 (15)	31.27 (30)
	0.798 (30)				
Tetradecamethyl-cyclohepta-siloxane		2.7 (20)			
Tetradecamethyl-hexasiloxane		2.5 (20)	1.58 (20) [lq]	17.42 (25)	
Tetradecane	2.131 (22)		0	28.30	0.086 9
Tetradecanoic acid		0.76 (25) [B]	33.90	0.093 2	
1-Tetradecanol		4.72 (38)	1.69 (25) [C]	32.72	0.070 3
		4.40 (48)			
Tetraethylene glycol	44.9 (25)		5.84 (20) [lq]	45 (25)	
Tetraethyllead			0.3 (20) [B]	30.50	0.096 9
Tetraethylsilane			0	25.22	0.107 9
Tetraethyl silicate		4.1 (20)	1.72 (32) [B]	23.63	0.097 9
Tetrahydrofuran	0.55 (20)	11.6 (−70)	1.75 (25) [B]	26.5 (25)	
	0.460 (25)	7.58 (25)			
2,5-Tetrahydro-furandi-methanol	225 (25)				
Tetrahydro-2-furanmethanol	6.24 (20)	13.61 (23)	2.12 (35) [lq]	39.96	0.100 8
1,2,3,4-Tetrahy-dronaphthalene	2.202 (20)	2.76 (20)	0.60 (25) [lq]	35.55	0.095 4
	2.003 (25)				
1,2,3,4-Tetrahydro-2-naphthol		11.7 (20)			
		6.7 (90)			

TABLE 4.10 Viscosity, Dielectric Constant, Dipole Moment, and Surface Tension of Selected Organic Substances (*continued*)

Substance	Viscosity η, mN·s·m^{-2}	Dielectric constant, ε	Dipole moment, D	Surface tension, dyn·cm^{-1}	
				a	b
Tetrahydropyran	0.826 (20) 0.764 (25)	5.61 (25)	1.55 (25) [B]		
Tetrahydropyran-2-methanol	11.0 (20)			34.1 (25)	
Tetrahydrothiophene-1,1-dioxide	9.87 (30)	43.3 (30)	4.81 (25) [B]	35.5 (30)	
Tetrahydrothiophene oxide	52 (30) 19 (80)	42.5 (30)			
1,1,2,2-Tetramethylurea		23.06	3.47 (25) [B]		
Tetranitromethane	1.76 (20)	2.32 (20)	0		
Tetrathiomethylmethane		2.82 (70)			
Thiacyclohexane				36.06 (20)	33.74 (40)
Thiacyclopentane	1.042 (20) 0.971 (25)		1.90 (25) [B]	38.44	0.134 2
Thioacetic acid		12.8 (20)			
2,2′-Thiodiethanol	65.2 (20)			53.8 (20)	
Thiophene	0.662 (20) 0.353 (82)	2.76 (16) 2.57 (25)	0.55 [g] 0.52 (25) [B]	34.00	0.132 8
Thymol			1.55 (25) [B]	33.95	0.082 1
Toluene	0.623 (15) 0.523 (30)	2.385 (20) 2.364 (30)	0.45 (20) [lq]	30.90	0.118 9
p-Toluenesulfonyl chloride				42.41	0.090 3
o-Toluidine	5.195 (15) 4.39 (20)	6.34 (18) 5.71 (58)	1.60 (25) [B]	42.87	0.109 4
m-Toluidine	4.418 (15) 2.741 (30)	5.95 (18) 5.45 (58)	1.45 (25) [B]	40.33	0.097 9
p-Toluidine	1.945 (45) 1.557 (60)	4.98 (54)	1.52 (25) [B]	39.58	0.095 7
m-Tolunitrile			4.21 (22) [B]	38.85	0.101 3
p-Tolunitrile			4.47 (20) [B]	39.79	0.110 0
Tribenzylamine			0.65 (20) [B]	42.41	0.095 3
Tributyl phosphite	1.9 (25)		1.92 (20) [C]	27.57	0.086 5
2,2,2-Tribromo-acetaldehyde		7.6 (20)	1.70 (20) [B]		
Tribromoethane	2.152 (15) 1.741 (30)	4.39 (20)	0.99 [g]	48.14	0.130 8
1,2,3-Tribromopropane		6.45 (20)	1.59 (25) [B]	47.99	0.126 7
Tributylamine	1.35 (25)		0.78 (25) [B]	26.47	0.083 1
Tributyl borate	1.776 (20) 1.601 (25)		0.78 (25) [C]	26.2 (20)	25.8 (25)

TABLE 4.10 Viscosity, Dielectric Constant, Dipole Moment, and Surface Tension of Selected Organic Substances (*continued*)

Substance	Viscosity η, mN·s·m^{-2}	Dielectric constant, ε	Dipole moment, D	Surface tension, dyn·cm^{-1}	
				a	b
Tributyl phosphate	111.1 (15)	3.39 (25)	7.96 (30) [B]	3.07 (25) 28.71	0.066 6
Trichloroace- taldehyde		7.6 (−40) 4.9 (20)	1.96 (25) [B]	27.66	9.119 7
Trichloroacetic acid		4.6 (60)	1.1 (25) [B, dimer]	35.4	0.089 5
Trichloro- acetonitrile		7.85 (19)	1.93 (19) [lq]		
1,1,1-Trichlo- roethane	0.903 (15) 0.725 (30)	7.1 (7) 7.52 (20)	1.79 [g] 1.6 (25) [B]	28.28	0.124 2
1,1,2-Trichlo- roethane	0.119 (20) 0.110 (25)	8.78 (23)	1.45 [g]	37.40	0.135 1
Trichloroethylene	0.566 (20) 0.532 (25)	3.42 (16)	0.77 (30) [lq] 0.95 (30) [B]	29.5 (20)	28.8 (25)
Trichlorofluoro- methane	0.42 (25) 0.011 (25) [g]	2.28 (29)	0.45 [g] 0.49 [lq]	18 (25)	
Trichloro- methylsilane	0.47 (20)		1.87 (25) [B]	20.3 (20)	
2,4,6-Trichloro- phenol			1.88 (25) [D]	43.13	0.095 5
1,2,3-Trichloro- propane		7.5 (20)	1.61 [g]	37.8 (20)	37.05 (25)
Trichlorosilane	0.332 (20) 0.316 (25)		0.86 [g] 0.98 (25) [B]	20.43	0.107 6
α,α,α-Trichloro- toluene	3.07 (10) 2.55 (17)	6.9 (21)	2.17 (20) [B]		
1,1,2-Trichloro- 1,2,2,-trifluoro- ethane	0.711 (20) 0.627 (30)	2.41 (25)		17.75 (20)	16.56 (30)
Tridecane	1.883 (20) 1.55 (23)		0	27.73	0.087 2
1-Tridecene				28.01	0.088 4
Triethanolamine	613.6 (25) 208.1 (40)	29.36 (25)	3.57 (25) [B]		
Triethylaluminum		2.9 (20)			
Triethylamine	0.394 (15) 0.363 (30)	2.42 (25)	0.66 [g] 0.9 (25) [B]	22.70	0.099 2
Triethylene glycol	49.0 (20) 8.5 (60)	23.7 (20)	5.58 (20) [lq]	47.33	0.088 0
Triethyl phosphate	1.684 (40) 1.376 (55)	13.43 (15) 10.93 (65)	3.08 (25) [B]	31.81	0.092 8
Triethyl phosphite	0.72 (25)	5.0	1.82 (25) [D]	25.73	0.087 8

TABLE 4.10 Viscosity, Dielectric Constant, Dipole Moment, and Surface Tension of Selected Organic Substances (*continued*)

Substance	Viscosity η, mN·s·m^{-2}	Dielectric constant, ε	Dipole moment, D	Surface tension, dyn·cm^{-1}	
				a	*b*
Trifluoroacetic acid	0.926 (20) 0.653 (40)	8.55 (20) 5.76 (50)	2.28 [g]	15.64	0.184 4
2,2,2-Trifluoro-ethanol	1.996 (20)		2.03 (25) [cH]	20.6 (33)	
α,α,α-Trifluoro-toluene		9.2 (30) 8.1 (60)			
Trimethylamine	0.321 (−33)	2.4 (25)		16.24	0.113 3
1,2,3-Trimethyl-benzene		2.636 (20) 2.609 (30)	0.56 (20) [lq]	30.91	0.104 0
1,2,4-Trimethyl-benzene	0.894 (15) 0.730 (30)	2.38 (20) 2.36 (30)	0.30 (20) [lq]	31.76	0.102 5
1,3,5-Trimethyl-benzene	1.154 (20)	2.28	0	29.79	0.089 7
Trimethyl borate		8 (20)	0.82 (25) [C]		
2,2,3-Trimethyl-butane	0.579 (20)	1.93 (20)	0	20.70	0.097 3
cis, cis-1,3,5-Trimethyl-cyclohexane	0.632 (20) 0.558 (30)				
trans-1,3,5-Trimethyl-cyclohexane	0.714 (20) 0.624 (30)				
Trimethylene sulfide	0.638 (20) 0.607 (25)		1.78 (25) [B]	36.3 (20)	35.0 (30)
3,5,5-Trimethyl-1-hexanol	11.06 (25)				
2,6,8-Trimethyl-4-nonanone	1.9 (20)				
1,3,5-Trimethyl-2-oxabicyclo-[2.2.2]octane		4.57 (24)	1.54 (25) [C]	32.1 (20)	31.1 (25)
2,2,3-Trimethyl-pentane	0.598 (20)	1.962 (20)	0	22.46	0.089 5
2,2,4-Trimethyl-pentane	0.502 (20)	1.940 (20)	0	20.55	0.088 8
Trimethyl phosphite	0.61 (20)		1.83 (20) [C]	27.18 (20)	24.88 (40)
2,4,6-Trimethyl-pyridine	1.498 (20) 1.496 (25)	6.6	1.95 (25) [B]		
Triphenylamine				46.2	0.095 5
Triphenyl phosphite	25.18 (15) 6.95 (45)	3.67 (45) 3.57 (65)	2.04 (25) [B]		
Tripropylamine			0.58 (20) [lq] 0.76 (20) [B]	24.58	0.087 8

TABLE 4.10 Viscosity, Dielectric Constant, Dipole Moment, and Surface Tension of Selected Organic Substances (*continued*)

Substance	Viscosity η, mN·s·m^{-2}	Dielectric constant, ε	Dipole moment, D	Surface tension, dyn·cm^{-1}	
				a	b
Tripropylene glycol	56.1 (25)			34 (25)	
Tripropylene glycol butyl ether	6.58 (25)			28.8 (25)	
Tripropylene glycol ethyl ether	5.17 (25)			28.2 (25)	
Tripropylene glycol isopropyl ether	7.7 (25)			27.4 (25)	
Tripropylene glycol methyl ether	5.96 (25)			30.0 (25)	
Tris(dimethy-lamino) phos-phine oxide	3.34 (30)	30 (20)			
Tris(4-ethyl-phenyl) phosphite	30.22 (15) 9.047 (45)	3.74 (15) 3.61 (45)	2.08 (25) [B]		
Tris(*m*-tolyl) phosphite	37.55 (15) 9.132 (45)	3.67 (15) 3.53 (45)	1.62 (25) [B]		
Tris(*p*-tolyl) phosphite	35.52 (15) 8.794 (45)	3.88 (15) 3.74 (45)	1.77 (25) [B]		
Tritolyl phosphate	38.8 (35) 16.8 (55)	6.92 (40)	2.84 (40) [C]	40.9 (20)	
Undecane	1.186 (20) 0.761 (50)	2.00 (20) 1.84 (150)	0	26.26	0.090 1
2-Undecanone	1.61 (30)		2.71 (15) [B]		
Urea			4.59 (25) [D]		
Vinyl acetate	0.421 (20)		1.79 (25) [B]	23.95 (20)	22.54 (30)
o-Xylene	0.809 (20) 0.627 (40)	2.57 (20) 2.54 (30)	0.62 [g] 0.52 (25) [lq]	32.51	0.110 1
m-Xylene	0.617 (20) 0.497 (40)	2.37 (20) 2.35 (30)	0.33 (20) [lq] 0.37 (20) [B]	31.23	0.110 4
p-Xylene	0.644 (20) 0.513 (40)	2.26 (20) 2.22 (50)	0	30.69	0.107 4
Xylitol		40 (20)			

TABLE 4.11 Viscosity, Dielectric Constant, Dipole Moment, and Surface Tension of Selected Inorganic Substance

Temperature in degree celsius are indicated in parentheses. The physical state of the substance is indicated in square brackets

Substance	Viscosity, $mN \cdot s \cdot m^{-2}$	Dielectric constant, ε	Dipole moment, D	Surface tension, $dyn \cdot cm^{-1}$	
				a	b
Air (20°C)	0.018 2	1.000 536 4			
AlBr$_3$		3.38^{100}	5.2		
Ar					
[g] (20°C)	0.022 3	1.000 517 2			
[lq]		1.538^{-191}	0	34.28	0.249 3
AsBr$_3$		8.83^{35}	1.61	54.51	0.1043
AsCl$_3$		12.6^{20}	1.59	41.67	0.097 81
AsH$_3$ (arsine)		2.05^{20}	0.20		
BBr$_3$		2.58^0	0	31.90	0.128 0
BCl$_3$			0		
BF$_3$			0	−2.92	0.230 0
B$_2$H$_6$ (diborane)		1.872$^{-92.5}$	0	−3.13	0.178 5
B$_5$H$_9$			2.13		
B$_3$H$_6$N$_3$ (triborotriazine)			0		
Br$_2$					
[g] (20°C)		1.012 8			
[lq]	1.03^{16}	3.09^{20}	0	45.5	0.182 0
BrF$_3$	2.22^{20}		1.1	38.30	0.099 9
BrF$_5$	0.62^{24}	7.91$^{24.5}$	1.51	25.24	0.109 8
Cl$_2$					
[g] (20°C)	0.013 2		0		
[lq]		1.91^{14}			
ClF$_3$	0.48^{12}	4.29^{25}	0.554	26.9	0.166 0
ClO$_3$F (perchloryl fluoride)			0.023	12.24	0.157 6
Co					
[g]	0.017 5^{20}	1.000 70^0	0.112		
[lq]				−30.20	0.207 3
CO$_2$					
[g] (20°C)	0.014 7	1.000 922	0		
[lq]	0.071^{20}	1.60$^{0 °C, 50 atm}$			
COCl$_2$		4.34^{22}	1.17	22.59	0.145 6
COF$_2$			0.95		
COS			0.712	12.12	0.177 9
COSe		3.47^{10}	0.73		
CS			1.98		
CS$_2$					
[g]		1.002 9^0	0		
[lq]	0.375^{20}	2.6^{20}			
CrO$_2$Cl$_2$ [chromyl(VI) chloride]		2.6^{20}	0.47		
D$_2$ (deuterium)		1.277^{-253}			
DH				6.537	0.188 3
D$_2$O	1.098^{25}	78.25^{25}	1.87	(71.72^{20})*	(68.38^{40})*

*Actual values of surface tension.

TABLE 4.11 Viscosity, Dielectric Constant, Dipole Moment, and Surface Tension of Selected Inorganic Substances (*continued*)

Substance	Viscosity, $mN \cdot s \cdot m^{-2}$	Dielectric constant, ε	Dipole moment, D	Surface tension, $dyn \cdot cm^{-1}$	
				a	*b*
F_2		1.54^{-202}		-16.10	0.164 6
$GaCl_3$			0.85	35.0	0.100 0
$GeCl_4$		2.430^{25}	0	(22.44^{30})*	
H_2					
[g] (20 °C)	0.008 8	1.000 253 8	0		
[lq]		$1.228^{20.4 K}$			
HBr					
[g]		$1.003\ 13^0$	0.82		
[lq]	0.83^{-67}	3.82^{25}		13.10	0.207 9
HCl					
[g]		$1.004\ 6^0$	1.08		
[lq]	0.51^{-95}	4.60^{28}			
HCN	0.206^{18}	116^{20}	2.98	(19.45^{10})*	(18.33^{20})*
HCNO (isocyanate)			1.6		
HCNS (isothiocyanate)			1.7		
HF	0.256^0	83.6^0	1.82	10.41	0.078 67
HI					
[g]		$1.002\ 34^0$	0.44		
[lq]		2.90^{22}			
NH_3 (azide)			0.8		
H_2O (see Table 4.12)					
H_2O_2	1.25^{20}	84.2^0	2.2	78.97	0.154 9
HNO_3			2.17		
H_2S					
[g]		$1.00\ 4\ 0^0$	0.97		
[lq]	0.412^0	5.93^{10}		48.95	0.175 8
H_2Se			0.24	22.32	0.148 2
H_2SO_4	24.54^{25}	100^{25}			
HSO_3Cl (chlorosulfonic acid)	2.43^{20}	60^{20}			
HSO_2F (fluorosulfonic acid	1.56^{25}	$\sim120^{25}$			
H_2Te			<0.2	29.03	0.261 9
He					
[g] (20 °C)	0.019 6	1.000 065 0	0		
Hg	1.552^{20}		0	490.6	0.204 9
I_2	1.98^{116}	11.1^{118}	0		
IF_5			2.18	33.16	0.131 8
Kr					
[g] (20 °C)	0.025 0		<0.05		
[lq]				40.576	0.289 0
Ne[g] (20 °C)	0.031 3	1.000 063 9	0		
N_2					
[g] (20 °C)	0.017 6	1.000 548 0	0		
[lq]		1.454^{-203}		26.42	0.226 5

*Actual values of surface tension.

TABLE 4.11 Viscosity, Dielectric Constant, Dipole Moment, and Surface Tension of Selected Inorganic Substances (*continued*)

Substance	Viscosity, $mN \cdot s \cdot m^{-2}$	Dielectric constant, ε	Dipole moment, D	Surface tension, $dyn \cdot cm^{-1}$ a	b
NH$_3$					
[g]		$1.007\ 2^{0}$	1.47		
[lq]	$0.254^{-33.5}$	$22.4^{-33.4}$		$(37.91^{-50})*$	$(35.38^{-40})*$
N$_2$H$_4$ (hydrazine)	0.97^{20}	52.9^{20}	1.75		
NO			0.153	-67.48	0.585 3
N$_2$O					
[g]	$0.014\ 6^{20}$	$1.001\ 13^{0}$	0.167		
[lq]		1.52^{15}		5.09	0.203 2
NO$_2$			0.316		
N$_2$O$_4$		2.56^{15}	0.5		
NOBr (nitrosyl bromide)		13.4^{15}	1.8		
NOCl		18.2^{12}	1.9	29.49	0.149 3
NOF			1.81	14.00	0.116 5
NO$_2$F (nitryl fluoride)			0.47	8.26	0.185 4
O$_2$					
[g] (20 °C)	0.020 4	$1.000\ 494\ 7$	0		
[lq]		1.507^{-193}		-33.72	0.256 1
O$_3$			0.53	$(38.1^{-183})*$	
OF$_2$ (oxygen difluoride)			0.297		
OsO$_4$			0		
PBr$_3$		3.9^{20}	0.5	45.34	0.128 3
PCl$_3$		3.43^{25}	0.78	31.14	0.126 6
PCl$_5$		2.7^{165}	0.9		
PF$_5$			0		
PH$_3$		2.9^{15}	0.58		
PI$_3$		4.12^{65}	0	61.66	0.067 71
POCl$_3$	1.065^{25}	13.7^{25}	2.41	35.22	0.127 5
POF$_3$			1.76		
PSCl$_3$		5.8^{22}	1.42	37.00	0.127 2
PbCl$_4$		2.78^{20}			
S$_2$Cl$_2$ dimer		4.79^{15}	1.0	46.23	0.146 4
S$_2$F$_2$					
FSSF isomer			1.45		
S$=$SF$_2$ isomer			1.03		
SF$_4$			0.632	12.87	0.173 4
SF$_6$			0	5.66	0.119 0
S$_2$F$_{10}$		2.020^{20}	0		
SO$_2$					
[g]	$0.012\ 6^{29}$	$1.009\ 3^{0}$	1.63		
[lq]		15.0^{0}		26.58	0.194 8
SO$_3$		3.11^{18}	0		
SOBr$_2$ (thionyl bromide)		9.06^{20}	9.11		
SOCl$_2$		9.25^{20}	1.45	36.10	0.141 6

*Actual values of surface tension.

TABLE 4.11   Viscosity, Dielectric Constant, Dipole Moment, and Surface Tension of Selected Inorganic Substances (*continued*)

Substance	Viscosity, $mN \cdot s \cdot m^{-2}$	Dielectric constant, ε	Dipole moment, D	Surface tension, $dyn \cdot cm^{-1}$	
				a	b
SO_2Cl_2 (sulfuryl chloride)		9.15^{20}	1.81	32.10	0.132 8
$SbCl_3$		33.2^{75}	3.93	47.87	0.123 8
$SbCl_5$		3.22^{20}	0		
SbF_5				49.07	0.193 7
SbH_3			0.12		
SeF_4				38.61	0.127 4
SeF_6			0		
$SeOCl_2$		55^{25}	2.64		
$SiCl_4$		2.40^{16}	0	20.78	0.099 62
SiF_4			0		
SiH_4			0		
$SiHCl_3$			0.86	20.43	0.107 6
$SnBr_4$			0		
$SnCl_4$		2.89^{20}	0	29.92	0.113 4
TeF_6			0		
$TiCl_4$		2.80^{20}	0	$(33.54^{20})*$	$(31.06^{40})*$
UF_6					
[g]		$1.002\ 92^{67}$	0		
[lq]		2.18^{65}		25.5	0.124 0
VCl_4		3.05^{25}	0		
$VOBr_3$		3.6^{25}			
$VOCl_3$		3.4^{25}	0.3	$(36.36^{20})*$	$(33.60^{40})*$
Xe [g] (20°C)	0.022 8	1.001 23	0		

*Actual values of surface tension.

TABLE 4.12 Refractive Index, Viscosity, Dielectric Constant, and Surface Tension of Water at Various Temperatures

Temperature, °C	Refractive index, n_D	Viscosity, $mn \cdot s \cdot m^{-2}$	Dielectric constant, ε	Surface tension, $dyn \cdot cm^{-1}$
0	1.333 95	1.770 2	87.74	75.83
5	1.333 88	1.510 8	85.76	75.09
10	1.333 69	1.303 9	83.83	74.36
15	1.333 39	1.137 4	81.95	73.62
20	1.333 00	1.001 9	80.10	72.88
21	1.332 90	0.976 4	79.73	72.73
22	1.332 80	0.953 2	79.38	72.58
23	1.332 71	0.931 0	79.02	72.43
24	1.332 61	0.910 0	78.65	72.29
25	1.332 50	0.890 3	78.30	72.14
26	1.332 40	0.870 3	77.94	71.99
27	1.332 29	0.851 2	77.60	71.84
28	1.332 17	0.832 8	77.24	71.69
29	1.332 06	0.814 5	76.90	71.55
30	1.331 94	0.797 3	76.55	71.40
35	1.331 31	0.719 0	74.83	70.66
40	1.330 61	0.652 6	73.15	69.92
45	1.329 85	0.597 2	71.51	69.18
50	1.329 04	0.546 8	69.91	68.45
55	1.328 17	0.504 2	68.35	67.71
60	1.327 25	0.466 9	66.82	66.97
65	1.326 16	0.434 1	65.32	66.23
70	1.325 11	0.405 0	63.86	65.49
75	1.323 99	0.379 2	62.43	64.75
80		0.356 0	61.03	64.01
85		0.335 2	59.66	63.28
90		0.316 5	58.32	62.54
95		0.299 5	57.01	61.80
100		0.284 0	55.72	61.80

COMBUSTIBLE MIXTURES

TABLE 4.13 Properties of Combustible Mixtures in Air

Additional compounds can be found in National Fire Protection Association, *Fire Protection Handbood*, 14th ed., 1976.

| Substance | Autoignition temperature, °C | Flammable limits, percent by volume of fuel (25°C, 760 mm) | |
		Lower	Upper
Acetaldehyde	175	4.0	6.0
Acetic acid, glacial	465	5.4	16.0
Acetic anhydride	390	2.9	10.3
Acetone	465	2.6	12.8
Acetonitrile	524	4.4	16.0
Acetylene	305	2.5	100
Acrolein	235*	2.8	31.0
Acrylonitrile	481	3.0	17
Allyl alcohol	378	2.5	18.0
Allylamine	374	2.2	22
Ammonia, anhydrous	651	16	25
Aniline	615	1.3	
Benzene	560	1.3	7.1
Biscyclohexyl	245	0.7 (100°C)	5.1 (150°C)
1-Bromobutane	265	2.6 (100°C)	6.6 (100°C)
3-Bromopropene	295	4.4	7.3
Butane	405	1.9	8.5
Butanol	365	1.4	11.2
2-Butanone	516	1.8	10
1-Butene	385	1.6	10.0
3-Buten-1-ol		4.7	34
Butyl acetate	425	1.7	7.6
Butylamine	312	1.7	9.8
Butylbenzene	410	0.8	5.8
Butylene oxide		1.5	18.3
Butyl formate	322	1.7	8.2
Butyraldehyde	230	2.5	12.5
Butyric acid	450	2.0	10.0
Carbon disulfide	90	1.3	50.0
Carbon monoxide	609	12.5	74
Carbonyl sulfide		12	29
Chlorobenzene	640	1.3	7.1
2-Chloro-1,3-butadiene		4.0	20.0
1-Chlorobutane		1.8	10.1
2-Chloro-2-butene		2.3	9.3
1-Chloro-1,1-difluoroethane		6.2	17.9
2-Chloroethanol	425	4.9	15.9
Chloromethane	632	10.7	17.4

*Unstable.

TABLE 4.13 Properties of Combustible Mixtures in Air (*continued*)

Substance	Autoignition temperature, °C	Flammable limits, percent by volume of fuel (25°C, 760 mm)	
		Lower	Upper
1-Chloropentane	260	1.6	8.6
2-Chloropropane	593	2.8	10.7
1-Chloro-1-propene		4.5	16
3-Chloro-1-propene	485	2.9	11.1
Chlorotrifluoroethylene		8.4	38.7
Crotonaldehyde	232	2.1	15.5
Cumene	425	0.9	6.5
Cyanogen		6.6	42.6
Cyclohexane	245	1.3	8
Cyclopropane	500	2.4	10.4
Decahydronaphthalene	250	0.7	4.9
Decane	210	0.8	5.4
Diborane	38–52†	0.8	88
Dibutyl ether	194	1.5	7.6
o-Dichlorobenzene	648	2.2	9.2
1,2-Dichloroethylene		9.7	12.8
Dichloropropane	557	3.4	14.5
Diisopropyl ether	443	1.4	7.9
Diethylamine	312	1.8	10.1
Diethyl ether	160	1.9	36.0
2,2-Dimethylbutane	425	1.2	7.0
Dimethyl ether		3.4	27.0
N,N-Dimethylformamide	445	1.2	7.0
1,1-Dimethylhydrazine	249	2	95
2,3-Dimethylpentane	335	1.1	6.7
2,2-Dimethylpropane	450	1.4	7.5
Dimethyl sulfide	206	2.2	19.7
Dimethyl sulfoxide	215	2.6	28.5
1,4-Dioxane	180	2.0	22.0
Divinyl ether	360	1.7	27
Ethane	515	3.0	12.5
Ethanol	365	3.3	19
2-Ethoxyethanol	235	1.8	14.0
1-Ethoxypropane		1.7	9.0
Ethyl acetate	427	2.2	11.0
Ethylamine	385	3.5	14.0
Ethylbenzene	432	1.0	6.7
Ethylcyclobutane	210	1.2	7.7
Ethylene	490	2.7	36.0
Ethyleneimine	320	3.6	46
Ethylene oxide	429	3.6	100
Ethyl formate	455	2.8	16.0
1,3-Ethylidene dichloride	440	6.2	16
Ethyl nitrite	90	3.0	50

†Ignites in moist air.

TABLE 4.13 Properties of Combustible Mixtures in Air (*continued*)

Substance	Autoignition temperature, °C	Flammable limits, percent by volume of fuel (25 °C, 760 mm)	
		Lower	Upper
Ethyl propionate	440	1.9	11
Ethyl vinyl ether	202	1.7	28
Formaldehyde	429	7.0	73
2-Furaldehyde	316	2.1	19.3
Furan		2.3	14.3
Furfuryl alcohol	491	1.8	16.3
Gasoline, 92 octane	~280	1.4	7.6
Heptane	215	1.0	6.7
Hexane	225	1.1	7.5
2-Hexanone	533	1.2	8
Hydrocyanic acid, 96%	538	5.6	40.0
Hydrogen	400	4.0	75
4-Hydroxy-4-methyl-2-pentanone	603	1.8	6.9
Isobutyl acetate	421	2.4	10.5
Isobutylbenzene	430	0.8	6.0
Isopentane	420	1.4	7.6
Isopentyl acetate	360	1.0	7.5
Isoprene	220	2	9
Isopropyl acetate	460	1.8	8
Isopropyl alcohol	399	2.0	12
Methane	540	5.4	15.0
Methanethiol		3.9	21.8
Methanol	385	6.7	36.0
2-Methoxyethyl acetate		1.7	8.2
Methyl acetate	502	3.1	16
Methyl acrylate		2.8	25
Methylamine	430	4.9	20.6
2-Methyl-2-butanol	437	1.2	9.0
3-Methyl-1-butene	365	1.5	9.1
Methylcyclohexane	250	1.2	6.7
Methyl formate	465	5.0	23
2-Methylpropene	465	1.8	9.6
4-Methyl-2-pentanone	460	1.4	7.5
2-Methylpropene	465	1.8	9.6
α-Methylstyrene	574	1.9	6.1
Methyl propionate	469	2.5	13
Nicotine	244	0.7	4.0
Nitrobenzene	482	1.8 (93 °C)	
Nonane	205	0.8	2.9
Octane	220	1.0	6.5

TABLE 4.13   Properties of Combustible Mixtures in Air (*continued*)

Substance	Autoignition temperature, °C	Flammable limits, percent by volume of fuel (25°C, 760 mm)	
		Lower	Upper
Pentanamine		2.2	22
Pentane	260	1.5	7.8
2-Pentanone	505	1.5	8.2
Pentyl acetate	360	1.1	7.5
Petroleum ether	550	1.1	5.9
Propane	450	2.2	9.5
1,3-Propanediol	371	2.6	12.5
Propanol	440	2.1	13.5
Propene	460	2.0	11.1
Propanamine	318	2.0	10.4
Propionaldehyde	207	2.9	17.0
Propyl acetate	450	2.0	8
Propylene oxide		2.8	37.0
Propyl nitrate	175	2	100
Pyridine	482	1.8	12.4
Styrene	490	1.1	6.1
Tetrahydrofuran	321	2	11.8
Tetrahydrofurfuryl alcohol	282	1.5	9.7
Tetrahydronaphthalene	385	0.8	5.0
Toluene	480	1.2	7.1
Trichlorothylene	420	12.5	90
Triethylamine		1.2	8.0
Triethylene glycol	371	0.9	9.2
Trimethylamine	190	2.0	11.6
Trioxane	414	3.6	29
Vinyl acetate	427	2.6	13.4
Vinyl butyrate		1.4	8.8
Vinyl chloride	461	3.6	33.0
Vinyl fluoride		2.6	21.7
Xylene, *m*- and *p*-	530	1.1	7.0
Xylene, *o*-	465	1.0	6.0

SECTION 5

THERMODYNAMIC PROPERTIES

ENTHALPIES AND GIBBS (FREE) ENERGIES OF FORMATION, ENTROPIES, AND HEAT CAPACITIES

The tables in this section contain values of the enthalpy (ΔHf) and Gibbs (ΔGf, free) energy of formation, entropy (S), and heat capacity (C_p) at 298.15 K (25 °C). The tables cover common organic compounds. No values are given in these tables for metal alloys or other solid solutions, for fused salts, or for substances of undefined chemical composition.

For a more complete listing of compounds see the tables in *Selected Values of Chemical Thermodynamical Properties*, by D. D. Wagman *et al.*, *National Bureau of Standards Technical Notes* 270-3, 270-4, 270-5, 270-6, 270-7, and 270-8, Washington; *JANAF Thermochemical Tables*, by D. R. Stull and H. Prophet, *National Bureau of Standards Publication 37*, Washington; supplements to JANAF appearing in *J. Phys. Chem. Ref. Data*; D. R. Stull, E. F. Westrum, Jr., and G. C. Sinke, *The Chemical Thermodynamics of Organic Compounds*, Wiley-Interscience, New York, 1969; and I. Barin and O. Knacke, *Thermochemical Properties of Inorganic Substances*, Springer-Verlag, Berlin, 1973.

The values of the thermodynamic properties of the pure substances given in these tables are, for the substances in their standard states, defined as follows:

Pure solid (c) *or liquid* (liq).   The substance is in the condensed phase under a pressure of 1 atm.

Gas (g).   The standard state is the hypothetical ideal gas at unit fugacity, in which state the enthalpy is that of the real gas at the same temperature and at zero pressure.

The values of $\Delta Hf°$ and $\Delta Gf°$ given in the tables represent the change in the appropriate thermodynamic quantity when one gram formula weight of the substance in its standard state is formed, isothermally at the indicated temperature, from the elements, each in its appropriate standard reference state. The standard reference state at 25 °C for each element has been chosen to be the standard state that is thermodynamically stable at 25 °C and 1 atm pressure. The standard reference states are indicated in the tables by the fact that the values of $\Delta Hf°$ and $\Delta Gf°$ are exactly zero.

The values of $S°$ represent the virtual or "thermal" entropy of the substance in the standard state at 298.15 K, omitting contributions from nuclear spins. Isotope mixing effects are also excluded except in the case of the ^1H–^2H system.

The physical state of each substance is indicated in the column headed "State" as crystalline solid (c), liquid (liq), gaseous (g), or amorphous (amorp). Solutions in water are listed as aqueous (aq). Solutions in water are designated as aqueous, and the concentration of the solution is expressed in terms of the number of moles of solvent associated with 1 mol of the solute. If no concentration is indicated, the solution is assumed to be dilute. The standard state for a solute in aqueous solution is taken as the hypothetical ideal solution of unit molality (indicated as std state, $m = 1$). In this state the partial molal enthalpy and the heat capacity of the solute are the same as in the infinitely dilute real solution (aq. m).

TABLE 5.1 Enthalpies and Gibbs (Free) Energies of Formation, Entropies, and Heat Capacities of Organic Compounds

Substance	State	$\Delta Hf°$, kcal·mol^{-1}	$\Delta Gf°$, kcal·mol^{-1}	$S°$, cal·deg^{-1}·mol^{-1}	$C_p°$, cal·deg^{-1}·mol^{-1}
Acenaphthene	c	16.8			
Acenaphthylene	c	44.7			
Acetaldeyde	liq	−45.96	−30.64	38.3	65.6
	g	−39.76	−31.86	63.15	13.06
Acetaldoxime	c	−18.6			
	liq	−19.5			
Acetamide	c	−76.0			
Acetamidoguanidine nitrate	c	−118.1			
1-Acetamido-2-nitroguanidine	c	−46.3			
5-Acetamidotetrazole	c	−1.2			
Acetanilide	c	−50.3			
Acetic acid	liq	−115.71	−93.2	38.2	29.7
	g	−103.93	−90.03	67.52	15.90
Ionized; std state, $m = 1$	aq	−116.16	−88.29	20.7	−1.5
Nonionized; std state, $m = 1$	aq	−116.70	−94.78	42.7	
Acetic anhydride	liq	−149.14	−116.82	64.2	
	g	−137.60	−113.93	93.20	23.78
Acetone	liq	−59.18	−37.22	47.9	30.22
	g	−51.78	−36.58	70.49	17.90
Acetone glyceraldehyde	liq	−180			
Acetonitrile	liq	12.8	23.7	35.76	21.86
	g	21.00	25.24	58.19	12.48
Acetophenone	liq	−34.07	−4.06	59.62	
	g	−20.76	0.44	89.12	
Acetyl radical	g	−4.0			
N-Acetylbenzidine	c	−38.0			
Acetyl bromide	liq	−53.5			
Acetyl chloride	liq	−65.44	−49.73	48.0	28
	g	−58.30	−46.29	70.47	16.21
Acetylene	g	54.19	50.00	48.00	10.50
Std state, $m = 1$	aq	50.54	51.88	29.5	
Acetylenedicarbonitrile	liq	119.6			
	g	127.50	122.10	69.31	20.53
Acetylene dicarboxylic acid	c	−138.1			
Acetyl fluoride	liq	−112.4			
N-Acetylhydrazobenzene	c	−2.0			
o-Acetylhydroxybenzoic acid	c	−194.93			
N-Acetylimidazole	c	−28.6			
Acetyl iodide	liq	−39.3			
4-Acetylresorcinol	c	−137.1			

TABLE 5.1   Enthalpies and Gibbs (Free) Energies of Formation, Entropies, and Heat Capacities of Organic Compounds (*continued*)

Substance	State	$\Delta Hf°$, kcal·mol^{-1}	$\Delta Gf°$, kcal·mol^{-1}	$S°$, cal·deg^{-1}· mol^{-1}	$C°_p$, cal·deg^{-1}· mol^{-1}
N-Acetyltetrazole	c	19.49			
Acridine	c	44.8			
Acrolein	liq	−29.97	−16.17		
	g	−20.50	−15.45		
Acrylic acid	liq	−91.8			
	g	−80.36	−68.37	75.29	18.59
Acrylonitrile	liq	36.1			
	g	44.20	46.68	65.47	15.24
Adenine	c	23.21	71.58	36.1	
Adipic acid	c	−237.60			
	liq	−235.51	−177.17		
Aetioporphyrin I	c	−6.0			
Aetioporphyrin II	c	0.4			
α-Alanine					
D	c	−134.03	−88.23	31.6	
L	c	−133.96	−88.49	30.88	
DL	c	−134.55	−88.92	31.6	
Alanine anhydride	c	−128.0			
α-Alanylglycine					
DL	c	−185.64	−117.00	51.0	
L	c	−197.52	−127.30	46.62	
DL-Alanylphenylalanine	c	−170.2			
Alanylphenylalanyl anhydride	c	−89.3			
Allantoin (5-ureidohydantoin)	c	−171.50	−106.65	46.6	
Allomucic acid	c	−142			
Alloxan monohydrate	c	−239.08	−182.08	44.6	
Alloxantin dihydrate	c	−510.3			
Allyl radical	g	38			
1-Allyl-5-allylamino-tetrazole	c	83.7			
1-Allyl-5-aminotetrazole	c	63.4			
2-Allyl-5-aminotetrazole	c	67.6			
Allyl chloride	g	−0.15	10.42	73.29	18.01
Allylcyclopentane	liq	−15.74			
Allyl ethyl sulfoxide	liq	−41.83			
Allyl trichloroacetate	liq	−94.5			
Amalic acid	c	−367.0			
Amarine	c	63			
p-Aminoacetophenone	c	70.2			
3-Aminoacridine	c	39.8			
5-Aminoacridine	c	38.1			
2-Aminobenzoic acid	c	−95.8			
3-Aminobenzoic acid	c	−98.2			

TABLE 5.1 Enthalpies and Gibbs (Free) Energies of Formation, Entropies, and Heat Capacities of Organic Compounds (*continued*)

Substance	State	$\Delta Hf°$, kcal·mol^{-1}	$\Delta Gf°$, kcal·mol^{-1}	$S°$, cal·deg^{-1}· mol^{-1}	$C_p°$, cal·deg^{-1}· mol^{-1}
4-Aminobenzoic acid	c	−98.8			
2-Aminobiphenyl	c	26.8			
4-Aminobiphenyl	c	19.4			
1-Aminobutane (butylamine)	liq	−30.52			
	g	−22.00	11.76	86.76	28.33
2-Aminobutane (*sec-* butylamine)	g	−24.90	9.71	83.90	27.99
4-Aminobutanoic acid	c	−138.1			
2-Aminoethanesulfonic acid	c	−187.7	−134.3	36.8	33.6
Ionized; std state, $m = 1$	aq	−171.92	−121.76	47.8	
Nonionized; std state, $m = 1$	aq	−181.92	−134.12	55.7	
2-Aminohexanoic acid (norleucine)	c	−152.7			
4-Aminohexanoic acid	c	−154.5			
5-Aminohexanoic acid	c	−153.7			
6-Aminohexanoic acid	c	−152.7			
3-Amino-2-methylpropane (2-butylamine)	liq	−31.68			
5-Aminopentanoic acid	c	−144.5			
5-Aminotetrazole	c	49.7			
5-Aminotetrazole nitrate	c	−6.6			
3-Amino-1,2,4-triazole	c	18.4			
Amygdalin	c	−455			
1,2-Anyhydroglucose-3,5,6-triacetate	c	−411.7			
Aniline	liq	7.55	35.63	45.72	45.90
	g	20.76	39.84	76.28	25.91
Anisine	c	−51			
Anisoyl glycine	c	−180.9			
Anthracene	c	29.0	68.30	49.58	49.7
9,10-Anthracenedione	c	−49.6			
β-D-Arabinose	c	−252.84			
β-L-Arabinose	c	−252.84			
D-Arabonic acid-γ-lactone	c	−238.2			
L-Arginine	c	−148.66			
D-Arginine	c	−149.05	−57.43	59.9	
L-Ascorbic acid (vitamin C)	c	−278.34			
L-Asparagine	c	−188.50	−126.73	41.7	
L-Aspartic acid	c	−232.47	−174.53	40.66	
Azobenzene					
cis	c	86.7			
trans	c	76.6			

TABLE 5.1 Enthalpies and Gibbs (Free) Energies of Formation, Entropies, and Heat Capacities of Organic Compounds (*continued*)

Substance	State	$\Delta Hf°$, kcal·mol⁻¹	$\Delta Gf°$, kcal·mol⁻¹	$S°$, cal·deg⁻¹· mol⁻¹	$C_p°$, cal·deg⁻¹· mol⁻¹
Azodicarbamide	c	−69.90			
Azulene	g	66.90	84.10	80.75	30.69
Barbituric acid	c	−152.2			
Benzaldehyde	liq	−21.23	2.24		
	g	−9.57	5.85		
Benzamide	c	−48.42			
Benzanilide	c	−22.3			
1,2-Benzanthracene	c	41			
2,3-Benzanthracene	c	38.3	85.79	51.48	
1,2-Benzanthra-9,10- quinone	c	−55.4			
Benzene	liq	11.71	29.72	41.41	19.52
	g	19.82	30.99	64.34	
Benzenethiol (thiophenol)	liq	15.32	32.02	53.25	41.40
	g	26.66	35.28	80.51	25.07
Benzidine	c	16.9			
Benzil	c	−36.8			
Benzoic acid	c	−92.03	−58.62	40.05	34.97
Benzoic anhydride	c	−103.0			
Benzonitrile	g	52.30	62.33	76.73	26.07
Benzophenone	c	−8.0	33.5	58.6	
p-Benzoquinone	c	−44.33			
Benzotriazole	c	59.74			
DL-Benzoylalanine	c	−147.9			
Benzoyl bromide	liq	−25.58			
Benzoyl chloride	liq	−39.17			
Benzoyl iodide	liq	−12.31			
Benzoylphenylalanine	c	−129.6			
Benzoyl sarcosine	c	−135.7			
3,4-Benzphenanthrene	c	44.2			
Benzyl radical	g	45			
Benzyl alcohol	liq	−38.49	−6.57	51.8	
Benzyl bromide (2-bromotoluene)	liq	5.6			
Benzyl chloride	liq	−7.8			
N-Benzyldiphenylamine	c	44.2			
Benzyl ethyl sulfide	liq	−1.17			
Benzyl iodide	liq	13.8			
Benzyl mercaptan	liq	10.4			
Benzyl methyl ketone	liq	−36.30			
Benzyl methyl sulfide	liq	6.27			
Bicyclo[4.1.0]heptane	g	0.33			
Bicyclo[3.1.0]hexane	g	9.09			
Bicyclo[4.2.0]octane	g	−6.39			
Bicyclo[5.1.0]octane	g	−3.85			

TABLE 5.1 Enthalpies and Gibbs (Free) Energies of Formation, Entropies, and Heat Capacities of Organic Compounds (*continued*)

Substance	State	$\Delta Hf°$, kcal·mol^{-1}	$\Delta Gf°$, kcal·mol^{-1}	$S°$, cal·deg^{-1}· mol^{-1}	$C_p°$, cal·deg^{-1}· mol^{-1}
Bicyclopropyl	g	30.9			
Biphenyl	c	24.02	60.75	49.2	38.80
	liq	28.5	62.07	59.8	
Biphenylene	liq	84.4			
N,N'-Bisuccinimide	c	−169.5			
Brassidic acid	c	−214			
Bromal	liq	−31.13			
Bromal hydrate	c	−112			
Bromobenzene	liq	14.5	30.12	52.0	37.17
4-Bromobenzoic acid	c	−90.4			
1-Bromobutane	g	−25.65	−3.08	88.39	26.13
2-Bromobutane	liq	−37.2	−4.60		
	g	−28.70	−6.16	88.50	26.48
Bromochlorodi-fluoromethane	g	−112.7	−107.18	76.14	
Bromochloro-fluoromethane	g	−70.5	−66.58	72.88	
Bromochloromethane	g	−12.0	−9.39	68.67	
Bromodichloro-fluoromethane	g	−64.4	−58.98	78.87	
Bromodichloromethane	g	−14.0	−10.16	75.56	
Bromodifluoromethane	g	−110.8	−106.90	70.51	
Bromoethane	liq	−21.99	−6.64	47.5	24.1
	g	−15.30	−6.29	68.71	15.45
Bromoethene (vinyl bromide)	g	18.73	19.30	65.83	13.26
Bromofluoromethane	g	−60.4	−57.71	65.97	
1-Bromoheptane	liq	−52.21			
1-Bromohexane	liq	−46.42			
Bromoiodomethane	g	12.0	9.36	73.49	
Bromomethane	g	−9.02	−6.75	58.76	10.15
2-Bromo-2-methylpropane	liq	−39.3			
	g	−32.00	−6.73	79.34	27.85
1-Bromooctane	liq	−58.57			
1-Bromopentane	liq	−40.68			
	g	−30.87	−1.37	97.70	31.60
1-Bromopropane	g	−21.00	−5.37	79.08	20.66
2-Bromopropane	g	−23.20	−6.51	75.53	21.37
N-Bromosuccinimide	c	−80.35			
Bromotrichloromethane	g	−8.9	−2.96	79.55	
Bromotrifluoromethane	g	−155.1	−148.8	71.16	16.57
Brucine	c	−188.6			
1,2-Butadiene	g	38.77	47.43	70.03	19.15
1,3-Butadiene	g	26.33	36.01	66.62	19.01
Butadiyne (biacetylene)	g	113.00	106.11	59.76	17.60

TABLE 5.1 Enthalpies and Gibbs (Free) Energies of Formation, Entropies, and Heat Capacities of Organic Compounds (*continued*)

Substance	State	$\Delta Hf°$, kcal·mol^{-1}	$\Delta Gf°$, kcal·mol^{-1}	$S°$, cal·deg^{-1}· mol^{-1}	$C_p°$, cal·deg^{-1}· mol^{-1}
Butane	liq	−35.29	−3.60	55.2	
	g	−30.15	−4.10	74.12	23.29
1,2-Butanediamine	liq	−28.74			
2,3-Butanedione (diacetyl)	liq	−87.44			
1,4-Butanedithiol	liq	−25.11			
1-Butanethiol	liq	−29.79	0.97	65.96	
(butyl mercaptan)					
	g	−21.05	2.64	89.68	28.24
2-Butanethiol	liq	−31.13	−0.04	64.87	
	g	−23.00	1.29	87.65	28.51
1-Butanol	liq	−78.18	−38.84	54.1	42.31
	g	−65.65	−36.04	86.7	26.29
2-Butanol	liq	−81.88	−42.31	53.8	47.5
	g	−69.94	−40.06	85.6	27.08
2-Butanone	liq	−65.29	−36.18	57.08	37.98
(methyl ethyl ketone)					
	g	−56.26	−34.91	80.81	24.59
1-Butene	g	−0.03	17.04	73.04	20.47
2-Butene					
cis	g	−1.67	15.74	71.90	18.86
trans	g	−2.67	15.05	70.86	20.99
1-Buten-3-yne	g	72.80	73.13	66.77	17.49
tert-Butoxy radical	g	−24.7			
tert-Butyl radical	g	6.7			
N-Butylacetamide	liq	−91.02			
Butyl acetate	liq	−126.52			
tert-Butylamine	liq	−35.97			
	g	−28.65	6.90	80.76	28.67
Butylbenzene	liq	−18.67 $^{18°C}$	27.50		
	g	−3.30	34.58	105.04	41.85
sec-Butylbenzene	liq	−15.87			
tert-Butylbenzene	liq	−16.90			
sec-Butyl butyrate	liq	−141.6			
Butyl chloroacetate	liq	−128.7			
Butyl 2-chlorobutyrate	liq	−156.6			
Butyl 3-chlorobutyrate	liq	−146.0			
Butyl 4-chlorobutyrate	liq	−147.7			
Butyl 2-chloropropionate	liq	−136.7			
Butyl 3-chloropropionate	liq	−133.4			
Butyl crotonate	liq	−111.8			
Butylcyclohexane	g	−50.95	13.49	109.58	49.50
Butylcyclopentane	g	−40.22	14.67	109.04	42.42
Butyl dichloroacetate	liq	−131.5			
Butyl ether	liq	−156.1			
	g	−87.2	114.96	48.82	
tert- Butyl hydroperoxide	liq	−70.2			

TABLE 5.1 Enthalpies and Gibbs (Free) Energies of Formation, Entropies, and Heat Capacities of Organic Compounds (*continued*)

Substance	State	$\Delta Hf°$, kcal·mol^{-1}	$\Delta Gf°$, kcal·mol^{-1}	$S°$, cal·deg^{-1}·mol^{-1}	$C_p°$, cal·deg^{-1}·mol^{-1}
Butyllithium	liq	−31.6			
Butyl trichloroacetate	liq	−130.6			
1-Butyne (ethyl acetylene)	g	39.48	48.30	69.51	19.46
2-Butyne (dimethylacetylene)	g	34.97	44.32	67.71	18.63
Butyraldehyde	g	−49.00	−27.43	82.44	24.52
Butyramide	c	−87.5			
Butyric acid	liq	−127.59	−90.27	54.1	42.1
Butyronitrile	g	8.14	25.97	77.98	23.19
Caffeine (methyl theobromine)	c	−76.2			
Capric acid (decanoic acid)	c	−170.59			
Caproic acid (hexanoic acid)	liq	−139.71			
ε-Caprolactam	c	−78.54	−22.72	40.3	
Caprylic acid (octanoic acid)	liq	−151.93			
Carbazole	c	30.3			
Carboxyl radical	g	−54			
CCH radical	g	114	105	49.6	8.87
Cellobiose	c	−532.5			
Chloroacetamide	c	−80.9			
Chloroacetic acid	c, l	−122.3			
Ionized	aq	−119.81			
Nonionized; std state, $m = 1$	aq	−118.92			
Chloroacetyl chloride	liq	−68.0			
2-Chlorobenzaldehyde	liq	−28.4			
3-Chlorobenzaldehyde	liq	−30.2			
4-Chlorobenzaldehyde	c	−35.1			
Chlorobenzene	liq	2.58	21.32	50.0	35.9
2-Chlorobenzoic acid	c	−95.3			
3-Chlorobenzoic acid	c	−101.2			
4-Chlorobenzoic acid	c	−102.19			
Chlorobenzoquinone	c	−52.7			
1-Chlorobutane	g	−35.20	−9.27	85.58	25.71
2-Chlorobutane	g	−38.60	−12.78	85.94	25.93
2-Chlorobutyric acid	liq	−137.6			
3-Chlorobutyric acid	liq	−133.0			
4-Chlorobutyric acid	liq	−135.4			
Chlorocyclohexane	liq	−49.54			
2-Chloro-1,1-difluoroethylene	g	−79.2	−72.90	72.28	
Chlorodifluoromethane	g	−115.6	−108.1	67.12	13.35

TABLE 5.1 Enthalpies and Gibbs (Free) Energies of Formation, Entropies, and Heat Capacities of Organic Compounds (*continued*)

Substance	State	$\Delta Hf°$, kcal·mol^{-1}	$\Delta Gf°$, kcal·mol^{-1}	$S°$, cal·deg^{-1}· mol^{-1}	$C_p°$, cal·deg^{-1}· mol^{-1}
Chloroethane (ethyl chloride)	g	−26.83	−14.46	65.91	14.97
Chloroethylene (vinyl chloride)	g	8.5	12.4	63.07	12.84
Chloroethyne	g	51	47	57.81	12.98
Chlorofluoromethane	g	−63.2	−57.11	63.16	11.24
Chloroform	liq	−31.6	−17.17	48.5	
	g	−24.60	−16.76	70.63	15.63
Chloroiodomethane	g	3.0	3.69	70.78	
Chloromethane (methyl chloride)	g	−19.59	−13.97	55.97	9.74
Chloromethyloxirane	liq	−35.48			
1-Chloro-2-methylpropane	g	−38.10	−11.87	84.56	25.93
2-Chloro-2-methylpropane	g	−43.80	−15.32	77.00	27.30
1-Chloronaphthalene	liq	13.0			
2-Chloronaphthalene	c	13.2			
1-Chloropentane	g	−41.80	−8.94	94.89	31.18
3-Chorophenol	c	−49.4			
4-Chlorophenol	c	−47.3			
1-Chloropropan-2,3-diol	liq	−125.58			
2-Chloropropan-1,3-diol	liq	−123.71			
1-Chloropropane	g	−31.10	−12.11	76.27	20.23
2-Chloropropane	g	−35.00	−14.94	72.70	20.87
3-Chloro-1-propene (allyl chloride)	g	−0.15	10.42	73.29	18.01
2-Chloropropionic acid	liq	−125.0			
3-Chloropropionic acid	c	−131.4			
N-Chlorosuccinimide	c	−85.58			
Chlorotrifluoromethane	g	−169.20	−159.38	68.16	15.98
Chlorotrinitromethane	liq	−6.54			
Chrysene	c	34.7			
Cinchonamine	c	−10.4			
Cinchonidine	c	7.1			
Cinchonine	c	7.4			
Cinnamic acid					
cis	c	−72.0			
trans	c	−80.53			
Cinnamic anhydride	c	−83.1			
Citraconic acid	c	−197.04			
Citric acid	c	−369.0	−295.5	39.73	
Citric acid monohydrate	c	−439.4	−352.0	67.74	1.276
Codeine monohydrate	c	−151.2			
Coniine	liq	−57.6			
Creatine	c	−128.16	−63.32	45.3	
Creatine hydrate	c	−199.1			
Creatinine	c	−56.77	−6.97	40.10	

TABLE 5.1 Enthalpies and Gibbs (Free) Energies of Formation, Entropies, and Heat Capacities of Organic Compounds (*continued*)

Substance	State	$\Delta Hf°$, kcal·mol⁻¹	$\Delta Gf°$, kcal·mol⁻¹	$S°$, cal·deg⁻¹· mol⁻¹	$C_p°$, cal·deg⁻¹· mol⁻¹
o-Cresol (2-methylphenol)	g	−30.74	−8.86	85.47	31.15
m-Cresol (3-methylphenol)	g	−31.63	−9.69	85.27	29.27
p-Cresol (4-methylphenol)	g	−29.97	−7.38	83.09	29.75
m-Cresol acetate	liq	−89.41			
Crotonic acid					
cis	liq	−83			
trans	c	−102.9			
trans-Crotononitrile	g	35.77	46.22	71.31	19.62
Cyanamide	c	14.05			
1-Cyanoguanidine	c	5.4	42.9	30.90	28.40
3-Cyanopyridine	c	46.23			
5-Cyanotetrazole	c	96.1			
4-Cyanothiazole	c	52.63			
Cyclobutane	g	6.37	26.30	63.43	17.26
Cyclobutene	g	31.00	41.76	62.98	16.03
Cyclododecane	c	−73.29			
Cycloheptane	liq	−37.47	12.92	57.97	29.42
Cycloheptanone	liq	−71.5			
1,3,5-Cycloheptatriene	liq	34.22	58.09	51.30	38.90
1,3-Cyclohexadien-5-yl radical	g	49.4			
Cyclohexane	liq	−37.34	6.37	48.84	37.4
	g	−29.43	7.59	71.28	25.40
Cyclohexane- 1,2-dicarboxylic acid					
cis	c	−229.7			
trans	c	−232.0			
Cyclohexanethiol	g	−22.80			
Cyclohexanol	liq	−83.22	−31.87	47.7	
Cyclohexanone	g	−55.00	−21.69	77.00	26.21
Cyclohexene	liq	−9.28	24.28	51.67	34.9
	g	−1.28	25.54	74.27	25.10
Cyclohexen-3-yl radical	g	29			
1-Cyclohexenylmethanol	liq	−91.4			
Cyclohexyl radical	g	13			
Cyclooctane	liq	−40.58	18.60	62.62	
Cyclooctanone	liq	−77.9			
1,3,5,7-Cyclooctatetraene	liq	60.93	85.70	52.65	
Cyclopentadiene	g	32.00	42.86	64.00	
Cyclopentane	liq	−25.28	8.70	48.82	30.80
	g	−18.46	9.23	70.00	19.84
Cyclopentane-1,2-diol					
cis	c	−115.9			
trans	c	−117.1			

TABLE 5.1   Enthalpies and Gibbs (Free) Energies of Formation, Entropies, and Heat Capacities of Organic Compounds (*continued*)

Substance	State	$\Delta Hf°$, kcal·mol^{-1}	$\Delta Gf°$, kcal·mol^{-1}	$S°$, cal·deg^{-1}· mol^{-1}	$C_p°$, cal·deg^{-1}· mol^{-1}
Cyclopentanethiol	g	−11.45	13.63	86.38	25.79
Cyclopentanol	liq	−71.74	−30.55	49.2	
Cyclopentanone	liq	−56.24			
	g	−46.03			
Cyclopentene	liq	1.02	25.93	48.10	29.24
	g	7.87	26.48	69.23	17.95
1-Cyclopentenylmethanol	liq	8.2			
Cyclopentyl-1-thiaethane	g	−15.41			
Cyclopropane	g	12.74	24.95	56.75	13.37
Cyclopropene	g	66.0	68.42	58.38	
Cyclopropyl radical	g	55			
L-Cysteine	c	−124.5			
L-Cystine	c	−245.7			
Decahydronaphthalene (decalin)					
cis	liq	−52.45	16.47	63.34	55.45
trans	liq	−55.14	13.79	63.32	54.61
Decanal	g	−79.09	−15.90	138.28	57.29
Decane	liq	−71.95	−4.19	101.70	75.16
1,10-Decanediol	c	−165.74			
1-Decanethiol	liq	−66.07			
	g	−50.54	14.68	145.82	61.08
1-Decanoic acid	c	−170.59			
1-Decanol	liq	−114.6	−31.6	10.2.9	
	g	−96.0	−24.9	142.8	59.1
1-Decene	liq	−41.73	25.10	101.58	
1-Decyne	g	9.85	60.28	125.36	52.51
Deoxybenzoin	c	−16.96			
Desoxyamalic acid	c	−285.7			
Diacetamide	c	−117			
Diacetyl peroxide	liq	−127.9			
o-Diallyl phthalate	liq	−131.6			
Dialuric acid	c	−314.4			
2,6-Diaminopyridine	c	−1.56			
Diamylose	c	−850			
Diazomethane	g	46.0	52.06	58.02	12.55
Dibenzoylethane	c	−61.1			
Dibenzoylethylene	c	−27.4			
Dibenzoylmethane	c	−53.6			
Dibenzoyl peroxide	c	−100			
Dibenzyl	c	10.53	62.15	64.4	61.0
Dibenzyl ketone	c	−20.1			
Dibenzyl sulfide	c	23.74			
Dibenzyl sulfone	c	−42.1			

TABLE 5.1 Enthalpies and Gibbs (Free) Energies of Formation, Entropies, and Heat Capacities of Organic Compounds (*continued*)

Substance	State	$\Delta Hf°$, kcal·mol^{-1}	$\Delta Gf°$, kcal·mol^{-1}	$S°$, cal·deg^{-1}· mol^{-1}	$C_p°$, cal·deg^{-1}· mol^{-1}
1,2-Dibromobutane	g	−23.70	−3.14	97.70	30.38
Dibromochlorofluoro-methane	g	−55.4	−53.40	81.99	
Dibromochloromethane	g	−5.0	−4.50	78.31	
1,2-Dibromocycloheptane	liq	−37.67			
1,2-Dibromocyclohexane	liq	−38.8			
1,2-Dibromocyclooctane	liq	−41.41			
Dibromodichloromethane	g	−7.0	−4.67	83.23	
Dibromodifluoromethane	g	−102.7	−100.16	77.66	
1,2-Dibromoethane	liq	−19.4	−5.0	53.37	32.51
Dibromofluoromethane	g	−53.4	−52.84	75.70	
Dibromomethane	g	−3.53	−3.87	70.10	13.04
1,2-Dibromopropane	g	−17.40	−4.22	89.90	24.57
Dibutylborinic acid	liq	−146.3			
Dibutyl ether	g	−79.80	−21.16	119.60	48.76
Dibutylmercury	liq	−23.4			
Di-*tert*-butyl peroxide	liq	−91.0			
Dibutyl *o*-phthalate	c	−201			
Dibutyl sulfate	liq	−216.1			
Dibutyl sulfite	liq	−165.6			
Dibutyl sulfone	c	−145.76			
Dichloroacetic acid	liq	−119.0			
Ionized	aq	−122.4			
Nonionized	aq	−120.4			
Dichloroacetylene	g	50	47	65	15.67
1,2-Dichlorobenzene	g	7.16	19.76	81.61	27.12
1,3-Dichlorobenzene	g	6.32	18.78	82.09	27.20
1,4-Dichlorobenzene	g	5.50	18.44	80.47	27.22
Dichlorodifluoromethane	g	−117.90	−108.51	71.91	17.31
1,1-Dichloroethane	liq	−38.3	−18.1	50.61	30.18
	g	−31.10	−17.52	72.91	18.25
1,2-Dichloroethane	liq	−39.49	−19.03	49.84	30.9
	g	−31.00	−17.65	73.66	18.80
1,1-Dichloroethylene	liq	−5.8	5.85	48.17	26.60
	g	0.30	5.78	68.85	16.02
cis-1,2-Dichloroethylene	liq	−6.6	5.27	47.42	27
	g	0.45	5.82	69.20	15.55
trans-1,2-Dichloroethylene	g	1.00	6.35	69.29	15.93
Dichlorofluoromethane	g	−68.10	−60.77	70.04	14.58
Dichloromethane	liq	−29.7	−16.83	42.7	
	g	−22.80	−16.46	64.61	12.16
1,2-Dichloropropane	g	−39.60	−19.86	84.80	23.47
1,3-Dichloropropane	g	−38.60	−19.74	87.76	23.81
2,2-Dichloropropane	g	−42.00	−20.21	77.92	25.30
Dicyanoacetylene	liq	119.6			

TABLE 5.1   Enthalpies and Gibbs (Free) Energies of Formation, Entropies, and Heat Capacities of Organic Compounds (*continued*)

Substance	State	$\Delta Hf°$, kcal·mol^{-1}	$\Delta Gf°$, kcal·mol^{-1}	$S°$, cal·deg^{-1}· mol^{-1}	$C_p°$, cal·deg^{-1}· mol^{-1}
1,4-Dicyano-2-butyne	c	87.6			
Dicyclohexadiene	liq	6.3			
Dicyclopentadiene	c	27.9			
Dicyclopentyl	liq	−41.8			
2,2-Diethoxypropane	liq	−128.83			
Diethylamine	g	−17.30	17.23	84.18	27.66
Diethylbarbituric acid (veronal)	c	−178.7			
1,2-Diethylbenzene	g	−4.53	33.72	103.81	43.63
1,3-Diethylbenzene	g	−5.22	32.67	104.99	42.27
1,4-Diethylbenzene	g	−5.32	32.95	103.73	42.10
Diethylenediamine	c	−3.2	57.4	20.5	
Diethylene glycol	liq	−150.2			
	g	−136.5		105.4	32.3
Diethyl ether (ethyl ether)	liq	−65.30	−27.88	60.5	40.8
	g	−60.26	−29.24	81.90	26.89
Diethylmercury	liq	7.1			
Diethylmethyl phos- phonate	liq	−245.3			
Diethylnitramine	liq	−25.4			
Diethyl oxalate	liq	−192.51			
Diethyl peroxide	liq	−53.4			
Diethyl *o*-phthalate	liq	−186			
Diethyl selenide	liq	−23.0			
Diethyl sulfate	liq	−194.28			
Diethyl sulfite	liq	−143.50			
Diethyl sulfone	c	−123.13			
Diethyl sulfoxide	liq	−63.97			
Diethylzinc	liq	4.0			
1,2-Difluorobenzene	liq	−79.04	−59.41	53.20	38.01
1,3-Difluorobenzene	g	−74.09	−61.43	76.57	25.40
1,4-Difluorobenzene	g	−73.43	−60.43	75.43	25.55
2,2′-Difluorobiphenyl	c	−70.73			
4,4-Difluorobiphenyl	c	−70.91			
2,2-Difluorochloroethylene	g	−75.4	−69.1	72.39	17.23
1,1-Difluoroethane	g	−119.70	−105.87	67.50	16.24
1,1-Difluoroethylene	g	−82.50	−76.84	63.38	14.14
Difluoromethane	g	−108.24	−101.66	58.94	10.25
9,10-Dihydroanthracene	c	15.87			
1,2-Dihydronaphthalene	liq	18.0			
1,4-Dihydronaphthalene	liq	21.0			
4*H*-Dihydropyran	liq	−37.5			
5,12-Dihydrotetracene	c	25.44			
2,3-Dihydrothiophene	liq	12.73			

TABLE 5.1 Enthalpies and Gibbs (Free) Energies of Formation, Entropies, and Heat Capacities of Organic Compounds (*continued*)

Substance	State	$\Delta Hf°$, kcal·mol^{-1}	$\Delta Gf°$, kcal·mol^{-1}	$S°$, cal·deg^{-1}· mol^{-1}	$C_p°$, cal·deg^{-1}· mol^{-1}
2,5-Dihydrothiophene	liq	11.31			
1,2-Dihydroxybenzene	c	−86.3	−50.20	35.9	31.6
1,3-Dihydroxybenzene	c	−87.95	−50.00	35.3	31.3
1,2-Diiodobenzene	c	41.2			
1,3-Diiodobenzene	c	44.7			
1,4-Diiodobenzene	c	38.4			
1,2-Diiodoethane	g	15.90	18.76	83.30	19.67
Diiodomethane	g	28.30	24.24	73.95	13.80
Diisopropyl ether	liq	−83.94	−21.1	70.4	
	g	−76.20	−29.13	93.27	37.83
Diisopropyl ketone	g	−74.40			
Diisopropylmercury	liq	−3.1			
1,2-Dimethoxybenzene	liq	−69.4			
Dimethoxyborane	liq	−144.5			
1,2-Dimethoxyethane	liq	−90.02			
2,2-Dimethoxypropane	liq	−108.92			
cis-α,β-Dimethylacrylic acid	c	−117.3			
Dimethyl adipate	liq	−211.9			
Dimethylamine	g	−4.50	16.25	65.24	16.50
Std state, $m=1$	aq	−16.88	13.85	31.8	
$(CH_3)_2NH_2^+$; std state, $m=1$	aq	−28.74	−0.80	41.2	
Dimethylaminotrimethyl-silane	liq	−66.8			
N,N-Dimethylaniline	liq	8.2			
2,2-Dimethylbutane	g	−44.35	−2.20	85.62	33.91
2,3-Dimethylbutane	g	−42.49	−0.98	87.42	33.59
2,3-Dimethyl-1-butene	g	−13.32	18.89	87.39	34.29
2,3-Dimethyl-2-butene	g	−14.15	18.18	87.15	29.54
3,3-Dimethyl-1-butene	g	−10.31	23.46	82.16	30.23
2,3-Dimethyl-2-butenoic acid	c	−108.9			
Dimethylcadmium	g	9.528		72.40	31.5
Dimethylchlorosilane	liq	−79.8			
1,1-Dimethylcyclohexane	liq	−52.31	6.34	63.87	
	g	−43.26	8.42	87.24	36.90
1,2-Dimethylcyclohexane					
cis	g	−41.15	9.85	89.51	37.40
trans	g	−43.02	8.24	88.65	38.00
1,3-Dimethylcyclohexane					
cis	g	−44.16	7.13	88.54	37.60
trans	g	−42.20	8.68	89.92	37.60
1,4-Dimethylcyclohexane					
cis	g	−42.22	9.07	88.54	37.60
trans	g	−44.12	7.58	87.19	37.70

TABLE 5.1 Enthalpies and Gibbs (Free) Energies of Formation, Entropies, and Heat Capacities of Organic Compounds (*continued*)

Substance	State	$\Delta Hf°$, kcal·mol^{-1}	$\Delta Gf°$, kcal·mol^{-1}	$S°$, cal·deg^{-1}·mol^{-1}	$C_p°$, cal·deg^{-1}·mol^{-1}
1,1-Dimethylcyclopentane	g	−33.05	9.33	85.87	31.86
1,2-Dimethylcyclopentane					
cis	g	−30.96	10.93	87.51	32.06
trans	g	−32.67	9.17	87.67	32.14
1,3-Dimethylcyclopentane					
cis	g	−32.47	9.37	87.67	32.14
trans	g	−31.93	9.91	87.67	32.14
Dimethyldichlorosilane	g	−110.2		80.16	24.17
cis-2,4-Dimethyl-1,3-dioxane	liq	−111.79			
4,5-Dimethyl-1,3-dioxane	liq	−108.32			
5,5-Dimethyl-1,3-dioxane	liq	−110.53			
4,4′-Dimethyldiphenyl-amine	c	−2.8			
Dimethyl ether	g	−43.99	−26.99	63.83	15.73
N,N-Dimethylformamide	liq	−57.2		28.5	37.45
Dimethylfulvene	liq	21.5			
Dimethyl fumarate	liq	−174.3			
Dimethyl glutarate	liq	−205.9			
Dimethylglyoxime	c	−42.51			
2,2-Dimethylhexane	liq	−62.63	0.71	79.33	
	g	−53.71	2.56	103.06	
2,3-Dimethylhexane	liq	−60.40	2.17	81.91	
2,3-Dimethylhexane	g	−51.13	4.23	106.11	
2,4-Dimethylhexane	liq	−61.47	0.89	82.62	
	g	−52.44	2.80	106.51	
2,5-Dimethylhexane	liq	−62.26	0.60	80.96	
	g	−53.21	2.50	104.93	
3,3-Dimethylhexane	liq	−61.58	1.23	81.12	
	g	−52.61	3.17	104.70	
3,4-Dimethylhexane	liq	−60.23	2.03	82.97	
	g	−50.91	4.14	107.15	
2,2-Dimethyl-3-hexene					
cis	liq	−30.22			
trans	liq	−34.64			
5,5-Dimethylhydantoin	c	−126.4			
1,1-Dimethylhydrazine	liq	11.8	49.4	47.32	39.21
1,2-Dimethylhydrazine	liq	13.3	50.8	47.60	40.88
Dimethyl maleate	liq	−168.2			
Dimethylmaleic anhydride	c	−139.0			
Dimethyl malonate	liq	−190.2			
Dimethylmercury	liq	14.0			
Dimethylnitramine	c	−16.9			
Dimethyl oxalate	liq	−181.0			
2,2-Dimethylpentane	g	−49.27	0.02	93.90	39.67

TABLE 5.1 Enthalpies and Gibbs (Free) Energies of Formation, Entropies, and Heat Capacities of Organic Compounds (*continued*)

Substance	State	$\Delta Hf°$, kcal·mol^{-1}	$\Delta Gf°$, kcal·mol^{-1}	$S°$, cal·deg^{-1}· mol^{-1}	$C_p°$, cal·deg^{-1}· mol^{-1}
2,3-Dimethylpentane	g	−47.62	0.16	98.96	39.67
2,4-Dimethylpentane	g	−48.28	0.74	94.80	39.67
3,3-Dimethylpentane	g	−48.17	0.63	95.53	39.67
2,7-Dimethylphenanthrene	c	8.70			
4,5-Dimethylphenanthrene	c	21.26			
9,10-Dimethyl-phenanthrene	c	11.4			
Dimethyl *m*-phthalate	c	−171			
Dimethyl *o*-phthalate	liq	−162			
Dimethyl *p*-phthalate	c	−170			
2,2-Dimethylpropane	g	−39.67	−0.364	73.23	29.07
2,3-Dimethylpyridine	liq	4.62			
2,4-Dimethylpyridine	liq	3.85			
2,5-Dimethylpyridine	liq	4.45			
2,6-Dimethylpyridine	liq	3.02			
3,4-Dimethylpyridine	liq	4.36			
3,5-Dimethylpyridine	liq	5.36			
Dimethyl succinate	liq	−199.6			
1,1-Dimethylsuccinic acid	c	−236.08			
1,2-Dimethylsuccinic acid					
cis	c	−233.6			
trans	c	−235.1			
Dimethyl sulfate	liq	−175.23			
Dimethyl sulfite	liq	−125.07			
Dimethyl sulfone	c	−107.8	−72.3	34.77	
Dimethyl sulfoxide	liq	−48.6	−23.7	45.0	35.2
3,3-Dimethyl-2-thiabutane	liq	−37.49			
2,2-Dimethylthia-cyclopropane	liq	−5.78			
2,2-Dimethyl-3-thiapentane	liq	−44.7			
2,4-Dimethyl-3-thiapentane	g	−33.76	6.48	99.30	40.45
2,3-Dinitroaniline	c	−2.8			
2,4-Dinitroaniline	c	−16.3			
2,5-Dinitroaniline	c	−10.6			
2,6-Dinitroaniline	c	−12.1			
3,4-Dinitroaniline	c	−7.8			
3,5-Dinitroaniline	c	−9.3			
2,4-Dinitroanisole	c	−44.6			
2,6-Dinitroanisole	c	−45.2			
1,2-Dinitrobenzene	c	2.06	50.56	51.7	
1,3-Dinitrobenzene	c	−4.04	44.13	52.8	
2,4-Dinitrophenol	c	−55.6			
2,6-Dinitrophenol	c	−50.2			
2,4-Dinitroresorcinol	c	−99.3			

TABLE 5.1   Enthalpies and Gibbs (Free) Energies of Formation, Entropies, and Heat Capacities of Organic Compounds (*continued*)

Substance	State	$\Delta Hf°$, kcal·mol^{-1}	$\Delta Gf°$, kcal·mol^{-1}	$S°$, cal·deg^{-1}· mol^{-1}	$C_p°$, cal·deg^{-1}· mol^{-1}
4,6-Dinitroresorcinol	c	−105.1			
2,4-Dinitrotoluene	c	−17.1			
2,6-Dinitrotoluene	c	−12.2			
1,4-Dioxane	liq	−84.47	−44.96	46.67	
	g	−75.30	−43.21	71.65	22.48
1,3-Dioxane	liq	−89.99			
1,4-Dioxatetralin	liq	−60.9			
Dioxindole	c	−76.9			
1,3-Dioxolane	g	−71.1			
Dipentene	liq	−12.1			
N,N-Diphenylacetamide	c	−10.3			
Diphenylamine	c	31.07			
1,4-Diphenylbutadiene					
cis,cis	c	47.51			
trans,trans	c	42.73			
Diphenylbutadiyne	c	123.91			
1,4-Diphenylbutane	c	−2.36			
1,4-Diphenyl-1,4-butanedione	c	−61.24	1.87	77.6	
1,4-Diphenyl-2-butene-1,4-dione	c	−27.55	26.64	76.3	
Diphenylcarbinol	c	−25.04			
Diphenyl carbonate	c	−95.93	−42.05	66.54	
Diphenyldichlorosilane	liq	−66.5			
Diphenyl disulfide	c	35.8			
Diphenyl disulfone	c	−153.59			
1,1-Diphenylethane	liq	11.7	58.58	80.28	
1,2-Diphenylethane	liq	12.31	63.87	64.6	
1,1-Diphenylethene	liq	41.21			
Diphenyl ether	liq	−3.48	34.47	69.62	
Diphenylethyne	c	74.66			
Diphenylfulvene	c	7.1			
Diphenylmercury	c	66.8			
Diphenylmethane	liq	21.25	66.19	57.2	55.7
Diphenyl sulfide	liq	39.1			
Diphenyl sulfone	c	−53.71			
Diphenyl sulfoxide	c	2.40			
Dipropyl ether	g	−70.00	−25.23	100.98	37.83
Dipropylmercury	liq	−5.0			
Dipropyl sulfate	liq	−205.22			
Dipropyl sulfite	liq	−154.52			
Dipropyl sulfone	liq	−130.94			
Dipropyl sulfoxide	liq	−78.65			
2,3-Dithiabutane	liq	−14.82	1.67	56.26	34.92
5,6-Dithiadecane	g	−37.86	12.87	136.91	55.23

TABLE 5.1   Enthalpies and Gibbs (Free) Energies of Formation, Entropies, and Heat Capacities of Organic Compounds (*continued*)

Substance	State	$\Delta Hf°$, kcal·mol^{-1}	$\Delta Gf°$, kcal·mol^{-1}	$S°$, cal·deg^{-1}·mol^{-1}	$C_p°$, cal·deg^{-1}·mol^{-1}
3,4-Dithiahexane	liq	−28.69	2.28	72.90	
1,3-Dithian-2-thione	c	−3.1			
4,5-Dithiaoctane	liq	−40.95	4.56	89.28	
N,N-Dithiodiethylamine	liq	−29.1			
1,3-Dithiolan-2-thione	c	3.1			
Di-*p*-tolyl sulfone	c	−74.32			
Divinyl ether	g	−9.53			
Divinyl sulfone	liq	−49.5			
Dodecane	liq	−84.16	6.71	117.26	89.86
Dodecanoic acid	c	−185.14			
1-Dodecene	g	−39.52	32.96	147.78	64.43
1-Dodecyne	g	−0.01	64.22	143.98	63.44
Dulcitol	c	−321.9			
Eicosane	g	−108.93	28.04	223.26	110.73
Eicosanoic acid (arachidic acid)	c	−241.9			
1-Eicosene	g	−78.93	49.03	222.26	108.15
Ergosterol	c	−188.8			
meso-Erythritol	c	−127.56	−152.12	39.9	
Ethane	g	−20.24	−7.84	54.76	12.54
1,2-Ethanedithiol	liq	−12.83			
Ethanethiol	g	−11.02	−1.12	70.77	17.37
Ethanol	liq	−66.20	−41.63	38.49	26.76
	g	−56.03	−40.13	67.54	15.64
Ethoxy radical	g	−6			
Ethyl radical	g	26.0	31	59.2	
Ethyl acetate	liq	−114.49	−79.52	62.0	
	g	−105.86	−78.25	86.70	27.16
Ethyl allyl sulfone	liq	−96.95			
Ethylamine	g	−11.00	8.91	68.08	17.36
N-Ethylaniline	liq	0.9	45.10	57.2	
Ethylbenzene	liq	−2.98	28.61	60.99	
	g	7.12	31.21	86.15	30.69
2-Ethyl-1-butene	g	−12.32	19.11	90.01	31.92
Ethyl carbamate (urethane)	c	−124.4			
Ethyl chloride	g	−26.83	−14.46	65.91	14.97
Ethyl crotonate	liq	−100.4			
Ethylcyclohexane	liq	−50.72	6.95	67.14	
1-Ethylcyclohexene	liq	−25.50			
Ethylcyclopentane	liq	−39.08	8.92	67.00	
	g	−30.37	10.65	90.42	31.49
Ethyldiethylcarbamate	liq	−141.6			
Ethylene	g	12.50	16.31	52.39	10.24
Ethylene carbonate	c	−138.9			

TABLE 5.1 Enthalpies and Gibbs (Free) Energies of Formation, Entropies, and Heat Capacities of Organic Compounds (*continued*)

Substance	State	$\Delta Hf°$, kcal·mol^{-1}	$\Delta Gf°$, kcal·mol^{-1}	$S°$, cal·deg^{-1}· mol^{-1}	$C_p°$, cal·deg^{-1}· mol^{-1}
Ethylene chlorohydrin	liq	−70.6			
1,2-Ethylenediamine	liq	−15.06		50	
	aq, 200	−13.32			
Ethylenediaminetetraacetic acid (EDTA)	c	−420.5			
Ethylenediammonium chloride	c	−122.7			
	aq, 5000	−115.92			
Ethylene glycol (2,1-ethanediol)	liq	−108.70	−77.25	39.9	35.8
	g	−93.05	−72.77	77.33	23.20
	aq, 1	−109.01			
Ethyleneimine (azirane)	g	29.50	42.54	59.90	12.55
Ethylene oxide	g	−12.58	−3.13	57.94	11.54
2-Ethyl-1-hexanal	liq	−83.30			
2-Ethyl-2-hexanal	liq	−62.44			
3-Ethylhexane	liq	−59.88	1.79	84.95	
Ethylidenecyclohexane	liq	−21.19			
Ethyl isovalerate	liq	−136.5			
Ethyllithium	c	−14.0			
Ethylmercury bromide	c	−25.7			
Ethylmercury chloride	c	−33.7			
Ethylmercury iodide	c	−15.7			
Ethyl methyl ether	g	−51.73	−28.12	74.24	21.45
Ethyl nitrate	g	−36.80	−8.81	83.25	23.27
Ethyl nitrite	g	−24.9		24.74	23.71
3-Ethylpentane	g	−45.33	2.63	98.35	39.67
Ethyl pentanoate	liq	−132.2			
Ethyl peroxyl radical	g	(−2)			
2-Ethylphenol	c	−49.91			
3-Ethylphenol	c	−51.21			
4-Ethylphenol	c	−53.63			
Ethylphosphonic acid	c	−251.3			
Ethyl propanoate	liq	−122.16	−79.16		
2-Ethylpyridine	liq	−1.2			
Ethylsuccinic acid	c	−236.4			
Ethyl thioacetate	liq	−64.01			
Ethyl β-vinylacrylate	liq	−80.8			
Ethyl vinyl ether	g	−33.63			
Ethynylbenzene (phenyl-acetylene)	g	78.22	86.46	76.88	27.46
Fluoranthene	c	45.75	82.60	55.09	
Fluoroacetamide	c	−118.7			
Fluoroacetic acid	c	−164.5			
Fluorobenzene	g	−27.86	−16.50	72.33	22.57
2-Fluorobenzoic acid	c	−135.67			

TABLE 5.1 Enthalpies and Gibbs (Free) Energies of Formation, Entropies, and Heat Capacities of Organic Compounds (*continued*)

Substance	State	$\Delta Hf°$, kcal·mol^{-1}	$\Delta Gf°$, kcal·mol^{-1}	$S°$, cal·deg^{-1}· mol^{-1}	$C_p°$, cal·deg^{-1}· mol^{-1}
3-Fluorobenzoic acid	c	−139.13			
4-Fluorobenzoic acid	c	−140.00			
Fluoroethane	g	−62.90	−50.44	63.34	14.21
2-Fluoroethanol	liq	−111.3			
Fluoromethane	g	−56.80	−51.09	53.25	8.96
1-Fluoropropane	g	−67.20	−47.87	72.71	19.75
2-Fluoropropane	g	−69.00	−48.81	69.82	19.60
4-Fluorotoluene	liq	−44.80	−19.06	56.67	
Fluorotrinitromethane	liq	−52.8			
Formaldehyde	g	−27.70	−26.27	52.29	8.46
unhydrolyzed	aq	−35.9	−31.02		
Formamide	liq	−60.7			
	g	−44.5	−33.71	59.41	10.84
Formanilide	c	−36.2			
Formic acid	liq	−101.51	−86.38	30.82	23.67
	g	−90.49	−83.89	59.45	10.81
Ionized; std state, $m = 1$	aq	−101.71	−83.9	22	−21.0
Nonionized; std state, $m = 1$	aq	−101.68	−89.0	39	
Dimer	g	−195.08			
Formyl					
HCO	g	10.4	6.76	53.66	8.27
HCO$^+$	g	204	201	48.3	8.62
Formyl fluoride	g	−90	−88	59.0	9.66
N-Formyl-DL-leucine	c	−222.1			
Formyl urea	c	−118			
β-D-Fructose	c	−302.2			
D-Fucose	c	−262.7			
Fumaric acid	c	−193.84	−156.70	39.7	
Fumaronitrile	c	64.11			
Furan	g	−8.23	0.21	63.86	15.64
Furfural	liq	−47.8			
Furfuryl alcohol	liq	−66.05	−36.85	51.50	
2-Furoic acid (pyromucic acid)	c	−119.12			
Furylacrylic acid	c	−109.7			
Furylethylene	liq	−2.5			
D-Galactonic acid	c	−384.8			
D-Galactose	c	−304.1	−219.60	49.1	
D-Glucaric acid-1,4-lactone	c	−343.2			
D-Glucaric acid-3,6-lactone	c	−343.6			
D-Gluconic acid	c	−379.3			
D-Gluconic acid-δ-lactone	c	−300.3			

TABLE 5.1 Enthalpies and Gibbs (Free) Energies of Formation, Entropies, and Heat Capacities of Organic Compounds (*continued*)

Substance	State	$\Delta Hf°$, kcal·mol^{-1}	$\Delta Gf°$, kcal·mol^{-1}	$S°$, cal·deg^{-1}· mol^{-1}	$C_p°$, cal·deg^{-1}· mol^{-1}
D-Glucose					
α	c	−304.26	−217.6	50.7	
β	c	−302.76			
D-Glutamic acid	c	−240.19	−173.87	45.7	
L-Glutamic acid	c	−241.32	−174.78	44.98	
L-Glutamine	c	−197.3			
Glutaric acid	c	−229.44			
Glyceraldehyde	liq	−143			
Glycerol	liq	−159.76	−114.01	48.87	35.9
Glyceryl-1-acetate	liq	−217.5			
Glyceryl-l-benzoate	c	−185.80			
Glyceryl-2-benzoate	c	−184.71			
Glyceryl-1-caprate	c	−265.05			
Glyceryl-2-caprate	c	−261.90			
Glyceryl-1,3-diacetate	liq	−268.2			
Glyceryl-1-laurate	c	−277.46			
Glyceryl-2-laurate	c	−275.48			
Glyceryl-2-myristate	c	−292.31			
Glyceryl-1-palmitate	c	−306.28			
Glyceryl-1-stearate	c	−319.64			
Glyceryl triacetate	liq	−318.3			
Glyceryl trilaurate	c	−489			
Glyceryl trimyristate	c	−520			
Glyceryl trinitrate	liq	−88.6			
Glycine	c	−126.22	−88.09	24.74	23.71
Ionized; std state, $m = 1$	aq	−112.28	−75.28	26.54	
Nonionized; std state, $m = 1$	aq	−122.85	−88.62	37.84	
NH$_3$$^+CH_2$COOH; std state, $m = 1$	aq	−123.78	−91.82	45.46	
Glycol acetal	liq	−91.1			
Glycolic acid (hydroxyacetic acid)	c	−158.6			
Glycylglycine	c	−178.51	−117.25	45.4	
Glycylphenylalanine	c	−163.9			
Glycylvaline	c	−200.0			
Glyoxal	g	−50.66			
Glyoxime	c	−21.63			
Glyoxylic acid	c	−199.7			
Guanidine	c	−13.39			
Guanidine carbonate	c	−232.10	−133.23	70.6	61.87
Guanidine nitrate	c	−92.5			
Guanidine sulfate	c	−288.0			
Guanine	c	−43.72	11.33	38.3	

TABLE 5.1   Enthalpies and Gibbs (Free) Energies of Formation, Entropies, and Heat Capacities of Organic Compounds (*continued*)

Substance	State	$\Delta Hf°$, kcal·mol^{-1}	$\Delta Gf°$, kcal·mol^{-1}	$S°$, cal·deg^{-1}· mol^{-1}	$C_p°$, cal·deg^{-1}· mol^{-1}
Guanylurea nitrate	c	−102.1			
Heptadecane	g	−94.15	22.01	195.33	94.33
Heptadecanoic acid	c	−220.9			
1-Heptadecene	g	−64.15	43.00	194.33	91.76
1-Heptanal	g	−63.10	−20.71	110.34	40.89
Heptane	liq	−53.63	0.42	77.92	53.76
Heptanedioic acid	g	−44.88	1.91	102.27	39.67
1-Heptanethiol	g	−35.76	8.65	117.89	44.68
	liq	−145.75			
Heptanoic acid (enanthic acid)	c	−241.75			
1-Heptanol	liq	−95.8	−34.0	76.5	66.5
	g	−79.3	−28.9	114.8	42.7
1-Heptene	liq	−23.41	21.22	78.31	50.62
	g	−14.89	22.90	101.24	37.10
1-Heptyne	g	24.62	54.18	97.44	36.11
Hexachlorobenzene	c	−31.30	0.25	62.20	48.11
	g	−8.10	10.56	105.45	41.40
Hexachloroethane	g	−33.20	−13.13	95.30	32.68
Hexadecafluoroethyl-cyclohexane	liq	−799.1			
Hexadecafluoroheptane	liq	−817.6	−739.24	134.28	
	g	−808.9	−737.87	158.88	
Hexadecane	g	−89.23	20.00	186.02	88.86
Hexadecanoic acid (palmitic acid)	c	−213.3	−75.54	108.12	
1-Hexadecanol (cetyl alcohol)	c, II	−163.4	−23.6	108.0	104.8
	liq	−151.86	−23.08	145.0	
1-Hexadecene	g	−59.23	40.99	185.02	86.29
Hexafluorobenzene	liq	−237.27	−211.43	66.90	52.96
	g	−228.64	−210.18	91.59	37.43
Hexafluoroethane	g	−320.90	−300.15	79.30	25.43
Hexahydroindane					
cis	g	−30.4			
trans	g	−31.4			
Hexamethylbenzene	c	−39.19	28.06	71.66	61.5
Hexamethyldisiloxane	liq	−194.7	−129.5	103.69	74.42
Hexamethylenetetramine (urotropine)	c	30.0	103.92	39.05	
	liq	−18.7	28.65	73.28	
Hexanal	g	−59.37	−23.93	101.07	35.43
Hexanamide	c	−101.48			
Hexane	liq	−47.52	−0.91	70.76	45.2
	g	−39.96	−0.06	92.83	34.20
1-Hexanethiol	g	−30.83	6.65	108.58	39.21
Hexanoic acid	liq	−139.71			

TABLE 5.1 Enthalpies and Gibbs (Free) Energies of Formation, Entropies, and Heat Capacities of Organic Compounds (*continued*)

Substance	State	$\Delta Hf°$, kcal·mol^{-1}	$\Delta Gf°$, kcal·mol^{-1}	$S°$, cal·deg^{-1}· mol^{-1}	$C_p°$, cal·deg^{-1}· mol^{-1}
1-Hexanol	liq	−90.7	−36.4	69.2	56.6
	g	−75.9	−32.4	105.5	37.2
1-Hexene	liq	−17.30	19.93	70.55	43.81
	g	−9.96	20.90	91.93	31.63
2-Hexene					
cis	g	−12.51	18.22	92.37	30.04
trans	g	−12.27	18.27	90.97	31.64
3-Hexene					
cis	g	−11.38	19.84	90.73	29.55
trans	g	−13.01	18.55	89.59	31.75
1-Hexyne	g	29.55	52.24	88.13	30.65
Hippuric acid (benzoylglycine)	c	−145.63	−88.33	57.2	
Hydantoic acid	c	−179			
Hydantoin	c	−107.2			
Hydrazobenzene	c	52.9			
Hydroquinone	c	−87.08	−49.48	33.5	33.9
Hydrosorbic acid	liq	−110.2			
Hydroxyacetic acid	c	−158.6			
o-Hydroxybenzoic acid	c	−140.64	−100.7	42.6	38.03
m-Hydroxybenzoic acid	c	−139.8	−99.74	42.3	37.59
p-Hydroxybenzoic acid	c	−139.7	−99.55	42.0	37.08
β-Hydroxybutyric acid	liq	−162.3			
Hydroxyisobutyric acid	c	−177.9			
L-Hydroxyproline	c	−158.1			
8-Hydroxyquinoline	c	−19.9			
Hypoxanthene (6-oxypurine)	c	−26.47	18.39	34.8	
Imidazole	c	14.5			
Indane	liq	2.56	36.04	56.01	45.47
Indene	liq	26.39	52.00	51.19	44.68
Indole	c	29.8			
Iodobenzene	g	38.85	44.88	79.84	24.08
2-Iodobenzoic acid	c	−72.2			
3-Iodobenzoic acid	c	−75.7			
4-Iodobenzoic acid	c	−75.5			
Iodocyclohexane	liq	−23.5			
Iodoethane	liq	−9.6	3.5	50.6	27.5
	g	−2.00	5.10	70.82	15.76
Iodomethane	liq	−3.29	3.61	38.9	
	g	3.29	3.72	60.64	10.54
2-Iodo-2-methylpropane	g	−17.60	5.65	81.79	28.27
1-Iodonaphthalene	liq	38.6			
2-Iodonaphthalene	c	34.5			
2-Iodophenol	c	−22.9			

TABLE 5.1 Enthalpies and Gibbs (Free) Energies of Formation, Entropies, and Heat Capacities of Organic Compounds (*continued*)

Substance	State	$\Delta Hf°$, kcal·mol^{-1}	$\Delta Gf°$, kcal·mol^{-1}	$S°$, cal·deg^{-1}· mol^{-1}	$C_p°$, cal·deg^{-1}· mol^{-1}
3-Iodophenol	c	−22.6			
4-Iodophenol	c	−22.8			
1-Iodopropane	g	−7.30	6.68	80.32	21.48
2-Iodopropane	g	−10.00	4.80	77.55	21.53
3-Iodopropene (allyl iodide)	liq	13.7			
3-Iodopropionic acid	c	−109.9			
2-Iodotoluene	liq	18.7			
3-Iodotoluene	liq	18.9			
4-Iodotoluene	liq	16.1			
Isatin	c	−62.7			
Isobutylbenzene	liq	−16.68			
Isobutyl dichloracetate	liq	−132.4			
Isobutyl phenyl ketone	liq	−52.63			
Isobutyl trichloroacetate	liq	−132.4			
Isobutyronitrile	g	6.07	24.76	74.88	23.04
L-Isoleucine	c	−151.8	−82.97	49.71	45.00
Isopropenyl acetate	liq	−92.31			
Isopropyl radical	g	17.6			
Isopropyl acetate	liq	−124.01			
Isopropylbenzene (cumene)	liq	−9.85	29.70	66.87	
	g	0.94	32.74	92.87	36.26
Isopropyl nitrate	g	−45.65	−9.72	89.20	28.84
Isopropyl thiolacetate	liq	−71.26			
Isopropyl trichloroacetate	liq	−128.2			
Isoquinoline	c	37.9			
L-Isoserine	c	−177.8			
Isothiocyanic acid	g	30.50	26.98	59.28	11.09
Itaconic acid	c	−201.06			
Ketene	g	−14.60	−14.41	57.79	12.37
α-Ketoglutaric acid	c	−245.35			
D-Lactic acid	c	−165.88		34.3	
L-Lactic acid	c	−165.89	−124.98	34.00	
	liq	−161.2	−123.84	45.9	
β-Lactose	c	−534.1	−374.52	92.3	
Lauric acid (dodecanoic acid)	c	−185.14			
D-Leucine	c	−152.36	−82.97	49.71	
L-Leucine	c	−154.6	−82.76	50.62	48.03
DL-Leucine	c	−153.14	−83.54	49.5	
DL-Leucylglycine	c	−205.7	−112.14	67.2	
Leucylglycylglycine	c	−259.6			
Levulinic acid	c	−166.6			
Levulinic lactone	liq	−76.2			
(+)-Limonene	liq	−13.0			
DL-Lysine	c	−162.2			

TABLE 5.1   Enthalpies and Gibbs (Free) Energies of Formation, Entropies, and Heat Capacities of Organic Compounds (*continued*)

Substance	State	$\Delta H f°$, kcal·mol^{-1}	$\Delta G f°$, kcal·mol^{-1}	$S°$, cal·deg^{-1}· mol^{-1}	$C_p°$, cal·deg^{-1}· mol^{-1}
Maleic acid	c	−188.94	−149.40	38.1	32.36
Maleic anhydride	c	−112.08			
L-Malic acid	c	−263.78	−211.45		
DL-Malic acid	c	−264.27			
Malonamide	c	−130.5			
Malonic acid	c	−212.96			
Malonic diamide	c	−130.52			
Malononitrile	c	44.6			
Maltose	c	−530.8	−412.60		
L-Mandelic acid	c	−138.8			
D-Mannitol	c	−139.61	−225.20	57.0	
D-Mannose	c	−301.9			
Melamine	c	−17.3	44.10	35.63	
(triaminotraizine)					
Melezitose	c	−815			
2-Mercaptopropionic acid	liq	−111.9	−82.19	54.70	
Mesaconic acid	c	−197			
Mesoxalic acid	c	−290.7			
2,2-Metacyclophane	g	40.8			
Methane	g	−17.89	−12.15	44.52	8.54
Methanethiol	g	−5.49	−2.37	60.96	12.01
(methyl mercaptan)					
Methanol	liq	−57.13	−39.87	30.41	19.40
	g	−48.06	−38.82	57.29	10.49
Std state, $m = 1$	aq	−58.78			
L-Methionine	c	−180.4	−120.88	55.32	
Methoxyl radical	g	(2)			
2-Methoxybenzaldehyde	c	−63.7			
3-Methoxybenzaldehyde	liq	−66.0			
4-Methoxybenzaldehyde	liq	−63.9			
Methoxybenzene (anisole)	g	−17.3			
Methoxymethyl radical	g	(−4)			
2-Methoxytetrahydropyran	liq	−105.7			
5-Methoxytetrazole	c	16.6			
Methyl (CH_3) radical	g	34.82	35.35	46.38	9.25
Methyl acetate	liq	−106.4			
Methyl acrylate	g	−70.10	−56.78		
Methyl allantoin (pyvurile)	c	−177.0			
Methyl allyl sulfone	liq	−91.95			
Methylamido radical	g	35			
(CH_3NH)					
Methylamine	g	−5.50	7.71	57.98	11.97
Std state, $m = 1$	aq	−16.77	4.94	29.5	
Methylaminolithium	c	−22.92			
N-Methylaniline	liq	7.7			

TABLE 5.1 Enthalpies and Gibbs (Free) Energies of Formation, Entropies, and Heat Capacities of Organic Compounds (*continued*)

Substance	State	$\Delta Hf°$, kcal·mol^{-1}	$\Delta Gf°$, kcal·mol^{-1}	$S°$, cal·deg^{-1}· mol^{-1}	$C_p°$, cal·deg^{-1}· mol^{-1}
Methyl benzoate	liq	−79.8			
Methyl benzyl sulfone	c	−88.65			
2-Methylbiphenyl	liq	25.8			
3-Methylbiphenyl	liq	20.4			
4-Methylbiphenyl	c	13.2			
2-Methyl-1,3-butadiene (isoprene)	g	18.10	34.86	75.44	25.00
3-Methyl-1,2-butadiene	g	31.00	47.47	76.40	25.20
2-Methylbutane	g	−36.92	−3.54	82.12	28.39
3-Methyl-1-butanethiol	g	−27.44			
2-Methyl-2-butanethiol	liq	−38.90	0.56	69.34	
	g	−30.36	2.20	92.48	34.30
2-Methyl-1-butanol	liq	−85.2			52.6
3-Methyl-1-butanol	liq	−85.2			50.3
2-Methyl-2-butanol	liq	−90.7	−41.9	54.8	59.2
	g	−78.8	−39.5	86.7	
3-Methyl-2-butanol	liq	−87.5			55.5
2-Methyl-1-butene	g	−8.68	15.68	81.15	26.28
3-Methyl-1-butene	g	−6.92	17.87	79.70	28.35
2-Methyl-2-butene	g	−10.17	14.26	80.92	25.10
Methyl butyl sulfone	liq	−128.00			
Methyl *tert*-butyl sulfone	c	−132.8			
3-Methyl-1-butyne	g	32.60	49.12	76.23	25.02
Methyl caprate	liq	−153.07			
Methyl caproate (methyl hexanoate)	liq	−129.10			
N-Methylcaprolactam	liq	−73.3			
5-Methylcaprolactam	c	−86.9			
7-Methylcaprolactam	c	−86.5			
Methyl caprylate (methyl octanoate)	liq	−141.07			
Methyl chloride	g	−19.59	−13.97	55.97	9.74
Methyl crotonate	liq	−91.5			
Methylcyclohexane	liq	−45.45	4.86	59.26	
	g	−36.99	6.52	82.06	32.27
2-Methylcyclohexanol					
cis	liq	−93.3			
trans	liq	−99.4			
3-Methylcyclohexanol					
cis	liq	−99.5			
trans	liq	−94.3			
4-Methylcyclohexanol					
cis	liq	−98.8			
trans	liq	−103.6			
2-Methylcyclohexanone	liq	−68.8			
Methylcyclopentane	g	−25.50	8.55	* 81.24	26.24

TABLE 5.1 Enthalpies and Gibbs (Free) Energies of Formation, Entropies, and Heat Capacities of Organic Compounds (*continued*)

Substance	State	$\Delta Hf°$, kcal·mol^{-1}	$\Delta Gf°$, kcal·mol^{-1}	$S°$, cal·deg^{-1}· mol^{-1}	$C_p°$, cal·deg^{-1}· mol^{-1}
1-Methylcyclopentanol	liq	−82.3			
2-Methylcyclopentanone	liq	−63.4			
1-Methylcyclopentene	g	−1.30	24.41	78.00	24.10
3-Methylcyclopentene	g	2.07	27.48	79.00	23.90
4-Methylcyclopentene	g	3.53	29.06	78.60	23.90
Methyldichlorosilane	liq	−105.9			
2-Methyl-1,3-dioxane	liq	−104.60			
4-Methyl-1,3-dioxane	liq	−99.80			
N-Methyldiphenylamine	liq	28.8			
4-Methyldiphenylamine	c	11.7			
Methylene	g	92.35	88.25	46.32	8.27
2-Methylenecyclohexanol	liq	−66.3			
2-Methylenecyclopentanol	liq	11.2			
β-Methylene-β-propio-lactone (diketene)	liq	−55.72			
Methylene sulfate	c	−164.6			
1-Methyl-2-ethylbenzene	g	0.29	31.33	95.42	37.74
1-Methyl-3-ethylbenzene	g	−0.46	30.22	96.60	36.38
1-Methyl-4-ethylbenzene	g	−0.78	30.28	95.34	36.22
2-Methyl-3-ethylpentane	liq	−59.69	3.03	81.41	
	g	−50.48	5.08	105.43	
3-Methyl-3-ethylpentane	liq	−60.46	2.69	79.97	
	g	−51.28	4.76	103.48	
2-Methyl-3-ethyl-1-pentene	g	−23.97			
Methyl ethyl sulfite	liq	−135.55			
Methyl ethyl sulfone	c	−116.17			
Methyl formate	liq	−90.60	−71.53	29	
	g	−83.70	−71.03	72.00	15.90
Methylglyoxal	g	−64.8			
Methylglyoxime	c	−30.3			
2-Methylheptane	liq	−60.98	0.92	84.16	
	g	−51.50	3.05	108.81	
3-Methylheptane	liq	−60.34	1.12	85.66	
	g	−50.82	3.28	110.32	
4-Methylheptane	liq	−60.17	1.87	83.72	
	g	−50.69	4.00	108.35	
Methyl heptanoate	liq	−135.54			
2-Methylhexane	liq	−54.93	−0.69	77.28	53.28
	g	−46.59	0.77	100.38	39.67
3-Methylhexane	liq	−54.35	−0.39	78.23	
	g	−45.96	1.10	101.37	39.67
Methyl hexanoate	liq	−129.11			
5-Methylhydantoin	c	−116.3			
Methylhydrazine	liq	12.9	43.0	39.66	32.25
	g	22.55	44.66	66.61	17.0

TABLE 5.1 Enthalpies and Gibbs (Free) Energies of Formation, Entropies, and Heat Capacities of Organic Compounds (*continued*)

Substance	State	$\Delta Hf°$, kcal·mol^{-1}	$\Delta Gf°$, kcal·mol^{-1}	$S°$, cal·deg^{-1}· mol^{-1}	$C_p°$, cal·deg^{-1}· mol^{-1}
Methylidyne					
CH	g	142.00	134.02	43.72	6.97
CH$^+$	g	388.8	380.1	41.00	6.97
α-Methylindole	c	14.5			
Methyl isocyanide	g	35.6	39.6	58.99	12.65
1-Methyl-2-isopropyl-benzene (*o*-cymene)	liq	−18.19			
1-Methyl-3-isopropyl-benzene	liq	−18.69			
Methyl isopropyl ether	g	−60.24	−28.89	80.86	26.55
Methyl isopropyl ketone	g	−62.76			
Methyl isopropyl sulfone	liq	−120.44			
Methyl isothiocyanate (CH$_3$NCS)	g	31.3	34.5	69.29	15.65
3-Methylisoxazole	liq	−5.0			
5-Methylisoxazole	liq	−6.4			
Methyl laurate	liq	−165.66			
Methylmercury bromide	c	−20.6			
Methylmercury chloride	c	−27.8			
Methylmercury iodide	c	−10.4			
Methyl myristate	liq	−177.80			
1-Methylnaphthalene	liq	13.43	46.26	60.90	53.63
2-Methylnaphthalene	c	10.72	46.03	52.58	46.84
Methyl nitrate	liq	−38.0	−10.4	51.9	37.6
	g	−29.8	−9.4	76.1	
Methyl nitrite	g	−15.30	0.24	67.95	15.11
Methyl oleate	liq	−174.2			
Methyl pelargonate	liq	−147.29			
2-Methylpentane	g	−41.66	−1.20	90.95	34.46
3-Methylpentane	g	−41.02	−0.51	90.77	34.20
Methyl pentanoate	liq	−122.90			
2-Methyl-1-pentene	g	−12.49	18.55	91.34	32.41
3-Methyl-1-pentene	g	−10.76	20.66	90.06	34.04
4-Methyl-1-pentene	g	−10.54	21.52	87.89	30.23
2-Methyl-2-pentene	g	−14.28	17.02	90.45	30.26
3-Methyl-2-pentene					
cis	g	−13.80	17.50	90.45	30.26
trans	g	−14.02	17.04	91.26	30.26
4-Methyl-2-pentene					
cis	g	−12.03	19.63	89.23	31.92
trans	g	−12.99	19.03	88.02	33.80
Methyl pentanoate	liq	−122.89			
Methyl phenyl sulfone	c	−82.49			
(2-Methyl phenol)	g	−30.74	−8.86	85.47	31.15

TABLE 5.1 Enthalpies and Gibbs (Free) Energies of Formation, Entropies, and Heat Capacities of Organic Compounds (*continued*)

Substance	State	$\Delta Hf°$, kcal·mol^{-1}	$\Delta Gf°$, kcal·mol^{-1}	$S°$, cal·deg^{-1}· mol^{-1}	$C_p°$, cal·deg^{-1}· mol^{-1}
(3-Methyl phenol)	g	−31.63	−9.69	85.27	29.27
(4-Methyl phenol)	g	−29.97	−7.38	83.09	29.75
Methylphosphonic acid	c	−252			
2-Methylpropanal	g	−52.25			
2-Methylpropane	g	−32.15	−4.99	70.42	23.14
2-Methyl-1,2-propanediamine	liq	−32.00			
2-Methyl-1-propanethiol	g	−23.24	1.33	86.73	28.28
2-Methyl-2-propanethiol	g	−26.17	0.17	80.79	28.91
2-Methyl-1-propanol	g	−67.69	−39.99	85.81	26.6
2-Methyl-2-propanol	liq	−85.86	−44.14	46.10	52.61
	g	−74.67	−42.46	77.98	27.10
2-Methylpropene	g	−4.04	13.88	70.17	21.30
Methyl propyl ether	g	−56.82	−26.27	83.52	26.89
7-Methylpurine	c	51.3			
2-Methylpyridine (2-picoline)	liq	13.83	39.80	52.07	37.86
	g	24.05	42.32	77.68	23.90
3-Methylpyridine	liq	15.57	41.16	51.70	37.93
4-Methylpyridine	liq	13.58			
N-Methylpyrrolidone	liq	−62.64			
Methyl salicylate	liq	−127.1			
α-Methylstyrene	liq	16.8			
	g	27.00	49.84	91.70	34.70
β-Methylstyrene					
cis	g	29.00	51.84	91.70	34.70
trans	g	28.00	51.08	90.90	34.90
Methylsuccinic acid	c	−229.02			
3-Methyl-2-thiabutane	g	−21.61	3.21	85.87	28.00
2-Methylthiacyclopentane	g	−15.12			
2-Methyl-3-thiapentane	liq	−37.3			
4-Methylthiazole	liq	16.31			
2-Methylthiophene	liq	10.75	27.35	52.22	29.43
3-Methylthiophene	liq	10.38	27.00	52.19	29.38
4-Methyluracil	c	−109.2			
Methyl valerate (methyl pentanoate)	liq	−122.89			
Morphine monohydrate	c	−170.1			
Mucic acid	c	−423			
Murexide	c	−289.7			
Myrcene	liq	3.5			
Myristic acid (tetradecanoic acid)	c	−199.21			
Naphthalene	c	18.0	48.05	39.89	
	g	35.6	53.44	80.22	31.68
1-Naphthol	g	−5.1			

TABLE 5.1 Enthalpies and Gibbs (Free) Energies of Formation, Entropies, and Heat Capacities of Organic Compounds (*continued*)

Substance	State	$\Delta Hf°$, kcal·mol^{-1}	$\Delta Gf°$, kcal·mol^{-1}	$S°$, cal·deg^{-1}· mol^{-1}	$C_p°$, cal·deg^{-1}· mol^{-1}
2-Naphthol	g	−10.1			
1,4-Naphthoquinone	c	−43.83			
1-Naphthyl acetate	c	−68.89			
2-Naphthyl acetate	c	−72.72			
1-Naphthylamine	c	16.2			
2-Naphthylamine	c	14.4			
Narceine dihydrate	c	−421.2			
Narcotine	c	−210.9			
Nicotine	liq	9.4			
Nitrilotriacetic acid	c		−312.5		
2-Nitroaniline	c	−3.45	42.60	42.1	39.3
3-Nitroaniline	c	−4.46	41.60	42.1	40.2
4-Nitroaniline	c	−9.91	36.10	42.1	40.4
Nitrobenzene	liq	3.80	34.95	53.6	44.4
2-Nitrobenzoic acid	c	−94.25	−46.95	49.8	
3-Nitrobenzoic acid	c	−100.25	−52.71	49.0	
4-Nitrobenzoic acid	c	−101.25	−53.07	50.2	43.3
3-Nitrobiphenyl	c	15.6			
4-Nitrobiphenyl	c	9.7			
1-Nitrobutane	g	−34.40	2.42	94.28	29.85
2-Nitrobutane	g	−39.10	−1.49	91.62	29.51
3-Nitro-2-butanol	liq	−93.2			
2-Nitrodiphenylamine	c	15.4			
Nitroethane	g	−24.4	−1.17	75.39	18.69
aci form	aq	−30.7			
nitro form	aq	−32			
2-Nitroethanol	liq	−83.8			
Nitroguanidine	c	−22.1			
Nitromethane	liq	−27.03	−3.47	41.05	25.33
	g	−17.86	−1.66	65.73	13.70
1-Nitronaphthalene	c	10.2			
1-Nitropropane	g	−30.00	0.08	85.00	24.41
2-Nitropropane	g	−33.21	−3.06	83.10	24.26
4-Nitrosodiphenylamine	c	50.9			
Nonadecane	g	−104.00	26.03	213.95	105.26
1-Nonadecene	g	−74.00	47.02	212.95	102.69
1-Nonanal	g	−74.16	−17.91	128.97	51.82
Nonane	liq	−65.84	2.81	94.09	
	g	−54.74	5.93	120.86	50.60
1-Nonanethiol	g	−45.61	12.67	136.51	55.61
Nonanoic acid	liq	−157.68			
1-Nonanol	liq	−109.2	−32.4	91.3	67.50
1-Nonene	g	−24.74	26.93	119.86	48.03
Octadecane	g	−99.08	24.02	204.64	99.80
Octadecanoic acid	c	−226.5			

TABLE 5.1 Enthalpies and Gibbs (Free) Energies of Formation, Entropies, and Heat Capacities of Organic Compounds (*continued*)

Substance	State	$\Delta Hf°$, kcal·mol^{-1}	$\Delta Gf°$, kcal·mol^{-1}	$S°$, cal·deg^{-1}· mol^{-1}	$C_p°$, cal·deg^{-1}· mol^{-1}
1-Octadecene	g	−69.08	45.01	203.64	97.22
Octafluorocyclobutane	g	−365.20	−334.33	95.69	37.32
1-Octanal	g	−69.23	−19.91	119.66	46.36
Octanamide	c	−113.1			
Octane	liq	−59.74	1.77	85.50	45.14
	g	−49.82	3.92	111.55	45.14
1-Octanethiol	g	−40.68	10.67	127.20	50.14
Octanoic acid (caprylic acid)	liq	−151.93			
1-Octanol	liq	−101.6	−34.2	90.2	77.7
2-Octanone	liq	−91.9	−33.54	89.35	65.31
1-Octene	liq	−29.52	22.49	86.15	57.65
	g	−19.82	24.91	110.55	42.56
1-Octyne	g	19.70	56.26	106.75	41.58
Oleic acid	c	−187.2			
DL-Ornithine	c	−156.0			
Oxacyclobutane (trimethylene oxide)	g	−19.25	−2.33	65.46	
Oxalic acid	c	−197.7	−166.8	28.7	
Std state, $m = 1$	aq	−197.2	−161.1	10.9	
Oxalic acid dihydrate	c	−341.0			
Oxalyl chloride	liq	−85.6			
Oxamic acid	c	−160.4			
Oxamide	c	−123.0	−81.9	28.2	
Oxindole	c	−41.2			
8-Oxypurine	c	−15.4			
Palmitic acid (hexadecanoic acid)	c	−213.10			
Papaverine	c	−120.2			
Parabanic acid	c	−138.0			
[1,8]-Paracyclophane	c	−19.6			
[2,2]-Paracyclophane	g	59.9			
[6,6]-Paracyclophane	c	−46.1			
Paraldehyde	liq	−164.2			
Pentachloroethane	g	−34.8	−16.79	91.17	28.22
Pentachlorofluoroethane	g	−75.8	−55.93	93.54	
Pentachlorophenol	c	−70.6	−34.44	60.21	48.27
Pentadecane	g	−84.31	17.98	176.71	83.40
1-Pentadecene	g	−54.31	38.97	175.71	80.82
1-Pentadecyne	g	−14.78	70.25	171.91	79.84
1,2-Pentadiene	g	34.80	50.29	79.70	25.20
1,3-Pentadiene					
cis	g	18.70	34.84	77.50	22.60
trans	g	18.60	35.07	76.40	24.70
1,4-Pentadiene	g	25.20	40.69	79.70	25.10

TABLE 5.1   Enthalpies and Gibbs (Free) Energies of Formation, Entropies, and Heat Capacities of Organic Compounds (*continued*)

Substance	State	$\Delta Hf°$, kcal·mol^{-1}	$\Delta Gf°$, kcal·mol^{-1}	$S°$, cal·deg^{-1}· mol^{-1}	$C_p°$, cal·deg^{-1}· mol^{-1}
2,3-Pentadiene	g	33.10	49.21	77.60	24.20
Pentaerythritol	c	−220.0	−146.73	47.34	45.51
Pentaerythritol tetranitrate	c	−128.8			
Pentafluorobenzoic acid	c	−296.34			
Pentafluoroethane	g	−264.00	−246.00	79.76	22.88
Pentafluorophenol	c	−244.86			
Pentamethylbenzene	liq	−32.33	25.64	70.22	51.74
Pentamethylbenzoic acid	c	−128.13			
1-Pentanal	g	−54.45	−25.88	91.53	29.96
Pentanamide	c	−90.70			
Pentan-2,4-dione	liq	−101.33			
(acetylacetone)	g	−90.47		95.1	28.7
Pentan-1,5-dithiol	liq	−30.99			
Pentane	g	−35.00	−2.00	83.40	28.73
1-Pentanethiol	liq	−35.72	2.28	74.18	
Pentanoic acid	liq	−133.71	−89.10	62.10	50.48
(valeric acid)					
1-Pentanol	liq	−85.0	−38.3	62.0	49.8
2-Pentanol	liq	−87.7			
3-Pentanol	liq	−88.5	−40.4	57.4	60.0
2-Pentanone	g	−61.82	−32.76	89.91	28.91
3-Pentanone	liq	−70.87			
1-Pentene	g	−5.00	18.91	82.65	26.19
2-Pentene					
cis	g	−6.71	17.17	82.76	24.32
trans	g	−7.59	16.71	81.36	25.92
2-Pentenoic acid	liq	−106.7			
3-Pentenoic acid	liq	−103.9			
4-Pentenoic acid	liq	−102.9			
1-Pentyne	g	34.50	50.25	78.82	25.50
2-Pentyne	g	30.80	46.41	79.30	23.59
Perfluoropiperidine	liq	−482.9	−422.67	94.02	70.93
Perylene	c	43.69			
α-Phellandrene	liq	−14.3			
Phenacetin	c	−101.1			
9,10-Phenanthraquinone	c	−55.18			
Phenanthrene	c	27.3	64.12	50.6	
Phenazine	c	56.4			
Phenol	c	−39.44	−12.05	34.42	32.2
	liq	−37.80	−11.02		30.46
	g	−23.03	−7.86	75.43	24.75
Phenoxy radical	g	10			
Phenoxyacetic acid	c	−122.8			
Phenyl radical	g	71			
Phenyl acetate	liq	−80.02			

TABLE 5.1 Enthalpies and Gibbs (Free) Energies of Formation, Entropies, and Heat Capacities of Organic Compounds (*continued*)

Substance	State	$\Delta Hf°$, kcal·mol^{-1}	$\Delta Gf°$, kcal·mol^{-1}	$S°$, cal·deg^{-1}·mol^{-1}	$C_p°$, cal·deg^{-1}·mol^{-1}
Phenylacetic acid	c	−95.3			
Phenylacetylene	g	78.22	86.46	76.88	27.46
β-Phenyl-1-alanine, DL- and L-	c	−111.9	−50.6	51.06	48.52
Phenyl benzoate	c	−57.7			
2-Phenylbenzoic acid	c	−83.4			
Phenylboronic acid	c	−172.0			
1-Phenylcyclohexene	liq	−4.0			
Phenylcyclopropane	liq	24.7			
N-Phenyldiacetimide	c	−86.63			
p-Phenylenediamine	c	0.73			
Phenyl ethyl sulfide	liq	5.29			
DL-Phenylglyceric acid	c	−178.5			
N-Phenylglycine	c	−96.2			
a-Phenylglycine	c	−103.2			
Phenylglyoxime					
α	c	−4.9			
β	c	10.1			
Phenylglyoxylic acid	c	−115.3			
Phenylhydrazine	liq	34.03			
Phenyl methyl sulfide	liq	11.5			
N-Phenyl-2-naphthylamine	c	38.2			
N-Phenylpyrrole	c	38.1			
2-Phenylpyrrole	c	34.5			
Phenyl salicylate	c	−104.3			
Phenyl thiolacetate	liq	−29.16			
Phosgene	g	−52.80	−49.42	67.82	13.79
Phthalamide	c	−104.4			
m-Phthalic acid	c	−191.91			
o-Phthalic acid	c	−186.91	−141.39	49.7	45.0
p-Phthalic acid	c	−195.05			
Phthalic anhydride	c	−110.1	−79.12	42.9	38.5
Phthalonitrile	c	65.82			
Pimelic acid (heptanedioic acid)	c	−241.25			
Pinene					
α	liq	−3.9			
β	liq	−1.8			
Piperazine	c	−10.90			
Piperidine	liq	−21.05			
α-Piperidone	c	−73.3	−26.79	39.4	
DL-Proline	c	−125.7			
Propadiene	g	45.92	48.37	58.30	14.10
Propane	g	−24.82	−5.63	64.58	17.59
1,2-Propanediamine	liq	−23.38			

TABLE 5.1 Enthalpies and Gibbs (Free) Energies of Formation, Entropies, and Heat Capacities of Organic Compounds (*continued*)

Substance	State	$\Delta Hf°$, kcal·mol^{-1}	$\Delta Gf°$, kcal·mol^{-1}	$S°$, cal·deg^{-1}· mol^{-1}	$C_p°$, cal·deg^{-1}· mol^{-1}
1,2-Propanediol	liq	−119.6			
1,3-Propanediol	liq	−124.4			
1,3-Propanedithiol	liq	−18.83			
2,3-Propanedithiol	liq	−18.82			
1-Propanethiol	g	−16.22	0.52	80.40	22.65
2-Propanethiol	g	−18.22	−0.61	77.51	22.94
1-Propanol	liq	−72.66	−40.78	46.5	33.7
	g	−61.28	−38.67	77.61	20.82
2-Propanol	liq	−75.97	−43.09	43.16	36.06
	g	−65.11	−41.44	74.07	21.21
1,2,3-Propenetricarboxylic acid					
cis	c	−292.7			
trans	c	−294.7			
2-Propen-1-ol (allyl alcohol)	g	−31.55	−17.03	73.51	18.17
Propionaldehyde	g	−45.90	−31.18	72.83	18.80
Propionamide	c	−81.7			
Propionic acid	liq	−122.07	−91.65		
Propionic anhydride	liq	−161.53	−113.66		
Propionitrile	liq	3.5	21.31	45.25	
	g	12.10	22.98	68.50	17.46
1-Propylamine	g	−17.30	9.51	77.48	22.89
2-Propylamine	liq	−26.83			
Propylbenzene	g	1.87	32.80	95.76	36.41
Propylcarbamate	c	−132.07			
Propyl chloroacetate	liq	−123.3			
Propylcyclohexane	g	−46.20	11.31	100.27	44.03
Propylcyclopentane	g	−35.39	12.57	99.73	36.96
Propylene (propene)	g	4.88	15.02	63.72	15.37
Propylene oxide	g	−22.17	−6.16	68.53	17.29
Propyl nitrate	g	−41.60	−6.53	92.10	28.99
Propyl phenyl ketone	liq	−45.14			
Propyl thiolacetate	liq	−70.29			
Propyl trichloroacetate	liq	−122.7			
Propyne (methyl acetylene)	g	44.32	46.47	59.30	14.50
Pyrazine	c	33.41			
Pyrazole	c	28.3			
Pyrene	c	27.44	64.40	53.75	56.4
Pyridazine	liq	53.74			
Pyridine	liq	23.96	43.34	42.52	31.72
	g	33.61	45.46	67.59	18.67
Pyrimidine	liq	35.04			
Pyrrole	liq	15.08			

TABLE 5.1 Enthalpies and Gibbs (Free) Energies of Formation, Entropies, and Heat Capacities of Organic Compounds (*continued*)

Substance	State	$\Delta Hf°$, kcal·mol^{-1}	$\Delta Gf°$, kcal·mol^{-1}	$S°$, cal·deg^{-1}· mol^{-1}	$C_p°$, cal·deg^{-1}· mol^{-1}
Pyrrole-2-aldehyde	c	−24.8			
Pyrrole-2-aldoxime	c	2.9			
Pyrrolidine	liq	−9.84	25.94	48.76	
	g	−0.86	27.41	73.97	19.39
2-Pyrrolidone	c	−68.3			
Pyruvic acid	liq	−139.7	−110.75	42.9	
Quinaldine	c	39.3			
Quinhydrone	c	−19.79	−77.19	77.9	66.2
Quinidine	c	−38.3			
Quinine	c	−37.1			
Quinoline	liq	37.33	65.90	51.9	
p-Quinone	c	−44.10	−20.0	38.9	
Raffinose	c	−761			
L-Rhamnose	c	−256.5			
Rhamnose triacetate	c	−455.4			
D-Ribose	c	−251.16			
Saccharinic acid lactone	c	−249.6			
Salicylaldehyde	liq	−66.9			
Salicylaldoxime	c	−43.91			
Salicyclic acid	c	−140.9	−99.93	42.6	
Sarcosine	c	−121.2			
Sebacic acid (decanedioic acid)	c	−258.8			
Semicarbazide, std state, $m=1$	aq	−39.9	−9.7	71.2	
L-Serine	c	−173.6			
Serylserine	c	−281.8			
Sorbic acid	c	−93.4			
L-Sorbose	c	−303.68	−217.10	52.8	
5,5′-Spirobis(1,3-dioxane)	c	−167.8			
Spiropentane	g	44.27	63.41	67.45	21.06
Stearic acid (octadecanoic acid)	c	−226.5			
Stilbene					
cis	liq	43.81			
trans	c	32.27	75.90	60.0	
Strychnine	c	−41.0			
Styrene	liq	24.83	48.37	56.78	43.64
	g	35.22	51.10	82.48	29.18
Suberic acid (octanedioic acid)	c	−248.1			
Succinamide	c	−138.9			
Succinic acid	c	−224.79	−178.64	42.0	35.8
Sucrose	c	2531.9	2369.18	86.1	

TABLE 5.1 Enthalpies and Gibbs (Free) Energies of Formation, Entropies, and Heat Capacities of Organic Compounds (*continued*)

Substance	State	$\Delta Hf°$, kcal·mol^{-1}	$\Delta Gf°$, kcal·mol^{-1}	$S°$, cal·deg^{-1}· mol^{-1}	$C_p°$, cal·deg^{-1}· mol^{-1}
L-Tartaric acid	c	−306.5			
DL-Tartaric acid	c	−308.5			
meso-Tartaric acid	c	−305.9			
Tetrabromomethane	g	19.00	15.61	85.53	21.78
Tetracene	c	37.95			
Tetrachlorobenzoquinone	c	−69.0			
1,1,1,2-Tetrachlorodi-fluoroethane	g	−117.1	−97.3	91.5	29.5
1,1,1,2-Tetrachloroethane	g	−35.7	−19.2	85.05	24.67
1,1,2,2-Tetrachloroethane	liq	−47.0	−22.7	59.0	39.6
	g	−36.50	−20.45	86.69	24.09
Tetrachloroethylene	g	−3.40	4.90	81.46	22.69
Tetrachloromethane	liq	−31.75	−14.97	51.67	
	g	−22.90	−12.80	74.07	19.94
1,1,2,2-Tetracyano-cyclopropane	c	141			
Tetracyanoethylene	c	149.1			
Tetradecane	g	−79.38	15.97	167.40	77.93
Tetradecanoic acid	c	−199.2			
1-Tetradecene	g	−49.36	36.99	166.40	75.36
Tetraethylene glycol	liq	−234.6			
Tetraethyllead	liq	12.7	80.4	112.92	
	g	26.3			
1,1,1,2-Tetrafluoroethane	g	−214.10	−197.46	75.58	20.62
Tetrafluoroethylene	g	−157.40	−149.07	71.69	19.24
Tetrafluoromethane	g	−223.0	−212.3	62.45	14.59
Tetrahydrofuran	liq	−51.67			
Tetrahydrofurfuryl alcohol	liq	−104.1			
1,2,3,4-Tetrahydro-naphthalene (Tetralin)	liq	−6.1			
Tetrahydropyran	liq	−61.1			
1,2,5,6-Tetrahydropyridine	liq	8.0			
Tetraiodomethane	g	62.84	51.89	93.60	22.91
1,2,3,4-Tetramethylbenzene	liq	−23.0	25.49	69.45	
1,2,3,5-Tetramethylbenzene	liq	−23.54	23.58	99.55	57.5
1,2,4,5-Tetramethylbenzene	liq	−29.48	24.20	71.83	51.6
2,2,3,3-Tetramethylbutane	g	−53.99	5.26	93.06	
Tetramethyllead	liq	23.5	62.8	76.5	
	g	32.6	64.7	100.5	34.42
Tetramethylsilane	g	−68.50	−23.92	86.30	31.12
Tetramethylsuccinic acid	c	−242.0			
Tetramethylthia-cyclopropane	c	−19.84			
Tetranitromethane	liq	8.9			
1,1,1,2-Tetraphenylethane	c	53.31			

TABLE 5.1 Enthalpies and Gibbs (Free) Energies of Formation, Entropies, and Heat Capacities of Organic Compounds (*continued*)

Substance	State	$\Delta Hf°$, kcal·mol^{-1}	$\Delta Gf°$, kcal·mol^{-1}	$S°$, cal·deg^{-1}· mol^{-1}	$C_p°$, cal·deg^{-1}· mol^{-1}
1,1,2,2-Tetraphenylethane	c	51.63			
Tetraphenylethene	c	74.46			
Tetraphenylhydrazine	c	109.4			
Tetraphenylmethane	c	59.1	137.20		
Tetrazole	c	56.7			
Thebaine	c	−63.0			
Theobromine	c	−86.4			
Thiaadamantane	c	−34.22			
2-Thiabutane	liq	−21.89	1.79	57.14	
	g	−14.25	2.73	79.62	22.73
Thiacyclobutane	g	14.61	25.69	68.17	16.57
Thiacycloheptane	g	−14.66	20.09	86.50	29.78
Thiacyclohexane	liq	−25.32	9.96	52.16	
	g	−15.12	12.68	77.26	25.86
Thiacyclopentane	liq	−17.39	8.97	49.67	
	g	−8.08	11.00	73.94	21.72
Thiacyclopropane	liq	12.41	22.52	38.84	
	g	19.65	23.16	61.01	12.83
4-Thia-5,5-dimethylhex-1-ene	liq	−21.68			
2-Thiaheptane	g	−29.34	8.39	107.73	39.10
3-Thiaheptane	g	−29.92	7.65	108.27	38.71
4-Thiaheptane	liq	−40.62	5.12	80.85	
	g	−29.96	7.94	107.16	38.53
2-Thiahexane	liq	−34.15	4.08	73.49	
	g	−24.42	6.37	98.43	33.64
3-Thiahexane	liq	−34.58	3.50	73.98	
	g	−25.00	5.63	98.97	33.25
5-Thianonane	liq	−52.74	7.66	96.82	
	g	−39.99	11.76	125.76	49.46
2-Thiapentane	liq	−28.21	2.79	65.14	
	g	−19.54	4.40	88.84	28.05
3-Thiapentane	liq	−28.43	2.81	64.36	40.97
	g	−19.95	4.25	87.96	27.97
2-Thiapropane	g	−8.97	1.66	68.32	17.71
6-Thiaundecane	liq	−63.61			
Thioacetic acid	g	−43.49	−36.81	74.86	19.33
Thiohydantoic acid	c	−132.6			
Thiohydantoin	c	−59.5			
Thiolacetic acid	liq	−52.39			
β-Thiolactic acid	liq	−111.6			
Thiophene	liq	19.24	28.97	43.30	
	g	27.66	30.30	66.65	17.42
Thiosemicarbazide	c	6.0			
Thiourea	c	−21.13	5.2	27.7	
	aq, 100	−15.6			

TABLE 5.1 Enthalpies and Gibbs (Free) Energies of Formation, Entropies, and Heat Capacities of Organic Compounds (*continued*)

Substance	State	$\Delta Hf°$, kcal·mol^{-1}	$\Delta Gf°$, kcal·mol^{-1}	$S°$, cal·deg^{-1}· mol^{-1}	$C_p°$, cal·deg^{-1}· mol^{-1}
Threonine, L- and DL-	c	−181.4			
Thymine	c	−111.9			
Thymol	c	−74.0			
Tiglic acid	c	−117.3			
Toluene	liq	2.87	27.19	52.81	37.58
	g	11.95	29.16	76.64	24.77
2-Toluenethiol	liq	10.57			
m-Toluic acid	c	−101.85			
o-Toluic acid	c	−99.55			
p-Toluic acid	c	−102.59			
o-Toluic anhydride	c	−127.5			
p-Toluic anhyride	c	−124.5			
Trehalose	c	−531.3			
2,4,6-Triamino-	c	−17.3	44.10	35.63	
1,3,5-triazine	g	−17.13	42.33	74.10	20.93
(triaminotriazine)					
2-Triazoethanol	liq	22.6			
Tribenzylamine	c	33.6			
Tribromochloromethane	g	3.0	2.17	85.36	
Tribromofluoromethane	g	−45.4	−46.14	82.65	
Tribromomethane	g	4.00	1.78	79.01	16.96
Tributylamine	liq	−67.32			
Tributyl borate	liq	−286.7			
Tributylboron	liq	−83.4			
Tributyl phosphate	liq	−348			
Tributylphosphine oxide	c	−110			
Trichloroacetaldehyde	liq	−56.1			
Trichloroacetamide	c	−85.6			
Trichloroacetic acid	c	−120.7			
Ionized	aq	−123.4			
Trichloroacetyl chloride	liq	−66.4			
Trichlorobenzoquinone	c	−64.5			
1,1,1-Trichloroethane	g	−34.01	−18.21	76.49	22.07
1,1,2-Trichloroethane	g	−33.10	−18.52	80.57	21.47
Trichloroethylene	g	−1.40	4.75	77.63	19.17
Trichlorofluoromethane	g	−68.10	−58.68	74.06	18.66
Trichloromethyl	g	19	22	70.9	15.21
1,2,3-Trichloropropane	g	−44.40	−23.37	91.52	26.82
1,1,1-Tricyanoethane	c	83.9			
Tricyanoethylene	c	105.0			
Tridecane	g	−74.45	13.97	158.09	72.47
Tridecanoic acid	c	−192.8			
1-Tridecene	g	−44.45	34.96	157.09	69.89
Triethylaluminum	liq	−56.6			

TABLE 5.1 Enthalpies and Gibbs (Free) Energies of Formation, Entropies, and Heat Capacities of Organic Compounds (*continued*)

Substance	State	$\Delta Hf°$, kcal·mol^{-1}	$\Delta Gf°$, kcal·mol^{-1}	$S°$, cal·deg^{-1}· mol^{-1}	$C_p°$, cal·deg^{-1}· mol^{-1}
Triethylamine	g	−23.80	26.36	96.90	38.46
Triethylaminoborane	liq	−47.47			
Triethyl arsenite	liq	−168.9			
Triethylarsine	liq	3.1			
Triethyl borate	liq	−250.4			
Triethylenediamine	c	−3.4	57.28	37.67	
Triethylene glycol	liq	−192.2			
Triethyl phosphate	liq	−297			
Triethylphosphine	liq	−21.3			
Triethyl phosphite	liq	−205.9			
Triethylstibine	liq	1.2			
Triethylsuccinic acid	c	−254.9			
Triethyl thionophosphate	liq	−232.5			
Trifluoroacetic acid	liq	−255.4			
Trifluoroacetonitrile	g	−118.4	−110.4	71.3	18.70
1,1,1-Trifluoroethane	g	−178.20	−162.11	68.67	18.76
2,2,2-Trifluoroethanol	liq	−207.4			
Trifluoroethylene	g	−118.50	−112.22	69.94	16.54
Trifluoroiodomethane	g	−141.0	−136.70	73.50	
Trifluoromethane	g	−165.71	−157.48	62.04	12.22
Trifluoromethyl					
CF$_3$·	g	−112.4	−109.2	63.3	11.90
CF$_3$$^+$	g	100.6	103.1	60.8	11.87
Trifluoromethylbenzene	liq	−152.40	−123.98	64.89	
	g	−143.42	−122.20	89.05	31.17
Trifluoromethylhypo- fluorite (CF$_3$OF)	g	−183	−169	77.06	18.97
DL-Trihydroxyglutaric acid	c	−356			
Triiodomethane	g	50.40	42.54	84.97	17.94
Trimethylacetic acid	liq	−134.9			
Trimethylacetic anhydride	liq	−186.4			
2,4,5-Trimethylaceto- phenone	liq	−60.3			
2,4,6-Trimethylaceto- phenone	liq	−63.9			
Trimethylaluminum	liq	−36.1		50.05	37.19
Trimethylamine	g	−5.70	23.64	69.02	21.93
Std state, molarity = 1	aq	−18.17	22.22	31.9	
Trimethylamine aluminum chloride adduct	c	−210.1			
Trimethylammonium ion Std state, molarity = 1	aq	−26.99	8.90	47.0	
Trimethyl arsenite	liq	−141.2			
Trimethylarsine	liq	−3.9			

TABLE 5.1 Enthalpies and Gibbs (Free) Energies of Formation, Entropies, and Heat Capacities of Organic Compounds (*continued*)

Substance	State	$\Delta Hf°$, kcal·mol^{-1}	$\Delta Gf°$, kcal·mol^{-1}	$S°$, cal·deg^{-1}· mol^{-1}	$C_p°$, cal·deg^{-1}· mol^{-1}
1,2,3-Trimethylbenzene	liq	−14.01	25.68	66.40	
1,2,4-Trimethylbenzene	liq	−14.79	24.46	67.93	
1,3,5-Trimethylbenzene	liq	−15.18	24.83	65.38	
Trimethyl borate	liq	−222.9			
Trimethylboron	liq	−34.1			
2,2,3-Trimethylbutane	g	−48.95	1.02	91.61	39.33
Trimethylchlorosilane	liq	−91.8			
cis,cis-1,3,5-Trimethyl-cyclohexane	g	−51.48	8.10	93.30	42.93
2,2,3-Trimethylpentane	liq	−61.44	2.21	78.30	
	g	−52.61	4.09	101.62	
2,2,4-Trimethylpentane	liq	−61.97	1.65	78.40	
	g	−53.57	3.27	101.15	
2,3,3-Trimethylpentane	liq	−60.63	2.54	79.93	
	g	−51.73	4.52	103.14	
2,3,4-Trimethylpentane	liq	−60.98	2.55	78.71	
	g	−51.97	4.52	102.31	
2,4,4-Trimethyl-1-pentene	liq	−35.21	20.66	73.2	
2,4,4-Trimethyl-2-pentene	liq	−34.44	21.04	74.5	
Trimethylphosphine	liq	−29.2			
Trimethylphosphine-N-ethylimine	liq	−35.8			
Trimethylphosphine oxide	c	−114.2			
Trimethyl phosphite	liq	−177.1			
Trimethylsilanol	liq	−130.3			
Trimethylstibine	liq	0.2			
Trimethylsuccinic acid	c	−239.2			
Trimethylsuccinic anhydride	c	−164.5			
Trimethylthiacyclopropane	liq	−14.47			
Trimethylurea	c	−79.0			
2,4,6-Trinitroanisole	c	−37.6			
1,3,5-Trinitrobenzene	c	−10.40			
Trinitromethane	c	−11.50			
1,4,5-Trinitronaphthalene	c	8.7			
1,3,8-Trinitronaphthalene	c	5.8			
2,4,6-Trinitrophenetole	c	−48.9			
2,4,6-Trinitrophenol	c	−51.23			
2,4,6-Trinitrophenyl-hydrazine	c	8.8			
2,4,6-Trinitrotoluene	c	−16.0			
2,4,6-Trinitro-*m*-xylene	c	−24.5			
Triphenylamine	c	58.70$^{18°C}$	120.50		
Triphenylarsine	c	74.1			
Triphenylcarbinol	c	−0.80	65.2	78.7	

TABLE 5.1 Enthalpies and Gibbs (Free) Energies of Formation, Entropies, and Heat Capacities of Organic Compounds (*continued*)

Substance	State	$\Delta Hf°$, kcal·mol^{-1}	$\Delta Gf°$, kcal·mol^{-1}	$S°$, cal·deg^{-1}· mol^{-1}	$C_p°$, cal·deg^{-1}· mol^{-1}
Triphenylene	c	33.72	78.68	60.87	
1,1,1-Triphenylethane	c	37.56			
1,1,2-Triphenylethane	c	31.11			
Triphenylethylene	c	55.8	123.00		
Triphenylmethane	c	38.71	98.60	74.6	70.5
Triphenyl phosphate	c	−181			
Triphenylphosphine	c	55.5			
Triphenylphosphine oxide	c	−14.4			
Tripropylamine	liq	−49.51			
Tris(acetylacetonato)-chromium	c	−366.4			
1,1,1-Tris(hydroxymethyl)-ethane	c	−177.96			
Tropolone	c	−57.18			
L-Tryptophan	c	−99.8	−28.54	60.00	56.92
L-Tyrosine	c	−163.4	−92.18	51.15	51.73
Undecane	liq	−78.05	5.44	109.49	
	g	−64.60	9.94	139.48	61.53
1-Undecene	g	−34.60	30.94	138.48	58.96
Urea	c	−79.71	−47.19	25.00	22.26
Std state, $m = 1$	aq	−75.95			
Urea nitrate	c	−134.8			
Urea oxalate	c	−365.3			
Uric acid	c	−147.73	−85.75	41.4	
Valeric acid	liq	−133.71	−89.10	62.10	50.48
Valine, L and DL-	c	−148.2	−85.80	42.75	40.35
Valylphenylalanine	c	−183.5			
Veronal	c	−178.7			
Vinyl radical	g	63			
Vinyl bromide	g	18.7	19.3	65.90	13.27
Vinyl chloride	g	8.5	12.4	63.07	12.84
Vinylcyclohexane	liq	−21.19			
Vinylcyclopropane	liq	29.3			
2-Vinylpyridine	liq	37.2			
Xanthine	c	−90.49	−39.64	38.5	
o-Xylene	liq	−5.84	26.37	58.91	44.9
	g	4.54	29.18	84.31	31.85
m-Xylene	liq	−6.08	25.73	60.27	43.8
	g	4.12	28.41	85.49	30.49
p-Xylene	liq	−5.84	26.31	59.12	
	g	4.29	28.95	84.23	30.32
2,3-Xylenol	g	−37.57			
2,4-Xylenol	g	−38.93			
2,5-Xylenol	g	−38.63			

TABLE 5.1 Enthalpies and Gibbs (Free) Energies of Formation, Entropies, and Heat Capacities of Organic Compounds (*continued*)

Substance	State	$\Delta Hf°$, kcal·mol^{-1}	$\Delta Gf°$, kcal·mol^{-1}	$S°$, cal·deg^{-1}·mol^{-1}	$C_p°$, cal·deg^{-1}·mol^{-1}
2,6-Xylenol	g	−38.66			
3,4-Xylenol	g	−37.42			
3,5-Xylenol	g	−38.61			
Xylitol	c	−267.32			
D-Xylose	c	−252.8			

TABLE 5.2 Heats of Melting and Vaporization (or Sublimation) and Specific Heat at Various Temperatures of Organic Compounds

Abbreviations Used in the Table

ΔHm, enthalpy of melting (at the melting point) in kcal·mol⁻¹
ΔHv, enthalpy of vaporization (at the boiling point) in kcal·mol⁻¹
ΔHs, enthalpy of sublimation (at 298 K) in kcal·mol⁻¹
C_p, specific heat (at temperature specified, measured on the Kelvin scale) for physical state in existence at that temperature, expressed in cal·K⁻¹·mol⁻¹
ΔHt, enthalpy of transition (at temperature specified, measured in degrees Celsius) in kcal·mol⁻¹

Substance	ΔHm	ΔHv	ΔHs	C_p 400 K	600 K	800 K	1000 K
Acenaphthene			20.6				
Acenaphthylene			17.0				
Acetaldehyde	0.770	6.24		15.73	20.52	24.20	29.96
Acetanilide			19.3				
Acetic acid	2.80	5.663		19.52	25.15	29.08	31.99
Acetic anhydride	2.51	9.85	11.54	30.86	41.62	48.91	54.11
Acetone	1.366	6.952		22.00	29.34	34.93	39.15
Acetonitrile	1.952	7.3	7.94	14.62	18.35	21.26	23.50
(ΔHt, 0.215 at −56°C)							
Acetophenone		9.275	13.4	18.86	23.18	26.30	28.60
Acetyl bromide			7.9				
Acetyl chloride			7.2				
Acetylene	0.900	4.05	5.1	11.97	13.73	14.93	15.92
Acetylenedicarbonitrile			6.88	22.66	25.37	27.26	28.62
Acetyl fluoride			6.0				
Acetyl iodide			7.9				
Acrylic acid		11.21	12.98	22.94	29.50	33.93	37.12
Acrylonitrile		7.8		18.36	23.11	26.43	28.88
Adenine			25.8				

Compound							
Adipic acid			30.8				
α-Alanine			33.0				
Allyl ethyl sulfoxide			17.1				
Allyl trichloroacetate			12.5				
1-Aminobutane			8.50	35.44	47.30	56.01	62.54
2-Aminobutane			7.5	35.40	47.55	56.42	62.54
Aniline	2.519	10.643	13.325	34.17	46.09	53.79	69.18
Anthracene		13.5	24.7				
9,10-Anthracenedione			26.8				
Azoisopropane			8.5				
Azulene	2.89	13.26	22.8	42.15	59.32	70.59	78.24
Benzaldehyde			12				
1,2-Benzanthra-9,10-quinone			19.8				
Benzene	2.358	7.352	8.090	26.74	37.73	45.06	50.16
Benzenethiol	2.736	9.53	11.64	32.76	44.13	51.59	56.79
Benzil			23.5				
Benzoic acid	4.32	12.10	22.70				
Benzoic anhydride			23				
Benzonitrile	2.60	11.0	13.26	33.65	44.80	52.08	57.08
Benzophenone			22.5				
1,4-Benzoquinone			15.00				
Benzoyl bromide			14.0				
Benzoyl chloride			13.1				
Benzoyl iodide			14.8				
3,4-Benzophenanthrene			25.4				
Benzyl bromide			11.3				
Benzyl chloride			12.3				
Benzyl ethyl sulfide			13.6				
Benzyl iodide			11.3				
Benzyl methyl ketone			12.78				
Benzyl methyl sulfide			12.8				
Bicyclo[4.1.0]heptane			9.14				
Bicyclo[3.1.0]hexane			7.85				

TABLE 5.2 Heats of Melting and Vaporization (or Sublimation) and Specific Heat at Various Temperatures of Organic Compounds (*continued*)

Substance	ΔH_m	ΔH_v	ΔH_s	C_p 400 K	600 K	800 K	1000 K
Bicyclo[4.2.0]octane			9.85				
Bicyclo[5.1.0]octane			10.42				
Bicyclopropyl			8.0				
Biphenyl	4.44	10.9		52.83	73.54	86.92	96.00
Biphenylene			30.8				
Bromobenzene	2.54	9.05	10.62	30.44	40.99	47.78	52.40
4-Bromobenzoic acid			21.0				
1-Bromobutane	1.6	7.78		32.64	43.00	50.48	56.03
2-Bromobutane				33.09	43.76	51.31	56.93
Bromoethane	1.4	6.41	8.45	18.93	24.56	28.58	31.59
Bromoethene			6.57	15.91	19.83	22.50	24.46
1-Bromoheptane			12.05				
1-Bromohexane			10.91				
Bromomethane ΔH_t, 0.113 at $-99.4°C$	1.429			11.94	14.98	17.26	19.01
2-Bromo-2-methylpropane ΔH_t, 1.35 at $-64.5°C$; 0.25 at $-41.6°C$	0.47	5.715	7.4	34.93	45.58	52.65	57.74
1-Bromooctane			13.14				
1-Bromopentane	2.74	8.24		39.58	52.34	61.55	68.36
1-Bromopropane	1.56	7.14		25.70	33.66	39.41	43.70
2-Bromopropane		6.79		26.34	34.42	40.09	44.26
1,2-Butadiene	1.665	5.82	5.71	23.54	30.72	36.01	40.02
1,3-Butadiene	1.908	5.42	5.03	24.29	31.84	36.84	40.52
n-Butadiene sulfone			14.7				
Butadiyne				20.17	23.14	25.11	26.61
Butane ΔH_t, 0.494 at $-165.60°C$	1.114	5.352	5.035	29.60	40.30	48.23	54.22
2,3-Butanedione			9.25				
1,4-Butanedithiol			13.22				
1-Butanethiol	2.500	7.702	8.73	34.95	46.54	55.68	62.95

Compound							
2-Butanethiol	1.548	7.312	8.14	35.38	46.42	54.29	60.02
1-Butanol	2.24	10.31	12.52	32.80	43.90	52.11	58.26
2-Butanol		9.75	11.87	33.70	44.72	52.68	58.62
2-Butanone	2.017	7.475	8.34	29.81	39.09	46.08	51.33
1-Butene	0.920	5.238	4.81	26.04	35.14	41.80	46.82
2-Butene							
cis	1.747	5.580	5.29	24.33	33.80	40.87	46.15
trans	2.332	5.439	5.10	26.02	34.80	44.20	46.58
1-Buten-3-yne				21.26	26.67	30.40	33.16
N-Butylacetamide			18.2				
Butyl acetate		8.58	10.42				
tert-Butylamine			7.10	36.46	48.87	57.49	63.79
Butylbenzene							
stable(I)	2.682(I)	9.38	11.98	54.75	75.20	89.37	99.49
metastable(II)	2.691(11)						
sec-Butylbenzene			11.72				
tert-Butylbenzene			11.50				
sec-Butyl butyrate			11.3				
Butyl chloroacetate			12.2				
Butyl 2-chlorobutyrate			12.6				
Butyl 3-chlorobutyrate			12.7				
Butyl 4-chlorobutyrate			13.0				
Butyl 2-chloropropionate			13.0				
Butyl 3-chloropropionate			13.3				
Butyl crotonate			12.4				
sec-Butyl crotonate			11.8				
Butylcyclohexane	3.384	9.20	11.96	66.00	93.10	112.30	125.70
Butylcyclopentane	2.704	8.69	11.00	57.77	80.38	97.35	114.80
N-Butyldiacetimide			15.4				
Butyl dichloroacetate			12.5				
tert-Butyl hydroperoxide			11.41				
Butylisobutylamine			10.73				
Butyl lithium			25.6				
Butyl trichloroacetate			12.8				

TABLE 5.2 Heats of Melting and Vaporization (or Sublimation) and Specific Heat at Various Temperatures of Organic Compounds (*continued*)

Substance	ΔH_m	ΔH_v	ΔH_S	C_p			
				400 K	600 K	800 K	1000 K
1-Butyne	1.441		5.67	23.87	30.83	35.95	39.84
2-Butyne	2.207	5.861	6.38	22.62	29.68	35.14	39.29
Butyraldehyde	2.654	6.340	8.05	30.20	39.60	46.60	51.70
Butyric acid	2.50		15.2				
Butyronitrile	1.2	10.04	9.53	28.39	37.07	43.48	48.22
D-Camphor	1.635	8.13					
ε-Caprolactam		14.22	19.9				
Carbazole			20.2				
Carbon disulfide	1.049	6.401					
Chloroacetic acid			18				
Chloroacetyl chloride			9.3				
2-Chlorobenzaldehyde			13.3				
Chlorobenzene	2.28	8.73	9.81	30.62	41.16	47.89	52.48
2-Chlorobenzoic acid			19.0				
3-Chlorobenzoic acid			19.6				
4-Chlorobenzoic acid			21.0				
Chlorobenzoquinone			16.5	32.30	42.77	50.31	55.92
1-Chlorobutane		7.38	8.0	32.52	43.18	50.84	56.60
2-Chlorobutane		6.98	7.60				
Chlorocyclohexane			10.4				
Chlorodifluoromethane	0.985	4.833		15.63	18.87	20.84	22.10
Chloroethane	1.064	5.892		18.54	24.28	28.39	31.48
1-Chloro-2-ethylbenzene			11.3				
1-Chloro-4-ethylbenzene			11.5				
Chloroethylene				15.56	19.61	22.35	24.35
Chloroethyne				14.39	15.97	16.98	17.75
Chlorofluoromethane				13.29	16.57	18.81	20.39

Compound							
Chloroform	2.28	7.08	7.48	17.75	20.38	21.87	22.83
Chloromethane	1.537	5.147		11.52	14.66	17.04	18.86
Chloromethyloxirane							
1-Chloro-2-methylpropane			9.7	32.52	43.18	50.84	56.60
2-Chloro-2-methylpropane	0.48	6.6	7.57	34.00	44.20	51.50	57.00
ΔH_t, 0.41 at −90.1°C; 1.39 at −53.6°C							
1-Chloronaphthalene			15.6				
2-Chloronaphthalene			19.6				
1-Chloropentane		7.93	9.1	39.24	52.11	61.38	68.25
3-Chlorophenol			12.7				
4-Chlorophenol			12.4				
1-Chloropropane		6.62	6.9	25.36	33.43	39.24	43.59
2-Chloropropane		6.34	6.47	25.99	34.20	39.94	44.16
3-Chloro-1-propene				22.12	28.43	32.93	36.30
Chlorotrifluoromethane				18.53	21.60	23.17	24.03
Chlorotrinitromethane			10.86				
Chrysene			28.1				
o-Cresol		10.20	18.17	39.74	52.77	61.55	68.82
m-Cresol		10.32	14.75	38.74	52.26	61.27	68.50
p-Cresol		10.32	17.67	38.65	52.10	61.11	68.48
m-Cresyl acetate			14.51				
Cubane			19.2				
4-Cyanothiazole			17.67				
Cyclobutane ΔH_t, 1.38 at −126.79°C	0.260	5.781	5.65	23.89	34.76	42.42	47.96
Cyclobutene				21.59	30.30	36.26	40.53
Cyclododecane			18.26				
Cycloheptane ΔH_t, 1.187 at −138.4°C; 0.069 at −75.0°C; 0.108 at −60.8°C	0.450	7.93	9.21	41.82	62.42	77.03	87.40
Cycloheptanone			12.4				
1,3,5-Cycloheptatriene ΔH_t, 0.561 at −118.19°C	0.277	9.250		37.13	50.07	58.58	64.58
Cyclohexane ΔH_t, 1.611 at −87°C	0.640	7.160	7.896	35.82	53.83	66.76	75.80
Cyclohexanol ΔH_t, 1.96 at −9.7°C	0.406	10.875	12.820	41.14	59.29	72.18	81.13
Cyclohexanone		9.00	10.77	36.00	52.90	65.00	73.00
Cyclohexene ΔH_t, 1.016 at −134.4°C	0.787	7.285	8.00	34.64	49.45	59.49	66.62

TABLE 5.2 Heats of Melting and Vaporization (or Sublimation) and Specific Heat at Various Temperatures of Organic Compounds (*continued*)

Substance	ΔH_m	ΔH_v	ΔH_s	C_p			
				400 K	600 K	800 K	1000 K
Cyclooctane ΔH_t, 1.507 at −106.7°C; 0.114 at −89.35°C	0.576	8.58	10.36	47.82	71.00	87.30	99.01
Cyclooctanone			13.0				
1,3,5,7-cyclooctatetraene	2.695	8.700	10.30	38.45	52.77	62.23	68.88
Cyclopentadiene			6.78				
Cyclopentane ΔH_t, 1.167 at −150.76°C; 0.823 at −135.08°C	0.1455	6.524	6.818	28.38	42.57	52.60	59.84
Cyclopentanethiol	1.872	8.443	9.93	34.53	48.65	58.61	65.84
Cyclopentanol			13.74				
Cyclopentanone			10.21				
Cyclopentene ΔH_t, 0.115 at −186.08°C	0.804			25.08	37.19	45.78	51.94
Cyclopropane	1.301	4.793	6.71	18.31	26.15	33.57	35.39
Decahydronaphthalene cis ΔH_t, 0.511 at −57.1°C	2.268	9.940	12.0	56.64	84.14	103.36	116.91
trans	3.455	9.260	11.6	56.78	84.20	103.40	116.93
Decanal	6.863	9.388	12.277	71.80	95.70	113.00	125.70
Decane	7.4			71.24	96.36	114.92	128.20
1-Decanethiol	7.0	11.1	15.5	76.63	102.63	122.10	136.98
Decanoic acid			28.4				
1-Decanol	9.0	11.9	18.6	74.44	99.94	118.53	132.24
1-Decene ΔH_t, 1.90 at −74.8°C	3.300	9.24	12.06	67.79	91.27	108.28	120.90
1-Decyne				65.64	86.96	102.42	113.90
Deoxybenzoin			22.3				
Dibenzilidene azine			22.3				
Dibenzyl ketone			21.3				
Dibenzyl sulfide			22.3				
Dibenzyl sulfone			27.8				

1,2-Dibromobutane			10.8	36.77	46.70	53.60	58.50
1,2-Dibromocycloheptane			12.43				
1,2-Dibromocyclohexane			12.07				
1,2-Dibromocyclooctane			13.04				
1,2-Dibromoethane	2.62	8.69	9.86	23.83	29.24	32.94	35.80
1,2-Dibromoheptane			13.01				
1,2-Dibromopropane				29.74	37.63	42.91	46.74
Dibutylborinic acid			15				
Dibutyl ether		8.83	10.5	60.78	81.29	96.52	107.86
Dibutyl mercury			15.6				
Di-tert-butyl peroxide			7.6				
Dibutyl o-phthalate			21.9				
Dibutyl sulfate			18.1				
Dibutyl sulfite			16.2				
Dibutyl sulfone			24.0				
Dichloroacetyl chloride			9.4				
1,2-Dichlorobenzene	3.19	9.7	11.56	34.12	44.07	50.28	54.42
1,3-Dichlorobenzene			11.44	34.18	44.09	50.29	54.42
1,4-Dichlorobenzene	4.34	9.5	15.5	34.24	44.16	50.35	54.46
2,6-Dichlorobenzoquinone			16.7				
2,2'-Dichlorobiphenyl			23.0				
4,4'-Dichlorobiphenyl			24.8				
Dichlorodifluoromethane				19.69	22.37	23.69	24.39
1,1-Dichloroethane	1.881	6.97	7.36	21.85	27.18	30.79	33.40
1,2-Dichloroethane	2.112	7.65	8.47	22.00	26.90	30.40	33.00
1,1-Dichloroethylene	1.557	6.26	6.328	18.80	22.44	24.71	26.29
1,2-Dichloroethylene							
cis	1.72	7.08	7.43	18.41	22.23	24.60	26.23
trans	1.72	6.65	6.92	18.58	22.28	24.62	26.24
Dichlorofluoromethane				16.78	19.70	21.41	22.51
Dichloromethane	1.1	6.74	6.94	14.24	17.30	19.32	20.76
1,2-Dichloropropane		7.59	8.68	28.60	36.47	41.97	46.08
1,3-Dichloropropane		8.10	9.66	28.69	36.22	41.56	45.50
2,2-Dichloropropane		7.0	7.8	30.56	38.06	43.00	46.56

TABLE 5.2 Heats of Melting and Vaporization (or Sublimation) and Specific Heat at Various Temperatures of Organic Compounds (*continued*)

Substance	ΔH_m	ΔH_v	ΔH_s	C_p 400 K	600 K	800 K	1000 K
Dicyanoacetylene			6.88	34.88	47.14	56.16	62.91
2,2-Diethoxypropane			7.61				
Diethylamine			7.6				
1,2-Diethylbenzene	4.01	9.42	12.61	56.01	75.66	89.54	99.49
1,3-Diethylbenzene	2.62	9.41	12.55	55.01	75.19	89.31	99.37
1,4-Diethylbenzene	2.53	9.41	12.54	54.68	74.84	89.04	99.16
Diethylene glycol		12.50	13.7				
Diethyl ether	1.745	6.38	6.516	33.01	43.92	52.26	58.51
Diethylmercury			10.7				
Diethylmethyl phosphonate			13.5				
Diethylnitramine			12.7				
Diethyl oxalate		10.04	15.2				
Diethyl peroxide			7.3				
Diethyl o-phthalate			21.1				
Diethyl selenide			9.3				
Diethyl sulfate			13.6				
Diethyl sulfite			11.6				
Diethyl sulfone			20.6				
Diethyl sulfoxide	2.640	7.699	14.9				
1,2-Difluorobenzene			8.65	32.76	43.33	50.12	54.72
1,3-Difluorobenzene			8.29	32.72	43.13	49.67	53.93
1,4-Difluorobenzene			8.51	32.84	43.20	49.68	53.99
2,2'-Difluorobiphenyl			22.7				
4,4'-Difluorobiphenyl			21.8				
1,1-Difluoroethane		5.1		19.93	25.70	29.70	32.57
1,1-Difluoroethylene				17.16	21.32	23.95	25.74
Difluoromethane				12.22	15.72	18.22	19.98

5.52

Compound							
9,10-Dihydroanthracene			22.3				
4H-Dihydropyran			7.7				
5,12-Dihydrotetracene			27.7				
2,3-Dihydrothiophene			9.02				
2,5-Dihydrothiophene			9.55				
1,2-Diiodobenzene			15.5				
1,2-Diiodoethane	3.02(I) 2.88(II)		15.7	22.94 / 15.74	27.92 / 18.37	31.37 / 20.06	33.84 / 21.29
Diiodomethane	2.635	6.95	12.2	46.90	62.61	74.39	83.17
Diisopropyl ether			7.75				
Diisopropyl ketone			9.93				
Diisopropylmercury			12.8				
1,2-Dimethoxybenzene			16.0				
Dimethoxyborane			6.14				
2,2-Dimethoxypropane	1.420		7.03				
Dimethylamine		6.330	6.07	20.89	28.41	33.94	38.19
Dimethylaminotrimethylsilane			7.6				
2,2-Dimethylbutane ΔHt, 1.289 at −147.34°C; 0.068 at −132.28°C	0.138	6.287	6.618	43.70	60.00	71.40	79.70
2,3-Dimethylbutane ΔHt, 1.552 at −137.08°C	0.194	6.519	6.96	43.30	59.20	75.20	79.10
2,3-Dimethyl-1-butene		6.55	6.97	42.60	55.40	65.00	72.20
2,3-Dimethyl-2-butene a. ΔHt, b. 0.844 at −76.34°C	1.542	7.083	7.776	37.48	51.78	62.78	71.14
3,3-Dimethyl-1-butene ΔHt, 1.037 at −148.3°C	0.261	6.13	6.36	38.90	53.40	63.60	71.00
Dimethylcadmium			9.07				
1,1-Dimethylcyclohexane ΔHt, 1.430 at −120.01°C	0.495	7.79	9.043	50.70	74.10	90.70	102.20
1,2-Dimethylcyclohexane cis ΔHt, 1.974 at −100.6°C	0.393	8.04	9.492	51.10	74.00	90.10	101.40
trans	2.491(I) 2.508(II)	7.86	9.168	51.90	74.60	90.50	101.70
1,3-Dimethylcyclohexane cis	2.586	7.84	9.137	51.20	74.20	90.50	102.00
trans	2.358	8.09	9.369	51.10	73.80	89.80	101.10

TABLE 5.2 Heats of Melting and Vaporization (or Sublimation) and Specific Heat at Various Temperatures of Organic Compounds (*continued*)

Substance	ΔH_m	ΔH_v	ΔH_s	C_p 400 K	600 K	800 K	1000 K
1,4-Dimethylcyclohexane							
cis	2.225	8.07	9.329	51.10	73.80	89.80	101.10
trans	2.947	7.79	9.053	51.60	74.60	90.60	101.90
1,1-Dimethylcyclopentane	0.258	7.239	8.079	43.55	62.78	76.18	85.83
ΔH_t, 1.551 at −126.36°C							
1,2-Dimethylcyclopentane							
cis ΔH_t, 1.594 at −131.66°C	0.396	7.576	8.549	43.67	62.72	75.98	85.57
trans	1.713	7.375	8.259	43.71	62.66	75.84	85.43
1,3-Dimethylcyclopentane							
cis	1.761	7.265	8.200	43.71	62.66	75.84	85.43
trans	1.738	7.361	8.248	43.71	62.66	75.84	85.43
Dimethyldichlorosilane			8.2				
cis-2,4-Dimethyl-1,3-dioxane			9.53				
4,5-Dimethyl-1,3-dioxane			10.16				
5,5-Dimethyl-1,3-dioxane			9.86				
Dimethyl ether	1.180	5.141		19.02	25.16	30.04	33.79
N,N-Dimethylformamide			11.4				
Dimethylfulvene			10.6				
Dimethylglyoxime			23.2				
2,2-Dimethylhexane	1.62	7.71	8.91				
2,3-Dimethylhexane		7.94	9.27				
2,4-Dimethylhexane		7.79	9.03				
2,5-Dimethylhexane	3.096	7.80	9.05				
3,3-Dimethylhexane	1.7	7.76	8.97				
3,4-Dimethylhexane		7.95	9.32				
2,2-Dimethyl-3-hexene							
cis			8.88				
trans			8.91				

1,1-Dimethylhydrazine			8.37				
1,2-Dimethylhydrazine			9.40				
Dimethylmercury			8.26				
Dimethylnitramine			16.7				
2,2-Dimethylpentane	1.392	6.97	7.75	50.42	68.33	81.43	91.20
2,3-Dimethylpentane		7.26	8.19	50.42	68.33	81.43	91.20
2,4-Dimethylpentane	1.636	7.05	7.86	50.42	68.33	81.43	91.20
3,3-Dimethylpentane	1.689	7.09	7.89	50.42	68.33	81.43	91.20
2,7-Dimethylphenanthrene			25.5				
4,5-Dimethylphenanthrene			25.0				
9,10-Dimethylphenanthrene			28.6				
2,2-Dimethylpropane	0.752	5.438	5.205	37.55	51.21	60.78	67.80
ΔH_t, 0.616 at $-133.14\,^{\circ}\mathrm{C}$							
2,3-Dimethylpyridine			11.70				
2,4-Dimethylpyridine			11.42				
2,5-Dimethylpyridine			11.43				
2,6-Dimethylpyridine			11.01				
3,4-Dimethylpyridine			12.38				
3,5-Dimethylpyridine			12.04				
Dimethyl sulfate			11.6				
Dimethyl sulfite			9.6				
Dimethyl sulfone			18.4				
Dimethyl sulfoxide	1.56	12.66	12.64				
3,3-Dimethyl-2-thiabutane	2.011(I) 1.83(II)	7.523	8.57				
2,2-Dimethylthiacyclopropane	1.69	8.00	8.55				
2,4-Dimethyl-3-thiapentane	2.49	8.04	9.4	50.64	66.22	77.12	85.24
2,4-Dimethyl-3-thiapentane			9.44				
1,3-Dinitrobenzene			14.3				
2,4-Dinitrophenol			25				
2,6-Dinitrophenol			26.8				
1,1-Dinitropropane			14.93				
1,4-Dioxane ΔH_t, 0.562 at $-0.3\,^{\circ}\mathrm{C}$	3.07		9.20	30.23	43.44	52.15	58.05
1,3-Dioxolan			8.5				

TABLE 5.2 Heats of Melting and Vaporization (or Sublimation) and Specific Heat at Various Temperatures of Organic Compounds (*continued*)

Substance	ΔH_m	ΔH_V	ΔH_S	C_p				
				400 K	600 K	800 K	1000 K	
Dipentene			11.5					
Diphenylamine			23.1					
Diphenylchlorosilane			16.6					
Diphenyl disulfide			22.7					
Dipfenyl disulfone			38.7					
1,2-Diphenylethane		12.3	20.1					
1,1-Diphenylethene		15.5[25]	17.5					
Diphenyl ether	4.115		19.6					
Diphenylfulvene			25					
Diphenylmercury			26.95					
Diphenylmethane			19.7					
Diphenyl sulfide			16.2					
Diphenyl sulfone			25.4					
Diphenyl sulfoxide			23.2					
Dipropyl ether			8.6	46.90	62.61	74.39	83.17	
Dipropylmercury			13.2					
Dipropyl sulfate			16.0					
Dipropyl sulfite			14.0					
Dipropyl sulfone			19.1					
Dipropyl sulfoxide			17.8					
2,3-Dithiabutane	2.197	8.05	9.17	26.36	32.83	37.66	41.31	
5,6-Dithiadecane		11.2	15.2	68.38	89.98	105.83	117.86	
3,4-Dithiahexane	2.248	9.01	10.89	40.90	52.24	60.19	65.97	
1,3-Dithian-2-thione			21.85					
4,5-Dithiaoctane	3.30	10.02	12.55	44.50	71.30	83.70	93.20	
N,N-Dithiodiethylamine			12.6					
1,3-Diothiolan-2-thione			19.56					
Di-p-tolyl sulfone			26.2					

Compound							
Divinyl ether							
Divinyl sulfone	8.57			85.13	115.04	136.76	152.90
Dodecane		10.43	6.26; 13.5				
Dodecanedioic acid			14.65; 36.6				
1-Dodecene	4.76	10.27	14.42	8.68	109.95	130.41	145.50
$\quad\Delta H_t$, 1.088 at -60.2°C							
Eicosane	16.70	13.74	24.1; 48	140.65	189.78	225.28	251.60
Eicosanoic acid	17.2						
1-Eicosene	8.2	13.35	23.86; 32.3	137.20	184.69	218.93	244.20
meso-Erythritol		3.517					
Ethane	0.683		1.200; 10.68	15.65	21.35	25.81	29.30
1,2-Ethanedithiol	1.189		6.526				
Ethanethiol	1.198	6.401	10.11	21.08	27.21	31.83	35.38
Ethanol		9.255	8.63	19.36	25.69	30.33	33.83
Ethyl acetate	2.505	7.720	20.0	32.84	43.65	51.01	56.05
Ethyl allyl sulfone							
Ethylamine			6.7; 12.5	21.65	28.68	33.89	37.88
N-Ethylaniline	6.7						
Ethylbenzene	2.195	8.50	10.10	40.76	56.44	67.15	74.77
3-Ethyl-1-butene		6.88	7.41	40.70	54.50	64.40	71.90
Ethyl crotonate			10.6				
Ethylcyclohexane	1.992	8.20	9.67	51.60	74.10	90.10	101.30
1-Ethylcyclohexene			10.34				
Ethylcyclopentane	1.642(I); 1.889(II)	7.715	8.72	43.89	61.70	75.22	85.16
Ethylene	0.801	3.237	17.5	12.67	17.87	20.03	22.43
Ethylene carbonate	2.41		15.68				
Ethylene glycol	2.78	11.86	7.55	27.06	32.72	36.90	39.88
Ethyleneimine		7.24	5.96	16.83	23.56	28.14	31.45
Ethylene oxide		6.101		14.95	20.62	24.60	27.47
Ethyl formate	1.236	7.201	11.70				
2-Ethyl-1-hexanal	2.20		9.48				
3-Ethylhexane		8.03	10.5				
Ethylisovalerate			27.9				
Ethyllithium	8.03						

TABLE 5.2 Heats of Melting and Vaporization (or Sublimation) and Specific Heat at Various Temperatures of Organic Compounds (*continued*)

Substance	ΔHm	ΔHv	ΔHs	C_p				
				400 K	600 K	800 K	1000 K	
Ethylmercury bromide			18.3					
Ethylmercury chloride			18.2					
Ethylmercury iodide			19.0					
Ethyl methyl ether	2.04	7.92	8.67	26.08	34.58	41.19	46.18	
Ethyl nitrate	2.282	7.40	8.42	28.73	37.07	42.72	46.69	
3-Ethylpentane			11.0	50.42	68.33	81.43	91.20	
Ethyl pentanoate			15.20					
2-Ethylphenol			16.30					
3-Ethylphenol			19.20					
4-Ethylphenol			12.1					
Ethylphosphonic acid		8.178	9.0					
Ethyl propanoate			11.6					
Ethyl β-vinylacrylate			6.35					
Ethyl vinyl ether				35.95	48.01	55.79	61.17	
Ethynylbenzene		7.457	24.65	29.99	40.86	47.83	52.58	
Fluoranthrene	2.702		8.27					
Fluorobenzene			21.8					
4-Fluorobenzoic acid				17.71	23.56	27.82	31.00	
Fluoroethane				10.56	13.83	16.45	18.44	
Fluoromethane				24.55	32.82	38.88	43.37	
1-Fluoropropane				24.72	33.14	39.14	43.55	
2-Fluoropropane		8.144	9.42	36.43	49.70	58.60	64.84	
4-Fluorotoluene	2.235		8.3					
Fluorotrinitromethane								
Formaldehyde		5.85		9.38	11.52	13.37	14.81	
Formic acid	3.035	5.24	11.03	12.85	16.02	18.35	19.95	
Formyl								
HCO·				8.73	9.79	10.75	11.49	
HCO⁺				9.39	10.39	11.14	11.78	

Compound							
Fumaric acid			32.5				
Fumaronitrile	0.909		17.2				
Furan ΔHt, 0.489 at −123.2°C	3.12	6.474	6.61				
Furfuryl alcohol			15.4	21.20	29.31	34.41	37.89
2-Furoic acid			25.92				
Furylethylene	4.416		9.1				
Glycerol			20.5				
Glyceryl triacetate			19.6				
Glyceryl trinitrate			23.9				
Heptadecane ΔHt, 2.62 at 11.1°C	9.67	12.64	20.6	119.83	161.75	192.08	214.60
Heptadecanoic acid	12.3						
1-Heptadecene	7.5	12.39	20.32	116.38	156.66	185.74	207.20
1-Heptanal	5.637		11.40	51.00	67.70	79.80	88.70
Heptane	3.359	7.575	8.74	50.42	68.33	81.43	91.20
1-Heptanethiol	6.067	9.5	12.06	55.81	74.60	88.91	99.98
Heptanoic acid	3.16		18.0	53.62	71.92	85.32	95.25
1-Heptanol	2.964(I); 3.021(II)	11.5	16.5	46.97	63.24	75.09	83.90
1-Heptene ΔHt, 0.07 at 1.36°C		7.43	8.52				
Hexachlorobenzene	6.1		23.2	48.08	55.78	59.96	62.34
Hexachloroethane ΔHt, 1.9 at 71.3°C	2.33	12.2	16.5	36.21	39.82	41.48	42.38
Hexadecafluoroethylcyclohexane			9.20				
Hexadecafluoroheptane			8.7				
Hexadecane	12.39	12.24	19.38	112.89	152.41	181.02	202.20
Hexadecanoic acid	12.8		36.9	116.09	156.00	184.90	206.30
1-Hexadecanol	7.8		40.5				
ΔHt, 4.8 at 44.0°C; 5.7 at 49.1°C							
1-Hexadecene	7.216	12.05	19.14	109.44	147.32	174.67	194.80
Hexafluorobenzene	2.770	7.571	8.61	43.88	52.55	57.62	60.63
Hexafluoroethane	0.642	3.860		30.01	35.60	38.40	39.87
ΔHt, 0.893 at −169.17°C							
Hexahydroindane							
cis			11.0				
trans			10.7				

TABLE 5.2 Heats of Melting and Vaporization (or Sublimation) and Specific Heat at Various Temperatures of Organic Compounds (*continued*)

Substance	ΔH_m	ΔH_v	ΔH_s	C_p			
				400 K	600 K	800 K	1000 K
Hexamethylbenzene	4.93		17.9	74.18	97.13	113.51	125.55
$\quad\Delta H_t$, 0.269 at $-156.67°C$; 0.422 at 110.7°C							
Hexmethyldisiloxane			8.9				
Hexanal			22.72	44.00	58.30	68.70	76.40
Hexanamide	3.126	6.896	7.54				
Hexane	4.305	8.9	11.14	43.47	58.99	70.36	78.89
1-Hexanethiol	6.98	15.45	17.3	48.87	65.26	77.84	87.65
Hexanoic acid		11.6	14.8				
1-Hexanol	3.68			46.68	62.58	74.25	82.92
1-Hexene	2.234	6.76	7.32	40.03	53.90	64.02	71.54
2-Hexene							
$\quad$*cis*		6.96	7.52	38.60	53.00	63.40	71.20
$\quad$*trans*		6.91	7.54	39.70	53.40	63.60	71.20
3-Hexene							
$\quad$*cis*		6.86	7.47	38.50	53.20	63.50	71.20
$\quad$*trans*		6.92	7.54	40.20	53.90	63.90	71.40
1-Hexyne				37.87	49.59	58.16	64.56
Hydroquinone			23.7				
8-Hydroxyquinoline			26.0				
Indane			11.8				
Indene			12.64				
Indole			16.7				
Iodobenzene	2.33	9.44	11.85	31.10	41.43	48.07	52.60
4-Iodobenzoic acid			21.0				
Iodocyclohexane			11.3				
Iodoethane		7.115	7.7	19.18	24.64	28.65	31.65
Iodomethane		6.52	6.63	12.33	15.28	17.47	19.17

2-Iodo-2-methylpropane	3.47		8.46	35.27	45.82	52.85	57.91
1-Iodonaphthalene			17.3	26.27	34.11	39.80	44.03
2-Iodonaphthalene			21.7	26.59	34.58	40.21	44.34
1-Iodopropane			8.6				
2-Iodopropane			8.14				
3-Iodopropene			9.1				
Iodotoluene, 3- and 4-			13.0				
Isobutylbenzene			11.54				
Isobutyl dichloroacetate			12.5				
Isobutyl phenyl ketone			14.22				
Isobutyl trichloroacetate			12.7				
Isobutyronitrile	7.754	8.99	28.56	37.39	43.74	48.40	
Isopropyl acetate	1.86	8.97	8.89	48.00	66.20	78.60	87.30
Isopropylbenzene		8.35	10.79	35.96	46.81	54.13	59.26
Isopropyl nitrate			9.27				
Isopropyl trichloroacetate			12.4				
Isothiocyanic acid				12.71	14.57	15.74	16.57
Ketene	8.8		4.18	14.22	16.89	18.80	20.25
Lauric acid			31.7				
Leucine			36.0				
(+)-Limonene			11.5				
Maleic acid			26.3				
Maleic anhydride			17.1				
Malononitrile			18.9				
D-Mannitol							
Melamine	5.39		29.7				
2,2-Metacyclophane			22.0				
Methane	0.225	1.953	5.7	9.71	12.55	15.18	17.40
ΔH_t, 0.0187 at −248 to −252.7°C							
Methanethiol ΔH_t, 0.0525 at −135.6°C	1.411	5.872	8.94	14.04	17.57	20.32	22.48
Methanol ΔH_t, 0.152 at −115.8°C	0.768	8.24	15.42	12.29	16.02	19.04	21.38
4-Methoxybenzaldehyde			11.18				
Methoxybenzene			10.2				
2-Methoxytetrahydropyran							
Methyl (CH$_3$)				10.05	11.54	12.89	14.09

TABLE 5.2 Heats of Melting and Vaporization (or Sublimation) and Specific Heat at Various Temperatures of Organic Compounds (*continued*)

Substance	ΔH_m	ΔH_v	ΔH_s	C_p 400 K	C_p 600 K	C_p 800 K	C_p 1000 K
Methyl allyl sulfone	1.466	6.169	19.0	14.38	18.86	22.44	25.26
Methylamine			5.80				
Methyl benzyl sulfone	1.155		23.7				
2.Methyl-1,3-butadiene		6.191	6.32	31.80	41.40	48.00	52.90
3-Methyl-1,2-butadiene		6.51	6.68	31.00	40.30	47.20	52.40
2-Methylbutane	1.231	5.901	5.94	36.49	49.89	59.71	67.12
2-Methyl-1-butanethiol	1.78						
3-Methyl-1-butanethiol		8.0					
2-Methyl-2-butanethiol	0.1454	7.50	8.51	42.79	56.58	66.28	73.30
$\quad \Delta H_t$, 1.907 at $-114.0\,°C$							
3-Methylbutanoic acid	1.750	10.32	12.9				
2-Methyl-1-butanol		10.5	13.0				
3-Methyl-1-butanol		10.54	11.9				
2-Methyl-2-butanol	1.06	9.6	12.4				
$\quad \Delta H_t$, 0.47 at $-127.2\,°C$							
3-Methyl-2-butanol	1.891	9.9					
2-Methyl-1-butene	1.281	6.094	6.181	33.20	44.72	53.15	59.43
3-Methyl-1-butene	1.816	5.750	5.70	35.26	45.90	53.85	59.83
2-Methyl-2-butene		6.287	6.468	31.93	43.42	52.05	58.55
Methyl butyl sulfone			18.2				
Methyl *tert*-butyl sulfone			19.7				
3-Methyl-1-butyne		6.25	6.16	31.10	40.60	47.40	52.40
Methyl crotonate			9.8				
Methylcyclohexane	1.614	7.44	8.45	44.35	64.46	78.74	88.79
2-Methylcyclohexanol, *cis*- and *trans*-			15.1				
3-Methylcyclohexanol							
$\quad$ *cis*			15.6				
$\quad$ *trans*			15.7				

Compound							
4-Methylcyclohexanol							
cis	1.656		15.7				
trans			15.8				
Methylcyclopentane		6.95	7.55	36.11	52.43	64.00	72.44
1-Methylcyclopentene			7.55	32.50	46.80	57.00	64.30
3-Methylcyclopentene			7.7	32.60	47.10	57.20	64.50
4-Methylcyclopentene			7.7	32.60	47.00	57.10	64.40
Methyldichlorosilane			6.7				
2-Methyl-1,3-dioxane			9.23				
4-Methyl-1,3-dioxane			9.36				
Methylene (CH$_2$)				8.64	9.37	10.14	10.89
1-Methyl-2-ethylbenzene	2.38 / 2.28	9.29	11.40	48.50	65.80	78.10	86.90
1-Methyl-3-ethylbenzene	1.82 / 1.79	9.21	11.21	47.50	65.40	77.80	86.80
1-Methyl-4-ethylbenzene	3.19	9.18	11.14	47.20	65.00	77.60	86.60
2-Methyl-3-ethylpentane	2.71	7.88	9.20				
3-Methyl-3-ethylpentane	2.59	7.84	9.08				
2-Methyl-3-ethyl-1-pentene			8.98				
Methyl ethyl sulfite			10.4				
Methyl ethyl sulfone			18.6				
Methyl formate	1.800	6.75		19.50	25.20	29.10	32.00
Methylglyoxal			9.1				
2-Methylheptane	2.839	8.08	9.48				
3-Methylheptane	2.779	8.10	9.52				
4-Methylheptane	2.59	8.10	9.48				
Methyl heptanoate			12.0				
2-Methylhexane	2.195	7.33	8.32	50.42	68.33	81.43	91.20
3-Methylhexane		7.36	8.39	50.42	68.33	81.43	91.20
Methyl hexanoate			11.1				
Methylhydrazine							
Methylidyne			9.65				
CH				6.98	7.11	7.40	7.78
CH$^+$				6.98	7.10	7.36	7.65
1-Methyl-2-isopropylbenzene	2.39	9.17	12.10				

TABLE 5.2 Heats of Melting and Vaporization (or Sublimation) and Specific Heat at Various Temperatures of Organic Compounds (*continued*)

Substance	ΔH_m	ΔH_v	ΔH_s	C_p				
				400 K	600 K	800 K	1000 K	
1-Methyl-3-isopropylbenzene	3.27	9.11	11.94					
1-Methyl-4-isopropylbenzene	2.31	9.12	12.02	32.97	44.17	52.67	59.08	
Methyl isopropyl ether			6.27					
Methyl isopropyl ketone			8.82					
Methyl isopropyl sulfone			16.8					
3-Methylisoxazole			9.8					
5-Methylisoxazole			10.0					
Methylmercury bromide			16.2					
Methylmercury chloride			15.5					
Methylmercury iodide			15.6					
1-Methylnaphthalene	1.160	11.0		50.74	69.79	82.48	91.21	
ΔH_t, 1.190 at $-32.37°C$								
2-Methylnaphthalene	2.808	11.0		50.50	69.31	82.03	90.86	
ΔH_t, 1.34 at 15.4°C								
Methyl nitrate	1.97	7.54	8.1	21.87	27.54	31.47	34.19	
Methyl nitrite		5.0	5.4	18.24	23.35	26.97	29.52	
2-Methylpentane	1.498	6.643	7.138	44.00	59.60	70.80	79.20	
3-Methylpentane		6.711	7.236	43.47	59.00	70.40	78.90	
Methyl pentanoate			10.2					
2-Methyl-1-pentene		6.71	7.29	40.80	54.40	64.40	71.80	
3-Methyl-1-pentene		6.43	6.83	42.50	55.60	65.20	72.30	
4-Methyl-1-pentene		6.47	6.86	38.90	52.90	63.10	70.70	
2-Methyl-2-pentene		6.93	7.55	39.00	53.20	58.60	71.10	
3-Methyl-2-pentene								
cis		6.89	7.49	39.00	53.20	63.40	71.10	
trans		7.00	7.67	39.00	53.20	63.40	71.10	
4-Methyl-2-pentene								
cis		6.59	7.04	40.05	54.10	64.00	71.50	
trans		6.68	7.16	41.90	54.80	64.50	71.80	

Methyl phenyl sulfone			22.0				
Methylphosphonic acid			11.5				
2-Methylpropanal			7.5	29.77	40.62	48.49	54.40
2-Methylpropane	1.085	5.089	4.57	35.31	46.26	53.77	59.17
2-Methyl-1-propanethiol	1.191	7.412	8.28	36.13	47.60	55.53	61.24
2-Methyl-2-propanethiol	0.593	6.80	7.36	34.16	45.37	53.28	59.16
ΔHt, 0.972 at −121.6°C; 0.155 at −116.2°C; 0.232 at −73.8°C							
2-Methyl-1-propanol	1.602	9.80	12.04	26.57	35.30	41.86	46.85
2-Methyl-2-propanol	1.418	9.33	12.73	33.01	43.92	52.26	58.51
ΔHt, 0.20 at 12.99°C							
2-Methylpropene		5.286	4.92				
Methyl propyl ether			6.6				
2-Methylpyridine	2.324	8.654	10.15	31.92	44.55	53.21	59.34
3-Methylpyridine	3.389	8.932	10.62	31.82	44.47	53.12	59.23
α-Methylstyrene				44.80	60.70	71.80	79.80
β-Methylstyrene							
cis				44.80	60.70	71.80	79.80
trans				45.20	61.20	72.20	80.00
3-Methyl-2-thiabutane	2.236	7.338	8.15	34.69	46.01	54.95	62.29
2-Methylthiacyclopentane			10.1				
2-Methyl-3-thiapentane	2.08	8.7	9.2				
4-Methylthiazole			10.48				
2-Methylthiophene	2.263	8.103	9.26	29.43	39.57	46.43	51.30
3-Methylthiophene	2.518	8.186	9.44	29.38	39.34	45.95	50.59
Naphthalene	4.536	10.34	17.6	42.83	59.67	70.77	78.38
1-Naphthol			21.9				
2-Naphthol			19.8				
1,4-Naphthoquinone			17.3				
1-Naphthylamine			21.5				
2-Naphthylamine			21.1				
p-Nitroaniline	5.04		26				
Nitrobenzene	2.78	9.744					
1-Nitrobutane		9.3	11.6	37.65	50.21	59.03	65.39
2-Nitrobutane		8.8	10.48	37.61	50.46	59.44	65.96

TABLE 5.2 Heats of Melting and Vaporization (or Sublimation) and Specific Heat at Various Temperatures of Organic Compounds (*continued*)

Substance	ΔHm	ΔHv	ΔHs	C_p				
				400 K	600 K	800 K	1000 K
Nitroethane		8.4	9.9	23.66	31.45	36.81	40.67
Nitromethane	2.319	8.12	9.17	16.80	21.92	25.56	28.17
1-Nitronaphthalene			25.6				
1-Nitropropane		8.8	10.37	30.72	40.87	47.96	53.06
2-Nitropropane		8.4	9.88	30.89	41.19	48.22	53.24
Nonadecane ΔHt, 3.30 at 22.8°C	10.95	13.39	22.9	133.71	180.43	214.21	239.20
1-Nonadecene	8.0	13.06	22.68	130.26	175.35	207.86	231.80
1-Nonanal			17.28	64.80	86.40	101.90	113.40
Nonane ΔHt, 1.50 at −55.97°C	3.72	8.82	11.10	64.30	87.01	103.56	115.90
1-Nonanethiol	8.0	10.6		69.69	93.28	111.04	124.65
Nonanoic acid			19.7				
1-Nonanol		13.0	18.6	67.50	90.60	107.46	119.91
1-Nonene	4.3	8.68	10.88	60.85	81.93	97.22	108.50
Octadecane	14.81	13.02	21.7	126.77	171.09	203.15	226.90
Octadecanoic acid	15.1		39.8				
1-Octadecene	7.8	12.74	21.50	123.32	166.00	196.80	219.50
Octafluorocyclobutane	0.662	5.58		44.50	53.85	58.65	61.50
1-Octanal			16.28	57.90	77.00	90.90	101.00
Octanamide			26.4				
Octane	4.957	8.225	9.916	57.35	77.67	92.50	103.60
1-Octanethiol	58	10.1		62.75	83.94	99.97	112.31
Octanoic acid	3.30	16.73	19.2				
1-Octanol	10.1	11.2	15.6	60.56	81.26	96.39	107.58
1-Octene	3.660	8.07	9.70	53.91	72.58	86.15	96.20
1-Octyne				51.75	68.28	80.30	89.20
Oxalic acid ΔHt, 0.3($\alpha \rightarrow \beta$)			23.4				
Oxalyl chloride			7.6				
Oxamide			26.8				

Compound							
Palmitic acid	10.30						
[1.8]-Paracyclophane			37				
[2.2]-Paracyclophane			26.5				
[6.6]-Paracyclophane			23.0				
Paraldehyde		8.9	27.5	31.96	36.35	38.71	40.17
Pentachloroethane	2.7		9.9				
Pentachlorofluoroethane	0.449		10.9				
Pentachlorophenol	8.31	11.82	16.1	105.95	143.07	169.95	189.90
Pentadecane ΔHt, 2.19 at −2.25°C		11.63	18.20	102.50	137.98	163.60	182.50
1-Pentadecene	6.9	6.59	17.96	31.40	40.80	47.70	52.80
1,2-Pentadiene			6.85				
1,3-Pentadiene							
cis	1.468	6.60	6.77	29.50	39.90	47.00	52.20
trans		6.46	6.64	31.20	40.90	47.70	52.60
1,4-Pentadiene		6.01	6.01	31.30	40.80	47.60	52.70
2,3-Pentadiene		6.75	7.05	29.90	39.40	46.60	52.00
Pentaerythritol			34.4				
Pentaerythritol tetranitrate			36.3				
Pentafluorobenzoic acid			21.9				
Pentafluoroethane	2.95		16.1	27.20	32.94	36.12	37.98
Pentafluorophenol				65.00	86.08	101.29	112.33
Pentamethylbenzene				37.10	49.00	57.70	64.00
ΔHt, 0.473 at 23.7°C							
1-Pentanal			21.34				
Pentanamide			10.82				
Pentan-2,4-dione			6.32				
Pentane	2.008	6.16	14.17	36.53	49.64	59.30	66.55
Pentan-1,5-dithiol							
Pentanenitrile	1.130	7.98	9.83				
1-Pentanethiol	4.19	8.34	16.6	41.93	55.92	66.78	75.32
Pentanoic acid	3.850	10.53	13.61	39.74	53.24	63.18	70.59
1-Pentanol	2.34	10.6	12.7				
2-Pentanol		10.3	12.8				
3-Pentanol		10.1					
2-Pentanone		7.98	9.89	36.42	48.32	57.13	63.61

TABLE 5.2 Heats of Melting and Vaporization (or Sublimation) and Specific Heat at Various Temperatures of Organic Compounds (*continued*)

Substance	ΔHm	ΔHv	ΔHs	C_p				
				400 K	600 K	800 K	1000 K
1-Pentene	1.388	6.02	6.09	33.10	44.56	52.95	59.21
2-Pentene							
cis	1.700	6.24	6.41	31.57	43.62	52.29	58.78
trans	1.996	6.23	6.38	32.67	44.02	52.45	58.81
1-Pentyne		6.63	6.79	31.10	40.40	47.10	52.20
2-Pentyne		6.99	7.35	29.20	38.70	45.90	51.40
Perylene			30.0				
α-Phellandrene			12.1				
9,10-Phenanthraquinone			21.9				
Phenanthrene	2.752	13.3	21.1				
Phenol		9.73	16.41	32.45	43.54	50.62	55.49
Phenyl acetate			13.0				
β-Phenyl-1-alanine, DL- and L-			36.8				
Phenyl benzoate			23.0				
N-Phenyldiacetimide			21.5				
Phenyl ethyl sulfide			13.2				
Phenylhydrazine			14.69				
1-Phenyl-2-methylpropane	2.99	9.04	11.82				
Phenyl methyl sulfide			12.1				
Phenyl salicylate			22.0				
Phosgene	1.372	5.832		15.28	16.98	17.92	18.49
	1.335						
	1.131						
m-Phthalic acid			25.5				
p-Phthalic acid			23.5				
Phthalic anhydride			21.19				
α-Pinene			10.7				

β-Pinene			11.1				
Propadiene		4.45	7.09	17.21	22.00	25.42	28.00
1-Propanal			3.605	23.09	30.22	35.45	39.27
Propane	0.842	4.487	11.87	22.47	30.76	36.99	41.73
Propane-2,3-dithiol							
1-Propanethiol ΔHt, 0.949 at −131.06°C	1.309	7.059	7.62	27.86	36.72	43.60	49.01
2-Propanethiol ΔHt, 0.013 at −160.6°C	1.371	6.670	7.039	28.35	37.02	43.26	47.92
1-Propanol	1.242	9.982	11.36	25.86	34.56	41.04	45.93
2-Propanol	1.293	9.510	10.85	26.78	35.76	42.13	46.82
2-Propen-1-ol			11.3	22.81	30.11	35.28	39.06
Propionic acid	1.800	7.716	13.7	21.18	27.42	32.14	35.70
Propionic anhydride			12.6				
Propionitrile ΔHt, 0.408 at −96.19°C	1.202	7.353	8.632	28.51	37.99	44.94	50.21
1-Propylamine			7.46				
Propylbenzene	2.215	9.14	11.05	47.82	65.86	78.30	87.16
	2.03						
Propyl carbamate			19.4				
Propyl chloroacetate			11.6				
Propylcyclohexane	2.479	8.62	10.78	59.10	83.80	101.20	113.40
Propylcyclopentane	2.398	8.15	9.82	50.83	71.04	86.28	97.50
Propylene	0.718	4.40	6.67	19.23	25.81	30.77	34.52
Propylene oxide		6.87	9.70	22.16	30.07	35.68	39.79
Propyl nitrate	1.561	8.58	14.51	35.79	46.49	53.87	59.08
Propyl phenyl ketone			12.7				
Propyl trichloroacetate							
Propyne		5.29	13.45	17.33	21.80	25.14	27.71
Pyrazine			22.5				
Pyrene							
Pyridazine			12.78				
Pyridine	1.979	8.39	9.61	25.42	35.72	42.49	47.17
Pyrimidine			11.95				
Pyrrole			10.80				
Pyrrolidine ΔHt, 0.129 at −66.01°C	2.050	7.89	8.98	27.33	40.31	49.35	55.84
Salicyclic acid			22.74				
Sebacic acid			38.4				

TABLE 5.2 Heats of Melting and Vaporization (or Sublimation) and Specific Heat at Various Temperatures of Organic Compounds (*continued*)

Substance	ΔH_m	ΔH_v	ΔH_s	C_p 400 K	C_p 600 K	C_p 800 K	C_p 1000 K
5,5'-Spirobis(1,3-dioxane)			17.4				
Spiropentane	1.538	6.39	6.58	28.55	40.10	47.91	53.51
cis-Stilbene			16.5				
Styrene	2.617	8.85	10.50	38.32	52.14	61.40	67.92
Suberic acid			34.2				
Succinic acid			28.1				
Tetrabromomethane				23.20	24.51	25.51	25.32
Tetracene			30				
Tetrachlorobenzoquinone			23.6				
1,1,2-Tetrachloroethane		9.24		28.36	33.28	36.24	38.17
1,1,2,2-Tetrachloroethane			10.7	27.90	32.91	35.85	37.76
Tetrachloroethylene	2.5	8.3	9.4	25.10	27.86	29.29	30.07
Tetrachloromethane							
$\quad \Delta H_t$, 1.095 at −47.9°C	0.601	7.16	7.79	21.92	23.82	24.64	25.05
Tetracyanoethylene			19.4				
Tetradecane	10.90	11.38	17.01	99.01	133.72	158.89	177.60
Tetradecanoic acid			33.4				
1-Tetradecene	6.6	11.21	16.78	95.56	128.64	152.54	170.20
Tetraethylene glycol			24				
Tetraethyllead			13.6				
1,1,1,2-Tetrafluoroethane	1.844			24.90	30.76	34.20	36.36
Tetrafluoroethylene		4.02		21.97	25.53	27.61	28.86
Tetrafluoromethane							
$\quad \Delta H_t$, 0.353 at −196.92°C	0.167						
Tetrahydrofuran			7.65	17.30	20.74	22.58	23.61
Tetrahydrofurfuryl alcohol			15.9				
1,2,3,4-Tetrahydronaphthalene			13.4				
Tetrahydropyran			8.35				

Compound							
Tetraiodomethane	2.684	10.76	13.66	24.00	24.94	25.31	25.49
1,2,3,4-Tetramethylbenzene	2.561	10.47	13.34	56.81	75.68	89.42	99.47
1,2,3,5-Tetramethylbenzene	5.02	10.88	18	55.76	74.81	88.79	99.01
1,2,4,5-Tetramethylbenzene	1.802	7.51	10.24	55.50	74.38	88.41	98.71
2,2,3,3-Tetramethylbutane ΔH_t, 0.478 at $-120.66°C$							
Tetramethyllead			9.1				
Tetranitromethane			10.3				
Tetrazole			23				
2-Thiabutane	2.333	7.06	7.61	27.81	36.41	42.93	47.94
Thiacyclobutane ΔH_t, 0.160 at $-96.45°C$	1.971	7.7	8.56	21.89	30.45	36.40	40.67
Thiacycloheptane			11.30	42.0	65.0	79.0	88.0
Thiacyclohexane ΔH_t, 0.262 at $-71.75°C$; 1.858 at $-33.14°C$	0.585	8.60	10.22	35.71	52.37	64.00	72.34
Thiacyclopentane	1.757	8.28	9.28	28.95	40.04	47.66	53.14
Thiacyclopropane		6.98	7.24	16.53	21.99	25.61	28.21
4-Thia-5,5'-dimethyl-1-hexene			10.6				
2-Thiaheptane	2.96	8.78	10.88	48.67	65.02	77.59	87.41
3-Thiaheptane	2.90	8.76	10.74	48.37	64.96	77.74	87.75
4-Thiaheptane	2.976	8.2	10.64	48.21	65.13	78.45	89.05
2-Thiahexane	2.529	8.3	9.8	41.73	55.68	66.53	75.08
3-Thiahexane			9.58	41.43	55.62	66.68	75.42
5-Thianonane	4.64		12.75	62.09	83.81	100.58	113.71
2-Thiapentane	2.369	7.62	8.65	34.64	45.86	54.45	61.14
3-Thiapentane	2.845	7.59	8.55	34.65	46.11	54.91	61.79
2-Thiapropane	1.908	6.45	6.61	21.12	27.01	31.58	35.17
6-Thiaundecane			14.7				
Thioacetic acid	1.216	7.52	8.27	22.25	26.72	30.41	32.62
Thiophene ΔH_t, 0.152 at $-101.6°C$			21.8	23.02	30.95	36.01	39.54
Thymol			9.08				
Toluene	1.586	7.93	12.3	33.48	47.20	56.61	63.32
2-Toluenethiol			29.5				
2,4,6-Triamino-1,3,5-triazine							

TABLE 5.2 Heats of Melting and Vaporization (or Sublimation) and Specific Heat at Various Temperatures of Organic Compounds (*continued*)

Substance	ΔH_m	ΔH_v	ΔH_s	C_p 400 K	600 K	800 K	1000 K
Tribromomethane			17.2	18.80	21.03	22.29	23.12
Tributyl phosphate			9.8				
Trichloroacetyl chloride			21.2				
Trichlorobenzoquinone							
1,1,1-Trichloroethane							
ΔH_t, 1.79 at −48.95°C	0.45	7.96	7.76	25.72	30.68	33.73	35.81
1,1,2-Trichloroethane	2.7	8.3	9.4	25.03	30.13	33.28	35.42
Trichloroethylene		7.52	8.2	21.80	25.06	26.94	28.15
Trichlorofluoromethane				20.84	23.13	24.19	24.74
Trichloromethyl (CCl₃)				16.66	18.16	18.83	19.18
1,2,3-Trichloropropane			11.22	31.71	38.87	43.79	47.34
Tricyanoethylene			19.4				
Tridecane ΔH_t, 1.831 at −18.2°C	6.81	10.91	15.83	92.07	124.38	147.82	165.20
Tridecanoic acid	8.2		35.0				
1-Tridecene	6.2	10.75	15.60	88.62	119.29	141.48	157.80
Trimethylaluminum			17.5				
Triethylamine			8.29	48.70	66.10	78.56	87.80
Triethylaminoborane			14.5				
Triethyl arsenite			12.1				
Triethylarsine			10.3				
Triethyl borate			10.5				
Triethylenediamine							
ΔH_t, 2.30 at 79.8°C	1.45	17.07	14.8				
Triethylene glycol			18.9				
Triethyl phosphate			13.7				
Triethylphosphine			9.5				
Triethyl phosphite			10.0				
Triethylstibine			10.4				

1,1,1-Trifluoroethane	1.480	4.58		22.75	28.38	31.98	34.44
Trifluoroethylene	0.970	3.99		19.39	23.30	25.69	27.23
Trifluoromethane				14.61	18.16	20.35	21.76
Trifluoromethyl							
CF_2				13.74	16.17	17.50	18.25
CF_3^+				13.62	16.00	17.35	18.13
Trifluoromethylbenzene	3.29	7.80	8.98	40.59	54.20	62.75	68.45
Triiodomethane	3.9		16.7	19.60	21.52	22.64	23.38
2,4,5-Trimethylacetophenone			15.1				
2,4,6-Trimethylacetophenone			14.9				
Trimethylaluminum	1.564	5.48	15.1				
Trimethylamine			5.26	28.08	38.34	45.62	50.98
Trimethyl arsenite			10.1				
Trimethylarsine	1.955	9.57	6.9	46.90	64.00	76.70	85.90
1,2,3-Trimethylbenzene			11.73				
$\Delta H t$, 0.157 at $-54.46\,°C$; 0.319 at $-42.89\,°C$							
1,2,4-Trimethylbenzene	3.153	9.38	11.46	46.96	64.29	76.93	86.10
1,3,5-Trimethylbenzene	2.274	9.33	11.35	46.41	64.08	76.84	86.07
	1.932						
	1.892						
Trimethyl borate			8.3				
Trimethylboron			4.83				
2,2,3-Trimethylbutane	0.540	6.92	7.65	50.83	69.61	82.73	92.32
$\Delta H t$, 0.586 at $-157.8\,°C$							
Trimethylchlorosilane			7.2				
cis,cis-1,3,5-Trimethylcyclohexane				58.05	83.94	102.20	115.21
2,2,3-Trimethylpentane	2.06	7.65	8.82				
2,2,4-Trimethylpentane	2.20	7.41	8.40				
2,3,3-Trimethylpentane	0.205	7.73	8.90				
$\Delta H t$, 1.850 at $-109.01\,°C$							
2,3,4-Trimethylpentane	2.215	7.82	9.01				
2,4,4-Trimethyl-1-pentene		7.5	8.5				
2,4,4-Trimethyl-2-pentene		7.8	8.9				
Trimethylphosphine			6.7				
Trimethylphosphine oxide			12.0				

TABLE 5.2 Heats of Melting and Vaporization (or Sublimation) and Specific Heat at Various Temperatures of Organic Compounds (*continued*)

Substance	ΔH_m	ΔH_V	ΔH_S	C_p			
				400 K	600 K	800 K	1000 K
Trimethyl phosphite			8.8				
Trimethylsilanol			10.9				
Trimethylstibine			7.5				
Trimethylsuccinic anhydride			17.7				
Trimethylthiacyclopropane			9.40				
2,4,6-Trinitroanisole			31.8				
1,3,5-Trinitrobenzene			23.8				
Trinitromethane			11.15				
2,4,6-Trinitrophenetole			28.8				
2,4,6-Trinitrotoluene			28.3				
Triphenylarsine			23.5				
Triphenylene			28.2				
Triphenylmethane			23.9				
Triphenylphosphine			23				
Tropolone			20.0				
Undecane ΔH_t, 1.64 at $-36.55\,°C$	5.28	9.92	13.47	78.18	105.80	125.69	140.60
Undecanoic acid	6.2		29.0				
1-Undecene ΔH_t, 2.202 at $-55.8\,°C$	4.06	9.77	13.24	74.74	100.61	119.34	133.20
Urea			21.0				
o-Xylene	3.25	8.80	10.38	41.03	55.98	66.64	74.35
m-Xylene	2.765	8.69	10.20	40.03	55.51	66.41	74.23
p-Xylene	4.09	8.60	10.13	39.70	55.16	66.14	74.02
2,3-Xylenol			20.1				
2,4-Xylenol			15.74				
2,5-Xylenol			20.31				
2,6-Xylenol			18.07				
3,4-Xylenol			20.49				
3,5-Xylenol			19.80				

CRITICAL PHENOMENA

The *critical temperature* T_c of a gas is the temperature above which the gas cannot be liquefied no matter how high the pressure.

The *critical pressure* P_c is the lowest pressure which will liquefy the gas at its critical temperature.

The *critical molar volume* V_c is the volume of 1 mol at the critical temperature and the critical pressure. It can be computed from the critical density ρ_c as follows:

$$\frac{\text{Molecular weight in g} \cdot \text{mol}^{-1}}{\rho_c \text{ in g} \cdot \text{cm}^{-3}} = V_c \text{ in cm}^3 \cdot \text{mol}^{-1}$$

The critical pressure, critical molar volume, and critical temperature are the values of the pressure, molar volume, and thermodynamic temperature at which the densities of coexisting liquid and gaseous phases just become identical. At this critical point the *critical compressibility factor Z_c* is

$$Z_c = \frac{P_c V_c}{RT_c}$$

Since pressure, volume, and temperature are related to the corresponding critical properties, the function connecting the reduced properties becomes the same for each substance. The reduced property is expressed as a fraction of the critical property.

$$P_r = \frac{P}{P_c} \quad V_r = \frac{V}{V_c} \quad T_r = \frac{T}{T_c}$$

TABLE 5.3 Critical Properties

Substance	T_c, K	P_c, atm	V_c, cm$^3 \cdot$mol^{-1}
Acetaldehyde	461	55	154
Acetic acid	594.4	57.1	171.3
Acetic anydride	569	46.2	290
Acetone	508.1	46.4	209
Acetonitrile	548	47.7	173
Acetophenone	701	38	376
Acetyl chloride	508	58	204
Acetylene	308.3	60.6	113
Acrylic acid	615	56	210
Acrylonitrile	536	45	210
Air	132.5	37.2	92.7
Allene	393		
Allyl alcohol	545	56.4	203
Allyl sulfide	653		
Aluminum trichloride	629	26	261
Aminoethanol	614	44	196
Ammonia	405.6	111.3	72.5
Aniline	699	52.4	270
Anisole	368	41.2	
Anthracene	883		
Antimony tribromide	904.5	56	
Antimony trichloride	794		270

TABLE 5.3 Critical Properties (*continued*)

Substance	T_c, K	P_c, atm	V_c, cm$^3 \cdot$ mol^{-1}
Argon	150.8	48.1	74.9
Arsine	373.0		
Benzaldehyde	695	46	
Benzene	562.1	48.3	259
Benzoic acid	752	45	341
Benzonitrile	699.4	41.6	
Benzyl alcohol	677	46	334
Biphenyl	789	38	502
Bismuth tribromide	1219		301
Bismuth trichloride	1179	118	261
Boron pentafluoride	470		
Boron tribromide	573		280
Boron trichloride	451.9	38.2	
Boron trifluoride	260.8	49.2	
Bromine	584	102	127
Bromobenzene	670	44.6	324
Bromoethane	503.8	61.5	215
Bromomethane	464	85	
Bromopentafluorobenzene	670	44.6	
Bromotrifluoromethane	340.2	39.2	200
1,2-Butadiene	443.7	44.4	219
1,3-Butadiene	425	42.7	221
Butane	425.2	37.5	255
1-Butanol	562.9	43.6	274
2-Butanol	536.0	41.4	268
2-Butanone	535.5	41.0	267
1-Butene	419.6	39.7	240
cis-2-Butene	435.6	41.5	234
trans-2-Butene	428.6	40.5	238
3-Butenenitrile	585	39	265
1-Buten-3-yne	455	49	202
Butyl acetate	579	31	400
1-Butylamine	524	41	288
N-Butylaniline	72	28	518
Butylbenzene	660.5	28.5	497
sec-Butylbenzene	664	29.1	
tert-Butylbenzene	660	29.3	
Butyl benzoate	723	26	561
Butylcyclohexane	667	31.1	
sec-Butylcyclohexane	669	26.4	
tert-Butylcyclohexane	659	26.3	
Butyl ethyl ether	531	30	390
1-Butyne	463.7	46.5	220
2-Butyne	488.6	502	221
Butyraldehyde	524	40	278
Butyric acid	628	52.0	292
Butyronitrile	582.2	37.4	285
Carbon dioxide	304.2	72.8	94.0
Carbon disulfide	552	78.0	170

TABLE 5.3   Critical Properties (*continued*)

Substance	T_c, K	P_c, atm	V_c, cm$^3 \cdot$ mol^{-1}
Carbon monoxide	132.9	34.5	93.1
Carbon tetrachloride	556.4	45.0	276
Carbon tetrafluoride	227.6	36.9	140
Carbonyl chloride (phosgene)	455	56	190
Carbonyl sulfide	375	58	140
Chlorine	417	76.1	124
Chlorine pentafluoride	415.7	51.9	230.9
Chlorine trifluoride	426.6		
Chlorobenzene	632.4	44.6	308
1-Chlorobutane	542	36.4	312
2-Chlorobutane	520.6	39	305
1-Chloro-1, 1-difluoroethane	410.2	40.7	231
2-Chloro-1,l-difluoroethylene	400.5	44.0	197
Chlorodifluoromethane	369.2	49.1	165
Chloroethane	460.4	52.0	199
Chloroform	536.4	54.0	239
Chloromethane	416.3	65.9	139
2-Chloro-2-methylpropane	507	39	295
Chloropentafluoroacetone	410.7	28.4	
Chloropentafluoroethane	353.2	31.2	252
1-Chloropropane	503	45.2	254
2-Chloropropane	485	46.6	230
3-Chloropropene	514	47	234
Chlorotrifluoromethane	302.0	38.7	180
Chlorotrifluorosilane	308.5	34.2	
o-Cresol	697.6	49.4	282
m-Cresol	705.8	45.0	310
p-Cresol	704.6	50.8	277
Cyanogen	400	59	
Cyclobutane	459.9	49.2	210
Cycloheptane	589	36.7	390
Cyclohexane	553.4	40.2	308
Cyclohexanol	625	37	327
Cyclohexanone	629	38	312
Cyclohexene	560.4	42.9	292
Cyclopentane	511.6	44.5	260
Cyclopentanone	626	53	268
Cyclopentene	506.0		
Cyclopropane	397.8	54.2	170
Cymene	658		
cis-Decalin	702.2	31	
trans-Decalin	690.0	31	
Decane	617.6	20.8	603
Decanenitrile	621.9	32.1	
1-Decanol	700	22	600
1-Decene	615	21.8	650
Decylcyclohexane	750	13.4	
Decylcyclopentane	723.8	15.0	

TABLE 5.3 Critical Properties (*continued*)

Substance	T_c, K	P_c, atm	V_c, cm$^3 \cdot$ mol^{-1}
Deuterium			
(equilibrium)	38.3	16.28	60.4
(normal)	38.4	16.43	60.3
Deuterium bromide	361.9		
Deuterium chloride	328.4		
Deuterium hydride	35.8	14.64	62.8
Deuterium iodide	421.7		
Deuterium oxide	644.0	213.8	55.6
Diborane	289.0	39.5	
1,2-Dibromoethane	582.9	70.6	
Dibromomethane	583	71	
Dibromotetrafluoroethane	487.6	34	329
Dibutylamine	596	25	517
Dibutyl ether	580	25	500
1,2-Dichlorobenzene	697.3	40.5	360
1,3-Dichlorobenzene	684	38	359
1,4-Dichlorobenzene	685	39	372
Dichlorodifluoromethane	385.0	40.7	217
1,1-Dichloroethane	523	50	240
1,2-Dichloroethane	561	53	220
1,1-Dichloroethylene	544		
1,2-Dichloroethylene	516.5	54.4	
Dichlorofluoromethane	451.6	51.0	197
Dichloromethane	510	60.0	193
1,2-Dichloropropane	577	44	226
Dichlorosilane	449	46.1	
1,1-Dichloro-1,2,2,2-tetrafluoroethane	418.6	32.6	294
1,2-Dichloro-1,1,2,2-tetrafluoroethane	418.9	32.6	293
Diethylamine	496.6	36.6	301
1,4-Diethylbenzene	657.9	27.7	480
Diethyl disulfide	642		
Diethylene glycol	681	46	316
Diethyl ether	466.7	35.9	280
3,3-Diethylpentane	610	26.4	
Diethyl sulfide	557	39.1	318
Difluoroamine (HNF$_2$)	403	93	
cis-Difluorodiazine (N$_2$F$_2$)	272	70	
trans-Difluorodiazine	260	55	
1,1-Difluoroethane	386.6	44.4	181
1,1-Difluoroethylene	302.8	44.0	154
Dihexyl ether	657	18	720
Dihydrogen disulfide	572	58.3	
Dihydrogen heptasulfide	1015	33	
Dihydrogen hexasulfide	980	36	
Dihydrogen octasulfide	1040	32	
Dihydrogen pentasulfide	930	38.4	
Dihydrogen tetrasulfide	855	43.1	
Dihydrogen trisulfide	738	50.6	
Diisopropyl ether	500	28.4	385

TABLE 5.3 Critical Properties (*continued*)

Substance	T_c, K	P_c, atm	V_c, cm^3·mol^{-1}
1,2-Dimethoxyethane	536	38.2	271
Dimethoxymethane	497		
Dimethylamine	437.6	52.4	187
N,N-Dimethylaniline	687	35.8	
2,2-Dimethylbutane	488.7	30.4	359
2,3-Dimethylbutane	499.9	30.9	358
2,3-Dimethyl-1-butene	501	32.0	343
2,3-Dimethyl-2-butene	524	33.2	351
3,3-Dimethyl-1-butene	490	32.1	340
1,1-Dimethylcyclohexane	591	29.3	416
cis-1,2-Dimethylcyclohexane	606	29.3	
trans-1,2-Dimethylcyclohexane	596	29.3	
cis-1,3-Dimethylcyclohexane	591	29.3	
trans-1,3-Dimethylcyclohexane	598	29.3	
1,1-Dimethylcyclopentane	547	34.0	360
cis-1,2-Dimethylcyclopentane	564.8	34.0	368
trans-1,2-Dimethylcyclopentane	553.2	34.0	362
Dimethyl ether	400.0	53.0	178
2,2-Dimethylhexane	549.8	25.0	478
2,3-Dimethylhexane	563.4	25.9	468
2,4-Dimethylhexane	553.5	25.2	472
2,5-Dimethylhexane	550.0	24.5	482
3,3-Dimethylhexane	562.0	26.2	443
3,4-Dimethylhexane	568.8	26.6	466
Dimethyl oxalate	628	39.2	
2,2-Dimethylpentane	520.4	27.4	416
2,3-Dimethylpentane	537.3	28.7	393
2,4-Dimethylpentane	519.7	27.0	418
3,3-Dimethylpentane	536.3	29.1	414
2,2-Dimethylpropane	433.8	31.6	303
2,2-Dimethyl-1-propanol	549	39	319
2,3-Dimethylpyridine	655.4		
2,4-Dimethylpyridine	644.2		
2,5-Dimethylpyridine	644		
2,6-Dimethylpyridine	623.7		
3,4-Dimethylpyridine	683.8		
3,5-Dimethylpyridine	667.2		
N,N-Dimethyl-*o*-toluidine	668	30.8	
1,4-Dioxane	587	51.4	238
Diphenyl ether	766	31	
Diphenylmethane	767	29.4	
Dipropylamine	550	31	407
Dodecane	658.3	18.0	713
1-Dodecanol	679	19	718
1-Dodecene	657	18.3	
Dodecylcyclopentane	750	12.8	
Ethane	305.4	48.2	148
Ethanethiol	498.6	54.2	207

TABLE 5.3   Critical Properties (*continued*)

Substance	T_c, K	P_c, atm	V_c, cm$^3 \cdot$ mol^{-1}
Ethanol	516.2	63.0	167
Ethoxybenzene	647.1	33.8	
Ethyl acetate	523.2	37.8	286
Ethyl acetoacetate	673		
Ethyl acrylate	552	37.0	320
Ethylamine	456	55.5	178
Ethylbenzene	617.1	35.6	374
Ethyl benzoate	697	32	451
2-Ethyl-1-butanol	418.8		
Ethyl butyrate	565.9	30.2	395
Ethyl crotonate	599		
Ethylcyclohexane	609	29.9	450
Ethylcyclopentane	369.5	33.5	375
Ethylene	282.4	49.7	129
Ethylenediamine	592.9	62.1	206
Ethylene glycol	645	76	186
Ethylene oxide	469	71.0	140
Ethyl formate	508.4	46.8	229
3-Ethylhexane	565.4	25.7	455
2-Ethylhexanol	613	27.2	494
2-Ethyl-1-methylbenzene	651	30.0	460
3-Ethyl-1-methylbenzene	637	28.0	490
4-Ethyl-1-methylbenzene	640	29.0	470
Ethyl 3-methylbutyrate	588.0		
1-Ethyl-1-methylcyclopentane	592	29.5	
Ethyl methyl ether	437.8	43.4	221
Ethyl methyl ketone	535.6	41.0	267
3-Ethyl-2-methylpentane	567.0	26.7	443
3-Ethyl-2-methylpentane	576.5	27.7	455
3-Ethyl-3-methylpentane	576.4	27.7	455
Ethyl-2-methylpropanoate	553	30	410
Ethyl methyl sulfide	533	42	
3-Ethylpentane	540.6	28.5	416
o-Ethylphenol	703.0		
m-Ethylphenol	716.4		
p-Ethylphenol	716.4		
Ethylpropanoate	546.0	33.2	345
Ethyl propyl ether	500.6	32.1	244
o-Ethyltoluene	653	31	461
m-Ethyltoluene	636	31	461
p-Ethyltoluene	636	31	461
Ethyl vinyl ether	475	40.2	260
Fluorine	144.3	51.5	66.2
Fluorobenzene	560.1	44.9	271
Fluoroethane	375.3	49.6	169
Fluoromethane	317.8	58.0	124
Fluorotrichloromethane	471.1	43.2	248
Formaldehyde	408	65	

TABLE 5.3 Critical Properties (*continued*)

Substance	T_c, K	P_c, atm	V_c, cm$^3 \cdot$ mol^{-1}
Formic acid	580		
Furan	490.2	54.3	218
Germanium tetrachloride	550.0	38	330
Glycerol	726	66	255
Hafnium tetrabromide	746		415
Hafnium tetrachloride	723	57.0	304
Hafnium tetraiodide	916		528
Helium-3	3.30	1.167	73.2
Helium-4	5.19	2.24	57.3
Heptadecane	733	13	1000
1-Heptadecanol	736	14	
Heptane	540.2	27.0	432
1-Heptanol	633	30	435
1-Heptene	537.2	28	440
Heptylcyclopentane	679	19.2	
Hexadecane	717	14	
1-Hexadecene	717	13.2	
Hexadecylcyclopentane	791	9.6	
1,5-Hexadiene	507	34	328
Hexafluoroethane	292.8	29.4	223.7
Hexamethylbenzene	767		
Hexane	507.4	29.3	370
1-Hexanol	610	40	381
1-Hexene	504.3	31.3	350
cis-2-Hexene	518	32.4	351
trans-2-Hexene	516	32.3	351
cis-3-Hexene	517	32.4	350
trans-3-Hexene	519.9	32.1	350
Hexylcyclopentane	660.1	21.1	
Hydrazine	653	145	96.1
Hydrogen			
(equilibrium)	32.9	12.77	65.4
(normal)	33.2	12.8	65.0
Hydrogen bromide	363.2	84.4	100.0
Hydrogen chloride	324.6	82.0	81.0
Hydrogen cyanide	456.8	53.2	139
Hydrogen deuteride,			
see Deuterium hydride			
Hydrogen fluoride	461	64	69
Hydrogen iodide	424.0	82.0	131
Hydrogen selenide	411	88	
Hydrogen sulfide	373.2	88.2	98.5
Icosane	767	11.0	
1-Icosanol	770	12.0	
Iodine	819	115	155
Iodobenzene	721	44.6	351
Iodomethane	528	65	190

TABLE 5.3   Critical Properties (*continued*)

Substance	T_c, K	P_c, atm	V_c, cm³·mol⁻¹
Isobutyl acetate	561	30	414
Isobutylamine	516	42	284
Isobutylbenzene	650	31	480
Isobutyl butyrate	611		
lsobutylcyclohexane	659	30.8	
Isobutyl formate	551	38.3	350
Isobutyl 3-methylbutyrate	621		
Isobutyl propanoate	592		
Isobutyric acid	609	40	292
Isopropylamine	476	50	229
lsopropylbenzene	631.0	31.7	428
lsopropylcyclohexane	640	28	
Isopropylcyclopentane	601	29.6	
2-Isopropyl-1-methylbenzene	670	28.6	
3-Isopropyl-l-methylbenzene	666	290	
4-Isopropyl-1-methylbenzene	653	27.9	
Isoquinoline	803		
Isoxazole	552.0		
Ketene	380	64	145
Krypton	209.4	54.3	91.2
Mercury	1173	180	
Methane	190.6	45.4	99.0
Methanethiol	470.0	71.4	145
Methanol	512.6	79.9	118
Methoxybenzene (anisole)	641	41.2	
Methyl acetate	506.8	46.3	228
Methyl acrylate	536	42	265
Methylamine	430	73.6	140
N-Methylaniline	701	51.3	
Methyl benzoate	692	36	396
2-Methyl-1,3-butadiene	484	38.0	276
3-Methyl-1,2-butadiene	496	40.6	267
2-Methylbutane	460.4	33.3	306
2-Methyl-1-butanol	571	38	322
3-Methyl-1-butanol	579.5	38	329
2-Methyl-2-butanol	545	39	319
3-Methyl-2-butanone	553.4	38.0	310
2-Methyl-1-butene	465	34.0	294
2-Methyl-2-butene	470	34.0	318
3-Methyl-1-butene	450	34.7	300
Methyl butyrate	554.4	34.3	340
3-Methylbutyric acid	634		
Methylcyclohexane	572.1	34.3	368
Methylcyclopentane	532.7	37.4	319
N-Methylethylamine	496.6	36.6	243
Methyl formate	487.2	59.2	172
2-Methylheptane	559.6	24.5	488

TABLE 5.3 Critical Properties (*continued*)

Substance	T_c, K	P_c, atm	V_c, cm$^3 \cdot$ mol^{-1}
3-Methylheptane	563.6	25.1	464
4-Methylheptane	561.7	25.1	476
2-Methylhexane	530.3	27.0	421
3-Methylhexane	535.2	27.8	404
Methylhydrazine	567	79.3	271
Methyl isobutyrate	540.8	33.9	339
Methyl isocyanate	491	55	
1-Methylnaphthalene	772	35.2	445
2-Methylnaphthalene	761	34.6	462
2-Methylpentane	497.5	29.7	367
3-Methylpentane	504.4	30.8	367
2-Methyl-2,4-pentanediol	678	33.9	
4-Methyl-2-pentanone	571	32.3	371
2-Methyl-2-pentene	518	32.4	351
cis-3-Methyl-2-pentene	518	32.4	351
trans-3-Methyl-2-pentene	521	32.3	350
cis-4-Methyl-2-pentene	490	30	360
trans-4-Methyl-2-pentene	493	30	360
Methyl phenyl ether	641	41.2	
2-Methylpropanal	513	41	274
2-Methylpropane	408.1	36.0	263
Methyl propanoate	530.6	39.5	282
2-Methyl-1-propanol (isobutyl alcohol)	547.7	42.4	273
2-Methyl-2-propanol	506.2	39.2	275
2-Methylpropene	417.9	39.5	239
2-Methylpyridine	621		
3-Methylpyridine	645		
4-Methylpyridine	646	44	311
α-Methylstyrene	654	33.6	397
Methyl vinyl ether	436	47	205
Morpholine	618	54	253
Naphthalene	748.4	40.0	410
Neon	44.4	27.2	41.7
Niobium pentabromide	1010		469
Niobium pentachloride	807		400
Niobium pentafluoride	737	62	155
Nitric oxide	180	64	58
Nitrobenzene	732		
Nitrogen-14	126.2	33.5	89.5
Nitrogen-15	126.3	33.5	90.4
Nitrogen dioxide (equilibrium)	431.4	100	170
Nitrogen trifluoride	234.0	44.7	
Nitromethane	588	62.3	173
Nitrosyl chloride	440	90	139
Nitrous oxide	309.6	71.5	97.4
Nitryl fluoride	349.4		
Nonadecane	756	11.0	
Nonane	594.6	22.8	548
1-Nonanol	677		546

TABLE 5.3 Critical Properties (*continued*)

Substance	T_c, K	P_c, atm	V_c, cm$^3 \cdot$mol^{-1}
1-Nonene	592	23.1	580
Nonylcyclopentane	710.5	16.3	
Octadecane	745	11.9	
1-Octadecanol	747	14	
1-Octadecene	739	11.2	
Octane	568.8	24.5	492
1-Octanol	658	34	490
2-Octanol	637	27	494
1-Octene	566.6	25.9	464
trans-2-Octene	580	27.3	
Octylcyclopentane	694	17.7	
Oxygen	154.6	49.8	73.4
Oxygen difluoride	215.2	48.9	97.7
Ozone	161.3	55.0	88.9
Paraldehyde	563		
Pentachloroethane	646.1		
Pentadecane	707	15	880
1-Pentadecene	704	14.4	
Pentadecylcyclopentane	780	10.1	
1,2-Pentadiene	503	40.2	276
trans-1,3-Pentadiene	496	39.4	275
1,4-Pentadiene	478	37.4	276
Pentafluorobenzene	532.0	34.7	
1,1,2*H*-Pentafluoropropane	380.11	31.0	273
Pentanal	554	35	333
Pentane	469.6	33.3	304
Pentanoic acid	651	38	340
1-Pentanol	586	38	326
2-Pentanone	564.0	38.4	301
3-Pentanone	561.0	36.9	336
1-Pentene	464.7	40.0	300
cis-2-Pentene	476	36.0	300
trans-2-Pentene	475	36.1	300
Pentyl formate	576		
1-Pentyne	493.4	40	278
Perchloryl fluoride	368.4	53.0	161
Perfluoroacetone	357.3	28.0	
Perfluorobenzene	516.7	32.6	
Perfluorobutane	386.4	22.9	378
Perfluoro-(2-butyltetrahydrofuran)	500.3	15.9	588
Perfluorocyclobutane	388.4	27.41	260
Perfluorocyclohexane	457.2	24	
Perfluorocyclohexene	461.8		
Perfluorodecene	542.3	14.3	
Perfluoroethane	292.8	29.4	223.7
Perfluoroheptane	474.8	16.0	664
Perfluoroheptene	478.1		
Perfluorohexane	451.7	18.8	442

TABLE 5.3   Critical Properties (*continued*)

Substance	T_c, K	P_c, atm	V_c, cm$^3 \cdot$ mol^{-1}
Perfluorohexene	454.3		
Perfluoromethylcyclohexane	486.8	23	
Perfluoronaphthalene	673.1		
Perfluorononane	524.0	15.4	
Perfluorooctane	502	16.4	
Perfluoropentane	422	20.1	
Perfluoropropane	345.1	26.5	299
Phenanthrene	878		
Phenetole	647	33.8	
Phenol	694.2	60.5	229
Phosgene	455	56	190
Phosphine	324.4	64.5	
Phosphonium chloride	322.2	72.7	
Phosphorus bromide difluoride	386		
Phosphorus chloride difluoride	362.32	44.6	
Phosphorus dibromide fluoride	527		
Phosphorus dichloride fluoride	463.0	49.3	
Phosphorus pentachloride	645		
Phosphorus trichloride	563		260
Phosphorus trifluoride	271.2	42.7	
Phosphoryl chloride difluoride	423.8	43.4	
Phosphoryl trichloride	602		
Phosphoryl trifluoride	346.5	41.8	
Phthalic anhydride	810	47	368
Piperidine	594.0	47	289
Propadiene	393	54.0	162
Propane	369.8	41.9	203
1,2-Propanediol	625	60	237
1,3-Propanediol	658	59	241
Propanoic acid	612	53.0	230
1-Propanol	536.7	51.0	218.5
2-Propanol	508.3	47.0	220
2-Propenal	506	51	
Propionaldehyde	496	47	223
Propionitrile	564.4	41.3	230
Propyl acetate	549.4	23.9	345
Propylamine	497.0	46.8	233
Propylbenzene	638.3	31.6	440
Propylcyclopentane	603	29.6	425
Propylcyclohexane	639	27.7	
Propylene	365.0	45.6	181
Propylene oxide	482.2	48.6	186
Propyl formate	538.0	40.1	285
Propyl propanoate	578		
1-Propyne	402.4	55.5	164
Pyridine	620.0	55.6	254
Pyrrole	639.6	56	
Pyrrolidine	568.6	55.4	249

TABLE 5.3 Critical Properties (*continued*)

Substance	T_c, K	P_c, atm	V_c, cm$^3 \cdot$ mol^{-1}
Quinoline	794.4		
Radon	376.9	62	139
Rhenium(VII) oxide	942		334
Selenium	1766		
Silane	269.6	47.8	
Silicon chloride trifluoride	307.6	34.2	
Silicon tetrachloride	507	37	326
Silicon tetrafluoride	259.1	36.7	
Silicon trichlorofluoride	438.5	35.3	
Styrene	647	39.4	
Sulfur	1314		
Sulfur dioxide	430.8	77.8	122
Sulfur hexafluoride	318.7	37.1	198
Sulfur tetrafluoride	364.0		
Sulfur trioxide	491.0	81	130
Tantalum pentabromide	974		461
Tantalum pentachloride	767		400
o-Terphenyl	891.0	38.5	769
m-Terphenyl	924.8	34.6	784
p-Terphenyl	926.0	32.8	779
1,1,2,2-Tetrachloro-1,2-difluoroethane	551	34	370
1,1,2,2-Tetrachloroethane	661.1		
Tetrachloroethylene	620	44	290
Tetradecane	694	16	830
1-Tetradecene	689	15.4	
Tetradecylcyclopentane	772	11.1	
Tetrafluoroethylene	306.4	38.9	175
Tetrafluorohydrazine	309.4	37	
Tetrahydrofuran	540.2	51.2	224
1,2,3,4-Tetrahydronaphthalene	719	34.7	
Tetrahydrothiophene	631.9		
1,2,4,5-Tetramethylbenzene	675	29	480
2,2,3,3-Tetramethylbutane	567.8	28.3	461
2,2,3,3-Tetramethylhexane	623.1	24.8	
2,2,5,5-Tetramethylhexane	581.5	21.6	
2,2,3,3-Tetramethylpentane	607.6	27.0	
2,2,3,4-Tetramethylpentane	592.7	25.7	
2,2,4,4-Tetramethylpentane	574.7	24.5	
2,3,3,4-Tetramethylpentane	607.6	26.8	
2-Thiapropane	503.1	54.6	201
Thiophene	579.4	56.2	219
Thymol	698		
Tin(IV) chloride	591.8	37.0	351
Titanium tetrachloride	638	46	340
Toluene	591.7	40.6	316
o-Toluidine	694	37	343
m-Toludine	709	41	343
p-Toluidine	667		
Toluonitrile	723		

TABLE 5.3 Critical Properties (*continued*)

Substance	T_c, K	P_c, atm	V_c, cm$^3 \cdot$mol^{-1}
Tributylamine	643	18	
1,1,2-Trichloroethane	602	41	294
Trichloroethylene	571	48.5	256
Trichlorofluoromethane	471.2	43.5	248
1,2,3-Trichloropropane	651	39	348
1,2,2-Trichloro-1,1,2-trifluoroethane	487.2	33.7	304
Tridecane	675.8	17.0	780
1-Tridecene	674	16.8	
Tridecylcyclopentane	761	11.9	
Triethanolamine	787.4	24.2	
Triethylamine	535	30	390
Trifluoroacetic acid	491.3	32.2	204
1,1,1-Trifluoroethane	346.2	37.1	221
Trifluoromethane	298.89	47.7	133.3
Trimethylamine	433.2	40.2	254
1,2,3-Trimethylbenzene	664.5	34.1	430
1,2,4-Trimethylbenzene	649.1	31.9	430
1,3,5-Trimethylbenzene	637.3	30.9	433
2,2,3-Trimethylbutane	531.1	29.2	398
2,2,3-Trimethyl-1-butene	533	28.6	400
Trimethylchlorosilane	497.7	31.6	
1,1,2-Trimethylcyclopentane	579.5	29.0	
1,1,3-Trimethylcyclopentane	569.5	27.9	
cis,cis,trans- 1,2, 4-Trimethylcyclopentane	579	28.4	
cis,trans,cis- 1,2, 4-Trimethylcyclopentane	571	27.7	
3,3,5-Trimethylheptane	609.6	22.9	
2,2,3-Trimethylhexane	588	24.6	
2,2,4-Trimethylhexane	573.7	23.4	
2,2,5-Trimethylhexane	567.9	23.0	519
2,2,3-Trimethylpentane	563.4	26.9	436
2,2,4-Trimethylpentane	543.9	25.3	468
2,3,3-Trimethylpentane	573.5	27.8	455
2,3,4-Trimethylpentane	566.3	26.9	461
2,2,4-Trimethyl-1,3-pentanediol	671	25.6	364.6
1*H*-Undecafluoropentane	443.9		
Undecane	638.8	19.4	660
1-Undecene	637	19.7	
Uranium hexafluoride	505.8	45.5	250
Vinyl acetate	525	43	265
Vinyl chloride	429.7	55.3	169
Vinyl fluoride	327.8	51.7	114
Vinyl formate	475	57	210
Water	647.3	217.6	56.0
Xenon	289.7	57.6	118
o-Xylene	630.2	36.8	369

TABLE 5.3 Critical Properties (*continued*)

Substance	T_c, K	P_c, atm	V_c, cm$^3 \cdot$ mol^{-1}
m-Xylene	617.0	35.0	376
p-Xylene	616.2	347	379
2,3-Xylenol	722.6	48	470
2,4-Xylenol	707.6	43	509
2,5-Xylenol	723.0	48	470
2,6-Xylenol	700.9	42	509
3,4-Xylenol	729.8	49	552
3,5-Xylenol	715.6	36	611
Zirconium tetrabromide	805		415
Zirconium tetrachloride	778	56.9	319
Zirconium tetraiodide	960		528
Zirconium tetraiodide	960		528

Estimation of Critical Properties

When the critical properties are unavailable, they may be estimated employing structural contributions to estimate T_c, P_c, and V_c. Lydersen's critical-property increments* provide good estimates for T_c and P_c; Vetere's group contributions† yield reasonable estimates for V_c. The units employed are kelvins, atmospheres, and cubic centimeters per mole. Typical errors in estimated values are less than 2% for T_c but may rise up to 5% for higher-molecular-weight (greater than 100) nonpolar materials; errors are uncertain for molecules with multifunctional polar groups. Errors for estimated values of P_c and V_c are about double those for T_c.

The relations are

$$T_c = T_b \left[0.567 + \sum \Delta_T - \left(\sum \Delta_T \right)^2 \right]^{-1}$$

$$P_c = M \left(0.34 + \sum \Delta \right)^{-2}$$

$$V_c = 33 + \left[\sum_i \left(M_i \Delta_v \right) \right]^{1.029}$$

where T_b is the normal boiling point and M is the molular weight. Group contributions are listed in Table 5.4.

TABLE 5.4 Group Contributions for the Estimation of Critical Properties

There are no increments for hydrogen. All bonds shown as free are connected with atoms other than hydrogen. Values in parentheses are based upon very few experimental values.

Group	Δ_T, K	Δ_P, atm	Δ_V, cm$^3 \cdot$ mol^{-1}
		Nonring increments	
—CH$_3$, —CH$_2$—	0.020	0.227	3.360 (linear chain) 2.888 (side chain)

---

* A. L. Lydersen, Univ. Wisconsin Coll. Eng., Eng. Exp. Stn, Rep 3, Madison, April 1955.
† A Vetere, cited in R. C. Reid, J. M. Prausnitz, and T. K Sherwood, *The Properties of Gases and Liquids*, 3d ed., McGraw-Hill, New York, 1977, p. 17.

TABLE 5.4 Group Contributions for the Estimation of Critical Properties (*continued*)

Group	Δ_T, K	Δ_P, atm	$_v$, cm$^3 \cdot$ mol^{-1}
$-$CH	0.012	0.210	3.360 (linear chain)
			2.888 (side chain)
$-$C$-$	0.0	0.210	3.360 (linear chain)
			2.888 (side chain)
$=$CH$_2$, $=$CH	0.018	0.198	2.940
$=$C$-$	0.0	0.198	2.940
$=$C$=$	0.0	0.198	2.908
$\equiv$CH, $\equiv$C$-$	0.005	0.153	2.648
$-$O$-$	0.021	0.16	1.075
$>$C$=$O	0.040	0.29	1.765
$>$NH	0.031	0.135	2.333
$>$N$-$	0.014	0.17	1.793
$-$S$-$	0.015	0.27	0.591

Ring increments			
$-$CH$_2$$-$	0.013	0.184	2.813
$-$CH$-$	0.012	0.192	2.813
$-$C$-$	($-$0.007)	(0.154)	2.813
$=$CH, $=$C, $=$C$=$	0.011	0.154	2.538
$-$O$-$	(0.014)	(0.12)	0.790
$>$C$=$O	(0.033)	(0.2)	1.500
$>$NH	(0.024)	(0.09)	1.736
$>$N$-$	(0.007)	(0.13)	1.883
$-$S$-$	(0.008)	(0.24)	0.911

General substituents			
$-$F	0.018	0.224	0.770
$-$Cl	0.017	0.320	1.237
$-$Br	0.010	(0.50)	0.899
$-$I	0.012	(0.83)	0.702
$-$OH			
Alcohols	0.082	0.06	0.704
Phenols	0.031	($-$0.02)	1.553
HC$=$O (aldehyde)	0.048	0.33	2.333
$-$COOH	0.085	(0.4)	1.652
$-$COO$-$ (ester)	0.047	0.47	1.607
$-$NH$_2$	0.031	0.095	2.184
$-$CN	(0.060)	(0.36)	2.784
$-$NO$_2$	(0.055)	(0.42)	1.559
$-$SH	0.015	0.27	1.537
$-$Si$-$	0.03	(0.54)	

SECTION 6
SPECTROSCOPY

For more than half a century, spectroscopy has been the key structural tool for organic chemistry and biochemistry. This remains the case today although the availability of enhanced methods of X-ray analysis has somewhat altered the balance in both disciplines. A vast array of specialized texts are available that survey spectroscopic techniques in general or individual methods in particular. It would be folly to try to duplicate those efforts here. A brief overview of the methods may prove helpful and serve as a reminder for the practitioner who consults this handbook.

ULTRAVIOLET–VISIBLE SPECTROSCOPY

Generally, spectroscopic methods involve the absorption of radiation at certain wavelengths (and therefore certain energies) by molecules. For the techniques of ultraviolet–visible (UV–VIS) and infrared spectroscopy, there is an inverse proportionality between the amount of energy absorbed and the structural information that is revealed. The UV–VIS wavelength range is generally considered to be 200–800 nm. At the lower end, oxygen absorbs energy and the lowest end of the range is called the "vacuum UV." Relatively limited structural information may be obtained from the typically broad bands that are observed. Even so, the electronic transitions that are observed are sensitive to structure. Individual absorbing groups or chromophores absorb light at a characteristic wavelength. The absorption usually shifts when two or more chromophores are linked or in conjugation.

The UV–VIS technique may be used quantitatively by application of Beer's law:

$$\varepsilon = abc.$$

In this relationship, ε is the extinction coefficient that is characteristic of a given compound. The values a and c represent absorption and concentration, respectively. The

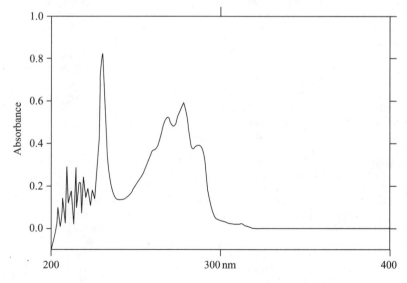

UV Sprectrum of Napthalene.

variable *b* is the pathlength of light through the sample, usually 1 cm. If the extinction coefficient is known, the measured absorption will give the compound's concentration.

The spectrum of naphthalene is shown above. Its broad bands reflect the absorption of energy by the extended pi-electron system.

The tables in this section present the essential data required for interpretation of UV–VIS spectra. Table 6.1 lists representative chromophores. These are identified by λ_{max}, the maximum height of any given peak, and by ε_{max}, the extinction coefficient at the maximum wavelength. Both the wavelength and absorption intensity are characteristic of individual chromophores.

Table 6.2 lists the ultraviolet cutoff for a variety of solvents commonly used in UV–VIS spectroscopy. The solvent chosen must dissolve the sample, yet be relatively transparent in the spectral region of interest. Typically, very low concentrations of sample will be present in the solvent. It is therefore important to avoid solvents that have even weak absorptions near the solute's bands of interest. Methanol and ethanol are two of the most commonly used solvents. Care must be exercised when using the latter that no benzene (an azeotropic drying agent) is present as this will alter the solvent's transparency. Normally, this will not be a problem in spectral grade solvents.

Tables 6.3–6.5 record data developed to undertake structural analysis in systems possessing chromophores that are conjugated or otherwise interact with each other. Chromophores within a molecule interact when linked directly to each other or when they are forced into proximity owing to structural constraints. Certain combinations of functional groups comprise chromophoric systems that exhibit characteristic absorption bands. In the era when UV–VIS was one of the principal spectral methods available to the organic chemist, sets of empirical rules were developed to extract as much information as possible from the spectra. The correlations referred to as Woodward's rules or the Woodward–Fieser rules, enable the absorption maxima of dienes (Table 6.3) and enones and dienones (Table 6.4) to be predicted. When this method is applied, wavelength increments correlated to structural features are added to the respective base values (absorption wavelength of parent compound). The data refer to spectra determined in methanol or ethanol. When other solvents are used, a numerical correction must be applied. These corrections are recorded in Table 6.5.

The benzene ring is a common structural element in organic chemistry. It is rigid and substituents are arranged in a fixed orientation. Shifts in the wavelength are expected for substituents that can interact electronically with the aromatic ring. Typically, shifts to longer wavelength (and intensification of the absorption band) are observed for any ring substitution. In the absence of conjugation, the shifts are small. Also, interposition of a single methylene group, or *meta* orientation within the aromatic ring, is sufficient to insulate chromophores almost completely from each other. With electron-withdrawing substituents, practically no change in the maximum position is observed. Directly conjugated groups may produce quite large spectral shifts. Examples include double and triple bonds and carbonyl groups. The spectra of heteroaromatics are related to their isocyclic analogs, but only in a general way. As with benzene, the magnitude of substituent shifts can be estimated, but tautomeric possibilities may invalidate the empirical method. Table 6.6 records data for substituents bonded directly to a benzene ring.

When electronically complementary groups are situated *para* to each other in disubstituted benzenes, there is a more pronounced shift to a longer wavelength than would be expected from the additive effect due to the extension of the chromophore from the electron-donating group through the ring to the electron-withdrawing group. When the *para* groups are not complementary, or when the groups are situated *ortho* or *meta* to each other, disubstituted benzenes show a more or less additive effect of the two substituents on the wavelength maximum. Calculation of the principal band of selected substituted benzenes is illustrated in Table 6.7.

TABLE 6.1 Electronic Absorption Bands for Representative Chromophores

Chromophore	System	λ_{max}	ε_{max}
Acetylene (ethynyl)	—C≡C—	175–180	6 000
Aldehyde	—CHO	210	strong
		280–300	11–18
Amine	—NH₂	195	2 800
Azido	>C=N—	190	5 000
Azo	—N=N—	285–400	3–25
Bromide	—Br	208	300
Carbonyl	>C=O	195	1 000
		270–285	18–30
Carboxyl	—COOH	200–210	50–70
Disulfide	—S—S—	194	5 500
		255	400
Ester	—COOR	205	50
Ether	—O—	185	1 000
Ethylene	—C=C—	190	8 000
Iodide	—I	260	400
Nitrate ester	—ONO₂	270 (shoulder)	12
Nitrile	—C≡N	160	—
Nitrite ester	—ONO	220–230	1 000–2 000
		300–400	10
Nitro	—NO₂	210	strong
Nitroso	—NO	302	100
Oxime	—NOH	190	5 000
Sulfone	—SO₂—	180	—
Sulfoxide	>S=O (>S→O)	210	1 500
Thiocarbonyl	>C=S	205	strong
Thioether	—S—	194	4 600
		215	1 600
Thiol	—SH	195	1 400
	—(C=C)₂— (acyclic)	210–230	21 000
	—(C=C)₃—	260	35 000
	—(C=C)₄—	300	52 000
	—(C=C)₅—	330	118 000
	—(C=C)₂— (alicyclic)	230–260	3 000–8 000
	C=C—C≡C	219	6 500
	C=C—C=N	220	23 000
	C=C—C=O	210–250	10 000–20 000
		300–350	weak
	C=C—NO₂	229	9 500
Benzene		184	46 700
		204	6 900
		255	170
Diphenyl		246	20 000
Naphthalene		222	112 000
		275	5 600
		312	175
Anthracene		252	199 000
		375	7 900
Phenanthrene		251	66 000
		292	14 000

TABLE 6.1 Electronic Absorption Bands for Representative Chromophores (*continued*)

Chromophore	System	λ_{max}	ε_{max}
Naphthacene		272	180 000
		473	12 500
Pentacene		310	300 000
		585	12 000
Pyridine		174	80 000
		195	6 000
		257	1 700
Quinoline		227	37 000
		270	3 600
		314	2 750
Isoquinoline		218	80 000
		266	4 000
		317	3 500

TABLE 6.2 Ultraviolet Cutoffs of Spectrograde Solvents

Absorbance of 1.00 in a 10.0 mm cell vs. distilled water

Solvent	Wavelength, nm	Solvent	Wavelength, nm
Acetic acid	260	Hexadecane	200
Acetone	330	Hexane	210
Acetonitrile	190	Isobutyl alcohol	230
Benzene	280	Methanol	210
1-Butanol	210	2-Methoxyethanol	210
2-Butanol	260	Methylcyclohexane	210
Butyl acetate	254	Methylene chloride	235
Carbon disulfide	380	Methyl ethyl ketone	330
Carbon tetrachloride	265	Methyl isobutyl ketone	335
1-Chlorobutane	220	2-Methyl-1-propanol	230
Chloroform (stabilized		N-Methylpyrrolidone	285
with ethanol)	245	Nitromethane	380
Cyclohexane	210	Pentane	210
1,2-Dichloroethane	226	Pentyl acetate	212
Diethyl ether	218	1-Propanol	210
1,2-Dimethoxyethane	240	2-Propanol	210
N,N-Dimethylacetamide	268	Pyridine	330
N,N-Dimethylformamide	270	Tetrachloroethylene	
Dimethylsulfoxide	265	(stabilized with thymol)	290
1,4-Dioxane	215	Tetrahydrofuran	220
Ethanol	210	Toluene	286
2-Ethoxyethanol	210	1,1,2-Trichloro-1,2,2-	
Ethyl acetate	255	trifluoroethane	231
Ethylene chloride	228	2,2,4-Trimethylpentane	215
Glycerol	207	o-Xylene	290
Heptane	197	Water	191

TABLE 6.3 Absorption Wavelength of Dienes

Heteroannular and acyclic dienes usually display molar absorptivities in the 8000–20000 range, whereas homoannular dienes are in the 5000–8000 range. Poor correlations are obtained for cross-conjugated polyene systems such as

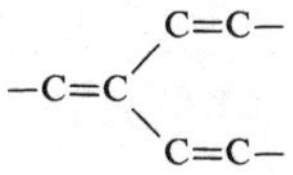

The correlations presented here are sometimes referred to as Woodward's rules or the Woodward-Fieser rules.

Base value for heteroannular or open chain diene, nm	214
Base value for homoannular diene, nm	253
Increment (in nm) for	
double bond extending conjugation	30
Alkyl substituent or ring residue	5
Exocyclic double bond	5
Polar groupings:	
-O-acyl	0
-O-alkyl	6
-S-alkyl	30
-Cl, -Br	5
-N(alkyl)$_2$	60
Solvent correction (see Table 6.5)	
Calculated wavelength =	total

TABLE 6.4 Absorption Wavelength of Enones and Dienones

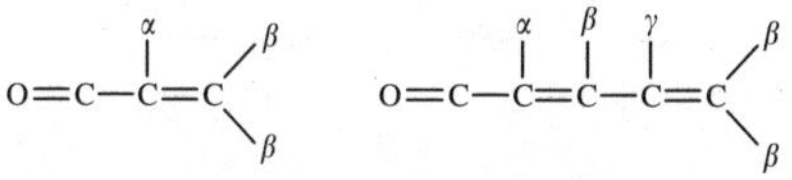

Base values, nm	
Acyclic α,β-unsaturated ketones	215
Acyclic α,β-unsaturated aldehyde	210
Six-membered cyclic α,β-unsaturated ketones	215
Five-membered cyclic α,β-unsaturated ketones	214
α,β-Unsaturated carboxylic acids and esters	195
Increments (in nm) for	
Double bond extending conjugation:	
Heteroannular	30
Homoannular	69
Alkyl group or ring residue:	
α	10
β	12
γ, δ	18

TABLE 6.4 Absorption Wavelength of Enones and Dienones (*continued*)

Polar groups:	
—OH	
α	35
β	30
γ	50
—O—CO—CH$_3$ and —O—CO—C$_6$H$_5$: $\alpha, \beta, \gamma, \delta$	6
—OCH$_3$	
α	35
β	30
γ	17
δ	31
—S—alkyl, β	85
—Cl	
α	15
β	12
—Br	
α	25
β	30
—N(alkyl)$_2$, β	95
Exocyclic double bond	5
Solvent correction (see Table 6.5)	———
	Calculated wavelength = total

TABLE 6.5 Solvent Correction for UV–VIS spectroscopy

Solvent	Correction, nm
Chloroform	+1
Diethyl ether	+11
1,4-Dioxane	+5
Ethanol	0
Hexane	+11
Methanol	0
Water	−8

TABLE 6.6 Primary Band of Substituted Benzene and Heteroaromatics

In methanol
Base value: 203.5 nm

Substituent	Wavelength shift, nm	Substituent	Wavelength shift, nm
—CH$_3$	3.0	—COOH	25.5
—CH=CH$_2$	44.5	—COO$^-$	20.5
—C≡CH	44	—CN	20.5
—C$_6$H$_5$	48	—NH$_2$	26.5
—F	0	—NH$_3^+$	−0.5
—Cl	6.0	—N(CH$_3$)$_2$	47.0
—Br	6.5	—NH—CO—CH$_3$	38.5
—I	3.5	—NO$_2$	57
—OH	7.0	—SH	32
—O$^-$	31.5	—SO—C$_6$H$_5$	28
—OCH$_3$	13.5	—SO$_2$CH$_3$	13
—OC$_6$H$_5$	51.5	—SO$_2$NH$_2$	14.0
—CHO	46.0	—CH=CH—C$_6$H$_5$	
—CO—CH$_3$	42.0	*cis* (*Z*)	79
—CO—C$_6$H$_5$	48	*trans* (*E*)	92.0
		—CH=CH—COOH, *trans*	69.5

Heteroaromatic	Base value, nm	Heteroaromatic	Base value, nm
Furan	200	Pyridine	257
Pyrazine	257	Pyrimidine	ca 235
Pyrazole	214	Pyrrole	209
Pyridazine	ca 240	Thiophene	231

TABLE 6.7 Wavelength Calculation of the Principal Band of Substituted Benzene Derivatives
In ethanol

Base value of parent chromophore, nm	
C$_6$H$_5$COOH or C$_6$H$_5$COO—alkyl	230
C$_6$H$_5$—CO—alkyl (or aryl)	246
C$_6$H$_5$CHO	250
Increment (in nm) for each substituent on phenyl ring	
—Alkyl or ring residue	
o-, *m-*	3
p-	10
—OH and —O— alkyl	
o-, *m-*	7
p-	25
—O$^-$	
o-	11
m-	20
p-	78*

* Value may be decreased markedly by steric hindrance to coplanarity.

TABLE 6.7   Wavelength Calculation of the Principal Band of Substituted Benzene Derivatives (*continued*)

—Cl	
o-, m-	0
p-	10
—Br	
o-, m-	2
p-	15
—NH$_2$	
o-, m-	13
p-	58
—NHCO—CH$_3$	
o-, m-	20
p-	45
—NHCH$_3$	
p-	73
—N(CH$_3$)$_2$	
o-, m-	20
p-	85

PHOTOLUMINESCENCE

Luminescence processes may be categorized by the excitation method used with any particular luminescent molecule. Photoluminescence is the excitation process that involves the interaction of electromagnetic radiation with photons. The process is termed chemiluminescence when the exciting energy results from a chemical reaction. Any luminescence arising from an organism is referred to as bioluminescence.

The most common application of photoluminescence is found in fluorescence spectroscopy. Fluorescence is the immediate release of electromagnetic energy from an excited molecule or release of the energy from the singlet state. If the emitted energy arises from the triplet state or is delayed, the process is referred to as phosphorescence.

A fluorescence spectrum is characteristic of a given compound. It is observed as a result of radiative emission of the energy absorbed by the molecule. The observed spectrum does not depend on the wavelength of the exciting light, except that the spectrum will be more intense if irradiation occurs at the absorption maximum. The spectral intensity is called the quantum efficiency and is usually abbreviated as Φ. The quantum yield or quantum efficiency, Φ, which is solvent dependent, is the ratio: Φ = number of quanta emitted/number of quanta absorbed. Approximate values of quantum efficiencies are as follows: naphthalene, ~0.1; anthracene, ~0.3; indole, ~0.5; and fluorescein, ~0.9.

An equation similar to Beer's law applies to fluorescence spectroscopy at dilute concentrations. In its most general form, it is given as

$$F = \Phi I_0 (1 - e^{-\varepsilon bc}).$$

In this equation, F is the observed fluorescence, Φ is the quantum efficiency (see above), I_0 is the intensity of the incident radiation, ε is the molar absorptivity, b is the cell's path length, and c is the compound's molar concentration.

The appearance of a fluorescence spectrum is reminiscent of a UV–VIS spectrum. The fluorescence spectrum for a 3-substituted indole derivative is shown in Figure 6.1. The

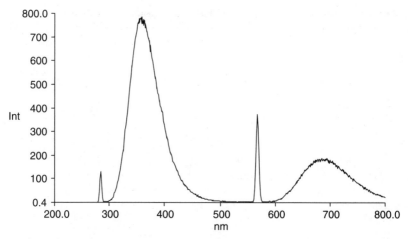

FIGURE 6.1 Fluorescence Spectrum for a 3-substituted indole derivative.

compound concentration was approximately 10^{-5} M in acetonitrile (CH_3CN). The sample was irradiated at a wavelength of 283 nm. The abscissa shows the wavelength in nanometers (nm) and the intensity ("int") is recorded on the ordinate. The maximum emission intensity is observed at 356 nm. The longest wavelength peak (λ_{max}) is observed at approximately 680 nm.

TABLE 6.8 Fluorescence Spectroscopy Data for Selected Organic Compounds

Compound	Solvent	pH	Excitation wavelength, nm	Emission wavelength, nm
Acenaphthene	Pentane		291	341
Acridine	CF₃COOH		358	475
Adenine	Water	1	280	375
Adenosine	Water	1	285	395
Adenosine triphosphate	Water	1	285	395
Adrenalin			295	335
p-Aminobenzoic acid	Water	8	295	345

TABLE 6.8 Fluorescence Spectroscopy Data for Selected Organic Compounds (*continued*)

Compound	Solvent	pH	Excitation wavelength, nm	Emission wavelength, nm
Aminopterin 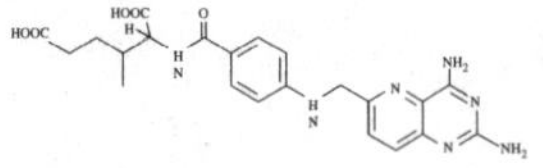	Water	7	280, 370	460
1-Aminopyrene	CF₃COOH		330, 342	415
p-Aminosalicylic acid	Water	11	300	405
Amobarbital	Water	14	265	410
Anilines (aminobenzenes)	Water	7	280, 291	344, 361
Anthracene 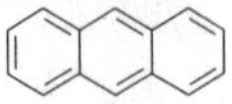	Pentane		420	430
Anthranilic acid (2-aminobenzoic acid)	Water	7	300	405
Azaindoles	Water	10	290, 299	310, 347
Benz[c]acridine	CF₃COOH		295, 380	480
Benz[a]anthracene	Pentane		284	382
1,2-Benzanthracene			280, 340	390, 410
Benzanthrone	CF₃COOH		370, 420	550
Benzo[b]chrysene	Pentane		283	398
11-H-Benzo[a]fluorene	Pentane		317	340
Benzoic acid	70% H₂SO₄		285	385
3,4-Benzopyrene (benzo[a]pyrene) 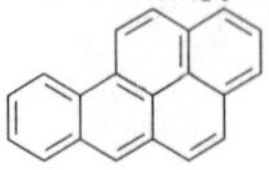	Benzene		365	390, 480
4,5-Benzopyrene (benzo[e]pyrene) 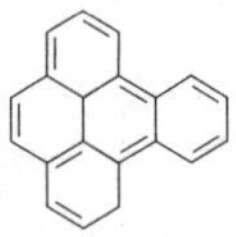	Pentane		329	389
Benzoquinoline	CF₃COOH		280	425
Benzoxanthane	Pentane		363	418
Bromolysergic acid diethylamide	Water	1	315	460
Brucine	Water	7	305	500

Note: The table includes structural diagrams for Aminopterin, Anthracene, 3,4-Benzopyrene (benzo[a]pyrene), and 4,5-Benzopyrene (benzo[e]pyrene).

TABLE 6.8 Fluorescence Spectroscopy Data for Selected Organic Compounds (*continued*)

Compound	Solvent	pH	Excitation wavelength, nm	Emission wavelength, nm
Carbazole 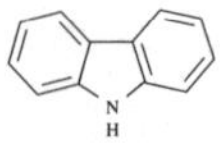	*N,N-* Dimethyl- formamide		291	359
Carboxyfluorescein 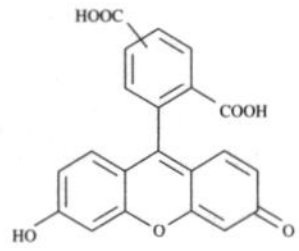	Solvent		000	000
Chlortetracycline			355	445
Chrysene 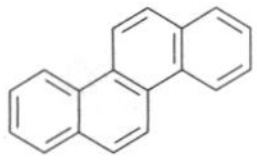	Pentane		250, 300, 310	260, 380
Cinchonine	Water	1	320	420
Coumarin 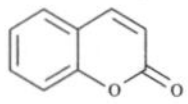	Ethanol		280	352
Dibenzo[*a,c*]anthracene 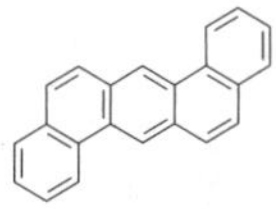	Pentane		280	381
Dibenzo[*b,k*]chrysene	Pentane		308	428
Dibenzo[*a,e*]pyrene	Pentane		370	401
3,4,8,9-Dibenzopyrene			370, 335, 390, 410	480, 510
5,12-Dihydronaphthacene	Pentane		282	340
1,4-Diphenylbutadiene	Pentane		328	370
Epinephrine	Water	7	295	335
Ethacridine	Water	2	370, 425	515
Fluoranthrene 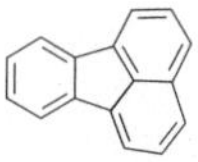	Pentane		354	464
Fluorene 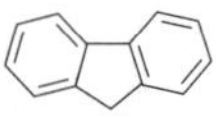	Pentane		300	321

TABLE 6.8 Fluorescence Spectroscopy Data for Selected Organic Compounds (*continued*)

Compound	Solvent	pH	Excitation wavelength, nm	Emission wavelength, nm
Fluorescein 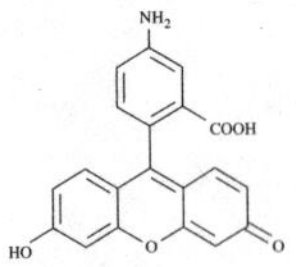	Water	7–11	490	515

Folic acid	Water	7	365	450
Gentisic acid	Water	7	315	440
Griseofulvin	Water	7	295, 335	450
Guanine 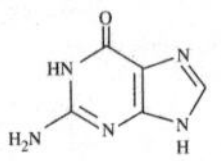	Water	1	285	365

| Harmine 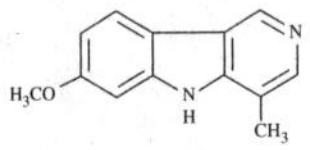 | Water | 1 | 300, 365 | 400 |

Hippuric acid	70% H_2SO_4		270	370
Homovanillic acid	Water	7	270	315
m-Hydroxybenzoic acid	Water	12	314	430
p-Hydroxycinnamic acid	Water	7	350	440
7-Hydroxycoumarin	Ethanol		325	441
5-Hydroxyindole	Water	1	290	355
5-Hydroxyindoleacetic acid	Water	7	300	355
3-Hydroxykynurenine	Water	11	365	460
p-Hydroxymandelic acid	Water	7	300	380
p-Hydroxyphenylacetic acid	Water	7	280	310
p-Hydroxyphenylpyruvic acid	Water	7	290	345
p-Hydroxyphenylserine	Water	1	290	320
5-Hydroxytryptophan	Water	7	295	340
Imipramine 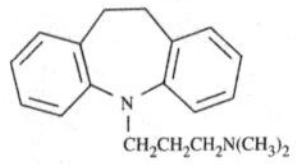	Water	14	295	415

| Indoleacetic acid | Water | 8 | 285 | 360 |
| Indoles 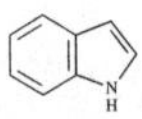 | Water | 7 | 269, 315 | 355 |

TABLE 6.8 Fluorescence Spectroscopy Data for Selected Organic Compounds (*continued*)

Compound	Solvent	pH	Excitation wavelength, nm	Emission wavelength, nm
Indomethacin	Water	13	300	410
Kynurenic acid	Water	7	325	405
		11	325	440
Lysergic acid diethylamide	Water	1	325	445
Menadione	Ethanol		335	480
9-Methylanthracene	Pentane		382	410
3-Methylcholanthrene	Pentane		297	392
7-Methyldibenzopyrene	Pentane		460	467
2-Methylphenanthrene	Pentane		257	357
3-Methylphenanthrene	Pentane		292	368
1-Methylpyrene	Pentane		336	394
4-Methylpyrene	Pentane		338	386
Naphthacene (2,3-benzanthracene)			290, 310	480, 515
1-Naphthol	0.1 M NaOH 20% ethanol		365	480
2-Naphthol	0.1 M NaOH 20% ethanol		356	426
Oxytetracycline			390	520
Phenanthrene	Pentane		252	362

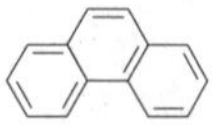 Compound	Solvent	pH	Excitation wavelength, nm	Emission wavelength, nm
Phenylalanine	Water		215, 260	282
o-Phenylenepyrene	Pentane		360	506
Phenylephrine			270	305
Picene	Pentane		281	398
Procaine	Water	11	275	345
Pyrene	Pentane		330	382

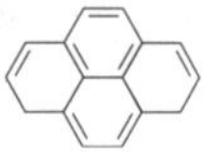 Compound	Solvent	pH	Excitation wavelength, nm	Emission wavelength, nm
Pyridoxal	Water	12	310	365

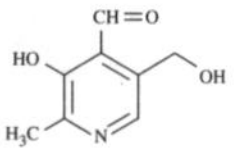 Compound	Solvent	pH	Excitation wavelength, nm	Emission wavelength, nm
Quinacrine	Water	11	285	420
Quinidine	Water	1	350	450
Quinine	Water	1	250, 350	450

TABLE 6.8 Fluorescence Spectroscopy Data for Selected Organic Compounds (*continued*)

Compound	Solvent	pH	Excitation wavelength, nm	Emission wavelength, nm
Reserpine	Water	1	300	375
Resorcinol	Water		265	315
Riboflavin	Water	7	270, 370, 445	520
Rutin	Water	1	430	520
Salicylic acid	Water	11	310	435
Scoparone	Water	10	350, 365	430
Scopoletin	Water	10	365, 390	460
Serotonin	3 M HCl		295	550
Skatole	Water		290	370
Streptomycin	Water	13	366	445
p-Terphenyl	Pentane		284	338
Thiopental			315	530
Thymol	Water	7	265	300
Tocopherol	Hexane–ethanol		295	340
Tribenzo[*a,e,i*]pyrene	Pentane		384	448
Triphenylene	Pentane		288	357
Tryptamine	Water	7	290	360
Tryptophan	Water	11	285	365
Tyramine	Water	1	275	310
Tyrosine	Water	7	275	310
Uric acid	Water	1	325	370
Vitamin A	1-Butanol		340	490
Vitamin B_{12}	Water	7	275	305
Warfarin	Methanol		290, 342	385
Xanthine	Water	1	315	435
2,6-Xylenol			275	305
3,4-Xylenol			280	310
Yohimbine	Water	1	270	360
Zoxazolamine	Water	11	280	320

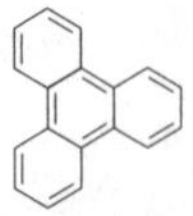

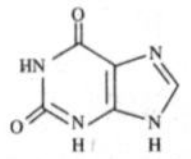

TABLE 6.9   Fluorescence Quantum Yield Values

Compound	Solvent	Q_F value vs. Q_F standard	
	Q_F standard		
9-Aminoacridine	Water	0.99	
Anthracene	Ethanol	0.30	
POPOP*	Toluene	0.85	
Quinine sulfate dihydrate	$1 N H_2SO_4$	0.55	
	Secondary standards		
Acridine orange hydrochloride	Ethanol	0.54	Quinine sulfate
		0.58	Anthracene
1,8-ANS† (free acid)	Ethanol	0.38	Anthracene
		0.39	POPOP
1,8-ANS (magnesium salt)	Ethanol	0.29	Anthracene
		0.31	POPOP
Fluorescein	0.1 N NaOH	0.91	Quinine sulfate
		0.94	POPOP
Fluorescein, ethyl ester	0.1 N NaOH	0.99	Quinine sulfate
		0.99	POPOP
Rhodamine B	Ethanol	0.69	Quinine sulfate
		0.70	Anthracene
2,6-TNS‡ (potassium salt)	Ethanol	0.48	Anthracene
		0.51	POPOP

* POPOP *p*-bis[2-(5-phenyloxazoyl)]benzene.
† ANS, anilino-8-naphthalenesulfonic acid.
‡ TNS, 2-*p*-toluidinylnaphthalene-6-sulfonate.

TABLE 6.10   Phosphorescence Spectroscopy of Some Organic Compounds

Abbreviation Used in the Table

EPA: diethyl ether, isopentane, and ethanol (5:5:2) volume ratio

Compound	Solvent	Lifetime, s	Excitation wavelength, nm	Emission wavelength, nm
Acenaphthene	Ethanol		300	515
3-Acetylpyridine	Ethanol	0.5	395	525
Adenine	Water–methanol (9:1)	2.9	278	406
Adenosine	Ethanol	0.8	280	422
p-Aminobenzoic acid	Ethanol		305	425
2-Aminofluorene	Ethanol	4.6	380	590
6-Amino-6-methylmercapto-purine	Water–methanol (9:1)	0.66	321	456
2-Amino-4-methylpyrimidine	Ethanol	2.1	302	438

TABLE 6.10 Phosphorescence Spectroscopy of Some Organic Compounds (*continued*)

Compound	Solvent	Lifetime, s	Excitation wavelength, nm	Emission wavelength, nm
2-Amino-5-nitrobenzothiazole	EPA	0.41	375	515
2-Amino-5-nitrobiphenyl	EPA	0.56	380	520
3-L-Aminotyrosine·2HCl	Ethanol	0.8	286	398
Anthracene	Ethanol		300	462
Aspirin	EPA	2.1	240	380
Atropine	Ethanol	1.4		410
8-Azaguanine	Ethanol	1.8	282	442
Benzaldehyde	Ethanol	3.4	254	433
1,2-Benzanthracene	Ethanol	2.2	310	510
Benzimidazole	Ethanol	2.3	280	406
Benzocaine	Ethanol	3.4	310	430
1,2-Benzofluorene	Ethanol		315	502
Benzoic acid	EPA	2.4	240	400
3,4-Benzopyrene	Ethanol		325	508
Benzyl alcohol	Ethanol		219	393
6-Benzylaminopurine	Water–methanol (9:1)	2.8	286	413
Biphenyl	Ethanol	1.0	270	385
6-Bromopurine	Water–methanol (9:1)	0.5	273	420
Brucine	Ethanol	0.9	305	435
Caffeine	Ethanol	2.0	285	440
Carbazole	Ethanol	7.8	341	436
2-Chloro-4-aminobenzoic acid	Ethanol	1.0	312	337
p-Chlorophenol	Ethanol	<0.2	290	505
o-Chlorophenoxyacetic acid	Ethanol	0.7	280	518
p-Chlorophenoxyacetic acid	Ethanol	<0.5	283	396
6-Chloropurine	Water–methanol (9:1)	0.64	273	419
Chlorpromazine·HCl	Ethanol	0.3	320	490
Chlorotetracycline	Ethanol	2.7	280	410
Cocaine·HCl	Ethanol	2.7	240	400
Codeine	Ethanol	0.3	270	505
Cytidine	Water–methanol (9:1)		290	420
Desoxypyridoxine·HCl	Ethanol	1.4	290	442
Diacetylsulfanilamide	Ethanol	1.3	280	405
2,6-Diaminopurine	Water–methanol (9:1)	2.7	288	410
2,6-Diaminopurine sulfate	Ethanol	1.7	294	424
1,2,5,6-Dibenzanthracene	Ethanol	1.3	340	550
2,6-Dichloro-4-nitroaniline	EPA	0.5	368	525
2,4-Dichlorophenoxyacetic acid	Ethanol	<0.5	289	490
2,6-Diethyl-4-nitroaniline	EPA	0.66	388	525
3,4-Dihydroxymandelic acid	Ethanol	1.1	294	412
3,4-Dihydroxyphenylacetic acid	Ethanol	0.9	295	430

TABLE 6.10 Phosphorescence Spectroscopy of Some Organic Compounds (*continued*)

Compound	Solvent	Lifetime, s	Excitation wavelength, nm	Emission wavelength, nm
2,5-Dimethoxy-4-methyl-amphetamine	Water–methanol (9:1)	3.9	289	411
5,7-Dimethyl-1,2-benzacridine	Ethanol	0.6	310	555
N,N-Dimethyl-4-nitroaniline	EPA	0.54	398	525
N,N-Dimethyltryptamine	Water-methanol (9:1)	6.9	286	434
Dopamine	Ethanol	0.9	285	430
Ephedrine	Ethanol	3.6	225	390
Epinephrine	Ethanol	1.0	283	425
N-Ethylcarbazole	Ethanol	7.8	340	437
Ethyl 3-indoleacetate	Ethanol	3.3	290	440
Folic acid	Ethanol		367	425
Hippuric acid	EPA	4.9	311	450
Homovanillic acid	Ethanol	0.8	289	435
DL-5-Hydroxytryptophan	Ethanol	6.3	315	435
Indole-3-acetic acid	Ethanol	<0.5	290	438
3-Indoleacetonitrile	Ethanol	7.1	285	438
Indole-3-butonoic acid	Ethanol	0.6	284	510
Indolecarboxylic acid	Ethanol	5.5	290	429
Indole-2-propanoic acid	Ethanol	0.6	290	440
D-Lysergic acid	Water–methanol (9:1)	0.1	310	518
2-Methylcarbazole	Ethanol	8.1	333	442
N-Methylcarbazole	Ethanol	8.4	336	437
6-Methylmercaptopurine	Water–methanol (9:1)	0.6	291	420
N-Methyl-4-nitroaniline	EPA	0.5	390	522
6-Methylpurine	Water–methanol (9:1)	3.2	272	405
Morphine	Ethanol	0.3	285	500
Naphthacene	Ethanol		300	518
Naphthalene	EPA	1.8	310	475
1-Naphthaleneacetic acid	Ethanol	2.8	295	510
1-Naphthol	Ethanol	1.1	320	475
2-Naphthoxyacetic acid	Ethanol	2.6	328	497
2-Naphthylamine	Ethanol	2.3	270	303
Niacinamide	Ethanol		270	410
Nicotine	Ethanol	5.2	270	390
5-Nitroacenaphthene	EPA		380	540
4-Nitroaniline	EPA	0.6	380	510
9-Nitroanthracene	EPA		248	488
1-Nitroanthraquinone	EPA	0.3	250	490
4-Nitrobiphenyl	EPA		330	480
3-Nitro-*N*-ethylcarbazole	EPA	0.4	315	475
2-Nitrofluorene	EPA	0.4	340	517
6-Nitroindole	EPA	0.4	372	520
1-Nitronaphthalene	EPA		340	520

TABLE 6.10   Phosphorescence Spectroscopy of Some Organic Compounds (*continued*)

Compound	Solvent	Lifetime, s	Excitation wavelength, nm	Emission wavelength, nm
2-Nitronaphthalene	EPA	0.4	260	500
4-Nitro-1-naphthylamine	EPA		400	578
4-Nitrophenol	Ethanol	< 0.2	355	520
4-Nitrophenylhydrazine	EPA	0.5	390	520
4-Nitro-2-toluidine	EPA	0.5	375	520
Papaverine·HCl	Ethanal	1.5	260	480
Phenacetin	EPA			410
Phenanthrene	EPA	2.6	340	465
Phenobarbital	Ethanol	1.8	240	380
Phenylalanine	Ethanol		270	385
DL-2-Phenyllactic acid	Ethanol	5.4	262	383
Phthalylsulfathiazole	Ethanol	0.9	305	405
Procaine·HCl	Ethanol	3.5	310	430
Purine	Water–methanol (9:1)	2.2	272	405
Pyrene	Ethanol		330	515
Pyridine	Ethanol	1.4	310	440
Pyridine-3-sulfonic acid	Ethanol	1.2	272	408
Pyridoxine·HCl	Ethanol		290	425
Quercetin	Ethanol	2.1	345	480
Quinidine sulfate	Ethanol	1.3	340	500
Quinine·HCl	Ethanol	1.3	340	500
Salicyclic acid	Ethanol	6.2	315	430
Strychnine phosphate	Ethanol	1.2	290	440
Sulfabenzamide	Ethanol	0.7	305	405
Sulfadiazine	Ethanol	0.7	275	410
Sulfanilamide	Ethanol	2.9	300	410
Sulfapyridine	Ethanol	1.4	310	440
Sulfathiazole	Ethanol	0.9	310	420
1,2,4,5-Tetramethylbenzene	EPA	4.5	275	390
2-Thiouracil	Ethanol	< 0.5	310	430
2,4,5-Trichlorophenol	Ethanol	<0.2	305	485
2,4,5-Trichlorophenoxyacetic acid	Ethanol	1.1	295	475
Triphenylene	Ethanol	15	290	460
Tryptophan	Ethanol	1.5	295	440
Tyrosine	Ethanol	2.8	290	390
Vitamin K$_1$	Hexane	0.4	345	570
Vitamin K$_3$	Hexane	0.5	335	510
Vitamin K$_5$	Water–methanol (9:1)	1.3	310	535
Warfarin	Ethanol	0.8	305	460
Yohimbine·HCl	Ethanol	7.4	290	410

INFRARED SPECTROSCOPY

Infrared (IR) and Raman spectroscopy rely on the interaction of a bond between two elements and IR radiation in the $400\text{--}4\,000\,\text{cm}^{-1}$ range. The two techniques are distinct but closely related. Historically, infrared analysis has been the more widely used in organic chemistry but much of the brief discussion that follows applies equally to both methods.

A chemical bond may be considered as a spring to which two weights are attached. The weights are atoms of different atomic masses. The length and strength of the spring may be correlated to the length and strength of the chemical bond. Each bond will vibrate at a frequency that is characteristic of the attached atoms and the type of bond (*i.e.*, single, double, or triple) between them. Radiation of an appropriate frequency will be absorbed by the bond and the wavelength at which this occurs will be detected and recorded by the instrument in the spectrum. Indeed, most molecules have many bonds so multiple peaks are observed in an IR spectrum.

An electrical dipole is required for IR energy to be efficiently absorbed by a molecule. Thus, bonds between different elements will give more prominent absorption peaks than will symmetrical bonds. This is because the bond's dipole moment changes as the bond stretches and contracts. Symmetrical bonds do not change dipole moment even if the bond distance changes. Usually, the most prominent peaks are observed when the electronegativity difference between the bound elements is greatest. Thus, a C—O bond will usually be more prominent than a C—H bond.

The position (frequency) of the absorption depends on the strength of the bond linking them. Thus, single, double, and triple bonds are observed in characteristic ranges. In addition, the frequency is related to the masses of the attached atoms. The largest mass differences occur when an element is attached to hydrogen. Such bonds as C—H, O—H, and N—H typically are observed in the $2\,900\text{--}3\,600\,\text{cm}^{-1}$ range.

The remarkable versatility of NMR as an analytical method has diminished the importance of IR analysis in modern laboratories but it remains a very useful technique. The very small amount of sample required and the prominence of functional group absorption means that the progress of a reaction can be monitored very conveniently. For example, the reduction of an aldehyde to an alcohol will be accompanied by the disappearance of the prominent C=O peak and the appearance of a C—OH absorption. Because both peaks are so readily identifiable in a small sample, the reaction is easy to follow and its completeness can be assayed.

TABLE 6.11 Absorption Frequencies of Single Bonds to Hydrogen

Abbreviations Used in the Table		
m, moderately strong	var, of variable strength	
m–s, moderate to strong	w, weak	
s, strong	w–m, weak to moderately strong	

Group	Band, cm^{-1}	Remarks
Saturated C—H		
H | —C—H | H	2 975–2 950 (s) 2 885–2 865 (w)	Two or three bands usually; asymmetrical and symmetrical CH stretching, respectively. In

TABLE 6.11 Absorption Frequencies of Single Bonds to Hydrogen (*continued*)

Group	Band, cm^{-1}	Remarks
	Saturated C—H (*continued*)	
	1450–1260 (m)	presence of double bond adjacent to CH$_3$ group symmetrical band splits into two. Sensitive to adjacent negative substituents
H | —C— acyclic | H	ca 2930 (s) 2870–2840 (w) 1480–1440 (m) ca 720 (w)	Frequency increased in strained systems. Symmetrical band splits into two bands when double bond adjacent. Scissoring mode Rocking mode
	Alkane residues attached to carbon	
Cyclopropane	ca 3050 (w) 540–500 470–460 (s)	CH stretching Aliphatic cyclopropanes
Cyclobutanes Cyclopentanes	580–490 (s) 595–490 (s)	Alkyl derivatives: 550–530 cm^{-1} Alkyl derivatives: 585–530 cm^{-1}
$>$C(CH$_3$)$_2$	ca 1380 (m) 1175–1165 (m) 1150–1130 (m)	A roughly symmetrical doublet If no H on central carbon, then one band at ca 1190 cm^{-1}
—C(CH$_3$)$_3$	1395–1385 (m) 1365 (s)	Split into two bands
Aryl-CH$_3$ Aryl-C$_2$H$_5$ Aryl-C$_3$H$_7$ (or C$_4$H$_9$) —(CH$_2$)$_n$— $n = 1$ $n = 2$ $n = 3$ $n \geq 4$	390–260 (m) 565–540 (m–s) 585–565 (m) 785–770 (w–m) 745–735 (w–m) 735–725 (w–m) 725–720 (w–m)	 Two bands Rocking vibrations
	Alkane residues attached to miscellaneous atoms	
Epoxide C—H 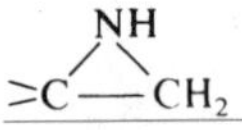 NH $>$C——CH$_2$	ca 3050 (m–s) ca 3050 (m–s)	
—CH$_2$—halogen	ca 3050 (m–s) 1435–1385 (m) 1300–1240 (s)	Halogens except fluorine

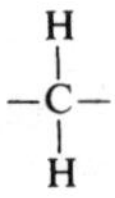

TABLE 6.11 Absorption Frequencies of Single Bonds to Hydrogen (*continued*)

Group	Band, cm^{-1}	Remarks
	Alkane residues attached to miscellaneous atoms (*continued*)	
—CHO	2900–2800 (w) 2775–2700 (w) 1420–1370 (m)	
—CO—CH$_3$	3100–2900 (w) 1450–1400 (s) 1360–1355 (s)	
—O—CH$_3$ ethers	2835–2810 (s) 1470–1430 (m–s) ca 1030 (w–m)	Two bands
—O—C(CH$_3$)$_3$	1200–1155 (s)	
—O—CH$_2$—O—	2790–2770 (m)	
—O—CH$_2$— esters	1475–1460 (m–s) 1470–1435 (m–s)	Acyclic esters. Frequency increased ca 30 cm^{-1} for cyclic and small ring systems.
—O—CO—CH$_3$	1450–1400 (s) 1385–1365 (s) 1360–1355 (s)	Acetate esters The high intensity of these bands often dominates this region of the spectrum.
—CH$_2$—$\overset{\mid}{C}$=C<	1445–1430 (m)	
—CH$_2$—SO$_2$—	ca 1250 (m)	
P—CH$_3$ Se—CH$_3$ B—CH$_3$ Si—CH$_3$ Sn—CH$_3$ Pb—CH$_3$ As—CH$_3$ Ge—CH$_3$ Sb—CH$_3$ Bi—CH$_3$ —CH$_2$—(Cd, Hg, Zn, Sn)	1320–1280 (s) ca 1280 (m) 1460–1405 (m) 1320–1280 (m) 1265–1250 (m–s) 1200–1180 (m) 1170–1155 (m) 1265–1240 (m) 1240–1230 (m) 1215–1195 (m) 1165–1145 (m) 1430–1415 (m)	
N—CH$_3$ and N—CH$_2$— N—CH$_2$—CH$_2$—N	2820–2780 (s) 1440–1390 (m) 1480–1450 (s)	 Ethylenediamine complexes Ethylenediamine complexes

TABLE 6.11 Absorption Frequencies of Single Bonds to Hydrogen (*continued*)

Group	Band, cm^{-1}	Remarks
	Alkane residues attached to miscellaneous atoms (continued)	
N—CH$_3$		
Amine · HCl	1 475–1 395 (m)	
Amino acid · HCl	1 490–1 480 (m)	
Amides	1 420–1 405 (s)	
N—CH$_2$— amides	ca 1 440 (m)	
S—CH$_3$	2 990–2 955 (m–s)	
	2 900–2 865 (m–s)	
	1 440–1 415 (m)	
	1 325–1 290 (m)	
	1 030–960 (m)	
	710–685 (w–m)	
S—CH$_2$—	2 950–2 930 (m)	
	2 880–2 845 (m)	
	1 440–1 415 (m)	
	1 270–1 220 (s)	
—C≡CH	ca 3 300 (s)	Sharp
	700–600	Bending
C=C (vinyl, one H trans)	3 040–3 010 (m)	
C=C (terminal =CH$_2$)	3 095–3 075 (m)	CH stretching sometimes
	2 985–2 970 (m)	obscured by much stronger
		bands of saturated CH groups
C=C (R, H / H, H)	995–980 (s)	
	940–900 (s)	
	ca 635 (s)	
	485–445 (m–s)	
C=C (H / H, H)	895–885 (s)	
	560–530 (s)	
	470–435 (m)	
C=C (R, H / H, R)	980–955 (s)	
	455–370 (m–s)	
C=C (H, H / R, R)	730–655 (m)	
	670–455 (s)	
C=C (R, H / R, R)	850–790 (m)	
	570–515 (s)	
	525–470 (s)	

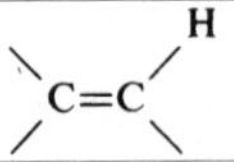

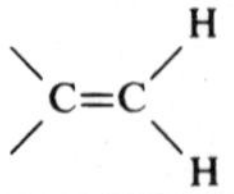

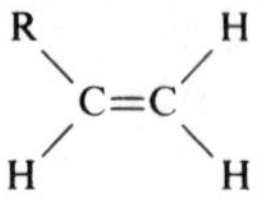

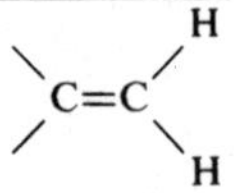

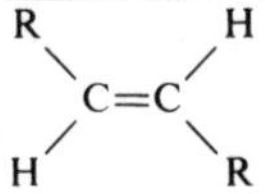

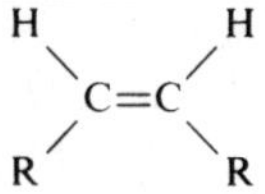

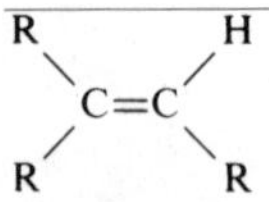

TABLE 6.11 Absorption Frequencies of Single Bonds to Hydrogen (*continued*)

Group	Band, cm^{-1}	Remarks
Alkane residues attached to miscellaneous atoms (*continued*)		
—O—CH=CH$_2$	965–960 (s) 945–940 (m) 820–810 (s)	
—S—CH=CH$_2$	ca 965 (s) ca 860 (s)	
—CO—CH=CH$_2$ —CO—OCH=CH$_2$ —CO—C=CH$_2$ —CO—OC=CH$_2$ —O—CH=CH— *trans* —CO—CH=CH— *trans*	995–980 (s) 965–955 (m) 950–935 (s) 870–850 (s) ca 930 (s) 880–865 940–920 (s) ca 990 (s)	
Hydroxyl group O—H compounds		
Primary aliphatic alcohols	3 640–3 630 (s) 1 350–1 260 (s) 1 085–1 030 (s)	Only in very dilute solutions in nonpolar solvents OH bending Also broad band at 700–600 cm^{-1}

Secondary aliphatic alcohols	3 625–3 620 (s)
	
	1 350–1 260 (s)
	1 125–1 085 (s)
		See comments under primary
		  aliphatic alcohols
		
		Also for α-unsaturated and cyclic
		  tertiary aliphatic alcohols

Tertiary aliphatic alcohols	3 620–3 610 (s) 1 410–1 310 (s) 1 205–1 125 (s)	See comments under primary aliphatic alcohols

Aryl—OH	ca 3 610 (s)
	
	1 410–1 310 (s)
	1 260–1 180 (s)
	1 085–1 030 (s)
		See comments under primary
		  aliphatic alcohols
		
		Also for unsaturated secondary
		  aliphatic alcohols

Carboxylic acids	3 300–2 500 (w–m) 995–915 (s)	Broad Broad diffuse band
Enol form of β-diketones	2 700–2 500 (var)	Broad

TABLE 6.11 Absorption Frequencies of Single Bonds to Hydrogen (*continued*)

Group	Band, cm^{-1}	Remarks
Hydroxyl group O—H compounds (*continued*)		
Free oximes	3 600–3 570(w–m)	Shoulder
Free hydroperoxides	3 560–3 530 (m)	
Peroxy acids	ca 3 280 (m)	
Phosphorus acids	2 700–2 560 (m)	Broad
Water in solution	3 710	When solution is damp
Intermolecular H bond Dimeric	3 600–3 500	Rather sharp. Absorptions arising from H bond with polar solvents also appear in this region.
Polymeric	3 400–3 200 (s)	Broad
Intramolecular H bond Polyvalent alcohols Chelation	3 600–3 500 (s) 3 200–2 500	Sharper than dimeric band above Broad and occasionally weak; the lower the frequency, the stronger the intramolecular bond
Water of crystallation (solid state spectra)	3 600–3 100 (w)	Usually a weak band at 1 640–1 615 cm^{-1} also. Water in trace amounts in KBr disks shows a broad band at 3 450 cm^{-1}.
Amine, imine, ammonium, and amide N—H		
Primary amines Aliphatic	3 550–3 300 (m) 1 650–1 560 (m) 1 090–1 020 (w–m) 850–810 (w–m)	Two bands in this range With α-carbon branching at 795 cm^{-1} and strong
	495–445 (m–s) ca 290 (s)	Broad Broad
Aromatic	1 350–1 260 (s) 445–345	Also for secondary aryl amines
Amino acids	3 100–3 030 (m)	Values for solid states; broad bands also (but not always) near 2 500 and 200 cm^{-1}
	2 800–2 400 (m) 1 625–1 560 (m) 1 550–1 550 (m)	Number of sharp bands; dilute solution
Amino salts	3 550–3 100 (m) ca 3 380 ca 3 280	Values for solid state Dilute solutions

TABLE 6.11 Absorption Frequencies of Single Bonds to Hydrogen (*continued*)

Group	Band, cm^{-1}	Remarks
	Amine, imine, ammonium, and amide N—H (*continued*)	
Secondary amines	3 550–3 400 (w)	Only one band, whereas primary amines show two bands
	1 580–1 490 (w)	Often too weak to be noticed
	1 190–1 170 (m)	
	1 145–1 130 (m)	
	455–405 (w–m)	
Salts	ca 2 500	Sharp; broad values for solid state
	ca 2 400	Sharp; broad values for solid state
	1 620–1 560 (m–s)	
Tertiary amines $R_1R_2R_3NH^+$	2 700–2 250	Group of relatively sharp bands; broad bands in solid state
Ammonium ion	3 300–3 030 (s)	Group of bands
	1 430–1 390 (s)	
Imines =N=H	3 350–3 310 (w)	Aliphatic
	3 490 (s)	Aryl
	3 490 (s)	Pyrroles, indoles; band sharp
Imine salts	2 700–2 330 (m–s)	Dilute solutions
	2 200–1 800 (m)	One or more bands; useful to distinguish from protonated tertiary amines
Primary amide —CONH$_2$	ca 3 500 (m)	Lowered ca 150 cm^{-1} in the solid state and on H bonding; often several bands 3 200–3 050 cm^{-1}
	ca 3 400 (m)	
Secondary amide —CONH—	3 460–3 400 (m)	Two bands; lowered on H bonding and in solid state. Only one band with lactams
	3 100–3 070 (w)	Extra band with bonded and solid-state samples
	Miscellaneous R—H	
—S—H	2 600–2 550 (w)	Weaker than OH and less affected by H bonding
P—H	2 440–2 350 (m)	Sharp
P(=O)OH	2 700–2 560 (m)	Associated OH
	100/137 times the	
R—D	corresponding RH frequency	Useful when assigning RH bands; deuteration leads to a known shift to lower frequency

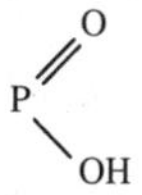

TABLE 6.12   Absorption Frequencies of Triple Bonds

Abbreviations Used in the Table

m, moderately strong	var, of variable strength
m–s, moderate to strong	w–m, weak to moderately strong
s, strong	

Group	Band, cm^{-1}	Remarks
Alkynes		
Terminal	3 300 (s)	CH stretching
	2 140–2 100 (w–m)*	C≡C stretching
	1 375–1 225 (w–m)	
	695–575 (m–s)	Two bands if molecule has axial symmetry
	ca 630 (s)	Alkyl monosubstituted
Nonterminal	2 260–2 150 (var)*	Symmetrical or nearly symmetrical substitution makes the C≡C stretching frequency inactive. When more than one C≡C linkage is present, and sometimes when there is only one, there are frequently more absorption bands in this region than there are triple bonds to account for them.
R$_1$—C≡C—R$_2$	540–465 (m)	The longer the chain, the lower the frequency
Aryl—C≡C—	ca 550 (m)	
	ca 350 (var)	
—C≡C—halogen (Cl, Br, I)	185–160 (var)	
Nitriles—C≡N	2 260–2 200 (var)	Stronger and toward the lower end of the range when conjugated; occasionally very weak or absent
Aliphatic	580–555 (m–s)	
	560–525 (m–s)	
	390–350 (s)	
Aromatic	580–540 (s)	
	430–380 (m)	
Isonitriles R—N⁺≡C⁻ or R—N=C:	2 175–2 150 (s)	Very sensitive to changes in substituents
	2 150–2 115 (s)	Not found for nitriles
	1 595	
Cyanamides >N—C≡N⇌>N⁺—C=N⁻	2 225–2 210 (s)	

* Conjugation with olefinic or acetylenic groups lowers the frequency and raises the intensity. Conjugation with carbonyl groups usually has little effect on the position of absorption.

TABLE 6.12 Absorption Frequencies of Triple Bonds (*continued*)

Group	Band, cm^{-1}	Remarks
Thiocyanates R—S—C≡N	2 175–2 140 (s)	Aryl thiocyanates at the upper end of the range, alkyl at the lower end
	404–400 (s) ca 600 (m–s)	Aliphatic derivatives
Nitrile *N*-oxides —C≡N→O	2 305–2 285 (s) 1 395–1 365 (s)	Aryl derivatives
Diazonium salts R—N$^+$≡N	2 300–2 230 (m–s)	
Selenocyanates R—Se—C≡N	ca 2 160 (m–s) 545–520 ca 390 ca 350	

TABLE 6.13 Absorption Frequencies of Cumulated Double Bonds

Abbreviations Used in the Table

m–s, moderate to strong   vs, very strong
s, strong   w, weak

Group	Band, cm^{-1}	Remarks
Carbon dioxide O=C=O	2 349 (s)	Appears in many spectra as a result of inequalities in path length
Isocyanates —N=C=O	2 275–2 250 (vs)	Position unaffected by conjugation
Isoselenocyanates —N=C=Se	2 200–2 000 (s) 675–605	Broad; usually two bands
Azides —N$_3$ or —N=$\overset{+}{N}$=$\overset{-}{N}$	2 140–2 030 (s) 1 340–1 180 (w)	Not observed for ionic azides
—N=C=N—	2 155–2 130 (s)	Split into unsymmetrical doublet by conjugation with aryl groups: 2 145–2 125 (vs) and 2 115–2 105 (vs)

TABLE 6.13 Absorption Frequencies of Cumulated Double Bonds (*continued*)

Group	Band, cm^{-1}	Remarks
Isothiocyanates —N=C=S	2140–1990 (vs) 649–600 (m–s) 565–510 (m–s) 470–440 (m–s)	Broad; usually a doublet
Ketenes >C=C=O	ca 2150 (s)	
Ketenimines C=C=N—	2050–2000 (s)	
Allenes >C=C=C<	2000–1915 (m–s)	Two bands when terminal allene or when bonded to electron-attracting groups
Thionylamines —N=S=O	1300–1230 (s) 1180–1110 (s)	
Diazoalkanes R$_2$C=$\overset{+}{N}$=$\overset{-}{N}$ —CH=$\overset{+}{N}$=$\overset{-}{N}$	2030–2000 (s) 2050–2035 (s)	
Diazoketones —CO—CH=$\overset{+}{N}$=$\overset{-}{N}$	2100–2080 2075–2050	Monosubstituted Disubstituted

Position of Carbonyl Absorption

Because the carbonyl absorption is one of the most prominent and identifiable bands in the IR spectrum, it is often used diagnostically. The general trends of structural variation on the position of C=O stretching frequencies are summarized in six principles, as follows. Details of carbonyl absorptions are recorded in Table 6.14 and for other double bonds in Table 6.15, for aromatic bonds in Table 6.16 and other miscellaneous bonds in Table 6.17.

1. The more electronegative the group X in the system R—CO—X—, the higher is the frequency.
2. α,β-Unsaturation lowers the frequency by 15–40 cm^{-1}, except in amides, where little shift is observed; if present it is usually to higher frequency.
3. Further conjugation has relatively little effect.

4. Ring strain in cyclic compounds causes a relatively large shift to higher frequency. This phenomenon provides a remarkably reliable test of ring size, distinguishing clearly between four-, five-, and larger-membered-ring ketones, lactones, and lactams. Six-membered-ring and larger ketones, lactones, and lactams show the normal frequency found for the open-chain compounds.

5. Hydrogen bonding to a carbonyl group causes a shift to lower frequency of 40–$60\,cm^{-1}$. Acids, amides, enolized β-keto carbonyl systems, and o-hydroxyphenol and o-aminophenyl carbonyl compounds show this effect. All carbonyl compounds tend to give slightly lower values for the carbonyl stretching frequency in the solid state compared with the value for dilute solutions.

6. Where more than one of the structural influences on a particular carbonyl group is operating, the net effect is usually close to additive.

An especially convenient aspect of IR spectroscopy is its practice. A small amount of sample can be pressed between two NaCl or KBr (Table 6.19) disks and the spectrum can be determined without further preparation. A spectrum so obtained is recorded as "neat" or "between salts." If the sample is a solid, it may be mixed in a mortar and pestle with KBr and then pressed into a disk. The salt disk may be placed directly in the IR beam. In neither case is there a concern about solvent peaks. Of course, solvents may be used. Carbon tetrachloride and chloroform are the most commonly used solvents when the compound requires dissolution. Alternately, the sample may be intimately mixed (mulled) with mineral oil (a hydrocarbon oil). The thick slurry may then be smeared on a salt disk and placed in the spectrometer. The brand of mineral oil used historically is Nujol and such slurries are still called "Nujol mulls." The transmission characteristics of potential solvents for IR spectroscopy may be found in Table 6.20.

Traditional analog spectrometers were calibrated by taking a second spectrum of polystyrene. The sharp $1\,641\,cm^{-1}$ band was recorded on the same sheet as the original spectrum. Modern Fourier transform instruments do not usually require this step but still require calibration.

TABLE 6.14 Absorption Frequencies of Carbonyl Bands

All bands quoted are strong.

Groups	Band, cm^{-1}	Remarks
Acid anhydrides —CO—O—CO— Saturated	1 850–1 800 1 790–1 740	Two bands usually separated by about $60\,cm^{-1}$. The higher-frequency band is more intense in acyclic anhydrides, and the lower-frequency band is more intense in cyclic anhydrides.
Aryl and α,β-unsaturated	1 830–1 780 1 700–1 710	
Saturated five-ring	1 870–1 820 1 800–1 750	
All classes	1 300–1 050	One or two strong bands due to CO stretching

TABLE 6.14 Absorption Frequencies of Carbonyl Bands (*continued*)

Groups	Band, cm^{-1}	Remarks
Acid chlorides—COCl		
Saturated	1815–1790	Acid fluorides higher, bromides and iodides lower
Aryl and α,β-unsaturated	1790–1750	
Acid peroxide		
CO—O—O—CO—		
Saturated	1820–1810	
	1800–1780	
Aryl and α,β-unsaturated	1805–1780	
	1785–1755	
Esters and lactones		
—CO—O—		
Saturated	1750–1735	
Aryl and α,β-unsaturated	1730–1715	
Aryl and vinyl esters		
C=C—O—CO—alkyl	1800–1750	The C=C stretching band also shifts to higher frequency.
Esters with electronegative α substituents; *e.g.*,		
>CCl—CO—O—	1770–1745	
α-Keto esters	1755–1740	
Six-ring and larger lactones	Similar values to the corresponding open-chain esters	
Five-ring lactone	1780–1760	
α,β-Unsaturated five-ring lactone	1770–1740	When α-CH is present, there are two bands, the relative intensity depending on the solvent.
β,γ-Unsaturated five-ring lactone, vinyl ester type	ca 1800	
Four-ring lactone	ca 1820	
β-Keto ester in H bonding enol form	ca 1650	Keto from normal; chelate-type H bond causes shift to lower frequency than the normal ester. The C=C band is strong and is usually near 1630 cm^{-1}.
All classes	1300–1050	Usually two strong bands due to CO stretching
Aldehydes —CHO		
(See also Table 6.39 for C—H.) All values given below are lowered in liquid-film or solid-state spectra by about 10–20 cm^{-1}. Vapor-phase spectra have values raised about 20 cm^{-1}.		

TABLE 6.14 Absorption Frequencies of Carbonyl Bands (*continued*)

Groups	Band, cm^{-1}	Remarks
Aldehydes —CHO (*continued*)		
Saturated	1 740–1 720	
Aryl	1 715–1 695	*o*-Hydroxy or amino groups shift this value to 1 655–1 625 cm^{-1} because of intramolecular H bonding.
α,β-Unsaturated	1 705–1 680	
$\alpha,\beta,\gamma,\delta$-Unsaturated	1 680–1 660	
β-Ketoaldehyde in enol form	1 670–1 645	Lowering caused by chelate-type H bonding
Ketones $>$C$=$O		
All values given below are lowered in liquid-film or solid-state spectra by about 10–20 cm^{-1}. Vapor-phase spectra have values raised about 20 cm^{-1}.		
Saturated	1 725–1 705	
Aryl	1 700–1 680	
α,β-Unsaturated	1 685–1 665	
$\alpha,\beta,\alpha',\beta'$-Unsaturated and diaryl	1 670–1 660	
Cyclopropyl	1 705–1 685	
Six-ring ketones and larger	Similar values to the corresponding open-chain ketones	
Five-ring ketones	1 750–1 740	α,β Unsaturation, $\alpha,\beta,\alpha',\beta'$ unsaturation, etc., have a similar effect on these values as on those of open-chain ketones.
Four-ring ketones	ca 1 780	
α-Halo ketones	1 745–1 725	Affected by conformation; highest values are obtained when both halogens are in the same plane as the C$=$O.
α,α'-Dihaloketones	1 765–1 745	
1,2-Diketones, *syn-trans-* open chains	1 730–1 710	Anti-symmetrical stretching frequency of both C$=$O's. The symmetrical stretching is inactive in the infrared but active in the Raman.
*syn-cis-*1,2-Diketones, six-ring	1 760 and 1 730	
*syn-cis-*1,2-Diketones, five ring	1 775 and 1 760	
o-Amino-aryl or *o*-hydroxy-aryl ketones	1 655–1 635	Low because of intramolecular H bonding. Other substituents and steric hindrance affect the position of the band.

TABLE 6.14   Absorption Frequencies of Carbonyl Bands (*continued*)

Groups	Band, cm^{-1}	Remarks
Ketones $>$C$=$O (*continued*)		
Quinones	1 690–1 660	C$=$C band is strong and is usually near 1 600 cm^{-1}.
Extended quinones	1 655–1 635	
Tropone	1 650	Near 1 600 cm^{-1} when lowered by H bonding as in tropolones
Carboxylic acids —CO$_2$H		
All types	3 000–2 500	OH stretching; a characteristic group of small bands due to combination bands
Saturated	1 725–1 700	The monomer is near 1 760 cm^{-1}, but is rarely observed. Occasionally both bands, the free monomer, and the H-bonded dimer can be seen in solution spectra. Ether solvents give one band near 1 730 cm^{-1}.
α,β-Unsaturated	1 715–1 690	
Aryl	1 700–1 680	
α-Halo-	1 740–1 720	
Carboxylate ions —CO$_2^-$		
Most types	1 610–1 550 1 420–1 300	Anti-symmetrical and symmetrical stretching, respectively
Amides —CO—N$<$ (See also Table 6.39 for NH stretching and bending.)		
Primary —CONH$_2$		
In solution	ca 1 690	Amide I; C$=$O stretching
Solid state	ca 1 650	
In solution	ca 1 600	Amide II: mostly NH bending
Solid state	ca 1 640	
		Amide I is generally more intense than amide II. (In the solid state, amides I and II may overlap.)
Secondary —CONH—		
In solution	1 700–1 670	Amide I
Solid state	1 680–1 630	
In solution	1 550–1 510	Amide II; found in open-chain amides only
Solid state	1 570–1 515	Amide I is generally more intense than amide II.
Tertiary	1 670–1 630	Since H bonding is absent, solid and solution spectra are much the same.
Lactams		
Six-ring and larger rings	ca 1 670	
Five-ring	ca 1 700	Shifted to higher frequency when the N atom is in a bridged system
Four-ring	ca 1 745	

TABLE 6.14 Absorption Frequencies of Carbonyl Bands (*continued*)

Groups	Band, cm^{-1}	Remarks
R—CO—N—C=C		Shifted $+15$ cm^{-1} by the additional double bond
C=C—CO—N		Shifted by up to $+15$ cm^{-1} by the additional double bond. This is an unusual effect by α,β unsaturation. It is said to be due to the inductive effect of the C=C on the well-conjugated CO—N system, the usual conjugation effect being less important in such a system.
Imides —CO—N—CO—		
Cyclic six-ring	ca 1 710 and ca 1 700	Shift of $+15$ cm^{-1} with α,β unsaturation
Cyclic five-ring	ca 1 770 and ca 1 700	
Ureas N—CO—N		
RNHCONHR	ca 1 660	
Six-ring	ca 1 640	
Five-ring	ca 1 720	
Urethanes R—O—CO—N	1 740–1 690	Also shows amide II band when nonsubstituted on N
Thioesters and Acids RCO—S—R′		
RCOSH	ca 1 720	α,β-Unsaturated or aryl acid or ester shifted about -25 cm^{-1}
RCOS—alkyl	ca 1 690	
RCOS—aryl	ca 1 710	

TABLE 6.15 Absorption Frequencies of Other Double Bonds

Abbreviations Used in the Table

m, moderately strong
m-s, moderate to strong
var, of variable strength

vs, very strong
w, weak

Group	Band, cm^{-1}	Remarks
Alkenes >C=C<		
Nonconjugated	1 680–1 620 (w-m)	May be very weak if symmetrically substituted
Conjugated with aromatic ring	1 640–1 610 (m)	More intense than with unconjugated double bonds

TABLE 6.15 Absorption Frequencies of Other Double Bonds

Alkenes $>$C$=$C$<$ (continued)		
Internal (ring)	3 060–2 995 (m)	Highest frequencies for smallest ring
Carbons: $n = 3$	ca 1 665 (w-m)	
$n = 4$	ca 1 565 (w-m)	
$n = 5$	ca 1 610 (w-m)	
	1 370–1 340 (s)	Characteristic
$n \geq 6$	1 650–1 645 (w-m)	
Exocyclic C$=$C(CH$_2$)$_n$ $n = 2$	1 780–1 730 (m)	
$n = 3$	ca 1 680 (m)	
$n \geq 4$	1 655–1 650 (m)	
Fulvene	1 645–1 630 (m)	
	1 370–1 340 (s)	
	790–765 (s)	
Dienes, trienes, etc.	1 650 (s) and 1 600 (s)	Lower-frequency band usually more intense and may hide or overlap the higher-frequency band
α,β-Unsaturated carbonyl compounds	1 640–1 590 (m)	Usually much weaker than the C$=$O band
Enol esters, enol ethers, and enamies	1 700–1 650 (s)	

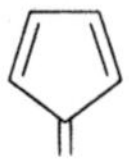

Imines, oximes, and amidines $>$C$=$N$-$		
Imines and oximes		
Aliphatic	1 690–1 640 (w)	
α,β-Unsaturated and aromatic	1 650–1 620 (m)	
Conjugated cyclic systems	1 660–1 480 (var)	
	960–930 (s)	NO stretching of oximes
Imino ethers $-$O$-$C$=$N$-$	1 690–1 640 (var)	Usually a strong doublet
Imino thioethers $-$S$-$C$=$N$=$	1 640–1 605 (var)	
Imine oxides $>$C$=$$\overset{+}{N}-\overset{-}{O}$	1 620–1 550 (s)	
Amidines $>$N$-$C$-$N$-$	1 685–1 580 (var)	
Benzamidines Aryl$-$C$=$N$=$N	1 630–1 590	

TABLE 6.15 Absorption Frequencies of Other Double Bonds (continued)

Group	Band, cm^{-1}	Remarks
Imines, oximes, and amidines $>$C$=$N$-$ (continued)		
Guanidine $>$N$-$C$=$N$-$ $\underset{\text{N}}{\vert}$	1 725–1 625 (s)	
Azines $>$C$=$N$-$N$=$C$<$	1 670–1 600	
Hydrazoketones $-$CO$-$C$=$N$-$N	1 600–1 530 (vs)	
Azo compounds $-$N$=$N$-$		
Azo $-$N$=$N$-$ Aliphatic Aromatic *cis* (Z) *trans* (E)	ca 1 575 (var) ca 1 510 (w) 1 440–1 410 (w)	Very weak or inactive
Azoxy $-$N^{+}$=$N$-$ $\underset{\text{O}^-}{\vert}$ Aliphatic Aromatic	1 590–1 495 (m-s) 1 345–1 285 (m-s) 1 480–1 450 (m-s) 1 340–1 315 (m-s)	
Azothio $-$N$=$N^{+}$-$S^{-}$-$	1 465–1 445 (w) 1 070–1 055 (w)	
Nitro compounds N$=$O		
Nitro C$-$NO$_2$ Aliphatic	ca 1 560 (s) 1 385–1 350 (s)	The two bands are due to asymmetrical and symmetrical stretching of the N$=$O bond. Electron-withdrawing substituents adjacent to nitro group increase the frequency of the asymmetrical band and decrease that of the symmetrical frequency.
Aromatic	1 570–1 485 (s) 1 380–1 320 (s) 865–835 (s)	See above remark; also bulky orthosubstituents shift band to higher frequencies. Strong H bonding shifts frequency to lower end of range. Strong and sometimes at ca 750 cm^{-1}

TABLE 6.15 Absorption Frequencies of Other Double Bonds

Group	Band, cm^{-1}	Remarks
Nitro compounds N$=$O (*continued*)		
α,β-Unsaturated Nitroalkenes	580–520 (var) 1 530–1 510 (s) 1 360–1 335 (s)	
Nitrates —O—NO$_2$	1 650–1 625 (vs) 1 285–1 275 (vs) 870–855 (vs) 760–755 (w-m) 710–695 (w-m)	
Nitramines $>$N—NO$_2$ Nitrates —O—N$=$O	1 630–1 550 (s) 1 300–1 250 (s) 1 680–1 610 (vs) 815–750 (s) 850–810 (s) 690–615 (s)	Two bands *Trans* (*E*) form *Cis* (*Z*) form
Thionitrites —S—N$=$O	730–685 (m-s)	
Nitroso $\geqq$C—N$=$O	1 600–1 500 (s)	
N—$\overset{+}{N}$$=$$\overset{-}{O}$ Aliphatic Aromatic	1 530–1 495 (m-s) 1 480–1 450 (m-s) 1 335–1 315 (m-s)	
Nitrogen oxides N$\rightarrow$O Pyridine Pyrazine	1 320–1 230 (m-s) 1 190–1 150 (m-s) 1 380–1 280 (m-s) 1 040–990 (m-s) ca 850 (m)	Affected by ring substituents

TABLE 6.16 Absorption Frequencies of Aromatic Bands

Abbreviations Used in the Table

m, moderately strong   var, of variable strength
m-s, moderate to strong   w-m, weak to moderately strong
s, strong

Group	Band, cm^{-1}	Remarks
Aromatic rings	ca 1 600 (m)	
	ca 1 580 (m)	Stronger when ring is further conjugated
	ca 1 470 (m)	When substituent on ring is electron acceptor
	ca 1 510 (m)	When substituent on ring is electron donor
Five adjacent H	900–860 (w-m)	
	770–730 (s)	
	720–680 (s)	
	625–605 (w-m)	
	ca 550 (w-m)	Substituents: $C{=}C, C{\equiv}C, C{\equiv}N$
1,2-Substitution	770–735 (s)	
	555–495 (w-m)	
	470–415 (m-s)	
1,3-Substitution	810–750 (s)	
	560–505 (m)	
	460–415 (m-s)	490–460 cm^{-1} when substituents are electron-accepting groups
1,4-Substitution	860–800 (s)	
	650–615 (w-m)	
	520–440 (m-s)	520–490 cm^{-1} when substituents are electron-donating groups
1,2,3-Trisubstitution	800–760 (s)	
	720–685 (s)	
	570–535 (s)	
	ca 485	
1,2,4-Trisubstitution	900–885 (m)	
	780–760 (s)	
	475–425 (m-a)	
1,3,5-Trisubstitution	950–925 (var)	
	865–810 (s)	
	730–680 (m-s)	
	535–495 (s)	
	470–450 (w-m)	
Pentasubstitution	900–860 (m-s)	
	580–535 (s)	
Hexasubstitution	415–385 (m-s)	

TABLE 6.17 Absorption Frequencies of Miscellaneous Bands

Abbreviations Used in the Table

m, moderately strong	vs, very strong
m-s, moderate to strong	w, weak
s, strong	w-m, weak to moderately strong
var, of variable strength	

Group	Band, cm^{-1}	Remarks
	Ethers	
Saturated aliphatic $\geq$C—O—C$\leq$	1 150–1 060 (vs)	Two peaks may be observed for branched chain, usually 1 140–1 110 cm^{-1}.
	1 140–900 (s)	Usually 930–900 cm^{-1}; may be absent for symmetric ethers
Alkyl-aryl =C—O—C$\leq$	1 270–1 230 (vs)	=CO stretching
	1 120–1 020 (s)	CO stretching
Vinyl	1 225–1 200 (s)	Usually about 1 205 cm^{-1}
Diaryl =C—O—C=	1 200–1 120 (s)	
	1 100–1 050 (s)	
Cyclic	1 270–1 030 (s)	
Epoxides 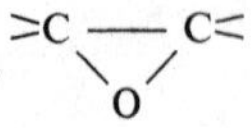 $>$C——C$<$ O	1 260–1 240 (m-s)	
	880–805 (m)	Monosubstituted
	950–860 (var)	*Trans* (*E*) form
	865–785 (m)	*Cis* (*Z*) form
	770–750 (m)	Trisubstituted
Ketals and acetals	1 190–1 140 (s)	
	1 195–1 125 (s)	
	1 100–1 000 (s)	Strongest band
	1 060–1 035 (s)	Sometimes obscured
Phthalanes	915–895 (s)	
Aromatic methylenedioxy	1 265–1 235 (s)	
	Peroxides	
—O—O—	900–830 (w)	
	1 150–1 030 (m-s)	Alkyl
	ca 1 000 (m)	Aryl

TABLE 6.17 Absorption Frequencies of Miscellaneous Bands (*continued*)

Group	Band, cm^{-1}	Remarks
	Sulfur compounds	
Thiols —S—H —CO—SH —CS—SH	2600–2450 (w) 840–830 (m) ca 860 (s)	Broad
Thiocarbonyl >C=S >N—C̣=S —S—C̣=S	1200–1050 (s) 1570–1395 1420–1260 1140–940 ca 580 (s)	Behaves generally in a manner similar to carbonyl band
Sulfoxides >S=O	1075–1040 (vs) 730–690 (var) 395–360 (var)	Halogen or oxygen atom bonded to sulfur increases the frequency.
Sulfones >SO$_2$	1360–1290 (vs) 1170–1120 (vs) 610–545 (m-s) 525–495 (m-s)	Halogen or oxygen atom bonded to sulfur increases the frequency.
Sulfonamides —SO$_2$—N<	1380–1330 (vs) 1170–1140 (vs) 950–860 (m) 715–700 (w-m)	
Sulfonates —SO$_2$—O—	1420–1330 (s) 1200–1145 (s)	May appear as doublet
Thiosulfonates —SO$_2$—S—	ca 1340 (vs)	
Sulfates —O—SO$_2$—O— Primary alkyl salts	1415–1380 (s) 1200–1185 (s) 1315–1220 (s) 1140–1075 (m)	Electronegative substituents increase frequencies. Strongly influenced by metal ion

TABLE 6.17 Absorption Frequencies of Miscellaneous Bands (*continued*)

Group	Band, cm^{-1}	Remarks
Sulfur compounds (*continued*)		
Sulfates —O—SO$_2$—O (*continued*) Secondary alkyl salts	1 270–1 210 (vs) 1 075–1 050 (s)	Doublet; both bands strongly influenced by metal ion
Stretching frequencies of C—S and S—S bonds —S—CH$_3$ —S—CH$_2$— —S—CH< —S—C≦ —S—aryl	710–685 (w-m) 660–630 (w-m) 630–600 (w-m) 600–570 (w-m) 1 110–1 070 (m) 710–685 (w-m)	
R—S—S—R	705–570 (w) 520–500 (w)	
Aryl—S—S—aryl Polysulfides CH$_2$—S—CH$_2$— (R—S)$_2$C=O —CO—S— —CS—S	500–430 (w-m) 500–470 (w-m) 695–655 (w-m) 880–825 (s) 570–560 (var) 1 035–935 (s) ca 580 (s)	CSC stretching
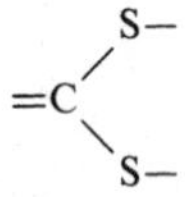 =C⟨ S— / S—	1 050–900 (m-s) 980–850 (m-s) 900–800 (m-s)	Monoionic Ionic 1,1-dithiolates
Phosphorus compounds		
P—H	2 455–2 265 (m) 1 150–965 (w-m)	Sharp. Phosphines lie in the region 2 285–2 265 cm^{-1}.
—PH$_2$	1 100–1 085 (m) 1 065–1 040 (w-m) 940–910 (m)	
P—alkyl	795–650 (m-s)	
P—aryl	1 130–1 090 (s) 750–680 (s)	
P—O—alkyl	1 050–970 (s)	Broad
P—O—aryl	1 240–1 190 (s)	
P—O—P	970–910	Broad

TABLE 6.17 Absorption Frequencies of Miscellaneous Bands (*continued*)

Group	Band, cm^{-1}	Remarks
Phosphorus compounds (*continued*)		
P=O	1 350–1 150 (s)	May appear as doublet
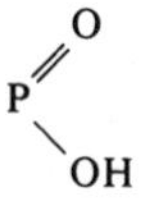 P⟋O ⟍OH	2 725–2 520 (w-m) 2 350–2 080 (w-m) 1 740–1 600 (w-m) 1 335 (s) 1 090–910 (s) 540–450 (w-m)	H-bonded; broad Broad; may be doublet for aryl acids P=O stretching
P=S	865–655 (m-s) 595–530 (var)	
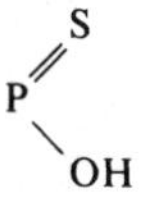 P⟋S ⟍OH	3 100–3 000 (w) 2 360–2 200 (w) 935–910 (s) 810–750 (m-s) 655–585 (var)	 PO stretching P=S stretching P=S stretching
Silicon compounds		
Si—H	2 250–2 100 (s) 985–800	SiH$_3$ has two bands.
Si—C≡	860–760	Accompanied by CH$_2$ rocking
Si—CH$_3$	1 280–1 250 (s)	Sharp
Si—C$_2$H$_5$	1 250–1 220 (m) 1 020–1 000 (m) 970–945 (m)	
Si—Aryl	1 125–1 090 (vs)	Splits into two bands when two aryl groups are attached to one silicon atom, but has only one band when three aryl groups attached
≥Si—OH	870–820	OH deformation band
≥Si—O—Si≤	1 100–1 000	
≥Si—N—Si≤	940–870 (s)	
≥Si—Cl	550–470 (s) 250–150	

TABLE 6.17 Absorption Frequencies of Miscellaneous Bands (*continued*)

Group	Band, cm^{-1}	Remarks
	Silicon compounds (*continued*)	
$>$SiCl$_2$	595–535 (s) 540–460 (m)	
—SiCl$_3$	625–570 (s) 535–450 (m)	
	Boron compounds	
Boranes $>$BH or —BH$_2$	2 640–2 450 (m-s) 2 640–2 570 (m-s) 2 535–2 485 (m-s) 2 380–2 315 (s) 2 285–2 265 (s) 2 140–2 080 (w-m) 2 580–2 450 (m)	Free H in BH Free H in BH$_2$ plus second band In complexes; second band for BH$_2$ Bridged H Borazoles and borazines
BH$_4^-$	2 310–2 195 (s)	Two bands
B—N	1 550–1 330 750–635	Borazines and borazoles
B—O	1 390–1 310 (s) 1 280–1 200	BO stretching Metal orthoborates
B—Cl B—Br	1 090–890 (s)	Plus other bands at lower frequencies for BX$_2$ and BX$_3$
B—F	1 500–840 (var)	Isotope splitting present
XBF$_2$	1 500–1 410 (s) 1 300–1 200 (s)	
X$_2$BF	1 360–1 300 (s)	
BF$_3$ complexes	1 260–1 125 (s) 1 030–800 (s)	Band splitting may be added to isotopic splittings.
BF$_4^-$	ca 1 030 (vs)	

TABLE 6.17 Absorption Frequencies of Miscellaneous Bands (*continued*)

Group	Band, cm^{-1}	Remarks
	Halogen compounds	
C—F		
Aliphatic, mono-F	1 110–1 000 (vs)	
	780–680 (s)	
Aliphatic, di-F	1 250–1 050 (vs)	Two bands
Aliphatic, poly-F	1 360–1 090 (vs)	Number of bands
Aromatic	1 270–1 100 (m)	
	680–520 (m-s)	
	420–375 (var)	
	340–240 (s)	
—CF$_3$		
Aliphatic	1 350–1 120 (vs)	
	780–680 (s)	
	680–590 (s)	
	600–540 (s)	
	555–505 (s)	
Aromatic	1 330–1 310 (m-s)	
	600–580 (s)	
C—Cl		
Primary alkanes	730–720 (s)	
	685–680 (s)	
	660–650 (s)	
Secondary alkanes	ca 760 (m)	
	675–655 (m-s)	
	615–605 (s)	
Tertiary alkanes	635–610 (m-s)	
	580–560 (m-s)	
Poly-Cl	800–700 (vs)	
Aryl:		
1,2-	1 060–1 035 (m)	
1,3-	1 080–1 075 (m)	
1,4-	1 100–1 090 (m)	
Chloroformates	ca 690 (s)	
	485–470 (s)	
Axial Cl	730–580 (s)	
Equatorial Cl	780–740 (s)	
C—Br		
Primary alkanes	645–635 (s)	
	565–555 (s)	
	440–430 (var)	
Secondary alkanes	620–605 (s)	
	590–575 (m-w)	
	540–530 (s)	

TABLE 6.17 Absorption Frequencies of Miscellaneous Bands (*continued*)

Group	Band, cm^{-1}	Remarks
	Halogen compounds (*continued*)	
C—Br (*continued*)		
Tertiary alkanes	600–595 (m-s)	
	525–505 (s)	
Axial	690–550 (s)	
Equatorial	750–685 (s)	
Aryl:		
1,2-	1 045–1 025 (m)	
1,3-; 1,4-	1 075–1 065 (m)	
Other bands	400–260 (s)	
	325–175 (m-s)	
	290–225 (m-s)	
C—I		
Primary alkanes	600–585 (s)	
	515–500 (s)	
Secondary alkanes	ca 575 (s)	
	550–520 (s)	
	490–480 (s)	
Tertiary alkanes	580–560 (s)	
	510–485 (m)	
	485–465 (s)	
Aromatic	1 060–1 055 (m-s)	
	310–160 (s)	
	265–185	
Axial	ca 640 (s)	
Equatorial	ca 655 (s)	
	Inorganic ions	
Ammonium	3 300–3 030	Several bands, all strong
Cyanate	2 220–2 130 (s)	
Cyanide	2 200–2 000	
Carbonate	1 450–1 410	
Hydrogen sulfate	1 190–1 160 (s)	
	1 180–1 000 (s)	
	880–840 (m)	
Nitrate	1 410–1 350 (vs)	
	860–800 (m)	
Nitrite	1 275–1 230 (s)	
	835–800 (m)	Shoulder

TABLE 6.17 Absorption Frequencies of Miscellaneous Bands (*continued*)

Group	Band, cm^{-1}	Remarks
	Inorganic ions (*continued*)	
Phosphate	1 100–1 000	
Sulfate	1 130–1 080 (s)	
Thiocyanate	ca 2 050 (s)	

TABLE 6.18 Absorption Frequencies in the Near Infrared

Values in parentheses are molar absorptivity

Class	Band, cm^{-1}	Remarks
Acetylenes	9 800–9 430 6 580–6 400 (1.0)	Overtone of ≡CH stretching
Alcohols (nonhydrogen-bonded)	7 140–7 010 (2.0)	Overtone of OH stretching

Aldehydes
  Aliphatic
	4 640–4 520 (0.5)	Combination of C=O and CH
		  stretchings

Aromatic	ca 8 000 ca 4 525 ca 4 445	
Formate	4 775–4 630 (1.0)	
Alkanes —CH$_3$	9 000–8 350 (0.02) 5 850–5 660 (0.1) 4 510–4 280 (0.3)	
—CH$_2$—	9 170–8 475 (0.02) 5 830–6 640 (0.1) 4 420–4 070 (0.25)	
≧CH	8 550–8 130 7 000–6 800 5 650–5 560	All bands very weak
Cyclopropane	6 160–6 060 4 500–4 400	
Alkenes ＼C=C／ （C=C with H)	6 850–6 370 (1.0)	
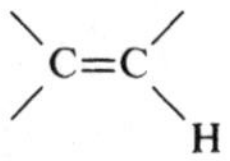 >C=CH$_2$ and —CH=CH$_2$	7 580–7 300 (0.02) 6 140–5 980 (0.2) 4 760–4 700 (1.2)	

TABLE 6.18 Absorption Frequencies in the Near Infrared (*continued*)

Class	Band, cm^{-1}	Remarks
Alkenes (continued) H H $\backslash$ / C=C / $\backslash$ —O—CH=CH$_2$ —CO—CH=CH$_2$	4760–4660 (0.15) 6250–6040 (0.3) 7580–7410 (0.02) 6190–5990 (0.3) 4820–4750 (0.2–0.5)	*Trans* (*E*) isomers have no unique bands.
Amides Primary	7400–6540 (0.7) 5160–5060 (3.0) 5040–4990 (0.5) 4960–4880 (0.5)	Two bands; overtone of NH stretch Second overtone of C=O stretch; second overtone of NH deformation; combination of C=O and NH
Secondary	7330–7140 (0.5) 5050–4960 (0.4)	Overtone of NH stretch Combination of NH stretch and NH bending
Amines, aliphatic Primary	9710–9350 6670–6450 (0.5) 5075–4900 (0.7)	Second overtone of NH stretch Two bands; overtone of NH stretch Two bands; combination of NH stretch and NH bending
Secondary	9800–9350 6580–6410 (0.5)	Second overtone of NH stretch Overtone of NH stretch
Amines, aromatic Primary	9950–9520 (0.4) 7040–6850 (0.2) 6760–6580 (1.4) 5140–5040 (1.5)	
Secondary	10000–9710 6800–6580 (0.5)	
Aryl-H	7660–7330 (0.1) 6170–5880 (0.1)	Overtone of CH stretch
Carbonyl	5200–5100	
Carboxylic acids	7000–6800	
Epoxide (terminal)	6135–5960 (0.2) 4665–4520 (1.2)	Cyclopropane bands in same region

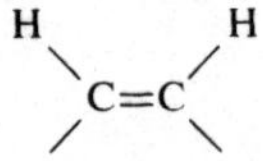

TABLE 6.18 Absorption Frequencies in the Near Infrared (*continued*)

Class	Band, cm^{-1}	Remarks
Glycols	7140–7040	
Hydroperoxides		
Aliphatic	6940–6750 (2.0)	
	4960–4880 (0.8)	
Aromatic	7040–6760 (1.0)	Two bands
	4950–4850 (1.3)	
Imides	9900–9620	
	6540–6370	
Nitriles	5350–5200 (0.1)	
Oximes	7140–7050	
Phosphines	5350–5260 (0.2)	
Phenols		
Nonbonded	7140–6800 (3.0)	
	5000–4950	
Intramolecularly bonded	7000–6700	
Thiols	5100–4950 (0.05)	

TABLE 6.19 Infrared Transmitting Materials

Material	Wavelength range, μm	Wavenumber range, cm^{-1}	Refractive index at 2 μm
NaCl, rock salt	0.25–17	40000–590	1.52
KBr, potassium bromide	0.25–25	40000–400	1.53
KCl, potassium chloride	0.30–20	33000–500	1.5
AgCl, silver chloride*	0.40–23	25000–435	2.0
AgBr, silver bromide*	0.50–35	20000–286	2.2
CaF$_2$, calcium fluoride (Irtran-3)	0.15–9	66700–1110	1.40
BaF$_2$, barium fluoride	0.20–11.5	50000–870	1.46
MgO, magnesium oxide (Irtran-5)	0.39–9.4	25600–1060	1.71
CsBr, cesium bromide	1–37	10000–270	1.67
CsI, cesium iodide	1–50	10000–200	1.74
TlBr–TlI, thallium bromide–iodide (KRS-5)*	0.50–35	20000–286	2.37
ZnS, zinc sulfide (Irtran-2)	0.57–14.7	17500–680	2.26

* Useful for internal reflection work.

TABLE 6.19   Infrared Transmitting Materials (*continued*)

Material	Wavelength range, μm	Wavenumber range, cm^{-1}	Refractive index at 2 μm
ZnSe, zinc selenide* (vacuum deposited) (Irtran-4)	1–18	10 000–556	2.45
CdTe, cadmium telluride (Irtran-6)	2–28	5 000–360	2.67
Al$_2$O$_3$, sapphire*	0.20–6.5	50 000–1 538	1.76
SiO$_2$, fused quartz	0.16–3.7	62 500–2 700	
Ge, germanium*	0.50–16.7	20 000–600	4.0
Si, silicon*	0.20–6.2	50 000–1 613	3.5
Polyethylene	16–300	625–33	1.54

* Useful for internal reflection work.

TABLE 6.20 Infrared Transmission Characteristics of Selected Solvents

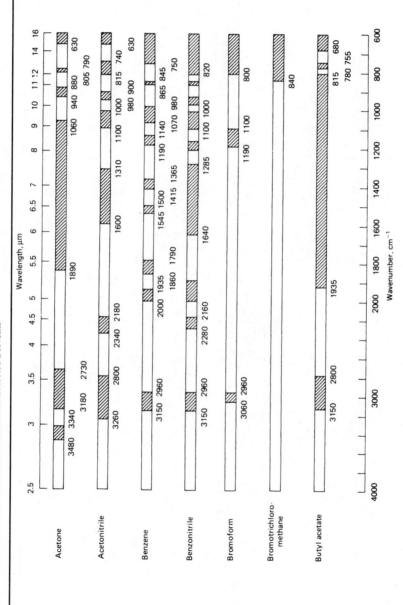

TABLE 6.20 Infrared Transmission Characteristics of Selected Solvents (*continued*)

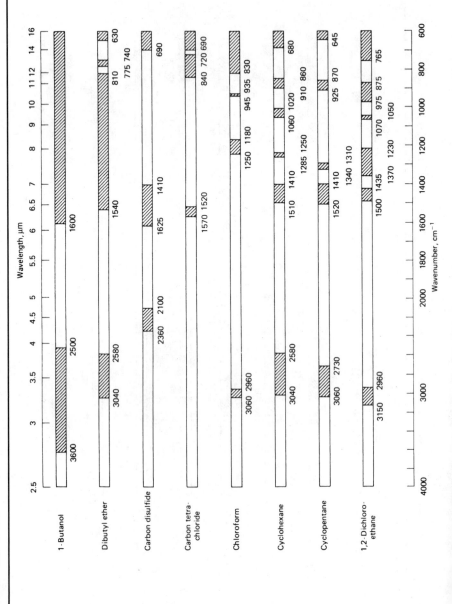

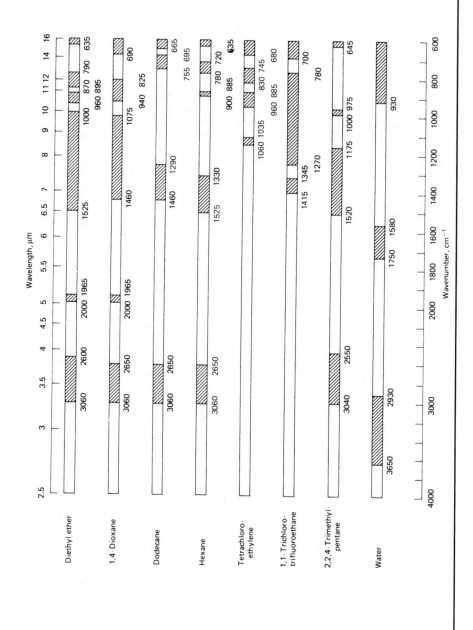

RAMAN SPECTROSCOPY

Infrared and Raman spectroscopy are related by the fact that both permit the detection of bond vibrations. Like IR spectroscopy, the spectral bands are reported in cm^{-1}. An important difference is that the wavelength and intensity of inelastically scattered light is measured in the Raman spectroscopic method. The "Raman effect" causes the scattered radiation to shift according to the energies of molecular vibrations. Although Raman spectroscopy involves a physical principle different from that in IR spectroscopy, the two techniques are complementary.

Infrared spectroscopy relies on a changing dipole during a bond vibration for absorption of energy to occur. In Raman, it is a change in polarizability in the bond that permits absorption. The simple molecule carbon dioxide, $O=C=O$, is an instructive example. Both $C=O$ bonds have dipoles but they oppose each other and the net dipole is 0 Debye (0 D). The symmetrical stretch in which both $C=O$ bonds simultaneously extend and contract does not change the dipole but is detectable by Raman because the polarizability of the system alters.

Raman scattering is not a very efficient process and an energy source of considerable power is required. This is typically an argon (Ar) laser. A variety of chemical bonds and systems can be detected by modern Raman spectrometers and typical data are summarized in Tables 6.21–6.30.

TABLE 6.21   Raman Frequencies of Single Bonds to Hydrogen and Carbon

Abbreviations Used in the Table

m, moderately strong	vw, very weak
m-s, moderate to strong	w, weak
m-vs, moderate to very strong	w-m, weak to moderately strong
s, strong	w-m, weak to moderately strong
vs, very strong	w-vs, weak to very strong

Group	Band, cm^{-1}	Remarks
	Saturated C—H and C—C	
—CH$_3$	2 969–2 967 (s)	
	2 884–2 883 (s)	
	ca 1 205 (s)	In aryl compounds
	1 150–1 135	In unbranched alkyls
	1 060–1 056	In unbranched alkyls
	975–835 (s)	Terminal rocking of methyl group
	280–220	CH$_2$—CH$_3$ torsion
—CH$_2$—	2 949–2 912 (s)	
	2 861–2 849 (s)	
	1 473–1 443 (m-vs)	Intensity proportional to
	1 305–1 295 (s)	number of CH$_2$ groups
	1 140–1 070 (m)	Often two bands; see above
	888–837 (w)	
	425–150	
	500–490	Substituent on aromatic ring

TABLE 6.21 Raman Frequencies of Single Bonds to Hydrogen and Carbon (*continued*)

Group	Band, cm^{-1}	Remarks
	Saturated C—H and C—C (*continued*)	
—CH(CH$_3$)$_2$	1350–1330 (m) 835–750 (s)	If attached to C=C bond, 870–800 cm^{-1}. If attached to aryl ring, 740 cm^{-1}
—C(CH$_3$)$_3$	1265–1240 (m) 1220–1200 (m) 760–685 (vs)	Not seen in *tert*-butyl bromide Not seen in *tert*-butyl bromide If attached to C=C or aromatic ring, 760–720 cm^{-1}
Internal tertiary carbon atom	855–805 (w) 455–410	
Internal quaternary carbon atom	710–680 (vs) 490–470	
Two adjacent tertiary carbon atoms	730–920 770–725	Often a band at 530–524 cm^{-1} indicates presence of adjacent tertiary and quaternary carbon atoms.
Dialkyl substitution at α-carbon atom	800–700 (m-s) 680–650 (vs) 605–550	
Cyclopropane	3101–3090 3038–3019 1210–1180 (s)	Shifts to 1200 cm^{-1} for monoalkyl or 1,2-dialkyl substitution and to 1320 cm^{-1} for *gem*-1,1-dialkyl substitution
Cyclobutane	1001–960 (vs)	Shifts to 933 cm^{-1} for monoalkyl, to 887 cm^{-1} for *cis*-1,3-dialkyl, and to 891 cm^{-1} plus 855 cm^{-1} (doublet) for *trans*-1,3-dialkyl subsition
Cyclopentane	900–800 (s)	
Cyclohexane	825–815 (vs) 810–795 (vs)	Boat configuration Chair configuration

TABLE 6.21 Raman Frequencies of Single Bonds to Hydrogen and Carbon (*continued*)

Group	Band, cm^{-1}	Remarks
	Saturated C—H and C—C (*continued*)	
Cycloheptane	ca 733	
Cyclooctane	ca 703	
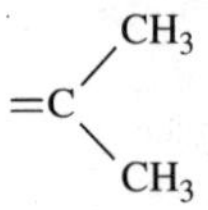 $=C\begin{smallmatrix}CH_3\\CH_3\end{smallmatrix}$	1392–1377 450–400 (vw) 270–250 (m)	
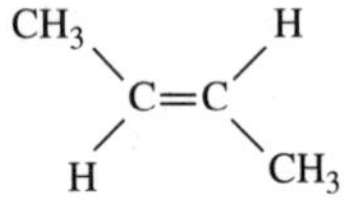 $CH_3,H / C=C / H, CH_3$	1380–1379 492–455 (vw) 220–200 (m)	
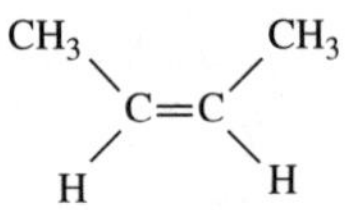 $CH_3, CH_3 / C=C / H, H$	1372–1368 970–952 (m) 592–545 (vw) 420–400 (m) 310–290 (m)	
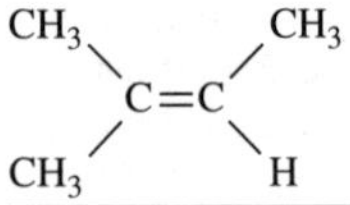 $CH_3, CH_3 / C=C / CH_3, H$	1385–1375 522–488 (w)	
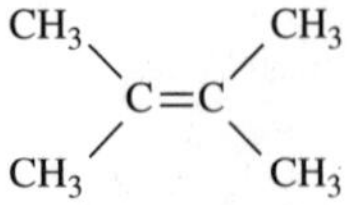 $CH_3, CH_3 / C=C / CH_3, CH_3$	1392–1386 690–678 (m-s) 510–485 (m) 424–388 (w)	
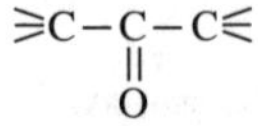 $\geqslant$C–C–C$\leqslant$, ‖ , O	1170–1100 (w-m) 600–580 (m-s)	
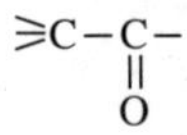 $\geqslant$C–C– , ‖ , O	1120–1090 (m-vs) 600–510 (w-m)	Tertiary or quaternary carbon adjacent to carbonyl group lowers the frequency 300 cm^{-1}
—CH$_2$—CO—	1420–1410 (s)	
—CHO	2850–2810 (m) 2720–2695 (vs)	Often appears as a shoulder

TABLE 6.21 Raman Frequencies of Single Bonds to Hydrogen and Carbon (*continued*)

Group	Band, cm^{-1}	Remarks
Unsaturated C—H		
—C≡C—H	3340–3270 (w-m)	Alkyl substituents at higher frequencies; unsaturated or aryl substituents at lower frequencies
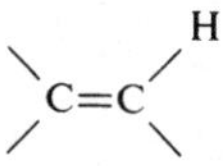	3040–2995 (m)	
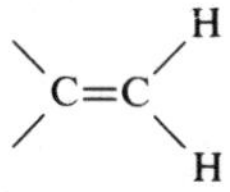	3095–3050 (m) 2990–2983 (s)	Asymmetric =CH$_2$ stretch Symmetric =CH$_2$ stretch
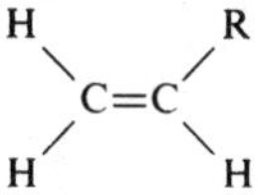	1419–1415 (m) 1309–12888 (m)	Plus =CH and =CH stretching bands
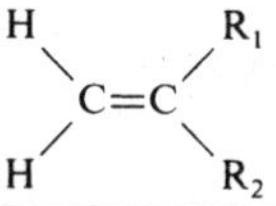	1413–1399 (m) 909–885 (m) 711–684 (w)	Plus =CH$_2$ stretching bands
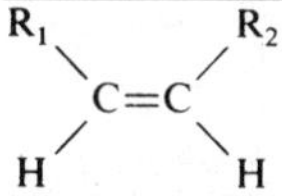	1270–1251 (m)	Plus =CH stretching band
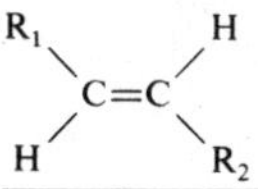	1314–1290 (m)	Plus =CH stretching band
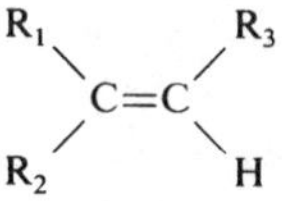	1360–1322 (w) 830–800 (vw)	Plus =CH stretching band
Hydroxy O—H		
Free —OH Intermolecularly bonded Aromatic —OH	3650–3250 (w) 3400–3300 (w) ca 3160 (s)	
—OH	1460–1320 (w) 1276–1205 (w-m) 1260 (w-m)	Common to all OH substituents Primary Secondary

TABLE 6.21 Raman Frequencies of Single Bonds to Hydrogen and Carbon (*continued*)

Group	Band, cm^{-1}	Remarks
Hydroxy O—H (*continued*)		
C—C—OH primary	1 070–1 050 (m-s) 1 030–960 (m-s) 480–430 (w-m)	CCO stretching CCO deformation

C—C—OH
  Secondary
 	
  Tertiary
	1 135–1 120 (m-s)
	825–815 (vs)
	500–490 (w-m)
	1 210–1 200 (m-s)
	755–730 (vs)
	360–350 (w-m)	

—CO—O—H	1 305–1 270	CO stretching
N—H and C—N bonds		

Amine >N—H
  Associated
  Nonbonded
  Salts
 	
  —NH$_2$
	3 400–3 250 (s)
	3 550–3 250 (s)
	2 986–2 974
	
	1 650–1 590 (w-vs)
		Primary amines show two bands.
		
		Often obscured by intense CH
		  stretching bands
		Bending

Amides
  Primary
 	
  Secondary
	3 540–3 500 (w)
	3 400–3 380 (w)
	1 310–1 250 (s)
	
	1 150–1 095 (m)
	3 491–3 404 (m-s)
	
	1 190–1 130 (m)
	931–865 (m-s)
	430–395 (w-m)
		Both bands lowered ca 150 cm^{-1}
		  in solid state and H bonding
		Interaction of NH bending and
		  CN stretching; lowered 50 cm^{-1}
		  in nonbonded state
		Rocking of NH$_2$
		Two bands; lowered in frequency
		  on H bonding and in solid state

—CO—N	607–555 (m)	O=CN bending

C—N—C
  |
  C
	1 070–1 045 (m)	Stretching

≡C—N<
  Primary carbon
  Secondary α carbon
 	
  Tertiary α carbon
	1 090–1 060 (m)
	1 140–1 035 (m)
	
	1 240–1 020 (m)
		CN stretching
		Two bands but often obscured.
		  Strong band at 800 cm^{-1}
		Two bands. Strong band also at
		  745 cm^{-1}

TABLE 6.22 Raman Frequencies of Triple Bonds

Abbreviations Used in the Table

m, moderately strong	s-vs, strong to very strong
m-s, moderate to strong	vs, very strong
s, strong	

Group	Band, cm^{-1}	Remarks
R—C≡CH	2160–2100 (vs)	Monoalkyl substituted; C≡C stretch
	650–600 (m)	C≡CH deformation
	356–335 (s)	C≡C—C bending of monoalkyls
R$_1$—C≡C—R$_2$	2300–2190 (vs)	C≡C stretching of disubstituted alkyls; sometimes two bands
—C≡C—C≡C—	2264–2251 (vs)	
—C≡N	2260–2240 (vs)	Unsaturated nonaryl substituents lower the frequency and enhance the intensity.
	2234–2200 (vs)	Lowered ca 30 cm^{-1} with aryl and conjugated aliphatics
	840–800 (s-vs)	CCCN symmetrical stretching
	385–350 (m-s)	
	200–160 (vs)	Aliphatic nitriles
H—C≡N	2094 (vs)	
Azides —$\overset{+}{\text{N}}$—$\overset{-}{\text{N}}$≡N	2170–2080 (s)	Asymmetric NNN stretching
	1258–1206 (s)	Symmetric NNN stretching; HN$_3$ at 1300 cm^{-1}
Diazonium salts R—$\overset{+}{\text{N}}$≡N	2300–2240 (s)	
Isonitriles —$\overset{+}{\text{N}}$≡C$^-$	2146–2134	Stretching of aliphatics
	2124–2109	Stretching of aromatics
Thiocyanates —S—C≡N	2260–2240 (vs)	Stretching of C≡N
	650–600 (s)	Stretching of SC bond

TABLE 6.23 Raman Frequencies of Cumulated Double Bonds

Abbreviations Used in the Table

s, strong	vw, very weak
vs, very strong	w, weak

Group	Band, cm^{-1}	Remarks
Allenes C=C=C	2000–1960 (s) 1080–1060 (vs) 356	Pseudo-asymmetric stretching Symmetric stretching C=C=C bending

Carbodiimides (cyanamides)
—N=C=N—
	2140–2125 (s)
	
	2150–2100 (vs)
	
	1460
	
	1150–1140 (vs)
		Asymmetric stretching of
		  aliphatics
		Asymmetric stretching of
		  aromatics; two bands
		Symmetrical stretching of
		  aliphatics
		Symmetric stretching of aryls

Cumulenes (trienes) C=C=C=C	2080–2030 (vs) 878	
Isocyanates —N=C=O	2300–2250 (vw) 1450–1400 (s)	Asymmetric stretching Symmetric stretching
Isothiocyanates —N=C=S	2220–2100 690–650	Two bands Alkyl derivatives
Ketenes C=C=O	2060–2040 (vs) 1130 (s) 1374 (s) 1120 (s)	Pseudo-asymmetric stretching Pseudo-symmetric stretching Alkyl derivatives Aryl derivatives
Sulfinylamines R—N=S=O	1306–1214 (w) 1155–989 (s)	Asymmetric stretching Symmetric stretching

TABLE 6.24 Raman Frequencies of Carbonyl Bonds

Abbreviations Used in the Table

m, moderately strong	s-vs, strong to very strong
m-s, moderate to strong	vs, very strong
s, strong	w, weak

Group	Band, cm^{-1}	Remarks
Acid anhydrides —CO—O—CO—		
Saturated	1 850–1 780 (m) 1 771–1 770 (m)	
Conjugated, noncyclic	1 775 1 720	
Acid fluorides —CO—F		
Alkyl	1 840–1 835	
Aryl	1 812–1 800	
Acid chlorides —CO—Cl		
Alkyl	1 810–1 770 (s)	
Aryl	1 774 1 731	
Acid bromides —CO—Br		
Alkyl	1 812–1 788	
Aryl	1 775–1 754	
Acid iodides —CO—I		
Alkyl	ca 1 806	
Aryl	ca 1 752	
Lactones	1 850–1 730 (s)	
Esters		
Saturated	1 741–1 725	Alkyl branching on carbon adjacent to C=O lowers frequency by 5–15 cm^{-1}.
Aryl and α,β-unsaturated	1 727–1 714	
Diesters		
Oxalates	1 763–1 761	
Phthalates	1 738–1 728	
C≡C—CO—O—	1 716–1 708	
Carbamates	1 694–1 688	
Aldehydes	1 740–1 720 (s-vs)	
Ketones		
Saturated	1 725–1 700 (vs)	
Aryl	1 700–1 650 (m)	

TABLE 6.24   Raman Frequencies of Carbonyl Bonds (*continued*)

Group	Band, cm^{-1}	Remarks
Ketones (*continued*)		
Alicyclic		
$n = 4$	1 782 (m)	
$n = 5$	1 744 (m)	
$n \geq 6$	1 725–1 699 (m)	
Carboxylic acids		
Mono-	1 686–1 625 (s)	These α-substituents increase the frequency: F, Cl, Br, OH.
Poly-	1 782–1 645	Solid state; often two bands
	1 750–1 710	In solution; very broad band
Amino acids	1 743–1 729	
Carboxylate ions	1 690–1 550 (w)	
	1 440–1 340 (vs)	
Amino acid anion	1 743–1 729	
	1 600–1 570 (w)	Often masked by water deformation band near 1 630 cm^{-1}
Amides (see also Table 6.21)		
Primary		
Associated	1 686–1 576 (m-s)	
	1 650–1 620 (m)	
Nonbonded	1 715–1 675 (m)	
	1 620–1 585 (m)	
Secondary		
Associated	1 680–1 630 (w)	Both *cis* (*Z*) and *trans* (*E*) forms
	1 570–1 510 (w)	*Trans* (*E*) form
	1 490–1 440	*Cis* (*Z*) form
Nonbonded	1 700–1 650	Both *cis* (*Z*) and *trans* (*E*) forms
	1 550–1 500	*Trans* (*E*) form (no *cis* band)
Tertiary	1 670–1 630 (m)	
Lactams	1 750–1 700 (m)	

TABLE 6.25 Raman Frequencies of Other Double Bonds

Abbreviations Used in the Table

m, moderately strong — vs, very strong
m-s, moderate to strong — w, weak
s, strong — s-vs, strong to very strong
w-m, weak to moderately strong

Group	Band, cm^{-1}	Remarks
Alkenes $>C=C<$		
$>C=C<$	1 680–1 576 (m-s)	General range
(H, H)C=C(R$_1$, H)	1 648–1 638 (vs)	C=C stretching
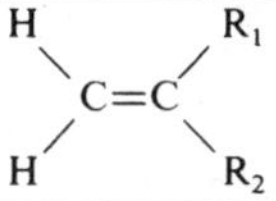 (H, H)C=C(R$_1$, R$_2$)	ca 1 650 (vs) 270–252 (w)	C=C stretching C=C—C skeletal deformation
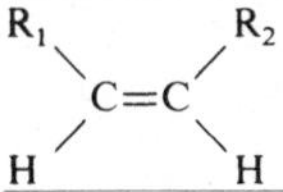 (R$_1$, H)C=C(R$_2$, H)	ca 1 660 (vs) 970–952 (w)	C=C stretching Asymmetric CC stretching
(R$_1$, H)C=C(H, R$_2$)	1 676–1 665 (s)	C—C stretching
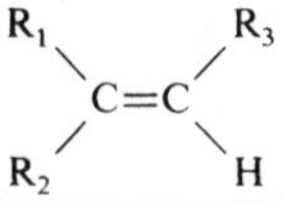 (R$_1$, R$_2$)C=C(R$_3$, H)	1 678–1 664 (vs) 522–488 (w)	C=C stretching C=C—C skeletal deformation
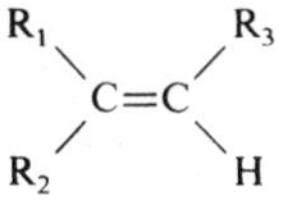 (R$_1$, R$_2$)C=C(R$_3$, H)	1 680–1 665 (s) 690–678 (m-s) 510–485 (m) 424–388 (w)	C=C stretching Symmetrical CC stretching Skeletal deformation Skeletal deformation

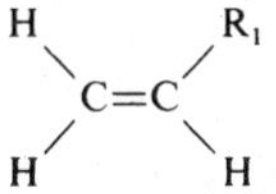

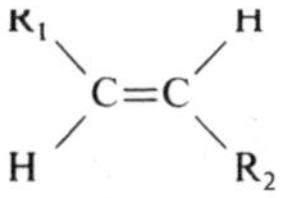

Haloalkene	X = fluorine	X = chlorine	X = bromine	X-iodine
$>C=C<$ stretch of haloalkanes				
$H_2C=CHX$	1 654	1 603–1 601	1 596–1 593	1 581
$HXC=CHX$				
cis (Z)	1 712	1 590–1 587	1 587–1 583	1 543
trans (E)	1 694	1 578–1 576	1 582–1 581	1 537
$H_2C=CX_2$	1 728	1 616–1 611	1 593	
$X_2C=CHX$	1 792	1 589–1 582	1 552	
$X_2C=CX_2$	1 872	1 577–1 571	1 547	1 465 (solid)

TABLE 6.25 Raman Frequencies of Other Double Bonds (*continued*)

Group	Band, cm^{-1}	Remarks
	>C—N—bonds	
Aldimines (azomethines) 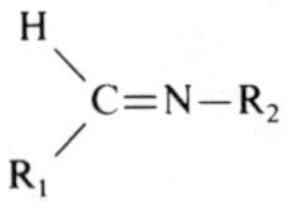 H \ C=N—R$_2$ / R$_1$	1 673–1 639 1 405–1 400 (s)	Dialkyl substituents at higher frequency; diaryl substituents at lower end of range
Aldoximines and Ketoximes >C—N—OH	1 680–1 617 (vs) 1 335–1 330 (w)	
Azines >C=N—N=C<	1 625–1 608 (s)	
Hydrazones H H \ / C=N—N / \ R$_1$ R$_2$ 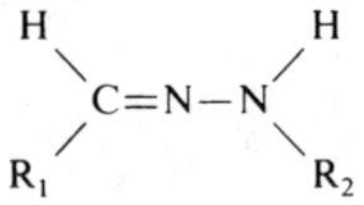	1 660–1 610 (s-vs)	
Imido ethers 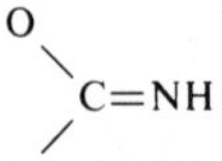 O \ C=NH /	1 658–1 648	NH stretching at 3 360–3 327 cm^{-1}

Semicarbazones and
  thiosemicarbazones 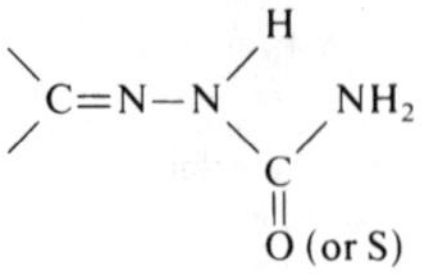
 	
\   H
    \   /
C=N—N   NH$_2$
/   \   /
        C
       ‖
      O (or S)
	1 665–1 642 (vs)
	
	1 620–1 610 (vs)
		Aliphatic. Thiosemicarbazones fall in lower end of range. Aromatic derivatives

	Azo compounds —N=N—	
—N=N—	1 580–1 570 (vs) 1 442–1 380 (vs) 1 060–1 030 (vs)	Nonconjugated Conjugated to aromatic ring CN stretching in aryl compounds
	Nitro compounds N=O	
Alkyl nitrites	1 660–1 620 (s)	N=O stretching
Alkyl nitrates	1 635–1 622 (w-m) 1 285–1 260 (vs) 610–562 (m)	Asymmetric NO$_2$ stretching Symmetric NO$_2$ stretching NO$_2$ deformation

TABLE 6.25 Raman Frequencies of Other Double Bonds (*continued*)

Group	Band, cm^{-1}	Remarks
	Nitro compounds N$=$O (*continued*)	
Nitroalkanes Primary	1 560–1 548 (m-s) 1 395–1 370 (s)	Sensitive to substituents attached to CNO$_2$ group
	915–898 (m-s) 894–873 (m-s) 618–609 (w)	
	640–615 (w)	Shoulder
	494–472 (w-m)	Broad; useful to distinguish from secondary nitroalkanes
Secondary	1 553–1 547 (m) 1 375–1 360 (s) 908–868 (m) 863–847 (s) 625–613 (m)	
	560–516 (s)	Sharp band
Tertiary	1 543–1 533 (m) 1 355–1 345 (s)	
Nitrogen oxides $\geqslant\overset{+}{N}\rightarrow\overset{-}{O}$	1 612–1 602 (s) 1 252 (m) 1 049–1 017 (s) 835 (s) 541 (w) 469 (w)	

TABLE 6.26 Raman Frequencies of Aromatic Compounds

Abbreviations Used in the Table

m, moderately strong	var, of variable strength
m-s, moderate to strong	vs, very strong
m-vs, moderate to very strong	w, weak
s, strong	w-m, weak to moderately strong
s-vs, strong to very strong	

Group	Band, cm^{-1}	Remarks
	Common features	
Aromatic compounds	3070–3020 (s) 1630–1570 (m-s)	CH stretching C—C stretching
	Substitution patterns of the benzene ring	
Monosubstituted	1180–1170 (w-m) 1035–1015 (s) 1010–990 (vs) 630–605 (w)	Characteristic feature; found also with 1,3- and 1,3,5-substitutions
1,2-Disubstituted	1230–1215 (m) 1060–1020 (s) 740–715 (m)	Characteristic feature Lowered 60 cm^{-1} for halogen substituents
1,3-Disubstituted	1010–990 (vs) 750–640 (s)	Characteristic feature
1,4-Disubstituted	1230–1200 (s-vs) 1180–1150 (m) 830–750 (vs) 650–630 (m-w)	Lower frequency with Cl substituents
Isolated hydrogen	1379 (s-vs) 1290–1200 (s) 745–670 (m-vs) 580–480 (s)	Characteristic feature
1,2,3-Trisubstituted	1100–1050 (m) 670–500 (vs) 490–430 (w)	The lighter the mass of the substituent, the higher the frequency
1,2,4-Trisubstituted	750–650 (vs) 580–540 (var) 500–450 (var)	Lighter mass at higher frequencies

TABLE 6.26 Raman Frequencies of Aromatic Compounds (*continued*)

Group	Band, cm^{-1}	Remarks
Substitution patterns of the benzene ring (continued)		
1,3,5-Trisubstituted	1 010–990 (vs)	
Completely substituted	1 296 (s)	
	550 (vs)	
	450 (m)	
	361 (m)	
Other aromatic compounds		
Naphthalenes	1 390–1 370	Ring breathing
	1 026–1 012	α or β substituents
	767–762	β substituents
	535–512	α substituents
	519–512	β substituents
Disubstituted naphthalenes	773–737 (s)	1,2-; 1,3-; 2,3-; 2,6-; 2,7-
	726–705 (s)	1,3-; 1,4-(two bands); 1,6-; 1,7- (two bands)
	690–634 (s)	1,2-; 1,4-(two bands); 1,5-; 1,8- (two bands)
	608	1,3-
	575–569	1,2-; 1,3-; 1,6-
	544–537	1,2-; 1,7-; 1,8-
Anthracenes	1 415–1 385	Ring breathing

TABLE 6.27 Raman Frequencies of Sulfur Compounds
Abbreviations Used in the Table

m, moderately strong	s-vs, strong to very strong
m-s, moderate to strong	vs, very strong
s, strong	w-m, weak to moderately strong

Group	Band, cm^{-1}	Remarks
—S—H	2 590–2 560 (s)	SH stretching for both aliphatic and aromatic
>C=S	1 065–1 050 (m)	
	735–690 (vs)	Solid state
>S=O		
In (RO$_2$)$_2$SO	1 209–1 198	One or two bands
In (R$_2$N)$_2$SO	1 108	

TABLE 6.27   Raman Frequencies of Sulfur Compounds (*continued*)

Group	Band, cm^{-1}	Remarks
$>$S$=$O (*continued*) In R$_2$SO SOF$_2$ SOCl$_2$ SOBr$_2$	1070–1010 (w-m) 1308 1233 1121	Broad
—SO$_2$—	1330–1260 (m-s) 1155–1110 (s) 610–540 (m) 512–485 (m)	Asymmetric SO$_2$ stretching Symmetric SO$_2$ stretching Scissoring mode of aryls Scissoring mode of alkyls
—SO$_2$—N$<$	ca 1322 (m) 1163–1138 (s) 524–510 (s)	Asymmetric SO$_2$ stretching Symmetric SO$_2$ stretching Scissoring mode
—SO$_2$—O	1363–1338 (w-m) 1192–1165 (vs) 589–517 (w-m)	SO$_2$ stretching. Aryl substituents occur at higher range. Scissoring (two bands). Aryl substituents occur at higher range of frequencies.
—SO$_2$—S—	1334–1305 (m-s) 1128–1126 (s) 559–553 (m-s)	
X—SO$_2$—X	1412–1361 (w-m) (F) (Cl) 1263–1168 (s) (F) (Cl) 596–531 (s)	
—O—SO$_2$—O—	1388–1372 (s) 1196–1188 (vs)	
—O—C—S— $\parallel$ S	670–620 (vs) 480–450 (vs)	C$=$S stretching CS stretching
$\equiv$C—SH	920 (m) 850–820 (m)	C—SH deformation of aryls
$\equiv$C—S—	752 (vs), 731 (vs) 742–722 (m-s) 698 (w), 678 (s) 693–639 (s) 651–610 (s-vs) 589–585 (vs)	With vinyl group attached With CH$_3$ attached With allyl group attached Ethyl or longer alkyl chain Isopropyl group attached *tert*-Butyl group attached

TABLE 6.27 Raman Frequencies of Sulfur Compounds (*continued*)

Group	Band, cm^{-1}	Remarks
$\gtreqqless$C—S—(*continued*)		
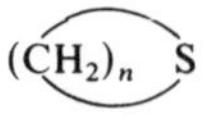 (CH$_2$)$_n$ S		
$n = 2$	1112	
$n = 4$	688	
$n = 5$	659	
$\gtreqqless$C—(S—S)$_n$—C$\lesseqqgtr$	715–620 (vs)	Two bands; CS stretching
	525–510 (vs)	Two bands; SS stretching
Didi-*n*-alkyl disulfides	576 (s)	CS stretching
Di-*tert*-butyl disulfide	543 (m)	SS stretching
Trisulfides	510–480 (s)	SS stretching

TABLE 6.28 Raman Frequencies of Ethers

Abbreviations Used in the Table

m, moderately strong	var, of variable strength
s, strong	vs, very strong

Group	Band, cm^{-1}	Remarks
$\gtreqqless$C—O—C$\lesseqqgtr$		
Aliphatic	1200–1070 (m)	Asymmetrical COC stretching. Symmetrical substitution gives higher frequencies
	930–830 (s)	Symmetrical COC stretching
	800–700 (s)	Branching at α carbon gives higher frequencies.
	550–400	
Aromatic	1310–1210 (m)	
	1050–1010 (m)	
$\gtreqqless$C—O—C—O—C$\lesseqqgtr$	1145–1129 (m)	
	900–800 (vs)	
	537–370 (s)	
	396–295	
$\gtrless$C——C$\lessgtr$ O	1280–1240 (s)	Ring breathing
—O—O—	800–770 (var)	
(CH$_2$)$_n$ O $n = 3$	1040–1010 (s)	
$n = 4$	920–900 (s)	
$n = 5$	820–800 (s)	

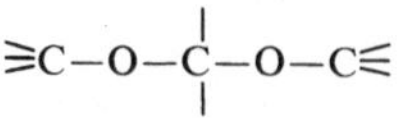

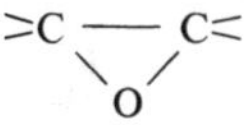

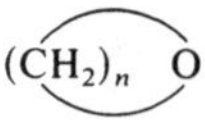

TABLE 6.29 Raman Frequencies of Halogen Compounds

Abbreviations Used in the Table

m-s, moderate to strong   var, of variable strength
s, strong   vs, very strong

Group	Band, cm^{-1}	Remarks
C—F	1400–870	Correlations of limited applicability because of vibrational coupling with stretching
C—Cl	350–290 (s)	CCCl bending; general
Primary	660–650 (vs)	
Secondary	760–605 (s)	May be one to four bands
Tertiary	620–540 (var)	May be one to three bands
=C—Cl	844–564	
	438–396	
	381–170	
=CCl$_2$	601–441	
	300–235	
C—Br	690–490 (s)	Often several bands; primary at higher range of frequencies. Tertiary has very strong band at ca 520 cm^{-1}.
	305–258 (m-s)	
=C—Br	745–565	
	356–318	
	240–115	
=CBr$_2$	467–265	
	185–145	
C—I	663–595	
	309	
	154–85	
=C—I	ca 180	Solid state
=CI$_2$	ca 265	Solid state
	ca 105	Solid state

TABLE 6.30   Raman Frequencies of Miscellaneous Compounds

Abbreviations Used in the Table

m, moderately strong	vs, very strong
s, strong	vvs, very very strong

Group	Band, cm^{-1}	Remarks
C—As	570–550 (vs)	CAs stretching
	240–220 (vs)	CAsC deformation
C—Pb	480–420 (s)	CPb stretching
C—Hg	570–510 (vvs)	CHg stretching
C—Si	1300–1200 (s)	CSi stretching
C—Sn	600–450 (s)	CSn stretching
P—H	2350–2240 (m)	PH stretching

Heterocyclic rings		
Trimethylene oxide	1029	
Trimethylene imine	1026	
Tetrahydrofuran	914	
Pyrrolidine	899	
1,3-Dioxolane	939	
1,4-Dioxane	834	
Piperidine	815	
Tetrahydropyran	818	
Morpholine	832	
Piperazine	836	
Furan	1515–1460	2-Substituted
	1140	
Pyrazole	1040–990	
Pyrrole	1420–1360 (vs)	
	1144	
Thiophene	1410 (s)	
	1365 (s)	
	1085 (vs)	
	1035 (s)	
	832 (vs)	
	610 (s)	
Pyridine	1030 (vs)	
	990 (vs)	

NUCLEAR MAGNETIC RESONANCE SPECTROSCOPY

Nuclear Magnetic Resonance (NMR) Spectroscopy is by far the most widely used analytical technique in the modern organic chemistry lab. Numerous monographs have been written on this subject. It would be impossible to cover all of the significant points here. The reader who is interested in knowing what the proton (^{1}H) or carbon (^{13}C) spectrum of a particular compound is directed to the Aldrich Library of NMR Spectra or the Sadtler Library.

A number of resources are also available online. These include software for the prediction and analysis of spectral data and databases. Resources include:

Proton NMR basics:
http://jchemed.chem.wisc.edu/JCESoft/Programs/PNMRB/
NMR database:
www.acornnmr.com/database.htm
NMR prediction:
www.acdlabs.com/products/spec_lab/predict_nmr/
NMR library:
www.acdlabs.com/products/spec_lab/exp_spectra/spec_libraries/aldrich.html
NMR Periodic Table for half-integer quadrupole spins:
www.pascal-man.com/periodic-table/periodictable.html

Nuclear Magnetic Resonance

Table 6.31 presents the nuclear properties of the elements. Hydrogen (^{1}H) is an almost ideal nucleus for NMR spectroscopy. First, its natural abundance is high so most of the nuclei present in the sample will be detected in the NMR experiment. Second, its sensitivity is high meaning that its signal is readily detected. The nuclei that have been most generally used in organic chemistry are ^{1}H and ^{13}C although the natural abundance of the latter is low. Advances in instruments have made the acquisition of ^{13}C-NMR spectra routine. Several other nuclei have high natural abundance and occur frequently in organic compounds. These include ^{7}Li, ^{11}B, ^{14}N, ^{19}F, ^{23}Na, and ^{35}Cl, which are shown in bold type in Table 6.31. Modern NMR spectrometers permit the acquisition of NMR spectra from many nuclei depending on the probe. The most favorable nuclei are those that have spin 1/2, high natural abundance, high sensitivity, and no quadrupole moment. Of course, the importance of the problem under study will ultimately dictate whether the investigator will invest the time and effort to obtain the spectrum when the experiment is difficult.

Table 6.31 Nuclear properties of the elements
In the following table the magnetic moment μ is in multiples of the nuclear magneton μ_N ($eh/4\pi Mc$) with diamagnetic correction, the spin I is in multiples of $h/2\pi$, and the electric quadrupole moment Q is in multiples of 10^{-28} square meters. Nuclei with spin $\frac{1}{2}$ have no quadrupole moment. The sign of μ and Q is uncertain for those nuclides for which no sign is given. Sensitivity is for equal number of nuclei at constant field. NMR frequency at any magnetic field is the entry for column 5 multiplied by the value of the magnetic field in kilogauss. For example, in a magnetic field of 14.0924 kG, protons (^{1}H) will precess at a frequency of $4.25760 \times 14.0924\,\text{kG} = 60.000\,\text{MHz}$. In a magnetic field of 23.4924 kG, protons will precess at $4.25760 \times 23.4924\,\text{kG} = 100.00\,\text{MHz}$.

TABLE 6.31 Nuclear Properties of the Elements

Nuclide	Natural abundance, %	Spin I	Sensitivity at constant field relative to ^{1}H	NMR frequency for a 1000 G field, MHz	Magnetic moment μ/μ_N, J·T^{-1}	Electric quadrupole moment Q, 10^{-28} m^2
^{1}n	—	−1/2	0.322	2.91670	−1.91312	—
^{1}**H**	**99.985**	**1/2**	**1.000**	**4.25760**	**+2.79278**	**—**
^{2}H	0.015	1	0.00964	0.65357	+0.85742	+0.00228
^{3}H	—	1/2	1.21	4.54131	+2.9789	—
^{3}He	0.00013	−1/2	0.443	3.24338	−2.1276	—
^{6}Li	7.42	1	0.00851	0.62655	+0.82203	−0.0008
^{7}**Li**	**92.58**	**3/2**	**0.294**	**1.65465**	**+3.25636**	**−0.04**
^{9}Be	100	−3/2	0.0139	0.59827	−1.17745	0.05
^{10}B	19.7	3	0.0199	0.4574	+1.8006	+0.111
^{11}**B**	**80.3**	**3/2**	**0.165**	**1.36595**	**+2.6885**	**+0.041**
^{13}**C**	**1.108**	**1/2**	**0.0159**	**1.07054**	**+0.7024**	**—**
^{14}N	99.635	1	0.00101	0.3076	+0.40375	+0.01
^{15}N	0.365	−1/2	0.00104	0.4315	−0.2831	—
^{17}O	0.037	−5/2	0.0291	0.57739	−1.8937	−0.004
^{19}**F**	**100**	**1/2**	**0.834**	**4.00543**	**+2.6288**	**—**
^{21}Ne	0.257	−3/2	0.0272	0.33611	−0.66176	+0.09
^{22}Na	—	3	0.0181	0.4434	1.746	—
^{23}**Na**	**100**	**3/2**	**100**	**1.12621**	**+2.21740**	**+0.10**
^{24}Na	—	4	0.00115	0.322	1.690	—
^{25}Mg	10.11	−5/2	0.0268	0.2606	−0.8554	+0.22
^{27}**Al**	**100**	**5/2**	**0.207**	**1.10940**	**+3.6413**	**+0.15**
^{29}Si	4.71	−1/2	0.0785	0.8458	−0.55526	—
^{31}**P**	**100**	**1/2**	**0.0664**	**1.7238**	**+1.1317**	**—**
^{33}S	0.76	3/2	0.00226	0.3266	+0.6435	−0.055
^{35}S	—	3/2	0.00850	0.508	—	+0.038
^{35}**Cl**	**75.53**	**3/2**	**0.00471**	**0.4171**	**+0.82181**	**−0.080**
^{36}Cl	—	2	0.0121	0.4893	+1.2853	−0.10
^{37}Cl	24.47	3/2	0.00272	0.3472	+0.68407	−0.0062
^{39}K	93.22	3/2	0.000508	0.19864	+0.39143	+0.049
^{40}K	0.0118	4	0.00521	0.2470	−1.2981	−0.061
^{41}K	6.77	3/2	0.0000839	0.10903	+0.2149	+0.060
^{43}Ca	0.145	7/2	0.0639	0.28654	−1.3172	—
^{45}Sc	100	7/2	0.301	1.03434	+4.7559	−0.22
^{47}Ti	7.32	−5/2	0.00210	0.23997	−0.78846	+0.29
^{49}Ti	5.46	−7/2	0.00376	0.24004	−1.10414	+0.24
^{50}V	0.25	6	0.0553	0.4243	+3.3470	0.06
^{51}V	99.75	7/2	0.383	1.11922	+5.1485	−0.05
^{53}Cr	9.55	3/2	0.00010	0.24063	−0.4735	+0.03
^{55}Mn	100	5/2	0.178	1.05542	+3.449	+0.4
^{57}Fe	2.17	1/2	0.0000333	0.138	+0.09042	—
^{59}Co	100	7/2	0.281	1.0072	+4.616	+0.38
^{61}Ni	1.25	3/2	0.00350	0.38048	−0.7498	+0.16
^{63}Cu	69.1	3/2	0.0938	1.1285	+2.2228	−0.211
^{65}Cu	30.9	3/2	0.116	1.2090	+2.3812	−0.195
^{67}Zn	4.11	5/2	0.00286	0.2663	+0.87524	+0.16
^{69}Ga	60.2	3/2	0.0693	1.02188	+2.0145	+0.19

TABLE 6.31 Nuclear Properties of the Elements 9 (*continued*)

Nuclide	Natural abundance, %	Spin I	Sensitivity at constant field relative to ^1H	NMR frequency for a 1000 G field, MHz	Magnetic moment μ/μ_N, J·T^{-1}	Electric quadrupole moment Q, 10^{-28}m^2
^{71}Ga	39.8	3/2	0.142	1.29840	+2.5597	+0.12
^{75}As	100	3/2	0.0251	0.7292	+1.439	+0.29
^{77}Se	7.58	1/2	0.00697	0.8118	+0.534	—
^{79}Br	50.52	3/2	0.0786	1.0669	+2.1055	+0.37
^{81}Br	49.48	3/2	0.0984	1.1498	+2.2696	+0.31
^{87}Rb	27.85	3/2	0.177	1.2923	+2.7500	+0.13
^{93}Nb	100	9/2	0.482	1.04048	+6.167	−0.22
^{113}In	4.23	−1/2	0.345	0.9312	−0.6225	—
^{119}Sn	8.58	−1/2	0.0518	1.5868	−1.0461	—
^{121}Sb	57.25	5/2	0.160	1.0192	+3.3592	−0.28
^{123}Sb	42.75	7/2	0.0457	0.5519	+2.5466	−0.36
^{125}Te	6.99	−1/2	0.0316	1.3453	−0.8872	—
^{127}I	100	5/2	0.0935	0.8517	+2.8091	−0.79
^{129}Xe	26.44	−1/2	0.0212	1.17779	−0.7768	—
^{195}Pt	33.8	1/2	0.00994	0.91523	+0.6022	—
^{199}Hg	16.84	1/2	0.00572	0.7612	+0.50415	—
^{203}Tl	29.50	1/2	0.187	2.4332	+1.6115	—
207Pb	21.7	1/2	0.00913	0.8898	10.5783	—

Chemical Shifts

In essence, the chemical shift of a nucleus such as proton (^1H) is its resonance frequency. It is usually expressed in parts per million (ppm) relative to a standard. The most common standard is tetramethylsilane [$(CH_3)_4Si$, TMS] which defines 0 on the delta (δ) scale and 10 on the older, less used τ scale. A small amount of TMS is typically added to the NMR solution to be examined. The presence of an internal standard minimizes experimental variations. This is particularly important because the chemical shift is typically a change of only a few hertz per megahertz, hence the part per million (ppm) scale. The separation of peaks will be greater in hertz at higher field but spectra obtained at different field strengths are comparable on the ppm scale. Common reference standards are listed in Table 6.32.

TABLE 6.32 Proton Chemical Shifts of Reference Compounds Relative to Tetramethylsilane

Compound	Chemical shift δ, ppm*	Solvent(s)
Tetramethylsilane, $(CH_3)_4Si$	0.0	CDCl$_3$, CCl$_4$
3-(Trimethylsilyl)-1-propanesulfonic acid, sodium salt (DSS), $(CH_3)_3SiCH_2CH_2COONa$	0.0	D$_2$O
Sodium acetate	1.90	D$_2$O
1,2-Dibromoethane	3.63	CDCl$_3$

TABLE 6.32 Proton Chemical Shifts of Reference Compounds Relative to Tetramethylsilane (*continued*)

Compound	Chemical shift δ, ppm*	Solvent(s)
1,1,2,2-Tetrachloroethane, $Cl_2HCCHCl_2$	5.95	$CDCl_3$, CCl_4
1,4-Benzoquinone	6.78	$CDCl_3$, CCl_4
1,4-Dichlorobenzene	7.23	CCl_4
Chloroform, $CHCl_3$	7.27	$CDCl_3$, CCl_4
Benzene	7.37	$CDCl_3$, CCl_4
1,3,5-Trinitrobenzene	9.21	DMSO-d_6† $CDCl_3$

*Shift relative to TMS, first entry in table; † Dimethylsulfoxide-d_6

A typical solution prepared for NMR analysis rarely contains more than a few percent of solute. Thus, protons on the solvent could significantly distort the spectrum. When ¹H-NMR are desired, solvents having no protons (CS_2 or CCl_4) or deuterated solvents are used. Table 6.33 gives the common NMR solvents used. Although modern manufacturing methods typically produce NMR solvents having high isotopic purity, incomplete deuteration of a protonic solvent will result in a residual signal.

TABLE 6.33 Common NMR Solvents

Solvent	Detail	Group*	δ (ppm)
Acetic acid-d_4	$D_3C-COOD$	CD_2H	2.05
		OH	11.5†
Acetone-d_6	CD_3COCD_3	CD_2H	2.057
Acetonitrile-d_3	$CD_3C\equiv N$	CD_2H	1.95
Benzene-d_6	C_6D_6	$C-H$	6.78
tert-Butanol-d_1	$(CH_3)_3COD$	CD_2H	1.28
Chloroform-d_1	Cl_3CD	Cl_3CH	7.25
Cyclohexane-d_{12}	C_6D_{12}	CHD	1.40
Deuterium oxide	D_2O	HOD	4.7†
Dimethylformamide-d_7	$(CD_3)_2N-CD=O$	CD_2H	2.75; 2.95
		$-CH=O$	8.05
Dimethylsulfoxide-d_6 (DMSO-d_6)	CD_3SOCD_3	CD_2H	2.51
		Absorbed H_2O	3.3†
1,4-Dioxane-d_8	$\begin{matrix} D_2C \diagdown^O\diagup CD_2 \\ D_2C \diagup_O \diagdown CD_2 \end{matrix}$	Methylene	3.55
Hexamethylphosphoramide-d_{18}, HMPA-d_{18}, HMPT-d_{18}	$(D_3C)_2N-\overset{\overset{O}{\|}}{\underset{\underset{N(CD_3)_2}{\|}}{P}}-N(CD_3)_2$	Methyl	2.60
Methanol-d_4	CD_3OD	CD_2H	3.35
		OH	4.8†

*Impurity peak resulting from incomplete deuteration or exchange.
†These values may vary greatly depending on the solute and its concentration.

TABLE 6.33 Common NMR Solvents (*continued*)

Solvent	Detail	Group*	δ(ppm)
Dichloromethane-d_2	Cl_2CD_2	Cl_2CDH	5.35
Pyridine-d_5	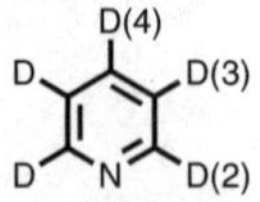 D(4) ... D(3), D(2), N	C-2—H	8.5
		C-3—H	7.0
		C-4—H	7.35
Tetrahydrofuran-d_6	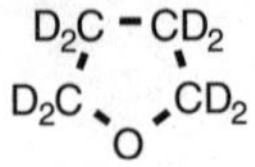 D_2C—CD_2 / D_2C CD_2 / O	CD-2—H	3.58
		CD-3—H	1.73
Toluene-d_8	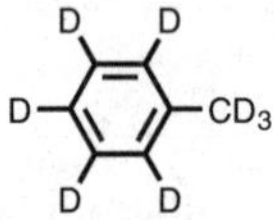 D ring with CD_3	CD_2H	2.3
		Ring CD	7.1
Trifluoroacetic acid-d_1	F_3C—COOD	Hydroxyl	11.3†

*Impurity peak resulting from incomplete deuteration or exchange. †These values may vary greatly depending on the solute and its concentration.

TABLE 6.34 Proton Chemical Shifts

Values are given on the δ scale; $\tau = 10.00 - \delta$.

Abbreviations Used in the Table
R, alkyl group   Ar, aryl group

Substituent group	Methyl protons	Methylene protons	Methine proton
HC—C—CH_2	0.95	1.20	1.55
HC—C—NR_2	1.05	1.45	1.70
HC—C—C=C	1.00	1.35	1.70
HC—C—C=O	1.05	1.55	1.95
HC—C—NRAr	1.10	1.50	1.80
HC—C—H(C=O)R	1.10	1.50	1.90
HC—C—(C=O)NR_2	1.10	1.50	1.80
HC—C—(C=O)Ar	1.15	1.55	1.90
HC—C—(C=O)OR	1.15	1.70	1.90
HC—C—Ar	1.15	1.55	1.80
HC—C—OH	1.20	1.50	1.75
HC—C—OR	1.20	1.50	1.75
HC—C—C≡CR	1.20	1.50	1.80
HC—C—C≡N	1.25	1.65	2.00
HC—C—SR	1.25	1.60	1.90
HC—C—OAr	1.30	1.55	2.00
HC—C—O(C=O)R	1.30	1.60	1.80
HC—C—SH	1.30	1.60	1.65
HC—C—(S=O)R and HC—C—SO_2R	1.35	1.70	
HC—C—NR_3^+	1.40	1.75	2.05

TABLE 6.34 Proton Chemical Shifts (*continued*)

Substituent group	Methyl protons	Methylene protons	Methine proton
HC—C—O—N=O	1.40		
HC—C—O(C=O)CF$_3$	1.40	1.65	
HC—C—Cl	1.55	1.80	1.95
HC—C—F	1.55	1.85	2.15
HC—C—NO$_2$	1.60	2.05	2.50
HC—C—O(C=O)Ar	1.65	1.75	1.85
HC—C—I	1.75	1.80	2.10
HC—C—Br	1.80	1.85	1.90
HC—CH$_2$	0.90	1.30	1.50
HC—C=C	1.60	2.05	
HC—C≡C	1.70	2.20	2.80
HC—(C=O)OR	2.00	2.25	2.50
HC—(C=O)NR$_2$	2.00	2.25	2.40
HC—SR	2.05	2.55	3.00
HC—O—O	2.10	2.30	2.55
HC—(C=O)R	2.10	2.35	2.65
HC—C≡N	2.15	2.45	2.90
HC—I	2.15	3.15	4.25
HC—CHO	2.20	2.40	
HC—Ar	2.25	2.45	2.85
HC—NR$_2$	2.25	2.40	2.80
HC—SSR	2.35	2.70	
HC—(C=O)Ar	2.40	2.70	3.40
HC—SAr	2.40		
HC—NRAr	2.60	3.10	3.60
HC—SO$_2$R and HC—(SO)R	2.60	3.05	
HC—Br	2.70	3.40	4.10
HC—NR$_3^+$	2.95	3.10	3.60
HC—NH(C=O)R	2.95	3.35	3.85
HC—SO$_3$R	2.95		
HC—Cl	3.05	3.45	4.05
HC—OH and HC—OR	3.20	3.40	3.60
HC—PAr$_3$	3.20	3.40	
HC—NH$_2$	3.50	3.75	4.05
HC—O(C=O)R	3.65	4.10	4.95
HC—OAr	3.80	4.00	4.60
HC—O(C=O)Ar	3.80	4.20	5.05
HC—O(C=O)CF	3.95	4.30	
HC—F	4.25	4.50	4.80
HC—NO$_2$	4.30	4.35	4.60
Cyclopropane		0.20	0.40
Cyclobutane		2.45	
Cyclopentane		1.65	
Cyclohexane		1.50	1.80
Cycloheptane		1.25	

TABLE 6.34 Proton Chemical Shifts (*continued*)

Substituent group	Proton shift	Substituent group	Proton shift
HC≡CH	2.35	HO—C=O	10–12
HC≡CAr	2.90	HO—SO$_2$	11–12
HC≡C—C=C	2.75	HO—Ar	4.5–6.5
HAr	7.20	HO—R	0.5–4.5
HCO—O	8.1	HS—Ar	2.8–3.6
HCO—R	9.4–10.0	HS—R	1–2
HCO—Ar	9.7–10.5	HN—Ar	3–6
HO—N=C (oxime)	9–12	HN—R	0.5–5

Saturated heterocyclic ring systems

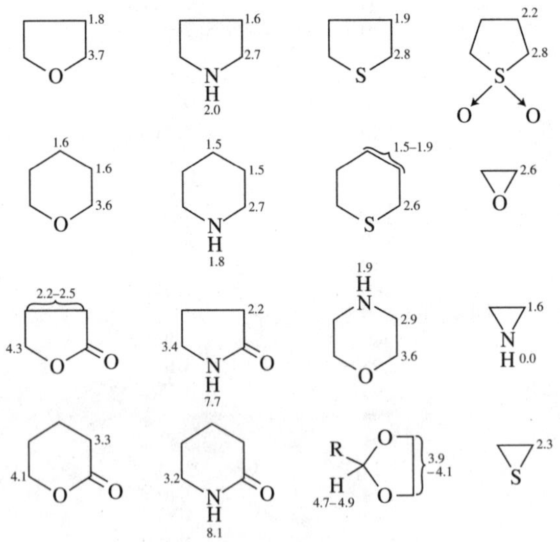

TABLE 6.34 Proton Chemical Shifts (*continued*)

Substituent group	Methyl protons	Methylene protons	Methine proton
		Unsaturated cyclic systems	

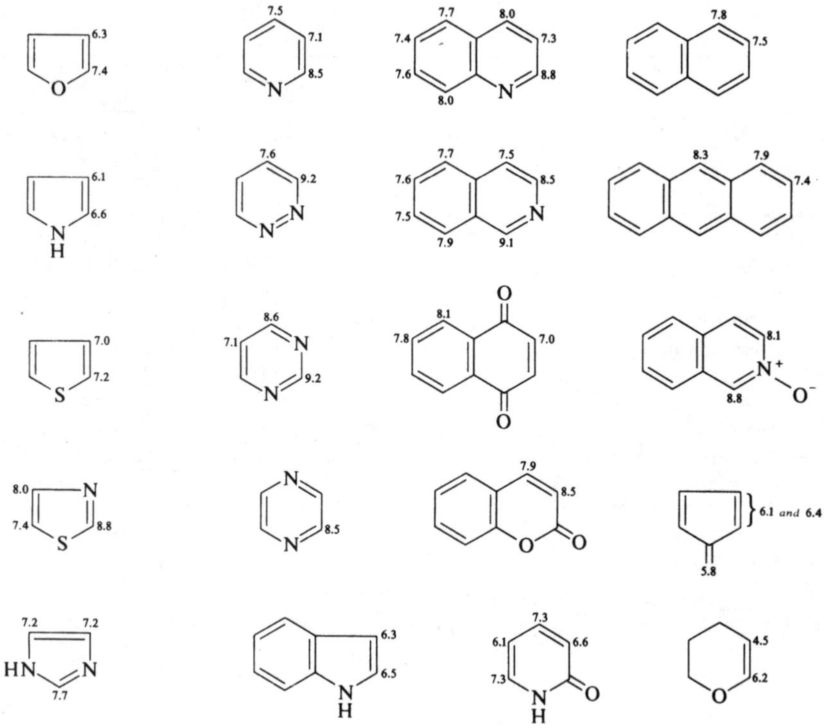

TABLE 6.35 Estimation of Chemical Shift for Protons of —CH_2— and >CH— Groups

$\delta_{CH_2} = 0.23 + C_1 + C_2$ $\delta_{CH} = 0.23 + C_1 + C_2 + C_3$

X*	C	X*	C	X*	C
—CH_3	0.5	—SR	1.6	—OR	2.4
—CF_3	1.1	—C≡C—Ar	1.7	—Cl	2.5
>C—C<	1.3	—CN	1.7	—OH	2.6
—C≡C—R	1.4	—CO—R	1.7	—N=C=S	2.9
—COOR	1.5	—I	1.8	—OCOR	3.1
—NR_2	1.6	—Ph	1.8	—OPh	3.2
—$CONR_2$	1.6	—Br	2.3		

*R, alkyl group; Ar, aryl group; Ph, phenyl group.

TABLE 6.36 Estimation of Chemical Shift of Proton Attached to a Double Bond

Positive Z values indicate a downfield shift, and an arrow indicates the point of attachment of the substituent group to the double bond.

$$\delta_{C=C_{\substack{\\ H}}} = 5.25 + Z_{gem} + Z_{cis} + Z_{trans}$$

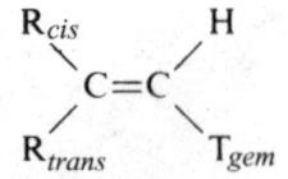

$$\begin{array}{c} R_{cis} \qquad H \\ \diagdown \quad \diagup \\ C=C \\ \diagup \quad \diagdown \\ R_{trans} \qquad T_{gem} \end{array}$$

R	Z_{gem}, ppm	Z_{cis}, ppm	Z_{trans}, ppm
→H	0	0	0
→alkyl	0.45	−0.22	−0.28
→alkyl—ring (5- or 6-member)	0.69	−0.25	−0.28
→CH$_2$O—	0.64	−0.01	−0.02
→CH$_2$S—	0.71	−0.13	−0.22
→CH$_2$X (X: F, Cl, Br)	0.70	0.11	−0.04
→CH$_2$N<	0.58	−0.10	−0.08
$\diagup$C=C (isolated)	1.00	−0.09	−0.23
$\diagup$C=C (conjugated)	1.24	0.02	−0.05
→C≡N	0.27	0.75	0.55
→C≡C—	0.47	0.38	0.12
$\diagup$C=O (isolated)	1.10	1.12	0.87
$\diagup$C=O (conjugated)	1.06	0.91	0.74
→COOH (isolated)	0.97	1.41	0.71
→COOH (conjugated)	0.80	0.98	0.32
→COOR (isolated)	0.80	1.18	0.55
→COOR (conjugated)	0.78	1.01	0.46
$\begin{array}{c} H \\ \mid \\ →C=O \end{array}$	1.02	0.95	1.17
$\begin{array}{c} \diagdown N \diagup \\ \mid \\ →C=O \end{array}$	1.37	0.98	0.46
$\begin{array}{c} Cl \\ \mid \\ →C=O \end{array}$	1.11	1.46	1.01
→OR (R: aliphatic)	1.22	−1.07	−1.21
→OR (R: conjugated)	1.21	−0.60	−1.00
→OCOR	2.11	−0.35	−0.64
→CH$_2$—C̷=O; →CH$_2$—C≡N	0.69	−0.08	−0.06
→CH$_2$—aromatic ring	1.05	−0.29	−0.32
→F	1.54	−0.40	−1.02
→Cl	1.08	0.18	0.13
→Br	1.07	0.45	0.55
→I	1.14	0.81	0.88
→N—R (R: aliphatic)	0.80	−1.26	−1.21

TABLE 6.36 Estimation of Chemical Shift of Proton Attached to a Double Bond (*continued*)

R	Z_{gem}, ppm	Z_{cis}, ppm	Z_{trans}, ppm
$\rightarrow$N—R (R: conjugated)	1.17	−0.53	−0.99
$\rightarrow$N—C=O	2.08	−0.57	−0.72
$\rightarrow$aromatic	1.38	0.36	−0.07
$\rightarrow$CF$_3$	0.66	0.61	0.32
$\rightarrow$aromatic (*o*-substituted)	1.65	0.19	0.09
$\rightarrow$SR	1.11	−0.29	−0.13
$\rightarrow$SO$_2$	1.55	1.16	0.93

TABLE 6.37 Chemical Shifts in Monosubstituted Benzene

$\delta = 7.27 + \Delta_i$

Substituent	Δ_{ortho}	Δ_{meta}	Δ_{para}
NO$_2$	0.94	0.18	0.39
CHO	0.58	0.20	0.26
COOH	0.80	0.16	0.25
COOCH$_3$	0.71	0.08	0.20
COCl	0.82	0.21	0.35
CCl$_3$	0.8	0.2	0.2
COCH$_3$	0.62	0.10	0.25
CN	0.26	0.18	0.30
CONH$_2$	0.65	0.20	0.22
$\overset{+}{N}H_3$	0.4	0.2	0.2
CH$_2$X*	0.0–0.1	0.0–0.1	0.0–0.1
CH$_3$	−0.16	−0.09	−0.17
CH$_2$CH$_3$	−0.15	−0.06	−0.18
CH(CH$_3$)$_2$	−0.14	−0.09	−0.18
C(CH$_3$)$_2$	−0.09	0.05	−0.23
F	−0.30	−0.02	−0.23
Cl	0.01	−0.06	−0.08
Br	0.19	−0.12	−0.05
I	0.39	−0.25	−0.02
NH$_2$	−0.76	−0.25	−0.63
OCH$_3$	−0.46	−0.10	−0.41
OH	−0.49	−0.13	−0.2
OCOR	−0.2	0.1	−0.2
NHCH$_3$	−0.8	−0.3	−0.6
N(CH$_3$)$_2$	−0.60	−0.10	−0.62

*X = Cl, alkyl, OH, or NH$_2$.

TABLE 6.38 Proton Spin Coupling Constants

Structure	*J*, Hz	Structure	*J*, Hz
H—C—H	12–15	aziridine (H—N) cis (Z)	2
		trans (E)	6
		gem	4
CH—CH (free rotation)	6–8	furan 2–3	1.8
>CH—OH (no exchange)	5	3–4	3.5
>CH—NH	4–8	2–4	0–1
CH—SH	6–8	2–5	1–2
CH—C(H)=O	1–3	thiophene 2–3	5–6
		3–4	3.5–5.0
—N=C(H)(H)	8–16	2–4	1.5
	0–3	2–5	3.4
H$_t$, H$_g$ gem	0–3	benzene—F *o*	6–12
C=C cis (Z)	6–14	*m*	4–8
H$_c$, H trans (E)	11–18	*p*	1.5–2.5
H$_c$, CH cis (Z)	0.5–3	benzene—CH$_3$ *o*	2.5
C=C trans (E)	0.5–3	(F) *m*	1.5
H$_t$, H$_g$ gem	4–10	*p*	0
>C=CH—CH=C<	10–13	cyclohexane *a–a*	8–10
=CH—C(H)=O	6	*a–e*	2–3
—CH$_2$—C≡C—CH	0–3	*e–e*	2–3
>CH—C≡CH	0–3	Cyclopentane cis (Z)	4–6
H, H C=C (ring) 3-member	0–2	trans (E)	4–6
4-member	2–4	Cyclobutane cis (Z)	8
5-member	5–7	trans (E)	8
6-member	6–9	Cyclopropane cis (Z)	9–11
7-member	10–13	trans (E)	6–8
O (epoxide) cis (Z)	4–5	gem	4–6
trans (E)	3	benzene *o*	6–10
gem	5–6	*m*	1–3
S (thiirane) cis (Z)	0	*p*	0–1
trans (E)	7	naphthalene 1–2	8–9
gem	6	2–3	6
		pyridine 2–3	5–6
		3–4	7–9
		2–4	1–2
		3–5	1–2
		2–5	0–1
		2–6	0–1

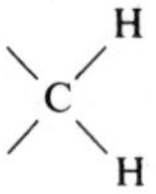

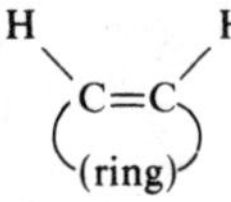

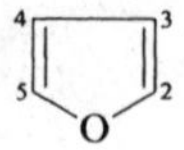

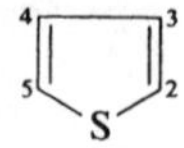

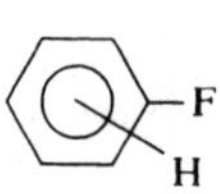

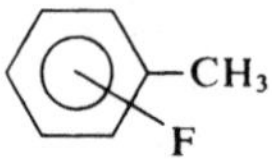

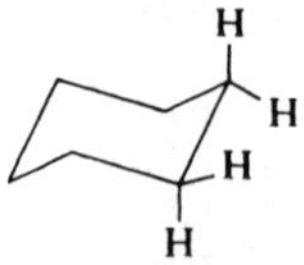

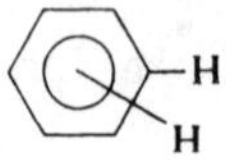

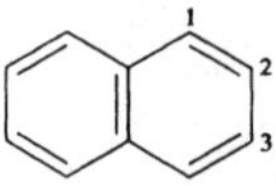

TABLE 6.38 Proton Spin Coupling Constants (*continued*)

Structure	J, Hz	Structure	J, Hz
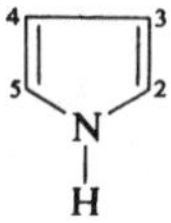 1–2 / 1–3 / 2–3 / 3–4 / 2–4 / 2–5	2–3 / 2–3 / 2–3 / 3–4 / 1–2 / 1–3	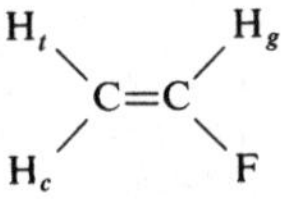 H_g gem / cis (Z) / trans (E)	72–90 / −3 to 20 / 12–40
			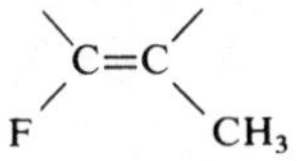 2–4
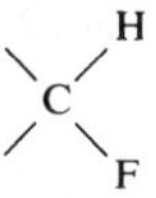	45–52	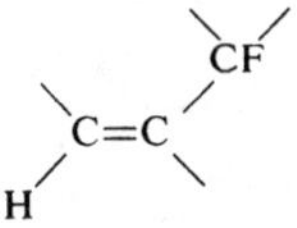	0–6
		HC≡CF	21
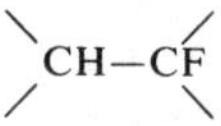	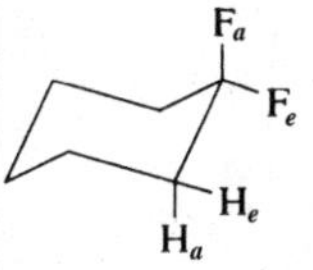	F_a F_e a–a / a–e / e–e} e–e} H_e H_a	34 / 12 / <5–8
CH—CF gauche / trans (Z)	0–12 / 10–45		

TABLE 6.39 Carbon-13 Chemical Shifts

Values given in ppm on the δ scale, relative to tetramethylsilane

Substituent group	Primary carbon	Secondary carbon	Tertiary carbon	Quaternary carbon
Alkanes				
C—C	5–30	25–45	23–58	28–50
C—O	45–60	42–71	62–78	73–86
C—N	13–45	44–58	50–70	60–75
C—S	10–30	22–42	55–67	53–62
C—halide (I to Cl)	3–25	3–40	34–58	35–75

Substituent group	δ, ppm	Substituent group	δ, ppm
Cyclopropane	−5–5	Alcohols R—OH	45–87
Cycloalkane C_4–C_{10}	5–25	Ethers R—O—R	57–87
Mercaptanes	5–70	Nitro R—NO_2	60–78
Amines		Alkynes	
R$_2$N—C	20–70	HC≡CR	63–73
Aryl—N	128–138	RC≡CR	72–95
Sulfoxides, sulfones	35–55	Acetals, ketals	88–112
Thiocyanates R—SCN	96–118	Esters	
Alkenes		Saturated	158–165
H_2C=	100–122	α,β-Unsaturated	165–176
R_2C=	110–150	Isocyanides R—NC	162–175

TABLE 6.39 Carbon-13 Chemical Shifts (*continued*)

Substituent group	δ, ppm	Substituent group	δ, ppm
Heteroaromatics		Carboxylic acids	
C=N	100–152	Nonconjugated	162–165
C$_\alpha$	142–160	Conjugated	165–184
Cyanates R—OCN	105–120	Salts (anion)	175–195
Isocyanates R—NCO	115–135	Ketones	
Isothiocyanates R—NCS	115–142	α-Halo	160–200
Nitriles, cyanides	117–124	Nonconjugated	192–202
Aromatics		α,β-Unsaturated	202–220
Aryl-C	125–145	Imides	165–180
Aryl-P	119–128	Acyl chlorides R—CO—Cl	165–183
Aryl-N	128–138	Thioureas	165–185
Aryl-O	133–152	Aldehydes	
Azomethines	145–162	α-Halo	170–190
Carbonates	159–162	Nonconjugated	182–192
Ureas	150–170	Conjugated	192–208
Anhydrides	150–175	Thioketones R—CS—R	190–202
Amides	154–178	Carbonyl M(CO)$_n$	190–218
Oximes	155–165	Allenes =C=	197–205

Saturated heterocyclic ring systems

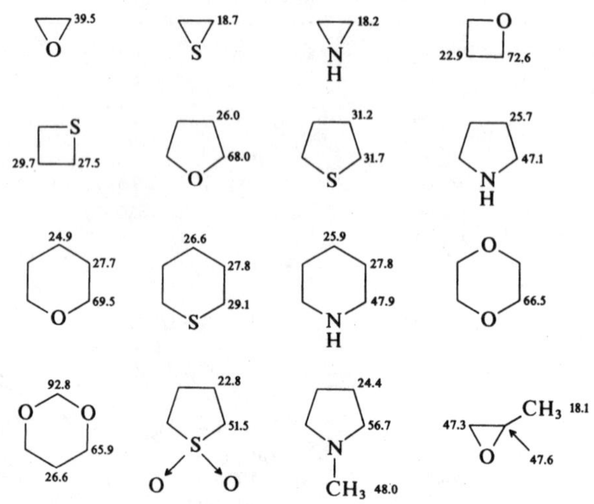

TABLE 6.39 Carbon-13 Chemical Shifts (*continued*)

Unsaturated cyclic systems

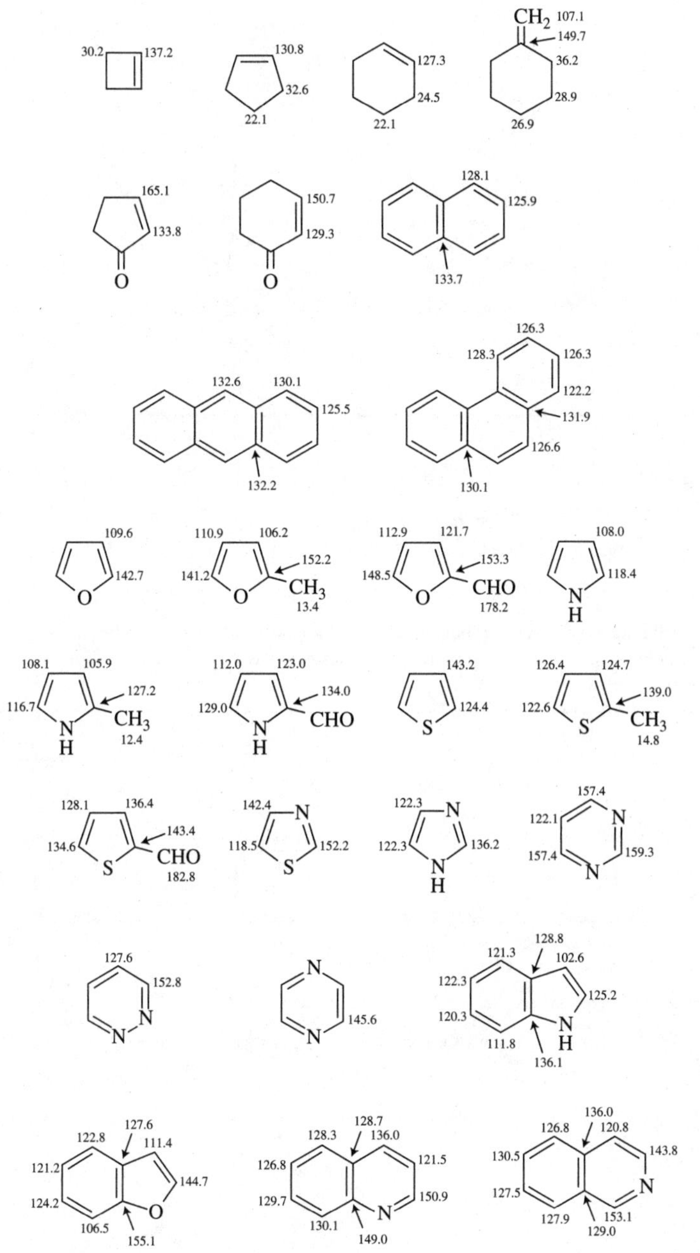

TABLE 6.39   Carbon-13 Chemical Shifts (*continued*)

Saturated alicyclic ring systems

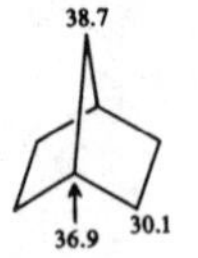

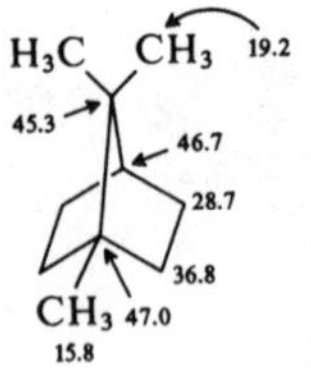

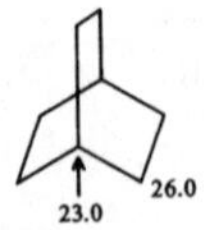

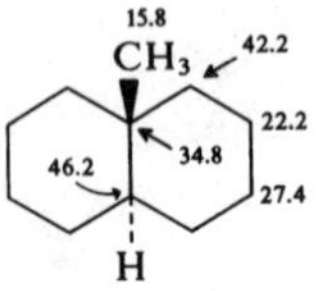

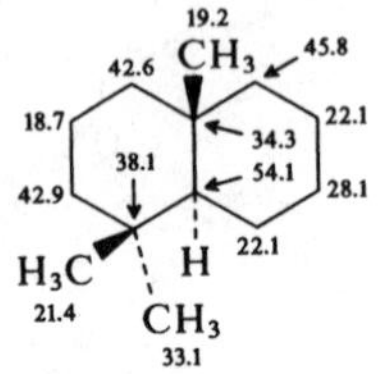

TABLE 6.40   Estimation of Chemical Shifts of Alkane Carbons

Relative to tetramethylsilane

Positive terms indicate a downfield shift.

$$\delta_C = -2.6 + 9.1n_\alpha + 9.4n_\beta - 2.5n_\gamma + 0.3n_\delta + 0.1n_\varepsilon \text{ (plus any correction factors)}$$

where n_α is the number of carbons bonded directly to the ith carbon atom and n_β, n_γ, n_δ, and n_ε are the number of carbon atoms two, three, four, and five bonds removed. The constant is the chemical shift for methane.

Chain branching*	Correction factor	Chain branching*	Correction factor
1°(3°)	−1.1	4°(1°)	−1.5
1°(4°)	−3.4	2°(4°)	−7.2
2°(3°)	−2.5	3°(3°)	−9.5
3°(2°)	−3.7	4°(2°)	−8.4

*1° signifies a CH_3— group; 2°, a —CH_2— group; 3°, a $>$CH— group; and 4°, a $>$C$<$ group. 1°(3°) signifies a methyl group bound to a $>$CH— group, and so on.

Examples: For 3-methylpentane, CH_3—CH_2—$CH(CH_3)$—CH_2—CH_3,

$$\delta_{C=2} = -2.6 + 9.1(2) + 9.4(2) - 2.5 - 1(1)[2°(3°)] = 29.4$$
$$\delta_{C=3} = -2.6 + 9.1(3) + 9.4(2) + (2)[3°(2°)] = 36.2$$

TABLE 6.41 Effect of Substituent Groups on Alkyl Chemical Shifts

These increments are added to the shift value of the appropriate carbon atom as calculated from Table 6.40.

Straight: $Y-\underset{\alpha}{CH_2}-\underset{\beta}{CH_2}-CH_3$ Branched: $-\underset{\gamma}{CH_2}-\underset{\beta}{CH_2}-\overset{\overset{\displaystyle Y}{\displaystyle |}}{\underset{\alpha}{CH}}-\underset{\beta}{CH_2}-\underset{\gamma}{CH_2}-$

Substituent group Y*	α carbon		β carbon		γ carbon
	Straight	Branched	Straight	Branched	
—CO—OH	20.9	16	2.5	2	−2.2
—COO⁻ (anion)	24.4	20	4.1	3	−1.6
—CO—OR	20.5	17	2.5	2	−2
—CO—Cl	33	28		2	
—CO—NH₂	22	2.5			−0.5
—CHO	31		0		−2
—CO—R	30	24	1	1	−2
—OH	48.3	40.8	10.2	7.7	−5.8
—OR	58	51	8	5	−4
—O—CO—NH₂	51		8		
—O—CO—R	51	45	6	5	−3
—C—CO—Ar	53				
—F	68	63	9	6	−4
—Cl	31.2	32	10.5	10	−4.6
—Br	20.0	25	10.6	10	−3.1
—I	−8	4	11.3	12	−1.0
—NH₂	29.3	24	11.3	10	−4.6
—NH₃⁺	26	24	8	6	−5
—NHR	36.9	31	8.3	6	−3.5
—NR₂	42		6		−3
—NR₃⁺	31		5		−7
—NO₂	63	57	4	4	
—CN	4	1	3	3	−3
—SH	11	11	12	11	−6
—SR	20		7		−3
—CH=CH₂	20		6		−0.5
—C₆H₅	23	17	9	7	−2
—C≡CH	4.5		5.5		−3.5

*R, alkyl group; Ar, aryl group.

TABLE 6.42   Estimation of Chemical Shift of Carbon Attached to a Double Bond

The olefinic carbon chemical shift is calculated from the equation

$$\delta_C = 123.3 + 10.6n_\alpha + 7.2n_\beta - 7.9n_{\alpha'} - 1.8n_{\beta'} \text{ (plus any steric correction terms)}$$

where n is the number of carbon atoms at the particular position, namely,

$$\begin{array}{cccc} \beta & \alpha & \alpha' & \beta' \\ C-C & = & C-C \end{array}$$

Substituents on both sides of the double bond are considered separately. Additional vinyl carbons are treated as if they were alkyl carbons. The method is applicable to alicyclic alkenes; in small rings carbons are counted twice, that is, from both sides of the double bond where applicable. The constant in the equation is the chemical shift for ethylene. The effect of other substituent groups is tabulated below.

Substituent group	β	α	α'	β'
—OR	2	29	−39	−1
—OH	6			−1
—O—CO—CH$_3$	−3	18	−27	4
—CO—CH$_3$		15	6	
—CHO		13.6	13.2	
—CO—OH		5.2	9.1	
—CO—OR		6	7	
—CN		−15.4	14.3	
—F		24.9	−34.3	
—Cl	−1	3.3	−5.4	2
—Br	0	−7.2	−0.7	2
—I		−37.4	7.7	
—C$_6$H$_5$		12	−11	

Substituent pair		Steric correction term
α,α'	trans (E)	0
α,α'	cis (Z)	−1.1
α,α	gem	−4.8
α',α'		+2.5
β,β		+2.3

TABLE 6.43 Carbon-13 Chemical Shifts in Substituted Benzenes

$\delta_C = 128.5 + \Delta$

Substituent group	Δ_{C-1}	Δ_{ortho}	Δ_{meta}	Δ_{para}
—CH$_3$	9.3	0.8	−0.1	−2.9
—CH$_2$CH$_3$	15.6	−0.4	0	−2.6
—CH(CH$_3$)$_2$	20.2	−2.5	0.1	−2.4
—C(CH$_3$)$_3$	22.4	−3.1	−0.1	−2.9
—CH$_2$O—CO—CH$_3$	7.7	0	0	0
—C$_6$H$_5$	13.1	−1.1	0.4	−1.2
—CH=CH$_2$	9.5	−2.0	0.2	−0.5
—C≡CH	−6.1	3.8	0.4	−0.2
—CH$_2$OH	12.3	−1.4	−1.4	−1.4
—CO—OH	2.1	1.5	0	5.1
—COO$^-$ (anion)	8	1	0	3
—CO—OCH$_3$	2.1	1.1	0.1	4.5
—CO—CH$_3$	9.1	0.1	0	4.2
—CHO	8.6	1.3	0.6	5.5
—CO—Cl	4.6	2.4	1	6.2
—CO—CF$_3$	−5.6	1.8	0.7	6.7
—CO—C$_6$H$_5$	9.4	1.7	−0.2	3.6
—CN	−15.4	3.6	0.6	3.9
—OH	26.9	−12.7	1.4	−7.3
—OCH$_3$	31.4	−14.0	1.0	−7.7
—OC$_6$H$_5$	29.2	−9.4	1.6	−5.1
—O—CO—CH$_3$	23.0	−6.4	1.3	−2.3
—NH$_2$	18.0	−13.3	0.9	−9.8
—N(CH$_3$)$_2$	22.4	−15.7	0.8	−11.5
—N(C$_6$H$_5$)$_2$	19	−4	1	−6
—NHC$_6$H$_5$	14.6	−10.7	0.7	−7.7
—NH—CO—CH$_3$	11.1	−9.9	0.2	−5.6
—NO$_2$	20.0	−4.8	0.9	5.8
—F	34.8	−12.9	1.4	−4.5
—Cl	6.2	0.4	1.3	−1.9
—Br	−5.5	3.4	1.7	−1.6
—I	−32.2	9.9	2.6	−1.4
—CF$_3$	−9.0	−2.2	0.3	3.2
—NCO	5.7	−3.6	1.2	−2.8
—SH	2.3	1.1	1.1	−3.1
—SCH$_3$	10.2	−1.8	0.4	−3.6
—SO$_2$—NH$_2$	15.3	−2.9	0.4	3.3
—Si(CH$_3$)$_3$	13.4	4.4	−1.1	−1.1

TABLE 6.44 Carbon-13 Chemical Shifts in Substituted Pyridines*

$\delta_c(k) = C_k + \Delta_i$

Substituent group	$C_2 = C_6 = 149.6$ Δ_{C-2} or Δ_{C-6}	Δ_{23}	Δ_{24}	Δ_{25}	Δ_{26}
—CH$_3$	9.1	−1.0	−0.1	−3.4	−0.1
—CH$_2$CH$_3$	14.0	−2.1	0.1	−3.1	0.2
—CO—CH$_3$	4.3	−2.8	0.7	3.0	−0.2
—CHO	3.5	−2.6	1.3	4.1	0.7
—OH	14.9	−17.2	0.4	−3.1	−6.8
—OCH$_3$	15.3	−13.1	2.1	−7.5	−2.2
—NH$_2$	11.3	−14.7	2.3	10.6	−0.9
—NO$_2$	8.0	−5.1	5.5	6.6	0.4
—CN	−15.8	5.0	−1.7	3.6	1.9
—F	14.4	−14.7	5.1	−2.7	−1.7
—Cl	2.3	0.7	3.3	−1.2	0.6
—Br	−6.7	4.8	3.3	−0.5	1.4

Substituent group	Δ_{32}	$C_3 = C_5 = 124.2$ Δ_{C-3} or Δ_{C-5}	Δ_{34}	Δ_{35}	
—CH$_3$	1.3	9.0	0.2	−0.8	−2.3
—CH$_2$CH$_3$	0.3	15.0	−1.5	−0.3	−1.8
—CO—CH$_3$	0.5	−0.3	−3.7	−2.7	4.2
—CHO	2.4	7.9	0	0.6	5.4
—OH	−10.7	31.4	−12.2	1.3	−8.6
—NH$_2$	−11.9	21.5	−14.2	0.9	−10.8
—CN	3.6	−13.7	4.4	0.6	4.2
—Cl	−0.3	8.2	−0.2	0.7	−1.4
—Br	2.1	−2.6	2.9	1.2	−0.9
—I	7.1	−28.4	9.1	2.4	0.3

Substituent group	$\Delta_{42} = \Delta_{46}$	$\Delta_{43} = \Delta_{45}$	$C_4 = 136.2$ Δ_{C-4}
—CH$_3$	0.5	0.8	10.8
—CH$_2$CH$_3$	0	−0.3	15.9
—CH=CH$_2$	0.3	−2.9	8.6
—CO—CH$_3$	1.6	−2.6	6.8
—CHO	1.7	−0.6	5.5
—NH$_2$	0.9	−13.8	19.6
—CN	2.1	2.2	−15.7
—Br	3.0	3.4	−3.0

*May be used for disubstituted, polyheterocyclic, and polynuclear systems if deviations due to steric and mesomeric effects are allowed for.

TABLE 6.45 Carbon-13 Chemical Shifts of Carbonyl Group

$$\underset{X-\overset{\displaystyle O}{\overset{\displaystyle \|}{C}}-Y}{}$$

X	Y	δ_C	X	Y	δ_C
H—	—CH₃	199.7	CH₃—	—CH=CH₂	196.9
H—	—CCl₃	175.3	CH₃—	—C₆H₅	197.6
H—	—NH₂	165.5	CH₃—	—CH₂—CO—CH₃	201.9 (keto)
H—	—N(CH₃)₂	162.4			191.4 (enol)
H—	2-Furyl	153.3	CH₃—	—CH₂CHO	167.7
H—	2-Pyrrolyl	134.0	CH₃—	—C₆H₅—CH₃	196 (m, p)
H—	2-Thienyl	143.3			199 (o)
(CH₃)₂CH—	—OH	184.8	CH₃—	—2,6-(CH₃)₂C₆H₅	206
C₆H₅—	—OH	172.6	CH₃—	—OH	178
CF₃—	—OH	163.0	CH₃—	—O⁻ (anion)	181.5
CCl₃—	—OH	168.0	CH₃—	—OCH₃	170.7
CH₃CH(NH₂)—	—OH	176.5	CH₃—	—O—CH=CH₂	167.7
CF₃—	—OCH₂CH₃	158.1	CH₃—	—O—CH(CH₃)₂	170.3
H₂N—	—OCH₂CH₃	157.8	CH₃—	—O—CO—CH₃	167.3
2-Furyl	—OCH₃	159.1	CH₃—	—NH₂	172.7
(CH₃)₂N—	—C₆H₅	170.8	CH₃—	—NHCH₃	172
CH₂=CHCH₂O-			CH₃—	—N(CH₃)₂	169.5
CO—	—OCH₂CH=CH₂	157.6	CH₃—	—Cl	169.6
CH₃CH₂—	—CH₂CH₃	211.4	CH₃—	—Br	165.6
CH₃—CH₂—	—O—CO—CH₂CH₃	170.3	CH₃—	—I	158.9
CH₃—	—CH₃	205.8			
CH₃—	—CH₂CH₃	207			

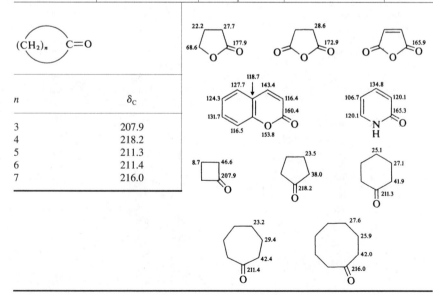

n	δ_C
3	207.9
4	218.2
5	211.3
6	211.4
7	216.0

TABLE 6.46 One-Bond Carbon–Hydrogen Spin Coupling Constants

Structure	J_{CH}, Hz	Structure	J_{CH}, Hz
H—CH$_3$	125.0	H$_t$ H$_g$ *gem*	177
H—CH$_2$CH$_3$	124.9	C=C *cis (Z)*	163
CH$_3$—$\underline{CH}_2$—CH$_3$	119.2	H$_c$ CN *trans (E)*	165
H—C(CH$_3$)$_2$	114.2		
H—CH$_2$CH$_2$OH	126.9	H OH *cis (Z)*	163
H—CH$_2$CH=CH$_2$	122.4	C=N *trans (E)*	177
H—CH$_2$C$_6$H$_5$	129.4	CH$_3$	
H—CH$_2$C≡CH	132.0		
H—CH$_2$CN	136.1	H—CH=O; CH$_3$—$\underline{CH}$=O	172
H—CH(CN)$_2$	145.2	H$_2$N—CH=O	188.3
H—CH$_2$—halogen	149–152	(CH$_3$)$_2$N—$\underline{CH}$=O	191
H—CHF$_2$	184.5	H—COOH	222
H—CHCl$_2$	178.0	H—COO$^-$ (anion)	195
H—CH$_2$NH$_2$	133.0	H—CO—OCH$_3$	226
H—CH$_2$NH$_3^+$	145.0	H—CO—F	267
H—CH$_2$OH (or H—CH$_2$OR)	140–141	CH$_3$CH$_2$—O—$\underline{CHO}$	225.6
H—CH(OR)$_2$	161–162	Cl$_3$—CHO	207
H—C(OR)$_3$	186	H—C≡CH	249
H—C(OH)R$_2$	143	H—C≡CCH$_3$	248
H—CH$_2$NO$_2$	146.0	H—C≡CC$_6$H$_5$	251
H—CH(NO$_2$)$_2$	169.4	H—C≡CCH$_2$OH	241
H—CH$_2$COOH	130.0	H—CN	269
H—CH(COOH)$_2$	132.0	Cyclopropane	161
H—CH=CH$_2$	156.2	Cyclobutane	136
H—C(CH$_3$)=C(CH$_3$)$_2$	148.4	Cyclopentane	131
H—CH=C(*tert*-C$_4$H$_9$)$_2$	152	Cyclohexane	123
H—C(*tert*-C$_4$H$_9$)=		Tetrahydrofuran 2,5	149
C(*tert*-C$_4$H$_9$)$_2$	143	3,4	133
Methylenecycloalkane		1,4-Dioxane	145
C$_4$–C$_7$	153–155	Benzene	159
H—CH=C=CH$_2$	168	Fluorobenzene 2,6	155
H—C(C$_6$H$_5$)=CH(C$_6$H$_5$)		3,5	163
cis (Z)	155	4	161
trans (E)	151	Bromobenzene 2,6	171
Cyclopropene	220	3,5	164
		4	161
H$_t$ H$_g$ *gem*	200	Benzonitrile 2,6	173
C=C *cis (Z)*	159	3,6	166
H$_c$ F *trans (E)*	162	4	163
		Nitrobenzene 2,6	171
H$_t$ H$_g$ *gem*	195	3,5	167
C=C *cis (Z)*	163	4	163
H$_c$ Cl *trans (E)*	161	Mesitylene	154
H$_t$ H$_g$ *gem*	162	2,6	170
C=C *cis (Z)*	157	3,5	163
H$_c$ CHO *trans (E)*	162	4	152

TABLE 6.46 One-Bond Carbon–Hydrogen Spin Coupling Constants (*continued*)

Structure	J_{CH}, Hz	Structure	J_{CH}, Hz
2,4,6-Trimethylpyridine	158		
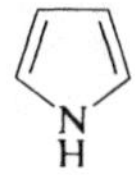 pyrrole 2,5 3,4	183 170	 imidazole 2 4	208 199
 furan 2,5 3,4	201 175 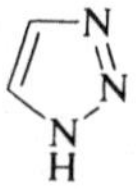	1,2,3-triazole (N–N)	205
 thiophene 2,5 3,4	185 167 	1,2,3-triazole (N–N–N)	216
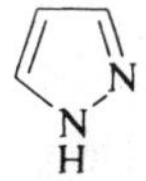 pyrazole 3,5 4	190 178		

TABLE 6.47 Two-Bond Carbon–Hydrogen Spin Coupling Constants

Structure	$^2J_{CH}$, Hz	Structure	$^2J_{CH}$, Hz
$\underline{C}H_3-CH_2-H$	−4.5	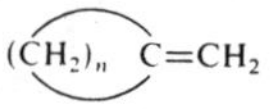 $(\underline{C}H_2)_n\ C{=}CH_2$ $n=4$	4.2
$\underline{C}Cl_3-CH_2-H$	5.9	$n=5$	5.2
$Cl\underline{C}H_2-CH_2Cl$	−3.4	$n=6$	5.5
$Cl_2\underline{C}H-CHCl_2$	1.2	$\overset{H}{\underset{Cl}{}} C{=}C \overset{H}{\underset{Cl}{}}$ *cis* (Z)	16.0
$\underline{C}H_3-\underline{C}HO$	26.7	*trans* (E)	0.8
$CH_2{=}CH_2$	−2.4	$H\underline{C}{\equiv}CH$	49.3
$(\underline{C}H_3)_2\underline{C}{=}O$	5.5	$C_6H_5O-\underline{C}{\equiv}CH$	61.0
$CH_2{=}CH-\underline{C}H{=}O$	26.9	$HC{\equiv}\underline{C}-\underline{C}HO$	33.2
$(C_2H_5)_2\underline{C}H-\underline{C}HO$	26.9	$Cl\underline{C}H_2-\underline{C}HO$	32.5
$H_2NCH{=}CH-\underline{C}HO$	6.0	$Cl_2\underline{C}H-\underline{C}HO$	35.3
$H_2NCH-\underline{C}H-\underline{C}HO$	20.0	$Cl_3\underline{C}-\underline{C}HO$	46.3
C_6H_6	1.0	$C_6H_5-\underline{C}{\equiv}\underline{C}{\equiv}CH_3$	10.8

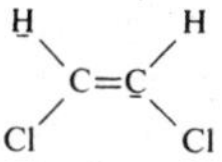

TABLE 6.48 Carbon–Carbon Spin Coupling Constants

Structure*	J_{CC}, Hz	Structure*	J_{CC}, Hz
H_3C-CH_3	35	$H_3C-CH_2NH_2$	37
H_3C-CHR_2	37	$C-C{=}O$	38–40
H_3C-CH_2Ar	34	$\underline{C}-\underline{C}-C{=}O$	36
H_3C-CH_2CN	33	$\underline{C}-\underline{C}-Ar$	43
$H_3C-CH_2-CH_2OH$		$C-CO-O^-$ (anion)	52
C-1, C-2	38	$C-CO-N$	52
C-2, C-3	34	$C-CO-OH$	57

TABLE 6.48 Carbon–Carbon Spin Coupling Constants (*continued*)

Structure*	J_{CC}, Hz	Structure*	J_{CC}, Hz
C—CO—OR	59	$C_6H_5NH_2$	
C—CN	52–57	1–2	61
C—C≡C $^2J_{CC}$ = 11.8	67	2–3	58
H_2C=CH_2	68	3–4	57
>C̲=C̲—CO—OH	70–71	$^3J_{2-5}$	7.9
>C̲=C̲—CN	71	$C_6H_5CH_3$	44
>C̲=C̲—Ar	67–70	Pyridine	
C_6H_6	57	2–3	54
$C_6H_5NO_2$		3–4	56
1–2	55	$^3J_{2-5}$	14
2–3, 3–4	56	Furan	69
$^3J_{2-5}$	7.6	Pyrrole	69
C_6H_5I		Thiophene	64
1–2	60	$H_2C̲$=C̲=$C(CH_3)_2$	100
2–3	53	—C≡C—	170–176
3–4	58		

Structure	$^2J_{CC}$, Hz
C̲H$_3$—CO—C̲H$_3$	16
C̲H$_3$—C≡C̲H	11.8
C̲H$_3$CH$_2$—C̲N	33

(continuation of left column)

Structure*	J_{CC}, Hz
$^3J_{2-5}$	8.6
C_6H_5—OCH_3	
2–3	58
3–4	56

* R, alkyl group; Ar, aryl group.

TABLE 6.49 Carbon–Fluorine Spin Coupling Constants

Structure*	J_{CF}, Hz	Structure*	J_{CF}, Hz
F,H / C / H,H	−158	F,F / C / F,CH$_3$	−271
F,H \ C / F,H	−235	F,H \ C / H,Ar	−165
F,F \ C / F,H	−274	F—CH$_2$CH$_2$— or F—CR$_3$	−167
		p-F—C_6H_4—OR	−237
		p-F—C_6H_4—R	−241
F,F \ C / F,F	−259	p-F—C_6H_4—CF$_3$	−252
		p-F—C_6H_4—CO—CH$_3$	−253
		p-F—C_6H_4—NO$_2$	−257
		F—C_6H_5 $^2J_{CF}$ = 21.0 $^3J_{CF}$ = 7.7 $^4J_{CF}$ = 3.4	−244

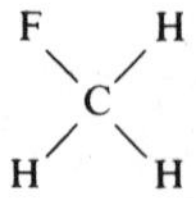

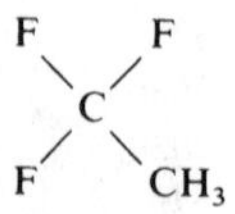

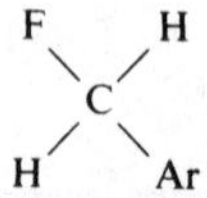

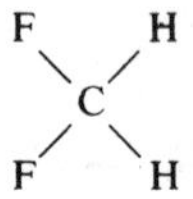

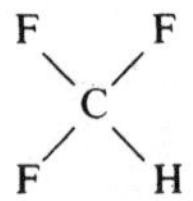

TABLE 6.49 Carbon–Fluorine Spin Coupling Constants (*continued*)

Structure[*]	J_{CF}, Hz	Structure[*]	J_{CF}, Hz

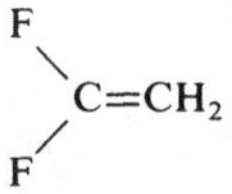 F⟍
  C=CH₂
 F⟋
	−287	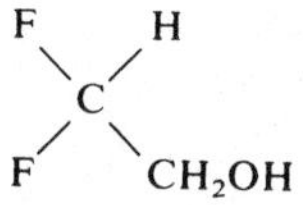 F⟍ ⟋H
		  C
		 F⟋ ⟍CH₂OH
			−241

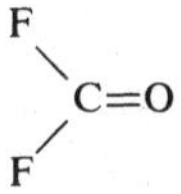 F⟍
  C=O
 F⟋
	−308	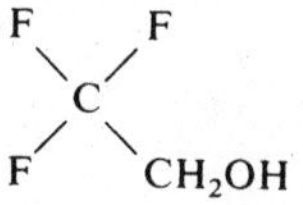 F⟍ ⟋F
		  C
		 F⟋ ⟍CH₂OH
			−278

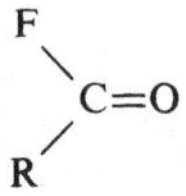 F⟍
  C=O
 R⟋
	−353	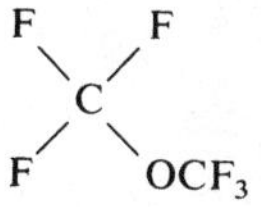 F⟍ ⟋F
		  C
		 F⟋ ⟍OCF₃
			−265

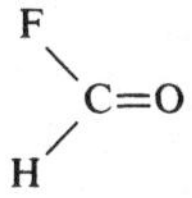 F⟍
  C=O
 H⟋
	−369	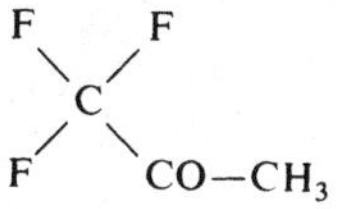 F⟍ ⟋F
		  C
		 F⟋ ⟍CO−CH₃
			−289

* Ar, aryl group; R, alkyl group.

TABLE 6.50 Carbon-13 Chemical Shifts of Deuterated Solvents

Relative to tetramethylsilane

Solvent	Group	δ, ppm
Acetic-d_3 acid-d_1	Methyl	20.0
	Carbonyl	205.8
Acetone-d_6	Methyl	28.1
	Carbonyl	178.4
Acetonitrile-d_3	Methyl	1.3
	Carbonyl	117.7
Benzene-d_6		128.5
Carbon disulfide		193
Carbon tetrachloride		97
Chloroform-d_1		77
Cyclohexane-d_{12}		25.2
Dimethyl sulfoxide-d_6		39.5
1,4-Dioxane-d_6		67
Formic acid-d_1	Carbonyl	165.5
Methanol-d_4		47–49
Methylene chloride-d_2		53.8
Nitromethane-d_3		57.3
Pyridine-d_5	C₃, C₅	123.5
	C₄	135.5
	C₂, C₆	149.9

TABLE 6.51   Carbon-13 Spin Coupling Constants with Various Nuclei

Nuclei	Structure	^1J, Hz	^2J, Hz	^3J, Hz	^4J, Hz
^{2}H	$CDCl_3$	32			
	$CD_3{-}CO{-}CD_3$	20			
	$(CD_3)_2SO$	22			
	C_6D_6	26			
^{7}Li	CH_3Li	15			
^{11}B	$(C_6H_5)_4B^-$	49		3	
^{14}N	$(CH_3)_4N^+$	10			
	CH_3NC	8			
^{29}Si	$(CH_3)_4Si$	52			
^{31}P	$(CH_3)_3P$	14			
	$(C_4H_9)_3P$	11	12	5	
	$(C_6H_5)_3P$	12	20	7	0
	$(CH_3)_4P^+$	56			
	$(C_4H_9)_4P^+$	48	4	15	
	$(C_6H_5)_4P^+$	88	11	13	3
	$R(RO)_2P=O$	142	5–7		
	$(C_4H_9O)_3P{=}O$		6	7	
^{77}Se	$(CH_3)_2Se$	62			
	$(CH_3)_3Se^+$	50			
^{113}Cd	$(CH_3)_2Cd$	513, 537			
^{119}Sn	$(CH_3)_4Sn$	340			
	$(CH_3)_3SnC_6H_5$	474	37	47	11
^{125}Te	$(CH_3)_2Te$	162			
^{199}Hg	$(CH_3)_2Hg$	687			
	$(C_6H_5)_2Hg$	1186	88	102	18
^{207}Pb	$(CH_3)_2Pb$	250			
	$(C_6H_5)_4Pb$	481	68	81	20

TABLE 6.52   Boron-11 Chemical Shifts

Values given in ppm on the δ scale, relative to $B(OCH_3)_3$

Structure	δ, ppm	Structure	δ, ppm
R_3B	−67 to −68	$C_6H_5BCl_2$	−36
Ar_3B	−43	$C_6H_5B(OH)_2$	−14
BF_3	24	$C_6H_5B(OR)_2$	−10
BCl_3	−12	$M(BH_4)$	55–61
BBr_3	−6	$B(BF_4)$	19–20
BI_3	41		
$B(OH)_3$	36	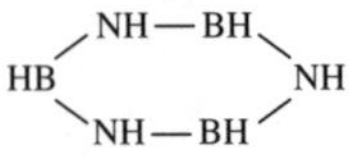	
$B(OR)_3$	0–1	NH—BH	
$B(NR_2)_3$	−13	HB⟨ ⟩NH NH—BH	−12

TABLE 6.52 Boron-11 Chemical Shifts (*continued*)

Structure	δ, ppm	Structure	δ, ppm
	37	$R_2O(\text{or ROH}) \cdot BCl_3$ $R_2O(\text{or ROH}) \cdot BBr_3$ $R_2O(\text{or ROH}) \cdot BI_3$	-7 to -8 23–24 74–82
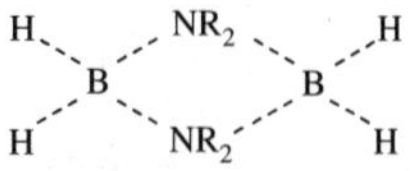	15	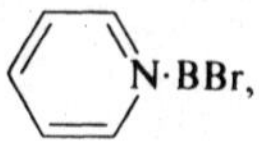	24
$(CH_3)_2N\!-\!B(CH_3)_2$	62	Boranes B_2H_6	1
Addition complexes $R_2O \cdot BH_3$ $R_3N \cdot BH_3$ $R_2NH \cdot BH_3$	18–19 25 33	B_4H_{10} (BH_2) (BH)	25 60

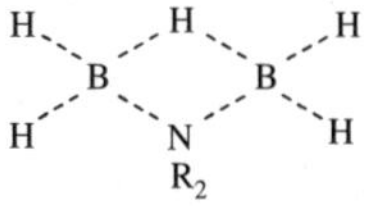

			Base	Apex
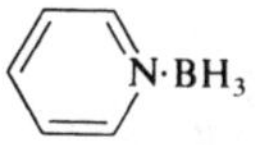	31	B_5H_9	31	70
		B_5H_{11}	-16	50
$R_2O(\text{or ROH}) \cdot BF_3$	17–19	$B_{10}H_{14}$	7	54

TABLE 6.53 Nitrogen-15 (or Nitrogen-14) Chemical Shifts

Values given in ppm on the δ scale, relative to NH_3 liquid

Substituent group	δ, ppm	Substituent group	δ, ppm
Aliphatic amines		Ureas	
Primary	1–59	Aliphatic	63–84
Secondary	7–81	Aryl	105–108
Tertiary	14–44	Sulfonamides	79–164
Cyclo, primary	29–44	Amides	
Aryl amines	40–100	HCO—NHR	
Aryl hydrazines	40–100	R = primary	100–115
Piperidines,		R = secondary	104–148
decahydroquinolines	30–82	R = tertiary	96–133
Amine cations		HCO—NH—Aryl	138–141
Primary	19–59	RCO—NHR or	103–130
Secondary	40–74	RCO—NR$_2$	
Tertiary	30–67	RCO—NH—Aryl	131–136
Quaternary	43–70	Aryl—CO—H—Aryl	ca 126
Enamines, tertiary type		Guanidines	
Alkyl	29–82	Amino	30–60
Cycloalkyl	55–104	Imino	166–207
Aminophosphines	59–100	Thioureas	85–111
Amine N-oxides	95–122	Thioamides	135–154

TABLE 6.53 Nitrogen-15 (or Nitrogen-14) Chemical Shifts (*continued*)

Substituent group	δ, ppm	Substituent group	δ, ppm
Cyanamides		Diazo	
R_2N—	−12 to −38	Internal	226–303
—CN	175–200	Terminal	315–440
Carbodiimides	95–120	Nitrilium ions	123–150
Isocyanates		Azinium ions	185–220
Alkyl, primary	14–32	Azine *N*-oxides	230–300
Alkyl, secondary and tertiary	54–57	Nitrones	270–285
Aryl	ca 46	Imides	170–178
Isothiocyanates	90–107	Imimes	310–359
Azides	52–80	Oximes	340–380
	108–122	Nitramines	
	240–260	Amine	252–280
Lactams	113–122	—NO_2	328–355
Hydrazones		Nitrates	310–353
Amino	141–167	*gem*-Polynitroalkanes	310–353
Imino	319–327	Nitro	
Cyanates	155–182	Aryl	350–382
Nitrile *N*-oxides, fulminates	195–225	Alkyl	372–410
Isonitriles		Hetero, unsaturated	354–367
Alkyl, primary	162–178	Azoxy	330–356
Alkyl, secondary	191–199	Azo	504–570
Aryl	ca 180	Nitrosamines	222–250
Nitriles			525–550
Alkyl	235–241	Nitrites	555–582
Aryl	258–268	Thionitrites	720–790
Thiocyanates	265–280	Nitroso	
Diazonium		Aliphatic amines, NO	535–560
Internal	222–230	Aryl	804–913
Terminal	315–322		

TABLE 6.53 Nitrogen-15 (or Nitrogen-14) Chemical Shifts (*continued*)

Substituent group	δ, ppm	Substituent group	δ, ppm
Saturated cyclic systems			

$(CH_2)_n$ N—H

$n = 2$	−8.5
$n = 3$	25.3
$n = 4$	36.7
$n = 5$	37.7

Morpholine structure — 32.1

Piperazine structure — 35.5

Quinuclidine structure — 7.5 (in C_6H_6), 18.0 (in H_2O)

Decahydroquinoline structure:
cis (Z) — 42.4
trans (E) — 52.9

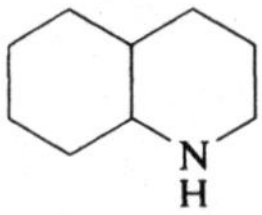

Unsaturated cyclic systems			

Pyrrole — N 145

Pyrazole — N 245

Imidazole — N 205, —N

1,2,4-Triazole — N, N, N 240

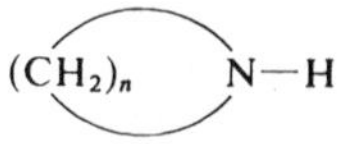

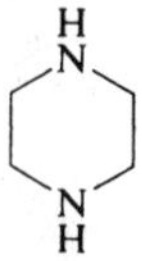

1,2,3-Triazole — N 316, N 316, 244 N

Oxazole — N 251

Isoxazole — N 383

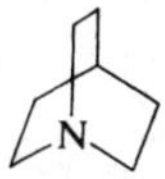

Thiazole — N 323

Isothiazole — N 298

1,2,3-Triazole — 270 N—N 351, N 207, 351 N

Pyridine — N 317

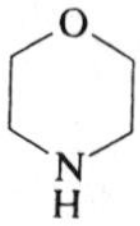

Pyridazine — N, N 396

Pyrimidine — N, N 295

Pyrazine — N 331

1,3,5-Triazine — N, N, N 383

1,2,4,5-Tetrazine — N, N, N, N 381

Indole — N 133

Benzimidazole — N, N 191

Indazole — N 301, N 179

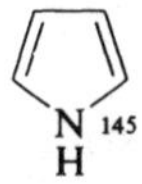

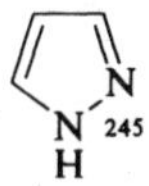

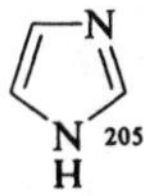

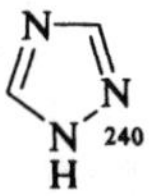

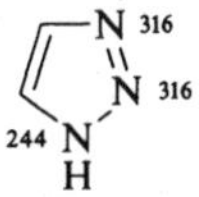

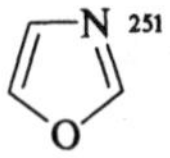

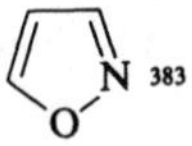

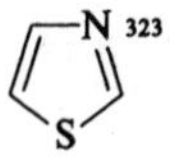

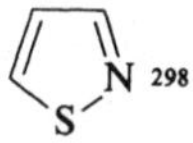

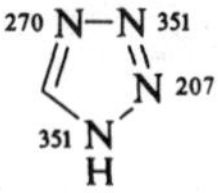

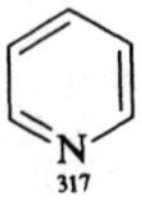

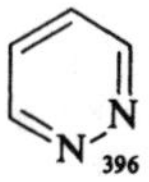

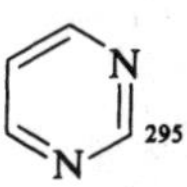

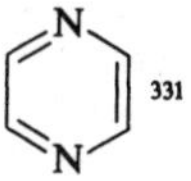

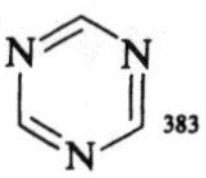

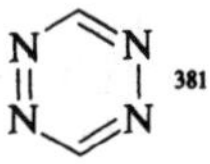

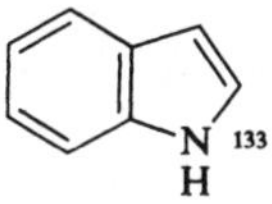

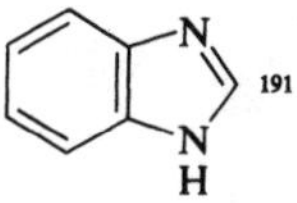

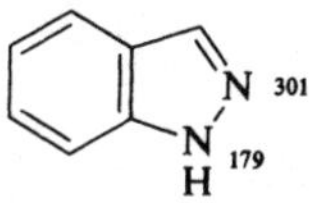

TABLE 6.53 Nitrogen-15 (or Nitrogen-14) Chemical Shifts (*continued*)

Unsaturated cyclic systems (*continued*)

X	δ,ppm
O	517
S	331
Se	373

TABLE 6.54 Nitrogen-15 Chemical Shifts in Monosubstituted Pyridine

$\delta = 317.3 + \Delta_i$

Substituent	$\Delta_{C\text{-}2}$	$\Delta_{C\text{-}3}$	$\Delta_{C\text{-}4}$
—CH₃	-0.4	0.3	-8.0
—CH₂CH₃	-1.8		-6.6
—CH(CH₃)₂	-5.1		-5.9
—C(CH₃)₃	-2.5		-5.8
—CN	-0.9	-0.8	10.6
—CHO	10	11	29
—CO—CH₃	-9	15	11
—CO—OCH₂CH₃	11.8		-5
—OCH₃	-49	0	-23
—OH	-126	-2	-118
—NO₂	-23	1	22
—NH₂	-45	10	-46
—F	-42	-18	
—Cl	-4	4	-6
—Br	2	8	7

TABLE 6.55 Nitrogen-15 Chemical Shifts for Standards

Values given in ppm, relative to NH$_3$ liquid at 23 °C

Substance	δ, ppm	Conditions
Nitromethane (neat)	380.2	For organic solvents and acidic aqueous solutions
Potassium (or sodium) nitrate (saturated aqueous solution)	376.5	For neutral and basic aqueous solutions
C(NO$_2$)$_4$	331	For nitro compounds
(CH$_3$)$_2$—CHO (neat)	103.8	For organic solvents and aqueous solutions
(C$_2$H$_5$)$_4$N$^+$Cl$^-$	64.4	Saturated aqueous solution
(CH$_3$)$_4$N$^+$Cl$^-$	43.5	Saturated aqueous solution
NH$_4$Cl	27.3	Saturated aqueous solution
NH$_4$NO$_3$	20.7	Saturated aqueous solution
NH$_3$	0.0	Liquid, 25 °C
	-15.9	Vapor, 5 atm

TABLE 6.56 Nitrogen-15 to Hydrogen-1 Spin Coupling Constants

Structure	J, Hz	Structure	J, Hz
R—NH$_2$ and R$_2$NH	61–67	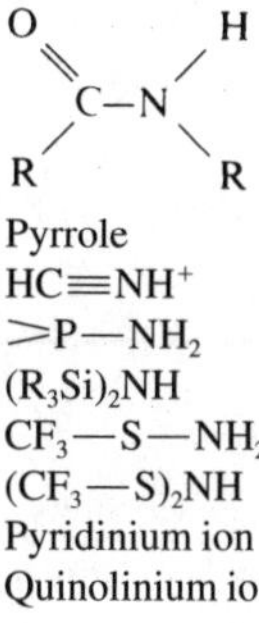 $\begin{array}{c}\text{O}\diagdown\quad\diagup\text{H}\\[-2pt]\text{C—N}\\[-2pt]\diagup\quad\diagdown\\[-2pt]\text{R}\qquad\text{R}\end{array}$	88–92
Aryl—NH$_2$	78		
p-CH$_3$O—aryl—NH$_2$	79		
p-O$_2$N—aryl—NH$_2$	90–93		
Amine salts (alkyl and aryl)	73–76	Pyrrole	97
Aryl—NHOH	79	HC≡NH$^+$	133–136
Aryl—NHCH$_3$	87	⪢P—NH$_2$	82–90
Aryl—NHCH$_2$F	90	(R$_3$Si)$_2$NH	67
Aryl—NHNH$_2$	90	CF$_3$—S—NH$_2$	81
p-O$_2$N—aryl—NHNH$_2$	99	(CF$_3$—S)$_2$NH	99
Aryl—SO$_2$—NH$_2$	81	Pyridinium ion	90
Aryl—SO$_2$—NHR	86	Quinolinium ion	96
$\begin{array}{c}\text{O}\diagdown\qquad\text{H}_{syn}\ (\text{to —CO—})\\[-2pt]\text{C—N}\\[-2pt]\diagup\qquad\diagdown\\[-2pt]\text{H}\qquad\text{H}_{anti}\end{array}$	88 92–93		

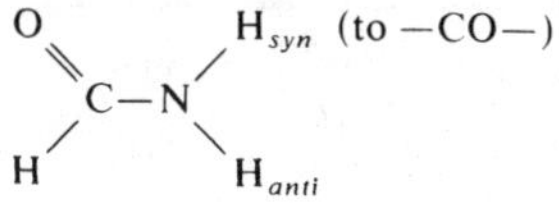

TABLE 6.57 Nitrogen-15 to Carbon-13 Spin Coupling Constants

Structure	J, Hz	Structure	J, Hz
Alkyl amines	4–4.5	Alkyl—NO_2	11
Cyclic alkyl amines	2–2.5	R—CN	18
Alkyl amines protonated	4–5	CH_3—$\overset{+}{N}\equiv\overset{-}{C}$	
Aryl amines	10–14	H_3C—N	10
Aryl amines protonated	9	—N≡C	9
CH_3CO—NH_2	14–15	Diaryl azoxy	
H_2N—CO—NH_2	20	*anti*	18
Aryl—NO_2	15	*syn*	13

TABLE 6.58 Nitrogen-15 to Fluorine-19 Spin Coupling Constants

Structure	J, Hz	Structure	J, Hz
NF_3	155	Pyridine	
F_4N_2	164	2-F	52
FNO_2	158	3-F	4
F_3NO	190	2,6-di-F	37
F_3C—O—NF_2	164–176	Pyridinium ion	
FCO—NF_2	221	2-F	23
$(NF_4)^+SbF_6^-$	323	3-F	3
$(NF_4)^+AsF_6^-$	328	Quinoline, 8-F	3
$(N_2F)^+AsF_6^-$	459	Aniline	
F_3C—NO_2	215	2-F	0
		3-F	0
$\underset{N=N}{F\diagdown}$ $(^2J = 10)$	190	4-F	1.5
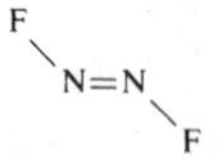		Anilinium ion	
		2-F	1.4
$\underset{N=N}{F\diagup\diagdown F}$ $(^2J = 52)$		3-F	0.2
		4-F	0
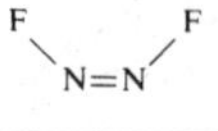			203

TABLE 6.59 Fluorine-19 Chemical Shifts
Values given in ppm on the δ scale, relative to CCl_3F

Substituent group	δ, ppm	Substituent group	δ, ppm
—SO_2—F	−67 to −42	R—CF_2Cl	61–71
	(aryl) (alkyl)	$>$C—CF_3 and aryl—CF_3	56–73
—CO—F	−29 to −20	—CS—CF_3	70
$>$N—CO—F	−5	$>$CF—CF_3	71–73
Aryl—CF_2Cl	49	—S—CF_3	41
—CF_2I	56	—S—CF_2—S—	39
—CF_2Br	63	$>$P—CF_3	46–66

TABLE 6.59 Fluorine-19 Chemical Shifts (*continued*)

Substituent group	δ, ppm	Substituent group	δ, ppm
$>$N—CF$_3$	40–58	Perfluorocycloalkane	131–138
$>$N—CF$_2$—C	85–127	$>$CF—CF$_3$	163–198
—O—CF$_2$—R	70–91	$>$CF(CF$_3$)$_2$	180–191
—O—CF$_2$—CF$_3$	70–91	—CFH—	198–231
—CH$_2$—CF$_3$	76–77	—CFH$_2$	235–244
HO—CO—CF$_3$	77	F$_2$C=CF$_2$	133
—CHF—CF$_3$	81		
—CF$_2$—CF$_3$	78–88	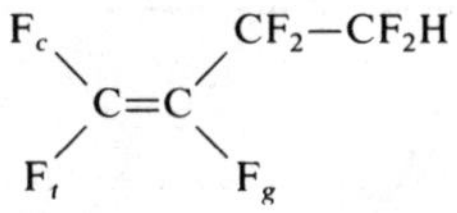	
—CS—F	81		
CF$_3$—C—N$<$	84–96		
—CO—CF$_2$—CF$_3$	83	*cis (Z)*	108
—CF$_2$—	86–126	*trans (E)*	92
—CF$_2$Br	91	*gem*	192
—C—CF$_2$—S—	91–98		
—CF=	180–192		
—CF$_2$—CF$_3$	111	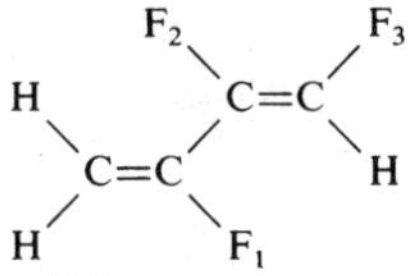	
—CO—CF$_2$—	116–131		
—C(halide)—CF$_2$—	119–128		
—CF$_2$—CF$_3$	121–125	F-1	126
—CF$_2$—CF$_2$—	121–129	F-2	155
—CF$_2$—CH$_2$—	122–133	F-3	162
—CF$_2$—CHF$_2$	128–132	ClFC=CH—CF$_3$	61
—CF$_2$H	136–143	Cycloalkenes	
		=CF—CF$_2$—	
$\triangleright$F$_2$	151–156	C(CF$_3$ or H)—	101–113
		—CF$_2$—CF$_2$—	
$\diamondsuit$F$_2$	147	C(CF$_3$ or CH$_3$)=	110–114
		—CF$_2$—CF$_2$—CH=	113–116
 F$_2$	96–133	—CF$_2$—CF$_2$—CF=	119–122
		Aryl—F	113
		C$_{10}$H$_7$—F	
F		F-1	127
	159	F-2	114
		C$_6$H$_5$—C$_6$H$_4$—F	
Cyclohexane-F		F-2	117
	210	F-3	113
	(axial)	F-4	109
	to	C$_6$F$_6$	163
	240		
	(equatorial)		

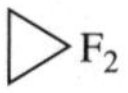

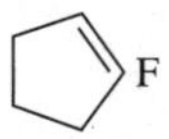

F$_c$ ... CF$_2$—CF$_2$H

C=C

F$_t$ ... F$_g$

F$_2$ ... F$_3$

H ... C=C

C=C ... H

H ... F$_1$

TABLE 6.60 Fluorine-19 Chemical Shifts for Standards

Substance	Formula	δ, ppm
Trichlorofluoromethane	$CFCl_3$	0.0
α,α,α-Trifluorotoluene	$C_6H_5CF_3$	63.8
Trifluoroacetic acid	CF_3COOH	76.5
Carbon tetrafluoride	CF_4	76.7
Fluorobenzene	C_6H_5F	113.1
Perfluorocyclobutane	C_4F_8	138.0

TABLE 6.61 Fluorine-19 to Fluorine-19 Spin Coupling Constants

Structure	J_{FF}, Hz
$F_2C\diagdown$ cycloalkane	
gem	212–260
Unsaturated compounds $>$C=C$<$	
gem	30–90
trans	115–130
cis	9–58
Aromatic compounds, monocyclic	
ortho	18–22
meta	0–7
para	12–15
Alkanes	
C$\underline{F}$Cl$_2$—C$\underline{F}_2$—CFCl$_2$	6
C$\underline{F}$Cl$_2$—C$\underline{F}_2$—CCl$_3$	5
C$\underline{F}_2$Cl—C$\underline{F}_2$—CF$_2$Cl	1
C$\underline{F}_3$—C$\underline{F}_2$—CF$_2$Cl (or —CF$_3$)	<1
CF$_3$—C$\underline{F}_2$—C$\underline{F}_2$Cl	2
C$\underline{F}_3$—CF$_2$—C$\underline{F}_2$Cl	9
C$\underline{F}_3$—CF$_2$—C$\underline{F}_3$	7

TABLE 6.62 Silicon-29 Chemical Shifts
Values given in ppm on the δ scale relative to tetramethylsilane

Substituent group X in $(CH_3)_{4-n}SiX_n$	n			
	1	2	3	4
—F	35	9	−52	−109
—Cl	30	32	13	−19
—Br	26	20	−18	−94
—I	9	−34	−18	−346
—H	−19	−42	−65	−93
—C_2H_5	2	5	7	8

TABLE 6.62 Silicon-29 Chemical Shifts (*continued*)

Substituent group X in $(CH_3)_{4-n}SiX_n$	n			
	1	2	3	4
—C_6H_5	−5	−9	−12	
—CH=CH_2	−7	−14	−21	−23
—Oalkyl	14–17	−3 to −6	−41 to −45	−79 to −83
—Oaryl	17	−6	−54	−101
—O—CO—alkyl	22	4	−43	−75
—$N(CH_3)_2$	6	−2	−18	−28

Structure	δ, ppm	Structure	δ, ppm
Hydrides		$CH_3\overset{\displaystyle O-}{\underset{\displaystyle O-}{Si}}-O-$ (branching)	−65 to −66
$H_3Si—$	−39 to −60		
—$H_2Si—$	−5 to −37		
HSi≦	−2 to −39		
Silicates		$-O-\overset{\displaystyle O-}{\underset{\displaystyle O-}{Si}}-O-$ (cross-linked)	−105 to −110
Orthosilicate anions	−69 to −72		
Silicon in end position	−77 to −81		
Silicon in middle	−85 to −89	**Polysilanes**	
Branching silicons	−93 to −97	$F_3Si—SiF_3$ −74	
Cross-linked silicons	−107 to −120	$Cl_3Si—SiCl_3$	−8
Methyl siloxanes		$(CH_3O)_3Si—Si(OCH_3)_3$	−53
$(CH_3)_2Si—O—$(end position)	6–8	$(CH_3)_3Si—Si(CH_3)_3$	−20
$(CH_3)_2Si\overset{\displaystyle O-}{\underset{\displaystyle O-}{\big\langle}}$ (middle)	−18 to −23	$(CH_3)_2\underline{Si}[Si(CH_3)_3]_2$	−48
		$H\underline{Si}[Si(CH_3)_3]_3$	−117
		$\underline{Si}[Si(CH_3)_3]_4$	−135
$CH_3Si(H)\overset{\displaystyle O-}{\underset{\displaystyle O-}{\big\langle}}$ (middle)	−35 to −36		

TABLE 6.63 Phosphorus-31 Chemical Shifts

Values given in ppm on the δ scale, relative to 85% H_3PO_4

Structure	Identical atoms attached directly to phosphorus	Non-identically substituted phosphorus		
		$R = CH_3$	$R = C_2H_5$	$R = C_6H_5$
P_4	461			
PR_3		62	20	6
PHR_2		99	56	41
PH_2R		164	128	122
PH_3	241			
PF_3	−97			
PRF_2			−168	−207

TABLE 6.63 Phosphorus-31 Chemical Shifts (*continued*)

Structure	Identical atoms attached directly to phosphorus	Non-identically substituted phosphorus		
		$R = CH_3$	$R = C_2H_5$	$R = C_6H_5$
PCl_3	-220			
$PRCl_2$		-192	-196	-162
PR_2Cl		-94	-119	-81
PBr_3	-227			
$PRBr_2$		-184	-194	-152
PR_2Br		-91	-116	-71
PI_3	-178			
$P(CN)_3$	136			
$P(SiR_3)_3$		251		
$P(OR)_3$		-141	-139	-127
$P(OR)_2Cl$		-169	-165	-157
$P(OR)Cl_2$		-114	-177	-173
$P(SR)_3$		-125	-115	-132
$P(SR)_2Cl$		-188	-186	-183
$P(SR)Cl_2$		-206	-211	-204
$P(SR)_2Br$				-184
$P(SR)Br_2$		-204		
$P(NR_2)_3$		-123	-118	
$P(NR_2)Cl_2$		-166	-162	-151
$PR(NR_2)_2$		-86	-100	-100
$PR_2(NR_2)$		-39	-62	
F_2P-PF_2	-226			
Cl_2P-PCl_2	-155			
I_2P-PI_2	-170			
$PH_2^- K^+$	255			
$P(CF_3)_3$	3			
P_4O_6	-113			

Structure	Identical atoms attached directly to phosphorus	Non-identically substituted phosphorus		
		$X = F$	$X = Cl$	$X = Br$
$P(NCO)_3$	-97			
$P(NCO)_2X$		-128	-128	-127
$P(NCO)X_2$		-131	-166	
$P(NCS)_3$	-86			
$P(NCS)_2X$			-114	-112
$P(NCS)X_2$			-155	-153

Structure	Identical atoms attached directly to phosphorus	Non-identically substituted phosphorus		
		$R = CH_3$	$R = C_2H_5$	$R = C_6H_5$
$O=PR_3$		-36	-48	-25
$O=PHR_2$		-63		-23
$O=PF_3$	36			

TABLE 6.63 Phosphorus-31 Chemical Shifts (*continued*)

Structure	Identical atoms attached directly to phosphorus	Non-identically substituted phosphorus		
		$R = CH_3$	$R = C_2H_5$	$R = C_6H_5$
$O{=}PRF_2$		-27	-29	-11
$O{=}PCl_3$	-2			
$O{=}PRCl_2$		-45	-53	-34
$O{=}PR_2Cl$		-65	-77	-43
$O{=}P(OR)_3$		-1	1	18
$O{=}P(OR)_2Cl$		-6	-3	6
$O{=}P(OR)Cl_2$		-6	-6	-2
$O{=}PH(OR)_2$		-19	-15	
$O{=}PR_2(OC_2H_5)$		-50	-52	-31
$O{=}PR(OC_2H_5)_2$		-30	-33	-17
$O{=}P(NR_2)_3$		-23	-24	-2
$O{=}PR_2(NR_2)$		-44		-26
$O{=}P(OR)_2NH_2$		-15	-12	-3
$O{=}P(OR)_2(NCS)$			19	29
$O{=}P(SR)_3$		-66	-61	-55
$O{=}PBr_3$	103			
$O{=}P(NCO)_3$	41			
$O{=}P(NCS)_3$	62			
$O{=}P(NH_2)_3$	-22			

Structure	Identical atoms attached directly to phosphorus	Structure	Identical atoms attached directly to phosphorus
PF_5	35	$\begin{array}{c} O \\ | \\ -O-P-O- \\ | \\ OR \end{array}$ (middle group)	ca 18
$PF_6{}^-H^+$	144		
PBr_5	101		
$P(OC_2H_5)_5$	71		
$PO_4{}^{3-}$	-6		
$O{=}P[OSi(CH_3)_3]_3$	33	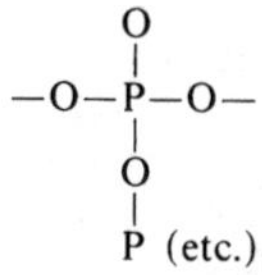	
$H_4P_2O_7$	11	$\begin{array}{c} O \\ | \\ -O-P-O- \\ | \\ O \\ | \\ P \text{ (etc.)} \end{array}$ (branch group)	ca 30
Phosphonates	-24 to -2		
Phosphonium cations			
Alkyl	-43 to -32		
Aryl	-35 to -18		
$(O_3P{-}PO_3)^{4-}$	-9		
Polyphosphates			
$\begin{array}{c} O{=}P-O- \\ | \\ (OR)_2 \end{array}$ (end group)	ca 6		

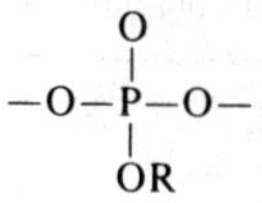

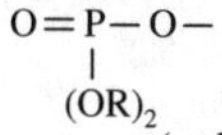

TABLE 6.63 Phosphorus-31 Chemical Shifts (*continued*)

Structure	Identical atoms attached directly to phosphorus	Non-identically substituted phosphorus		
		R = CH$_3$	R = C$_2$H$_5$	R = C$_6$H$_5$
S=PR$_3$		−59	−55	−43
S=PCl$_3$	−29			
S=PRCl$_2$		−80	−94	−75
S=PR$_2$Cl		−87	−109	−80
S=PBr$_3$	112			
S=PRBr$_2$		−21	−42	−20
S=PR$_2$Br		−64	−98	
S=P(OR)$_3$		−73	−68	−53
S=P(OR)Cl$_2$		−59	−56	−54
S=P(OR)$_2$Cl		−73	−68	−59
S=PH(OR)$_2$		−74	−69	−59
S=P(SR)$_3$		−98	−92	−92
S=P(NH$_2$)$_3$	−60			
S=P(NR$_2$)$_3$		−82	−78	
Se=P(OR)$_3$		−78	−71	−58
Se=P(SR)$_3$		−82	−76	
P(OR)$_5$			71	86
PRF$_4$		30	30	42
PR$_2$F$_3$		−9	−6	

TABLE 6.64 Phosphorus-31 Spin Coupling Constants

Substituent group	J_{PH}, Hz	Substituent group	J_{PH}, Hz
>PH	180–225	>P—N—CH	8–25
—PH$_2^-$	134	>P—C—CH	0–4
RPH$_2$	160–210	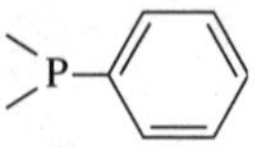	
>P—CH$_3$	1–6		
>P—CH$_2$—	14		
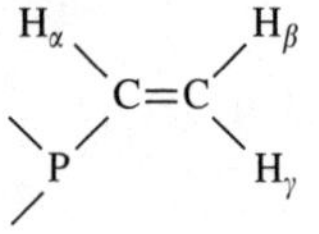		*ortho*	7–10
		meta	2–4
α	12–22	O=PHR$_2$	210–500
β	30–40	O—PH(S)R	490–540
γ	14–20	O$_2$PHR	500–575
(Halogen)$_2$P=CH	16–20	O$_2$PH(N)	560–630
>P—NH	10–28	O$_2$PH(S or Se)	630–655
>P—O—CH$_3$	11–15	O$_3$PH	630–760
>P—O—CH$_2$—R	6–10	S(or Se)=P—H	490–650
>P—O—CHR$_2$	3–7		
>P—SCH	5–20	S(or Se)=PHR$_2$	420–454

TABLE 6.64 Phosphorus-31 Spin Coupling Constants

Substituent group	J_{PH}, Hz	Substituent group	J_{PH}, Hz
O=P—CH$_3$	7–15	—P F axial	600–860
		equatorial	800–1000
O=P—CH=C	15–30	O=P—CF	110–113
O=P—CH—Aryl (or C=O)	15–30	O=P—F	980–1190
(Halogen)$_2$P—N—CH	9–18	P—O—P—F	2
S=P—CH	11–15	**Substituent group**	**J_{PB}, Hz**
≡P—CH$_3{}^+$	12–17	H$_3$B—P—N	80
≡P—H$^+$	490–600		

Substituent group	J_{PP}, Hz	Substituent group	J_{PP}, Hz
>P—F	1320–1420 (1F) (3F)	>P—P<	220–400
RPF$_2$	1140–1290	O=P—P=O	330–500
R$_2$PF	1020–1110	S=P—P=S	15–500
RP(N)F	920–985 (alkyl) (aryl)		
—O\ PF —O/	1225–1305	\P—C—P/	ca 70
(OCN)PF	1310	>P—O—P<	20–40
\N—P/ F	1100–1200	>P—S—P<	86–90
>P—CF	60–90	O=P—O=P=O	15–25
>P—⟨ ⟩—F		O=P—N=P=O H	8–30
ortho	0–60	P—N N P P—N	5–66
meta	1–7	>P=N—P=N—	5–65
para	0–3		

ELECTRON SPIN RESONANCE

Electron paramagnetic resonance (EPR) is also referred to as electron spin resonance (ESR). In many respects, it is similar to NMR and the corresponding principles, discussed in the previous section, apply. The critical difference is that an unpaired electron spin is detected in this method instead of a nuclear spin. The method applies only to paramagnetic systems. The electron spin is more readily detected than is a nuclear spin and magnets on EPR instruments are correspondingly smaller and less expensive.

Certain transition metals such as cobalt (Co), copper (Cu), iron (Fe), manganese (Mn), nickel (Ni), and vanadium (V) have unpaired spins and are readily detectable by EPR. Likewise, organic radicals can be detected and studied. Numerous organic radicals have been studied by this method. When an unpaired electron spin is present, it will be observed at a characteristic resonance position. The resonance line is split by nuclei such as protons resulting in spin–spin coupling as observed in NMR. The coupling constant is referred to as "g" in EPR rather than "J" as in NMR. It is referred to as the hyperfine splitting or hyperfine coupling constant.

Certain radicals are especially stable and are of biological consequence or have been used as antioxidants. TEMPO radicals (TEMPO = 2,2,6,6-tetramethylpiperidin-1-oxyl) have been particularly well studied because they are readily prepared and extremely stable.

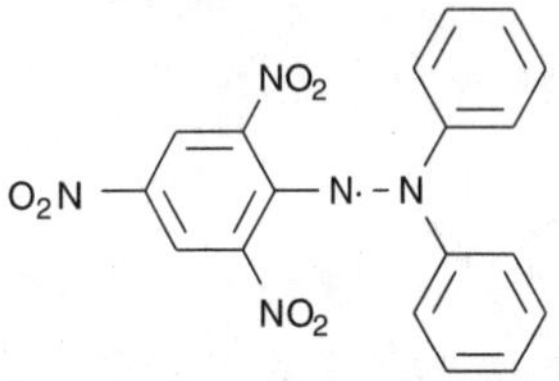

diphenylpicrylhydrazyl
DPPH radical

nitrous oxide
radical

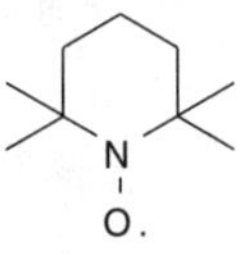

TEMPO

A noticeable difference between NMR and EPR is that NMR is typically presented in absorption mode whereas EPR normally is presented in derivative mode. The EPR spectrum may be analyzed mathematically to determine the coupling pattern and thus the structural relationships. A database of software useful in EPR research is available at http://epr.niehs.nih.gov/software.html.

TABLE 6.65   Spin–Spin Coupling (Hyperfine Splitting Constants)

Values of coupling constant a_i given in gauss
Involves protons unless otherwise indicated.

H˙	Li˙	Na˙	K˙	Cs˙	HĊ—H	H₂Ċ—H	(^{1}H) 22.8
508	81	632	165	3280	15		(^{13}C) 41

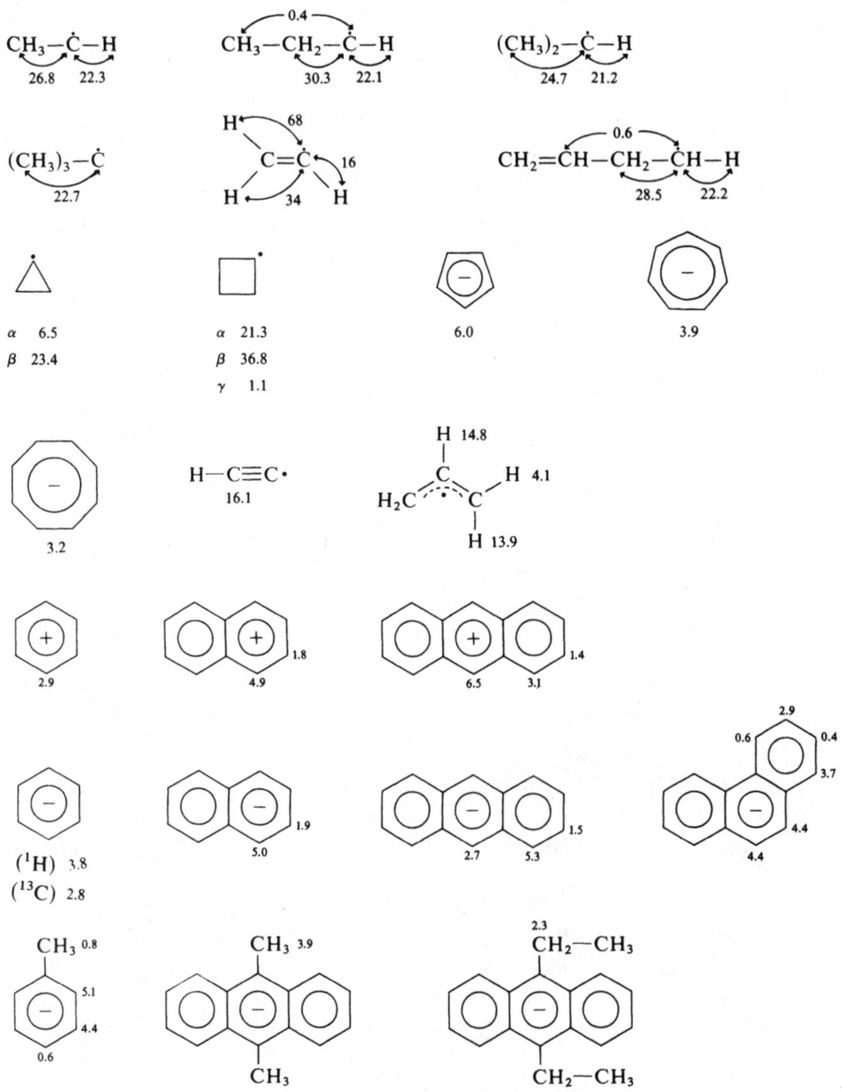

CH₃—Ċ—H
26.8   22.3

CH₃—CH₂—Ċ—H   (0.4)
30.3   22.1

(CH₃)₂—Ċ—H
24.7   21.2

(CH₃)₃—Ċ
22.7

C=C
H 68
16
H 34 H

CH₂=CH—CH₂—ĊH—H   (0.6)
28.5   22.2

△
α   6.5
β 23.4

▢˙
α 21.3
β 36.8
γ   1.1

(pentagon) 6.0

(heptagon) 3.9

(octagon) 3.2

H—C≡C˙
16.1

H 14.8
C
H₂C   C   H 4.1
H 13.9

(benzene +) 2.9

(naphthalene +) 4.9 1.8

(anthracene +) 6.5 3.1 1.4

(benzene −)
(^{1}H) 3.8
(^{13}C) 2.8

(naphthalene −) 5.0 1.9

(anthracene −) 2.7 5.3 1.5

(phenanthrene −) 2.9 0.6 0.4 3.7 4.4 4.4

CH₃ 0.8
5.1
4.4
0.6

CH₃ 3.9
CH₃

2.3
CH₂—CH₃
CH₂—CH₃

TABLE 6.65 Spin–Spin Coupling (Hyperfine Splitting Constants) (*continued*)

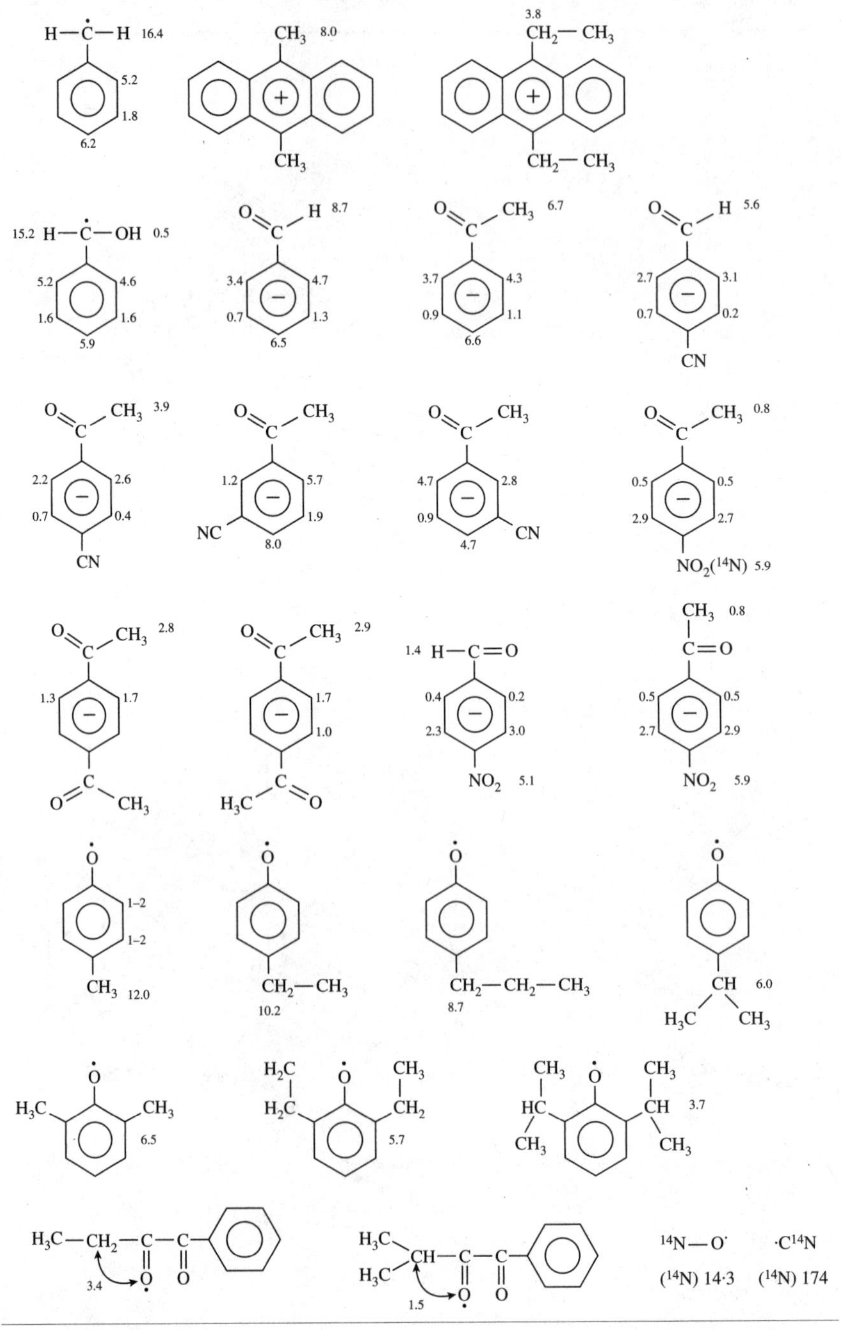

H—Ċ—H 16.4 5.2 1.8 6.2

CH₃ 8.0 (+) CH₃

3.8 CH₂—CH₃ (+) CH₂—CH₃

15.2 H—Ċ—OH 0.5 5.2 4.6 1.6 1.6 5.9

O=C—H 8.7 3.4 4.7 0.7 1.3 6.5

O=C—CH₃ 6.7 3.7 4.3 0.9 1.1 6.6

O=C—H 5.6 2.7 3.1 0.7 0.2 CN

O=C—CH₃ 3.9 2.2 2.6 0.7 0.4 CN

O=C—CH₃ 1.2 5.7 1.9 NC 8.0

O=C—CH₃ 4.7 2.8 0.9 CN 4.7

O=C—CH₃ 0.8 0.5 0.5 2.9 2.7 NO₂(¹⁴N) 5.9

O=C—CH₃ 2.8 1.3 1.7 O=C—CH₃

O=C—CH₃ 2.9 1.7 1.0 H₃C—C=O

1.4 H—C=O 0.4 0.2 2.3 3.0 NO₂ 5.1

CH₃ 0.8 C=O 0.5 0.5 2.7 2.9 NO₂ 5.9

Ȯ 1–2 1–2 CH₃ 12.0

Ȯ CH₂—CH₃ 10.2

Ȯ CH₂—CH₂—CH₃ 8.7

Ȯ CH 6.0 H₃C CH₃

H₃C Ȯ CH₃ 6.5

H₂C Ȯ CH₃ H₂C CH₂ 5.7

CH₃ Ȯ CH₃ HC CH 3.7 CH₃ CH₃

H₃C—CH₂—C—C— 3.4 Ö O

H₃C CH—C—C— H₃C 1.5 Ö O

¹⁴N—Ȯ (¹⁴N) 14·3

·C¹⁴N (¹⁴N) 174

TABLE 6.65 Spin–Spin Coupling (Hyperfine Splitting Constants) (*continued*)

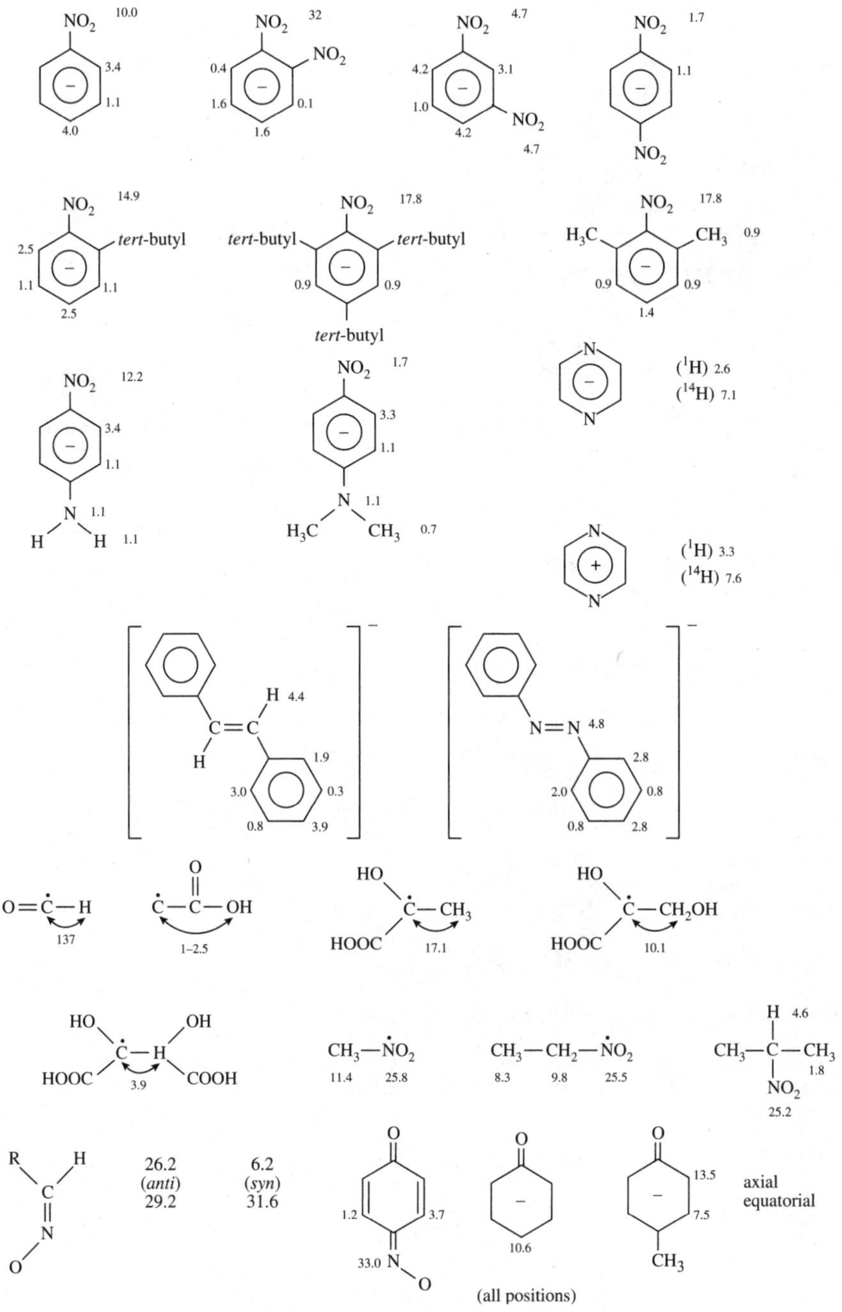

TABLE 6.65 Spin–Spin Coupling (Hyperfine Splitting Constants) (*continued*)

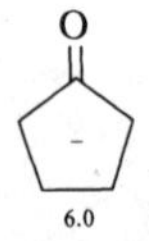

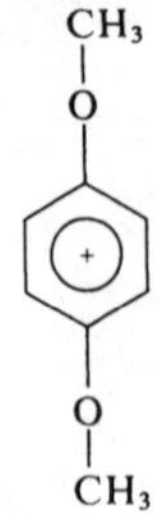

		3.5	3.3
		trans (E)	*cis* (Z)
		3.0	2.7
		1.5	1.8
(all positions)		Ring H	

6.0

IONIZATION POTENTIALS

The ionization potential is the energy required to remove an electron from an element or compound. Table 6.66A presents the ionization potentials for molecular species. The values are given in electron volts (eV). The values in parentheses are uncertainties in the final figure(s). Smaller numbers indicate lower energies or greater ease of electron removal. Table 6.66B is arranged alphabetically by element. Within an element, the compounds are arranged by increasing molecular weight.

TABLE 6.66A Ionization Potentials of Molecular Species

$1\,eV = 23.061\,kcal\cdot mol^{-1}$
Values in parentheses are uncertainties in the final figure(s).

Species	Ionization potential, eV	Species	Ionization potential, eV
Diborane(6)	12.0	2-Methyl-1-propene	9.23(2)
Pentaborane(9)	10.5	Cyclobutane	10.58
Hexaborane(10)	9.3(1)	Butane	10.63(3)
Trimethylborane	8.8(2)	Isobutane	10.57
Triethylborane	9.0(2)	Cyclopentadiene	8.97
Methane	12.6	1,2-Pentadiene	9.42
CD_4	12.888	1,3-Pentadiene	8.68
Acetylene	11.4	1,4-Pentadiene	9.58
C_2D_2	11.416(6)	2,3-Pentadiene	8.68
Ethylene	10.5	2-Methyl-1,4-butadiene	8.845(5)
Ethane	11.5	Cyclopentene	9.01(1)
Propyne	10.36	1-Pentene	9.50(2)
Allene	10.16(2)	*cis*-2-Pentene	9.11
Cyclopropene	9.95	*trans*-2-Pentene	9.06
Cyclopropane	10.09(2)	2-Methyl-1-butene	9.12(2)
Propane	11.1	3-Methyl-1-butane	9.51(3)
1,2-Butadiene	9.57(2)	3-Methyl-2-butene	8.69(2)
1,3-Butadiene	9.07	Cyclopentane	10.53(5)
1-Butyne	10.18(1)	Pentane	10.35
2-Butyne	9.9(1)	Isopentane	10.32
1-Butene	9.6	Neopentane	10.35

TABLE 6.66A Ionization Potentials of Molecular Species (*continued*)

Species	Ionization potential, eV	Species	Ionization potential, eV
cis-2-Butene	9.13	Benzene	9.24
trans-2-Butene	9.13	Hexa-1,3-diene-5-yne	9.50
1,3-Hexadiyne	9.25	2,2,4-Trimethylpentane	9.86
1,4-Hexadiyne	9.75	2,2,3,3-Tetramethylbutane	9.79
1,5-Hexadiyne	10.35	Indene	8.81
2,4-Hexadiyne	9.75	β-Methylstyrene	8.35(1)
1-Methylcyclopentadiene	8.43(5)	Propylbenzene	8.72(1)
2-Methylcyclopentadiene	8.46(5)	Isopropylbenzene	8.69(1)
Cyclohexene	8.72	1,2,3-Trimethylbenzene	8.48
1-Hexene	9.45(2)	1,2,4-Trimethylbenzene	8.27
2,3-Dimethyl-2-butene	8.30	1,3,5-Trimethylbenzene	8.4
Cyclohexane	9.8	Naphthalene	8.12
Hexane	10.18	Azulene	7.42
2-Methylpentane	10.12	Butylbenzene	8.69(1)
3-Ethylbutane	10.08	*sec*-Butylbenzene	8.68(1)
2,2-Dimethylbutane	10.06	*tert*-Butylbenzene	8.68(1)
2,3-Dimethylbutane	10.02	1,2,3,5-Tetramethylbenzene	8.47(5)
Toluene	8.82(1)	1,2,4,5-Tetramethylbenzene	8.03
Cycloheptatriene	8.5	*cis*-Decalin	9.61(2)
Bicyclo[2.2.1]heptane	8.67	*trans*-Decalin	9.61(2)
Bicyclo[3.2.0]heptane	9.37	1-Methylnaphthalene	7.96(1)
1,2-Dimethylcyclopentadiene	8.1(1)	2-Methylnaphthalene	7.955(10)
5,5-Dimethylcyclopentadiene	8.22(5)	Pentamethylbenzene	7.92(2)
1,3-Cycloheptadiene	8.55	Hexamethylcyclopentadiene	7.74(5)
Norbornene	8.95(15)	Biphenyl	8.27(1)
4-Methylcyclohexene	8.91(1)	Hexamethylbenzene	7.85(2)
Methylcyclohexane	9.85(3)	Fluorene	8.63
Heptane	9.90(5)	Diphenylacetylene	8.85(5)
Phenylacetylene	8.815(5)	Anthracene	7.55
Styrene	8.47(2)	Phenanthrene	8.1
Cyclooctatetraene	8.0	1,2-Benzanthracene	8.01
Cubane	8.74(15)	1-Phenyldodecane	9.05(10)
Ethylbenzene	8.76(1)	3-Phenyldodecane	8.95(10)
o-Xylene	8.56	7-Phenyltridecane	8.91(10)
m-Xylene	8.58	1-Phenylicosane	9.34(10)
p-Xylene	8.44	2-Phenylicosane	9.22(10)
7-Methylcycloheptatriene	8.39(10)	3-Phenylicosane	8.95(10)
1-Methylspiroheptadiene	8.02(10)	4-Phenylicosane	9.01(10)
6-Methylspiroheptadiene	8.4(1)	5-Phenylicosane	9.04(10)
1,2,3-Trimethylcyclo-pentadiene	7.96(5)	7-Phenylicosane	8.97(10)
1,5,5-Trimethylcyclo-pentadiene	8.0(1)	9-Phenylicosane	9.06(10)
		N_2	15.576
4-Vinylcyclohexene	8.93(2)	NH_3	10.2
cis-1,2-Dimethylcyclohexane	10.08(2)	N_2H_2	9.85(10)
trans-1,2-Dimethylcyclo-hexane	10.08(3)	N_2H_4	8.74(6)
		HCN	13.8
		C_2N_2	13.6

TABLE 6.66A   Ionization Potentials of Molecular Species (*continued*)

Species	Ionization potential, eV	Species	Ionization potential, eV
Methylamine	8.97	N,N-Dimethyl-p-toluidine	7.33
Acetonitrile	12.2	Tripropylamine	7.23
Ethyleneimine	9.94(15)	N-Butylaniline	7.53
Ethylamine	8.86(2)	N,N-Diethylaniline	6.99
Dimethylamine	8.24(2)	N,N-Dimethyl-4-ethylaniline	7.38
Acrylonitrile	10.91(1)	N,N-2,4-Tetramethylaniline	7.17
Propionitrile	11.84(2)	N,N-2,6-Tetramethylaniline	7.22
Propylamine	8.78(2)	N,N-3,5-Tetramethylaniline	7.25
Isopropylamine	8.72(3)	N,N-Diethyl-4-toluidine	6.93
Trimethylamine	7.82(2)	N,N-Dimethyl-4-	
3-Butenonitrile	10.39(1)	isopropylaniline	7.41
Pyrrole	8.20(1)	Diphenylamine	7.25(3)
Butyronitrile	11.67(5)	N,N-Dipropylaniline	6.96
Pyrrolidine	8.41	N,N-Dimethyl-4-*tert*-	
Butylamine	8.71(3)	butylaniline	7.43
sec-Butylamine	8.70	N,N-Dibutylaniline	6.95
Isobutylamine	8.70	Triphenylamine	6.86(3)
tert-Butylamine	8.64	Diazirine	10.18(5)
Diethylamine	8.01(1)	Diazomethane	8.999(1)
Pyridine	9.3	Methylhydrazine	8.00(6)
Aniline	7.7	1,1-Dimethylhydrazine	7.67(5)
2-Methylpyridine	9.02(3)	1,2-Dimethylhydrazine	7.75(10)
3-Methylpyridine	9.04(3)	o-Diazine	9.9
4-Methylpyridine	9.04(3)	m-Diazine	9.9
Cyclohexylamine	8.86	p-Diazine	9.8
Dipropylamine	7.84(2)	1,1-Diethylhydrazine	7.59(5)
Diisopropylamine	7.73(3)	1-Butyl-1-methylhydrazine	7.62(5)
Triethylamine	7.50(2)	p-Bis(dimethylamino)benzene	6.9
Benzonitrile	9.705(10)	Methyl azide	9.5(1)
N-Methylaniline	7.32	O_2	12.063(1)
m-Toluidine	7.50(2)	O_3	12.3(1)
2,3-Dimethylpyridine	8.85(2)	Water (and D_2O)	12.6
2,4-Dimethylpyridine	8.85(3)	H_2O_2	11.0
2,6-Dimethylpyridine	8.85(2)	CO	14.013(4)
Phenylacetonitrile	9.4(5)	CO_2	13.769(30)
3-Methylbenzonitrile	9.66(5)	NO	9.25
4-Methylbenzonitrile	9.76	N_2O	12.894
N-Ethylcyclohexylamine	7.56	NO_2	9.79
N,N-Dimethylcyclo-		Formaldehyde	10.88
hexylamine	7.12	Methanol	10.84
Dibutylamine	7.69(3)	Acetaldehyde	10.2
N-Propylaniline	7.54	Ethylene oxide	10.6
N-Ethyl-N-methylaniline	7.37	Ethanol	10.49
N,N-Dimethyl-o-toluidine	7.37	Dimethyl ether	9.98
N,N-Dimethyl-m-toluidine	7.35	Propenal	10.10(1)
Propionaldehyde	9.98	Diphenyl ether	8.82(5)

TABLE 6.66A Ionization Potentials of Molecular Species (*continued*)

Species	Ionization potential, eV	Species	Ionization potential, eV
Acetone	9.69	Benzophenone	9.4
Allyl alcohol	9.67(5)	4-Methylbenzophenone	9.13(5)
Methyl vinyl ether	8.93(2)	Formic acid	11.05(1)
Propylene oxide	10.22(2)	Acetic acid	10.69(3)
Trimethylene oxide	9.667(5)	Methyl formate	10.815(5)
1-Propanol	10.1	Propionic acid	10.24(1)
2-Propanol	10.15	Ethyl formate	10.61(1)
Furan	8.89	Methyl acetate	10.27(2)
2-Butenal	9.73(1)	Dimethoxymethane	10.00(5)
Butyraldehyde	9.86(2)	Vinyl acetate	9.19(5)
2-Methylpropionaldehyde	9.74(3)	2,3-Butanedione	9.24(3)
2-Butanone	9.5	Butanoic acid	10.16(5)
Tetrahydrofuran	9.42	Isobutyric acid	10.02(5)
1-Butanol	10.04	Propyl formate	10.54(1)
Diethyl ether	9.6	Ethyl acetate	10.11(2)
Cyclopentanone	9.26(1)	Methyl propionate	10.15(3)
Dihydropyran	8.34(1)	1,4-Dioxane	9.13(3)
Pentanal	9.82(5)	1,1-Dimethoxyethane	9.65(3)
3-Methylbutyraldehyde	9.71(5)	2-Furaldehyde	9.21(1)
2-Pentanone	9.37(2)	2,4-Pentanedione	8.87(3)
3-Methyl-2-butanone	9.30(2)	Butyl formate	10.50(2)
3-Pentanone	9.32(1)	Isobutyl formate	10.46(2)
Cyclopentanone	9.25(1)	Propyl acetate	10.04(3)
Phenol	8.51	Isopropyl acetate	9.99(1)
4-Methyl-3-penten-2-one	9.08(3)	Ethyl propionate	10.00(2)
Cyclohexanone	9.14(1)	Methyl butyrate	10.07(3)
2-Hexanone	9.35	Methyl isobutyrate	9.98(2)
4-Methyl-2-pentanone	9.30	Diethoxymethane	9.70(5)
3,3-Dimethyl-2-butanone	9.17(3)	1,4-Quinone	9.67(2)
Dipropyl ether	9.27(5)	Butyl acetate	9.56(3)
Diisopropyl ether	9.20(5)	Isobutyl acetate	9.97
Benzaldehyde	9.52	sec-Butyl acetate	9.91(3)
Tropone	9.68(2)	Benzoic acid	9.73(9)
Benzyl alcohol	9.14(5)	p-Hydroxybenzaldehyde	9.32(2)
Methoxybenzene	8.21(2)	α-Hydroxyacetophenone	9.33(5)
m-Cresol	8.52(5)	Methyl benzoate	9.35(6)
2-Heptanone	9.33(3)	p-Methoxybenzaldehyde	8.60(3)
Acetophenone	9.27(3)	m-Hydroxyacetophenone	8.67(5)
4-Methylbenzaldehyde	9.33(5)	p-Hydroxyacetophenone	8.70(3)
Benzyl methyl ether	8.85(3)	α-Methoxyacetophenone	8.60(5)
Ethyl phenyl ether	8.13(2)	m-Methoxyacetophenone	8.53(5)
3-Methylanisole	8.31(5)	p-Methoxyacetophenone	8.62(5)
Propiophenone	9.27(5)	Methyl p-methylbenzoate	8.94(4)
3-Methylacetophenone	9.15(5)	p-Hydroxybenzophenone	8.59(5)
Phenyl benzoate	8.98(5)	N_2F_4	12.04(10)
Benzil	8.78(5)	OF_2	13.6
Methyl methoxyacetate	9.56(5)	XeF_2	11.5(2)
Methyl p-methoxybenzoate	8.43(4)	Fluoromethane	12.85(1)

TABLE 6.66A Ionization Potentials of Molecular Species (*continued*)

Species	Ionization potential, eV	Species	Ionization potential, eV
Diphenyl carbonate	9.01(5)	Fluoroethylene	10.37
Acetamide	9.77(2)	Fluorobenzene	9.2
N,N-Dimethylformamide	9.12(2)	1,2-Difluorobenzene	9.31
N-Methylacetamide	8.90(2)	1,4-Difluorobenzene	9.15
NN-Dimethylacetamide	8.81(3)	Trifluoroethylene	10.14
N,N-Diethylformamide	8.89(2)	3,3,3-Trifluoro-1-propene	10.9
2-Pyridinecarboxaldehyde	9.75(5)	*o*-Fluorophenol	8.66(1)
4-Pyridinecarboxaldehyde	10.12(5)	PH$_3$	9.98
N,N-Diethylacetamide	8.60(2)	PF$_3$	9.71
Phenyl isocyanate	8.77(2)	Methylphosphine	9.72(15)
Benzamide	9.4(2)	Ethylphosphine	9.47(50)
p-Aminobenzaldehyde	8.25(2)	Trimethylphosphine	8.6(2)
p-Methoxyaniline	7.82	Triphenylphosphine	7.36(5)
Acetanilide	8.39(10)	S$_6$	9.7
m-Aminoacetophenone	8.09(5)	S$_7$	9.2(3)
p-Aminoacetophenone	8.17(2)	Hydrogen sulfide	10.4
α-Cyanoacetophenone	9.56(5)	Carbon disulfide	10.080
Nitromethane	11.1	Sulfur dioxide	12.34(2)
Nitroethane	10.88(5)	Methanethiol	9.440(5)
1-Nitropropane	10.81(3)	Ethylene sulfide	8.87(15)
2-Nitropropane	10.71(5)	Ethanethiol	9.285(5)
Nitrobenzene	9.92	Dimethyl sulfide	8.685(5)
m-Nitrotoluene	9.65(5)	Propylene sulfide	8.6(2)
p-Nitrotoluene	9.87	1-Propanethiol	9.195
o-Nitroaniline	8.66	Ethyl methyl sulfide	8.55(1)
m-Nitroaniline	8.7	Thiophene	8.860(5)
p-Nitroaniline	8.85	Methyl 1-propenyl sulfide	8.7(2)
Ethyl nitrate	11.22	1-Butanethiol	9.14(2)
Propyl nitrate	11.07(2)	Diethyl sulfide	8.430(5)
p-Nitrophenol	9.52	Methyl propyl sulfide	8.80(15)
p-Nitrobenzaldehyde	10.27(1)	Isopropyl methyl sulfide	8.7(2)
m-Nitroacetophenone	9.89(5)	Thiophenol	8.32(1)
p-Nitroacetophenone	10.07(2)	2-Ethylthiophene	8.8(2)
Methyl *p*-nitrobenzoate	10.20(3)	Dipropyl sulfide	8.5
F$_2$	15.7	Methyl phenyl sulfide	8.9
HF	15.77(2)	2-Propylthiophene	8.6(2)
BF$_3$	15.5	2-Butylthiophene	8.5(2)
C$_2$F$_4$	10.12	Dimethyl disulfide	8.46(3)
Hexafluorobenzene	9.97	Diethyl disulfide	8.27(3)
trans-N$_2$F$_2$	13.1(1)	COS	11.17(1)
NF$_3$	13.2(2)	SO$_2$F$_2$	13.3(1)
Methyl isothiocyanate	9.25(3)	*p*-Dichlorobenzene	8.95
Methyl thiocyanate	10.065(10)	Chloroform	11.42(3)
Ethyl isothiocyanate	9.14(3)	Trichloroethylene	9.45
Ethyl thiocyanate	9.89(1)	1,1,2,2-Tetrachloroethane	11.10(5)
Phenyl isothiocyanate	8.520(5)	CNCl	12.49(4)
Tolyl thiocyanate	9.06(5)	CF$_3$Cl	12.91(3)
Thiourea	8.50(5)	Chlorotrifluoroethylene	10.4(2)

TABLE 6.66A Ionization Potentials of Molecular Species (*continued*)

Species	Ionization potential, eV	Species	Ionization potential, eV
1-Methylthiourea	8.29(5)	Chloropentafluorobenzene	10.4(1)
1-Vinylthiourea	8.29(5)	Dichlorodifluoromethane	12.31(5)
1,1-Dimethylthiourea	8.34(5)	$CF_3CCl=CClCF_3$	10.36(1)
1,3-Dimethylthiourea	8.17(5)	Trichlorofluoromethane	11.77(2)
1,1,3-Trimethylthiourea	7.93(5)	CF_3CCl_3	11.78(3)
Tetramethylthiourea	7.95(5)	$CFCl_2CF_2Cl$	11.99(2)
CH_3COSH	10.00(2)	ClO_3F	13.6(2)
Cl_2	11.48(1)	1-Bromo-1-propene	9.30(5)
HCl	12.74	1-Bromopropane	10.18(1)
CCl_4	11.47(1)	2-Bromopropane	10.075(10)
Tetrachloroethylene	9.32(1)	1-Bromobutane	10.125(10)
PCl_3	9.91	2-Bromobutane	9.98(1)
Chloromethane	11.3	1-Bromo-2-methylpropane	10.09(2)
Chloroethane	10.97	2-Bromo-2-methylpropane	9.89(3)
Chloroethylene	9.996	1-Bromopentane	10.10(2)
1-Chloro-1-propyne	9.9(1)	Bromobenzene	8.98(2)
1-Chloropropane	10.82(3)	*o*-Bromotoluene	8.78(1)
2-Chloropropane	10.78(2)	*m*-Bromotoluene	8.81(2)
1-Chlorobutane	10.67(3)	*p*-Bromotoluene	8.67(2)
2-Chlorobutane	10.65(3)	Dibromomethane	10.49(2)
1-Chloro-2-methylpropane	10.66(3)	*cis*-1,2-Dibromoethylene	9.45
2-Chloro-2-methylpropane	10.61(3)	*trans*-1,2-Dibromoethylene	9.46
Chlorobenzene	9.07	1,1-Dibromoethane	10.19(3)
α-Chlorotoluene	9.19(5)	1,3-Dibromopropane	10.07(2)
o-Chlorotoluene	8.83(2)	Bromoform	10.51(2)
m-Chlorotoluene	8.83(2)	Tribromoethylene	9.27
p-Chlorotoluene	8.69(2)	Cyanogen bromide	11.95(8)
endo-5-Chloro-2-norbornene	9.10(15)	Bromotrifluoromethane	11.89
exo-5-Chloro-2-norbornene	9.15(15)	2-Bromopyridine	9.65(5)
Dichloromethane	11.35(2)	4-Bromopyridine	9.94(5)
cis-1,2-Dichloroethylene	9.65	Acetyl bromide	10.55(5)
trans-1,2-Dichloroethylene	9.64	Methyl bromoacetate	10.37(5)
1,2-Dichloroethane	11.12(5)	CF_2BrCH_2Br	10.83(1)
2,3-Dichloro-1-propene	9.82(3)	Bromochloromethane	10.77(1)
1,2-Dichloropropane	10.87(5)	1-Bromo-2-chloroethane	10.63(3)
1,3-Dichloropropane	10.85(5)	Bromodichloromethane	10.88(5)
o-Dichlorobenzene	9.06	Bromotrimethylsilane	10.24(2)
m-Dichlorobenzene	9.12(1)	I_2	9.28(2)
HI	10.39	$CF_3CF_2CF_2CH_2Cl$	11.84(2)
ICl	10.31(2)	Dichlorofluoromethane	12.39(20)
IBr	9.98(3)	Chlorotrimethylsilane	10.58(4)
Iodomethane	9.54	Trichloromethylsilane	11.36(3)
Iodoethane	9.33	Trichlorovinylsilane	10.79(2)
1-Iodopropane	9.26(1)	Trichloroethylsilane	10.74(4)
2-Iodopropane	9.17(2)	Trichloroisopropylsilane	10.28(10)
1-Iodobutane	9.21(1)	$C_2H_5V(CO)_4$	8.2(3)
2-Iodobutane	9.09(2)	$Cr(CO)_6$	8.03(3)
1-Iodo-2-methylpropane	9.18(2)	$C_2H_5Mn(CO)_3$	8.3(4)

TABLE 6.66A   Ionization Potentials of Molecular Species

Species	Ionization potential, eV	Species	Ionization potential, eV
2-Iodo-2-methylpropane	9.02(2)	$Fe(CO)_5$	7.95(3)
1-Iodopentane	9.19(1)	$Ni(CO)_4$	8.28(3)
Iodobenzene	8.73	$Mo(CO)_6$	8.12(3)
o-Iodotoluene	8.62(1)	$W(CO)_6$	8.18(3)
m-Iodotoluene	8.61(3)	As_4	9.07(7)
p-Iodotoluene	8.50(1)	Arsine	10.03
RuO_4	12.33(23)	$AsCl_3$	11.7(1)
2-Chloropyridine	9.91(5)	Trimethylarsine	8.3(1)
4-Chloropyridine	10.15(5)	Triphenylarsine	7.34(7)
Acetyl chloride	11.02(5)	Br_2	10.54(3)
1-Chloro-2-propanone	9.99	HBr	11.62(3)
2-Chlorophenol	9.28	BrCl	11.1(2)
4-Chlorophenol	9.07	Bromomethane	10.53
Benzoyl chloride	9.70(1)	Bromoethylene	9.80
4-Chlorobenzaldehyde	9.61(1)	Bromoethane	10.29
α-Chloroacetophenone	9.5	1-Bromo-1-propyne	10.1(1)
p-Chloroacetophenone	9.47(5)	OsO_4	12.97(12)
Methyl chloroacetate	10.53(5)	Dimethylmercury	9.0
4-Methoxybenzoyl chloride	8.87(5)	Diethylmercury	8.5(1)
4-Chlorobenzoyl chloride	9.58(3)	Diisopropylmercury	7.6(1)
cis-Chlorofluoroethylene	9.86	CH_3HgCl	11.5(2)
trans-Chlorofluoroethylene	9.87	Triphenylbismuth	7.3(1)
o-Chlorofluorobenzene	9.155(10)	Stibine	9.58
m-Chlorofluorobenzene	9.21(1)	Triphenylstibine	7.3(1)
p-Chlorofluorobenzene	9.43(2)	Tetramethylstannane	8.25(15)
Chlorodifluoromethane	12.45(5)	Tetramethylplumbane	8.0(4)
1-Chloro-1,1-difluoroethane	11.98(1)	Tetramethylgermane	9.2(2)

TABLE 6.66B   Alphabetical Listing of Ionization Potentials of Molecular Species

Species	IP (eV)	Species	IP (eV)
Acetylene	11.4	Butane	10.63(3)
Acetylene-d_2; C_2D_2	11.416	1-Butene	9.6
Allene	10.16	cis-2-Butene	9.13
Ammonia (NH_3)	10.2	trans-2-Butene	9.13
Anthracene	7.55	1-Butyne	10.18(1)
Azine (N_2H_2)	9.85(10)	2-Butyne	9.9(1)
Azulene	7.42	Butylbenzene	8.69(1)
1,2-Benzanthracene	8.01	sec-Butylbenzene	8.68(1)
Benzene	9.24	tert-Butylbenzene	8.68(1)
Bicyclo[2.2.1]heptane	8.67	Carbon dioxide, CO_2	12.888
Bicyclo[3.2.0]heptane	9.37	Cubane	8.74(15)
Biphenyl	8.27(1)	Cyclobutane	10.58
1,2-Butadiene	9.57(2)	1,3-Cycloheptadiene	8.55
1,3-Butadiene	9.07	Cycloheptatriene	8.5

TABLE 6.66B Alphabetical Listing of Ionization Potentials of Molecular Species (*continued*)

Species	IP (eV)	Species	IP (eV)
Cyclohexane	9.8	4-Methylcyclohexene	8.91(1)
Cyclohexene	8.72	1-Methylcyclopentadiene	8.43(5)
Cyclooctatetraene	8	2-Methylcyclopentadiene	8.46(5)
Cyclopentadiene	8.97	1-Methylnaphthalene	7.96(1)
Cyclopentane	10.53(5)	2-Methylnaphthalene	7.955(10)
Cyclopentene	9.01(1)	2-Methylpentane	10.12
Cyclopropane	10.09	2-Methyl-l-propene	9.23(2)
Cyclopropene	9.95	1-Methylspiroheptadiene	8.02(10)
cis-Decalin	9.61(2)	6-Methylspiroheptadiene	8.4(1)
trans-Decalin	9.61(2)	β-Methylstyrene	8.35(1)
Diazomethane (C_2N_2)	13.6	Naphthalene	8.12
Diborane	12	Neopentane	10.35
2,2-Dimethylbutane	10.06	Nitrogen (N_2)	15.576
2,3-Dimethylbutane	10.02	Norbornene	8.95(15)
2,3-Dimethyl-2-butene	8.3	Pentaborane(9)	10.5
cis-1,2-Dimethylcyclohexane	10.08(2)	1,2-Pentadiene	9.42
trans-1,2-Dimethylcyclohexane	10.08(3)	1,3-Pentadiene	8.68
1,2-Dimethylcyclopentadiene	8.1(1)	1,4-Pentadiene	9.58
5,5-Dimethylcyclopentadiene	8.22(5)	2,3-Pentadiene	8.68
Diphenylacetylene	8.85(5)	Pentamethylbenzene	7.92(2)
Ethane	11.5	Pentane	10.35
Ethylbenzene	8.76(1)	1-Pentene	9.50(2)
3-Ethylbutane	10.08	*cis*-2-Pentene	9.11
Ethylene	10.5	*trans*-2-Pentene	9.06
Fluorene	8.63	Phenanthrene	8.1
Heptane	9.90(5)	Phenylacetylene	8.815(5)
Hexaborane	9.3	1-Phenyldodecane	9.05(10)
Hexa-1,3-diene-5-yne	9.5	3-Phenyldodecane	8.95(10)
1,3-Hexadiyne	9.25	1-Phenylicosane	9.34(10)
1,4-Hexadiyne	9.75	2-Phenylicosane	9.22(10)
1,5-Hexadiyne	10.35	3-Phenylicosane	8.95(10)
2,4-Hexadiyne	9.75	4-Phenylicosane	9.01(10)
Hexamethylbenzene	7.85(2)	5-Phenylicosane	9.04(10)
Hexamethylcyclopentadiene	7.74(5)	7-Phenylicosane	8.97(10)
Hexane	10.18	9-Phenylicosane	9.06(10)
1-Hexene	9.45(2)	7-Phenyltridecane	8.9100)
Hydrazine (N_2H_4)	8.74(6)	Propane	11.1
Hydrocyanic acid (HCN)	13.8	Propylbenzene	8.72(1)
Indene	8.81	Propyne	10.36
Isobutane	10.57	Styrene	8.47(2)
Isopentane	10.32	1,2,3,5-Tetramethylbenzene	8.47(5)
Isopropylbenzene	8.69(1)	1,2,4,5-Tetramethylbenzene	8.03
Methane	12.6	2,2,3,3-Tetramethylbutane	9.79
2-Methyl-1,4-butadiene	8.845(5)	Toluene	8.82(1)
3-Methyl-1-butane	9.51(3)	Triethylborane	9
2-Methyl-1-butene	9.12(2)	1,2,3-Trimethylbenzene	8.48
3-Methyl-2-butene	8.69(2)	1,2,4-Trimethylbenzene	8.27
7-Methylcycloheptatriene	8.39(10)	1,3,5-Trimethylbenzene	8.4
Methylcyclohexane	9.85(3)	Trimethylborane	8.8

TABLE 6.66B Alphabetical Listing of Ionization Potentials of Molecular Species (*continued*)

Species	IP (eV)	Species	IP (eV)
1,2,3-Trimethylcyclopentadiene	7.96(5)	*m*-Xylene	8.58
1,5,5-Trimethylcyclopentadiene	8.0(1)	*o*-Xylene	8.56
2,2,4-Trimethylpentane	9.86	*p*-Xylene	8.44
4-Vinylcyclohexene	8.93(2)		

TABLE 6.67 Ionization potentials of radical species

$$1 \, \text{eV} = 23.061 \, \text{kcal·mol}^{-1}$$

Values in parentheses are uncertainties in the final figure(s).

Species	Ionization potential, eV	Species	Ionization potential, eV
BH	9.77(5)	*tert*-Pentyl	7.1(1)
BH_2	11.4(2)	Neopentyl	8.3(1)
BF	11.3	Benzyne	9.6
C_2	12.0(6)	Cyclohexyl	7.7
C_3	12.6	Benzyl	7.76(8)
CH	11.1(2)	Cycloheptatrienyl	6.24(1)
CH_2	10.396(3)	1-Methylnaphthyl	7.35
CH_3	9.83	2-Methylnaphthyl	7.56(5)
CD_3	9.832(2)	$(CH_3)_2CCN$	9.15(10)
C_2H_3	9.4	*m*-Nitrobenzyl	8.56(10)
C_2H_5	8.4	OH	13.17(10)
$HC{\equiv}CCH_2$	8.25	HO_2	11.53(2)
Allyl	8.15	CHO	9.8
Cyclopropyl	8.05	CH_3CO	10.3
C_3H_6	9.73	C_6H_5O	8.84
Propyl	8.1	CF_2	11.8
Isopropyl	7.5	NF_2	11.9
C_4H_2	10.2(1)	CH_2F	9.35
C_4H_4	9.87	CHF_2	9.45
Cyclobutyl	7.88(5)	HS	10.5(1)
$CH_3CH{=}CHCH_2$	7.71(5)	CH_3S	8.06(10)
$CH_2{=}C(CH_3)CH_2$	8.03(5)	C_6H_5S	8.63(10)
Butyl	8.64(5)	CCl_3	8.78(5)
sec-Butyl	7.93(5)	CH_2Cl	9.32
Isobutyl	8.35(5)	$CHCl_2$	9.30
tert-Butyl	7.42(7)	NH_2	11.3
Cyclopentyl	7.79(2)		

X-RAY DIFFRACTION

The X-ray diffraction method utilizes a monochromatic beam of X-rays to which a solid material is exposed. The beam of radiation interacts with the solid, and is both reflected and diffracted. The reflection pattern is recorded by a detector system sensitive to the X-radiation. Until recently, this involved an intricate mechanical device whose complex motion permitted X-rays to be recorded over a range of positions, as the detector position changed. Automated instruments using this technology are often referred to as "4-circle diffractometers," a term that refers to this complex detector motion. Newer X-ray instruments use charge coupled devices (CCDs) to simultaneously detect X-ray diffraction position and intensity over a much broader area. This enhanced detection technology reduces, often dramatically, the time required for data acquisition.

The Bragg equation describes the relationship between the impinging X-radiation, the diffraction angle, and the separation between lattice planes in the crystal under study. The Bragg equation is generally written as

$$\theta = \sin^{-1}(\lambda/2d)$$

where θ is the angle of the diffracted beam (usually called a reflection), λ is the wavelength of the incident X-ray beam, and d is the inter-planar spacing. From the diffraction pattern, both position and intensity, one can obtain structural information about the crystal under study.

Two types of X-ray studies are commonplace: X-ray powder analysis and crystal structure determination. Even in powders, the regular arrangement of atoms within the solid leads to characteristic diffraction patterns. X-ray powder patterns may therefore be used to characterize solids in much the same way that a UV or IR spectrum will give useful information but not necessarily a definitive structure.

The X-ray powder pattern obtained for the sodium salt of 2-propylpentanoic acid $(CH_3CH_2CH_2)_2CHCOO^- Na^+$ is shown in Figure 6.2. The X-ray powder pattern was detected over a range 2θ, in this case $2-40°$ from the incident beam. This is a typical range although other ranges are used as well. The peak intensities are expressed in counts per second (cps) and may vary from experiment to experiment. However, the ratios of the peak heights are characteristic. Thus, X-ray powder patterns obtained from different samples of the same compound should give very similar, if not identical, patterns. Because the X-ray powder patterns are complex, the identity of two spectra suggests that the compounds producing them are also identical.

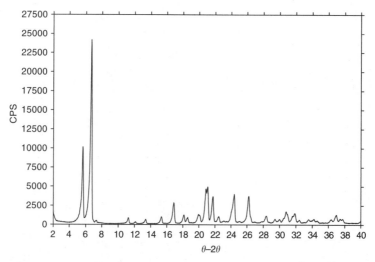

FIGURE 6.2 X-ray powder pattern of 2-propylpentanoic acid, $(CH_3CH_2CH_2)_2CHCOO^- Na^+$.

The determination of a molecular structure from X-ray diffraction data is of critical importance to modern chemical and biological sciences. For small molecules, the structure is normally determined by direct methods. The X-ray diffraction pattern resulting from the interaction of X-rays with the electron clouds of different elements gives a pattern from which the elements present and their connectivity may be deduced. The expected diffraction pattern calculated for an apparent structure is then compared with the observed data to refine the result. Refinement factors (usually expressed as Rw) of 1–3% are common in modern small molecule structure determinations.

The Cambridge Structural Database is a repository for more than 250 000 (as of 2002) small molecule crystal structures. It is accessible at http:// www.ccdc.cam.ac.uk/ by subscription.

The process is more complex for such large molecules as proteins. Typically, a model of the amino acid backbone will be constructed first to obtain a general sense of the overall structure. Amino acid sidechains will then be added and the experimental data are again compared with the calculated diffraction pattern. This process is repeated until the complete structure is obtained. Because the uncertainties are larger in these systems, the resolution of the structure is typically reported in Ångstroms. Structures that have a resolution of 3 Å can and do give important information, especially when the gross structure of a protein was previously unknown. Recent improvements in chemical and biological techniques, computers, and X-ray instrumentation (especially in detectors) have made resolutions in the 1–2 Å range more common.

A database, called the Protein Data Bank or "PDB," is a repository for protein structures. The database may be consulted at no charge and gives access to structures obtained by X-ray methods as well as by NMR and theoretical techniques. The Internet address is http://www.rcsb.org/pdb.

SECTION 7
PHYSICOCHEMICAL RELATIONSHIPS

LINEAR FREE ENERGY RELATIONSHIPS

Organic chemists have studied the influence of substituents on various reactions for the better part of a century. Linear free energy relationships have played an important role in this pursuit by correlating equilibrium and rate processes. One of the earliest examples is now known as the Hammett equation. It emerged from the observation that the acidities of benzoic acids correlated with the rates at which ethyl esters of benzoic acids hydrolyzed. The relationship was expressed as follows in which K represents an equilibrium constant and k is a rate constant. The proportionality constant, m, is the slope of the log–log data plot for the two processes.

$$m \log \frac{K}{K_0} = \log \frac{k}{k_0}$$

When ΔH and ΔS vary linearly for two processes or when ΔH or ΔS is constant, the free energy relationship will be linear. The common form of the relationship is either

$$\log \frac{K}{K_0} = \sigma \rho \quad \text{or} \quad \log \frac{k}{k_0} = \sigma \rho$$

for equilibrium or rate processes, respectively. The Greek letters rho (ρ) and sigma (σ) symbolize the reaction and substituent constants, respectively. These equations may be used to describe and understand the influence of substituents on a reaction. Separate sigma values are defined by this reaction for *meta* and *para* substituents and provide a measure of the total electronic influence (polar, inductive, and resonance effects) in the absence of conjugation effects. The correlation is not as useful for *ortho*-substituted aromatic compounds because steric or other proximity effects intercede.

Typically in aromatic systems, the inductive effect is transmitted about equally to the *meta* and *para* positions. Consequently, σ_m is an approximate measure of a substituent's inductive effect whereas σ_p gives an approximate measure of a substituent's resonance effect. Consider the dissociation of benzoic acids in water. This process is assigned a reaction constant ρ of 1. The reaction is illustrated using "Sub" to represent a substituent.

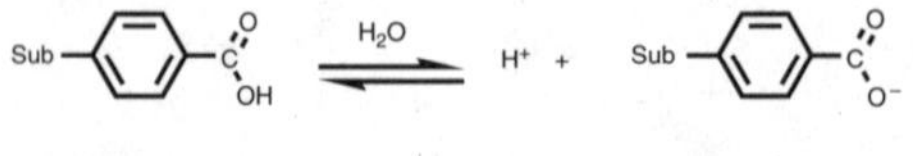

We may compare three *para*-substituted benzoic acids. The reference point is *para*-hydrogen, which has a σ_p constant of 0, by definition. The methoxy group is electron donating and has a σ_p constant of -0.27. Adding electrons to the benzoate anion (structure at right, above) should make the anion less stable. Thus, the ability to dissociate a cation (the acidity) should be diminished. The pK_A is $-\log K_A$ so the higher the pK_A, the lower the acidity (the weaker the acid). In contrast, the nitro group is electron withdrawing. Its σ_p constant is $+0.78$ and its presence in the benzoate anion should be stabilizing. A more stable conjugate base implies a stronger acid and, indeed, the pK_A for 4-nitrobenzoic acid is 3.44. Because these values are logarithmic, there is an order of magnitude difference in the acidities as a result of these substituents.

Comparison of acidities and sigma constants for three benzoic acids

Compound	COOH (OCH₃)	COOH (H)	COOH (NO₂)
Substituent	OCH_3	H	NO_2
σ_p	−0.27	0.00	0.78
pK_A	4.49	4.20	3.44

Values of Hammett sigma constants are listed in Table 7.1. Taft sigma* (σ^*) values may be used similarly with respect to aliphatic and alicyclic systems. Values of σ^* constants are also listed in Table 7.1.

The reaction constant ρ is related to the reaction process rather than to the substituents present. A somewhat oversimplified way of considering ρ is to say that it indicates the demand the process makes on the substituents. The acidity of a benzoic acid, C_6H_5COOH, derivative is affected directly by substituents in the aromatic ring. Substituents exert somewhat less influence in phenylacetic acids, $C_6H_5CH_2COOH$, because the methylene group between carboxylate and the aromatic ring tends to insulate the latter from the former. This "insulation" is even greater for phenyipropanoic acids, $C_6H_5CH_2CH_2COOH$. The reaction constant ρ for dissociation of benzoic acid in water is set at 1.0. The reaction constants for dissociation of phenylacetic and phenylpropanoic acids are 0.49 and 0.21, respectively, under the same conditions. Values of the reaction parameter for some aromatic and aliphatic systems are given in Tables 7.2 and 7.3.

Since substituent effects in aliphatic systems and in *meta* positions in aromatic systems are essentially inductive in character, σ^* and σ_m values are related by the expression $\sigma_m = 0.217\sigma^* - 0.106$. Substituent effects fall off with increasing distance from the reaction center. The decline is generally a factor of 0.36 for the interposition of a $-CH_2-$ group. This enables σ^* values to be estimated for $R-CH_2-$ groups not otherwise available.

Modified sigma constants have been formulated for situations in which the substituent enters into resonance with the reaction center in an electron-demanding transition state (σ^+) or for an electron-rich transition state (σ^-). Generally, σ^- constants give better correlations in reactions involving phenols, anilines, pyridines, and in nucleophilic substitutions. Values for some modified sigma constants are given in Table 7.4.

TABLE 7.1 Hammett and Taft Substituent Constants

Substituent	Hammett constants		Taft constant σ^*
	σ_m	σ_p	
$-AsO_3H$	−0.09	−0.02	0.06
$-B(OH)_2$	0.01	0.45	
$-Br$	0.39	0.23	2.84
$-CH_2Br$			1.00
m-BrC_6H_4-		0.09	

TABLE 7.1 Hammett and Taft Substituent Constants (*continued*)

Substituent	Hammett constants		Taft constant σ^*
	σ_m	σ_p	
p-BrC$_6$H$_4$—		0.08	
—CH$_3$	−0.07	−0.17	0.0
—CH$_2$CH$_3$	−0.07	−0.15	−0.10
—CH$_2$CH$_2$CH$_3$	−0.05	−0.15	−0.12
—CH(CH$_3$)$_2$ (isopropyl)	−0.07	−0.15	−0.19
—CH$_2$CH$_2$CH$_2$CH$_3$	−0.07	−0.16	−0.13
—CH$_2$CH(CH$_3$)$_2$ (isobutyl)	−0.07	−0.12	−0.13
—CH(CH$_3$)CH$_2$CH$_3$ (*sec*-butyl)		−0.12	−0.19
—C(CH$_3$)$_3$ (*t*-butyl)	−0.10	−0.20	−0.30
—CH$_2$CH$_2$CH$_2$CH$_2$CH$_3$ (*n*-pentyl)			−0.25
—CH$_2$CH$_2$CH(CH$_3$)$_2$ (isopentyl)			−0.17
—CH$_2$C(CH$_3$)$_3$ (*t*-amyl)		−0.23	−0.12
—CH$_2$CH$_2$CH$_2$CH$_2$CH$_2$CH$_3$			−0.37
—CH(CH$_2$)$_2$ (cyclopropyl)	−0.07	−0.21	
—CH(CH$_2$)$_5$ (cyclohexyl)			−0.15
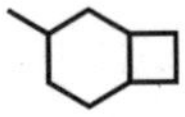		−0.26	
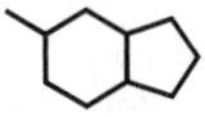		−0.48	
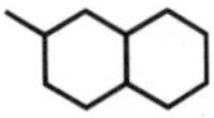	0.06	0.04	
—CH=CH$_2$ (vinyl, ethenyl)	0.02		0.56
—CH=C(CH$_3$)$_2$			0.19
—CH=CHCH$_3$, *trans*			0.36
—CH$_2$CH=CH$_2$			0.0
—CH=CHC$_6$H$_5$	0.14	−0.05	0.41
—C≡CH	0.21	0.23	2.18
—C≡CC$_6$H$_5$	0.14	0.16	1.35
—CH$_2$C≡CH			0.81
—C$_6$H$_5$ (phenyl)	0.06	−0.01	0.60
p—CH$_3$C$_6$H$_4$ (*p*-tolyl)		−0.5	
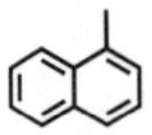 (1-naphthyl)			0.75
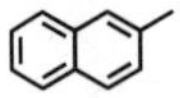 (2-naphthyl)			0.75
—CH$_2$C$_6$H$_5$ (benzyl)		0.46	0.22
—CH$_2$CH$_2$C$_6$H$_5$ (2-phenylethyl)			−0.06
—CH(CH$_3$)C$_6$H$_5$ (α-phenylethyl)			0.37
—CH(C$_6$H$_5$)$_2$ (benzhydryl)			0.41
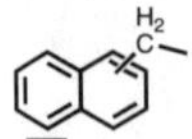			0.44
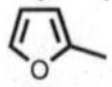 (2-furoyl)			0.25
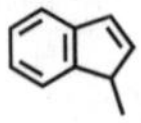 (3-indolyl)			−0.06

TABLE 7.1 Hammett and Taft Substituent Constants (*continued*)

Substituent	Hammett constants		Taft constant σ^*
	σ_m	σ_p	
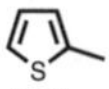 (2-thienyl)			1.31
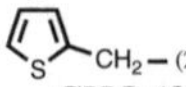 CH$_2$— (2-thienylmethylene)			0.31
—CHO (formyl)	0.36	0.22	
—COCH$_3$ (acetyl)	0.38	0.50	1.65
—COCH$_2$CH$_3$ (propionyl)		0.48	
—COCH(CH$_3$)$_2$		0.47	
—COC(CH$_3$)$_3$		0.32	
—COCF$_3$ (trifluoroacetyl)	0.65		3.7
—COC$_6$H$_5$ (benzoyl)	0.34	0.46	2.2
—CONH$_2$	0.28	0.36	1.68
—CONHC$_6$H$_5$			1.56
—CH$_2$COCH$_3$ (acetonyl)			0.60
—CH$_2$CONH$_2$ (acetamido)			0.31
—CH$_2$CH$_2$CONH$_2$			0.19
—CH$_2$CH$_2$CH$_2$CONH$_2$			0.12
—CH$_2$CONHC$_6$H$_5$			0.0
—COO$^-$ (carboxylate)	−0.1	0.0	−1.06
—COOH (carboxyl)	0.36	0.43	2.08
—CO—OCH$_3$ (carbomethoxy)	0.32	0.39	2.00
—CO—OCH$_2$CH$_3$ (carbethoxy)	0.37	0.45	2.12
—CH$_2$CO—OCH$_3$			1.06
—CH$_2$CO—OCH$_2$CH$_3$			0.82
—CH$_2$COOH			−0.06
—CH$_2$CH$_2$COOH	−0.03	−0.07	
—Cl	0.37	0.23	2.96
—CCl$_3$ (trichloromethyl)	0.47		2.65
—CHCl$_2$ (dichloromethyl)			1.94
—CH$_2$Cl (chloromethyl)	0.12	0.18	1.05
—CH$_2$CH$_2$Cl			0.38
—CH$_2$CCl$_3$			0.75
—CH$_2$CH$_2$CCl$_3$			0.25
—CH=CCl$_2$			1.00
—CH$_2$CH=CCl$_2$			0.19
p-ClC$_6$H$_4$— (p-chlorophenyl)		0.08	
—F	0.34	0.06	3.21
—CF$_3$ (trifluoromethyl)	0.43	0.54	2.61
—CHF$_2$ (difluoromethyl)			2.05
—CH$_2$F (fluoromethyl)			1.10
—CH$_2$CF$_3$			0.90
—CH$_2$CF$_2$CF$_2$CF$_3$			0.87
—C$_6$F$_5$ (pentafluorophenyl)	−0.12	−0.03	
—Ge(CH$_3$)$_3$ (trimethylgermyl)		0.0	
—Ge(CH$_2$CH$_3$)$_3$ (triethylgermyl)		0.0	
—H	0.00	0.00	0.49
—I	0.35	0.28	2.46
—CH$_2$I (iodomethyl)			0.85
—N$_2^+$ (diazonio)	1.76	1.91	

TABLE 7.1 Hammett and Taft Substituent Constants (*continued*)

Substituent	Hammett constants		Taft constant σ^*
	σ_m	σ_p	
—N$_3$ (azido)	0.33	0.08	2.62
—NH$_2$ (amino)	−0.16	−0.66	0.62
—NH$_3^+$	1.13	1.70	3.76
—CH$_2$—NH$_2$ (aminomethyl)			0.50
—CH$_2$—NH$_3^+$			2.24
—NH—CH$_3$ (methylamino)	−0.30	−0.84	
—NH—C$_2$H$_5$ (ethylamino)	−0.24	−0.61	
—NH—C$_4$H$_9$ (butylamino)	−0.34	−0.51	
—NH(CH$_3$)$_2^+$			4.36
—NH$_2^+$—CH$_3$	0.96		3.74
—NH$_2^+$—C$_2$H$_5$	0.96		3.74
—N(CH$_3$)$_3^+$ (trimethylammonium)	0.88	0.82	4.55
—N(CH$_3$)$_2$ (dimethylamino)	−0.2	−0.83	0.32
—CH$_2$—N(CH$_3$)$_3^+$			1.90
—N(CF$_3$)$_2$ [bis(trifluoromethyl)amino]	0.45	0.53	
p-H$_2$N—C$_6$H$_5$— (*p*-aminophenyl)		−0.30	
—NH—CO—CH$_3$	0.21	0.00	1.40
—NH—CO—C$_2$H$_5$			1.56
—NH—CO—C$_6$H$_5$	0.22	0.08	1.68
—NH—CHO	0.25		1.62
—NH—CO—NH$_2$	0.18		1.31
—NH—OH (hydroxylamino)	−0.04	−0.34	
—NH—CO—OC$_2$H$_5$	0.33		1.99
—CH$_2$—NH—CO—CH$_3$			0.43
—NH—SO$_2$—C$_6$H$_5$			1.99
—NH—NH$_2$ (hydrazido)	−0.02	−0.55	
—C≡N (cyano)	0.56	0.66	3.30
—CH$_2$—CN (cyanomethyl)	0.17	0.01	1.30
—N=O (nitroso)		0.12	
—NO$_2$ (nitro)	0.71	0.78	4.0
—CH$_2$—NO$_2$ (nitromethyl)			1.40
—CH$_2$—CH$_2$—NO$_2$ (2-nitroethyl)			0.50
—CH=CHNO$_2$ (2-nitroethylenyl)	0.33	0.26	
m-O$_2$N—C$_6$H$_4$— (*m*-nitrophenyl)		0.18	
p-O$_2$N—C$_6$H$_4$— (*p*-nitrophenyl)		0.24	
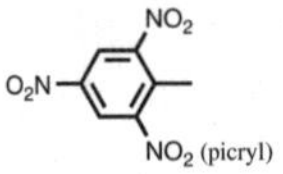 (picryl)	0.43	0.41	
			1.37
			1.65
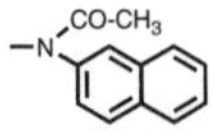			
—O$^-$	−0.71	−0.52	
—OH (hydroxy)	0.12	−0.37	1.34
—O—CH$_3$ (methoxy)	0.12	−0.27	1.81
—O—C$_2$H$_5$ (ethoxy)	0.10	−0.24	1.68

TABLE 7.1 Hammett and Taft Substituent Constants (*continued*)

Substituent	Hammett constants		Taft constant σ^*
	$\sigma\mu$	σp	
$-O-C_3H_7$ (propoxy)	0.00	−0.25	1.68
$-O-CH(CH_3)_2$ (isopropoxy)	0.05	−0.45	1.62
$-O-C_4H_9$ (butoxy)	−0.05	−0.32	1.68
$-O-C_5H_9$ (cyclopentyloxy)			1.62
$-O-C_6H_{11}$ (cyclohexyloxy)	0.29		1.81
$-O-CH_2-C_6H_{11}$ (cyclohexylmethoxy)	0.18		1.31
$-O-C_6H_5$ (phenoxy)	0.25	−0.32	2.43
$-O-CH_2-C_6H_5$ (phenylmethoxy)		−0.42	
$-OCF_3$ (trifluoromethoxy)	0.40	0.35	
(3,4- methylenedioxyphenyl, piperonyl)		−0.27	
(3,4-ethylenedioxyphenyl)		−0.12	
$-O-CO-CH_3$ (acetoxy)	0.39	0.31	
$-ONO_2$ (nitrate ester)			3.86
$-O-N=C(CH_3)_2$			1.81
$-ONH_3^+$			2.92
$-CH_2-O^-$			0.27
$-CH_2-OH$	0.08	0.08	0.31
$-CH_2-O-CH_3$			0.52
$-CH(OH)-CH_3$			0.12
$-CH(OH)-C_6H_5$			0.50
$p\text{-}HO-C_6H_4-$ (*p*-hydroxyphenyl)		−0.24	
$p\text{-}CH_3O-C_6H_4-$ (*p*-methoxyphenyl)		−0.10	
$-CH_2-CH(OH)-CH_3$			−0.06
$-CH_2-C(OH)(CH_3)_2$			−0.25
$-P(CH_3)_2$ (dimethylphosphino)	0.1	0.05	
$-P(CH_3)_3^+$ (trimethylphosphino)	0.8	0.9	
$-P(CF_3)_2$	0.6	0.7	
$-PO_3H^{\ominus}$	0.2	0.26	
$-PO(OC_2H_5)_2$	0.55	0.60	
$-SH$ (thio, mercapto)	0.25	0.15	1.68
$-SCH_3$ (methylthio)	0.15	0.00	1.56
$-S(CH_3)_2^+$ (dimethylsulfonium)	1.0	0.9	
$-SCH_2CH_3$ (ethylthio)	0.23	0.03	1.56
$-SCH_2CH_2CH_3$ (propylthio)			1.49
$-SCH_2CH_2CH_2CH_3$ (butylthio)			1.44
$-SC_6H_{11}$ (cyclohexylthio)			1.93
$-SC_6H_5$ (phenylthio)	0.30		1.87
$-SC(C_6H_5)_3$ (triphenylmethylthio)			0.69
$-SCH_2C_6H_5$ (benzylthio)			1.56
$-SCH_2CH_2C_6H_5$ (phenethylthio)			1.44
$-CH_2SH$ (thiomethyl)	0.03		0.62
$-CH_2SCH_2C_6H_5$			0.37
$-SCF_3$ (trifluoromethylthio)	0.40	0.50	
$-SCN$ (thiocyanato)	0.63	0.52	3.43

TABLE 7.1 Hammett and Taft Substituent Constants (*continued*)

Substituent	Hammett constants		Taft constant σ^*
	$\sigma\mu$	σp	
—S—CO—CH₃	0.39	0.44	
—S—CO—NH₂	0.34		2.07
—SO—CH₃ (methylsulfoxy)	0.52	0.49	
—SO—C₆H₅ (phenylsulfoxy)			3.24
—CH₂—SO—CH₃			1.33
—SO₂—CH₃ (methylsulfonyl)	0.60	0.68	3.68
—SO₂—CH₂CH₃ (ethylsulfonyl)			3.74
—SO₂—CH₂CH₂CH₃ (propylsulfonyl)			3.68
—SO₂—C₆H₅ (phenylsulfonyl)	0.67		3.55
—SO₂—CF₃ (trifluoromethylsulfonyl)	0.79	0.93	
—SO₂—NH₂	0.46	0.57	
—CH₂—SO₂—CH₃			1.38
—SO₃⁻	0.05	0.09	0.81
—SO₃H		0.50	
—SeCH₃	0.1	0.0	
—Se—C₆H₁₁ (cyclohexylselenyl)			2.37
—SeCN	0.67	0.66	3.61
—Si (CH₃)₃	−0.04	−0.07	−0.81
—Si(CH₂CH₃)₃		0.0	
—Si(CH₃)₂C₆H₅			−0.87
—Si(CH₃)₂—O—Si(CH₃)₃			−0.81
—CH₂Si(CH₃)₃	−0.16	−0.22	−0.25
—CH₂CH₂Si(CH₃)₃			−0.25
—Sn(CH₃)₃		0.0	
—Sn(CH₂CH₃)₃		0.0	

TABLE 7.2 pK_A and Rho (ρ) Values for the Hammett Equation

Acid	pK_A	ρ
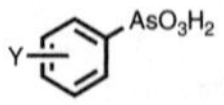 Y—⬡—AsO₃H₂ (arenearsonic acids)		
pK₁	3.54	1.05
pK₂	8.49	0.87
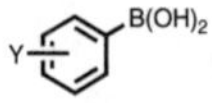 Y—⬡—B(OH)₂ (areneboronic acids, in aqueous 25% ethanol)	9.70	2.15
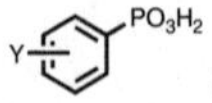 Y—⬡—PO₃H₂ (arenephosphonic acids)		
pK₁	1.84	0.76
pK₂	6.97	0.95
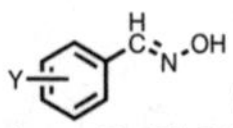 Y—⬡—C(H)=N—OH (α-arylaldoximes)	10.70	0.86
Ar—Se(O)OH (benzeneseleninic acids)	4.78	1.03
Ar—SO₂—NH₂ (benzenesulfonamides, 20°C)	10.00	1.06
Ar¹—SO₂—NHAr² (benzenesulfonanilides, 20°C)		
Y—C₆H₄—SO₂—NH—C₆H₅	8.31	1.16

TABLE 7.2   pK$_A$ and Rho (ρ) Values for the Hammett Equation (*continued*)

Acid	pK$_A$	ρ
$C_6H_5-SO_2-NH-C_6H_4-Y$	8.31	1.74
$Ar-CO-OH$ (benzoic acids)	4.21	1.00
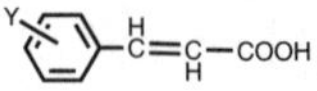 (cinnamic acids)	4.45	0.47
$Ar-OH$ (phenols)	9.92	2.23
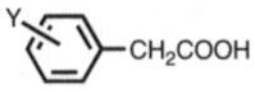 (phenylacetic acids)	4.30	0.49
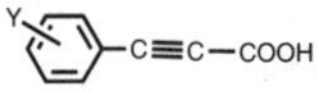 (phenylpropiolic acids, in aqueous 35% dioxane)	3.24	0.81
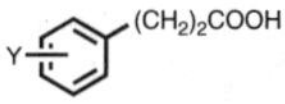 (phenylpropionic acids)	4.45	0.21
$Ar-CHOH-CF_3$ (phenyltrifluoromethylcarbinols)	11.90	1.01
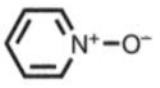 (pyridine-1-oxides, pyridine-N-oxides)	0.94	2.09
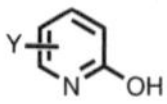 (2-pyridones, 2-hydroxypyridines)	11.65	4.28
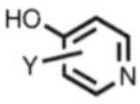 (4-pyridones, 4-hydroxypyridines)	11.12	4.28
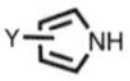 (pyrroles)	17.00	4.28
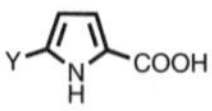 (5-substituted pyrrole-2-carboxylic acids)	2.82	1.40
$Ar-CO-SH$ (thiobenzoic acids)	2.61	1.0
$Ar-SH$ (thiophenols)	6.50	2.2
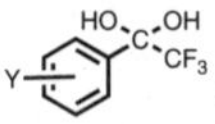 (trifluoroacetophenone hydrates)	10.00	1.11
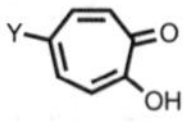 (5-substituted tropolones)	6.42	3.10
Cations resulting from protonation of		
$Ar-CO-CH_3$ (acetophenones)	−6.0	2.6
$Ar-NH_2$ (anilines)	4.60	2.90
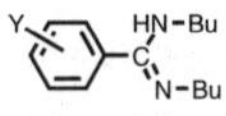 (C-aryl-N,N'-dibutylamidines, in aqueous 50% ethanol)	11.14	1.41
N,N-Dimethylanilines	5.07	3.46
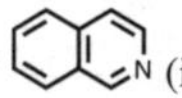 (isoquinolines)	5.32	5.90
1-Naphthylamines	3.85	2.81
2-Naphthylamines	4.29	2.81
Pyridines	5.18	5.90
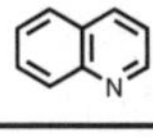 (quinolines)	4.88	5.90

TABLE 7.3 pK$_A$ and Rho (ρ) Values for the Taft Equation

Acid	pK$_A$	Rho (ρ)
RCOOH	4.66	1.62
RCH$_2$COOH	4.76	0.67
RC≡C—COOH	2.39	1.89
H$_2$C=C(R)—COOH	4.39	0.64
(CH$_3$)$_2$C=C(R)—COOH	4.65	0.47
Z-C$_6$H$_5$—CH=C(R)—COOH	3.77	0.63
E-C$_6$H$_5$—CH=C(R)—COOH	4.61	0.47
R—CO—CH$_2$—COOH	4.12	0.43
HO—N=CR—COOH	4.84	0.34
RCH$_2$OH	15.9	1.42
RCH(OH)$_2$	14.4	1.42
R^{1}CO—NHR^2	22.0	3.1*
CH$_3$CO—CR=C(OH)CH$_3$	9.25	1.78
CH$_3$CO—CHR—CO—OC$_2$H$_5$	12.59	3.44
R—CO—NHOH	9.48	0.98
R^{1}R^{2}C=N—OH (R^1, R^2 are not acyl groups)	12.35	1.18
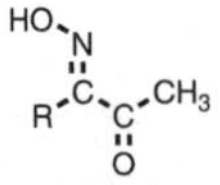 HO—N‖ R—C·C·CH$_3$ ‖ O	9.00	0.94
RCH(NO$_2$)$_2$	5.24	3.60
RSH	10.22	3.50
RCH$_2$SH	10.54	1.47
R—CO—SH	3.52	1.62
Cations resulting from protonation of		
RNH$_2$	10.15	3.14
R^{1}R^{2}NH	10.59	3.23
R^{1}R^{2}R^{3}N	9.61	3.30
R^{1}R^{2}PH	3.59	2.61
R^{1}R^{2}R^{3}P	7.85	2.67

σ for R^{1}CO and R^2

TABLE 7.4 Special Hammett Sigma Constants

Substituent	σ_m^+	σ_p^+	σ_p^-
—CH$_3$	−0.07	−0.31	−0.17
—C(CH$_3$)$_3$	−0.06	−0.26	
—C$_6$H$_5$	0.11	−0.18	
—CF$_3$	0.52	0.61	0.74
—F	0.35	−0.07	0.02
—Cl	0.40	0.11	0.23
—Br	0.41	0.15	0.26
—I	0.36	0.14	
—C≡N	0.56	0.66	0.88
—CH=O			1.13

TABLE 7.4 Special Hammett Sigma Constants (*continued*)

Substituent	σ_m^{+}	σ_p^{+}	σ_p^{-}
—CO—NH$_2$			0.63
—CO—CH$_3$			0.85
—COOH	0.32	0.42	0.73
—CO—OCH$_3$	0.37	0.49	0.66
—CO—OCH$_2$CH$_3$	0.37	0.48	0.68
—N$_2^{+}$			3.2
—NH$_2$	0.16	−1.3	−0.66
—N(CH$_3$)$_2$		−1.7	
—N(CH$_3$)$_3^{+}$	0.36	0.41	
—NH—CO—CH$_3$		−0.60	
—NO$_2$	0.67	0.79	1.25
—OH		−0.92	
—O^{-}			−0.81
—OCH$_3$	0.05	−0.78	−0.27
—SF$_5$			0.70
—SCF$_3$			0.57
—SO$_2$CH$_3$			1.05
—SO$_2$CF$_3$			1.36

SECTION 8

ELECTROLYTES, ELECTROMOTIVE FORCE, AND CHEMICAL EQUILIBRIUM

EQUILIBRIUM CONSTANTS

The acidities of organic compounds are typically expressed by citing their pK_A values. These are defined as $-\log_{10} K_A$ for the reaction

$$HA \rightleftharpoons H^{\oplus} + A^{\ominus}$$

The equilibrium constant K_A is defined as

$$K_A = \frac{[H^+][A^-]}{[HA]}$$

Thus, for example, the pK_A of water is $-\log_{10}([H^{\oplus}][HO^{\ominus}]/[H_2O])$ or $(10^{-7}) \cdot (10^{-7})/55.5$. The concentration of protons or hydroxide ions in water is $10^{-7} M$ and the concentration of water in water is $55.5 M$. The equilibrium constant K_A is therefore $10^{-15.74}$. The operator "p" means "$-\log$" so the pK_A of water is 15.7. The equilibrium constant K_W for water is 10^{-14} and is simply the product of $[H^{\oplus}] \cdot [HO^{\ominus}]$.

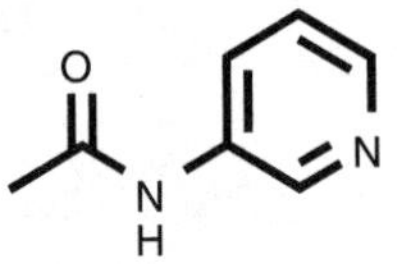

3-acetamidopyridine

Acidity constants are given for a range of compounds in Table 8.1. When more than one ionizable proton is present, pK_1, pK_2, etc. values are given. Cations formed from the indicated compound by protonation are indicated by "(+1)" or "(+2)" for a dication. For example, the dissociation of 3-acetamidopyridine is reported in Table 8.1 as "4.37(+1)." This means dissociation of the compound that is protonated (at the pyridine nitrogen atom).

Temperature values different from 25°C are given in parentheses as are other relevant variations. For example, the dissociation constant for acetic acid-d_1 is reported in D_2O.

TABLE 8.1 pK_A Values of Organic Materials in Water at 25 °C

Ionic strength μ is zero unless otherwise indicated. The protonation state of cations is designated by a value $(+1)$, $(+2)$, etc. that follows the pK_A value. Neutral species are indicated by (0), if it is not obvious otherwise. The charge state of anionic species is designated by (-1), (-2), etc.

Substance	pK_1	pK_2	pK_3	pK_4
Abietic acid	7.62			
Acetamide	−0.37(+1)			
Acetamidine	1.60(+1)			
N-(2-Acetamido)-2-aminoethanesulfonic acid (20 °C)	6.88			
2-Acetamidobenzoic acid	3.63			
3-Acetamidobenzoic acid	4.07			
4-Acetamidobenzoic acid	4.28			
2-(Acetamido)butanoic acid	3.716			
N-(2-Acetamido)iminodiacetic acid (20 °C)	6.62			
3-Acetamidopyridine	4.37(+1)			
Acetanilide	0.4(+1)	$13.39(0)^{40\,°C}$		
Acetic acid	4.756			
Acetic acid-d (in D_2O)	5.32			
Acetoacetic acid (18 °C)	3.58			
Acetohydrazine	3.24(+1)			
Acetone oxime	12.2			
2-Acetoxybenzoic acid (acetylsalicylic acid)	3.48			
3-Acetoxybenzoic acid	4.00			
4-Acetoxybenzoic acid	4.38			
Acetylacetic acid (18 °C)	3.58			
N-Acetyl-α-alanine	3.715			
N-Acetyl-β-alanine	4.455			
2-Acetylaminobutanoic acid	3.72			
3-Acetylaminopropionic acid	4.445			
2-Acetylbenzoic acid	4.13			
3-Acetylbenzoic acid	3.83			
4-Acetylbenzoic acid	3.70			
2-Acetylcyclohexanone	14.1			
N-Acetylcysteine (30 °C)	9.52			
Acetylenedicarboxylic acid	1.75	4.40		
N-Acetylglycine	3.670			

Abietic acid

TABLE 8.1 pK$_A$ Values of Organic Materials in Water at 25 °C (*continued*)

Substance	pK$_1$	pK$_2$	pK$_3$	pK$_4$
N-Acetylguanidine	8.23(+1)			
N-α-Acetyl-L-histidine	7.08			
Acetylhydroxamic acid (20 °C)	9.40			
N-Acetyl-2-mercaptoethylamine	9.92(SH)			
4-Acetyl-β-mercaptoisoleucine				
(30 °C)	10.30			
2-Acetyl-1-naphthol (30 °C)	13.40			
N-Acetylpenicillamine (30 °C)	9.90			
2-Acetylphenol	9.19			
4-Acetylphenol	8.05			
2-Acetylpyridine	2.643(+1)			
3-Acetypyridine	3.256(+1)			
4-Acetylpyridine	3.505(+1)			
Aconitine	8.11(+1)			
Acridine	5.60(+1)			
Acrylic acid	4.26			
Adenine	4.17(+1)	9.75(0)		
Adeninedeoxyriboside-5′-				
phosphoric acid		4.4	6.4	
Adenine-*N*-oxide	2.69(+1)	8.49(0)		
Adenosine	3.5(+1)	12.34(0)		
Adenosine-5′-diphosphoric acid		4.2(−1)	7.20(−2)	
Adenosine-2′-phosphoric acid	3.81(+1)	6.17(0)		
Adenosine-3′-phosphoric acid	3.65(0)	5.88(−1)		
Adenosine-5′-phosphoric acid	3.74(0)	6.05(−1)	13.06(−2)	
Adenosine-5′-triphosphoric acid		4.00(−1)	6.48(−2)	
Adipamic acid (adipic acid				
monoamide)	4.629			
Adipic acid	4.418	5.412		
α-Alanine	2.34(+1)	9.87(0)		
β-Alanine	3.55(+1)	10.238(0)		
α-Alanine, methyl ester (μ = 0.10)	7.743(+1)			
β-Alanine, methyl ester (μ = 0.10)	9.170(+1)			
N-D-Alanyl-α-D-alanine (μ = 0.1)	3.32(+1)	8.13(0)		
N-L-Alanyl-α-L-alanine (μ = 0.1)	3.32(+1)	8.13(0)		
N-L-Alanyl-α-D-alanine	3.12(+1)	8.30(0)		
N-α-Alanylglycine	3.11(+1)	8.11(0)		

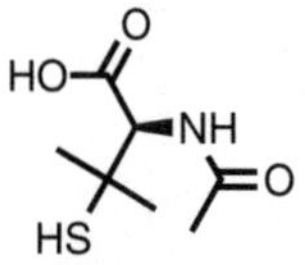

N-Acetylpenicillamine

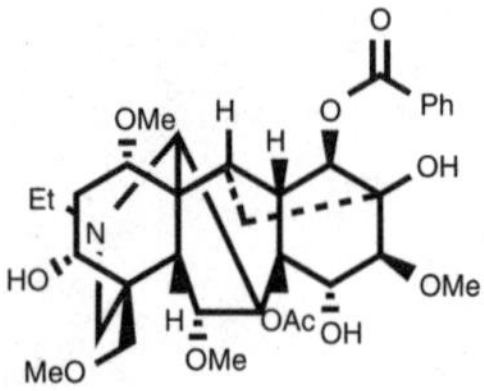

Aconitine

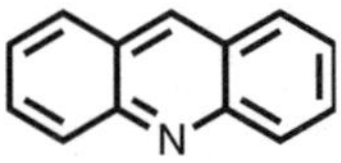

Acridine

TABLE 8.1 pK_A Values of Organic Materials in Water at 25 °C (*continued*)

Substance	pK_1	pK_2	pK_3	pK_4
Alanylglycylglycine	3.190(+1)	8.15(0)		
β-Alanylhistidine	2.64	6.86	9.40	
Albumin (bovine serum, $\mu = 0.15$)	10–10.3			
2-Aldoxime pyridine	3.42(+1)	10.22(0)		
Alizarin Black SN	5.79	12.8		
Alizarin-3-sulfonic acid	5.54	11.01		
Allantoin	8.96			
Allothreonine	2.108(+1)	9.096(0)		
Alloxanic acid	6.64			
Allylacetic acid	4.68			
Allylamine	9.69(+1)			
5-Allylbarbituric acid	4.78(+1)			
5-Allyl-5-(-methylbutyl)barbituric acid	8.08			
2-Allylphenol	10.28			
1-Allylpiperidine	9.65(+1)			
2-Allylpropionic acid	4.72			
3-Amidotetrazoline	3.95(+1)			
2-Aminoacetamide	7.95(+1)			
Aminoacetonitrile	5.34(+1)			
9-Aminoacridine (20 °C)	9.95(+1)			
4-Aminoantipyrine	4.94(+1)			
2-Aminobenzenesulfonic acid	2.459(0)			
3-Aminobenzenesulfonic acid	3.738(0)			
4-Aminobenzenesulfonic acid	3.227(0)			
2-Aminobenzoic acid	2.09(+1)	4.79(0)		
3-Aminobenzoic acid	3.07(+1)	4.79(0)		
4-Aminobenzoic acid	2.41(+1)	4.85(0)		
2-Aminobenzoic acid, methyl ester	2.36(+1)			
3-Aminobenzoic acid, methyl ester	3.58(+1)			
4-Aminobenzoic acid, methyl ester	2.45(+1)			

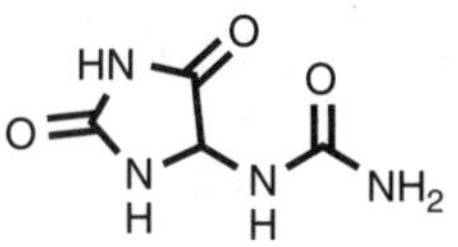

Allantoin 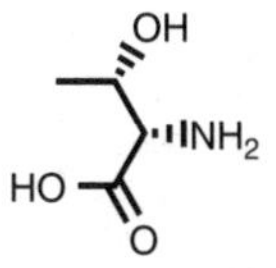   Allothreonine 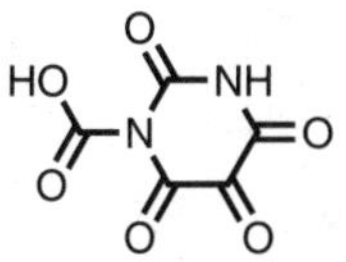   Alloxanic acid 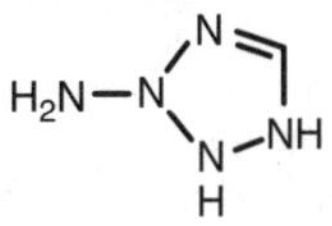   3-Aminotetrazoline

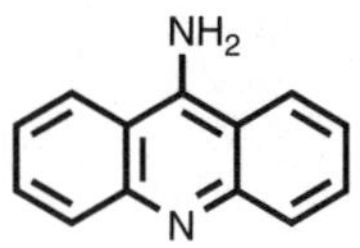

9-Aminoacridine 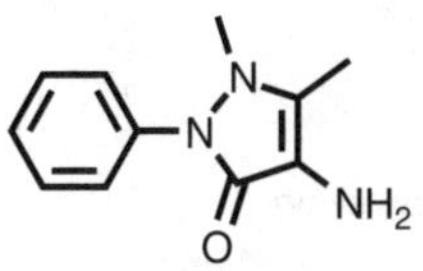   4-Aminoantipyrine

TABLE 8.1 pK$_A$ Values of Organic Materials in Water at 25 °C (*continued*)

Substance	pK$_1$	pK$_2$	pK$_3$	pK$_4$
3-Aminobenzonitrile	2.75(+1)			
4-Aminobenzonitrile	1.74(+1)			
4-Aminobenzophenone	2.15(+1)			
2-Aminobenzothiazole (20 °C)	4.48(+1)			
2-Aminobenzoylhydrazide	1.85	3.47	12.80	
2-Aminobiphenyl	3.78(+1)			
3-Aminobiphenyl	4.18(+1)			
4-Aminobiphenyl	4.27(+1)			
4-Amino-3-bromomethylpyridine	7.47(+1)			
4-Amino-3-bromopyridine (20 °C)	7.04(+1)			
2-Aminobutanoic acid	2.286(+1)	9.830(0)		
3-Aminobutanoic acid		10.14(0)		
4-Aminobutanoic acid	4.031(+1)	10.556(0)		
2-Aminobutanoic acid, methyl ester				
($\mu = 0.1$)	7.640(+1)			
4-Aminobutanoic acid, methyl ester				
($\mu = 0.1$)	9.838(+1)			
D-(+)-2-Amino-1-butanol	9.52(+1)			
3-Amino-N-butyl-3-methyl-2-				
butanone oxime	9.09(+1)			
4-Aminobutylphosphonic acid	2.55	7.55	10.9	
2-Amino-N-carbamoylbutanoic acid	3.886(+1)			
4-Amino-N-carbamoylbutanoic acid	4.683(+1)			
2-Amino-N-carbamoyl-2-				
methylpropanoic acid	4.463			
1-Amino-1-cycloheptanecarboxylic	2.59(+1)	10.46(0)		
acid				
1-Amino-1-cyclohexanecarboxylic	2.65(+1)	10.03(0)		
acid				
2-Amino-1-cyclohexanecarboxylic	3.56(+1)	10.21(0)		
acid				
1-Aminocyclopentane	10.65(+1)			
1-Aminocyclopropane	9.10(+1)			
10-Aminodecylphosphonic acid		8.0	11.25	
10-Aminodecylsulfonic acid	2.65(+1)			
1-Amino-2-di(aminomethyl)butane	3.58(+3)	8.59(+2)	9.66(+1)	
2-Amino-N,N-dihydroxyethyl-				
2-hydroxyl-1,3-propanediol	6.484(+1)			
2-Amino-N,N-dimethylbenzoic acid	1.63(+1)	8.42(0)		
4-Amino-2,5-dimethylphenol	5.28(+1)	10.40(0)		
4-Amino-3,5-dimethylpyridine (20 °C)	9.54(+1)			
12-Aminododecanoic acid	4.648(+1)			
2-Aminoethane-1-phosphoric acid	5.838	10.64		
1-Aminoethanesulfonic acid	−0.33	9.06		
2-Aminoethanesulfonic acid	1.5	9.061		
2-Aminoethanethiol (cysteamine)				
($\mu = 0.01$)	8.23(+1)			
2-Aminoethanol (ethanolamine)	9.50(+1)			
2-[2-(2-Aminoethyl)	3.50	6.59	9.51	
aminoethyl] pyridine				

TABLE 8.1   pK_A Values of Organic Materials in Water at 25 °C (*continued*)

Substance	pK$_1$	pK$_2$	pK$_3$	pK$_4$
2-Amino-2-ethyl-1-butanol	9.82(+1)			
3-(2-Aminoethyl)indole		10.2		
3-Amino-*N*-ethyl-3-methyl-2-butanone oxime	9.23(+1)			
N-(2-Aminoethyl)morpholine	4.06(+2)	9.15(+1)		
p-(2-Aminoethyl)phenol	9.3	10.9		
2-Aminoethylphosphonic acid	2.45(+1)	7.0(0)	10.8(−1)	
N-(2-Aminoethyl)piperidine (30 °C)	6.38	9.89		
2-(2-Aminoethyl)pyridine ($\mu = 0.5$)	4.24(+2)	9.78(+1)		
4-Amino-3-ethylpyridine (20 °C)	9.51(+1)			
N-(2-Aminoethyl)pyrrolidine (30 °C)	6.56(+2)	9.74(+1)		
2-Aminofluorine	10.34(+1)			
2-Amino-D-β-glucose ($\mu = 0.05$)	2.20(+1)	9.08(0)		
2-Amino-*N*-glycylbutanoic acid	3.155(+1)	8.331(0)		
7-Aminoheptanoic acid	4.502			
2-Aminohexanoic acid	2.335(+1)	9.834(0)		
6-Aminohexanoic acid	4.373(+1)	10.804(0)		
C-Amino-*C*-hydrazine carbonylmethane	2.38(+2)	7.69(+1)		
2-Amino-3-hydroxybenzoic acid	2.5(+1)	5.192(0)	10.118(OH)	
L-2-Amino-3-hydroxybutanoic acid (threonine)	2.088(+1)	9.100(0)		
DL-2-Amino-4-hydroxybutanoic acid ($\mu = 0.1$)	2.265(+1)	9.257(0)		
DL-4-Amino-3-hydroxybutanoic acid ($\mu = 0.1$)	3.834(+1)	9.487(0)		
2-Amino-2′-hydroxydiethyl sulfide	9.27(+1)			
4-Amino-2-hydroxypyrimidine (cytosine)	4.58(+1)	12.15(0)		
3-Amino-*N*-isopropyl-3-methyl-2-butanone oxime	9.09(+1)			
4-Amino-3-isopropylpyridine (20 °C)	9.54(+1)			
1-Aminoisoquinoline (20 °C, $\mu = 0.01$)	7.62(+1)			
3-Aminoisoquinoline (20 °C, $\mu = 0.005$)	5.05(+1)			
4-Aminoisoxazolidine-3-one	7.4(+1)			
Aminomalonic acid	3.32(+1)	9.83(0)		
DL-2-Amino-4-mercaptobutanoic acid	2.22(+1)	8.87(0)	10.86(SH)	

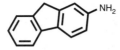

2-Aminofluorene

4-Amino-2-hydroxypyrimidine (cytosine)

TABLE 8.1   pK$_A$ Values of Organic Materials in Water at 25 °C (*continued*)

Substance	pK$_1$	pK$_2$	pK$_3$	pK$_4$
2-Amino-3-mercapto- 3-Methylbutanoic acid	1.8(+1)	7.9(0)	10.5(SH)	
2-Amino-6-methoxybenzothiazole	4.50(+1)			
3-Amino-4-methylbenzenesulfonic acid	3.633			
4-Amino-3-methylbenzenesulfonic acid	3.125			
2-Amino-4-methylbenzothiazole	4.7(+1)			
1-Amino-3-methylbutane	10.64(+1)			
3-Amino-3-methyl-2-butanone oxime	9.09(+1)			
3-Amino-*N*-methyl-3-methyl-2- butanone oxime	9.23(+1)			
2-Amino-3-methylpentanoic acid	2.320(+1)	9.758(0)		
3-Aminomethyl-6-methylpyridine (30 °C)	8.70(+1)			
Aminomethylphosphonic acid	2.35	5.9	10.8	
2-Amino-2-methyl-1,3-propanediol	8.801			
2-Amino-2-methyl-1-propanol	9.694(+1)			
2-Amino-2-methylpropanoic acid	2.357(+1)	10.205(0)		
(2-Aminomethyl)pyridine ($\mu = 0.5$)	2.31(+2)	8.79(+1)		
2-Amino-3-methylpyridine	˙7.24(+1)			
4-Amino-3-methylpyridine	9.43(+1)			
2-Amino-4-methylpyridine	7.48(+1)			
2-Amino-5-methylpyridine	7.22(+1)			
2-Amino-6-methylpyridine	7.41(+1)			
2-Amino-4-methylpyrimidine (20 °C)	4.11(+1)			
Aminomethylsulfonic acid	5.75(+1)			
N-Aminomorpholine	4.19(+1)			
4-Amino-1-naphthalenesulfonic acid	2.81			
1-Amino-2-naphthalenesulfonic acid	1.71			
1-Amino-3-naphthalenesulfonic acid	3.20			
1-Amino-5-naphthalenesulfonic acid	3.69			
1-Amino-6-naphthalenesulfonic acid	3.80			
1-Amino-7-naphthalenesulfonic acid	3.66			
1-Amino-8-naphthalenesulfonic acid	5.03			
2-Amino-1-naphthalenesulfonic acid	2.35			
2-Amino-4-naphthalenesulfonic acid	3.79			
2-Amino-6-naphthalenesulfonic acid	3.79	8.94		
2-Amino-8-naphthalenesulfonic acid	3.89			
3-Amino-1-naphthoic acid	2.61	4.39		
4-Amino-2-naphthoic acid	2.89	4.46		
8-Amino-2-naphthol	4.20(+1)			
DL-2-Aminopentanoic acid (DL-norvaline)	2.318(+1)	9.808		
3-Aminopentanoic acid	4.02(+1)	10.399(0)		
4-Aminopentanoic acid	3.97(+1)	10.46(0)		
5-Aminopentanoic acid	4.20(+1)	9.758(0)		
5-Aminopentanoic acid, ethyl ester	10.151			
2-Aminophenol	9.28	9.72		

TABLE 8.1 pK_A Values of Organic Materials in Water at 25 °C (*continued*)

Substance	pK_1	pK_2	pK_3	pK_4
3-Aminophenol	9.83	9.87		
4-Aminophenol	8.50	10.30		
4-Aminophenylacetic acid (20 °C)	3.60	5.26		
2-Aminophenylarsonic acid	ca 2	3.77	8.66	
3-Aminophenylarsonic acid	ca 2	4.02	8.92	
4-Aminophenylarsonic acid	ca 2	4.02	8.62	
3-Aminophenylboric acid	4.46	8.81		
4-Aminophenylboric acid	3.71	9.17		
4-Aminophenyl (4-chlorophenyl) sulfone	1.38			
2-Aminophenylphosphonic acid		4.10	7.29	
3-Aminophenylphosphonic acid			7.16	
4-Aminophenylphosphonic acid			7.53	
1-Amino-1,2,3-propanetricarboxylic acid ($\mu = 2.2$)	2.10(+1)	3.60(0)	4.60(−1)	9.82(−2)
3-Aminopropanoic acid	3.551(+1)	10.235(0)		
1-Amino-1-propanol	9.96(+1)			
DL-2-Amino-1-propanol	9.469(+1)			
3-Amino-1-propanol	9.96(+1)			
3-Aminopropene	9.691(+1)			
3-Amino-N-propyl-3-methyl-2-butanone oxime	9.09(+1)			
2-Aminopropylsulfonic acid		9.15		
2-Aminopyridine	6.71(+1)			
3-Aminopyridine	6.03(+1)			
4-Aminopyridine	9.114(+1)			
2-Aminopyridine-1-oxide	2.58(+1)			
3-Aminopyridine-1-oxide	1.47(+1)			
4-Aminopyridine-1-oxide	3.54(+1)			
8-Aminoquinaldine	4.86(+1)			
2-Aminoquinoline (20 °C, $\mu = 0.01$)	7.34(+1)			
3-Aminoquinoline (20 °C, $\mu = 0.01$)	4.95(+1)			
4-Aminoquinoline (20 °C, $\mu = 0.01$)	9.17(+1)			
5-Aminoquinoline (20 °C, $\mu = 0.01$)	5.46(+1)			
6-Aminoquinoline (20 °C, $\mu = 0.01$)	5.63(+1)			
8-Aminoquinoline (20 °C, $\mu = 0.01$)	3.99(+1)			
4-Aminosalicylic acid	1.991(+1)	3.917(0)	13.74	
5-Aminosalicylic acid	2.74(+1)	5.84(0)		
2-Amino-3-sulfopropanoic acid	1.89(+1)	8.70(0)		
4-Amino-2,3,5,6-tetramethylpyridine (20 °C)	10.58(+1)			

2-Aminopyridine-1-oxide   8-Aminoquinaldine

TABLE 8.1 pK$_A$ Values of Organic Materials in Water at 25 °C (*continued*)

Substance	pK$_1$	pK$_2$	pK$_3$	pK$_4$
5-Amino-1,2,3,4-tetrazole (20 °C)	1.76	6.07		
2-Aminothiazole (20 °C)	5.36(+1)			
1-Amino-3-thiobutane (30 °C)	9.18(+1)			
5-Amino-3-thio-1-pentanol (30 °C)	9.12(+1)			
2-Aminothiophenol	<2(+1)	7.90(0)		
2-Amino-4,4,4-trifluorobutanoic acid		8.171(0)		
3-Amino-4,4,4-trifluorobutanoic acid		5.831(0)		
3-Amino-2,4,6-trinitrotoluene		9.5(+1)		
Angiotensin II	10.37			
Anhydroplatynecine	9.40			
Aniline	4.60(+1)			
2-Anilinoethylsulfonic acid	3.80(+1)			
3-Anilinoethylsulfonic acid	4.85(+1)			
Anthracene-1-carboxylic acid	3.68			
Anthracene-2-carboxylic acid	4.18			
Anthracene-9-carboxylic acid	3.65			
Anthraquinone-1-carboxylic acid (20 °C)	3.37			
Anthraquinone-2-carboxylic acid (20 °C)	3.42			
9,10-Anthraquinone monoxime	9.78			
9,10-Anthraquinone-1-sulfonic acid	0.27			
9,10-Anthraquinone-2-sulfonic acid	0.38			
Antipyrine	1.45(+1)			
Apomorphine (15 °C)		8.92		
D-(−)-Arabinose	12.34			
L-(+)-Arginine		8.994(+1)	12.47(−1)	
Arsenazo III [pK$_5$ = 10.5(−4); pK$_6$ = 12.0(−5)]		1.2	2.7	7.9(−3)
Arsenoacetic acid		4.67	7.68	
Arsenoacrylic acid		4.23	8.60	
Arsenobutanoic acid		4.92	7.64	
2-Arsenocrotonic acid		4.61	8.75	
3-Arsenocrotonic acid		4.03	8.81	
Arsenopentanoic acid		4.89	7.75	
L-(+)-Ascorbic acid (vitamin C)	4.17	11.57		
L-(+)-Asparagine		8.80(0)		
L-Asparaginylglycine		4.53	9.07	
D-Aspartic acid		3.87(0)	10.00(−)	
Aspartic diamide (μ = 0.2)	7.00			

Anhydroplatynecine

Antipyrine

TABLE 8.1 pK_A Values of Organic Materials in Water at 25 °C (*continued*)

Substance	pK_1	pK_2	pK_3	pK_4
Aspartylaspartic acid		3.40	4.70	8.26
α-Aspartylhistidine (38 °C, $\mu = 0.1$)		3.02	6.82	7.98
β-Aspartylhistidine (38 °C, $\mu = 0.1$)		2.95	6.93	8.72
N-Aspartyl-p-tyrosine ($\mu = 0.01$)		3.57	8.92	10.23(OH)
Aspidospermine	7.65			
Atropine (17 °C)	4.35(+1)			
1-Azacycloheptane	11.11(+1)			
1-Azacyclooctane	11.1(+1)			
Azetidine	11.29(+1)			
Aziridine	8.04(+1)			
Barbituric acid		8.372(0)		
m-Benzbetaine	3.217(+1)			
p-Benzbetaine	3.245(+1)			
Benzenearsonic acid (22 °C)		8.48(−1)		
Benzene-1-arsonic acid-4-carboxylic				
acid		4.22	5.59	
		(COOH)		
Benzeneboronic acid	13.7			
Benzene-1-carboxylic acid-				
2-phosphoric acid		3.78	9.17	
Benzene-1-carboxylic acid-				
3-phosphoric acid		4.03	7.03	
Benzene-1-carboxylic acid-				
4-phosphoric acid	1.50	3.95	6.89	
Benzenediazine	11.08(+1)			
1,3-Benzenedicarboxylic acid				
(isophthalic acid)	3.62(0)	4.60(−1)		
1,4-Benzenedicarboxylic acid				
(terephthalic acid)	3.54(0)	4.46(−1)		
1,3-Benzenedicarboxylic acid				
mononitrile	3.60(0)			
1,4-Benzenedicarboxylic acid	3.55(0)			
mononitrile				
Benzenehexacarboxylic acid				
($pK_5 = 6.32$; $pK_6 = 7.49$)	0.68	2.21	3.52	5.09
Benzenepentacarboxylic acid				
($pK_5 = 6.46$)	1.80	2.73	3.96	5.25
Benzenesulfinic acid	1.50			
Benzenesulfonic acid	2.554			
1,2,3,4-Benzenetetracarboxylic acid	2.05	3.25	4.73	6.21

Azetidine

Aziridine (ethyleneimine)

TABLE 8.1 pK$_A$ Values of Organic Materials in Water at 25 °C (*continued*)

Substance	pK$_1$	pK$_2$	pK$_3$	pK$_4$
1,2,3,5-Benzenetetracarboxylic acid	2.38	3.51	4.44	5.81
1,2,4,5-Benzenetetracarboxylic acid	1.92	2.87	4.49	5.63
1,2,3-Benzenetricarboxylic acid	2.88	4.75	7.13	
1,2,4-Benzenetricarboxylic acid	2.52	3.84	5.20	
1,3,5-Benzenetricarboxylic acid	2.12	4.10	5.18	
Benzil-α-dioxime	12.0			
Benzilic acid	3.09			
Benzimidazole	5.53(+1)	12.3(0)		
Benzohydroxamic acid (20 °C)	8.89(0)			
Benzoic acid	4.204			
5,6-Benzoquinoline (20 °C)	5.00(+1)			
7,8-Benzoquinoline (20 °C)	4.15(+1)			
1,4-Benzoquinone monoxime	6.20			
Benzosulfonic acid	0.70			
1,2,3-Benzotriazole	8.38(+1)			
1-Benzoylacetone	8.23			
Benzoylamine	9.34(+1)			
2-Benzoylbenzoic acid	3.54			
Benzoylglutamic acid	3.49	4.99		
N-Benzoyglycine (hippuric acid)	3.65			
Benzoylhydrazine	3.03(+2)	12.45(+1)		
Benzoylpyruvic acid	6.40	12.10		
3-Benzoyl-1,1,1-trifluoroacetone	6.35			
Benzylamine	9.35(+1)			
Benzylamine-4-carboxylic acid	3.59	9.64		
2-Benzyl-2-phenylsuccinic acid (20 °C)	3.69	6.47		
2-Benzylpyridine	5.13(+1)			
4-Benzylpyridine-1-oxide	−1.018(+)			
1-Benzylpyrrolidine	9.51(+1)			
2-Benzylpyrrolidine	10.31(+1)			
Benzylsuccinic acid (20 °C)	4.11	5.65		
3-(Benzylthio)propanoic acid	4.463			
Berberine (18 °C)	11.73(+1)			
Betaine	1.832(+1)			
Biguanide	2.96(+2)	11.51(+1)		
2,2′-Biimidazolyl (μ = 0.3)	5.01(+1)			
2-Biphenylcarboxylic acid	3.46			

Benzilic acid   5,6-Benzoquinoline   Biguanide

TABLE 8.1 pK_A Values of Organic Materials in Water at 25 °C (*continued*)

Substance	pK_1	pK_2	pK_3	pK_4
(1,1'-Biphenyl)-4,4'-diamine	3.63(+2)	4.70(+1)		
Bis(2-aminoethyl) ether (30 °C)	8.62(+2)	9.59(+1)		
N,N'-Bis(2-aminoethyl)- ethylenediamine (20 °C)	3.32(+4)	6.67(+3)	9.20(+2)	9.92(+1)
N,N-Bis(2-hydroxyethyl)-2- aminoethane sulfonic acid (BES) (20 °C)	7.15			
N,N-Bis(2-hydroxyethyl)glycine (bicine) (20 °C)	8.35			
Bis(2-hydroxyethyl)iminotris (hydroxymethyl)- methane (bis-tris)	6.46(+1)			
1,3-Bis[tris(hydroxymethyl) methylamino]propane (20 °C)	6.80(+1)			
Bromoaetic acid	2.902			
2-Bromoaniline	2.53(+1)			
3-Bromoaniline	3.53(+1)			
4-Bromoaniline	3.88(+1)			
2-Bromobenzoic acid	2.85			
3-Bromobenzoic acid	3.810			
4-Bromobenzoic acid	3.99			
2-Bromobutanoic acid (35 °C)	2.939			
erythro-2-Bromo-3-chlorosuccinic acid (19 °C, $\mu = 0.1$)	1.4	2.6		
threo-2-Bromo-chlorosuccinic acid (19 °C, $\mu = 0.1$)	1.5	2.8		
trans-2-Bromocinnamic acid	4.41			
3-Bromo-4-(dimethylamino)pyridine (20 °C)	6.52(+1)			
2-Bromo-4,6-dinitroaniline	−6.94(+1)			
3-Bromo-2-hydroxymethylbenzoic acid (20 °C)	3.28			
6-Bromo-2-hydroxymethylbenzoic acid (20 °C)	2.25			
7-Bromo-8-hydroxyquinoline- 5-sulfonic acid	2.51	6.70		
3-Bromomandelic acid	3.13			
3-Bromo-4-methylaminopyridine (20 °C)	7.49(+1)			
(2-Bromomethyl)butanoic acid	3.92			
Bromomethylphosphonic acid	1.14	6.52		

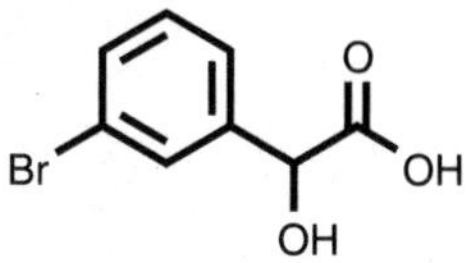

3-Bromomandelic acid

TABLE 8.1 pK$_A$ Values of Organic Materials in Water at 25 °C (*continued*)

Substance	pK$_1$	pK$_2$	pK$_3$	pK$_4$
2-Bromo-6-nitrobenzoic acid	1.37			
2-Bromophenol	8.452			
3-Bromophenol	9.031			
4-Bromophenol	9.34			
2-(2′-Bromophenoxy)acetic acid	3.12			
2-(3′-Bromophenoxy)acetic acid	3.09			
2-(4′-Bromophenoxy) acetic acid	3.13			
2-Bromo-2-phenylacetic acid	2.21			
2-(Bromophenyl)acetic acid	4.054			
4-(Bromophenyl)acetic acid	4.188			
4-Bromophenylarsonic acid	3.25	8.19		
4-Bromophenylphosphinic acid				
(17 °C)	2.1			
2-Bromophenylphosphonic acid	1.64	7.00		
3-Bromophenylphosphonic acid	1.45	6.69		
4-Bromophenylphosphonic acid	1.60	6.83		
3-Bromophenylselenic acid	4.43			
4-Bromophenylselenic acid	4.50			
2-Bromopropanoic acid	2.971			
3-Bromopropanoic acid	3.992			
Bromopropynoic acid	1.855			
2-Bromopyridine	0.71(+1)			
3-Bromopyridine	2.85(+1)			
4-Bromopyridine	3.71(+1)			
3-Bromoquinoline	2.69(+1)			
Bromosuccinic acid	2.55	4.41		
2-Bromo-*p*-tolylphosphonic acid	1.81	7.15		
Brucine (15 °C)	2.50(+2)	8.16(+1)		
2-Butanamine (*sec*-butylamine)	10.56(+1)			
1,2-Butanediamine	6.399(+2)	9.388(+1)		
1,4-Butanediamine	9.35(+2)	10.82(+1)		
2,3-Butanediamine	6.91(+2)	10.00(+1)		
1,2,3,4-Butanetetracarboxylic acid	3.43	4.58	5.85	7.16
cis-2-Butenoic acid				
(isocrotonic acid)	4.44			
trans-2-Butenoic acid (*trans*-				
crotonic acid) (35 °C)	4.676			
3-Butenoic acid (vinylacetic acid)	4.68			
3-Butoxybenzoic acid (20 °C)	4.25			
Butylamine	10.64(+1)			
tert-Butylamine	10.685(+1)			
4-*tert*-Butylaniline	3.78(+1)			
N-*tert*-Butylaniline	7.10(+1)			
Butylarsonic acid (18 °C)	4.23	8.91		
2-*tert*-Butylbenzoic acid	3.57			
3-*tert*-Butylbenzoic acid	4.199			
4-*tert*-Butylbenzoic acid	4.389			
N-Butylethylenediamine	7.53(+2)	10.30(+1)		
N-Butylglycine	2.35(+1)	10.25(0)		

TABLE 8.1 pK_A Values of Organic Materials in Water at 25 °C (*continued*)

Substance	pK_1	pK_2	pK_3	pK_4
tert-Butylhydroperoxide	12.80			
1-(*tert*-Butyl)-2-hydroxybenzene	10.62			
1-(*tert*-Butyl)-3-hydroxybenzene	10.119			
1-(*tert*-Butyl)-4-hydroxybenzene	10.23			
Butylmethylamine	10.90(+1)			
2-Butyl-1-methyl-2-pyrroline	11.84(+1)			
4-*tert*-Butylphenylactic acid	4.417			
Butylphosphinic acid	3.41			
tert-Butylphosphinic acid	4.24			
tert-Butylphosphonic acid	2.79	8.88		
1-Butylpiperidine ($\mu = 0.02$)	10.43(+1)			
2-*tert*-Butylpyridine	5.76(+1)			
3-*tert*-Butylpyridine	5.82(+1)			
4-*tert*-Butylpyridine	5.99(+1)			
2-*tert*-Butylthiazole ($\mu = 0.1$)	3.00(+1)			
4-*tert*-Butylthiazole ($\mu = 0.1$)	3.04(+1)			
2-Butyn-1,4-dioic acid	1.75	4.40		
2-Butynoic acid (tetrolic acid)	2.620			
Butyric acid	4.817			
4-Butyrobetaine (20 °C)	3.94(+1)			
Caffeine (40 °C)	10.4			
Calcein ($pK_5 > 12$)	<4	5.4	9.0	10.5
Calmagite	8.14	12.35		
D-Camphoric acid	4.57	5.10		
Canaline	2.40	3.70	9.20	
Canavanine	2.50(+2)	6.60(+1)	9.25(0)	
N-Carbamoylacetic acid	3.64			
N-Carbamoyl-α-D-alanine	3.89(+1)			
N-Carbamoyl-β-alanine	4.99(+1)			
DL-*N*-Carbamoylalanine	3.892(+1)			
N-Carbamoylglycine	3.876			
2-Carbamoylpyridine (20 °C)	2.10(+1)			
3-Carbamoylpyridine	3.328(+1)			
4-Carbamoylpyridine (20 °C)	3.61(+1)			
β-Carboxymethylaminopropanoic acid	3.61(+1)	9.46(0)		
Chloroacetic acid	2.867			
N-(2′-Chloroacetyl)glycine	3.38(0)			
cis-3-Chloroacrylic acid (18 °C, $\mu = 0.1$)	3.32			
trans-3-chloroacrylic acid (18 °C, $\mu = 0.1$)	3.65			
2-Chloroaniline	2.64(+1)			
3-Chloroaniline	3.52(+1)			
4-Chloroaniline	3.99(+1)			
2-Chlorobenzoic acid	2.877			
3-Chlorobenzoic acid	3.83			
4-Chlorobenzoic acid	3.986			
2-Chlorobutanoic acid	2.86			

TABLE 8.1   pK$_A$ Values of Organic Materials in Water at 25 °C (*continued*)

Substance	pK$_1$	pK$_2$	pK$_3$	pK$_4$
3-Chlorobutanoic acid	4.05			
4-Chlorobutanoic acid	4.50			
2-Chloro-3-butenoic acid	2.54			
3-Chlorobutylarsonic acid (18 °C)	3.95	8.85		
trans-2'-Chlorocinnamic acid	4.234			
trans-3'-Chlorocinnamic acid	4.294			
trans-4'-Chlorocinnamic acid	4.413			
2-Chlorocrotonic acid	3.14			
3-Chlorocrotonic acid	3.84			
Chlorodifluoroacetic acid	0.46			
1-Chloro-1,2-dihydroxybenzene	8.522			
1-Chloro-2,6-dimethyl- 4-hydroxybenzene	9.549			
4-Chloro-2,6-dinitrophenol	2.97			
2-Chloroethylarsonic acid	3.68	8.37		
3-Chlorohexyl-1-arsonic acid (18 °C)	3.51	8.31		
2-Chloro-3-hydroxybutanoic acid	2.59			
3-Chloro-2-(hydroxymethyl)benzoic acid (20 °C)	3.27			
6-Chloro-2-(hydroxymethyl)benzoic acid (20 °C)	2.26			
7-Chloro-8-hydroxyquinoline- 5-sulfonic acid	2.92	6.80		
2-Chloroisocrotonic acid	2.80			
3-Chloroisocrotonic acid	4.02			
3-Chlorolactic acid	3.12			
3-Chloromandelic acid	3.237			
3-Chloro-4-methoxyphenyl- phosphonic acid	2.25	6.7		
3-Chloro-4-methylaniline	4.05(+1)			
4-Chloro-*N*-methylaniline	3.9(+1)			
4-Chloro-3-methylphenol	9.549			
Chloromethylphosphonic acid	1.40	6.30		
2-Chloro-2-methylpropanoic acid	2.975			
2-Chloro-6-nitroaniline	−2.41(+1)			
4-Chloro-2-nitroaniline	−1.10(+1)			
2-Chloro-3-nitrobenzoic acid	2.02			
2-Chloro-4-nitrobenzoic acid	1.96			
2-Chloro-5-nitrobenzoic acid	2.17			
2-Chloro-6-nitrobenzoic acid	1.342			
4-Chloro-2-nitrophenol	6.48			

2-Chloroisocrotonic acid

TABLE 8.1   pK$_A$ Values of Organic Materials in Water at 25 °C (*continued*)

Substance	pK$_1$	pK$_2$	pK$_3$	pK$_4$
2-Chlorophenol	8.55			
3-Chlorophenol	9.10			
4-Chlorophenol	9.43			
(4-Chloro-3-nitrophenoxy)acetic acid	2.959			
2-Chloro-4-nitrophenylphosphonic acid	1.12	6.14		
3-Chloropentyl-l-arsonic acid (18 °C)	3.71	8.77		
2-Chlorophenoxyacetic acid	3.05			
3-Chlorophenoxyacetic acid	3.07			
4-Chlorophenoxyacetic acid	3.10			
4-Chlorophenoxy-2-methylacetic acid	3.26			
2-Chlorophenylacetic acid	4.066			
3-Chlorophenylacetic acid	4.140			
4-Chlorophenylacetic acid	4.190			
2-Chlorophenylalanine	2.23(+1)	8.94(0)		
3-Chlorophenylalanine	2.17(+1)	8.91(0)		
DL-4-Chlorophenylalanine	2.08(+1)	8.96(0)		
4-Chlorophenylarsonic acid	3.33	8.25		
2-Chlorophenylphosphonic acid	1.63	6.98		
3-Chlorophenylphosphonic acid	1.55	6.65		
4-Chlorophenylphosphonic acid	1.66	6.75		
3-(2'-Chlorophenyl)propanoic acid	4.577			
3-(3'-Chlorophenyl)propanoic acid	4.585			
3-(4'-Chlorophenyl)propanoic acid	4.607			
3-Chlorophenylselenic acid	4.47			
4-Chlorophenylselenic acid	4.48			
4-Chloro-1,2-phthalic acid	1.60			
2-Chloropropanoic acid	2.84			
3-Chloropropanoic acid	3.992			
2-Chloropropylarsonic acid (18 °C)	3.76	8.39		
3-Chloropropylarsonic acid (18 °C)	3.63	8.53		
Chloropropynoic acid	1.845			
2-Chloropyridine	0.49(+1)			
3-Chloropyridine	2.84(+1)			
4-Chloropyridine	3.83(+1)			
7-Chlorotetracycline	3.30(+1)	7.44	9.27	
4-Chloro-2-(2'-thiazolylazo)phenol	7.09			
4-Chlorothiophenol	5.9			
N-Chloro-p-toluenesulfonamide	4.54(+1)			
3-Chloro-o-toluidine	2.49(+1)			
4-Chloro-o-toluidine	3.385(+1)			
5-Chloro-o-toluidine	3.85(+1)			
6-Chloro-o-toludine	3.62(+1)			
Chrome Azurol S	2.45	4.86	11.47	
Chrome Dark Blue	7.56	9.3	12.4	
Cinchonine	5.85(+2)	9.92(+1)		

TABLE 8.1   pK$_A$ Values of Organic Materials in Water at 25°C (*continued*)

Substance	pK$_1$	pK$_2$	pK$_3$	pK$_4$
cis-Cinnamic acid	3.879			
trans-Cinnamic acid	4.438			
Citraconic acid	2.29(0)	6.15(−1)		
Citric acid	3.128	4.761	6.396	
L-(+)-Citrulline	2.43(+1)	9.41(0)		
Cocaine	8.41(+1)			
Codeine	7.95(+1)			
Colchicine	1.65(+1)			
Coniine ($\mu = 0.5$)	11.24(+1)			
Creatine (40°C)	3.28(+1)			
Creatinine	3.57(+1)			
o-Cresol	10.26			
m-Cresol	10.00			
p-Cresol	10.26			
Cumene hydroperoxide	12.60			
Cupreine	7.63(+1)			
Cyanamide	10.27			
Cyanoacetic acid	2.460			
Cyanoacetohydrazide	2.34(+2)	11.17(+1)		
2-Cyanobenzoic acid	3.14			
3-Cyanobenzoic acid	3.60			
4-Cyanobenzoic acid	3.55			
4-Cyanobutanoic acid	4.44			
trans-1-Cyanocyclohexane-2-carboxylic acid	3.865			
4-Cyano-2,6-dimethylphenol	8.27			
4-Cyano-3,5-dimethylphenol	8.21			
2-Cyanoethylamine	7.7(+1)			
N-(2-Cyano)ethylnorcodeine	5.68(+1)			
Cyanomethylamine	5.34(+1)			
2-Cyano-2-methyl-2-phenylacetic acid	2.290			
1-Cyanomethylpiperidine	4.55(+1)			
2-Cyano-2-methylpropanoic acid	2.422			
3-Cyanophenol	8.61			
o-Cyanophenoxyacetic acid	2.98			
m-Cyanophenoxyacetic acid	3.03			
p-Cyanophenoxyacetic acid	2.93			
2-Cyanopropanoic acid	2.37			
3-Cyanopropanoic acid	3.99			
2-Cyanopyridine	−0.26(+1)			

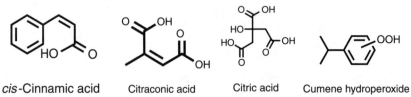

cis-Cinnamic acid   Citraconic acid   Citric acid   Cumene hydroperoxide

TABLE 8.1   pK$_A$ Values of Organic Materials in Water at 25 °C (*continued*)

Substance	pK$_1$	pK$_2$	pK$_3$	pK$_4$
3-Cyanopyridine	1.45(+1)			
4-Cyanopyridine	1.90(+1)			
Cyanuric acid	6.78			
Cyclobutanecarboxylic acid	4.785			
1,1-Cyclobutanedicarboxylic acid	3.13	5.88		
cis-1,2-Cyclobutanedicarboxylic acid	3.90	5.89		
trans-1,2-Cyclobutanedicarboxylic acid	3.79	5.61		
cis-1,3-Cyclobutanedicarboxylic acid	4.04	5.31		
trans-1,3-Cyclobutanedicarboxylic acid	3.81	5.28		
Cyclohexanecarboxylic acid	4.90			
1,1-Cyclohexanediacetic acid	3.49	6.96		
cis-1,2-Cyclohexanediacetic acid (20 °C)	4.42	5.45		
trans-1,2-Cyclohexanediacetic acid (20 °C)	4.38	5.42		
cis-1,2-Cyclohexanediamine	6.43(+2)	9.93(+1)		
trans-1,2-Cyclohexanediamine	6.34(+2)	9.74(+1)		
1,1-Cyclohexanedicarboxylic acid	3.45	4.11		
cis-1,2-Cyclohexanedicarboxylic acid (20 °C)	4.34	6.76		
trans-1,2-Cyclohexanedicarboxylic acid (20 °C)	4.18	5.93		
cis-1,3-Cyclohexanedicarboxylic acid (16 °C)	4.10	5.46		
trans-1,3-Cyclohexanedicarboxylic acid (19 °C)	4.31	5.73		
trans-1,4-Cyclohexanedicarboxylic acid (16 °C)	4.18	5.42		
1,3-Cyclohexanedione	5.26			
cis,cis-1,3,5-Cyclohexanetriamine	6.9(+3)	8.7(+2)	10.4(+1)	
Cyclohexanonimine	9.15			
cis-4-Cyclohexene-1,2-dicarboxylic acid (20 °C)	3.89	6.79		
trans-4-Cyclohexene-1,2-dicarboxylic acid (20 °C)	3.95	5.81		
Cyclohexylacetic acid	4.51			
Cyclohexylamine	10.64(+1)			
2-(Cyclohexylamino)ethanesulfonic acid (CHES) (20 °C)	9.55			

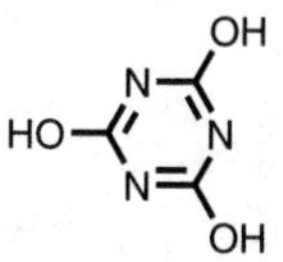

Cyanuric acid

TABLE 8.1   pK$_A$ Values of Organic Materials in Water at 25 °C (*continued*)

Substance	pK$_1$	pK$_2$	pK$_3$	pK$_4$
3-Cyclohexylamino- 1-propanesulfonic acid (CAPS) (20 °C)	10.40			
4-Cyclohexylbutanoic acid	4.95			
Cyclohexylcyanoacetic acid	2.367			
1,2-Cyclohexylenedinitriloacetic acid ($\mu = 0.1$)	2.4	3.5	6.16	12.35
3-Cyclohexylpropanoic acid	4.91			
2-Cyclohexylpyrrolidine	10.76(+1)			
2-Cyclohexyl-2-pyrroline	7.91(+1)			
Cyclohexylthioacetic acid	3.488			
Cyclopentanecarboxylic acid	4.905			
cis-Cyclopentane-1-carboxylic acid- 2-acetic acid	4.40	5.79		
trans-Cyclopentane-1-carboxylic acid-2-acetic acid	4.39	5.67		
Cyclopentane-1,2-diamine-*N,N′,N′*- tetraacetic acid ($\mu = 0.1$)				10.20
Cyclopentane-1,1-dicarboxylic acid	3.23	4.08		
cis-Cyclopentane-1,2-dicarboxylic acid	4.43	6.67		
trans-Cyclopentane-1,2-dicarboxylic acid	3.96	5.85		
cis-Cyclopentane-1,3-dicarboxylic acid	4.26	5.51		
trans-Cyclopentane-1,3-dicarboxylic acid	4.32	5.42		
Cyclopentylamine	10.65(+1)			
1,1-Cyclopentyldiacetic acid	3.80	6.77		
cis-Cyclopentyl-1,2-diacetic acid	4.42	5.42		
trans-Cyclopentyl-1,2-diacetic acid	4.43	5.43		
Cyclopropanecarboxylic acid	4.827			
Cyclopropane-1,1-dicarboxylic acid	1.82	5.43		
cis-Cyclopropane-1,2-dicarboxylic acid	3.33	6.47		
trans-Cyclopropane-1,2-dicarboxylic acid	3.65	5.13		
Cyclopropylamine	9.10(+1)			
5-Cyclopropyl-1,2,3,4-tetrazole	4.90(+1)			
L-Cysteic acid (3-sulfo-L-alanine)	1.89(+1)	8.7(0)		
L-(+)-Cysteine	1.71(+1)	8.39(0)	10.70(SH)	

5-Cyclopropyl-1,2,3,4-tetrazole

TABLE 8.1 pK_A Values of Organic Materials in Water at 25 °C (*continued*)

Substance	pK_1	pK_2	pK_3	pK_4
L-(+)-Cysteine, ethyl ester	6.69 (NH_3^+)	9.17(SH)		
L-(+)-Cysteine, methyl ester	6.56 (NH_3^+)	8.99(SH)		
L-Cysteinyl-L-asparagine	2.97	7.09	8.47	
L-Cystine (35 °C)	1.6(+2)	2.1(+1)	8.02(0)	8.71(−1)
Cystinylglycylglycine (35 °C)	3.12	3.21	6.01	6.87
Cytidine	4.08(+1)	12.24(0)		
Cytidine-2′-phosphoric acid	0.8(+1)	4.36(0)	6.17(−1)	
Cytidine-3′-phosphoric acid	0.80(+1)	4.31(0)	6.04(−1)	13.2(sugar)
Cytidine-5′-phosphoric acid		4.39(0)	6.62(−1)	
Cytosine	4.58(+1)	12.15(0)		
Decanedioic acid (sebacic acid)	4.59	5.59		
Dehydroascorbic acid (20 °C)	3.21	7.92	10.3	
2′-Deoxyadenosine ($\mu = 0.1$)	3.8(+1)			
Deoxycholic acid	6.58			
2-Deoxyglucose	12.52			
2-Deoxyguanosine ($\mu = 0.1$)	2.5(+1)			
5-Desoxypyridoxal ($\mu = 0$)	4.17(+1)	8.14(OH)		
1,1-Diacetic acid semicarbazide (30 °C, $\mu = 0.1$)	2.96	4.04		
Diacetylacetone	7.42			
Diallylamine ($\mu = 0.02$)	9.29(+1)			
5,5-Diallylbarbituric acid	7.78(0)			
1,3-Diamino-2-aminomethylpropane	6.44(+3)	8.56(+2)	10.38(+1)	
3,5-Diaminobenzoic acid	5.30			
1,3-Diamino-N,N′-bis- (2-aminoethyl)propane ($\mu = 0.5$)	6.01(+4)	7.26(+3)	9.49(+2)	10.23(+1)
2,4-Diaminobutanoic acid (20 °C)	1.85(+2)	8.24(+1)	10.40(0)	
2,2′-Diaminodiethyl sulfide (30 °C)	8.84(+2)	9.64(+1)		
1,8-Diamino-3,6-dithiooctane (30 °C)	8.43(+2)	9.31(+1)		
2,7-Diaminooctanedioic acid (20 °C, $\mu = 0.1$)	1.84(+2)	2.64(+1)	9.23(0)	9.89(−1)
1,8-Diamino-3,6-octanedione (30 °C)	8.60(+2)	9.57(+1)		
1,8-Diamino-3-oxa-6-thiooctane	8.54(+2)	9.46(+1)		
2,3-Diaminopropanoic acid ($\mu = 0.1$)	1.33(+2)	6.674(+1)	9.623(0)	
2,3-Diaminopropanoic acid, methyl ester ($\mu = 0.1$)	4.412(+1)	8.250(0)		

Cytidine

Deoxycholic acid

TABLE 8.1 pK$_A$ Values of Organic Materials in Water at 25°C (*continued*)

Substance	pK$_1$	pK$_2$	pK$_3$	pK$_4$
1,3-Diamino-2-propanol (20°C)	7.93(+2)	9.69(+1)		
2,5-Diaminopyridine (20°C)	2.13(+2)	6.48(+1)		
1,4-Diazabicyclo[2.2.2]octane	2.90(+2)	8.60(+1)		
Dibenzylamine	8.52(+1)			
Dibenzylsuccinic acid (20°C)	3.96	6.66		
Dibromoacetic acid	1.39			
3,5-Dibromoaniline	2.35(+1)			
3,5-Dibromophenol	8.056			
2,2-Dibromopropanoic acid	1.48			
2,3-Dibromopropanoic acid	2.33			
rac-2,3-Dibromosuccinic acid				
(20°C)	1.43	2.24		
meso-2,3-Dibromosuccinic acid				
(20°C)	1.51	2.71		
3,5-Dibromo-*p*-L-tyrosine	2.17(+1)	6.45(0)	7.60(−1)	
Dibutylamine	11.25(+1)			
Di-*sec*-butylamine	10.91(+1)			
2,6-Di-*tert*-butylpyridine	3.58(+1)			
rac-2,3-Di-*tert*-butylsuccinic acid				
($\mu = 0.1$)	3.58	10.2		
1,12-Dicarboxydodecaborane	9.07	10.23		
Dichloroacetic acid	1.26			
Dichloroacetylacetic acid	2.11			
3,5-Dichloroaniline	2.37(+1)			
1,3-Dichloro-2,5-dihydroxybenzene				
($\mu = 0.65$)	7.30	9.99		
2,5-Dichloro-3,6-dihydroxy-				
p-benzoquinone	1.09	2.42		
Dichloromethylphosphonic acid	1.14	5.61		
2,4-Dichloro-6-nitroaniline	−3.00(+1)			
2,5-Dichloro-4-nitroaniline	−1.74(+1)			
2,6-Dichloro-4-nitroaniline	−3.31(+1)			
2,3-Dichlorophenol	7.44			
2,4-Dichlorophenol	7.85			
2,6-Dichlorophenol	6.78			
3,4-Dichlorophenol	8.630			
3,5-Dichlorophenol	8.179			
2,4-Dichlorophenoxyacetic acid				
(2,4-D)	2.64			
4,6-Dichlorophenoxy-2-methylacetic				
acid	3.13			
3,6-Dichlorophthalic acid	1.46			
2,2-Dichloropropanoic acid	2.06			
2,3-Dichloropropanoic acid	2.85			
rac-2,3-Dichlorosuccinic acid (20°C)	1.43	2.81		
meso-2,3-Dichlorosuccinic acid	1.49	2.97		
3,5-Dichloro-*p*-tyrosine	2.12	6.47	7.62	
2-Dicyanoethylamine	5.14(+1)			
2,2-Dicyanopropanoic acid	−2.8			

TABLE 8.1   pK$_A$ Values of Organic Materials in Water at 25 °C (*continued*)

Substance	pK$_1$	pK$_2$	pK$_3$	pK$_4$
Dicyclohexylamine	11.25(+1)			
Dicyclopentylamine	10.93(+1)			
Didodecylamine	10.99(+)			
Diethanolamine	8.88(+1)			
Di(ethoxyethyl)amine	8.47(+1)			
3,5-Diethoxyphenol	9.370			
3-(Diethoxyphosphinyl)benzoic acid	3.65			
4-(Diethoxyphosphinyl)benzoic acid	3.60			
3-(Diethoxyphosphinyl)phenol	8.66			
4-(Diethoxyphosphinyl)phenol	8.28			
Diethylamine	10.8(+1)			
2-(Diethylamino)ethyl- 4-aminobenzoate	8.85(+1)			
α-(Diethylamino)toluene	9.44(+1)			
N,N-Diethylaniline	6.56(+1)			
5,5-Diethylbarbituric acid (veronal)	8.020(0)			
N,N-Diethylbenzylamine	9.48(+1)			
Diethylbiguanide (30 °C)	2.53(+1)	11.68(0)		
Diethylenetriamine	4.42(+3)	9.21(+2)	10.02(+1)	
Diethylenetriaminepentaacetic acid (pK$_5$ = 10.58)	1.80(0)	2.55(−1)	4.33(−2)	8.60(−3)
N,N-Diethylethylenediamine	7.70(+2)	10.46(+1)		
2,2-Diethylglutaric acid	3.62	7.12		
N,N-Diethylglycine	2.04(+1)	10.47(0)		
Diethylglycolic acid (18 °C)	3.804			
Diethylmalonic acid	2.151	7.417		
Diethylmethylamine	10.43(+1)			
rac-2,3-Diethylsuccinic acid	3.63	6.46		
meso-2,3-Diethylsuccinic acid	3.54	6.59		
N,N-Diethyl-o-toluidine	7.18(+1)			
Difluoroacetic acid	1.33			
3,3-Difluoroacrylic acid	3.17			
Diglycolic acid	2.96			
Diguanidine	12.8			
Dihexylamine	11.0(+1)			
Dihydroarecaidine	9.70			
Dihydroarecaidine, methyl ester	8.39			
Dihydrocodeine	8.75(+1)			
Dihydroergonovine	7.38(+1)			
α-Dihydrolysergic acid	3.57	8.45		

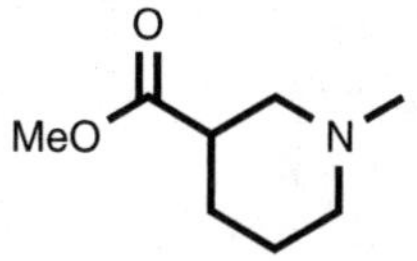

Dihydroarecaidine

TABLE 8.1   pK$_A$ Values of Organic Materials in Water at 25 °C (*continued*)

Substance	pK$_1$	pK$_2$	pK$_3$	pK$_4$
γ-Dihydrolysergic acid	3.60	8.71		
α-Dihydrolysergol	8.30			
β-Dihydrolysergol	8.23			
Dihydromorphine	9.35			
3,4-Dihydroxyalanine	2.32(+1)	8.68(0)	9.87(−1)	

1,2-Dihydroxyanthraquinone-
  3-sulfonic acid
  (alizarin-3-sulfonic acid)
		5.54(−1)	11.01(−2)	

3,4-Dihydroxybenzaldehyde	7.55			
1,2-Dihydroxybenzene (pyrocatechol) (μ = 0.1)	9.356(0)	12.98(−1)		
1,3-Dihydroxybenzene (resorcinol)	9.44(0)	12.32(−1)		
1,4-Dihydroxybenzene (hydroquinone)	9.91(0)	12.04(−1)		
4,5-Dihydroxybenzene-1,3-disulfonic acid			7.66(−2)	12.6(−3)
2,3-Dihydroxybenzoic acid (30 °C)	2.98	10.14		
2,4-Dihydroxybenzoic acid (β-resorcyclic acid)	3.29	8.98		
2,5-Dihydroxybenzoic acid	2.97	10.50		
2,6-Dihydroxybenzoic acid	1.30			
3,4-Dihydroxybenzoic acid	4.48	8.67	11.74	
3,5-Dihydroxybenzoic acid	4.04			
2,5-Dihydroxy-p-benzoquinone	2.71	5.18		
3,4-Dihydroxy-3-cyclobutene- 1,2-dione	0.541	3.480		
2,3-Dihydroxy-2-cyclopenten-1-one (20 °C)	4.72			
1,4-Dihydroxy-2,6-dinitrobenzene	4.42	9.14		
Di(2,2′-hydroxyethyl)amine	8.8(+1)			
N,N-Di(2-hydroxyethyl)glycine	8.333			
Dihydroxymaleic acid	1.10			
Dihydroxymalic acid	1.92			
1,3-Dihydroxy-2-methylbenzene (μ = 0.65)	10.05	11.64		
2,2-Di(hydroxymethyl)- 3-hydroxypropanoic acid	4.460			
2,4-Dihydroxy-5-methylpyrimidine	9.90			
2,4-Dihydroxy-6-methylpyrimidine	9.52			
1,4-Dihydroxynaphthalene (26 °C, μ = 0.65)	9.37	10.93		
1,2-Dihydroxy-3-nitrobenzene	6.68			
1,2-Dihydroxy-4-nitrobenzene (μ = 0.1)	6.701			
2,4-Dihydroxy-1-phenylazobenzene (μ = 0.1)	11.98			
2,4-Dihydroxyoxazolidine	6.11(+1)			
2,4-Dihydroxypteridine	<1.3	7.92		
2,6-Dihydroxypurine	7.53(0)	11.84(−1)		

TABLE 8.1 pK$_A$ Values of Organic Materials in Water at 25 °C (*continued*)

Substance	pK$_1$	pK$_2$	pK$_3$	pK$_4$
2,4-Dihydroxypyridine (20 °C)	1.37(+1)	6.45(0)	13(−1)	
Dihydroxytartaric acid	1.95	4.00		
1,4-Dihydroxy-2,3,5,6-tetramethylbenzene ($\mu = 0.65$)	11.25	12.70		
3,5-Diiodoaniline	2.37(+1)			
2,5-Diiodohistamine	2.31(+2)	8.20(+1)	10.11(0)	
2,5-Diiodohistidine ($\mu = 0.1$)	2.72	8.18	9.76	
3,5-Diiodophenol	8.103			
3,5-Diiodotyrosine	2.117(+1)	6.479(0)	7.821(−1)	
Diisopropylmalonic acid	2.124	8.848		
Dilactic acid	2.955			
threo-1,4-Dimercapto-2,3-butanediol	8.9			
meso-2,3-Dimercaptosuccinic acid	2.71	3.48	8.89(SH)	10.79(SH)
3,5-Dimethoxyaniline	3.86(+1)			
2,6-Dimethoxybenzoic acid	3.44			
1,10-Dimethoxy-3,8-dimethyl-4,7-phenanthroline	7.21			
Di(2-methoxyethyl)amine	9.51(+1)			
3,5-Dimethoxyphenol	9.345			
(3,4-Dimethoxy)phenylacetic acid	4.333			
Dimethylamine	10.77(+1)			
4-Dimethylaminobenzaldehyde	1.647(+1)			
N,N-Dimethylaminocyclohexane	10.72(+1)			
4-Dimethylamino-2,3-dimethyl-1-phenyl-3-pyrazolin-5-one	4.18(+1)			
4-Dimethylamino-3,5-dimethylpyridine (20 °C)	8.15(+1)			
2-(Dimethylamino)ethanol	9.26(+1)			
2-[2-(Dimethylamino)ethyl]pyridine	3.46(+2)	8.75(+1)		
3-(Dimethylaminoethyl)pyridine	4.30(+2)	8.86(+1)		
4-(Dimethylaminoethyl)pyridine	4.66(+2)	8.70(+1)		
4-(Dimethylamino)-3-ethylpyridine (20 °C)	8.66(+1)			
4-(Dimethylamino)-3-isopropylpyridine (20 °C)	8.27(+1)			
2-(Dimethylaminomethyl)pyridine	2.58(+2)	8.12(+1)		
3-(Dimethylaminomethyl)pyridine	3.17(+2)	8.00(+1)		
4-(Dimethylaminomethyl)pyridine	3.39(+2)	7.66(+1)		
4-(Dimethylamino)-3-methylpyridine (20 °C)	8.68(+1)			
4-(Dimethylaminophenyl)phosphonic acid	2.0(+1)	4.2	7.35	
3-(Dimethylamino)propanoic acid	9.85(+1)			
4-(Dimethylamino)pyridine (20 °C)	6.09(+1)			
N,N-Dimethylaniline	5.15(+1)			
2,3-Dimethylaniline	4.70(+1)			
2,4-Dimethylaniline	4.89(+1)			
2,5-Dimethylaniline	4.53(+1)			
2,6-Dimethylaniline	3.95(+1)			

TABLE 8.1   pK$_A$ Values of Organic Materials in Water at 25 °C (*continued*)

Substance	pK$_1$	pK$_2$	pK$_3$	pK$_4$
3,4-Dimethylaniline	5.17(+1)			
3,5-Dimethylaniline	4.765(+1)			
N,N-Dimethylaniline-4-phosphonic				
acid (17 °C)	2.0(+1)	4.2	7.39	
Dimethylarsinic acid (cacodylic acid)	6.273			
1,3-Dimethylbarbituric acid	4.68(+1)			
2,3-Dimethylbenzoic acid	3.771			
2,4-Dimethylbenzoic acid	4.217			
2,5-Dimethylbenzoic acid	3.990			
2,6-Dimethylbenzoic acid	3.362			
3,4-Dimethylbenzoic	4.41			
3,5-Dimethylbenzoic acid	4.302			
N,N-Dimethylbenzylamine	9.02(+1)			
Dimethylbiguanide	2.77(+1)	11.52		
2,2-Dimethylbutanoic acid (18 °C)	5.03			
Dimethylchlorotetracycline ($\mu = 0.01$)	3.30(+1)			
2,6-Dimethyl-4-cyanophenol	8.27			
3,5-Dimethyl-4-cyanophenol	8.21			
5,5-Dimethyl-1,3-cyclohexanedione	5.15			
cis-3,3-Dimethyl-1,2-				
cyclopropanedicarboxylic acid	2.34	8.31		
trans-3,3-Dimethyl-				
1,2-cyclopropanedicarboxylic acid	3.92	5.32		
3,5-Dimethyl-4-(dimethylamino)-				
pyridine (20 °C)	8.12(+1)			
2,2-Dimethyl-1,3-dioxane-4,6-dione	5.1			
1,1-Dimethylethanethiol ($\mu = 0.1$)	11.22			
N,N-Dimethylethylenediamine-				
N,N-diacetic acid	6.63	9.53		
N,N'-Dimethylethylenediamine-				
N,N'-diacetic acid	7.40	10.16		
N,N-Dimethylethylenediamine-				
N,N'-diacetic acid	5.99	9.97		
N,N-Dimethylglycine	2.146(+1)	9.940(0)		
Dimethylglycolic acid (18 °C)	4.04			
N,N-Dimethylglycylglycine	3.11(+1)	8.09(0)		
Dimethylglyoxime	10.60			
5,5-Dimethyl-2,4-hexanedione	10.01			
5,5-Dimethylhydantoin	9.19			
2,4-Dimethyl-8-hydroxyquinoline	6.20(+1)	10.60(0)		
3,4-Dimethyl-8-hydroxyquinoline	5.80(+1)	10.05(0)		

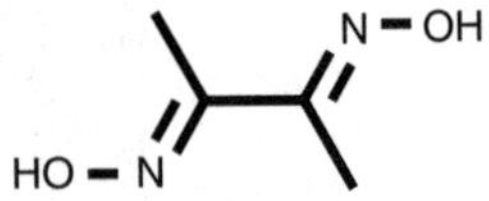

Dimethylglyoxime

TABLE 8.1   pK$_A$ Values of Organic Materials in Water at 25 °C (*continued*)

Substance	pK$_1$	pK$_2$	pK$_3$	pK$_4$
2,4-Dimethyl-8-hydroxyquinoline- 7-sulfonic acid	3.20 (NH$^+$)	10.14(OH)		
Dimethylhydroxytetracycline	7.5	9.4		
2,4-Dimethylimidazole	8.38(+1)			
Dimethylmalic acid	3.17	6.06		
2,2-Dimethylmalonic acid	3.17	6.06		
3,5-Dimethyl-4-(methylamino) pyridine (20 °C)	9.96(+1)			
2,3-Dimethylnaphthalene- 1-carboxylic acid	3.33			
2,6-Dimethyl-4-nitrophenol	7.190			
3,5-Dimethyl-4-nitrophenol	8.245			
α,α-Dimethyloxaloacetic acid	1.77	4.62		
3,3-Dimethylpentanedioic acid	3.70	6.34		
2,2-Dimethylpentanoic acid	4.969			
4,4-Dimethylpentanoic acid (18 °C)	4.79			
2,3-Dimethylphenol	10.50			
2,4-Dimethylphenol	10.58			
2,5-Dimethylphenol	10.22			
2,6-Dimethylphenol	10.59			
3,4-Dimethylphenol	10.32			
3,5-Dimethylphenol	10.15			
2,6-Dimethylphenoxyacetic acid	3.356			
Dimethylphenylsilylacetic acid	5.27			
N,N'-Dimethylpiperazine	4.630(+2)	8.539(+1)		
1,2-Dimethylpiperidine	10.22			
cis-2,6-Dimethylpiperidine	11.07(+1)			
2,2-Dimethylpropanoic acid (pivalic acid)	5.031			
2,2'-Dimethylpropylphosphonic acid	2.84	8.65		
2,4-Dimethylpyridine (2,4-lutidine)	6.74(+1)			
2,5-Dimethylpyridine (2,5-lutidine)	6.43(+1)			
2,6-Dimethylpyridine (2,6-lutidine)	6.71(+1)			
3,4-Dimethylpyridine (3,4-lutidine)	6.47(+1)			
3,5-Dimethylpyridine (3,5-lutidine)	6.09(+1)			
2,4-Dimethylpyridine-1-oxide	1.627(+1)			
2,5-Dimethylpyridine-1-oxide	1.208(+1)			
2,6-Dimethylpyridine-1-oxide	1.366(+1)			
3,4-Dimethylpyridine-1-oxide	1.493(+1)			
3,5-Dimethylpyridine-1-oxide	1.181(+1)			
2,3-Dimethylquinoline	4.94(+1)			
2,6-Dimethylquinoline	5.46(+1)			
meso-2,2-Dimethylsuccinic acid	3.77	5.936		
rac-2,2-Dimethylsuccinic acid	3.93	6.20		
D-2,3-Dimethylsuccinic acid	3.82	5.93		
meso-2,3-Dimethylsuccinic acid	3.67	5.30		
rac-2,3-Dimethylsuccinic acid	3.94	6.20		
2,4-Dimethylthiazole ($\mu = 0.1$)	3.98			

TABLE 8.1 pK$_A$ Values of Organic Materials in Water at 25 °C (*continued*)

Substance	pK$_1$	pK$_2$	pK$_3$	pK$_4$
2,5-Dimethylthiazole ($\mu = 0.1$)	3.91			
4,5-Dimethylthiazole ($\mu = 0.1$)	3.73			
N,N-Dimethyl-o-toluidine	5.86(+1)			
N,N-Dimethyl-p-toluidine	7.24(+1)			
2,4-Dinitroaniline	−4.25(+1)			
2,6-Dinitroaniline	−5.23(+1)			
3,5-Dinitroaniline	0.229(+1)			
2,3-Dinitrobenzoic acid	1.85			
2,4-Dinitrobenzoic acid	1.43			
2,5-Dinitrobenzoic acid	1.62			
2,6-Dinitrobenzoic acid	1.14			
3,4-Dinitrobenzoic acid	2.82			
3,5-Dinitrobenzoic acid	2.85			
1,1-Dinitrobutane (20 °C)	5.90			
1,1-Dinitrodecane	3.60			
1,1-Dinitroethane (20 °C)	5.21			
Dinitromethane (20 °C)	3.60			
1,1-Dinitropentane	5.337			
2,4-Dinitrophenol	4.08			
2,5-Dinitrophenol	5.216			
2,6-Dinitrophenol	3.713			
3,4-Dinitrophenol	5.424			
3,5-Dinitrophenol	6.732			
2,4-Dinitrophenylacetic acid	3.50			
1,1-Dinitropropane (20 °C)	5.5			
2,6-Dioxo-1,2,3,6-tetrahydro- 　4-pyrimidinecarboxylic acid 　(orotic acid)	1.8(+1)	9.55(0)		
Diphenylacetic acid	3.939			
Diphenylamine	0.9(+1)			
2,2-Diphenylglutaric acid (20 °C)	3.91	5.38		
1,3-Diphenylguanidine	10.12			
2,2-Diphenylheptanedioic acid 　(20 °C)	4.28	5.39		
2,2-Diphenylhexanedioic acid (20 °C)	4.17	5.40		
3,3-Diphenylhexanedioic acid	4.22	5.19		
Diphenylhydroxyacetic acid (35 °C)	3.05			
Diphenylketimine	6.82			
2,2-Diphenylnonanedioic acid (20 °C)	4.33	5.38		

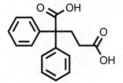

2,2-Diphenylglutaric acid

Diphenylketimine

TABLE 8.1   pK$_A$ Values of Organic Materials in Water at 25 °C (*continued*)

Substance	pK$_1$	pK$_2$	pK$_3$	pK$_4$
meso-2,2-Diphenylsuccinic acid	3.48			
rac-2,2-Diphenylsuccinic acid	3.58			
2,2-Diphenylsuccinic acid, 1-methyl ester (20 °C)	4.47			
2,2-Diphenylsuccinic acid, 4-methyl ester (20 °C)	3.900			
Diphenylthiocarbazone	4.50	15		
Dipropylamine	10.91(+1)			
Dipropylenetriamine	7.72(+3)	9.56(+2)	10.65(+1)	
2,2-Dipropylglutaric acid	3.688	7.31		
Dipropylmalonic acid	2.04	7.51		
2,2'-Dipyridyl	−0.52(+2)	4.352(+1)		
2,3'-Dipyridyl (20 °C)	1.52(+2)	4.42(+1)		
2,4'-Dipyridyl (20 °C)	1.19(+2)	4.77(+1)		
3,3'-Dipyridyl (20 °C, $\mu = 0.2$)	3.0(+2)	4.60(+1)		
3,4'-Dipyridyl (20 °C, $\mu = 0.2$)	3.0(+2)	4.85(+1)		
4,4'-Dipyridyl	3.17(+2)	4.82(+1)		
Dithiodiacetic acid (18 °C)	3.075	4.201		
1,4-Dithioerythritol	9.5			
Dithiooxamide (rubeanic acid)	10.89			
Dulcitol	13.46			
Ecgonine	10.91			
Emetine	7.36(+1)	8.23(0)		
Epinephrine enantiomorph	9.39(+1)			
Epinephrine, pseudo	9.53(+1)			
Ergometrinine	7.32(+1)			
Ergonovine	6.73(+1)			
Eriochrome Black T	6.3	11.55		
1,2-Ethanediamine	6.85(+2)	9.92(+1)		
Ethane-1,2-diamino-N,N'-dimethyl-N,N'-diacetic acid (20 °C)	6.047(0)	10.068(−1)		
1,2-Ethanedithiol	8.96	10.54		
Ethanethiol ($\mu = 0.015$)	10.61			
Ethoxyacetic acid (18 °C)	3.65			
2-Ethoxyaniline (*o*-phenetidine)	4.47(+1)			
3-Ethoxyaniline	4.17(+1)			
4-Ethoxyaniline	5.25(+1)			
2-Ethoxybenzoic acid (20 °C)	4.21			
3-Ethoxybenzoic acid (20 °C)	4.17			
4-Ethoxybenzoic acid (20 °C)	4.80			
Ethoxycarbonylethylamine	9.13(+1)			
2-Ethoxyethanethiol	9.38			
2-Ethoxyethylamine	6.26(+1)			
2-Ethoxyphenol	10.109			
3-Ethoxyphenol	9.655			
(4-Ethoxyphenyl)phosphonic acid	2.06	7.28		
4-Ethoxypyridine	6.67(+1)			
Ethyl acetoacetate	10.68			
3-Ethylacrylic acid	4.695			

TABLE 8.1 pK_A Values of Organic Materials in Water at 25 °C (*continued*)

Substance	pK_1	pK_2	pK_3	pK_4
N-Ethylalanine	2.22(+1)	10.22(0)		
Ethylamine	10.63(+1)			
(3-Ethylamino)phenylphosphonic acid	1.1(+1)	4.90(0)	7.24(−1)	
N-Ethylaniline	5.11(+1)			
2-Ethylaniline	4.42(+1)			
3-Ethylaniline	4.70(+1)			
4-Ethylaniline	5.00(+1)			
Ethylarsonic acid (18 °C)	3.89	8.35		
Ethylbarbituric acid	3.69(+1)			
2-Ethylbenzimidazole ($\mu = 0.16$)	6.27(+1)			
2-Ethylbenzoic acid	3.79			
4-Ethylbenzoic acid	4.35			
Ethylbiguanide	2.09(+1)	11.47(0)		
2-Ethylbutanoic acid (20 °C)	4.710			
S-Ethyl-L-cysteine ($\mu = 0.1$)	2.03(+1)	8.60(0)		
Ethylenebiguanide (30 °C)	1.74	2.88	11.34	11.76
Ethylenebis(thioacetic acid) (18 °C)	3.382(0)	4.352(−1)		
Ethylenediamine-N,N'-diacetic acid	6.42	9.46		
Ethylenediamine-N,N-dimethyl-N',N'-diacetic acid	6.047	10.068		
Ethylenediamine-N',N-dipropanoic acid (30 °C)	6.87	9.60		
Ethylenediamine-N,N,N',N'-tetraacetic acid ($\mu = 0.1$)	1.99	2.67	6.16	10.26
Ethylenediamine-N,N,N',N'-tetrapropanoic acid (30 °C)	3.00	3.43	6.77	9.60
Ethylene glycol	14.22			
Ethyleneimine	8.04(+1)			
cis-Ethylene oxide dicarboxylic acid	1.93	3.92		
trans-Ethylene oxide dicarboxylic acid	1.93	3.25		
N-Ethylethylenediamine	7.63(+2)	10.56(+1)		
N-Ethylglycine ($\mu = 0.1$)	2.34(+1)	10.23(0)		
3-Ethylglutaric acid	4.28	5.33		
Ethyl hydroperoxide	11.80			
Ethyl hydrogen malonate	3.55			
3-Ethyl-2-hydroxypyridine	5.00(+1)			
Ethylmalonic acid	2.90(0)	5.55(−1)		

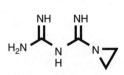

Ethylenebiguanide   3-Ethylglutaric acid

TABLE 8.1 pK_A Values of Organic Materials in Water at 25 °C (*continued*)

Substance	pK_1	pK_2	pK_3	pK_4
N-Ethyl mercaptoacetamide	8.14(SH)			
Ethyl 2-mercaptoacetate	7.95(SH)			
Ethyl 3-mercaptopropanoate	9.48(SH)			
3-Ethyl-4-(methylamino)pyridine (20 °C)	9.90(+1)			
5-Ethyl-5-(1-methylbutyl)barbituric acid	8.11(0)			
Ethyl methyl ketoxime	12.45			
Ethylmethylmalonic acid	2.86(0)	6.41(−1)		
1-Ethyl-2-methylpiperidine	10.66(+1)			
3-Ethyl-6-methylpyridine (20 °C)	6.51(+1)			
3-Ethyl-4-methylpyridine-1-oxide	−1.534(+1)			
5-Ethyl-2-methylpyridine-1-oxide	−1.288(+1)			
1-Ethyl-2-methyl-2-pyrroline	11.84(+1)			
Ethylmorphine (15 °C)	8.08			
Ethyl nitroacetate	5.85			
3-Ethylpentane-2,4-dione	11.34			
2-Ethylpentanoic acid (18 °C)	4.71			
5-Ethyl-5-pentylbarbituric acid	7.960			
2-Ethylphenol	10.2			
3-Ethylphenol	10.07			
4-Ethylphenol	10.0			
4-Ethylphenylacetic acid	4.373			
5-Ethyl-5-phenylbarbituric acid	7.445			
Ethylphosphinic acid	3.29			
Ethylphosphonic acid	2.43	8.05		
1-Ethylpiperidine ($\mu = 0.01$)	10.45(+1)			
2,2-Ethylpropylglutaric acid	3.511			
Ethylpropylmalonic acid	3.14	7.43		
2-Ethylpyridine	5.89(+1)			
3-Ethylpyridine (20 °C)	5.80(+1)			
4-Ethylpyridine	5.87(+1)			
Ethyl 3-pyridinecarboxylate	3.35(+1)			
Ethyl 4-pyridinecarboxylate	3.45(+1)			
2-Ethylpyridine-1-oxide	−1.19(+1)			
3-Ethylpyridine-1-oxide	−0.965(+1)			
Ethylpyrrolidine	10.43(+1)			
2-Ethyl-2-pyrroline	7.87(+1)			
Ethylsuccinic acid	4.08(0)			
S-Ethylthioacetic acid	5.06			
N-Ethyl-o-toluidine	4.92(+1)			
N-Ethylveratramine	7.40(+1)			
β-Eucaine	9.35(+1)			
Fluoroacetic acid	2.586			
2-Fluoroacrylic acid	2.55			
2-Fluoroaniline	3.20(+1)			
3-Fluoroaniline	3.58(+1)			
4-Fluoroaniline	4.65(+1)			
2-Fluorobenzoic acid	3.27			

TABLE 8.1 pK$_A$ Values of Organic Materials in Water at 25 °C (*continued*)

Substance	pK$_1$	pK$_2$	pK$_3$	pK$_4$
3-Fluorobenzoic acid	3.865			
4-Fluorobenzoic acid	4.14			
Fluoromandelic acid	4.244			
2-Fluorophenol	8.73			
3-Fluorophenol	9.29			
4-Fluorophenol	9.89			
2-Fluorophenoxyacetic acid	3.08			
3-Fluorophenoxyacetic acid	3.08			
4-Fluorophenoxyacetic acid	3.13			
4-Fluorophenylacetic acid	4.25			
2′-Fluorophenylalanine	2.14(+1)	9.01(0)		
3′-Fluorophenylalanine	2.10(+1)	8.98(0)		
4-Fluorophenylalanine	2.13(+1)	9.05(0)		
2-Fluorophenylphosphonic acid	1.64	6.80		
3-Fluorophenylselenic acid	4.34			
4-Fluorophenylselenic acid	4.50			
2-Fluoropyridine	−0.44(+1)			
3-Fluoropyridine	2.97(+1)			
5-Fluorouracil	8.00(0)	ca 13(−1)		
Folic acid (pteroylglutamic acid)	8.26			
Formic acid	3.751			
N-Formylglycine	3.43			
2-Formyl-3-hydroxypyridine (20 °C)	3.40(+1)	6.95(OH)		
4-Formyl-3-hydroxypyridine	4.05(+1)	6.77(OH)		
2-Formyl-3-methoxypyridine (20 °C)	3.89(+1)	12.95		
Formyl-3-methoxypyridine (20 °C)	4.45(+1)	11.7		
D-(−)-Fructose	12.03			
Fumaric acid	3.10	4.60		
2-Furancarboxylic acid (2-furoic acid)	3.164			
D-(+)-Galactose	12.35			
Galactose-1-phosphoric acid	1.00	6.17		
Glucoascorbic acid	4.26	11.58		
D-Gluconic acid	3.86			
α-D-(+)-Glucose	12.28			
α-D-Glucose-1-phosphate	1.11(0)	6.504(−1)		
trans-Glutaconic acid	3.77	5.08		
D-(−)-Glutamic acid	2.162(+1)	4.272(0)	9.358(−1)	
L-Glutamic acid	2.13(+1)	4.31(0)	9.76(−1)	
Glutamic acid, 1-ethyl ester	3.85(+1)	7.84(0)		

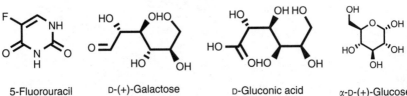

5-Fluorouracil   D-(+)-Galactose   D-Gluconic acid   α-D-(+)-Glucose

TABLE 8.1 pK_A Values of Organic Materials in Water at 25 °C (*continued*)

Substance	pK_1	pK_2	pK_3	pK_4
Glutamic acid, 5-ethyl ester	2.15(+1)	9.19(0)		
L-Glutamine ($\mu = 0.2$)	2.15(+1)	9.00(0)		
Glutaric acid	3.77	6.08		
Glutaric acid monoamide	4.600(0)			
Glutarimide	11.43			
Glutathione	2.12(+1)	3.53(0)	8.66	9.12
DL-Glyceric acid	3.64			
Glycerol	14.15			
Glyceryl-1-phosphoric acid		6.656(−1)		
Glyceryl-2-phosphoric acid	1.335(0)	6.650(−1)		
Glycine	2.351(+1)	9.70(0)		
Glycine amide	8.03(+1)			
Glycine, ethyl ester	7.66(+1)			
Glycine hydroxamic acid	7.10	9.10		
Glycine, methyl ester	7.59(+1)			
Glycine-O-phenylphosphorylserine	2.96	8.07		
Glycolic acid	3.831			
N-Glycyl-α-alanine	3.15(+1)	8.33(0)		
Glycylalanylalanine	3.38(+1)	8.10(0)		
N-Glycylasparagine	2.942			
Glycylaspartic acid	2.81(+1)	4.45(0)	8.60(−1)	
Glycyl-DL-glutamine (18 °C)	2.88(+1)	8.33(0)		
N-Glycylglycine	3.126(+1)	8.252(0)		
Glycylglycylcysteine (35 °C)	2.71	2.71	7.94	7.94
Glycylglycylglycine	3.225(+1)	8.090(0)		
Glycyl-L-histidine ($\mu = 0.16$)	6.79	8.20		
Glycylisoleucine	8.00			
N-Glycyl-L-leucine	3.180(+1)	8.327(0)		
Glycyl-O-phosphorylserine	2.90	6.02	8.43	
L-Glycylproline ($\mu = 0.1$)	2.81(+1)	8.65(0)		
N-Glycylsarcosine ($\mu = 0.1$)	2.98(+1)	8.55(0)		
N-Glycylserine	2.98(+1)	8.38(0)		
Glycylserylglycine	3.32	7.99		
Glycyltyrosine	2.93	8.45	10.49	
Glycylvaline	3.15	8.18		
Glyoxaline	7.03(+1)			
Glyoxylic acid	3.30(0)			
Guanidineacetic acid	2.82(+1)			

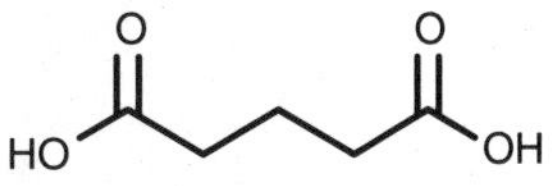

Glutaric acid

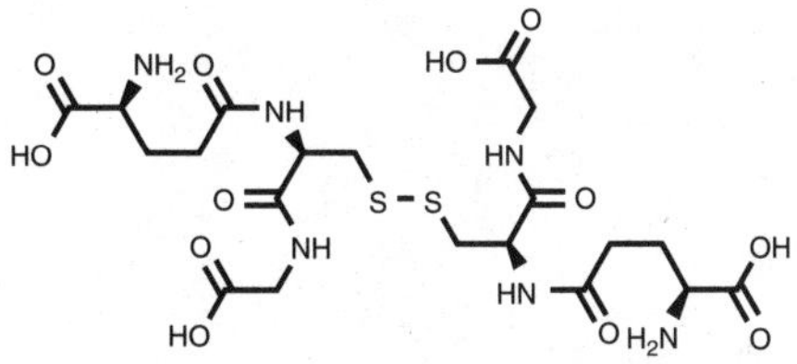

Glutathione

TABLE 8.1 pK$_A$ Values of Organic Materials in Water at 25 °C (*continued*)

Substance	pK$_1$	pK$_2$	pK$_3$	pK$_4$
Guanine	3.3(+1)	9.2	12.3	
Guanine deoxyriboside- 3′-phosphoric acid		2.9	6.4	9.7
Guanosine	1.9(+1)	9.25(0)	12.33 (OH)	
Guanosine-5′-diphosphoric acid ($\mu = 0.1$; pK$_5 = 59.6$)			2.9	6.3
Guanosine-3′-phosphoric acid	0.7	2.3	5.92	9.38
Guanosine-5′-phosphoric acid ($\mu = 0.1$)		2.4	6.1	9.4
Guanosine-5′-triphosphoric acid [$\mu = 0.1$; pK$_5 = 7.10(-3)$; pK$_6 = 9.3(-4)$]				3.0(−2)
Guanylurea	1.80	8.20		
Harmine (20 °C)	7.61(+1)			
Heptafluorobutanoic acid	0.17			
4,4,5,5,6,6,6-Heptafluorohexanoic acid	4.18			
4,4,5,5,6,6,6-Heptafluoro-2-hexenoic acid	3.23			
Heptanedioic acid (pimelic acid)	4.484	5.424		
2,4-Heptanedione	8.43(keto): 9.15(enol)			
Heptanoic acid	4.893			
Heroin	7.6(+1)			
2,4-Hexadienoic acid (sorbic acid)	4.77			
1,1,1,3,3,3-Hexafluoro- 2,2-propanediol	8.801			
1,1,1,3,3,3-Hexafluoro-2-propanol	9.42			
Hexahydroazepine	11.07			
Hexamethyldisilazine	7.55			
1,2,3,8,9,10-Hexamethyl- 4,7-phenanthroline (20 °C)	7.26			
1,6-Hexanediamine	9.830(+2)	10.930(+1)		
1,6-Hexanedioic acid	4.418	5.412		
2,4-Hexanedione	8.49 (enol); 9.32 (keto)			
2,2′,4,4′,6,6′-Hexanitrodiphenylamine	5.42 (+1)			
Hexanoic acid (20 °C)	4.849			
trans-2-Hexenoic acid	4.74			
trans-3-Hexenoic acid	4.72			
3-Hexen-4-oic acid	4.58			
4-Hexen-5-oic acid	4.74			
Hexylamine	10.64(+1)			
Hexylarsonic acid	4.16	9.19		
Hexylphosphonic acid	2.6	7.9		
DL-Histidine	1.82(+2)	6.00(+1)	9.16(0)	

TABLE 8.1 pK$_A$ Values of Organic Materials in Water at 25 °C (*continued*)

Substance	pK$_1$	pK$_2$	pK$_3$	pK$_4$
Histidine amide ($\mu = 0.2$)	5.78(+2)	7.64(+1)		
Histidine, methyl ester ($\mu = 0.1$)	5.01(+2)	7.23(+1)		
Histidylglycine	2.40(+2)	5.80(+1)	7.82(0)	
Histidylhistidine ($\mu = 0.16$)	5.40(+2)	6.80(+1)	7.95(0)	
DL-Homatropine	9.7(+1)			
DL-Homocysteine	2.222(+1)	8.87	10.86	
Homocysteine ($\mu = 0.1$)	1.593(+2)	2.523(+1)	8.676(0)	9.413(−1)
Hydantoin	9.12			
Hydrastine	6.23(+1)			
Hydrazine-*N,N*-diacetic acid	<0.1	2.8	3.8	
Hydrazine-*N′,N′*-diacetic acid	2.40	3.12	7.32	
4-Hydrazinocarbonylpyridine (20 °C)	1.82	3.52	10.79	
N-Hydroxyacetamide	9.40			
2′-Hydroxyacetophenone	9.90			
3′-Hydroxyacetophenone	9.19			
4′-Hydroxyacetophenone	8.05			
1-Hydroxyacridine (15 °C)	5.72			
2-Hydroxyacridine (15 °C)	5.62			
3-Hydroxyacridine (15 °C)	5.30			
α-Hydroxyasparagine	2.28(+1)	7.20(0)		
β-Hydroxyasparagine	2.09(+1)	8.29(0)		
Hydroxyaspartic acid	1.91(+1)	3.51(0)	9.11(−1)	
2-Hydroxybenzaldehyde (salicylaldehyde)	8.34			
3-Hydroxybenzaldehyde	9.00			
4-Hydroxybenzaldehyde	7.620			
2-Hydroxybenzaldehyde oxime	1.37(+1)	9.18	12.11	
2-Hydroxybenzamide	8.36			
2-Hydroxybenzenemethanol (2-hydroxybenzyl alcohol)	9.92			
3-Hydroxybenzenemethanol	9.83			
4-Hydroxybenzenemethanol	9.82			
4-Hydroxybenzenesulfonic acid		9.055(−1)		
2-Hydroxybenzohydroxamic acid	5.19			
2-Hydroxybenzoic acid (salicylic acid)	2.98	12.38		
3-Hydroxybenzoic acid	4.076	9.85		
4-Hydroxybenzoic acid	4.582	9.23		
4-Hydroxybenzonitrile	7.95			
2-Hydroxy-5-bromobenzoic acid	2.61			
2-Hydroxybutanoic acid (30 °C)	3.65			

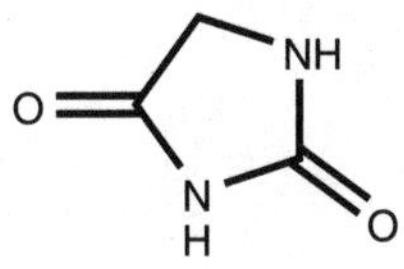

Hydantoin

TABLE 8.1 pK$_A$ Values of Organic Materials in Water at 25°C (*continued*)

Substance	pK$_1$	pK$_2$	pK$_3$	pK$_4$
L-3-Hydroxybutanoic acid (30°C)	4.41			
4-Hydroxybutanoic acid (30°C)	4.71			
2-Hydroxy-5-chlorobenzoic acid	2.63			
trans-2′-Hydroxycinnamic acid	4.614			
trans-3′-Hydroxycinnamic acid	4.40			
10-Hydroxycodeine	7.12			
cis-2-Hydroxycyclohexane- 1-carboxylic acid	4.796			
trans-2-Hydroxycyclohexane- 1-carboxylic acid	4.682			
cis-3-Hydroxycyclohexane- 1-carboxylic acid	4.602			
trans-3-Hydroxycyclohexane- 1-carboxylic acid	4.815			
cis-4-Hydroxycyclohexane- 1-carboxylic acid	4.836			
trans-4-Hydroxycyclohexane- 1-carboxylic acid	4.687			
1-Hydroxy-2,4- dihydroxymethylbenzene	9.79			
N-(Hydroxyethyl)biguanide	2.8(+2)	11.53(+1)		
N-(2-Hydroxyethyl)ethylenediamine	7.21(+2)	10.12(+1)		
N′-(2-Hydroxyethyl)ethylene- diamine-N,N,N′-triacetic acid	2.39	5.37	9.93	
N-(2-Hydroxyethyl)iminodiacetic acid ($\mu = 0.1$)	2.2	8.65		
N-(2-Hydroxyethyl)piperazine-N′- ethanesulfonic acid (20°C)	7.55			
4′-(2-Hydroxyethyl)-1′-piperazine- propanesulfonic acid (20°C)	8.00			
2-Hydroxyethyltrimethylamine	8.94(+1)			
L-β-Hydroxyglutamic acid	2.09	4.18	9.20	
1-Hydroxy-4-hydroxymethylbenzene	9.84			
5-Hydroxy-2-(hydroxymethyl)- 4H-pyran-4-one	7.90	8.03		
3-Hydroxy-2-hydroxymethylpyridine (20°C, $\mu = 0.2$)	5.00(+1)	9.07(OH)		
3-Hydroxy-4-hydroxymethylpyridine (20°C, $\mu = 0.2$)	5.00(+1)	8.95(OH)		
8-Hydroxy-7-iodoquinoline- 5-sulfonic acid	2.51(0)	7.417(−1)		
Hydroxylysine (38°C, $\mu = 0.1$)	2.13(+2)	8.62(+1)	9.67(0)	
2-Hydroxy-3-methoxybenzaldehyde	7.912			
3-Hydroxy-4-methoxybenzaldehyde (isovanillin)	8.889			
4-Hydroxy-3-methoxybenzaldehyde (vanillin)	7.396			
4-Hydroxy-3-methoxybenzoic acid	4.355			
1-Hydroxy-2-methoxybenzylamine	8.70(+1)	10.52(0)		

TABLE 8.1 pK_A Values of Organic Materials in Water at 25 °C (*continued*)

Substance	pK_1	pK_2	pK_3	pK_4
2-Hydroxy-1-methoxybenzylamine	8.89(+1)	10.52(0)		
3-Hydroxy-2-methoxybenzylamine	8.94(+1)	10.42(0)		
2-Hydroxymethyl-2-benzeneacetic acid	4.12			
(2-Hydroxy-5-methylbenzene)- methanol	10.15			
2-Hydroxy-3-methylbenzoic acid	2.99			
2-Hydroxy-4-methylbenzoic acid	3.17			
2-Hydroxy-5-methylbenzoic acid	4.08			
2-Hydroxy-6-methylbenzoic acid	3.32			
2-Hydroxy-2-methylbutanoic acid (18 °C)	3.991			
3-Hydroxy-2-methylbutanoic acid (18 °C)	4.648			
4-Hydroxy-4-methylpentanoic acid (18 °C)	4.873			
1-Hydroxymethylphenol	9.95			
Hydroxymethylphosphoric acid	1.91	7.15		
2-Hydroxy-2-methylpropanoic acid ($\mu = 0.1$)	3.717			
2-Hydroxy-4-methylpyridine	4.529(+1)			
8-Hydroxy-2-methylquinoline	5.55(+1)	10.31(0)		
8-Hydroxy-4-methylquinoline	5.56(+1)	10.00(0)		
8-Hydroxy-2-methylquinoline- 5-sulfonic acid	4.80(0)	9.30(−1)		
8-Hydroxy-4-methylquinoline- 7-sulfonic acid	4.78(0)	10.01(−1)		
8-Hydroxy-6-methylquinoline- 5-sulfonic acid	4.20(0)	8.7(−1)		
2-Hydroxy-1-naphthoic acid (20 °C)	3.29	9.68		
2-Hydroxy-2-nitrobenzoic acid	2.23			
2-Hydroxy-3-nitrobenzoic acid	1.87			
2-Hydroxy-5-nitrobenzoic acid	2.12			
2-Hydroxy-6-nitrobenzoic acid	2.24			
2-Hydroxy-4-nitrophenylphosphonic acid	1.22	5.39		
8-Hydroxy-7-nitroquinoline- 5-sulfonic acid	1.94(0)	5.750(−1)		
3-Hydroxy-4-nitrotoluene ($\mu = 0.1$)	7.41			
4-Hydroxypentanoic acid (18 °C)	4.686			
4-Hydroxy-3-pentenoic acid	4.30			
3-Hydroxyphenazine (15 °C)	2.67			
4-Hydroxyphenylarsonic acid	3.89	8.37	10.05 (phenol)	
3-Hydroxyphenylboric acid	8.55	10.84		
2-Hydroxy-2-phenylpropanoic acid	3.532			
2-(2-Hydroxyphenyl)pyridine (20 °C)	4.19(+1)	10.64		

TABLE 8.1 pK$_A$ Values of Organic Materials in Water at 25 °C (*continued*)

Substance	pK$_1$	pK$_2$	pK$_3$	pK$_4$
trans-4-Hydroxyproline	1.818(+1)	9.662(0)		
Hydroxypropanedioic acid (tartronic acid)	2.37	4.74		
2-Hydroxypropanoic acid	3.858			
1-Hydroxy-2-propylbenzene	10.50			
4-Hydroxypteridine	1.3(+1)	7.89(0)		
2-Hydroxypyridine	1.25(+1)	11.62(0)		
3-Hydroxypyridine	4.80(+1)	8.72(0)		
4-Hydroxypyridine	3.23(+1)	11.09(0)		
2-Hydroxypyridine-*N*-oxide	−0.62(+1)	5.97(0)		
2-Hydroxypyrimidine	2.24(+1)	9.17(0)		
4-Hydroxypyrimidine	1.85(+1)	8.59(0)		
8-Hydroxyquinazoline	3.41(+1)	8.65(0)		
2-Hydroxyquinoline (20 °C)	−0.31(+1)	11.74		
3-Hydroxyquinoline (20 °C)	4.30(+1)	8.06(0)		
4-Hydroxyquinoline (20 °C)	2.27(+1)	11.25(0)		
5-Hydroxyquinoline (20 °C)	5.20(+1)	8.54(0)		
6-Hydroxyquinoline (20 °C)	5.17(+1)	8.88(0)		
7-Hydroxyquinoline (20 °C)	5.48(+1)	8.85(0)		
8-Hydroxyquinoline (20 °C)	4.91(+1)	9.81(0)		
8-Hydroxyquinoline-5-sulfonic acid	4.092(+1)	8.776(0)		
DL-Hydroxysuccinic acid (malic acid)	3.458	5.097		
L-Hydroxysuccinic acid	3.40	5.05		
Hydroxytetracycline	3.27(+1)	7.32(0)	9.11(−1)	
5-Hydroxy-1,2,3,4-tetrazole	3.32			
4-Hydroxy-3-(2′-thiazolyazo)toluene	8.36			
2-Hydroxytoluene	10.33			
3-Hydroxytoluene	10.10			
4-Hydroxytoluene	10.276			
4-Hydroxy-α,α,α-trifluorotoluene	8.675			
1-Hydroxy-2,4,6-trihydroxymethylbenzene	9.56			
Hydroxyuracil	8.64			
Hydroxyvaline	2.55(+1)	9.77(0)		
Hyoscyamine	9.68(+1)			
Hypoxanthene	1.79(+1)	8.91(0)	12.07(−1)	
Hypoxanthine	5.3			
Imidazole	6.993(+1)	10.58(0)		
Imidazolidinetrione (parabanic acid)	6.10			

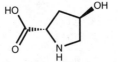

trans-4-Hydroxyproline

Imidazole

TABLE 8.1 pK$_A$ Values of Organic Materials in Water at 25 °C (*continued*)

Substance	pK$_1$	pK$_2$	pK$_3$	pK$_4$
4-(4-Imidazolyl)butanoic acid				
($\mu = 0.1$)	4.26(+1)	7.62(0)		
2-(4-Imidazolyl)ethylamine	5.784(+2)	9.756(+1)		
3-(4-Imidazolyl)propanoic acid				
($\mu = 0.16$)	3.96(+1)	7.57(0)		
3,3'-Iminobispropanoic acid	4.11(0)	9.61(−1)		
3,3'-Iminobispropylamine (30 °C)	8.02(+2)	9.70(+1)	10.70(0)	
2,2'-Iminodiacetic acid (diglycine)				
(30 °C, $\mu = 0.1$)	2.54(0)	9.12(−1)		
4-Indanol	10.32			
Indole-3-acetic acid	4.75			
Inosine	ca 1.5(+1)	8.96(0)	12.36	
Inosine-5'-phosphoric acid	1.54(0)	6.66(−1)		
Inosine-5'-triphosphoric acid				
[pK$_5 = 7.68(−4)$]			2.2(−2)	6.92(−3)
Iodoacetic acid	3.175			
2-Iodoaniline	2.54(+1)			
3-Iodoaniline	3.58(+1)			
4-Iodoaniline	3.82(+1)			
2-Iodobenzoic acid	2.86			
3-Iodobenzoic acid	3.86			
4-Iodobenzoic acid	4.00			
5-Iodohistamine	4.06(+2)	9.20(+1)	11.88(0)	
	(imidazole)	(NH$_3^+$)	(imino)	
7-Iodo-8-hydroxyquinoline-5-sulfonic				
acid	2.514	7.417		
Iodomandelic acid	3.264			
Iodomethylphosphoric acid	1.30	6.72		
2-Iodophenol	8.464			
3-Iodophenol	8.879			
4-Iodophenol	9.200			
2-Iodophenoxyacetic acid	3.17			
3-Iodophenoxyacetic acid	3.13			
4-Iodophenoxyacetic acid	3.16			
2-Iodophenylacetic acid	4.038			
3-Iodophenylacetic acid	4.159			
4-Iodophenylacetic acid	4.178			
2-Iodophenylphosphoric acid	1.74	7.06		
2-Iodopropanoic acid	3.11			
3-Iodopropanoic acid	4.08			
2-Iodopyridine	1.82(+1)			

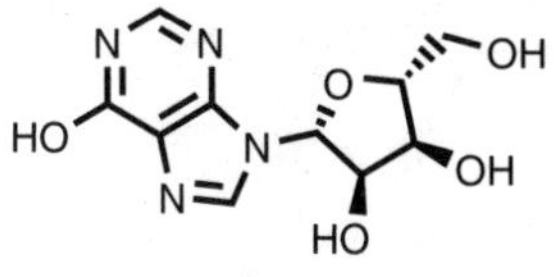

Inosine

TABLE 8.1 pK$_A$ Values of Organic Materials in Water at 25 °C (*continued*)

Substance	pK$_1$	pK$_2$	pK$_3$	pK$_4$
3-Iodopyridine	3.25(+1)			
4-Iodopyridine (20 °C)	4.02(+1)			
Isoasparagine	2.97(+1)	8.02(0)		
Isobutylacetic acid (18 °C)	4.79			
Isobutylamine	10.41(+1)			
Isochlorotetracycline	3.1(+1)	6.7(0)	8.3(−1)	
Isocreatine	2.84(+1)			
Isoglutamine	3.81(+1)	7.88(0)		
Isohistamine (μ = 0.1)	6.036(+2)	9.274(+1)		
L-Isoleucine	2.318(+1)	9.758(0)		
Isolysergic acid	3.33(0)	8.46(NH)		
Isopilocarpine (15 °C)	7.18(+1)			
2-(Isopropoxy)benzoic acid (20 °C)	4.24			
3-(Isopropoxy)benzoic acid (20 °C)	4.15			
4-(Isopropoxy)benzoic acid (20 °C)	4.68			
Isopropylamine	10.64(+1)			
N-Isopropylaniline	5.50(+1)			
5-Isopropylbarbituric acid	4.907(+1)			
2-Isopropylbenzene acid	3.64			
4-Isopropylbenzoic acid	4.36			
N-Isopropylglycine (μ = 0.1)	2.36(+1)	10.06(0)		
Isopropylmalonic acid	2.94	5.88		
Isopropylmalonic acid mononitrile	2.401			
3-Isopropyl-4-(methylamino)pyridine (20 °C)	9.96(+1)			
3-Isopropylpentanedioic acid	4.30	5.51		
4-Isopropylphenylacetic acid	4.391			
Isopropylphosphinic acid	3.56			
Isopropylphosphonic acid	2.66	8.44		
2-Isopropylpyridine	5.83(+1)			
3-Isopropylpyridine (20 °C)	5.72(+1)			
4-Isopropylpyridine	6.02(+1)			
DL-Isoproterenol	8.64(+1)			
Isoquinoline	5.40(+1)			
Isoretronecanol	10.83			
L-Isoserine (μ = 0.16)	2.72(+1)	9.25(0)		
Isothiocyanatoacetic acid	6.62			
L-(+)-Lactic acid	3.858			
L-Leucine	2.328(+1)	9.744(0)		
Leucine amide	7.80(+1)			

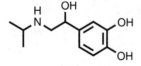

DL-Isoproterenol

Isoquinoline

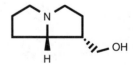

Isoretronecanol

TABLE 8.1 pK_A Values of Organic Materials in Water at 25 °C (*continued*)

Substance	pK_1	pK_2	pK_3	pK_4
Leucine, ethyl ester ($\mu = 0.1$)	7.57(+1)			
L-Leucyl-L-asparagine	3.00(+1)	8.12(0)		
L-Leucyl-L-glutamine	2.99(+1)	8.11(0)		
DL-Leucylglycine	3.25(+1)	8.28(0)		
Leucylisoserine (20 °C)	3.188(+1)	8.207(0)		
D-Leucyl-L-tyrosine	3.12(+1)	8.38(0)	10.35(−1)	
L-Leucyl-L-tyrosine	3.46(+1)	7.84(0)	10.09(−1)	
Lysergic acid	3.44(+1)	7.68(0)		
L-(+)-Lysine	2.18(+2)	8.95(+1)	10.53(0)	
Lysine, methyl ester ($\mu = 0.1$)	6.965(+1)	10.251(0)		
L-Lysyl-L-alanine	3.22(+1)	7.62(0)	10.70(−1)	
L-Lysyl-D-alanine	3.00(+1)	7.74(0)	10.63(−1)	
Lysylglutamic acid	2.93(+2)	4.47(+1)	7.75(0)	10.50(+1)
L-Lysyl-L-lysine ($\mu = 0.1$)	3.01(+2)	7.53(+1)	10.05(0)	10.01(−1)
L-Lysyl-D-lysine ($\mu = 0.1$)	2.85(+2)	7.53(+1)	9.92(0)	10.89(−1)
L-Lysyl-L-lysyl-L-lysine ($\mu = 0.1$)	3.08(+2)	7.34(+1)	9.80(0)	10.54(−1)
L-Lysyl-D-lysyl-L-lysine ($\mu = 0.1$)	2.91(+2)	7.29(+1)	9.79(0)	10.54(−1)
L-Lysyl-D-lysyl-lysine ($\mu = 0.1$)	2.94(+2)	7.15(+1)	9.60(0)	10.38(−1)
α-D-Lyxose	12.11			
Maleic acid	1.910	6.33		
Malonamic acid	3.641(0)			
Malonic acid	2.826	5.696		
Malonitrile (cyanoacetic acid)	2.460			
Mandelic acid	3.411			
D-(+)-Mannose	12.08			
Mercaptoacetic acid (thioglycolic acid)	3.60(0)	10.56(SH)		
2-Mercaptobenzoic acid (20 °C)	4.05(0)			
2-Mercaptobutanoic acid	3.53(0)			
Mercaptodiacetic acid	3.32	4.29		
2-Mercaptoethanesulfonic acid (20 °C)		9.5(−1)		
2-Mercaptoethanol	9.88			
2-Mercaptoethylamine	8.27(+1)	10.53(0)		
2-Mercaptohistidine	1.84(+1)	8.47(0)	11.4(SH)	
Mercapto-S-phenylacetic acid ($\mu = 0.1$)	3.9			
2-Mercaptopropane ($\mu = 0.1$)	10.86			
3-Mercapto-1,2-propanediol ($\mu = 0.5$)	9.43			
2-Mercaptopropanoic acid	4.32(0)	10.20(SH)		
3-Mercaptopropanoic acid		10.84(SH)		

Mandelic acid   D-(+)-Mannose

TABLE 8.1   pK$_A$ Values of Organic Materials in Water at 25 °C (*continued*)

Substance	pK$_1$	pK$_2$	pK$_3$	pK$_4$
2-Mercaptopyridine (20 °C)	−1.07(+1)	10.00(0)		
3-Mercaptopyridine (20 °C)	2.26(+1)	7.03(0)		
4-Mercaptopyridine (20 °C)	1.43(+1)	8.86(0)		
2-Mercaptoquinoline (20 °C)	−1.44(+1)	10.21(0)		
3-Mercaptoquinoline (20 °C)	2.33(+1)	6.13(0)		
4-Mercaptoquinoline (20 °C)	0.77(+1)	8.83(0)		
Mercaptosuccinic acid	3.30(0)	4.94(−1)	10.94(SH)	
Mesitylenic acid	4.32			
Mesoxaldialdehyde	3.60			
Methacrylic acid	4.66			
Methanethiol	10.70			
DL-Methionine	2.13(+1)	9.28(0)		
2-(*N*-Methoxyacetamido)pyridine	2.01(+1)			
3-(*N*-Methoxyacetamido)pyridine	3.52(+1)			
4-(*N*-Methoxyacetamido)pyridine	4.62(+1)			
Methoxyacetic acid	3.570			
3-Methoxy-D-α-alanine	2.037(+1)	9.176(0)		
2-Methoxyaniline	4.53(+1)			
3-Methoxyaniline	4.20(+1)			
4-Methoxyaniline	5.36(+1)			
2-Methoxybenzoic acid	4.09			
3-Methoxybenzoic acid	4.08			
4-Methoxybenzoic acid	4.49			
N,N-Methoxybenzylamine	9.68(+1)			
2-Methoxycarbonylaniline	2.23(+1)			
3-Methoxycarbonylaniline	3.64(+1)			
4-Methoxycarbonylaniline	2.38(+1)			
Methoxycarbonylmethylamine	7.66(+1)			
2-Methoxycarbonylpyridine	2.21(+1)			
3-Methoxycarbonylpyridine	3.13(+1)			
4-Methoxycarbonylpyridine	3.26(+1)			
trans-2-Methoxycinnamic acid	4.462			
trans-3-Methoxycinnamic acid	4.376			
trans-4-Methoxycinnamic acid	4.539			
2-Methoxyethylamine	9.45(+1)			
2-Methoxy-4-nitrophenylphosphonic acid	1.53	6.96		
2-Methoxyphenol	9.99			
3-Methoxyphenol	9.652			
4-Methoxyphenol	10.20			
(2′-Methoxy)phenoxyacetic acid	3.231			

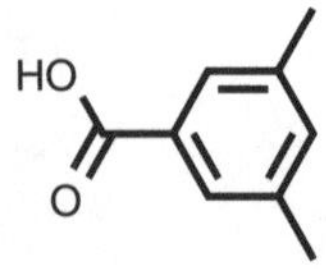

Mesitylenic acid

TABLE 8.1 pK_A Values of Organic Materials in Water at 25 °C (*continued*)

Substance	pK_1	pK_2	pK_3	pK_4
(3′-Methoxy)phenoxyacetic acid	3.141			
(4′-Methoxy)phenoxyacetic acid	3.213			
4′-Methoxyphenylacetic acid	4.358			
(4-Methoxyphenyl)phosphinic acid				
(17 °C)	2.35			
(2-Methoxyphenyl)phosphonic acid	2.16	7.77		
(4-Methoxyphenyl)phosphonic acid				
(17 °C)	2.4	7.15		
3-(2′-Methoxyphenyl)propanoic acid	4.804			
3-(3′-Methoxyphenyl)propanoic acid	4.654			
3-(4′-Methoxyphenyl)propanoic acid	4.689			
3-Methoxyphenylselenic acid	4.65			
4-Methoxyphenylselenic acid	5.05			
2-Methoxy-4-(2-propenyl)phenol	10.0			
2-Methoxypyridine	3.06(+1)			
3-Methoxypyridine	4.91(+1)			
4-Methoxypyridine	6.47(+1)			
4-Methoxy-2-(2′-thiazoylazo)phenol	7.83			
2-Methylacrylic acid (18 °C)	4.66			
N-Methylalanine	2.22(+1)	10.19(0)		
O-Methylallothreonine ($\mu = 0.1$)	1.92(+1)	8.90(0)		
Methylamine	10.62(+1)			
2-(N-Methylamino)benzoic acid	1.93(+1)	5.34(0)		
3-(N-Methylamino)benzoic acid		5.10(0)		
4-(N-Methylamino)benzoic acid		5.05		
Methylaminodiacetic acid (20 °C)	2.146	10.088		
2-(Methylamino)ethanol	9.88(+1)			
2-(2-Methylaminoethyl)pyridine				
(30 °C)	3.58(+2)	9.65(+1)		
2-(Methylaminomethyl)-6-methyl-				
pyridine				
($\mu = 0.5$)	3.03(+2)	9.15(+1)		
2-(Methylaminomethyl)pyridine				
(30 °C)	2.92(+2)	8.82(+1)		
4-Methylamino-3-methylpyridine				
(20 °C)	9.83(+1)			
(3-Methylamino)phenylphosphonic				
acid	1.1(+1)	4.72(+1)	7.30(−1)	
(4-Methylamino)phenylphosphonic				
acid			7.85(−1)	
3-(Methylamino)pyridine (30 °C)	8.70(+1)			
4-(Methylamino)pyridine (20 °C)	9.65(+1)			
4-(Methylamino)-2,3,5,6-tetramethyl-				
pyridine (20 °C)	10.06(+1)			
N-Methylaniline	4.85(+1)			
Methylarsonic acid (18 °C)	3.41	8.18		
1-Methylbarbituric acid	4.35(+1)			
5-Methylbarbituric acid	3.386(+1)			
2-(N-Methylbenzamido)pyridine	1.44(+1)			

TABLE 8.1 pK$_A$ Values of Organic Materials in Water at 25 °C (*continued*)

Substance	pK$_1$	pK$_2$	pK$_3$	pK$_4$
3-(*N*-Methylbenzamido)pyridine	3.66(+1)			
4-(*N*-Methylbenzamido)pyridine	4.68(+1)			
2-Methylbenzimidazole ($\mu = 0.16$)	6.29(+1)			
2-Methylbenzoic acid (*o*-toluic acid)	3.90			
3-Methylbenzoic acid	4.269			
4-Methylbenzoic acid	4.362			
N-Methyl-1-benzoylecgonine	8.65			
Methylbiguanidine	3.00(+2)	11.44(+1)		
2-Methyl-2-butanethiol	11.35			
2-Methylbutanoic acid	4.761			
3-Methylbutanoic acid (20 °C)	4.767			
(*E*)-2-Methyl-2-butendioic acid (mesaconic acid)	3.09	4.75		
3-Methyl-2-butenoic acid	5.12			
(*E*)-2-Methyl-2-butenoic acid (tiglic acid)	4.96			
(*Z*)-2-Methyl-2-butenoic acid (angelic acid)	4.30			
4-Methylcarboxylphenol	8.47			
(*E*)-2-Methylcinnamic acid	4.500			
(*E*)-3-Methylcinnamic acid	4.442			
(*E*)-4-Methylcinnamic acid	4.564			
1-Methylcyclohexane-1-carboxylic acid	5.13			
cis-2-Methylcyclohexane-1-carboxylic acid	5.03			
trans-2-methylcyclohexane-1-carboxylic acid	5.73			
cis-3-methylcyclohexane-1-carboxylic acid	4.88			
trans-3-Methylcyclohexane-1-carboxylic acid	5.02			
cis-4-Methylcyclohexane-1-carboxylic acid	5.04			
trans-4-Methylcyclohexane-1-carboxylic acid	4.89			
2-Methylcyclohexyl-1,1-diacetic acid	3.53	6.89		
3-Methylcyclohexyl-1,1-diacetic acid	3.49	6.08		
4-Methylcyclohexyl-1,1,1-diacetic acid	3.49	6.10		
3-Methylcyclopentyl-1,1-diacetic acid	3.79	6.74		
S-Methyl-L-cysteine	8.97			
N-Methylcytidine	3.88			
5-Methylcytidine	4.21			
N-Methyl-2'-deoxycytidine	3.97			
5-Methyl-2'-deoxycytidine	4.33			
2-Methyl-3,5-dinitrobenzoic acid	2.97			
5-Methyldipropylenetriamine (30 °C)	6.32(+3)	9.19(+2)	10.33(+1)	
2,2'-Methylenebis(4-chlorophenol)	7.6	11.5		

TABLE 8.1 pK_A Values of Organic Materials in Water at 25 °C (*continued*)

Substance	pK_1	pK_2	pK_3	pK_4
2,2′-Methylenebis(4,6-dichloro- phenol)	5.6	10.56		
Methylenebis(thioacetic acid) (18 °C)	3.310	4.345		
3,3′-(Methylenedithio)dialanine	2.200(+1)	8.16(0)		
Methylenesuccinic acid	3.85	5.45		
N-Methylethylamine	4.23(+1)			
N-Methylethylenediamine	6.86(+1)	10.15(+1)		
α-Methylglucoside	13.71			
3-Methylglutaric acid	4.24	5.41		
N-Methylglycine (sarcosine)	2.12(+1)	10.20(0)		
5-Methyl-2,4-heptanedione	8.52(enol); 9.10(keto)			
5-Methyl-2,4-hexanedione	8.66(enol); 9.31(keto)			
5-Methyl-4-hexenoic acid	4.80			
3-Methylhistamine	5.80(+1)	9.90(0)		
1-Methylhistidine	1.69	6.48	8.85	
2-Methylhistidine (18 °C)	1.7	7.2	9.5	
2-Methyl-8-hydroxyquinoline ($\mu = 0.005$)	4.58(+1)	11.71(0)		
4-Methyl-8-hydroxyquinoline	4.67(+1)	11.62(0)		
1-Methylimidazole	7.06(+1)			
4-Methylimidazole	7.55(+1)			
N-Methyliminodiacetic acid	2.15	10.09		
S-Methylisothiourea	9.83(+1)			
O-Methylisourea	9.72(+1)			
Methylmalonic acid	3.07	5.87		
2-(N-Methylmethane- sulfonamido)pyridine	1.73(+1)			
3-(N-Methylmethane- sulfonamido)pyridine	3.94(+1)			
4-(N-Methylmethane- sulfonamido)pyridine	5.14(+1)			
2-Methyl-6-methyl- aminopyridine (20 °C)	3.17(+1)	8.84(0)		
3-Methyl-4-methyl- aminopyridine (20 °C)		9.84(0)		
4-Methyl-2,2′- (4-methylpyridyl)pyridine	5.32(+1)			
N-Methylmorpholine	7.13(+1)			
2-Methyl-1-naphthoic acid	3.11			
N-Methyl-1-naphthylamine	3.70(+1)			
2-Methyl-4-nitrobenzoic acid	1.86			
2-Methyl-6-nitrobenzoic acid	1.87			
1-Methyl-2-nitroterephthalic acid	3.11			
4-Methyl-2-nitroterephthalic acid	1.82			
3-Methylpentanedioic acid	4.25	5.41		
3-Methylpentane-2,4-dione	10.87			
2-Methylpentanoic acid	4.782			

TABLE 8.1 pK$_A$ Values of Organic Materials in Water at 25 °C (*continued*)

Substance	pK$_1$	pK$_2$	pK$_3$	pK$_4$
3-Methylpentanoic acid	4.766			
4-Methylpentanoic acid	4.845			
cis-3-Methyl-2-pentenoic acid	5.15			
trans-3-Methyl-2-pentenoic acid	5.13			
4-Methyl-2-pentenoic acid	4.70			
4-Methyl-3-pentenoic acid	4.60			
6-Methyl-1,10-phenanthroline	5.11(+1)			
(2-Methylphenoxy)acetic acid	3.227			
(3-Methylphenoxy)acetic acid	3.203			
(4-Methylphenoxy)acetic acid	3.215			
(2-Methylphenyl)acetic acid (18 °C)	4.35			
(4-Methylphenyl)acetic acid	4.370			
5-Methyl-5-phenylbarbituric acid	8.011(0)			
3-(2-Methylphenyl)propanoic acid	4.66			
3-(3-Methylphenyl)propanoic acid	4.677			
3-(4-Methylphenyl)propanoic acid	4.684			
1-Methyl-2-phenylpyrrolidine	8.80			
5-Methyl-1-phenyl-1,2,3-triazole-4-carboxylic acid	3.73			
Methylphosphinic acid	3.08			
Methylphosphonic acid	2.38	7.74		
3-Methyl-*o*-phthalic acid	3.18			
4-Methyl-*o*-phthalic acid	3.89			
N-Methylpiperazine ($\mu = 0.1$)	4.94(+2)	9.09(+1)		
2-Methylpiperazine	5.62(+2)	9.60(+1)		
N-Methylpiperidine	10.19(+1)			
2-Methylpiperidine	10.95(+1)			
3-Methylpiperidine	11.07(+1)			
4-Methylpiperidine ($\mu = 0.5$)	11.23(+1)			
2-Methyl-1,2-propanediamine	6.178(+2)	9.420(+1)		
2-Methyl-2-propanethiol	11.2			
2-Methylpropanoic acid	4.853			
2-Methyl-2-propylamine	10.682(+1)			
2-Methyl-2-propylglutaric acid	3.626			
2-Methylpyridine	5.96(+1)			
3-Methylpyridine	5.68(+1)			
4-Methylpyridine	6.00(+1)			
Methyl 4-pyridinecarboxylate	3.26(+1)			
6-Methylpyridine-2-carboxylic acid	5.83			
2-Methylpyridine-1-oxide	1.029(+1)			
3-Methylpyridine-1-oxide	10.921(+1)			
4-Methylpyridine-1-oxide	1.258(+1)			
O-Methylpyridoxal ($\mu = 0.16$)	4.74			
Methyl-2-pyridyl ketoxime	9.97			
1-Methyl-2-(3-pyridyl)pyrrolidine	3.41	7.94		
1-Methylpyrrolidine	10.46(+1)			
1-Methyl-3-pyrroline	9.88(+1)			
5-Methylquinoline	4.62(+1)			
Methylsuccinic acid	4.13	5.64		
Methylsulfonylacetic acid	2.36			

TABLE 8.1   pK$_A$ Values of Organic Materials in Water at 25 °C (*continued*)

Substance	pK$_1$	pK$_2$	pK$_3$	pK$_4$
3-Methylsulfonylaniline	2.68(+1)			
4-Methylsulfonylaniline	1.48(+1)			
3-Methylsulfonylbenzoic acid	3.52			
4-Methylsulfonylbenzoic acid	3.64			
4-Methylsulfonyl-3,5-dimethylphenol	8.13			
3-Methylsulfonylphenol	9.33			
4-Methylsulfonylphenol	7.83			

1-Methyl-1,2,3,4-tetrahydro-
  3-pyridinecarboxylic acid
  (arecaidine; isoguvacine)
	9.07			

5-Methyl-1,2,3,4-tetrazole	3.32			
2-Methylthiazole ($\mu = 0.1$)	3.40(+1)			
4-Methylthiazole ($\mu = 0.1$)	3.16(+1)			
5-Methylthiazole ($\mu = 0.1$)	3.03(+1)			
Methylthioacetic acid	3.72			
4-Methylthioaniline	4.40(+1)			
2-Methylthioethylamine (30 °C)	9.18(+1)			
Methylthioglycolic acid	7.68			
3-(S-Methylthio)phenol	9.53			
4-(S-Methylthio)phenol	9.53			
2-Methylthiopyridine (20 °C)	3.59(+1)			
3-Methylthiopyridine (20 °C)	4.42(+1)			
4-Methylthiopyridine (20 °C)	5.94(+1)			
5-Methylthio-1,2,3,4-tetrazole	4.00(+1)			
O-Methylthreonine	2.02(+1)	9.00(0)		
O-Methyltyrosine	2.21(+1)	9.35(0)		
1-Methylxanthine	7.70	12.0		
3-Methylxanthine	8.10	11.3		
7-Methylxanthine	8.33	ca 13		
9-Methylxanthine	6.25			
Morphine (20 °C)	7.87(+1)	9.85(0)		
Morpholine	8.492(+1)			
2-(N-Morpholino)ethanesulfonic acid (MES) (20 °C)	6.15			
3-(N-Morpholino)-2-hydroxy- propanesulfonic acid (37 °C)	6.75			
3-(N-Morpholino)propanesulfonic acid (20 °C)	7.20			
Murexide	0.0	9.20	10.50	
Myosmine	5.26			
1-Naphthalenecarboxylic acid (1-naphthoic acid)	3.695			

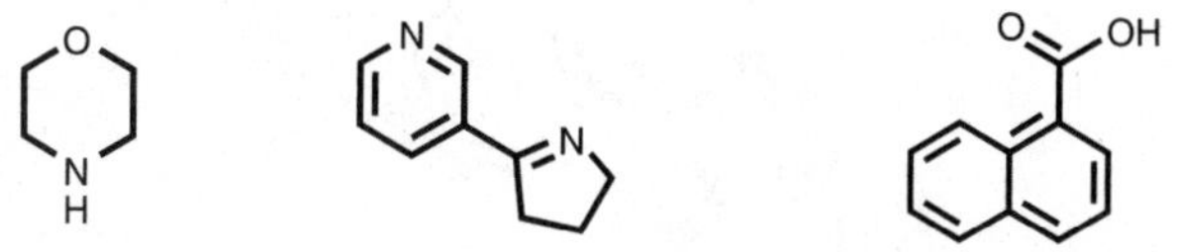

Morpholine

Myosmine

1-Naphthalenecarboxylic acid

TABLE 8.1 pK$_A$ Values of Organic Materials in Water at 25 °C (*continued*)

Substance	pK$_1$	pK$_2$	pK$_3$	pK$_4$
2-Naphthalenecarboxylic acid	4.161			
1-Naphthol (20 °C)	9.30			
2-Naphthol (20 °C)	9.57			
Naphthoquinone monoxime	8.01			
1-Naphthylacetic acid	4.236			
2-Naphthylacetic acid	4.256			
1-Naphthylamine	3.92(+1)			
2-Naphthylamine	4.11(+1)			
1-Naphthylarsonic acid	3.66	8.66		
1-Naphthylsulfonic acid	0.57			
Narceine (15 °C)	3.5(+1)	9.3		
Narcotine	6.18(+1)			
Nicotine	3.15(+1)	7.87(0)		
Nicotyrine	4.76(+1)			
Nitrilotriacetic acid (NTA) (20 °C)	1.65	2.94	10.33	
Nitroacetic acid	1.68			
2-Nitroaniline	−0.28(+1)			
3-Nitroaniline	2.46(+1)			
4-Nitroaniline	1.01(+1)			
2-Nitrobenzene-1,4-dicarboxylic acid	1.73			
3-Nitrobenzene-1,2-dicarboxylic acid	1.88			
4-Nitrobenzene-1,2-dicarboxylic acid	2.11			
2-Nitrobenzoic acid	2.18			
3-Nitrobenzoic acid	3.46			
4-Nitrobenzoic acid	3.441			
trans-2-Nitrocinnamic acid	4.15			
trans-3-Nitrocinnamic acid	4.12			
trans-4-Nitrocinnamic acid	4.05			
Nitroethane	8.57			
2-Nitrohydroquinone	7.63	10.06		
N-Nitroiminodiacetic acid	2.21	3.33		
3-Nitromesitol	8.984			
Nitromethane	10.21			
1-Nitro-6,7-phenanthroline ($\mu = 0.2$)	3.23(+1)			
5-Nitro-1,10-phenanthroline	3.232(+1)			
6-Nitro-1,10-phenanthroline	3.23(+1)			
2-Nitrophenol	7.222			
3-Nitrophenol	8.360			
4-Nitrophenol	7.150			
(2-Nitrophenoxy)acetic acid	2.896			

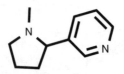

Nicotine

TABLE 8.1 pK_A Values of Organic Materials in Water at 25 °C (*continued*)

Substance	pK_1	pK_2	pK_3	pK_4
(3-Nitrophenoxy)acetic acid	2.951			
(4-Nitrophenoxy)acetic acid	2.893			
2-Nitrophenylacetic acid	4.00			
3-Nitrophenylacetic acid	3.97			
4-Nitrophenylacetic acid	3.85			
2-Nitrophenylarsonic acid	3.37	8.54		
3-Nitrophenylarsonic acid	3.41	7.80		
4-Nitrophenylarsonic acid	2.90	7.80		
7-(4-Nitrophenylazo)-8-hydroxy-5-quinolinesulfonic acid	3.14(0)	7.495(−1)		
3-Nitrophenylphosphonic acid	1.30	6.27		
4-Nitrophenylphosphonic acid	1.24	6.23		
3-(2′-Nitrophenyl)propanoic acid	4.504			
3-(4′-Nitrophenyl)propanoic acid	4.473			
3-Nitrophenylselenic acid	4.07			
4-Nitrophenylselenic acid	4.00			
1-Nitropropane	8.98			
2-Nitropropane	7.675			
2-Nitropropanoic acid	3.79			
2-Nitropyridine ($\mu = 0.02$)	−2.06(+1)			
3-Nitropyridine ($\mu = 0.02$)	0.79(+1)			
4-Nitropyridine ($\mu = 0.02$)	1.23(+1)			
N-Nitrosoiminodiacetic acid	2.28	3.38		
4-Nitrosophenol	6.48			
Nitrourea	4.15(+1)			
1,9-Nonanedioic acid (azelaic acid)	4.53	5.40		
Nonanoic acid (pelargonic acid)	4.95			
DL-Norleucine	2.335(+1)	9.834(0)		
Novocaine	8.85(+1)			
2,2,3,3,4,4,5,5-Octafluoropentanoic acid	2.65			
1,8-Octanedioic acid (suberic acid)	4.512	5.404		
Octanoic acid (caprylic acid)	4.895			
Octopine-DD	1.35	2.30	8.68	11.25
Octopine-LD	1.40	2.30	8.72	11.34
Octylamine	10.65(+1)			
L-(+)-Ornithine	1.94(+2)	8.65(+1)	10.76(0)	
Oxalic acid	1.271	4.272		
3,6-Oxaoctanedioic acid ($\mu = 1.0$)	3.055	3.676		
Oxoacetic acid	3.46			
2-Oxabutanedioic acid (oxaloacetic acid)	2.56	4.37		
2-Oxobutanoic acid	2.50			
5-Oxohexanoic acid (5-ketohexanoic acid) (18 °C)	4.662			
3-Oxo-1,5-pentanedioic acid	3.10			
4-Oxopentanoic acid (levulinic acid)	4.59			
2-Oxopropanoic acid (pyruvic acid)	2.49			

TABLE 8.1 pK$_A$ Values of Organic Materials in Water at 25 °C (*continued*)

Substance	pK$_1$	pK$_2$	pK$_3$	pK$_4$
Oxytetracycline	3.10(+1)	7.26	9.11	
Papaverine	5.90(+1)			
Pentamethylenebis(thioacetic acid) (18 °C)	3.485	4.413		
3,3-Pentamethylenepentanedioic acid	3.49	6.96		
1,5-Pentanediamine	10.05(+2)	10.916(+1)		
2,4-Pentanedione	8.24(enol); 8.95(keto)			
1-Pentanoic acid (valeric acid)	4.842			
2-Pentenoic acid	4.70			
3-Pentenoic acid	4.52			
4-Pentenoic acid	4.677			
Pentylarsonic acid	4.14	9.07		
N-Pentylveratramine	7.28(+1)			
Perhydrodiphenic acid (20 °C)	4.96	6.68		
Perlolidine (18 °C)	4.01	11.39		
Peroxyacetic acid	8.20			
1,7-Phenanthroline	4.30(+1)			
1,10-Phenanthroline	4.857(+1)			
6,7-Phenanthroline	4.857(+1)			
Phenazine	1.2(+1)			
Phenethylthioacetic acid	3.795			
Phenol	9.99			
Phenol-3-phosphoric acid	1.78	7.03	10.2	
Phenol-4-phosphoric acid	1.99	7.25	9.9	
Phenolphthalein	9.4			
3-Phenolsulfonic acid		9.05(−1)		
Phenolsulfonephthalein	7.9			
Phenoxyacetic acid	3.171			
2-Phenoxybenzoic acid	3.53			
3-Phenoxybenzoic acid	3.95			
4-Phenoxybenzoic acid	4.52			
5-Phenoxy-1,2,3,4-tetrazole	3.49(+1)			
Phenylacetic acid	4.312			
L-3-Phenyl-α-alanine	2.16(+1)	9.31(0)		
3-Phenyl-α-alanine, methyl ester	7.05(+1)			
Phenylalanylarginine ($\mu = 0.01$)	2.66(+1)	7.57(0)	12.40(−1)	
Phenylalanylglycine ($\mu = 0.01$)	3.10(+1)	7.71(0)		
7-Phenylazo-8-hydroxy-5-quinolinesulfonic acid	3.41(0)	7.850(−1)		

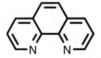

1,10-Phenanthroline

TABLE 8.1 pK_A Values of Organic Materials in Water at 25 °C (*continued*)

Substance	pK_1	pK_2	pK_3	pK_4
5-Phenylbarbituric acid	2.544(+1)			
2-Phenyl-2-benzylsuccinic acid	3.69	6.47		
1-Phenylbiguanide	2.13(+2)	10.76(+1)		
4-Phenylbutanoic acid	4.757			
Phenylbutazone	4.5(+1)			
2-Phenylenediamine	<2(+2)	4.47(+1)		
3-Phenylenediamine	2.65(+2)	4.88(+1)		
4-Phenylenediamine	3.29(+2)	6.08(+1)		
2-Phenylethylamine	9.83(+1)			
β-Phenylethylboronic acid	10.0			
DL-α-Phenylglycine	1.83(+1)	4.39(0)		
Phenylguanidine	10.77(+1)			
Phenylhydrazine	5.20(+1)			
2-Phenyl-3-hydroxypropanoic acid	3.53			
3-Phenyl-3-hydroxypropanoic acid	4.40			
Phenyliminodiacetic acid (20 °C)	2.40	4.98		
Phenylmalonic acid	2.58	5.03		
Phenylmethanethiol	10.70			
2-Phenyl-2-phenethylsuccinic acid (20 °C)	3.74	6.52		
2-Phenylphenol	9.55			
3-Phenylphenol	9.63			
4-Phenylphenol	9.55			
Phenylphosphinic acid (17 °C)	2.1			
Phenylphosphonic acid	1.83	7.07		
O-Phenylphosphorylserine	2.13(+1)	8.79		
O-Phenylphosphorylserylglycine	3.18(+1)	6.95(0)		
O-Phenylphosphoryl-L-seryl-L-leucine	3.16(+1)	7.12(0)		
N-Phenylpiperazine ($\mu = 0.1$)	8.71(+1)			
2-Phenylpropanoic acid	4.38			
3-Phenylpropanoic acid (35 °C)	4.664			
3-Phenyl-1-propylamine	10.39(+1)			
Phenylpropynoic acid (35 °C)	2.269			
Phenylselenic acid	4.79			
Phenylselenoacetic acid ($\mu = 0.1$)	3.75			
β-Phenylserine ($\mu = 0.16$)	8.79(0)			
Phenylsuccinic acid (20 °C)	3.78	5.55		
Phenylsulfenylacetic acid	2.66			
Phenylsulfonylacetic acid	2.44			
5-Phenyl-1,2,3,4-tetrazole	4.38(+1)			
1-Phenyl-1,2,3-triazole-4-carboxylic acid	2.88			
1-Phenyl-1,2,3-triazole-4,5-dicarboxylic acid	2.13	4.93		
Phosphoramidic acid	3.08	8.63		
O-Phosphorylethanolamine	5.838(+1)	10.638(0)		
O-Phosphorylserylglycine	3.13	5.41	8.01	

TABLE 8.1   pK$_A$ Values of Organic Materials in Water at 25 °C (*continued*)

Substance	pK$_1$	pK$_2$	pK$_3$	pK$_4$
O-Phosphoryl-L-seryl-L-leucine	3.11	5.47	8.26	
Phosphoserine	2.08	5.65	9.74	
Phthalamide	3.79(0)			
Phthalazine	3.47(+1)			
o-Phthalic acid	2.950	5.408		
Phthalimide	9.90(0)			
Physostigmine	1.76(+1)	7.88(0)		
Picric acid (2,4,6-trinitrophenol)				
(18°C)	0.419			
Pilocarpine	1.3(+1)	6.85(0)		
Piperazine	5.333(+2)	9.781(+1)		
1,4-Piperazinebis(ethanesulfonic				
acid) (20°C)	6.80			
Piperazine-2-carboxylic acid	1.5	5.41	9.53	
Piperidine	11.123(+1)			
2-Piperidinecarboxylic acid	2.12(+1)	10.75(0)		
3-Piperidinecarboxylic acid	3.35(+1)	10.64(0)		
4-Piperidinecarboxylic acid	3.73(+1)	10.72(0)		
1-(2-Piperidinyl)-2-propanone (15°C)	9.45			
Piperine (15°C)	1.98(+1)			
Proline	1.952(+1)	10.640(0)		
1,2-Propanediamine	6.607(+2)	9.720(+1)		
1,3-Propanediamine	8.49(+2)	10.47(+1)		
1-Propanethiol	10.86			
1,2,3-Propanetriamine	3.72(+3)	7.95(+2)	9.59(+1)	
1,2,3-Propanetricarboxylic acid	3.67	4.87	6.38	
Propanoic acid	4.874			
Propenoic acid	4.247			
N-Propionylglycine	3.718(0)			
2-Propoxybenzoic acid (20°C)	4.24			
3-Propoxybenzoic acid (20°C)	4.20			
4-Propoxybenzoic acid (20°C)	4.78			
N-Propylalanine	2.21(+1)	10.19(0)		
Propylamine	10.568(+1)			
Propylarsonic acid (18°C)	4.21	9.09		
Propylenimine	8.18(+1)			
N-Propylglycine ($\mu = 0.1$)	2.38(+1)	10.03(0)		
L-Propylglycine	3.19(+1)	8.97(0)		
Propylmalonic acid	2.97	5.84		
Propylphosphinic acid	3.46			

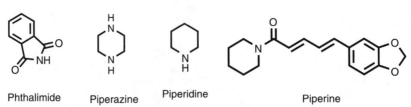

Phthalimide   Piperazine   Piperidine   Piperine

TABLE 8.1 pK$_A$ Values of Organic Materials in Water at 25 °C (*continued*)

Substance	pK$_1$	pK$_2$	pK$_3$	pK$_4$
Propylphosphonic acid	2.49	8.18		
2-Propylpyridine	6.30(+1)			
N-Propylveratramine	7.20(+1)			
2-Propynoic acid	1.887			
Pseudoecgonine	9.70			
Pseudoisocyanine ($\mu = 0.2$)	4.59(+2)			
Pseudotropine	9.86(+1)			
Pteroylglutamic acid	8.26			
Purine	2.52(+1)	8.92(0)		
Pyrazine	0.6(+1)			
Pyrazinecarboxamide	0.5(+1)			
Pyrazole	2.61(+1)			
Pyridazine	2.33(+1)			
Pyridine	5.17(+1)			
Pyridine-d_5	5.83(+1)			
2-Pyridinealdoxime	3.56(+1)	10.17(0)		
3-Pyridinealdoxime	4.07(+1)	10.39(0)		
4-Pyridinealdoxime	4.73(+1)	10.03(0)		
2-Pyridinecarbaldehyde	3.84(+1)			
3-Pyridinecarbaldehyde	3.80(+1)			
4-Pyridinecarbaldehyde	4.74(+1)			
3-Pyridinecarbamide (nicotinamide)	3.33(+1)			
3-Pyridinecarbonitrile	1.35(+1)			
Pyridine-2-carboxylic acid (picolinic acid)	1.01(+1)	5.29(0)		
Pyridine-3-carboxylic acid (nicotinic acid)	2.07(+1)	4.75(0)		
Pyridine-4-carboxylic acid (isonicotinic acid)	1.84(+1)	4.86(0)		
Pyridine-2,3-dicarboxylic acid	2.36(+1)	7.08(0)		
Pyridine-2,4-dicarboxylic acid	2.23(+1)	7.02(0)		
Pyridine-2,6-dicarboxylic acid	2.16(+1)	6.92(0)		
Pyridine-1-oxide	0.688(+1)			
Pyridoxal	4.20(+1)	8.66(ring OH)		
Pyridoxal-5-phosphate ($\mu = 0.15$)	<2.5	4.14	6.20	8.69
Pyridoxamine ($\mu = 0.1$)	3.37(+2)	8.01(+1)	10.13(ring OH)	
Pyridoxamine-5-phosphate ($\mu = 0.15$; pK$_5$ = 10.92)	2.5	3.69	5.76	8.61
Pyridoxine (vitamin B$_6$) (18 °C)	5.00(+1)	8.96(ring OH)		

Purine

Pyrazine

Pyrazole

Pyridazine

Pyridine

TABLE 8.1   pK$_A$ Values of Organic Materials in Water at 25 °C (*continued*)

Substance	pK$_1$	pK$_2$	pK$_3$	pK$_4$
3-(2′-Pyridyl)alanine	1.37(+2)	4.02(+1)	9.22(0)	
3-(3′-Pyridyl)alanine	1.77(+2)	4.64(+1)	9.10(0)	
2-(2′-Pyridyl)benzimidazole				
($\mu = 0.16$)	5.58(+1)			
2-(2′-Pyridyl)imidazole ($\mu = 0.005$)	8.98(+1)			
4-(2′-Pyridyl)imidazole ($\mu = 0.1$)	5.49(+1)			
Pyrimidine	1.30(+1)			
2,4(1*H*,3*H*)-Pyrimidinedione				
(uracil)	0.6(+1)	9.46(0)		
2,4,5,6(1*H*,3*H*)-Pyrimidinetetrone-				
5-oxime	4.57(0)			
Pyrocatecholsulfonephthalein	7.82	9.76	11.73	
Pyroxilidine	11.11(+1)			
Pyrrole-1-carboxylic acid	4.45			
Pyrrole-2-carboxylic acid	4.45			
Pyrrole-3-carboxylic acid	4.453			
Pyrrolidine	11.305(+1)			
Pyrrolidine-2-carboxylic acid				
(proline)	1.952(+1)	10.640(0)		
2-[2-(*N*-Pyrrolidinyl)ethyl]pyridine	3.60(+2)	9.39(+1)		
3-[2-(*N*-Pyrrolidinyl)ethyl]pyridine	4.28(+2)	9.28(+1)		
4-[2(*N*-Pyrrolidinyl)ethyl]pyridine	4.65(+2)	9.27(+1)		
2-(1-Pyrrolidinylmethyl)pyridine	2.54(+1)	8.56(+1)		
3-(1-Pyrrolidinylmethyl)pyridine	3.14(+2)	8.36(+1)		
4-(1-Pyrrolidinylmethyl)pyridine	3.38(+2)	8.16(+1)		
3-Pyrroline	−0.27(+1)			
Quinidine	4.0(+1)	8.54(0)		
Quinine	4.11(+1)	8.52(0)		
Quinoline	4.80(+1)			
Quinoxaline	0.72(+1)			
D-Raffinose	12.74			
Riboflavin (vitamin B$_2$) ($\mu = 0.01$)	ca −0.2	9.69		
α-D-Ribofuranose	12.11			
D-Ribose-5′-phosphonic acid		6.70(−1)	13.05(−2)	

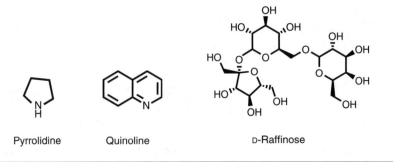

Pyrrolidine   Quinoline   D-Raffinose

TABLE 8.1 pK_A Values of Organic Materials in Water at 25 °C (*continued*)

Substance	pK_1	pK_2	pK_3	pK_4
D-Saccharic acid	5.00(0)			
Saccharin (*o*-benzoic sulfimide)	2.32			
Sarcosine	2.12(+1)	10.20(0)		
Sarcosine amide	8.35(+1)			
Sarcosine dimethylamide	8.86(+1)			
Sarcosine methylamide	8.28(+1)			
Sarcosylglycine ($\mu = 0.16$)	3.15(+1)	8.56(0)		
Sarcosylleucine	3.15(+1)	8.67(0)		
Sarcosylsarcosine	2.92(+1)	9.15(0)		
Sarcosylserine	3.17(+1)	8.63(0)		
3-Selenosemicarbazide ($\mu = 0.1$)	0.8(+1)			
Semicarbazide ($\mu = 0.1$)	3.53(+1)			
L-Serine	2.186(+1)	9.208(0)		
Serine, methyl ester ($\mu = 0.1$)	7.03(+1)			
Serylglycine ($\mu = 0.15$)	2.10(+1)	7.33(0)		
L-Seryl-L-leucine	3.08(+1)	7.45(0)		
Solanine	7.34(+1)			
D-Sorbitol (17.5 °C)	13.60			
L-(−)-Sorbose (18 °C)	11.55			
Sparteine	4.49(+1)	11.76(0)		
Spinaceamine ($\mu = 0.1$)	4.895(+2)	8.90(+1)		
Spinacine	1.649(+2)	4.936(+1)	8.663(0)	
L-Strychnine (15 °C)	2.50	8.20		
Succinamic acid (succinic acid monoamide)	4.39(0)			
Succinic acid	4.207	5.635		
DL-Succinimide	9.623			
β-(4′-Sulfaminophenyl)alanine	1.99(+1)	8.64(0)	10.26(−1)	
3-Sulfamylbenzoic acid	3.54			
4-Sulfamylbenzoic acid	3.47			
4-Sulfamylphenylphosphoric acid	1.42	6.38	10.0	
Sulfanilamide	10.43(+1)			
Sulfoacetic acid		4.0		
3-Sulfobenzoic acid		3.78		
4-Sulfobenzoic acid		3.72		
3-Sulfophenol	0.39	9.07		
4-Sulfophenol	0.58	8.70		
2-Sulfopropanoic acid	1.99			
5-Sulfosalicylic acid	2.49	12.00		
Sylvic acid	7.62			
D-Tartaric acid	3.036	4.366		
meso-Tartaric acid	3.22	4.81		

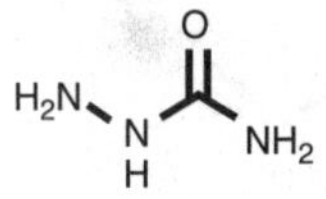

Semicarbazide 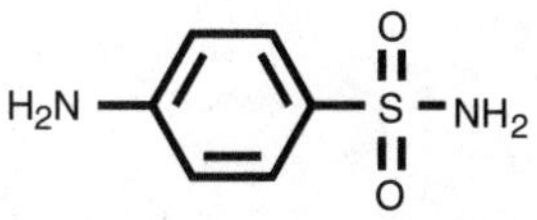   Sulfanilamide 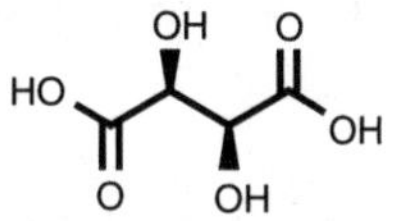   D-Tartaric acid

TABLE 8.1 pK$_A$ Values of Organic Materials in Water at 25 °C (*continued*)

Substance	pK$_1$	pK$_2$	pK$_3$	pK$_4$
Tetracycline ($\mu = 0.005$)	3.30(+1)	7.68	9.69	
Tetradehydroyohimbine	10.59(+1)			
Tetraethylenepentamine				
[$\mu = 0.1$; pK$_5 = 9.67$(+1)]	2.98(+5)	4.72(+4)	8.08(+3)	9.10(+2)
1,4,5,6-Tetrahydro-				
1,2-dimethylpyridine	11.38(+1)			
1,4,5,6-Tetrahydro-2-methylpyridine	9.53(+1)			
cis-Tetrahydronaphthalene-				
2,3-dicarboxylic acid (20 °C)	3.98	6.47		
trans-Tetrahydronaphthalene-				
2,3-dicarboxylic acid (20 °C)	4.00	5.70		
5,6,7,8-Tetrahydro-1-naphthol	10.28			
5,6,7,8-Tetrahydro-2-naphthol	10.48			
Tetrahydroserpentine	10.55(+1)			
2,3,5,6-Tetramethylbenzoic acid	3.415			
Tetramethylenebis(thioacetic acid)				
(18 °C)	3.463	4.423		
Tetramethylenediamine	9.22(+2)	10.75(+1)		
N,N,N′,N′-				
Tetramethylethylenediamine	2.20(+2)	6.35(+1)		
2,3,5,6-Tetramethyl-				
4-methylaminopyridine	0.07(+1)			
2,2,6,6-Tetramethylpiperidine				
($\mu = 0.5$)	1.24(+1)			
2,3,5,6-Tetramethylpyridine (20 °C)	7.90(+1)			
Tetramethylsuccinic acid	3.50	7.28		
1,2,3,4-Tetrazole	4.90			
Thebaine	7.95(+1)			
2-Thenoyltrifluoroacetone	5.70(0)			
Theobromine	0.68(+1)	7.89		
Theophylline	<1(+1)	8.80		
Thiazoline	2.53(+1)			
Thioacetic acid	3.33			
o-Thiocresol	6.64			
m-Thiocresol	6.58			
p-Thiocresol	6.52			
Thiocyanatoacetic acid	2.58			
2,2′-Thiodiacetic acid	3.32	4.29		
4,4′-Thiodibutanoic acid (18 °C)	4.351	5.275		

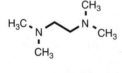

N,N,N,N-Tetramethylethylenediamine   1,2,3,4-Tetrazole   Thiazoline

TABLE 8.1 pK_A Values of Organic Materials in Water at 25 °C (*continued*)

Substance	pK_1	pK_2	pK_3	pK_4
3,3'-Thiodipropanoic acid (18 °C)	4.085	5.075		
3-Thio-*S*-methylcarbazide ($\mu = 0.1$)	7.563(+1)			
1-Thionylcarboxylic acid	3.53			
2-Thionylcarboxylic acid	4.10			
2-Thiophenecarboxylic acid (30 °C)	3.529			
3-Thiophenecarboxylic acid				
(3-thenoic acid)	4.10			
Thiophenol	6.50			
3-Thiosemicarbazide ($\mu = 0.1$)	1.5(+1)			
3-Thiosemicarbazide-1,1-diacetic				
acid (30 °C)	2.94	4.07		
Thiourea	2.03(+1)			
Thorin	3.7	8.3	11.8	
Thymidine	9.79	12.85		
p-Toluenesulfinic acid	1.7			
Toluhydroquinone	10.03	11.62		
o-Toluidine	4.45(+1)			
m-Toluidine	4.71(+1)			
p-Toluidine	5.08(+1)			
o-Tolylacetic acid (18 °C)	4.36			
p-Tolylacetic acid (18 °C)	4.36			
o-Tolylarsonic acid	3.82	8.85		
m-Tolylarsonic acid	3.82	8.60		
p-Tolylarsonic acid	3.70	8.68		
o-Tolylphosphonic acid	2.10	7.68		
m-Tolylphosphonic acid	1.88	7.44		
p-Tolylphosphonic acid	1.84	7.33		
3-Tolylselenic acid	4.80			
4-Tolylselenic acid	4.88			
Triacetylmethane	5.81			
Triallylamine	8.31(+1)			
1,3,5-Triazine-2,4,6-triol	7.20	11.10		
1*H*-1,2,3-Triazole		9.26		
1*H*1,2,4-Triazole	2.386(+1)	9.972		
1,2,3-Triazole-4-carboxylic acid	3.22	8.73		
1,2,3-Triazole-4,5-dicarboxylic acid	1.86	5.90	9.30	
1,2,4-Triazolidine-3,5-dione (urazole)	5.80			
Tribromoacetic acid	−0.147			
2,4,6-Tribromobenzoic acid	1.41			
Trichloroacetic acid	0.52			
Trichloroacrylic acid	1.15			

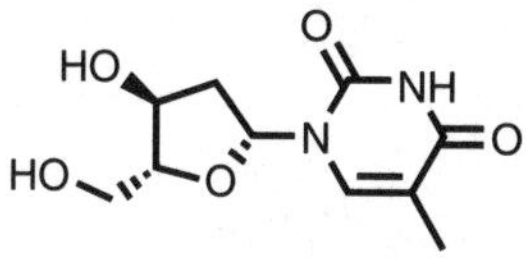

Thymidine

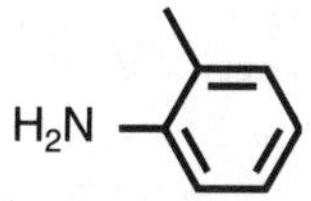

o-Toluidine

TABLE 8.1   pK$_A$ Values of Organic Materials in Water at 25 °C (*continued*)

Substance	pK$_1$	pK$_2$	pK$_3$	pK$_4$
3,3,3-Trichlorolactic acid	2.34			
Trichloromethylphosphonic acid	1.63	4.81		
2,4,5-Trichlorophenol	7.37			
3,4,5-Trichlorophenol	7.839			
Tricine (20 °C)	8.15			
Triethanolamine	7.76(+1)			
Triethylamine	10.72(+1)			
Triethylenediamine	4.18(+2)	8.19(+1)		
Triethylenetetramine (20 °C)	3.32(+4)	6.67(+3)	9.20(+2)	9.92(+1)
Triethylsuccinic acid	2.74			
Trifluoroacetic acid	0.50			
Trifluoroacrylic acid	1.79			
4,4,4-Trifluoro-2-aminobutanoic acid	1.600(+1)	8.169(0)		
4,4,4-Trifluoro-3-aminobutanoic acid	2.756(+1)	5.822(0)		
4,4,4-Trifluorobutanoic acid	4.16			
α,α,α-Trifluoro-*m*-cresol	8.950			
4,4,4-Trifluorocrotonic acid	3.15			
5,5,5-Trifluoroleucine	2.045(+1)	8.942(0)		
3-(Trifluoromethyl)aniline	3.5(+1)			
4-(Trifluoromethyl)aniline	2.6(+1)			
3-Trifluoromethylphenol	8.950			
5-Trifluoromethyl-1,2,3,4-tetrazole	1.70			
6,6,6-Trifluoronorleucine	2.164(+1)	9.463(0)		
5,5,5-Trifluoronorvaline	2.042(+1)	8.916(0)		
5,5,5-Trifluoropentanoic acid	4.50			
3,3,3-Trifluoropropanoic acid	3.06			
4,4,4-Trifluorothreonine	1.554(+1)	7.822(0)		
4,4,4-Trifluorovaline	1.537(+1)	8.098(0)		
1,2,3-Trihydroxybenzene				
(pyrogallol)	9.03(0)	11.63(−1)		
1,3,5-Trihydroxybenzene	8.45(0)	8.88(−1)		
(phloroglucinol)				
2,4,6-Trihydroxybenzoic acid	1.68(0)			
3,4,5-Trihydroxybenzoic acid	4.19(0)	8.85(−1)		
3,4,5-Trihydroxycyclohex-1-ene-				
1-carboxylic acid				
[D-(−)-shikimic acid]	4.15			
2,4,6-Tri(hydroxymethyl)phenol	9.56			
Triisobutylamine	10.42(+1)			
Trimethylamine	9.80(+1)			
3-(Trimethylamino)phenol	8.06			
4-(Trimethylamino)phenol	8.35			
2,4,6-Trimethylaniline	4.38(+1)			
2,4,6-Trimethylbenzoic acid	3.448			
Trimethylenebis(thioacetic acid)	3.435	5.383		
(18 °C)				
2,3,4-Trimethylphenol	10.59			
2,4,5-Trimethylphenol	10.57			
2,4,6-Trimethylphenol	10.88			
3,4,5-Trimethylphenol	10.25			
2,3,6-Trimethylpyridine ($\mu = 0.5$)	7.60(+1)			

TABLE 8.1 pK_A Values of Organic Materials in Water at 25 °C (*continued*)

Substance	pK_1	pK_2	pK_3	pK_4
2,4,6-Trimethylpyridine	7.43(+1)			
2,4,6-Trimethylpyridine-1-oxide	1.990(+1)			
3-(Trimethylsilyl)benzoic acid	4.089			
4-(Trimethylsilyl)benzoic acid	4.192			
2,4,5-Trimethylthiazole ($\mu = 0.1$)	4.55			
2,4,6-Trinitroaniline (picramide)	$-10.23(+1)$			
2,4,6-Trinitrobenzene acid	0.654			
2,2,2-Trinitroethanol	2.36			
Trinitromethane (20 °C)	0.17			
Triphenylacetic acid	3.96			
Tripropylamine	10.66(+1)			
Tris(2-hydroxyethyl)amine	7.762(+1)			
Tris(hydroxymethyl)aminomethane (TRIS)	8.08(+1)			
2-[Tris(hydroxymethyl)methyl amino]-1-ethanesulfonic acid (TES)	7.50			
3-[Tris(hydroxymethyl)methyl amino]-1-propanesulfonic acid (TAPS) (20 °C)	8.4			
N-[Tris(hydroxymethyl)methyl]-glycine (tricine)	2.023(+1)	8.135		
Tris(trimethylsilyl)amine	4.70(+1)			
Trithiocarbonic acid (20 °C)	2.64			
Tropacocaine (15 °C)	9.88(+1)			
3-Tropanol (tropine)	10.33(+1)			
Trypsin ($\mu = 0.1$)	6.25			
L-Tryptophan	2.38(+1)	9.39(0)		
DL-Tyrosine	2.18(+1)	9.21(0)	10.47(OH)	
Tyrosine amide	7.48	9.89		

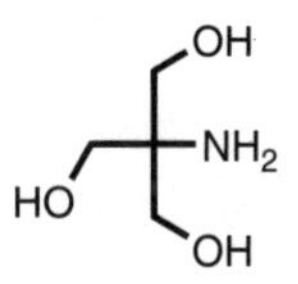

Tris(hydroxymethyl)aminomethane

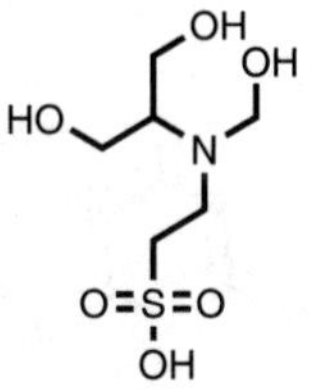

2-[Tris(hydroxymethyl)methylamino]-1-ethanesulfonic acid

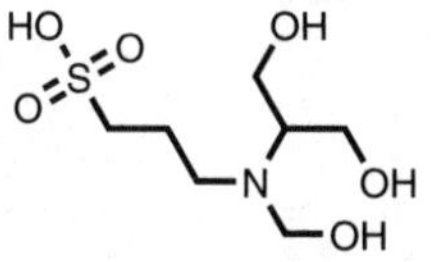

3-[Tris(hydroxymethyl)methylamino]-1-propanesulfonic acid

TABLE 8.1   pK$_A$ Values of Organic Materials in Water at 25 °C (*continued*)

Substance	pK$_1$	pK$_2$	pK$_3$	pK$_4$
Tyrosine, ethyl ester	7.33	9.80		
Tyrosylarginine ($\mu = 0.01$)	2.65(+1)	7.39(0)	9.36(−1)	11.62(−2)
Tyrosyltyrosine	3.52(+1)	7.68(0)	9.80(−1)	10.26(−2)
α-Ureidobutanoic acid	3.886(0)			
γ-Ureidobutanoic acid	4.683(0)			
β-Ureidopropanoic acid	4.487(0)			
Uric acid	5.40	5.53		
Uridine	9.30			
Uridine-5′-diphosphoric acid	7.16			
Uridine-5′-phosphoric acid				
(5′-uridylic acid)	6.63			
Uridine-5′-triphosphoric acid	7.58			
DL-Valine	2.286(+1)	9.719(0)		
L-Valine	2.296(+1)	9.79(0)		
Valine amide ($\mu = 0.2$)	8.00			
L-Valine, methyl ester	7.49(+1)			
L-Valylglycine	3.23(+1)	8.00(0)		
Vetramine	7.49(+1)			
Veratrine	8.85(+1)			
Vinylmethylamine	9.69(+1)			
2-Vinylpyridine	4.98(+1)			
4-Vinylpyridine	5.62(+1)			
Vitamin B$_{12}$	7.64(+1)			
Xanthine (40 °C)	0.68(+1)			
Xanthosine	<2.5(+1)	5.67(0)	12.00(−1)	
Xylenol Orange [pK$_5$ = 10.46(−4);				
pK$_6$ = 12.28(−5)]		2.58(−1)	3.23(−2)	6.37(−3)
D-(+)-Xylose	12.15(0)			
Zincon		4	7.85	15

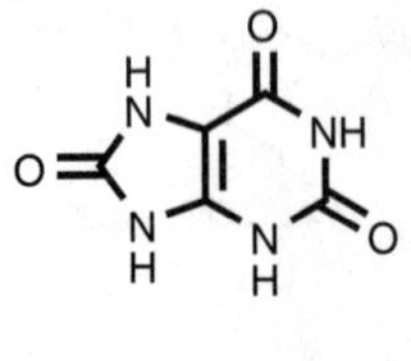

Uric acid 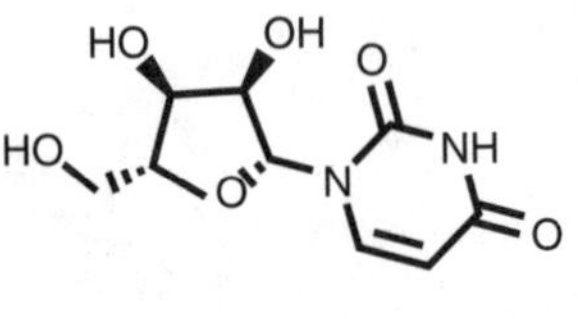   Uridine

Table 8.2 records the acidities of inorganic compounds expressed as their pK$_A$ values (see page 8.2 for a discussion of pK$_A$). When more than one ionizable proton is present, pK$_1$, pK$_2$, etc. values are given. Cations formed from the indicated compound by protonation are indicated by "(+1)" or "(+2)" for a dication.

Temperature values different from 25 °C are given in parentheses as are other relevant variations. For example, the dissociation constant for acetic acid-d_l is reported in D$_2$O.

TABLE 8.2 Proton-Transfer Reactions of Inorganic Materials in Water at 25 °C

The protonation states of cations are designated by values $(+1)$, $(+2)$, etc. that follow pK_A values.

Substance	Formula	pK_1	pK_2	pK_3	pK_4
Aluminic acid (alumina)	H_3AlO_3	11.2			
Amidophosphoric acid	$H_2NPO(OH)_2$	3.3	8.28		
Aminodisulfonic acid	$HN(SO_3H)_2$			8.50	
Ammonium ion	NH_4^+	9.24			
Arsenic acid	H_3AsO_4	2.25	6.77	11.53	
Arsenous acid	$HAsO_2$ or $HAs(OH)_4$	9.23			
Boric acid, ortho-	H_3BO_3	9.236	12.74		
Boric acid, etra-	$H_2B_4O_7$	4	9		
Carbonic acid	$CO_2 + H_2O$	6.35	10.53		
	(without including				
	dehydration constant)	3.76	10.329		
	$CO_2 + D_2O$ (solvent)	6.77	11.076		
Chloric acid	$HClO_3$	-1.58			
Chlorous acid	$HClO_2$	2.021			
Chlorosulfonic acid	$HOSO_2Cl$	-10.43			
Chromic acid	H_2CrO_4	-0.98	6.50		
Cyanic acid	$HOCN$	3.47			
Deuterium oxide	D_2O (solvent)	14.87			
Diamidophosphoric acid	$(H_2N)_2PO_2H$	4.83			
Dithionic acid	$H_2S_2O_6$	-3.4	-0.2		
Dithionous acid	$H_2S_2O_4$	0.35	2.45		
Ferricyanic acid	$H_3Fe(CN)_6$	<1			
Ferrocyanic acid	$H_2(Fe(CN)_6)^{2-}$			2.57	4.35
Fluorophosphoric acid	$FPO(OH)_2$		4.79		
Hexapolyphosphoric acid	$H_8P_6O_{19}$	ca 2.1	2.19	5.98	8.13
Hydrazinium$(2+)$ ion	$^+H_3NNH_3^+$	-0.88	7.956		
(20 °C)		$(+2)$	$(+1)$		
Hydrazinosulfuric acid	H_2NNHSO_3H	3.85			
Hydrazoic acid	HN_3	4.64			
Hydrocyanic acid	HCN	9.21			
Hydrogen bromide	HBr	-20.68			
Hydrogen chloride	HCl	-6.1			
Hydrogen fluoride	HF	3.17			
Hydrogen iodide	HI	-9.5			
Hydrogen peroxide	H_2O_2	11.58			
Hydrogen polysulfide					
(20 °C)	H_2S_4	3.8	6.3		
Hydrogen selenide	H_2Se	3.89	11.0		
Hydrogen sulfide	H_2S	6.96	12.90		
Hydrogen telluride (20 °C)	H_2Te	2.64	$11-12$		
Hydroperoxy radical	$HO_2 = H^+ + O_2^-$	4.45			
Hydroxide radical	$OH^.$	11.9			
Hydroxylamine-N,N-di					
sulfonic acid	$HON(SO_3H)_2$			11.85	
Hydroxylamine-N-sulfonic					
acid	$HONH—OSO_2H$		ca 12.5		
Hydroxylammonium ion	$HONH_3^-$	5.98			

TABLE 8.2 Proton-Transfer Reactions of Inorganic Materials in Water at 25 °C (*continued*)

Substance	Formula	pK_1	pK_2	pK_3	pK_4
Hypobromous acid	HBrO	8.597			
Hypochlorous acid	HClO	7.54			
Hypoiodous acid	HIO	10.64			
Hyponitrous acid	HON=NOH	7.05	11.54		
Hypophosphoric acid					
(20 °C)	$H_4P_2O_6$	2	2.19		
Hypophosphorus acid	HPH_2O_2	1.23			
Hyposulfurous acid	$H_2S_2O_4$	0.35	2.45		
Imidodiphosphoric acid	(HO)$_2$PO—NH—				
	PO(OH)$_2$	ca 2	2.85	7.08	9.72
Iodic acid (30 °C)	HIO_3	0.815			
Nitramide	O_2NNH_2	6.48			
Nitric acid	HNO_3	1.38			
Nitrous acid	HNO_2	3.14			
Osmic acid	H_2OsO_5 (mainly OsO$_4$)	12.0	14.5		
Perchloric acid	$HClO_4$ (completely				
	dissociated up to				
	10 M)				
Periodic acid, para-	H_5IO_6	1.55	8.27		
Permanganic acid	$HMnO_4$	−2.25			
Peroxide radical	$HO_2^{\cdot}$	4.90			
Peroxoboric acid	$H_3BO_3 + H_2O_2 =$				
	$(H_2BO_3 \cdot H_2O_2)^- + H^+$	7.91			
Peroxochromic acid	H_2CrO_5	4.30			
Peroxomonosulfuric acid	H_2SO_5	1.0	9.3		
Perxenic acid	H_4XeO_6	ca 2	ca 6	ca 10	
Phosphoric acid, ortho-	H_3PO_4	2.148	7.198	12.38	
Deuterated	D_3PO_3	2.420	7.201		
Phosphoric acid, di-	$H_4P_2O_7$	0.91	2.10	6.70	9.38
Phosphorous acid (20 °C)	H_2PHO_3	1.20	6.70		
Selenic acid	H_2SeO_4	−3	1.74		
Selenous acid	H_2SeO_3	2.27	7.78		
Silicic acid	H_2SiO_3	9.77	11.80		
Sulfamic acid	$HOSO_2NH_2$	0.988			
Sulfuric acid	H_2SO_4	ca −3	1.987		
Sulfurous acid	SO$_2$ + H$_2$O (includes				
	dehydration constant)	1.89	7.20		
Telluric acid	H_6TeO_6	7.70	10.99		
Tellurous acid	H_2TeO_3	2.46	7.7		
Tetraperoxochromic acid					
(30 °C)	H_3CrO_8	7.16			
Tetrapolyphosphoric acid					
($pK_5 = 6.63$; $pK_6 = 8.34$)	$H_6P_4O_{13}$			1.3	2.23
Thiocyanic acid	HSCN	0.95			
Thiosulfuric acid	$H_2S_2O_3$	0.60	1.5–1.7		
Trimetaphosphoric acid	$H_3P_3O_4$			2.0	
Tripolyphosphoric acid	$H_5P_3O_{10}$	−0.51	1.20	2.30	6.61
($\mu > 1$)* ($pK_5 = 9.26$)					
Trithiocarbonic acid					
(20 °C)	H_2CS_3	2.68	8.18		

TABLE 8.2 Proton-Transfer Reactions of Inorganic Materials in Water at 25 °C (*continued*)

Substance	Formula	pK_1	pK_2	pK_3	pK_4
Tungstic acid (20 °C)	H_2WO_4	ca 3.5	ca 4.6		
Vanadic acid	H_3VO_4	3.78	7.8	13.0	
Water	H_2O	14.003			
Xenon trioxide	XeO_3 (aqueous) = $HXeO_4^- + H^+$	10.8			

* Ionic strength.

TABLE 8.3 Selected Equilibrium Constants in Aqueous Solution at Various Temperatures

Abbreviations Used in the Table

(+1), monoprotonated cation   (−1), monoanion pK_{auto}, negative logarithm (base 10) of autoprotolysis constant
(0), neutral molecule   (−2), dianion pK_{sp}, negative logarithm (base 10) of solubility product

Substance	Temperature, °C									
	0	5	10	15	20	25	30	35	40	50
Acetic acid (0)	4.780	4.770	4.762	4.758	4.757	4.756	4.757	4.762	4.769	4.787
DL-N-Acetylalanine (+1)		3.699	3.699	3.703	3.708	3.715	3.725	3.733	3.745	3.774
β-Acetylaminopropionic (+1)		4.479	4.465	4.465	4.449	4.445	4.444	4.443	4.445	4.457
N-Acetylglycine (+1)		3.682	3.676	3.673	3.667	3.670	3.673	3.678	3.685	3.706
α-Alanine										
(+1)	2.42		2.39		2.35	2.34	2.33	2.33	2.33	2.33
(0)	10.59		10.29		10.01	9.87	9.74	9.62	9.49	9.26
2-Aminobenzenesulfonic acid (0), pK₂	2.633	2.591	2.556	2.521	2.448	2.459	2.431	2.404	2.380	2.338
3-Aminobenzenesulfonic acid (0), pK₂	4.075	4.002	3.932	3.865	3.799	3.738	3.679	3.622	3.567	3.464
4-Aminobenzenesulfonic acid (0), pK₂	3.521	3.457	3.398	3.338	3.283	3.227	3.176	3.126	3.079	2.989
3-Aminobenzoic acid (0)					4.90	4.79	4.75		4.68	4.60
4-Aminobenzoic acid (0)					4.95	4.85	4.90		4.95	5.10
2-Aminobutyric acid										
(+1)			2.334			2.286		2.289$^{37.5°C}$		2.297
(0)			10.530			9.380		9.518$^{37.5°C}$		9.234
4-Aminobutyric acid										
(+1)			4.057	4.046	4.038	4.031	4.027	4.025	4.027	4.032
(0)			11.026	10.867	10.706	10.556	10.409	10.269	10.114	9.874
2-Aminoethylsulfonic acid (0)			9.452	9.316	9.186	9.061	8.940	8.824	8.712	9.499
2-Amino-3-methylpentanoic acid										
(+1)	2.365$^{1°C}$		2.338$^{12.5°C}$			2.320		2.317$^{37.5°C}$		2.332
(0)	10.460$^{1°C}$		10.100$^{12.5°C}$			9.758		9.439$^{37.5°C}$		9.157

Name	0	5	10	15	20	25	30	35	40	45
2-Amino-2-methyl-1,3-propanediol	9.612	9.433	9.266	9.104	8.951	8.801	8.659	8.519	8.385	8.132
2-Amino-2-methylpropionic acid (+1)	2.419$^{1°C}$		2.380$^{12.5°C}$			2.357		2.351$^{37.5°C}$		2.356
(0)	10.960$^{1°C}$		10.580$^{12.5°C}$			10.205		9.872$^{37.5°C}$		9.561
2-Aminopentanoic acid (+1)	2.376$^{1°C}$		2.347			2.318			2.309	2.313
(0)	10.508$^{1°C}$			10.154$^{12.5°C}$		9.808		9.490$^{37.5°C}$		9.198
3-Aminopropionic acid (+1)	3.656	3.627		3.583		3.551		3.524	3.517	
(0)	11.000	10.830		10.526		10.235		9.963	9.842	
4-Aminopyridine (+1)	9.873	9.704	9.549	9.398	9.252	9.114	8.978	8.846	8.717	8.477
Ammonium ion (+1)	10.081	9.904	9.731	9.564	9.400	9.245	9.093	8.947	8.805	8.539
Arginine (+1)	1.914	1.885	1.870	1.849	1.837	1.823	1.814	1.801	1.800	1.787
(0)	9.718	9.563	9.407	9.270	9.123	8.994	8.859	8.739	8.614	8.385
Barbituric acid (+1)				3.969	3.980	4.02	4.00	4.008	4.017	4.032
(0)				8.493	8.435	8.372	8.302	8.227	8.147	7.974
Benzoic acid (0)		4.231	4.220	4.215	4.206	4.204	4.203	4.207	4.219	4.223
Boric acid (0)	9.508	9.439	9.380	9.327	9.280	9.236	9.197	9.161	9.132	9.080
Bromoacetic acid (0)				2.875	2.887	2.902	2.918	2.936		
3-Bromobenzoic acid (0)				3.818	3.813	3.810	3.808	3.810	3.813	
4-Bromobenzoic acid (0)				4.011	4.005	3.99	4.001	4.001	4.003	
Bromopropynoic acid (0)			1.786	1.814	1.839	1.855	1.879	1.900	1.919	
3-tert-Butylbenzoic acid (0)				4.266	4.231	4.199	4.170	4.143	4.119	
4-tert-Butylbenzoic acid (0)				4.463	4.425	4.389	4.354	4.320	4.287	
2-Butynoic acid (0)			2.618	2.626	2.611	2.620	2.618	2.621	2.631	
Butyric acid (0)	4.806	4.804	4.803	4.805	4.810	4.817	4.827	4.840	4.854	4.885
DL-N-Carbamoylalanine (+1)		3.898	3.894	3.891	3.890	3.892	3.896	3.902	3.908	3.931
N-Carbamoylglycine (+1)		3.911	3.900	3.889	3.879	3.876	3.874	3.873	3.875	3.888

TABLE 8.3 Selected Equilibrium Constants in Aqueous Solution at Various Temperatures (*continued*)

Substance	Temperature, °C									
	0	5	10	15	20	25	30	35	40	50
Carbon dioxide + water										
(0)	6.583	6.517	6.465	6.429	6.382	6.365	6.327	6.31	6.296	6.297
(−1)	10.627	10.558	10.499	10.431	10.377	10.33	10.290	10.25	10.220	10.172
Chloroacetic acid (0)				2.845	2.856	2.867	2.883	2.900		
3-Chlorobenzoic acid (0)				3.838	3.831	3.83	3.825	3.826	3.829	
4-Chlorobenzoic acid (0)				4.000	3.991	3.986	3.981	3.980	3.981	
Chloropropynoic acid (0)			1.766	1.796	1.820	1.845	1.864	1.879	1.893	
Citric acid										
(0)	3.220	3.200	3.176	3.160	3.142	3.128	3.116	3.109	3.099	3.095
(−1)	4.837	4.813	4.797	4.782	4.769	4.761	4.755	4.751	4.750	4.757
(−2)	6.393	6.386	6.383	6.384	6.388	6.396	6.406	6.423	6.439	6.484
Cyanoacetic acid (0)		2.445	2.447	2.452	2.460	2.460	2.482	2.496	2.511	
2-Cyano-2-methylpropionic acid		2.342	2.360	2.379	2.400	2.422	2.446	2.471	2.498	
5,5-Diethylbarbituric acid (0)	8.40	8.30	8.22	8.169	8.094	8.020	7.948	7.877	7.808	7.673
Diethylmalonic acid										
(0)			2.129	2.136	2.144	2.151	2.160	2.172	2.187	
(−1)			7.400	7.401	7.408	7.417	7.428	7.441	7.457	
2,3-Dimethylbenzoic acid (0)				3.663	3.687	3.771	3.726	3.762	3.788	
2,4-Dimethylbenzoic acid (0)				4.154	4.187	4.217	4.244	4.268	4.290	
2,5-Dimethylbenzoic acid (0)				3.911	3.954	3.990	4.020	4.045	4.065	
2,6-Dimethylbenzoic acid (0)				3.234	3.304	3.362	3.409	3.445	3.472	
3,5-Dimethylbenzoic acid (0)				4.292	4.299	4.302	4.304	4.306	4.306	
N,N'-Dimethylethyleneamine-N,N'-diacetic acid										
(0)	6.294		6.169		6.047		5.926		5.803	
(−1)	10.446		10.268		10.068		9.882		9.684	

Dissociation (ionization) constants — pK_a values at successive temperature columns. (Superscripts give the actual measurement temperature where it differs from the nominal column.)

Acid / base (charge)										
N,N-Dimethylglycine (0)			9.76		9.94		10.14		10.34	
3,5-Dinitrobenzoic acid (0)	3.07	2.96		2.85		2.73		2.60		
2-Ethylbutyric acid (0)	4.869	4.812		4.758	4.751	4.710		4.664		4.623
5-Ethyl-5-phenylbarbituric acid (0)	7.130	7.248	7.311	7.377	7.445	7.517	7.592			
Fluoroacetic acid (0)			2.624	2.604	2.586	2.571	2.555			
Formic acid (0)	3.782	3.766	3.758	3.752	3.751	3.753	3.757	3.762	3.772	3.786
2-Furancarboxylic acid (0)		3.239	3.216	3.200	3.164					
Glucose-1-phosphate (0)	6.561	6.531	6.519	6.510	6.504	6.500	6.499	6.500	6.506	
Glycerol-1-phosphoric acid (−1)	6.733	6.695	6.679	6.666	6.656	6.648	6.643	6.641	6.642	
Glycerol-2-phosphoric acid (0)	1.554	1.457	1.413	1.372	1.335	1.301	1.271	1.245	1.223	
(−1)	6.712	6.679	6.666	6.657	6.650	6.646	6.646	6.650	6.657	
Glycine (+1)	2.32	2.327	2.33	2.34	2.351	2.36	2.380	2.397		
(0)	9.19	9.412	9.53	9.65	9.780	9.91	10.044	10.193	10.34	
Glycolic acid (0)	3.849		3.833$^{37.5°C}$		3.831			3.844$^{12.5°C}$		3.875
Glycylasparagine (+1)	2.959	2.947	2.944	2.942	2.942	2.943	2.952	2.958	2.968	
N-Glycylglycine (+1)	3.159				3.126					3.201
(second pK)	7.668		7.948$^{37.5°C}$		8.252			8.594$^{12.5°C}$		
Hexanoic acid (0)	4.920	4.890		4.865		4.849		4.839		4.840
Hydrogen cyanide (0)		8.88	8.99	9.11	9.21	9.36	9.49	9.63		
Hydrogen peroxide (0)	11.21		11.45	11.55	11.65	11.75	11.86			12.23
Hydrogen sulfide (0)	6.69	6.79	6.82	6.90	6.97	7.05	7.13	7.24	7.33	
(−1)			12.6	12.75	12.90		13.2		13.5	
4-Hydroxybenzoic acid (0)		4.578	4.576	4.577	4.582	4.586	4.596			
Hydroxylamine (0)			5.730		5.948	6.063	6.186			
2-Hydroxy-1-naphthoic acid (0)	3.26	3.19				3.29				
(−1)	9.58	9.61				9.68				
4-Hydroxyproline (+1)	1.796		1.798$^{37.5°C}$		1.818			1.850$^{12.5°C}$		1.900$^{1°C}$
(0)	9.138		9.394$^{37.5°C}$		9.662			9.958$^{12.5°C}$		10.274$^{1°C}$

TABLE 8.3 Selected Equilibrium Constants in Aqueous Solution at Various Temperatures (*continued*)

Substance	Temperature, °C									
	0	5	10	15	20	25	30	35	40	50
2-Hydroxypropionic acid (0)	3.880	3.873	3.868	3.861	3.857	3.858	3.861	3.867	3.873	3.895
DL-2-Hydroxysuccinic acid										
(0)	3.537	3.520	3.494	3.482	3.472	3.458	3.452	3.446	3.444	3.445
(−1)	5.119	5.108	5.098	5.096	5.096	5.097	5.099	5.104	5.117	5.149
Hypobromous acid (0)				8.83		8.60		8.47	$8.37^{45°C}$	
Hypochlorous acid (0)	7.82	7.75	7.69	7.63	7.58	7.54	7.50	7.46		7.05
Imidazole (+1)	7.581	7.467	7.334	7.216	7.103	6.993	6.887	6.784	6.685	6.497
Iodoacetic acid (0)				3.143	3.158	3.175	3.193	3.213		
DL-Isoleucine										
(+1)	2.365		$2.338^{12.5°C}$			2.318		$2.317^{37.5°C}$		2.332
(0)	10.460		$10.100^{12.5°C}$			9.758		$9.439^{37.5°C}$		9.157
Isopropylmalonic acid, mononitrile (0)		2.299	2.320	2.343	2.365	2.401	2.427	2.452	2.481	
Lactic acid (0)	3.880	3.873	3.868	3.862	3.857	3.858	3.861	3.867	3.873	3.895
Lead sulfate, pK_{sp}	8.01			7.87		7.80		7.73		7.63
DL-Leucine										
(+1)	$2.383^{1°C}$		$2.348^{12.5°C}$			2.328		$2.327^{37.5°C}$		2.333
(0)	$10.458^{1°C}$		$10.095^{12.5°C}$			9.744		$9.434^{37.5°C}$		9.142
Malonic acid (−1)	5.670	5.665	5.667	5.673	5.683	5.696	5.710	5.730	5.753	5.803
Mannose (0)			12.45			12.08			11.81	
Mercury(I) chloride, pK_{sp}			18.65	18.48	18.27	17.88		16.79		
Methanol (solvent), pK_{auto}		17.12		16.84		16.71	16.65	16.53		
Methylamine (+1)	11.496		11.130		10.787	10.62	10.466		10.161	9.876
Methylaminodiacetic acid										
(0)	2.138		2.142		2.146		2.150		2.154	
(−1)	10.474		10.287		10.088		9.920		9.763	

Substance (charge)	0	5	10	15	20	25	30	35	40	50
3-Methylbenzoic acid (0)	4.726			4.303	4.285	4.269	4.256	4.244	4.235	4.871
4-Methylbenzoic acid (0)	4.827			4.390	4.376	4.362	4.349	4.336	4.322	4.908
3-Methylbutyric acid (0)					4.742		4.767		4.794	4.831
4-Methylpentanoic acid (0)					4.827		4.837		4.853	4.879
5-Methyl-5-phenylbarbituric acid (0)	8.104		8.057		8.011	7.966	7.922		7.879	7.797
2-Methylpropionic acid (0)	4.825		4.827		4.840	4.853	4.886		4.918	4.955
2-Methyl-2-propylamine (+1)	11.439		11.240		11.048	10.862	10.682		10.511	10.341
Nitric acid (0)	−1.65					−1.38				−1.20
Nitrilotriacetic acid (0)	1.69		1.65		1.65		1.66		1.67	
(−1)	2.95		2.95		2.94		2.96		2.98	
(−2)	10.59		10.45		10.33		10.23			
4-Nitrobenzoic acid (0)			3.448		3.444	3.441	3.441		3.442	3.445
Nitrous acid (0)			3.244		3.177	3.138			3.100	
DL-Norleucine (+1)	2.394			$2.356^{12.5°C}$		2.335			$2.324^{37.5°C}$	2.328
(0)	10.564			$10.190^{12.5°C}$		9.834			$9.513^{37.5°C}$	9.224
Oxalic acid (−1)	4.210	4.216	4.227	4.240	4.254	4.272	4.295	4.318	4.349	4.409
2,4-Pentanedione (0)	9.07					8.95			8.90	
Pentanoic acid (0)	4.823				4.763	4.835	4.842	4.851	4.861	4.906
Phenylalanine (0)	9.75			9.31		8.96			8.90	8.95
Phosphoric acid (0)	2.056	2.073	2.088	2.107	2.127	2.148	2.171	2.196	2.224	2.277
(−1)	7.313	7.282	7.254	7.231	7.213	7.198	7.189	7.185	7.181	7.183
o-Phthalic acid (0)	2.925	2.927	2.931	2.937	2.943	2.950	2.958	2.967	2.978	3.001
(−1)	5.432	5.418	5.410	5.405	5.405	5.408	5.416	5.427	5.442	5.485
Piperidine (+1)	11.963	11.786	11.613	11.443	11.280	11.123	10.974	10.818	10.670	10.384
Proline (+1)	2.011			$1.964^{12.5°C}$		1.952			$1.950^{37.5°C}$	1.958
(0)	11.296			$10.972^{12.5°C}$		10.640			$10.342^{37.5°C}$	10.064

TABLE 8.3 Selected Equilibrium Constants in Aqueous Solution at Various Temperatures (*continued*)

Substance	Temperature, °C									
	0	5	10	15	20	25	30	35	40	50
Propenoic acid (0)				4.267	4.250	4.247	4.249	4.267	4.301	3.750
N-Propionylglycine (+1)		3.728	3.723	3.718	3.716	3.718	3.721	3.725	3.731	
Propynoic acid (0)			1.791	1.829	1.867	1.887	1.940	1.932	1.963	
Pyrrolidine (+1)	12.17	11.98	11.81	11.63	11.43	11.30	11.15	10.99	10.84	11.56
Serine										
(+1)	$2.296^{1°C}$		$2.232^{12.5°C}$			2.186		$2.154^{37.5°C}$		2.132
(0)	$9.880^{1°C}$		$9.542^{12.5°C}$			9.208		$8.904^{37.5°C}$		8.628
Silver bromide, pK$_{sp}$		13.33		12.83	12.57	12.30	12.07	11.83	11.61	11.19
Silver chloride, pK$_{sp}$		10.595		10.152		9.749		9.381	9.21	8.88
Succinic acid										
(0)	4.285	4.263	4.245	4.232	4.218	4.207	4.198	4.191	4.188	4.186
(−1)	5.674	5.660	5.649	5.642	5.639	5.635	6.541	5.647	5.654	5.680
Sulfuric acid (−1)	1.778	$1.812^{4.3°C}$		1.894		1.987	2.05	2.095	2.17	2.246
Sulfurous acid (0)	1.63		1.74			1.89		1.98		2.12
D-Tartaric acid										
(0)	3.118	3.095	3.075	3.057	3.044	3.036	3.025	3.019	3.018	3.021
(−1)	4.426	4.407	4.391	4.381	4.372	4.366	4.365	4.367	4.372	4.391
2,3,5,6-Tetramethylbenzoic acid (0)				3.310	3.367	3.415	3.453	3.483	3.505	
Threonine										
(+1)	$2.200^{1°C}$		$2.132^{12.5°C}$			2.088		$2.070^{37.5°C}$		2.055
(0)	$9.748^{1°C}$		$9.420^{12.5°C}$			9.100		$8.812^{37.5°C}$		8.548
o-Toluidine (0)				4.58	4.495	4.45	4.345	4.28	4.20	
1,2,4-Triazole										
(+1)				2.451	2.418	2.386	2.327			
(0)				10.205	10.083	9.972	9.768			
3,4,5-Trihydroxybenzoic acid (0)					4.19		4.30		4.38	4.53

Tris(2-hydroxyethyl)amine (+1)	8.290	8.173	8.067	7.963	7.861	7.762	7.666	7.570	7.477	7.299
2,4,6-Trimethylbenzoic acid (0)				3.325	3.391	3.448	3.498	3.541	3.577	
3-Trimethylsilylbenzene acid (0)				4.142	4.116	4.089	4.060	4.029	3.996	
4-Trimethylsilylbenzoic acid (0)				4.270	4.230	4.192	4.155	4.119	4.084	
β-Ureidopropionic acid (0)		4.514	4.505	4.497	4.490	4.487	4.486	4.486	4.488	4.500
DL-Valine										
(+1)	2.320		$2.297^{12.5\,°C}$			2.296		$2.292^{37.5\,°C}$		2.310
(0)	10.413		$10.064^{12.5\,°C}$			9.719		$9.405^{37.5\,°C}$		9.124

TABLE 8.4 Indicators for Aqueous Acid–Base Titrations

Table 8.4 lists selected common indicators. The table is arranged according to function over increasing pH range or transition interval given (third column). Note that this range may vary appreciably from one observer to another, and that it is also affected by ionic strength, temperature, and illumination. The values given should therefore be considered to be approximate. These values refer to solutions having low ionic strengths and a temperature of about $25\,°C$. In the fourth column the pK_A ($-\log K_A$) of the indicator as determined spectrophotometrically is listed. In the fifth and sixth columns, the wavelength of maximum absorption is given for the acidic and basic forms of the indicator, respectively. The lower to higher pH color change is given in the last column. The abbreviations used to describe the colors of the two forms of the indicator are as follows:

B, Blue   P, Purple

C, Colorless   R, Red

G, Green   V, Violet

O, Orange   Y, Yellow

OBr, Orange-brown

Indicator	Chemical name	pH range	pK_A	λ_{max}, nm Acid	λ_{max}, nm Base	Color change
Cresol red (acid range)	o-Cresolsulfonephthalein	0.2–18				R–Y
Cresol purple (acid range)	m-Cresolsulfonephthalein	1.2–2.8	1.51	533		R–Y
Thymol blue (acid range)	Thymolsulfonephthalein	1.2–2.8	1.65	544	430	R–Y
Tropeolin OO	Diphenylamino-p-benzene sodium sulfonate	1.3–3.2	2.0	527		R–Y
2,6-Dinitrophenol	2,6-Dinitrophenol	2.4–4.0	3.69			C–Y
2,4-Dinitrophenol	2,4-Dinitrophenol	2.5–4.3	3.90			C–Y
Methyl yellow	Dimethylaminoazobenzene	2.9–4.0	3.3	508		R–Y
Methyl orange	Dimethylaminoazobenzene sodium sulfonate	3.1–4.4	3.40	522	464	R–O
Bromophenol blue	Tetrabromophenolsulfonephthalein	3.0–4.6	3.85	436	592	Y–BV
Bromocresol green	Tetrabromo-m-cresol-sulfonephthalein	4.0–5.6	4.68	444	617	Y–B
Methyl red	o-Carboxybenzeneazo-dimethylaniline	4.4–6.2	4.95	530	427	R–Y
Chlorophenol red	Dichlorophenolsulfonephthalein	5.4–6.8	6.0		573	Y–R
Bromocresol purple	Dibromo-o-cresolsulfonephthalein	5.2–6.8	6.3	433	591	Y–P
Bromophenol red	Dibromophenolsulfonephthalein	5.2–6.8			574	Y–R
p-Nitrophenol	p-Nitrophenol	5.3–7.6	7.15	320	405	C–Y
Bromothymol blue	Dibromothymolsulfonephthalein	6.2–7.6	7.1	433	617	Y–B
Neutral red	Aminodimethylaminotoluphen-azonium chloride	6.8–8.0	7.4			R–Y
Phenol red	Phenolsulfonephthalein	6.4–8.0	7.9	433	558	Y–R

TABLE 8.4 Indicators for Aqueous Acid–base Titrations (*continued*)

Indicator	Chemical name	pH range	pK_A	λ_{max}, nm Acid	λ_{max}, nm Base	Color change
m-Nitrophenol	*m*-Nitrophenol	6.4–8.8	8.3		570	C–Y
Cresol red	*o*-Cresolsulfonephthalein	7.2–8.8	8.2	434	572	Y–R
m-Cresol purple	*m*-Cresolsulfonephthalein	7.6–9.2	8.32		580	Y–P
Thymol blue	Thymolsulfonephthalein	8.0–9.6	8.9	430	596	Y–B
Phenolphthalein	Phenolphthalein	8.0–10.0	9.4		553	C–R
α-Naphtholbenzein	*α*-Naphtholbenzein	9.0–11.0				Y–B
Thymolphthalein	Thymolphthalein	9.4–10.6	10.0		598	C–B
Alizarin yellow	5-(*p*-Nitrophenylazo)salicylic acid, Na salt	10.0–12.0	11.16			Y–V
Tropeolin O	*p*-Sulfobenzeneazoresorcinol	11.0–13.0				Y–OBr
Nitramine	2,4,6-Trinitrophenyl-methylnitroamine	10.8–13.0				C–OBr

BUFFER SOLUTIONS

TABLE 8.5 National Institute of Standards and Technology (formerly National Bureau of Standards U.S.). Reference PH Buffer Solutions.

Temperature °C	Secondary standard 0.05 M Potassium tetraoxalate	Potassium hydrogen tartrate (saturated at 25 °C)	0.05 M Potassium dihydrogen citrate	0.05 M Potassium hydrogen phthalate	0.025 M KH$_2$PO$_4$, 0.025 M Na$_2$HPO$_4$	0.0087 M KH$_2$PO$_4$, 0.0302 M Na$_2$HPO$_4$	0.01 M Na$_2$B$_4$O$_7$	0.025 M NaHCO$_3$, 0.025 M Na$_2$CO$_3$	Secondary standard Ca(OH)$_2$ (saturated at 25 °C)
0	1.666		3.860	4.003	6.984	7.534	9.464	10.317	13.423
5	1.668		3.840	3.999	6.951	7.500	9.395	10.245	13.207
10	1.670		3.820	3.998	6.923	7.472	9.332	10.179	13.003
15	1.672		3.802	3.999	6.900	7.448	9.276	10.118	12.810
20	1.675		3.788	4.002	6.881	7.429	9.225	10.062	12.627
25	1.679	3.557	3.776	4.008	6.865	7.413	9.180	10.012	12.454
30	1.683	3.552	3.766	4.015	6.853	7.400	9.139	9.966	12.289
35	1.688	3.549	3.759	4.024	6.844	7.389	9.102	9.925	12.133
38	1.691	3.548		4.030	6.840	7.384	9.081		12.043
40	1.694	3.547	3.753	4.035	6.838	7.380	9.068	9.889	11.984
45	1.700	3.547		4.047	6.834	7.373	9.038		11.841
50	1.707	3.549	3.749	4.060	6.833	7.367	9.011	9.828	11.705
55	1.715	3.554		4.075	6.834		8.985		11.574
60	1.723	3.560		4.091	6.836		8.962		11.449
70	1.743	3.580		4.126	6.845		8.921		
80	1.766	3.609		4.164	6.859		8.885		
90	1.792	3.650		4.205	6.877		8.850		
95	1.806	3.674		4.227	6.886		8.833		
Dilution value $\Delta pH_{1/2}$	+0.186	+0.049	0.024	+0.052	+0.080	+0.070	+0.01	0.079	−0.28

Source: R. G. Bates, J. Res. Natl. Bur. Stand. (U.S.), **66A**:179 (1962) and B. R. Staples and R. G. Bates, ibid, **73A**: 37 (1969).

TABLE 8.6   Compositions of National Institute of Standards and Technology. Standard pH Buffer Solutions

Air weight of material per liter of buffer solution

Standard	Weight, g
$KH_3(C_2O_4)_2 \cdot 2H_2O$, 0.05 M	12.61
Potassium hydrogen tartrate, about 0.034 M	Saturated at 25 °C
Potassium hydrogen phthalate, 0.05 M	10.12
Phosphate (solution 1)	
$\quad KH_2PO_4$, 0.025 M	3.39
$\quad Na_2HPO_4$, 0.025 M	3.53
Phosphate (solution 2)	
$\quad KH_2PO_4$, 0.008665 M	1.179
$\quad Na_2HPO_4$, 0.03032 M	4.30
$Na_2B_4O_7 \cdot 10H_2O$, 0.01 M	3.80
Carbonate	
$\quad NaHCO_3$, 0.025 M	2.10
$\quad Na_2CO_3$, 0.025 M	2.65
$Ca(OH)_2$, about 0.0203 M	Saturated at 25 °C

Standard Reference pH Buffer Solutions

The buffer value for the National Institute of Standards and technology (U.S.) reference pH buffer solutions is given below:

Buffer solution	KH tartrate	0.05 M KH$_2$ citrate	0.05 M KH phthalate	0.025 M KH$_2$PO$_4$, 0.025 M Na$_2$HPO$_4$	0.0087 M KH$_2$PO$_4$, 0.0302 M Na$_2$HPO$_4$	0.01 M Na$_2$B$_4$O$_7$	0.025 M NaHCO$_3$, 0.025 M Na$_2$CO$_3$
Buffer value β	0.027	0.034	0.016	0.029	0.016	0.020	0.029

For the secondary pH reference standards, the buffer value is 0.070 for potassium tetraoxalate and 0.09 for calcium hydroxide.

To prepare the standard pH buffer solutions recommended by the National Bureau of Standards (U.S.), the indicated weights of the pure materials in Table 8.6 should be dissolved in water of specific conductivity not greater than 5 micromhos. The tartrate, phthalate, and phosphates can be dried for 2 h at 110 °C before use. Potassium tetraoxalate and calcium hydroxide need not be dried. Fresh-looking crystals of borax should be used. Before use, excess solid potassium hydrogen tartrate and calcium hydroxide must be removed. Buffer solutions pH 6 or above should be stored in plastic containers and should be protected from carbon dioxide with soda-lime traps. The solutions should be replaced within 2 to 3 weeks, or sooner if formation of mold is noticed. A crystal of thymol may be added as a preservative.

Buffer Solutions other than Standards

The range of the buffering effect of a single weak acid group is approximately one pH unit on either side of the pK_A. The ranges of some useful buffer systems are collected in Table 8.7. After all the components have been brought together, the pH of the resulting solution should be determined at the temperature to be employed with reference to standard reference solutions. Buffer components should be compatible with other components in the system under study; this is particularly significant for buffers employed in biological studies. Check tables of formation constants to ascertain whether metal-binding character exists.

When there are two or more acid groups per molecule, or a mixture is composed of several overlapping acids, the useful range is larger. Universal buffer solutions consist of a mixture of acid groups which overlap such that successive pK_A values differ by 2 pH units or less. The Prideaux–Ward mixture comprises phosphate, phenyl acetate, and borate plus HCl and covers the range from 2 to 12 pH units. The McIlvaine buffer is a mixture of citric acid and Na_2HPO_4 that covers the range from pH 2.2 to 8.0. The Britton–Robinson system consists of acetic acid, phosphoric acid, and boric acid plus NaOH and covers the range from pH 4.0 to 11.5. A mixture composed of Na_2CO_3, NaH_2PO_4, citric acid, and 2-amino-2-methyl-1,3-propanediol covers the range from pH 2.2 to 11.0.

TABLE 8.7   pH Values of Buffer Solutions for Control Purposes

Materials*	pH range
Glycine and HCl	1.0–3.7
Citrate and HCl	1.3–4.7
p-Toluenesulfonate and p-toluenesulfonic acid	1.1–3.3
Formate and HCl	2.8–4.6
Succinic acid and borax	3.0–5.8
Phenyl acetate and HCl	3.5–5.0
Acetate and acetic acid	3.7–5.6
Succinate and succinic acid	4.8–6.3
2-(N-Morpholino)ethanesulfonic acid and NaOH	5.2–7.1
2,2-Bis(hydroxymethyl)-2,2′,2″-nitrilotriethanol and HCl	5.8–7.2
KH_2PO_4 and borax	5.8–9.2
N-Tris(hydroxymethyl)methyl-2-aminoethanesulfonic acid and NaOH	6.8–8.2
KH_2PO_4 and Na_2HPO_4	6.1–7.5
N-2-Hydroxyethylpiperazine-N′-2-ethanesulfonic acid and NaOH	6.9–8.3
Triethanolamine and HCl	6.9–8.5
Diethylbarbiturate (veronal) and HCl	7.0–8.5
Tris(hydroxymethyl)aminomethane and HCl	7.2–9.0
N-Tris(hydroxymethyl)methylglycine and HCl	
N,N-Bis(2-hydroxyethyl)glycine and HCl	
Borax and HCl	7.6–8.9
Glycine and NaOH	8.2–10.1
Ammonia (aqueous) and NH_4Cl	8.3–9.2
Ethanolamine and HCl	8.6–10.4

Borax and NaOH	9.4–11.1
Carbonate and hydrogen carbonate	9.2–11.0
Na$_2$HPO$_4$ and NaOH	11.0–12.0

General directions for the preparation of buffer solutions of varying pH but fixed ionic strength are given by Bates.* Preparation of McIlvaine buffered solutions at ionic strengths of 0.5 and 1.0 and Britton–Robinson solutions of constant ionic strength have been described by Elving et al.† and Frugoni,‡ respectively.

* Bates, *Determination of pH, Theory and Practice*, Wiley, New York, 1964, pp. 121–122.
† Elving, Markowitz, and Rosenthal, *Anal. Chem.*, **28**:1179 (1956).
‡ Frugoni, *Gazz. Chim. Ital.*, **87**:403 (1957).

REFERENCE ELECTRODES

TABLE 8.8   Potentials of Reference Electrodes (in volts) as a Function of Temperature

Liquid-junction potential included

Temp., °C	0.1 M KCl, calomel*	1.0 M KCl, calomel*	3.5 M KCl, calomel*	Saturated KCl, calomel*	1.0 M KCl, Ag/AgCl†	1.0 M KBr, Ag/AgBr‡	1.0 M KI, Ag/AgI§
0	0.3367	0.2883		0.25918	0.23655	0.08128	−0.14637
5				0.23413		0.07961	−0.14719
10	0.3362	0.2868	0.2556	0.25387	0.23142	0.07773	−0.14822
15	0.3361			0.2511	0.22857	0.07572	−0.14942
20	0.3358	0.2844	0.2520	0.24775	0.22557	0.07349	−0.15081
25	0.3356	0.2830	0.2501	0.24453	0.22234	0.07106	−0.15244
30	0.3354	0.2815	0.2481	0.24118	0.21904	0.06856	−0.15405
35	0.3351			0.2376	0.21565	0.06585	−0.15590
38	0.3350		0.2448	0.2355			
40	0.3345	0.2782	0.2439	0.23449	0.21208	0.06310	−0.15788
45					0.20835	0.06012	−0.15998
50	0.3315	0.2745		0.22737	0.20449	0.05704	−0.16219
55					0.20056		
60	0.3248	0.2702		0.2235	0.19649		
70					0.18782		
80				0.2083	0.1787		
90					0.1695	0.0251	

* Bates et al., *J. Res. Natl. Bur. Stand.*, **45**:418 (1950).
† Bates and Bower, *J. Res. Natl. Bur. Stand.*, **53**:283 (1954).
‡ Hetzer, Robinson, and Bates, *J. Phys. Chem.*, **66**:1423 (1962).
§ Hetzer, Robinson, and Bates, *J. Phys. Chem.*, **68**:1929 (1964).

Temp., °C	125	150	175	200	225	250	275
1.0 M KCl, Ag/AgCl*	0.1330	0.1032	0.0708	0.0348	−0.0051	−0.054	−0.090
1.0 M KBr, Ag/AgBr†	−0.0048	−0.0312	−0.0612	−0.0951			

* Greeley et al., *J. Phys. Chem.*, **64**:652 (1960).
† Towns et al., *J. Phys. Chem.*, **64**:1861 (1960).

The values of several additional reference electrodes at 25 °C are listed:

Reference electrode	Potential, V
Ag/AgCl, saturated KCl	0.198
Ag/AgCl, 0.1 M KCl	0.288
Hg/HgO, 1.0 M NaOH	0.140
Hg/HgO, 0.1 M NaOH	0.165
Hg/Hg$_2$SO$_4$, saturated K$_2$SO$_4$ (22 °C)	0.658
Hg/HgSO$_4$, saturated KCl	0.655

TABLE 8.9 Potentials of Reference Electrodes (in volts) at 25°C for water–organic solvent mixtures

Electrolyte solution of 1 M HCl

Solvent, wt %	Methanol, Ag/AgCl	Ethanol, Ag/AgCl	2-Propanol, Ag/AgCl	Acetone, Ag/AgCl	Dioxane, Ag/AgCl	Ethylene glycol, Ag/AgCl	Methanol, calomel	Dioxane, calomel
5	0.2153	0.2146	0.2180	0.2190		0.2190		
10	0.2090	0.2075	0.2138	0.2156		0.2160		
20		0.2003	0.2063	0.2079	0.2031	0.2101	0.255	0.2501
30		0.1945				0.2036		
40	0.1968	0.1859		0.1859	0.1635	0.1972	0.243	0.2104
45								
50		0.173		0.158				
60	0.1818	0.158				0.1807		
70	0.1492	0.136			0.0659		0.216	0.1126
80								
82					−0.0614			−0.0014
90	0.1135	0.196		−0.034				
94.2	0.0841	0.0215						
98								
99							0.103	
100	−0.0099	−0.0081		−0.53				

ELECTRODE POTENTIALS

TABLE 8.10 Potentials of Selected Half-Reactions at 25 °C

This table is a summary of oxidation–reduction half-reactions arranged in order of decreasing oxidation strength and is useful for selecting reagent systems.

Abbreviations Used in the Table
g, gas   liq, liquid   s, solid

Half-reaction		$E°$, V
$F_2(g) + 2H^+ + 2e^-$	$= 2HF$	3.06
$O_3 + 2H^+ + 2e^-$	$= O_2 + H_2O$	2.07
$S_2O_8^{2-} + 2e^-$	$= 2SO_4^{2-}$	2.01
$Ag^{2+} + e^-$	$= Ag^+$	2.00
$H_2O_2 + 2H^+ + 2e^-$	$= 2H_2O$	1.77
$MnO_4^- + 4H^+ + 3e^-$	$= MnO_2(s) + 2H_2O$	1.70
$Ce(IV) + e^-$	$= Ce(III)$ (in 1 M $HClO_4$)	1.61
$H_5IO_6 + H^+ + 2e^-$	$= IO_3^- + 3H_2O$	1.6
$Bi_2O_4(bismuthate) + 4H^+ + 2e^-$	$= 2BiO^+ + 2H_2O$	1.59
$BrO_3^- + 6H^+ + 5e^-$	$= \frac{1}{2}Br_2 + 3H_2O$	1.52
$MnO_4^- + 8H^+ + 5e^-$	$= Mn^{2+} + 4H_2O$	1.51
$PbO_2 + 4H^+ + 2e^-$	$= Pb^{2+} + 2H_2O$	1.455
$Cl_2 + 2e^-$	$= 2Cl^-$	1.36
$Cr_2O_7^{2-} + 14H^+ + 6e^-$	$= 2Cr^{3+} + 7H_2O$	1.33
$MnO_2(s) + 4H^+ + 2e^-$	$= Mn^{2+} + 2H_2O$	1.23
$O_2(g) + 4H^+ + 4e^-$	$= 2H_2O$	1.229
$IO_3^- + 6H^+ + 5e^-$	$= \frac{1}{2}I_2 + 3H_2O$	1.20
$Br_2(liq) + 2e^-$	$= 2Br^-$	1.065
$ICl_2^- + e^-$	$= \frac{1}{2}I_2 + 2Cl^-$	1.06
$VO_2^+ + 2H^+ + e^-$	$= VO^{2+} + H_2O$	1.00
$HNO_2 + H^+ + e^-$	$= NO(g) + H_2O$	1.00
$NO_3^- + 3H^+ + 2e^-$	$= HNO_2 + H_2O$	0.94
$2Hg^{2+} + 2e^-$	$= Hg_2^{2+}$	0.92
$Cu^{2+} + I^- + e^-$	$= CuI$	0.86
$Ag^+ + e^-$	$= Ag$	0.799
$Hg_2^{2+} + 2e^-$	$= 2Hg$	0.79
$Fe(III) + e^-$	$= Fe^{2+}$	0.771
$O_2(g) + 2H^+ + 2e^-$	$= H_2O_2$	0.682
$2HgCl_2 + 2e^-$	$= Hg_2Cl_2(s) + 2Cl^-$	0.63
$Hg_2SO_4(s) + 2e^-$	$= 2Hg + SO_4^{2-}$	0.615
$H_3AsO_4 + 2H^+ + 2e^-$	$= HAsO_2 + 2H_2O$	0.581
$Sb_2O_5 + 6H^+ + 4e^-$	$= 2SbO^+ + 3H_2O$	0.559
$I_3^- + 2e^-$	$= 3I^-$	0.545
$Cu^+ + e^-$	$= Cu$	0.52
$VO^{2+} + 2H^+ + e^-$	$= V^{3+} + H_2O$	0.337
$Fe(CN)_6^{3-} + e^-$	$= Fe(CN)_6^{4-}$	0.36
$Cu^{2+} + 2e^-$	$= Cu$	0.337
$UO_2^{2+} + 4H^+ + 2e^-$	$= U^{4+} + 2H_2O$	0.334
$BiO^+ + 2H^+ + 3e^-$	$= Bi + H_2O$	0.32
$Hg_2Cl_2(s) + 2e^-$	$= 2Hg + 2Cl^-$	0.2676
$AgCl(s) + e^-$	$= Ag + Cl^-$	0.2223

TABLE 8.10 Potentials of Selected Half-Reactions at 25 °C (*continued*)

Half-reaction		$E°$, V
$SbO^+ + 2H^+ + 3e^-$	$= Sb + H_2O$	0.212
$CuCl_3^{2-} + e^-$	$= Cu + 3Cl^-$	0.178
$SO_4^{2-} + 4H^+ + 2e^-$	$= SO_2(aq) + 2H_2O$	0.17
$Sn^{4+} + 2e^+$	$= Sn^{2+}$	0.154
$S + 2H^+ + 2e^-$	$= H_2S(g)$	0.141
$TiO^{2+} + 2H^+ + e^-$	$= Ti^{3+} + H_2O$	0.10
$S_4O_6^{2-} + 2e^-$	$= 2S_2O_3^{2-}$	0.08
$AgBr(s) + e^-$	$= Ag + Br^-$	0.071
$2H^+ + 2e^-$	$= H_2$	0.0000
$Pb^{2+} + 2e^-$	$= Pb$	−0.126
$Sn^{2+} + 2e^-$	$= Sn$	−0.136
$AgI(s) + e^-$	$= Ag + I^-$	−0.152
$Mo^{3+} + 3e^-$	$= Mo$	ca−0.2
$N_2 + 5H^+ + 4e^-$	$= H_2NNH_3^+$	−0.23
$Ni^{2+} + 2e^-$	$= Ni$	−0.246
$V^{3+} + e^-$	$= V^{2+}$	−0.255
$Co^{2+} + 2e^-$	$= Co$	−0.277
$Ag(CN)_2^- + e^-$	$= Ag + 2CN^-$	−0.31
$Cd^{2+} + 2e^-$	$= Cd$	−0.403
$Cr^{3+} + e^-$	$= Cr^{2+}$	−0.41
$Fe^{2+} + 2e^-$	$= Fe$	−0.440
$2CO_2 + 2H^+ + 2e^-$	$= H_2C_2O_4$	−0.49
$H_3PO_3 + 2H^+ + 2e^-$	$= H_3PO_2 + H_2O$	−0.50
$U^{4+} + e^-$	$= U^{3+}$	−0.61
$Zn^{2+} + 2e^-$	$= Zn$	−0.763
$Cr^{2+} + 2e^-$	$= Cr$	−0.91
$Mn^{2+} + 2e^-$	$= Mn$	−1.18
$Zr^{4+} + 4e^-$	$= Zr$	−1.53
$Ti^{3+} + 3e^-$	$= Ti$	−1.63
$Al^{3+} + 3e^-$	$= Al$	−1.66
$Th^{4+} + 4e^-$	$= Th$	−1.90
$Mg^{2+} + 2e^-$	$= Mg$	−2.37
$La^{3+} + 3e^-$	$= La$	−2.52
$Na^+ + e^-$	$= Na$	−2.714
$Ca^{2+} + 2e^-$	$= Ca$	−2.870
$Sr^{2+} + 2e^-$	$= Sr$	−2.89
$K^+ + e^-$	$= K$	−2.925
$Li^+ + e^-$	$= Li$	−3.045

TABLE 8.11 Half-Wave Potentials (vs. Saturated Calomel Electrode) of Organic Compounds at 25 °C

The solvent systems in this table are listed below:

A, acetonitrile and a perchlorate salt such as $LiClO_4$ or a tetraalkyl ammonium salt
B, acetic acid and an alkali acetate, often plus a tetraalkyl ammonium iodide
C, 0.05 to 0.175 M tetraalkyl ammonium halide and 75% 1,4-dioxane
D, buffer plus 50% ethanol (EtOH)

Abbreviations Used in the Table

Bu, butyl	M, molar	MeOH, methanol
Et, ethyl	Me, methyl	PrOH, propanol
EtOH, ethanol		

Compound	Solvent system	$E_{1/2}$
Unsaturated aliphatic hydrocarbons		
Acrylonitrile	C but 30% EtOH	−1.94
Allene	C	−2.29
1,3-Butadiene	A	−2.03
	C	−2.59
1,3-Butadiyne	C	−1.89
1-Buten-2-yne	C	−2.40
1,4-Cyclohexadiene	A	−1.6
Cyclohexene	A	−1.89
1,3,5,7-Cyclooctatetraene	B	−1.42
	C	−1.51
Diethyl fumarate	B, pH 4.0	−0.84
Diethyl maleate	B, pH 4.0	−0.95
2,3-Dimethyl-1,3-butadiene	A	−1.83
Dimethylfulvene	C	−1.89
Diphenylacetylene	C	−2.20
1,1-Diphenylethylene	B	−1.52
	C	−2.19
Ethyl methacrylate	0.1 N LiCl + 25% EtOH	−1.9
2-Methyl-1,3-butadiene	A	−1.84
2-Methyl-1-butene	A	−1.97
1-Piperidino-4-cyano-4-phenyl-1,3-butadiene	$LiClO_4$ in dimethylformamide	−0.16
trans-Stilbene	B	−1.51
Tetrakis(dimethylamino)ethylene	A	−0.75
Aromatic hydrocarbons		
Acenaphthene	A	−0.95
	B	−1.36
	C	−2.58
Anthracene	A	−0.84
	B	−1.20
	C	−1.94

TABLE 8.11 Half-Wave Potentials (vs. Saturated Calomel Electrode) of Organic Compounds at 25 °C (*continued*)

Compound	Solvent system	$E_{1/2}$
	Aromatic hydrocarbons (*continued*)	
Azulene	A	−0.71
	C	−1.66, −2.26, −2.56
1,2-Benzanthracene	C	−2.03, −2.54
2,3-Benzanthracene	A	−0.54, −1.20
Benzene	A	−2.08
1,2-Benzo[*a*]pyrene	A	−0.76
Biphenyl	A	−1.48
	B	−1.91
	C	−2.70
Chrysene	A	−1.22
1,2,5,6-Dibenzanthracene	A	−1.00, −1.26
1,2-Dihydronaphthalene	C	−2.57
9,10-Dimethylanthracene	A	−0.65
2,3-Dimethylnaphthalene	A	−1.08, −1.34
9,10-Diphenylanthracene	A	−0.92
Fluorene	A	−1.25
	B	−1.65
	C	−2.65
Hexamethylbenzene	A	−1.16
	B	−1.52
Indan	A	−1.59, −2.02
Indene	A	−1.23
	C	−2.81
1-Methylnaphthalene	A	−1.24
	B	−1.53
	C	−2.46
2-Methylnaphthalene	A	−1.22
	B	−1.55
	C	−2.46
Naphthalene	A	−1.34
	B	−1.72
Pentamethylbenzene	A	−1.28
	B	−1.62
Phenanthrene	A	−1.23
	B	−1.68
	C	−2.46, −2.71
Phenylacetylene	C	−2.37
Pyrene	A	−1.06, −1.24
trans-Stilbene	B	−1.51
	C	−2.26
Styrene	C	−2.35
1,2,3,5-Tetramethylbenzene	A	−1.50, −1.99
1,2,4,5-Tetramethylbenzene	A	−1.29
Tetraphenylethylene	C	−2.05

TABLE 8.11 Half-Wave Potentials (vs. Saturated Calomel Electrode) of Organic Compounds at 25 °C (*continued*)

Compound	Solvent system	$E_{1/2}$
Aromatic hydrocarbons (continued)		
1,4,5,8-Tetraphenylnaphthalene	A	−1.39
Toluene	A	−1.98
1,2,3-Trimethylbenzene	A	−1.58
1,2,4-Trimethylbenzene	A	−1.41
1,3,5-Trimethylbenzene	A	−1.50
	B	−1.90
Triphenylene	A	−1.46, −1.55
Triphenylmethane	C	−1.01, −1.68, −1.96
o-Xylene	A	−1.58, −2.04
m-Xylene	A	−1.58
p-Xylene	A	−1.56
Aldehydes		
Acetaldehyde	B, pH 6.8–13	−1.89
Benzaldehyde	McIlvaine buffer, pH 2.2	−0.96, −1.32
Bromoacetaldehyde	pH 8.5	−0.40
	pH 9.8	−1.58, −1.82
Chloroacetaldehyde	Ammonia buffer, pH 8.4	−1.06, −1.66
Cinnamaldehyde	Buffer + EtOH, pH 6.0	−0.9, −1.5, −1.7
Crotonaldehyde	B, pH 1.3–2.0	−0.92
	Ammonia buffer, pH 8.0	−1.30
Dichloroacetaldehyde	Ammonia buffer, pH 8.4	−1.03, −1.67
3,7-Dimethyl-2,6-octadienal	0.1 M Et$_4$NI	−1.56, −2.22
Formaldehyde	0.05 M KOH + 0.1 M KCl, pH 12.7	−1.59
2-Furaldehyde	pH 1–8	−0.86, −0.07 pH
	pH 10	−1.43
Glucose	Phosphate buffer, pH 7	−1.55
Glyceraldehyde	Britton–Robinson buffer, pH 5.0	−1.47
	Britton–Robinson buffer, pH 8.0	−1.55
Glycolaldehyde	0.1 M KOH, pH 13	−1.70
Glyoxal	B, pH 3.4	−1.41
4-Hydroxybenzaldehyde	Britton–Robinson buffer, pH 1.8	−1.16
	Britton–Robinson buffer, pH 6.8	−1.45
4-Hydroxy-2-methoxybenzaldehyde	McIlvaine buffer, pH 2.2	−1.05
	McIlvaine buffer, pH 5.0	−1.16, −1.36
	McIlvaine buffer, pH 8.0	−1.47
o-Methoxybenzaldehyde	Britton–Robinson buffer, pH 1.8	−1.02
	Britton–Robinson buffer, pH 6.8	−1.49
p-Methoxybenzaldehyde	Britton–Robinson buffer, pH 1.8	−1.17
	Britton–Robinson buffer, pH 6.8	−1.48
Methyl glyoxal	A, pH 4.5	−0.83

TABLE 8.11 Half-Wave Potentials (vs. Saturated Calomel Electrode) of Organic Compounds at 25 °C (*continued*)

Compound	Solvent system	$E_{1/2}$
	Aldehydes (*continued*)	
m-Nitrobenzaldehyde	Buffer + 10% EtOH, pH 2.0	−0.28, −1.20
Phthalaldehyde	Buffer, pH 3.1	−0.64, −1.07
	Buffer, pH 7.3	−0.89, −1.29
2-Propenal (acrolein)	pH 4.5	−1.36
	pH 9.0	−1.1
Propionaldehyde	0.1 M LiOH, pH 13	−1.93
Pyrrole-2-carbaldehyde	0.1 M HCl + 50% EtOH	−1.25
Salicylaldehyde	McIlvaine buffer, pH 2.2	−0.99, −1.23
	McIlvaine buffer, pH 5.0	−1.20, −1.30
	McIlvaine buffer, pH 8.0	−1.32
Trichloroacetaldehyde	Ammonia buffer, pH 8.4	−1.35, −1.66
	0.1 M KCl + 50% EtOH	−1.55
	Ketones	
Acetone	B, pH 9.3	−1.52
	C	−2.46
Acetophenone	D + McIlvaine buffer, pH 4.9	−1.33
	D + McIlvaine buffer, pH 7.2	−1.58
	D + McIlvaine buffer, pH 1.3	−1.08
7*H*-Benz[*de*]anthracen-7-one	0.1 N H_2SO_4 + 75% MeOH	−0.96
Benzil	D + McIlvaine buffer, pH 1.3	−0.27
	D + McIlvaine buffer, pH 4.9	−0.50
Benzoin	D + McIlvaine buffer, pH 1.3	−0.90
	D + McIlvaine buffer, pH 8.6	−1.49
Benzophenone	D + McIlvaine buffer, pH 1.3	−0.94
	D + McIlvaine buffer, pH 8.6	−1.36
Benzoylacetone	Buffer, pH 2.6	−1.60
	Buffer, pH 5.3 and pH 7.6	−1.68
	Buffer, pH 9.7	−1.72
Bromoacetone	0.1 M LiCl	−0.29
2,3-Butanedione	0.1 M HCl	−0.84
3-Buten-2-one	0.1 M KCl	−1.42
Butyrophenone	0.1 M NH₄Cl + 50% EtOH	−1.55
D-Carvone	0.1 M Et₄NI + 80% EtOH	−1.71
Chloroacetone	0.1 M LiCl	−1.18
Coumarin	McIlvaine buffer, pH 2.0	−0.95
	McIlvaine buffer, pH 5.0	−1.11, −1.44
Cyclohexanone	C	−2.45
cis-Dibenzoylethylene	D, pH 1	−0.30
	D, pH 11	−0.62, −1.65
trans-Dibenzoylethylene	D, pH 1	−0.12
	D, pH 11	−0.57, −1.52
Dibenzoylmethane	D, pH 1.3	−0.59
	D, pH 11.3	−1.30, −1.62

TABLE 8.11   Half-Wave Potentials (vs. Saturated Calomel Electrode) of Organic Compounds at 25 °C (*continued*)

Compound	Solvent system	$E_{1/2}$
	Ketones (continued)	
9,10-Dihydro-9-oxoanthracene	D, pH 2.0	−0.93
1,5-Diphenyl-1,5-pentanedione	A	−2.10
1,5-Diphenylthiocarbazone	D, pH 7.0	−0.6
Flavanone	Acetate buffer + Me$_4$NOH + 50% 2-PrOH, pH 6.1	−1.30
	Acetate buffer + Me$_4$NOH + 50% 2-PrOH, pH 9.6	−1.51
Fluorescein	Acetate buffer, pH 2.0	−0.50
	Phthalate buffer, pH 5.0	−0.65
	Borate buffer, pH 10.1	−1.18, −1.44
Fructose	0.02 M LiCl	−1.76
Girard derivatives of aliphatic ketones	pH 8.2	−1.52
o-Hydroxyacetophenone	D, pH 5	−1.36
p-Hydroxyacetophenone	D, pH 5	−1.46
1,2,3-Indantrione (ninhydrin)	Britton–Robinson buffer, pH 2.5	−0.67, −0.83
	Britton–Robinson buffer, pH 4.5	−0.73, −1.01
	Britton–Robinson buffer, pH 6.8	−0.10, −0.90, −1.20
	Britton–Robinson buffer, pH 9.2	−1.35
α-Ionone	C	−1.59, −2.08
Isatin	Phosphate buffer + citrate buffer, pH 2.9	−0.3, −0.5
	Phosphate buffer + citrate buffer, pH 4.3	−0.3, −0.5, −0.8
	Phosphate buffer + citrate buffer, pH 5.4	−0.8
4-Methyl-3,5-heptadien-2-one	A	−0.64
4-Methyl-2,6-heptanedione	A	−1.28
4-Methyl-3-penten-2-one	D + McIlvaine buffer, pH 1.3	−1.01
	D + McIlvaine buffer, pH 11.3	−1.60
4-Phenyl-3-buten-2-one	D, pH 1.3	−0.72
	D, pH 8.6	−1.27
Phthalide	0.1 M Bu$_4$NI + 50% dioxane	−0.20
Phthalimide	pH 4.2	−1.1, −1.5
	pH 9.7	−1.2, −1.4
Pulegone	C	−1.74
Quinalizarin	Phosphate buffer + 1% EtOH, pH 8.0	−0.56
Testosterone	D + Britton–Robinson buffer, pH 2.6	−1.20
	D + Britton–Robinson buffer, pH 5.8	−1.40
	D + Britton–Robinson buffer, pH 8.8	−1.53, −1.79

TABLE 8.11   Half-Wave Potentials (vs. Saturated Calomel Electrode) of Organic Compounds at 25 °C (*continued*)

Compound	Solvent system	$E_{1/2}$
	Quinones	
Anthraquinone	Acetate buffer + 40% dioxane, pH 5.6	−0.51
	Phosphate buffer + 40% dioxane, pH 7.9	−0.71
o-Benzoquinone	Britton–Robinson buffer, pH 7.0	+0.20
	Britton–Robinson buffer, pH 9.0	+0.08
2,3-Dimethylnaphthoquinone	D, pH 5.4	−0.22
1,2-Naphthoquinone	Phosphate buffer, pH 5.0	−0.03
	Phosphate buffer, pH 7.0	−0.13
1,4-Naphthoquinone	Britton–Robinson buffer, pH 7.0	−0.07
	Britton–Robinson buffer, pH 9.0	−0.19
	Acids	
Acetic acid	A	−2.3
Acrylic acid	pH 5.6	−0.85
Adenosine-5′-phosphoric acid	$HClO_4 + KClO_4$, pH 2.2	−1.13
4-Aminobenzenesulfonic acid	0.05 M Me_4NI	−1.58
3-Aminobenzoic acid	pH 5.6	−0.67
Anthranilic acid	pH 5.6	−0.67
Ascorbic acid	Birtton–Robinson buffer, pH 3.4	+0.17
	Britton–Robinson buffer, pH 7.0	−0.06
Barbituric acid	Borate buffer, pH 9.3	−0.04
Benzoic acid	A	−2.1
Benzoylformic acid	Britton–Robinson buffer, pH 2.2	−0.48
	Britton–Robinson buffer, pH 5.5	−0.85, −1.26
	Britton–Robinson buffer, pH 7.2	−0.98, −1.25
	Britton–Robinson buffer, pH 9.2	−1.25
Bromoacetic acid	pH 1.1	−0.54
2-Bromopropionic acid	pH 2.0	−0.39
Crotonic acid	C	−1.94
Dibromoacetic acid	pH 1.1	−0.03, −0.59
Dichloroacetic acid	pH 8.2	−1.57
5,5-Diethylbarbituric acid	Borate buffer, pH 9.3	0.00
Flavanol	D, pH 5.6	−1.25
	D, pH 7.7	−1.40
Folic acid	Britton–Robinson buffer, pH 4.6	−0.73
Formic acid	0.1 M KCl	−1.66
Fumaric acid	HCl + KCl, pH 2.6	−0.83
	Acetate buffer, pH 4.0	−0.93
	Acetate buffer, pH 5.9	−1.20
2,4-Hexadienedioic acid	Acetate buffer, pH 4.5	−0.97
Iodoacetic acid	pH 1	−0.16
Maleic acid	Britton–Robinson buffer, pH 2.0	−0.70
	Britton–Robinson buffer, pH 4.0	−0.97
	Britton–Robinson buffer, pH 6.0	−1.11, −1.30
	Britton–Robinson buffer, pH 10.0	−1.51

TABLE 8.11 Half-Wave Potentials (vs. Saturated Calomel Electrode) of Organic Compounds at 25 °C (*continued*)

Compound	Solvent system	$E_{1/2}$
	Acids (*continued*)	
Mercaptoacetic acid	B, pH 6.8	−0.38
Methacrylic acid	D + 0.1 M LiCl	−1.69
Nitrobenzoic acids	Buffer + 10% EtOH, pH 2.0	−0.2, −0.7
Oxalic acid	B, pH 5.4–6.1	−1.80
2-Oxo-1,5-pentanedioic acid	HCl + KCl, pH 1.8	−0.59
	Ammonia buffer, pH 8.2	−1.30
2-Oxopropionic acid	Britton–Robinson buffer, pH 5.6	−1.17
	Britton–Robinson buffer, pH 6.8	−1.22, −1.53
	Britton–Robinson buffer, pH 9.7	−1.51
Phenolphthalein	Phthalate buffer, pH 2.5	−0.67
	Phthalate buffer, pH 4.7	−0.80
	D, pH 9.6	−0.98, −1.35
Picric acid	pH 4.2	−0.34
	pH 11.7	−0.36, −0.56, −0.96
1,2,3-Propenetricarboxylic acid	pH 7.0	−2.1
Trichloroacetic acid	Ammonia buffer, pH 8.2	−0.84, −1.57
	Phosphate buffer, pH 10.4	−0.9, −1.6
3,4,5-Trihydroxybenzoic acid	Phosphate buffer, pH 2.9	+0.50
	Phosphate buffer, pH 8.8	+0.1
p-Aminophenol	Britton–Robinson buffer, pH 6.3	+0.14
	Britton–Robinson buffer, pH 8.6	−0.04
	Britton–Robinson buffer, pH 12.0	−0.16
o-Chlorophenol	pH 5.6	−0.63
m-Chlorophenol	pH 5.6	−0.73
p-Chlorophenol	pH 5.6	−0.65
o-Cresol	pH 5.6	−0.56
m-Cresol	pH 5.6	−0.61
p-Cresol	pH 5.6	−0.54
1,2-Dihydroxybenzene	pH 5.6	−0.35
1,3-Dihydroxybenzene	pH 5.6	−0.61
1,4-Dihydroxybenzene	pH 5.6	−0.23
o-Methoxyphenol	pH 5.6	−0.46
m-Methoxyphenol	pH 5.6	−0.62
p-Methoxyphenol	pH 5.6	−0.41
1-Naphthol	A	−0.74
2-Naphthol	A	−0.82
1,2,3-Trihydroxybenzene	Britton–Robinson buffer, pH 3.1	+0.35
	Britton–Robinson buffer, pH 6.5	+0.10
	Britton–Robinson buffer, pH 9.5	−0.10
	Halogen compounds	
Bromobenzene	A	−1.98
	C	−2.32
1-Bromobutane	C	−2.27

TABLE 8.11 Half-Wave Potentials (vs. Saturated Calomel Electrode) of Organic Compounds at 25 °C (*continued*)

Compound	Solvent system	$E_{1/2}$
	Halogen compounds (*continued*)	
Bromoethane	C	−2.08
Bromomethane	C	−1.63
1-Bromonaphthalene (also 2-bromonaphthalene)	A	−1.55, −1.60
3-Bromo-1-propene	C	−1.29
p-Bromotoluene	A	−1.72
Carbon tetrachloride	C	−0.78, −1.71
Chlorobenzene	A	−2.07
Chloroform	C	−1.63
Chloromethane	C	−2.23
3-Chloro-1-propene	C	−1.91
α-Chlorotoluene	C	−1.81
p-Chlorotoluene	A	−1.76
N-Chloro-p-toluenesulfonamide	0.5 M K_2SO_4	−0.13
9,10-Dibromoanthracene	A	−1.15, −1.47
p-Dibromobenzene	C	−2.10
1,2-Dibromobutane	D + 1% Na_2SO_3	−1.45
Dibromoethane	C	−1.48
meso-2,3-Dibromosuccinic acid	Acetate buffer, pH 4.0	−0.23, −0.89
Dichlorobenzenes	C	−2.5
Dichloromethane	C	−1.60
Diiodomethane	C	−1.12, −1.53
Hexabromobenzene	C	−0.8, −1.5
Hexachlorobenzene	C	−1.4, −1.7
Iodobenzene	A	−1.72
Iodoethane	C	−1.67
Iodomethane	A	−2.12
	C	−1.63
Tetrabromomethane	C	−0.3, −0.75, −1.49
Tetraiodomethane	C	−0.45, −1.05, −1.46
Tribromomethane	C	−0.64, −1.47
α,α,α-Trichlorotoluene	C	−0.68, −1.65, −2.00
	Nitro and nitroso compounds	
1,2-Dinitrobenzene	Phthalate buffer, pH 2.5	−0.12, −0.32, −1.26
	Borate buffer, pH 9.2	−0.38, −0.74
1,3-Dinitrobenzene	Phthalate buffer, pH 2.5	−0.17, −0.29
	Borate buffer, pH 9.2	−0.46, −0.68
1,4-Dinitrobenzene	Phthalate buffer, pH 2.5	−0.12, −0.33
	Borate buffer, pH 9.2	−0.35, −0.80

TABLE 8.11 Half-Wave Potentials (vs. Saturated Calomel Electrode) of Organic Compounds at 25 °C (*continued*)

Compound	Solvent system	$E_{1/2}$
	Nitro and nitroso compounds (*continued*)	
Methyl nitrobenzoates	Buffer + 10% EtOH, pH 2.0	−0.20 to −0.25 −0.68 to −0.74
p-Nitroacetophenone	Britton–Robinson buffer, pH 2.2	−0.16, −0.61, −1.09
	Britton–Robinson buffer, pH 10.0	−0.51, −1.40, −1.73
o-Nitroaniline	0.03 M LiCl + 0.02 M benzoic acid in EtOH	−0.88
m-Nitroaniline	Britton–Robinson buffer, pH 4.3	−0.3, −0.8
	Briton-Robinson buffer, pH 7.2	−0.5
	Britton–Robinson buffer, pH 9.2	−0.7
p-Nitroaniline	pH 2.0	−0.36
	Acetate buffer, pH 4.6	−0.5
o-Nitroanisole	Buffer + 10% EtOH, pH 2.0	−0.29, −0.58
p-Nitroanisole	Buffer + 10% EtOH, pH 2.0	−0.35, −0.64
1-Nitroanthraquinone	Britton–Robinson buffer, pH 7.0	−0.16
Nitrobenzene	HCl + KCl + 8% EtOH, pH 0.5	−0.16, −0.76
	Phthalate buffer, pH 2.5	−0.30
	Borate buffer, pH 9.2	−0.70
Nitrocresols	Britton–Robinson buffer, pH 2.2	−0.2 to −0.3
	Britton–Robinson buffer, pH 4.5	−0.4 to −0.5
	Britton–Robinson buffer, pH 8.0	−0.6
Nitroethane	Britton–Robinson buffer + 30% MeOH, pH 1.8	−0.7
	Britton–Robinson buffer + 30% MeOH, pH 4.6	−0.8
2-Nitrohydroquinone	Phosphate buffer + citrate buffer, pH 2.1	−0.2
	Phosphate buffer + citrate buffer, pH 5.2	−0.4
	Phosphate buffer + citrate buffer, pH 8.0	−0.5
Nitromethane	Britton–Robinson buffer + 30% MeOH, pH 1.8	−0.8
	Britton–Robinson buffer + 30% MeOH, pH 4.6	−0.85
o-Nitrophenol	Britton–Robinson buffer + 10% EtOH, pH 2.0	−0.23
	Britton–Robinson buffer + 10% EtOH, pH 4.0	−0.4
	Britton–Robinson buffer + 10% EtOH, pH 8.0	−0.65
	Britton–Robinson buffer + 10% EtOH, pH 10.0	−0.80

TABLE 8.11   Half-Wave Potentials (vs. Saturated Calomel Electrode) of Organic Compounds at 25 °C (*continued*)

Compound	Solvent system	$E_{1/2}$
Nitro and nitroso compounds (continued)		
m-Nitrophenol	Britton–Robinson buffer + 10% EtOH, pH 2.0	−0.37
	Britton–Robinson buffer + 10% EtOH, pH 4.0	−0.40
	Britton–Robinson buffer + 10% EtOH, pH 8.0	−0.64
	Britton–Robinson buffer + 10% EtOH, pH 10.0	−0.76
p-Nitrophenol	Britton–Robinson buffer + 10% EtOH, pH 2.0	−0.35
	Britton–Robinson buffer + 10% EtOH, pH 4.0	−0.50
	Britton–Robinson buffer + 10% EtOH, pH 8.0	−0.82
1-Nitropropane	Britton–Robinson buffer + 30% MeOH, pH 1.8	−0.73
	Britton–Robinson buffer + 30% MeOH, pH 8.6	−0.88
	Britton–Robinson buffer + 30% MeOH, pH 8.0	−0.95
2-Nitropropane	McIlvaine buffer, pH 2.1	−0.53
	McIlvaine buffer, pH 5.1	−0.81
Nitrosobenzene	McIlvaine buffer, pH 6.0	−0.03
	McIlvaine buffer, pH 8.0	−0.14
1-Nitroso-2-naphthol	D + buffer, pH 4.0	+0.02
	D + buffer, pH 7.0	−0.20
	D + buffer, pH 9.0	−0.31
N-Nitrosophenylhydroxylamine	pH 2.0	−0.84
o-Nitrotoluene	Phthalate buffer, pH 2.5	−0.35, −0.66
	Phthalate buffer, pH 7.4	−0.60, −1.06
m-Nitrotoluene (also *p*-nitrotoluene)	Phthalate buffer, pH 2.5	−0.30, −0.53
	Phthalate buffer, pH 7.4	−0.58, −1.06
Tetranitromethane	pH 12.0	−0.41
1,3,5-Trinitrobenzene	Phthalate buffer, pH 4.1	−0.20, −0.29, −0.34
	Borate buffer, pH 9.2	−0.34, −0.48, −0.65
Heterocyclic compounds containing nitrogen		
Acridine	D, pH 8.3	−0.80, −1.45
Cinchonine	B, pH 3	−0.90
2-Furanmethanol	Britton–Robinson buffer, pH 2.0	−0.96
	Britton–Robinson buffer, pH 5.8	−1.38, −1.70
2-Hydroxyphenazine	Britton–Robinson buffer, pH 4.0	−0.24

TABLE 8.11 Half-Wave Potentials (vs. Saturated Calomel Electrode) of Organic Compounds at 25 °C (*continued*)

Compound	Solvent system	$E_{1/2}$
Heterocyclic compounds containing nitrogen (*continued*)		
8-Hydroxyquinoline	B, pH 5.0	−1.12
	Phosphate buffer, pH 8.0	−1.18, −1.71
3-Methylpyridine	D + 0.1 M LiCl	−1.76
4-Methylpyridine	D + 0.1 M LiCl	−1.87
Phenazine	Phosphate buffer + citrate buffer, pH 7.0	−0.36
Pyridine	Phosphate buffer + citrate buffer, pH 7.0	−1.75
Pyridine-2-carboxylic acid	B, pH 4.1	−1.10
	B, pH 9.3	−1.48, −1.94
Pyridine-3-carboxylic acid	0.1 M HCl	−1.08
Pyridine-4-carboxylic acid	Britton–Robinson buffer, pH 6.1	−1.14
	pH 9.0	−1.39, −1.68
Pyrimidine	Citrate buffer, pH 3.6	−0.92, −1.24
	Ammonia buffer, pH 9.2	−1.54
Quinoline-8-carboxylic acid	pH 9	−1.11
Quinoxaline	Phosphate buffer + citrate buffer, pH 7.0	−0.66, −1.52
Azo, hydrazine, hydroxylamine, and oxime compounds		
Azobenzene	D, pH 4.0	−0.20
	D, pH 7.0	−0.50
Azoxybenzene	Buffer + 20% EtOH, pH 6.3	−0.30
Benzoin-1-oxime	Buffer, pH 2.0	−0.88
	Buffer, pH 5.6	−1.08
	Buffer, pH 8.2	−1.67
Benzoylhydrazine	0.13 M NaOH, pH 13.0	−0.30
Dimethylglyoxime	Ammonia buffer, pH 9.6	−1.63
Hydrazine	Britton–Robinson buffer, pH 9.3	−0.09
Hydroxylamine	Britton–Robinson buffer, pH 4.6	−1.42
	Britton–Robinson buffer, pH 9.2	−1.65
Oxamide	Acetate buffer	−1.55
Phenylhydrazine	McIlvaine buffer, pH 2	+0.19
	0.13 M NaOH, pH 13.0	−0.36
Phenylhydroxylamine	McIlvaine buffer + 10% EtOH, pH 2	−0.68
	McIlvaine buffer + 10% EtOH, pH 4–10	−0.33
		0.061 pH
Salicylaldoxime	Phosphate buffer, pH 5.4	−1.02
Thiosemicarbazide	Borate buffer, pH 9.3	−0.26
Thiourea	0.1 M sulfuric acid	+0.02

TABLE 8.11 Half-Wave Potentials (vs. Saturated Calomel Electrode) of Organic Compounds at 25 °C (*continued*)

Compound	Solvent system	$E_{1/2}$
	Indicators and dyestuffs	
Brilliant Green	HCl + KCl, pH 2.0	$-0.2, -0.5$
Indigo carmine	pH 2.5	-0.24
Indigo disulfonate	pH 7.0	-0.37
Malachite Green G	HCl + KCl, pH 2.0	$-0.2, -0.5$
Metanil yellow	Phosphate buffer + 1% EtOH, pH 7.0	-0.51
Methylene blue	Britton–Robinson buffer, pH 4.9	-0.15
	Britton–Robinson buffer, pH 9.2	-0.30
Methylene green	Phosphate buffer + 1% EtOH, pH 7.0	-0.12
Methyl orange	Phosphate buffer + 1% EtOH, pH 7.0	-0.51
Morin	D, pH 7.6	-1.7
Neutral red	Britton–Robinson buffer, pH 2.0	-0.21
	Britton–Robinson buffer, pH 7.0	-0.57
	Peroxide	
Ethyl peroxide	0.02 M HCl	-0.2

SECTION 9
DATA USEFUL IN LABORATORY MANIPULATION AND ANALYSIS

COOLING MIXTURES

Convenient cooling mixtures can be prepared in several ways. First, an inorganic salt may be mixed with finely shaved dry ice. Such a mixture can be used to maintain temperatures as shown in Table 9.1.

TABLE 9.1   Cooling Mixtures Made from Dry Ice and Salts

Salt	Dry ice, g/100g	Minimum temperature, °C
$CaCl_2 \cdot 6H_2O$	41	−9.0
	81	−21.5
	123	−40.3
	143	−55
NH_4Cl	25	−15.4
NaBr	66	−28
$MgCl_2$	85	−34

A more common method for preparing a low temperature bath is to mix an organic substance with either dry ice or liquid nitrogen. Dry ice (CO_2, −78 °C) can be added in small lumps to the solvent until a slight excess of dry ice remains. Alternately, liquid nitrogen (N_2, −196 °C) can be poured into the solvent until a slush is formed that consists of the solid–liquid mixture at its melting point.

TABLE 9.2   Dry Ice or Liquid Nitrogen Slush Baths

Substance	Temperature, °C	Substance	Temperature, °C
Ethylene glycol	−13	Acetone–CO_2	−77
1,2-Dichlorobenzene	−17	Ethyl acetate	−84
Carbon tetrachloride	−22.9	2-Butanone	−87
Bromobenzene	−31	Hexane	−95
Methoxybenzene	−37	Methanol	−98
Chlorobenzene	−45	Carbon disulfide	−112
Bis(2-ethoxyethyl) ether	−44	Bromoethane	−119
N-Methylaniline	−57	Pentane	−130
p-Cymene	−68	2-Methylbutane	−160

HUMIDIFICATION AND DRYING

A saturated aqueous solution in contact with an excess of a definite solid phase at a given temperature will maintain constant humidity in an enclosed space. Table 9.3 identifies a number of salts suitable for this purpose. The aqueous tension (in millimeters of Hg) of a solution at a given temperature is found by multiplying the decimal fraction of the humidity by the aqueous tension at 100% humidity for the specific temperature. For example, the aqueous tension of a saturated solution of NaCl at 20 °C is $0.757 \times 17.54 = 13.28$ mmHg and at 80 °C is $0.764 \times 355.1 = 271.3$ mmHg.

TABLE 9.3 Humidity (%) Maintained by Saturated Solutions of Various Salts at Specified Temperatures

Solid phase	Temperature, °C						
	10	20	25	30	40	60	80
$K_2Cr_2O_7$			98.0				
K_2SO_4	98	97	97	96	96	96	
KNO_3	95	93	92.5	91	88	82	
KCl	88	85.0	84.3	84	81.7	80.7	79.5
KBr	86	84	80.7		79.6	79.0	79.3
$NaCl$	76	75.7	75.3	74.9	74.7	74.9	76.4
$NaNO_3$	77	75	73.8	72.8	71.5	67.5	65.5
KI					66.8	63.1	60.8
$NaNO_2$		66	65	63.0	61.5	59.3	58.9
$Na_2CrO_4 \cdot 4H_2O$				64.6	61.8	55.6	56.2
$NaBr \cdot 2H_2O$	58	57.9	57.7		52.4	49.9	50.0
$Na_2Cr_2O_7 \cdot 2H_2O$	58	55	54		53.6	55.2	56.0
$Mg(NO_3)_2 \cdot 6H_2O$	57	55	52.9	52	49	43	
$K_2CO_3 \cdot 2H_2O$	47	44	42.8	42	40		
$NaI \cdot 2H_2O$		47		36.4	32.3	25.3	23.2
$MgCl_2 \cdot 6H_2O$	34	33	33.0	33	32	30	
$CaCl_2 \cdot 6H_2O$	38	32.6	29	26			
$KF \cdot 2H_2O$				27.4	22.8	21.0	22.8
$KC_2H_3O_2 \cdot 1.5H_2O$	24	23	22.5	22	20		
$LiCl \cdot H_2O$	13	12	11.1	12	11	11	
KOH	13	9	8	7	6	5	
Aqueous tension at 100% humidity, mmHg	9.21	17.54	23.76	31.82	55.32	149.4	355.1

TABLE 9.4 Humidity (%) Maintained by Saturated Solutions of Common Salts at Specified Temperatures

Solid phase	Temperature, °C	Humidity, %
KF	100	22.9
KI	100	56.2
$(NH_4)_2SO_4$	20–30	81.1
	108	75
$BaCl_2 \cdot 2H_2O$	25	90.2
NaF	100	96.6

TABLE 9.5 Drying Agents

Drying agent	Most useful for	Residual water, mg H_2O per liter of dry air (25°C)	Grams water removed per gram of desiccant	Regeneration, °C
Al_2O_3	Hydrocarbons	0.002–0.005	0.2	175 (24h)
$Ba(ClO_4)_2$[a]	Inert gas streams	0.6–0.8	0.17	140
BaO	Basic gases: hydrocarbons, aldehydes, alcohols	0.0007–0.003	0.12	1000
CaC_2[b]	Ethers		0.56	Impossible
$CaCl_2$[c]	Inert organics	0.1–0.2	0.15 (1 H_2O) / 0.30 (2 H_2O)	250
CaH_2[d]	Hydrocarbons, ethers, amines, esters, higher alcohols	1×10^{-5}	0.85	Impossible
CaO	Ethers, esters, alcohols, amines	0.01–0.003	0.31	Difficult, 1000
$CaSO_4$	Most organic substances	0.005–0.07	0.07	225
Dow Desiccant 812[e]	Most materials	(5–200 ppm)	0.16	No
K_2CO_3	Most materials except acids and phenols	0.01–0.9		158
KOH	Amines		1.9	Impossible
$LiAlH_4$[f]	Hydrocarbons	0.0005–0.002	0.24	Impossible
$Mg(ClO_4)_2$[a]	Gas streams	0.008	0.45	250 (high vacuum)
MgO	All but acidic compounds	1–12	0.15–0.75	800
$MgSO_4$	Most organic compounds			Not feasible
Molecular sieves				
4X	Molecules with effective diameter >4Å	0.001	0.18	250
5X	Molecules with effective diameter >5Å	0.001	0.18	250
9.5% Na–Pb alloy[d]	Hydrocarbons, ethers	(For solvents only)	0.08	Impossible
Na_2SO_4	Ketones, acids, alkyl and aryl halides	12	1.25	150
P_2O_5	Gas streams; not suitable for alcohols, amines, ketones, or amines	2×10^{-5}	0.5	Not feasible
Silica gel	Most organic amines	0.002–0.07	0.2	200–350
Sulfuric acid	Air and inert gas streams	0.003–0.008	Indefinite	Not feasible

[a]May form explosive mixtures on contact with organic material.
[b]Explosive C_2H_2 formed.
[c]Drying action slow.
[d]H_2 formed.
[e]Used for column drying of organic liquids.
[f]Strong reductant.

TABLE 9.6 Solvents of Chromatographic Interest (arranged in order of increasing solvent strength)

Solvent	Boiling point, °C	Solvent strength parameter $e°$ (SiO_2)	Solvent strength parameter $e°$ (Al_2O_3)	Viscosity, mN·s·m^{-2} (20°C)	Refractive index (20°C)	UV cutoff, nm
Fluoroalkanes			−0.25		1.25	
Pentane	36	0.0	0.0	0.24$^{15°C}$	1.358	210
Hexane	69	0.0	0.0	0.31	1.375	210
2,2,4-Trimethylpentane	99		0.01	0.50	1.392	215
Decane	174		0.04	0.93	1.412	210
Cyclohexane	81	−0.05	0.04	0.98	1.426	210
Cyclopentane	49		0.05	0.44	1.407	210
Diisobutylene	101		0.06		1.411	
1-Pentene	30		0.08	0.24$^{0°C}$	1.371	
Carbon disulfide	46	0.14	0.15	0.36	1.626	380
Carbon tetrachloride	77	0.14	0.18	0.97	1.466	265
1-Chlorobutane	78		0.26	0.43	1.402	220
1-Chloropentane	98		0.26	0.58	1.412	225
o-Xylene	144		0.26	0.81	1.505	290
Diisopropyl ether	68		0.28	0.38$^{25°C}$	1.369	220
2-Chloropropane	35		0.29	0.33	1.378	225
Toluene	111		0.29	0.59	1.497	286
1-Chloropropane	47		0.30	0.35	1.389	225
Chlorobenzene	132		0.40	0.80	1.525	
Benzene	80	0.25	0.32	0.65	1.501	280
Bromoethane	38		0.37	0.40	1.424	
Diethyl ether	35	0.38	0.38	0.25	1.353	218
Diethyl sulfide	92		0.38	0.45	1.443	290
Chloroform	62	0.26	0.40	0.57	1.443	245

9.5

TABLE 9.6 Solvents of Chromatographic Interest (*continued*)

Solvent	Boiling point, °C	Solvent strength parameter $e°$ (SiO$_2$)	Solvent strength parameter $e°$ (Al$_2$O$_3$)	Viscosity, mN·s·m^{-2} (20°C)	Refractive index (20°C)	UV cutoff, nm
Dichloromethane	41		0.42	0.44	1.425	235
4-Methyl-2-pentanone	116		0.43	0.42$^{15°C}$	1.396	335
Tetrahydrofuran	66		0.45	0.55	1.407	220
1,2-Dichloroethane	84		0.49	0.80	1.445	228
2-Butanone	80		0.51	0.42$^{15°C}$	1.379	330
1-Nitropropane	131		0.53	0.80$^{25°C}$	1.402	380
Acetone	56		0.56	0.32	1.359	330
1,4-Dioxane	101	0.47	0.56	1.44$^{15°C}$	1.420	215
Ethyl acetate	77	0.49	0.58	0.45	1.372	255
Methyl acetate	56	0.38	0.60	0.48$^{15°C}$	1.362	260
1-Pentanol	138		0.61	4.1	1.410	210
Dimethyl sulfoxide	189		0.62	2.47	1.478	265
Aniline	184		0.62	4.40	1.586	
Diethylamine	56		0.63	0.33	1.386	275
Nitromethane	101		0.64	0.67	1.394	380
Acetonitrile	82	0.50	0.65	0.37	1.344	190
Pyridine	115		0.71	0.97	1.510	330
2-Butoxyethanol	170		0.74	3.15$^{25°C}$	1.420	220
1-Propanol	97		0.82	2.25	1.386	210
2-Propanol	82		0.82	2.50	1.377	210
Ethanol	78		0.88	1.20	1.361	210
Methanol	65		0.95	0.59	1.328	210
Ethylene glycol	198		1.11	21.8	1.432	210
Acetic acid	118		large	1.23	1.372	260
Water	100		large	1.00	1.333	191

TABLE 9.7   Solvents having the Same Refractive Index and the Same Density at 25 °C

		Refractive index		Density, g/mL	
Solvent 1	Solvent 2	1	2	1	2
Acetone	Ethanol	1.357	1.359	0.788	0.786
Ethyl formate	Methyl acetate	1.358	1.360	0.916	0.935
Ethanol	Propionitrile	1.359	1.363	0.786	0.777
2,2-Dimethylbutane	2-Methylpentane	1.366	1.369	0.644	0.649
2-Methylpentane	Hexane	1.369	1.372	0.649	0.655
Isopropyl acetate	2-Chloropropane	1.375	1.376	0.868	0.865
3-Butanone	Butyraldehyde	1.377	1.378	0.801	0.799
Butyraldehyde	Butyronitrile	1.378	1.382	0.799	0.786
Dipropyl ether	Butyl ethyl ether	1.379	1.380	0.753	0.746
Propyl acetate	Ethyl propionate	1.382	1.382	0.883	0.888
Propyl acetate	1-Chloropropane	1.382	1.386	0.883	0.890
Butyronitrile	2-Methyl-2-propanol	1.382	1.385	0.786	0.781
Ethyl propionate	1-Chloropropane	1.382	1.386	0.888	0.890
1-Propanol	2-Pentanone	1.383	1.387	0.806	0.804
Isobutyl formate	1-Chloropropane	1.383	1.386	0.881	0.890
1-Chloropropane	Butyl formate	1.386	1.387	0.890	0.888
Butyl formate	Methyl butyrate	1.387	1.391	0.888	0.875
Methyl butyrate	2-Chlorobutane	1.392	1.395	0.875	0.868
Butyl acetate	2-Chlorobutane	1.392	1.395	0.877	0.868
4-Methyl-2-pentanone	Pentanonitrile	1.394	1.395	0.797	0.795
4-Methyl-2-pentanone	1-Butanol	1.394	1.397	0.797	0.812
2-Methyl-1-propanol	Pentanonitrile	1.394	1.395	0.798	0.795
2-Methyl-1-propanol	2-Hexanone	1.394	1.395	0.798	0.810
2-Butanol	2,4-Dimethyl-3-pentanone	1.395	1.399	0.803	0.805
2-Hexanone	1-Butanol	1.395	1.397	0.810	0.812
Pentanonitrile	2,4-Dimethyl-3-pentanone	1.395	1.399	0.795	0.805
2-Chlorobutane	Isobutyl butyrate	1.395	1.399	0.868	0.860
Butyric acid	2-Methoxyethanol	1.396	1.400	0.955	0.960
1-Butanol	3-Methyl-2-pentanone	1.397	1.398	0.812	0.808
1-Chloro-2-methylpropane	Isobutyl butyrate	1.397	1.399	0.872	0.860
1-Chloro-2-methylpropane	Pentyl acetate	1.397	1.400	0.872	0.871
Methyl methacrylate	3-Methyl-2-pentanone	1.398	1.398	0.795	0.808
Triethylamine	2,2,3-Trimethylpentane	1.399	1.401	0.723	0.712
Butylamine	Dodecane	1.399	1.400	0.736	0.746
Isobutyl butyrate	1-Chlorobutane	1.399	1.401	0.860	0.875
1-Nitropropane	Propionic anhydride	1.399	1.400	0.995	1.007
Pentyl acetate	1-Chlorobutane	1.400	1.400	0.871	0.881
Pentyl acetate	Tetrahydrofuran	1.400	1.404	0.871	0.885
Dodecane	Dipropylamine	1.400	1.400	0.746	0.736
1-Chlorobutane	Tetrahydrofuran	1.401	1.404	0.871	0.885
Isopentanoic acid	2-Ethoxyethanol	1.402	1.405	0.923	0.926
Dipropylamine	Cyclopentane	1.403	1.404	0.736	0.740
2-Pentanol	4-Heptanone	1.404	1.405	0.804	0.813
3-Methyl-1-butanol	Hexanonitrile	1.404	1.405	0.805	0.801
3-Methyl-1-butanol	4-Heptanone	1.404	1.405	0.805	0.813
Hexanonitrile	4-Heptanone	1.405	1.405	0.801	0.813

TABLE 9.7 Solvents having the Same Refractive Index and the Same Density at 25 °C (*continued*)

Solvent 1	Solvent 2	Refractive index 1	2	Density, g/mL 1	2
Hexanonitrile	1-Pentanol	1.405	1.408	0.801	0.810
Hexanonitrile	2-Methyl-1-butanol	1.405	1.409	0.801	0.815
4-Heptanone	1-Pentanol	1.405	1.408	0.813	0.810
2-Ethoxyethanol	Pentanoic acid	1.405	1.406	0.926	0.936
2-Heptanone	1-Pentanol	1.406	1.408	0.811	0.810
2-Heptanone	2-Methyl-1-butanol	1.406	1.409	0.811	0.815
2-Heptanone	Dipentyl ether	1.406	1.410	0.811	0.799
2-Pentanol	3-Isopropyl-2-pentanone	1.407	1.409	0.804	0.808
1-Pentanol	Dipentyl ether	1.408	1.410	0.810	0.799
2-Methyl-1-butanol	Dipentyl ether	1.409	1.410	0.815	0.799
Isopentyl isopentanoate	Allyl alcohol	1.410	1.411	0.853	0.847
Dipentyl ether	2-Octanone	1.410	1.414	0.799	0.814
2,4-Dimethyldioxane	3-Chloropentene	1.412	1.413	0.935	0.932
2,4-Dimethyldioxane	Hexanoic acid	1.412	1.415	0.935	0.923
Diethyl malonate	Ethyl cyanoacetate	1.412	1.415	1.051	1.056
3-Chloropentene	Octanoic acid	1.413	1.415	0.932	0.923
2-Octanone	1-Hexanol	1.414	1.416	0.814	0.814
2-Octanone	Octanonitrile	1.414	1.418	0.814	0.810
3-Octanone	3-Methyl-2-heptanone	1.414	1.416	0.830	0.818
3-Methyl-2-heptanone	1-Hexanol	1.415	1.416	0.818	0.814
3-Methyl-2-heptanone	Octanonitrile	1.415	1.418	0.818	0.810
1-Hexanol	Octanonitrile	1.416	1.418	0.814	0.810
Dibutylamine	Allylamine	1.416	1.419	0.756	0.758
Allylamine	Methylcyclohexane	1.419	1.421	0.758	0.765
Butyrolactone	1,3-Propanediol	1.434	1.438	1.051	1.049
Butyrolactone	Diethyl maleate	1.434	1.438	1.051	1.064
2-Chloromethyl-2-propanol	Diethyl maleate	1.436	1.438	1.059	1.064
N-Methylmorpholine	Dibutyl decanedioate	1.436	1.440	0.924	0.932
1,3-Propanediol	Diethyl maleate	1.438	1.438	1.049	1.064
Methyl salicylate	Diethyl sulfide	1.438	1.442	0.836	0.831
Methyl salicylate	1-Butanethiol	1.438	1.442	0.836	0.837
1-Chlorodecane	Mesityl oxide	1.441	1.442	0.862	0.850
Diethylene glycol	Formamide	1.445	1.446	1.128	1.129
Diethylene glycol	Ethylene glycol diglycidyl ether	1.445	1.447	1.128	1.134
Formamide	Ethylene glycol diglycidyl ether	1.446	1.447	1.129	1.134
2-Methylmorpholine	Cyclohexanone	1.446	1.448	0.951	0.943
2-Methylmorpholine	1-Amino-2-propanol	1.446	1.448	0.951	0.961
Dipropylene glycol monoethyl ether	Tetrahydrofurfuryl alcohol	1.446	1.450	1.043	1.050
1-Amino-2-methyl-2-pentanol	2-Butylcyclohexanone	1.449	1.453	0.904	0.901
2-Propylcyclohexanone	4-Methylcyclohexanol	1.452	1.454	0.923	0.908

TABLE 9.7 Solvents having the Same Refractive Index and the Same Density at 25 °C (*continued*)

Solvent 1	Solvent 2	Refractive index		Density, g/mL	
		1	2	1	2
Carbon tetrachloride	4,5-Dichloro-1,3-dioxolan-2-one	1.459	1.461	1.584	1.591
N-Butyldiethanolamine	Cyclohexanol	1.461	1.465	0.965	0.968
D-α-Pinene	trans-Decahydro-naphthalene	1.464	1.468	0.855	0.867
Propylbenzene	p-Xylene	1.490	1.493	0.858	0.857
Propylbenzene	Toluene	1.490	1.494	0.858	0.860
Phenyl 1-hydroxyphenyl ether	1,3-Dimorpholyl-2-propanol	1.491	1.493	1.081	1.094
Phenetole	Pyridine	1.505	1.507	0.961	0.978
2-Furanmethanol	Thiophene	1.524	1.526	1.057	1.059
m-Cresol	Benzaldehyde	1.542	1.544	1.037	1.041

McReynolds' Constants

The *Kovats retention indices* (R.I.) indicate where compounds will appear on a chromatogram with respect to unbranched alkanes injected with the sample. By definition, the R.I. for pentane is 500, for hexane is 600, for heptane is 700, and so on, regardless of the column used or the operating conditions, although the exact conditions and column must be specified, such as liquid loading, particular support used, and any pretreatment. For example, suppose that on a 20% squalane column at 100 °C, the retention times for hexane, benzene, and octane are found to be 15, 16, and 25 min, respectively. On a graph of ln t'_R (naperian logarithm of the adjusted retention time) of the alkanes versus their retention indices, a R.I. of 653 for benzene is read off the graph. The number 653 for benzene (see the last line of Table 9.8 in the column headed "1" under "Reference compounds") means that it elutes halfway between hexane and heptane on a logarithmic time scale. If the experiment is repeated with a dinonyl phthalate column, the R.I. for benzene is found to be 736 (lying between heptane and octane), which implies that dinonyl phthalate will retard benzene slightly more than squalane will; that is, dinonyl phthalate is slightly more polar than squalane by $\Delta I = 83$ units (the entry in Table 9.8 for dinonyl phthalate in the column headed "1" under "Reference compounds"). The difference gives a measure of solute–solvent interaction due to all intermolecular forces other than London dispersion forces. The latter are the principal solute–solvent effects with squalane.

Now the overall effects due to hydrogen bonding, dipole moment, acid–base properties, and molecular configuration can be expressed as

$$\Sigma \, \Delta I = ax' + by' + cz' + du' + es'$$

where $x' = \Delta I$ for benzene (the column headed "1" in Table 9.8, intermolecular forces typical of aromatics and olefins), $y' = \Delta I$ for 1-butanol (the column headed "2" in Table 9.8, electron attraction typical of alcohols, nitriles, acids, and nitro and alkyl monochlorides, dichlorides and trichlorides), $z' = \Delta I$ for 2-pentanone (the column headed "3" in Table 9.8, electron repulsion typical of ketones, ethers, aldehydes, esters, epoxides, and dimethylamino derivatives), $u' = \Delta I$ for 1-nitropropane (the column headed "4" in Table 9.8, typical of nitro and nitrile derivatives), and $s' = \Delta I$ for pyridine (or dioxane) (the column headed "5" in Table 9.8).

TABLE 9.8 McReynolds' Constants for Stationary Phases in Gas Chromatography

The McReynolds' constants listed are differences in retention index units between the reference compound run on squalane and on the other phases listed. The last entry in the table shows the absolute retention indices for the reference compounds on squalane. Reference compounds are (1) benzene, (2) 1-butanol, (3) 2-pentanone, (4) 1 nitropropane, and (5) pyridine. (Note that Rohrschneider's constants are based on these reference compounds and may differ slightly from the McReynolds' constants. The reference compounds for Rohrschneider's constants are (1) benzene, (2) ethanol, (3) 2-butanone, (4) nitromethane, and (5) pyridine.) The minimum temperature is that at which normal gas–liquid chromatography (GLC) behavior is expected. Below that temperature, the phase will be a solid or an extremely viscous gum. The maximum temperature is that above which the bleed rate will be excessive.

Liquid phase	Chemical type	Similar liquid phases	Temperature, °C Minimum	Maximum	Reference compounds 1	2	3	4	5	Sum
Squalane	(2,6,10,15,19,23-Hexamethyl)tetracosane		20	150	0	0	0	0	0	0
Paraffin oil	(24,24-Diethyl-19,29-dioctadecyl)heptatetracontane		30	280	9	5	2	6	11	33
Apolane-87					21	10	3	12	25	71
Apiezon L	Poly(dimethylsiloxane)	SP-2100, SF 96	50	250	32	22	15	32	42	143
SE 30		OV-1, DC 200, DC 410	50	350	15	53	44	64	41	217
OV-101			50	350	17	57	45	67	43	229
OV-73	Poly(diphenyldimethylsiloxane), 5%:95%	SE 52	0	325	32	72	65	98	67	334
SE 54	Poly(diphenylvinyldimethylsiloxane), 5%:1%:94%		50	300	33	72	66	99	67	337
OV-3	Poly(diphenyldimethylsiloxane), 10%:90%		0	350	44	86	81	124	88	423
Dexsil 300	Poly(carboranemethylsiloxane)		50	500	47	80	103	148	96	474
Kel F Wax				150	55	67	114	143	116	495
Apiezon H				300	59	86	81	151	129	506

Stationary phase	Commercial name	Chemical type							
Dexsil 400		Carborane and methylphenyl-silicone	50	500	72	108	118	166	123	587
OV-7	DC 550	Poly(diphenyldimethylsiloxane), 20%:80%	20	350	69	113	111	171	128	592
Di(2-ethylhexyl) sebacate			0		72	168	108	180	125	653
Diisodecyl adipate				125	71	171	113	185	128	668
Decyl octyl adipate				175	79	179	119	193	134	704
Bis(2-ethylhexyl)-tetrachlorophthalate			0	150	112	150	123	168	181	734
Diisodecyl phthalate			0	175	84	173	137	218	155	767
Dinonyl phthalate			20	150	83	183	147	231	159	803
OV-11	DC 710	Poly(diphenyldimethylsiloxane), 35%:65%	0	350	107	149	153	228	190	827
Dioctyl phthalate			20	125	92	186	150	236	167	831
Hallcomid M-18			40	150	79	268	130	222	146	845
OV-17		Poly(diphenyldimethylsiloxane), 50%:50%	0	325	119	158	162	243	202	884
Dexsil 410		Carborane and methylcyanoethylsilicone	50	500	72	286	174	249	171	952
UCON LB-550-X			0	200	118	271	158	243	206	996
Span 80			15	150	97	266	170	216	268	1017
OV-22		Poly(diphenyldimethylsiloxane), 65%:35%	0	350	160	188	191	283	253	1075
Polypropylene glycol			0	150	128	294	173	264	226	1085
Didecyl phthalate			10	175	136	255	213	320	235	1159
OV-25		Poly(diphenyldimethylsiloxane), 75%:25%	0	350	178	204	208	305	280	1175
Polyphenyl ether OS-138 (6 rings)			0	225	182	233	228	313	293	1249
Neopentyl glycol sebacate	HI-EFF-3CP		50	225	172	327	225	344	326	1394
Squalene			0	100	152	341	328	329	344	1404
UCON 50-HB-280X			0	200	177	362	227	351	302	1419
Tricresyl phosphate			20	125	176	321	250	374	299	1420

TABLE 9.8 McReynolds' Constants for Stationary Phases in Gas Chromatography (*continued*)

Liquid phase	Chemical type	Similar liquid phases	Temperature, °C		Reference compounds					
			Minimum	Maximum	1	2	3	4	5	Sum
Sucrose acetate isobutyrate			0	200	172	330	251	378	295	1426
QF-1	Poly(trifluoropropylsiloxane)	SP-2401, FS 1265	0	250	144	233	355	463	305	1500
OV-210	Poly(trifluoropropylmethyl-siloxane)		0	275	146	238	358	468	310	1520
OV-215		XE 6O	0	275	149	240	363	478	315	1545
UCON 50-HB-2000	Emulphor ON-870		0	200	202	394	253	392	341	1582
Triton X-100			0	200	203	399	268	402	362	1634
UCON 50-HB-5100			0	200	214	418	278	421	375	1706
Siponate DS-10			0	150	99	569	320	344	388	1720
Tween 80			0		227	430	256	438	396	1747
XE-60	Poly(cyanoethylphenyl-methylsiloxane)		0	250	204	381	340	493	367	1785
OV-225	Poly(cyanopropylphenyl-methylsiloxane)	HI-EFF-3AP	0	265	228	369	338	492	386	1813
Neopentyl glycol adipate			50	225	232	421	311	461	424	1849
UCON 75-H-90000	Igepal CO-880		100	200	255	452	299	470	406	1882
Triton X-305		HI-EFF-3BP	0	200	262	467	314	488	430	1961
Neopentyl glycol succinate			50	230	272	469	366	539	474	2120
Igepal CO 990	Poly(ethylene glycol)		100	200	298	508	345	540	475	2166
Carbowax 20M		FFAP, SP-2300	25	275	322	536	368	572	510	2308
Epon 1001			50	225	284	489	406	539	601	2319
Carbowax 4000			60	200	325	551	375	582	520	2353
Ethylene glycol isophthalate		HI-EFF-2EP	100	225	326	508	425	607	561	2427
Ethylene glycol adipate		HI-EFF-2AP	100	225	372	576	453	655	617	2673
Butane-1,4-diol succinate		HI-EFF-4BP	50	225	369	591	457	661	629	2207
Phenyldiethanolamine succinate		HI-EFF-10BP	0	200	386	555	472	674	654	2741

Diethylene glycol adipate	HI-EFF-1AP, LAC-1-R-296, SP-2330	25	275	378	603	460	665	658	2764
Carbowax 1540		50	175	371	639	453	666	641	2770
Hyprose SP-80		0	175	336	742	492	639	727	2936
SILAR-7CP		0	250	440	638	605	844	673	3200
ECNSS-M		30	200	421	690	581	803	732	3227
EGSS-X		90	200	484	710	585	831	778	3388
Ethylene glycol phthalate	HI-EFF-2GP	100	200	453	697	602	816	872	3410
SILAR-9CP		0	250	489	725	631	910	778	3536
SILAR-10C	SP-2340	25	275	523	757	659	942	801	3682
Diethylene glycol succinate	HI-EFF-1BP, LAC-3-R-728	20	200	499	751	593	840	860	3543
Tetrahydroxyethylethylenediamine	THEED	0	150	463	942	626	801	893	3725
Tetracyanoethylated pentaerythritol		30	175	526	782	677	920	837	3742
Ethylene glycol succinate	HI-EFF-2BP	100	200	537	787	643	903	889	3759
1,2,3,4-Tetrakis-(2-cyanoethoxy)butane		110	200	617	860	773	1048	941	4239
1,2,3,4,5,6-Hexakis(2-cyanoethoxy)cyclohexane		125	150	567	825	713	978	901	3984
1,2,3-Tris-(2-cyanoethoxy)propane		0	175	593	857	752	1028	915	4145
N,N-Bis(2-cyanoethyl)-formamide		0	125	690	991	853	1110	1000	4644
OV-275			250	781	1006	885	1177	1089	4938
Dicyanoallylsilicone		25							
Absolute retention index values on squalane for reference compounds				653	590	627	652	699	

SECTION 10
POLYMERS, RUBBERS, FATS, OILS, AND WAXES

POLYMERS

General

Polymers are macromolecules that result from combinations of individual building blocks called monomer molecules. Most polymers are regular structures in which a single unit repeats many times. This produces a range of macromolecules that have similar structures and molecular weights. The ensemble of molecules typically exhibit average molecular weights but have characteristic properties. Polymerization of ethylene, $CH_2 = CH_2$, results in polyethylene, $(-CH_2-CH_2-)_n$ such that "n" is a range of values. If the average chain possesses 1000 monomer units, chains having values of n such as 998, 999, 1001, 1002, etc. will be present as well. There will be more such chains near the average value and fewer or none when n is far from the average value.

In some polymers, long segments of linear polymer chains are oriented in a regular manner with respect to one another. Such polymers have many of the physical characteristics of crystals and are said to be *crystalline*. Polymers that have polar functional groups show a greater tendency to be crystalline. Orientation is aided by alignment of dipoles on different chains. Van der Waals' interactions between long hydrocarbon chains may provide sufficient total attractive energy to account for a high degree of regularity within the polymers.

Irregularities such as branch points, co-monomer units, and cross-links lead to *amorphous* polymers. These have less regular structures and typically do not have true melting points. Instead, they have glass transition temperatures at which the rigid and glass-like material becomes a viscous liquid as the temperature is raised.

Elastomers. Elastomer is a generic name for polymers that exhibit rubber-like elasticity. Elastomers are soft yet sufficiently elastic that they can usually be stretched several hundred percent under tension. When the stretching force is removed, they quickly retract and recover their original dimensions.

Polymers that soften or melt and then solidify and regain their original properties on cooling are called *thermoplastic*. A thermoplastic polymer is usually a single strand of linear polymer with few if any cross-links.

Thermosetting Polymers. Polymers that soften or melt on warming and then become infusible solids are called *thermosetting*. The term implies that thermal decomposition has not taken place. Thermosetting plastics contain a cross-linked polymer network that extends through the final material, making it stable to heat and insoluble in organic solvents. Many molded plastics are shaped while molten and are then heated further to become rigid solids of desired shapes.

Synthetic Rubbers. Synthetic rubbers are polymers with rubber-like characteristics that are prepared from dienes or olefins. Rubbers with special properties can also be prepared from other polymers, such as polyacrylates, fluorinated hydrocarbons, and polyurethanes.

Structural Differences. Polymers exhibit structural differences resulting from the type of monomer used, the polymerization method employed, and other factors. A *linear* polymer consists of long segments of single strands that are oriented in a regular manner with respect to one another. *Branched* polymers have substituents attached to the repeating units that extend the polymer laterally. When these units participate in chain propagation and link together chains, a *cross-linked* polymer is formed. A *ladder* polymer results when repeating units have a tetravalent structure such that a polymer consists of two backbone chains regularly cross-linked at short intervals.

Generally polymers involve bonding of the most substituted carbon of one monomeric unit to the least substituted carbon atom of the adjacent unit in a *head-to-tail* arrangement. An example is the formation of polypropylene from propylene. This is shown for three monomer units. The wavy lines indicate that more monomers would lead to extended chains.

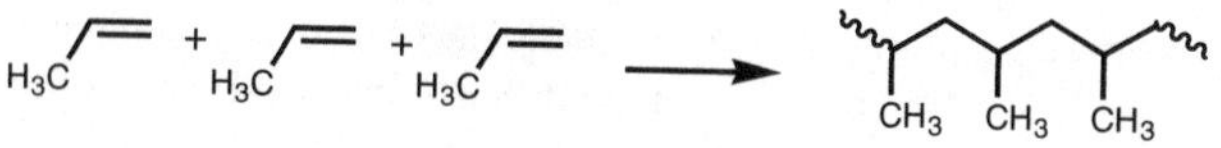

Substituents appear on alternate carbon atoms. *Tacticity* refers to the configuration of substituents relative to the backbone axis. In an *isotactic* arrangement, substituents are on the same plane of the backbone axis; that is, the configuration at each chiral center is identical.

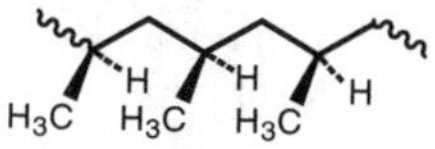

In a *syndiotactic* arrangement, the substituents are in an ordered alternating sequence, appearing alternately on one side and then on the other side of the chain, as shown for a segment of a vinyl chloride polymer. If the sidechains are not in any particular order with respect to each other (random), the polymer is said to be *atactic*.

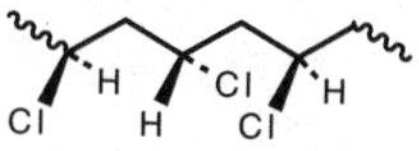

Copolymerization. Copolymerization occurs when a mixture of two or more monomer types polymerizes so that each kind of monomer enters the polymer chain. The fundamental structure resulting from copolymerization depends on the nature of the monomers and the relative rates of monomer reactions with the growing polymer chain. A tendency toward alternation of monomer units is common. Random copolymerization is known but it is rather unusual.

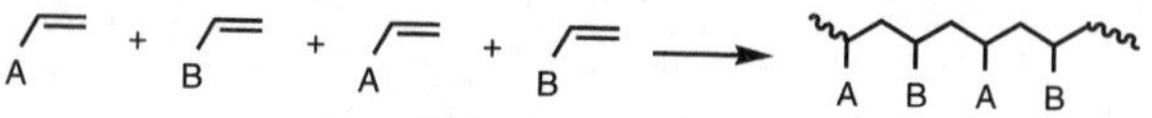

In *graft copolymers* the chain backbone is composed of one kind of monomer and the branches are made up of another kind of monomer. The structure of a *block copolymer* consists of a homopolymer attached to chains of another homopolymer. In either case, *cis* or *trans* (*Z* or *E*) double bond configurations around any double bond not involved in the polymerization will normally be unaltered.

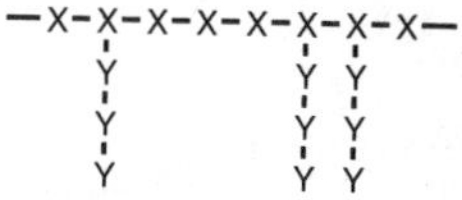

Schematic of a graft copolymer 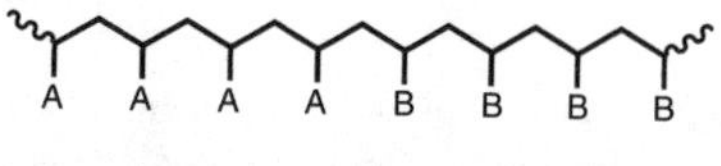   Schematic of a block copolymer

Dendrimers

A relatively recent development in polymer chemistry is the family of compounds known as dendrimers. The term derives from the Greek *dendra* meaning tree. Tree-like structures of this general type have also been referred to as arborols. Dendrimers differ from typical polymers in that they radiate from a central unit or core rather than being either linear or planar. Dendrimers are built up using a "generational" structure in discrete synthetic steps. As a result, the product is nearer to being a single compound than is a typical polymer. A typical dendrimer consists of a multifunctional core unit. Each functional group of the core unit is elaborated by a further molecular unit, often referred to as a "dendritic wedge." The core unit is usually designated "generation 0" and additional units radiating from it or prior units are designated as higher generations, that is, generation 1, generation 2, etc. When the synthesis begins at the core and radiates outward, it is called a divergent synthesis. Alternately, synthesis may begin at the outside and terminate with a core unit.

The more common divergent method is illustrated in the following scheme. 1,3,5-Tricarboxybenzene (trimesic acid) serves as the core or generation 0. It could be functionalized by converting it into the tris(acid chloride). Reaction with the secondary amine of diethanolamine would give the tris(amide) hexahydroxy compound shown here. This structure represents generations 0 and 1. Treatment with an appropriately substituted benzyl chloride could lead to the hexaether that comprises generations 0, 1, and 2. There are now a dozen "Y" groups that could be further functionalized in going to generation 3. A difficulty is that the functional or protecting groups present at each stage must be compatible with the chemistry used to make the connections.

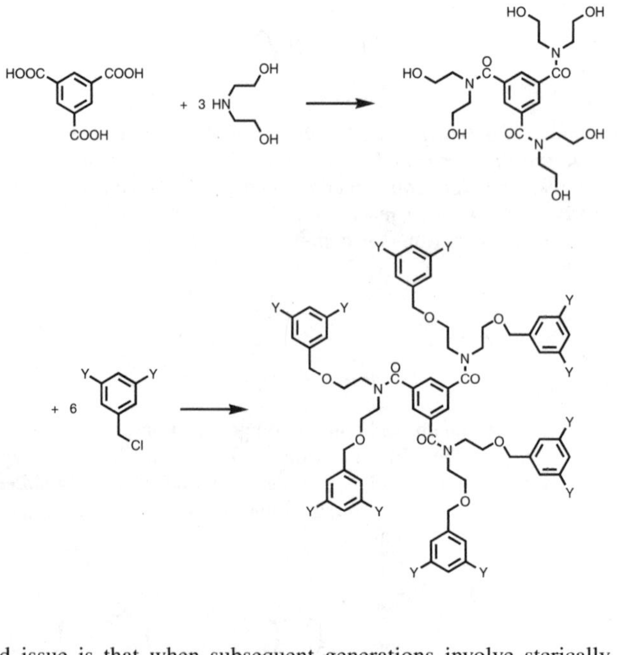

A second issue is that when subsequent generations involve sterically demanding structural units, incomplete substitution may occur. Thus, in the final structure shown, five benzyl ethers might form, rather than six, owing to steric crowding. This would introduce a defect in the dendrimer structure. Indeed, some molecules might possess all six benzyl ethers whereas others might lack one or even two. It must be possible to manipulate the substituents designated "Y" in the presence of ether and amide groups to further extend the dendrimer.

Novel dendrimers have novel and unique properties that make them promising candidates for use in the development of nanoscale devices and in drug delivery systems.

Additives to Polymers

Antioxidants. Antioxidants markedly retard the rate of autoxidation throughout the useful life of the polymer. Chain-terminating antioxidants have a reactive —NH or —OH functional group and include compounds such as secondary aryl amines or hindered phenols. They function by transfer of hydrogen to free radicals, principally to peroxy radicals. Butylated hydroxytoluene is a widely used example.

Peroxide-decomposing antioxidants destroy hydroperoxides, the sources of free radicals in polymers. Phosphites and thioesters such as tris(nonylphenyl) phosphite, distearyl pentaerythritol diphosphite, and dialkyl thiodipropionates are examples of peroxide-decomposing antioxidants.

Antistatic Agents. External antistatic agents are usually quaternary ammonium salts of fatty acids and ethoxylated glycerol esters of fatty acids that are applied to the plastic surface. Internal antistatic agents are compounded into plastics during processing. Carbon blacks provide a conductive path through the bulk of the plastic. Other types of internal agents must bloom to the surface after compounding in order to be active. These latter materials are ethoxylated fatty amines and ethoxylated glycerol esters of fatty acids, which often must be individually selected to match chemically each plastic type.

Antistatic agents require ambient moisture to function. Consequently their effectiveness is dependent on the relative humidity. They provide a broad range of protection at 50% relative humidity. Much below 20% relative humidity, only materials that provide a conductive path through the bulk of the plastic to ground (such as carbon black) will reduce electrostatic charging.

Chain-Transfer Agents. Chain-transfer agents are used to regulate the molecular weight of polymers. These agents react with the developing polymer and interrupt the growth of a particular chain. The products, however, are free radicals that are capable of adding to monomers and initiating the formation of new chains. The overall effect is to reduce the average molecular weight of the polymer without reducing the rate of polymerization. Branching may occur as a result of chain transfer between a growing but rather short chain with another and longer polymer chain. Branching may also occur if the radical end of a growing chain abstracts a hydrogen atom from a carbon atom four or five carbons removed from the end. Thiols are commonly used as chain-transfer agents.

Coupling Agents. Coupling agents are molecular bridges between the interface of an inorganic surface (or filler) and an organic polymer matrix. Titanium-derived coupling agents interact with the free protons at the inorganic interface to form organic monomolecular layers on the inorganic surface. The titanate-coupling-agent molecule has six functions:

$$(RO)_m—Ti—(O—Y—R^1—Z)_n$$

where

Type	m	n
Monoalkoxy	1	3
Coordinate	4	2
Chelate	1	2

Function 1 is the attachment of the hydrolyzable portion of the molecule to the surface of the inorganic (or proton-bearing) species.

Function 2 is the ability of the titanate molecule to transesterify.

Function 3 affects performance as determined by the chemistry of alkylate, carboxyl, sulfonyl, phenolic, phosphate, pyrophosphate, and phosphite groups.

Function 4 provides van der Waals' entanglement via long carbon chains.

Function 5 provides thermoset reactivity via functional groups such as methacrylates and amines.

Function 6 permits the presence of two or three pendent organic groups. This allows all functionality to be controlled to the first-, second-, or third-degree levels.

Silane coupling agents are represented by the formula

$$Z - R - SiY_3$$

where Y represents a hydrolyzable group (typically alkoxy); Z is a functional organic group, such as amino, methacryloxy, epoxy; and R typically is a short aliphatic linkage that serves to attach the functional organic group to silicon in a stable fashion. Bonding to surface hydroxy groups of inorganic compounds is accomplished by the $-SiY_3$ portion, either by direct bonding of this group or more commonly via its hydrolysis product $-Si(OH)_3$. Subsequent reaction of the functional organic group with the organic matrix completes the coupling reaction and establishes a covalent chemical bond from the organic phase through the silane coupling agent to the inorganic phase.

Flame Retardants.   Flame retardants are thought to function via several mechanisms, dependent upon the class of flame retardant used. Halogenated flame retardants are thought to function principally in the vapor phase either as a diluent and heat sink or as a free-radical trap that stops or slows flame propagation. Phosphorus compounds are thought to function in the solid phase by forming a glaze or coating over the substrate that prevents the heat and mass transfer necessary for sustained combustion. With some additives, as the temperature is increased, the flame retardant acts as a solvent for the polymer, causing it to melt at lower temperatures and flow away from the ignition source.

Mineral hydrates, such as alumina trihydrate and magnesium sulfate heptahydrate, are used in highly filled thermosetting resins.

Foaming Agents (Chemical Blowing Agents).   Foaming agents are added to polymers during processing to form minute gas cells throughout the product. Physical foaming agents include liquids and gases. Compressed nitrogen is often used in injection molding. Common liquid foaming agents are short-chain aliphatic hydrocarbons in the C_5 to C_7 range and their chlorinated or fluorinated analogs.

The chemical foaming agent used varies with the temperature employed during processing. At relatively low temperatures (15–200 °C), the foaming agent is often 4,4'-oxybis(benzenesulfonylhydrazide) or *p*-toluenesulfonylhydrazide. In the midrange (160–232 °C), either sodium hydrogen carbonate or 1,1'azobisformamide is used. For the high range (200–285 °C), there are *p*-toluenesulfonylsemicarbazide, 5-phenyltetrazole and analogs, and trihydrazinotriazine.

Inhibitors. Inhibitors slow or stop polymerization by reacting with the initiator or the growing polymer chain. The free radical formed from an inhibitor must be sufficiently unreactive that it does not function as a chain-transfer agent and begin another growing chain. Benzoquinone is a typical free-radical chain inhibitor. The resonance-stabilized free radical usually dimerizes or disproportionates to produce inert products and end the chain process.

Lubricants. Materials such as fatty acids are added to reduce the surface tension and improve the handling qualities of plastic films.

Plasticizers. Plasticizers are relatively nonvolatile liquids which are blended with polymers to alter their properties by intrusion between polymer chains. Diisooctyl phthalate is a common plasticizer. A plasticizer must be compatible with the polymer to avoid bleeding out over long periods of time. Products containing plasticizers tend to be more flexible and workable.

Ultraviolet Stabilizers. 2-Hydroxybenzophenones represent the largest and most versatile class of ultraviolet stabilizers that are used to protect materials from the degradative effects of ultraviolet radiation. They function by absorbing ultraviolet radiation and by quenching electronically excited states.

Hindered amines, such as 4-(2,2,6,6-tetramethylpiperidinyl) decanedioate, serve as radical scavengers and will protect thin films under conditions in which ultraviolet absorbers are ineffective. Metal salts of nickel, such as dibutyldithiocarbamate, are used in polyolefins to quench singlet oxygen or electronically excited states of other species in the polymer. Zinc salts function as peroxide decomposers.

Vulcanization and Curing. Originally, vulcanization implied heating natural rubber with sulfur, but the term is now also employed for curing polymers. When sulfur is employed, sulfide and disulfide cross-links form between polymer chains. This provides sufficient rigidity to prevent *plastic flow*. Plastic flow is a process in which coiled polymers slip past each other under an external deforming force; when the force is released, the polymer chains do not completely return to their original positions.

Organic peroxides are used extensively for the curing of unsaturated polyester resins and the polymerization of monomers having vinyl unsaturation. The —O—O— bond is split into free radicals which can initiate polymerization or cross-linking of various monomers or polymers.

TABLE 10.1 Plastic Families

Acetals	**Allyls**
Acrylics	Allyl-diglycol-carbonate polymer
Poly(methyl methacrylate) (PMMA)	Diallyl phthalate (DAP) polymer
Poly(acrylonitrile)	**Cellulosics**
Alkyds	Cellulose acetate resin
	Cellulose-acetate–propionate resin
Alloys	Cellulose-acetate–butyrate resin
Acrylic-poly(vinyl chloride) alloy	Cellulose nitrate resin
Acrylonitrile–butadiene–styrene–	Ethyl cellulose resin
poly(vinyl chloride) alloy (ABS–PVC)	Rayon
Acrylonitrile–butadiene–styrene–	
polycarbonate alloy (ABS–PC)	**Chlorinated polyether**
	Epoxy

TABLE 10.1   Plastic Families (*continued*)

Fluorocarbons
  Poly (tetrafluoroethylene) (PTFE)
  Poly (chlorotrifluoroethylene) (PCTFE)
  Perfluoroalkoxy (PFA) resin
  Fluorinated ethylene–propylene (FEP)
    resin
  Poly(vinylidene fluoride) (PVDF)
  Ethylene–chlorotrifluoroethylene
    copolymer
  Ethylene–tetrafluoroethylene copolymer
  Poly(vinyl fluoride) (PVF)

Melamine formaldehyde

Melamine phenolic

Nitrile resins

Phenolics

Polyamides
  Nylon 6
  Nylon 6/6
  Nylon 6/9
  Nylon 6/12
  Nylon 11
  Nylon 12
  Aromatic nylons

Poly(amide–imide)

Poly(aryl ether)

Polycarbonate (PC)

Polyesters
  Poly(butylene terephthalate) (PBT)
    [also called polytetramethylene
    terephthalate (PTMT)]
  Poly(ethylene terephthalate) (PET)
  Unsaturated polyesters (SMC, BMC)
    Butadiene–maleic acid copolymer (BMC)
    Styrene–maleic acid copolymer (SMC)

Polyimide

Poly(methylpentene)

Polyolefins (PO)
  Low-density polyethylene (LDPE)
  High-density polyethylene (HDPE)
  Ultrahigh-molecular-weight polyethylene
    (UHMWPE)
  Polypropylene (PP)
  Polybutylene (PB)
  Polyallomers

Poly(phenylene oxide)

Poly(phenylene sulfide) (PPS)

Polyurethanes

Silicones

Styrenics
  Polystyrene (PS)
  Acrylonitrile–butadiene–styrene (ABS)
    copolymer
  Sytrene–acrylonitrile (SAN) copolymer
  Styrene–butadiene copolymer

Sulfones
  Polysulfone (PSF)
  Poly(ether sulfone)
  Poly(phenyl sulfone)

Thermoplastic elastomers
  Polyolefin
  Polyester
  Block copolymers
    Styrene–butadiene block copolymer
    Styrene–isoprene block copolymer
    Styrene–ethylene block copolymer
    Styrene–butylene block copolymer

Urea formaldehyde

Vinyls
  Poly(vinyl chloride) (PVC)
  Poly(vinyl acetate) (PVAC)
  Poly(vinylidene chloride)
  Poly(vinyl butyrate) (PVB)
  Poly(vinyl formal)
  Poly(vinyl alcohol) (PVAL)

FORMULAS AND KEY PROPERTIES OF PLASTIC MATERIALS

Acetals

Homopolymer. Acetal homopolymers are prepared from formaldehyde and consist of high-molecular-weight linear polymers of formaldehyde. The trimer of formaldehyde is shown to the left and the structure of the polymer is shown at the right, below.

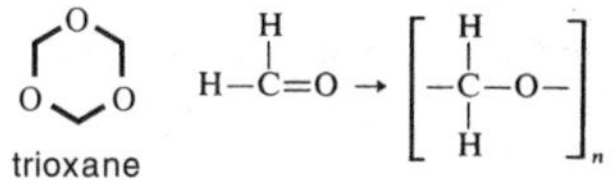

$$
\text{trioxane} \qquad H-\underset{\underset{H}{|}}{\overset{\overset{H}{|}}{C}}=O \;\rightarrow\; \left[-\underset{\underset{H}{|}}{\overset{\overset{H}{|}}{C}}-O- \right]_n
$$

The good mechanical properties of this homopolymer result from the ability of the oxymethylene chains to pack together into a highly ordered crystalline configuration as the polymers change from the molten to the solid state.

Key properties include high melt point, strength and rigidity, good frictional properties, and resistance to fatigue. Higher molecular weight increases toughness but reduces melt flow.

Copolymer. Acetal copolymers are prepared by copolymerization of 1,3,5-trioxane with small amounts of a co-monomer. Carbon–carbon bonds are distributed randomly in the polymer chain. These carbon–carbon bonds help to stabilize the polymer against thermal, oxidative, and acidic attack.

Acrylics

Poly(methyl methacrylate). Acrylic acid is $H_2C{=}CH{-}COOH$ and methacrylic acid is $H_2C{=}C(CH_3)COOH$. These compounds and their methyl esters are both quite reactive and difficult to store and handle. The monomer used to form poly(methyl methacrylate), 2-hydroxy-2-methylpropanenitrile, is prepared by the following reaction:

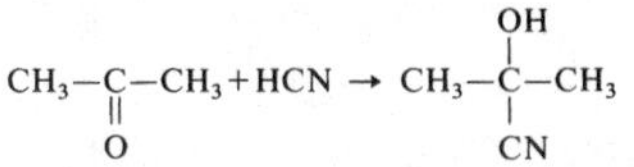

$$
CH_3-\underset{\underset{O}{\|}}{C}-CH_3 + HCN \;\rightarrow\; CH_3-\underset{\underset{CN}{|}}{\overset{\overset{OH}{|}}{C}}-CH_3
$$

2-Hydroxy-2-methylpropanenitrile is then reacted with methanol (or other alcohol) to yield methacrylate ester. Free-radical polymerization is initiated by peroxide or azo catalysts and produce poly(methyl methacrylate) resins having the following formula:

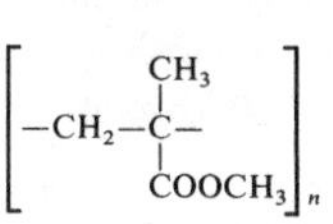

$$
\left[-CH_2-\underset{\underset{COOCH_3}{|}}{\overset{\overset{CH_3}{|}}{C}}- \right]_n
$$

Key properties are good resistance to heat, light, and weathering. This polymer is unaffected by most detergents, cleaning agents, and solutions of inorganic acids, alkalies, and aliphatic hydrocarbons. Poly(methyl methacrylate) has light transmittance of 92% with a haze of 1–3% and its clarity is equal to glass.

Poly(methyl acrylate).   The structure of methyl acrylate is $H_2C{=}CH{-}COOCH_3$. The monomer used to prepare poly(methyl acrylate) is produced by the oxidation of propylene. The resin is made by free-radical polymerization initiated by peroxide catalysts and has the following formula:

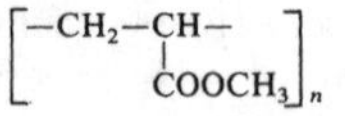

$$\left[\begin{array}{c} -CH_2-CH- \\ \mid \\ COOCH_3 \end{array} \right]_n$$

Poly(methyl acrylate) resins vary from soft, elastic, film-forming materials to hard plastics.

Poly(acrylic acid) and Poly(methacrylic acid).   Glacial acrylic acid and glacial methacrylic acid can be polymerized to produce water-soluble polymers having the following structures:

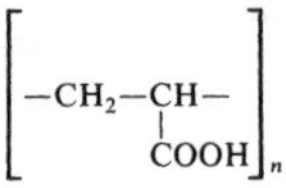

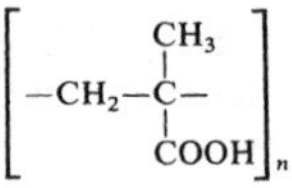

$$\left[\begin{array}{c} -CH_2-CH- \\ \mid \\ COOH \end{array} \right]_n \quad \left[\begin{array}{c} CH_3 \\ \mid \\ -CH_2-C- \\ \mid \\ COOH \end{array} \right]_n$$

These monomers provide a means for introducing carboxyl groups into copolymers. In copolymers these acids can improve adhesion properties, improve freeze–thaw and mechanical stability of polymer dispersions, provide stability in alkalies (including ammonia), increase resistance to attack by oils, and provide reactive centers for cross-linking by divalent metal ions, diamines, or epoxides.

Functional Group Methacrylate Monomers.   Hydroxyethyl methacrylate and dimethylaminoethyl methacrylate produce polymers having the following formulas:

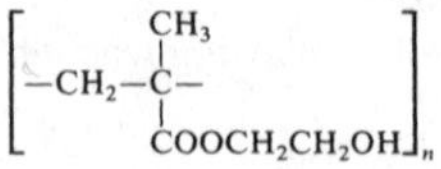

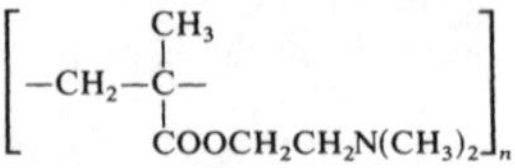

$$\left[\begin{array}{c} CH_3 \\ \mid \\ -CH_2-C- \\ \mid \\ COOCH_2CH_2OH \end{array} \right]_n \quad \left[\begin{array}{c} CH_3 \\ \mid \\ -CH_2-C- \\ \mid \\ COOCH_2CH_2N(CH_3)_2 \end{array} \right]_n$$

The use of hydroxyethyl (also hydroxypropyl) methacrylate as a monomer permits the introduction of reactive hydroxyl groups into the copolymers. This offers the possibility for subsequent cross-linking with an HO-reactive difunctional agent (diisocyanate, diepoxide, or melamineformaldehyde resin). Hydroxyl groups promote adhesion to polar substrates.

Use of dimethylaminoethyl (also *tert*-butylaminoethyl) methacrylate as a monomer permits the introduction of pendent amino groups which can serve as sites for secondary cross-linking, provide a way to make the copolymer acid-soluble, and provide anchoring sites for dyes and pigments.

Poly(acrylonitrile).   Acrylonitrile has the formula $H_2C{=}CH{-}C{\equiv}N$. Poly(acrylonitrile) polymers have the following formula:

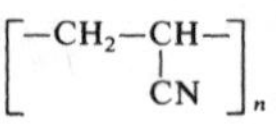

$$\left[\begin{array}{c} -CH_2-CH- \\ \mid \\ CN \end{array} \right]_n$$

Alkyds

Alkyds are formulated from polyester resins, cross-linking monomers, and fillers of mineral or glass. The unsaturated polyester resins used for thermosetting alkyds are the reaction products of polyfunctional organic alcohols (glycols) and dibasic organic acids. Key properties of alkyds are dimensional stability, colorability, and arc track resistance. Chemical resistance, however, is generally poor.

Alloys

Polymer alloys are physical mixtures of structurally different homopolymers or copolymers. The mixture is held together by secondary intermolecular forces such as dipole interaction, hydrogen bonding, or van der Waals' forces.

Homogeneous alloys have a single glass transition temperature which is determined by the ratio of the components. The physical properties of these alloys are averages based on the composition of the alloy. Heterogeneous alloys can be formed when graft or block copolymers are combined with a compatible polymer. Alloys of incompatible polymers can be formed if an interfacial agent can be found.

Allyls

Diallyl Phthalate (and Diallyl 1,3-Phthalate).   Phthalic acid is 1,2-dicarboxybenzene. The 1,3-isomer is generally referred to as isophthalic acid. These allyl polymers are prepared from

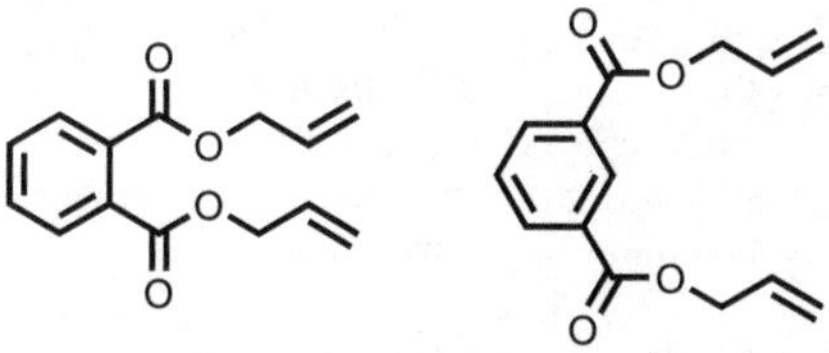

These resulting polymers are solid, linear, internally cyclized, thermoplastic structures containing unreacted allylic groups spaced at regular intervals along the polymer chain. Compounds derived from these polymers that are molded with mineral, glass, or synthetic fiber filling exhibit good electrical properties under high humidity and high temperature conditions. They also show stable low-loss factors, high surface and volume resistivity, and high arc and track resistance.

Cellulosics

Cellulose Triacetate.   Cellulose triacetate is prepared according to the following reaction:

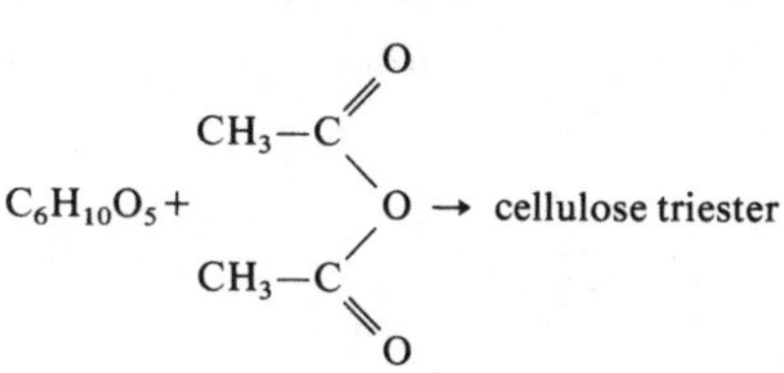

$$C_6H_{10}O_5 + \begin{matrix} CH_3-C \overset{\displaystyle O}{\diagup} \\ O \\ CH_3-C \diagdown_O \end{matrix} \rightarrow \text{cellulose triester}$$

Because cellulose triacetate has a high softening temperature, it must be processed in solution. A mixture of dichloromethane and methanol is a common solvent.

Cellulose triacetate sheeting and film have good gauge uniformity and good optical clarity. Cellulose triacetate products have good dimensional stability and resistance to water and have good folding endurance and burst strength. It is highly resistant to solvents such as acetone. Cellulose triacetate products have good heat resistance and a high dielectric constant.

Cellulose Acetate, Propionate, and Butyrate. Cellulose acetate is prepared by hydrolyzing the triester to remove some of the acetyl groups; the plastic-grade resin contains 38–40% acetyl. The propionate and butyrate esters are made by substituting propionic acid and its anhydride (or butyric acid and its anhydride) for some of the acetic acid and acetic anhydride. Plastic grades of cellulose-acetate–propionate resin contain 39–47% propionyl and 2–9% acetyl; cellulose-acetate–butyrate resins contain 26–39% butyryl and 12–15% acetyl.

These cellulose esters form tough, strong, stiff, hard plastics with almost unlimited color possibilities. Articles made from these plastics have a high gloss and are suitable for use in contact with food.

Cellulose Nitrate. Cellulose nitrate is prepared according to the following reaction:

$$C_6H_{10}O_5 + HNO_3 \rightarrow [-C_6H_7O_2(OH)(ONO_2)_2-]_n$$

The nitrogen content for plastics is usually about 11%, for lacquers and cement base it is 12%, and for explosives it is 13%. The standard plasticizer added is camphor.

Key properties of cellulose nitrate are good dimensional stability, low water absorption, and toughness. Its disadvantages are its flammability and lack of stability to heat and sunlight.

Ethyl Cellulose. Ethyl cellulose is prepared by reacting cellulose with caustic to form caustic cellulose, which is then reacted with chloroethane to form ethyl cellulose. Plastic-grade material contains 44–48% ethoxyl.

Although not as resistant as cellulose esters to acids, it is much more resistant to bases. An outstanding feature is its toughness at low temperatures.

Rayon. Viscose rayon is obtained by reacting the hydroxy groups of cellulose with carbon disulfide in the presence of alkali to give xanthates. When this solution is poured (spun) into an acid medium, the reaction is reversed and the cellulose is regenerated (coagulated).

Epoxy

Epoxy resin is prepared by the following condensation reaction:

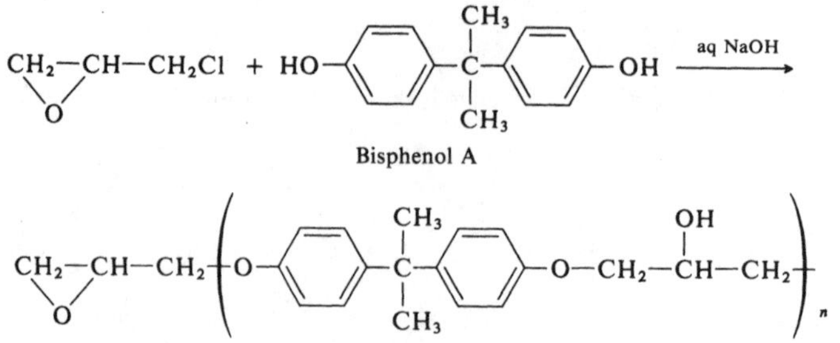

Bisphenol A

The condensation leaves epoxy end groups that are then reacted in a separate step with nucleophilic compounds (alcohols, acids, or amines). For use as an adhesive, the epoxy resin and the curing resin (usually an aliphatic polyamine) are packaged separately and mixed together immediately before use.

Epoxy novolac resins are produced by glycidation of the low-molecular-weight reaction products of phenol (or cresol) with formaldehyde. Highly cross-linked systems are formed that have superior performance at elevated temperatures.

Fluorocarbon

Poly(tetrafluoroethylene). Poly(tetrafluoroethylene) is prepared from tetrafluoroethylene and consists of repeating units in a predominantly linear chain:

$$F_2C = CF_2 \rightarrow [-CF_2 - CF_2 -]_n$$

Tetrafluoroethylene polymer has the lowest coefficient of friction of any solid. It has remarkable chemical resistance and a very low brittleness temperature ($-100\,°C$). Its dielectric constant and loss factor are low and stable across a broad temperature and frequency range. Its impact strength is high.

Fluorinated Ethylene–Propylene Resin. Polymer molecules of fluorinated ethylene-propylene consist of predominantly linear chains with this structure:

$$\left[-CF_2 - CF_2 - CF_2 - \underset{\underset{CF_3}{|}}{CF} - \right]_n$$

Key properties are its flexibility, translucency, and resistance to all known chemicals except molten alkali metals, elemental fluorine and fluorine precursors at elevated temperatures, and concentrated perchloric acid. It withstands temperatures from $-270°$ to $250\,°C$ and may be sterilized repeatedly by all known chemical and thermal methods.

Perfluoroalkoxy Resin. Perfluoroalkoxy resin has the following formula:

$$\left[-CF_2 - CF_2 - \underset{\underset{\underset{R}{|}}{\overset{\overset{|}{O}}{CF}}}{} - CF_2 - CF_2 - \right]_n \qquad \text{where R is} -C_n F_{2n+1}$$

It resembles polytetrafluoroethylene and fluorinated ethylene propylene in its chemical resistance, electrical properties, and coefficient of friction. Its strength, hardness, and wear resistance are about equal to the former plastic and superior to that of the latter at temperatures above $150\,°C$.

Poly(vinylidene fluoride). Poly(vinylidene fluoride) consists of linear chains in which the predominant repeating unit is

$$[-CH_2 - CF_2 -]_n$$

It has good weathering resistance and does not support combustion. It is resistant to most chemicals and solvents and has greater strength, wear resistance, and creep resistance than the preceding three fluorocarbon resins.

Poly(1-chloro-1,2,2-trifluoroethylene).   Poly(1-chloro-1,2,2-trifluoroethylene) consists of linear chains in which the predominant repeating unit is

$$\left[-CF_2-\underset{\underset{Cl}{|}}{CF}- \right]_n$$

It possesses outstanding barrier properties to gases, especially water vapor. It is surpassed only by the fully fluorinated polymers in chemical resistance. A few solvents dissolve it at temperatures above 100°C, and it is swollen by a number of solvents, especially chlorinated solvents. It is harder and stronger than perfluorinated polymers, and its impact strength is lower.

Ethylene–Chlorotrifluoroethylene Copolymer.   Ethylene–chlorotrifluoroethylene copolymer consists of linear chains in which the predominant 1:1 alternating copolymer is

$$\left[-CH_2-CH_2-CF_2-\underset{\underset{Cl}{|}}{CF}- \right]_n$$

This copolymer has useful properties from cryogenic temperatures to 180°C. Its dielectric constant is low and stable over a broad temperature and frequency range.

Ethylene–Tetrafluoroethylene Copolymer.   Ethylene–tetrafluoroethylene copolymer consists of linear chains in which the repeating unit is

$$[-CH_2-CH_2-CF_2-CF_2-]_n$$

Its properties resemble those of ethylene–chlorotrifluoroethylene copolymer.

Poly(vinyl fluoride).   Poly(vinyl fluoride) consists of linear chains in which the repeating unit is

$$[-CH_2-CHF-]_n$$

It is used only as a film, and it has good resistance to abrasion and resists staining. It also has outstanding weathering resistance and maintains useful properties from -100 to 150°C.

Nitrile Resins

The principal monomer of nitrile resins is acrylonitrile (see "Polyacrylonitrile"), which constitutes about 70% by weight of the polymer and provides the polymer with good gas barrier and chemical resistance properties. The remainder of the polymer is 20–30% methyl acrylate (or styrene), with 0–10% butadiene to serve as an impact-modifying termonomer.

Melamine Formaldehyde

The monomer used for preparing melamine formaldehyde is formed as follows:

$$H_2N-C\underset{N=C}{\overset{N=C}{\underset{\underset{NH_2}{|}}{}}}C-NH_2 \;+\; 6\; HC\overset{H}{=}O \;\longrightarrow\; (HOCH_2)_2N-C\underset{N-C}{\overset{N-C}{\underset{\underset{N(CH_2OH)_2}{|}}{}}}C-N(CH_2OH)_2$$

Hexamethylolmelamine

Hexamethylolmelamine can further condense in the presence of an acid catalyst; ether linkages can also form (see "Urea Formaldehyde"). A wide variety of resins can be obtained by careful selection of pH, reaction temperature, reactant ratio, amino monomer, and extent of condensation. Liquid coating resins are prepared by reacting methanol or butanol with the initial methylolated products. These can be used to produce hard, solvent-resistant coatings by heating with a variety of hydroxy, carboxyl, and amide functional polymers to produce a cross-linked film.

Phenolics

Phenol–formaldehyde resin.   Phenol–formaldehyde resin is prepared from phenol by reaction with formaldehyde. Phenol is an enol, the 2-, 4-, and 6-positions of which are activated for reaction with an electrophile. Phenol is sequentially hydroxymethylated approximately as illustrated below. Dehydration of the phenolic hydroxymethyl groups affords a benzyl cation, a new electrophile that can react with another substituted or unsubstituted molecule of phenol. Both linear polymerization and cross-linking are possible, depending on the ratio of the reactants and the polymerization conditions.

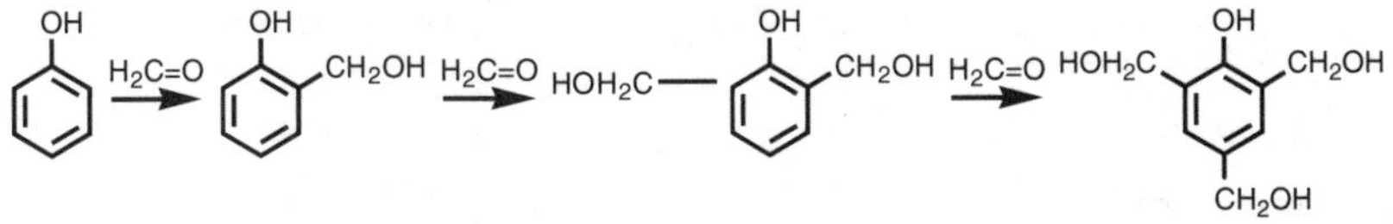

One-Stage Resins.   The ratio of formaldehyde to phenol is high enough to allow the thermosetting process to take place without the addition of other sources of cross-links.

Two-Stage Resins   The ratio of formaldehyde to phenol is low enough to prevent the thermosetting reaction from occurring during manufacture of the resin. At this point the resin is termed *novolac* resin. Subsequently, hexamethylenetetramine is incorporated into the material to act as a source of chemical cross-links during the molding operation (and conversion to the thermoset or cured state).

Polyamides

Nylon 6, 11, and 12.   This class of polymers is polymerized by addition reactions of ring compounds that contain both acid and amine groups on the monomer.

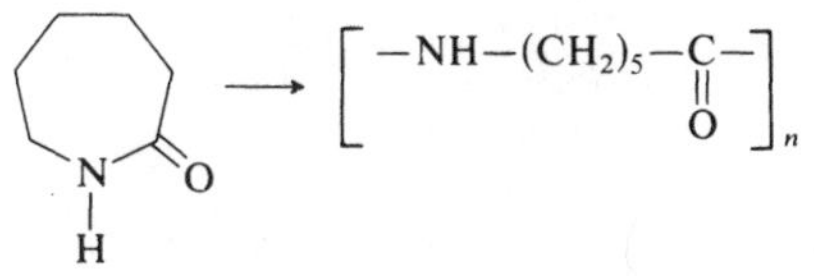

Nylon 6 is polymerized from 2-oxohexamethyleneimine (6 carbons); nylon 11 and 12 are made this way from 11- and 12-carbon rings, respectively.

Nylon 6/6, 6/9, and 6/12.   As illustrated below, nylon 6/6 is polymerized from 1,6-hexanedioic acid (six carbons) and 1,6-hexanediamine (six carbons).

$$\text{HOOC}-(\text{CH}_2)_4-\text{COOH} + \text{H}_2\text{N}-\text{CH}_2-(\text{CH}_2)_4-\text{CH}_2-\text{NH}_2 \rightarrow$$

1,6-Hexanedioic acid   1,6-Hexanediamine

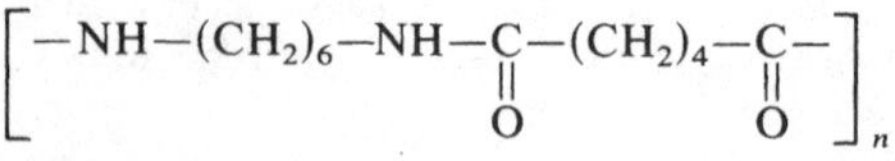

$$\left[-\text{NH}-(\text{CH}_2)_6-\text{NH}-\underset{\text{O}}{\overset{\text{O}}{\underset{\|}{\text{C}}}}-(\text{CH}_2)_4-\underset{\text{O}}{\overset{\text{O}}{\underset{\|}{\text{C}}}}-\right]_n$$

Poly(hexamethylene 1,6- hexanediamide)

Other nylons are made this way from different combinations of monomers to produce types 6/9, 6/10, and 6/12.

Nylon 6 and 6/6 possess the maximum stiffness, strength, and heat resistance of all the types of nylon. Type 6/6 has a higher melt temperature, whereas type 6 has a higher impact resistance and better processibility. At a sacrifice in stiffness and heat resistance, the higher analogs of nylon are useful primarily for improved chemical resistance in certain environments (acids, bases, and zinc chloride solutions) and for lower moisture absorption.

Aromatic nylons, $[-\text{NH}-\text{C}_6\text{H}_4-\text{CO}-]_n$, (also called aramids) have specialty uses because of their improved clarity.

Poly (amide-imide)

Poly(amide-imide) is the condensation polymer of 1,2,4-benzenetricarboxylic anhydride and various aromatic diamines and has the general structure:

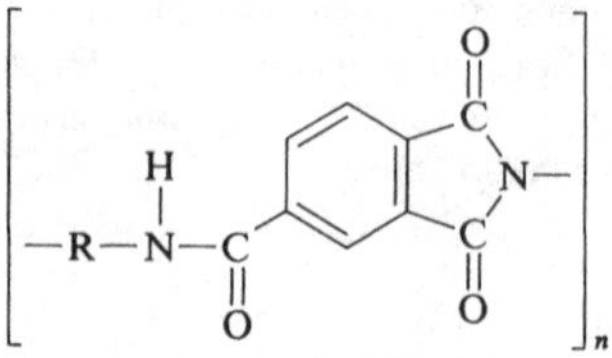

It is characterized by high strength and good impact resistance, and retains its physical properties at temperatures up to to 260°C. Its radiation (gamma) resistance is good.

Polycarbonate

Polycarbonate is a polyester in which dihydric (or polyhydric) phenols are joined through carbonate linkages. The general-purpose type of polycarbonate is based on 2,2-bis(4′-hydroxybenzene) propane (bisphenol A) and has the general structure:

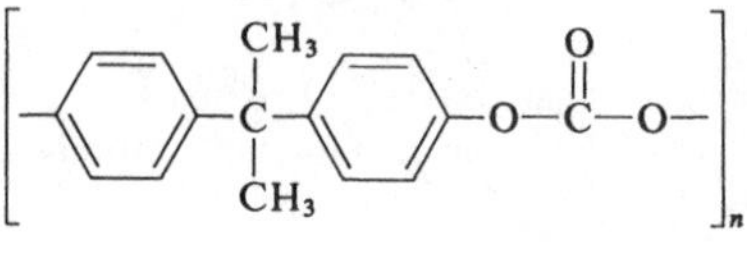

Polycarbonates are the toughest of all thermoplastics. They are window-clear, amazingly strong and rigid, autoclavable, and nontoxic. They have a brittleness temperature of $-135\,°C$.

Polyester

Poly(butylene terephthalate).   Poly(butylene terephthalate) is prepared in a condensation reaction between dimethyl terephthalate and 1,4-butanediol and its repeating unit has the general structure

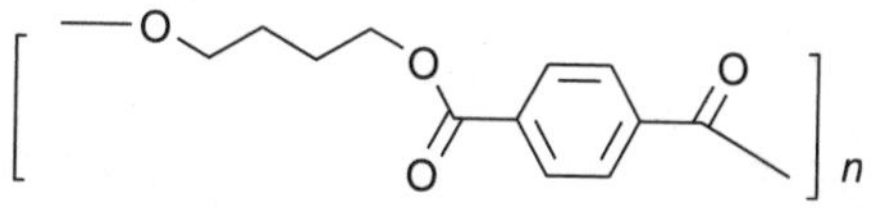

This thermoplastic shows good tensile strength, toughness, low water absorption, and good frictional properties, plus good chemical resistance and electrical properties.

Poly(ethylene terephthalate).   Poly(ethylene terephthalate) is prepared by the reaction of either terephthalic acid or dimethyl terephthalate with ethylene glycol, and its repeating unit has the general structure

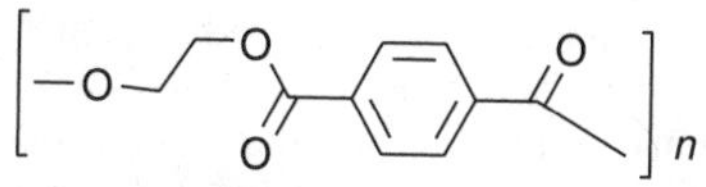

The resin has the ability to be oriented by a drawing process and crystallized to yield a high-strength product.

Unsaturated Polyesters.   Unsaturated polyesters are produced by reaction between two types of dibasic acids, one of which is unsaturated, and an alcohol to produce an ester. Double bonds in the body of the unsaturated dibasic acid are obtained by using maleic anhydride or fumaric acid.

PCTA Copolyester.   Poly(1,4-cyclohexanedimethylene terephthalic acid) (PCTA) copolyester is a polymer of cyclohexanedimethanol and terephthalic acid, with another acid substituted for a portion of the terephthalic acid otherwise required. It has the following formula:

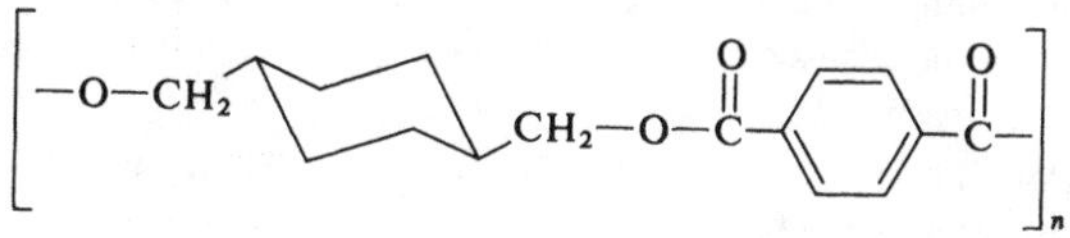

Polyimides.   Polyimides have the following formula:

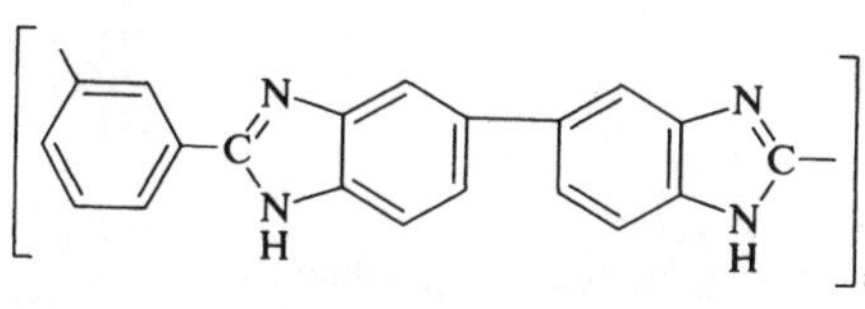

They are used as high-temperature structural adhesives since they become rubbery rather than melt at about 300°C.

Poly(methylpentene)

Poly(methylpentene) is obtained by a Ziegler-type catalytic polymerization of 4-methyl-1-pentene.

Its key properties are its excellent transparency, rigidity, and chemical resistance, plus its resistance to impact and to high temperatures. It withstands repeated autoclaving, even at 150°C.

Polyolefins

Polyethylene. Polymerization of ethylene results in an essentially straight-chain high-molecular-weight hydrocarbon.

$$CH_2{=}CH_2 \rightarrow [-CH_2-CH_2-]_n$$

Branching occurs to some extent and can be controlled. Minimum branching results in a "high-density" polyethylene because of its closely packed molecular chains. More branching gives a less compact solid known as "low-density" polyethylene.

A key property is its chemical inertness. Strong oxidizing agents eventually cause some oxidation, and some solvents cause softening or swelling, but there is no known solvent for polyethylene at room temperature. The brittleness temperature is $-100°C$ for both types. Polyethylene has good low-temperature toughness, low water absorption, and good flexibility at subzero temperatures.

Polypropylene. The polymerization of propylene results in a polymer with the following structure:

$$CH_2{=}CH{-}CH_3 \rightarrow \left[\begin{array}{c} -CH_2-CH- \\ | \\ CH_3 \end{array}\right]_n$$

The desired form in homopolymers is the isotactic arrangement (at least 93% is required to give the desired properties). Copolymers have a random arrangement. In block copolymers a secondary reactor is used where active polymer chains can further polymerize to produce segments that use ethylene monomer.

Polypropylene is translucent and autoclavable and has no known solvent at room temperature. It is slightly more susceptible to strong oxidizing agents than polyethylene.

Polybutylene. Polybutylene is composed of linear chains having an isotactic arrangement of ethyl side groups along the chain backbone.

$$CH_2{=}CH{-}CH_2{-}CH_3 \rightarrow \left[\begin{array}{c} -CH_2-CH- \\ | \\ CH_2 \\ | \\ CH_3 \end{array}\right]_n$$

It has a helical conformation in the stable crystalline form.

Polybutylene exhibits high tear, impact, and puncture resistance. It also has low creep, excellent chemical resistance, and abrasion resistance with coilability.

Ionomer. Ionomer is the generic name for polymers based on sodium or zinc salts of ethylene–methacrylic acid copolymers in which interchain ionic bonding, occurring randomly between the long-chain polymer molecules, produces solid-state properties.

The abrasion resistance of ionomers is outstanding, and ionomer films exhibit optical clarity. In composite structures ionomers serve as a heat-seal layer.

Poly(phenylene sulfide)

Poly(phenylene sulfide) has the following formula:

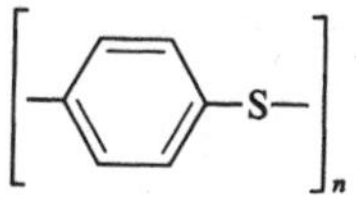

The recurring *para*-substituted benzene rings and sulfur atoms form a symmetrical rigid backbone.

The high degree of crystallization and the thermal stability of the bond between the benzene ring and sulfur are the two properties responsible for the polymer's high melting point, thermal stability, inherent flame retardance, and good chemical resistance. There are no known solvents of poly(phenylene sulfide) that can function below 205 °C.

Polyurethane

Foams. Polyurethane foams are prepared by the polymerization of polyols with isocyanates. One of the most commonly used reactive isocyanates toluenediisocyanate, TDI. It is made from toluene by nitration and then reduction followed by treatment with phosgene. The isocyanate residue reacts readily with alcohols to give carbamates (urethanes) or amines to give ureas.

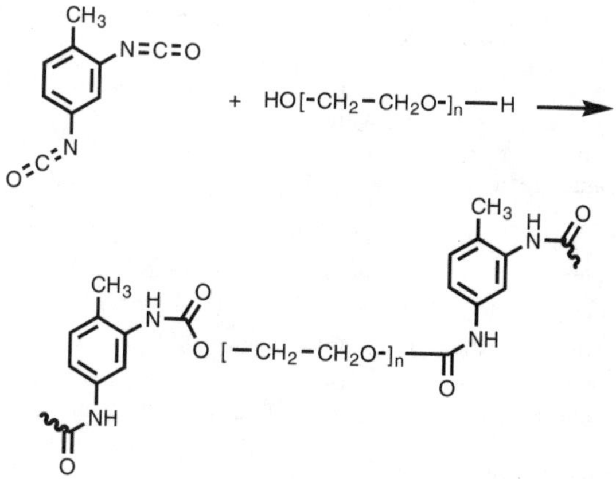

Commonly used isocyanates are toluenediisocyanate, methylenediphenylisocyanate, and polymeric isocyanates. Polyols used are macroglycols based on either polyester or polyether. The former [poly(ethylene phthalate) or poly(ethylene 1,6-hexanedioate)] have hydroxyl groups that are free to react with the isocyanate. Most flexible foam is made from 80/20 toluene diisocyanate (which refers to the ratio of 2,4-toluenediisocyanate to 2,6-toluene diisocyanate). High-resilience foam contains about 80% 80/20 toluenediisocyanate and 20% poly(methylene diphenyl isocyanate), while semiflexible foam is almost always 100% poly(methylene diphenyl isocyanate). Much of the latter reacts by trimerization to form isocyanurate rings.

Flexible foams are used in mattresses, cushions, and safety applications. Rigid and semiflexible foams are used in structural applications and to encapsulate sensitive components to protect them against shock, vibration, and moisture. Foam coatings are tough, hard, flexible, and chemically resistant.

Elastomeric Fiber. Elastomeric fibers are prepared by the polymerization of polymeric polyols with diisocyanates.

toluenediisocyanate

The structure of elastomeric fibers is similar to that illustrated for polyurethane foams.

SILICONES

Silicones are formed in the following multistage reaction:

$$R_2SiCl_2 + 2H_2O \rightarrow R_2Si(OH)_2 + 2HCl$$
$$\downarrow$$
$$[-Si(R)_2-O-]_n$$

The silanols formed above are unstable and undergo dehydration. On polycondensation, they give polysiloxanes (or silicones) which are characterized by their three-dimensional branched-chain structure. Various organic groups introduced within the polysiloxane chain impart certain characteristics and properties to these resins.

Methyl groups impart water repellency, surface hardness, and noncombustibility.

Phenyl groups impart resistance to temperature variations, flexibility under heat, resistance to abrasion, and compatibility with organic products.

Vinyl groups strengthen the rigidity of the molecular stucture by creating easier cross-linkage of molecules.

Methoxy and alkoxy groups facilitate cross-linking at low temperatures.

Oils and gums are nonhighly branched or straight-chain polymers whose viscosity increases with the degree of polycondensation.

Styrenics

Polystyrene.   Polystyrene has the following formula:

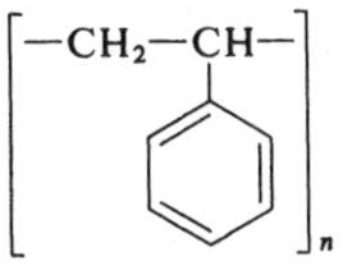

Polystyrene is rigid with excellent dimensional stability, has good chemical resistance to aqueous solutions, and is an extremely clear material.

Impact polystyrene contains polybutadiene added to reduce brittleness. The polybutadiene is usually dispersed as a discrete phase in a continuous polystyrene matrix. Polystyrene can be grafted onto rubber particles, which assures good adhesion between the phases.

Acrylonitrile–Butadiene–Styrene (ABS) Copolymers.   This basic three-monomer system can be tailored to yield resins with a variety of properties. Acrylonitrile contributes heat resistance, high strength, and chemical resistance. Butadiene contributes impact strength, toughness, and retention of low-temperature properties. Styrene contributes gloss, processibility, and rigidity. ABS polymers are composed of discrete polybutadiene particles grafted with the styrene–acrylonitrile copolymer; these are dispersed in the continuous matrix of the copolymer.

Styrene–Acrylonitrile (SAN) Copolymers.   SAN resins are random, amorphous copolymers whose properties vary with molecular weight and copolymer composition. An increase in molecular weight or in acrylonitrile content generally enhances the physical properties of the copolymer but at some loss in ease of processing and with a slight increase in polymer color.

SAN resins are rigid, hard, transparent thermoplastics which process easily and have good dimensional stability—a combination of properties unique in transparent polymers.

Sulfones

Below are the formulas for three polysulfones.

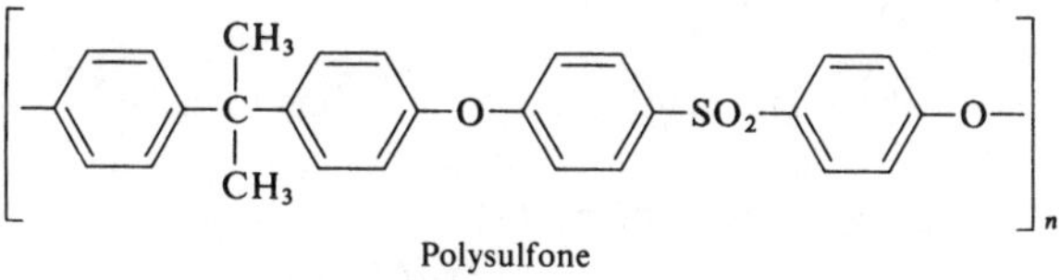

Polysulfone

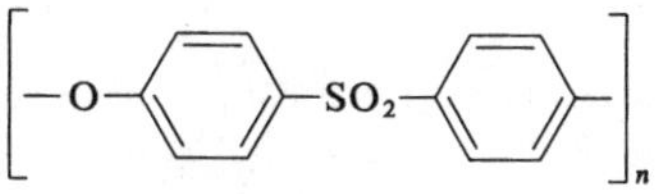

Poly(ester sulfone)

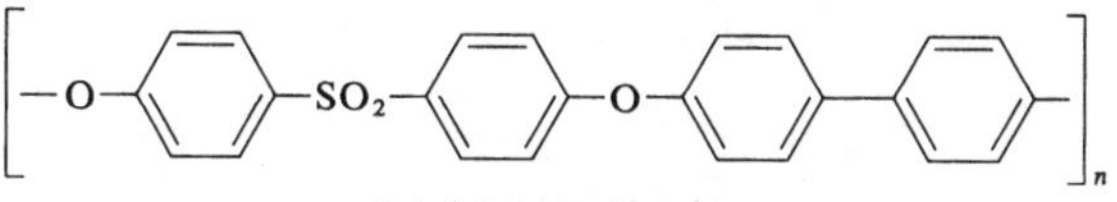

Poly(phenyl sulfone)

The isopropylidene linkage imparts chemical resistance, the ether linkage imparts temperature resistance, and the sulfone linkage imparts impact strength. The brittleness temperature of polysulfones is $-100\,°C$. Polysulfones are clear, strong, nontoxic, and virtually unbreakable. They do not hydrolyze during autoclaving and are resistant to acids, bases, aqueous solutions, aliphatic hydrocarbons, and alcohols.

Thermoplastic Elastomers

Polyolefins. In these thermoplastic elastomers the hard component is a crystalline polyolefin, such as polyethylene or polypropylene, and the soft portion is composed of ethylene–propylene rubber. Attractive forces between the rubber and resin phases serve as labile cross-links. Some contain a chemically cross-linked rubber phase that imparts a higher degree of elasticity.

Styrene–Butadiene–Styrene Block Copolymers. Styrene blocks associate into domains that form hard regions. The midblock, which is normally butadiene, ethylene–butene, or isoprene blocks, forms the soft domains. Polystyrene domains serve as cross-links.

Polyurethanes. The hard portion of polyurethane consists of a chain extender and polyisocyanate. The soft component is composed of polyol segments.

Polyesters. The hard portion consists of copolyester, and the soft portion is composed of polyol segments.

Vinyl

Poly(vinyl chloride) (PVC). Polymerization of vinyl chloride results in the formation of a polymer with the following formula:

$$CH_2\!=\!CHCl \;\rightarrow\; \left[-CH_2-\underset{\underset{Cl}{|}}{CH}-\right]_n$$

When blended with phthalate ester plasticizers, PVC becomes soft and pliable.

Its key properties are good resistance to oils and a very low permeability to most gases.

Poly(vinyl acetate). Poly(vinyl acetate) has the following formula:

$$\left[-CH_2-\underset{\underset{O-CO-CH_3}{|}}{CH}-\right]_n$$

Poly(vinyl acetate) is used in latex water paints because of its weathering, quick-drying, recoatability, and self-priming properties. It is also used in hot-melt and solution adhesives.

Poly(vinyl alcohol). Poly(vinyl alcohol) has the following formula:

$$\left[-CH_2-\underset{\underset{OH}{|}}{CH}-\right]_n$$

It is used in adhesives, paper coating and sizing, and textile warp size and finishing applications.

Poly(vinyl butyral). Poly(vinyl butyral) is prepared according to the following reaction:

$$\left[-CH_2-\underset{\underset{OH}{|}}{CH}-\right]_n + CH_3CH_2CH_2CHO \rightarrow \left[\begin{array}{c}-CH_2-CH-CH_2-CH- \\ \underset{|}{O}-CH\underline{\quad}O \\ CH_2-CH_2-CH_3\end{array}\right]_n$$

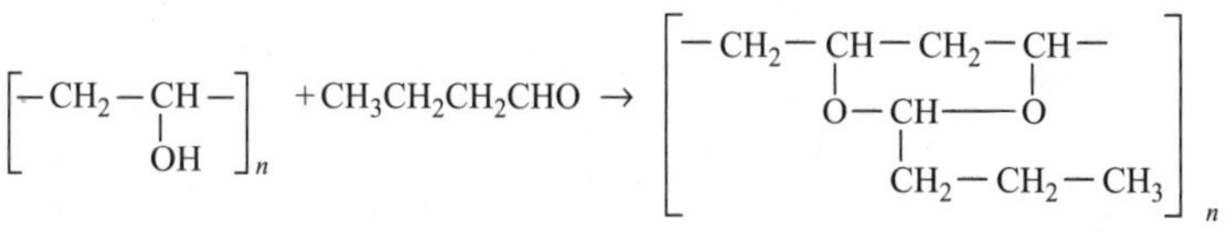

Its key characteristics are its excellent optical and adhesive properties. It is used as the interlayer film for safety glass.

Poly(vinylidene chloride). Poly(vinylidene chloride) is prepared according to the following reaction:

$$CH_2=CCl_2 + CH_2=CHCl \rightarrow [-CH_2-CCl_2-CH_2-CHCl-]_n$$

Random copolymer

Urea Formaldehyde

The reaction of urea with formaldehyde yields the following products, which are used as monomers in the preparation of urea formaldehyde resin.

$$H_2N-CO-NH_2 + H_2CO \rightarrow H_2N-CO-NH-CH_2OH$$

$$+ HOCH_2-NH-CO-NH-CH_2OH$$

The reaction conditions can be varied so that only one of these monomers is formed. 1-Hydroxymethylurea and 1,3-bis(hydroxymethyl)urea condense in the presence of an acid catalyst to produce urea formaldehyde resins. A wide variety of resins can be obtained by careful selection of the pH, reaction temperature, reactant ratio, amino monomer, and degree of polymerization. If the reaction is carried far enough, an infusible polymer network is produced.

Liquid coating resins are prepared by reacting methanol or butanol with the initial hydroxymethylureas. Ether exchange reactions between the amino resin and the reactive sites on the polymer produce a cross-linked film.

TABLE 10.2 Properties of Commercial Plastics

| Properties | Acetal | | | | 21% poly(tetrafluoroethylene)-filled homopolymer |
	Homopolymer	Copolymer	20% glass-reinforced homopolymer	25% glass-reinforced copolymer	
Physical					
Melting temperature, °C					
Crystalline	175	175	181	175	181
Amorphous					
Specific gravity	1.42	1.41	1.56	1.61	1.54
Water absorption (24 h), %	0.25–0.40	0.22	0.25	0.29	0.20
Dielectric strength, KV · mm^{-1}	19.7	19.7	19.3	22.8	15.7
Electrical					
Volume (dc) resistivity, ohm-cm	10^{15}	10^{15}	5×10^{14}		3×10^{16}
Dielectric constant (60 Hz)	3.7	3.7	3.9		3.1
Dielectric constant (10^6 Hz)	3.7	3.7	3.9		3.1
Dissipation (power) factor (60 Hz)					
Dissipation factor (10^6 Hz)	0.005	0.005	0.005		0.005
Mechanical					
Compressive modulus, 10^3 lb · in^{-2}	670	450			

Property					
Compressive strength, rupture or 1% yield, $10^3\,\text{lb}\cdot\text{in}^{-2}$	5.29	16 (10% yield)	18 (10% yield)	17 (10% yield)	13 (10% yield)
Elongation at break, %	25–75	40–75	7	3	15–22
Flexural modulus at 23°C, $10^3\,\text{lb}\cdot\text{in}^{-2}$	380–430	375	730	1100	340–350
Flexural strength, rupture or yield, $10^3\,\text{lb}\cdot\text{in}^{-2}$	14	13	15	28	
Hardness, Rockwell (or Shore)	M94	M78	M90	M79	M78
Impact strength (Izod) at 23°C, $\text{J}\cdot\text{m}^{-1}$	69–123	53–80	43	96	37–64
Tensile modulus, $10^3\,\text{lb}\cdot\text{in}^{-2}$	520	410	1000	1250	
Tensile strength at break, $10^3\,\text{lb}\cdot\text{in}^{-2}$	10	10	8.5	18.5	7.6
Tensile yield strength, $10^3\,\text{lb}\cdot\text{in}^{-2}$	9.5–12	8.5			6.9–7.6
Thermal					
Burning rate, $\text{mm}\cdot\text{min}^{-1}$	27.9				
Coefficient of linear thermal expansion, $10^{-6}°\text{C}$	100	85	36–81		75
Deflection temperature under flexural load ($264\,\text{lb}\cdot\text{in}^{-2}$), °C	124	110	157	163	
Maximum recommended service temperature, °C	84				100
Specific heat, $\text{cal}\cdot\text{g}^{-1}$	0.35				
Thermal conductivity, $\text{W}\cdot\text{m}^{-1}\cdot\text{K}^{-1}$	0.23	0.23			

TABLE 10.2 Properties of Commercial Plastics (*continued*)

Properties	Acrylic				Alkyd, molded	Alloy	
	Poly(methyl methacrylate)	Cast sheet	Impact-modified	Heat-resistant		Acrylic poly(vinyl chloride) alloy	Acrylonitrile–butadiene–styrene poly(vinyl chloride) alloy
Physical							
Melting temperature, °C							
Crystalline							
Amorphous	90–105	90–105	80–100	100–125		105	
Specific gravity	1.17–1.20	1.18–1.20	1.11–1.18	1.16–1.19	2.22–2.24		
Water absorption (24h), %	0.1–0.4	0.2–0.4	0.2–0.8	0.2–0.3		0.06	
Dielectric strength, KV·mm^{-1}	15.7–19.9	17.7–21.7	15.0–19.9	15.7–19.9		>15.7	19.7
Electrical							
Volume (dc) resistivity, ohm-cm	>10^{14}	>10^{14}					
Dielectric constant (60 Hz)	3.3–4.5	3.5–4.5			3.8–5.0		
Dielectric constant (10^6 Hz)		3.0–3.5			3.6–4.7		
Dissipation (power) factor (60 Hz)		0.04–0.06			0.012–0.026		
Dissipation factor (10^6 Hz)		0.02–0.03			0.01–0.016		
Mechanical							
Compressive modulus, 10^4 lb·in^{-2}	370–460	390–475	240–370	350–460		330–400	

Property							
Compressive strength, rupture or 1% yield, 10^3 lb·in^{-2}	12–18	11–19	4–14	17	16–20	8.4	
Elongation at break, %	2–10	2–7	20–70	3–5		100	
Flexural modulus at 23°C, 10^3 lb·in^{-2}	420–460	390–475	200–380	460–500		330–400	340
Flexural strength, rupture or yield, 10^3 lb·in^{-2}	13–19	12–17	7–13	12–16		10.7	9.6
Hardness, Rockwell (or Shore)	M85–M105	M80–M100	R105–R120	M95–M105	E76	R99–R105	R100
Impact strength (Izod) at 23°C, J·m^{-1}					27–240		
Tensile modulus, 10^3 lb·in^{-2}	16–27	16–21	43–133	16–21		800	560
Tensile strength at break, 10^3 lb·in^{-2}	380–450	350–450	200–400	350–460	4.5–6.5	330–335	330
Tensile yield strength, 10^3 lb·in^{-2}	7–11	8–11	5–9	10	10–13	6.5	5.8
Thermal Burning rate, mm·min^{-1}		0.5–2.2			Self-extinguishing		
Coefficient of linear thermal expansion, 10^{-6}°C	50–90	50–90	50–80	50–60	40–55		46
Deflection temperature under flexural load (264 lb·in^{-2}), °C	74–99	71–102	74–95	88–104	177–204	71	
Maximum recommended service temperature, °C		60–71			220		
Specific heat, cal·g^{-1}	0.36	0.35					
Thermal conductivity, W·m^{-1}·K^{-1}	0.17–0.25	0.17–0.25	0.17–0.21	0.19			

TABLE 10.2 Properties of Commercial Plastics (*continued*)

Properties	Alloy	Allyl			Cellulosic		
	Polycarbonate acrylonitrile-butadiene-styrene alloy	Allyl-diglycol-carbonate polymer	Diallyl phthalate molding		Cellulose acetate		Cellulose-acetate-butyrate resin
			Glass-filled	Mineral-filled	Sheet	Molding	Sheet
Physical							
Melting temperature, °C							
Crystalline							
Amorphous	150	Thermoset	Thermoset	Thermoset	230	230	140
Specific gravity	1.12–1.20	1.3–1.4	1.7–2.0	1.65–1.85	1.27–1.34	1.29–1.34	1.15–1.22
Water absorption (24 h), %	0.21–0.24	0.2	0.12–0.35	0.2–0.5	2–7	1.7–6.5	0.9–2.2
Dielectric strength, kV·mm^{-1}	17.7	15.0	15.7–17.7	15.7–17.7	11–24	9–24	9–18
Electrical							
Volume (dc) resistivity, ohm-cm					10^{10}–10^{13}	10^{10}–10^{13}	10^{10}–10^{12}
Dielectric constant (60 Hz)					3.4–7.4	3.5–7.5	3.7–4.3
Dielectric constant (10^6 Hz)					3.2–7.0	3.2–7.0	3.3–3.8
Dissipation (power) factor (60 Hz)					0.01–0.06	0.01–0.06	0.01–0.04
Dissipation factor (10^6 Hz)					0.01–0.06	0.01–0.10	0.01–0.04
Mechanical							
Compressive modulus, 10^3 lb·in^{-2}		300					

Property							
Compressive strength, rupture or 1% yield, 10^3 lb·in^{-2}	11	21–23	25–35	20–32	22–33	25–36	50–100
Elongation at break, %	10–15		3–5	3–5	17–40	6–40	
Flexural modulus at 23°C, 10^3 lb·in^{-2}	300–400	250–330	1200–1500	1000–1400			740–1300
Flexural strength, rupture or yield, 10^3 lb·in^{-2}	13.0–13.7	6–13	9–20	8.5–11	6–10	2–16	4–9
Hardness, Rockwell (or Shore)	R117	M95–M100	E80–E87	E61	R85–R120	R100–R123	R50–R95
Impact strength (Izod) at 23°C, J·m^{-1}	560	11–21	21–800	16–43	107–454	53–214	133–288
Tensile modulus, 10^3 lb·in^{-2}	370–380	300	1400–2200	1200–2200			200–250
Tensile strength at break, 10^3 lb·in^{-2}	7.0–7.3	5–6	6–11	5–8	4.5–8.0	1.9–9.0	2.6–6.9
Tensile yield strength, 10^3 lb·in^{-2}	8.5				2.2–7.4	4.1–7.6	
Thermal							
Burning rate, mm·min^{-1}	63–67	5.4–9.6	0.68–2.4	2.8		1.3–3.8	1.3–3.8
Coefficient of linear thermal expansion, 10^{-6}°C	104–116	60–88			100–150	80–180	110–170
Deflection temperature under flexural load (264 lb·in^{-2}), °C			165–288+	160–288	44–91	51–98	49–58
Maximum recommended service temperature, °C							
Specific heat, cal·g^{-1}					0.3–0.4	0.3–0.42	0.3–0.4
Thermal conductivity, W·m^{-1}·K^{-1}	0.25–0.38	0.20–0.21	0.21–0.63	0.30–1.04	0.17–0.34	0.17–0.34	0.17–0.34

TABLE 10.2 Properties of Commercial Plastics (*continued*)

Properties	Cellulosic				Chlorinated polyether	Epoxy Bisphenol	
	Cellulose-acetate–butyrate resin, molding	Cellulose-acetate–propionate resin, molding	Ethyl cellulose	Cellulose nitrate		Glass-fiber-reinforced	Mineral-filled
Physical							
Melting temperature, °C							
Crystalline						Thermoset	Thermoset
Amorphous	140	190	135		125		
Specific gravity	1.15–1.22	1.17–1.24	1.09–1.17	1.35–1.40	1.4	1.6–2.0	1.6–2.1
Water absorption (24h), %	0.9–2.2	1.2–2.8	0.8–1.8			0.04–0.20	0.03–0.20
Dielectric strength, kV·mm^{-1}	9–13	12–17.7	13.8–19.7			9.8–15.7	9.8–15.7
Electrical							
Volume (dc) resistivity, ohm-cm	10^{10}–10^{12}			10^{10}			
Dielectric constant (60 Hz)	3.5–6.4			7.0–7.5			
Dielectric constant (10^6 Hz)	3.2–6.2		3.01	6.6			
Dissipation (power) factor (60 Hz)	0.01–0.04						
Dissipation factor (10^6 Hz)	0.01–0.04						
Mechanical							
Compressive modulus, 10^3 lb·in^{-2}						3000	

Compressive strength, rupture or 1% yield, 10^3 lb·in^{-2}	2.1–7.5	2.4–7.0	5–40	2.1–8.0	600–800	18000–40000	18000–40000
Elongation at break, %	40–88	29–100		40–45		4	
Flexural modulus at 23°C, 10^3 lb·in^{-2}	90–300	120–350				2–4.5	
Flexural strength, rupture or yield, 10^3 lb·in^{-2}	1.8–9.3	2.9–11.4	4–12	9–11	5	8–30	6–18
Hardness, Rockwell (or Shore)	R31–R116	R10–R122	R50–R115	R95–R115	R100	M100–M112	M100–M112
Impact strength (Izod) at 23°C, J·m^{-1}	53–582	27 to no break	21	267–374	21	16–533	16–22
Tensile modulus, 10^3 lb·in^{-2}	50–200	60–215		190–220		3	
Tensile strength at break, 10^3 lb·in^{-2}	2.6–6.9	2.0–7.8	2–8	7–8	1.5–1.8	5–20	4–10
Tensile yield strength, 10^3 lb·in^{-2}							
Thermal							
Burning rate, mm·min^{-1}	1.3–3.8				Self-extinguishing		
Coefficient of linear thermal expansion, 10^{-6}°C	110–170	110–170	100–200	80–120	6.6	11–50	20–60
Deflection temperature under flexural load (264 lb·in^{-2}), °C	44–94	44–109	45–88	60–71	185	107–260	107–260
Maximum recommended service temperature, °C					255		
Specific heat, cal·g^{-1}	0.3–0.4			0.31–0.41			
Thermal conductivity, W·m^{-1}·K^{-1}	0.17–0.30	0.17–0.30	0.16–0.30	0.23		0.17–0.42	0.17–1.48

TABLE 10.2 Properties of Commercial Plastics (*continued*)

Properties	Epoxy		Novolac resin	Poly(tetrafluoroethylene)		Fluorocarbon	
	Casting resin						
	Unfilled	Flexible	Mineral-filled	Granular	Glass-fiber-reinforced	Poly(chlorotrifluoroethylene)	Perfluoroalkoxy
Physical							
Melting temperature, °C							
Crystalline	Thermoset	Thermoset	Thermoset	327	327	220	310
Amorphous							
Specific gravity	1.11–1.40	1.05–1.35	1.7–2.1	2.14–2.20	2.2–2.3	2.1–2.2	2.12–2.17
Water absorption (24h), %	0.08–0.15	0.27–0.50	0.05–0.2	0.01		0.03	
Dielectric strength, kV·mm^{-1}	11.8–19.7	9.3–15.8	11.8–13.8	18.9	12.6	19.7–23	19.7
Electrical							
Volume (dc) resistivity, ohm·cm	10^{12}–10^{17}			10^{18}		10^{18}	
Dielectric constant (60 Hz)	3.5–5.0			2.1		2.3–2.7	
Dielectric constant (10^6 Hz)	3.5–5.0			2.1		2.3–2.5	
Dissipation (power) factor (60 Hz)				0.0002		0.001	
Dissipation factor (10^6 Hz)				0.0002		0.005	
Mechanical							
Compressive modulus, 10^3 lb·in^{-2}				60			

Compressive strength, rupture or 1% yield, 10^3 lb·in^{-2}	15–25	1–14	30	1.7	200–300	4.6–7.4	300
Elongation at break, %	3–6	20–70	2–4	200–400		80–250	
Flexural modulus at 23°C, 10^3 lb·in^{-2}			2000	80	235	120	
Flexural strength, rupture or yield, 10^{-3} lb·in^{-2}	13–21	1–13	16–20		2	7.4–9.3	
Hardness, Rockwell (or Shore)	M80–M110			(D50–D55)	(D60–D70)	R75–R95	(D64)
Impact strength (Izod) at 23°C, J·m^{-1}	10.7–53	187–267	21	160	144	133–160	No break
Tensile modulus, 10^3 lb·in^{-2}	350	1–350		58–80		150–300	
Tensile strength at break, 10^3 lb·in^{-2}	4–13	2–10	6–12	2–5	2–2.7	4.5–6	4–4.3
Tensile yield strength, 10^3 lb·in^{-2}			30				
Thermal							
Burning rate, mm·min^{-1}				Self-extinguishing	Self-extinguishing	Self-extinguishing	
Coefficient of linear thermal expansion, 10^{-6}°C	45–65	20–100	22–30	100	77–100	70	
Deflection temperature under flexural load (264 lb·in^{-2}), °C	46–288	23–121	149–260	121 (66 lb·in^{-2})		126 (66 lb·in^{-2})	74 (66 lb·in^{-2})
Maximum recommended service temperature, °C				260		200	
Specific heat, cal·g^{-1}				0.25		0.22	
Thermal conductivity, W·m^{-1}·K^{-1}	0.17–0.21			0.25	0.34–0.40	0.19–0.22	0.25

TABLE 10.2 Properties of Commercial Plastics (*continued*)

Properties	Fluorocarbon					Melamine formaldehyde	
	Fluorinated ethylene-propylene resin	Poly(vinylidene fluoride)	Ethylene–tetrafluoroethylene copolymer		Ethylene–chlorotrifluoro-ethylene copolymer	Cellulose-filled	Glass-fiber-reinforced
			Unfilled	Glass-fiber-reinforced			
Physical							
Melting temperature, °C							
Crystalline	275	156	270	270	245	Thermoset	Thermoset
Amorphous							
Specific gravity	2.14–2.17	1.75–1.78	1.7	1.8	1.68	1.47–1.52	1.5–2.0
Water absorption (24h), %	<0.01	0.04–0.06	0.03	0.02	0.01	0.1–0.8	0.09–1.3
Dielectric strength, kV·mm^{-1}	20–24	10	16	17	19	11–16	5–15
Electrical							
Volume (dc) resistivity, ohm-cm							
Dielectric constant (60 Hz)	2.1	8–9	2.6		2.6		
Dielectric constant (10^6 Hz)	2.1	8–9	2.6		2.6		
Dissipation (power) factor (60 Hz)		High					
Dissipation factor (10^6 Hz)		High					
Mechanical							
Compressive modulus, 10^3 lb·in^{-2}		120	120	1 200	240		

Property							
Compressive strength, rupture or 1% yield, 10^3 lb·in^{-2}	2.2	8.7–10	7.1	10	200–300	33–45	20–35
Elongation at break, %	250–330	25–500	100–400	8	240	0.6–1.0	0.6
Flexural modulus at 23°C, 10^3 lb·in^{-2}	80–95	200	200	950		1100	
Flexural strength, rupture or yield, 10^3 lb·in^{-2}		8.6–11	5.5	10.7	7	9–16	14–23
Hardness, Rockwell (or Shore)	(D60–D65)	(D80)	R50 (D75)	R74	R95	M115–M125	M115
Impact strength (Izod) at 23°C, J·m^{-1}	No break 50	192–214 120	No break 120	480 1200	No break 240	11–21 1.1–1.4	32–961 1.6–2.4
Tensile modulus, 10^3 lb·in^{-2}							
Tensile strength at break, 10^3 lb·in^{-2}	2.7–3.1	5.5–7.4	6.5	12	7	5–13	5–10.5
Tensile yield strength, 10^3 lb·in^{-2}							
Thermal							
Burning rate, mm·min^{-1}	Not combustible	Not combustible	Not combustible	Not combustible	Not combustible	Self-extinguishing	Self-extinguishing
Coefficient of linear thermal expansion, 10^{-6}°C	83–105	85	59	10–32	80	40–45	15–28
Deflection temperature under flexural load (264 lb·in^{-2}), °C	70 (66 lb·in^{-2})	80–90	71	210	77	177–199	190–204
Maximum recommended service temperature, °C	205	150				210	
Specific heat, cal·g^{-1}	0.28						
Thermal conductivity, W·m^{-1}·K^{-1}	0.25	0.19–0.24	0.24		0.16	0.27–0.41	0.41–0.49

TABLE 10.2 Properties of Commercial Plastics (*continued*)

Properties	Melamine phenolic, woodflour- and cellulose-filled	Nitrile	Phenolic				
			Unfilled	Woodflour-filled	Glass-fiber-reinforced	Cellulose-filled	Mineral-filled
	Thermoset		Thermoset	Thermoset	Thermoset	Thermoset	Thermoset
Physical							
Melting temperature, °C							
Crystalline							
Amorphous		95					
Specific gravity	1.5–1.7	1.15	1.24–1.32	1.37–1.46	1.69–2.0	1.38–1.42	1.42–1.84
Water absorption (24 h), %	0.3–0.65	0.28	0.1–0.36	0.3–1.2	0.03–1.2	0.5–0.9	0.1–0.3
Dielectric strength, kV·mm^{-1}	8.7–12.8	8.7–9.5	9.8–15.8	10.2–15.8	5.5–15.8	11.8–15	7.9–13.8
Electrical							
Volume (dc) resistivity, ohm-cm		1.9×10^{15}	1×10^{12} to 7×10^{12}				
Dielectric constant (60 Hz)			6.5–7.5				
Dielectric constant (10^6 Hz)			4.0–5.5				
Dissipation (power) factor (60 Hz)			0.10–0.15				
Dissipation factor (10^6 Hz)			0.04–0.05				
Mechanical							
Compressive modulus, 10^3 lb·in^{-2}							

Compressive strength, rupture or 1% yield, 10^3 lb·in^{-2}	26–30	12	18–32	25–31	26–70	22–31	22.5–34.6
Elongation at break, %	0.4–0.8	3–4	1.5–2.0	0.4–0.8	0.2	1–2	0.1–0.5
Flexural modulus at 23°C, 10^3 lb·in^{-2}	1000–1200	500–590	700–1500	1000–1200	2000–33000	900–1300	1000–2000
Flexural strength, rupture or yield, 10^3 lb·in^{-2}	8–10	14	11–17	7–14	15–60	5.5–11	11–14
Hardness, Rockwell (or Shore)	E95–E100	M72–M76	M93–M120	M100–M115	E54–E101	M95–M115	E88
Impact strength (Izod) at 23°C, J·m^{-1}	11–21	80–256	13–21	11–32	27–960	21–59	14–19
Tensile modulus, 10^3 lb·in^{-2}	800–1700	510–580	700–1500	800–1700	1900–3300		2400
Tensile strength at break, 10^3 lb·in^{-2}	6–8	9	6–9	5–9	7–18	3.5–6.5	6–9.7
Tensile yield strength, 10^3 lb·in^{-2}			12–15				
Thermal							
Burning rate, mm·min^{-1}			Self-extinguishing				
Coefficient of linear thermal expansion, 10^{-6}°C	10–40	66	68	30–45	8–21	20–31	19–26
Deflection temperature under flexural load (264 lb·in^{-2}), °C	140–154	73	74–80	149–188	177–316	149–177	320–246
Maximum recommended service temperature, °C							
Specific heat, cal·g^{-1}							
Thermal conductivity, W·m^{-1}·K^{-1}	0.17–0.30	0.26	0.15	0.17–0.34	0.34–0.59	0.25–0.38	0.42–0.57

TABLE 10.2 Properties of Commercial Plastics (*continued*)

Properties	Polyamide						
	Nylon 6		High-impact copolymer	Molding	Nylon 6/6		Nylon 6/6-nylon 6 copolymer
	Molding and extrusion	30–35% glass-fiber-reinforced			33% glass-fiber-reinforced	Molybdenum disulfide-filled	
Physical							
Melting temperature, °C							
Crystalline	216	216	216	265	265	265	240
Amorphous							
Specific gravity	1.12–1.14	1.35–1.42	1.08–1.17	1.13–1.15	1.38	1.15–1.17	1.08–1.14
Water absorption (24h), %	2.9	1.2	1.3–1.5	1.0–1.3	1.0	0.8–1.1	1.5–2.0
Dielectric strength, kV·mm^{-1}	15.8	15.8	22	24		14	15.8
Electrical							
Volume (dc) resistivity, ohm-cm	10^{12}			10^{12}–10^{15}			10^{10}
Dielectric constant (60 Hz)	9.8			4.0			16
Dielectric constant (10^6 Hz)	3.7			3.6			4
Dissipation (power) factor (60 Hz)	0.14			0.01–0.02			0.4
Dissipation factor (10^6 Hz)	0.12			0.02–0.03			0.1
Mechanical							
Compressive modulus, 10^3 lb·in^{-2}	250						

Property							
Compressive strength, rupture or 1% yield, 10^3 lb·in^{-2}	13–16	19	150–270	15 (yield)	24.9	12.5	
Elongation at break, %	30–100	3–6		60	3	15	40
Flexural modulus at 23°C, 10^3 lb·in^{-2}	390	1500	110–320	420	1300	450	150–410
Flexural strength, rupture or yield, 10^3 lb·in^{-2}	14	33	5–12	17	41	17	
Hardness, Rockwell (or Shore)	R119	M101	R81–R110	R120	M100	R119	R119
Impact strength (Izod) at 23°C, J·m^{-1}	32–53	160	96 to no break	43–53	117	240	37
Tensile modulus, 10^3 lb·in^{-2}	380	1450				550	150–410
Tensile strength at break, 10^3 lb·in^{-2}	11.8	25	7.5–11	12	28	13.7	7.4–12.4
Tensile yield strength, 10^3 lb·in^{-2}	8			8			
Thermal							
Burning rate, mm·min^{-1}	Self-extinguishing	Self-extinguishing	Self-extinguishing	Self-extinguishing	Self-extinguishing	Self-extinguishing	Self-extinguishing
Coefficient of linear thermal expansion, 10^{-6}°C	80–90	20–30	30–40	80	15–20	54	
Deflection temperature under flexural load (264 lb·in^{-2}), °C	68–85	210	45–54	75	249	127	77
Maximum recommended service temperature, °C	107			135			
Specific heat, cal·g^{-1}	0.4			0.4			
Thermal conductivity, W·m^{-1}·K^{-1}	0.24	0.24		0.24	0.22		

TABLE 10.2 Properties of Commercial Plastics (*continued*)

		Polyamide					
		Nylon 6/12					
Properties	Nylon 6/9, molding and extrusion	Molding	30–35% glass-fiber-reinforced	Nylon 11, molding and extrusion	Nylon 12, molding and extrusion	Aromatic nylon (aramid), molded and unfilled	Poly(amide-imide), unfilled
Physical							
Melting temperature, °C							
Crystalline	205	217	217	194	179	275	275
Amorphous							
Specific gravity	1.08–1.10	1.06–1.08	1.31–1.38	1.03–1.05	1.01–1.02	1.30	1.40
Water absorption (24h), %	0.5	0.4	0.2	0.3	0.25	0.6	0.28
Dielectric strength, kV·mm^{-1}	24	16	21	17	18	31	24
Electrical							
Volume (dc) resistivity, ohm-cm		10^{15}			10^{14}		
Dielectric constant (60 Hz)		4.0			3.8		
Dielectric constant (10^6 Hz)		3.5			3.0		
Dissipation (power) factor (60 Hz)		0.02			0.07		
Dissipation factor (10^6 Hz)		0.02			0.04		
Mechanical							
Compressive modulus, 10^3 lb·in^{-2}				180		290	413

Property							
Compressive strength, rupture or 1% yield, 10^3 lb·in^{-2}	1125	2.4			7.5	30	40
Elongation at break, %		150	4	300	300	5	12–18
Flexural modulus at 23°C, 10^3 lb·in^{-2}	290	290	1120	150	165	640	664
Flexural strength, rupture or yield, 10^3 lb·in^{-2}					1.5	25.8	30
Hardness, Rockwell (or Shore)	R111	R114	E40–E50	R108	R106–R109	E90	E78
Impact strength (Izod) at 23°C, J·m^{-1}	59	53	139	96	107–300	75	133
Tensile modulus, 10^3 lb·in^{-2}	275	290	1200	185	180		730
Tensile strength at break, 10^3 lb·in^{-2}	8.5	8.8	24	8	8–9	17.5	26.9
Tensile yield strength, 10^3 lb·in^{-2}	8.5	8.8					
Thermal Burning rate, mm·min^{-1}				Self-extinguishing			
Coefficient of linear thermal expansion, 10^{-6}°C	57–60	90		55–100	67–100	40	36
Deflection temperature under flexural load (264 lb·in^{-2}), °C	82	82	93–218	54	54	260	274
Maximum recommended service temperature, °C			260	100–120			260
Specific heat, cal·g^{-1}	0.4	0.4		0.58			
Thermal conductivity, W·m^{-1}·K^{-1}	0.22	0.22		0.34	0.22	0.22	0.25

TABLE 10.2 Properties of Commercial Plastics (*continued*)

| Properties | Poly(aryl ether), unfilled | Polycarbonate | | Thermoplastic polyester | | | |
| | | | | Poly(butylene terephthalate) | | Poly(ethylene terephthalate) | |
		Low viscosity	30% glass-fiber-reinforced	Unfilled	30% glass-fiber-reinforced	Unfilled	30% glass-fiber-reinforced
Physical							
Melting temperature, °C							
Crystalline				232–267	232–267	245	245
Amorphous	160	140	150				
Specific gravity	1.14	1.2	1.4	1.31–1.38	1.52	1.34–1.39	1.27
Water absorption (24h), %	0.25	0.15	0.14	0.08–0.09	0.06–0.08	0.1–0.2	0.05
Dielectric strength, $kV \cdot mm^{-1}$	17	15	19	16–22	18–22		22
Electrical							
Volume (dc) resistivity, ohm-cm		2×10^{16}	$> 10^{16}$		10^{16}	10^{16}	
Dielectric constant (60 Hz)		3.17	3.35				
Dielectric constant (10^6 Hz)		2.96	3.31			3.25	
Dissipation (power) factor (60 Hz)		0.0009	0.011				
Dissipation factor (10^6 Hz)		0.010	0.007				
Mechanical							
Compressive modulus, $10^3 \, lb \cdot in^{-2}$		350	1 300				

Property							
Compressive strength, rupture or 1% yield, 10^3 lb·in^{-2}	80	12.5	18	8.6–14.5	18–23.5	11–15	25
Elongation at break, %		110	3–5	50–300	2–4	50–300	3
Flexural modulus at 23°C, 10^3 lb·in^{-2}	300	340	1100	330–400	1100–1200	35–450	1440
Flexural strength, rupture or yield, 10^3 lb·in^{-2}	11	13.5	23	12–16.7	26–29	14–18	33.5
Hardness, Rockwell (or Shore)	R117	M70	M92	M68–M78	M90	M94–M101	M100
Impact strength (Izod) at 23°C, J·m^{-1}	427	14	107	43–53	69–85	13–32	101
Tensile modulus, 10^3 lb·in^{-2}	320	345	1250	280	1300	400–600	1440
Tensile strength at break, 10^3 lb·in^{-2}	7.5	9.5	19	8.2	17–19	8.5–10.5	23
Tensile yield strength, 10^3 lb·in^{-2}		9.0					
Thermal							
Burning rate, mm·min^{-1}		Self-extinguishing	Self-extinguishing				
Coefficient of linear thermal expansion, 10^{-6}°C	65	68	22	60–95	25	65	29
Deflection temperature under flexural load (264 lb·in^{-2}), °C	149	138–145	146	50–85	220	38–41	224
Maximum recommended service temperature, °C		143					
Specific heat, cal·g^{-1}		0.3				0.27	
Thermal conductivity, W·m^{-1}·K^{-1}	0.30	0.20	0.22	0.18–0.30	0.30	0.15	

TABLE 10.2 Properties of Commercial Plastics (*continued*)

| | Thermoplastic polyester | | Thermosetting and alkyd polyester | | | | |
| | Aromatic polyester | | Unsaturated polyester | | Alkyd molding compounds | | |
Properties	Extrusion-transparent	Injection molding	Styrene–maleic acid copolymer, low-shrink	Butadiene–maleic acid copolymer	Putty, mineral-filled	Glass-fiber-reinforced	Polyimide, unfilled
Physical							
Melting temperature, °C							
Crystalline							310–365
Amorphous	81		Thermoset	Thermoset	Thermoset	Thermoset	
Specific gravity		1.39					1.36–1.43
Water absorption (24 h), %		0.01					0.24
Dielectric strength, kV·mm^{-1}		14					22
Electrical							
Volume (dc) resistivity, ohm-cm							$>10^{16}$
Dielectric constant (60 Hz)							3–4
Dielectric constant (10^6 Hz)							
Dissipation (power) factor (60 Hz)							
Dissipation factor (10^6 Hz)							
Mechanical							
Compressive modulus, 10^3 lb·in^{-2}					2000–3000		
Compressive strength, rupture or 1% yield, 10^3 lb·in^{-2}	10		15–30	14–30	12–38	15–36	30–40

Property							
Elongation at break, %	225	7–10	3–5				8–10
Flexural modulus at 23°C, 10^3 lb·in^{-2}	290	700	1000–2500		2000	2000	450–500
Flexural strength, rupture or yield, 10^3 lb·in^{-2}	10.6	12	9–35	16–24	6–17	8.5–26	19–28.8
Hardness, Rockwell (or Shore)	R105		40–70 (Barcol)	50–60 (Barcol)	E98	E95	E52–E99
Impact strength (Izod) at 23°C, J·m^{-1}	101	300	133–800	214–694	16–27	27–854	80
Tensile modulus, 10^3 lb·in^{-2}			1000–2500	1500–2500	500–3000		300
Tensile strength at break, 10^3 lb·in^{-2}	6	11	4.5–20	5–10	3–9	4–9.5	10.5–17.1
Tensile yield strength, 10^3 lb·in^{-2}	7						12.5
Thermal							
Burning rate, mm·min^{-1}							
Coefficient of linear thermal expansion, 10^{-6}°C		29	6–30		20–50	15–33	45–56
Deflection temperature under flexural load (264 lb·in^{-2}), °C	63	282	190–260	160–177	177–260	204–260	277–360
Maximum recommended service temperature, °C							
Specific heat, cal·g^{-1}							0.27
Thermal conductivity, W·m^{-1}·K^{-1}	0.29	0.29		0.76–0.93	0.51–0.89	0.6–0.89	0.10–0.11

TABLE 10.2 Properties of Commercial Plastics (*continued*)

Properties	Poly(methyl pentene), unfilled	Polyolefin					Ethylene–vinyl acetate copolymer
		Polyethylene			Ultra high-molecular-weight	Glass-fiber-reinforced, high-density	
		Low-density	Medium-density	High-density			
Physical							
Melting temperature, °C							
Crystalline	230–240	95–130	120–140	120–140	125–135	120–140	65–90
Amorphous							
Specific gravity	0.84	0.910–0.925	0.926–0.94	0.941–0.965	0.94	1.28	0.92–0.95
Water absorption (24h), %	0.01	<0.01	<0.01	<0.01	<0.01	0.02	0.05–0.13
Dielectric strength, kV·mm^{-1}		18–39	18–39	18–39	28	20	24–30
Electrical							
Volume (dc) resistivity, ohm-cm		$>10^{15}$	$>10^{15}$	$<10^{15}$			
Dielectric constant (60 Hz)		2.3	2.3	2.3			
Dielectric constant (10^6 Hz)		2.3	2.3	2.3			
Dissipation (power) factor (60 Hz)		<0.0005	<0.0005	<0.0005			
Dissipation factor (10^6 Hz)		<0.0005	<0.0005	<0.0005			
Mechanical							
Compressive modulus, 10^3 lb·in^{-2}	114–171						

Property							
Compressive strength, rupture or 1% yield, 10^3 lb·in^{-2}	5–6.6			2.7–3.6		7	
Elongation at break, %	10–50	90–800	50–600	20–130		1.5	550–900
Flexural modulus at 23°C, 10^3 lb·in^{-2}	110–260	8–60	60–115	100–260	450–525	800	1–20
Flexural strength, rupture or yield, 10^3 lb·in^{-2}	4–6.5				130–140	11	
Hardness, Rockwell (or Shore)	L67–L74	(D40–D51)	(D50–D60)	R30–R50	R50	R75	
Impact strength (Izod) at 23°C, J·m^{-1}	16–64	No break	27–854	27–1068	No break	59	No break
Tensile modulus, 10^3 lb·in^{-2}	160–280	14–38	25–55	60–180			20–120
Tensile strength at break, 10^3 lb·in^{-2}	3.5–4	0.6–2.3	1.2–3.5	3.1–5.5	5.6	9	1.4–2.8
Tensile yield strength, 10^3 lb·in^{-2}		0.8–1.2	1.0–2.2	3–4	3.1–4.0		
Thermal							
Burning rate, mm·min^{-1}		1.0	1.0	1.0			
Coefficient of linear thermal expansion, 10^{-6}°C	117	100–220	140–160	110–130	130	48	160–200
Deflection temperature under flexural load (264 lb·in^{-2}), °C	41	32–41	41–49	43–54	43–49	121	34
Maximum recommended service temperature, °C	175	70	93	200			
Specific heat, cal·g^{-1}		0.55	0.55	0.46–0.55			
Thermal conductivity, W·m^{-1}·K^{-1}	0.17	0.34	0.34–0.42	0.46–0.51		0.46	

TABLE 10.2 Properties of Commercial Plastics (continued)

Properties	Polyolefin — Polybutylene extrusion	Polyolefin — Polypropylene Homopolymer	Polyolefin — Polypropylene Copolymer	Polyolefin — Polypropylene Impact copolymer	Polyolefin — Polyallomer	Poly(phenylene sulfide) Injection molding	Poly(phenylene sulfide) 40% glass-fiber-reinforced
Physical							
Melting temperature, °C							
Crystalline	126	168	160–168		120–135	290	290
Amorphous							
Specific gravity	0.91–0.925	0.90–0.91	0.89–0.905	0.90	0.90	1.3	1.6
Water absorption (24h), %	0.01–0.02	0.01–0.03	0.03	<0.03	<0.01	<0.02	0.05
Dielectric strength, kV·mm^{-1}	18	24	24	24	31	15	18
Electrical							
Volume (dc) resistivity, ohm-cm		10^{17}	10^{17}	10^{17}			
Dielectric constant (60 Hz)		2.2–2.6	2.3	2.3			
Dielectric constant (10^6 Hz)		2.2–2.6	2.3				
Dissipation (power) factor (60 Hz)		<0.0005	0.0001–0.0005	0.0003			
Dissipation factor (10^6 Hz)		0.0005–0.002	0.0001–0.002				
Mechanical							
Compressive modulus, 10^3 lb·in^{-2}	31	150–300					

Compressive strength, rupture or 1% yield, 10^3 lb·in^{-2}		5.5–8.0	3.5–8.0		400–500	16	21
Elongation at break, %	300–380	100–600	200–700	8–20	400–500	1–2	1
Flexural modulus at 23°C, 10^3 lb·in^{-2}	45–50	170–250	130–200	130–190	70–110	550	1700
Flexural strength, rupture or yield, 10^3 lb·in^{-2}	2–2.3	6–8	5–7			14	29
Hardness, Rockwell (or Shore)		R80–R102	R50–R96	R40–R90	R50–R85	R123	R123
Impact strength (Izod) at 23°C, J·m^{-1}	No break	21–53	53–1068	80–900		<27	75
Tensile modulus, 10^3 lb·in^{-2}	30–40	165–225	100–170		91–203	480	1100
Tensile strength at break, 10^3 lb·in^{-2}	3.8–4.4	4.5–6	4–5.5	2.5–3.1	3–3.8	9.5	19.5
Tensile yield strength, 10^3 lb·in^{-2}	1.7–2.5	4.5–5.4	3.5–4.3		3–3.4		
Thermal							
Burning rate, mm·min^{-1}		81–100	68–95				
Coefficient of linear thermal expansion, 10^{-6}°C	128–150	48–57	45–57	60–90	83–100	49	22
Deflection temperature under flexural load (264 lb·in^{-2}), °C	54–60	160	240	90–105 (66 lb·in^{-2})	51–56	135	249
Maximum recommended service temperature, °C				140–160			
Specific heat, cal·g^{-1}		0.44–0.46	0.45–0.50	0.45–0.50			
Thermal conductivity, W·m^{-1}·K^{-1}	0.22	0.12	0.15–0.17	0.12–0.17	0.09–0.17	0.29	0.29

TABLE 10.2 Properties of Commercial Plastics (*continued*)

Properties	Polyurethane			Silicone			Styrenic
	Casting resin		Thermoplastic elastomer	Cast resin, flexible	Mineral- and/or glass-filled	Epoxy molding and encapsulating compound	Polystyrene
	Liquid	Unsaturated					Crystal
Physical							
Melting temperature, °C							
Crystalline	Thermoset	Thermoset	120–160	Thermoset	Thermoset	Thermoset	
Amorphous							85–105
Specific gravity	1.1–1.5	1.05	1.05–1.25	0.99–1.5	1.8–1.94	1.84	1.04–1.05
Water absorption (24 h), %	0.02–1.5	0.1–0.2	0.7–0.9	22	8–15	10	0.03–0.10
Dielectric strength, kV·mm^{-1}	12–20		13–25				24
Electrical							
Volume (dc) resistivity, ohm-cm	10^{11}–10^{15}		10^{11}–10^{13}	10^{14}–10^{15}			> 10^{16}
Dielectric constant (60 Hz)	4.0–7.5		5.4–7.6	2.7–4.2			2.5
Dielectric constant (10^6 Hz)							
Dissipation (power) factor (60 Hz)							
Dissipation factor (10^6 Hz)							
Mechanical							
Compressive modulus, 10^3 lb·in^{-2}	10–100		4–9				

Property							
Compressive strength, rupture or 1% yield, 10^3 lb·in^{-2}	20		20		10–16	28	11.5–16
Elongation at break, %	100–1000	3–6	100–1100	100–700			1–2
Flexural modulus at 23°C, 10^3 lb·in^{-2}	10–100	610	10–350		1000–2500		380–450
Flexural strength, rupture or yield, 10^3 lb·in^{-2}	0.7–4.5	19	0.7–9		9–14	17	8–14
Hardness, Rockwell (or Shore)			(A65–D80)	(A15–A65)	M80–M90		M60–M75
Impact strength (Izod) at 23°C, J·m^{-1}	1334 to flexible	21	No break		13–427	16	13–21
Tensile modulus, 10^3 lb·in^{-2}	10–100		10–350				350–485
Tensile strength at break, 10^3 lb·in^{-2}	0.175–10	10–11	1.5–8.4	0.35–1.0	4–6.5	6–8	5.3–7.9
Tensile yield strength, 10^3 lb·in^{-2}							
Thermal							
Burning rate, mm·min^{-1}	100–200		100–200	300–800	0–78		
Coefficient of linear thermal expansion, 10^{-6}°C					20–50	30	
Deflection temperature under flexural load (264 lb·in^{-2}), °C	Varies over wide range	87–93	Varies over wide range		260	74–100	70–80
Maximum recommended service temperature, °C					371		93
Specific heat, cal·g^{-1}	0.43		0.43				0.3
Thermal conductivity, W·m^{-1}·K^{-1}	0.21		0.07–0.31	0.15–0.31	0.30	0.68	0.09–0.13

TABLE 10.2 Properties of Commercial Plastics (*continued*)

Properties	Polystyrene	Styrenic — Acrylonitrile–butadiene–styrene copolymer					
	Heat-resistant	Extrusion	Molding				
			Heat-resistant	High-impact	Flame-retarded	Platable	20% glass-reinforced
Physical							
Melting temperature, °C							
Crystalline							
Amorphous	110–125	88–120	110–125	100–110	110–125	100–110	
Specific gravity	1.05–1.09	1.02–1.06	1.05–1.08	1.01–1.04	1.16–1.21	1.06–1.07	1.22
Water absorption (24 h), %	0.03–0.12	0.20–0.45	0.20–0.45	0.20–0.45	0.2–0.6		
Dielectric strength, kV·mm^{-1}	20	14–20	14–20	14–20	14–20	16–22	18
Electrical							
Volume (dc) resistivity, ohm-cm							
Dielectric constant (60 Hz)				2.4–5.0			
Dielectric constant (10^6 Hz)				2.4–3.8			
Dissipation (power) factor (60 Hz)				0.003–0.008			
Dissipation factor (10^6 Hz)				0.007–0.015			
Mechanical							
Compressive modulus, 10^3 lb·in^{-2}		150–390	190–440	140–300	130–310		

Compressive strength, rupture or 1% yield, $10^3\,lb \cdot in^{-2}$	11.5–16	5.2–10	7.2–10	4.5–8	6.5–7.5		14
Elongation at break, %	2–60	20–100	3–20	5–70	5–25		
Flexural modulus at 23°C, $10^3\,lb \cdot in^{-2}$	340–470	130–420	300–400	250–350	300–400	340–390	710
Flexural strength, rupture or yield, $10^3\,lb \cdot in^{-2}$	8.9–14	4–14	10–13	8–11	9–14	10.5–11.5	15.5
Hardness, Rockwell (or Shore)	L80–L108	R75–R115	R100–R115	R85–R105	R100–R120	R103–R109	M85
Impact strength (Izod) at 23°C, $J \cdot m^{-1}$	21–181	133–640	107–347	347–400	160–640	267–283	64
Tensile modulus, $10^3\,lb \cdot in^{-2}$	320–460	130–380	300–350	230–330	320–400	330–380	740
Tensile strength at break, $10^3\,lb \cdot in^{-2}$	5–7.8	2.5–8.0	6–7.5	4.8–6.3	5–8	6–6.4	11
Tensile yield strength, $10^3\,lb \cdot in^{-2}$			5.5–7	4–5.5	4–6		
Thermal							
Burning rate, $mm \cdot min^{-1}$		1.3		1.3			
Coefficient of linear thermal expansion, 10^{-6}°C	60–70	60–130	60–93	95–110	65–95	47–53	21
Deflection temperature under flexural load (264 lb·in⁻²), °C	93–120 annealed	77–104 annealed	104–116 annealed	96–102 annealed	90–107 annealed	96–102 annealed	99
Maximum recommended service temperature, °C				110			
Specific heat, $cal \cdot g^{-1}$							
Thermal conductivity, $W \cdot m^{-1} \cdot K^{-1}$			0.19–0.34	0.3–0.4			

TABLE 10.2 Properties of Commercial Plastics (*continued*)

| | Styrenic | | | Polysulfone | | Sulfone | |
| | Styrene-acrylonitrile copolymer | | Styrene-butadiene copolymer, high-impact | | | | |
Properties	Unfilled	20% glass-fiber-reinforced		Unfilled	20% glass-fiber-reinforced	Poly(ether sulfone)	Poly(phenyl sulfone)
Physical							
Melting temperature, °C							
Crystalline	115–125	115–125	90–110	200	200	230	220
Amorphous							
Specific gravity	1.07–1.08	1.22	1.03–1.06	1.24	1.46	1.37	1.29
Water absorption (24 h), %	0.2–0.3	0.15–0.20	0.05–0.10	0.22	0.23	0.43	1.1–1.3 (saturated)
Dielectric strength, kV·mm⁻¹	16–20	20	18	17	17	17	16
Electrical							
Volume (dc) resistivity, ohm-cm				10^{15}			
Dielectric constant (60 Hz)				3.14	3.7		
Dielectric constant (10^6 Hz)				3.26	3.7		
Dissipation (power) factor (60 Hz)				0.004	0.002		
Dissipation factor (10^6 Hz)				0.008	0.009		
Mechanical							
Compressive modulus, 10^3 lb·in⁻²	530			370			

Property							
Compressive strength, rupture of 1% yield, 10^3 lb·in⁻²	14–17	19	4–9	13.9	22		
Elongation at break, %	1–4	1–2	13–50	50–100	2	30–80	60
Flexural modulus at 23°C, 10^3 lb·in⁻²	550	100–1100	280–450	390	1000	375	330
Flexural strength, rupture or yield, 10^3 lb·in⁻²	14–17	20	5.3–9.4	15.4	23	18.7	12.4
Hardness, Rockwell (or Shore)	M80–M90	R122	M10–M68	M69, R120	R123	M88	
Impact strength (Izod) at 23°C, J·m⁻¹	19–27	53	32–192	64	59	85	
Tensile modulus, 10^3 lb·in⁻²	400–560	1150–1200	280–465	360	1200	350	640
Tensile strength at break, 10^3 lb·in⁻²	9–12	15.8–18	3.2–4.9				310
Tensile yield strength, 10^3 lb·in⁻²			2.9–4.9	10.2	17	12.2	10.4
Thermal							
Burning rate, mm·min⁻¹	36–38	38–40	70–101				
Coefficient of linear thermal expansion, 10^{-6}°C	88–104	99	74–93	52–56	25	55	31
Deflection temperature under flexural load (264 lb·in⁻²), °C				174	182	203	204
Maximum recommended service temperature, °C				149			
Specific heat, cal·g⁻¹							
Thermal conductivity, W·m⁻¹·K⁻¹	0.12	0.26–0.28	0.12–0.21	0.12	0.38	0.14–0.19	

TABLE 10.2 Properties of Commercial Plastics (*continued*)

| Properties | Thermoplastic elastomers | | | | Urea formaldehyde, alpha-cellulose filled | Vinyl | |
| | Polyolefin | Polyester | Block copolymers of styrene and butadiene or styrene and isoprene | Block copolymers of styrene and ethylene or styrene and butylene | | Poly(vinyl chloride) and poly(vinyl acetate) | |
						Rigid	Flexible and unfilled
Physical							
Melting temperature, °C					Thermoset		
Crystalline		168–206					
Amorphous						75–105	75–105
Specific gravity	0.88–0.90	1.17–1.25	0.9–1.2	0.9–1.2	1.47–1.52	1.30–1.58	1.16–1.35
Water absorption (24 h), %	0.01		0.19–0.39		0.4–0.8	0.04–0.4	0.15–0.75
Dielectric strength, kV·mm^{-1}	24–26		16–21		12–16	14–20	12–16
Electrical							
Volume (dc) resistivity, ohm-cm					0.5–5.0	10^{12}–10^{15}	10^{11}–10^{14}
Dielectric constant (60 Hz)					7.7–9.5	3.2–4.0	5.0–9.0
Dielectric constant (10^6 Hz)					6.7–8.0	3.0–4.0	3.0–4.0
Dissipation (power) factor (60 Hz)					0.036–0.043	0.01–0.02	0.03–0.05
Dissipation factor (10^6 Hz)					0.025–0.035	0.006–0.02	0.06–0.1
Mechanical							
Compressive modulus, 10^3 lb·in^{-2}			3.6–120				

Property							
Compressive strength, rupture or 1% yield, $10^3\,\text{lb}\cdot\text{in}^{-2}$	150–300	350–450	500–1350	600–800	25–45	8–13	0.9–1.7
Elongation at break, %	1.5–2.0	7–75	4–150	4–100	<1	40–80	200–450
Flexural modulus at 23°C, $10^3\,\text{lb}\cdot\text{in}^{-2}$					1300–1600	300–500	
Flexural strength, rupture or yield, $10^3\,\text{lb}\cdot\text{in}^{-2}$					10–18	10–16	
Hardness, Rockwell (or Shore)	(A65–A92)	(D40–D72)	(A40–A90)	(A50–A90)	M110–M120	(D65–D95)	(A50–A100)
Impact strength (Izod) at 23°C, $\text{J}\cdot\text{m}^{-1}$	No break	208 to no break	No break	No break	13–21	21–1068	Varies over wide range
Tensile modulus, $10^3\,\text{lb}\cdot\text{in}^{-2}$		1.1–2.5	0.8–50		1000–1500	350–600	
Tensile strength at break, $10^3\,\text{lb}\cdot\text{in}^{-2}$	0.65–2.0	3.7–5.7	0.6–3.0	1–3	5.5–13	6–75	1.5–3.5
Tensile yield strength, $10^3\,\text{lb}\cdot\text{in}^{-2}$							
Thermal							
Burning rate, $\text{mm}\cdot\text{min}^{-1}$					Self-extinguishing	Self-extinguishing	Slow to self-extinguishing
Coefficient of linear thermal expansion, 10^{-6}°C	130–170		130–137		22–36	50–100	70–250
Deflection temperature under flexural load (264 $\text{lb}\cdot\text{in}^{-2}$), °C			<0–49		127–143	60–77	
Maximum recommended service temperature, °C					77	70–74	80–105
Specific heat, $\text{cal}\cdot\text{g}^{-1}$					0.6	0.2–0.28	0.36–0.5
Thermal conductivity, $\text{W}\cdot\text{m}^{-1}\cdot\text{K}^{-1}$	0.19–0.21		0.15		0.30–0.42	0.15–0.21	0.13–0.17

TABLE 10.2 Properties of Commercial Plastics (*continued*)

Properties	Vinyl					
	Poly(vinyl chloride) and poly(vinyl acetate) Flexible and filled	Poly(vinyl chloride), 15% glass-fiber-reinforced	Poly(vinylidene chloride)	Poly(vinyl formal)	Chlorinated poly(vinyl chloride)	Poly(vinyl butyral), flexible
Physical						
Melting temperature, °C						
Crystalline						
Amorphous	75–105	75–105	210	105	110	49
Specific gravity	1.3–1.7	1.54	1.65–1.72	1.2–1.4	1.49–1.56	1.05
Water absorption (24 h), %	0.5–1.0	0.01	0.1	0.5–3.0	0.02–0.15	1.0–2.0
Dielectric strength, kV·mm^{-1}	9.8–12	24–31	16–24	19		14
Electrical						
Volume (dc) resistivity, ohm-cm			10^{14}–10^{16}			
Dielectric constant (60 Hz)			4.5–6.0			
Dielectric constant (10^6 Hz)						
Dissipation (power) factor (60 Hz)						
Dissipation factor (10^6 Hz)						
Mechanical						
Compressive modulus, 10^3 lb·in^{-2}					335–600	

Property						
Compressive strength, rupture or 1% yield, 10^3 lb·in^{-2}	1.0–1.8	9	2–2.7	5–20	9–22	150–450
Elongation at break, %	200–400	2–3	50–250		4–65	
Flexural modulus at 23°C, 10^3 lb·in^{-2}		750			380–450	
Flexural strength, rupture or yield, 10^3 lb·in^{-2}		13.5	4.2–6.2	17–18	14.5–17	
Hardness, Rockwell (or Shore)	(A50–A100)	R118	M50–M65	M85	R117–R122	A10–A100
Impact strength (Izod) at 23°C, J·m^{-1}	Varies over wide range	53	16–53	43–75	53–299	Varies over wide range
Tensile modulus, 10^3 lb·in^{-2}		870	50–80	350–600	360–475	
Tensile strength at break, 10^3 lb·in^{-2}	1–3.5	9.5	3–5	10–12	7.5–9	
Tensile yield strength, 10^3 lb·in^{-2}						0.5–3.0
Thermal						
Burning rate, mm·min^{-1}			Self-extinguishing			Slow
Coefficient of linear thermal expansion, 10^{-6}°C			190			
Deflection temperature under flexural load (264 lb·in^{-2}), °C	68		54–71	64	68–78	
Maximum recommended service temperature, °C			100	71–77	94–112	
Specific heat, cal·g^{-1}			0.32			
Thermal conductivity, W·m^{-1}·K^{-1}	0.13–0.17		0.13	0.16	0.14	

FORMULAS AND ADVANTAGES OF RUBBERS

Gutta Percha

Gutta percha is a natural polymer of isoprene (3-methyl-1,3-butadiene) in which the configuration around each double bond is *trans*. It is hard and horny and has the following formula:

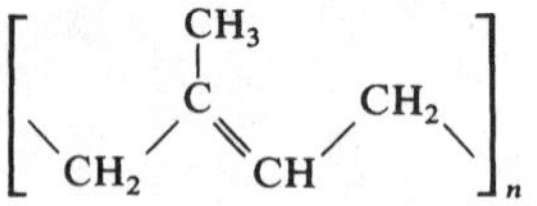

$$\left[\begin{array}{c} \overset{\displaystyle CH_3}{\underset{\displaystyle CH_2}{\overset{|}{C}}} \diagdown \\ CH_2 \diagup \quad \diagdown_{CH} \overset{CH_2}{\diagdown} \end{array} \right]_n$$

Natural Rubber

Natural rubber is a polymer of isoprene in which the configuration around each double bond is *cis* (or Z):

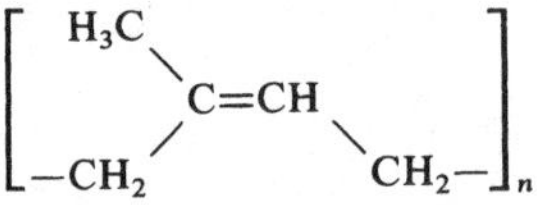

$$\left[\begin{array}{c} H_3C \diagdown \\ \qquad\quad C{=}CH \diagdown \\ {-}CH_2 \diagup \qquad CH_2{-} \end{array} \right]_n$$

Its principal advantages are high resilience and good abrasion resistance.

Chlorosulfonated Polyethylene

Chlorosulfonated polyethylene is prepared as follows:

$$[-CH_2-CH_2-]_n + HSO_3Cl \rightarrow \left[-CH_2-\underset{SO_3H}{\overset{|}{CH}}- \right]_n + HCl$$

Cross-linking, which can occur as a result of side reactions, causes an appreciable gel content in the final product.

The polymer can be vulcanized to give a rubber with very good chemical (solvent) resistance, excellent resistance to aging and weathering, and good color retention in sunlight.

Epichlorohydrin

Epichlorohydrin is a product of covulcanization of epichlorohydrin (epoxy) polymers with rubbers, especially *cis*-polybutadiene.

Its advantages include impermeability to air, excellent adhesion to metal, and good resistance to oils, weathering, and low temperature.

Nitrile Rubber (NBR, GRN, Buna N)

Nitrile rubber can be prepared as follows:

$$CH_2{=}CH-CH{=}CH_2 + CH_2{=}CH-CN \rightarrow$$
$$\quad\text{2 parts} \qquad\qquad \text{1 part}$$

$$\left[-CH_2-CH{=}CH-CH_2-CH_2-\underset{CN}{\overset{|}{CH}}-CH_2-CH{=}CH-CH_2- \right]_n$$

Nitrile rubber is also known as nitrile–butadiene rubber (NBR), government rubber nitrile (GRN), and Buna N.

It possesses resistance to oils up to 120 °C and excellent abrasion resistance and adhesion to metal.

Polyacrylate

Polyacrylate has the following formula:

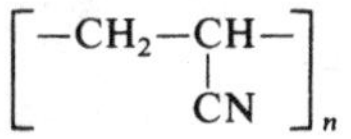

$$\left[\begin{array}{c} -CH_2-CH- \\ | \\ CN \end{array} \right]_n$$

It possesses oil and heat resistance to 175 °C and excellent resistance to ozone.

cis-Polybutadiene Rubber (BR)

cis-Polybutadiene is prepared by polymerization of butadiene by mostly 1,4-addition.

$$CH_2=CH-CH=CH_2 \rightarrow [-CH_2-CH=CH-CH_2-]_n$$

The polybutadiene produced is in the Z (or *cis*) configuration.

cis-Polybutadiene has good abrasion resistance, is useful at low temperature, and has excellent adhesion to metal.

Polychloroprene (Neoprene)

Polychloroprene is prepared as follows:

$$CH_2=CH-\underset{\underset{Cl}{|}}{C}=CH_2 \rightarrow [-CH_2-CH=C(Cl)-CH_2-]_n$$

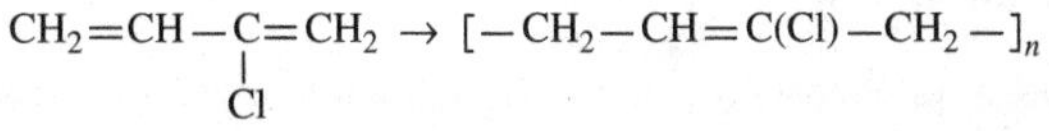

It has very good weathering characteristics, is resistant to ozone and to oil, and is heat-resistant to 100 °C.

Ethylene–Propylene–Diene Rubber (EPDM)

Ethylene–propylene–diene rubber is polymerized from 60 parts ethylene, 40 parts propylene, and a small amount of nonconjugated diene. The nonconjugated diene permits sulfur vulcanization of the polymer instead of using peroxide.

It is a very lightweight rubber and has very good weathering and electrical properties, excellent adhesion, and excellent ozone resistance.

Polyisobutylene (Butyl Rubber)

Polyisobutylene is prepared as follows:

$$\underset{\text{98 parts}}{H_3C-\underset{\overset{|}{CH_3}}{C}=CH_2} + \underset{\text{2 parts}}{\underset{\overset{|}{CH_3}}{CH_2=C}-CH=CH_2} \rightarrow \left[\left(-\underset{\overset{|}{CH_3}}{\overset{\overset{CH_3}{|}}{C}}-CH_2- \right)_n -CH_2-\underset{\overset{|}{}}{\overset{\overset{CH_3}{|}}{C}}=CH-CH_2- \right]$$

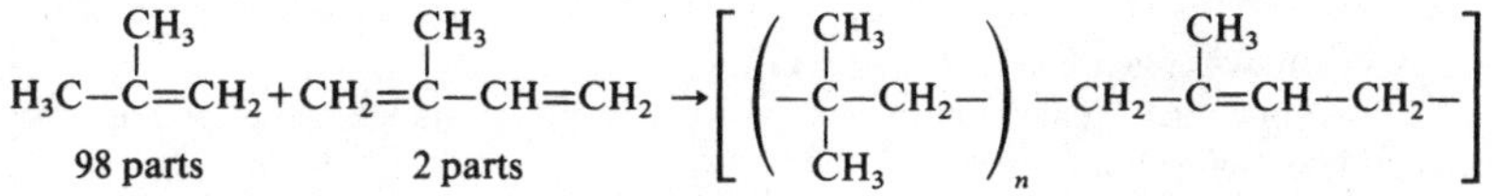

It possesses excellent ozone resistance, very good weathering and electrical properties, and good heat resistance.

(Z)-Polyisoprene (Synthetic Natural Rubber)

Polymerization of isoprene by 1,4-addition produces polyisoprene that has a *cis* (or *Z*) configuration.

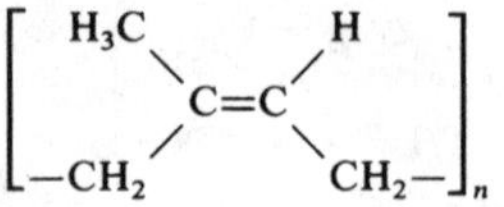

$$\left[\begin{array}{c} H_3C \diagdown \qquad \diagup H \\ \qquad C{=}C \\ {-}CH_2 \diagup \qquad \diagdown CH_2{-} \end{array} \right]_n$$

Polysulfide Rubbers

Polysulfide rubbers are prepared as follows:

$$Cl-R-Cl + Na-S-S-S-S-Na \rightarrow HS\,[-R-S-S-S-S-]_n R-SH$$

where R can be

$$-CH_2CH_2-, \quad -CH_2CH_2-O-CH_2CH_2-,$$

or

$$-CH_2CH_2-O-CH_2-O-CH_2CH_2-.$$

Polysulfide rubbers possess excellent resistance to weathering and oils and have very good electrical properties.

Poly(vinyl chloride) (PVC)

Poly(vinyl chloride) as previously discussed under "Formulas and Key Properties of Plastic Materials" has the following structures:

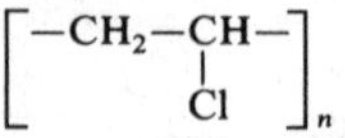

$$\left[\begin{array}{c} -CH_2-CH- \\ \qquad | \\ \qquad Cl \end{array} \right]_n$$

PVC polymer plus special plasticizers are used to produce flexible tubing which has good chemical resistance.

Silicone Rubbers

Silicone rubbers are prepared as follows:

$$Cl-\underset{\underset{CH_3}{|}}{\overset{\overset{CH_3}{|}}{Si}}-Cl \xrightarrow{H_2O} HO-\underset{\underset{CH_3}{|}}{\overset{\overset{CH_3}{|}}{Si}}-OH \xrightarrow{polymerize} \left[-\underset{\underset{CH_3}{|}}{\overset{\overset{CH_3}{|}}{Si}}-O- \right]_n$$

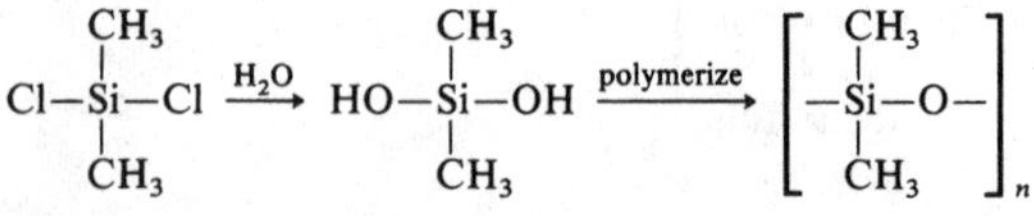

Other groups may replace the methyl groups.
Silicone rubbers have excellent ozone and weathering resistance, good electrical properties, and good adhesion to metal.

Styrene–Butadiene Rubber (GRS, SBR, Buna S)

Styrene–butadiene rubber is prepared from the free-radical copolymerization of one part by weight of styrene and three parts by weight of 1,3-butadiene. The butadiene is incorporated by both 1,4-addition (80%) and 1,2-addition (20%). The configuration around the double bond of the 1,4-adduct is about 80% *trans*. The product is a random copolymer with these general features:

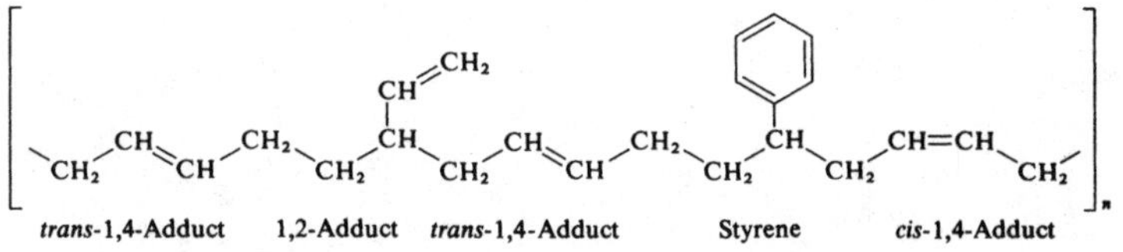

trans-1,4-Adduct   1,2-Adduct   *trans*-1,4-Adduct   Styrene   *cis*-1,4-Adduct

Styrene–butadiene rubber (SBR) is also known as government rubber styrene (GRS) and Buna S.

Urethane

See Table 10.3.

TABLE 10.3 Properties of Natural and Synthetic Rubbers

Rubber	Specific gravity	Durometer hardness (or Shore)	Ultimate elongation % (23°C)	Tensile strength, lb·in^{-2} (23°C)	Service temperature, °C	
					Minimum	Maximum
Gutta percha (hard rubber)	1.2–1.95	(65–95)	3–8	4000–10,000		104
Natural rubber (NR)	0.93	20–100	750–850	3000–4500	–56	82
Chlorosulfonated polyethylene	1.10	50–95	100–500	500–3000	–54	121
Epichlorohydrin	1.27	60–90	100–400	1000–2500	–46	121
Fluoroelastomers	1.4–1.95	60–90	100–350	2000–3000	–40	232
Isobutene–isoprene rubber (IIR) [also known as government rubber I(GR-I)]	0.91	(40–70)	750–950	2300–3000		121
Nitrile rubber (butadiene–acrylonitrile rubber) (also known as Buna N and NBR)	1.00	30–100	100–600	500–4000	–54	121
Polyacrylate	1.10	40–100	100–400	1000–2200	–18	149
Polybutadiene rubber (BR)	0.93	30–100	100–700	2500–3000	–62	79–100
Polychloroprene (neoprene)	1.23	20–90	800–1000	2000–3500	–54	121
Poly(ethylene–propylene–diene) (EPDM)	0.85	30–100	100–300	1000–3000	–40	149
Polyisobutylene (butyl rubber)	0.92	30–100	100–700	1000–3000	–54	100
Polyisoprene	0.94	20–100	100–750	2000–3000	–54	79–82
Polysulfide (Thiokol ST)	1.34	20–80	100–400	700–1250	–54	82–100
Poly(vinyl chloride) (Koroseal)	1.32	(80–90)		2400–3000		71
Silicone, high-temperature				700–800		316
Silicone	0.98	20–95	50–800	500–1500	–84	232
Styrene–butadiene rubber (SBR) (also known as Buna S)	0.94	40–100	400–600	1600–3700	–60	107
Urethane	0.85	62–95	100–700	1000–8000	–54	100

CHEMICAL RESISTANCE

TABLE 10.4 Resistance of Selected Polymers and Rubbers to Various Chemicals at 20 °C

The information in this table is intended to be used only as a general guide. The chemical resistance classifications are E = excellent (30 days of exposure causes no damage), G = good (some damage after 30 days), F = fair (exposure may cause crazing, softening, swelling, or loss of strength), N = not recommended (immediate damage may occur).

Polymers	Acids, dilute or weak	Acids, strong and concentrated	Alcohols, aliphatic	Aldehydes	Alkalies, concentrated	Esters	Ethers	Glycols	Hydrocarbons, aliphatic	Hydrocarbons, aromatic	Hydrocarbons, halogenated	Ketones	Oxidizing agents, strong
Acetals	F	N	F	N	N	N	N	G	N	N	N	N	N
Acrylics: poly(methyl methacrylate)	G	N	E		N	N	E	E	G	N	N	N	N
Allyls: diallyl phthalate	G			N	N				E	G	G	N	
Cellulosics: cellulose-acetate–butyrate and cellulose-acetate–propionate polymers	F	N	N			N	N	G	F	N	N	N	
Fluorocarbons	E	E	E	E	E					E	E	E	E
Polyamides	N	N	G	E	E	E	E	E	E	E	E	E	N
Polycarbonates	G	N	G	F	N	G		G	G	F	F	G	N
Polyesters	G	G	N		N	N	N	G	N	F	F	N	F
Poly(methyl pentene)	E	E	G	G	E	N	F	G	G	G	N	N	F
Low-density polyethylene	E	E	E	G	E	G	N	E	F	F	N	F	F
High-density polyethylene	E	E	E	E	E	G	N	E	F	G	N	G	F
Polybutadiene	G	F	E			G	N	E	G	E	E	G	F

TABLE 10.4 Resistance of Selected Polymers and Rubbers to Various Chemicals at 20°C (*continued*)

	Chemical												
	Acids, dilute or weak	Acids, strong and concentrated	Alcohols, aliphatic	Aldehydes	Alkalies, concentrated	Esters	Ethers	Glycols	Hydrocarbons, aliphatic	Hydrocarbons, aromatic	Hydrocarbons, halogenated	Ketones	Oxidizing agents, strong
Polymers (*continued*)													
APolypropylene and polyallomer	E	E	E	E	E	G	N	E	G	F	N	G	F
Polystyrene	N	N	E		E	N		E	N	N	N	N	N
Styrene–acrylonitrile copolymers			N		N			F	N				
Styrene–acrylonitrile–butadiene copolymers		N	G	F	G	N			F	N	N	N	
Sulfones: polysulfone	G	N	F	G	E	N	F	G	F	N	N	N	G
Vinyls: poly(vinyl chloride)	E	G	E		G	N	F	F	G	N	N	N	G
Rubbers													
Natural rubber			E			N	N	E	N	Z	Z	Z	
Nitrile rubber			E			Z	G	E	E	Z	Z	Z	
Polychloroprene			E			F	F	E	F	Z	Z	Z	
Polyisobutylene			E			E	E	E	N	Z	Z	Z	
Polysulfide rubbers: Thiokol			E			E	E	E	E	F	Z	Z	
Styrene–butadiene rubber			E			N	N	E	N	Z	Z	Z	

TABLE 10.5 Common Abbreviations Used in Polymer Chemistry

Acronym	Expansion
ABA	Acrylonitrile–butadiene–acrylate
ABS	Acrylonitrile–butadiene–styrene copolymer
ABS–PC	Acrylonitrile–butadiene–styrene–polycarbonate alloy
ABS–PVC	Acrylonitrile–butadiene–styrene–poly(vinyl chloride) alloy
ACM	Acrylic acid ester rubber
ACS	Acrylonitrile–chlorinated pe-styrene
AES	Acrylonitrile–ethylene–propylene–styrene
AMMA	Acrylonitrile–methyl methacrylate
AN	Acrylonitrile
APET	Amorphous polyethylene terephthalate
APP	Atactic polypropylene
ASA	Acrylic–styrene–acrylonitrile
BR	Butadiene rubber
BS	Butadiene styrene rubber
CA	Cellulose acetate
CAB	Cellulose acetate–butyrate
CAP	Cellulose acetate–propionate
CN	Cellulose nitrate
CP	Cellulose propionate
CPE	Chlorinated polyethylene
CPET	Crystalline polyethylene terephthalate
CPP	Cast polypropylene
CPVC	Chlorinated polyvinyl chloride
CR	Chloroprene rubber
CTA	Cellulose triacetate
DAM	Diallyl maleate
DAP	Diallyl phthalate
DMT	Terephthalic acid, dimethyl ester
ECTFE	Ethylene–chlorotrifluoroethylene copolymer
EEA	Ethylene–ethyl acrylate
EMA	Ethylene–methyl acrylate
EMAA	Ethylene methacrylic acid
EMAC	Ethylene–methyl acrylate copolymer
EMPP	Elastomer modified polypropylene
EnBA	Ethylene normal butyl acrylate
EP	Epoxy resin, also ethylene–propylene
EPM	Ethylene–propylene rubber
ESI	Ethylene–styrene copolymers
EVA(C)	Polyethylene–vinyl acetate
EVOH	Polyethylene–vinyl alcohol copolymers
FEP	Fluorinated ethylene–propylene copolymers
HDI	Hexamethylene diisocyanate
HDPE	High-density polyethylene
HIPS	High-impact polystyrene
HMDI	Diisocyanato dicyclohexylmethane
IPI	Isophorone diisocyanate
LDPE	Low-density polyethylene
LLDPE	Linear low-density polyethylene
MBS	Methacrylate–butadiene–styrene

TABLE 10.5 Common Abbreviations Used in Polymer Chemistry (*continued*)

Acronym	Expansion
MC	Methyl cellulose
MDI	Methylene diphenylene diisocyanate
MEKP	Methyl ethyl ketone peroxide
MF	Melamine formaldehyde
MMA	Methyl methacrylate
MPEG	Polyethylene glycol monomethyl ether
MPF	Melamine–phenol–formaldehyde
NBR	Nitrile butyl rubber
NDI	Naphthalene diisocyanate
NR	Natural rubber
OPET	Oriented polyethylene terephthalate
OPP	Oriented polypropylene
OSA	Olefin–modified styrene–acrylonitrile
PA	Polyamide
PAEK	Poly(aryl ether–ketone)
PAI	Poly(amide–imide)
PAN	Polyacrylonitrile
PB	Polybutylene
PBAN	Poly(butadiene–acrylonitrile)
PBI	Polybenzimidazole
PBN	Polybutylene naphthalate
PBS	Poly(butadiene–styrene)
PBT	Poly(butylene terephthalate)
PC	Polycarbonate
PCD	Polycarbodiimide
PCT	Poly(cyclohexylene–dimethylene terephthalate)
PCTFE	Polychlorotrifluoroethylene
PE	Polyethylene
PEC	Chlorinated polyethylene
PEG	Poly(ethylene glycol)
PEI	Poly(ether–imide)
PEK	Poly(ether–ketone)
PEN	Polyethylene naphthalate
PES	Polyether sulfone
PET	Polyethylene terephthalate
PF	Phenol–formaldehyde copolymer
PFA	Perfluoroalkoxy resin
PI	Polyimide
PIBI	Poly(isobutylene), Butyl rubber
PMDI	Polymeric methylene diphenylene diisocyanate
PMMA	Poly(methyl methacrylate)
PMP	Poly(methylpentene)
PO	Polyolefins
PP	Polypropylene
PPA	Polyphthalamide
PPC	Chlorinated polypropylene
PPO	Poly(phenylene oxide)
PPS	Poly(phenylene sulfide)
PPSU	Poly(phenylene sulfone)

TABLE 10.5 Common Abbreviations Used in Polymer Chemistry (*continued*)

Acronym	Expansion
PS	Polystyrene
PSF	Polysulfone (also PSU)
PSU	Polysulfone (also PSF)
PTFE	Polytetrafluoroethylene
PU	Polyurethane
PUR	Polyurethane
PVA	Poly(vinyl acetate)
PVAL	Poly(vinyl alcohol)
PVB	poly(vinyl butyrate)
PVC	Poly(vinyl chloride)
PVCA	Poly(vinyl chloride–acetate)
PVDA	Poly(vinylidene acetate)
PVDC	Poly(vinylidene chloride)
PVDF	Poly(vinylidene fluoride)
PVF	Poly(vinyl fluoride)
PVOH	Poly(vinyl alcohol)
SAN	Styrene–acrylonitrile copolymer
SB	Styrene–butadiene copolymer
SBC	Styrene block copolymer
SBR	Styrene butadiene rubber
SMA	Styrene–maleic anhydride (also SMC)
SMC	Styrene–maleic anhydride (also SMA)
TA	Terephthalic acid (also TPA)
TDI	Toluene diisocyanate
TEFE	Ethylene–tetrafluoroethylene copolymer
TPA	Terephthalic acid (also TA)
UF	Urea formaldehyde
ULDPE	Ultralow-density polyethylene
UP	Unsaturated polyester resin
UR	Urethane
VLDPE	Very low-density polyethylene
ZNC	Ziegler-Natta catalyst

GAS PERMEABILITY

TABLE 10.6 Gas Permeability Constants ($10^{10}P$) at 25°C for Polymers and Rubbers

The gas permeability constant P is defined as

$$P = \frac{\text{amount of permeant}}{(\text{area}) \times (\text{time}) \times (\text{driving forced across the film})}$$

The gas permeability constant is the amount of gas expressed in cubic centimeters passed in 1 s through a 1-cm^2 area of film when the pressure across a film thickness of 1 cm is 1 cmHg and the temperature is 25°C. All tabulated values are multiplied by 10^{10} and are in units of seconds^{-1} (centimeters of Hg)$^{-1}$. Other temperatures are indicated by exponents and are expressed in degrees Celsius.

Polymer or rubber	Gas						
	He	N_2	H_2	O_2	CO_2	H_2O	Other
Cellulose (cellophane)	0.005^{20}	0.0032	0.0065	0.0021	0.0047	1900	0.006^{45} (H_2S); 0.0017 (SO_2)
Cellulose acetate	13.6^{20}	0.28^{30}	3.5^{20}	0.78^{30}	22.7^{30}	5500	3.5^{30} (H_2S); 17^0 (ethylene oxide); 6.8^{60} (bromomethane)
Cellulose nitrate	6.9	0.12	2.0^{20}	1.95	2.12	6290	57.1 (NH_3); 1.76 (SO_2)
Ethyl cellulose	400^{30}	8.4^{30}	87^{20}	26.5^{30}	41.0^{30}	$12\,000^{20}$	705 (NH_3); 204 (SO_2); 420^0 (ethylene oxide)
Gutta percha		2.17	14.4	6.16	35.4	510	15.7 (CO); 30.1 (CH_4);
Natural rubber		9.43	52.0	23.3	15.3	2290	1.68 (C_3H_8); 98.9 (C_2H_2); 550 ($CH_3C{\equiv}CH$); 3.59 (SF_6);
Nylon 6	0.53^{20}	0.0095^{30}		0.038^{30}	0.10^{30}	177	0.33^{30} (H_2S); 1.2^{20} (NH_3); 0.84^{60} (CH_3Br)
Nylon 11	1.95^{30}		1.78^{30}		1.00^{40}		0.344^{30} (Ne); 0.189^{40} (Ar); 13.6^{50} (propyne)
Poly(acrylonitrile)				0.0002	0.0008	300	

Polymer							Permeability to other gases
Acrylonitrile–styrene copolymer (66:34)							
Poly(1,3-butadiene)		6.42	41.9	0.048	0.21	2000	
Poly(cis-1,4-butadiene)		19.2		19.0	138.0	5070	
Butadiene–acrylonitrile copolymer (80:20)	32.6						19.2 (Ne); 41.0 (Ar)
Butadiene–styrene copolymer (80:20)	12.2	1.06	15.9	3.85	30.8		24.8 (C_2H_2); 7.7 (propyne); 5.01 (Ne); 4.49 (Ar)
Butadiene–styrene copolymer (92:8)	13.4	1.71					9.70 (Ne); 12.7 (Ar); 3.79 (Ar); 3.27 (CH_4)
Polychloroprene	22.9	5.11	13.6	4.0	25.8		2.88 (CH_4); 6.81 (C_2H_6); 9.43 (C_3H_8); 1.48 (CO); 49^0 (ethylene oxide); 14.4 (propene); 42.2 (propyne);
Polyethylene, low-density	4.9	1.2; 0.969	12.0^{30}	2.88	12.6	90	0.170 (SF_6); 472^{60} (CH_3Br); 0.388 (CH_4); 0.590 (C_2H_6);
Polyethylene, high-density	1.14	0.143	3.0^{20}	0.403	0.36	12.0	0.537 (C_3H_8); 0.0083 (SF_6); 1.69 (Ar); 4.01 (propene)
Poly(ethylene terephthalate) Crystalline	1.32	0.0065	3.70^{20}	0.035	0.17		0.0032 (CH_4); 0.08^{60} (CH_3Br); 0.009 (CH_4)
Amorphous	3.28	0.013		0.059	0.30	130	
Poly(ethyl methacrylate)	6.82	0.220		1.15	5.00	3200	2.98 (Ne); 0.565 (Ar); 0.370 (Kr); 3.83 (H_2S); 0.00000165 (SF_6)
Isobutene–isoprene copolymer (98:2)	8.38	0.324	7.20	1.30	5.16	110^{38}	13.6^{50} (C_3H_8)
Isoprene–acrylonitrile copolymer (76:24)	7.77	0.181	7.41	0.852	4.32		

TABLE 10.6 Gas Permeability Constants ($10^{10}P$) at 25°C for Polymers and Rubbers (*continued*)

Polymer or rubber	Gas						
	He	N_2	H_2	O_2	CO_2	H_2O	Other
Isoprene–methacrylonitrile copolymer (76:24)		0.596	13.6	2.34	14.1		
Methacrylonitrile–styrene–butadiene copolymer (88:7:5)				0.0048	0.014	600	0.33^{20} (H_2S); 9.2^{20} (NH_3)
Poly(methylpentene)	101	7.83	136	32.0	92.6		
Polypropylene	38^{20}	0.44^{30}	41^{20}	2.3^{30}	9.2^{30}	51	191^{0} (Ne); 550^{0} (Ar);
Silicone rubber, 10% filler	233^{0}	227^{0}	464^{0}	489^{0}	3240	43000^{35}	1020^{0} (Kr); 2550^{0} (Xe); 19000^{0} (butane)
Polystyrene	18.7	0.788	23.3	2.63	10.5	1200	15.7 (NO_2); 37.5 (N_2O_4);
Poly(tetrafluoroethylene)	6.8^{20}	1.4	9.8	4.2	11.7		1.2^{0} (ethylene oxide); 4.6^{0} (CH_3Br)
Poly(trifluoroethylene)	12.6^{30}	0.003	0.94^{20}	0.025^{40}	0.048^{40}	0.29	2.64^{30} (Ne); 0.19^{30} (Ar); 0.078^{30} (Kr); 0.050^{30} (CH_4)
Poly(vinyl acetate)	0.001^{30}		89^{30}	0.50^{30}			0.007 (H_2S); 0.002^{0} (ethylene oxide)
Poly(vinyl alcohol)		$<0.001^{14}$	0.009	0.0089	0.001^{23}		
Poly(vinyl chloride)	2.05	0.0118	1.70	0.0453	0.157	275	3.92 (Ne); 0.0115 (Ar); 0.0286 (CH_4)
Poly(vinylidene chloride)	0.31^{34}	0.00094^{30}		0.0053^{30}	0.03^{30}	0.5	0.03^{30} (H_2S); 0.008^{60} (CH_3Br)

TABLE 10.7 Vapor Permeability Constants ($10^{10}P$) at 35 °C for Polymers

All tabulated values are multiplied by 10^{10} and are in units of seconds^{-1} (centimeters of Hg)$^{-1}$.

Polymer	Vapor				
	Benzene	Hexane	Carbon tetrachloride	Ethanol	Ethyl acetate
Cellulose	1.4	0.912	0.836	85.8	13.4
Cellulose acetate	512	2.80	3.74	2980	3595
Poly(acrylonitrile)	2.61	1.59	1.47	0	1.34
Polyethylene, low-density	5300	2910	3810	55.9	513
Polystyrene	10600		6820	0	soluble
Poly(vinyl alcohol)	3.58	2.34	1.61	32.7	2.53

FATS, OILS, AND WAXES

TABLE 10.8 Constants of Fats and Oils

Fat or oil	Solidification point, °C	Specific gravity (15°C/15°C)	Refractive index	Acid value	Saponification value	Iodine value
Animal origin						
Butterfat	20–23	$0.91^{40°C}_{15°C}$	1.455	0.5–35	210–230	26–38
Chicken fat	21–27	0.924		1.2	193–205	66–72
Cod-liver oil		0.92–0.93	$0.925^{25°C}$	5.6	171–189	137–166
Deer fat	−3	0.96–0.97		0.8–5.3	195–200	26–36

TABLE 10.8 Constants of Fats and Oils (*continued*)

Fat or oil	Solidification point, °C	Specific gravity (15°C/15°C)	Refractive index	Acid value	Saponification value	Iodine value
Animal origin (continued)						
Dolphin	−3 to +5	0.91–0.93		2–12	203 (body); 290 (jaw)	127 (body); 33 (jaw)
Goat butter		0.91 − 0.94$^{38°C}_{38°C}$			233–236	25–37
Goose fat	22–24	0.92–0.93		0.6	191–193	58–67
Herring oil		0.92–0.94	0.900$^{60°C}$	1.8–44	170–194	102–149
Horse fat	20–45	0.92–0.93		0–2.4	195–200	75–86
Human fat	15	0.903	1.460		193–200	57–73
Lard oil	−2 to +4	0.913–0.915	1.462	0.1–2.5	193–198	63–79
Lard oil, fatty tissue	27–30	0.93–0.94	1.462	0.5–0.8	195–203	47–67
Menhaden oil	−5	0.92–0.93	1.465$^{60°C}$	3–12	189–193	148–185
Neat's-foot oil	−2 to +10	0.91–0.92	1.464$^{25°C}$	0.1–0.6	193–199	58–75
Porpoise, body oil	−16	0.926		1.2	203	127
Rabbit fat	17–23	0.93–0.94		1.4–7.2	199–203	70–100
Sardine oil	20–22	0.92–0.93	1.466$^{60°C}$	4–25	188–196	130–152
Seal	3	0.915–0.926		1.9–40	188–196	130–152
Shark		0.916–0.919			157–164	115–139
Sperm oil	15.5	0.878–0.884		13	120–137	80–84
Tallow, beef	31–38	0.895		0.25	196–200	35–42
Tallow, mutton	32–41	0.937–0.953	1.457$^{40°C}$	2–14	195–196	48–61
Whale oil	−2 to 0	0.917–0.924	1.460$^{60°C}$	1.9	160–202	90–146
Plant origin						
Acorn	−10	0.916			199	100
Almond	−20 to −15	0.914–0.921		0.5–3.5	183–208	93–103

Oil						
Babassu oil	22–26	$0.893^{60°C}$	$1.443^{60°C}$		247	16
Beechnut oil	−17	0.922			191–196	97–111
Castor oil	−18 to −17	0.960–0.967	1.477	0.1–0.8	175–183	84
Chaulmoogra oil, USP	<−25	$0.950^{25°C}$			196–213	98–110
Chinese vegetable tallow	24–34	0.918–0.922		2.4	179–206	23–41
Cocoa butter	21.5–23	0.964–0.974	$1.457^{40°C}$	1.1–1.9	193–195	33–42
Coconut oil	14–22	0.926	$1.449^{40°C}$	2.5–10	153–262	6–10
Corn (maize) oil	−20 to −10	0.921–0.928	$1.473^{40°C}$	1.4–2.0	187–193	111–128
Cottonseed oil	−13 to +12	$0.918^{25°C}$	$1.474^{40°C}$	0.6–0.9	194–196	103–111
Hazelnut oil	−18 to −17	0.917			191–197	87
Hemp-seed oil	−28 to −15	0.928–0.934		0.45	190–195	145–162
Linseed oil	−27 to −19	0.930–0.938	$1.478^{25°C}$	1–3.5	188–195	175–202
Mustard oil, black	16	0.918–0.921	$1.475^{40°C}$	5.7–7.3	173–175	99–110
Neem oil	−3	0.917	$1.462^{40°C}$		195	71
Niger-seed oil		0.925	$1.471^{40°C}$		190	129
Oiticica oil		$0.974^{25°C}$				140–180
Olive oil	−6	0.914–0.918	$1.468^{40°C}$	0.3–1.0	185–196	79–88
Palm oil	35–42	0.915	$1.458^{40°C}$	10	200–205	49–59
Palm kernel oil	24	0.918–0.925	$1.457^{40°C}$	0.3–0.6	220–231	26–32
Peanut oil	3	0.917–0.926	$1.469^{40°C}$	0.8	186–194	88–98
Perilla oil		0.930–0.937	$1.481^{25°C}$		188–194	185–206
Pistachio–nut oil	−10 to −5	0.913–0.919			191	83–87
Poppy-seed oil	−18 to −16	0.924–0.926	$1.469^{40°C}$	2.5	193–195	128–141
Pumpkin-seed oil	−15	0.923–0.925			188–193	121–130
Rapeseed oil	−10	0.913–0.917	$1.471^{40°C}$	0.36–1.0	168–179	94–105
Safflower oil	−18 to −13	0.925–0.928	$1.462^{60°C}$	0.6	188–203	122–141
Sesame oil	−6 to −4	$0.919^{25°C}$	$1.465^{40°C}$	9.8	188–193	103–117
Soybean oil	−16 to −10	0.924–0.927	$1.473^{40°C}$	0.3–1.8	189–194	122–134
Sunflower-seed oil	−17	0.924–0.926	$1.469^{40°C}$	11.2	188–193	129–136
Tung oil	−2.5	0.94–0.95	$1.517^{25°C}$	2	190–197	163–171
White-mustard-seed oil		0.912–0.916		5.4	171–174	94–98
Wheat-germ oil	−16 to −8					125

TABLE 10.9 Constants of Waxes

Wax	Melting point, °C	Specific gravity (15°C/15°C)	Refractive index	Acid value	Saponification value	Iodine value
Bamboo leaf	79–80	$0.961^{25°C}$	$1.436^{80°C}$	14–15	43–44	7.8
Bayberry (myrtle)	47–49	0.99		3–4	205–212	4–9.5
Beeswax, ordinary	62–66	0.95–0.97	1.44–$1.48^{40°C}$	17–21	88–100	8–11
Beeswax, East Indian	61–67	0.95–0.97	$1.44^{40°C}$	5–10.5	87–117	4–10.5
Beeswax, white, USP	61–69	0.95–0.98	1.45–$1.47^{65°C}$	17–24	90–96	7–11
Candelilla	73–77	0.98–0.99	1.45–$1.46^{85°C}$	19–24	55–64	14–20
Cape berry	40–45	1.01	$1.45^{45°C}$	2.5–4.0	211–215	0.5–2.5
Caranda	80–85	0.99–1.00		5.0–9.5	64–79	8–9
Carnauba, No. 1 yellow	86–88	0.99–1.00		1.5–2.5	75–86	
Carnauba, No. 3, crude	86–90	0.99–1.01		3.0–8.5	75–89	
Carnauba, No. 3, refined	86–89	0.96–0.97	$1.47^{40°C}$	3.0–5.0	76–85	7–13.5
Castor oil, hydrogenated	83–88	0.98–$0.99^{20°C}$	$1.46^{40°C}$	1.0–5.0	177–181	2.5–8.5
Chinese insect	80–85	0.95–0.97		2–9	78–93	1.0–2.5
Cotton	68–71	0.96		32	71	25
Cranberry	207–218	0.97–0.98		42–59	131–134	44–53
Esparto	75–79	0.985–0.995		22–27	58–73	7–15

Flax	61–70	0.91–0.99		17–48	37–102	22–29
Japan	49–56	0.97–1.00		4–15	210–235	4–15
Jojoba	11–12	0.86–0.90$^{25°C}$	1.465$^{25°C}$	0.2–0.6	92–95	82–88
Microcrystalline, amber	64–91	0.91–0.94	1.42–1.45$^{80°C}$	0	0	0
Microcrystalline, white	71–89	0.93–0.94	1.441$^{80°C}$	0	0	0
Montan, crude	76–86	1.01–1.02$^{25°C}$		22–31	59–92	14–18
Montan, refined	77–84	1.02–1.04		23–45	72–115	10–14
Ouricury	86–89	0.99–1.01		12–19	88–96	6.9–7.8
Ozokerite	56–82	0.90–1.00		0	0	4–8
Palm	74–86	0.99–1.05		5–11	64–104	9–17
Paraffin, American	49–63	0.896–0.925	1.44–1.48$^{80°C}$	0	0	0
Shellac	79–82	0.97–0.98		12–24	64–83	6–9
Sisal hemp	74–81	1.007–1.010		16–19	56–58	28–29
Spermaceti	41–49	0.905–0.960		0.5–3.0	121–135	2.5–8.5
Sugarcane, refined	76–82	0.96–0.98	1.51$^{25°C}$	8–23	55–70	13–29
Wool	38–40	0.97	1.48$^{40°C}$	6–22	82–130	15–47

SECTION 11

ABBREVIATIONS, CONSTANTS, AND CONVERSION FACTORS

PHYSICAL CONSTANTS

TABLE 11.1 Fundamental Physical Constants

A. Defined values

Name of unit	Symbol	Definition
		SI base units
Meter (metre) (preferred spelling in U.S. is meter)	m	1 650 763.73 wavelengths in vacuum of the orange-red line of the spectrum of krypton-86
Kilogram	kg	Mass of a cylinder of platinum–iridium alloy kept at Paris
Second	s	Duration of 9 192 631 770 cycles of the radiation associated with a specified transition of the cesium atom
Ampere	A	Magnitude of the current that, when flowing through each of two long parallel wires separated by one meter in free space, results in a force between the two wires 2×10^{-7} newton for each meter of length
Kelvin (degree Kelvin)	K	Defined in the thermodynamic scale by assigning 273.16 K to the triple point of water (freezing point, $273.15 \text{ K} = 0°\text{C}$)
Candela	cd	Luminous intensity of 1/600 000 of a square meter of a radiating cavity at the temperature of freezing platinum (2042 K)
Mole	mol	Amount of substance which contains as many specified entities (molecules, atoms, ions, electrons, photons, etc.) as there are atoms of carbon-12 in exactly 0.012 kg of that nuclide
		Supplementary SI units
Radian	rad	The plane angle between two radii of a circle which cut off on the circumference an arc equal in length to the radius
Steradian	sr	The solid angle which, having its vertex in the center of a sphere, cuts off an area of the surface of the sphere equal to that of a square with sides of length equal to the radius of the sphere

B. Derived SI units

Quantity and symbol	Name of SI unit	Symbol and definition
Capacitance (electric), C	farad	$F = C \cdot V^{-1}$
Charge (electric), quantity of electricity, Q	coulomb	$C = A \cdot s$
Conductance (electric), $G (= 1/R)$	siemens	$S = \Omega^{-1}$
Energy, work, quantity of heat, H	joule	$J = kg \cdot m^2 \cdot s^{-2}$

TABLE 11.1 Fundamental Physical Constants (*continued*)

Quantity and symbol	Name of SI unit	Symbol and definition
Force	newton	$N = kg \cdot m \cdot s^{-2}$
Frequency	hertz	$Hz = s^{-1}$
Illuminance, illumination	lux	$lx = lm \cdot m^{-2}$
Inductance, L	henry	$H = \Omega \cdot s$
Luminous flux	lumen	$lm = cd \cdot sr$
Magnetic flux	weber	$Wb = V \cdot s$
Magnetic flux density	tesla	$T = Wb \cdot m^{-2}$
Potential difference, E	volt	$V = kg \cdot m^2 \cdot s^{-3} \cdot A^{-1} = J \cdot A^{-1} \cdot s^{-1}$
Power, radiant flux	watt	$W = kg \cdot m^2 \cdot s^{-3} = J \cdot s^{-1}$
Pressure, stress	pascal	$Pa = N \cdot m^{-2} = kg \cdot m^{-1} \cdot s^{-2}$
Resistance (electric), R	ohm	$\Omega = V \cdot A^{-1} = kg \cdot m^2 \cdot s^{-3} \cdot A^{-2}$

C. Recommended Consistent Values of Constants

The digits in parentheses following a numerical value represent the standard deviation of that value in terms of the final listed digits.

Constant	Symbol and value
Anomalous electron moment correction	$(\mu_e/\mu_0) - 1 = 1.159\,615(15) \times 10^{-3}$
Atomic mass unit	$u = (10^{-3}\,kg \cdot mol^{-1})/N_A = 1.660\,566(9) \times 10^{-27}\,kg$
Avogadro constant	$N_A = 6.022\,045(31) \times 10^{23}\,mol^{-1}$
Bohr magneton	$\mu_B = e\hbar/2m_ec = 9.274\,078(36) \times 10^{-24}\,J \cdot T^{-1}$
Bohr radius	$a_0 = \alpha/4\pi R_\infty = 0.529\,177\,06(44) \times 10^{-10}\,m$
Boltzmann constant	$k = R/N_A = 1.380\,662(44) \times 10^{-23}\,J \cdot K^{-1}$
Charge-to-mass ratio for electron	$e/m_e = 1.758\,805(5) \times 10^{11}\,C \cdot kg^{-1}$
Compton wavelength of electron	$\lambda_c = \alpha^2/2R_\alpha = 2.426\,309(4) \times 10^{-12}\,m$
	$\lambdabar_c = \lambda_c/2\pi = \alpha a_0 = 3.861\,591(6) \times 10^{-13}\,m$
Compton wavelength of neutron	$\lambda_{c,n} = h/m_nc = 1.319\,591(2) \times 10^{-15}\,m$
Compton wavelength of proton	$\lambda_{c,p} = h/m_pc = 1.321\,410(2) \times 10^{-15}\,m$
Diamagnetic shielding factor, spherical H_2O molecule	$1 + \sigma(H_2O) = 1.000\,025\,64(7)$
Electron g-factor	$g_e/2 = \mu_e/\mu_B = 1.001\,159\,657(4)$
Electron magnetic moment	$\mu_e = 9.284\,832(36) \times 10^{-24}\,J \cdot T^{-1}$
Electron radius (classical)	$\alpha\lambdabar_c = \mu_0e^2/4\pi m_e = r_e = 2.817\,938(7) \times 10^{-15}\,m$
Electron rest mass	$m_e = 0.910\,953(5) \times 10^{-30}\,kg$
	$= 5.485\,803(2) \times 10^{-4}\,u$
Elementary charge	$e = 1.602\,189(5) \times 10^{-19}\,C$
Faraday constant	$N_Ae = F = 9.648\,456(27) \times 10^4\,C \cdot mol^{-1}$
Fine structure constant	$\mu_0ce^2/2h = \alpha = 0.007\,297\,351(6)$
	$1/\alpha = 1.370\,360(1)$
First radiation constant	$2\pi hc^2 = c_1 = 3.741\,83(2) \times 10^{-16}\,W \cdot m^2$
Gas constant (molar)	$R = P_0V_m/T_0 = 8.314\,41(26)\,J \cdot mol^{-1} \cdot K^{-1}$
	$= 82.0568(26)\,cm^3 \cdot atm \cdot mol^{-1} \cdot K^{-1}$
	$= 1.987\,19(6)\,cal \cdot mol^{-1} \cdot K^{-1}$
Gravitational constant	$G = 6.672(4) \times 10^{-11}\,N \cdot m^2 \cdot kg^{-2}$

TABLE 11.1 Fundamental Physical Constants (*continued*)

Constant	Symbol and value
Gyromagnetic ratio of proton (uncorrected for diamagnetism of H_2O)	$\gamma_p = 2.675\,199(8) \times 10^8\,\text{s}^{-1} \cdot \text{T}^{-1}$
	$\gamma_p' = 675\,130(8) \times 10^8\,\text{s}^{-1} \cdot \text{T}^{-1}$
Josephson frequency–voltage ratio	$2e/h = 4.835\,939(13) \times 10^{14}\,\text{Hz} \cdot \text{V}^{-1}$
Magnetic flux quantum	$\Phi_0 = h/2e = 2.067\,851(5) \times 10^{-15}\,\text{Wb}$
Molar standard volume, ideal gas	$V_m = RT_0/P_0 = 0.022\,413\,8(7)\,\text{m}^3 \cdot \text{mol}^{-1}$
Muon g-factor	$e\hbar/2m_\mu c = g_\mu/2 = 1.001\,166\,16(31)$
Muon magnetic moment	$\mu_\mu = 4.490\,474(18) \times 10^{-26}\,\text{J} \cdot \text{T}^{-1}$
Muon rest mass	$m_\mu = 1.883\,566(11) \times 10^{-28}\,\text{kg}$
Neutron rest mass	$m_n = 1.674\,954(9) \times 10^{-27}\,\text{kg}$
Normal volume, perfect gas	$V_0 = 2.241\,36(30) \times 10^4\,\text{cm}^3 \cdot \text{mol}^{-1}$
Nuclear magneton	$\mu_N = e\hbar/2m_pc = 5.050\,824(20) \times 10^{-27}\,\text{J} \cdot \text{T}^{-1}$
Permeability of vacuum	$\mu_0 = 4\pi \times 10^{-7}\,\text{H} \cdot \text{m}^{-1}$
Permittivity of vacuum	$\varepsilon_0 = (\mu_0 c^2)^{-1} = 8.854\,187\,82(7) \times 10^{-12}\,\text{F} \cdot \text{m}^{-1}$
Planck constant	$h = 6.626\,176(36) \times 10^{-34}\,\text{J} \cdot \text{s}$
	$\hbar = h/2\pi = 1.054\,589(6) \times 10^{-34}\,\text{J} \cdot \text{s}$
Proton magnetic moment:	$\mu_p = 1.410\,617(5) \times 10^{-26}\,\text{J} \cdot \text{T}^{-1}$
In Bohr magnetons	$\mu_p/\mu_B = 1.521\,032\,209(16) \times 10^{-3}$
In nuclear magnetons	$\mu_p/\mu_N = 2.792\,845\,6(11)$
Proton rest mass	$m_p = 1.672\,649(9) \times 10^{-27}\,\text{kg}$
Quantum–charge ratio	$h/e = 4.135\,701(11) \times 10^{-15}\,\text{J} \cdot \text{Hz}^{-1} \cdot \text{C}^{-1}$
Quantum of circulation	$h/m_e = 7.273\,89(1) \times 10^{-4}\,\text{J} \cdot \text{s} \cdot \text{kg}^{-1}$
Ratio, electron to proton magnetic moments	$\mu_e/\mu_p = 6.582\,106\,88(7) \times 10^2$
Ratio, kxu (Siegbahn) to angstrom	$= 1.000\,020\,5(56)$
Ratio, muon moment to proton moment	$\mu_\mu/\mu_p = 3.183\,340(7)$
Rydberg constant	$R_\infty = 1.097\,373\,18(8) \times 10^7\,\text{m}^{-1}$
Second radiation constant	$c_2 = hc/k = 1.438\,786(45) \times 10^{-2}\,\text{m} \cdot \text{K}$
Speed of light in vacuum	$c = 2.997\,924\,58(12) \times 10^8\,\text{m} \cdot \text{s}^{-1}$
Stefan–Boltzmann constant	$\sigma = (\pi^2/60)k^4/\hbar^3 c^2 = 5.670\,3(7) \times 10^{-8}\,\text{W} \cdot \text{m}^{-2} \cdot \text{K}^{-4}$
Thomson cross section	$\sigma_e = 8\pi r_e^2/3 = 6.652\,448(33) \times 10^{-28}\,\text{m}^2$
Voltage–wavelength product	$V\lambda = 1.239\,852(3) \times 10^{-6}\,\text{eV} \cdot \text{m}$
Wien displacement constant	$b = 0.289\,78(4)\,\text{cm} \cdot \text{K}$
Zeeman splitting constant	$\mu_B/hc = 4.668\,58(4) \times 10^{-5}\,\text{cm}^{-1} \cdot \text{G}^{-1}$
Energy equivalents:	
1 atomic mass unit	$u = 931.501\,6(26)\,\text{MeV}$
1 proton mass	$m_v = 938.279\,6(27)\,\text{MeV}$
1 neutron mass	$m_n = 939.573\,1(27)\,\text{MeV}$
1 muon mass	$m_\mu = 105.659\,48(35)\,\text{MeV}$
1 electron mass	$m_e = 0.511\,003\,4(14)\,\text{MeV}$
1 electronvolt	$1\,\text{eV}/k = 1.160\,450(36) \times 10^4\,\text{K}$
	$1\,\text{eV}/hc = 8.065\,479(21) \times 10^3\,\text{cm}^{-1}$
	$1\,\text{eV}/h = 2.417\,970(6) \times 10^{14}\,\text{Hz}$

Source: E.R. Cohen and B.N. Taylor, *J. Phys. Chem. Ref. Data*, **2**(4): 663 (1973)

GREEK ALPHABET

TABLE 11.2   Greek Alphabet

Capital letter	Lowercase letter	Letter name	Capital letter	Lowercase letter	Letter name
A	α	Alpha	N	ν	Nu
B	β	Beta	Ξ	ξ	Xi
Γ	γ	Gamma	O	o	Omicron
Δ	δ	Delta	Π	π	Pi
E	ε	Epsilon	P	ρ	Rho
Z	ζ	Zeta	Σ	σ	Sigma
H	η	Eta	T	τ	Tau
Θ	θ	Theta	Y	υ	Upsilon
I	ι	Iota	Φ	ϕ	Phi
K	κ	Kappa	X	χ	Chi
Λ	λ	Lambda	Ψ	ψ	Psi
M	μ	Mu	Ω	ω	Omega

PREFIXES

TABLE 11.3   Prefixes for Naming Multiples and Submultiples of Units

For example: 10^{-9} gram is one nanogram, or 1 ng.

Factor	Prefix	Symbol	Factor	Prefix	Symbol
10^{12}	tera	T	10^{-2}	centi	c
10^{9}	giga	G	10^{-3}	milli	m
10^{6}	mega	M	10^{-6}	micro	μ
10^{3}	kilo	k	10^{-9}	nano	n
10^{2}	hecto	h	10^{-12}	pico	p
10	deka	da	10^{-15}	femto	f
10^{-1}	deci	d	10^{-18}	atto	a

TABLE 11.4   Numerical Prefixes

Number	Prefix	Number	Prefix	Number	Prefix
$\frac{1}{2}$	hemi	6	hexa	13	trideca
1	mono	7	hepta	14	tetradeca
$1\frac{1}{2}$	sesqui	8	octa	15	pentadeca
2	di or bi	9	nona	16	hexadeca
3	tri	10	deca	17	heptadeca
4	tetra	11	undeca	18	octadeca
5	penta	12	dodeca	19	nonadeca

TABLE 11.4 Numerical Prefixes (*continued*)

Number	Prefix	Number	Prefix	Number	Prefix
20	icosa	34	tetratriaconta	48	octatetraconta
21	henicosa	35	pentatriaconta	49	nonatetraconta
22	docosa	36	hexatriaconta	50	pentaconta
23	tricosa	37	heptatriaconta	51	henpentaconta
24	tetracosa	38	octatriaconta	52	dopentaconta
25	pentacosa	39	nonatriaconta	53	tripentaconta
26	hexacosa	40	tetraconta	54	tetrapentaconta
27	heptacosa	41	hentetraconta	55	pentapentaconta
28	octacosa	42	dotetraconta	56	hexapentaconta
29	nonacosa	43	tritetraconta	57	heptapentaconta
30	triaconta	44	tetratetraconta	58	octapentaconta
31	hentriaconta	45	pentatetraconta	59	monapentaconta
32	dotriaconta	46	hexatetraconta	60	hexaconta
33	tritriaconta	47	heptatetraconta		

TRANSFORMATIONS

TABLE 11.5 Conversion Formulas for Solutions Having concentrations expressed in Various Ways

Abbreviations Used in the Table

wt %, weight percent of solute m, molality
MW_1, molecular weight of solute M, molarity
MW_2, molecular weight of solvent n, mole fraction
d, density of solution $(g \cdot mL^{-1})$ G, grams of solute per liter of solution

To obtain	From	Compute
molarity	weight per cent of solute	$M = \dfrac{10\,d(\text{wt \%})}{MW_1}$
molarity	molality	$M = \dfrac{1000\,dm}{1000 + (MW_1)m}$
molarity	grams of solute per liter of solution	$M = \dfrac{G}{MW_1}$
molarity	mole fraction	$M = \dfrac{1000\,dn}{n(NW_1) + MW_2(1-n)}$
mole fraction	weight per cent of solute	$n = \dfrac{(\text{wt \%})/MW_1}{(\text{wt \%})/MW_1 + (100 - \text{wt \%})MW_2}$
mole fraction	molality	$n = \dfrac{(MW_2)m}{(MW_2)m + 1000}$
mole fraction	molarity	$n = \dfrac{M(MW_2)}{M(MW_2 - MW_1) + 1000\,d}$

TABLE 11.5 Conversion Formulas for Solutions Having concentrations expressed in Various Ways (*continued*)

To obtain	From	Compute
mole fraction	grams of solute per liter of solution	$n = \dfrac{G(MW_2)}{G(MW_2 - MW_1) + 1000\,d(MW_1)}$
weight percent of solute	mole fraction	$wt\% = \dfrac{100\,n(MW_1)}{n(MW_1) + MW_2(1-n)}$
weight percent of solute	grams of solute per liter of solution	$wt\% = \dfrac{G}{10\,d}$
weight percent of solute	molarity	$wt\% = \dfrac{M(MW_1)}{10\,d}$
weight percent of solute	molality	$wt\% = \dfrac{100\,m(MW_1)}{1000 + m(MW_1)}$
molality	molarity	$m = \dfrac{1000\,M}{1000\,d - M(MW_1)}$
molality	grams of solute per liter of solution	$m = \dfrac{1000\,G}{MW_1(1000\,d - G)}$
molality	weight percent of solute	$m = \dfrac{1000(wt\%)}{MW_1(100 - wt\%)}$
molality	mole fraction	$m = \dfrac{1000\,n}{MW_2 - n(MW_2)}$

TABLE 11.6 Conversion Factors

The data have been compared with the *International Standard ISO* 31 (1979–80) and the *American Society for Testing and Materials Standard for Metric Practice E* 380-79. Relations which are exact are indicated by an asterisk (*). Factors in parentheses are also exact.

To convert	Into	Multiply by
ampere per square centimeter	ampere per square inch*	6.4516
ampere-hour	coulomb*	3600
ampere-turn	gilbert	1.256637
angstrom	meter*	1×10^{-10}
	nanometer*	0.1
apostib	candela per square meter	$0.3183099(1\,\pi)$
	lambert*	1×10^{-4}
atmosphere	bar*	1.01325
	inch of mercury	29.92126
	millimeters of mercury*	760
	millimeter of water	1.033227×10^4
	newton per square meter*	1.01325×10^5
	pascal*	1.01325×10^5
	torr*	760

TABLE 11.6 Conversion Factors (*continued*)

To convert	Into	Multiply by
bar	atmosphere	0.986 923
	dyne per square centimeter*	1×10^6
	millimeter of mercury	750.062
	pascal	1×10^5
barn	square meter*	1×10^{-28}
barrel (petroleum)	gallon (British)	34.9723
	gallon (U.S.)*	42
	liter	158.987
barrel (U.S., dry)	bushel (U.S.)	3.281 22
	liter	115.6271
barrel (U.S., liquid)	gallon (U.S.)	31.5
	liter	119.2405
becquerel	curie*	2.7×10^{-11}
British thermal unit (Btu)	calorie	251.996
	joule	1 055.056
	kilowatt-hour	2.93071×10^{-4}
	liter-atmosphere	10.4126
bushel (U.S.)	barrel (U.S., dry)	0.304 765
	cubic foot	1.244 456
	cubic inch*	2 150.42
	gallon (U.S.)	9.309 18
	liter	3.523 907
	pint (U.S., dry)	64
	quart (U.S., dry)	32
calorie	Btu	0.003 968 320
	joule*	4.1868
	liter-atmosphere	0.041 3205
calorie (thermochemical)	joule*	4.184
calorie per minute	watt*	0.069 78
calorie per second	watt*	4.1868
candela	Hefner unit	1.11
	lumen per steradian*	1
candela per square centimeter	candela per square foot*	929.3034
	lambert	$3.141593(\pi)$
carat (metric)	gram*	0.2
Celsius (Centigrade) temperature scale, °C	Fahrenheit temperature scale, °F	$\frac{9}{5}(°C + 32) = °F$
centimeter	foot	0.032 808 4
	inch	0.393 700 8
	mil	393.700 8
centimeter of mercury	pascal	1 333.22
centimeter per second	foot per second	0.032 808 4

TABLE 11.6 Conversion Factors (*continued*)

To convert	Into	Multiply by
centimeter-dyne	erg*	1
	joule*	1×10^{-7}
centipoise	pascal-second*	0.001
centistokes	square meter per second*	1×10^{-6}
coulomb	ampere-second*	1
cubic centimeter	cubic foot	3.53147×10^{-5}
	liter*	0.001
	ounce (U.S., fluid)	0.03381402
	quart (U.S., dry)	9.08083×10^{-4}
	quart (U.S., liquid)	0.001056688
cubic centimeter per second	liter per hour*	3.6
curie	becquerel*	3.7×10^{10}
cycle per second	hertz*	1
day (mean solar)	hour*	24
	minute*	1440
	second	8.64×10^{4}
Debye unit	coulomb-meter	3.33564×10^{-30}
decibel	neper	0.115129255
degree (angle)	circumference	0.00277778(1/360)
	minute (angle)*	60
	quadrant	0.0111111(1/90)
	radian	0.01745329(π/180)
degree Celcius (Centigrade) (temperature difference), °C	degree Fahrenheit, °F*	1.8
	degree Rankine*	1.8
	kelvin*	1
dram (apothecaries or troy)	dram (avoirdupois)	2.1942857
dram (avoirdupois)	grain*	27.34375
	gram	1.7718452
	ounce (avoirdupois)	0.0625(1/16)
dram (U.S., fluid)	cubic centimeter	3.6966912
	ounce (U.S., fluid)*	0.125(1/8)
	pint (U.S., liquid)*	0.0078125(1/128)
dyne	kilogram-force	1.019716×10^{-6}
	newton*	1×10^{-5}
dyne per square centimeter	bar*	1×10^{-6}
	millimeter of mercury	7.500617×10^{-4}
	pascal	0.1
dyne-centimeter	erg*	1
	joule*	1×10^{-7}
	newton-meter*	1×10^{-7}

TABLE 11.6 Conversion Factors (*continued*)

To convert	Into	Multiply by
dyne-second per square centimeter	poise*	1
	pascal-second	0.1
electronvolt	erg	1.60219×10^{-12}
	joule	1.60219×10^{-19}
em	millimeter	4.21752
erg	dyne-centimeter*	1
	joule*	1×10^{-7}
	watt-hour	2.77778×10^{-11}
Fahrenheit temperature, °F	Celsius temperature, °C	$\frac{5}{9}(°F - 32) = °C$
fathom	foot*	6
fermi	meter*	1×10^{-15}
foot	centimeter*	30.48
	inch	12
foot-candle	lumen per square foot*	1
	lumen per square meter	10.7639
foot-lambert	candela per square centimeter	3.42626×10^{-4}
	candela per square foot	0.3183099
	lambert	0.00107639
gallon (British, imperial)	gallon (U.S.)	1.20095
	liter*	4.54609
gallon (U.S.)	liter	3.785412
	ounce (U.S., fluid)*	128
	pint (U.S., liquid)*	8
gauss	tesla*	1×10^{-4}
	weber per square meter	1×10^{-4}
gilbert	ampere-turn	0.795775
grain	milligram*	64.79891
gram	carat (metric)*	5
	grain	15.432358
	ounce (avoirdupois)	0.035273962
	ounce (troy)	0.032150747
	pound	0.0022046226
	ton (metric)	1×10^{-6}
gram-force	dyne*	980.665
	newton*	0.00980665
gram-force per square centimeter	pascal*	98.0665
gram-force-centimeter	joule*	9.80665×10^{-5}
Hefner unit	candela	0.9
hertz	cycles per second*	1

TABLE 11.6 Conversion Factors (*continued*)

To convert	Into	Multiply by
hour (mean solar)	minute*	60
	second	3 600
inch	centimeter*	2.54
	foot	0.083 333 3(1/12)
	mil*	1 000
	millimeter*	25.4
joule	Btu	9.478170×10^{-4}
	calorie	0.238 845 9
	erg*	1×10^7
	liter-atmosphere	0.009 869 233
	newton-meter*	1
	watt-hour	$2.77778 \times 10^{-4}(1/3600)$
kelvin temperature scale, K	Celsius scale, °C	$°C + 273.1 = K$
kilocalorie per second	kilowatt*	4.186 8
kilogram	ounce (avoirdupois)	35.273 963
	ounce (troy)	32.150 747
	pound	2.204 622 6
	ton (long)	$9.842065 3 \times 10^{-4}$
	ton (metric)	0.001
	ton (short)	0.001 102 311 3
kilometer	foot	3 280.840
	light-year	1.05702×10^{-13}
	mile (statute)	0.621 371 192
kilowatt	Btu per hour	3 412.14
	horsepower (metric)	1.359 62
	joule per hour*	3.6×10^{-6}
	kilocalorie per hour	859.845
knot	foot per minute	101.268 6
	meter per minute	30.866 7
	mile (nautical) per hour*	1
	mile (statute) per hour	1.150 78
lambert	candela per square centimeter	0.318 310
liter	cubic centimeter*	1 000
	cubic decimeter*	1
	cubic inch	61.023 74
	gallon (U.S.)	0.264 172 1
	ounce (U.S., fluid)	33.814 02
	pint (U.S., liquid)	2.113 376
	quart (U.S., liquid)	1.056 688
liter per minute	gallon (U.S.) per hour	15.850 3
liter-atmosphere	Btu	0.096 037 6
	calorie	24.201 1
	joule*	101.325

TABLE 11.6 Conversion Factors (*continued*)

To convert	Into	Multiply by
lumen per square centimeter	lux*	1×10^4
lux	lumen per square meter*	1
maxwell	weber*	1×10^{-8}
megaohm	ohm*	1×10^6
meter	angstrom*	1×10^{10}
	foot	3.280 839 895
mho (ohm^{-1})	siemens*	1
micrometer (micron)	angstrom	1×10^4
	millimeter*	0.001
mil	inch*	0.001
	millimeter*	0.025 4
mile (statute)	foot*	5.280
	furlong*	8
	kilometer*	1.609 344
	mile (nautical)	0.868 976
milligram per assay ton	milligram per kilogram	34.285 714
	ounce (troy) per ton (short)*	1
milliliter	cubic centimeter*	1
millimeter	inch	0.039 370 8
millimeter of mercury	atmosphere	0.001 315 789(1/760)
	dyne per square centimeter	1 333.224
	pascal	133.322 4
	torr*	1
minute (angle)	circumference	$4.629 63 \times 10^{-5}$
	degree (angle)	0.016 666 7(1/60)
	radian	$2.908 88 \times 10^{-4}$
	second (angle)*	60
minute	day	$6.944 444 \times 10^{-4}$
	hour	0.016 666 7(1/60)
	second*	60
newton	dyne*	1×10^5
newton per square centimeter	pascal*	1×10^4
oersted	ampere per meter	79.577 5
ounce (avoirdupois)	dram*	16
	grain*	437.5
	gram*	28.349 523 125
	ounce (troy)	0.911 458 33
	pound*	0.062 5(1/16)
ounce (U.S., fluid)	cubic centimeter	29.573 530
	gallon (U.S.)*	0.007 812 5(1/128)
	milliliter	29.573 530

TABLE 11.6 Conversion Factors (*continued*)

To convert	Into	Multiply by
	pint (U.S., liquid)*	0.0625(1/16)
	quart (U.S., liquid)*	0.03125(1/32)
parsec	kilometer	3.08568×10^{13}
part per million	gram per ton (metric)*	1
	milligram per kilogram*	1
pascal	bar*	1×10^{-5}
	dyne per square centimeter*	10
	inch of mercury	2.95300×19^{-4}
	millimeter of mercury	7.50062×10^{-3}
	newton per square meter*	1
pascal-second	poise*	10
pica (printer's)	point*	12
pint (U.S., liquid)	cubic centimeter	473.1765
point (printer's, U.S.)	millimeter*	0.3514598
poise	pascal-second*	0.1
pound	dram*	256
	grain*	7000
	gram*	453.59237
	ounce (avoirdupois)*	16
	ton (long)	4.4622857×10^{-4}
	ton (metric)*	4.5359237×10^{-4}
	ton (short)*	$5 \times 10^{-4}(1/2000)$
poundal	gram-force	14.0981
	newton	0.138255
proof (U.S.)	percent alcohol by volume*	0.5
quart (U.S., dry)	cubic centimeter	1101.221
	cubic foot	0.03888925
	pint (U.S., dry)	2
quart (U.S., liquid)	gallon (U.S.)*	0.25
	liter	0.946353
	ounce (U.S., fluid)*	32
	pint (U.S., liquid)*	2
radian	degree (angle)	57.295780
	minute (angle)	3.437.75
	revolution	0.159155
ream	quire*	20
	sheet	480 or 500
revolution	degree (angle)*	360
revolution per minute	radian per second	0.140720
roentgen	coulomb per kilogram*	2.58×10^{-4}
second (angle)	degree	2.77778×10^{-4}
	radian	4.848137×10^{-6}

TABLE 11.6 Conversion Factors (*continued*)

To convert	Into	Multiply by
siemens	mho (ohm^{-1})*	1
steradian	sphere	0.079 5775
	spherical right angle	0.636 620
stokes	square meter per second*	1×10^{-4}
tablespoon (metric)	cubic centimeter*	15
teaspoon (metric)	cubic centimeter*	5
tesla	weber per square meter*	1
ton (long)	kilogram*	1016.046 908 8
	pound*	2240
	ton (metric)	1.016 0469
	ton (short)*	1.12
torr	millimeter of mercury	1
	pascal	133.322 4
volt-second	weber*	1
watt	Btu per hour	3.412 14
	calorie per second	0.238 846
	erg per second*	1×10^{7}
	joule per second*	1
weber	maxwell*	1×10^{8}
X unit	meter	$1.002 02 \times 10^{-13}$

STATISTICS

TABLE 11.7 Values of t

df	$t_{.60}$	$t_{.70}$	$t_{.80}$	$t_{.90}$	$t_{.95}$	$t_{.975}$	$t_{.99}$	$t_{.995}$
1	0.325	0.727	1.376	3.078	6.314	12.706	31.821	63.657
2	0.289	0.617	1.061	1.886	2.920	4.303	6.965	9.925
3	0.277	0.584	0.978	1.638	2.353	3.182	4.541	5.841
4	0.271	0.569	0.941	1.533	2.132	2.776	3.747	4.604
5	0.267	0.559	0.920	1.476	2.015	2.571	3.365	4.032
6	0.265	0.553	0.906	1.440	1.943	2.447	3.143	3.707
7	0.263	0.549	0.896	1.415	1.895	2.365	2.998	3.499
8	0.262	0.546	0.889	1.397	1.860	2.306	2.896	3.355
9	0.261	0.543	0.883	1.383	1.833	2.262	2.821	3.250
10	0.260	0.542	0.879	1.372	1.812	2.228	2.764	3.169

TABLE 11.7 Values of t (continued)

df	$t_{.60}$	$t_{.70}$	$t_{.80}$	$t_{.90}$	$t_{.95}$	$t_{.975}$	$t_{.99}$	$t_{.995}$
11	0.260	0.540	0.876	1.363	1.796	2.201	2.718	3.106
12	0.259	0.539	0.873	1.356	1.782	2.179	2.681	3.055
13	0.259	0.538	0.870	1.350	1.771	2.160	2.650	3.012
14	0.258	0.537	0.868	1.345	1.761	2.145	2.624	2.977
15	0.258	0.536	0.866	1.341	1.753	2.131	2.602	2.947
16	0.258	0.535	0.865	1.337	1.746	2.120	2.583	2.921
17	0.257	0.534	0.863	1.333	1.740	2.110	2.567	2.898
18	0.257	0.534	0.862	1.330	1.734	2.101	2.552	2.878
19	0.257	0.533	0.861	1.328	1.729	2.093	2.539	2.861
20	0.257	0.533	0.860	1.325	1.725	2.086	2.528	2.845
21	0.257	0.532	0.859	1.323	1.721	2.080	2.518	2.831
22	0.256	0.532	0.858	1.321	1.717	2.074	2.508	2.819
23	0.256	0.532	0.858	1.319	1.714	2.069	2.500	2.807
24	0.256	0.531	0.857	1.318	1.711	2.064	2.492	2.797
25	0.256	0.531	0.856	1.316	1.708	2.060	2.485	2.787
26	0.256	0.531	0.856	1.315	1.706	2.056	2.479	2.799
27	0.256	0.531	0.855	1.314	1.703	2.052	2.473	2.771
28	0.256	0.530	0.855	1.313	1.701	2.048	2.467	2.763
29	0.256	0.530	0.854	1.311	1.699	2.045	2.462	2.756
30	0.256	0.530	0.854	1.310	1.697	2.042	2.457	2.750
40	0.255	0.529	0.851	1.303	1.684	2.021	2.423	2.704
60	0.254	0.527	0.848	1.296	1.671	2.000	2.390	2.660
120	0.254	0.526	0.845	1.289	1.658	1.980	2.358	2.617
∞	0.253	0.524	0.842	1.282	1.645	1.960	2.326	2.576
df*	$-t_{.40}$	$-t_{.30}$	$-t_{.20}$	$-t_{.10}$	$-t_{.05}$	$-t_{.025}$	$-t_{.01}$	$-t_{.006}$

*When the table is read from the foot, the table values should be prefixed with a negative sign. Interpolation should be performed using the reciprocals of the degrees of freedom.

Source: Perry, Chilton, and Kirkpatrick, *Chemical Engineers' Handbook*, 4th ed., McGraw-Hill, New York (1963).

INDEX

ABOUT THE AUTHOR

George W. Gokel, Ph.D., is a professor of molecular biology and pharmacology and the director of the Chemical Biology Program at Washington University School of Medicine. He lives in Chesterfield, Missouri.